Irrigation Engineering

This textbook provides a comprehensive treatment of irrigation engineering for advanced undergraduates and graduate students. It does not require a background in calculus, hydrology, or hydraulics, offering a one-stop overview of the entire field of study. It includes everything a student of irrigation engineering needs to know: concepts of climate, soils, crops, water quality, hydrology, and hydraulics, as well as their application to design and environmental management. To demonstrate the practical applications of the theories discussed, there are over 500 worked examples and end-of-chapter exercises. The exercises allow readers to solve real-world problems and apply the information they've learned to a diverse range of scenarios. To further prepare students for their future careers, each chapter includes many illustrative diagrams and tables containing data to help design irrigation systems. For instructors' use when planning and teaching, a solutions manual can be found online alongside a suite of PowerPoint lecture slides.

VIJAY P. SINGH is University Distinguished Professor, Regents Professor, and Caroline and William N. Lehrer Distinguished Chair in Water Engineering at Texas A&M University. To date, he has published more than 1370 journal articles, 120 books, and 113 book chapters in hydrology, groundwater, hydraulics, irrigation, pollutant transport, copulas, entropy, and water resources. He has received more than 100 national and international awards, and is past President of the American Institute of Hydrology (AIH) and the American Academy of Water Resources Engineers (AAWRE).

QIONG SU is based in the Water Management and Hydrological Science Program at Texas A&M University. She is a member of the American Society of Agricultural and Biological Engineers (ASABE) and the American Geophysical Union (AGU). She has received several travel awards, graduate student scholarships, and an outstanding contributions in reviewing award. She has published peer-reviewed papers in the areas of hydrology, water quality, irrigation, water-energy nexus, and water resources. She was a teaching assistant for the irrigation engineering course at Texas A&M.

Irrigation Engineering

Principles, Processes, Procedures, Design, and Management

VIJAY P. SINGH
Texas A&M University

QIONG SU
Texas A&M University

CAMBRIDGE
UNIVERSITY PRESS

University Printing House, Cambridge CB2 8BS, United Kingdom

One Liberty Plaza, 20th Floor, New York, NY 10006, USA

477 Williamstown Road, Port Melbourne, VIC 3207, Australia

314–321, 3rd Floor, Plot 3, Splendor Forum, Jasola District Centre, New Delhi – 110025, India

103 Penang Road, #05–06/07, Visioncrest Commercial, Singapore 238467

Cambridge University Press is part of the University of Cambridge.

It furthers the University's mission by disseminating knowledge in the pursuit of education, learning, and research at the highest international levels of excellence.

www.cambridge.org
Information on this title: www.cambridge.org/highereducation/isbn/9781316511220
DOI: 10.1017/9781009049610

First published 2022

Printed in the United Kingdom by TJ Books Limited, Padstow Cornwall

A catalogue record for this publication is available from the British Library.

Library of Congress Cataloging-in-Publication Data
Names: Singh, V. P. (Vijay P.), author. | Su, Qiong, 1964– author.
Title: Irrigation engineering : principles, processes, procedures, design, and management / Vijay P. Singh, Qiong Su.
Description: First edition. | New York, NY : Cambridge University Press, 2022. | Includes bibliographical references and index.
Identifiers: LCCN 2022012599 | ISBN 9781316511220 (hardback)
Subjects: LCSH: Irrigation engineering. | Agricultural engineering. | Environmental hydraulics. | BISAC: SCIENCE / Earth Sciences / Hydrology
Classification: LCC TC805 .S498 2022 | DDC 631.5/87–dc23/eng/20220413
LC record available at https://lccn.loc.gov/2022012599

ISBN 978-1-316-51122-0 Hardback

VPS: Wife, Anita, who is no more; son, Vinay; daughter, Arti; daughter-in-law, Sonali; son-in-law, Vamsi; and grandsons: Ronin, Kayden, and Davin

QS: Parents and husband, Yun

Contents

Detailed Contents

Preface

Food security is a major concern, especially in developing countries, and this concern is being heightened by increasing population and climatic vagaries and change causing higher uncertainty in the space–time distribution of rainfall, rising standards of living leading to greater demand for food and greater wastage, inadequate facilities for grain storage and transport, and improper soil and water management. Productive agriculture is therefore fundamental to sustained food security, which is not possible without irrigation. Irrigation is also needed for transforming non-agricultural land, such as wasteland, into agricultural land, and less productive lands into more productive lands.

Irrigation has many aspects, such as agricultural, environmental, engineering, design, operation, and management, which together constitute a vast discipline. Over the years, irrigation engineering has branched out into canal irrigation, tubewell irrigation, and agricultural irrigation. Canal irrigation primarily entails design, operation, and management of canal systems that include canals, gates and diversion systems, and delivery systems. This is a discipline in civil engineering, but agricultural engineers also deal with some aspects of canals – especially small canals. It goes without saying that climatic, soil, and crop characteristics are considered in the design and operation of these canal systems. Tubewell irrigation involves design, operation, and management of tubewells. Since tubewells are constructed in agricultural lands, they primarily fall within the discipline of agricultural engineering, although civil engineering is also concerned with the design and operation of tubewells. Agricultural or farm irrigation is a major discipline in agricultural engineering, although many civil engineering aspects are involved therein. Because of increasing concern for the environment, diminishing availability of water, and growing incorporation of socio-political, economic, legal, and administrative issues, irrigation engineering has taken on a much broader scope and no single book can cover all aspects in complete detail.

There are many excellent authored and edited books on agricultural irrigation engineering. However, many of these books are at an advanced level or at the graduate level, and much work is needed to extract material that will be suitable for a one-semester undergraduate course in agricultural irrigation engineering. Furthermore, there are several aspects that an undergraduate student learning irrigation engineering for the first time needs to understand but which are not covered. Thus, a book primarily designed for an undergraduate course in irrigation engineering seems to be lacking. This book aims at filling this gap. To help the student appreciate the importance of irrigation engineering, a large number of solved problems are included in each chapter and a number of end-of-chapter problems are provided that are solved in the solutions manual that accompanies this book.

Introducing agricultural irrigation, including definitions, needs, benefits, and limitations of irrigation; food security; development of irrigation worldwide and in the United States; impacts of global warming and climate change; environmental concern; and the future of irrigation in Chapter 1, the subject matter of the book is divided into seven parts. Part I discusses fundamentals spanning six chapters. Chapter 2 deals with those aspects of climate that are pertinent to agricultural irrigation, such as weather and climate, types of climate, causes of climate variability, meteorological variables, the Earth system, and the greenhouse effect. Sources and availability of water are described in Chapter 3. Both surface water and subsurface water sources, availability of water, water sharing, and allocation are included.

In almost all countries, a significant portion of irrigation water is derived from groundwater using wells. Chapter 4 therefore discusses rudimentary aspects of groundwater and wells from the perspective of pumping water for irrigation. Since the quality of water used for irrigation has a significant impact on crop yield, degradation of soil, groundwater, and operation and life of irrigation systems, Chapter 5 discusses water quality from an agricultural irrigation point of view. Physical and chemical characteristics of water, and biological characteristics if wastewater is used for irrigation, that are important for irrigation are presented in the chapter. Irrigation of crops significantly depends on the type of soil, which is fundamental to our biosphere and requires proper management. Chapter 6 discusses basic soil properties relevant to crops and irrigation. Different types of crops require different types of climate and soil, and different crops and their optimum production have different irrigation requirements with respect to frequency and timing of irrigation, and the amount of water used per irrigation. Chapter 7 discusses the types of crops and their water requirements.

Hydraulic principles that are relevant to agricultural irrigation constitute the subject matter of Part II, which comprises four chapters. Water is conveyed from the source to the farm either by channels or by pipelines. Channels used in irrigation systems can be either erodible or non-erodible, and flow in these channels is governed by the principles of hydraulics. Chapter 8 discusses the rudimentary aspects of hydraulics and the design of open channels. The hydraulics of a pipe when it runs full is different from the hydraulics of open channels in which the upper surface of the flow is exposed to the atmosphere. Pipes or closed conduits are used in sprinkler and drip irrigation systems to carry water from the source to the individual sprinkler or emitter. Chapter 9 reviews the principles of pipeline hydraulics. Pumps are used to lift groundwater to the ground surface, raise water from a lower elevation to a higher elevation, transport water, overcome friction, or generate pressure for the operation of sprinkler and trickle irrigation systems. For many agricultural irrigation systems, pumps are therefore an integral part. Chapter 10 discusses the rudimentary aspects of pumps and their operation and selection. For irrigation management and maintenance, it is necessary to determine the volume of water that is applied to the field and the rate at which water is applied. Measurement devices are therefore included in irrigation systems. Chapter 11 discusses different methods and devices that are commonly employed for determining flow rates and volume.

Part III presents the principles of hydrology in three chapters. The purpose of irrigation is to supply water to plants, and this supply of water is made possible through infiltration into the soil. The infiltrated water is extracted by plant roots. Thus, infiltration plays a fundamental role in the design, management, and operation of irrigation systems. Chapter 12 discusses elementary aspects of infiltration and methods for computing it. The infiltrated water, called soil water, occurs in the unsaturated zone and is the only source of water for most agricultural crops. Since nutrients, including fertilizers, are dissolved in soil water, it is the only source from which plants can extract them. Different aspects of this water are discussed in Chapter 13. In cropland, water evaporates from soil and is transpired by plants – that is, water is transported to the atmosphere by evaporation plus transpiration, which together form evapotranspiration. Determination of evapotranspiration is vital for determining crop water requirements. Chapter 14 discusses the process of evaporation and some of the methods that are used in irrigation engineering for computing evaporation, including empirical and energy-balance-based methods, which constitute the basis of quasi-theoretical and theoretical methods.

Principles of irrigation science are discussed in Part IV, which comprises three chapters. The water used by a crop is not the same as its potential evapotranspiration, which is then modified for a specific crop using a crop coefficient, which is what constitutes the crop water use or consumptive use. Chapter 15 discusses the methods for converting estimated evapotranspiration to crop water use. When irrigating crops, one of the key considerations is to do irrigation efficiently so that crop yield is maximized. However, irrigation efficiency is not uniquely defined. Chapter 16 discusses the concept of irrigation efficiency and related aspects. Surface irrigation is commonly practiced around the world, more so in developing countries. There are different methods of surface irrigation and the selection of a particular method depends on a number of factors, including climate, soil, crop, water availability, landscape, availability of labor, energy, costs and benefits, and traditions. Chapter 17 discusses the preliminaries of the entire irrigation system.

Part V, consisting of five chapters, deals with methods of irrigation. Depending on the type, crops are irrigated by different methods. Basin irrigation is a common method for surface irrigation, especially in developing countries. A basin is an agricultural field with zero to little slope, which is diked from all sides. The usual method of water application is flooding. Chapter 18 briefly discusses the basin method of irrigation and its design. A simple but popular method is also border irrigation, which constitutes the subject matter of Chapter 19. Furrow irrigation is another popular method of irrigation, which is presented in Chapter 20. The availability of water for agricultural irrigation is diminishing for a variety of reasons. In order to use water in agriculture more efficiently, sprinkler irrigation and trickle irrigation are becoming popular all over the world. Chapter 21 discusses different aspects of sprinkler irrigation, and Chapter 22 covers trickle irrigation.

Irrigation design is the subject matter of Part VI, comprising four chapters. Irrigation system design begins with planning, which depends on the size of the system. Small systems may be owned by individual farmers, and farmers plan these systems on their own, with limited outside help. On the other hand, large systems are owned by government or groups of farmers and their planning is quite technical. Chapter 23 discusses the rudimentary aspects of irrigation planning. In irrigation, land surface characteristics play a significant role. Indeed, the type of irrigation system to be employed is determined by the land surface, which ideally should be such that the irrigation water moves as uniformly as possible, but the natural landscape is not always so. Therefore, the natural landscape or topography is altered by land leveling, which is discussed in Chapter 24. When irrigation is done, especially using border, furrow, or flooding methods, more water is applied than is actually used by plants. The excess water from the soil surface, as well as the soil profile or root zone, needs to be removed without eroding the soil and damaging crops. The orderly removal of this excess water is called drainage, which provides a suitable environment for the maximization of plant growth, keeping in mind financial constraints. Chapter 25 presents rudiments of agricultural drainage. Design of a farm irrigation system entails both technical and non-technical considerations. It is an integration of principles borrowed from agriculture, meteorology, hydrology, hydraulics, irrigation, and drainage engineering, as well as economic, environmental, and management sciences. Chapter 26 provides a snapshot of steps involved in designing a farm irrigation system.

The final part is on irrigation management, comprising five chapters. Considering the type of crop, soil, climate, method of irrigation, and agricultural practices, scheduling of irrigation is done. There are different methods for irrigation scheduling which are presented in Chapter 27.

Although irrigated agriculture is vital for food security, irrigation should be practiced such that it causes minimum environmental damage. Chapter 28 discusses important environmental considerations that should be kept in mind in planning, designing, operating, and managing irrigation systems. Irrigation systems are usually designed with the long-term objective that they are economically sustainable, although that is not always the case in many developing countries. Chapter 29 visits fundamental concepts of analyzing benefits and costs, which in the long term define the benefit–cost ratio. For large irrigation systems there is usually an organizational structure tasked with managing water, structures and equipment, and people. It engages in decision-making, resource mobilization, communication, and conflict resolution. Chapter 30 provides a snapshot of elements of irrigation management. The concluding Chapter 31 deals with irrigated agriculture for food security and the impact of climate change. Defining food and nutritional security, it discusses the factors affecting the security and the impacts of climate change.

Although the book is primarily designed for undergraduate or early graduate students in agricultural or irrigation engineering, it will also be useful for students in civil engineering, agriculture, horticulture, and water resources management. Faculty members engaged in teaching irrigation engineering and those engaged in farm management will also find the book useful.

Acknowledgments

This book grew out of a senior undergraduate course, "Irrigation and Drainage Engineering," being taught in the Department of Biological & Agricultural Engineering at Texas A&M University. Dr. Kyungtae Lee helped solve a number of problems in the book and prepare a number of figures. His help is deeply appreciated.

Like any other book, the authors have drawn from the works of a large number of irrigation engineers from the United States and abroad. Without their contributions, irrigation engineering would not be what it is today. The authors have tried to acknowledge these works as specifically as possible, and any omission on their part will be entirely inadvertent and they would like to express their apologies in advance. In the wake of growing population, rising demand for food, and increasing food wastage during the entire supply chain and consumption, irrigation is vital to ensure food security. The contributions of irrigation engineers to food security that we take for granted these days can hardly be over-emphasized. The authors are grateful that these engineers chose to pursue irrigation engineering.

Part I Fundamentals

1 Introduction

1.1 PRELIMINARY REMARKS

Irrigation is vital for productive agriculture and consequent food and nutritional security. Agriculture is either rainfed or is dependent on irrigation. In either case, it is greatly impacted by the vagaries of nature, especially weather, as well as by soil, crops to be irrigated, and the source, availability, and quality of water. Irrigation is needed for productive agriculture and consequent food and nutritional security, because rainfall is seldom adequate and timely for meeting agricultural needs, even in humid areas. Also, non-agricultural lands, such as wastelands, can be brought under agriculture by irrigation, and less productive land can be transformed into more productive land. Rainfall varies from place to place for a given month or year and from month (or year) to month (or year) for a given place, as illustrated in Figures 1.1–1.3 for the United States. Figure 1.1 depicts the monthly distribution of rainfall at different places, Figure 1.2 shows the annual precipitation from 1982 to 2010 at different places, and Figure 1.3 shows the percentage of normal precipitation in January 2015.

The requirement for irrigation is significantly affected by climate. In the United States, there are four main types of climate regions: arid, semi-arid, sub-humid, and humid, as shown in Figure 1.4. These climate regions are delineated by the Köppen–Geiger climate classification system based upon the annual and monthly averages of temperature and precipitation (Beck et al., 2018). In arid and semi-arid areas, irrigation is needed for growing crops. In humid and sub-humid areas, there may be enough annual rainfall, but it is not timely distributed when needed and hence irrigation may be required. Even in humid areas, periods between rainfall events may be too long and crops may suffer due to the lack of moisture. Crops in certain soil conditions begin to suffer if they do not receive water within five days or less. Droughts also occur, and their frequency seems to be increasing due to climate change and global warming, thus necessitating irrigation.

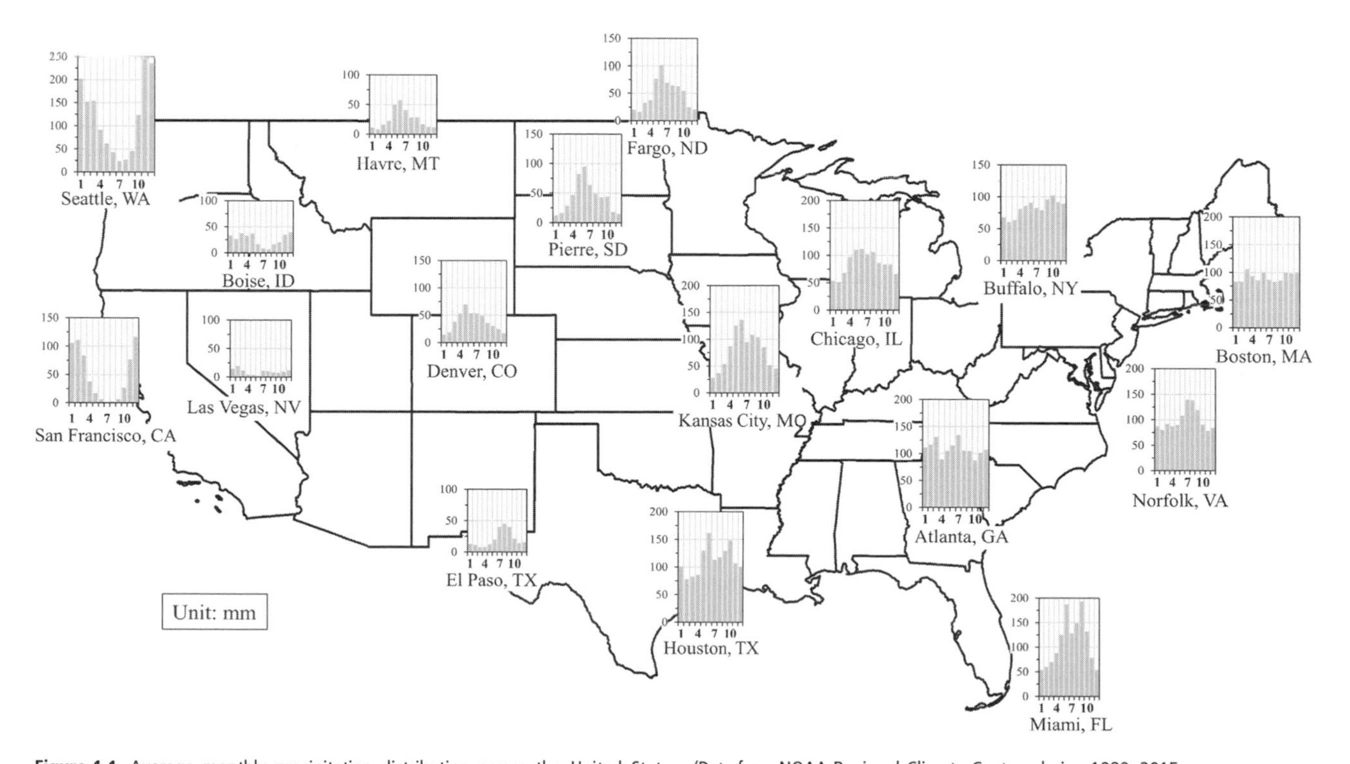

Figure 1.1 Average monthly precipitation distribution across the United States. (Data from NOAA Regional Climate Centers during 1980–2015, http://scacis.rcc-acis.org.)

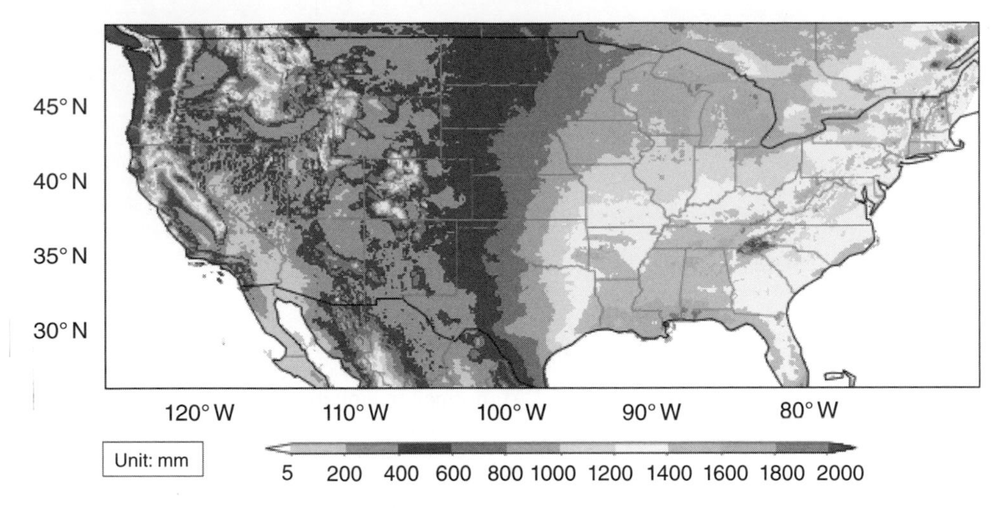

Figure 1.2 Average annual precipitation in millimeters during 1982–2010 across the contiguous United States. (Data from Daily Surface Weather and Climatological Summaries, DAYMET, https://daymet.ornl.gov.) A black and white version of this figure will appear in some formats. For the color version, please refer to the plate section.

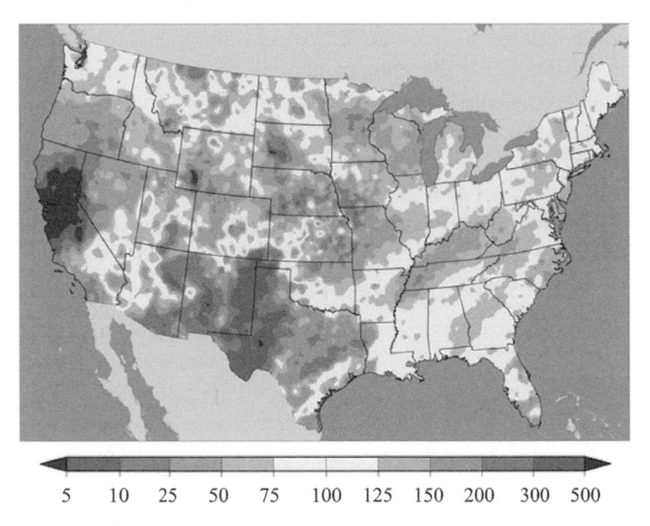

Figure 1.3 Percentage of the normal precipitation in January 2015 across the United States. (Data from NOAA National Climate Report, www.ncdc.noaa.gov/sotc/national/201501.) A black and white version of this figure will appear in some formats. For the color version, please refer to the plate section.

Irrigation of agricultural lands is now practiced all over the world. The purpose of irrigation is to provide water to crops where and when crop water requirements cannot be met by natural rainfall. In many areas, there is deficit rainfall and in some areas rainfall is not enough during the crop-growing season, and in other areas there is hardly any rainfall. Many areas without or with little rainfall are wastelands, but they can be transformed by irrigation into crop-producing areas.

Providing a snapshot of irrigation worldwide as well as in the United States, this chapter discusses different aspects of irrigation, including need, importance, and development. It also discusses the organization of the book, irrigation practices, and environmental concerns arising due to irrigation, and concludes with a reflection on the future of irrigation.

1.2 ORGANIZATION OF CONTENTS

Before discussing the scope of the book, it is desirable to point out the interaction or overlap of irrigation engineering with other disciplines, as shown in Figure 1.5, such as climate science, water resources (including water quality), soil science, crop science, hydraulics, hydrology, environmental impact assessment, economics, and management. These disciplines define the factors that influence the planning, design, and management of irrigation systems and in turn define the domain of irrigation engineering. The techniques developed in these disciplines are brought to bear on solutions to irrigation problems.

Irrigation engineering borrows techniques from solutions in non-engineering disciplines, such as mathematics, statistics, operations research, probability theory, social sciences, economics, and law. For farming, irrigation is almost always required and irrigation systems are therefore planned, designed, built, operated, and managed. There are a number of considerations that need to be kept in mind for planning, designing, operating, and maintaining an irrigation system, as shown in Figure 1.6. The first consideration is the selection of crops. However, before selecting crops for cultivation on an area of land, it is important to know five things: climate, soil, source and availability of water and energy, quality of water, and crop types. The crops selected for cultivation depend not only on these things, but also on irrigation.

The second consideration includes principles of hydraulics, which are needed for bringing water from its source to the field. Water is transported either by open channels or pipelines, and often it may need to be lifted using pumps. Also, for proper application of water, it is important to measure its flow. The third consideration comprises hydrologic principles that are needed for the application of water. Once water is applied to the land, it has four ways to go: vertically downward (infiltration and percolation), temporary storage in the pore spaces of soil (soil water), horizontal movement

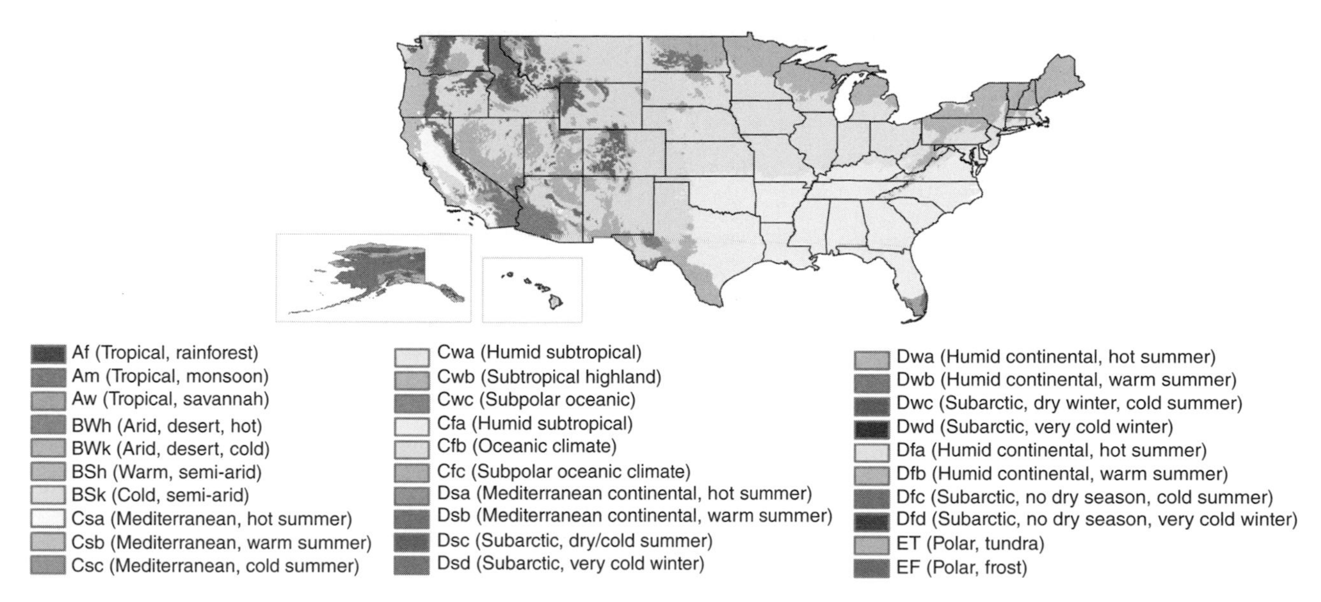

Figure 1.4 Köppen climate classification for the United States. (Data from Beck et al. [2018] using weather data from 1980–2016.) A black and white version of this figure will appear in some formats. For the color version, please refer to the plate section.

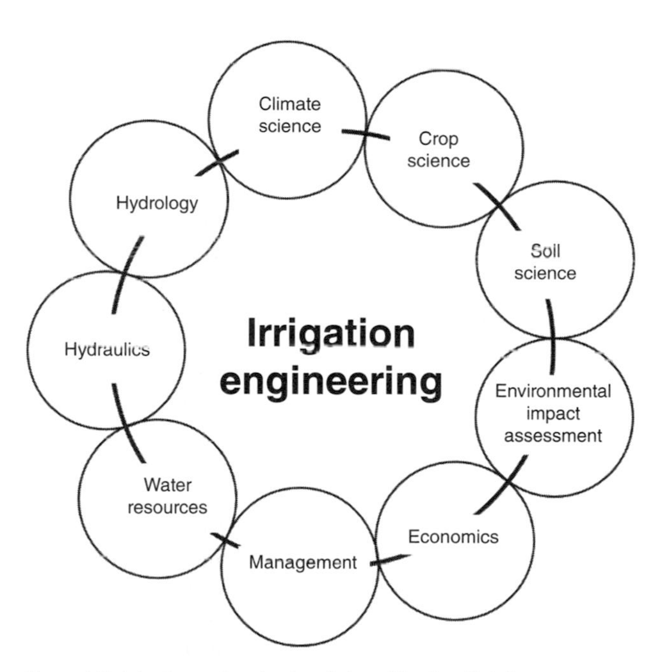

Figure 1.5 Irrigation engineering interfacing with other disciplines.

(drainage and seepage), and vertically upward (evapotranspiration), combining evaporation and transpiration.

The fourth consideration comprises irrigation principles, which include consumptive use, irrigation efficiency, and equations governing flow over porous beds. In general, crops do not consume water at the same rate or in the same amount, so crop water consumptive use becomes important. The fifth consideration includes methods of irrigation, including basin, border, furrow, sprinkler, and trickle (or drip). The sixth consideration entails design, operation, and management of irrigation systems, including land leveling, drainage, irrigation scheduling, environmental considerations, economic evaluation, and maintenance.

1.3 DEFINITION OF IRRIGATION

Irrigation is defined as artificial application of water to plants for overcoming the lack, insufficiency, or poor distribution of rainfall. It is the controlled application of water to croplands, with the primary objective of creating an optimal soil moisture regime for maximizing crop production and quality, and at the same time minimizing environmental degradation inherent in the irrigation of agricultural lands. Irrigation is critical for food as well as nutritional security, especially in semi-arid and arid areas, and will become even more so with increasing population and under the specter of climate change.

1.4 BENEFITS OF IRRIGATION

Irrigation has several benefits, including the following: (1) It increases the productivity of agricultural lands, as shown in Table 1.1 for Brazil and in Table 1.2 for different states in the United States (Figure 1.7 shows the percentage increase in different crop yields in the United States [US Department of Agriculture, 2002]); (2) it makes it possible to have more than one harvest per year; (3) it guarantees production by removing water deficit; (4) it enables harvesting of crops in the off-season; (5) it permits growing more crops in a year on the same land; (6) it leads to improved product quality; (7) it allows growing crops that are highly sensitive to moisture deficit but are economically more profitable; (8) it permits transformation of agriculturally unproductive areas into agriculturally productive areas; (9) it creates jobs; and

(10) it helps arrest migration of people from rural areas to urban centers.

1.5 LIMITATIONS OF IRRIGATION

Irrigation is not without limitations, some of which are as follows. Traditional methods of irrigation – such as flooding, border, and furrow – require large volumes of water, hence irrigation management is needed. Modern methods of irrigation have high implementation costs. They may need hand-specialized labor which may not be available or may be too expensive. If not properly done, irrigation may lead to salinization of soils, so soils need to be appropriately managed. Irrigation has environmental impacts, such as water logging, waste of water, generation of mosquitoes, and changes in ecosystems, that need to be minimized.

Table 1.1 *Increase in yield due to irrigation in Brazil*

Crop	Not-irrigated (kg/ha)	Irrigated (kg/ha)	Increase in yield (%)
Cotton	848	2,700	218
Rice	1,739	3,750	115
Beans	388	2,300	492
Corn	1,985	5,500	177
Soyabean	1,844	3,000	62
Oat	1,668	3,400	104

1.6 NEED FOR IRRIGATION

The demand for food is increasing each year because of the growing population, rising food requirements, and increasing standards of living. Further, a lot of food grains and other agricultural produce are wasted during harvesting, transportation, distribution, storage, and consumption, as shown in Figure 1.8. The per capita food loss in Europe and North America is 280–300 kg/year. In sub-Saharan Africa and South/Southeast Asia it is 120–170 kg/year (FAO, 2011). This means that more food will have to be produced to ensure food security. In the next 35–45 years, world food production will have to be doubled in order to meet the demands of increasing population. It may be noted that 90% of this increased food production will have to come from existing lands, and 70% of this increased food production will have to come from irrigated lands.

Nearly 20% of the cultivated area in the world uses irrigation, and this area generates 44% of the agricultural production of the world. For example, 5% of the total area in Brazil uses irrigation, and this area produces 18% of agricultural produce in the country. Irrigated and rainfed agricultural lands and their production in different countries are shown in Figure 1.9.

The increase in irrigated land has not kept pace with the increase in global population, as shown in Figure 1.10. The decreasing trend of irrigation withdrawal since the 1980s with a growing population in the United States, as shown in Figure 1.11, poses a challenge to design even more efficient irrigation techniques and systems.

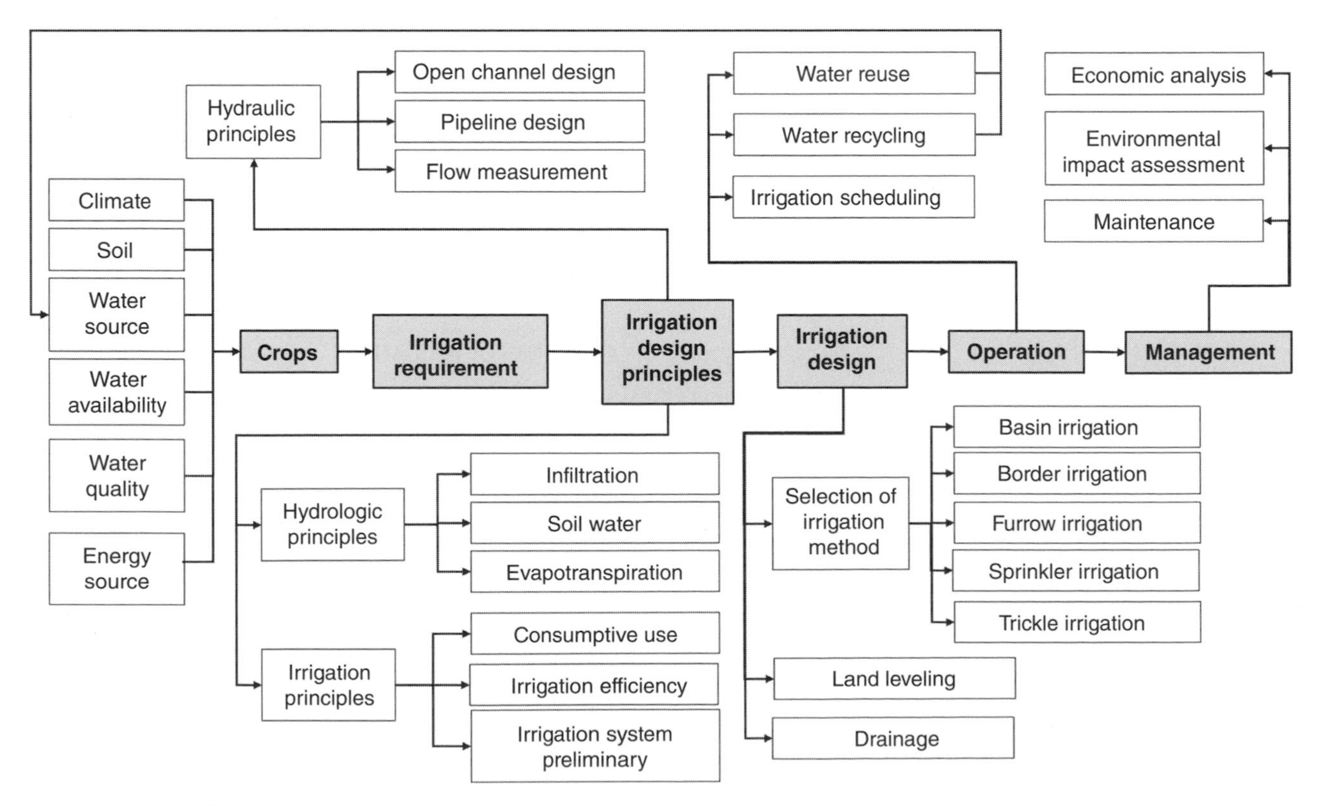

Figure 1.6 Planning, designing, operating, and managing an irrigation system.

Table 1.2 *Average yields of non-irrigated crops in comparison with irrigated crops*

	Yield per acre		
Crops, state, and year	Non-irrigated	Irrigated	Increase
Alfalfa			
North Dakota (1966)	2.0 tons	4.4 tons	2.4 tons
South Dakota (1966)	2.5 tons	5.3 tons	2.8 tons
Cabbage			
New Jersey (1955–9)	12.5 tons	18.9 tons	6.4 tons
Corn			
Florida (1971)	115 bu.	190 bu.	75 bu.
Nebraska (1966)	36 bu.	102 bu.	66 bu.
North Carolina (1963–8)	101 bu.	139 bu.	38 bu.
North Dakota (1966)	44.0 bu.	77.9 bu.	33.9 bu.
South Dakota (1949–55)	32.0 bu.	92.0 bu.	60 bu.
Virginia (1954–5)	83.3 bu.	109.2 bu.	28.9 bu.
Cotton (lint and seed)			
Arizona (1964–5)	NGWI	2,137 lb	2,137 lb
Arkansas (1950–2)	1,608 lb	2,083 lb	475 lb
Georgia (1949–53)	1,216 lb	1,902 lb	686 lb
Missouri (1953)	1,414 lb	2,683 lb	1,269 lb
North Carolina (1963–7)	1,836 lb	1,932 lb	96 lb
South Carolina (1954–5)	1,077 lb	1,668 lb	591 lb
Field beans (edible)			
Nebraska (1956)	27 bu.	54 bu.	27 bu.
Grain Sorghum			
Arizona (1964–5)	NGWI	72 bu.	72 bu.
Nebraska (1966)	39 bu.	87 bu.	48 bu.
Oklahoma (1958–62)	9.3 bu.	44.4 bu.	35.1 bu.
Grapefruit			
Florida (1960–7)	735 bo.	1,056 bo.	321 bo.
Oranges			
Florida (19600–67)	369 bo.	493 bo.	124 bo.
Peanuts			
North Carolina (1963–68)	2,632 lb	3,168 lb	536 lb
Oklahoma (1956–59)	1,014 lb	2,306 lb	1,292 lb
Peaches			
Maryland (1955–64)	300 lb	372 lb	72 lb
Pole beans			
Georgia (1950)	2,583 lb	6,025 lb	3,442 lb
Potatoes (Irish)			
Arizona (1964–5)	NGWI	6.5 tons	6.5 tons
California (1968)	350 sacks	450 sacks	100 sacks
New York (1946)	NP	NP	57 bu.
Texas	1.5 tons	7.0 tons	5.5 tons
Wisconsin (1946)	NP	NP	100 bu.
Silage			
Alabama (1966–8)	31 tons	47 tons	16 tons
Soybeans			
Arkansas (1966–8)	28.9 bu.	37.2 bu.	8.3 bu.
Georgia (1978)	30 bu.	53 bu.	23 bu.
Missouri (1959)	NP	NP	8.0 bu.
Sugar Beets			
Arizona (1964–85)	NGWI	20.5 tons	20.5 tons
North Dakota (1949–52)	NGWI	20 tons	20 tons
Wyoming (1956)	NGWI	16 tons	16 tons
Sweet Corn			
New Jersey (1955–8)	5,600 lb	11,900 lb	6,300 lb
Sweet Potatoes			
Louisiana (1953–6)	117.9 bu.	271.9 bu.	154 bu.
Tobacco			
South Carolina (1951–4)	1,183 lb	1,547 lb	364 lb
Virginia (1954–7)	2,699 lb	3,042 lb	343 lb
Tomatoes			
Georgia (1947–53)	17,430 lb	23,485 lb	6,055 lb
Wheat			
Kansas (1954–9)	21 bu.	48.6 bu.	27.6 bu.
Oklahoma (1954)	13 bu.	34 bu.	21.0 bu.
Texas (1966–7)	15.8 bu.	53.8 bu.	38 bu.

NGWI, not grown without irrigation; NP, not published; bu., bushels; bo., boxes.
Data from the US Department of Agriculture (www.ers.usda.gov/data-products/irrigated-agriculture-in-the-united-states)

1.7 PLANNING FOR AN IRRIGATION SYSTEM

Planning for an irrigation system entails answering a number of questions: (1) Should irrigation be done? (2) How much yield will irrigation increase? (3) How much will irrigation cost, and will it be affordable? (4) Is there sufficient water of good quality available? (5) What are the energy requirements, and can they be met? (6) What is the labor cost, and will labor be available? (7) How will the system be paid for? (8) What type of irrigation will be needed? (9) Will the irrigation system pay for itself in the long run? (10) Who will design the system? These and related questions need to be answered when planning for an irrigation system.

1.8 FOOD SECURITY

1.8.1 Population and Growth Pattern

The world population exceeded 6 billion in 2000 and it is projected to reach 9.7 billion by 2050 and 10.9 billion by the turn of 2100 (United Nations Population Division, 2019).

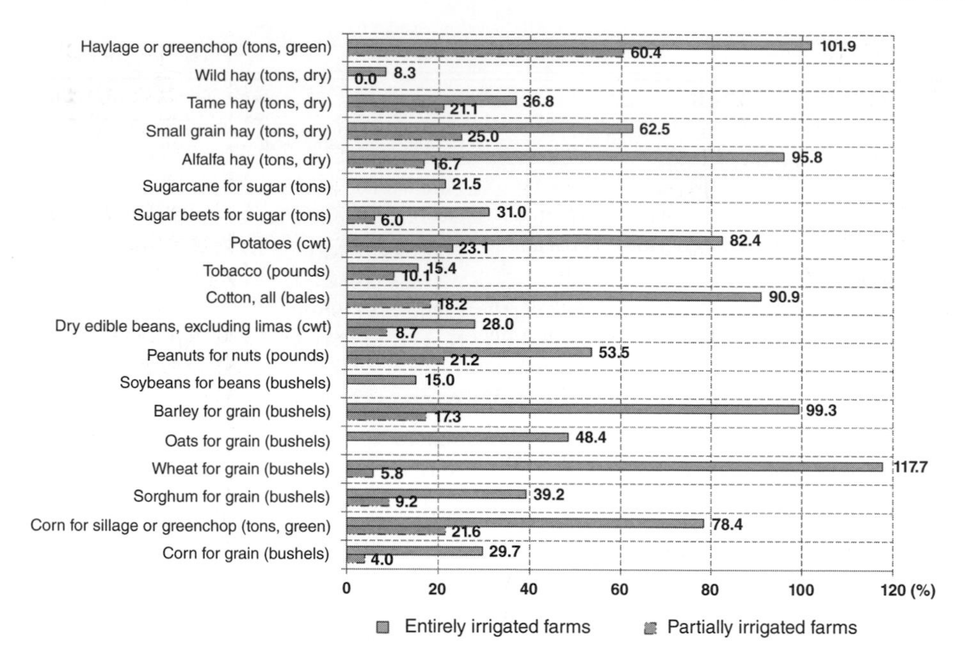

Figure 1.7 Percentage increase in average yield per acre of harvested crops in the United States (tons, cwt, pounds, bales, and bushels indicate different units used for crops yield in the United States). (Data from US Census of Agriculture, US Department of Agriculture, 2002.)

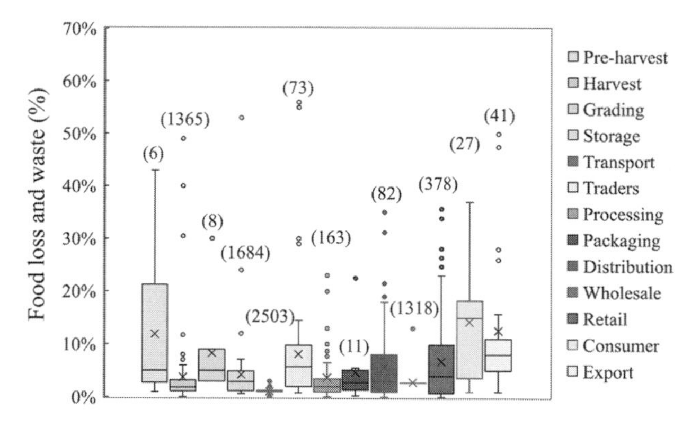

Figure 1.8 Range of reported global food loss and waste percentage during different processes (2001–2017). Number of observations is shown in brackets. (Data from FAO, 2019.) A black and white version of this figure will appear in some formats. For the color version, please refer to the plate section.

Interestingly, for most of human history the rate of increase of global population was relatively small until a century ago. However, the twentieth century witnessed a relatively rapid rise, resulting in the doubling of population, and population continues to grow at an alarming rate, as shown in Figure 1.12. The rapid growth is occurring in regions that are water stressed or are subject to droughts and extreme seasonal changes in precipitation and evaporation (Falkenmark and Widstrand, 1992). The pressure of the population in the top water-scarce areas in the world is shown in Table 1.3.

1.8.2 Food Requirement

The food requirements of a person depend on age, lifestyle, and work or level of activity. There is a large variation in the amount of food consumed daily from one country to another, and within the same country. In terms of calories, the US Department of Agriculture estimates that most women need 1600–2400 calories, while the majority of men need 2000–3000 calories each day to maintain a healthy weight. In the United States, the daily per capita consumption of food is 2730 grams, whereas on a global basis nearly 1880 grams of food is consumed per person on daily basis. Table 1.4 shows the temporal variation of per capita food consumption of different countries. This food comprises grains, dairy products, fruits, vegetables, meat, etc.

Food consumption has changed significantly over the past 100 years, as shown in Figure 1.13. These days, people eat more and waste more. To produce enough food to satisfy global food requirements, a huge amount of water is needed. For example, producing one ton of grains requires nearly 1000 m^3 of water. Much more water is needed to produce livestock products: for example, about 15,415 m^3 of water is needed to produce 1 ton of beef, as shown in Figure 1.14. There is a large variation in the water footprint of various foods and beverages, as illustrated in Figure 1.14.

Nearly one billion people, approximately one in six people globally, do not have access to adequate food at present. In India alone, nearly 195 million people are malnourished – about 25% of the global malnourished population. China has about 134 million malnourished people. Malnourishment is caused by poverty, inadequate supply chains, rampant food wastage, and poor farming. Ensuring food security requires good agricultural and management practices, advances in irrigation and technology, proper policies, and strong political will (Brabeck-Letmathe and Biswas, 2015).

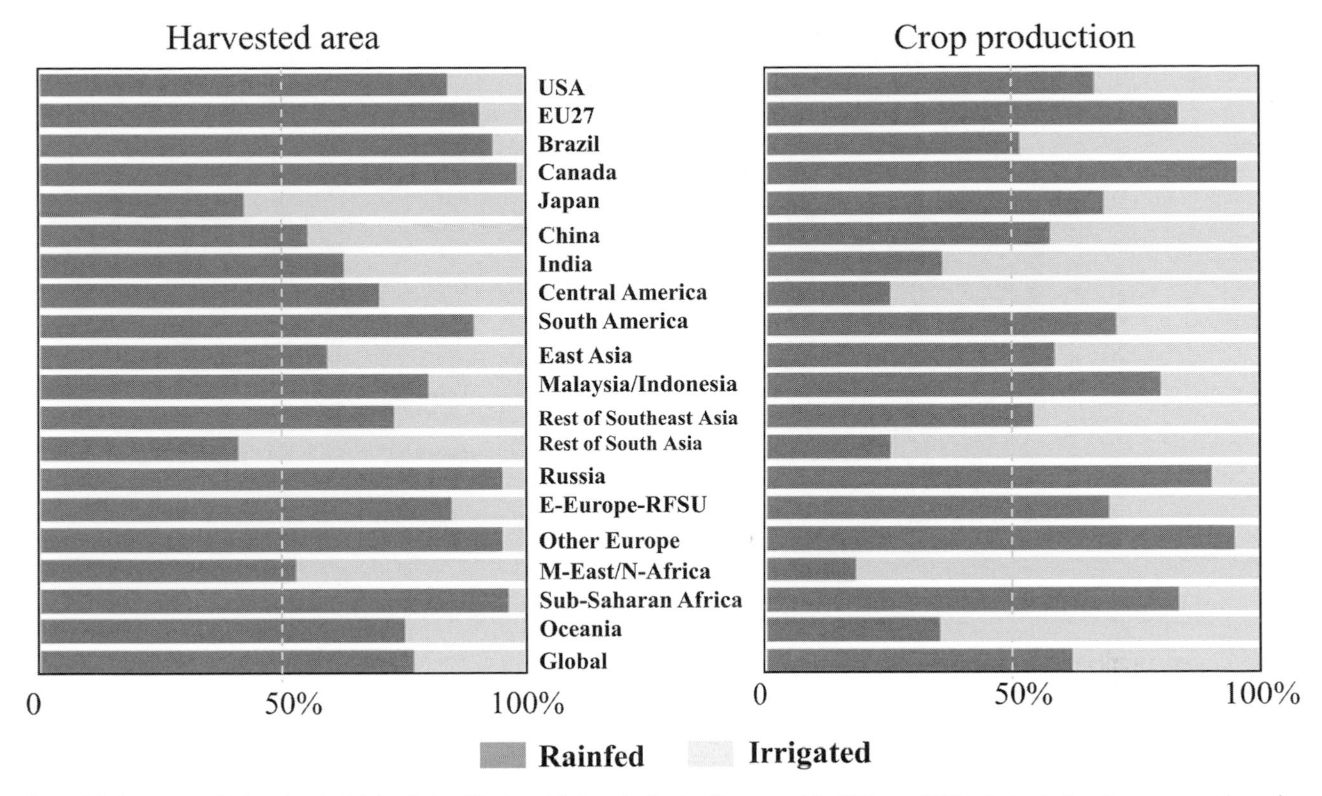

Figure 1.9 Percentage of irrigated and rainfed agricultural lands and their production in different countries (E-Europe-RFSU indicates Eastern European countries and a portion of former Soviet Union; M-East/N-Africa indicates the Middle East and North Africa). (Adapted from Taheripour et al., 2013.)

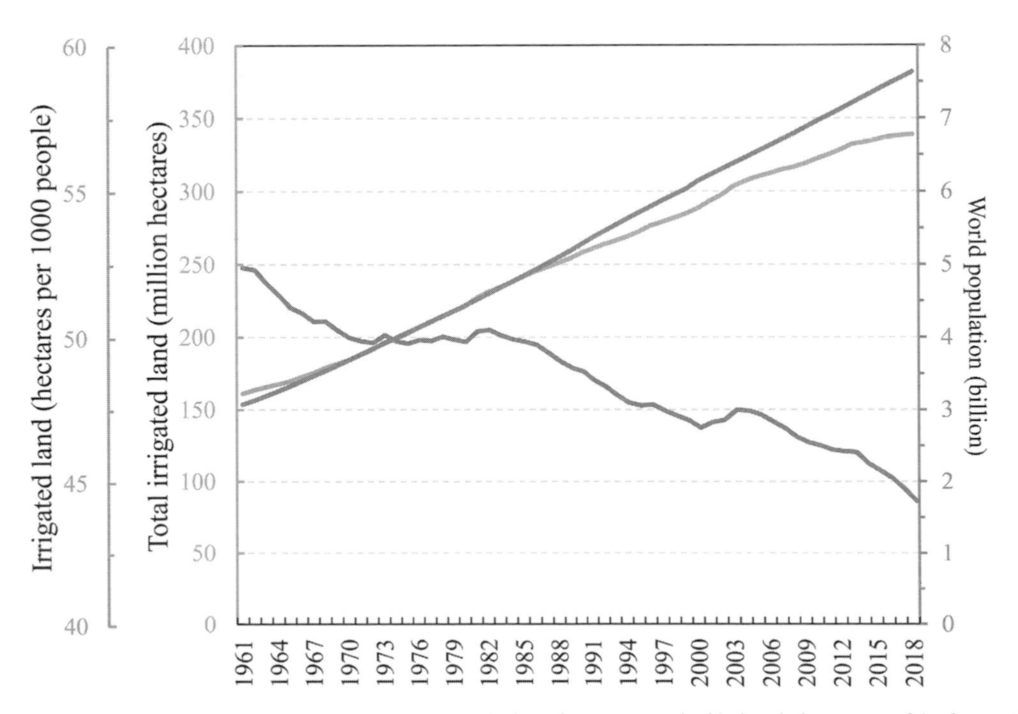

Figure 1.10 World population, world irrigated area, and hectares per 1000 people. (Data from FAO, 2020.) A black and white version of this figure will appear in some formats. For the color version, please refer to the plate section.

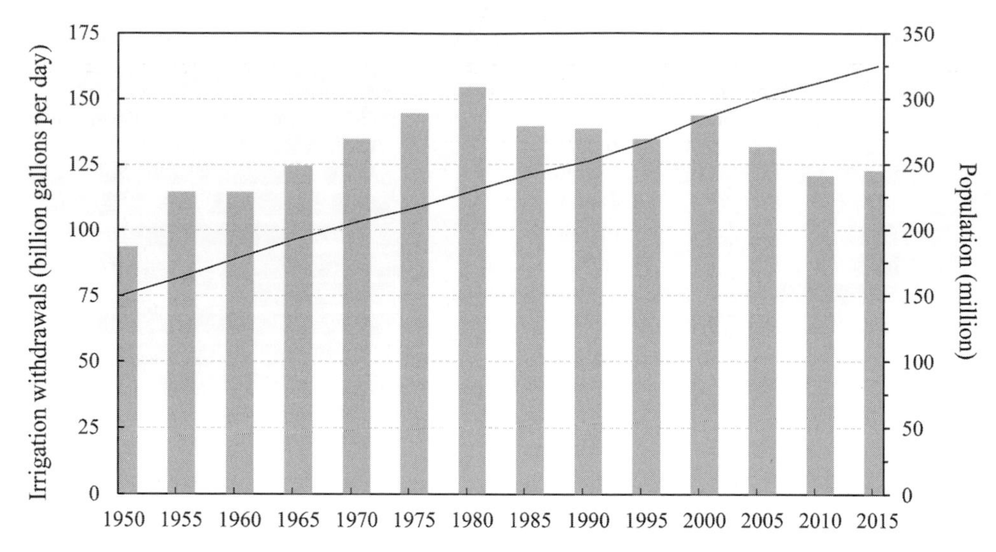

Figure 1.11 Trend of irrigation withdrawals with increasing population in the United States. (Data from US Geological Survey, http://water.usgs.gov/edu/wuir.html.)

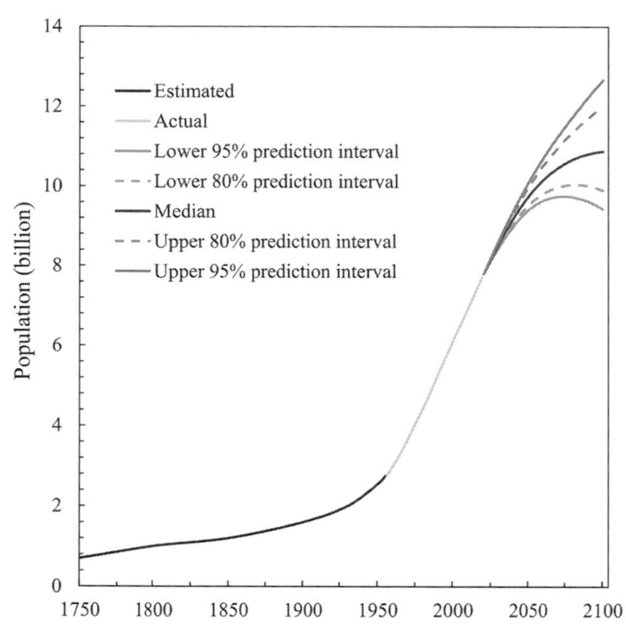

Figure 1.12 Population growth as a function of time. World population estimates from 1750 to 2100, based on the probabilistic median, and the upper and lower 80% and 95% prediction intervals of population projections. (Data from United Nations, DESA, Population Division, World Population Prospects 2019, http://population.un.org/wpp.) A black and white version of this figure will appear in some formats. For the color version, please refer to the plate section.

1.8.3 Rising Living Standards

The usual connotation of living standards has to do with the quality of life, which is measured by a number of factors, such as income, housing, healthcare, education, environmental quality, infrastructure, freedom, and social and cultural environment. There has been a substantial rise in the standard of living for the past 50 years all over the world, and this has translated into greater food and fiber requirements. Associated with this rise in the standard of living in rapidly developing economies is a steady increase in the demand for meat products and meat consumption (as shown in Table 1.5). For example, in China meat consumption rose from 20 kg/capita in 1995 to 99 kg/capita in 2017, increasing pressure on livestock production and water withdrawals. Consumption by an American is, on average, nearly the same as the consumption by nine Nepalis (Table 1.5).

1.8.4 Nutritional Security

Lack of food security is not the main or even the sole cause of malnutrition or lack of nutritional security. Many developing countries produce enough food to combat hunger, but a significant proportion of their population suffers from a lack of nutritional security, which encompasses malnutrition and obesity. This may partly be due to the lack of understanding of nutritional security, male domination resulting in gender discrimination, social taboos, lack of proper health education, corruption, and national pride. Nearly 3.1 million children under the age of five die each year because of malnutrition, accounting for about 45% of child mortality. About two-thirds of the world's malnourished people live in Asia and about one in four people living in sub-Saharan Africa is malnourished. The prevalence of undernourishment is the highest in Africa (as shown in Figure 1.15).

1.9 DEVELOPMENT OF IRRIGATION WORLDWIDE

Irrigation was introduced thousands of years ago in ancient civilizations. For example, irrigation started nearly 6000 years ago in Mesopotamia, 5000 years ago in the Nile River valley in Egypt, 4000 years ago in the Yellow

Table 1.3 *Population pressure in the top water-scarce areas of the world*

Country	Population 2010 (thousands)	Projected population 2035 (thousands)	Per capita water availability 2035 (m^3/person/year)
Kuwait	2,737	4,328	4.6
United Arab Emirates	7,512	11,042	13.6
Qatar	1,759	2,451	21.6
Bahamas	343	426	46.9
Saudi Arabia	27,448	40,444	59.3
Bahrain	1,262	1,711	67.8
Libya	6,355	8,081	74.3
Maldives	316	392	76.6
Yemen	24,053	46,196	88.8
Singapore	5,086	6,098	98.4

Data from Population Action International, http://pai.org/wp-content/uploads/2012/04/PAI-1293-WATER-4PG.pdf.

Table 1.4 *Per capita food consumption (kcal/person/day) across the world*

Region	1961–1970	1971–1980	1981–1990	1991–2000	2001–2010	2010–2013
World	2,301	2,413	2,585	2,663	2,778	2,876
Africa	2,042	2,133	2,248	2,380	2,526	2,619
Asia	1,945	2,094	2,356	2,512	2,634	2,768
Europe	3,153	3,292	3,350	3,200	3,329	3,366
North America	2,940	3,092	3,315	3,591	3,718	3,654
South America	2,401	2,535	2,609	2,728	2,889	3,022
Oceania	3,062	3,048	3,057	3,051	3,108	3,201

Data from FAO, 2020.

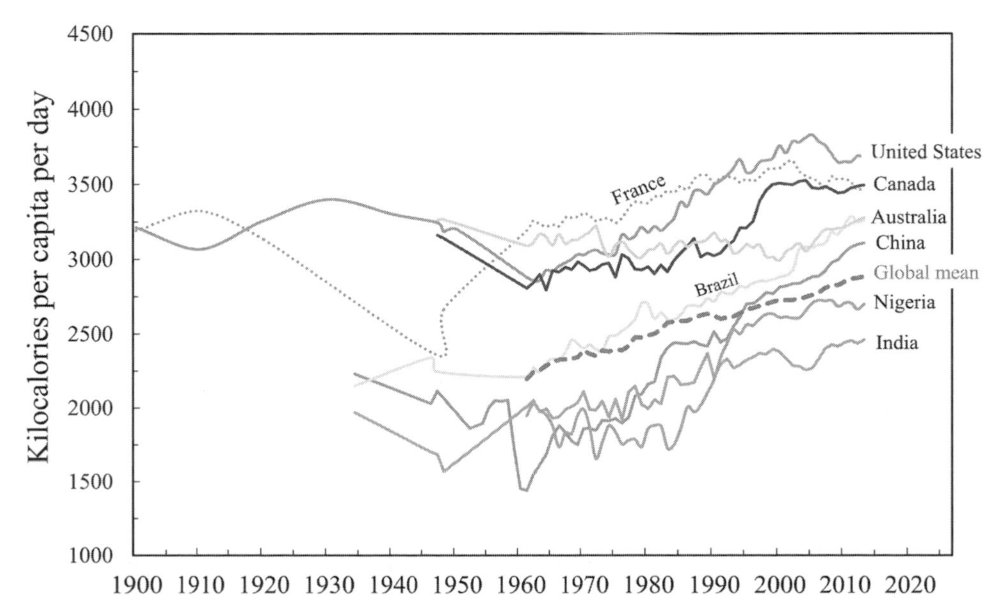

Figure 1.13. Country-level food consumption as a function of time. (Data from FAO, 2020.) A black and white version of this figure will appear in some formats. For the color version, please refer to the plate section.

Table 1.5 *Meat consumption in 2017*

Rank	Country	Meat consumption (kg/capita/year)	Rank	Country	Meat consumption (kg/capita/year)
1	Iceland	181.7	**39**	**China**	**98.8**
2	Macao	159.5	**65**	**Mexico**	**79.3**
3	Portugal	150.9	66	Fiji	78.7
4	Australia	147.5	67	Kuwait	78.6
5	**United States**	**146.5**	68	Estonia	78.4
6	French Polynesia	143.3	**103**	**Egypt**	**50.0**
7	Samoa	143.0	104	Moldova	49.7
8	Spain	142.7	105	Peru	49.4
9	South Korea	125.7	106	Iran	47.7
10	New Zealand	125.6	107	Jordan	43.8
11	Saint Lucia	125.0	108	Turkey	43.7
12	Antigua and Barbuda	124.6	**165**	**Nepal**	**16.4**
13	Israel	122.9	166	Nigeria	16.3
14	Norway	118.8	167	Tanzania	16.0
32	**Canada**	**105.1**	168	Niger	14.8
33	Ireland	102.1	169	India	10.7
34	Austria	101.3	170	Afghanistan	8.8
35	Vietnam	101.0	171	Ethiopia	5.8
36	Germany	100.5			
37	United Kingdom	99.6			

Bold text is for the purpose of highlighting specific countries and showing where the ranking is discontinuous.
Data from https://vegetarian.procon.org/view.resource.php?resourceID=004716.

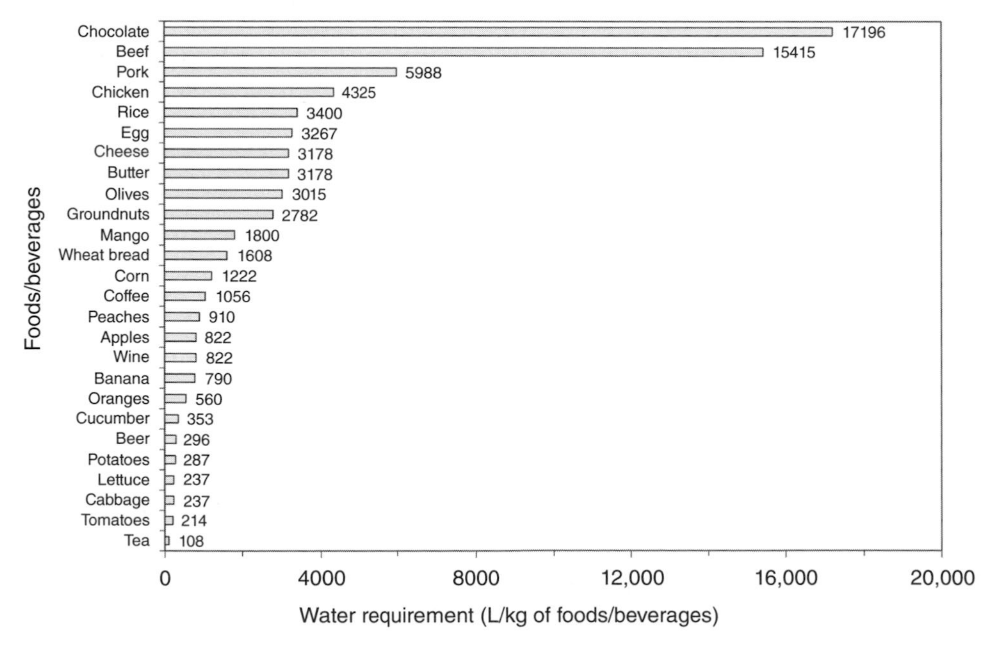

Figure 1.14 Global average water footprint of various foods/beverages. (Data from Water Footprint Network, https://waterfootprint.org/en/resources/interactive-tools/product-gallery.)

River valley in China, 3500 years ago in the Indus River valley in India and Pakistan, 2300 years ago in Mexico, 2000 years ago in Central America, and 1000 years ago in North America. Without irrigation it would have been almost impossible to grow enough food to feed the world, and this is even more true now and will be so in the future, as millions of people are added each year to the already large population.

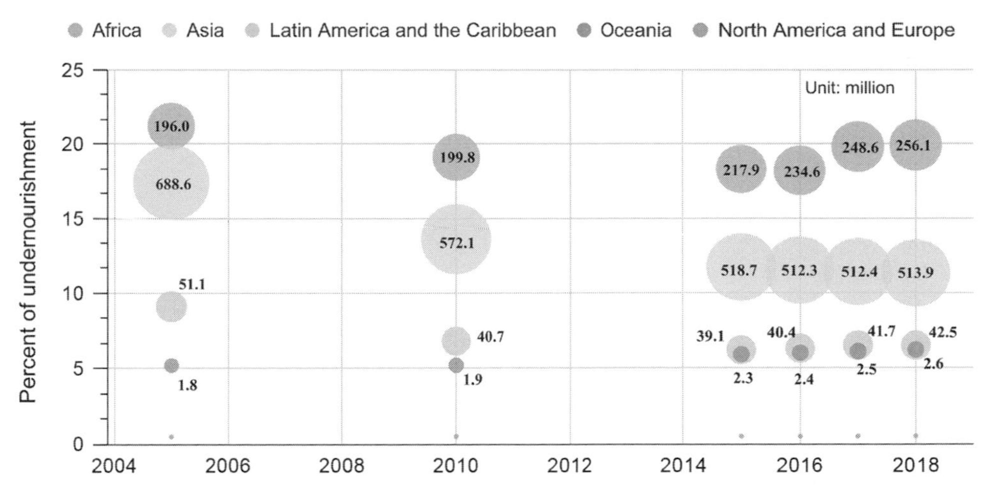

Figure 1.15 The absolute number of undernourished people. The size of the bubble indicates the total number of undernourished people in the unit of millions. (Data from FAO et al., 2019.) A black and white version of this figure will appear in some formats. For the color version, please refer to the plate section.

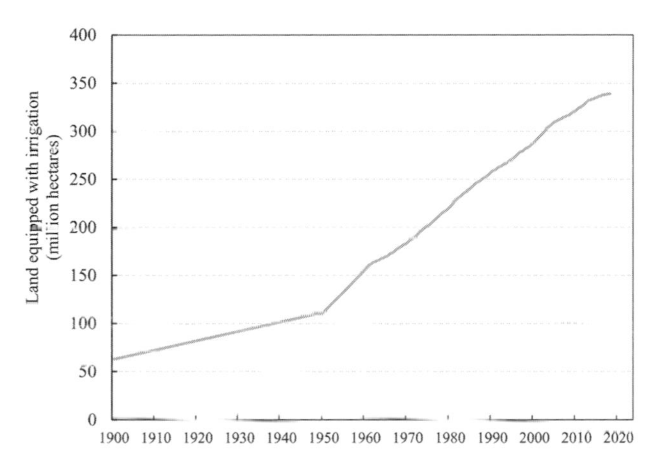

Figure 1.16 Increase in irrigated land during the twentieth century. (Data from FAO, 2020; estimated data for 1900 and 1950 are from Siebert et al., 2015.)

Currently, about 20% of the world's total cultivated land is irrigated, and this irrigated land produces about 40% of the food and fiber. Asia has about 32% of arable land irrigated and Africa has only about 7%. The increase in irrigated land occurred mainly during the twentieth century, as shown in Figure 1.16. The total global irrigated area was merely 8 million hectares (ha) in 1800, and increased to nearly 40 million ha in the last quarter of the nineteenth century, and to about 270 million ha during the twentieth century. In 2018, the most recent year for which global data are available from the UN Food and Agriculture Organization (FAO), 339 million ha of land in the world were equipped for irrigation.

The percentage of irrigated land in a given country depends on climate, availability of water, energy supply, amount of arable land, and population. In the United States, about 13% of the arable land is irrigated, which is about 22.3 million ha. Uzbekistan and Egypt are the only two countries where nearly all arable land is irrigated. Bangladesh, Iran, Iraq, Japan, and Pakistan have more than half of their arable land irrigated. China and India together represent about 40% of the global arable land. Figure 1.17 showcases the top 20 countries in total irrigated land in 2018.

1.10 IRRIGATION IN THE UNITED STATES

Although irrigation in the Southwestern United States existed by about 100 BCE, the expansion of irrigation occurred along with the settlement of the West, and much of the expansion occurred in the twentieth century with the support of the federal government. Irrigated land increased from one million ha in the 1880s to 8 million ha by the middle of the twentieth century, primarily in the Southwest, the Mountain States, and the Pacific Northwest (US Department of Commerce, 1983). In the second half of the twentieth century, irrigation expanded to the southern Great Plains, central Great Plains, and southeastern states, largely triggered by developments in irrigation technologies, such as sprinklers. Figure 1.18 shows irrigated areas in different states.

1.11 IRRIGATION PRACTICE IN THE UNITED STATES

The practice of irrigation has changed over time. In the beginning, water was diverted from streams by ditches dug by hand. Water was also withdrawn from open dug wells. Then, water storage reservoirs and canal systems were built. Thereafter, tubewells were developed. Then came sprinklers and microirrigation systems. In the latter half of the twentieth century, sprinkler technology along with low-cost aluminum and PVC pipe became popular. Now, in many areas, more cropland is irrigated by sprinklers than by surface irrigation methods. Table 1.6 shows cropland irrigated by various irrigation methods.

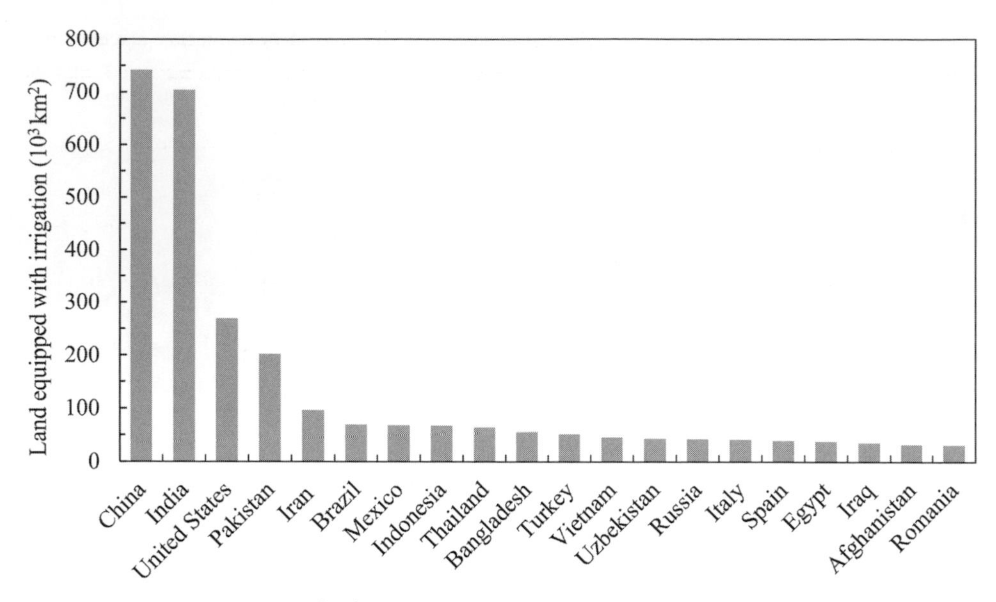

Figure 1.17 Top 20 nations by irrigated land in 2018 (10^3 km^2). (Data from FAO, 2020.)

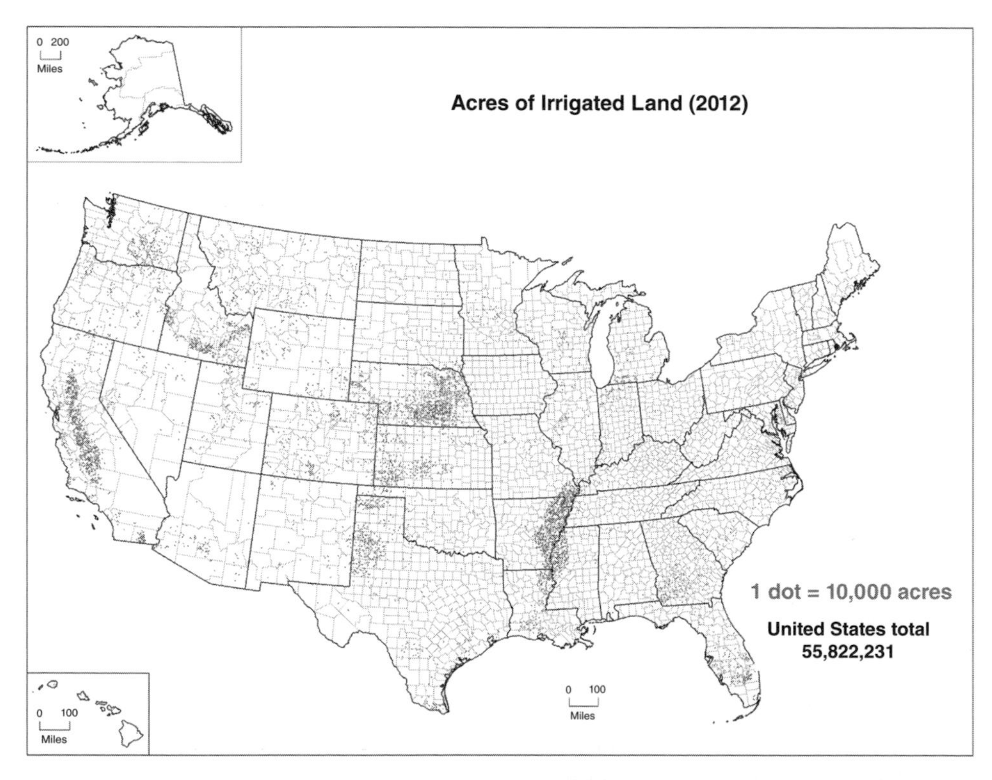

Figure 1.18 Map of irrigated cropland in the United States. (Adapted from US Department of Agriculture, www.nass.usda.gov/Publications/AgCensus/2012/Online_Resources/Ag_Atlas_Maps/Farms/Land_in_Farms_and_Land_Use/12-M080-RGBDot1-largetext.pdf.)

1.12 IMPACT OF GLOBAL WARMING AND CLIMATE CHANGE

It is now well accepted that the globe is warming and climate is changing, and will continue to change in the foreseeable future. From an agricultural point of view, the rise in temperature translates into more evaporation and evapotranspiration and changing patterns of precipitation and cropping patterns. Crop-growing seasons may also shift. The hydrologic cycle may be undergoing change. Hydrologic extremes,

Table 1.6 *Cropland irrigated by various methods in the United States: comparison of irrigation methods in 2013*

Irrigation method	Irrigated area (acre)	Percentage of total
Gravity systems		
Furrow	10,485,453	
Border/basin	8,487,054	
Uncontrolled flooding	1,801,259	
Other	730,918	
US total, gravity systems	21,504,684	35.1
Sprinkler systems		
Center pivot, pressures above 60 psi	1,172,234	
Center pivot, pressures 30–59 psi	13,396,454	
Center pivot, pressures <30 psi	12,770,489	
Linear move tower sprinklers (low pressure, <30 psi)	257,237	
Linear move tower sprinklers (low pressure, >30 psi)	368,329	
Solid set and permanent sprinklers (low pressure, <30 psi)	341,288	
Solid set and permanent sprinklers (low pressure, >30 psi)	1,145,451	
Side roll, wheel move, or other mechanical move	1,859,017	
Traveler or big gun	558,308	
Hand move	820,806	
Other sprinkler systems	1,664,496	
US total, sprinkler systems	34,894,109	56.9
Microirrigation systems		
Surface drip	2,583,201	
Subsurface drip	768,901	
Microsprinklers	1,269,483	
Other microsprinklers	270,327	
US total, microirrigation systems	4,889,912	8.0
Total US irrigation[a]	61,288,705	

[a] The US total irrigated area is larger than the 21.3 million ha quoted previously because more than one irrigation method may be used on some lands.

Adapted from US Department of Agriculture, 2014.

such as droughts and floods, will occur more frequently. This will pose a challenge for agriculture, agricultural irrigation, and the operation and management of irrigation systems.

1.13 ENVIRONMENTAL CONCERNS

Irrigated agriculture has both positive and negative impacts on the environment. On the positive side, it has led to increased wetlands, which serve a variety of useful purposes such as refuge for migratory and non-migratory birds, wildlife, recreation, reduction in pollution, and groundwater recharge. On the negative side, salinization of soil, water logging, declining water table, loss of aquatic and riparian habitats, decline in native species, increase in pollution, decline in fish spawning, etc. may be caused by irrigation.

Consider the case of the Aral Sea disaster in Central Asia, which is now regarded as the largest man-made disaster in the world. The Aral Sea was once the largest inland sea in the world. Figure 1.19 shows the sea in 2000 and 2011 and the approximate shoreline in 1960. In 2001 the sea had lost 60% of its area and 40% of its volume, and it continued shrinking in 2011. The Amudaz River (Amu Darya) (Figure 1.20), originating in the Hindu Kush mountains and flowing through five nations, was the largest river in Central Asia. In the 1960s, the (then) Soviet Union expanded irrigated cotton production that diverted so much water for irrigation that otherwise flowed into the Aral Sea that eventually the sea collapsed. This resulted in over 100 species of animals and fish becoming extinct and large areas of seabed contaminated with chemicals being exposed. Storms would pick up the dust and contaminants and deposit them over populated areas, increasing illnesses, especially in children and women. The fishing industry, dating back thousands of years, is now practically dead. Despite the ecological disaster, rice, which is a water-intensive crop, is still being grown in the Aral Sea region.

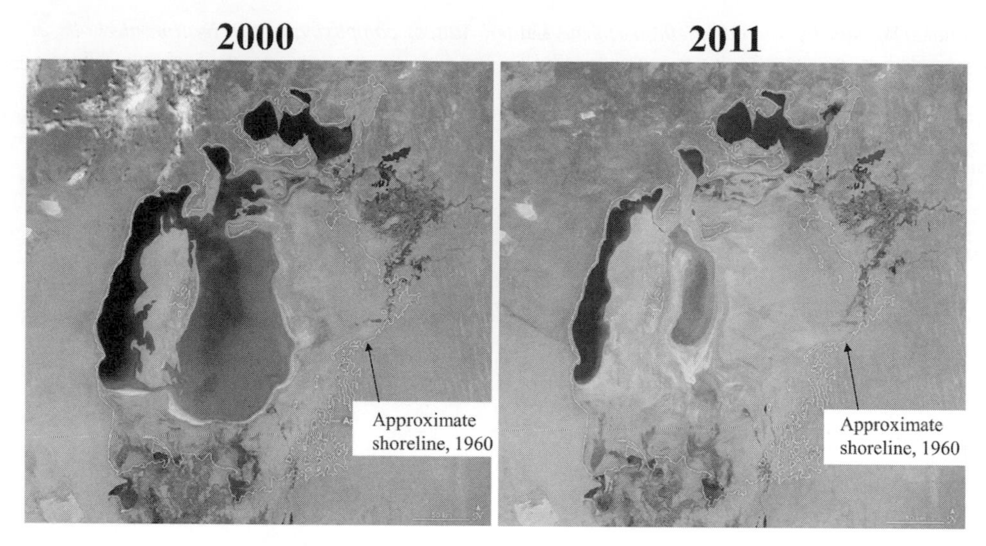

Figure 1.19 Aral Sea in 2000 and 2011. (Adapted from NASA, https://earthobservatory.nasa.gov/world-of-change/AralSea/show-all.) A black and white version of this figure will appear in some formats. For the color version, please refer to the plate section.

Figure 1.20 Amudaz River (Amu Darya), 100 miles away from the Aral Sea. A black and white version of this figure will appear in some formats. For the color version, please refer to the plate section.

1.14 FUTURE OF IRRIGATION

The population will continue to grow until it stabilizes at about 11 billion or so, and standards of living will also rise all across the world. Hence, food demand will increase. In order to meet the growing food and fiber demand, agricultural production will have to be increased. Production can be increased by developing higher-yield varieties, increasing irrigated agriculture, and improved irrigation technology. The only viable option – at least in the short term – to meet increasing food demand is by irrigation, entailing both increased irrigated land and improved irrigation technology and better management of irrigation systems.

In the future, the pressure on available water resources will increase, and other sectors, such as industry, domestic sectors, energy generation, environmental quality, ecosystem sustainability and integrity, and recreation, will demand more water. This will lead to a decline in the amount of water available for agriculture. For sustained agricultural productivity, irrigation technology will have to be more efficient and better managed and will have to compete with these sectors. Water allocated for irrigation will have to be justified, and this might involve water pricing.

QUESTIONS

Q.1.1 What is the percentage diversion of water for irrigation, industrial use, municipal use, livestock use, and other uses?

Q.1.2 What proportion of water for irrigation in the United States is supplied by groundwater aquifers and by surface streams and reservoirs?

Q.1.3 What is the yearly increase in population in Texas, United States?

Q.1.4 What was the annual population level in the United States from 2000 onwards?

Q.1.5 Texas has witnessed a significant growth of irrigated land. Comment on this growth.

Q.1.6 Provide the irrigated acreage since 1969 in the United States.

REFERENCES

Beck, H. E., Zimmermann, N. E., McVicar, T. R. et al. (2018). Present and future Köppen–Geiger climate classification maps at 1-km resolution. *Nature Scientific Data*, 5: 180214.

Brabeck-Letmathe, P., and Biswas, A. K. (2015). The nutritional security imperative. The Diplomat. https://thediplomat.com/2015/10/the-nutritional-security-imperative.

Falkenmark, M., and Widstrand, C. (1992). Population and water resources: a delicate balance. *Population Bulletin*, 47(3): 1–36.

FAO (2011). *Global Food Losses and Food Waste: Extent, Causes and Prevention*. Rome: FAO.

FAO (2019). Food Loss and Waste Database. www.fao.org/platform-food-loss-waste/flw-data/en.

FAO (2020). FAOSTAT Database. www.fao.org/faostat/en/#data.

FAO, IFAD, UNICEF, WFP, and WHO (2019). *The State of Food Security and Nutrition in the World 2019. Safeguarding Against Economic Slowdowns and Downturns*. Rome: FAO.

Siebert, S., Kummu, M., Porkka, M., et al. (2015). A global data set of the extent of irrigated land from 1900 to 2005. *Hydrology and Earth System Sciences*, 19: 1521–1545.

Taheripour, F., Hertel, T. W., and Liu, J. (2013). The role of irrigation in determining the global land use impacts of biofuels. *Energy, Sustainability and Society*, 3(1): 4.

United Nations Population Division (2019). World population prospects 2019. www.population.un.org/wpp.

US Department of Agriculture (2002). *Census of Agriculture*. Washington, DC: National Agricultural Statistics Service. www.agcensus.usda.gov/Publications/2002/Volume_1,_Chapter_1_US/st99_1_033_033.pdf.

US Department of Agriculture (2014). *Census of Agriculture*. Washington, DC: National Agricultural Statistics Service. www.nass.usda.gov/Publications/AgCensus/2012.

US Department of Commerce (1983). United States summary and state data. In *1982 Census of Agriculture*, Vol. 1. Washington, DC: US Department of Commerce.

2 Climate

Notation

A area (m^2)
c_p specific heat of air, $c_p = 0.2396$ cal/g·°C if $p = 1013$ mb or 1 atm
d mean distance between Earth and the sun (cm)
e vapor pressure (millibars, kPa, Pa)
e_a actual vapor pressure (millibars, kPa, Pa)
$elev$ elevation (m)
$elev_0$ base elevation (m)
e_s saturation vapor pressure (millibars, kPa, Pa)
E rate of energy production of the sun (cal/min)
F force exerted on an area (N/m^2)
g acceleration due to gravity (9.81 m/s^2)
I amount of radiation ($cal/cm^2/min$)
I_0 extraterrestrial radiation, solar radiation received at the outer Earth's surface or Earth's atmosphere (cal/cm^2·day)
L_{in} longwave radiation emitted by the atmosphere and clouds (cal/cm^2·day)
L_{out} outgoing longwave radiation from the surface (cal/cm^2·day)
m_d mass of dry air (kg)
n actual number of sunshine hours
N possible maximum number of sunshine hours
p atmospheric pressure (mb, kPa)
p_0 standard sea-level atmospheric pressure (mb)
p_d partial air pressure or dry air pressure (Pa)
p_{dry} precipitation of the driest month (mm)
P_{mean} mean annual precipitation (mm)
P_{sdry} precipitation of the driest month in summer (mm)
P_{swet} precipitation of the wettest month in summer (mm)
$P_{threshold}$ threshold of precipitation (mm)
P_{wdry} precipitation of the driest month in winter (mm)
P_{wwet} precipitation in the wettest month in winter (mm)
q specific humidity, defined as the mass of water vapor present in 1 g of moist air
r mixing ratio, defined as the mass of water vapor present per 1 g of dry air
r_e radius of the Earth (L)
R specific gas constant for dry air, 287.058 J/kg·K
R_E average longwave radiation ($cal/cm^2/min$)
RH relative humidity (percentage)
RH_{mean} mean relative humidity (percentage)
R_L net outgoing longwave radiation (cal/cm^2·day)
R_{L0} net outgoing clear sky longwave radiation (cal/cm^2·day)
R_s shortwave radiation(cal/cm^2·day)
R_{so} daily clear sky solar radiation at Earth's surface (cal/cm^2·day)
R_w specific gas constant for water vapor (461.495 J/kg·K)
S solar constant ($cal/cm^2/min$)
T absolute temperature (K)
T_0 standard sea-level temperature (K)
T_a air temperature or dry-bulb temperature (°C)
T_d dew-point temperature (°C)
T_E mean surface temperature of the Earth (°C)
T_{mean_a} mean annual air temperature (°C)
T_{mmax} temperature of the warmest month (°C)
T_{mmin} temperature of the coolest month (°C)
T_{mon10} number of months with air temperature > 10 °C (unitless)
T_s surface temperature (°C)
T_w wet-bulb temperature (°C)
u_2 wind velocity at a height of 2 m (L/T)
Uz wind velocity measured at the height of z (L/T)
v specific volume of moist air per mass unit of dry air and water vapor (m^3/kg)
v_{da} specific volume of moist air per unit mass of dry air (m^3/kg)
V total volume of moist air (m^3)
z height at which velocity U_Z is measured (m)
α albedo
β specific gravity of water vapor, which is the ratio of the molecular weight of water vapor (18) to the molecular weight of dry air (28.97), 0.622
γ psychrometric constant (the relationship between vapor pressure deficit and wet-bulb depression) (kPa/°C, mb/°C)

δ	standard lapse rate (K/m)	ρ_d	density of dry air (kg/m^3)
ε	effective emissivity	ρ_v	density of vapor (kg/m^3)
ε_a	emissivity of the atmosphere	ρ_w	density of water (kg/m^3)
ε_s	emissivity of Earth's surface	σ	Stefan–Boltzmann constant ($\sigma = 4.903 \times 10^{-9}$ MJ/m^2/day/K^4)
λ	latent heat of vaporization (cal/g, kJ/kg, MJ/kg)		
ρ	density of moist air (kg/m^3)		

2.1 INTRODUCTION

The crops that can be grown in a particular area depend on the climate and weather, which are not the same. Weather is an atmospheric condition that is described by various meteorological variables, such as temperature, pressure, humidity, wind, radiation, rain, snow, sunshine, evaporation, clouds, frost, and fog, for a short period of time – an hour, a day, a week, a month, or a year. These variables vary in both space and time and their continuous measurements are needed for understanding weather. Thus, weather is defined at a given location for a certain period of time. The weather is caused by the composite effect of several atmospheric variables. People are aware of the way it affects living organisms. The values of meteorological variables depend on the conditions within the atmosphere and the atmosphere's interactions with the land and ocean surfaces.

Weather is a short-term description of the air in the area. Weather may change from hour to hour, day to day, month to month, season to season, and year to year. Climate, on the other hand, is defined by long-term weather conditions and is regarded as the average weather of a place. Since weather is determined by meteorological variables, average weather implies averaging of measurements (or observations) of these variables over several years. The period of averaging usually ranges from 30 to 60 years. Climate also varies, but it does so on a long-term basis and this variation is small and is not monotonic. Currently, there is considerable discussion on climate change due to anthropogenic factors, as shown in Figure 2.1. Climate change, as being experienced presently, is sudden and significantly departs from the natural variability of climate. Crops that can be grown in an area depend on the type of climate, and the water requirements of crops are influenced by the climate and climate change. This chapter discusses those aspects of climate that are fundamental to agricultural farming and consequent irrigation.

2.2 VARIATION OF CLIMATE

Climate varies from one place to another, depending on the latitude, elevation, ocean and wind currents, terrain, and

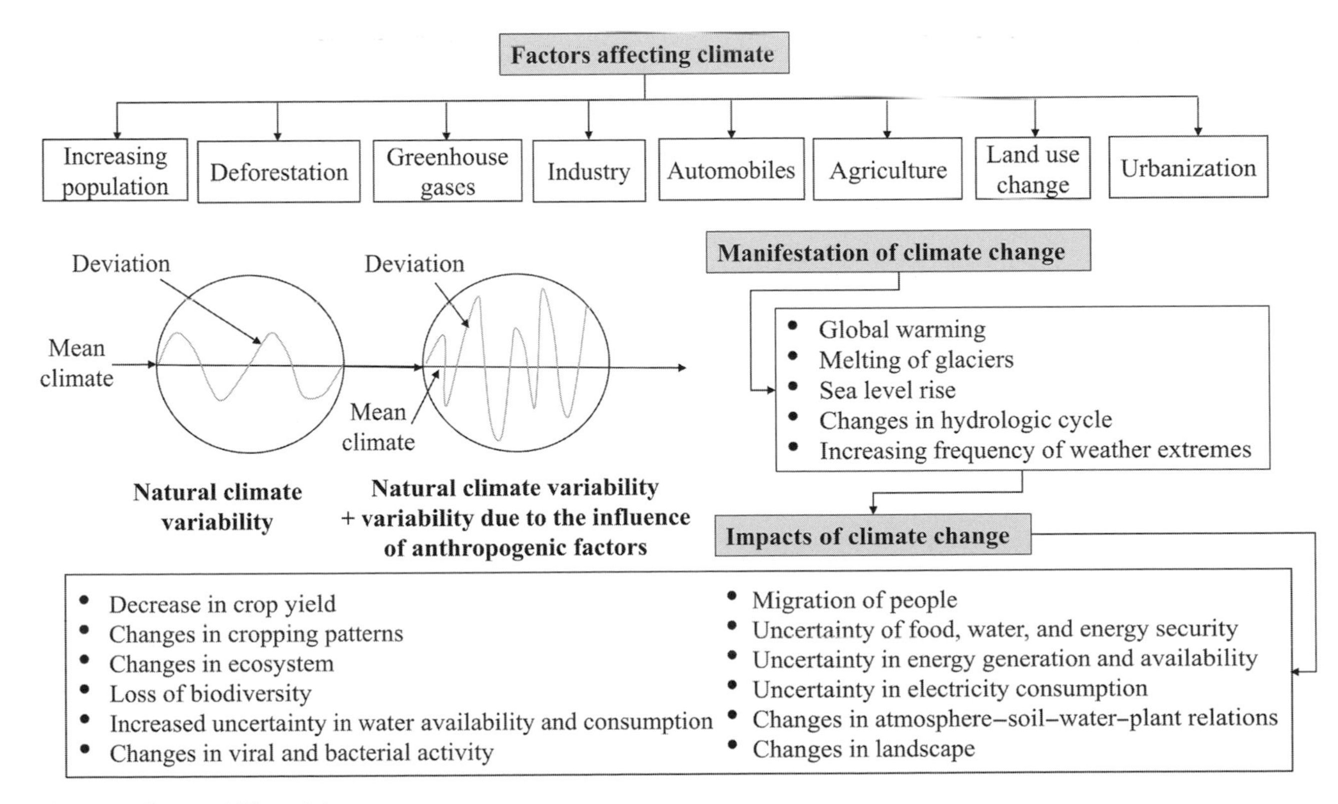

Figure 2.1 Climate variability and change.

closeness to large water bodies. The closeness to the equator is measured by latitude. The closer a place is to the equator, the warmer its climate is. The elevation of a place above the mean sea level affects temperature. The higher the elevation, the cooler the climate. Ocean and wind currents impact the climate of an area. When water is warm, warm air rises; when air is cool, it sinks. This creates wind currents that transport heat around the planet. If a place is located close to a large water body, it will have more precipitation and its temperature is affected. That is why places near seas or oceans are not extremely hot, because their climate is moderated by the seas. Mountains also have a strong influence on climate. They affect the amount of sunlight received and the movement of wind, and hence temperature and rains. The climate on the north side of a mountain is different from that on the south side.

2.3 CLASSIFICATION OF CLIMATES

There are five major types of climate in the world: tropical, dry, temperate, cold, and polar. Each climate type is further divided into subtypes, as will be seen in the discussion to follow. Likewise, there are three major climate zones in the world: tropical, temperate, and polar. The tropical climate zone extends from the Tropic of Cancer at 23.5° N; thus, tropical climate is found around the equator. The temperate climate zone extends from the southern edge of the Arctic Circle to the Tropic, so the temperate climate is found in mid-latitudes. The polar climate zone occupies areas between the Arctic and Antarctic Circles; this climate is extremely cold. There are different ways to classify climates. Here, three classifications are presented for brevity.

2.3.1 Köppen–Geiger Classification

One of the most popular systems of classification of climate is the Köppen–Geiger classification system, which classifies climates into five types, designated by upper case letters A, B, C, D, and E, as shown in Table 2.1. The criterion for defining climates A, C, D, and E is temperature, and that for climate B is vegetation. These climate types are further classified into various subtypes.

Climate A is the warmest climate, in which the temperature of the coolest month is 18 °C or higher. Based on the seasonality of precipitation, it is further divided into three subtypes, designated as Af, Am, and Aw. In subtype Af, there is no dry season even in the driest month. The precipitation is at least 60 mm. It is also called wet equatorial climate. Subtype Am corresponds to a short dry season in which the precipitation in the driest month is less than 60 mm but greater than [100 – (average annual precipitation in mm/25)]. This subtype is also referred to as tropical monsoon and trade wind littoral climate. In subtype Aw, also referred to as tropical wet climate, the winter season is dry and precipitation is less than 60 mm and less than [100 – (average annual precipitation in mm/25)]. Climates of type A are found at low altitudes, mostly between 15° north and south, and are controlled by trade winds, the Asian monsoon, and the Intertropical Convergence Zone. The difference between day and night temperatures is usually greater than the difference between temperatures of the warmest and coolest months.

Climate B corresponds to the dry climates in which vegetation is controlled by dryness or aridity, which is defined by a temperature–precipitation index. These climates encompass nearly one-quarter of the Earth's land surface, mostly between 50° N and 50° S, but mainly found in the 15–30° latitude in both hemispheres. Typical characteristics of these climates are low precipitation, high variability in yearly precipitation, intense solar radiation, clear skies, low humidity, and high evaporation. The B climate is further divided into two subtypes: arid, designated BW, and semi-arid, designated BS, which are further divided into hot (denoted by h) and cold (denoted by k), thus resulting in BWh, BWk, BSh, and BSk. Climates BWh and part of BWk are also referred to as tropical and subtropical desert climates. Climate BSh is known as mid-altitude steppe and desert climate. Climate BSk and part of BWk are regarded as tropical and subtropical steppe climate.

Climate types C and D occupy a major portion of mid- and high latitudes, mostly from 25° to 70° N and S, which lie below the upper level, mid-latitude westerlies throughout the year. These climates are further subdivided by adding a second letter: f for no dry season, w for dry winter, or s for dry summer. A third symbol – a, b, c, or d – is added to reflect the warmth of the summer or the coolness of the winter. Taken together, this classification results in six C climates (Cfa, Cfb, Cfc, Csa, Csb, and Cwa) and eight D climates (Dfa, Dfb, Dfc, Dfd, Dwa, Dwb, Dwc, and Dwd). Climates Cfa and Cwa are known as humid subtropical climate; climates Csa and Csb as Mediterranean climates; climates Cfb and Cfc as marine west coast climates; climates Dfa, Dfb, Dwa, and Dwb as humid continental climates; and climates Dfc, Dfd, Dwc, and Dwd as continental subarctic climates. In climate C the temperature of the warmest month is greater than or equal to 10 °C, and the temperature of the coldest month is greater than −3 °C. In climate D the temperature of the coldest month is less than −3 °C.

Climate E is characterized by low temperatures and precipitation and a large variety of subtypes, and is governed by polar and arctic air masses of high latitudes (60° N and S and higher). The temperature of the warmest month is less than 10 °C. This climate is further divided into two subtypes: ET, which is known as tundra climate, where the temperature of the warmest month is greater than 0 °C but less than 10 °C; and EF, which is known as frost, where the temperature of the warmest month is less than 0 °C.

Another climate, called H climate, is also included in the climate classification. It accounts for climate in regions with a high elevation greater than 1500 m.

Table 2.1 *Köppen–Geiger classification and its criterion for different climate classes*

First-level type	Second-level subtype	Third-level subtype	Description	Criterion
A			**Tropical**	Not (B) and $T_{mmin} \geq 18\,°C$
	f		Rain forest	$P_{dry} \geq 60$ mm
	m		Monsoon	$100 - P_{mean}/25 \leq P_{dry} < 60$ mm
	w		Savannah	$P_{dry} < 100 - P_{mean}/25$ and $P_{dry} < 60$ mm
B			**Arid**	$P_{mean} < 10 \times P_{threshold}$
	W		Desert	$P_{mean} < 5 \times P_{threshold}$
	S		Steppe	$P_{mean} \geq 5 \times P_{threshold}$
		h	Hot	$T_{mean\_a} \geq 18\,°C$
		k	Cold	$T_{mean\_a} < 18\,°C$
C			**Temperate**	Not (B) and $T_{mmax} > 10\,°C$ and $0 < T_{mmin} < 18\,°C$
	s		Dry summer	$P_{sdry} < 40$ mm and $P_{sdry} < P_{wwet}/3$
	w		Dry winter	$P_{wdry} < P_{swet}/10$
	f		No dry season	Not (Cs) or (Cw)
		a	Hot summer	$T_{mmax} \geq 22\,°C$
		b	Warm summer	Not (a) and $T_{mon10} \geq 4$
		c	Cold summer	Not (a or b) and $1 \leq T_{mon10} < 4$
D			**Cold**	Not (B) and $T_{mmax} > 10\,°C$ and $T_{mmin} \leq 0\,°C$
	s		Dry summer	$P_{sdry} < 40$ mm and $P_{sdry} < P_{wwet}/3$
	w		Dry winter	$P_{wdry} < P_{swet}/10$
	f		No dry season	Not (Cs) or (Cw)
		a	Hot summer	$T_{mmax} \geq 22\,°C$
		b	Warm summer	Not (a) and $T_{mon10} \geq 4$
		c	Cold summer	Not (a, b, or d)
		d	Very cold winter	Not (a or b) and $T_{mmin} \leq 38\,°C$
E			**Polar**	Not (B) and $T_{mmax} \leq 10\,°C$
	T		Tundra	$T_{mmax} > 0\,°C$
	F		Frost	$T_{mmax} \leq 0\,°C$
H			**Highland**	Elevation greater than 1500 m

Definitions of variables: T_{mean_a} = mean annual air temperature (°C); T_{mmin} = temperature of the coolest month (°C); T_{mmax} = temperature of the warmest month (°C); T_{mon10} = the number of months with air temperature > 10 °C (unitless); P_{dry} = precipitation of the driest month (mm); P_{mean} = mean annual precipitation (mm); P_{sdry} = precipitation of the driest month in summer (mm); P_{wdry} = precipitation of the driest month in winter (mm); P_{swet} = precipitation of the wettest month in summer (mm); P_{wwet} = precipitation of the wettest month in winter (mm); $P_{threshold} = 2 \times T_{mean\_a}$ if more than 70% of precipitation occurs in winter; $P_{threshold} = 2 \times T_{mean\_a} + 28$ if more than 70% of precipitation occurs in summer; otherwise $P_{threshold} = 2 \times T_{mean\_a} + 14$. Summer (winter) is the warmer (colder) six-month period of April–September and October–March. Adapted from Peel et al., 2007.

2.3.2 Types of Climate

On a global basis, there are 12 different types of climate observed on Earth, as shown in Figure 2.2. Eleven of these types can be grouped into five categories, and one does not fit any of these categories.

2.3.2.1 Tropical Wet Climate

The tropical wet areas are usually within 25° latitude of the equator. Large areas of Brazil, Indonesia, the Philippines, and the Democratic Republic of Congo possess this climate. There is only one season in this climate. Days receive direct sunlight and are warm throughout the year. The average temperature is about 80 °F (27 °C); the daytime temperature rarely goes over 93 °F (34 °C) and the nighttime temperature rarely goes below 68 °F (20 °C). Because of sunlight and high temperature, evaporation and humidity are high. There is regular rainfall throughout the year. Most tropical areas receive about 100 inches (2540 mm) of rainfall each year, but some may receive nearly 300 inches (7620 mm). Frequently it rains in the morning or early afternoon. Typical vegetation includes tropical rain forests. These forests cover about 6% of the Earth's surface, and yet produce 40% of oxygen and support nearly 40% of all plant and animal species on Earth.

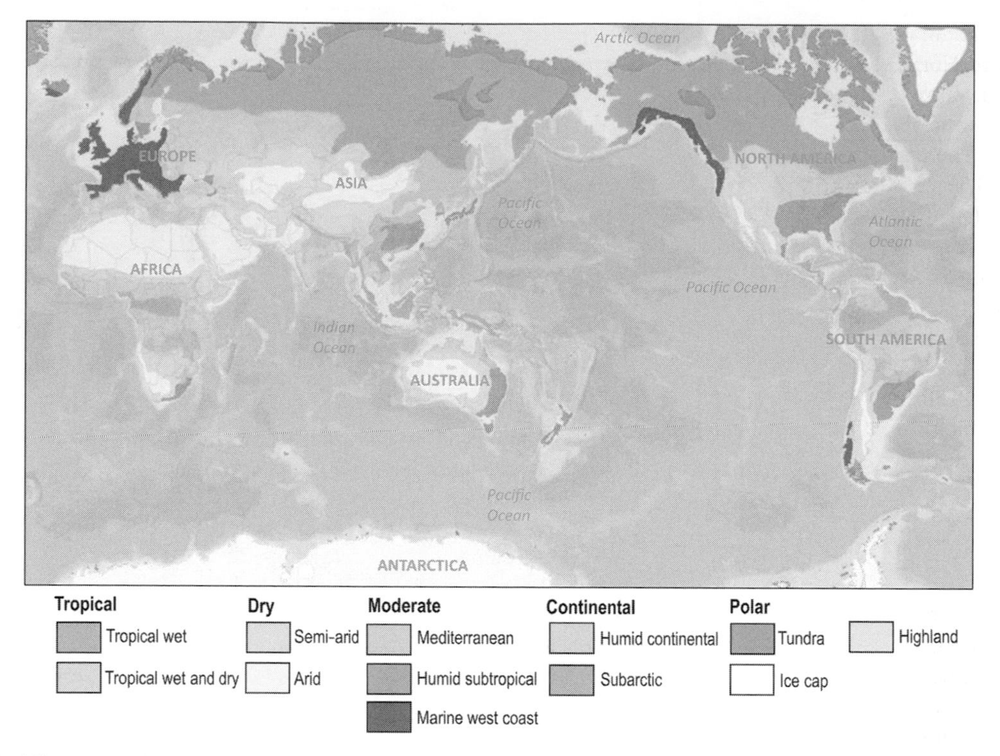

Figure 2.2 Twelve different types of climates on the Earth. (Data from www.nationalgeographic.org/mapmaker-interactive.) A black and white version of this figure will appear in some formats. For the color version, please refer to the plate section.

2.3.2.2 Tropical Wet and Dry Climate

The areas that have tropical wet and dry climates are located near the equator, on the outer fringes of tropical wet climates. The largest areas having this kind of climate are located in Brazil, India, and Africa. There are two seasons in this climate: wet season (summer) and dry season (winter). The variation in the climate is caused by changing wind and ocean currents. The average temperature is about 77 °F (25 °C) during the wet season and around 68 °F during the dry season. Rainfall occurs only during summer, usually from May through August or September, with June and July having the heaviest rainfall. At least 25 inches (635 mm) of rainfall occurs during the wet season. Sometimes unusually high rainfall occurs. Northern Madagascar once received 71 inches (1803 mm) of rainfall in one day. Cherrapunji in India received nearly 1000 inches (25 400 mm) of rain in one year. Typical vegetation includes grasses and shrubs with scattered trees.

2.3.2.3 Humid Subtropical Climate

The areas having humid subtropical climate are located at 20–40° N and S. They are located on the eastern sides of the continents, such as Florida in the southeastern United States. This climate has two seasons: summer and winter. Summers are hot and long and winters are mild and short. The coldest winter temperatures are around 40–50 °F (4–10 °C) and summer temperatures average 70–80 °F (21–27 °C). The warmest summer months are about 72 °F (22 °C). Large bodies of water keep the climate cool. Rain falls throughout the year and averages around 48 inches (1219 mm) per year. The areas with this kind of climate are known to have high evaporation and humidity. They experience strong storms, such as hurricanes and tornadoes. Typical vegetation includes trees, bushes and shrubs, palm trees, and ferns.

2.3.2.4 Mediterranean Climate

The areas with a Mediterranean climate are located between 30° and 45° latitude. They are on the western side of the continent. The climate is very mild with a few temperature extremes and has only two seasons: summer and winter. Summers are long, warm to hot, and dry, and winters are short, cool, and wet. Winter temperatures are around 30–65 °F (–1 to 18 °C) and summer temperatures over 50 °F (10 °C). The warmest summer months are about 72 °F. Large bodies of water keep the climate cool. There is almost no rainfall during summer, and around 20 inches (508 mm) during winter; snow may occur. Typical vegetation includes fruit trees and vines, such as grapes, figs, olives, and citrus fruits, pine and cypress trees, oaks, shrubs, grasses, and herbs.

2.3.2.5 Marine West Coast Climate

The areas with marine west coast climates are located in mid-latitudes – that is, midway between the Tropics and the Arctic/Antarctic Circles. The climate is influenced by the presence of mountains. Such areas are found more in Europe than in North America. In North America, mountains block the humid air from moving inland. There are two seasons: summer and winter. Temperatures do not have a large range. The climate is mild, with a few extremes in temperature. The coldest winter temperatures are seldom below 30 °F (–1 °C) and summer temperatures average around 72 °F (22 °C). It is similar to the Mediterranean climate, but is different in the amount of precipitation. The amount of rainfall varies with area – some areas receive around 30 inches (762 mm) each year, while others receive as much as 98 inches (2489 mm). In some places rain falls about 150 days out of 365 days. Typical vegetation includes thick forests and a wide variety of plant life, such as spruce, cedar, pine, and redwood.

2.3.2.6 Humid Continental Climate

The areas with humid continental climate are located between 30° and 60°, in the interior of continents, usually above the 40° line. They are found in the northern hemisphere. Northern Indiana in the United States is a typical example of this kind of climate. There are four seasons: warm and humid summer, cool and dry autumn (fall), cold and harsh winter, and warm and wet spring. There is a wide range of temperatures. Winter temperatures average around 25 °F (–4 °C), and summer temperatures average around 71 °F (22 °C). Summer months may be above 100 °F (38 °C). These areas are far away from the equator. For example, 40° latitude means about 2800 miles from the equator. The amount of rainfall varies between 20 and 50 inches (508–1270 mm) per year. Precipitation falls throughout the year and there can be plenty of snowfall. Evaporation from land is slow but regular. Typical vegetation includes a wide variety of plants and evergreen forests. This climate is excellent for farming.

2.3.2.7 Arid Climate

The areas with arid climate represent about 30% of the Earth and are located along the 30° latitude line, north and south of the equator. They are usually desert, found in the center of continents or in the rain shadow of mountain ranges. They have no regular seasons but usually have summer and winter. Some places are hot and dry. Temperatures can reach up to 130 °F (54 °C) and fall as low as 30 °F (–1 °C), depending on the latitude. Areas located farther from the equator have a colder climate. There is a lack of precipitation. The areas have less than 10 inches (254 mm) of rain each year, and some receive less than 10 inches (254 mm) in 10 years. An example is the Atacama Desert in Chile, which is the driest place on Earth, averaging around 0.04 inches (1 mm) of rain per year. Typical vegetation includes thorny and long root-scrub bushes, grasses, and cacti.

2.3.2.8 Semi-arid Climate

The areas with semi-arid arid climate are located on the outer edges of arid climate areas – they represent the transition between dry and wetter climate areas. The Sahel in sub-Saharan Africa is an example. Usually, they have two seasons: summer and winter, and winter is wet. The temperature depends on the latitude. Temperatures in the northern United States can be as low as –15 °F, but in northern Australia or African Sahel, temperatures remain high. The average precipitation varies between 10 inches (254 mm) and 20 inches (508 mm) per year. Some places have 5–10 inches (127–254 mm) of rain each year and some receive 20–40 inches (508–1016 mm). Typical vegetation includes major grasses and shrubs, thorny and long root-scrub bushes, grasses, and cacti.

2.3.2.9 Subarctic Climate

The areas with subarctic climate are located between 50° and 70° latitude and are in the interior (not coastal) of high-latitude continents. These areas are found in the northern hemisphere, because the southern hemisphere does not have any large continents in high latitudes. There are two seasons in this climate: summer and winter. Summer is cool to mild, lasting only 2–3 months, but winter is much longer and extremely cold. Sometimes the summer may be only one month long. Temperatures have a very wide range, reaching –40 °F (–4 °C) in the winter and going as high as 85 °F (29 °C) in the summer, resulting in a 125 °F (33 °C) temperature range. The reason is that these areas are away from the coast, so the ocean waters do not help warm the land in the winter. The temperatures are mainly determined by latitude. Because of the very low temperatures, there is little evaporation and very little precipitation. The average annual rainfall is 10–20 inches (254–508 mm), and much of this occurs in summer, when evaporation is high. For most of the year, these areas are covered in snow. During warm summer most of the snow is melted. Typical vegetation that can survive the harsh cold winters includes evergreen trees (conifers), such as pine and spruce; ferns, shrubs, and grasses are found in summer.

2.3.2.10 Tundra Climate

The areas having tundra climate are located between the 60–75° latitude lines. This climate is a transition climate between ice caps and subarctic, and is mainly found along the coast of the Arctic Ocean. There are two seasons: summer and winter, where summer is cool and winter is very harsh. Much of the snow and ice melts during summer, but some of the deeper layers of the soil remain frozen, even during summer, and these layers form what is called permafrost (permanent frost). The permafrost can be as thick as 10–35 inches (254–889 mm), and it prevents the melted snow and ice from draining into the groundwater, forming soggy marshes and bogs. Winter temperatures are usually between –50 and –18 °F (–46 to –28 °C) and summer temperatures range from 35 to 50 °F (2–10 °C). Because these areas are located in high latitudes, temperatures are low. These areas receive indirect sunlight, which provides light but little heat, and receive precipitation of 5–15 inches (127–381 mm) per year, usually in summer, and have different types of mosses, lichens, and algae, as well as grasses and low shrubs in summer. The amount of evaporation is low because of the low temperatures.

2.3.2.11 Ice Cap Climate

The areas having ice cap climate are located near the poles and cover nearly 20% of the Earth. They are located in Antarctica and the land around the Arctic Ocean, especially Greenland. The climate is the most extreme on Earth and has extreme seasons. There are two seasons – summer and winter – determined by the amount of light. These areas receive only indirect sunlight. Since the pole is pointed toward the sun during summer, there is nearly 24 hours of light daily. The opposite is true during winter, when the pole is facing away from the sun, causing nearly 24 hours of darkness. Antarctica is usually colder than the Arctic so the climate is the coldest climate on Earth. The average temperature during the warmest month is about –16 °F (–27 °C), and during the coldest month it is about –70 °F (–57 °C). Temperatures regularly reach –90 °F (–68 °C) during winter and the coldest temperature ever recorded was –128.6 °F (–89.2 °C). The areas receive less than 10 inches (254 mm) of precipitation. It is too cold to evaporate water, so the humidity is low. Plant life is in the form of moss and lichen – a mixture of fungus and algae.

2.3.2.12 Highland Climate

This climate is found in high mountain areas, hence term "highland." Examples of such highlands are the Tibetan Plateau and single mountains such as Mount Kilimanjaro. The Tibetan Plateau, the largest area of highland on Earth, is located at about 14,800 feet (4500 m) above sea level. Sometimes this climate is called alpine climate. The seasons are not well defined, but at low elevations near the bottom of a mountain seasonal differences are observed. The temperature in highland climates depends on elevation, because the temperature drops about 3 °F for every 1000 feet in elevation. It is likely that at the base (bottom) of a mountain it might be 80 °F and sunny, but up the mountain it will get colder and be rainy. Further up, it might be snowy and freezing cold. The amount of precipitation in highland climates depends on the elevation. The land around the base of a mountain is dry, but snow may cover the top of a mountain. This occurs because high mountains force warm air to rise, where it cools and causes precipitation. The type of vegetation that grows in highlands also depends on the elevation. At the base (bottom) of the mountain the vegetation will be the same as the surrounding climate type. For example, a mountain in the rain forest will have rain forest at the base. Farther up the elevation, however, it will change to plants that can survive colder weather, and at very high elevations there will be no vegetation. On high mountains one can witness a tree line or timberline, which is a clear elevation line that divides areas where trees are able to survive from areas that are too harsh for them to live. Most mountains exhibit different tree lines, depending on temperature, soil, and moisture.

2.3.3 Holdridge Life Zones System

The Holdridge life zones system is an empirically and objectively based global ecosystem classification system. Life zones represent different ecosystem functioning conditions, which are determined by precipitation (annual mean, logarithmic scale), potential evapotranspiration (PET) ratio, and biotemperature (annual mean, logarithmic scale). PET represents the evaporative demand of the atmosphere. The PET ratio (i.e., precipitation/PET) is the aridity index (AI), indicating a long-term climatic water deficit in a region. There are four degrees of aridity, with AI < 0.05 indicating hyperarid (7.5% of global land area), $0.05 <$ AI < 0.20 indicating arid (12.1% of global land area), $0.20 <$ AI < 0.50 indicating semi-arid (17.7% of global land area), and $0.50 <$ AI < 0.65 indicating dry humid (9.9% of global land area). Biotemperature is measured as the mean temperature during the growing season, with temperatures below 0 °C and above 30 °C adjusted to 0 °C, as it assumes that net productivity ceases at these temperatures. The climatic parameters and the life zones delineation has a logarithmic relationship, which reflects the limiting factor kinetics (Holdridge, 1967). As shown in Figure 2.3, life zones can be aggregated into larger regions/zones based on humidity provinces, belts of altitude, and latitude. The latitudinal regions show the utility of the system for global-scale classification, while the altitudinal belts reflect the differences in montane conditions.

In the United States, there are 38 life zones (Lugo et al., 1990). The life zones have five latitudinal regions, from boreal to tropical, and five altitudinal belts, from lower montane to Alvar, and range from rain forest to desert in terms of humidity provinces. The largest life zone type in the United States is the warm temperate moist forest, which accounts for 23% of the total area.

2.3.4 Climates in the United States

In the United States there are nine different types of climates: BSk (cold semi-arid steppe), Cfa (humid subtropical), Cfb (oceanic climate), Csa (Mediterranean hot summer), Dfa (humid continental, hot summer), Dfb (humid continental warm summer), H (highland climate), Aw (tropical savannah), and BWh (arid hot desert). The detailed climates in the United States can be found in Figure 1.4. In Texas, there are five different types of climate, as shown in Figure 2.4.

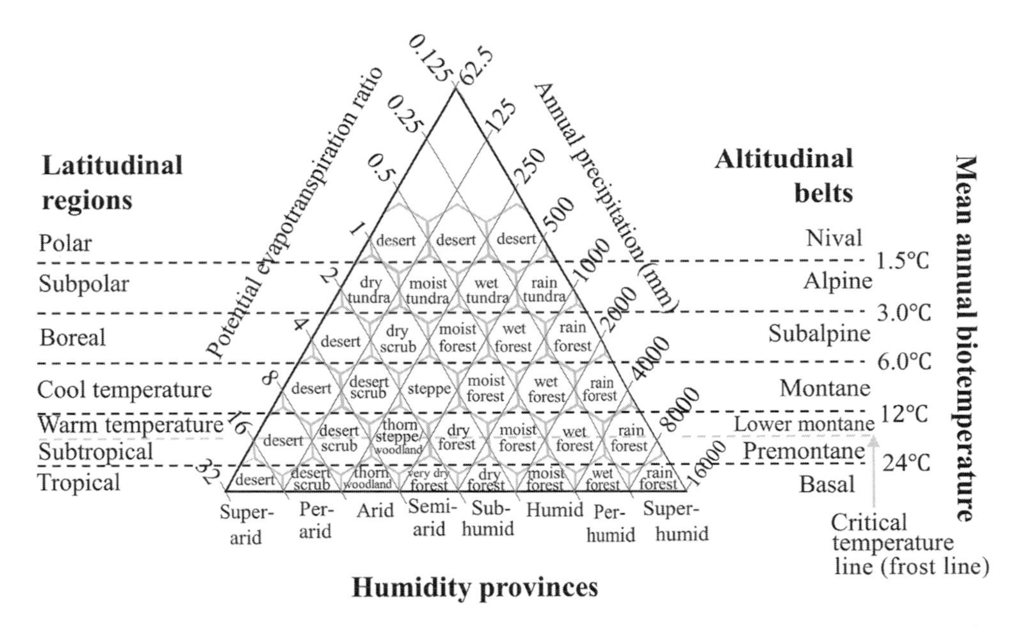

Figure 2.3 Holdridge life zone classification. (Adapted from Lugo et al., 1990.) A black and white version of this figure will appear in some formats. For the color version, please refer to the plate section.

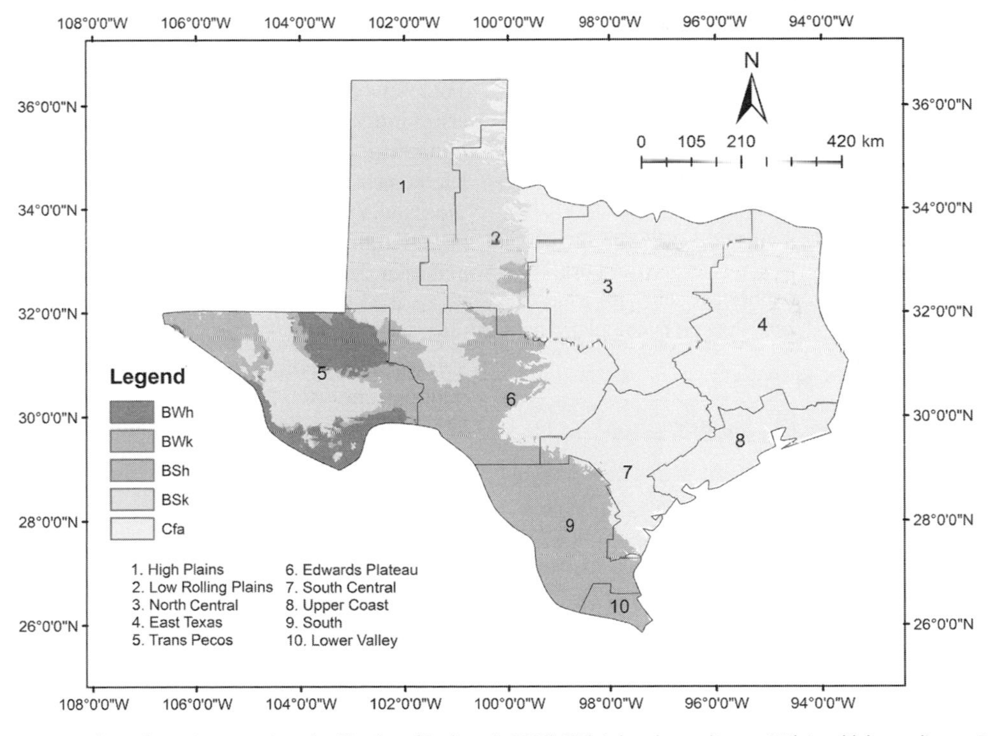

Figure 2.4 Climates in Texas. (Data from Köppen–Geiger classification of Beck et al., 2018). BWh is hot desert climate, BWk is cold desert climate, BSh is hot semi-arid, BSk is cold semi-arid, and Cfa is humid subtropical. Texas is divided into 10 climate regions. Some areas in Region 4 belong to a highland climate and are not shown in the figure. Cfa can be further divided into humid subtropical (Regions 4 and 8, part of Regions 3 and 7) and subhumid subtropical (other regions left in Cfa). The 1980–2016 global monthly datasets are used for the Köppen–Geiger classification. A black and white version of this figure will appear in some formats. For the color version, please refer to the plate section.

2.4 EARTH SYSTEM

Earth is a spheroid, slightly flattened at the poles. Its equatorial radius is 6378 km and its polar radius is 6357 km. Earth may be regarded as a sphere with a mean radius (r_e) of 6371 km as this has practically the same volume as the Earth. The equatorial circumference of Earth is: $2 \times \pi r_e = 2 \times 3.14 \times 6371 \approx 40{,}000$ km. The equator is the great circle whose plane is perpendicular to the axis of rotation of the Earth. The latitude circles are small circles whose planes are parallel to the equatorial plane. Latitudes are measured in degrees north or south of the equator from 0° at the equator to 90° at the poles. The longitude circles, also known as meridian circles, are great circles passing through the north and south poles. Longitudes are measured in degrees east or west of the meridian of Greenwich from 0° to 180° in either direction. The location of a point on Earth is specified by its latitude and longitude.

2.5 SUN

The sun is the principal source of energy (heat), providing 99.97% of the heat energy that drives the climate system. It is larger than one million Earths. Its radius is 6.96×10^5 km (6.96×10^{10} cm), which is about 109 times the radius of Earth. It is a gaseous sphere, where the gases are mostly hydrogen and helium. The sun gets its energy from the gradual conversion of hydrogen into helium by thermonuclear fusion in its deep interior under the extreme conditions of pressure and temperature that prevail there. The visible surface of the sun from which most of the radiation is emitted is called the photosphere. The temperature of the photosphere is approximately 6000 K.

Each minute the sun radiates about 56×10^{26} calories of energy. This amount of energy from the sun is usually expressed in terms of solar constant, which is defined as an amount of energy received in a unit time on a unit surface area perpendicular to the sun's rays at the outer boundary of the atmosphere at the mean distance between Earth and the sun. This distance (d) is about 1.5×10^8 km $(1.5 \times 10^{13}$ cm). The energy per unit area incident on a spherical cell with a radius of 1.5×10^{13} cm and concentric with the sun is calculated as:

$$S = \frac{E}{4\pi d^2} = \frac{56 \times 10^{26}(\text{cal/min})}{4 \times 3.14 \times (1.5 \times 10^{13})^2(\text{cm}^2)} \approx 2\ \text{cal/cm}^2/\text{min},$$

where S is the solar constant (cal/cm^2/min), E is the rate of energy production (cal/min), and d is the mean distance between Earth and the sun (cm). The solar constant is not a true constant, but fluctuates by as much as $\pm 1.5\%$ about its mean value. The rate of input of heat energy can also be expressed in terms of watts per square meter of Earth's surface (W/m^2) as:

$$S = 2\ \text{cal/cm}^2/\text{min} = \frac{2 \times 4.2 \times 100 \times 100}{60}\text{joules/s/m}^2 = 1390\ \text{joules/s/m}^2 = 1390\ \text{w/m}^2(1\ \text{joule/s} = 1\ \text{watt}).$$

The current best estimate is 1370 W/m^2.

Earth intercepts only a small fraction of the energy radiated by the sun. At any instant one half of the surface of Earth is exposed to the sun. The projected area normal to the solar rays which intercepts the radiation is πr_e^2, where r_e is the radius of Earth (6.37×10^5 km). Therefore, energy intercepted = $\pi r_e^2 S = 3.14 \times (6.37)^2 \times 10^{16} \times 2 = 2.55 \times 10^{18}$ cal/min. The energy intercepted during a complete rotation is distributed uniformly over the entire surface area of Earth $(4\pi r_e^2)$. Hence, the average solar energy that arrives at the top of the atmosphere is

$$\overline{Q_s} = \frac{\pi r_e^2 S}{4\pi r_e^2} = \frac{S}{4} = 0.50\ \text{cal/cm}^2/\text{min} = 263\ \text{kcal/cm}^2/\text{year} = 348\ \text{w/m}^2.$$

This energy flows down through the atmosphere and interacts with it, so that some energy is reflected back to space, some is absorbed and converted into heat, and some reaches the surface of Earth. The energy that arrives at the surface and is absorbed can heat the surface, evaporate water, melt snow, and heat the underlying soil by radiation processes.

2.6 SEASONS

Earth rotates from west to east about its axis in 23 hours and 56 minutes, and revolves around the sun in about 365.25 days. The rotation of the Earth produces day and night and its revolution produces seasons and different seasons in different places. For example, in the United States four seasons are commonly recognized: fall, winter, spring, and summer. There are countries where there are more seasons and there are countries where there are fewer seasons. For example, in northern India there are six seasons, each two months long: fall, winter, winter-end, spring, summer, and rainy season. In Thailand there are practically no seasons as the climate is humid tropical. Different crops are grown in different seasons and their irrigation requirements are different.

2.7 METEOROLOGICAL VARIABLES

In day-to-day language, the weather of an area is normally described by meteorological variables that include sunshine, temperature, precipitation, humidity, clouds, and wind. These variables are briefly discussed in what follows.

2.7.1 Temperature

Temperature is one of the most important meteorological variables and is a measure of the degree of hotness or coldness of the air. It is expressed in degrees and has three commonly used scales of measurement: (1) Celsius (C); (2) Fahrenheit (F); and (3) kelvin (K). In the United States we use the Fahrenheit scale and temperature is expressed in °F. In the rest of the world it is expressed in °C. When very low or very high temperature values are considered, it is convenient to use the kelvin scale. The absolute zero value corresponds to the temperature conversion of about −273 °C, beyond which no gas can be cooled any further. That is, the absolute scale commences at this scale. Then, the freezing point is 273 K and the boiling point 373 K.

Temperature is measured by a mercury thermometer. There are also other scales, such as Rankine, Newton,

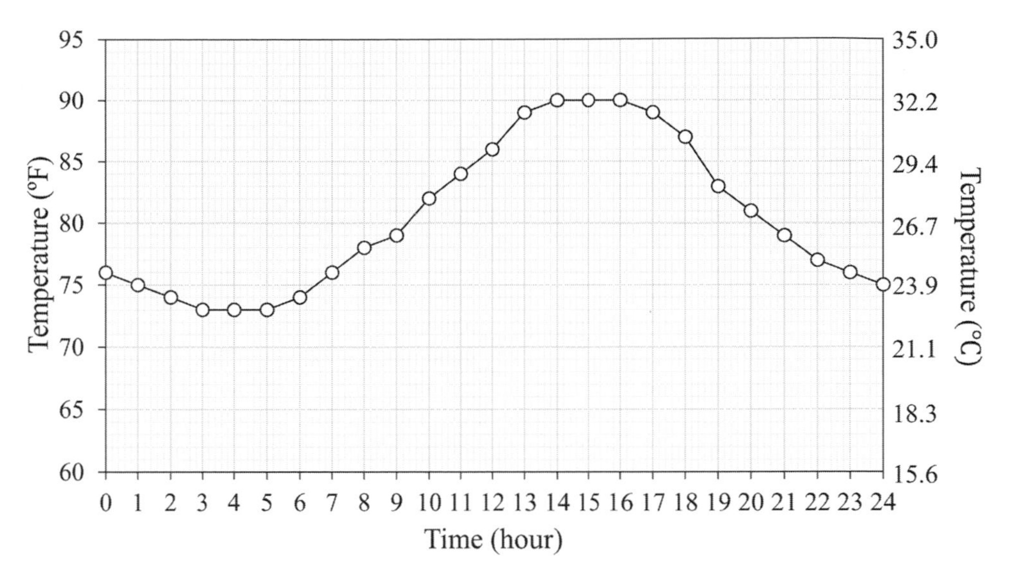

Figure 2.5 Hourly variation of temperature in College Station, Texas, November 18, 2018.

Romer, Delisle, and Reaumur. In the Rankine (R) scale, −459.67 °F is exactly equal to 0 °R. It may be noted that zero on both the kelvin scale and the Rankine scale is absolute zero. Thus, for absolute zero by definition, zero degree Rankine is equal to −273.15 °C, −459.67 °F, and 0 °R. The freezing point of water is 0 °C, 32 °F, 273.15 K, and 491.67 °R. Likewise, the boiling point of water occurs at 99.9839 °C, 211.9710 °F, 373.1339 K, and 671.6410 °R. If temperature is known at one scale, it can be transformed to another scale using the following equations:

$$\text{Between C and F}: \frac{C}{5} = \frac{F-32}{9} \tag{2.1}$$

$$\text{Between C and K}: K = C + 273.15 \tag{2.2a}$$

$$\text{Between C and R}: R = \frac{9}{5}C + 491.67 \tag{2.2b}$$

$$\text{Between F and K}: F = \frac{9}{5}K - 459.67 \tag{2.3a}$$

or

$$\frac{F-32}{9} \times 5 = K - 273.15 \tag{2.3b}$$

$$\text{Between F and R}: R = F + 459.67 \tag{2.4a}$$

$$\text{Between K and R}: R = \frac{9}{5}K \tag{2.4b}$$

The temperature of any place depends on radiation, altitude, proximity to the sea, wind, and vegetative cover. At any given place, temperature varies throughout the day. Figure 2.5 shows hourly variation of temperature in College Station, Texas. It is maximum during daytime and minimum during nighttime. The values of temperature that are useful for crops and irrigation are average, maximum, and minimum.

The US Weather Service publishes climatological data that contain normal and mean temperatures. The normal temperature for a particular timescale is the average value over a 30-year period, say 1971–2000. The timescale can be daily, weekly, monthly, seasonal, or yearly. The mean daily temperature in the United States is the average of daily maximum and daily minimum temperatures. Likewise, the mean monthly temperature is the average of the mean monthly maximum and mean monthly minimum temperatures. Mean seasonal and mean yearly temperatures are computed in a similar manner.

Temperature decreases with elevation, and this decrease defines the lapse rate. The average lapse rate in the atmosphere is about 3.8 °F/1000 ft (or 2.1 °C/328 m or 0.7 °C/100 m). The dry adiabatic lapse rate is about 5.4 °F/ft (3 °C/328 m). This is the rate at which the temperature of unsaturated air changes due to expansion or compression during descent or ascent of air. These lapse rates hold in normal atmospheric conditions.

Example 2.1: The temperature on a summer day in College Station is 95 °F. What is the value of temperature in Celsius (°C)? Determine the corresponding values of temperature in degrees kelvin and Rankine.

Solution: Given temperature = 95 °F,

temperature in Celsius (°C): $C = \frac{5}{9}(95 - 32) = 35$ °C

temperature in degrees kelvin (K): $K = C + 273.15 = 35 + 273.15 = 308.15$ K

temperature in degrees Rankine (°R): $R = F + 459.67 = 95 + 459.67 = 554.67$ °R

Example 2.2: The value of temperature on a given day is 313.15 K. What is the value of temperature in °F, °C, and °R?

Solution: Given temperature = 313.15 K

temperature in Celsius (°C): C = K − 273.15 = 313.15 − 273.15 = 40 °C

temperature in Fahrenheit (°*F*): $F = \frac{9C}{5} + 32 = \frac{9}{5}(40) + 32 = 104$ °F

temperature in degrees Rankine (°*R*): R = F + 459.67 = 104 + 459.67 = 563.67 °R

Example 2.3: A parcel of air has a temperature of 25 °C at an elevation of 500 m above the mean sea level. During its movement, the air comes across a mountain range with an elevation of 2000 m and rises above the mountain range and then descends 1000 m below. During the ascent the air reaches saturation at the elevation of 1500 m. Condensation occurs from 1500–2000 m. Compute the temperature of air if the average pseudo-adiabatic lapse rate is about half of the dry adiabatic lapse rate.

Solution: Since the air remains unsaturated from 500 to 1500 m, it cools at the dry adiabatic lapse rate of 0.7 °C/100 m. The temperature drop during a climb of 1000 m is 0.7 °C × 1000 m/100 m = 7 °C. Condensation occurs from 1500 to 2000 m so that the air cools at the pseudo-adiabatic lapse rate. The drop of the temperature during a climb of 500 m from 1500 m to 2000 m equals (0.7 °C/2)/100 m × 500 m = 1.75 °C. This air at an elevation of 2000 m has a temperature of 25 °C − 7 °C − 1.75 °C = 16.25 °C. Assuming the moisture has fallen out of the air during the condensation process, the air heats at the dry adiabatic lapse rate when it descends on the other side of the mountain. The air descends 1000 m so the temperature rise is 0.7 °C × 1000 m/100 m = 7 °C. Hence, the final air temperature is 16.25 °C + 7 °C = 23.25 °C.

2.7.2 Dew-Point Temperature

In irrigation engineering the dew-point temperature is also used. This may be needed in the computation of evaporation. It is the temperature at which air is cooled to reach saturation at constant pressure. At this temperature the saturation vapor pressure is the same as vapor pressure, since there is no change in water content. The dew-point temperature (T_d) can be computed from relative humidity (*RH*) and air temperature (T_a) as (Bosen, 1958):

$$T_d = \left(\frac{RH}{100}\right)^{\frac{1}{8}} (112 + 0.9T_a) - 112 + 0.1T_a, \tag{2.5}$$

where T_a is air temperature in °C, *RH* is a percentage, and T_d is the dew-point temperature in °C. The dew-point is also computed as:

$$T_d = T_a - (14.55 + 0.114T_a)\left(1 - \frac{RH}{100}\right) + \left[(2.5 + 0.0077T_a)\left(1 - \frac{RH}{100}\right)\right]^3 + (15.9 + 0.117T_a)\left(1 - \frac{RH}{100}\right)^{14}, \tag{2.6}$$

where T_a is air temperature in °C, and *RH* is a percentage.

2.7.3 Wet-Bulb Temperature

If a thermometer is covered with a cloth wetted with pure water in moving air then the thermometer is cooled until the heat gained from the air equals the latent heat of vaporization. At this point, the temperature reaches equilibrium; this is called the wet-bulb temperature. Also at this point the radiation energy balance is zero.

2.7.4 Atmospheric Pressure

Pressure (*P*) can be defined as the force exerted on a unit area. It is applied perpendicularly to the surface of the area, expressed as

$$P = \frac{F}{A}, \tag{2.7}$$

where *F* is the force and *A* is the area. In the metric system of units, *F* is expressed in Newton, so *P* is expressed in Newton per square meter (N/m^2).

Atmospheric air exerts pressure on the Earth's surface and this pressure is known as the atmospheric pressure. To calculate atmospheric pressure, consider a unit column of air standing above the Earth's surface. Then, the atmospheric pressure is the weight of this column. If the column is at the mean sea level and has a height h of air with an average density of ρ, then the atmospheric pressure p would be ρgh, where g is the acceleration due to gravity.

Pressure is measured with a mercury barometer and is expressed by the height of the mercury column in a barometer balancing the atmospheric pressure. The height of the mercury column at the average sea-level pressure is about 76 cm. Since the density of mercury at 15 °C is 13.59 g/cm^3 and the acceleration due to gravity (g) is 981 cm/s^2, the average sea-level pressure (p) is

$76 \times 13.59 \times 981 = 1,013,216\ \text{dynes/cm}^2$. The unit dynes/cm^2 is centimeter-gram per square second (CGS) and is an exceedingly small unit. Therefore, millibar (mb), which is 1000 times larger, has been introduced and is the unit of pressure often used in meteorology: 1 millibar = 1000 dynes/cm^2. Thus, the average sea-level pressure is about 1013 mb. The unit of pressure in the SI system is 1 N/m^2. This unit is also known as the pascal (Pa) and is equal to 10 dynes/cm^2. Hence, 1 mb = 100 Pa (= 1 hectopascal, or 1 hPa). One bar (*b*) is equal to 100 kPa (kilopascals). The atmospheric pressure of 1013 mb is 1013 hPa in the SI unit system. Thus, units of measurement of pressure include atmosphere (atm), bar (b), millibar (mb), kilogram per square cm (kg/cm^2), kilogram per square meter (kg/m^2), kilopascal (kPa), millimeter of mercury (mmHg), inches of mercury (inHg), pascal (Pa), hectopascal (hPa), pounds per square foot (psf), and pounds per square inch (psi). One atmosphere (atm) equals 1.01325 bars = 101.325 kilopascals. At sea level, the atmospheric pressure is about 100 kPa. The difference between 1 bar and 1 atm is small, so often they are used interchangeably.

For the US standard atmospheric pressure the relation between atmospheric pressure (p) in mb and elevation ($elev$) in meters can be expressed as

$$p = 1013 - 0.1055 \times elev. \tag{2.8}$$

A more general relation, which is applicable to any standard atmosphere, can be written as

$$p = p_0\left[1 - \frac{\delta}{T_0}(elev - elev_0)\right]^{\frac{g}{\delta R}}, \tag{2.9}$$

in which p_0 is the standard sea-level atmospheric pressure (mb), T_0 is the standard sea-level temperature (K), δ is the standard lapse rate (0.00976 K/m), R is the specific gas constant for dry air (287.058 J/kg·K), g is the acceleration due to gravity (9.81 m/s^2), $elev_0$ is the base elevation (m), and $elev$ is the elevation (m). The lapse rate for the US standard atmosphere is considered as 6.5 K/km, the standard sea-level pressure is 1013 mb, and the standard sea-level temperature is 288 K.

Example 2.4: If the pressure is 10 atm, then compute the equivalent value of pressure in bar, hectopascal, kilogram per square cm, kilogram per square meter, kilopascal, millimeter of mercury, inches of mercury, pascal, pounds per square foot, and pounds per square inch.

Solution: Given pressure = 10 atmosphere (atm)

1 atmosphere (atm) = 1.01325 bar, then 10 atm = 10 × 1.01325 = 10.1325 bar

1 bar = 1000 hPa, then 10.1325 bar = 10.1325 × 1000 = 10,132.5 hPa

1 bar = 1.0197 kg/cm^2, then 10.1325 bar = 10.1325 × 1.0197 = 10.332 kg/cm^2

1 bar = 10,197 kg/m^2, then 10.1325 bar = 10.1325 × 10,197 = 10,3321 kg/m^2

1 atm = 101.325 kPa, then 10 atm = 10 × 101.325 = 1013.25 kPa

1 atm = 760 mmHg, then 10 atm = 10 × 760 = 7600 mmHg

1 atm = 29.92 inHg, then 10 atm = 10 × 29.92 = 299.2 inHg

1 atm = 101,325 Pa, then 10 atm = 101,325 × 10 = 1,013,250

1 atm = 2116.21662 pounds per sq foot, then 10 atm = 10 × 2116.217 = 21,162.17 pounds per sq foot

1 atm = 14.6959 psi, then 10 atm = 10 × 14.6959 = 146.959 psi

2.7.5 Atmospheric Water Vapor and Its Indices

Atmospheric water vapor or atmospheric moisture, although small in comparison with other gases present in the atmosphere, is the source of precipitation received on land. Its importance can be further seen by recognizing that it absorbs and reradiates terrestrial radiation and hence stabilizes the Earth's temperature and is also one of the principal determinants of evaporation which, in turn, together with transpiration, is the source of water vapor. The amount of water vapor present in the atmosphere is not uniformly distributed in time and place.

2.7.5.1 Vaporization or Evaporation

Vaporization is a process by which water is transformed to vapors; this transformation entails the removal of the heat energy of water. The rate of vaporization increases with the increase of temperature. If water vapor is transformed to liquid, the process of transformation is called condensation, which adds heat energy to water. Condensation and vaporization always occur simultaneously in places in contact with water. In unsaturated environments, such as irrigation fields, evaporation is greater than condensation; in saturated environments under the same air and water temperature, evaporation and condensation are equal; in supersaturated environments condensation is higher than evaporation. There is another process by which a solid, such as snow or ice, can directly change into water vapor: sublimation, during which the intermediate process of conversion of solid to liquid is bypassed.

2.7.5.2 Latent Heat of Vaporization or Condensation

The amount of heat required to convert a unit mass of water to vapor without change in temperature is called the latent heat of vaporization. Likewise, the amount of heat released that is required to convert water vapor to a unit mass of water is called the latent heat of condensation. Numerically the latent heat of vaporization is the same as the latent heat of condensation and is approximately 597 cal/g at 0 °C. The unit of measurement is usually calories per gram (cal/g). The relation between latent heat of vaporization (λ) and air temperature (T_a) in °C can be expressed as

$$\lambda = 597.3 - 0.564T_a, \tag{2.10}$$

where λ is in cal/g and T_a is in °C. The latent heat of vaporization can also be computed as

$$\lambda = 2500.78 - 2.360T_a, \tag{2.11}$$

where λ is in kJ/kg and T_a is in °C.

2.7.5.3 Latent Heat of Fusion

The amount of heat required to convert 1 g of ice to water at the same temperature is called the latent heat of fusion. If 1 g of water is converted to ice at 0 °C (32 °F), the amount of heat released would be the same. The latent heat of fusion is 79.7 cal/g.

2.7.5.4 Latent Heat of Sublimation

The amount of heat needed to directly convert 1 g of ice to water vapor at the same temperature is called the latent heat of sublimation. Conversely, if water vapor is directly converted to ice an equivalent amount of heat would be released. The latent heat of sublimation is the sum of latent heat of fusion and the latent heat of vaporization, and is approximately 677 cal/g at 0 °C.

Example 2.5: Compute the latent heat of vaporization in cal/g for water at 5 °C, 10 °C, 15 °C, 20 °C, 25 °C, 30 °C, 35 °C, and 40 °C.

Solution: Using eq. (2.10),

λ at 5 °C = 597.3 − 0.564(5 °C) = 594.5 cal/g
λ at 10 °C = 597.3 − 0.564(10 °C) = 591.7 cal/g
λ at 15 °C = 597.3 − 0.564(15 °C) = 588.8 cal/g
λ at 20 °C = 597.3 − 0.564(20 °C) = 586.0 cal/g
λ at 25 °C = 597.3 − 0.564(25 °C) = 583.2 cal/g
λ at 30 °C = 597.3 − 0.564(30 °C) = 580.4 cal/g
λ at 35 °C = 597.3 − 0.564(35 °C) = 577.6 cal/g
λ at 40 °C = 597.3 − 0.564(40 °C) = 574.7 cal/g

Example 2.6: Compute the amount of calories needed to evaporate 10 liters, 20 liters, and 50 liters of water at 15 °C, 20 °C, 25 °C, 30 °C, and 35 °C.

Solution: This involves two steps. First, the latent heat of vaporization is computed using eq. (2.10) at the given temperatures. Then, the amount of calories needed to evaporate the given amount of water is computed for each temperature: 1 L = 1000 g, 10 L = 10,000 g, 20 L = 20,000 g, and 50 L = 50,000 g.

For 10 L:

$$\lambda \text{ at } 15\,^\circ\text{C} = 588.8\frac{\text{cal}}{\text{g}} \rightarrow 10\,\text{liters} \rightarrow \left(588.8\frac{\text{cal}}{\text{g}} \times 10{,}000\,\text{g}\right) = 5888\,\text{kcal},$$

$$\lambda \text{ at } 20\,^\circ\text{C} = 586.0\frac{\text{cal}}{\text{g}} \rightarrow 10\,\text{liters} \rightarrow \left(586.0\frac{\text{cal}}{\text{g}} \times 10{,}000\,\text{g}\right) = 5860\,\text{kcal},$$

$$\lambda \text{ at } 25\,^\circ\text{C} = 583.2\frac{\text{cal}}{\text{g}} \rightarrow 10\,\text{liters} \rightarrow \left(583.2\frac{\text{cal}}{\text{g}} \times 10{,}000\,\text{g}\right) = 5832\,\text{kcal},$$

$$\lambda \text{ at } 30\,^\circ\text{C} = 580.4\frac{\text{cal}}{\text{g}} \rightarrow 10\,\text{liters} \rightarrow \left(580.4\frac{\text{cal}}{\text{g}} \times 10{,}000\,\text{g}\right) = 5804\,\text{kcal},$$

$$\lambda \text{ at } 35\,^\circ\text{C} = 577.6\frac{\text{cal}}{\text{g}} \rightarrow 10\,\text{liters} \rightarrow \left(577.6\frac{\text{cal}}{\text{g}} \times 10{,}000\,\text{g}\right) = 5776\,\text{kcal}.$$

For 20 liters,

$$\lambda \text{ at } 15\,^\circ\text{C} = 588.8\frac{\text{cal}}{\text{g}} \rightarrow 20\,\text{liters} \rightarrow \left(588.8\frac{\text{cal}}{\text{g}} \times 20{,}000\text{g}\right) = 11{,}777\,\text{kcal},$$

$$\lambda \text{ at } 20\,^\circ\text{C} = 586.0\frac{\text{cal}}{\text{g}} \rightarrow 20\,\text{liters} \rightarrow \left(586.0\frac{\text{cal}}{\text{g}} \times 20{,}000\,\text{g}\right) = 11{,}720\,\text{kcal},$$

$$\lambda \text{ at } 25\,^\circ\text{C} = 583.2\frac{\text{cal}}{\text{g}} \rightarrow 20\,\text{liters} \rightarrow \left(583.2\frac{\text{cal}}{\text{g}} \times 20{,}000\,\text{g}\right) = 11{,}664\,\text{kcal},$$

$$\lambda \text{ at } 30\,^\circ\text{C} = 580.4\frac{\text{cal}}{\text{g}} \rightarrow 20\,\text{liters} \rightarrow \left(580.4\frac{\text{cal}}{\text{g}} \times 20{,}000\,\text{g}\right) = 11{,}608\,\text{kcal},$$

$$\lambda \text{ at } 35\,^\circ\text{C} = 577.6\frac{\text{cal}}{\text{g}} \rightarrow 20\,\text{liters} \rightarrow \left(577.6\frac{\text{cal}}{\text{g}} \times 20{,}000\,\text{g}\right) = 11{,}551\,\text{kcal}.$$

For 50 liters,

$$\lambda \text{ at } 15\,^\circ\text{C} = 588.8\frac{\text{cal}}{\text{g}} \rightarrow 50\,\text{liters} \rightarrow \left(588.8\frac{\text{cal}}{\text{g}} \times 50{,}000\,\text{g}\right) = 29{,}442\,\text{kcal},$$

$$\lambda \text{ at } 20\,^\circ\text{C} = 586.0\frac{\text{cal}}{\text{g}} \rightarrow 50\,\text{liters} \rightarrow \left(586.0\frac{\text{cal}}{\text{g}} \times 50{,}000\,\text{g}\right) = 29{,}301\,\text{kcal},$$

$$\lambda \text{ at } 25\,^\circ\text{C} = 583.2\frac{\text{cal}}{\text{g}} \rightarrow 50\,\text{liters} \rightarrow \left(583.2\frac{\text{cal}}{\text{g}} \times 50{,}000\,\text{g}\right) = 29{,}160\,\text{kcal},$$

$$\lambda \text{ at } 30\,^\circ\text{C} = 580.4\frac{\text{cal}}{\text{g}} \rightarrow 50\,\text{liters} \rightarrow \left(580.4\frac{\text{cal}}{\text{g}} \times 50{,}000\,\text{g}\right) = 29{,}019\,\text{kcal},$$

$$\lambda \text{ at } 35\,^\circ\text{C} = 577.6\frac{\text{cal}}{\text{g}} \rightarrow 50\,\text{liters} \rightarrow \left(577.6\frac{\text{cal}}{\text{g}} \times 50{,}000\,\text{g}\right) = 28{,}878\,\text{kcal}.$$

2.7.6 Atmospheric Humidity

The atmosphere always contains water vapor in varying proportions. The maximum amount of water vapor present in the atmosphere depends on the air temperature. The higher the temperature the more water vapor the air can hold. Water vapor is produced by evaporation from water bodies and land surfaces. Humidity represents the amount of water vapor present in the atmosphere and is indicated by several parameters, including vapor pressure, relative humidity, specific humidity, mixing ratio, absolute humidity, and dew-point. Each of these parameters is briefly presented here.

Vapor pressure is a very useful measure of water content in the atmosphere. Various gases are present in the atmosphere, and each gas exerts its pressure, called partial pressure. Likewise, the pressure exerted by water vapor, independent of the presence of other gases, is referred to as the vapor pressure and is usually denoted by lower case letter "e." When the atmospheric air holds the maximum possible water vapor at a given temperature and pressure, then the air is said to be saturated. The pressure exerted by the water vapor in the saturated air at a given temperature is called the saturation vapor pressure. For all practical purposes the saturation vapor pressure is the largest vapor pressure and is denoted by "e_s" and is measured in millibars or in millimeters of mercury. The difference "$e_s - e$" is called the saturation deficit. If more vapor is added to the atmosphere or if the air is cooled below the saturation point, the excess vapor condenses.

2.7.6.1 Absolute Humidity and Vapor Density

Absolute humidity (ρ_V) is defined as the mass of water vapor per unit volume of space and is the same as the density of vapor. In other words, the vapor density is the mass of water vapor present in a unit volume of moist air. This is sometimes called the absolute humidity. Its unit is g/cm^3 or kg/m^3. It can be determined from the equation of state for an ideal gas and is given by:

$$\rho_V = \frac{e}{R_w T}, \tag{2.12}$$

where R_w is the specific gas constant for water vapor (461.495 J/kg·K), e is the partial pressure of water vapor or vapor pressure (Pa), and T is the absolute temperature (K).

2.7.6.2 Relative Humidity

The relative humidity of air at a given temperature is defined as the percentage ratio of the actual vapor pressure (e_a) to the maximum possible vapor pressure, that is, the saturation vapor pressure (e_s):

$$RH = \frac{e_a}{e_s} \times 100. \tag{2.13a}$$

When air is saturated with moisture, the relative humidity is a percentage.

The relative humidity is a function of temperature, because vapor pressure is a function of air temperature (T_a). Bosen (1958) computed relative humidity as a percentage directly from air temperature (T_a; °C) and dew-point temperature (T_d; °C) as:

$$RH = \left(\frac{112 - 0.1T_a + T_d}{112 + 0.9T_a}\right)^8 \times 100. \tag{2.13b}$$

If T is given in °F then eq. (2.13b) can be expressed as

$$RH = \left(\frac{96 - 0.056T_a + 0.56T_d}{96 + 0.5T_a}\right)^8 \times 100 \tag{2.13c}$$

or

$$RH = \left(\frac{173 - 0.1T_a + T_d}{173 + 0.9T_a}\right)^8 \times 100. \tag{2.13d}$$

The saturation vapor pressure (e_s) in millibars (mb) is related to the temperature of air (T_a) in °C as

$$e_s = 6.108 \times \exp\left[\frac{17.27T_a}{T_a + 237.3}\right]. \tag{2.14}$$

The saturation vapor pressure can be computed for a given value of temperature using eq. (2.14). The saturation vapor pressure in millibars can also be computed as

$$e_s = 33.8639\left[(0.00738T_a + 0.8072)^8 - 0.000019|1.8T_a + 0.48| + 0.001316\right]. \tag{2.15}$$

The values of saturation vapor pressure for various values of temperature are given in Table 2.2.

Another term used in irrigation engineering is saturation vapor pressure deficit, which is the difference between saturation vapor pressure (e_s) and actual vapor pressure (e_a).

Table 2.2 *Saturation vapor pressure as a function of temperature (mb)*

	Tenth of °C									
	0.0	0.1	0.2	0.3	0.4	0.5	0.6	0.7	0.8	0.9
°C	mb	mb	mb	mb	mb	mb	mb	mb	mb	mb
0	6.11	6.15	6.20	6.24	6.29	6.33	6.38	6.43	6.47	6.52
1	6.57	6.61	6.66	6.71	6.76	6.81	6.86	6.91	6.96	7.01
2	7.06	7.11	7.16	7.21	7.26	7.31	7.37	7.42	7.47	7.52
3	7.58	7.63	7.69	7.74	7.80	7.85	7.91	7.96	8.02	8.08
4	8.13	8.19	8.25	8.31	8.36	8.42	8.48	8.54	8.60	8.66
5	8.72	8.78	8.85	8.91	8.97	9.03	9.10	9.16	9.22	9.29
6	9.35	9.42	9.48	9.55	9.61	9.68	9.75	9.81	9.88	9.95
7	10.02	10.09	10.16	10.23	10.30	10.37	10.44	10.51	10.58	10.65
8	10.73	10.80	10.87	10.95	11.02	11.10	11.17	11.25	11.33	11.40
9	11.48	11.56	11.64	11.72	11.79	11.87	11.95	12.03	12.12	12.20
10	12.28	12.36	12.45	12.53	12.61	12.70	12.78	12.87	12.95	13.04
11	13.13	13.21	13.30	13.39	13.48	13.57	13.66	13.75	13.84	13.93
12	14.03	14.12	14.21	14.31	14.40	14.49	14.59	14.69	14.78	14.88
13	14.98	15.08	15.17	15.27	15.37	15.47	15.58	15.68	15.78	15.88
14	15.99	16.09	16.19	16.30	16.41	16.51	16.62	16.73	16.84	16.94
15	17.05	17.16	17.27	17.39	17.50	17.61	17.72	17.84	17.95	18.07
16	18.18	18.30	18.42	18.53	18.65	18.77	18.89	19.01	19.13	19.25
17	19.38	19.50	19.62	19.75	19.87	20.00	20.13	20.25	20.38	20.51
18	20.64	20.77	20.90	21.03	21.16	21.30	21.43	21.57	21.70	21.84
19	21.97	22.11	22.25	22.39	22.53	22.67	22.81	22.95	23.09	23.24
20	23.38	23.53	23.67	23.82	23.97	24.12	24.27	24.42	24.57	24.72
21	24.87	25.02	25.18	25.33	25.49	25.64	25.80	25.96	26.12	26.28
22	26.44	26.60	26.76	26.93	27.09	27.26	27.42	27.59	27.76	27.92
23	28.09	28.26	28.44	28.61	28.78	28.96	29.13	29.31	29.48	29.66
24	29.84	30.02	30.20	30.38	30.56	30.75	30.93	31.12	31.30	31.49
25	31.68	31.87	32.06	32.25	32.44	32.63	32.83	33.02	33.22	33.42
26	33.61	33.81	34.01	34.22	34.42	34.62	34.83	35.03	35.24	35.44
27	35.65	35.86	36.07	36.29	36.50	36.71	36.93	37.14	37.36	37.58
28	37.80	38.02	38.24	38.46	38.69	38.91	39.14	39.37	39.60	39.83
29	40.06	40.29	40.52	40.76	40.99	41.23	41.47	41.71	41.95	42.19
30	42.43	42.67	42.92	43.17	43.41	43.66	43.91	44.16	44.42	44.67
31	44.93	45.18	45.44	45.70	45.96	46.22	46.48	46.75	47.01	47.28
32	47.55	47.82	48.09	48.36	48.63	48.91	49.18	49.46	49.74	50.02
33	50.30	50.58	50.87	51.15	51.44	51.73	52.02	52.31	52.60	52.90
34	53.19	53.49	53.79	54.09	54.39	54.69	55.00	55.30	55.61	55.92
35	56.23	56.54	56.85	57.17	57.48	57.80	58.12	58.44	58.76	59.08
36	59.41	59.74	60.07	60.39	60.73	61.06	61.39	61.73	62.07	62.41
37	62.75	63.09	63.43	63.78	64.13	64.48	64.83	65.18	65.53	65.89
38	66.25	66.61	66.97	67.33	67.69	68.06	68.43	68.80	69.17	69.54
39	69.91	70.29	70.67	71.05	71.43	71.81	72.20	72.58	72.97	73.36
40	73.76	74.15	74.55	74.94	75.34	75.74	76.15	76.55	76.96	77.37
41	77.78	78.19	78.61	79.02	79.44	79.86	80.28	80.71	81.13	81.56
42	81.99	82.42	82.85	83.29	83.73	84.17	84.61	85.05	85.50	85.95
43	86.40	86.85	87.30	87.76	88.21	88.67	89.14	89.60	90.07	90.53
44	91.01	91.48	91.95	92.43	92.91	93.39	93.87	94.36	94.84	95.33
45	95.82	96.32	96.81	97.31	97.81	98.32	98.82	99.33	99.84	100.35
46	100.86	101.38	101.90	102.42	102.94	103.47	103.99	104.52	105.06	105.59
47	106.13	106.67	107.21	107.75	108.30	108.85	109.40	109.95	110.51	111.07
48	111.63	112.19	112.76	113.33	113.90	114.47	115.04	115.62	116.20	116.79
49	117.37	117.96	118.55	119.14	119.74	120.34	120.94	121.54	122.15	122.76
50	123.37	123.98	124.60	125.22	125.84	126.46	127.09	127.72	128.35	128.99

To obtain the correct saturation vapor pressure value from the table corresponding to a given temperature value, the values of temperature in the first column should be added to the values in the first row. For example, for temperature 20.6 °C, go to the row corresponding to the value of 20 and the column corresponding to the value of 0.6 and the common element of 24.27 mb is obtained, which is the value of the saturation vapor pressure.

Knowing the values of e_s and e_a , one can compute the value of saturation vapor pressure deficit $(e_s - e_a)$. The value of actual vapor pressure for a certain period can be determined by knowing the minimum relative humidity and maximum relative humidity (RH_{min} and RH_{max}), corresponding to the minimum and maximum temperatures during the same period, respectively, and computing the mean relative humidity (RH_{mean}) as

$$e_a = e_s \frac{RH_{mean}}{100}. \tag{2.16}$$

If the dry-bulb temperature or the air temperature (T_a) corresponding to maximum temperature and wet-bulb temperature (T_w) corresponding to minimum temperature are known, then the actual vapor pressure can be computed as

$$e_a = e_s - \gamma(T_a - T_w) = e_s - \frac{pc_p}{0.622\lambda}(T_a - T_w), \tag{2.17a}$$

in which e_s is the saturated vapor pressure corresponding to T_w, c_p is the specific heat of air, λ is the latent heat of vaporization, and γ is the psychrometric constant. If $p = 1013$ mb or 1 atm, $c_p = 0.2396$ cal/g·°C, and $\lambda = 590$ cal/g at 10 °C then $pc_p/(0.622\lambda) = 0.662$ mb/°C (or 0.496 mmHg/°C or 0.276 mmHg/°F).

The actual vapor pressure can also be computed as

$$e_a = e_s - 10^{-4} \times 3.67p(T_a - T_w) \times \left(1 + \frac{T_w - 32}{1571}\right), \tag{2.17b}$$

in which p is the atmospheric pressure in mb, T_a is the dry-bulb temperature in °F, T_w is the wet-bulb temperature in °F, and e_s is the saturation vapor pressure in mb corresponding to T_w.

2.7.6.3 Specific Humidity and Density of Dry Air

The specific humidity (q) is defined as the mass of water vapor present in 1 g of moist air. It hardly ever exceeds 0.04. If p is the total pressure of air and e is the partial pressure of water vapor in it, then the specific humidity (q) can be expressed as

$$q = \beta\frac{e}{p} = \frac{0.622e}{p} = \frac{0.622e}{\rho_d RT}, \tag{2.18}$$

where $\beta = 0.622$ is the ratio of the molecular weight of water vapor (which is 18.02 g/mol) to the molecular weight of dry air (which is 28.97 g/mol), R is the specific gas constant for dry air (287.058 J/kg·K), and T is the absolute temperature (K). Here, ρ_d is in g/cm^3 and e is 10 mb or 10^3 Pa. The density of dry air (ρ_d; kg/m^3) can be written as

$$\rho_d = \frac{p}{RT}, \tag{2.19}$$

where R is the specific gas constant for dry air (287.058 J/kg·K), p is the air pressure (Pa), and T is the temperature (K).

2.7.6.4 Mixing Ratio

The mixing ratio (r) is defined as the mass of water vapor present per 1 g of dry air. The specific humidity and the mixing ratio are usually expressed as g/kg. The partial pressure of dry air is ($p - e$). The mixing ratio (r) can be defined as:

$$r = \frac{\beta e}{p - e} = \frac{0.622e}{p - e}. \tag{2.20}$$

Example 2.7: On a given winter day in College Station, Texas, the temperature is 5 °C, air pressure is 1000 mb, and humidity is 40%. Compute the actual vapor pressure and the saturated vapor pressure. Also compute water vapor density, dry air density, air density, and mixing ratio.

Solution: From Table 2.2, at temperature 5 °C, saturated vapor pressure is 8.72 mb.

Actual vapor pressure: $e_a = \frac{RH}{100} \times e_s = \frac{40}{100} \times 8.72 =$ 3.49 mb

Water vapor density: $\rho_V = \frac{P_w}{R_w T} = \frac{3.49 \times 100}{461.495 \times (5 + 273.15)} =$ 0.0027 kg/m^3 (R_w is the specific gas constant for water vapor [461.495 J/kg·K], P_w is the partial pressure of water vapor [Pa]).

Dry air density: $\rho_d = \frac{p_d}{RT} = \frac{p - e_a}{RT} = \frac{(1000 - 3.49) \times 100}{287.058 \times (5 + 273.15)} =$ 1.2481 kg/m^3 (R is the specific gas constant for dry air [287.058 J/kg·K], p is the air pressure [Pa], and p_d is the partial pressure of dry air [Pa]).

Note that the temperature used in the above formula is in kelvin.

Air density: $\rho = \rho_V + \rho_d = 0.0027 + 1.2481 =$ 1.251 kg/m^3.

Mixing ratio: $r = \frac{\beta e}{p - e} = \frac{0.622e}{p - e} = \frac{0.622 \times 3.49}{1000 - 3.49} = 0.002$ g/g, $r = 2$ g/kg.

2.7.7 Air Density

The density of moist air (ρ) can be defined as the mass of water vapor and mass of dry air per unit volume of moisture per unit volume of air. If p is the pressure of moist air, then $p - e$ is the partial pressure of dry air. Thus, air density is the sum of density of water vapor and density of dry air. Compared with the density of water, the air density (ρ) near the ground is quite small, about the order of 1/1000 of that of $r = 2$ g/kg water, and is seldom measured. Rather, it is computed from the gas equation ($\rho = p/RT$) and is given in g/cm^3 or kg/m^3. The air density is 0.0012 g/cm^3 for a surface pressure of 1000 mb and temperature of 290 K. Moist air is lighter than dry air. The specific gravity of the water vapor (β) is about 0.622 of that of dry air at the same temperature and pressure.

Example 2.8: Show mathematically that moist air is lighter than dry air.

Solution: Specific volume is defined as the total volume of dry air and water vapor mixture per kilogram of dry air. It can be expressed as:

$$v_{da} = \frac{V}{m_d}, \tag{2.21}$$

where v_{da} is the specific volume of moist air per unit mass of dry air (m^3/kg), V is the total volume of moist air (m^3), and m_d is the mass of dry air (kg). When dry air and water vapor with the same temperature occupy the same volume the equation for an ideal gas can be applied as

$$p_d V = m_d RT, \tag{2.22}$$

where p_d is partial air pressure (Pa), R is the dry air gas constant (287.058 J/kg·K), and T is the absolute temperature of moist air (K). By combining eqs (2.21) and (2.22):

$$v_{da} = \frac{RT}{p_d}. \tag{2.23}$$

The dry air pressure can be expressed as:

$$p_d = p - e, \tag{2.24}$$

where p is the pressure of the moist air and e is the partial pressure of water vapor. From eqs (2.23) and (2.24):

$$v_{da} = \frac{RT}{(p - e)}. \tag{2.25}$$

The ideal gas law for water vapor:

$$eV = m_w R_w T, \tag{2.26}$$

where $R_w = 461.495$, the gas constant for water vapor (J/kg·K).

Let the specific humidity be supposed to be equal to x. Then, the mass of water vapor can be expressed as:

$$m_w = x m_d. \tag{2.27}$$

From eqs (2.26) and (2.27),

$$eV = x m_d R_w T. \tag{2.28}$$

Combining eqs (2.28) and (2.21):

$$v_{da} = \frac{x R_w T}{e}, \tag{2.29}$$

which leads to

$$e = \frac{x R_w T}{v_{da}}. \tag{2.30}$$

Combining eqs (2.29) and (2.25):

$$v_{da} = \frac{RT}{\left[p - \left(\frac{x R_w T}{v_{da}}\right)\right]}. \tag{2.31}$$

Equation (2.31) can be transformed as:

$$v_{da} = \frac{RT}{p}\left(1 + \frac{x R_w}{R}\right). \tag{2.32}$$

As discussed earlier,

$$v = \frac{V}{m_d + m_w}, \tag{2.33}$$

where v is the specific volume of moist air per mass unit of dry air and water vapor (m^3/kg). Using eqs (2.27) and (2.33):

$$v = \frac{V}{m_d(1 + x)}. \tag{2.34}$$

Combining eqs (2.21) and (2.34):

$$v = \frac{v_{da}}{(1 + x)}. \tag{2.35}$$

Using eqs (2.32) and (2.35), the specific volume of moist air per unit mass of dry air and water vapor can be expressed as:

$$v = \frac{\frac{RT}{p}\left(1 + \frac{x R_w}{R}\right)}{(1 + x)}. \tag{2.36}$$

The inverse of eq. (2.36) can be used to express the density of moist air: $\rho = \frac{1}{v}$ as

$$\rho = \frac{\left(\frac{p(1 + x)}{RT}\right)}{\left(1 + \frac{x R_w}{R}\right)}. \tag{2.37}$$

If it is dry air, the density of air can be expressed as:

$$\rho_{da} = \frac{p}{RT}, \tag{2.38}$$

where ρ_{da} is the density of dry air (kg/m^3) and $\frac{R_W}{R} = 1.608$. Combining eqs (2.37) and (2.38) and using the value of $\frac{R_W}{R}$, we get the final equation for the density of moist air:

$$\rho = \frac{\rho_{da}(1 + x)}{(1 + 1.608x)}. \tag{2.39}$$

It is seen from eq. (2.39) that increased moisture content reduces the density of moist air. Hence, dry air is denser than moist air.

One can also show more simply as follows. The specific gravity of water vapor is 0.622 of that of dry air at the same temperature and pressure. The density of water vapor (ρ_w) can be expressed as

$$\rho_w = 0.622 \frac{e}{RT}, \tag{2.40}$$

where T is the absolute temperature in K, e is the vapor pressure in Pa, and R is the dry gas constant = 287.058 (J/kg·K). Here, the density is in kg/m^3.

Likewise, the density of dry air in kg/cm^3 can be written as

$$\rho_d = \frac{p_d}{RT}, \tag{2.41}$$

in which p_d is the partial air pressure in Pa. Noting that $\rho = \rho_w + \rho_d$, eqs (2.40) and (2.41) can be combined to yield

$$\rho = 0.622\frac{e}{RT} + \frac{p_d}{RT} = \frac{1}{RT}(0.622e + p_d). \tag{2.42}$$

Equation (2.42) can be expressed with p_d replaced by $p - e$ as

$$\rho = 0.622\frac{e}{RT} + \frac{p_d}{RT} = \frac{1}{RT}(0.622e + p - e) = \frac{p}{RT}\left(1 - 0.378\frac{e}{p}\right). \tag{2.43}$$

Equation (2.43) shows that moist air is lighter than dry air, because (e/P) is a positive quantity.

Example 2.9: What will be the weight of 0.5 m^3 of dry air if the air temperature is 15 °C and pressure is 800 mb?

Solution: Equation (2.19) can be used to compute the density of the dry air:

$$T = 273.15 + 15 = 288.15 \text{ K},$$

$$\rho_d = \frac{p}{RT} = \frac{800 \times 100}{287.058 \times 288.15} = 0.967 \text{ kg/m}^3,$$

$$0.967\frac{\text{kg}}{\text{m}^3} \times 0.5 \text{ m}^3 = 0.484 \text{ kg}.$$

Example 2.10: What will be the density of dry air at a temperature of 25 °C and at a pressure of 1050 mb, and the density of moist air with *RH* of 70% at the same temperature and pressure?

Solution: The dry air density is computed using eq. (2.19) and the moist air density using eq. (2.43). Therefore,

$$\rho_d = \frac{P}{RT} = \frac{1050 \times 100}{287.058 \times (273.15 + 25)} = 1.227 \text{ kg/m}^3.$$

The partial vapor pressure is computed using eq. (2.16). The saturation pressure is calculated using eq. (2.14) as

$$e_s = 6.108 \times \exp\left[\frac{17.27T_a}{T_a + 237.3}\right] = 6.108 \times \exp\left[\frac{17.27 \times 25}{25 + 237.3}\right] = 31.68 \text{ mb}.$$

The partial pressure is obtained as: $70\% \times e_s = 0.7 \times 31.68 \text{ mb} = 22.18$ mb.

From eq. (2.43),

$$\rho_a = \frac{1050 \times 100}{287.058 \times (273.15 + 25)}\left[1 - 0.378\frac{22.18}{1050}\right] = 1.227\frac{\text{kg}}{\text{m}^3} \times (0.9920) = 1.217 \text{ kg/m}^3.$$

Example 2.11: What will be the dew-point temperature, relative humidity, saturation vapor pressure, and actual vapor pressure if the dry-bulb temperature is 30 °C and the wet-bulb temperature is 25 °C?

Solution: The actual vapor pressure can be computed using eq. (2.17a) or (2.17b). Using atmospheric pressure as 1013 mb, the actual vapor pressure is:

$$e_a = e_s - \gamma(T_a - T_w) = e_s - 0.662(T_a - T_w),$$

where e_s is the saturated vapor pressure corresponding to $T_w = 25$ °C. Here, eq. (2.15) is used to calculate e_s, which can be compared with the value calculated using eq. (2.14) in Example 2.10:

$$e_s\,(\text{at } T_w = 25\,°\text{C}) = 33.8639\left[(7.38 \times 10^{-3} \times 25 + 0.8072)^8 - 1.9 \times 10^{-5}|1.8 \times 25 + 0.48| + 1.316 \times 10^{-3}\right] = 31.69 \text{ mb}.$$

Then, the actual vapor pressure is:

$$e_a = e_s - 0.662(30 - 25) = 31.69 - (0.662)(30 - 25) = 28.38 \text{ mb}.$$

The saturated vapor pressure at current temperature is

$$e_s(\text{at}\,T_a=30\,°\text{C})=(33.8639)\left[(7.38\times10^{-3}\times30+0.8072)^8-(1.9\times10^{-5})|1.8\times30+0.48|+1.316\times10^{-3}\right]=42.44\,\text{mb}.$$

The relative humidity is:

$$RH=\frac{e_a}{e_s}\times100=\frac{28.38}{42.44}\times100\%=66.9\%.$$

The dew-point temperature can also be computed using eq. (2.13b) as:

$$RH=\left(\frac{112-0.1T_a+T_d}{112+0.9T_a}\right)^8\times100,$$

$$T_d=0.669^{\frac{1}{8}}\times(112+0.9\times30)-112+0.1\times30=23.2\,°\text{C}.$$

Example 2.12: What is the relative humidity if the air temperature is 25 °C and dew-point temperature is 20 °C?

Solution: Using eq. (2.13b),

$$RH=\left(\frac{112-0.1\times25+20}{112+0.9\times25}\right)^8\times100=\left(\frac{129.5}{134.5}\right)^8=74\%.$$

Example 2.13: What will be the dew-point temperature if the air temperature is 25 °C and the relative humidity is 80%?

Solution: From eq. (2.13b), the dew-point temperature is computed as

$$(0.8)^{\frac{1}{8}}=\frac{112-0.1\times25+T_d}{112+0.9\times25},$$

$$T_d=0.8^{\frac{1}{8}}\times(112+0.9\times25)-112+0.1\times25=21.3\,°\text{C}.$$

Equation (2.6) can also be used to compute the dew-point temperature as

$$T_d=25-(14.55+0.114\times25)(1-0.8)+[(2.5+0.0077\times25)(1-0.8)]^3+(15.9+0.117\times25)(1-0.8)^{14}=21.7\,°\text{C}$$

2.7.8 Wind

Wind is given in terms of its direction and speed and is defined by the horizontal motion of air over the Earth's surface. The direction is taken as the direction from which the wind blows. Wind directions are given in terms of cardinal directions, such as N, NW, NNW, etc. – for example, the north (N) wind is that which comes from the north. Wind speed is measured in knots (nautical miles per hour), kilometers per hour (km/h) or miles per hour (mph), or meters per second (m/s). The direction of the wind is measured by a wind vane and the speed by an anemometer. The conversion factor for wind speed is:

$$1\text{ knot}=1.852\text{ km/h}=0.514\text{ m/s}=1.151\text{ mph}.$$

In irrigation engineering, the average wind velocity (in km/h) is often needed at a height of 2 m, which can be computed from observations of wind velocity at a height z (in m) as

$$u_2=U_z\left(\frac{2.0}{z}\right)^{0.2}, \tag{2.44}$$

in which u_2 is the wind velocity at a 2-meter height, and z is the height in meters at which velocity U_z is measured.

2.7.9 Clouds and Sunshine Hours

Clouds form when the ascending air becomes supersaturated at high altitudes. Cooling is caused by the expansion of air as it rises gradually to the upper atmosphere. Caused by evaporation from water bodies and evapotranspiration from vegetated lands, water vapor, being light, rises upward, where it expands adiabatically and is cooled below the saturation point. The excess moisture is condensed on suspended dust particles as minute globules of water which form clouds.

The solar radiation received at the Earth's surface is the duration of sunshine. The mean monthly values of possible sunshine hours depend on the month and the latitude, as shown in Table 2.3.

2.7.10 Radiation

Radiation is the process by which energy is emitted by all objects having a temperature above absolute zero in the form of electromagnetic waves. All bodies (solid, liquid, and gas) absorb and emit radiation, depending on the nature of the body and its temperature. A black body completely absorbs the radiation in all wavelengths falling on it, and the

Table 2.3 *Mean daily duration of maximum possible sunshine hours (N) for different months and latitudes*

Northern latitude	Jan.	Feb.	Mar.	Apr.	May	June	July	Aug.	Sept.	Oct.	Nov.	Dec.
Southern latitude	July	Aug.	Sept.	Oct.	Nov.	Dec.	Jan.	Feb.	Mar.	Apr.	May	June
50	8.5	10.1	11.8	13.8	15.4	16.3	15.9	14.5	12.7	10.8	9.1	8.1
48	8.8	10.2	11.8	13.6	15.2	16	15.6	14.3	12.6	10.9	9.3	8.3
46	9.1	10.4	11.9	13.5	14.9	15.7	15.4	14.2	12.6	10.9	9.5	8.7
44	9.3	10.5	11.9	13.4	14.7	15.4	15.2	14	12.6	11	9.7	8.9
42	9.4	10.6	11.9	13.4	14.6	15.2	14.9	13.9	12.6	11.1	9.8	9.1
40	9.6	10.7	11.9	13.3	14.4	15	14.7	13.7	12.5	11.2	10	9.3
35	10.1	11	11.9	13.1	14	14.5	14.3	13.5	12.4	11.3	10.3	9.8
30	10.4	11.1	12	12.9	13.6	14	13.9	13.2	12.4	11.5	10.6	10.2
25	10.7	11.3	12	12.7	13.3	13.7	13.5	13	12.3	11.6	10.9	10.6
20	11	11.5	12	12.6	13.1	13.3	13.2	12.8	12.3	11.7	11.2	10.9
15	11.3	11.6	12	12.5	12.8	13	12.9	12.6	12.2	11.8	11.4	11.2
10	11.6	11.8	12	12.3	12.6	12.7	12.6	12.4	12.1	11.8	11.6	11.5
5	11.8	11.9	12	12.2	12.3	12.4	12.3	12.3	12.1	12	11.9	11.8
0	12.1	12.1	12.1	12.1	12.1	12.1	12.1	12.1	12.1	12.1	12.1	12.1

After Doorenbos and Pruitt, 1977. Reprinted with permission from FAO.

radiation emitted is known as black body radiation. The term black has nothing to do with the color of the body. The sun, the Earth, and the clouds radiate very nearly as black bodies over a wide range of wavelengths.

The temperature of a radiating body and the radiation emitted by it can be expressed using the Stefan–Boltzmann law that states that the amount of energy radiated per unit time and wavelength from a unit surface area of an ideal black body is proportional to the fourth power of the absolute temperature of the body, expressed as

$$I = \sigma T^4, \tag{2.45}$$

where I is the amount of radiation in cal/cm^2/min, T is the absolute temperature of the black body in absolute degrees, and σ is the Boltzmann constant. The value of σ depends on the units used. Its value in CGS units is 5.67×10^{-5} erg cm^2/s^1/k^4 (1 erg = 10^{-7}J) or 5.67×10^{-8} J/m^2/s^1/k^4. Using eq. (2.45), the solar energy emitted at the sun's surface can be calculated. Solar energy arriving at the outer edge of Earth's atmosphere is called the solar constant, with an average rate of 1364 W/m^2 or 1.98 cal/cm^2/min. The solar radiation received by the entire Earth's surface area of $4\pi r^2$ (r is the radius of Earth) is, on average, equal to one-quarter of the solar constant (341 W/m^2), or about 0.4855 cal/cm^2/min.

The solar energy in Earth's atmosphere system involves absorption, scattering, and diffuse reflection. Figure 2.6 shows the average radiation balance in the northern hemisphere based on the data in Trenberth et al. (2009). To understand the relative importance of the various processes, the energy-flux values are normalized so that the incoming solar radiation from the sun is equal to 100 units: 100 units = 341 W/m^2.

On average, 23% (78 W/m^2) of solar radiation is absorbed in the troposphere, including 19% absorbed by the atmosphere and 4% by clouds. Another 30% (102 W/m^2) of the incoming solar radiation is returned to space during its journey from the top of the atmosphere to the surface of the Earth by scattering in the atmosphere and by diffused reflection from clouds, and also at the Earth's surface. When the incoming solar radiation encounters small particles (air molecules, dust, aerosols, and water droplets) in the atmosphere, the radiation energy is scattered in all directions. About 5% of the energy is scattered back into space. Clouds reflect much incoming solar radiation to space. About 18% of the total radiation is reflected from clouds and a further 7% is reflected from the Earth's surface. Thus, after absorption, scattering, and reflection, about 47% (161 W/m^2) of the solar radiation is absorbed by the Earth's surface, and most of the energy is in the visible range (i.e., incident radiation). It is this part of solar energy that is available for physical processes associated with weather and climate, such as heating land and water surfaces, evaporation, and melting. As shown in Figure 2.6, this amount of energy absorbed by the Earth is balanced by the net upward transport of terrestrial longwave radiation, latent heat, and sensible heat.

The fraction of the total incoming radiation that is reflected back to space is called albedo. It has been seen that 30% of the total radiation is reflected from clouds (18%) and the Earth's surface (7%), and a further 5% from scattered radiation. Under average conditions, the albedo (reflectivity) of the Earth as a whole (planetary albedo) is about 30%, with an estimated uncertainty of 3% (Trenberth et al., 2009). Of

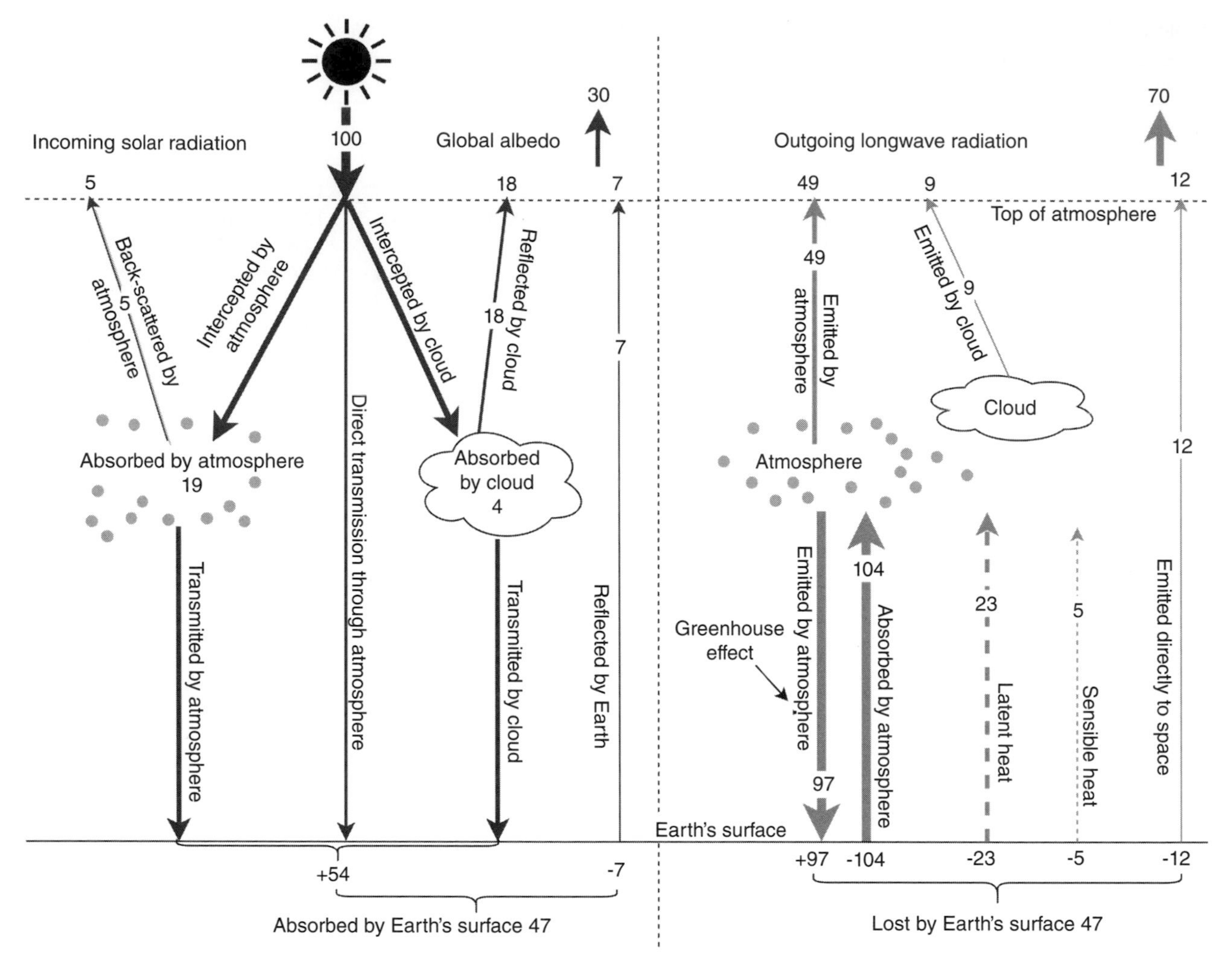

Figure 2.6 Average global energy balance. (Data from Trenberth et al., 2009.) A black and white version of this figure will appear in some formats. For the color version, please refer to the plate section.

the remaining 70%, a small part (23%) is absorbed in the atmosphere, but the bulk (47%) is absorbed by the Earth's surface.

The Earth–atmosphere system receives energy from the sun in the form of shortwave radiation and returns the same amount of energy to space in the form of longwave radiation. The energy absorbed by Earth's surface includes 47 units from the incoming radiation and 97 units of longwave radiation from gases in the atmosphere. Some of this energy, about 23 units (80 W/m^2), is transferred to the atmosphere by evaporation (latent heat), and 5 units (17 W/m^2) is transferred to the atmosphere by direct heating (sensible heat). The net energy changes on Earth's surface result in a mean surface temperature of about 288 K (15 °C). The radiation emitted (I) per unit area of the surface per unit time is given by the Stefan–Boltzmann formula (eq. 2.45) as $I = \sigma T^4$, where $\sigma = 8.12 \times 10^{-11}$ cal/cm^2/min/K^4. The Boltzmann constant is also expressed as $\sigma = 5.675 \times 10^{-8}$ J/m^2/s/K^4 $= 4.903 \times 10^{-9}$ MJ/m^2/day/K^4. For 288 K, I is obtained as $I = 8.12 \times 10^{-11} \times (288)^4 = 0.56$ cal/cm^2/min. Expressed on the same scale as for the incident solar radiation, this is equal to 116 units ($0.56/0.4855 \sim 116$). This is roughly 2.5 times the 47 units of solar radiation absorbed by Earth's surface. Of this emitted radiation, 104 units are absorbed by the water vapor and CO_2 in the troposphere, while 12 units escape to space (Figure 2.6).

It is estimated from the radiative transfer that 155 units of longwave radiation are emitted by the troposphere. Of this, 97 units are directed downward and absorbed by Earth. The balance of 58 units escapes to space at the tropopause (Table 2.4). The total longwave radiation emitted to space by the Earth and the troposphere together is 70 units (12 + 58). This is equal to the 70 (47 + 23) units of shortwave radiation absorbed by the Earth and the troposphere. Thus, the overall energy balance of the Earth–troposphere system is maintained.

Table 2.4 *Average global energy balance at the top of the atmosphere, of the atmosphere itself, and of Earth's surface*

Incoming energy			Outgoing energy		
	Normalized units	W/m^2		Normalized units	W/m^2
Top of the atmosphere					
Solar radiation (SW)	+100	341	Reflected by clouds (SW)	−18	62
			Scattered by atmosphere (SW)	−5	17
			Reflected by Earth's surface (SW)	−7	23
			Emitted by atmosphere (LW)	−49	167
			Emitted by cloud (LW)	−9	31
			Emitted by Earth's surface (LW)	−12	41
Total	**+100**	**341**		**−100**	**341**
Atmosphere					
Absorbed by atmosphere (SW)	+19	65	Emitted to space by atmosphere (LW)	−49	−167
Absorbed by cloud (SW)	+4	14	Emitted to space by cloud (LW)	−9	−31
Absorbed from Earth's surface (LW)	+104	355	Emitted to Earth's surface by atmosphere (LW)	−97	−331
Latent heat release	+23	78			
Sensible heat from the Earth	+5	17			
Total	**+155**	**529**		**−155**	**529**
Earth's surface					
Solar radiation (SW)	+54	184	Reflected to space (SW)	−7	−24
Emitted by atmosphere (LW)	+97	331	Emitted to atmosphere (LW)	−104	−355
			Emitted to Earth's surface (LW)	−12	−41
			Latent heat (evaporation)	−23	−78
			Sensible heat (removal of heat)	−5	−17
Total	**+151**	**515**		**−151**	**−515**

LW, longwave radiation; SW, shortwave radiation.
Adapted from US Department of Commerce, National Oceanic and Atmospheric Administration, and National Weather Service, www.weather.gov/jetstream/energy.

Example 2.14: Compute the amount of radiation emitted per unit surface area per unit time if the temperature is 20 °C, 25 °C, and 35 °C.

Solution: Converting the given temperatures of 20 °C, 25 °C, and 35 °C to kelvin we get, respectively, 293.15 K, 298.15 K, and 308.15 K. The radiation emitted (I) per unit area of the surface per unit time is given by the Stefan–Boltzmann formula as: $I = \sigma T^4$:

For $T = 293.15$ K, $I = 8.12 \times 10^{-11} \times (293.15)^4 = 0.60\ \text{cal/cm}^2/\text{min}$.

For $T = 298.15$ K, $I = 8.12 \times 10^{-11} \times (298.15)^4 = 0.64\ \text{cal/cm}^2/\text{min}$.

For $T = 308.15$ K, $I = 8.12 \times 10^{-11} \times (308.15)^4 = 0.73\ \text{cal/cm}^2/\text{min}$.

The shortwave radiation can be computed as

$$R_s = R_{so}\left(a + b\frac{n}{N}\right), \tag{2.46a}$$

$$R_s = I_0\left(a + b\frac{n}{N}\right), \tag{2.46b}$$

where R_{so} is the daily clear sky solar radiation at Earth's surface (cal/cm^2·day), I_0 is the extraterrestrial radiation, which is the solar radiation received at the Earth's outer surface or Earth's atmosphere (cal/cm^2·day), n is the actual number of sunshine hours, N is the possible maximum number of sunshine hours, a is a constant varying from 0.18 to 0.24, and b is a constant varying from 0.54 to 0.56. Table 2.5 gives the daily clear sky solar radiation (R_{so}) at the Earth's surface, and Table 2.6 gives the values of extraterrestrial radiation (I_0) in millimeters of evaporable water per

Table 2.5 *Total daily clear sky solar radiation at the surface of the Earth (cal/cm^2)*

	Month											
Latitude	Jan.	Feb.	Mar.	Apr.	May	June	July	Aug.	Sept.	Oct.	Nov.	Dec.
60° N	58	152	319	533	671	763	690	539	377	197	87	35
55	100	219	377	558	690	780	706	577	430	252	133	74
50	155	290	429	617	716	790	729	616	480	313	193	126
45	216	365	477	650	729	797	748	648	527	371	260	190
40	284	432	529	677	742	800	755	674	567	426	323	248
35	345	496	568	700	742	800	761	697	603	474	380	313
30	403	549	600	713	742	793	755	703	637	519	437	371
25	455	595	629	720	742	780	745	703	660	561	486	423
20	500	634	652	720	726	760	729	697	680	597	537	474
15	545	673	671	713	706	733	706	684	697	623	580	519
10	584	701	681	707	684	700	681	665	707	648	617	565
5	623	722	690	700	652	663	645	645	710	665	650	606
0	652	740	694	680	623	627	616	623	707	684	680	619
5° S	648	758	690	663	590	587	577	590	693	690	727	677
10	710	772	681	640	571	543	526	558	680	690	727	710
15	729	779	665	610	516	497	497	519	657	687	747	739
20	748	779	645	573	474	447	445	481	630	677	753	761
25	761	779	626	533	419	400	406	439	600	665	767	777
30	771	772	600	497	384	353	358	390	567	648	767	793
35	774	754	568	453	335	300	310	342	530	629	767	806
40	774	729	529	407	281	243	261	290	477	603	760	813
45	774	704	490	357	229	183	203	235	447	571	747	813
50	761	669	445	307	174	127	148	177	400	535	727	806
55	748	630	397	250	123	77	97	123	343	497	707	794
60	729	588	348	187	77	33	52	74	283	455	700	787

From Cuenca, 1989. Copyright © 1989 by Prentice Hall. Reprinted with permission from Prentice Hall.

Table 2.6 *Mean monthly solar radiation incident at Earth's outer surface (extraterrestrial radiation), I_0 in millimeters of evaporable water per day in the northern and southern hemispheres*

	Northern latitude (°N)						Southern latitude (°N)				
Month	50°	40°	30°	20°	10°	0°	50°	40°	30°	20°	10°
January	3.8	6.4	8.8	11.2	13.2	14.5	17.5	17.9	17.8	17.3	16.4
February	6.1	8.6	10.7	12.7	14.2	15.0	14.7	15.7	16.4	16.5	16.3
March	9.4	11.4	13.1	14.4	15.3	15.2	10.9	12.5	14.0	15.0	15.5
April	12.7	14.3	15.2	15.6	15.5	14.7	7.0	9.2	11.3	13.0	14.2
May	15.8	16.4	16.5	16.3	15.5	13.9	4.2	6.6	8.9	11.0	12.8
June	17.1	17.3	17.0	16.4	15.3	13.4	3.1	5.3	7.8	10.0	12.0
July	16.4	16.7	16.8	16.3	15.3	13.5	3.5	5.9	8.1	10.4	12.4
August	14.1	15.2	15.7	15.9	15.5	14.2	5.5	7.9	10..1	12.0	13.5
September	10.9	12.5	13.9	14.8	15.3	14.9	8.9	11.0	12.7	13.9	14.8
October	7.4	9.6	11.6	13.3	14.7	15.0	12.9	14.2	15.3	15.8	15.9
November	4.5	7.0	9.5	11.6	13.6	14.6	16.5	16.9	17.3	17.0	16.2
December	3.2	5.7	8.3	10.7	12.9	14.3	18.2	18.3	18.1	17.4	16.2

After Doorenbos and Pruitt, 1977. Reprinted with permission from FAO.

day. Fritz and MacDonald (1949) used $a = 0.35$ and $b = 0.61$ to compute R_s using eq. (2.46a), whereas Penman (1948) used $a = 0.18$ and $b = 0.55$ and Black et al. (1954) used $a = 0.23$ and $b = 0.48$ to compute R_s using eq. (2.46b).

Example 2.15: What will be the solar radiation incident at Earth's surface in the month of July for a place at a latitude of 30° N. The mean observed number of sunshine hours is 12.

Solution: The maximum number of sunshine hours obtained from Table 2.3 is 13.9. The value of I_0 can be obtained directly from Table 2.6 as 16.8 mm of water. Then, using eq. (2.46) with $a = 0.18$ and $b = 0.55$, the value of R_s in millimeters of water is computed as

$$R_s = \left(0.18 + 0.55 \frac{12}{13.9}\right)(16.8) = 11.0 \text{ mm of water/day}.$$

The net outgoing longwave radiation, R_L, can be computed as

$$R_L = R_{L0}\left(a_1 + \frac{R_s}{R_{so}} b_1\right), \tag{2.47}$$

in which R_{L0} is the net outgoing clear sky longwave radiation (cal/cm^2·day), R_{so} is the solar radiation under clear sky conditions (cal/cm^2·day) and is the same as R_{so} before, and a_1 and b_1 are constants. Penman (1948) used $a_1 = 0.1$ and $b_1 = 0.9$. The value of R_{L0} is computed as the difference between the outgoing longwave radiation from the surface (L_{out}) and the longwave radiation emitted by the atmosphere and clouds (L_{in}):

$$L_{out} = \varepsilon_s \,\sigma T_s^4 + (1 - \varepsilon_s) L_{in}, \tag{2.48a}$$

$$L_{in} = \varepsilon_a \,\sigma T_a^4, \tag{2.48b}$$

$$R_{L0} = L_{out} - L_{in} = \varepsilon_s \,\sigma T_s^4 - \varepsilon_s \varepsilon_a \,\sigma T_a^4, \tag{2.48c}$$

in which ε_s and ε_a are the emissivity of Earth's surface and that of the atmosphere, respectively; and T_s and T_a are the surface temperature and air temperature (K), respectively. The value of emissivity varies from 0.9056 to 0.985 for water surfaces. For grass surfaces, emissivity varies from 0.97 and 0.98. Since for grass surfaces ε_s is very close to 1, eq. (2.48c) can be simplified to

$$R_{L0} = \varepsilon_s \,\sigma T_s^4 \left(1 - \frac{\varepsilon_a \,\sigma T_a^4}{\varepsilon_s \sigma T_s^4}\right) = \sigma T_s^4 \left(1 - \frac{\varepsilon_a \,\sigma T_a^4}{\varepsilon_s \sigma T_s^4}\right) \tag{2.49}$$
$$= \sigma T_s^4 (1 - \varepsilon_0) = \varepsilon \sigma T_s^4,$$

in which ε_0 is defined as the ratio of longwave radiation down from the atmosphere to longwave radiation upward from the Earth, and ε is the effective emissivity of the surface – that is, $\varepsilon = 1 - \varepsilon_0$. ε depends on ε_a, T_a, and T_s, or equivalently, for clear sky, it depends largely on humidity as

$$\varepsilon = a_2 + b_2 \,(e_a)^{0.5}, \tag{2.50}$$

where e_a is the actual vapor pressure of the air (mb) at a height of 2 m, and a_2 and b_2 are constants.

ε can also be computed from mean temperature (°C) as (Burman et al., 1983):

$$\varepsilon = -0.02 + 0.261 \exp\left[-7.77 \times 10^{-4} (T_{mean})^2\right]. \tag{2.51}$$

Penman (1948) expressed R_L in terms of millimeters of water as

$$R_L = \sigma T^4 \left[0.56 - 0.08 (e_a)^{0.5}\right]\left(0.1 + 0.9 \frac{n}{N}\right), \tag{2.52}$$

where the temperature is in K, e_a is actual vapor pressure in mb. As shown here, $\varepsilon = 0.56 - 0.08(e_a)^{0.5}$. The term $0.1 + 0.9 \frac{n}{N}$ is similar to eq. (2.47). If the Stefan–Boltzmann constant, σ, is in 4.903×10^{-9} MJ/m^2/day/K^4, R_L can be converted to mm/day with latent heat of vaporization as 2.46 MJ/kg and water density as 1000 kg/m^3.

Doorenbos and Pruitt (1977) computed net outgoing longwave radiation from measured sunshine hours as

$$R_L = \sigma T^4 \left[0.34 - 0.044 (e_a)^{0.5}\right]\left(0.1 + 0.9 \frac{n}{N}\right), \tag{2.53}$$

where the temperature T is in K, e_a is actual vapor pressure in mb. As shown here, $\varepsilon = 0.34 - 0.044(e_a)^{0.5}$. The term $0.1 + 0.9 \frac{n}{N}$ is similar to eq. (2.47). If the Stefan–Boltzmann constant, σ, is in 4.903×10^{-9} MJ/m^2/day/K^4, R_L can be converted to mm/day with latent heat of vaporization as 2.46 MJ/kg and water density as 1000 kg/m^3.

Example 2.16: Determine the net outgoing longwave radiation for a day in July for a place at 30° N. The mean number of sunshine hours is 12, the mean air temperature is 35 °C, and the actual vapor pressure is 30 mb.

Solution: Given 4.903×10^{-9} MJ/m^2/day/K^4, $T = 35 + 273.15 = 308.15$ K, $n = 12$ hours, $e_a = 30$ mb, and $N = 15.9$ hours. From eq. (2.52),

$$R_L = \sigma T^4 \left[0.56 - 0.08(e_a)^{0.5}\right]\left(0.1 + 0.9\,\frac{n}{N}\right)$$
$$= \left(4.903 \times 10^{-9}\ \mathrm{MJ/m^2/day/K^4}\right) \times (308.15\ \mathrm{K})^4$$
$$\left[0.56 - 0.08(30)^{0.5}\right]\left(0.1 + 0.9\,\frac{12}{15.9}\right)$$
$$= 4196.7\ \mathrm{kJ/m^2/day}.$$

Converting from energy values to equivalent evaporation:

$$\text{Radiation (depth of water)} = \frac{\text{Radiation (energy)}}{\lambda \rho_w},$$

where ρ_w is the density of water (1000 kg/m^3), λ is the latent heat of vaporization in kJ/kg:

$$\lambda = 2500.78 - 2.360T_a = 2500.78 - 2.360 \times 35$$
$$= 2418.18\,\frac{\mathrm{kJ}}{\mathrm{kg}},$$

$$R_L = 4196.7\,\frac{\mathrm{kJ}}{\mathrm{m^2{\cdot}day}} \times \frac{1000\,\frac{\mathrm{mm}}{\mathrm{m}}}{(2418.18)\,\frac{\mathrm{kJ}}{\mathrm{kg}} \times \frac{1000\ \mathrm{kg}}{1\ \mathrm{m^3}}}$$
$$= 1.74\ \mathrm{mm/day}.$$

Example 2.17: What will be the value of ε if the air temperature is 25 °C and relative humidity is 50%? Compute using eq. (2.50) with the value of a_2 as 0.34 and b_2 as –0.044, and eq. (2.51).

Solution: The saturation vapor pressure is 31.68 mb at the air temperature of 25 °C, from Table 2.2. The vapor pressure is to be computed as

$$e_a = e_s \times RH = 31.68 \times 50\% = 15.84\ \mathrm{mb}.$$

Using eq. (2.50),

$$\varepsilon = a_2 + b_2(e_a)^{0.5} = 0.34 - 0.044(15.84)^{0.5} = 0.165.$$

Using eq. (2.51),

$$\varepsilon = -0.02 + 0.261\ \exp\left[-7.77 \times 10^{-4}(T_{mean})^2\right] = 0.141.$$

2.8 THE GREENHOUSE EFFECT

The Earth–atmosphere system receives energy from the sun in the form of shortwave radiation and returns the same amount of energy to space in the form of longwave radiation. The amount of solar energy absorbed by the Earth–atmosphere system is $(1 - \alpha)S/4$ cal/cm^2/min, where S is the solar constant and α is the planetary albedo – that is, the percentage of the incoming radiation that is reflected back to space, which amounts to approximately 30%. Because of the law of conservation of energy for Earth, the incoming radiation must be balanced by an amount equal to $(1 - \alpha)S/4$ cal/cm^2/min. This loss must be recouped in the longwave radiation that goes from the Earth–atmosphere system back to space. Assuming the mean surface temperature of Earth as T_E, then, from the Stefan–Boltzmann law, the average longwave radiation, R_E, per unit area of Earth's surface can be expressed as

$$R_E = \sigma T_E^4, \tag{2.54}$$

where σ is the Stefan–Boltzmann constant. The incoming solar radiation is balanced by the longwave terrestrial radiation:

$$\sigma T_E^4 = (1 - \alpha)\frac{S}{4}, \tag{2.55}$$

where $\sigma = 5.67 \times 10^{-8}\ \mathrm{W/m^2/k^4}$, $S = 1359\ \mathrm{W/m^2}$, and $\alpha = 0.3$. Introducing numerical values of quantities σ, S, and α, we obtain

$$T_E^4 = (1 - 0.3) \times \frac{1359}{4} \times \frac{10^8}{5.67}. \tag{2.56}$$
$$T_E^4 = 255\ \mathrm{K}(-18\ °\mathrm{C})$$

The result is much lower than the global mean temperature of Earth's surface. Temperature measurements averaged over all latitudes and all seasons yield a mean temperature of the Earth as 15 °C. The difference of 33 degrees is due to the fact that much of the radiation emitted by Earth's surface does not go straight into space, because certain gases in the atmosphere absorb and redirect it back to Earth, adding to its warmth. This process of warming is called the greenhouse effect and makes the Earth and the adjoining atmosphere warmer than they would have been if the atmosphere did not have greenhouse gases.

The greenhouse gases include water vapor, carbon dioxide, nitrous oxide, methane, and ozone. If our atmosphere did not absorb or trap the sun's heat in this manner, Earth would be as lifeless as the moon. The atmospheric concentration of each greenhouse gas is a result of the interaction of the biological systems on Earth with hydrologic and other biogeochemical cycles that involve various sinks and sources of greenhouse gases. Changes in the amounts of greenhouse gases throughout Earth's history, whether natural or recently human-induced, have coincided with the changes in Earth's climate. The most popular paradigm for the correlation between greenhouse gas concentrations and climate variations is that as concentrations of greenhouse gases increase, the atmosphere becomes an increasingly efficient trap for the redirected energy, and global atmospheric warming could occur (IPCC, 2001).

QUESTIONS

Q.2.1 Compute the amount of calories required to evaporate 5 liters of water at 20 °C. Compute the amount of ice that these calories will melt at 0 °C. The latent heat of fusion is 79.7 cal/g.

Q.2.2 Compute the weight of 2 m^3 of dry air having a temperature of 20 °C and pressure of 800 mb.

Q.2.3 Compute the density of dry air having a temperature of 20 °C and pressure of 800 mb and the density of moist air at the same temperature and pressure.

Q.2.4 Calculate the dew-point temperature, saturation vapor pressure, actual vapor pressure, and relative humidity if the dry-bulb temperature is 15 °C and the wet-bulb temperature is 10 °C.

Q.2.5 Compute the relative humidity if the air temperature is 15 °C and the dew-point temperature is 10 °C.

Q.2.6 Calculate the dew-point temperature if the air temperature is 35 °C and the relative humidity is 80%.

Q.2.7 A parcel of air at an elevation of 1000 m has a temperature of 25 °C. The air rises to an elevation of 2500 m and then descends to 2000 m. During the ascent, the air reaches saturation at an elevation of 1500 m. Condensation occurs from 1500 m to 2500 m. Compute the air temperature if the saturated adiabatic lapse rate is about half of the dry adiabatic lapse rate.

Q.2.8 If the atmospheric pressure is 1013 mb, the air temperature is 18 °C, and the wet-bulb temperature is 14 °C, calculate vapor pressure, vapor pressure deficit, relative humidity, and dew-point temperature in °C.

Q.2.9 Compute the saturation vapor pressure deficit in mb at an elevation of 2500 m where the dry-bulb temperature is 30 °C and the wet-bulb temperature is 25 °C.

Q.2.10 What is the net radiation if the incoming solar radiation is 12.5 mm/day, longwave radiation is 0.8 mm/day, and the terrestrial longwave radiation is 1.5 mm/day?

Q.2.11 Compute the net radiation in equivalent millimeters of evaporated water at a site having a latitude of 35° N in August and an elevation of 800 m if T_{max} is 32 °C and RH_{max} is 80% and T_{min} is 27 °C and RH_{min} is 60%, albedo (α) is 0.25, and $n = 11$ hours.

REFERENCES

Beck, H. E., Zimmermann, N. E., McVicar, T. R., et al. (2018). Present and future Köppen–Geiger climate classification maps at 1-km resolution. *Scientific Data*, 5: 180–214.

Black, J. N., Bonython, C. W., and Prescott, J. A. (1954). Solar radiation and duration of sunshine. *Proceedings of Royal Meteorological Society*, 80: 231–235.

Bosen, J. F. (1958). An appropriate formula compute relative humidity from dry-bulb and dew-point temperature. *Monthly Weather Review*, 86: 486.

Burman, R. D., Cuenca, R. H., and Weiss, A. (1983). Techniques for estimating irrigation water requirements. In *Advances in Irrigation*, Vol. 2, edited by D. Hillel, New York: Academic Publishers, pp. 335–394.

Cuenca, R. H. (1989). *Irrigation System Design: An Engineering Approach*. Englewood Cliffs, NJ: Prentice Hall.

Doorenbos, J., and Pruitt, W. O. (1977). Crop water requirements. Irrigation and Drainage Paper 24. FAO.

Fritz, S., and MacDonald, J. H. (1949). Average solar radiation in the United States. *Heat and Ventilation*, 46(7): 61–64.

Holdridge, L. R. (1967) *Life Zone Ecology*. San Jose: Tropical Science Center.

IPCC (Intergovernmental Panel on Climate Change) (2001). *Climate Change: The Scientific Basis*. Cambridge: Cambridge University Press.

Lugo A. E., Brown, S. L., Dodson, R., Smith, T. S., and Shugart, H. H. (1990). The Holdridge life of the conterminous United States in relation to ecosystem mapping. *Journal of Biogeography*, 26: 1025–1038.

Peel, M. C., Finlayson B. L., and McMahon T. A. (2007). Updated world map of the Köppen–Geiger climate classification. *Hydrology and Earth System Sciences*, 11: 1633–1644.

Penman, H. L. (1948). Natural evaporation for open water, bare soil, and grass. *Proceedings of the Royal Society of London*, A193: 120–146.

Trenberth, K. E., Fasullo, J. T., and Kiehl, J. (2009). Earth's global energy budget. *Bulletin of the American Meteorological Society*, 90: 311–324.

3 Sources and Availability of Water

Notation

a	parameter	p	probability (p) of each value
b	parameter	S_x	standard deviation of x
$f(y)$	probability density function	x	random variable which is the minimum flow rate
$F(y)$	cumulative probability distribution function	$\bar{x}$	mean value of x
m	rank of each value	y	transformed variate (also random)
n	number of value		

3.1 INTRODUCTION

Water for irrigation is derived either from surface sources or from below geologic formations. The surface water sources include primarily three sources: (1) streams; (2) lakes, ponds, tanks, or reservoirs; and (3) rainwater harvesting. Subsurface water sources include groundwater aquifers – confined and unconfined – from which water is extracted through wells and pumping. From both surface and subsurface sources, water is conveyed to the farm by pipelines, open channels, or canals. The distribution of water quantity and quality is variable in space and time.

Many streams are dependable for water supply, but some are not, and even for dependable streams the amount of water available in a given season may vary. Since streamflow has significant random variability, it may be necessary to provide for water impoundment as a lake, pond, or reservoir. However, sustainable withdrawal of surface water should be considered when designing storage facilities so that they hold enough water for dry seasons. This gives rise to the concept of safe yield, because there is always a risk associated with sourcing a river or reservoir. Reservoirs are designed based on past records, which do not guarantee that a worse drought will not occur in the future. Therefore, it is important to design such impoundments using risk analysis to evaluate the reservoir size and its cost. There is also the risk of pollution. Pipelines and canals are developed for large areas on a community basis, and water is received from a large reservoir or stream.

The other source of water is groundwater extracted through wells. Wells are either open-pit wells or are dug, usually in unconfined aquifers. Open dug wells are dug to a depth several feet or meters below the water table. Usually they are 9–24 meters (30–80 feet) deep, but do not provide enough water for irrigation. Deep wells are dug into confined aquifers and can be dependable sources of water. They can produce water at a rate of 3 liters per second (50 gallons per minute). Here, extraction of groundwater causes the lowering of the water table, which may cause social and environmental problems that have already started occurring, especially in developing countries. Chapter 4 provides more information on groundwater and wells. Figure 3.1a depicts the principal aquifers and the percentage of groundwater withdrawals to total water withdrawals in the United States. The figure shows that groundwater is the major water supply source for many regions in the United States. Wells are constructed using rotary drills, and in some wells gravel packing around a sand screen may be necessary to prevent pumping of sand. A good source of information when considering a well is the US Geological Survey, which has one or more offices in each state. A local driller can also be consulted, who can provide an estimate of the amount of water that can be expected and how deep the well can be drilled. A local water agency should be consulted to check if a permit will be required.

This chapter provides a snapshot of the sources of water – surface or subsurface – and the supply of water from reservoirs or rivers converted to farms through canals or from aquifers extracted through wells.

3.2 SURFACE WATER

A major source of water for irrigation is the withdrawal from rivers or surface water impoundments (like reservoirs, ponds, or lakes), but groundwater is becoming equally important, if it is not already. The main reason for surface water development is the cost and easy accessibility, as well as its wider prevalence. However, this is a limited resource, especially in the current environment of competing users, and has to be managed properly. Its management should encompass watershed

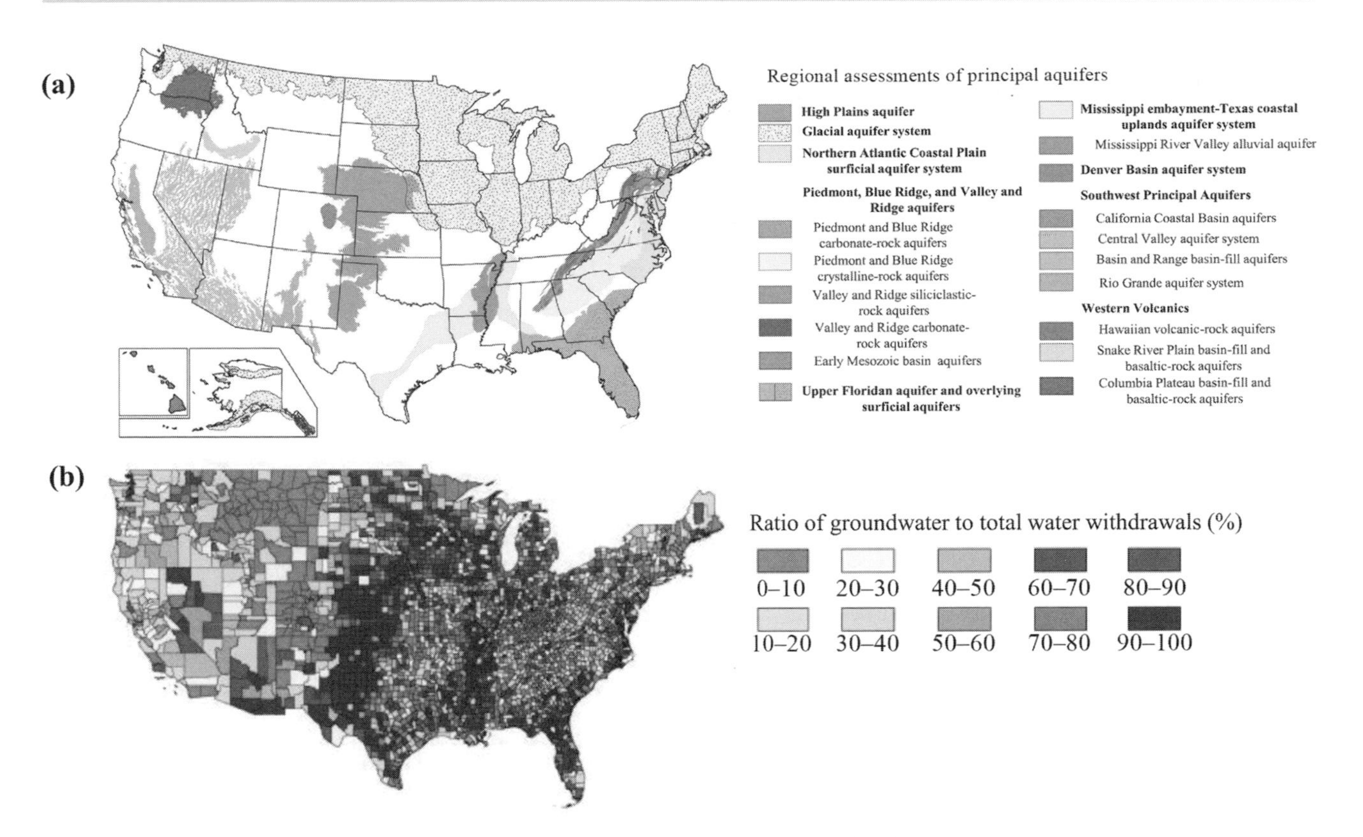

Figure 3.1 (a) Principal aquifers and (b) the percentage of groundwater withdrawals to total water withdrawals in the United States. (Adapted from US Geological Survey, www.usgs.gov/media/images/regional-assessments-principal-aquifers.) A black and white version of this figure will appear in some formats. For the color version, please refer to the plate section.

management and there may be legal issues, such as water rights and water sharing, that have to be dealt with.

To overcome the effect of seasonal variability of river flow, water is stored in reservoirs, and withdrawal of water should be managed properly. This water is not pristine and may contain sediment and other pollutants that may require sediment management.

Texas has approximately 191,000 miles of streams, 15 major river basins, 8 coastal basins, and 196 major reservoirs, as shown in Figure 3.2. Of these reservoirs, 175 are used for water supply, irrigation, and industry. More than half of the available surface water is derived from reservoirs: about 8.9 million acre-feet per year out of a total of 13.3 million. These reservoirs are able to capture and store flood waters for use during times of drought, when rivers are low or dry. The main regions that use streams are the Lower Rio Grande River valley, the Colorado River basin, and the Pecos River basin. Most of the water in these areas is considered good for irrigation, although the sodium content is slightly higher in the waters of the more western streams.

The Texas Water Development Board (2007) reports that approximately 90,000 acre-feet of water is lost per year from the major reservoirs in Texas due to sedimentation. This amount is approximately 0.27% of the total capacity per year. By 2060 the total loss would be 4.5 million acre-feet, which is greater than the capacity that can be gained from the construction of new reservoirs. This suggests that reservoirs will have to be managed more efficiently and sedimentation must be reduced. Surface water in Texas is the property of the state government, which permits individuals, cities, industries, businesses, and other public and private interests to use it.

The availability of water varies from basin to basin because of variations in precipitation and basin characteristics. Table 3.1 shows the average flow in acre-feet per year of major river basins in Texas. This flow is the average of the flow volume measured at the most downstream streamflow gage on the river in Texas. When the average annual flow is divided by the basin area the result is the average watershed yield. Existing water supplies correspond to the maximum amount of surface water available from existing sources that can be used during a drought of record conditions, and that is physically and legally available. Analogously, firm yield corresponds to the maximum amount of water that can be taken from a reservoir under a repeat of the drought of record. Extending further, safe yield is the firm yield in addition to an amount of water supply for an additional period of time. Run-of-river diversion corresponds to the amount of water that the permit holder can divert directly from the river under his water-right permit.

3.3 GROUNDWATER

Groundwater is an important source of water in most countries of the world. In Texas, groundwater represents

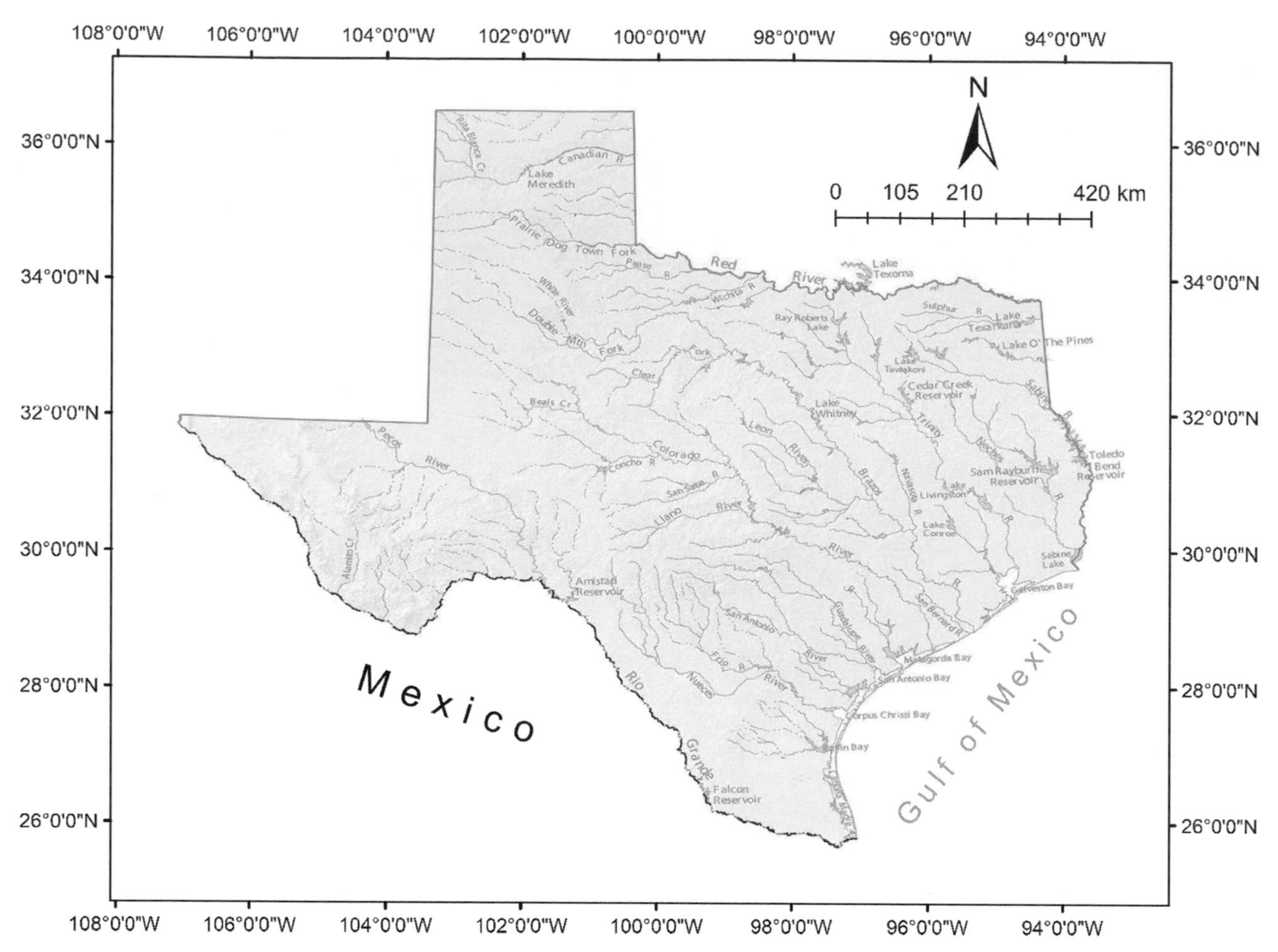

Figure 3.2 Lakes and reservoirs in Texas. (Adapted from Texas Water Development Board, 2007.)

approximately 59% of water used in the state and is pumped from 9 major and 21 minor recognized aquifers. Since the 1950s, groundwater production has been about 10 million acre-feet per year, which in 2003 represented 59% of 15.6 million acre-feet of total water use. About 79% of this groundwater was used for irrigation. It is projected that the amount of groundwater that can be produced with the current permits and existing infrastructures will decline by 32% between 2010 and 2060, from 8.5 million acre-feet per year to 5.8 million acre-feet per year. Likewise, the amount of water from an aquifer that is available for use will decrease by 22% from 12.7 million acre-feet per year in 2010 to 9.9 million acre-feet per year by 2060 (TWDB, 2007).

The nine major aquifers in Texas, as shown in Figure 3.3, are Ogallala, Pecos Valley, Hueco-Messila Bolsons, Edwards-Trinity (Plateau), Edwards (Balcones Fault Zone), Gulf Coast, Carrizo-Wilcox, Trinity, and Seymour. The 21 minor aquifers are shown in Figure 3.4. These aquifers are a critical source of water for Texas. Of these aquifers, Ogallala is the largest, supplying 82% of all groundwater used for irrigation, which is about 6.0 million acre-feet per year. Key aquifer characteristics are given in Table 3.2.

The TWDB (2007) has inventoried and entered records of nearly 140,000 water wells (including ~2000 springs) in the TWDB Groundwater Database (GWDB). In Texas, groundwater is the major source of irrigation water, especially in the High Plains, the El Paso valley (also partly supplied by the Rio Grande), the Winter Garden district, and the Gulf Coast area. Eighty percent of the total irrigated acreage comes from water pumped from wells. Important groundwater-supplied irrigated areas include the Winter Garden Region and adjacent lands below the Balcones Escarpment, the Trans-Pecos farms of Reeves, Pecos, and Ward counties, the Marfa, Van Horn, and Dell City vicinities, and part of the Gulf Coast, as well as several north central Texas localities (where, especially during the 1960s, there was an increased use of sprinkler systems for growing peanuts). The Southern High Plains still accounts for the largest portion at 68%. Of the total irrigated acres in Texas, 34% is watered by sprinkler irrigation and 70% of all irrigation is from wells. The use of drip irrigation had increased, somewhat, to 55,000 acres. The decline in irrigated acreage is attributed to decreasing groundwater supplies and higher fuel prices.

Table 3.1 *Average flow of river basins in Texas*

	Basin area		River length		Flow
River basin	Total area (mi^2)	In Texas (mi^2)	Total (miles)	In Texas (miles)	Average flow (acre-feet/year)
Brazos	45,573	42,865	840	840	6,074,000
Canadian	47,705	12,865	906	213[a]	196,000
Colorado	42,318	39,428	865	865	1,904,000
Cypress	3,552[b]	2,929[c]	90[d]	75[d]	493,700
Guadalupe	5,953[b]	5,953[b]	409[a]	409[a]	1,422,000
Lavaca	2,309[b]	2,309[b]	117[a]	117[a]	277,000
Neches	9,937[b]	9,937[b]	416	416	4,323,000
Nueces	16,700[b]	16,790[b]	315	315	539,700
Red	93,490[c]	24,297	1,360	695[a]	3,464,000
Rio Grande	182,215	49,387	1,896	889	645,500[f]
Sabine	9,756	7,570	360	360	5,864,000
San Antonio	4,180[b]	4,180[b]	238[a]	238[a]	562,700
San Jacinto	3,936[b]	3,936[b]	85	85	1,365,000[e]
Sulphur	3,767[b]	3,580[c]	227[d]	222[a]	932,200
Trinity	17,913[b]	17,913[b]	550	550	5,727,000

The source for total basin area and river length data is the USGS (1985) unless otherwise specified. The source for basin area within Texas data is TWDB unless otherwise specified. The source for average flow data is USGS (2004) unless otherwise specified.
[a] The source for these data is TCEQ (2004) and does not include length of in-channel reservoirs or lakes.
[b] The source for these data is TWDB.
[c] The Sulphur and Cypress River basins in Texas are sub-basins of the entire Red River basin.
[d] The sources for these data are TCEQ (2004) and ADEQ (2004) or LDEQ (2004), and do not include length of in-channel reservoirs or lakes.
[e] This value combines data from USGS (2004) for several tributaries.
[f] The source for these data is IBWC (2003).

3.4 WASTEWATER SOURCES

Owing to the shortage of water resources in many parts of the world, there is increasing attention on the use of wastewater generated domestically and industrially. Some countries, such as Israel (Kellis et al., 2013), have been emphasizing the greater use of wastewaters. However, there are issues related to human health and the likely effects on crops and soils when wastewaters are used in irrigation. On the other hand, wastewaters may contain beneficial elements, such as nitrogen, phosphates, and potassium, which may reduce the application of fertilizers and may help farmers with some cost savings. The application of wastewaters may require some treatment and satisfaction of local and state laws, if there are any.

3.5 WATER RIGHTS

In order to control the use of water in many states with limited supplies, water rights laws have been passed. There are two doctrines: (1) riparian and (2) appropriations. These doctrines apply mostly to streamflow. The use of water from lakes and ponds is also controlled in some states. In the eastern United States, the riparian doctrine states that a person can make reasonable use of water from a stream if (1) the stream borders his or her land, and (2) the water is used on the land. In general, a person can use water from a stream as long as someone downstream who depends on the stream is not deprived of using it. However, a person can legally limit the use of water by claiming a need or intention of use in the future.

In the western United States the appropriations doctrine states that the first user of water has the best lawful right, whether he owns the land next to the stream or not (Figure 3.5). No one can use the water from the stream if it reduces the water needed by the first user. The only way a person can use the water from this stream is by purchasing the water rights from the prior user or his or her heirs, or by purchasing both the land and the water rights.

The drilling of wells is also controlled in some western states. In the western United States almost all waters in existence are subject to grandfather clauses that appropriate water rights to those who used the water first, prior to the passage of new laws.

The water use from lakes may be regulated in a similar manner. The water from the lakes may not be withdrawn if it is reserved for recreational purposes. The water level in the lake must not be lowered below a specified level.

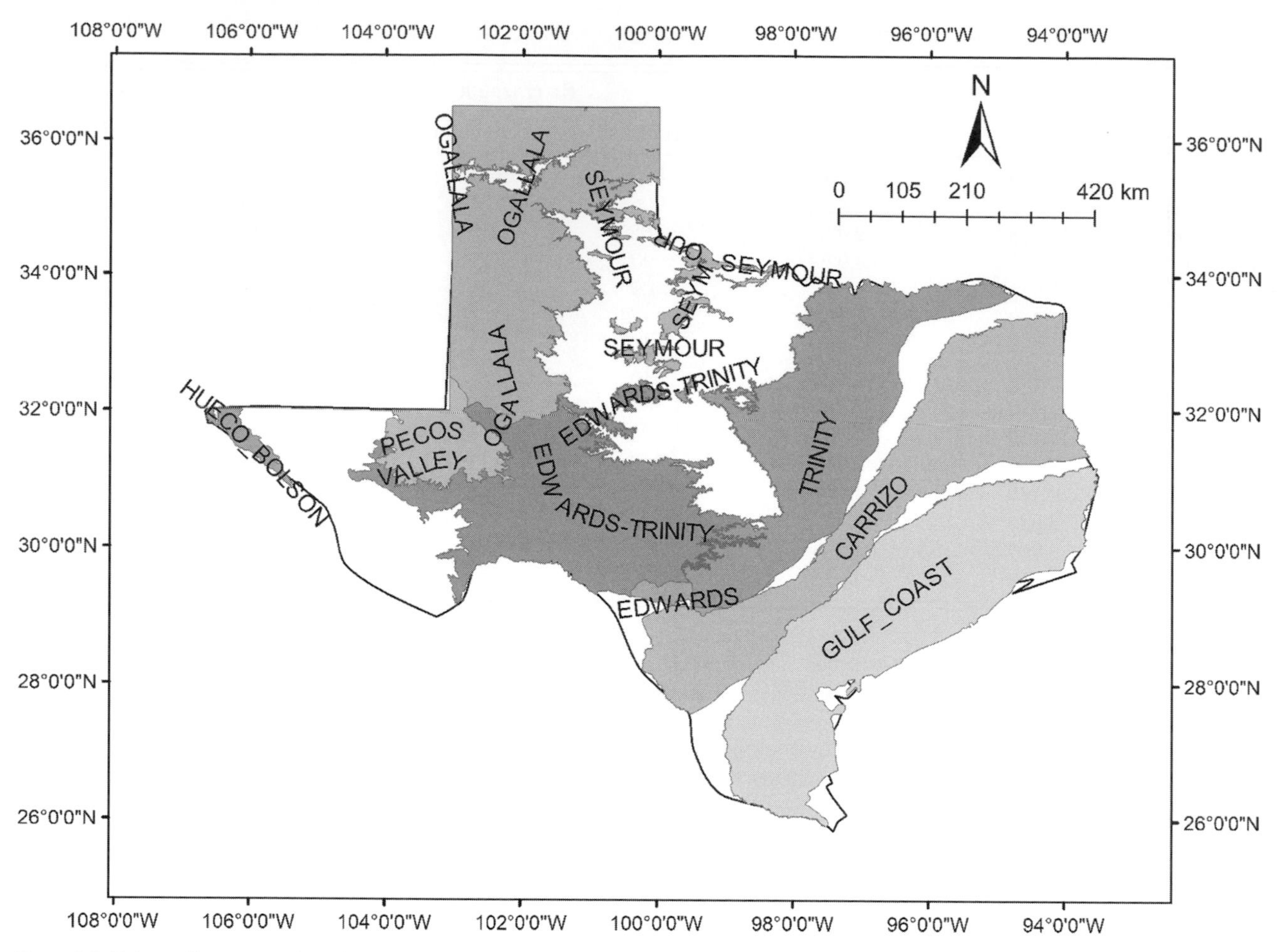

Figure 3.3 Major aquifers in Texas. (Data from Texas Water Development Board, www.twdb.texas.gov/mapping/gisdata.asp.) A black and white version of this figure will appear in some formats. For the color version, please refer to the plate section.

3.6 RAINWATER HARVESTING

Rainwater harvesting can supplement domestic supplies and water for gardening and small-scale irrigation. It is, however, limited by the size of tank that will provide a sustained water supply during periods of little or no rainfall. In South India tanks are widespread and irrigation has been practiced for centuries. For domestic purposes, rainwater collected from rooftops may contain significant amounts of debris, sediment, and other pollutants. There should therefore be some kind of filtration or treatment of water before use.

3.7 WASTEWATER AND RECYCLING

A significant amount of wastewater is generated in municipalities and agricultural feedlots. With some treatment, this wastewater can be used for irrigation, especially for irrigation of lawns and grasses, and for washing and cleaning, which will reduce competition for water for crop irrigation.

3.8 LOCATION OF WATER SOURCE

The distance between the source of water and the field is an important determinant for evaluating the adequacy of the source. The cost of bringing the water to the field determines the viability of the source. If the cost is high then other sources of water should be explored. If the cost is too high then it may not be profitable for irrigation.

3.9 HEIGHT OF WATER FOR PUMPING

The height of water to be pumped is an important consideration in determining the suitability of a source. It also affects the cost of irrigation. If the source of water is a surface source then one must consider (1) the height from the water surface to the pump, called suction lift; and (2) the height from the pump to the field. For centrifugal pumps, the suction lift should not be greater than 5 m (15 ft) for maximum efficiency.

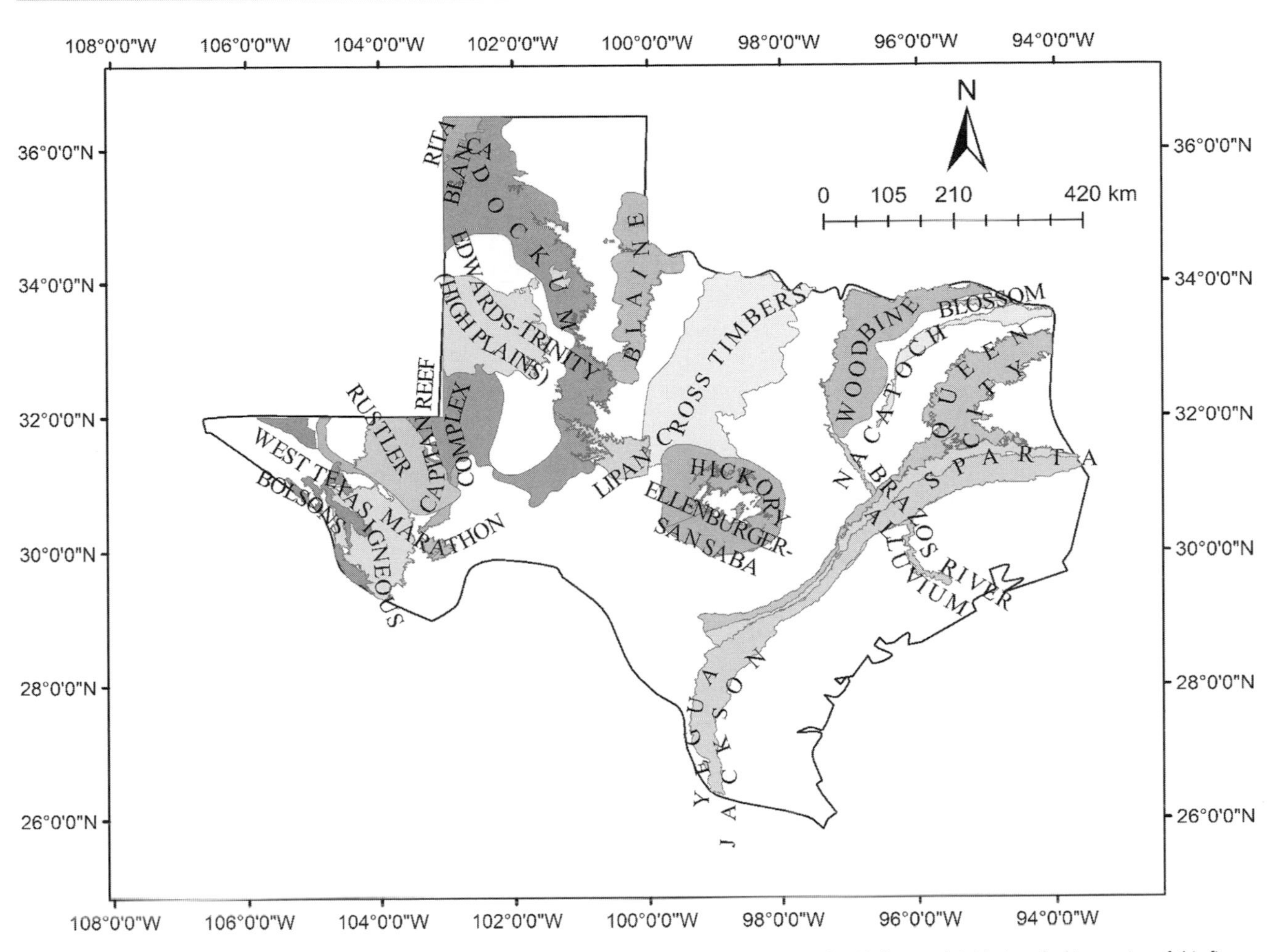

Figure 3.4 Minor aquifers in Texas. (Data from Texas Water Development Board, www.twdb.texas.gov/mapping/gisdata.asp.) A black and white version of this figure will appear in some formats. For the color version, please refer to the plate section.

For a groundwater source, the water is pumped from a drilled well. The total height is the sum of lift to the pump plus the height from the pump to the field. Different types of pumps are available that can pump water from as deep as 305 m (1000 ft.) or greater. There seems no reasonable limit to the height from the pump to the field.

3.10 AVAILABILITY OF WATER

It is important to know the amount of water available and the rate of flow from a given source. This information is important for determining the design capacity of irrigation systems. However, since irrigation needs must be met year after year it is important to determine variations in water supply sources as well as in irrigation needs, especially under climate change. These days, sophisticated measurement techniques are available that greatly help to determine the water availability of a source. However, there is considerable uncertainty about the availability of streamflow and consequent reservoir storage, and this uncertainty is being heightened by global warming and climate change. The impact of climate change is being manifested by the intensification of the hydrologic cycle, unseasonal changes in the precipitation regime, increased temperatures, and increased frequency of droughts, affecting agricultural irrigation.

3.10.1 Flow of Water in a Stream

There are three commonly used methods to determine flow in a stream: (1) the average velocity method; (2) a weir notch; and (3) the Parshall flume. The average velocity method involves the following steps:

1. Select a straight reach of the stream.
2. Make several measurements of the depth at several points in a cross-section in the transverse direction and then obtain the average cross-sectional depth.
3. Take several measurements of stream width and compute the average width of the stream.
4. Determine the average cross-section of the stream.
5. Determine the average velocity of flow.
6. Determine the discharge of the stream.

Table 3.2 *Key aquifer characteristics*

Aquifer	Area of aquifer (sq. miles)	Area of outcrop (sq. miles)	Area in subsurface (sq. miles)	Availability of water (acre-feet) per year
Blaine	–	3,443	2,203	315,183 (2010) to 313,933 (2060)
Blossom	–	182	95	2,270 (2010 to 2060)
Bone Spring-Victorio Peak	710	–	–	63,000 (2010 to 2060)
Brazos River Alluvium	1,053	–	–	99,632 (2010 to 2060)
Capital Reef Complex	1,842	–	–	52,150 (2010 to 2060)
Carrizo-Wilcox	–	11,186	25,409	1,014,753 (2010) to 1,010,793 (2060)
Dockum	–	3,519	21,992	408,138 (2010) to 246,720 (2060)
Edwards (Balcones Fault Zones)	–	1,560	2,314	373,811 (2010 to 2060)
Edwards-Trinity (High Plains)	7,889	–	–	4,160 (2010) to 2,066 (2060)
Edwards-Trinity Plateau	–	32,294	2,294	572,515 (2010) to 572,517 (2060)
Ellenburger-San Saba	–	1,147	4,262	45,672 (2010 to 2060)
Gulf Coast	–	271	8,193	1,825,976 (2010) to 1,661,736 (2060)
Hickory	41,679	–	–	278,316 (2010 to 2060)
Hueco-Messila Bolsons	1,370	–	–	183,000 (2010 to 2060)
Igneous	6,075	–	–	14,600 (2010 to 2060)
Lipan	–	1,565	422	48,536 (2010 to 2060)
Marathon	390	–	–	200 (2010 to 2060)
Marble Falls	213	–	–	22,637 (2010 to 2060)
Nacatoch	–	889	936	10,453 (2010 to 2060)
Ogalala	36,515	–	–	5,968,260 (2010) to 3,534,124 (2060)
Pecos Valley	6,829	–	–	200,690 (2010 to 2060)
Queen City	–	7,702	6,989	295,791 (2010 to 2060)
Rita Blanca	922	–	–	5,419 (2010 to 2060)
Rustler	–	309	4,660	2,492 (2010 to 2060)
Seymour	4,042	–	–	242,226 (2010) to 227,580 (2060)
Sparta	–	1,543	6,926	50,511 (2010 to 2060)
Trinity	–	10,652	21,306	205,799 (2010) to 202,603 (2060)
West Texas Bolsons	1,895	–	–	62,325 (2010 to 2060)
Woodbine	–	1,557	5,766	37,712 (2010) to 38,072 (2060)
Yegua-Jackson	10,904	–	–	24,720 (2010 to 2060)

From Texas Water Development Board, 2007.

Example 3.1: Consider a stream reach 30 m long and 1.5 m wide. In a selected cross-section the flow depth is measured at five points across the width. These depths are: 5 cm, 20 cm, 25 cm, 30 cm, and 50 cm. Using a float it takes 2.2 minutes to travel a distance of 30 m. Determine the discharge of water in the stream.

Solution:

Average depth $= \frac{5+20+25+30+50}{5} = 26$ cm or 0.26 m.

Using a float it takes 2.2 minutes to travel a distance of 30 m. Then, in 1 minute:

$$\text{float travels} = \frac{30}{2.2} = 13.64 \text{ m},$$

$$\text{velocity} = \frac{13.64}{60} \text{ m/s} = 0.227 \text{ m/s},$$

$$\text{cross-sectional area} = 0.26 \text{ m} \times 1.5 \text{ m} = 0.39 \text{ m}^2,$$

$$\text{discharge} = \text{velocity} \times \text{area} = 0.39 \frac{\text{m}}{\text{s}} \times 0.227 \text{ m}^2 = 0.089 \text{ m}^3/\text{s}.$$

3.10.2 Water in a Lake, Pond, or Reservoir

The amount of water available from a lake, pond, or reservoir can be computed as follows:

1. Determine the surface area using a rectangular method. This involves computing the average length and average width of the reservoir. Multiplying the width and length yields the surface area.

2. The surface area can also be determined more accurately by dividing the reservoir into a number of sections and computing the surface area of each section and then summing the section areas.
3. Determine the average depth of the water. This can be approximated by measuring the greatest depth. For small reservoirs the average depth is approximately 40% of the greatest depth.
4. Determine the volume of water by multiplying the surface area by the average depth.
5. Determine the volume of water available for irrigation by taking 80% of the volume of water in the lake. The remaining 20% accounts for the reduction in lake capacity due to evaporation, seepage, and sediment.

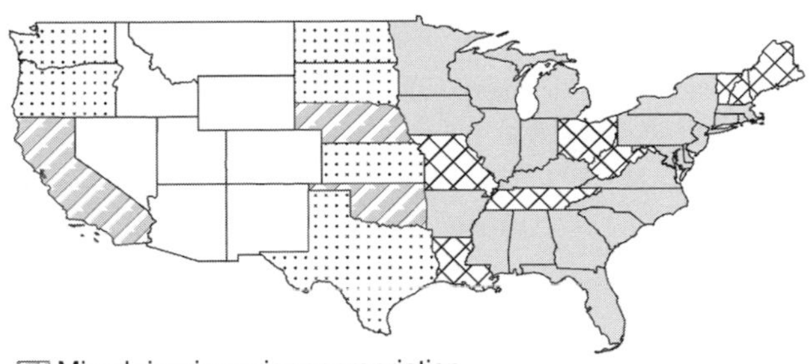

Mixed riparian prior appropriation
Prior appropriation formerly riparian
Pure prior appropriation
Pure riparian
Regulated riparianism

Figure 3.5 Water rights in the United States. Persons living in the eastern half of the United States under the "riparian doctrine" (either pure riparian or regulated riparianism) are entitled to reasonable use of water from a stream or lake. Those living in most of the western states must have earned the right by being the first to use the water under the "prior appropriations doctrine." (Data source: Christian-Smith et al., 2012.)

Example 3.2: Determine the amount of water available for irrigation from a lake 5 ha in area, as shown in Figure 3.6. The greatest depth in the lake is 5 m. Approximate the surface area of the lake by a rectangular shape. Take the average depth of water as 0.4 times the greatest depth.

Solution: Area = 5 ha (given), greatest depth = 5 m,

average depth = 0.4 times the greatest depth
= 0.4 × 5 = 2 m,

volume of water = 5 ha × 2 m
= 10 ha × m or 100,000 m^3,

volume of water available for irrigation
(assuming 80% efficiency of lake) = 10 × 0.8
= 8 ha × m = 100,000 × 0.8 = 80,000 m^3.

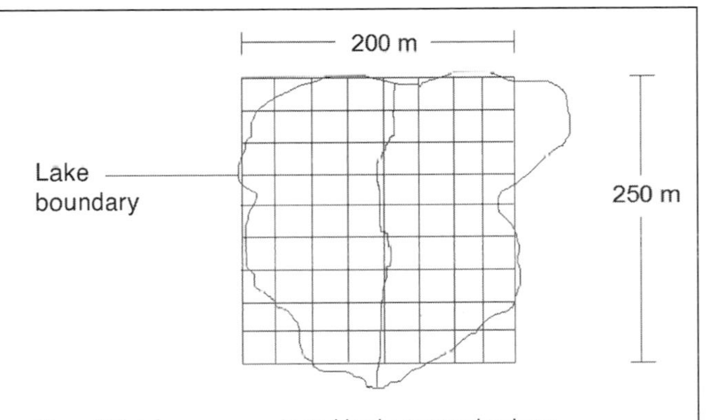

Figure 3.6 Lake area approximated by the rectangular shape.

3.10.3 Minimum Expected Supply Rate

Determination of the minimum expected supply rate requires frequency analysis of minimum daily flow data from stream or groundwater springs for the irrigation season. Frequency analysis involves the following steps:

1. Obtain the minimum flow from the irrigation season (say, April to October) for each year for several years (e.g., 30 or more years). The data for a large stream can be obtained from the US Geological Survey.
2. Arrange the minimum flow data in descending order.
3. Assign rank (m) to each value beginning with 1 ($m = 1$) for the highest value and 2 ($m = 2$) for the second highest value, and so on. The last value will assume the rank of $m = n$.
4. Compute the probability (p) of each value using an appropriate plotting position formula, such as

$$p = 1 - \frac{m}{n+1}. \tag{3.1}$$

If eq. (3.1) is multiplied by 100 then this will yield the probability of occurrence in percent. The inverse of the probability of occurrence multiplied by 100 defines the return period or recurrence interval (R).
5. Plot the value of the return period for each ranked value.
6. Plot the minimum values on the ordinate and the probabilities or return period on the abscissa on a probability paper. For simplicity, an extreme type I (or Gumbel) distribution can be fitted to the plotted data using the Gumbel distribution probability paper. The probability paper essentially is designed for the distribution so that

it transforms p in order to obtain a linear relation between p and minimum flow values. The Gumbel distribution can be expressed as

$$f(y) = a \exp[y - \exp(y)] \quad (3.2)$$

and

$$F(y) = 1 - \exp(-\exp(y)), \quad (3.3)$$

where

$$y = a(x - b), \quad (3.4)$$

where x is the random variable that is the minimum flow rate, y is the transformed variate (also random), $f(y)$ is the probability density function, $F(y)$ is the cumulative probability distribution function, and a and b are parameters estimated as

$$a = \frac{1.28255}{S_x}, \quad (3.5)$$

$$b = \bar{x} + 0.450041 S_x, \quad (3.6)$$

in which $\bar{x}$ is the mean value of x and S_x is the standard deviation of x computed respectively as

$$\bar{x} = \frac{\sum_{i=1}^{n} x_i}{n}, \quad (3.7)$$

$$S_x = \left[\frac{\sum_{i=1}^{n}(x_i - \bar{x})^2}{n-1}\right]^{0.5}. \quad (3.8)$$

It is noted that eq. (3.3) can be expressed as

$$y = \ln[-\ln(1 - F(y))], \quad (3.9)$$

where $F(y)$ expresses the non-exceedance probability of value y. Now the frequency analysis of minimum daily flow is illustrated with an example.

Example 3.3: The minimum daily flow rates for the Lower Colorado River at the gaging station Colorado River below Palo Verde Dam (USGS 09429100) for the irrigation season from April to October are given in Table 3.3. Compute the daily minimum flow rates that will be exceeded 5, 10, 20, 25, 30, 40, and 50% of the time, which will correspond to return periods of 20, 10, 5, 4, 3.33, 2.5, and 2 years, respectively.

Solution: The solution involves the following steps.

1. Arrange the minimum daily flow values in descending order.
2. Assign ranks to the ordered values.
3. Compute p for each value using eq. (3.1) and tabulate the p values.

   The p values are the non-exceedance probabilities of the distribution based on the observed data. Assuming these data follow the Gumbel distribution (minimum), one can use eqs (3.5) and (3.6) to estimate the parameters of the Gumbel distribution.

Table 3.3 *Minimum daily flow rates for the irrigation season (units: cfs)*

Year	Flow rate	Year	Flow rate	Year	Flow rate
1989	8700	2000	5330	2011	4200
1990	7890	2001	4950	2012	3390
1991	6650	2002	4870	2013	1330
1992	1900	2003	4450	2014	2780
1993	6470	2004	3050	2015	2350
1994	7390	2005	2980	2016	2420
1995	7080	2006	2980	2017	3250
1996	8740	2007	4280	2018	2150
1997	9990	2008	4450	2019	2070
1998	9690	2009	4100		
1999	5630	2010	3620		

4. Compute the mean and standard deviation of the daily minimum flow values. Using eqs (3.7) and (3.8), the standard deviation $S_x = 1268.543$ cfs and the daily minimum flow $\bar{x} = 3676.774$ cfs.
5. Compute the y values and tabulate them. The value of $F(y)$ corresponds to the probabilities. Using eqs (3.5) and (3.6),

$$a = \frac{1.28255}{S_x} = \frac{1.28255}{1268.543} = 0.001011,$$

$$b = \bar{x} + 0.450041 S_x = 3676.774 + 0.450041 \times 1268.543 = 4247.671.$$

Using eq. (3.4),

$$y = 0.001011(x - 4247.671).$$

Calculate the probability using eq. (3.3):

$$F(y) = 1 - \exp(-\exp(y)).$$

Flow rate (cfs)	Rank	p	y	$F(y)$
6250	1	0.9688	2.0244	0.9995
5650	2	0.9375	1.4178	0.9839
5330	3	0.9063	1.0943	0.9496
5330	4	0.8750	1.0943	0.9496
5260	5	0.8438	1.0235	0.9381
4950	6	0.8125	0.7101	0.8692
4870	7	0.7813	0.6292	0.8468
4820	8	0.7500	0.5786	0.8320
4450	9	0.7188	0.2046	0.7068
4450	10	0.6875	0.2046	0.7068
4280	11	0.6563	0.0327	0.6441
4200	12	0.6250	–0.0482	0.6144
4100	13	0.5938	–0.1493	0.5774
3780	14	0.5625	–0.4728	0.4638
3620	15	0.5313	–0.6346	0.4115

(*cont.*)

Flow rate (cfs)	Rank	*p*	*y*	*F*(*y*)
3540	16	0.5000	−0.7155	0.3867
3520	17	0.4688	−0.7357	0.3807
3390	18	0.4375	−0.8671	0.3431
3250	19	0.4063	−1.0087	0.3056
3200	20	0.3750	−1.0592	0.2930
3050	21	0.3438	−1.2109	0.2576
2980	22	0.3125	−1.2817	0.2424
2980	23	0.2813	−1.2817	0.2424
2780	24	0.2500	−1.4839	0.2029
2420	25	0.2188	−1.8479	0.1458
2350	26	0.1875	−1.9186	0.1365
2150	27	0.1563	−2.1208	0.1130
2070	28	0.1250	−2.2017	0.1047
1900	29	0.0938	−2.3736	0.0889
1730	30	0.0625	−2.5455	0.0754
1330	31	0.0313	−2.9499	0.0510

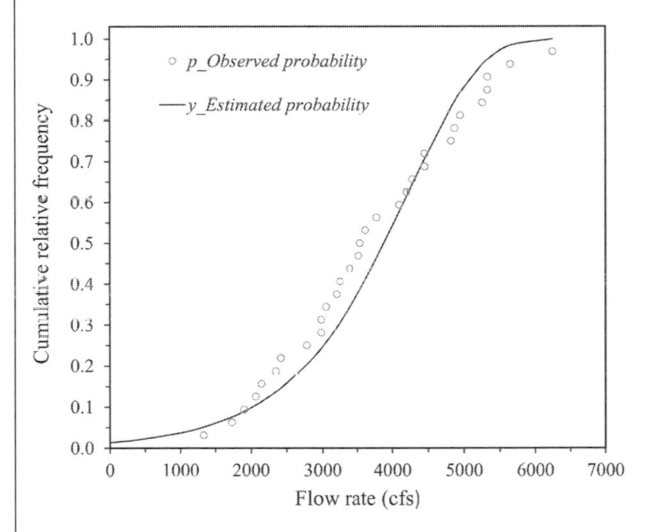

Figure 3.7 Cumulative probability distribution of daily minimum flow rate during the irrigation season (observed vs. estimated).

6. Plot the flow rate and $F(y)$ to show the cumulative distribution of the minimum flow rate, as seen in Figure 3.7. Compare it with the observed probability p.
7. Obtain the values of minimum daily flows for the desired recurrence intervals or probabilities.

The values of minimum daily flows for the desired recurrence intervals or probabilities can be calculated using eqs (3.9) and (3.4). Also, it can be directly read from Figure 3.7. For daily minimum flow rates that will be exceeded 10% of the time, $F(y) = 10\% = 0.1$. Using eq. (3.9),

$$y = \ln[-\ln(1 - F(y))] = \ln[-\ln(1 - 0.1)] = -2.25.$$

The minimum daily flow is computed using eq. (3.4) as

$$y = 0.001011(x - 4247.671) = -2.25,$$

Then, $x = 2021.9$ cfs.

The daily minimum flow rate that will be exceeded 10% of the time is 2021.9 cfs. The other results can be calculated using the same method:

Probability of occurrence (%)	Return period	Flow rate (cfs)
5	20.0	1309.9
10	10.0	2021.9
15	6.7	2450.6
20	5.0	2764.1
25	4.0	3015.4
30	3.3	3228.0
40	2.5	3583.3
50	2.0	3885.2

3.11 FLOW OF WATER FROM A WELL

The flow of water from a well can be accurately determined by a pumping test if the well has a pump. The flow from the well can be measured using a flow meter or an orifice plate.

3.12 COMBINATION OF SOURCES

Often, water for irrigation is drawn from a combination of sources. For example, for part of the growing season the water is withdrawn from a lake and part of the time from a well. A lake should be supplied with water before it gets dry or depleted. The effect of using combined sources can be enumerated using the following two relations:

1. Minimum lake capacity = (irrigation rate × hours pumped per day − recharge rate × hours pumped per day) / percent efficiency.
2. Minimum recharge rate = irrigation rate × $\frac{\text{hours pumped per day}}{\text{24 hours pumping per day}}$.

These relations must be satisfied when a combination of water sources is used.

The question arises: What is the capacity needed by the other sources? This can be answered in two ways, depending on the starting source of water: (a) a well or stream, or (b) a lake. Each case is illustrated using an example.

a. Consider the case of using a well or stream first, but it does not have enough capacity, so it can be used to build a store of water in the lake. Then, the minimum dimension of the lake can be determined as follows:

1. Determine the minimum recharge rate required for irrigation.
2. Determine the minimum capacity of the lake.
3. Determine the area of the lake.
4. Determine the length and width of the lake.

Example 3.4: It is assumed that an irrigation system requires 500 gallons per minute and the maximum time used in reaching the peak use demand rate is 12 hours. Assume the recharge from the well or stream is continuous, 24 hours a day, and the percent efficiency of the lake is 75% due to evaporation, seepage, and sediment losses. The lake surface area can be assumed to be rectangular. Determine the area of the lake. Do the calculations in SI units also.

Solution: Given Irrigation rate = 500 gallons/min or 1.89 m^3/min or 1890 liters/min, the minimum recharge rate required for irrigation is

$$1.89 \times \frac{12}{24} = 0.945\ \text{m}^3/\text{min or } 945\ \text{L/min or } 250\ \text{gal/min}.$$

The minimum capacity of the lake is

$$\frac{1890 \times 12 \times 60 - 945 \times 12 \times 60}{0.75} = 907,200\ \text{L or } 907.2\ \text{m}^3 \text{ or } 239,000\ \text{gal}.$$

To determine the area of the lake, let us assume that the average depth for the lake is 2 m. Then, area = 907.2/2 = 453.6 m^2. From the area, the length and width can be determined if one or the other is given in the question.

b. Now the lake does not have enough water but satisfies the minimum lake capacity requirement for irrigation. Additional water can be added from a well or stream. Then, the minimum capacity of the stream or well can be determined as follows:

1. Determine the recharge rate of the stream or well.
2. Determine the minimum lake capacity requirement for the above recharge rate, assuming some lake efficiency.

Example 3.5: It is assumed that an irrigation system requirement is 450 gallons per minute, and the maximum time used in reaching the peak use demand rate is 10 hours. Assume the recharge from the well or stream is continuous, 24 hours a day, and the percent efficiency of the lake is 80% due to evaporation, seepage, and sediment losses. The lake surface area can be assumed to be rectangular. Determine the minimum capacity of the lake and stream or well capacity. Do the calculations in SI units also.

Solution: The minimum recharge rate of the stream or well is:

$$450 \times \frac{10}{24} = 187.5\ \text{gal/min or } 0.71\ \text{m}^3/\text{min}.$$

The minimum capacity of the lake is: $\frac{450 \times 12 \times 60 - 187.5 \times 12 \times 60}{0.80} = 236,250$ gallons or 894.30 m^3.

If the lake has this capacity or greater, then a combination of water sources will work. The required stream or well capacity should be 0.71 m^3/min or 710 L/min on a continuous basis.

QUESTIONS

Q.3.1 A 5 m wide canal is being used to deliver water for irrigation. It has a flow depth of 2 m and an average flow velocity of 2 m/s. What is the discharge of water in the canal?

Q.3.2 A river is 100 m wide and depth measurements at every 10 m beginning from the left bank to the right bank are 0.5 m, 1.8 m, 2.4 m, 2.8 m, 3.1 m, 3.5 m, 3.8 m, 3.25 m, 3.0 m, 2.8 m, and 0.4 m. The average flow velocity is 1.5 m/s. What is the river discharge?

Q.3.3 A reservoir with a surface area of 0.5 km^2 is being used for irrigation. The average depth of water in the lake is 6 m. How much water does the lake have?

Q.3.4 The lake in Q.3.3 loses water by evaporation at a rate of 5 mm/day. If there were no source of recharge within a period of six months, how much water will there be in the lake and what will be the average depth of water? Suppose there is rainfall of 200 mm in the next six months. How much water will there be at the end of the year?

Q.3.5 The amount of water being withdrawn for irrigation from the lake in Q.3.4 during the crop-growing season of four months is 50 mm, 120 mm, 150 mm, and 100 mm, respectively. The crop season is during the rainy season. How much water will there be at the end of the four-month crop season?

Q.3.6 The minimum daily flow rate for the irrigation season is given for the Guadalupe River at the gaging station of Sattler, Texas (USGS 08167800) in Table 3.4. Compute the minimum flow rate that exceeds 5, 10, 20, and 50% of the time.

Table 3.4 *The minimum daily flow rates for the irrigation season (units: cfs)*

Year	Flow rate	Year	Flow rate	Year	Flow rate
1989	195	2000	47	2011	48.5
1990	53	2001	138	2012	55.5
1991	134	2002	133	2013	53.9
1992	203	2003	168	2014	54.6
1993	128	2004	182	2015	79.8
1994	104	2005	208	2016	134
1995	105	2006	58.4	2017	108
1996	25	2007	219	2018	51.9
1997	186	2008	151	2019	71.8
1998	121	2009	52		
1999	100	2010	185		

REFERENCES

ADEQ (Arizona Department of Environmental Quality) (2004). Surface water reports. https://azdeq.gov/node/4908.

Christian-Smith, J., Gleick, P. H., Cooley, H., et al. (2012). *A Twenty-First Century U.S. Water Policy*. Oxford: Oxford University Press.

IBWC (2003). Flow of the Rio Grande and related data. Water Bulletin 73. International Boundary and Water Commission the United States and Mexico.

Kellis, M., Kalavrouziotisi, K., and Gikas, P. (2013). Review of wastewater reuse in the Mediterranean countries, focusing on regulations and policies for municipal and industrial applications. *Global NEST Journal*, 15(3): 333–350.

LDEQ (2004). Louisiana water quality inventory: integrated (305(b)/303(d)). www1.deq.louisiana.gov/portal/DIVISIONS/WaterPermits/WaterQualityAssessment/WaterQualityInventorySection305b/2004305bReport.aspx.

TCEQ (2004). Science Advisory Committee report on water for environmental flows. www.tceq.texas.gov/assets/public/permitting/watersupply/water_rights/txefsac8132008article4.pdf.

Texas Water Development Board (TWDB) (2007). *Water for Texas: 2007*, Vol. 2. Austin, TX: State Government.

USGS (1985). *Estimated Use of Water in the United States in 1985*. Denver, CO: USGS.

USGS (2004). *Estimated Use of Water in the United States in 2000*. Washington, DC: United States Government Printing Office.

4 Groundwater and Wells

Notation

A	cross-sectional area of a porous bed (L^2 [length])
b	thickness of a saturated porous medium (L)
c	clogging factor
C	dimensionless constant in eq. (4.4)
d	mean grain diameter (L)
d_{40}	grain size (d_{40}) for which 40% of the material is coarser (L)
d_{90}	effective grain size (L)
d_s	screen diameter (cm)
D	hydraulic diffusivity of an aquifer (L^2/T [time])
h	piezometric head (L)
h_0	piezometric head when the radial distance is r_0 (L)
h_1, h_2	respective piezometric heads of two observation wells (L)
k	intrinsic or specific permeability (L^2)
k_1	a function of fluid properties
k_2	a function of porous medium properties
K	hydraulic conductivity (L/T)
L_s	screen length (m)
M_s	mass of solids (kg)
n	porosity (fraction)
P	percentage of screen open area (percentage)
q	specific discharge (L/T)
Q	discharge from the well (L^3/T)
Q_i	flow rate from the ith well (L^3/T)
r	radial distance measured from the center of the well (L)
r_0	radial distance from the well to the point where the decline in the piezometric head is negligible (L)
r_{0i}	radial distance of the ith well to a location where the water table is almost horizontal or drawdown is negligible (L)
r_1, r_2	distances of two observation wells (L)
r_i	radial distance from well i to the point at which the drawdown is to be determined (L)
r_v	void ratio (fraction)
r_{vf}	final void ratio (fraction)
r_{vi}	initial void ratio (fraction)
r_w	well radius (L)
s	drawdown of well (L)
S_c	storage coefficient of an aquifer (fraction)
S_r	specific retention of an aquifer, also known as field capacity or water-holding capacity (L^3/L^3, or L/L)
S_s	specific storage of an aquifer (L)
S_y	specific yield of an aquifer
S_{ya}	apparent specific yield of an aquifer
t	time (T)
T	transmissivity of an aquifer (L^2/T)
u	Boltzmann variable
UC	uniformity coefficient (UC) of an aquifer
v	specific discharge or flux through a porous bed (L/T)
v_s	screen entrance velocity (cm/s)
V	volume of an aquifer (L^3)
V_b	bulk volume (L^3)
V_i	ratio of the volume of pore spaces or interstices (fraction)
V_r	volume of water retained in an aquifer (L^3)
V_s	volume of solid particles (L^3)
V_w	volume of interstices (L^3)
V_y	volume of water that the aquifer will yield (L^3)
$W(u)$	Theis well function
z	elevation head (L)
α	compressibility of an aquifer (m^2/N)
β	compressibility of water (m^2/N)
γ	weight density (N/m^3)
γ_a	specific weight of water per unit area (i.e., ρg [N/m^3])
Δb	amount of subsidence of a porous medium (L)
Δh	difference in the piezometric head in the porous bed (L)
Δp	change in hydraulic pressure (N/m^2)
Δx	length of a porous bed (L)
θ	angle of water table (degree)
ρ	mass density (kg/m^3)
ρ_b	mass density of a dry sample (bulk density) (kg/m^3)
ρ_s	density of sediment or mineral particles (grain density) (kg/m^3)
μ	dynamic viscosity (L/T)

4.1 INTRODUCTION

Groundwater is of vital importance for agricultural irrigation, especially in arid and semi-arid regions, where groundwater is the main source of water supply. In almost all countries, a significant portion of irrigation water is derived from groundwater using wells. In developing countries where farm holdings are small, one well may suffice, but a number of wells are used in large farms. However, groundwater supply is being threatened by frequent occurrence of droughts and water shortages, rising demand, dwindling supplies of surface water, the specter of global warming and lurking climate change, egregious lack of properly thought-out strategies for the protection and management of water resources, and growing uncertainty about the availability of good-quality water. Each year there are scores of places that virtually run out of water to meet even basic needs. For example, in India tens of thousands of people, along with their herds, temporarily migrate during summer months from west to north in search of water. In 2012, two towns in Texas, United States, practically ran out of water due to a severe and prolonged drought, and water had to be transported to sustain the towns. The scarcity of water is becoming alarming in many parts of the world.

To meet growing water demand, groundwater is being pumped far in excess of its replenishment. For example, in the North Indian Plains, which are inhabited by nearly 600 million people, water levels are declining at an alarming rate. In many places, freshwater unconfined aquifers have already been depleted, and in some cases the water levels have gone down by as much as 10–20 m. Some investigators estimate the rate of decline of the water level at about 10 cm per year. This rate of decline is being further compounded by global warming and climate change. On top of this frightening scenario, groundwater is being indiscriminately contaminated by pollution from agriculture, industry, oil and gas, mining, and energy generation. Groundwater contamination directly impacts surface water and aquatic ecosystems. Yet systematic monitoring and measurement of groundwater contamination are lacking in many parts of the world. There must therefore be a national commitment in order to manage, monitor, protect, and remediate groundwater. It must begin with a groundwater information network providing information on mapping, characterization, and monitoring. There must be a repository of data on groundwater and contamination. Leading-edge science and engineering technologies must be employed to model groundwater, recharge, the fate of contaminants, saltwater intrusion, surface water–groundwater interaction, and optimal management. This should be supplemented by public policy, education, stakeholder participation, water pricing, and so on. Seemingly endless government organizations must be brought together or at least work together for the common good.

This chapter discusses rudimentary aspects of groundwater and wells from the perspective of irrigation. Although the material covered in this chapter is based on gross simplifications, it has been found to be useful nonetheless for irrigation purposes.

4.2 DEFINITIONS AND NOTATIONS

4.2.1 Aquifers

Groundwater is stored in unconsolidated porous rocks made up of particles of varying sizes, located below the land surface, and which may comprise alluvial and colluvial deposits. An aquifer is a geologic formation that is capable of yielding significant quantities of water under ordinary hydraulic gradients (Hantush, 1964). If a geologic formation contains sufficient quantities of water but yields water slowly because of low hydraulic conductivity and may not be suitable for the development of wells, it is called an aquitard and is considered semipermeable. If a geologic formation contains water but is not capable of yielding significant quantities of water, such as clay or shale, then it is called an aquiclude and is considered impermeable.

Aquifers can generally be classified as unconfined, confined, perched, and leaky, as shown in Figure 4.1. An unconfined aquifer – also called a phreatic, water table, or free aquifer – is bounded by the water table at the upper surface and by a confining layer at the bottom surface. The water table can rise or fall in response to recharge or pumping.

A confined aquifer is confined at the top and the bottom by impervious or semipervious strata, and has pressure greater than the atmospheric pressure. When a well penetrates a confined aquifer, the piezometric head (or water table) will rise above the upper confining layer but may not reach the land surface; if it does it is called artesian and the well becomes a free-flowing well. A confined aquifer has a limited recharge zone.

A leaky aquifer, also called a semi-confined aquifer, loses or gains water through adjacent layers and can be a confined aquifer or an unconfined one. A perched aquifer is an unconfined aquifer in which the water table rests over a relatively small impervious or semipervious stratum, and this water table is above the main water table and occurs over clay lenses in sedimentary deposits. Perched aquifers yield small quantities of water and are often exploited for domestic water supply and small-scale gardening.

4.2.2 Porosity

The ratio of the volume of pore spaces or interstices (V_i) to the bulk volume (V_b) is defined as porosity (n). In an aquifer the pore spaces are occupied by water whose volume is the same as the volume of interstices (V_w). If the volume of solid particles is V_s, then $V_b = V_i + V_s$. Then, porosity can be expressed as

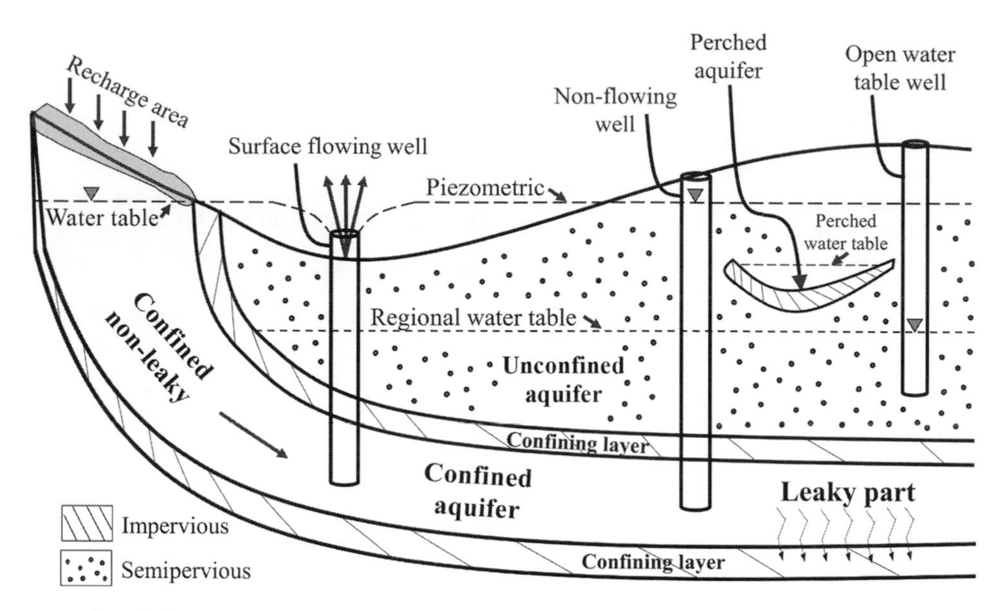

Figure 4.1 Different types of aquifer formations: confined, unconfined, perched, and leaky.

$$n = \frac{V_i}{V_b} = \frac{V_w}{V_b} = \frac{V_b - V_s}{V_b} = 1 - \frac{V_s}{V_b} = 1 - \frac{\frac{M_s}{\rho_s}}{\frac{M_s}{\rho_b}} = 1 - \frac{\rho_b}{\rho_s} \tag{4.1}$$

where ρ_b is the mass density of a dry sample (bulk density), ρ_s is the density of sediment or mineral particles (grain density), and M_s is the mass of solids. Typical values of porosity are 25–40% for sand and gravel, 35–50% for silt, and 40–70% for clay. Porosity can be 5–50% for fractured basalt and karst limestone, 5–30% for sandstone, 0–20% for limestone and dolomite, 1–10% for shale, 0–10% for fractured crystalline rock, and 0–5% for dense crystalline rock.

Example 4.1: Consider 1.5 m^3 of sand with a porosity of 0.35, and 2.5 m^3 of clay with a porosity of 0.45. Sand and clay are now mixed. Determine the resulting porosity.

Solution: For sand, the volume of pore space is 1.5 m^3 × 0.35 = 0.525 m^3 and the volume of solids is 1.5 m^3 × (1 − 0.35) = 0.975 m^3.

For clay, the volume of pore space is 2.5 m^3 × 0.45 = 1.125 m^3 and the volume of particles is 2.5 m^3 × (1 − 0.45) = 1.375 m^3.

Since sand has larger pore spaces, the volume of clay that will fill the void space within sand is 0.525 m^3.

The volume of clay left over is 1.375 m^3 − 0.525 m^3 = 0.85 m^3. Since all the pore space of sand is filled with clay, the total pore space of the mixed soil is the pore space of clay: 1.125 m^3.

The total volume of solids is: 0.975 m^3 (sand) + 1.375 m^3 (clay) = 2.35 m^3.

Total volume: 2.35 m^3 + 1.125 m^3 = 3.475 m^3.

Therefore,

$$n = \frac{(3.475 - 2.35)\ \text{m}^3}{3.475\ \text{m}^3} = 0.32.$$

4.2.3 Void Ratio

The void ratio (r_v) is defined as the ratio of the volume of voids or pores to the volume of solids, and is related to porosity as:

$$r_v = \frac{V_i}{V_s} = \frac{V_w}{V_s} = \frac{V_b - V_s}{V_b} = 1 - \frac{V_s}{V_b} \\ = 1 - \frac{M_s/\rho_s}{M_s/\rho_b} = \frac{n}{1-n}. \tag{4.2}$$

Because of excessive pumping for irrigation, grounds may subside. Since the void ratio is related to the strength of the porous medium, a decrease in the void ratio corresponds to an increase in the compaction of the material. If a saturated porous medium of thickness b and initial void ratio r_{vi} subsides, meaning its void ratio reduces to r_{vf}, then the amount of subsidence, Δb, can be determined as (Green, 1962):

$$\Delta b = \frac{r_{vi} - r_{vf}}{1 + r_{vi}} b. \tag{4.3}$$

Example 4.2: Compute the void ratio of the sand–clay mixture in Example 4.1.

Solution: The porosity of the mixture computed in Example 4.1 is 0.32. Therefore, using eq. (4.2), the void ratio is

$$r_v = \frac{0.32}{1 - 0.32} = 0.47.$$

Example 4.3: Consider a confined aquifer of thickness 50 m and porosity 40%. Due to excessive pumping, the aquifer porosity decreases to 30%. Determine the amount of subsidence.

Solution: First, the initial void ratio and the final void ratio are to be computed:

$$r_{vi} = \frac{n}{1-n} = \frac{0.4}{1-0.4} = 0.67,$$

$$r_{vf} = \frac{0.3}{1-0.3} = 0.43.$$

The amount of subsidence is

$$\Delta b = \frac{r_{vi} - r_{vf}}{1 + v_{vi}} b = \frac{0.67 - 0.43}{1 + 0.67} \times (50\text{ m}) = 7.19\text{ m}.$$

The aquifer thickness now reduces to $50 - 7.19 =$ 42.81 m.

4.2.4 Permeability and Hydraulic Conductivity

The permeability of a porous medium indicates its ability to transmit fluid under energy gradient and does not depend on fluid properties. It is often expressed as a function of mean grain diameter:

$$k = Cd^2, \tag{4.4}$$

where k (L^2) is the intrinsic or specific permeability, d is the mean grain diameter (L), and C is a dimensionless constant, depending on the porosity range.

Hydraulic conductivity (K), on the other hand, is a function of both porous medium properties and fluid properties. It is a fundamental parameter for determining the movement of water in porous media and can be expressed as a product of two factors, k_1 and k_2, where k_1 is a function of fluid properties and k_2 is a function of porous medium properties. Thus, K can be written as

$$K = k_1 k_2, \quad k_1 = \frac{\rho g}{\mu}, \quad k_2 = k. \tag{4.5a}$$

Therefore,

$$K = \frac{k\rho g}{\mu}, \tag{4.5b}$$

where $\gamma = \rho g$ is the weight density, g is the acceleration due to gravity, ρ is the mass density, and μ is the dynamic viscosity. It has dimensions of length per unit time (L/T). The values of K span over 13 orders of magnitude from a very low value of 10^{-13} m/s for metamorphic and igneous rocks to 1 m/s for gravel (Freeze and Cherry, 1979).

4.2.5 Anisotropy and Heterogeneity

An aquifer is characterized by the change in its hydraulic conductivity in space (or location) and direction. The property of homogeneity corresponds to location and the property of isotropy corresponds to direction. If the hydraulic conductivity (K) does not vary with location and with direction, the aquifer is regarded as homogeneous and isotropic. If K varies with direction as well as with location the aquifer is regarded as heterogeneous and anisotropic. If K is constant from location to location but varies from one direction to another at the same location, the aquifer is homogeneous but anisotropic. If K varies from location to location but is constant from one direction to another at the same location the aquifer is heterogeneous but isotropic. Aquifers are usually heterogeneous and anisotropic, but for simplicity they are assumed homogeneous and isotropic. Representative values of hydraulic conductivity are given in Table 4.1.

4.2.6 Transmissivity

Transmissivity (T) of an aquifer characterizes its ability to transmit water. For a confined aquifer of thickness b and hydraulic conductivity K, it is defined as the product of b and K:

$$T = bK. \tag{4.6}$$

Transmissivity T has the dimensions of L^2/T and defines the rate of transmission of water through its unit width under a unit hydraulic gradient.

4.2.7 Specific Yield

The specific yield (S_y) of an aquifer is the ratio of the volume of water that the aquifer will yield (V_y) under gravity to its own volume (V) (Meinzer, 1923):

$$S_y = \frac{V_y}{V}. \tag{4.7}$$

It is a dimensionless quantity. Because of pumping, there will be a change in the aquifer volume below the water table. Hence, apparent specific yield (S_{ya}) is defined as the ratio of the volume of water added to or extracted from the aquifer to the change in the aquifer volume. Usually, $S_{ya} \leq S_y$. Reasonable values of S_y are 0.2 for gravel and coarse-grained sand and 0.1 for medium- to fine-grained sand. Representative values of specific yield are given in Table 4.2.

4.2.8 Specific Retention

The specific retention (S_r) of an aquifer is defined as the ratio of the volume of water retained (V_r) against the force of gravity to the aquifer volume (V):

$$S_r = \frac{V_r}{V} = n - S_y. \tag{4.8}$$

For a soil, S_r is also known as the field capacity or water-holding capacity.

Table 4.1 *Representative values of hydraulic conductivity*

Material	Hydraulic conductivity (m/day)	Type of measurement	Material	Hydraulic conductivity (m/day)	Type of measurement
Coarse gravel	150	R	Dune sand	20	V
Medium gravel	270	R	Loess	0.08	V
Fine gravel	450	R	Peat	5.7	V
Coarse sand	45	R	Schist	0.2	V
Medium sand	12	R	Slate	0.00008	V
Fine sand	2.5	R	Predominantly sand till	0.49	R
Silt	0.08	H	Predominantly gravel till	30	R
Clay	0.0002	H	Tuff	0.2	V
Fine-grained sandstone	0.2	V	Basalt	0.01	V
Medium-grained sandstone	3.1	V	Weathered gabbro	0.2	V
Limestone	0.94	V	Weathered granite	1.4	V
Dolomite	0.001	V			

H, horizontal hydraulic conductivity; R, repacked sample; V, vertical hydraulic conductivity.
After Morris and Johnson, 1967.

Table 4.2 *Representative values of specific yield*

Material	Specific yield	Material	Specific yield
Coarse gravel	0.21	Limestone	0.14
Medium gravel	0.24	Dune sand	0.38
Fine gravel	0.28	Loess	0.18
Coarse sand	0.30	Peat	0.44
Medium sand	0.32	Schist	0.26
Fine sand	0.33	Siltstone	0.12
Silt	0.20	Predominantly silt till	0.06
Clay	0.06	Predominantly sand till	0.16
Fine-grained sandstone	0.21	Predominantly gravel till	0.16
Medium-grained sandstone	0.27	Tuff	0.21

After Morris and Johnson, 1967.

Example 4.4: An unconfined aquifer about 0.5 km^2 in area with a porosity of 0.35 is being pumped, and pumping causes a decline of 2 m in the water table. The specific retention of the aquifer is 0.25. Compute the specific yield and the change in storage of the aquifer.

Solution: From eq. (4.8),

$$S_y = n - S_r = 0.35 - 0.25 = 0.1.$$

The change in storage is the product of area multiplied by the specific yield and the drop in the water table:

$$S_y = \frac{V_y}{V} = 0.1.$$

The change in storage volume is equal to the aquifer area times the change in the water table:

$$\Delta V_y = 0.1 \times \Delta V = 0.1 \times \left(0.5\ \text{km}^2 \times 0.002\ \text{km}\right) = 0.0001\ \text{km}^3.$$

4.2.9 Specific Storage

The specific storage S_s (L^{-1}) of an aquifer is the volume of water that is released under a unit decline in the average hydraulic gradient by a unit volume of an aquifer due to the expansion of water and compression of an aquifer, and can be expressed as (Hantush, 1964):

$$S_s = \rho g n \beta \left(1 + \frac{\alpha}{n\beta}\right) = \rho g n \beta + \rho g \alpha. \tag{4.9}$$

Equation (4.9) contains two parts. $\rho g n \beta$ corresponds to the fraction of storage derived from the expansion of water; and $\rho g \alpha$ corresponds to the fraction of storage derived from the compression of the aquifer. Here, β is the compressibility of water and α is the compressibility of the aquifer.

4.2.10 Storativity

Storativity (S_c), or the storage coefficient of an aquifer, is the volume of water that a vertical column of a unit cross-sectional area releases or takes in storage under a unit change in the average hydraulic head. For unconfined aquifers, $S_c \approx S_y$, because most of the water is released under gravity and only a small portion is attributed to the expansion of water and compression of the aquifer. The value of S_c ranges from 0.1 to 0.3 for most unconfined aquifers.

For confined aquifers, the release of water is entirely due to the compression of the aquifer and expansion of water. For a confined aquifer of thickness b,

$$S_c = bS_s \tag{4.10a}$$

Therefore, the storage coefficient can also be expressed as

$$S_c = \rho g b n \beta \left(1 + \frac{\alpha}{n\beta}\right) = \rho g b n \beta + \rho g b \alpha. \tag{4.10b}$$

The value of S_c ranges from 0.00005 to 0.005. A value of 0.0001 or 0.0005 is reasonable for confined aquifers. For unconsolidated and loosely consolidated geologic formations the storage coefficient can be expressed as

$$S_c = S_y + bS_s. \tag{4.10c}$$

The aquifer is made up of sediment and has water that provides the hydraulic pressure. It is the sediment and the hydraulic pressure that provide the support to the overburden on the top of the aquifer. When the water is released from the confined aquifer, the hydraulic pressure is reduced because water is forced from the pores, even though the aquifer is still filled with water. This means that the reduction in support for the overburden provided by the hydraulic pressure will have to be compensated for by the aquifer skeleton, which will result in aquifer compaction and reduction in pore spaces. The water forced from the pores represents part of the storage coefficient due to compression. The amount of compaction or land subsidence for a confined aquifer can then be defined as

$$\Delta b = \Delta p \left(\frac{S_c}{\gamma_a} - n\beta\right), \tag{4.10d}$$

in which Δp is the change in hydraulic pressure (N/m^2) and γ_a is the specific weight of water per unit area: ρg (N/m^3).

Example 4.5: A confined aquifer with an area of 2 km^2 and a storage coefficient of 0.0025 experiences a drop in the hydraulic head of 2 m. Compute the change in storage of the aquifer.

Solution: The change in storage is given as

$$0.0025 \times 2\ \text{m} \times 2{,}000{,}000\ \text{m}^2 = 10{,}000\ \text{m}^3$$

Example 4.6: Compute the storage coefficient of an aquifer that has a porosity of 0.35, thickness of 150 m, and compressibility of 10^{-8} m²/N. The value of β can be taken as 5.0×10^{-10} m²/N.

Solution: From eq. (4.9),

$$S_s = \rho g n\beta\left(1 + \frac{\alpha}{n\beta}\right)$$

$$= 1000\frac{\text{kg}}{\text{m}^3} \times 9.81\frac{\text{N}}{\text{kg}} \times 0.35 \times 5 \times 10^{-10}\frac{\text{m}^2}{\text{N}} \times \left(1 + \frac{10^{-8}\frac{\text{m}^2}{\text{N}}}{0.35 \times 5 \times 10^{-10}\frac{\text{m}^2}{\text{N}}}\right) = \frac{0.0001}{\text{m}},$$

and from eq. (4.10a),

$$S_c = b \times S_s = 150\text{ m} \times \frac{0.0001}{\text{m}} = 0.015.$$

Example 4.7: A confined aquifer with porosity of 0.30, storativity of 0.0005, and thickness of 100 m experiences a pressure drop of 4000 N/m² because of pumping. What will be the amount of subsidence? The value of β can be taken as 4.4×10^{-10} m²/N.

Solution: The compressibility of water (β) can be taken as 5.0×10^{-10} m²/N. Using eq. (4.10d), the amount of subsidence is

$$\Delta b = \Delta p\left(\frac{S_c}{\gamma_a} - nb\beta\right)$$

$$= 4000\frac{\text{N}}{\text{m}^2}\left(\frac{0.0005}{1000 \times 9.81\frac{\text{N}}{\text{m}^3}} - 0.3 \times 100\,\text{m} \times 4.4 \times \frac{10^{-10}\text{m}^2}{\text{N}}\right)$$

$$= 0.00015\text{ m}.$$

4.2.11 Hydraulic Diffusivity

The hydraulic diffusivity (D) of an aquifer is defined as

$$D = \frac{T}{S_c} = \frac{K}{S_s}. \tag{4.11}$$

Hydraulic diffusivity has the dimension of L²/T.

4.3 DARCY'S LAW

Consider a porous bed of cross-sectional area A, as shown in Figure 4.2. Darcy's law states that the specific discharge or flux v through the bed of length Δx (or L) is directly proportional to the difference in the piezometric head ($-\Delta h$) and inversely proportional to Δx, where h is the piezometric head:

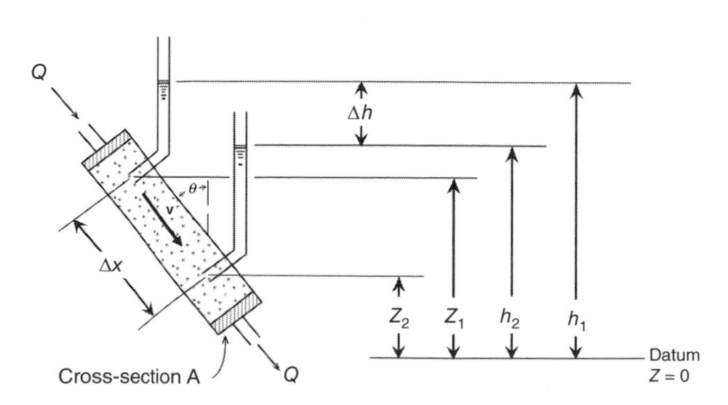

Figure 4.2 Experimental device illustrating Darcy's law.

$$v = \frac{Q}{A} = -K\frac{\Delta h}{\Delta x}. \tag{4.12}$$

For flow toward a well with radial distance r, the specific discharge q can be written as

$$q = -K\frac{dh}{dr}, \tag{4.13}$$

where K is the coefficient of proportionality or hydraulic conductivity, and dh/dr is the hydraulic gradient, as shown in Figure 4.3.

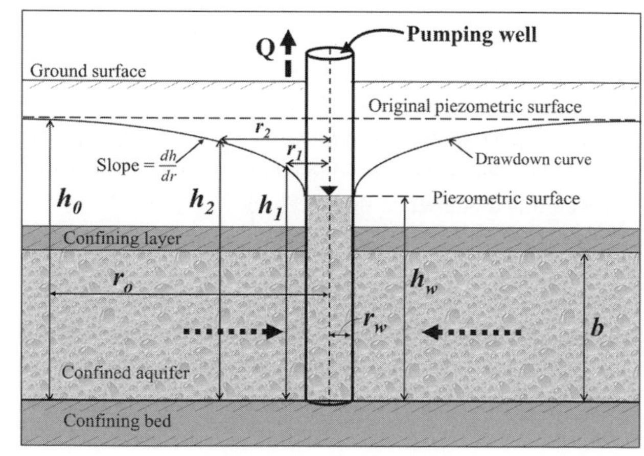

Figure 4.3 Radial flow to a well in a confined aquifer.

4.4 STEADY-STATE WELL HYDRAULICS

In steady-state well hydraulics the piezometric head does not vary with time, which is an assumption. Under this assumption, a short discussion of well hydraulics is given for both confined and unconfined aquifers.

4.4.1 Confined Aquifers

For confined aquifers, the discharge from the well under steady-state conditions can be written as

$$Q = 2\pi r b K \frac{dh}{dr}, \tag{4.14}$$

where r is the radial distance measured from the center of the well. Equation (4.14) can be integrated with the limits as $h = h_w$ at $r = r_w$ and $h = h_0$ at $r = r_0$, where r_w is the well radius and r_0 is the radial distance from the well to the point where the decline in the piezometric head is negligible, as shown in Figure 4.3. For integration, eq. (4.14) can be written as

$$Q\frac{dr}{r} = 2\pi b K dh. \tag{4.15}$$

Integrating eq. (4.15),

$$Q\int_{r_w}^{r_0} \frac{dr}{r} = 2\pi b K \int_{h_w}^{h_0} dh. \tag{4.16}$$

Integration of eq. (4.16) leads to

$$Q[\ln(r_0) - \ln(r_w)] = 2\pi b K[h_0 - h_w] \tag{4.17}$$

or

$$Q = 2\pi b K\left[\frac{h_0 - h_w}{\ln\left(\frac{r_0}{r_w}\right)}\right]. \tag{4.18}$$

If there are two observation wells located at distances r_1 and r_2 where the respective piezometric heads are h_1 and h_2, as shown in Figure 4.3, then one can write

$$Q = 2\pi b K\left[\frac{h_2 - h_1}{\ln\left(\frac{r_2}{r_1}\right)}\right]. \tag{4.19}$$

If drawdown s at any radius r defined as $h_0 - h$ is measured, then eq. (4.19) can be expressed as

$$Q = 2\pi b K\left[\frac{s_1 - s_2}{\ln\left(\frac{r_2}{r_1}\right)}\right]. \tag{4.20}$$

Equation (4.20) can also be written in terms of transmissivity ($T = bK$) as

$$Q = 2\pi T\left[\frac{s_1 - s_2}{\ln\left(\frac{r_2}{r_1}\right)}\right]. \tag{4.21}$$

Example 4.8: Consider a confined aquifer 50 m thick and 1000 m wide. Two wells 500 m apart are located in the aquifer. The difference in the hydraulic heads of these wells is 2.5 m. The hydraulic conductivity of the aquifer is 10 m/day. Compute the groundwater flow passing through a unit cross-section of the aquifer.

Solution: The hydraulic gradient is 2.5 m/500 m = 0.005. The aquifer area is: 50 m × 1000 m = 50,000 m^2.

Therefore, groundwater discharge is

$$q = K\frac{dh}{dr} = 10\frac{\text{m}}{\text{day}} \times (0.005) = 0.05\frac{\text{m}}{\text{day}},$$

$$Q = q \times A = 0.05\frac{\text{m}}{\text{day}} \times 50{,}000\ \text{m}^2 = 2500\ \frac{\text{m}^3}{\text{day}}.$$

Example 4.9: A confined aquifer is 50 m thick and its hydraulic conductivity is 12 m/day. A pumping well is installed penetrating the entire thickness of the aquifer. Two observation wells are installed at radial distances of 10 m and 80 m away from the pumping well. The well is being pumped at a constant rate of 1500 m^3/day. Compute the drawdown in the well 10 m away if the drawdown in the well 80 m away is 0.5 m.

Solution: Using eq. (4.20),

$$Q = 2\pi b K\left[\frac{s_1 - s_2}{\ln\left(\frac{r_2}{r_1}\right)}\right]$$

$$= 2\pi(50\ \text{m})\left(12\frac{\text{m}}{\text{day}}\right)\left[\frac{s_1 - 0.5\ \text{m}}{\ln\left(\frac{80\ \text{m}}{10\ \text{m}}\right)}\right] = 1500\frac{\text{m}^3}{\text{day}},$$

$$s_1 - 0.5\ \text{m} = \frac{1500\frac{\text{m}^3}{\text{day}}}{2\pi(50\ \text{m})\left(12\frac{\text{m}}{\text{day}}\right)} \times \ln\left(\frac{80\ \text{m}}{10\ \text{m}}\right) = 0.83\ \text{m},$$

$$s_1 = 0.83\ \text{m} + 0.5\ \text{m} = 1.33\ \text{m}.$$

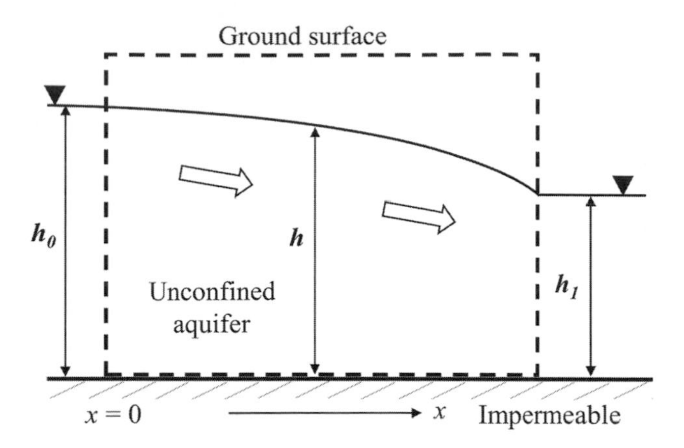

Figure 4.4 Steady flow in an unconfined aquifer with vertical boundaries.

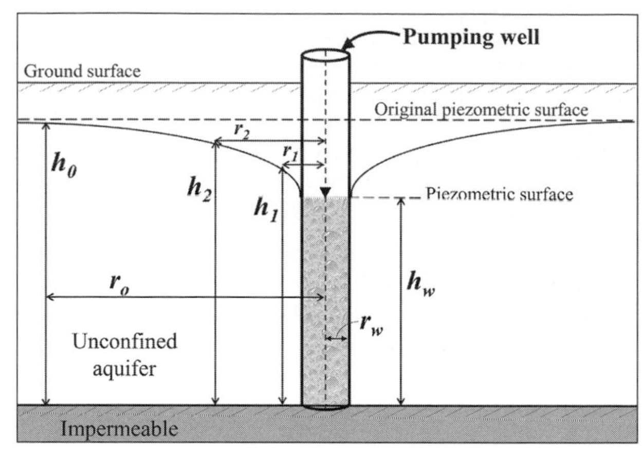

Figure 4.5 Radial flow to well in an unconfined aquifer.

4.4.2 Unconfined Aquifers

In unconfined aquifers, the water table, where the pressure is atmospheric ($p = 0$), is the piezometric surface, as shown in Figure 4.4. Therefore, the piezometric head equals the elevation head z. The flow in the unconfined aquifer is not actually horizontal but converges toward the downstream vertical boundary of the flow system. At a distance x along the water table, the specific discharge q is a function of piezometric head h and can be expressed as

$$q = -K\frac{dh}{dx} = -K\frac{dz}{dx}. \tag{4.22}$$

If the water table makes an angle of θ with the horizontal, then $dz/dx = \sin\theta$, then eq. (4.22) can be written as

$$q = -K\sin\theta. \tag{4.23}$$

It was shown by Dupuit (1863) that the slope of the water table in most cases is quite small, varying from 1/100 to 1/1000, and stated that for flow in unconfined aquifers the water table slope of less than 1/10 would correspond to a horizontal angle of less than 6°. He assumed that sin θ may be replaced by tan θ and the flow would be horizontal and have uniform velocity with depth. Considering radial flow, eq. (4.23) can be written for pump discharge Q as

$$Q = 2\pi rKh\frac{dh}{dr}. \tag{4.24}$$

Equation (4.24) can be integrated as

$$Q\int_{r_w}^{r_0}\frac{dr}{r} = 2\pi K\int_{h_w}^{h_0} h\,dh. \tag{4.25}$$

Integration of eq. (4.25) yields

$$Q = \pi K\left[\frac{(h_0)^2 - (h_w)^2}{\ln\left(\frac{r_0}{r_w}\right)}\right]. \tag{4.26}$$

Equation (4.26) can be recast for any two measuring wells located at distances r_1 and r_2, as shown in Figure 4.5, as

$$Q = \pi K\left[\frac{(h_2)^2 - (h_1)^2}{\ln\left(\frac{r_2}{r_1}\right)}\right]. \tag{4.27}$$

Equation (4.26) or (4.27) is reasonably accurate (Marino and Luthin, 1982) and can provide a reasonable distribution of piezometric head for radial distance greater than 1.5 times the height of the flow system (i.e., $r > 1.5\,h_0$). For computing transmissivity, the average thickness of the aquifer can be used.

Example 4.10: A well is pumping water at a rate of 2000 m^3 per day from an unconfined aquifer whose thickness is 40 m and hydraulic conductivity is 80 m/day. The pumping well penetrates the entire aquifer thickness. There are two observation wells located 20 m and 100 m away from the pumping well. The drawdown in the observation well at a distance of 100 m is 1 m. Compute the drawdown in the well 20 m away.

Solution: The head in terms of drawdown is $h = h_0 - s$. Therefore, $h_2 = h_0 - s = 40\text{ m} - 1\text{ m} = 39\text{ m}$. Then, the discharge equation can be expressed as

$$Q = \pi K\left[\frac{(h_2)^2 - (h_1)^2}{\ln\left(\frac{r_2}{r_1}\right)}\right],$$

$$2000\frac{\text{m}^3}{\text{day}} = \pi \times 80\frac{\text{m}}{\text{day}} \times \left[\frac{(39\text{ m})^2 - (h_1)^2}{\ln\left(\frac{100\text{ m}}{20\text{ m}}\right)}\right].$$

This can be expressed as

$$(39\text{ m})^2 - (h_1)^2 = \left[\frac{2000\frac{\text{m}^3}{\text{day}}}{\pi \times 80\frac{\text{m}}{\text{day}}}\right] \times \ln\left(\frac{100\text{ m}}{20\text{ m}}\right)$$

$$(39\text{ m})^2 - (h_1)^2 = 12.81\text{ m}^2.$$

Therefore, $h_1 = 38.8$ m.

The drawdown in the well 20 m away is: $s = h_0 - h = 40\text{ m} - 38.8\text{ m} = 1.2\text{ m}$.

4.4.3 Well Fields

On large farms it is common to have more than one well, and pumping from some of the wells may impact the drawdown in other wells; that is, the drawdown curves will overlap, resulting in a composite effect, as shown in Figure 4.6 for pumping from two wells. Likewise, on small farms, pumping from one well on one farm may impact the drawdown in a well located on the adjacent farm. The drawdown will depend on the rates at which water is being pumped at different wells and the distances between wells. First, consider an unconfined aquifer.

The head at any radial distance r from a single well can be obtained from eq. (4.27), which can be recast for constant rate of pumping Q as

$$h_0{}^2 - h^2 = \frac{Q \ln(r_0/r)}{\pi K}. \tag{4.28}$$

Equation (4.29) can be generalized for N pumping wells. For pumping from N wells, the total drawdown s_T at any point will be the sum of drawdowns caused by individual pumping, which is the composite effect. If r_i is the radial distance from well i to the point at which the drawdown is to be determined, r_{0i} is the radial distance of the ith well to a location where the water table is almost horizontal or drawdown is negligible, h_0 is the original height of the water table, h is the height due to the combined effect of N pumping wells, and Q_i is the flow rate from the ith well, then one can write

$$h_0^2 - h^2 = \sum_{i=1}^{N} \frac{Q_i}{\pi K} \ln\left(\frac{r_{0i}}{r_i}\right). \tag{4.29}$$

The value of r_0 can in practice be taken as 500–1000 ft (Todd, 1959). Equation (4.29) should be used when drawdowns are relatively small and can simply be expressed as

$$s_T = \sum_{i=1}^{N} s_i. \tag{4.30}$$

The composite drawdown at any point in a confined aquifer due to N pumping wells can be computed in a similar manner as:

$$h_0 - h = \sum_{i=1}^{N} \frac{Q_i}{2\pi K b} \ln\left(\frac{r_{0i}}{r_i}\right). \tag{4.31}$$

4.5 UNSTEADY WELL HYDRAULICS

In the preceding discussion it was assumed that the piezometric head did not change with time. In reality, groundwater flow is unsteady. As before, the flow is assumed to be radial and symmetric. The drawdown s is then expressed (Theis, 1935) as

$$s = \frac{Q}{4\pi T} W(u), \tag{4.32}$$

where $W(u)$ is referred to as the Theis well function, whose values have been tabulated as shown in Table 4.3, and u is the Boltzmann variable defined as

$$u = \frac{r^2}{4\alpha t}, \quad \alpha = \frac{T}{S_c}, \quad \text{or} \quad \alpha = \frac{T}{S_y}, \tag{4.33}$$

in which t is time.

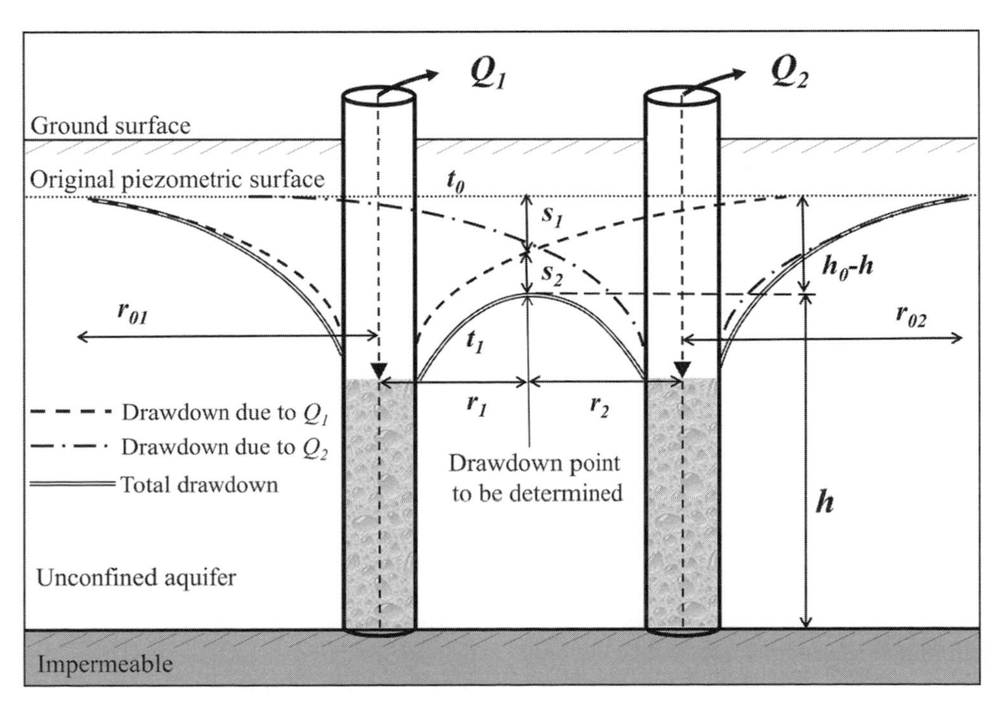

Figure 4.6 Drawdown due to simultaneous pumping from two wells.

Table 4.3 *Value of W(u) for different values of u*

u	1.0	2.0	3.0	4.0	5.0	6.0	7.0	8.0	9.0
$\times 1$	0.219	0.049	0.013	0.0038	0.0011	0.00036	0.00012	0.000038	0.000012
$\times 10^{-1}$	1.82	1.22	0.91	0.7	0.56	0.45	0.37	0.31	0.26
$\times 10^{-2}$	4.04	3.35	2.96	2.68	2.47	2.3	2.15	2.03	1.92
$\times 10^{-3}$	6.33	5.64	5.23	4.95	4.73	4.54	4.39	4.26	4.14
$\times 10^{-4}$	8.63	7.94	7.53	7.25	7.02	6.84	6.69	6.55	6.44
$\times 10^{-5}$	10.94	10.24	9.84	9.55	9.33	9.14	8.99	8.86	8.74
$\times 10^{-6}$	13.24	12.55	12.14	11.85	11.63	11.45	11.29	11.16	11.04
$\times 10^{-7}$	15.54	14.85	14.44	14.15	13.93	13.75	13.6	13.46	13.34
$\times 10^{-8}$	17.84	17.15	16.74	16.46	16.23	16.05	15.9	15.76	15.65
$\times 10^{-9}$	20.15	19.45	19.05	18.76	18.54	18.35	18.2	18.07	17.95
$\times 10^{-10}$	22.45	21.76	21.35	21.06	20.84	20.66	20.5	20.37	20.25
$\times 10^{-11}$	24.75	24.06	23.65	23.36	23.14	22.96	22.81	22.67	22.55
$\times 10^{-12}$	27.05	26.36	25.96	25.67	25.44	25.26	25.11	24.97	24.86
$\times 10^{-13}$	29.36	28.66	28.26	27.97	27.75	27.56	27.41	27.28	27.16
$\times 10^{-14}$	31.66	30.97	30.56	30.27	30.05	29.87	29.71	29.58	29.46
$\times 10^{-15}$	33.96	33.27	32.86	32.58	32.35	32.17	32.02	31.88	31.76

The well function can be expressed as an infinite series

$$W(u) = -0.5772 - \ln u + u - \frac{u^2}{2.2!} + \frac{u^3}{3.3!} - \ldots. \quad (4.34)$$

If $u \leq 0.01$, which it will be for small r and/or large time, then the well function can be reasonably approximated (Jacob, 1950) as

$$W(u) = -0.5772 - \ln u. \quad (4.35)$$

Inserting eq. (4.35) in eq. (4.32), the drawdown can be given as

$$s = \frac{Q}{4\pi T}\left[\ln\left(\frac{1}{u}\right) - 0.5772\right]. \quad (4.36)$$

This approximation is referred to as the Jacob method.

Example 4.11: Two wells are pumping water from a confined aquifer that has $T = 0.0055\ \mathrm{m^2/s}$ and $S_c = 0.00045$. The wells are 0.5 m in diameter and are 50 m apart. The rate of pumping from both wells is 10 L/s. Compute the drawdown curve for each well separately and then compute the composite drawdown when $t = 150$ s. Plot the drawdown curves.

Solution: $T = 0.0055\ \mathrm{m^2/s}$, $S_c = 0.00045$, $Q_1 = Q_2 = 10\ \mathrm{L/s} = 0.01\ \mathrm{m^3/s}$.

The drawdown curve for each well when $t = 150$ s:

The Boltzmann variable u can be calculated using eq. (4.33):

$$u_1 = u_2 = \frac{r^2 S_c}{4Tt} = \frac{r^2 \times 0.00045}{4 \times 0.0055\,\frac{\mathrm{m^2}}{\mathrm{s}} \times (150\ \mathrm{s})}$$

$$= 0.000136 \times r^2.$$

The drawdown curve for well 1:

$$s_1 = \frac{Q_1}{4\pi T}\left[\ln\left(\frac{1}{u_1}\right) - 0.5772\right]$$

$$= \frac{0.01\,\frac{\mathrm{m^3}}{\mathrm{s}}}{4\pi \times 0.0055\,\frac{\mathrm{m^2}}{\mathrm{s}}}\left[\ln\left(\frac{1}{0.000136 \times r^2}\right) - 0.5772\right],$$

$$s_1 = 0.1447\ \mathrm{m} \times \left[\ln\left(\frac{1}{0.000136 \times r^2}\right) - 0.5772\right]$$

for $(-64\ \mathrm{m} \leq r \leq 64\ \mathrm{m})$.

If r is considered the radial distance to well 1, then the drawdown curve for well 2 is

$$s_2 = 0.1447\ \mathrm{m} \times \left[\ln\left(\frac{1}{0.000136 \times (r - 50)^2}\right) - 0.5772\right]$$

for $(-14\ \mathrm{m} \leq r \leq 114\ \mathrm{m})$.

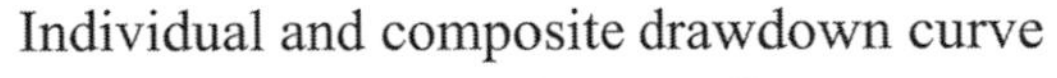

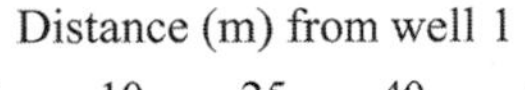

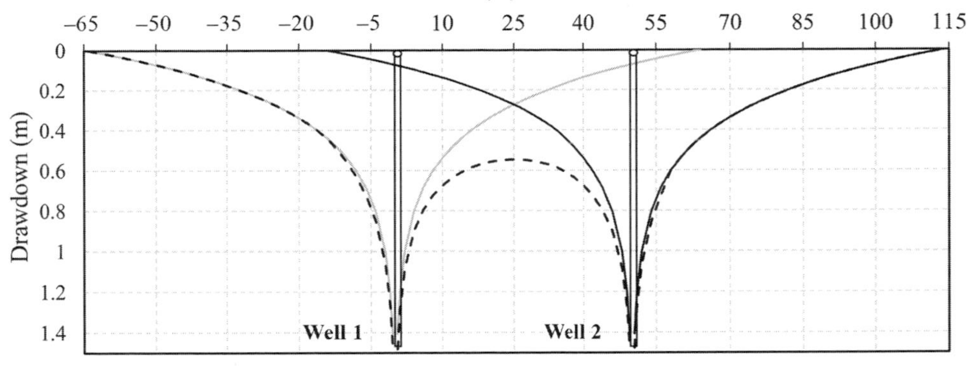

Figure 4.7 Individual and composite drawdowns due to simultaneous pumping from two wells.

The composite drawdown curve:

$$s_T = \sum_{i=1}^{N} s_i = s_1 + s_2,$$

$$s_T = s_1$$

$$= 0.1447\left[\ln\left(\frac{1}{0.000136 \times r^2}\right) - 0.5772\right]$$

for $(-64 \text{ m} \le r < -14 \text{ m})$,

$$s_T = s_1 + s_2$$

$$= 0.1447\left[\ln\left(\frac{1}{0.000136 \times r^2}\right) - 0.5772\right]$$

$$+ 0.1447\left[\ln\left(\frac{1}{0.000136 \times (r - 50)^2}\right) - 0.5772\right]$$

for $(-14 \text{ m} \le r \le 64 \text{ m})$,

$$s_T = s_2$$

$$= 0.1447\left[\ln\left(\frac{1}{0.000136 \times (r - 50)^2}\right) - 0.5772\right]$$

for $(64 \text{ m} < r \le 114 \text{ m})$.

The computed drawdown is shown in Figure 4.7.

Example 4.12: Consider the problem in Example 4.11, but the two wells are pumping water at different rates. One well is pumping at a rate of 10 L/s and the other is pumping at a rate of 15 L/s. Compute the drawdown curve for each well separately and then compute the composite drawdown when $t = 150$ s. Plot the drawdown curves.

Solution:

$T = 0.0055 \text{ m}^2/\text{s}, S_c = 0.00045,$
$Q_1 = 10 \text{ L/s} = 0.01 \text{ m}^3/\text{s},$
and $Q_2 = 15 \text{ L/s} = 0.015 \text{ m}^3/\text{s}.$

The drawdown curve for each well when $t = 150$ s:

The Boltzmann variable u can be calculated using eq. (4.33):

$$u_1 = u_2 = \frac{r^2 S_c}{4Tt} = \frac{r^2 \times 0.00045}{4 \times 0.0055 \frac{\text{m}^2}{\text{s}} \times (150 \text{ s})}$$

$$= 0.000136 \times r^2.$$

The drawdown curve for well 1 is:

$$s_1 = \frac{Q_1}{4\pi T}\left[\ln\left(\frac{1}{u_1}\right) - 0.5772\right]$$

$$= \frac{0.01 \frac{\text{m}^3}{\text{s}}}{4\pi \times 0.0055 \frac{\text{m}^2}{\text{s}}}\left[\ln\left(\frac{1}{0.000136 \times r^2}\right) - 0.5772\right],$$

$$s_1 = 0.1447 \text{ m} \times \left[\ln\left(\frac{1}{0.000136 \times r^2}\right) - 0.5772\right]$$

for $(-64 \text{ m} \le r \le 64 \text{ m})$.

If r is considered the radial distance to well 1, then the drawdown curve for well 2 is:

$$s_2 = \frac{Q_2}{4\pi T}\left[\ln\left(\frac{1}{u_1}\right) - 0.5772\right] = \frac{0.015 \frac{\text{m}^3}{\text{s}}}{4\pi \times 0.0055 \frac{\text{m}^2}{\text{s}}}$$

$$\left[\ln\left(\frac{1}{0.000136 \times (r - 50 \text{ m})^2}\right) - 0.5772\right],$$

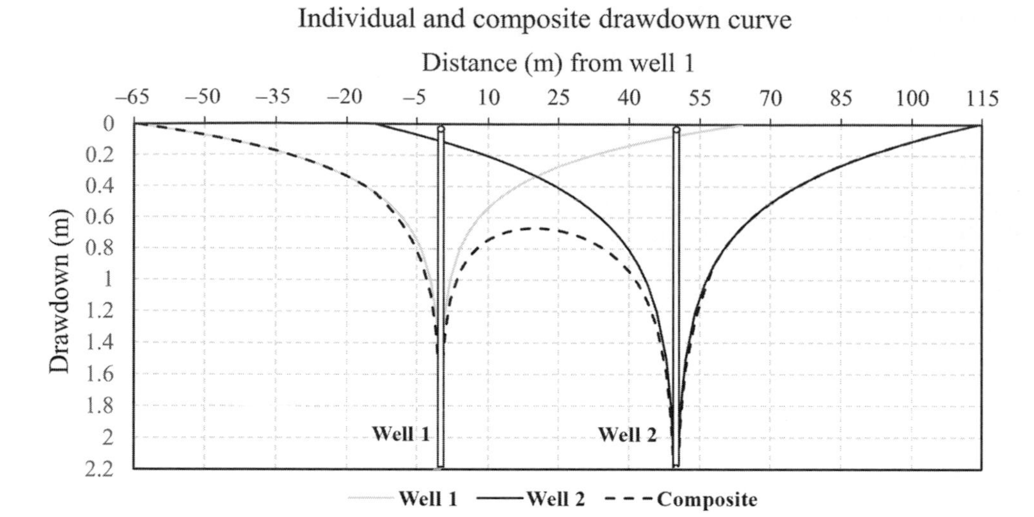

Figure 4.8 Individual and composite drawdowns due to simultaneous pumping from two wells.

$$s_2 = 0.217\ \text{m} \times \left[\ln\left(\frac{1}{0.000136 \times (r - 50\ \text{m})^2}\right) - 0.5772\right]$$

for $(-14\ \text{m} \le r \le 114\ \text{m})$.

The composite drawdown curve is:

$$s_T = \sum_{i=1}^{N} s_i = s_1 + s_2,$$

$$s_T = s_1$$
$$= 0.1447\left[\ln\left(\frac{1}{0.000136 \times r^2}\right) - 0.5772\right]$$

for $(-64\ \text{m} \le r < -14\ \text{m})$,

$$s_T = s_1 + s_2$$
$$= 0.1447\left[\ln\left(\frac{1}{0.000136 \times r^2}\right) - 0.5772\right]$$
$$+ 0.217\left[\ln\left(\frac{1}{0.000136 \times (r - 50)^2}\right) - 0.5772\right]$$

for $(-14\ \text{m} \le r \le 64\ \text{m})$,

$$s_T = s_2$$
$$= 0.217\left[\ln\left(\frac{1}{0.000136 \times (r - 50)^2}\right) - 0.5772\right]$$

for $(64\ \text{m} < r \le 114\ \text{m})$.

The drawdown curves are sketched in Figure 4.8.

4.5.1 Determination of Aquifer Parameters by Pumping Test

The aquifer parameters T and S_c can be estimated graphically using drawdown data. The graphical method (or type curve method) involves the following steps:

1. Plot on log-log paper drawdown s against t, where t is the time of drawdown. This is the data plot.
2. Plot on log-log paper $W(u)$ versus $1/u$. This plot is often referred to as the type curve. The values of $W(u)$ can be obtained from Table 4.3. The point to be noted here is that the length of each log cycle of the paper must be exactly the same as used in step 1.
3. Superimpose the data plot on the type curve, making sure that the drawdown axis is parallel to the $W(u)$ axis. Do the adjusting until most points fall on the type curve, ensuring that the coordinate axis is parallel.
4. Select an arbitrary point and note the values of $W(u)$ and $1/u$ coordinates and the corresponding s and t. It is not necessary to choose the match point on the type curve. For calculation simplicity, the point where $W(u) = 1$ and $1/u = 10$ is usually chosen.
5. The transmissivity is obtained as

$$T = \frac{QW(u)}{4\pi s}, \tag{4.37}$$

and the storativity or apparent specific yield is obtained as

$$S_c = \frac{4Ttu}{r^2}. \tag{4.38}$$

The aquifer parameters can also be determined using the Jacob method (Jacob, 1940), which involves the following steps:

1. Plot the measured drawdown against time on semilog paper (with time on the log axis). The plot should be approximately a straight line.

2. Compute the slope of the line in step 1, Δs.
3. Compute the transmissivity as

$$T = \frac{2.303\,Q}{4\pi\Delta s}, \quad (4.39)$$

in which Δs is the drawdown per log cycle.

4. Extrapolate the straight line to the time t_0 when $s = 0$. This yields the value of storativity as

$$S_c = \frac{2.246Tt_0}{r^2}. \quad (4.40)$$

Example 4.13: The US Geological Survey conducted a pumping test on a confined aquifer and reported the drawdown data. The radial distance where the drawdown was measured was 61 m, the rate of pumping was 1.88 m^3/min. The drawdown data are reported in Table 4.4.

Compute the transmissivity and storativity of the aquifer using both the type curve method and the Jacob method.

Solution: Type curve method:

1. Plot on log-log paper drawdown s against t using the pumping test data, as shown in Figure 4.9.
2. Plot on log-log paper $W(u)$ versus $1/u$. Then, make sure that the lengths of each log cycle are the same as used in step 1, as shown in Figure 4.10.
3. For calculation simplicity, the matching point where $W(u) = 1$ and $1/u = 10$ is chosen.

From the match point, the coordinates are:

$W(u) = 1,$

Table 4.4 *Drawdown data*

Time (min)	Drawdown (m)	Time (min)	Drawdown (m)	Time (min)	Drawdown (m)
1	0.200	12	0.600	80	0.925
2	0.300	14	0.635	100	0.965
3	0.370	18	0.670	120	1.000
4	0.415	24	0.720	150	1.045
5	0.450	30	0.760	180	1.070
6	0.485	40	0.810	210	1.100
8	0.530	50	0.850	240	1.120
10	0.570	60	0.875		

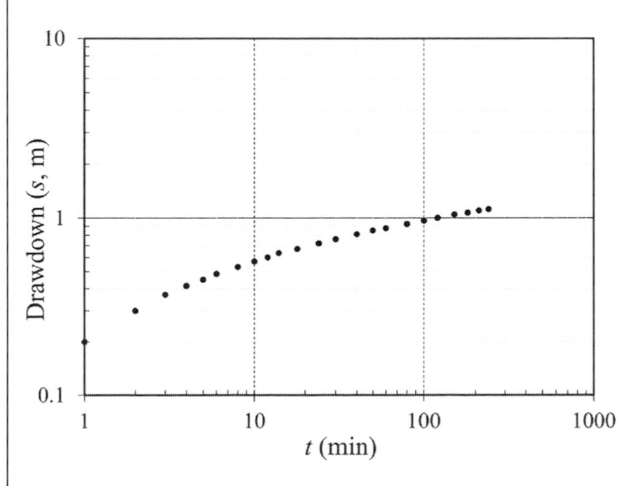

Figure 4.9 Plot of drawdown against time.

$\frac{1}{u} = 10,$

$u = 0.1,$

$s = 0.19$ m,

$t = 2.4$ min,

$$T = \frac{QW(u)}{4\pi s} = \frac{\left(1.88\,\frac{m^3}{min}\cdot\frac{1440\ min}{1\ day}\right)\times 1}{4\pi(0.19\ m)} = 1134\,\frac{m^2}{day}$$

$$S_c = \frac{4uTt}{r^2}$$
$$= \frac{4\times 0.1\times 1134\,\frac{m^2}{day}\times\left(2.4\ min\times\frac{0.000694\ day}{1\ min}\right)}{(61\ m)^2}$$
$$= 0.0002.$$

Solution: Jacob method:

1. Plot the measured drawdown against time on semilog paper (with time on the log axis). The plot should approximately be a straight line, as shown in Figure 4.11.
2. Compute the slope of the line in step 1, Δs, which is the drawdown per log cycle.

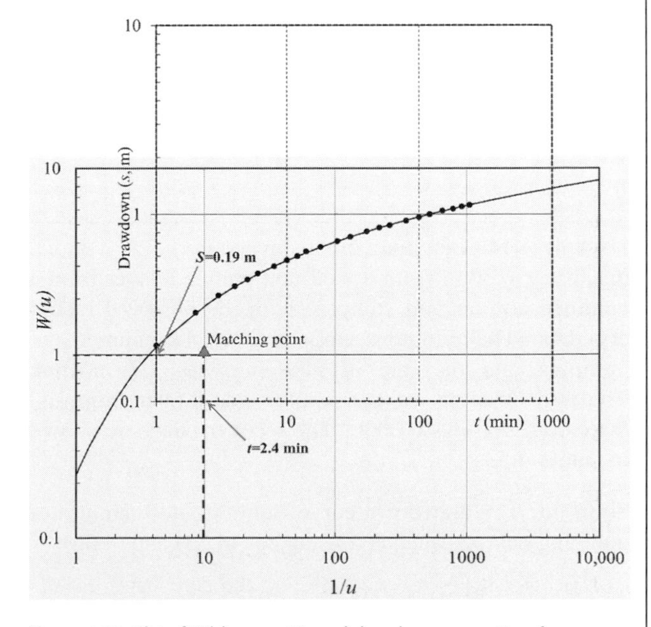

Figure 4.10 Plot of $W(u)$ versus $1/u$ and drawdown versus time for determining the match point.

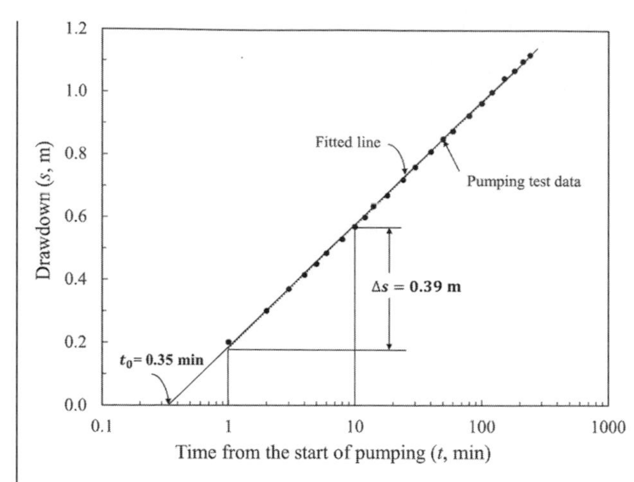

Figure 4.11 Plot of drawdown against time.

3. Compute the transmissivity as

$$T = \frac{2.303\ Q}{4\pi\Delta s} = 2.303 \times \frac{1.88\ \frac{\mathrm{m}^3}{\mathrm{min}} \cdot \frac{1440\ \mathrm{min}}{1\ \mathrm{day}}}{4\pi(0.39\ \mathrm{m})}$$

$$= 1272\frac{\mathrm{m}^2}{\mathrm{day}}.$$

4. Extrapolate the straight line to the time t_0 when $s = 0$. As shown in Figure 4.11, $t_0 = 0.35$. This yields the value of storativity as

$$S_c = \frac{2.246Tt_0}{r^2}$$

$$= 2.246 \times 1272\frac{\mathrm{m}^2}{\mathrm{day}} \times \left(\frac{0.35\ \mathrm{min}}{(61\ \mathrm{m})^2} \cdot \frac{0.000694\ \mathrm{day}}{1\ \mathrm{min}}\right)$$

$$= 0.0002.$$

4.5.2 Recovery of a Well

When pumping has ceased, the water level begins to increase or the well begins to recover, as shown in Figure 4.12. The drawdown equation (eq. 4.32) can be applied in this case also. If t is the time since pumping and t_* is the time since the cessation of pumping, then the drawdown at a radial distance r can be expressed as

$$s = h_0 - h = \frac{Q}{4\pi T}[W(u_1) - W(u_2)], \tag{4.41}$$

where

$$u_1 = \frac{r^2 S_c}{4Tt} \text{ and } u_2 = \frac{r^2 S_c}{4Tt_*}. \tag{4.42}$$

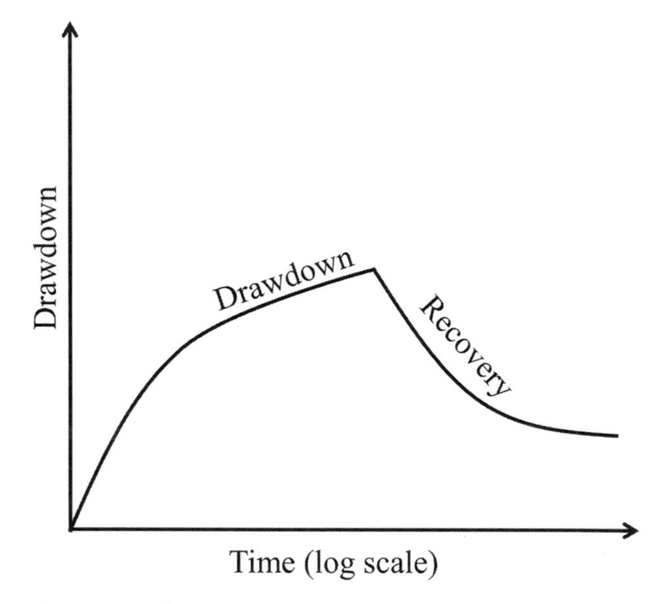

Figure 4.12 Well recovery.

Example 4.14: Compute the recovery curve at a radial distance $r = 30$ m from a well that pumped water from a confined aquifer and compare with the observed recovery data. The pumping stopped after 443 minutes of pumping and the rate of pumping was 1.79 m^3/min. The transmissivity of the aquifer was 0.60 m^2/min and the storativity was 0.0006. The recovery data are shown in Table 4.5.

Solution: The drawdown curve from 0 to 443 minutes of pumping can be calculated using eqs (4.33) and (4.36):

$$u = \frac{r^2 S_c}{4Tt} = \frac{(30\ \mathrm{m})^2 \times 0.0006}{4 \times 0.6\frac{\mathrm{m}^2}{\mathrm{min}} \times t} = \frac{0.225}{t},$$

$$s = \frac{Q}{4\pi T}\left[\ln\left(\frac{1}{u}\right) - 0.5772\right]$$

$$= \frac{1.79\frac{\mathrm{m}^3}{\mathrm{s}}}{4\pi \times 0.6\frac{\mathrm{m}^2}{\mathrm{min}}}\left[\ln\left(\frac{t}{0.225}\right) - 0.5772\right].$$

The recovery curve after 443 minutes of pumping is calculated using eqs (4.41) and (4.42):

$$s_r = \frac{Q}{4\pi T}\left(\ln\frac{4Tt}{r^2 S_c} - \ln\frac{4Tt_*}{r^2 S_c}\right).$$

The calculated and observed recovery data are shown in Table 4.6 and Figure 4.13.

Table 4.5 *Drawdown data*

Time (min)	Drawdown (m)	Time (min)	Drawdown (m)
443.5	1.640	451	1.060
444	1.595	455	0.930
444.5	1.535	459	0.845
445	1.490	464	0.755
445.5	1.445	469	0.700
446	1.400	479	0.590
447	1.305	489	0.521
447.5	1.235	499	0.451
448.5	1.200	514	0.384

Table 4.6 *Computed and observed recovery data of Example 4.14*

t (min)	t_* (min)	$t + t_*$	$\ln(4Tt/r^2S_c)$	$\ln(4Tt_*/r^2S_c)$	$Q/(4\pi T)$	Recovery (s_r)	Observed
443.5	0.5	444	7.586	0.799	0.237	1.611	1.640
444.0	1.0	445	7.587	1.492	0.237	1.447	1.595
444.5	1.5	446	7.589	1.897	0.237	1.351	1.535
445.0	2.0	447	7.590	2.185	0.237	1.283	1.490
445.5	2.5	448	7.591	2.408	0.237	1.230	1.445
446.0	3.0	449	7.592	2.590	0.237	1.187	1.400
447.0	4.0	451	7.594	2.878	0.237	1.120	1.305
447.5	4.5	452	7.595	2.996	0.237	1.092	1.235
448.5	5.5	454	7.598	3.196	0.237	1.045	1.200
451.0	8.0	459	7.603	3.571	0.237	0.957	1.060
455.0	12.0	467	7.612	3.977	0.237	0.863	0.930
459.0	16.0	475	7.621	4.264	0.237	0.797	0.845
464.0	21.0	485	7.632	4.536	0.237	0.735	0.755
469.0	26.0	495	7.642	4.750	0.237	0.687	0.700
479.0	36.0	515	7.663	5.075	0.237	0.614	0.590
489.0	46.0	535	7.684	5.320	0.237	0.561	0.521
499.0	56.0	555	7.704	5.517	0.237	0.519	0.451
514.0	71.0	585	7.734	5.754	0.237	0.470	0.384

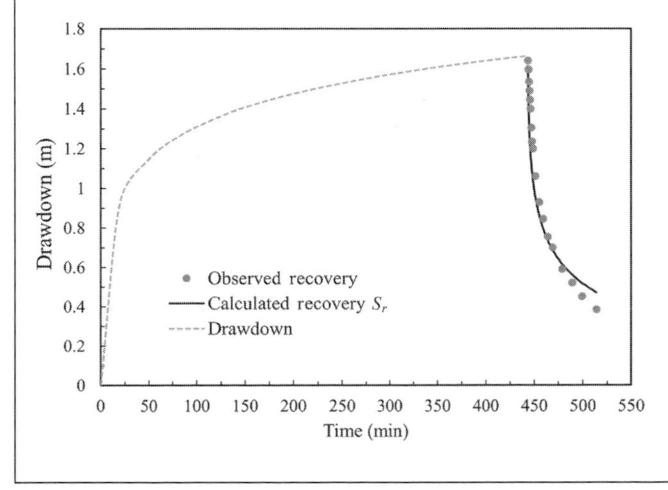

Figure 4.13 Computed and observed recovery data.

4.6 WELL CONSTRUCTION

Well construction involves drilling, well sizing, screen selection, and development. Each step is now briefly discussed.

4.6.1 Drilling

There are various methods of drilling whose applicability depends on the type of geologic formation, normal depths and diameters, and well yields. The Soil Conservation Service of the US Department of Agriculture (1969) has reported different well construction methods and their applicability, as shown in Table 4.7. Since time immemorial wells have been dug by hand, sometimes with the use of animal power, and this practice continues today in many developing countries. When wells are dug by hand, they are usually shallow and large in diameter, providing low yields. They are often used for domestic water supply or irrigating small fields or vegetable gardens. For moderate- to high-capacity irrigation wells, the dominant methods are cable tool, mud rotary, and air rotary. Driscoll (1986) compared the relative performance of different drilling methods applied to various types of geologic formations. Based on the rate of penetration, these methods can be classified into impossible, difficult, slow, medium, rapid, and very rapid. Augers have a slow penetration rate in clay and silt formations, and it is difficult to penetrate firm shale and sticky shale formations in this way. This method is impossible in unconsolidated materials such as dune sand, loose sand, and loose boulder. Driven wells are only recommended in dune sand, loose sand, and gravel, and have a slow penetration rate in these types of formations. Jetted wells have a rapid penetration rate in dune sand, loose sand and gravel, and quicksand formations, and a slow penetration rate in clay and silt formations, but are impossible or difficult in other types. Cable tool has a rapid rate of penetration in firm shale, brittle shale, chert nodules, limestone, dolomite, and basalts with thin layers in sedimentary rock formations, and a slow rate of penetration in clay and silt, sticky shale, sandstone, basalt with thick layers or highly fractured, metamorphic rocks, and granite formations. This method is less efficient in unconsolidated materials such as dune sand, loose sand, and quicksand. The direct rotary with fluid method, also called direct mud rotary, has a rapid penetration rate in dune sand, loose sand and gravel, quicksand, clay and silt, firm shale, sticky shale, brittle shale, limestone, and dolomite formations; a medium penetration rate in poorly cemented sandstone formations; and a slow penetration rate in well-cemented sandstone, chert nodules, thin- and thick-layer basalts, metamorphic rocks, and granite. Direct rotary with air is efficient in consolidated materials such as well-cemented sandstone, limestone, dolomite, thin layers of basalt, and granite (rapid rate of penetration), and thick layers of basalt (medium rate of penetration), but is not recommended for dune sand, loose sand and gravel, quicksand, clay and silt, firm shale, sticky shale, brittle shale, and poorly cemented sandstone. The reverse-circulation rotary method is considered a useful tool for drilling large-diameter wells in unconsolidated formations, for which the reverse dual-wall rotary is most used (Boulding, 1993). The reverse dual-wall rotary has very rapid penetration rates in dune sand, loose sand and gravel, and quicksand, and rapid penetration rates in firm shale, brittle shale, poorly and well-cemented sandstone, and limestone. This method has slow to medium penetration rates in other types of formations. More details can be found in the supplementary information.

4.6.2 Well Installations

Well installations can be of three types, as discussed by Bear (1979) and shown in Figure 4.14. These include uncased hole, screened well, and gravel-packed well. The uncased hole is employed in consolidated formations where the aquifer material does not collapse into the well and well radius equals the hole radius. The screened well is installed in unconsolidated formations wherein the screen prevents the aquifer material from collapsing into the well. Here also, the well radius equals the hole radius. The gravel-packed well is also installed in unconsolidated formations wherein a blanket of gravel about 5 cm thick is placed surrounding the well screen. The gravel increases the permeability of the aquifer in the vicinity of the well and hence leads to reduced drawdown for the same well discharge.

4.6.3 Screen Selection

Different types of screens are employed in wells, including slots cut into a pipe by a welding torch, perforations punched into the casing material by multi-bladed devices, slots machined into metal casing, louvered screens, and screens developed by winding wire of uniform thickness around a longitudinal wire frame. The screen reduces drawdown, which in turn reduces the cost of pumping. On the other hand, there is head loss due to friction when water flows past the well screen, which depends on the type and design of well screen.

The dimensions of well casing and screen depend on the size of the aquifer, expected drawdown in an unconfined aquifer, particle size distribution, hydraulic conductivity, and required discharge from the well. The US Bureau of Reclamation has recommended minimum diameters of well casings and screens, as shown in Table 4.8, which depend on well discharge. The screen length depends on the entrance velocity, aquifer thickness, open area per unit length of screen, and required well discharge. The entrance velocity is chosen to minimize blockage and encrustation due to contaminants. Walton (1962) has given recommended entrance velocities through well screens as a function of hydraulic conductivity, as shown in Table 4.9. The National Water Well Association (1981) has recommended

Table 4.7 *Well construction methods and applications*

Method	Material for which the method is best suited	Water table depth for which the method is best suited (m)	Usual maximum depth (m)	Usual diameter range (cm)	Usual casing material	Customary use	Yield (m^3/day)	Remarks
1. Augering (a) Hand auger	Clay, silt, sand, gravel less than 2 m	2–9	10	5–20	Sheet metal	Domestic drainage	15–250	Most effective for penetrating and removing clay. Limited by gravel over 2 m. Casing required if material is loose.
(b) Power auger	Clay, silt, sand, gravel less than 5 m	2–15	25	15–90	Concrete, steel, or wrought-iron	Domestic, irrigation, drainage	15–500	Limited by gravel over 5 m, otherwise same as for hand augur.
2. Driven wells (hand, air hammer)	Silt, sand, gravel less than 5 m	2–5	15	3–10	Standard weight pipe	Domestic, drainage	15–200	Limited to shallow water table, no large gravel.
3. Jetted wells (light, portable rig)	Silt, sand, gravel less than 2 m	2–5	15	4–8	Standard weight pipe	Domestic, drainage	15–150	Limited to shallow water table, no large gravel.
4. Drilled wells (a) Cable tools	Unconsolidated and consolidated medium hard and hard rock	Any depth	450	8–60	Steel or wrought-iron pipe	All uses	15–15,000	Effective for water exploration. Requires casing in loose materials. Mudscow and hollow rod bits developed for drilling unconsolidated fine to medium sediments.
(b) Rotary	Silt, sand, gravel less than 2 m; soft to hard consolidated rock	Any depth	450	8–45	Steel or wrought-iron pipe	All uses	15–15,000	Fastest method for all except hardest rock. Casing usually not required during drilling. Effective for gravel envelope wells.
(c) Reverse-circulation rotary	Silt, sand, gravel, cobble	2–30	60	40–120	Steel or wrought-iron pipe	Irrigation, industrial, municipal	2500–20,000	Effective for large-diameter holes in unconsolidated and partially consolidated deposits. Requires large volume of water for drilling. Effective for gravel envelope wells.
(d) Rotary-percussion	Silt, sand, gravel less than 5 cm; soft to hard consolidated rock	Any depth	600	30–50	Steel or wrought-iron pipe	Irrigation, industrial, municipal	2500–15,000	Now used in oil exploration. Very fast drilling. Combines rotary and percussion methods (air drilling) cuttings removed by air. Would be economical for deep water wells.

After Soil Conservation Service, US Department of Agriculture, 1969.

Table 4.8 *Recommended minimum diameters for well casings and screens*

Well yield (m^3/day)	Nominal pump chamber casing diameter (cm)	Surface casing diameter (cm) Naturally developed wells	Gravel-packed wells	Nominal screen diameter (cm)
<270	15	25	45	5
270–680	20	30	50	10
680–1,900	25	35	55	15
1,900–4,400	30	40	60	20
4,400–7,600	35	45	65	25
7,600–14,000	40	50	70	30
14,000–19,000	50	60	80	35
19,000–27,000	60	70	90	40

After US Bureau of Reclamation, 1977.

Table 4.9 *Values of screen velocities through screen openings*

Aquifer hydraulic conductivity		Entrance velocity through screen opening	
(m/day)	(gpd/ft^2)	(cm/s)	(ft/min)
<20	<500	1	2
20	500	1.5	3
40	1000	2	4
80	2000	3	6
120	3000	4	8
160	4000	4.5	9
200	5000	5	10
240	6000	5.5	11
>240	>6000	6	12

After Walton (1962).

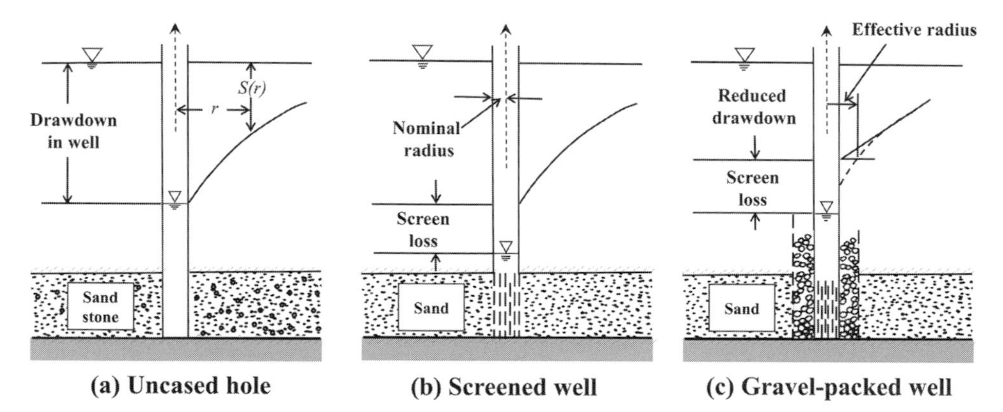

Figure 4.14 Three types of well installations in a confined aquifer. (Adapted from Bear (1979). Copyright © 1979 by Jacob Bear. Republished by Dover Publications in 2007.)

a maximum entrance velocity of 3 cm/s for any value of hydraulic conductivity equal to or greater than 120 m/day.

The screen slot size defined by the minimum width of the opening between successive windings of the screen wire is determined by sieve analysis of aquifer material. The effective diameter, defined by the grain size for which 90% of the aquifer material is coarser, is used as an index of the fineness of the material, and is about 0.025 cm based on sample data.

The uniformity coefficient (UC) of an aquifer is defined by the ratio of the grain size (d_{40}) for which 40% of the material is coarser to the effective grain size (d_{90}), expressed as

$$\mathrm{UC} = \frac{d_{40}}{d_{90}}. \tag{4.43}$$

For sample data, d_{40} is about 0.076. This then yields a value of UC as 3.0. For relatively fine and uniform material with UC less than or equal to 5, the screen opening should correspond to d_{40} for noncorrosive groundwater and to d_{50} for corrosive groundwater. For coarse sand and gravel, the opening should be between d_{30} and d_{50} of the sand fraction.

For nonuniform material with UC > 5, the screen opening should be d_{30} for stable material above the aquifer and d_{50} for unstable material. For wells with artificial gravel pack normally applied in sand aquifers, the slot size is d_{90} of the gravel grain size distribution. The diameter of the gravel material depends on the UC, as shown in Table 4.10. When screen size is selected, the length of the screen depends on the required well discharge, the percentage of open area for the screen as a function of slot size, and the thickness of the water-bearing formation within the aquifer, as shown in Table 4.11. Well screens are installed by the pull-back, bail-down, or wash-down method. The pull-back method is the simplest method; when it is not feasible one of the other two methods is used.

A functional relationship for well screen dimensions as a function of required well discharge and allowable entrance velocity has been reported as

Table 4.10 *Criteria for selection of gravel pack material*

Uniformity coefficient (UC)	Gravel pack criteria	Screen slot size
<2.5	(a) UC between 1 and 2.5 with 50% size not greater than 6 times the 50% size of the aquifer (b) If (a) is not available, UC between 2.5 and 5 with 50% size not greater than 9 times the 50% size of the aquifer	≤10% passing size of the gravel pack
2.5–5	(a) UC between 1 and 2.5 with 50% size not greater than 9 times the 50% size of the aquifer (b) If (a) is not available, UC between 2.5 and 5 with 50% size not greater than 12 times the 50% size of the aquifer	≤10% passing size of the gravel pack
>5	(a) Multiply the 30% passing size of the aquifer by 6–9 and locate the points on the grain size distribution graph on the same horizontal line (b) Through these points draw two parallel lines representing materials with UC ≤ 2.5 (c) Select gravel pack material that falls between the two lines.	≤10% passing size of the gravel pack

After US Bureau of Reclamation, 1977.

Table 4.11 *Intake areas for telescope V-slot wire-wound well screen (areas expressed as square centimeters per lineal meter of screen length)*

Nominal screen size (cm)	10-slot (0.01 in.) (0.25 mm)	20-slot (0.02 in.) (0.50 mm)	40-slot (0.04 in.) (1.0 mm)	60-slot (0.06 in.) (1.5 mm)	80-slot (0.08 in.) (2.0 mm)	100-slot (0.10 in.) (2.5 mm)	150-slot (0.15 in.) (3.7 mm)	250-slot (0.25 in.) (6.2 mm)
7.6	318	550	868	1,101	1,249	1,376	1,545	1,736
8.9	381	656	1,037	1,291	1,482	1,630	1,863	2,096
10.2	423	741	1,207	1,503	1,715	1,863	2,138	2,434
11.4	487	847	1,355	1,693	1,947	2,117	2,413	2,731
12.7	550	953	1,524	1,905	2,159	2,371	2,371	2,794
14.3	593	1,037	1,672	2,096	2,392	2,604	2,985	3,366
15.2	635	1,122	1,799	2,244	2,117	2,371	2,794	3,302
20.3	593	1,080	1,842	2,392	2,815	3,154	3,387	4,107
25.4	762	1,376	2,286	2,985	3,514	3,937	4,234	5,144
30.5	889	1,630	2,752	3,027	3,620	4,128	5,017	5,610
35.6	783	1,439	2,053	2,794	3,408	3,916	4,911	6,181
40.6	889	1,270	2,286	3,133	3,810	4,403	5,525	6,922
45.7	762	1,461	2,625	3,577	4,361	5,017	6,308	7,938
50.8	868	1,630	2,942	4,001	4,847	5,588	5,927	7,747
61.0	1,291	2,392	2,773	3,853	4,784	5,610	7,261	9,504
66.0	1,334	2,498	2,921	4,043	5,017	5,885	7,620	9,970
76.2	1,588	2,921	3,408	4,742	5,885	6,880	8,923	11,685
91.4	1,778	3,323	3,895	5,398	6,710	7,853	10,182	13,315

After Cuenca (1989).

$$Q = 0.0864 v_s (1 - c) \pi d_s L_s P, \tag{4.44}$$

where Q is the required well discharge (m^3/day), v_s is the screen entrance velocity (cm/s), c is the clogging factor, d_s is the screen diameter (cm), L_s is the screen length (m), and P is the percentage of screen open area. The value of c is taken as 0.5, indicating that 50% of the screen open area is blocked by aquifer material. The National Water Well Association (1981) has specified that at least 80% of aquifer thickness should be screened, with the screen centered in the water-bearing formation in the confined aquifer, and 33–50% of the aquifer thickness should be screened, with the screen placed at the bottom of the unconfined aquifer.

Example 4.15: Determine the casing diameter and screen dimensions and check the feasibility of installation in an aquifer 40 m thick having a hydraulic conductivity of 200 m/day and noncorrosive water. The grain size distribution of the aquifer material shows that d_{40} is 0.076 and d_{90} is 0.025, yielding a UC of 3.0. The required well discharge is 4500 m^3/day. The percentage of open area can be obtained from tabulated values.

Solution: The uniformity coefficient of 3.0 is less than 5.0 and the material is uniform and water supply is noncorrosive. Therefore, the required slot size is $s_s = d_{40} = 0.076\ \text{cm} = 0.03\ \text{in}$.

From Table 4.9, the allowable entrance screen velocity is 5 cm/s. From the National Water Well Association, the maximum allowable velocity which is more constraining is $v_{s\text{-}max} = 3.0\ \text{cm/s}$. For the required discharge of 4500 m^3/day, the nominal or minimum screen diameter from Table 4.8 is 25 cm.

The value of c can be taken as 0.5. From Table 4.11 for telescope-type well screens the required slot size is 0.76 mm (0.030 in.), the closest screen size is for a 20-slot screen ($s_s = 0.020$ in). For this screen with minimum screen diameter of 25 cm (10 in), the open (intake) area (A_{open}) is 1376 cm^2/m, in which per meter corresponds to per lineal meter of screen length. The total screen area per lineal meter can be calculated as

$$A_{total} = \pi d_s \times 100 \frac{\text{cm}}{\text{m}} = \pi(25\ \text{cm}) \times 100 \frac{\text{cm}}{\text{m}} = 7854\ \text{cm}^2/\text{m}.$$

The percentage open area is obtained as

$$P = \frac{A_{open}}{A_{total}} \times 100 = \frac{1376\ \text{cm}^2/\text{m}}{7854\ \text{cm}^2/\text{m}} \times 100 = 17.5\%.$$

Therefore,

$$Q = 0.0864 v_s (1 - c) \pi d_s L_s P,$$

$$4500 \frac{\text{m}^3}{\text{d}} = 0.0864 \times 3.0 \frac{\text{cm}}{\text{s}} \times (1 - 0.5) \times \pi \times 25\ \text{cm} \times L_s \times 17.5.$$

Solving for screen length L_s,

$$L_s = 25.26\ \text{m}.$$

The aquifer thickness is 40 m. Therefore,

$$\frac{L_s}{b} = \frac{25.26\ \text{m}}{40\ \text{m}} = 0.632.$$

It is not suitable for a confined aquifer ($\geq$80%) or unconfined aquifer (33–50%).

4.6.4 Well Development

Well development is needed for proper well operation and is undertaken for two reasons to conclude the well construction process: (1) The aquifer permeability decreases due to compaction and lining of the hole with drilling mud during drilling. The development step increases the permeability to the level that existed before drilling. (2) It removes the fine material that can pass through the screen. This material is in the vicinity of the well. Thus, the well is able to operate sand-free under normal conditions. The development process is regarded as completed when the well is able to discharge sand-free water at a rate equal to or greater than the maximum design discharge. Sand-free water means that a liter of water contains no more than a few sand grains when pumped at the maximum discharge. The main consideration for good development is to agitate the formation adjacent to the screen and allow the water to flow at high velocity through the screen into the well. The common methods of well development are: (1) surge block, (2) air displacement, and (3) jetting.

4.7 WELL TESTING

Well testing is the final step in the construction process and is done to determine the hydraulic characteristics as well as the operating characteristics of the well. Common methods for well testing are the specific capacity test and the constant-rate test. The specific capacity test involves pumping at different discharges for times equal to the expected time of well operation. Generally, testing is started at 20% of the design discharge and the drawdown is measured. Then, pumping discharge is increased at 20% increments up to 120%, and at each discharge level drawdown is measured. Then, measured drawdown is plotted against pumped discharge, as shown in Figure 4.15, and the inverse

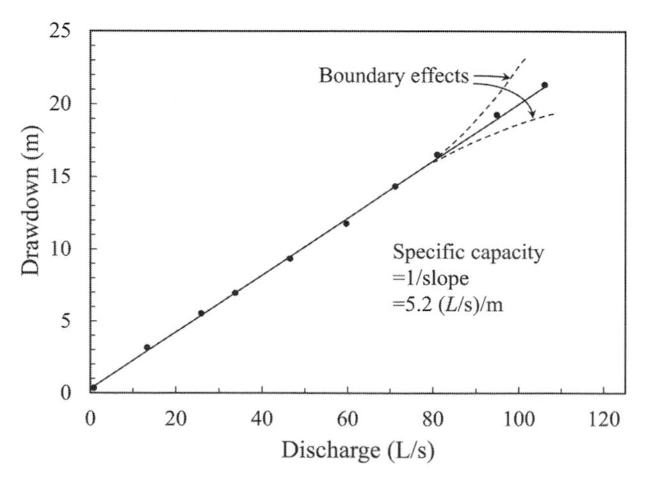

Figure 4.15 Well characteristic curve developed from well test data at different flow rates.

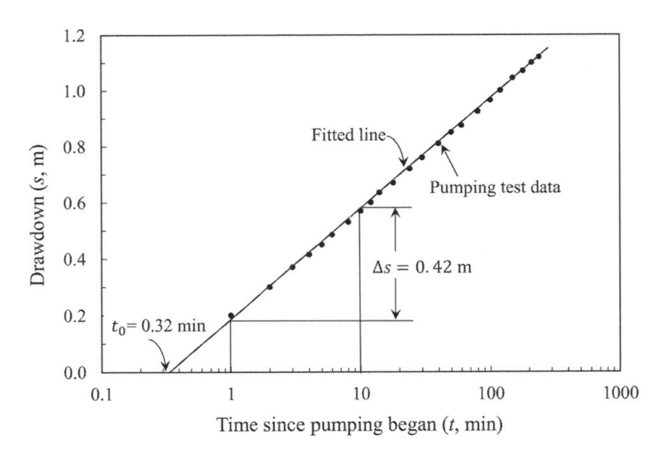

Figure 4.16 Example plot of the test data from a constant-rate well test.

of the slope of the linear plot defines the specific capacity of the aquifer, which is the ratio of well discharge to the drawdown.

The constant-rate test involves measuring the variation in the level of drawdown with time for a given discharge. If the test is done at a constant rate, then that should be the design discharge. For multiple constant rates, the discharge can be 60–125% of design discharge. The minimum time of operation should be 100 minutes and equal to or longer than 24 hours. The drawdown is then plotted against time since the beginning of pumping on semilog paper, as shown in Figure 4.16. Then, the hydraulic properties of the aquifer are determined. The transmissivity (T [m^2/day]) can also be determined using the Jacob method (Jacob, 1940):

$$T = \frac{2.303\ Q}{4\pi\Delta s}, \tag{4.45}$$

where Q is well discharge (m^3/day) and Δs is the drawdown (m) corresponding to one log cycle of pumping time.

The aquifer storativity (S_c) is determined as

$$S_c = 2.246T\frac{t_0}{r^2}. \tag{4.46}$$

Example 4.16: For a confined aquifer a pumping test was done at a constant discharge of 55 L/s, and the well test data are given in Figure 4.16. The drawdown was measured at a radial distance of 60 m away from the pumping well. Determine the transmissivity and storativity of the aquifer.

Solution: The transmissivity is computed as

$$T = \frac{2.303\ Q}{4\pi\Delta s} = \frac{2.303 \times 55\frac{\text{L}}{\text{s}}\cdot\frac{\text{m}^3}{1000\ \text{L}}\cdot\frac{86{,}400\ \text{s}}{1\ \text{day}}\text{m}^3/\text{day}}{4\times\pi\times 0.42\ \text{m}}$$

$$= 2073.5\frac{\text{m}^2}{\text{day}}.$$

The storativity is computed as

$$S_c = 2.246T\frac{t_0}{r^2}$$

$$= 2.246\times 2073.5\frac{\text{m}^2}{\text{day}}\times\left(\frac{0.32\ \text{min}}{(60\ \text{m})^2}\cdot\frac{1\ \text{day}}{1440\ \text{min}}\right)$$

$$= 0.000287.$$

QUESTIONS

Q.4.1 An aquifer has a porosity of 0.30 and the average flow velocity is 0.20 m/day, measured by introducing a tracer upstream and observing the time for it to reach downstream. The slope of the piezometric surface is found to be 0.25 m/km. What is the hydraulic conductivity of the aquifer?

Q.4.2 An unconfined aquifer with an area of 2 km^2 has a specific yield of 0.25. How much water can be withdrawn from the aquifer if the water table lowers by 5 m? If the water table is 20 m below the land surface, how much water will have to be added to raise the water table by 10 m?

Q.4.3 An unconfined aquifer has an area of 2 km^2 and a water table 10 m below the land surface. About 30 million m^3 of water is added to the aquifer, which increases the water table by 5 m. What is the specific yield of this aquifer?

Q.4.4 Compute the hydraulic conductivity and transmissivity of a confined aquifer having an average thickness of 40 m. A 40-cm diameter pumping well fully penetrates the aquifer. There are two observation wells, 40 m and 100 m away from the pumping well. The well is pumped at a discharge of 20 L/s until a steady-state condition is achieved. The drawdown in the nearest well is observed as 2.5 m and in the farthest well as 2.25 m.

Q.4.5 A confined aquifer has a thickness of 30 m and a fully penetrating 40-cm diameter pumping well. The discharge under steady-state conditions is 4 L/s and the drawdown in the well is 1.5 m. What will be the discharge from this well if the well diameter was 25 cm and 35 cm for the same drawdown of 1.5 m? The radius of influence can be assumed to 1000 m.

Q.4.6 A 50-m thick unconfined aquifer has a hydraulic conductivity of 5 m/day. A fully penetrating well is to be tested with a constant discharge of 2000 m^3/day. There are two observation wells located 50 m and 200 m away from the pumping well. If the drawdown in the nearest well is observed to be 2 m, what is the drawdown in the farthest well?

Q.4.7 An unconfined aquifer has a pumping well and three observation wells A, B, and C. Well A is located 20 m away from the pumping well; well B is 15 m away; and well C is 10 m away. Under steady-state conditions, dye travels from well A to well B in 50 hours. How much time will the dye take to travel from well B to well C?

Q.4.8 A confined aquifer has a transmissivity of 30 m^2/day and has two fully penetrating wells, A and B. The two wells are 50 m apart. Well A has a static water table 30 m above the lower confining bed and well B has a 25 m static water table. What is the aquifer thickness? If a tracer is injected into well A, how long will it take to reach well B?

Q.4.9 Compute the storage coefficient of an aquifer that has a porosity (n) of 0.31, thickness (b) of 200 m, and compressibility of confined aquifer (α) is 10^{-8} m^2/N. The value of β (compressibility of water) can be taken as 5.1×10^{-10} m^2/N.

Q.4.10 A confined aquifer is 60 m thick and its hydraulic conductivity is 14 m/day. A pumping well is installed penetrating the entire thickness of the aquifer. Two observation wells are installed at radial distances of 25 m (r_1) and 75 m (r_2) away from the pumping well. The well is being pumped at a constant rate of 1600 m^3/day. Compute the drawdown (s_1) in the well 25 m away if the drawdown in the well 75 m away is 0.8 m (s_2).

Q.4.11 A well is pumping water at a rate of 2500 m^3 per day from an unconfined aquifer whose thickness is 40 m and hydraulic conductivity is 70 m/day. The pumping well penetrates the entire aquifer thickness. There are two observation wells located 15 m and 120 m away from the pumping well. If the drawdown in the observation well at 120 m is 1 m, then compute the drawdown in the well 15 m away.

Q.4.12 A confined aquifer has a transmissivity of 0.0045 and storativity of 0.0008. There are three wells, A, B, and C, pumping water at rates of 10 L/s, 12 L/s, and 15 L/s. Well B is in the center; well A is located 40 m away on the left side; and well C is located 60 m away on the right side of well B. Well A has a diameter of 0.4 m, well B has a diameter of 0.5 m, and well C has a diameter of 0.6 m. Compute the drawdown curve for each well separately and the composite drawdown at a distance 20 m away from well B when $t = 150$ s.

Q.4.13 A pumping test was conducted on an aquifer 11 m thick and drawdown data were noted. The radial distance where the drawdown was measured was 4.4 m away from the pumped well. The rate of pumping was 1.872 m^3/min. Determine the transmissivity, apparent specific yield (or storativity), and horizontal hydraulic conductivity of the aquifer. The drawdown data were reported as shown in Table 4.12.

Table 4.12 *Drawdown data*

Time (min)	Drawdown (m)	Time (min)	Drawdown (m)	Time (m)	Drawdown (m)
0.5	0.340	11	0.765	61	0.995
2	0.635	13	0.770	94	1.060
3	0.705	15	0.800	156	1.140
4	0.715	18	0.805	253	1.230
5	0.705	20	0.820	400	1.315
6	0.715	22	0.845	482	1.340
7	0.720	30	0.870	509	1.355
9	0.745	32	0.885		

APPENDIX: SUPPLEMENTARY INFORMATION

Driscoll (1986) compared the relative performance of different drilling methods in various types of geologic formations. Based on the rate of penetration, these methods are classified into impossible, difficult, slow, medium, rapid, and very rapid.

1. Dune sand/loose sand and gravel. In these two formations, the drilling methods with reverse rotary (dual-wall) and direct rotary (drill through casing hammer) have very rapid penetration rates and are the most efficient methods. Direct rotary (with fluids), reverse rotary (with fluids, assuming sufficient hydrostatic pressure is available), hydraulic percussion, and jetted wells methods have rapid penetration rates. The driven method has a slow penetration rate. Cable tool and auger methods are difficult or impossible, and so are not suitable. The direct rotary with air method is not recommended.
2. Quicksand. In this formation, the drilling methods with reverse rotary (dual-wall) and direct rotary (drill through casing hammer) have very rapid penetration rates and are the most efficient methods. Direct rotary (with fluids), reverse rotary (with fluids, assuming sufficient hydrostatic pressure is available), hydraulic percussion, and jetted wells methods have rapid penetration rates. Cable tool and auger methods are difficult or impossible, and so are not suitable. Driven and direct rotary (with air) methods are not recommended.
3. Loose boulders in alluvial fans or glacial drift. In this formation, the drilling method with direct rotary (drill through casing hammer) has a rapid penetration rate and is the most efficient method. The drilling method with

reverse rotary (dual-wall) has a medium penetration rate. The drilling methods such as cable tool, direct rotary (with fluids), reverse rotary (with fluids), hydraulic percussion, auger, and jetted wells are difficult or impossible, and so are not suitable. Driven and direct rotary (with air) drilling methods are not recommended.

4. Clay and silt. In this formation, the drilling methods with direct rotary (with fluids or drill through casing hammer) and reverse rotary (with fluids or dual-wall) have rapid penetration rates and are the most efficient. Hydraulic percussion, cable tool, auger, and jetted wells methods have slow penetration rates. The drilling method with direct rotary (with air) is not recommended.
5. Firm shale. In this formation, the drilling methods with direct rotary (with fluids or drill through casing hammer), reverse rotary (with fluids or dual-wall), and cable tool have rapid penetration rates and are the most efficient. Hydraulic percussion has a slow penetration rate. Auger is difficult and is not suitable. Direct rotary (with air), driven, and jetted wells methods are not recommended.
6. Sticky shale. In this formation, the drilling methods with direct rotary (with fluids or drill through casing hammer) and reverse rotary (dual-wall) have rapid penetration rates and are the most efficient. Reverse rotary (with fluids), hydraulic percussion, and cable tool methods have slow penetration rates. Auger is difficult and is not suitable. Direct rotary (with air), driven, and jetted wells methods are not recommended.
7. Brittle shale. In this formation, the drilling methods with direct rotary (with fluids or drill through casing hammer), reverse rotary (with fluids or dual-wall), and cable tool have rapid penetration rates and are the most efficient methods. Hydraulic percussion has a slow penetration rate. Direct rotary (with air), auger, driven, and jetted wells methods are not recommended.
8. Sandstone, poorly cemented. In this formation, the drilling method with reverse rotary (dual-wall) has a rapid penetration rate and is the most efficient method. Direct rotary (with fluids), reverse rotary (with fluids), and hydraulic percussion methods have medium penetration rates. The cable tool drilling method has a slow penetration rate. Direct rotary (with air), driven, and jetted wells methods are not recommended.
9. Sandstone, well-cemented. In this formation, the drilling methods with direct rotary (with air) and reverse rotary (dual-wall) have rapid penetration rates and are the most efficient. Direct rotary (with fluids), reverse rotary (with fluids), cable tool, and hydraulic percussion methods have slow penetration rates. Direct rotary (down-the-hole air hammer), driven, and jetted wells methods are not recommended.
10. Chert nodules. In this formation, hydraulic percussion and cable tool are the most efficient drilling methods and have rapid penetration rates. Direct rotary (with fluids or air) and reverse rotary (with fluids or dual-wall) methods have slow penetration rates. Driven and jetted wells methods are not recommended.
11. Limestone/dolomite. In these two formations, the drilling method with direct rotary (down-the-hole air hammer) has a very rapid penetration rate and is the most efficient, followed by the direct rotary (with air or fluids), reverse rotary (with fluids or dual-wall), and hydraulic percussion methods (rapid penetration rates). Driven and jetted wells methods are not recommended.
12. Limestone with chert nodules. In this formation, the drilling method with direct rotary (down-the-hole air hammer) has a very rapid penetration rate and is the most efficient method, followed by the direct rotary (with air) and hydraulic percussion methods (rapid penetration rates). Direct rotary (with fluids) and reverse rotary (with fluids or dual-wall) methods have slow penetration rates. Driven and jetted wells methods are not recommended.
13. Limestone with small cracks or fractures. In this formation, the drilling method with direct rotary (down-the-hole air hammer) has a very rapid penetration rate and is the most efficient, followed by the direct rotary (with air) and hydraulic percussion methods (rapid penetration rates). Direct rotary (with fluids) and reverse rotary (with fluids or dual-wall) methods have slow penetration rates. Driven and jetted wells methods are not recommended.
14. Limestone, cavernous. In this formation, the drilling methods with direct rotary (down-the-hole air hammer), reverse rotary (dual-wall), and cable tool have rapid penetration rates and are the most efficient methods. Direct rotary (with fluids or air), reverse rotary (with fluids), and hydraulic percussion methods are difficult or impossible, and so are not suitable. Driven and jetted wells drilling methods are not recommended.
15. Basalts, thin layers in sedimentary rocks. In this formation, the drilling method with direct rotary (down-the-hole air hammer) has a very rapid penetration rate and is the most efficient method, followed by the direct rotary (with air), cable tool, reverse rotary (dual-wall), and hydraulic percussion methods (rapid penetration rates). Direct rotary (with fluids) and reverse rotary (with fluids) methods have slow penetration rates. Driven and jetted wells methods are not recommended.
16. Basalts, thick layers. In this formation, the drilling method with direct rotary (down-the-hole air hammer) has a rapid penetration rate and is the most efficient method. Direct rotary (with air) and reverse rotary (dual-wall) methods have medium penetration rates. Direct rotary (with fluids), reverse rotary (with fluids), cable tool, and hydraulic percussion methods have slow penetration rates. Driven and jetted wells methods are not recommended.
17. Basalts, highly fractured (lost circulation zones). In this formation, the drilling method with reverse rotary (dual-wall) has a medium penetration rate. Direct rotary (with air) and cable tool methods have slow penetration rates.

Direct rotary (with fluids), reverse rotary (with fluids), and hydraulic percussion methods are difficult and are not suitable. Driven and jetted wells methods are not recommended.

18. Metamorphic rocks. In this formation, the drilling method with direct rotary (down-the-hole air hammer) has a rapid penetration rate and is the most efficient method. Direct rotary (with air) and reverse rotary (dual-wall) methods have medium penetration rates. Direct rotary (with fluids), reverse rotary (with fluids), cable tool, and hydraulic percussion methods have slow penetration rates. Driven and jetted wells methods are not recommended.
19. Granite. In this formation, the drilling method direct rotary (with air) has a rapid penetration rate and is the most efficient method. The reverse rotary (dual-wall) method has a medium penetration rate. Direct rotary (with air), reverse rotary (with fluids), cable tool, and hydraulic percussion methods have slow penetration rates. Driven and jetted wells methods are not recommended.

REFERENCES

Bear, J. (1979). *Hydraulics of Groundwater*, McGraw-Hill, New York.

Boulding J. R. (1993). *Subsurface Characterization and Monitoring Techniques: A Desk Reference Guide.* Washington, DC: US Environmental Protection Agency, Office of Research and Development.

Cuenca, R. H. (1989). *Irrigation System Design: An Engineering Approach.* Englewood, Cliffs, NJ: Prentice Hall.

Driscoll, F. G. (1986). *Groundwater and Wells*, 2nd ed. St. Paul, MN: Johnson Screens.

Dupuit, J. (1863). *Etudes Theoriques et Pratiques sur le Mouvement des Eaux dans les Canaux Decouverts et a Travers les Terrains Perm eables*, 2nd ed. Paris: Dunod.

Freeze, R. A., and Cherry, J. A. (1979). *Groundwater.* Englewood Cliffs, NJ: Prentice Hall.

Green, J. H. (1962). Compaction of the aquifer system and land subsidence in the Santa Clara Valley, California. US Geological Survey professional paper 450-D, art. 172, pp. D175–D178.

Hantush, M. S. (1964). Hydraulics of wells. In *Advances in Hydroscience*, edited by V. T. Chow. New York: Academic Press, pp. 281–432.

Jacob, C. E. (1940). The flow of water in an elastic artesian aquifer: *Transactions of the American Geophysical Union*, 21: 574–586.

Jacob, C. E. (1950). Flow of ground water. In *Engineering Hydraulics*, edited by H. Rouse, New York: Wiley.

Marino, M. A. and Luthin, J. N. (1982). *Seepage and Groundwater.* New York: Elsevier.

McWhorter, D. B., and Sunada, D. K. (1977). *Ground Water Hydrology and Hydraulics.* Highlands Ranch, CO: Water Resources Publication.

Meinzer, O. E. (1923). Outline of groundwater hydrology, with definitions. US Geological Survey water-supply paper 494.

Morris, D. A., and Johnson, A. I. (1967). Summary of the hydrologic and physical properties of rock and soil materials as analyzed by the hydrologic laboratory of the US Geological Survey, 1948–1960. US Geological Survey water-supply paper 1839-D.

National Water Well Association (1981). *Water Well Specifications*, Berkeley, CA: Premier Press.

Soil Conservation Service (1969). *Engineering Field Manual for Conservation Practices.* Washington, DC: US Department of Agriculture.

Theis, C. V. (1935). The relation between the lowering of the piezometric surface and the rate and duration of discharge of a well using groundwater storage. *Transactions of the American Geophysical Union*, 16: 519–524.

Todd, D. K. (1959). *Ground Water Hydrology.* New York: Wiley.

Todd, D. K. (1980). *Groundwater Hydrology*, 2nd ed. New York: Wiley.

US Bureau of Reclamation (1977). *Groundwater Manual.* Washington, DC: US Department of the Interior.

Walton, W. C. (1962). Selected analytical methods for well and aquifer evaluation. Illinois State Water Survey Bulletin 49.

5 Irrigation Water Quality

Notation

Ca	calcium concentration (mmol/L or moles/m^3)
C_s	salinity concentration (mg/L)
D_d	depth of water (volume per unit area) draining from the root zone (mm, in.)
D_I	depth of water (volume per unit area) applied at the surface (mm, in.)
EC	electrical conductivity at 25 °C (mmhos/cm)
EC_{iw}	electrical conductivity of irrigation water (mmhos/cm)
EC_s	electrical conductivity of a saturation extract of the soil taken from the root zone (mmhos/cm)
EC_{sw}	electrical conductivity of soil water (mmhos/cm)
I_S	ionic strength
LF	leaching fraction
Mg	magnesium concentration (mmol/L or moles/m^3)
Na	sodium cation concentration (mmol/L or moles/m^3)
SAR	sodium absorption ratio
SAR_{final}	final concentration of total salt
SAR_{intial}	initial concentration of total salt
SC	sum of cations (meq/L)
SP	saturation percentage (%)
$(\theta_m)_{CEW}$	soil moisture content at the limit of crop-extractable water on a mass basis (g/g)
$(\theta_m)_{FC}$	soil moisture content at field capacity on a mass basis (g/g)
Ψ_{sw}	osmotic potential (kPa)

5.1 INTRODUCTION

The source of irrigation water is either surface or subsurface. Surface water contamination is caused by both point and non-point sources, as shown in Figure 5.1. Point sources are industrial and factory waste outlets, chemical dumping sites, waste disposal sites, waste outfall, animal feedlots, and chemical spills. Examples of non-point source pollution include runoff from urban, agricultural, and forest lands, acid precipitation, and highway runoff.

The quality of irrigation water is determined by the amount of sediment and chemical constituents in the water. The origin of sediment influences the fertility of soil and the quality of water; sediment is the most important pollutant from agricultural lands. Soil particles incorporate chemical ions and transport them, and chemical characteristics determine the suitability of irrigation water.

The quality of irrigation water has a significant impact on crop yield, degradation of soil, pollution of groundwater, and operation and life of irrigation systems. It also interacts with soil and its chemical and physical constituents. In irrigation engineering, water quality is evaluated by considering the physical and chemical characteristics of water, but biological characteristics may also be important if wastewater is used for irrigation. This chapter discusses water quality characteristics and their effect on agricultural productivity and irrigation systems.

5.2 WATER QUALITY CHARACTERISTICS

Water quality characteristics or parameters that are important from an agricultural irrigation point of view are total dissolved solids (TDS) (salinity), sodicity, acidity, alkalinity, dissolved oxygen (DO), biochemical oxygen demand (BOD), nitrate, arsenic, radionuclides, hardness, turbidity, color, concentration of selected cations and anions, and trace elements. The suitability of irrigation water can be evaluated by evaluating these characteristics, as well as their combinations, using appropriate tests. The most important chemical characteristics of irrigation water are the total concentration of soluble salts, the proportion of sodium to other ions, the concentration of potentially toxic elements, and bicarbonate concentration in relation to the concentration of calcium and magnesium. Table 5.1 summarizes various parameters for the assessment of irrigation water quality.

Table 5.1 *Various parameters for evaluating irrigation water quality*

Parameter	Symbol	Atomic weight	Unit of measurement	Usual range in irrigation water
Salinity				
Electrical conductivity	EC		dS/m	0–3
Total dissolved solids	TDS		mg/L	0–2000
Sodicity				
Sodium absorption ratio (SAR)	SAR		[a]	0–15
Adjusted SAR	ASAR		[a]	[b]
Concentration of selected cations				
Calcium	Ca	40.1	mg/L	0–800
Magnesium	Mg	24.3	mg/L	0–120
Sodium	Na	23	mg/L	0–900
Concentration of selected anions				
Carbonate	CO_3	60	mg/L	0–6
Bicarbonate	HCO_3	61	mg/L	0–600
Chloride	Cl	35.5	mg/L	0–1000
Sulfate	SO_4	96.1	mg/L	0–2000
Nitrate/nitrogen	NO_3–N	62	mg/L	0–10
Ammonium/nitrogen	NH_4–N	18	mg/L	0–5
Phosphate/phosphorous	PO_4	79	mg/L	0–2
Acidity or alkalinity	pH		[a]	6.0–8.5
Suspended materials				
Turbidity	T		NTU	[c]
Trace elements			mg/L	

mg/L = milligrams per liter; ppm = parts per million; $\frac{dS}{m}$ = deci-Siemens per meter at 25 °C.
[a] Usually reported without a unit.
[b] Depends on the pH value and concentration of CO_3 and HCO_3 ions.
[c] Maximum recommended values in Table 5.2.
After Ayers and Westcot, 1994.

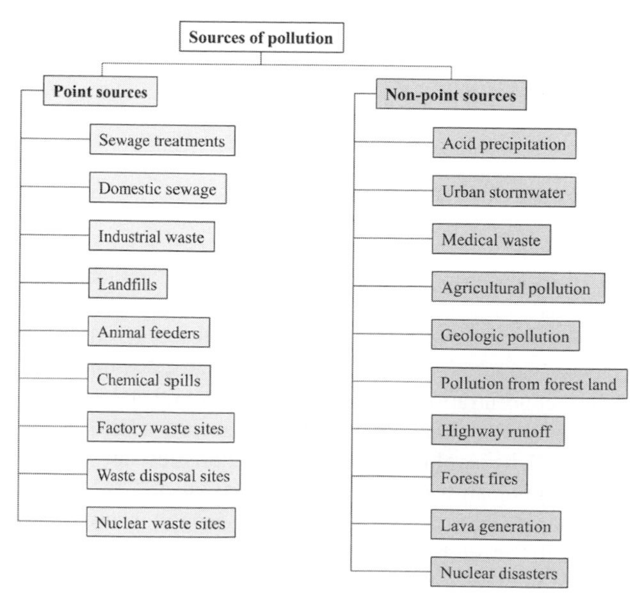

Figure 5.1 Point and non-point sources of pollution.

It may be useful to note different conversions for water quality analysis: 1 mg/L = 1 ppm
1 dS/m = 1 mmhos/cm = 1000 μmhos/cm
TDS (mg/L) = EC (dS/m) × 640 for EC < 5 dS/m
TDS (mg/L) = EC (dS/m) × 800 for EC > 5 dS/m
TDS (lb/ac-ft) = TDS (mg/L) × 2.72
concentration (ppm) = concentration (mol/m^3) × atomic weight
meq/L = EC (dS/m) × 10 for sum of cations/anions.

5.3 THE EFFECT OF POOR WATER QUALITY

Poor-quality irrigation water contains substances that are toxic to plants, detrimental to plant physiological activities, slow the extraction of water by plant roots, retard infiltration of water into the soil, and undermine the effectiveness of irrigation system operation and delivery. Not all crops have equal sensitivity to water quality – some are more sensitive than others. Further, the quality of water needed varies with crop, soil, fertilization, climate, and irrigation method. Table 5.2 gives information on hazards caused by sodium in irrigation water to crops. Table 5.3 shows the impact of water quality on soil permeability.

The influence of water quality on crop yield depends on soil characteristics, crop type, and crop variety. Soil texture and structure and organic matter influence infiltration, storage, and movement of water within the soil. The movement of water determines the leaching of pollutants below the root zone, the availability to plants, or the fixation to soil particles.

Table 5.2 *Sodium hazards of irrigation water*

Hazard level	SAR	Suitability of irrigation water
Low	1–10	Little problem with most soils and crops, but use of sodium crops, such as avocados, must be cautioned against.
Medium	10–18	Fine-textured soils may develop problems associated with excessive sodium accumulation; amendments, such as gypsum, and leaching may be needed.
High	18–26	Generally not suitable for irrigation purposes or continuous use.
Very high	>26	Generally unsuitable for use.

Adapted from Fipps, 2021.

Table 5.3 *Effect of irrigation water quality on soil permeability*

Evaluate using both SAR and *EC_{iw}	Effect		
	None	Moderate	Severe
SAR	EC_{iw} (mmhos/cm) *	EC_{iw} (mmhos/cm)	EC_{iw} (mmhos/cm)
0–3	>0.7	0.2–0.7	<0.2
3–6	>1.2	0.3–1.2	<0.3
6–12	>1.9	0.5–1.9	<0.5
12–20	>2.9	1.3–2.9	<1.3
20–40	>5.0	2.9–5.0	<2.9

* EC_{iw} is the electrical conductivity of irrigation water.
After Ayers and Westcot, 1994.

Climate characteristics such as temperature, cloud cover, and sunshine affect evapotranspiration, and rainfall affects the frequency of irrigation. These factors affect the availability of water to plants, infiltration and movement of water in the root zone, toxic effects on plant growth, and the quality of farm produce. Sprinkler and drip irrigation systems are impacted by the quality of water.

5.4 SALINITY

Almost all waters contain dissolved salts, which may be caused by natural processes and anthropogenic activities. Although ordinary table salt or sodium chloride characterizes salt, there are many types of salts that are observed in water resources. Table 5.4 shows different types of salts in the water resources of Texas. With evaporation, the salt concentration in water increases, and this is how salinization of soil in lowland areas is caused. For example, if there are 5 g of salt in 1 L of water, then the salt concentration is 5 g/L, or 0.5%. If half of the water is evaporated, the salt concentration will rise to 10 g/L, or 1%, which is double the original concentration. As water transpires from croplands, the salt accumulation grows. Salinity affects plants and may cause saline soil conditions.

Table 5.4 *Types of salts found in irrigation water*

Chemical name	Chemical symbol	Approximate proportion of total salt content
Sodium chloride	$NaCl$	Moderate to large
Sodium sulfate	Na_2SO_4	Moderate to large
Calcium chloride	$CaCl_2$	Moderate
Calcium sulfate (gypsum)	$CaSO_4{\cdot}2H_2O$	Moderate to small
Magnesium chloride	$MgCl_2$	Moderate
Magnesium sulfate	$MgSO_4$	Moderate to small
Potassium chloride	KCl	Small
Potassium sulfate	K_2SO_4	Small
Sodium bicarbonate	$NaHCO_3$	Small
Calcium carbonate	$CaCO_3$	Very small
Sodium carbonate	Na_2CO_3	Trace to none
Borates	BO^{-3}	Trace to none
Nitrates	NO^{-3}	Small to none

After Longenecker and Lyerly, 1974.

The salinity of water can be reflected by TDS or EC. TDS can be referred to as the total salinity, which is measured in mg/L or ppm. Based on TDS, the water can be characterized as: (1) low-saline, (2) moderately saline, (3) highly saline, and (4) very highly saline. This characterization gives an idea about the suitability of salt-affected water for irrigation. Of course, other factors are also considered when evaluating the suitability, such as rainfall, infiltration characteristics, depth of the water table, and salt transportability of soil, as shown in Figure 5.2. The movement of water is affected, in turn, by carbonate and bicarbonate concentrations and the pH value of water.

5.5 EFFECT OF SALINITY ON SOIL WATER

Soils may be affected by total salinity or a combination of both salinity and sodium. The salinity of irrigation water affects infiltration and the availability of soil water for plant extraction. Irrigation with saline water increases salt concentration in soil water, reducing osmotic pressure and consequently decreasing the availability of soil water for plant use. This affects crop growth and yield. In other words, high concentration of salt can cause physiological drought conditions despite the soil having plenty of moisture. On the other hand, low-salinity irrigation water is not always preferred. If water with EC < 0.5 dS/m is applied to a non-saline soil that has low to moderate SAR, infiltration may be hampered by the reduction in soil permeability due to plugging and sealing of surface pores by structural breakdown and dispersion. The minerals, especially calcium, may leach, reducing

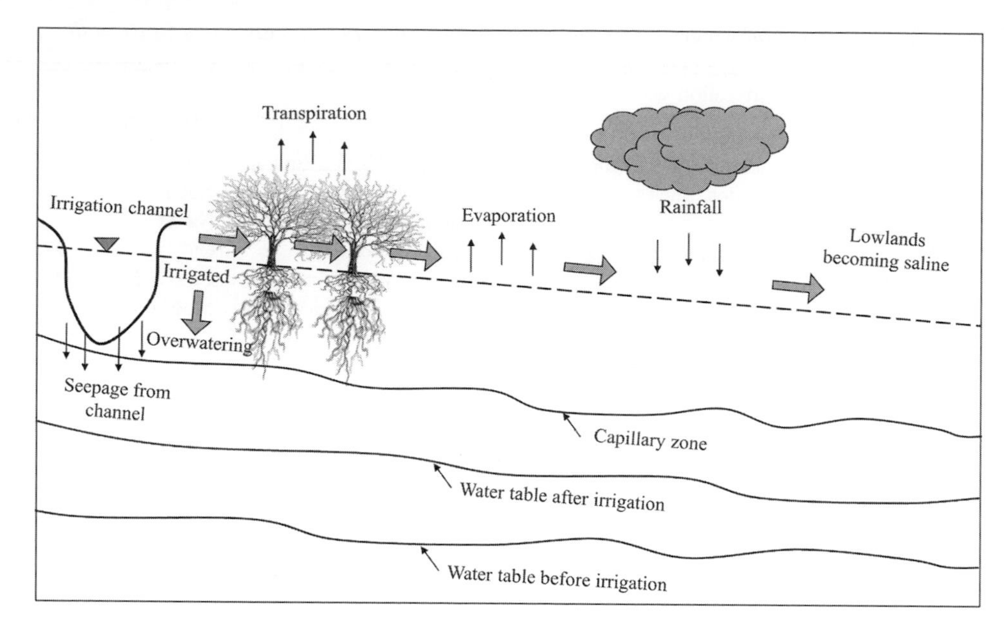

Figure 5.2 Hydrologic processes affecting salinity.

the stabilizing effect on soil structure and aggregate. In general, high total salinity increases infiltration. Although the increased level of salinity has possible effects on the aggregation and stabilization of soil, its potential impacts on plant health should be considered at the same time.

5.6 SODICITY

Sodicity refers to the amount of sodium in irrigation water. Sodium affects soils and may lead to sodic soil conditions. The sodicity of water is evaluated with the use of the SAR, defined as

$$\mathrm{SAR} = \frac{Na}{\sqrt{Ca + Mg}}, \tag{5.1}$$

where *Na*, *Ca*, and *Mg* are the sodium cation concentration, calcium concentration, and magnesium concentration, respectively, all measured in mmol/L or moles/m^3. For units in meq/L, the sum of Ca and Mg ions in eq. (5.1) should be divided by 2. The ions in the denominator are important, because their effect is to counterbalance the effect of sodium.

The value of SAR depends on the chemical reactions the water undergoes when moving through the soil, such as the pH value and concentration of carbonate and bicarbonate ions. Therefore, SAR is corrected by taking into account these reactions; then it is called the adjusted SAR (or ASAR). The value of ASAR will be lower than that of SAR if $CaCO_3$ or lime from soil dissolves, and will be higher if it precipitates.

The SAR can be determined as follows:

1. Calculate the sum of cations (*SC*) as

$$SC = Na[\mathrm{meq/L}] + Ca[\mathrm{meq/L}] + Mg[\mathrm{meq/L}]$$

and ionic strength (I_S) as

$$I_S = \frac{(1.3477 \times SC + 0.5355)}{1000}.$$

2. Calculate the factor of $\log_{10}(.)$ in the equation below as

$$\log_{10}(X) = \frac{1}{3}\Big[4.6629 + 0.6103\log_{10}(I_S) + 0.0844\{\log_{10}(I_S)\}^2 + 2\log_{10}\left(\frac{Ca[\mathrm{meq/L}]}{2 \times HCO_3[\mathrm{meq/L}]}\right)\Big].$$

3. Calculate equilibrated *Ca* concentration (Ca_{eq}) as

$$Ca_{eq}[\mathrm{meq/L}] = 10^{\log_{10}(X)} \times 0.17758.$$

4. Calculate the adjusted SAR by replacing Ca ions with equilibrated Ca concentration (Ca_{eq}). One can compute the percentage by which the ASAR is higher or lower than SAR.

In the ASAR calculation, the ions measurement needs to be converted from mg/L to meq/L:

$$I[\mathrm{meq/L}] = \frac{V}{AW} \times I[\mathrm{mg/L}],$$

where *I* denotes the ion measurement in respective units, *V* is the valence number of the ion, and *AW* is the atomic weight of the ion. The characteristics of ions in the water sample are as follows:

Ion	Atomic weight	Valence numbers
Na^+	22.99	1
Ca^{2+}	40.08	2
Mg^{2+}	24.31	2
HCO_3^-	61.02	1

Table 5.5 *Tolerance of crops to exchangeable sodium percentage*

Tolerance to ESP	Influencing range of ESP	Crop	Growth response under field conditions
Extremely sensitive	2–10	Deciduous fruits Nuts Citrus Avocados	Sodium toxicity symptoms even at low ESP values
Sensitive	10–20	Beans	Stunted growth at low ESP values even though the physical condition of the soil may be good
Moderately sensitive	20–40	Clover Oats Tall fescue Rice Dallisgrass	Stunted growth due to both nutritional factors and adverse soil conditions
Tolerant	40–60	Wheat Cotton Alfalfa Barley Tomatoes Beets	Stunted growth usually due to adverse physical conditions
Most tolerant	>60	Crested and fairway wheat grass Tall wheat grass Rhodes grass	Stunted growth usually due to adverse physical soil conditions

After Pearson, 1960.

5.7 EFFECT OF SODICITY

The sodicity of irrigation water affects plant physiology, soil structure, and infiltration. The effect of sodicity on soil physical properties is opposite to that of salinity. High sodium concentration degrades soil structure. Water with a high SAR causes soil to become hard and compact, making it less permeable to water and air upon wetting, especially when clay content is high. Recommended SAR values are given in Table 5.2. High SAR causes dispersion of soil particles, blocking of soil pores, and reduction of infiltration. Both salinity and sodicity act simultaneously. Carbonate (CO_3^-) and bicarbonate (HCO_3^-) in irrigation water will lead to the precipitation of calcium carbonate ($CaCO_3$) and magnesium carbonate ($MgCO_3$) when the concentration of dissolved salts increases in water. This will give rise to more sodium ions relative to calcium and sodium ions, thus increasing SAR. Even if SAR may be low, there is potential for sodium hazard as a result of high concentration of carbonate and bicarbonate. An increase in SAR will impact soil structure and infiltration.

The soluble sodium percentage (SSP), defined as the ratio of sodium in equivalents per million (epm) to the total cation epm multiplied by 100, is also employed to evaluate the effect of sodium. Irrigation water with SSP exceeding 60% is not desired, because it may lead to sodium accumulation and may cause a breakdown in the physical properties of the soil. The tolerance of crops to sodium is measured by the exchangeable sodium percentage (ESP), as shown in Table 5.5.

Table 5.6 *Classification of salt-affected soils based on* EC_s

Criteria	EC_s (mmhos/cm)
Very slightly saline	<4
Slightly saline	4–8
Moderately saline	8–16
Strongly saline	>16

After Soil Conservation Service, 1991.

5.8 CLASSIFICATION OF SALT-AFFECTED SOILS

Both SAR and EC_s can be used to classify salt-affected soils, in which EC_s is the electrical conductivity of a saturation extract of the soil taken from the root zone. Soils may be classified as normal ($EC_s < 4$ mmhos/cm and SAR < 13), saline ($EC_s > 4$ mmhos/cm and SAR < 13), sodic ($EC_s < 4$ mmhos/cm and SAR > 13), and saline–sodic ($EC_s > 4$ mmhos/cm and SAR < 13) (James et al., 1982). This classification serves as the basis for determining the type of soil reclamation and treatment. As shown in Table 5.6, soils can also be classified based on EC_s.

Saline soils normally have a pH value of less than 8.5. They are low in sodium and primarily comprise sodium, calcium, magnesium chlorides, and sulfates. The white crust that is often observed on the soil surface or salt streaks along furrows are caused by these compounds. The compounds that cause saline soils easily dissolve water and can therefore be removed by leaching.

Sodic soils have a pH value between 8.5 and 10, are dark in color, and are smooth. They are also called black alkali soils. Sodium destroys the permanent soil structure and makes the soil less pervious to water, implying that leaching alone will not be effective for treatment unless amendments are used.

The salt concentration affects the value of SAR. The change in SAR due to the change in salt concentration can be expressed as

$$SAR_{final} = (\Delta \text{ concentration })^{\frac{1}{2}} SAR_{initial}, \tag{5.2}$$

where Δ concentration denotes the ratio of the final concentration of total salt to the initial concentration. Equation (5.2) is useful when combined with the soil water system. For medium- to fine-textured soils, on a mass basis the water content at saturation is twice that at field capacity. If SAR_{intial} is assumed to be the same as that for the saturation abstract, then SAR at field capacity will be the square root of 2 multiplied by SAR_{intial}.

The saturation percentage can be defined by the soil water content on a mass basis for a saturated sample. For fine- and medium-textured soils, the relationship between saturation percentage (*SP*) and other moisture content levels can be defined as follows:

$$SP = 2(\theta_m)_{FC} \times 100, \tag{5.3}$$

$$(\theta_m)_{FC} = 2(\theta_m)_{CEW}, \tag{5.4}$$

$$SP = 4(\theta_m)_{CEW} \times 100, \tag{5.5}$$

where $(\theta_m)_{FC}$ is the soil moisture content at field capacity on a mass basis (g/g) and $(\theta_m)_{CEW}$ is the soil moisture content at the limit of crop-extractable water on a mass basis (g/g).

James et al. (1982) have given general relationships for practical applications; these, however, should be used with caution:

$$\text{Osmotic potential}: \psi_{sw} = -36EC, \tag{5.6}$$

$$\text{Salinity concentration}: C_s\left(\frac{\text{mg}}{\text{L}}\right) = 640EC;\ C_s\left(\frac{\text{meq}}{\text{L}}\right) = 10EC. \tag{5.7}$$

EC of soil water (EC_{sw}), irrigation water (EC_{iw}), and saturation extract (EC_s) are, respectively,

$$3EC_{iw} = EC_{sw}, \tag{5.8}$$

$$2EC_s = EC_{sw}, \text{ and} \tag{5.9}$$

$$EC_s = \left(\frac{3}{2}\right) EC_{sw}, \tag{5.10}$$

where Ψ_{sw} is the osmotic potential (kPa) and *EC* is the electrical conductivity at 25 °C (mmhos/cm).

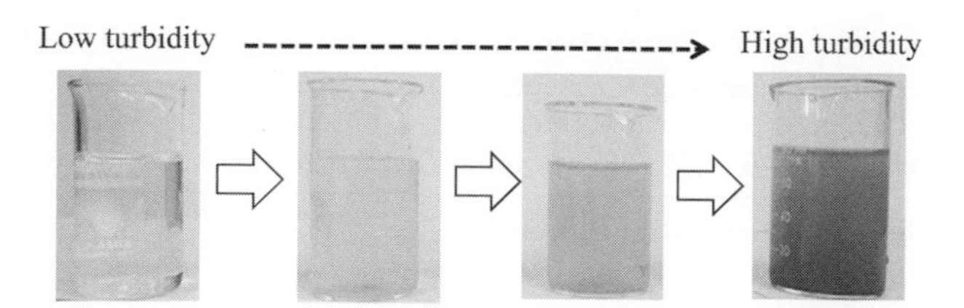

Figure 5.3 Turbidity levels. A black and white version of this figure will appear in some formats. For the color version, please refer to the plate section.

5.9 SUSPENDED MATTER

Surface water used for irrigation contains suspended matter that may stem from soil erosion, erosion caused within the water body, and/or material generated by human activities. A measure of suspended material is turbidity, which makes water muddy or discolored, as shown in Figure 5.3. Suspended matter affects the penetration of light whose measure, in turn, can be an indirect indicator of turbidity.

From an agricultural perspective, suspended matter may contain toxic heavy metals, pathogens, and nutrients that affect crop yield and the quality thereof. Further, sediments can decrease infiltration, clog sprinklers and drips, deposit in canals and channels, and increase maintenance and increase wear of pumps.

5.10 EFFECT OF TOXIC ELEMENTS

The main toxic ions in irrigation water are chloride, sodium, and boron, which can be toxic individually or in combination. Their uptake by plants may lead to excessive accumulation in leaves during transpiration and may damage leaves and affect yield. An element becomes toxic when its concentration reaches a certain threshold level. Further, some elements may also become inactive upon chemical reaction within the soil.

Chloride is useful for plant growth, but can be toxic if it is excessive. Burning of leaf tips or margins and premature yellowing of crops are symptoms of excessive chloride concentration. Fruit trees are generally more sensitive to chloride than are vegetables, grain, fiber, and fodder crops. Symptoms of excessive sodium concentration are burning of leaves, yellow discoloration, dead and drying areas between the veins in the leaves and on leaf margins, and lack of root development. Shrubs, trees, and woody ornamental plants are more sensitive to high values of SAR. For sprinkler irrigation, the sodium concentration should not exceed 70 mg/L. Irrigation water with SAR < 3 is not toxic for most crops. Boron toxicity is not as common as chloride or sodium toxicity, but in certain areas of Texas the boron concentration is excessively high, which makes the water unsuitable for irrigation.

5.11 TRACE ELEMENTS

All trace elements, such as iron, magnesium, and zinc, in small quantities are essential for plant growth; however,

Table 5.7 *Recommended limits for constituents in reclaimed water for irrigation*

Constituent	Long-term use (mg/L)	Short-term use (mg/L)	Remarks
Aluminum (Al)	5	20	Can cause acid soils to be non-productive, but soils at pH 5.5–8.0 will precipitate iron and eliminate toxicity.
Arsenic (As)	0.1	2	Toxicity to plants varies widely, ranging from 12 mg/L for Sudan grass to less than 0.05 mg/L for rice.
Beryllium (Be)	0.1	0.5	Toxicity to plants varies widely, ranging from 5 mg/L for kale to 0.5 mg/L for bush beans.
Boron (B)	0.75	2	Essential to plant growth, with optimum yields for many obtained at a few tenths of a mg/L in nutrient solutions. Toxic to many sensitive plants (e.g., citrus) at 1 mg/L. Most grasses are relatively tolerant at 2.0–10 mg/L.
Cadmium (Cd)	0.01	0.05	Toxic to beans, beets, and turnips at concentrations as low as 0.1 mg/L in nutrient solution. Conservative limits recommended.
Chromium (Cr)	0.1	1	Not generally recognized as an essential growth element. Conservative limits recommended due to a lack of knowledge on toxicity to plants.
Cobalt (Co)	0.05	5	Toxic to tomato plants at 0.1 mg/L in nutrient solution. Tends to be inactivated by neutral and alkaline soils.
Copper (Cu)	0.2	5	Toxic to a number of plants at 0.1–1.0 mg/L in nutrient solution.
Fluoride (F^-)	1	15	Inactivated by neutral and alkaline soils.
Iron (Fe)	5	20	Not toxic to plants in aerated soils, but can contribute to soil acidification and loss of essential phosphorus and molybdenum.
Lead (Pb)	5	10	Can inhibit plant cell growth at very high concentrations.
Lithium (Li)	2.5	2.5	Tolerated by most crops at up to 5 mg/L; mobile in soil; toxic to citrus at low doses; recommended limit of 0.075 mg/L.
Manganese (Mg)	0.2	10	Toxic to a number of crops at a few tenths to a few mg/L in acid soils.
Molybdenum (Mo)	0.01	0.05	Not toxic to plants at normal concentrations in soil and water. Can be toxic to livestock if forage is grown in soils with high levels of available molybdenum.
Nickel (Ni)	0.2	2	Toxic to a number of plants at 0.5–1.0 mg/L; reduced toxicity at neutral or alkaline pH.
Selenium (Se)	0.02	0.02	Toxic to plants at low concentrations and to livestock if forage is grown in soils with low levels of added selenium.
Vanadium (V)	0.1	1	Toxic to many plants at relatively low concentrations.
Zinc (Zn)	2	10	Toxic to many plants at widely varying concentrations; reduced toxicity at increased pH (6 or above) and in fine-textured or organic soils

After USEPA, 1992.

their excessive amounts can be toxic, as shown in Table 5.7. Cadmium, lead, and selenium can be toxic to humans and animals upon consumption of products generated with water containing an excessive concentration of trace elements. Trace elements in wastewater tend to be high. An imbalance of calcium and magnesium may cause toxic effects for crops. If the potassium concentration is high it may cause magnesium deficiency and iron chlorosis. An imbalance of magnesium and potassium may be toxic, but can be counterbalanced by high calcium concentrations. Sensitive crops are affected by sulfate salts, which limit their ability to take up calcium, and increase the adsorption of sodium and potassium, causing a cationic imbalance within the plant.

5.12 HEAVY METALS

Heavy metals include cadmium, chromium, copper, lead, nickel, and zinc. Their excessive concentration disrupts plant physiology. Crops produced with such irrigation water can be harmful to humans and animals. Heavy metals are released by the processing of ores and metals, industrial use of metal compounds (e.g., tanneries), and leaching from domestic and industrial dumps, as shown in Figure 5.4.

5.13 HARDNESS

Hardness corresponds to the availability of calcium and magnesium ions, such as chloride, bicarbonate, or sulfate.

Table 5.8 *Chloride tolerance of various crops in order of tolerance*

	Maximum Cl^{-a} concentration without loss in yield[b]	
Crop	Mol/m^3	ppm
Strawberry	10	350
Bean	10	350
Onion	10	350
Carrot	10	350
Radish	10	350
Lettuce	10	350
Turnip	10	350
Rice, paddy[c]	30[d]	1,050
Pepper	15	525
Clover, strawberry	15	525
Clover, red	15	525
Clover, alsike	15	525
Clover, ladino	15	525
Corn	15	525
Flax	15	525
Potato	15	525
Sweet potato	15	525
Broad bean	15	525
Cabbage	15	525
Foxtail, meadow	15	525
Celery	15	525
Clover, Berseem	15	525
Orchardgrass	15	525
Sugarcane	15	525
Trefoil, big	20	700
Lovegrass	20	700
Spinach	20	700
Alfalfa	20	700
Sesbania[c]	20	700
Cucumber	25	875
Tomato	25	875
Broccoli	25	875
Squash, scallop	30	1,050
Vetch, common	30	1,050
Wild rye, beardless	30	1,050
Sudan grass	30	1,050
Wheat grass, standard crested	35	1,225
Beet, red[c]	40	1,400
Fescue, tall	40	1,400
Squash, zucchini	45	1,575
Harding grass	45	1,575
Cowpea	50	1,750
Trefoil, narrow-leaf bird's foot	50	1,750
Ryegrass, perennial	55	1,925
Wheat, durum	55	1,925
Barley (forage)[c]	60	2,100
Wheat[c]	60	2,100
Sorghum	70	2,450
Bermudagrass	70	2,450
Sugar beet[c]	70	2,450
Wheat grass, fairway crested	75	2,625
Cotton	75	1,625
Wheat grass, tall	75	2,625
Barley	80	2,800

[a] These data serve only as a guideline to relative tolerances among crops. Absolute tolerances vary, depending on climate, soil conditions, and cultural practices.
[b] Cl^- concentrations in saturated extracts sampled in the root zone.
[c] Less tolerant during the seeding stage and emergence.
[d] Values for paddy rice refer to the Cl^- concentration in the soil water during the flooded growing conditions.
After Tanji, 1990.

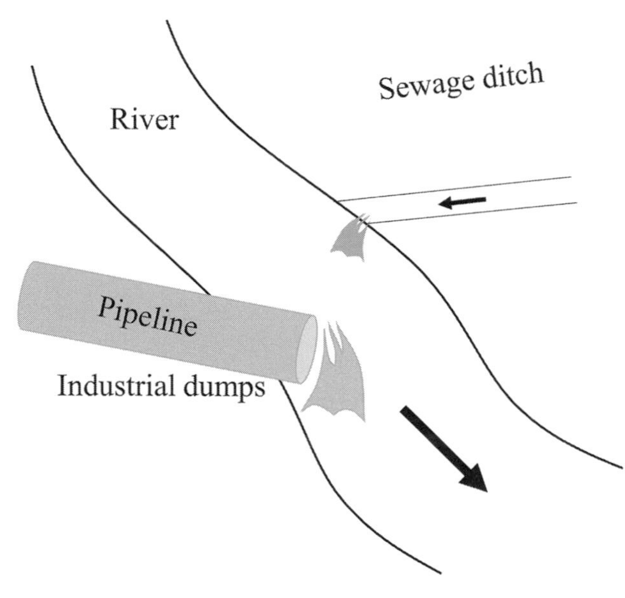

Figure 5.4 Industrial dumps.

It causes scale formation in pipelines and may increase friction loss, which increases loss of energy.

5.14 EFFECT OF CHEMICAL CONCENTRATIONS

Different crops have different degrees of sensitivity to different chemical constituents. Therefore, these constituents should be properly managed.

Chloride is regarded as toxic to crops when it exceeds a threshold concentration. Table 5.8 provides values of chloride tolerances of various crops (Tanji, 1990), Table 5.9 gives values of permissible limits of boron in irrigation water, and Table 5.10 gives values of boron tolerance by different crops. Table 5.11 gives values of tolerance of salt in soil by different crops.

Table 5.9 *Permissible limits of boron (ppm) in irrigation water*

	Crop		
Class of water	Sensitive crops	Semi-sensitive crops	Tolerant crops
Excellent	<0.33	<0.67	<1.00
Good	0.33–0.67	0.67–1.33	1.00–2.00
Permissible	0.67–1.00	1.33–2.00	2.00–3.00
Doubtful	1.00–1.25	2.00–2.50	3.00–3.75
Unusable	>1.25	>2.50	>3.75

After Wilcox, 1995.

5.15 CORROSION

Corrosion occurs because of low salinity and too high or too low pH. Encrustation occurs due to high salinity and low pH. Both corrosion and encrustation impact the life of irrigation equipment, pipelines, and concrete canals and channels. Equipment not resistant to corrosion and encrustation should be avoided.

5.16 CLASSIFICATION OF IRRIGATION WATER

The suitability of irrigation water for crops is often judged using TDS, EC_{iw}, and SAR. Fipps (2021) has indicated permissible limits for different classes of irrigation water, as shown in Table 5.12; sodium hazard of irrigation water based on SAR values is shown in Table 5.2.

The US Department of Agriculture has given guidelines for the suitability of waters for irrigation of crops on the basis of conductivity and SAR, as shown in Table 5.13.

5.17 EFFECT OF WATER QUALITY

Different crops have different degrees of tolerance to irrigation water salinity. Further, the tolerance of different crops may vary with their growth stage. For example, wheat, barley, and corn are more sensitive during the early growth stage than during germination and later growth stages, whereas sugar beets and safflower are more sensitive during germination. In the case of soybean, the sensitivity may increase or decrease during different growth stages.

5.18 WATER QUALITY GUIDELINES

Water quality guidelines are a set of criteria or standards to help with water quality management and must be enforced by law. There is a certain degree of value judgment involved in these guidelines. Irrigation water quality guidelines have been developed, taking into consideration the interactions of soil, crop, climate, and irrigation factors. These guidelines indicate safe values of water quality parameters and expected consequences if actual values deviate from them. It is not uncommon that the guidelines are subject to more than one interpretation and hence to misuse.

Table 5.10 *Limits of boron in irrigation water for different degrees of boron tolerance by crops*

Tolerant (4.0 to 2.0 ppm)	Semi-tolerant (2.0 to 1.0 ppm)	Sensitive (1.0 to 0.3 ppm)
Athel (*Tamarix aphylla*)	Sunflower (native)	Pecan
Asparagus	Potato	Walnut (black, Persian, or English)
Palm (*Phoenix canariensis*)	Cotton (acala and pina)	Jerusalem artichoke
Date palm (*P. dactylifera*)	Tomato	Navy bean
Sugar beet	Sweetpea	American elm
Mangel	Radish	Plum
Garden beet	Field pea	Apple
Alfalfa	Ragged-robin rose	Grape (sultanina and Malaga)
Gladiolus	Olive	Kadora fig
Broad bean	Barley	Persimmon
Onion	Wheat	Cherry
Turnip	Corn	Peach
Cabbage	Milo	Apricot
Lettuce	Oat	Thornless blackberry
Carrot	Zinnia	Orange
	Pumpkin	Avocado
	Bell pepper	Grapefruit
	Sweet potato	Lemon
	Lima bean	

After Wilcox, 1960.

5.19 MANAGING WITH POOR-QUALITY WATER

If poor-quality water is to be used for irrigation, then there are ways by which the effect on soil, crop growth, yield, quality of products, and irrigation equipment can be minimized by proper management. Some of the management steps are as follows:

1. Crops that are tolerant to salinity and toxicity can be grown.
2. Land can be graded or smoothed to improve water distribution and control.
3. Adequate leaching for the removal of excess salts and other undesirable substances can be provided.
4. Drainage can be improved, promoting leaching and removal of excessive salts.
5. Infusion of gypsum, lime, sulfur, and sulfuric acid can help improve soil structure, infiltration characteristics, and pH, and reduce the effect of sodium.
6. Tillage operations help improve soil characteristics.

Table 5.11 *Tolerances of salinity in irrigation water[a] by different crops (after Ayers and Westcot, 1994)*

	Yield potential, EC_{iw} (mmhos/cm)			
Crop	100%	90%	75%	50%
Field crops				
Barley	5.3	6.7	8.7	12.0
Bean (field)	0.7	1.0	1.5	2.4
Broad bean	1.1	1.8	2.0	4.5
Corn	1.1	1.7	2.5	3.9
Cotton	5.1	6.4	8.4	12.0
Cowpea	3.3	3.8	4.7	6.0
Flax	1.1	1.7	2.5	3.9
Groundnut	2.1	2.4	2.7	3.3
Rice (paddy)	2.0	2.6	3.4	4.8
Safflower	3.5	4.1	5.0	6.6
Sesbania	1.5	2.5	3.9	6.3
Sorghum	2.7	3.4	4.8	7.2
Soybean	3.3	3.7	4.2	5.0
Sugar beet	4.7	5.8	7.5	10.0
Wheat	4.0	4.9	6.3	8.7
Vegetable crops				
Bean	0.7	1.0	1.5	2.4
Beet	2.7	3.4	4.5	6.4
Broccoli	1.9	2.6	3.7	5.5
Cabbage	1.2	1.9	2.9	4.6
Cantaloupe	1.5	2.4	3.8	6.1
Carrot	0.7	1.1	1.9	3.1
Cucumber	1.7	2.2	2.9	4.2
Lettuce	0.9	1.4	2.1	3.4
Onion	0.8	1.2	1.8	2.9
Pepper	1.0	1.5	2.2	3.4
Potato	1.1	1.7	2.5	3.9
Radish	0.8	1.3	2.1	3.4
Spinach	1.3	2.2	3.5	5.7
Sweet corn	1.1	1.7	2.5	3.9
Sweet potato	1.6	1.6	2.5	4.0
Tomato	1.7	2.3	3.4	5.0
Forage crops				
Alfalfa	1.3	2.2	3.6	5.9
Barley hay	4.0	4.9	6.4	8.7
Bermudagrass	4.6	5.6	7.2	9.8
Clover, Berseem	1.0	2.1	3.9	6.8
Corn (forage)	1.2	2.1	3.5	5.7
Harding grass	3.1	3.9	5.3	7.4
Orchard grass	1.0	2.1	3.7	6.4
Perennial grass	3.7	4.6	5.9	8.1
Sudan grass	1.9	3.4	5.7	9.6
Tall fescue	2.6	3.6	5.2	7.8
Tall wheat grass	5.0	6.6	9.0	13.0
Trefoil, big	1.5	1.9	2.4	3.3
Trefoil, small	3.3	4.0	5.0	6.7
Wheat grass	5.0	6.0	7.4	9.8
Fruit crop				
Almond	1.0	1.4	1.9	2.8
Apple, pear	1.0	1.6	2.2	3.2
Apricot	1.1	1.3	1.8	2.5
Avocado	0.9	1.2	1.7	2.4
Date palm	2.7	4.5	7.3	12.0
Fig, olive, pomegranate	1.8	2.6	3.7	5.6
Grape	1.0	1.7	2.7	4.5
Grapefruit	1.2	1.6	2.2	3.3
Lemon	1.1	1.6	2.2	3.2
Orange	1.1	1.6	2.2	3.2
Peach	1.1	1.5	1.9	2.7
Plum	1.0	1.4	1.9	2.9
Strawberry	0.7	0.9	1.2	1.7
Walnut	1.1	1.6	2.2	3.2

[a] Based on the electrical conductivity of the irrigation (mmhos/cm).
After Ayers and Westcot, 1994. Reprinted with permission from FAO.

Table 5.12 *Permissible limits for classes of irrigation water*

	Concentration of TDS	
Classes of water	Electrical conductivity (μmhos[a])	Gravimetric ppm
Class 1, excellent	250	175
Class 2, good	250–750	175–525
Class 3, permissible[b]	750–2000	525–1400
Class 4, doubtful[c]	2000–3000	1400–2100
Class 5, unsuitable[c]	3000	2100

[a] Micromhos/cm at 25 °C
[b] Leaching needed if used.
[c] Good drainage needed and sensitive plants will have difficulty obtaining stands.
After Fipps, 2021.

7. Injecting acid into water helps improve the quality of the water, especially oxidization of iron.
8. Crop residues and other organic wastes can be added to improve soil structure.
9. Mixing of poor-quality and good-quality water can be done.
10. Proper irrigation methods can be selected. For example, the sprinkler method is not suitable for water with high sodium ion concentration.
11. Irrigation management can be strategized to reduce the effect of poor-quality water.

To maintain the targeted salinity level in the root zone one can compute the leaching fraction (LF) defined as the fraction of applied irrigation water to be leached through the root zone:

$$LF = \frac{EC_{iw}}{EC_{sw}} = \frac{D_d}{D_I}, \tag{5.11}$$

where EC_{iw} represents the electrical conductivity of the irrigation water, EC_{sw} represents the electrical conductivity of the soil water, which is equal to the electrical conductivity

Table 5.13 *Suitability of irrigation water*

Class	Salinity or conductivity	SAR
Low 1	Low-salinity water (C1) can be used for irrigation with most crops on most soils, with little likelihood that soil salinity will develop. Some leaching is required, but this occurs under normal irrigation practices except in soils of extremely low permeability.	Low-sodium water (S1) can be used for irrigation on almost all soils with little danger of the development of harmful levels of exchangeable sodium. However, sodium-sensitive crops such as stone-fruit trees and avocados may accumulate injurious concentrations of sodium.
Medium 2	Medium-salinity water (C2) can be used if a moderate amount of leaching occurs. Plants with moderate salt tolerance can be grown in most cases without special practices for salinity control.	Medium-sodium water (S2) will present an appreciable sodium hazard in fine-textured soils having high cation-exchange capacity, especially under low-leaching conditions, unless gypsum is present in the soil. This water may be used on coarse-textured or organic soils with good permeability.
High 3	High-salinity water (C3) cannot be used on soils with restricted drainage. Even with adequate drainage, special management for salinity control may be required and plants with good salt tolerance should be selected.	High-sodium water (S3) may produce harmful levels of exchangeable sodium in most soils and management is necessary: good drainage, high leaching, and organic matter additions. Gypsiferous soils may not develop harmful levels of exchangeable sodium from such waters. Chemical amendments may be required for replacement of exchangeable sodium, except that amendments may not be feasible with waters of very high salinity.
Very high 4	Very high-salinity water (C4) is not suitable for irrigation under ordinary conditions, but may be used occasionally under very special circumstances. The soils must be permeable, drainage must be adequate, irrigation water must be applied in excess to provide considerable leaching, and very salt tolerant crops should be selected.	Very high-sodium water (S4) is generally unsatisfactory for irrigation purposes except at low and perhaps medium salinity, where the solution of calcium from the soil or use of gypsum or other amendments may make the use of these waters feasible.

After US Salinity Laboratory (1954).

Table 5.14 *Leaching requirement as a function of electrical conductivities of irrigation and drainage water*

Electrical conductivity of irrigation water (mmhos/cm)	Leaching requirement based on the indicated maximum values of the drainage water at the bottom of the root zone (%)			
	4 mmhos/cm	8 mmhos/cm	12 mmhos/cm	16 mmhos/cm
0.75	13.3	9.4	6.3	4.7
1.00	25.0	12.5	8.3	6.3
1.25	31.3	15.6	10.4	7.8
1.50	37.5	18.7	12.5	9.4
2.00	50.0	25.0	16.7	12.5
2.50	62.5	31.3	20.8	15.6
3.00	75.0	37.5	25.0	18.7
5.00	–	62.5	41.7	31.2

After Fipps, 2021.

of the water draining below the root zone, D_d is the depth of water (volume per unit area) draining from the root zone (mm, in.), and D_I is the depth of water (volume per unit area) applied at the surface (mm, in.). This equation can also be used to determine the salinity level corresponding to the amount of available water for leaching. Table 5.14 gives the amount of leaching required to maintain the desired salinity level in the root zone for different classes of irrigation. Since irrigation systems are less than efficient, and salt already in the soil may need to be leached, additional water supply will be required, as shown in Table 5.15. Sodium ions get attached to soil particles and break the soil structure,

Table 5.15 *Typical overall on-farm efficiencies of different irrigation systems*

Irrigation system	Overall efficiency (%)
Surface	50–80
a. Average	50
b. Land leveling and delivery pipeline meeting design standards	70
c. Tailwater recovery with (b)	80
d. Surge	60–90[a]
Sprinkler (moving and fixed systems)	55–85
LEPA (low energy [pressure] precision application)	95–98
Drip	80–90[b]

[a] Surge has been found to increase efficiencies by 8–28% over non-surge systems.
[b] Drip systems are typically designed at 90% efficiency; short laterals (100 feet) or systems with pressure-compensating emitters may have higher efficiencies.

Table 5.16 *Amendments for reclaiming sodic soils and amount equivalent to gypsum*

Amendment	Physical description	Amount equivalent 100% gypsum
Gypsum[a]	White mineral	1.0
Sulfur[b]	Yellow element	0.2
Sulfuric acid[a]	Corrosive liquid	0.6
Lime sulfur[a]	Yellow-brown solution	0.8
Calcium carbonate[b]	White mineral	0.6
Calcium chloride[a]	White salt	0.9
Ferrous sulfate[a]	Blue-green salt	1.6
Pyrite[b]	Yellow-black mineral	0.5
Ferric sulfate[a]	Yellow-brown salt	0.6
Aluminum sulfate[a]	Corrosive granules	1.3

[a] Sulfate for use as a water or soil amendment.
[b] Suitable only for soil application.
After Fipps, 2021.

and in turn seal the soil. Chemical amendment can displace these sodium ions. These amendments include sulfur in its elemental form or related compounds, such as sulfuric acid and gypsum. Gypsum contains calcium, which helps replace the adsorbed sodium and hence restore the soil structure. Chemical amendment is only effective for sodium-affected soils. Table 5.16 gives common amendments.

Example 5.1: Determine the leaching requirement for corn if the salinity of irrigation water is 2 dS/m with 25 inches of water contributing to the crop water requirement. Assume that the corn has a yield potential of 100%.

Solution: Assume that the corn has a yield potential of 100%, and the salinity tolerance level is 1.1 mmhos/cm = 1.1 dS/m in the irrigation water from Table 5.10. From eq. 5.8, the soil salinity tolerance in the soil water is $EC_{sw} = 3EC_{iw} = 3 \times 1.1\ \text{dS/m} = 3.3\ \text{dS/m}$.

To maintain the targeted salinity level in the root zone, the leaching fraction is

$$LF = \frac{EC_{iw}}{EC_{sw}} = \frac{2}{3.3} = 0.61 \text{ or } 61\%.$$

The total crop water requirement is 25 in.

$$LF = \frac{D_d}{D_I} = \frac{D_d}{D_d + 25\text{ in.}} = 0.61, \quad \text{then } D_d = 38.45\text{ in.}$$

The leaching requirement is $D_d = 38.45$ in.

QUESTIONS

Q.5.1 Given here is an irrigation water quality sample obtained from a municipal wastewater treatment plant in Brazos County, Texas.

Cation (mg/L)			Anion (mg/L)
Na	Ca	Mg	HCO_3
175	95	20	168

a. Convert the ions measurement from mg/L to meq/L using eq. (1) and Table 5.17:

$$I[\text{meq/L}] = \frac{V}{AW} \times I[\text{mg/L}], \quad (1)$$

where I denotes the ion measurement in respective units, V is the valence number of the ion, and AW is the atomic weight of the ion.

b. Determine the SAR.

c. Use the following steps to determine the adjusted SAR value.

Table 5.17 *Characteristics of ions in the water sample*

Ion	Atomic weight	Valence numbers
Na^+	22.99	1
Ca^{2+}	40.08	2
Mg^{2+}	24.31	2
HCO_3^-	61.02	1

1. *Calculate the sum of cations (SC) and ionic strength (I_S) using eqs (2) and (3), respectively:*

$$SC = Na[\text{meq/L}] + Ca[\text{meq/L}] + Mg[\text{meq/L}], \quad (2)$$

$$I_S = \frac{(1.3477 \times SC + 0.5355)}{1000}. \quad (3)$$

2. *Calculate the factor of $log_{10}(X)$ using eq. (4):*

$$\log_{10}(X) = \frac{1}{3}\left[4.6629 + 0.6103\log_{10}(I_S) + 0.0844\{\log_{10}(I_S)\}^2 + 2\log_{10}\left(\frac{Ca[\text{meq/L}]}{2 \times HCO_3[\text{meq/L}]}\right)\right]. \quad (4)$$

3. *Calculate the equilibrated Ca concentration (Ca_{eq}) using eq. (5):*

$$Ca_{eq}[\text{meq/L}] = 10^{\log_{10}(X)} \times 0.17758. \quad (5)$$

4. *Calculate the ASAR by replacing Ca ions with the equilibrated Ca concentration (Ca_{eq}) in the SAR formula.*

d. By what percentage is the ASAR higher or lower than the SAR calculated in part (b)?

Q.5.2 An irrigation water has an electrical conductivity of 1.0 ds/m at 25 °C and a concentration of sodium of 36 meq/L, a concentration of calcium of 30 meq/L, and a concentration of magnesium of 10 meq/L. Determine whether the irrigation water will create a sodicity hazard.

Q.5.3 Determine the leaching requirement for tomatoes if the salinity of irrigation water is 3 dS/m with 16 inches of irrigation and 10 inches of rainfall contributing to the crop water requirement.

Q.5.4 The electrical conductivity of water entering the root zone is 0.5 dS/m. It is given that the crop salinity rating is moderately sensitive – that is, the electrical conductivity of water draining from the root zone is 3.0 dS/m. The amount of irrigation water supplied is 30 cm. Compute the depth of water entering the root zone and the depth of water draining from the root zone.

Q.5.5 The water season requirement of a crop is 90 cm, and the amount of irrigation water supplied is 100 cm. What is the depth of leaching and what is the fraction of leaching?

Q.5.6 The depth of irrigation water with an electrical conductivity of 0.6 dS/m applied to a corn crop field is 95 cm. What is the depth of leaching and what is the leaching fraction, assuming that the corn has a 75% yield potential?

REFERENCES

Ayers R. S., and Westcot D. W. (1994). Water quality for agriculture, revised. FAO irrigation and drainage paper 29.

Doorenbos, J., and Pruitt, W. O. (1977). Crop water requirements: FAO irrigation and drainage paper 24.

Fipps, G. (2021). *Irrigation Water Quality Standards and Salinity Management Strategies*. College Station, TX: Texas A&M University.

James, D. W., Hanks, R. J., and Jurinak, J. H. (1982). *Modern Irrigated Soils*. New York: Wiley.

Longenecker, D. E., and Lyerly, P. J. (1974). *Control of Soluble Salts in Farming and Gardening*. College Station, TX: Texas A&M University.

Pearson, G. A. (1960). Tolerance of crop to exchangeable sodium. Agricultural Research Service agriculture information bulletin 216.

Soil Conservation Service (1991). Soil–plant–water relationships. In *National Engineering Handbook*. Washington, DC: US Department of Agriculture.

Tanji, K. K. (1990). The nature and extent of agricultural salinity problems. In Tanji, K. K. (ed.), *Agricultural Salinity Assessment and Management*. Reston, VA: ASCE.

USEPA (1992). *Guidelines for Water Reuse, EPA/625/R-92/004*, Office of Water, Office of Wastewater Enforcement and Compliance. Washington, DC: USEPA.

US Salinity Laboratory (1954). *Diagnosis and Improvement of Saline and Alkali Soils*. Washington, DC: USDA.

Wilcox, L. V. (1960). Boron injury to plants. USDA bulletin 211.

Wilcox, L. V. (1995). Classification and use of irrigation waters. USDA circular 969.

6 Soils and Soil Management

Notation

A_s	apparent specific gravity of solids (dimensionless)	V_p	total pore volume (L^3)
M_s	mass of dry soil (kg, g)	V_s	volume of soil particles (L^3)
N	soil porosity (fraction)	V_w	volume of water (L^3)
N_a	air-filled porosity (fraction)	ρ_b	soil bulk density (kg/m³)
V_a	volume of air (L^3)	ρ_p	specific gravity of solids (dimensionless)
V_b	volume of bulk soil (L^3)	ρ_s	soil particle density (kg/m³)
		ρ_w	mass density of water (kg/m³)

6.1 INTRODUCTION

Soil is a thin surface layer of the Earth's pellicle, and this layer varies in thickness, depending on climate. For arid soils, the thickness may be 10–20 feet (3–6 m), but for humid soils it may be 6–18 inches (15–55 cm). The soil develops on the underlying rock as a result of climate and organic life upon or within it. The organic life includes plants, animals, and microorganisms that occur through different stages of decomposition. The thin layer has physical, mechanical, chemical, and biological characteristics that are different from those of lower layers and underlying rocks and rock ingredients. The soil is where plants grow and crops are produced, because it provides an environment where plant roots can absorb the water, oxygen, and nutrients needed to grow and develop. The ability of soil to supply the needed nutrients to plants is termed soil fertility. The science of soil is often referred to as pedology.

Crops are grown in certain types of soils – that is, not all soils are suitable for growing all crops, and some crops can be grown in only certain types of soils. Irrigation of crops significantly depends on the type of soil. Soil is fundamental to our biosphere and requires proper management. This chapter discusses the basic soil properties that are relevant to crops and farm irrigation.

6.2 DEVELOPMENT OF SOIL

Soils develop as a result of geological processes, both physical and chemical, in concert with biological, hydrological, and environmental processes. The time it takes for soils to develop is very long, up to millions of years.

6.2.1 Disintegration of Rocks

The origination of soil can be attributed to the slow processes of disintegration of rocks due to weathering enhanced by the movement of water, heat, cold, wind, other atmospheric factors, and biological factors. The disintegration results in an accumulation of fine material, such as sand and clay. During the process of disintegration or thereafter, vegetation takes root that also aids disintegration. Vegetation undergoes partial or complete decomposition by microorganisms and becomes incorporated in the soil. When it rains, water enters the soil and percolates down, removing soluble inorganic materials, such as carbonic material. The inorganic materials are also removed by vegetation. Some of these materials enter the soil again and become part of it.

6.2.2 Chemical Processes and Minerals

The chemical processes occurring during rock disintegration are hydrolysis, oxidation, hydration, solution, and carbonation. Soil minerals contain 21 of the known chemical elements. Eight of these elements – oxygen (47.33%), silicon (27.74%), aluminum (7.85%), iron (4.5%), calcium (3.47%), magnesium (2.24%), sodium (2.46%), and potassium (2.46%) – constitute about 98% of the total mineral content of Earth's crust, as shown in Figure 6.1. Interestingly, five of these elements – hydrogen, sulfur, carbon, titanium, and phosphorus – are found in many minerals and each comprises 0.1–0.5% of the inorganic matter. The remaining eight elements – fluorine, chlorine, zirconium, boron, nitrogen, barium, manganese, and chromium – together form only 0.35% of all soil mineral matter. Oxygen and silicon constitute 75% of the surface of the

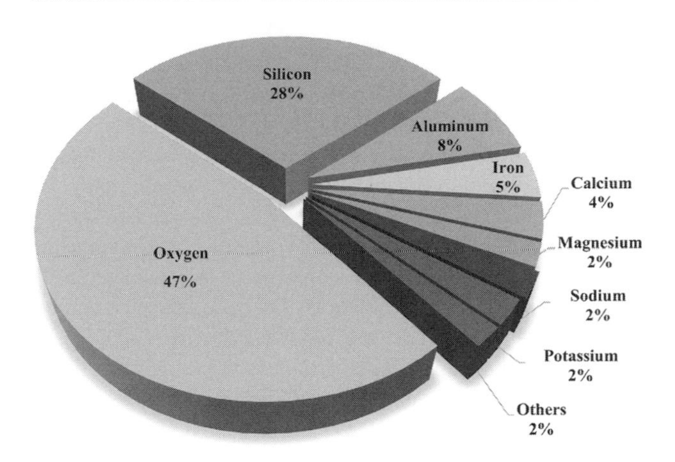

Figure 6.1 Chemical elements in soil. A black and white version of this figure will appear in some formats. For the color version, please refer to the plate section.

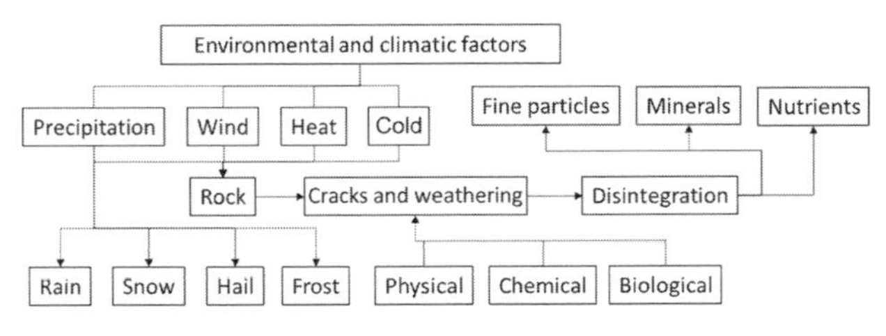

Figure 6.2 Formation of soil.

Earth, but silica as a compound and combined in silicates forms about 60% of the crust. This silica results from the disintegration of rocks and resistance to solution by water or dilute acids released by vegetation or generated by microorganisms in the soil.

It may be noted that although carbon is a source of fuel and is important in the synthesis of hundreds of compounds used by humans, it makes up only a small fraction of the Earth's surface and of the whole lithosphere, hydrosphere, and atmosphere. Carbon is in constant circulation in nature, and is essential to sustain life. Soil microbes cause certain phases of the transformation of carbon.

The changes in the chemical composition during rock weathering and the formation of the Earth's crust include the separation of silica and the bases, oxidation of the compounds of iron, removal of bases by leaching and replacement, general hydration of the remainder of the silicates, aluminum, and iron, and addition of organic matter due to vegetation and microorganisms.

6.2.3 Formation of Soil

Soil is formed in a series of steps, as shown in Figure 6.2. Initially, organisms weaken the rock constituents by removing some of their more soluble elements or compounds. The change in temperature causes swelling and shrinkage of the rock material, which in turn causes cracks and openings and small fragments. When these changes are combined with organic matter, the conditions become conducive for vegetation to establish a foothold and grow. Further disintegration of rock material occurs because of the physical and chemical forces exerted by plants. The organic material of the vegetation mixes with coarse and fine rock material, heralding the beginning of agricultural soils. Further disintegration occurs due to the action of air, water, and ice, making particles even finer. The repeated development of vegetation and inclusion of their residues in the soil leads to further disintegration, eventually producing a soil that has few characteristics in common with the rocks from which it was formed.

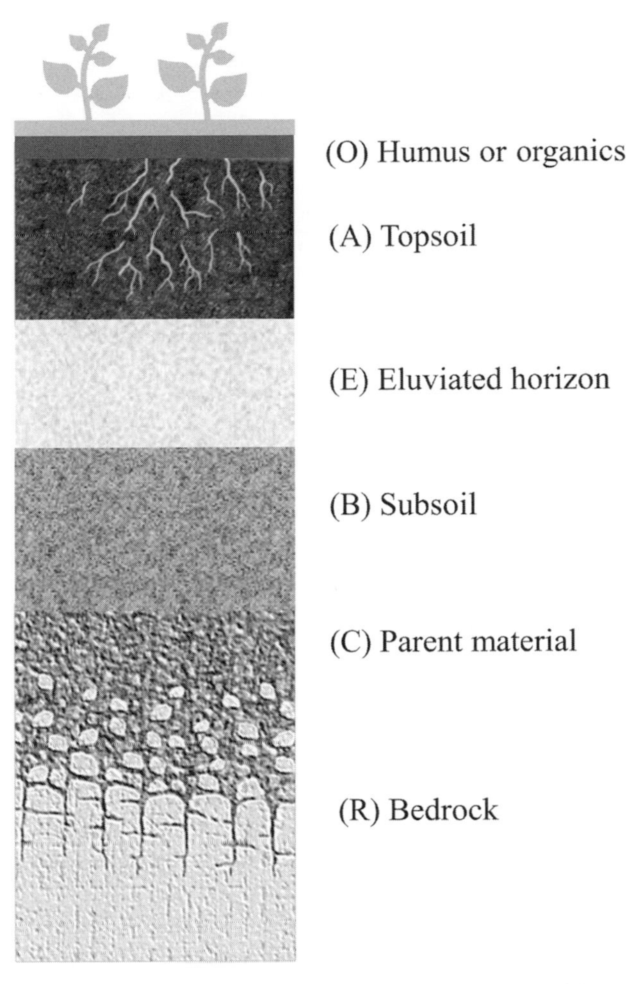

Figure 6.3 Soil horizons in a soil profile. A black and white version of this figure will appear in some formats. For the color version, please refer to the plate section.

6.3 SOIL CHARACTERIZATION AND PROFILE

The soil profile is a series arrangement, in vertical, of characteristic layers, called horizons, as shown in Figure 6.3. The arrangement of layers results from the way the soil has developed. A vertical cut through the soil shows its different layers or horizons that constitute the soil profile. Most soils have three major horizons, designated by the letters A, B, and C, as shown in Figure 6.3 (Soil Science Society of America, 2015), and may be subdivided as A_0, A_1, A_2, B_1, and B_2. These horizons are markedly different in the content of organic matter. Horizon A, also called the horizon of eluviation, is at the surface from which certain material has been mechanically or chemically removed. Eluviation denotes the washing out of clay and other minerals from the overlying horizon A. Much of the microbial activity occurs in this horizon, and most of the vegetation is incorporated into it. This horizon is the most fertile and has the greatest organic content; it is also referred to as topsoil. Some soils have an O layer, which is the layer of organic matter on the surface produced by decaying plants. However, this layer can also be buried.

Horizon B, also called the horizon of illuviation or subsurface, contains much less organic matter and comprises almost entirely mineral substances, some of which may have come from the surface and accumulated at deeper layers due to leaching. Illuviation denotes the accumulation in horizon B of the material removed from horizon A. The E layer is used for subsurface horizons that have a significant loss of eluviation.

Horizon C is the horizon of the parent material. The organic substances, which are mainly colloidal in nature, significantly determine the physical properties of soils. These substances have pronounced absorptive properties. There is little weathering and few plants grow in this zone. Layer R is the hard bedrock, which is not soil.

Irrespective of the way soils are formed, whether by chemical and physical weathering or stratification, they are not uniform in their characteristics, such as thicknesses of layers, distribution of chemical constituents, and physical properties, and they exhibit substantial spatial variability even over short distances of a few meters. This variability makes soil sampling extremely difficult.

Color, texture, structure, and chemical composition of the various horizons characterize the soil and vary from soil to soil. The color of soil depends on its geologic development, penetration of plant roots, activity of microorganisms, and the presence of organic matter at different stages of decomposition. The color of the upper horizon is darker because of the organic matter, and the color of lower horizons may become darker or lighter, depending on the movement of air and water, and the organic matter.

6.4 TYPES OF SOIL

The soil owes its composition to the rock from which it was formed. Rocks are heterogeneous and are aggregates of minerals that are themselves chemical substances varying in complexity from elements to complex substances. These minerals are found in varying proportions in a rock. The degree of abundance of any mineral or group of minerals in a rock, and the degree of their consolidation, essentially determines the nature of the rock and consequently the soil it forms.

6.4.1 Soil Texture

Soil texture denotes the size of particles that form the soil. The size of a soil particle corresponds to the diameter of a sphere whose volume is equivalent to that of the particle. Soil texture may be analyzed mechanically, and soil particles may be separated and classified into groups of various sizes, as shown for two typical soils in Table 6.1. When soils have significant percentages of various particle sizes, they are considered well graded. Soil texture, as a function of particle diameter, can be classified, as shown in Figure 6.4 (USDA-NRCS, 2001a). It provides qualitative information about soil compounds.

Depending on their size, soil particles are classified as sand, silt, or clay. Soil texture is determined by the type and amount of these soil particles; together they determine how fine or coarse the soil is. Figure 6.5 depicts these particles.

Sand and silt are formed by the weathering of rocks, and mostly comprise silica (SiO_3). Their particles do not stick to each other. Sand is the largest soil particle and is derived from sedimentary rock – mainly weathered quartz. As particles are large, pore spaces between them are also large and internal surface areas are small. Chemically, sand and silt are relatively inactive and provide few nutrients to plants.

Clay particles are the smallest and are crystalline and layered. They stick to each other, and clay swells and shrinks. Clay becomes sticky upon wetting. Clay particles comprise silica, aluminum, oxygen, hydrogen, potassium, magnesium, and phosphorus. Chemically, clay is the most active and provides the most nutrients to plants.

Combinations of the three soil particles represent agricultural soils. These are described by a textural triangle, as shown in Figure 6.6 (USDA-NRCS, 2001a). For example, a loam soil is 20% clay, 40% silt, and 40% sand. The proportions of these particles in a given soil are evaluated mechanically and govern the physical, chemical, and biological properties of soils.

The soil textural triangle can be used to characterize soil texture and hence to identify a soil for given percentages of sand, clay, and silt. To use the triangle, mark the point representing the percentage of clay on the left (clay) side of the triangle. Then, draw a horizontal line issuing from the point parallel to the base of the triangle (sand side). Now mark the point representing the percentage of sand on the

Table 6.1 *Classification of soil particles of two soils*

Fraction	Size (mm)	Fine sandy loam (%)	Clay (%)
Very coarse sand	2.00–1.00	1	1
Coarse sand	1.0–0.5	2	2
Medium sand	0.5–0.25	3	2
Fine sand	0.25–0.10	22	6
Very fine sand	0.10–0.05	35	11
Silt	0.05–0.005	24	41
Clay	<0.002	13	37

Adapted from Soil Conservation Service, 1991.

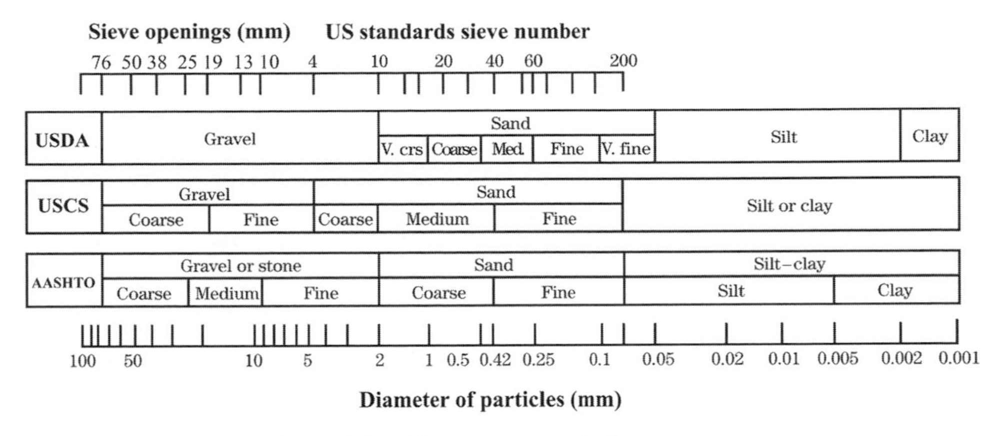

Figure 6.4 Systems for soil particle size-based classification. (Adapted from USDA-NRCS, 2001a.)

Figure 6.5 Three types of soil. A black and white version of this figure will appear in some formats. For the color version, please refer to the plate section.

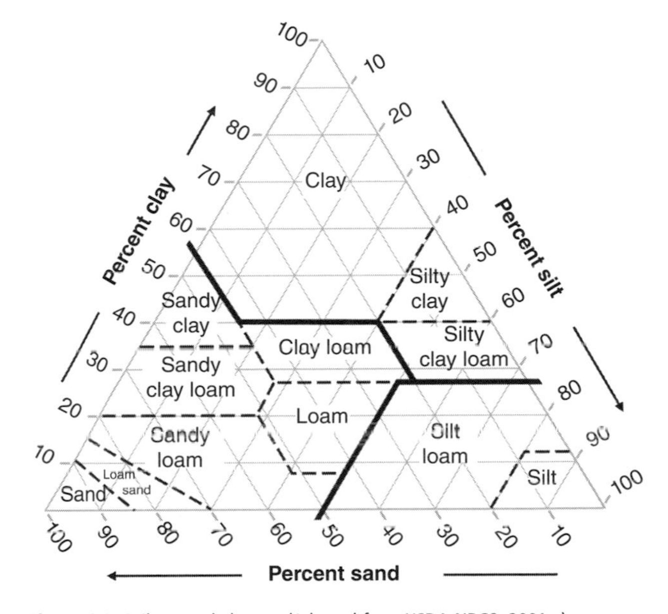

Figure 6.6 Soil textural classes. (Adapted from USDA-NRCS, 2001a.)

base of the triangle (sand side). Draw a line originating from this point parallel to the right side of the triangle (percent silt). Mark the intersection of the point, which will fall within one of the boxes. That box indicates the soil type. Alternatively, one can also mark the point representing the percentage of silt on the right side of the triangle (silt side) and draw a line issuing from this point parallel to the clay side of the triangle. This line will intersect the horizontal line drawn earlier. The intersection point will be exactly the same as the intersection point found earlier.

Example 6.1: Identify a soil that is 30% sand, 25% clay, and 45% silt. Also, identify a soil that contains 40% sand, 30% clay, and 30% silt.

Solution: Using Figure 6.6, the soil that is 30% sand, 25% clay, and 45% silt can be classified as loam. The soil that contains 40% sand, 30% clay, and 30% silt can be classified as clay loam.

6.4.2 Soil Structure

Soil structure qualitatively describes the soil condition. It affects the mechanical properties of the soil. Soils possess textural, structural, and chemical properties, depending upon the proportion of minerals, organic matter, and pore spaces. Soil is continually subject to a variety of influences that change its physical structure, chemical composition, and even its location. Although physical factors exercise a pronounced influence, biological factors, such as the development of vegetation and microorganisms, are, to a large degree, responsible for the creation of fertile agricultural lands from inorganic substances such as mixtures of sand and clay–loam.

Soil structure is represented by the clustering of soil particles in various shapes. Clusters of soil particles are bound together by organic matter and clay. A granular structure often represents a productive soil. Well-aggregated soils have good aeration and are good for crop growth. On the other hand, compacted soils are not good for crop production because they do not allow water to infiltrate freely and are hard to till.

There are three broad categories of soil structure: single-grained, massive-grained, and aggregated (Hillel, 1980). As the name suggests, single-grained soils are structureless and individual soil particles do not adhere to each other; an example of this is beach sand, which is preferred for sun bathing. Another example is desert sand. Massive structure represents the other extreme, in which clumps of soil as large as tens of centimeters are formed by physically and chemically bonded particles. An example of this structure is often observed in clayey soils during rainless summers. An aggregated structure represents clods that may be several centimeters or a fraction of a centimeter in diameter, and are often observed when soil is tilled some time after irrigation or rainfall. The degree of aggregation depends on the soil moisture content, physical, chemical, and biological characteristics, root depth, organic matter, and depth.

6.5 SPATIAL DISTRIBUTION OF SOILS

Soils are classified using soil order. There are 15 major soil orders in the world, as shown in Figure 6.7 (USDA-NRCS, 2005), and 12 major soil orders in the United States, as shown in Figure 6.8 (Natural Resources Conservation Service, 2007). A short description of these orders follows.

Alfisols represent about 14.5% of the United States and are dominant in the Great Lake states and in the Mississippi River basin. These are fertile soils but can be acidic.

Andisols represent about 1.7% of the United States and are found in the Pacific Northwest, Hawaii, and Alaska. They contain about 10–20% organic matter and are fertile.

Aridisols represent about 8.8% of the United States and are mostly desert soils found in the western and southwestern United States. They are used for rangeland.

Entisols represent about 12.2% of the United States, are relatively undeveloped, and do not have well-defined horizons. They are found throughout the United States.

Gelisols represent about 7.5% of the United States, are found in the tundra of Alaska, and have permanent frost within 100 cm of the surface.

Histosols represent about 1.3% of the United States and are found in Minnesota, Wisconsin, Florida, and Louisiana. They have more than 20% organic matter.

Inceptisols represent about 9.1% of the United States, are found in New England states, Appalachia, and the Pacific Northwest, and are used for cropland and forest land.

Mollisols represent about 22.4% of the United States, are found in Northern and Southern Plains states, and are fertile.

Oxisols represent about 0.01% of the United States, are found in Hawaii and Puerto Rico, and are used for sugarcane and forests.

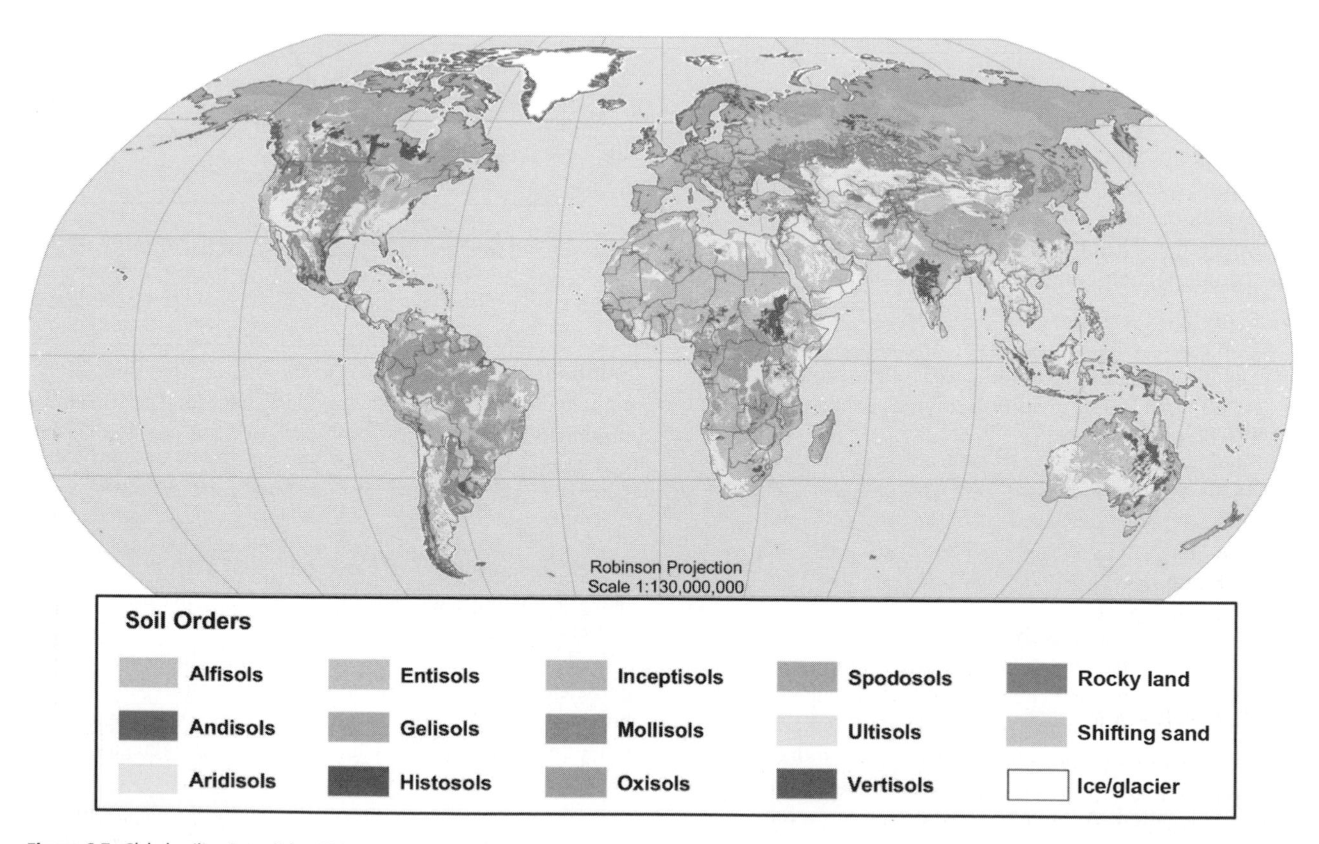

Figure 6.7 Global soil regions. (After USDA-NRCS, 2005.) A black and white version of this figure will appear in some formats. For the color version, please refer to the plate section.

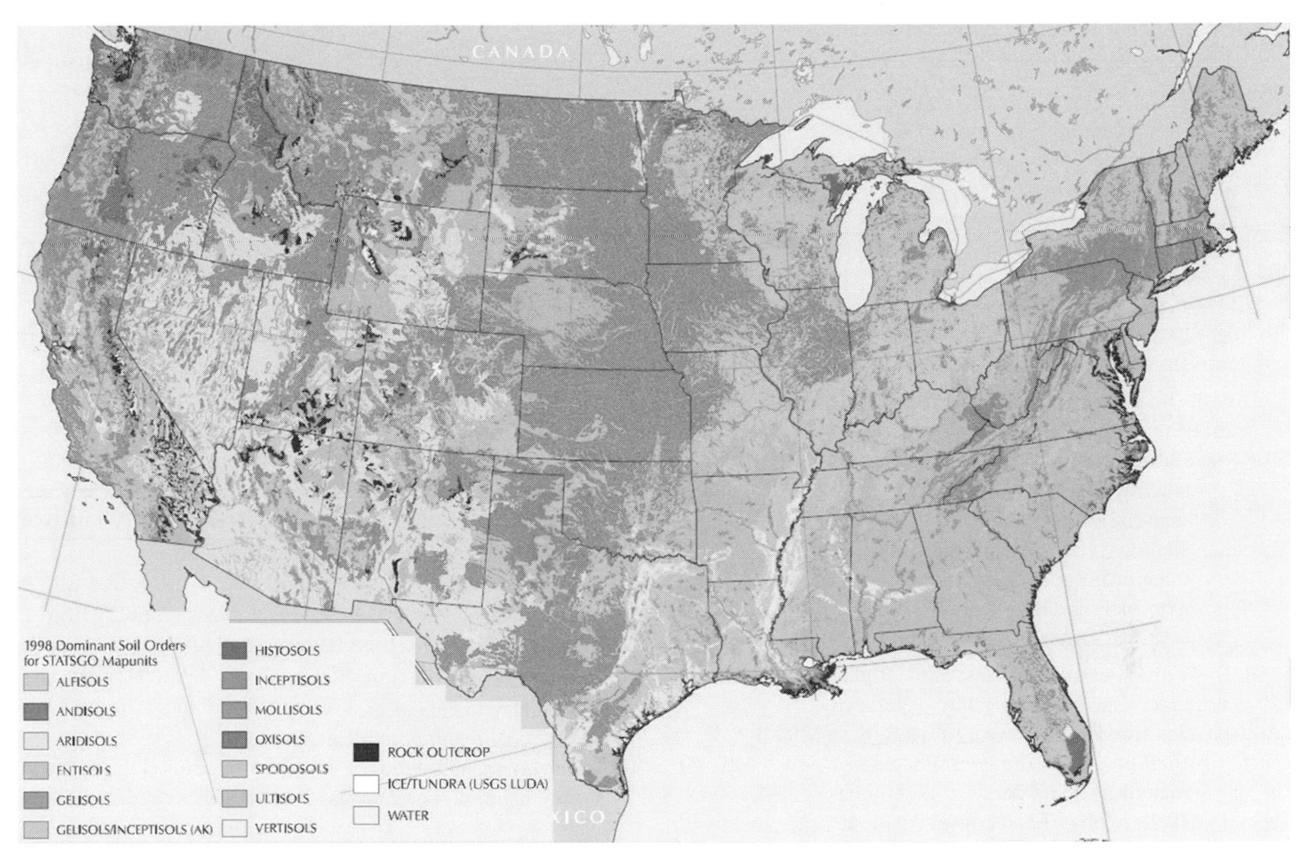

Figure 6.8 Twelve major soil orders in the United States. (After Natural Resources Conservation Service, 2007.) A black and white version of this figure will appear in some formats. For the color version, please refer to the plate section.

Spodosols represent about 3.3% of the United States, are found in Minnesota, Wisconsin, Michigan, and New England states, and are used for forestland and cropland along with adequate fertilizers.

Ultisols represent about 9.6% of the United States, are found from Pennsylvania to the southwestern United States, and are productive but can be acidic.

Vertisols represent about 1.7% of the United States, are found mostly in Louisiana and Arkansas, and are clayey. They are used for grass, savannah, open forest, and shrub.

Different places have different soil types or orders. On a regional scale, it is observed that the southeastern United States has predominantly ultisols, whereas the Midwest and Northwest, and parts of the southwestern United States have mollisols. China has gelisols, and central and parts of south India have vertisols. Parts of Russia and the British Isles have inceptisols.

6.6 CLASSIFICATION AND EVALUATION OF SOILS

Soils are evaluated and lands are classified for their agricultural potential in different ways. Storie (1978) developed an index, called the Storie Index, for rating the potential utilization and productive capacity of soil. The rating involves a product of four factors each expressed as a percentage, with a maximum value of 100%. The four factors are: (A) profile; (B) surface texture; (C) slope; and (X) other conditions. These are briefly discussed below, and then the Storie index is illustrated with an example.

A. Profile: This factor is rated based on the character of physical profile – that is, features of subsurface layers – and soil profiles are classified into nine groups, depending on soil or profile development, geologic evolution, and the source of the soil-forming material. Each group is further subdivided into several depths ranging from one to six, and each depth is then assigned a percentage. Finally, the factor is assigned a rating expressed as a percentage. Table 6.2 gives the rating for this factor.

B. Surface texture: Surface textures are classified into five groups: medium textured, heavy or fine textured, light or coarse textured, gravelly, and stony. Each texture is further subdivided into several textures, such as sand, silt, clay, loam, etc., and each is assigned a percentage. Fine sandy loams, loams, and silt loams are assigned a 100% rating, and gravelly and stony soils are assigned a lower percentage. Finally, the factor is assigned a percentage rating. Table 6.3 shows the rating of this factor.

C. Slope: Soils are rated based on slope. For example, A, nearly level (0–2%) is rated as 100%; AA, gently undulating (0–2%) is 95–100%; B, gently sloping (3–8%) is

Table 6.2 *Factor A: rating based on physical soil profile*

Soil profile	Percent
I: Soils on recent alluvial fans, flood plains, or other secondary deposits having undeveloped profiles	100
x – shallow phases (on consolidated material), 2 feet deep	50–60
x – shallow phases (on consolidated material), 3 feet deep	70
g – extremely gravelly subsoils	80–95
x – stratified clay subsoils	80–95
II. Soils on young alluvial fans, flood plains, or other secondary deposits having slightly developed profiles	95–100
x – shallow phases (on consolidated material), 2 feet deep	50–60
x – shallow phases (on consolidated material), 3 feet deep	70
g – extremely gravelly subsoils	80–95
x – stratified clay subsoils	80–95
III. Soils on older alluvial fans, alluvial plains, or terraces having moderately developed profiles (moderately dense subsoils)	80–95
x – shallow phases (on consolidated material), 2 feet deep	40–60
x – shallow phases (on consolidated material), 3 feet deep	60–70
g – extremely gravelly subsoils	60–90
IV. Soils on older plains or terraces having strongly developed profiles (dense clay subsoils)	40–80
V. Soils on older plains or terraces having hardpan subsoil layers	
– at less than 1 foot	5–20
– at 1–2 feet	20–30
– at 2–3 feet	30–40
– at 3–4 feet	40–50
– at 4–6 feet	50–80
VI. Soils on older terraces and upland areas having dense clay subsoils resting on moderately consolidated or consolidated material	40–80
VII. Soils on upland areas underlain by hard igneous bed rock	
– at less than 1 foot	10–30
– at 1–2 feet	30–50
– at 2–3 feet	50–70
– at 3–4 feet	70–80
– at 4–6 feet	80–100
– at more than 6 feet	100
VIII. Soils on upland areas underlain by consolidated sedimentary rocks	
– at less than 1 foot	10–30
– at 1–2 feet	30–50
– at 2–3 feet	50–70
– at 3–4 feet	70–80
– at 4–6 feet	80–100
– at more than 6 feet	100
IX. Soils on upland areas underlain by softly consolidated material	
– at less than 1 foot	20–40
– at 1–2 feet	40–60
– at 2–3 feet	60–80
– at 3–4 feet	80–90
– at 4–6 feet	90–100
– at more than 6 feet	100

Adapted from Storie, 1978.

Table 6.3 *Factor B: rating based on surface texture factor*

Soil texture	Percent
Medium textured:	
– fine sandy loam	100
– loam	100
– silt loam	100
– sandy loam	95
– silty clay loam, calcareous	95
– silty clay loam, non-calcareous	90
– clay loam, calcareous	95
– clay loam, non-calcareous	85–90
Heavy or fine textured:	
– silty clay, highly calcareous	70–90
– silty clay, non-calcareous	60–70
– clay, highly calcareous	70–80
– clay, non-calcareous	50–70
Light or coarse textured:	
– coarse sandy loam	90
– loamy sand	80
– very fine sand	80
– fine sand	65
– sand	60
– coarse sand	30–60
Gravelly:	
– gravelly fine sandy loam	70–80
– gravelly loam	60–80
– gravelly silt loam	60–80
– gravelly sandy loam	50–70
– gravelly clay loam	60–80
– gravelly clay	40–70
– gravelly sand	20–30
Stony:	
– stony fine sandy loam	70–80
– stony loam	60–80
– stony silt loam	60–80
– stony sandy loam	50–70
– stony clay loam	50–80
– stony clay	40–70
– stony sand	10–40

Adapted from Storie, 1978.

95–100%; BB, undulating (3–8%) is 85–100%; C, moderately sloping (9–15%) is 80–95%; CC, rolling (9–15%) is 80–95%; D, strongly sloping (16–30%) is 70–80%; DD, hilly (16–30%) is 70–80%; E, steep (30–45%) is 30–50%; and F–d, very steep (≥45%) is 5–30%. The slope factor is then assigned a percentage rating.

X. Other conditions: This factor comprises six conditions: drainage, alkalinity, nutrient level, acidity, erosion, and micro-relief. Each condition is further divided into several sub-conditions, and each is assigned a percentage rating. Finally, an overall rating expressed as a percentage is assigned. Table 6.4 gives the ratings of other conditions.

Example 6.2: Determine the Storie Index for rating a 5000-acre farm where 3000 acres of land is gently sloping (3–8%), has soils in group II with slightly developed profiles, loam texture, and is well drained. Another 1200 acres of land is moderately sloping (9–15%), has a silty loam texture, is underlain by softly consolidated material at 2–3 feet, and is fairly well drained. The remaining piece of acreage has a high nutrient level, is undulating (3–8%), comprises calcareous clay loam, and is underlain by consolidated sedimentary rocks at 1–2 feet.

Solution: In this case, the lower limits of ranges of percentages are used. One can also choose the midpoint values of ranges.

For the 3000 acres of land: factor A (from Table 6.2): 95%, B: 100%, C: 95%, and X: 100%. Thus, the rating is: $(95+100+95+100)/4=97.5\%$.

For the 1200 acres of land: factor A: 60%, B: 100%, C: 80%, and X: 80%. Thus, the rating is: $(60+100+95+100)/4=88.75\%$.

For the remaining 800 acres of land: A: 40%, B: 95%, C: 85%, and X: 100%. Thus, the rating is: $(40+95+85+80)/4=75.0\%$.

The index rating (weighted average) for the entire farm is: $((3000\times 97.5)+(1200\times 88.75)+(800\times 75))/5000=91.8\%$

6.7 SOIL GRADING

By combining soils having ranges in the index rating, Storie (1978) graded California soils into six grades:

Grade 1 (excellent): This characterizes the soils that are rated at 80–100%. These soils are suitable for a variety of crops, such as field crops, alfalfa, orchards, and truck crops.
Grade 2 (good): The soils in this grade are rated at 60–79% and are good for most crops, with generally good to excellent yields.
Grade 3 (fair): The soils in this grade are rated at 40–59% and are fair in quality, with wide-ranging suitability. They are good for specific crops.
Grade 4 (poor): This corresponds to soils that are rated at 20–39%. These soils are suitable for a limited number of crops.
Grade 5 (very poor): This grade includes soils that are rated at 10–19%, which are limited in their usefulness because of poor conditions. They can be used for pasture.
Grade 6 (non-agricultural): The soils in this grade are rated below 10% and cannot be used for agriculture. Examples of such soils are highly alkaline soils, tidelands, riverwash, and steep, broken land.

By a systematic appraisal of soil, topographic, and drainage conditions, the US Bureau of Reclamation evaluated the capability of lands to repay the cost of irrigation development through crop production and then divided soils or lands into six classes:

Class 1: Lands most suitable for crop development.
Class 2: Lands less desirable because of soil, topographic, or drainage conditions.
Class 3: Lands least suitable, with lowest potential for repayment.
Class 4: Lands with excessive deficiencies but prone to development with the potential for repayment.
Class 5: Lands not suitable for irrigation under current conditions.
Class 6: Lands not considered arable.

This classification provides limits for each class of soil texture, depth, water-holding capacity, permeability, salinity, and other characteristics of both surface and various horizons of the subsoil. These, in combination with topographic, natural drainage, vegetal cover, and other characteristics, influence the development of irrigation system and their economics.

6.8 SOIL PHYSICAL CHARACTERISTICS

The soil particles may consist of minerals and organic matter. Normally sand particles are 0.05–2.0 mm in diameter, silt particles are 0.002–0.05 mm, and clay particles are less than 0.002 mm.

6.8.1 Density

The soil bulk density ρ_b can be defined as the ratio of the mass of dry soil M_s to the volume of bulk soil V_b, expressed as

$$\rho_b=\frac{M_s}{V_b}=\frac{M_s}{V_s+V_a+V_w}, \tag{6.1}$$

Table 6.4 *Factor X: rating of other conditions*

Condition	Percent
Drainage:	
– well drained	100
– fairly well drained	80–90
– moderately waterlogged	40–80
– badly waterlogged	10–40
– subject to overflow	Variable
Alkali:	
– alkali-free	100
– slightly affected	60–95
– moderately affected	30–60
– moderately to strongly affected	15–30
– strongly affected	5–15
Nutrient (fertility) level:	
– high	100
– fair	95–100
– poor	80–95
– very poor	60–80
Acidity: according to degree	80–95
Erosion:	
– none to slight	100
– detrimental deposition	75–95
– moderate sheet gullies	80–95
– occasional shallow gullies	70–90
– moderate sheet erosion with shallow gullies	60–80
– deep gullies	10–70
– moderate sheet erosion with deep gullies	10–60
– severe sheet erosion	50–80
– severe sheet erosion with shallow gullies	40–50
– severe sheet erosion with deep gullies	10–40
– very severe erosion	10–40
– moderate wind erosion	80–95
– severe wind erosion	30–80
Micro-relief:	
– smooth	100
– channels	60–95
– hogwallows	60–90
– low hummocks	80–95
– high hummocks	20–60
– dunes	10–40

Adapted from Storie, 1978.

where V_s is the volume of soil particles, V_a is the volume of air, and V_w is the volume of water. The dry density is the mass of oven-dried (over 24 h at 105 °C) soil divided by the volume in its natural state. The relationship of soil bulk density to root growth of different soil textures is shown in Table 6.5. The soil bulk density is affected by soil organic matter, soil texture, the soil mineral density (sand, silt, and clay), and the packing arrangement of the soil mineral. An average value of 1.4 or 1.5 (10^3 kg/m³) is often considered for medium-textured soils.

Table 6.5 *Relationship of soil bulk density with root growth of different soil textures*

Soil texture	Ideal bulk densities (g/cm^3)	Bulk densities affect root growth (g/cm^3)	Bulk densities restrict root growth (g/cm^3)
Sands, loamy sands	<1.60	1.69	>1.80
Sandy loams, loams	<1.40	1.63	>1.80
Sandy clay loams, clay loams	<1.40	1.60	>1.75
Silts, silt loams	<1.40	1.60	>1.75
Silt loams, silty clay loams	<1.40	1.55	>1.65
Sandy clays, silty clays, clay loams	<1.10	1.49	>1.58
Clays (>45% clay)	<1.10	1.39	>1.47

Adapted from USDA-NRCS, www.nrcs.usda.gov/Internet/FSE_DOCUMENTS/nrcs142p2_053260.pdf.

Similarly, soil particle density ρ_s can be defined as

$$\rho_s = \frac{M_s}{V_s}. \tag{6.2}$$

The specific gravity of solids (ρ_p) can be written as

$$\rho_p = \frac{M_s}{V_s \rho_w}, \tag{6.3}$$

where ρ_w is the mass density of water. The specific gravity of minerals is about 2.65, of clay about 2.7, and of organic matter 2.60. The apparent specific gravity of solids can be defined as

$$A_s = \frac{\rho_b}{\rho_w} = \frac{M_s}{V_b \rho_w}. \tag{6.4}$$

6.8.2 Soil Porosity

Soil porosity, denoted by N, is defined as the ratio of total pore volume (V_p) to the bulk volume of soil (V_b), expressed as

$$N = \frac{V_p}{V_b} = \frac{V_a + V_w}{V_b} = \frac{V_a + V_w}{V_s + V_a + V_w}. \tag{6.5}$$

The bulk density consists of the volume of soil particles, volume of air, and volume of water.

6.8.3 Air-Filled Porosity

Denoted by N_a, the air-filled porosity is defined as the ratio of air-filled pore volume (V_a) to the bulk volume (V_b), expressed as

$$N_a = \frac{V_a}{V_b}. \tag{6.6}$$

6.9 SOIL SALINITY

The chemical properties of soils, such as cation exchange capacity, pH, and salinity affect the plant and production. Fertile soils have high cation exchange capacity, because they contain small crystalline clay particles and organic matter with a negative surface charge that attracts positively charged ions (cations), water, and chemical compounds, resulting in the holding of nutrients for plant use. Clay, sand, and humus soils have cation exchange capacities of 30, 9, and 200 milliequivalents per 100 g of soil, respectively. Some of the exchangeable cations in soils are aluminum (Al^{3+}), with the strongest attachment, followed by hydrogen (H^{+}), calcium (Ca^{2+}), magnesium (Mg^{2+}), potassium (K^{+}), ammonia (NH_4^{+}), and sodium (Na^{+}).

Soil reaction is described either by the concentration of hydrogen ions (H^{+}) or as pH. The value of pH can vary from 0 to 14, as shown in Figure 6.9, which shows that a pH of 0 to less than 7 is acidic, a pH of 7 is neutral, and a pH of greater than 7 to 14 is alkaline (Brady, 1974). The US Department of Agriculture Natural Resources Conservation Service (1998) has classified soil pH ranges as shown in Table 6.6. Low pH may cause toxicity due to aluminum and manganese. The availability of nutrients is limited by the soil pH. For example, a pH of 5 may limit the availability of phosphorus to plants, despite its adequate availability in the soil. Different plants need different pH values for their optimal growth. A pH value of 6–7 is good for most field crops, and a value of less than 5 may be good for some crops, such as potato and blueberry. Figure 6.10 shows the availability of minerals to plants as impacted by pH (USDA-NRCS, 1997).

Salinity is caused by the accumulation of soluble salts, including sodium, calcium, and magnesium, and is denoted by terms such as saline, sodic, and saline–sodic. Soil is impacted by salinity alone or in combination with sodium. Saline soils generally contain a pH value below 8.5 and low sodium, and mainly have calcium, sodium, and magnesium chloride and sulfates. In saline soils, salts rise to the surface during evaporation and accumulate there; this is found mostly in dry regions. These compounds form the white crust on the surface and are soluble in water. Therefore, leaching is effective in reclaiming these soils. This practice is often followed in developing countries such as India. Sodic soils, on the other hand, have a pH value of 8.5–10. Sodium has a tendency to debilitate the soil structure and makes it impervious to water. The soils turn dark and are often referred to as black alkaline soils. Depending on the level of soil salinity, crop yield is expected to decline. When measured by electrical conductivity under normal growing conditions, both electrical conductivity and sodium adsorption ratio can be employed to classify salt-affected soils, as shown in Table 6.7. Tolerance of selected crops is shown in

Table 6.6 *Classification of soil pH ranges*

Denomination	pH range
Ultra-acidic	<3.5
Extremely acidic	3.5–4.4
Very strongly acidic	4.5–5.0
Strongly acidic	5.1–5.5
Moderately acidic	5.6–6.0
Slightly acidic	6.1–6.5
Neutral	6.6–7.3
Slightly alkaline	7.4–7.8
Moderately alkaline	7.9–8.4
Strongly alkaline	8.5–9.0
Very strongly alkaline	>9.0

Data from USDA-NRCS, 1998.

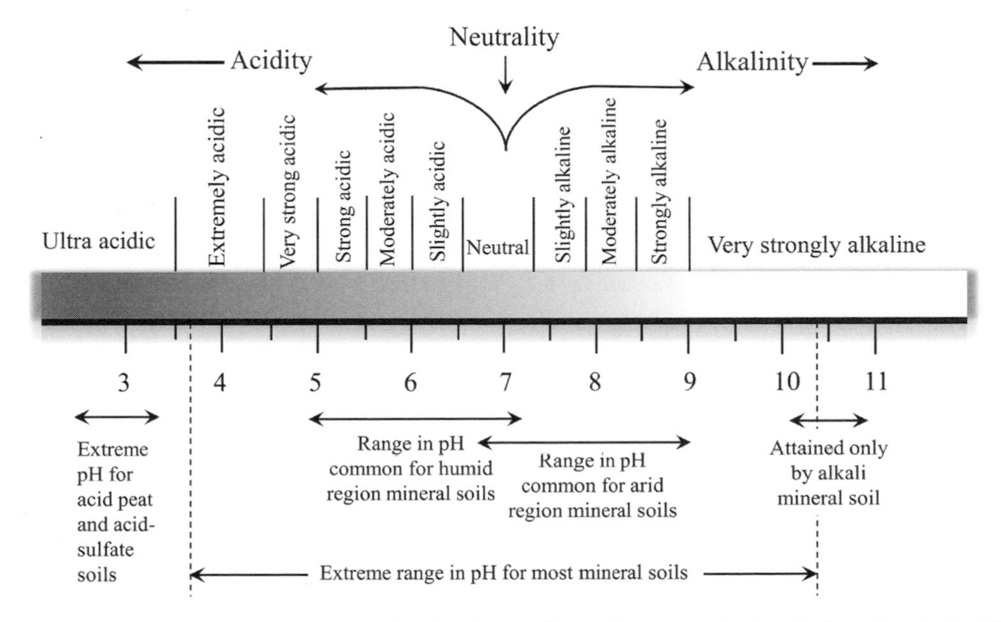

Figure 6.9 Range of pH in mineral soils. (Soil pH value ranges are based on data in Table 6.6; the pH ranges in mineral soils are from Brady, 1974.)

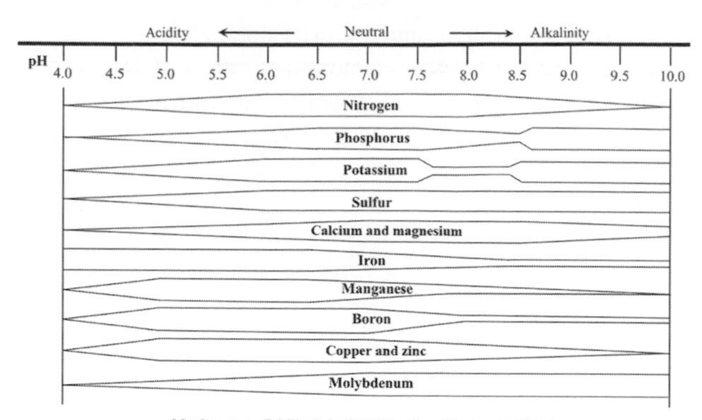

Figure 6.10 Nutrient availability in relation to soil pH. (USDA-NRCS, 1997.)

Table 6.8. Table 6.9 shows expected yield reductions for some crops. Forage crops are least sensitive to salinity, whereas fruit crops are most sensitive. Field crops and vegetable crops are in between.

6.10 SOIL DEGRADATION

Soil is a fundamental natural resource that must be protected, conserved, and managed. A number of activities, some natural but mostly human, cause the degradation of soil. Erosion is the primary mechanism of degradation and the primary agents are water and wind. Widespread agriculture has accelerated erosion and soil degradation, which may threaten the availability of productive soil and in turn food security. Table 6.10 indicates the average annual loss of topsoil caused by erosion in non-federal lands in the United States (Natural Resources Conservation Service, 2012). Annual soil losses vary from state to state and are greater in cultivated lands than in pasture and rangelands. Also, erosion by water is greater in high-rainfall regions, such as the eastern and southeastern United States, and erosion by wind is greater in dry windy areas, such as the western United States. Figure 6.11 shows both water (sheet and rill) and wind erosion on cropland by year in the United States (USDA-NRCS, 2007), Figure 6.12 shows water erosion in 1982 (the most serious year in the period 1982–2007) in the United States (USDA-NRCS, 2001b), and Figure 6.13 shows the locations of cropland in the United States with the highest potential for wind erosion (Nordstrom and Hotta, 2004).

6.11 SOIL CONSERVATION

Although soil erosion cannot be entirely prevented, appropriate management strategies can be put in place to minimize erosion and soil degradation. These include: (1) maintaining vegetative cover; (2) leaving crop residue on the field; (3) maintaining a high organic matter content; (4) conservation tillage; (5) crop rotation; (6) contour tillage; (7) strip cropping; (8) grass waterways; and (9) windbreaks.

Table 6.7 *Classification of salt-affected soils based on the analysis of saturated soil extracts*

Criteria	Normal	Saline	Saline–sodic	Sodic
pH		<8.5	<8.5	>8.5
EC_e (mmhos/cm)	<4	>4	>4	<4
SAR	<13	<13	>13	>13

Data from USDA, n.d.

Table 6.8 *Tolerances of crops to soil salinity*

Crop	Tolerant	Medium	Sensitive
Field crops	Barley Sugar beet Cotton	Corn Soybean Sorghum Wheat	Beans Flax Broad bean
Forage crops	Bermudagrass Wheat grass Tall fescue	Alfalfa Orchardgrass Perennial grass	Clovers
Vegetables	Beets Asparagus	Spinach Tomato Broccoli Cabbage Potato Sweet corn	Lettuce Bell pepper Onion Carrot Beans Celery
Fruits	Date palm	Grape Fig Olive	All others

Adapted from Peace Corps of the United States of America, 1986.

6.12 SALINITY MANAGEMENT

Management of salinity is always complicated. However, there are different techniques available, which are briefly discussed in this section.

6.12.1 Leaching

Leaching is the principal management tool for salt, as salts are soluble in water. By applying water in excess of what is needed by plants and consumed by evapotranspiration, salts can be leached below the root zone to maintain productivity. The amount of water needed is called the leaching requirement or leaching fraction. The frequency of leaching can vary from each irrigation to a few irrigations, to yearly irrigation, depending on the degree of salinity.

Table 6.9 *Soil salinity tolerance levels for different crops*

Crop	Yield potential 100%	90%	75%	50%	Maximum EC_s
Field crops					
Barley[a]	8.0	10	13	18	28
Bean (field)	1.0	1.5	2.3	3.6	6.3
Broad bean	1.5	2.6	4.2	6.8	12
Corn	1.7	2.5	3.8	5.9	10
Cotton	7.7	9.6	13	17	27
Cowpea	4.9	5.7	7.0	9.1	13
Flax	1.7	2.5	3.8	5.9	10
Groundnut	3.2	3.5	4.1	4.9	6.6
Rice (paddy)	3.0	3.8	5.1	7.2	11
Safflower	5.3	6.2	7.6	9.9	15
Sesbania	2.3	3.7	5.9	9.4	17
Sorghum	6.8	7.4	8.4	9.8	13
Soybean	5.0	5.5	6.3	7.5	10
Sugar beet	7.0	8.7	11	15	24
Wheat[a]	6.0	7.4	9.5	13	20
Vegetable crops					
Bean	1.0	1.5	2.3	3.6	6.3
Beet[b]	4.0	5.1	6.8	9.6	15
Broccoli	2.8	3.9	5.5	8.2	14
Cabbage	1.8	2.8	4.4	7.0	12
Cantaloupe	2.2	3.6	5.7	9.1	16
Carrot	1.0	1.7	2.8	4.6	8.1
Cucumber	2.5	3.3	4.4	6.3	10
Lettuce	1.3	2.1	3.2	5.1	9.0
Onion	1.2	1.8	2.8	4.3	7.4
Pepper	1.5	2.2	3.3	5.1	8.6
Potato	1.7	2.5	3.8	5.9	10
Radish	1.2	2.0	3.1	5.0	8.9
Spinach	2.0	3.3	5.3	8.6	15
Sweet corn	1.7	2.5	3.8	5.9	10
Sweet potato	1.5	2.4	3.8	6.0	11
Tomato	2.5	3.5	5.0	7.6	13
Forage crops					
Alfalfa	2.0	3.4	5.4	8.8	16
Barley hay	6.0	7.4	9.5	13	20
Bermudagrass	6.9	8.5	11	15	23
Clover, Berseem	1.5	3.2	5.9	10	19
Corn (forage)	1.8	3.2	5.2	8.6	15
Harding grass	4.6	5.9	7.9	11	18
Orchardgrass	1.5	3.1	5.5	9.6	18
Perennial rye	5.6	6.9	8.9	12	19
Sudan grass	2.8	5.1	8.6	14	26
Tall fescue	3.9	5.5	7.8	12	20
Tall wheat grass	7.5	9.9	13	19	31
Trefoil, big	2.3	2.8	3.6	4.9	7.6
Trefoil, small	5.0	6.0	7.5	10.0	15
Wheat grass	7.5	9.0	11.0	15.0	22
Fruit crops					
Almond	1.5	2.0	2.8	4.1	6.8
Apple pear	1.7	2.3	3.3	4.8	8
Apricot	1.6	2.0	2.6	3.7	5.8
Avocado	1.3	1.8	2.5	3.7	6

Table 6.9 (*cont.*)

Crop	Yield potential 100%	90%	75%	50%	Maximum EC_s
Date palm	4.0	6.8	10.9	17.9	32
Fig, olive, pomegranate	2.7	3.8	5.5	8.4	14
Grape	1.5	2.5	4.1	6.7	12
Grapefruit	1.8	2.4	3.4	4.9	8.0
Lemon	1.7	2.3	3.3	4.8	8.0
Orange	1.7	2.3	3.3	4.8	8.0
Peach	1.7	2.2	2.9	4.1	6.5
Plum	1.5	2.1	2.9	4.3	7.1
Strawberry	1.0	1.3	1.8	2.5	4.0
Walnut	1.7	2.3	3.3	4.8	8.0

Based on electrical conductivity of the saturated extract taken from a root zone soil sample (EC_s) measured in mmhos/cm.
[a] During germination and the seeding stage EC_s should not exceed 4–5 mmhos/cm, except for certain semi-dwarf varieties.
[b] During germination EC_s should not exceed 3 mmhos/cm.
After Ayers and Westcot, 1994. Reprinted with permission from FAO.

Table 6.10 *Estimated average annual loss of topsoil in tonnes/hectare caused by erosion in non-federal lands of the United States*

Land	Water erosion (tonnes/hectare/year) 1982	Water erosion 1997	Wind erosion (tonnes/hectare/year) 1982	Wind erosion 1997
Cultivated cropland	10.1	7.9	8.3	6.5
Pastureland	2.5	2.2	0.2	0.2
Rangeland	2.7	2.7	2.7	9.9

After Natural Resources Conservation Service, 2012.

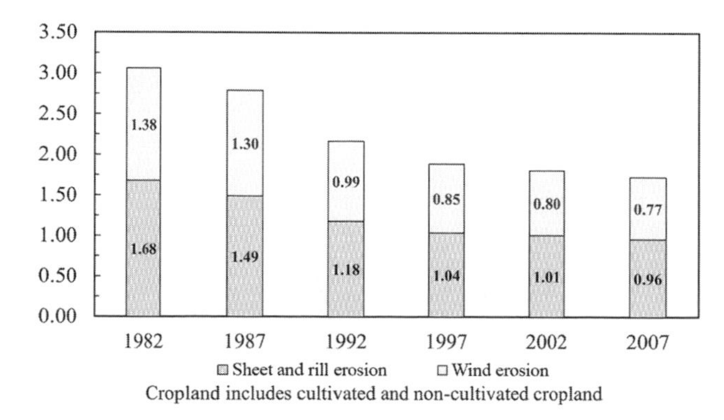

Figure 6.11 Erosion of cropland, billions of tons. (Data from USDA-NRCS, 2007.)

The leaching fraction can be determined by taking the ratio of the electrical conductivity of the irrigation water (EC_{iw}) to the electrical conductivity of the water draining below the root zone, which is equal to the electrical conductivity of the soil water (EC_{sw}). The amount of leaching required for different classes of irrigation and drainage water in order to maintain the soil salinity level in the root zone at the desired level is given in Table 5.14.

6.12.2 Subsurface Drainage

In many areas shallow saline water tables exist, in which case water carrying dissolved salts may move upward into the root zone. When this water is used up by plants and consumed by evapotranspiration, the salts are left behind. The shallow water tables also limit the downward leaching of salts. To overcome this problem, subsurface drainage is recommended.

6.12.3 Seed Placement

Appropriate planting procedures can be adopted to bring down the salinity in the soil around the germinating seeds.

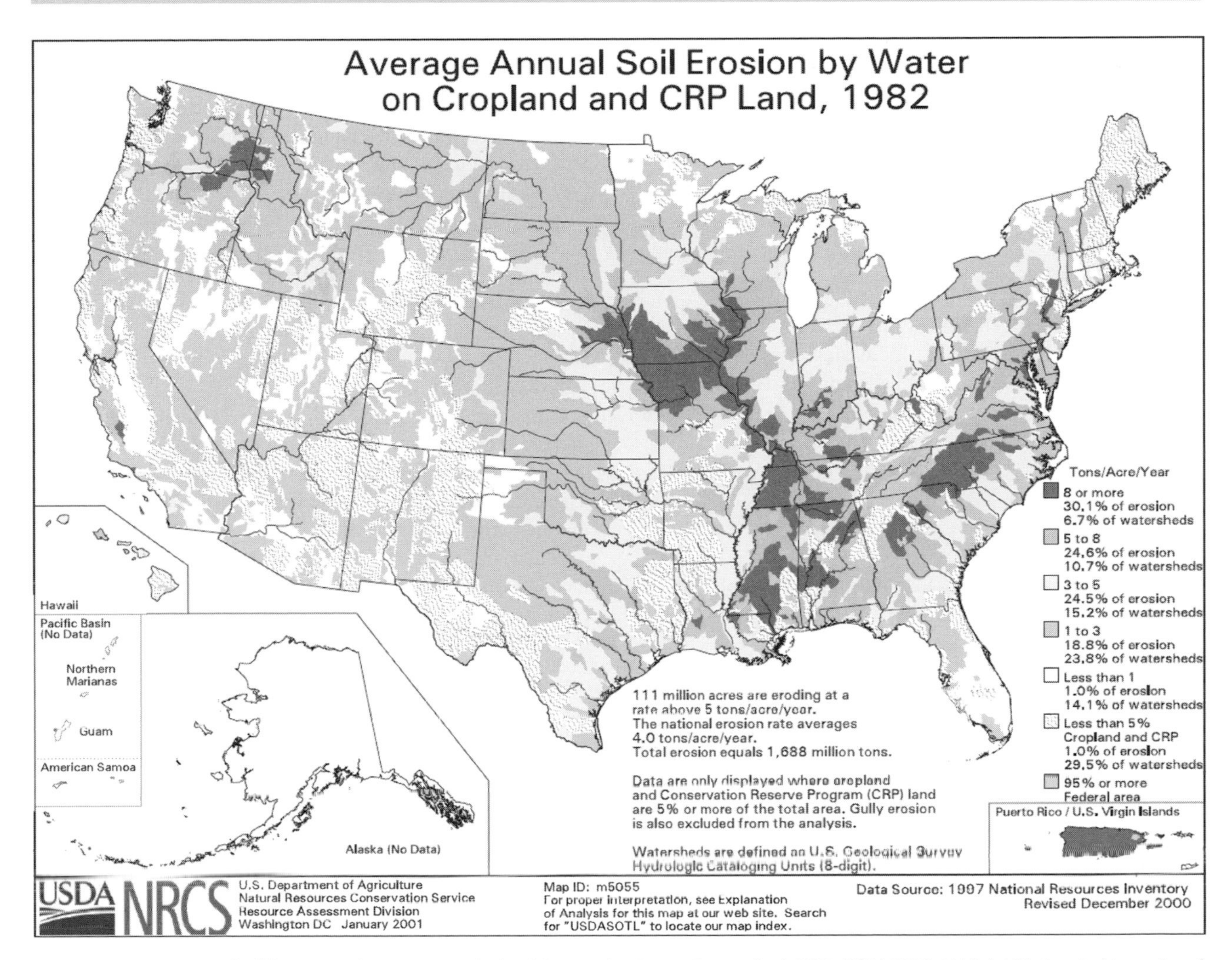

Figure 6.12 Average annual soil loss erosion by water on cropland and Conservation Reserve Program land, 1982. (USDA-NRCS, 2001b.) A black and white version of this figure will appear in some formats. For the color version, please refer to the plate section.

Figure 6.13 Wind erosion areas in the United States. (Adapted from Nordstrom and Hotta, 2004. Copyright © 2004, Reprinted with permission from Elsevier.)

For example, in the case of furrows, seeds can be planted near the shoulders, away from where the salt accumulation would be greatest. Sometimes more seeds are planted to compensate for the poor germination. Planting seeds on the sloping side above the water table can also be considered in the case of sloping beds.

6.12.4 Frequent Irrigation

In general, salt concentrations are greater before irrigation and lower during irrigation. Increasing the frequency of irrigation may help leaching of salts and to maintain a more consistent moisture content in the soil. This can be accomplished with sprinkler or drip irrigation systems.

6.12.5 Pre-planting Irrigation

When off-season rainfall is low and the water table is high, salts accumulate near the soil surface during fallow periods. It is therefore recommended that soil is properly leached or else seed germination and seeding growth will be severely impacted.

6.12.6 Change in Irrigation Method

Border, furrow, flood, and basin irrigation use a lot of water for irrigation and their irrigation efficiency is low. Therefore, methods of surface irrigation may not be used for frequent irrigation where water supply is a limiting factor. Sprinkler or drip irrigation systems may be the only option for frequent irrigation.

6.12.7 Use of Chemicals

For sodic soils, gypsum or sulfuric acid can be employed to dislodge the sodium ions attached and adsorbed onto the soil particles. Gypsum contains calcium, which can replace adsorbed sodium and help restore the infiltration capacity of soils. Polymers are also an option for treating sodic soils. However, practices do not obviate the need for leaching.

6.12.8 Water Delivery by Pipelines

The salinity content of irrigation water increases due to evaporation when delivered by open channels. The delivery of water by pipelines eliminates evaporation and prevents the increase of salinity of irrigation water, and hence stabilizes the salinity level. The pipes also minimize the loss of water to seepage and hence increase the availability of water for leaching.

6.13 EFFECT OF CLIMATE

Climate determines rock disintegration and the development of soil. It determines the course of mechanical and chemical transformation, the type and amount of vegetation, and the type of organic matter forming and accumulating in the soil. The two most important climatic factors are temperature and precipitation. Arid conditions permit accumulation of more soluble substances than do humid conditions. That is the reason that with the availability of soil moisture dry deserts turn to green areas. In humid regions, soluble substances such as chlorides and sulfates of the alkalis are rapidly removed from the soil.

Climate exercises an even greater influence on organic matter. In general, soils in cool and humid regions are rich in organic matter compared to those in dry and warm regions. The nitrogen content of soils in cool regions is higher than that in warmer regions. There is a close connection between nitrogen and organic matter, and higher nitrogen content means more organic matter in the soil. Further, climate also influences the chemical nature of organic matter, as reflected by the relationship between nitrogen and carbon.

QUESTIONS

Q.6.1 Demarcate zones of different soil types on a Texas map. Indicate on the map where College Station is located.

Q.6.2 Consider a 1 m^3 sample of a soil that contains 40% pore space.

a. What is the mass of water required to saturate it?
b. What is the mass of air when the soil is completely dry?
c. What is the mass of the solid phase?
d. What is the mass of the soil when saturated?
e. What would be the mass of the soil if the volume fraction of air were 18%?

Q.6.3 Calculate ρ_b of 1 m^3 of soil. Assume the volume fractions of solid, water, and air of 0.5, 0.34, and 0.16, respectively. Assume $\rho_w = 1.0\ g/cm^3$ and $\rho_a = 1.3\ kg/m^3$.

Q.6.4 A sharp-edged cylinder of 14 cm diameter is driven into the soil so that negligible compaction occurs. A 24 cm sample is secured. The net weight of the soil sample is 6225 g and the dry weight of the soil sample is 5400 g.

a. What is the bulk density of the soil sample (g/cm^3)?
b. What is the moisture content of the soil sample on a dry weight basis?
c. What is the moisture content of the soil sample on a volumetric basis?
d. What is the apparent specific gravity of the soil sample?

Q.6.5 Use the following table to answer the questions.

Depth of sample (cm)	Weight of wet sample (g)	Weight of oven-dry sample (g)
0–25	129	108
25–50	131	106
50–75	133	108
75–100	133	109

a. Compute the soil moisture content (dry weight and volumetric) in each soil layer. Assume a bulk density of 1.28 g/cm^3.

b. If the permanent wilting point of this soil is 14% (dry weight basis) and the soil has an average available water-holding capacity of 1.80 mm per centimeter of soil, how much water must be applied to bring the soil to field capacity?

Q.6.6 A soil sample taken from the field has a volume of 250 cm^3 and a weight of 450 g. The oven-dry weight of this sample is 375 g. Assuming a water density of 1000 kg/m^3 and soil density of 2650 kg/m^3, compute the following: (1) volume of solid or soil particles; (2) volume of water; (3) volume of air; (4) mass of water; (5) mass of solid particles; (6) volume of pore spaces; (7) porosity; (8) air-filled porosity; (9) bulk density; (10) moisture content on a mass basis; and (11) moisture content on a volume basis.

Q.6.7 A soil sample taken from the field has a volume of 400 cm^3 and a weight of 750 g. The oven-dry weight of this sample is 580 g. It is given that water density = 1000 kg/m^3 and soil density = 2900 kg/m^3. Compute the following parameters: (1) volume of soil particles; (2) volume of water; (3) volume of air; (4) mass of water; (5) volume of pore spaces; (6) porosity; (7) air-filled porosity; (8) bulk density; (9) moisture content on a mass basis; and (10) moisture content on a volume basis.

Q.6.8 Derive the formula of porosity in soil sample (N) in terms of bulk density (ρ_b) and soil density (ρ_s).

Q.6.9 Identify (a) a soil that is 40% sand, 20% clay, and 40% silt; and (b) a soil that contains 65% sand, 15% clay, and 20% silt.

Q.6.10 Determine the Storie Index for rating a 6000-acre farm with the following characteristics: 3500 acres of land is nearly level (0–2%), has soils in group I with undeveloped profiles, coarse sandy loam texture, and is well drained; another 1500 acres of land is gently sloping (3–8%), has loamy sand texture, is underlain by softly consolidated material at 3–4 feet, and has a very poor nutrient level; and the remaining acreage has a high nutrient level, is undulating (0–2%), has stony loam texture, and is underlain by consolidated sedimentary rock at 2–3 feet.

REFERENCES

Ayers, R. S., and Westcot D. W. (1994). Water quality for agriculture, revised. FAO irrigation and drainage paper 29.

Brady, N. C. (1974). *The Nature and Property of Soils*. New York: Macmillan.

Hillel, D. (1980). *Fundamentals of Soil Physics*. New York: Academic Press.

Natural Resources Conservation Service (2007). Distribution maps of dominant soil orders. www.nrcs.usda.gov/wps/portal/nrcs/detail/soils/survey/class/maps/?cid=nrcs142p2_053589.

Natural Resources Conservation Service (2012). *Summary Report: 2012 National Resources Inventory*. Washington, DC: USDA.

Nordstrom, K. F., and Hotta, S. (2004) Wind erosion from cropland in the USA: a review of problems, solutions and prospects. *Geoderma*, 121(3–4): 159–167.

Peace Corps of the United States of America (1986). Salinity and alkalinity problems. In *Soils, Crops and Fertilizer Use*. Washington, DC: Peace Corps of the United States of America.

Soil Conservation Service (1991). Soil–plant–water relationships. In *National Engineering Handbook*. Washington, DC: USDA.

Soil Science Society of America (2015) Dig Deeper. www.soils4kids.org.

Storie, R. E. (1978). *Storie Index Soil Rating*. Berkeley, CA: University of California.

USDA (n.d.) *New Mexico Integrated Water Management Handbook (AGRO-76)*. Washington, DC: USDA. www.nrcs.usda.gov/wps/portal/nrcs/detail/nm/technical /?cid=nrcs144p2_068965.

USDA-NRCS (1997). Soil. In *National Engineering Handbook*. Washington, DC: USDA.

USDA-NRCS (1998). Soil quality indicators: pH. USDA soil quality information sheet. www.nrcs.usda.gov/Internet/FSE_DOCUMENTS/nrcs142p2_052208.pdf.

USDA-NRCS (2001a). Water management (drainage). In *National Engineering Handbook*. Washington, DC: USDA.

USDA-NRCS (2001b). Average annual soil erosion by water on cropland and CRP land, 1982. USDA, Natural Resources Conservation Service. www.nrcs.usda.gov/wps/portal/nrcs/detail/national/technical/nra/nri/results/?cid=nrcs143_013800.

USDA-NRCS (2005). Global soil regions map. USDA, Natural Resources Conservation Service. www.nrcs.usda.gov/wps/portal/nrcs/detail/soils/use/?cid=nrcs142p2_054013.

USDA-NRCS (2007). National Resources Inventory. Soil erosion on cropland. Washington, DC: USDA, Natural Resources Conservation Service. https://www.nrcs.usda.gov/wps/portal/nrcs/detail/national/technical/nra/nri/results/?cid=nrcs143_013800.

7 Crops and Crop Production

Notation

ET_a	actual evapotranspiration (mm/day)	Y	crop yield (kg/ha); relative crop yield ($Y = 1.0$ for maximum yield)
ET_m	maximum evapotranspiration (mm/day)	Y_a	actual harvested yield (kg/ha)
F	fertilizer application (kg/ha)	Y_m	maximum potential yield (kg/ha)
k_y	coefficient or yield response factor for a crop		
X	water required ($X = 1.00$ for maximum yield of each crop)		

7.1 INTRODUCTION

The purpose of irrigation is to achieve a higher level of productivity. However, irrigation should be practiced in concert with other factors that affect crop production. It is the integration of irrigation with agricultural technology that has led to the doubling or even quadrupling of the yield of some crops. The factors that affect crop productivity and yield include water availability and distribution, water quality, crop variety, tillage operations, soil fertility and fertilizers, pesticides and weedicides, aeration and drainage, climatic factors, and availability of energy. These factors will be briefly reviewed here, but first it is necessary to know crops and their cultivation.

Different types of crops require different types of climate and soil. Further, different crops and their optimum production have different irrigation requirements with respect to the frequency of irrigation, timing of irrigation, and amount of irrigation water per irrigation. The objective of this chapter is to briefly discuss the types of crops and their water requirements.

7.2 CLASSIFICATION OF CROPS

Crops are grown in different seasons. Some crops are grown in fall, some in winter, some in spring, and some in summer. Except for California, parts of Texas, Louisiana, Florida, Puerto Rico, and Hawaii, only one crop per year is grown in most of the United States because the weather is too cold outside of the growing season. Crops can be classified based on a number of criteria, such as agronomic and forage crops, vegetable crops, and fruit or garden crops.

Agronomic crops can be categorized into cereals, grains, pulses, forages, fiber crops, root and tuber crops, cover crops, companion crops, green manure crops, and recreational crops.

Cereal crops: These crops are grasses that produce edible grains. Examples of such crops include wheat, barley, rice, corn, and sorghum. Cereal crops comprise about 75% starch, 12% protein, 2% fat, and smaller amounts of vitamins and minerals. Figure 7.1 shows some cereal crops.

Grain crops: For grain crops, only the portion that produces grain – which is the fruit or seed – is harvested, such as wheat, corn, soybean, kidney bean, flax, and buckwheat. Some of the grain crops produce small grains, such as wheat, rye, barley, and oat. Three grain crops are shown in Figure 7.2.

Pulses: Pulse crops are soybean, edible field beans, chickpeas, cowpeas, lupines, field peas, lentils, and peanuts. These are legumes and their seeds are rich in protein, containing up to one-third of their dry matter as protein. Figure 7.3 shows some pulses.

Forages: Forage crops are grasses that are used for livestock feed. Their vegetative parts include stem leaves, and sometimes attached seeds constitute the feed. These crops are grown on pastures, rangelands, and cropland pastures, but are stored for future use. Hay is dried up to about 20% and stored in the air. Silage is harvested at about 50–80% moisture and sealed in a silo. Forages are sometimes used for crop rotation. They are often grown on erodible lands that are unsuitable for tillage. They are natural feeds for ruminant animals, such as goats, sheep, cattle, deer, and other wildlife.

Depending on soil and climatic conditions, a wide variety of perennial and annual crops are grown in the United States. Shrubs and native grass, such as little bluestem and buffalo grass, are grazed in the dry regions of the western Great Plains. Annual clovers, Bermudagrass, and tall fescue are

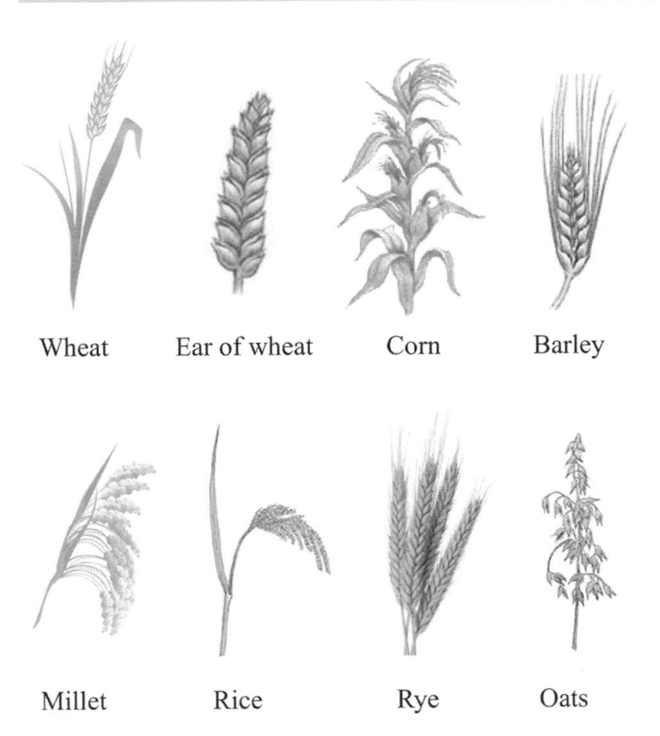

Figure 7.1 Cereal crops. A black and white version of this figure will appear in some formats. For the color version, please refer to the plate section.

Figure 7.2 Grain crops. A black and white version of this figure will appear in some formats. For the color version, please refer to the plate section.

Figure 7.3 Pulses. A black and white version of this figure will appear in some formats. For the color version, please refer to the plate section.

grazed and used for hay in the southern United States. Legumes, such as alfalfa and clovers, and grasses, such as bromegrass and timothy grass, are used for pasture, hay, and silage in cool temperate northern United States and Canada. Small grain crops and corn are harvested for silage when they are immature throughout the United States.

Fiber crops: Cotton, flax, and hemp are examples of fiber crops. Fiber plants are used for clothing, paper, rope, and baskets. Cotton is the most common fiber crop. Its fiber is white and surrounds the seed. Flax fiber is produced from the stem, is soft, and is called bast fiber. It is used for cloth production, such as for linen. Hemp fiber, also extracted from the stem, is coarse and brown. It is extremely strong and is used for rope, bags, and canvas.

Root and tuber crops: These crops are harvested for their underground storage organs, which are rich in starch. Such crops are potatoes, sweet potatoes, cassava, and Jerusalem artichoke. Potatoes are enlarged stems, called tubers, whereas sweet potatoes and cassavas are enlarged roots, not tubers. They are an important source of energy for human consumption.

Cover crops: These crops are used to prevent soil erosion via water or wind when the soil is barren. Consider, for example, the Midwest. Crops, such as corn and soybean, are grown from May through October. That means the fields remain uncovered, excepting crop residue, until the following spring. A cover crop, such as winter rye, is often grown in late fall to provide soil cover during winter.

Companion crops: Companion crops are seeded with main crops, such as wheat and oat. Such crops are alfalfa and mustard. Companion crops have three main functions. First, they help control soil erosion; second, they suppress weeds; and third, they provide some income if the main crop fails or underperforms. The legumes may regrow upon harvest.

Green manure crops: The purpose of seeding green crops is to add nutrients and organic matter to the soil and enhance soil quality. Examples of green manure crops are alfalfa, winter rye, and sweet clover. Sometimes they are grown for the purpose of plowing down.

Recreational turf grass: There are many kinds of grasses that are grown entirely for recreational uses, such as lawns, golf courses, soccer fields, football fields, and baseball fields. Examples of such grasses are Kentucky bluegrass, perennial ryegrass, and bentgrass.

Bioenergy crops: Bioenergy crops are divided into biomass crops that are utilized for energy, and biofuel crops, which are grown only for the production of liquid fuel. These crops constitute a renewable source of energy. Only a small portion of the total energy used in the United States comes from renewable sources, as shown in Table 7.1. Examples of biomass energy crops are corn, sugarcane, and canola. Bioenergy can be produced from starch or cellulose. Corn is used to produce starch, which is then converted to ethanol. Cellulose is also used to produce ethanol. Cellulosic ethanol sources are crop residues, such as corn stoves, rice strands, and wheat straw left after harvest; forest residue left after wood harvest and processing; energy crops, such as switchgrass, alfalfa, hybrid poplar, and willow; and surplus or damaged hay crops of alfalfa and smooth bromegrass.

Nutritional-use crops: The basic components of our diet comprise proteins, fats, and carbohydrates, including sugar and starch. Sugar crops are sugar beet and sugarcane. Corn provides starch that is a source of high-fructose corn syrup.

Oil crops: Many crops produce oil seeds that are used for the production of oil. Examples of soil crops are soybean, sunflower, flax, mustard, canola, peanuts, and cotton. Soybean represents about half of oil production in the United States.

Protein crops: Legumes are rich in protein. Examples are soybean, peas, field beans, and peanuts. Non-legumes used for seed production are quinoa, cotton, and sunflower.

Starchy seed crops: Cereal crops produce seeds that are high in starch. These include corn, wheat, oat, barley, rye, millet, and sorghum. Amaranth and buckwheat are non-grasses that also produce starchy seeds.

Starchy root and tuber crops: These crops also produce starch. They provide edible starch material in roots, subterranean stems, rhizomes, tubers, and corms. These include potatoes, yams, taro, cocoyams, cassava, sweet potatoes, canna, and arrowroots. Asia and Africa are the major producers.

Table 7.1 *US energy production/consumption by major sources in 2019*

Fuel/energy source	Share of total production	Share of total consumption
Petroleum	31%	37%
Natural gas	35%	32%
Renewables	12%	11%
Coal	14%	11%
Nuclear	8%	8%
Total energy consumption	101.04 quadrillion Btu	100.17 quadrillion Btu

After US Energy Information Administration, 2020.

7.3 CROPPING SYSTEMS

Farmers grow crops in order to get maximum returns. To achieve this objective it is desirable to efficiently use soil and climatic conditions for managing crops in concert with environmental, socioeconomic, and political considerations. There are two systems for growing crops on a farm: continuous crops and crop rotation. In the continuous cropping system, as shown in Figure 7.4, the same crop is grown year after year on the same land for at least two or three years. Examples are wheat in the Central and Southern Plains and the Northwest, cotton in the Southwest, and corn in the Midwest. These systems lack biodiversity and are more prone to crop failure due to extreme weather, pests, and other catastrophes. The advantage is in terms of specialization in management, equipment, and marketing.

In the crop rotation system, crops are grown in sequence for a period of time over the years on the same farm (Figure 7.4). The length and diversity in crop rotation can be different. Examples of crop rotation systems are corn–soybean (two-year rotation), wheat–fallow (two-year rotation), potato–soybean–wheat (three-year rotation), oat/red clover (or alfalfa)–corn–soybean (three-year rotation), and sugar beet–soybean (or dry bean)–wheat (three-year rotation).

Rotation cropping systems have a number of advantages, which depend on the number, type, and sequence of crops in the rotation, as shown in Figure 7.5. The systems are more

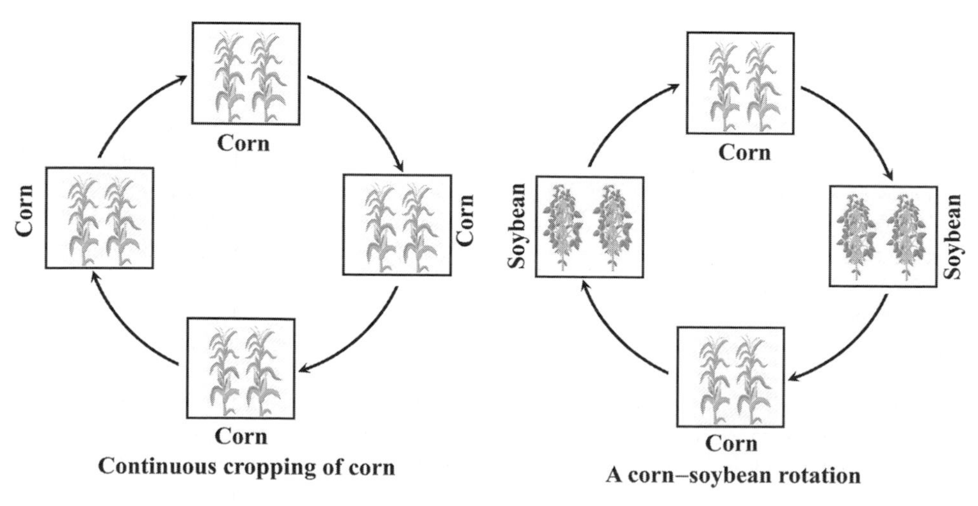

Figure 7.4 Continuous cropping and rotation cropping systems.

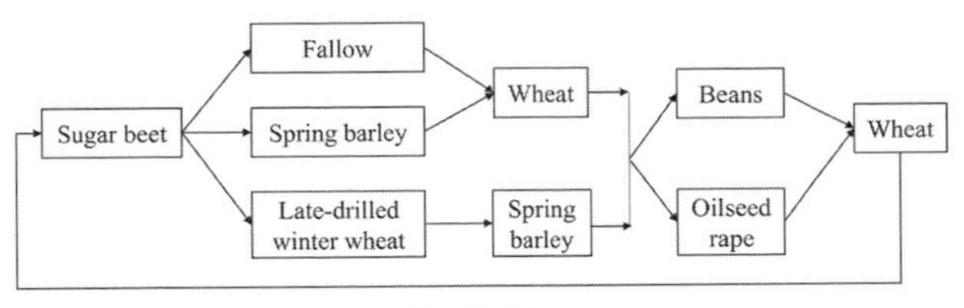

Figure 7.5 Crop rotation system. (Adapted from Castellazzi et al., 2008.)

advantageous if they have a greater diversity of crops, and if the crops are from different plant families and have different life cycles. The advantages of such systems include weed, disease, and insect control, increase in soil fertility, reduced soil erosion, increased yield and reduced risk, and recycling of nutrients.

Crops are grown as a single crop in a field during the growing season or two or more crops in the same field during the season. Important crops grown in monocultures are wheat, corn, rice, soybean, and cotton. Polyculture promotes spatial diversity of crops and more efficient use of soil and environmental resources. Polyculture is also known as intercropping. Polycultures may involve mixed intercropping, strip cropping, and living mulches. Multiple cropping systems involve production of two or more crops from the same land each year. They include cover crops and green manure crops.

7.4 FACTORS AFFECTING CROP PRODUCTION

The factors affecting crop production are both natural and man-made. Natural factors include climate, diseases, and crop types, but the latter two can be controlled by technology.

7.4.1 Climatic Factors

A number of climatic factors influence crop production and yield; these were discussed in Chapter 2. Here, the focus is on the effect of temperature, radiation, and evaporative potential on crop yield. Since temperature is the easiest to visualize and is related to radiation and evaporative potential, the focus will be on how crop yield is related to temperature. Daily temperature range (minimum and maximum temperatures) can be used to compute solar radiation and relative humidity. Plant density is directly related to solar radiation and optimum mean temperatures are related to plant growth.

Potential crop yields are related to evaporative potential within a certain temperature range which, in turn, varies with crop and its variety. For example, a low temperature range is more suitable for rice, soybean, and potato. A mean temperature of 28 °C is good for maize provided the soil is well aerated, but the yield declines for temperatures greater than 21 °C if the soil is poorly aerated.

Following Hargreaves and Samani (1987), FAO (1978), and Lorenz and Maynard (1988), crops can be divided into five groups for optimum and operative temperature ranges, as shown in Table 7.2.

It may be noted that some crops are included in more than one group and different crops have different temperature requirements. Similarly, different varieties of the same crop may have different temperature requirements, which may be significantly modified by relative humidity. Warm but not sunny hot days, but not above 25 °C and not below 4 °C, are preferable for bush berries, blackberries, dewberries, loganberries, youngberries, and raspberries during the period from leaf to harvest. Apples and apricots do well under a maximum temperature of 30 °C or less and cherries under 20 °C during fruit production. For almonds, figs, peaches, pears, and prunes the temperature should not exceed 30 °C. For citrus crops, the temperature can range from 5 °C to 38 °C. The crop growth rate is strongly correlated with mean temperature, as shown in Figure 7.6, which indicates that after the optimum temperature range is attained, the relative growth rate levels off with increasing temperature (after Hargreaves and Merkley, 1998).

7.4.2 Soil Fertility and Fertilizers

The economic viability of irrigation projects may depend on the proficiency of fertilizer management as optimal application of fertilizers in conjunction with irrigation has the potential to double crop yield. Doorenbos and Kassam (1979) provided ranges of fertilizer requirements for 26 crops. The nitrogen (N) requirement can be 0–40 kg/ha for leguminous crops and 40–300 kg/ha for non-leguminous crops. Similarly, the phosphorus (K) requirement can be 15–110 kg/ha for leguminous crops and 24–480 kg/ha for non-leguminous crops. Based on field experiments, crop yield (Y) has been observed to be related to fertilizer application (F) as a quadratic function:

$$Y = a + bF + cF^2, \tag{7.1}$$

where a, b, and c are parameters. If no fertilizer is applied (i.e., $F = 0$), then $a = Y$ (yield without fertilizer). Parameter b is the initial slope of the yield curve (increase in Y divided by the increase in F for the first 25–50 kg/ha of F applied). Parameter c is evaluated from the amount of F needed when Y reaches the maximum.

The fertilizer in eq. (7.1) can be N, P, or K. Often, interaction of fertilizer and moisture content is more useful. Figure 7.7 shows isoquants (contours of equal yield) of average cotton lint production in kg/ha in the San Joaquin Valley, California (after Hargreaves and Merkley, 1998). The figure depicts the interaction of moisture w and N; this kind of interaction is typical for many crops. It shows that the optimum average application of

Table 7.2 *Division of crops based on optimum and operative temperature ranges*

Crop group	Optimum temperature range (°C)	Operative temperature range (°C)	Crops
I	15–20	5–30	Arabica coffee, artichoke, asparagus, barley, beet, broccoli, Brussels sprouts, cabbage, carnations, carrots, cauliflower, celery, chard, chayote, chickpea, chrysanthemum, crucifers, cucumber, French bean, garlic, gladiola, grapes, green onion, lima bean, lentils, lettuce, linseed (flax), muskmelon, mustard, potato, rape, roses, rye, snap bean, southern pea, spinach, squash, strawberries, sugar beet, sunflower, sweet corn, sweet pepper, tomatoes, and wheat
II	25–30	10–35	Avocadoes, banana, cassava, castor bean, cocoa, coconut, cotton, cowpea, eggplant, fig, French bean, grape, greater yam, groundnut, hot pepper, hyacinth bean, kenaf, mango, okra, oil palm, olive, para rubber, rice, robusta coffee, roselle, safflower, sesame, sunflower, sweet potato, tobacco, tomato, watermelon, and white yarn
III	30–35	15–45	Corn (maize), hungry rice, millet, sorghum, and sugarcane
IV	20–30	10–35	Maize, millet, and sorghum
V	25–35	10–45	Pineapple and sisal

After Hargreaves and Samani (1987), FAO (1978), and Lorenz and Maynard (1988).

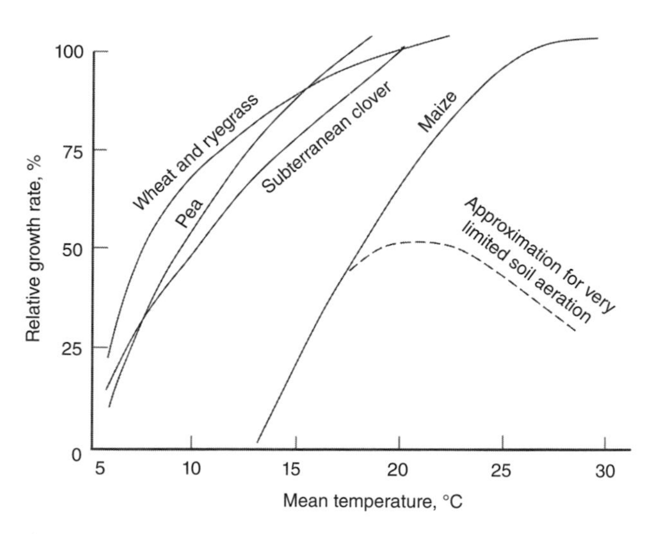

Figure 7.6 The effect of temperature on crop growth. (After Hargreaves and Merkley, 1998. Copyright © 1998 by Water Resource Publications, Highlands Ranch, Colorado. Reprinted with permission from Water Resource Publications.)

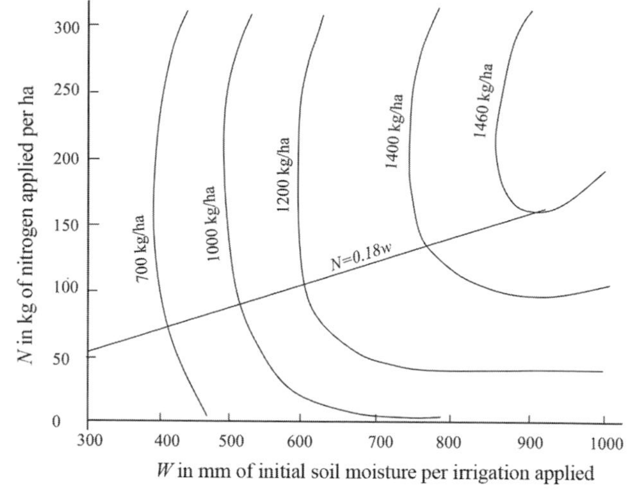

Figure 7.7 Isoquants of cotton lint production in kg/ha. (After Hargreaves and Merkley, 1998. Copyright © 1998 by Water Resource Publications, Highlands Ranch, Colorado. Reprinted with permission from Water Resource Publications.)

N in kg/ha is about 0.18 times the moisture availability in millimeters per crop season.

When crop yield is plotted as a function of nitrogen application and evapotranspiration, the graph looks similar to Figure 7.7, as shown in Figure 7.8, which is based on field trials using N (kg/ha) (after Hargreaves and Merkley, 1998). It is seen that N is about 0.32 times seasonal evapotranspiration (mm). This graph is typical for other crops, but the value, 0.32 here, will be different for different crops. Hargreaves and Merkley (1998) reported 0.12 for wheat, 0.18 for cotton, and 0.18 for sugar beets.

If two fertilizers are applied their interactions need to be considered. The average interaction of nitrogen and P_2O_5 on dryland maize production is shown in Figure 7.9. It is seen that $P_2O_5 = 15 + 0.20$ N and N = 150 kg/ha leads to the optimum interaction on yield and profit (after Hargreaves and Merkley, 1998).

7.4.3 Water Availability

Irrigation applies water to soil, and plants use this water through evapotranspiration (ET). Crop growth and production have been found to be approximately proportional to ET (the amount of evapotranspiration). Using yield data for various crops Hargreaves (1975) plotted relative yield

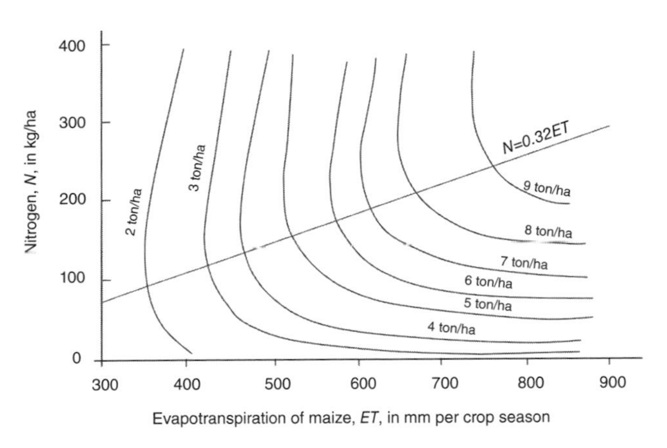

Figure 7.8 Isoquants of maze production. (After Hargreaves and Merkley, 1998. Copyright © 1998 by Water Resource Publications, Highlands Ranch, Colorado. Reprinted with permission from Water Resource Publications.)

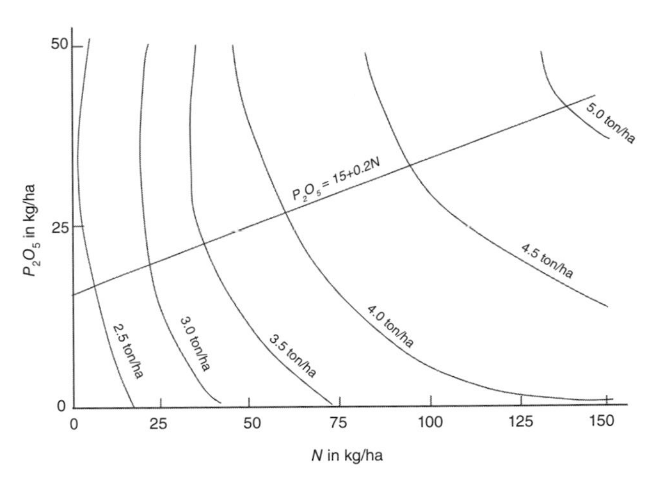

Figure 7.9 Isoquants from eight trials of dryland maize production. (After Hargreaves and Merkley, 1998. Copyright © 1998 by Water Resource Publications, Highlands Ranch, Colorado. Reprinted with permission from Water Resource Publications.)

(Y; $Y = 1.0$ for maximum yield) and water required (X) for crop production ($X = 100$ for maximum yield of each crop) and found

$$Y = 0.8X + 1.3X^2 - 1.1X^3. \tag{7.2}$$

The data used for eq. (7.2) covered the range of X from 0.30 to 1.20.

When actual ET is below maximum, the crop yield will be below maximum. Doorenbos and Kassam (1979) estimated crop yield as a function of water using the Stewart equation:

$$1 - \frac{Y_a}{Y_m} = k_y\left(1 - \frac{ET_a}{ET_m}\right), \tag{7.3}$$

where k_y is the coefficient or yield response factor for a crop, Y_a is the actual harvested yield, Y_m is the maximum potential yield, ET_a is the actual ET, and ET_m is the maximum ET. They developed values of k_y for 23 crops for the total growing period and various stages of growth. Table 7.3 gives values of k_y for these crops (after Doorenbos and Kassam, 1979). The values here should be taken as a guide, reflecting the relative significance of water during different crop growth stages. The values indicate that the flowering period is the most critical period for most crops, so irrigation should be adequate. To illustrate, if actual ET is 10% less than the maximum ET, then the actual yield will be 15% less than the maximum yield. Also, the moisture deficiency in one period conditions the plant such that the deficiency in the next period is less effective. However, there are also other factors that influence crop yield.

The calculation procedure to determine actual harvest yield is:

1. Determine the maximum potential yield (Y_m) of a specific crop variety, based on the assumption that water and nutrients are non-limiting and pests and diseases are well controlled.
2. Estimate the maximum ET (ET_m) as determined by multiplying the crop coefficient and reference evapotranspiration. The crop water requirement is fully satisfied with this condition. The calculation of maximum ET will be discussed in Chapter 15.
3. Calculate actual ET (ET_a) under specific conditions dependent on water availability. The calculation of actual ET will be discussed in Chapter 15.
4. Determine actual yield (Y_a) using eq. (7.3) with proper yield response factor k_y for the total growing period or over different growing stages.

Example 7.1: A farmer wants to apply deficit irrigation in a cotton field. The cotton yield in the past 10 years with sufficient irrigation is estimated to be 1200 kg/ha. What will be the cotton yield if the farmer uses 75%, 80%, and 90% ET replacement?

Solution: For 75% ET replacement, $\frac{ET_a}{ET_m} = 75\%$. Given $Y_m = 1200$ kg/ha, k_y for cotton from Table 7.3 is 0.85; then, using eq. 7.3, one gets

$$1 - \frac{Y_a}{1200\ \text{kg/ha}} = (0.85)(1 - 75\%).$$

The yield $Y_a = 945$ kg/ha.

When using 80% ET replacement, the actual yield is 996 kg/ha; when using 90% ET replacement, the actual yield is 1098 kg/ha.

Table 7.3 *Yield response factor (k_y)*

Crop	Vegetative period Early (la)	Vegetative period Late (lb)	Vegetative period Total	Flowering period (2)	Yield formation (3)	Ripening (4)	Total growing period
Alfalfa			0.7–1.1				0.7–1.1
Banana							1.2–1.35
Bean			0.2	1.1	0.75	0.2	1.15
Cabbage	0.2				0.45	0.6	0.95
Citrus							0.8–1.1
Cotton			0.2	0.5		0.25	0.85
Grape							0.85
Groundnut			0.2	0.8	0.6	0.2	0.7
Maize			0.4	1.5	0.5	0.2	1.25
Onion			0.45		0.8	0.3	1.1
Pea	0.2			0.9	0.7	0.2	1.15
Pepper							1.1
Potato	0.45	0.8			0.7	0.2	1.1
Safflower		0.3		0.55	0.6		0.8
Sorghum			0.2	0.55	0.45	0.2	0.9
Soybean			0.2	0.8	1.0		0.85
Sugar beet							
Beet							0.6–1.0
Sugar							0.7–1.1
Sugarcane			0.75		0.5	0.1	1.2
Sunflower	0.25	0.5		1.0	0.8		0.95
Tobacco	0.2	1.0			0.5		0.9
Tomato			0.4	1.1	0.8	0.4	1.05
Watermelon	0.45	0.7		0.8	0.8	0.3	1.1
Wheat							
Winter			0.2	0.6	0.5		1.0
Spring			0.2	0.65	0.55		1.15

After Doorenbos and Kassam, 1979.

Example 7.2: A corn field is located in Houston, Texas. The yield in the past 10 years with sufficient irrigation is estimated to be 6900 kg/ha. The planting day is April 20. The growing period, maximum crop ET, and yield response factor are given below. If the irrigation efficiency is 85%, what will be the total irrigation water requirement to obtain the maximum crop yield? If the farmer wants to apply deficit irrigation in this field, he has two choices. First, he wants to apply 90% ET replacement for the whole growing season. Second, he wants to apply 80% ET replacement in the initial/mid-/late season and 100% ET replacement in the crop development period. What will be the crop yield and total irrigation water demand for the two choices?

Growth period	Number of days	ET_m (mm/day)	k_y
Initial	15	2.7	0.4
Crop development	30	5.3	1.5
Mid-season	35	6.8	0.5
Late season	25	5.2	0.2

Solution: The total ET_m for the growing season is:

$$ET_m = 2.7 \times 15 + 5.3 \times 30 + 6.8 \times 35 + 5.2 \times 25 = 567.5 \text{ mm}.$$

Given the irrigation efficiency is 85%, the total irrigation water demand to obtain the maximum crop yield is $567.5 \text{ mm}/(85\%) = 667.6 \text{ mm}$.

For the first choice:

Total irrigation water demand is $667.6 \text{ mm} \times 90\% = 600.9 \text{ mm}$.

The total yield is

$$Y_a = Y_m \times [1 - (0.4 + 1.5 + 0.5 + 0.2) \times (1 - 0.9)] = 6900 \times (1 - 0.26) = 5106 \text{ kg/ha}.$$

For the second choice:

Total irrigation water demand is

$$[(2.7 \times 15 + 6.8 \times 35 + 5.2 \times 25)(0.8)] + 5.3 \times 30 = 485.8 \text{ mm}.$$

Given irrigation efficiency is 85%, the total irrigation water demand is $485.8 \text{ mm}/(85\%) = 571.5 \text{ mm}$.

The total yield is

$$Y_a = Y_m \times [1 - (0.4 + 0.5 + 0.2) \times (1 - 0.8)] = 6900 \times (1 - 0.22) = 5382 \text{ kg/ha}.$$

7.4.4 Soil Aeration and Drainage

The degree of aeration or the availability of oxygen to plant roots has a significant influence on plant growth. Respiration is vital for plant roots to extract water and nutrients. Well-aerated plants may take up 2–4 times more water than poorly aerated plants. Poor aeration leads to a number of undesirable consequences: (1) it reduces transpiration; (2) anaerobic respiration causes incomplete oxidation of organic matter and may form toxic products for plants; (3) it makes plants susceptible to fungi and other organisms; (4) NO_3 may be lost through denitrification; (5) it reduces root permeability for water; and (6) it reduces shoot growth and crop yield.

Respiration allows the conversion of carbohydrates to H_2O and CO_2 and releases energy which is used for plant growth. With increasing temperature, the need for respiration and energy from respiration increase exponentially. If the soil is waterlogged and the temperature is high even for a day, the crop production may be severely impacted. As seen from Figure 7.6, an optimum temperature for maize is about 28 °C for well-aerated conditions and is about 21 °C for poorly aerated soil. The crop growth decreases rapidly if temperatures are higher than these values. The tolerance to poor aeration varies with temperature, daily temperature range, plant species, and the variety within species.

7.4.5 Plant Density, Spacing, and Leaf Area Index

Plant spacing depends on cultivation and cultural practices, such as spacing between and within rows. It is important for plants to intercept enough light and grow adequately. Crop production depends on plant growth and crop yield increases with the increase in plant density. The optimum plant density depends on the crop variety and availability of resources. The plant density may be double or more in irrigated agriculture than in dryland farming. The amount of solar energy received by the plant influences the leaf area index (LAI), which is the ratio of the total leaf area of plants to the area of soil producing the plants. Figure 7.10 shows the relative yield of dry matter as a function of LAI and incident solar radiation. The maximum dry matter production was achieved when LAI was about 2.6 plus 0.0007 solar radiation (Hargreaves and Merkley, 1998).

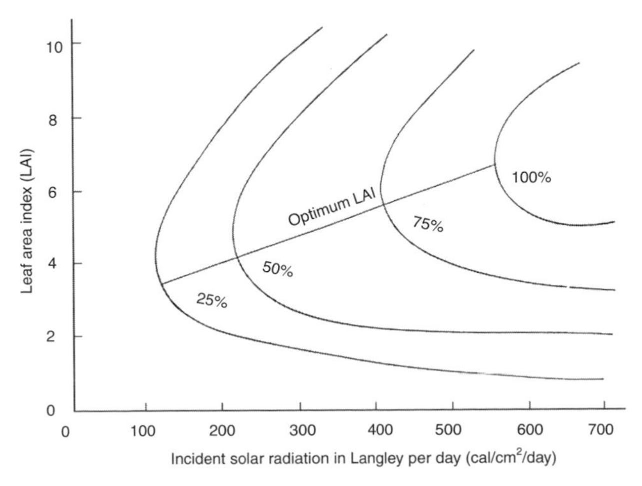

Figure 7.10 Relative yield of dry matter in isoquants. (After Hargreaves and Merkley, 1998). Copyright © 1998 by Water Resource Publications, Highlands Ranch, Colorado. Reprinted with permission from Water Resource Publications.)

7.4.6 Crop Variety

Crop production varies greatly with the variety of crops for given soil, nutrients, climate, cultural practices, and irrigation. Jones and Kiniry (1986) reported genetic coefficients for crop yields predicted by the CERES–maize nitrogen model and the genetic coefficients, as shown in Table 7.4. The field yield data representing 15 cultivars and 14 locations showed that grain yields ranged from 3475 to 17,106 kg/ha (51 datasets).

Table 7.4 *Selected genetic coefficients by cultivar and region used in the CERES–maize nitrogen model*

Cultivar name	Region	P1	P2	P5	G2	G3
B56 X OH 43	1	162	0.80	685	784	6.90
B60 X R 71	1	172	0.80	685	710	7.70
B59 X C103	1	172	0.80	685	825	10.15
PIO 3382	2	200	0.70	800	650	8.50
PIO 3901	2	215	0.76	600	560	9.00
PIO 3780	2	200	0.76	685	600	9.60
PIO 3720	2	180	0.80	685	825	10.00
PIO 511A	3	220	0.30	685	645	10.50
PIO 3183	3	260	0.50	750	600	8.50
W64A X W117	3	245	0.00	685	825	8.00
B14 X OH43	3	265	0.80	665	780	6.90
B8 X 153R	3	218	0.30	760	595	8.80
PIO 3147	4	255	0.76	685	834	10.00
PV82S	4	260	0.50	750	600	8.50
B73 X MO17	4	220	0.52	880	730	10.00
B56 X C131A	4	318	0.50	700	805	6.40
MCCYRDY 67–14	4	265	0.30	825	825	9.80
H 610	5	340	0.52	900	520	6.50
PIO X 304C	5	360	0.52	900	550	5.60

Regions: region 1: Northern United States; region 2: Northern Nebraska, Iowa, Illinois, and Indiana; region 3: Southern Nebraska, South Iowa, South Illinois, and South Indiana; region 4: Central Missouri, Kansas to North Carolina southward; region 5: Tropical.
Genetic coefficients: P1: growing degree days (based on a minimum of 8 °C) from seedling emergence to the end of the juvenile phase (*d* °C); P2: photoperiod sensitivity coefficient (1/h); P5: growing degree days (based on a minimum of 8 °C) from silking to physiological maturity (*d* °C); G2: potential kernel number (kernels/plant); and G3: potential kernel growth rate (mg/kernel/day)
Adapted from Jones and Kiniry, 1986. Copyright © 1986 by Texas A&M University Press, College Station, Texas. Reprint with permission from Texas A&M University Press.

Hargreaves and Merkley (1998) also reported that the predicted grain yields and biomass production using the CERES–maize nitrogen model with 14 cultivars under rainfed condition was 1995–5435 kg/ha and 10,753–12,190 kg/ha, respectively, which shows that yields of grains varied within a range of 2.7 times the lowest predicted. This shows that the best cultivar for the available resources should be used.

QUESTIONS

Q.7.1 Demarcate zones of different dominant crops on a Texas map. Indicate on the map where College Station is located.

Q.7.2 Using Figures 7.8 and 7.9, calculate how much nitrogen and P_2O_5 are required to maintain the maximum dryland maize production if the ET of maize is 550 mm per season.

Q.7.3 A farmer wants to apply deficit irrigation in his field. The crop type is winter wheat. The yield in the past 10 years with sufficient irrigation is estimated to be 2100 kg/ha. What will be the yield if the farmer uses 80%, 85%, and 90% ET replacement?

Q.7.4 Wheat was grown for grain planted on March 1 and was harvested on July 31. The growing period, maximum crop ET, and yield response factor are given below. The yield in the past 10 years with sufficient irrigation is estimated to be 2100 kg/ha. If the irrigation efficiency is 90%, what will be the total irrigation water demand to obtain the maximum crop yield? If the farmer wants to apply deficit irrigation during the initial, mid-, and late season with 80% ET replacement, how much water can he save, and what will be the yield at the end of the season?

Growth period	Number of days	ET_m (mm/day)	k_y
Initial	17	1.2	0.2
Crop development	50	4.1	0.7
Mid-season	43	6.5	0.6
Late season	43	1.2	–

REFERENCES

Castellazzi, M. Z, Wood, G. A., Burgess, P. J., et al. 2008. A systematic representation of crop rotations. *Agricultural Systems*, 97(1–2): 26–33.

Doorenbos, J., and Kassam, A. H. (1979). Yield response to water. FAO irrigation and drainage paper 33.

FAO (1978). *Report on the Agroecological Zones Project. Vol. 1. Methodology and Results for Africa*. Rome: FAO.

Hargreaves, G. H. (1975). Moisture availability and crop production. *Transactions of the ASAE*, 18(5): 980–984.

Hargreaves, G. H., and Merkley, G. P. (1998). *Irrigation Fundamentals: An Applied Technology Text for Teaching Irrigation at the Intermediate Level*. Highlands Ranch, CO: Water Resource Publications.

Hargreaves, G. H., and Samani, Z. A. (1987). Simplified irrigation scheduling and crop selection for El Salvador. *Journal of Irrigation and Drainage Engineering*, 113(2): 224–232.

Jones, C. A., and Kiniry, J. R. (1986). *CERES Maize: A Simulation Model for Maize Growth and Development*. College Station, TX: Texas A&M University Press.

Lorenz, O. A., and Maynard, D. N. (1988). *Knotts Handbook for Vegetable Growers*. New York: Wiley.

US Energy Information Administration (2020). Monthly energy review. www.eia.gov/energyexplained/us-energy-facts.

Part II Principles of Hydraulics

8 Channel Design

Notation

A	flow cross-sectional area (m^2, ft^2)
C	Chezy's roughness coefficient
d	depth of flow (m, ft)
D	hydraulic depth (m, ft)
E	specific energy head (m, ft)
f	friction factor
F	tractive force (lb/ft^2, N/m^2)
FB	freeboard (m, ft)
F_f	friction force on the surface of contact between water and channel (N, lbf)
Fr	Froude number
f_s	silt factor
g	acceleration due to gravity (9.81 m/s^2, 32.2 ft/s^2)
H	total water head (m, ft)
h_f	energy head loss (m, ft)
K	conveyance factor
K_s	shear stress reduction factor
L	length of flow (m, ft)
P	wetted perimeter (m, ft)
P_1, P_2	pressure force at sections 1 and 2 (N, lbf)
q	specific discharge, the discharge per unit width (L^2/T)
Q	discharge of the cross-sectional area (m^3/s, ft^3/s)
R	hydraulic radius (m, ft)
Re	Reynolds number, dimensionless
S	the slope of energy line, equal to the bed slope for uniform flow (L/L)
S_c	critical slope (L/L)
S_f	energy slope for gradually varied flow (L/L)
S_o	slope of the bottom (L/L)
S_w	slope of water surface (L/L)
T	top width (m, ft)
u	cross-sectional mean velocity (for critical flow, it is denoted as u_c) (m/s, ft/s)
W	weight of water between sections 1 and 2 (N, lbf)
y	depth of flow section, approximately equal to d (m, ft)
y_c	critical depth (m, ft)
y_o	normal depth (m, ft)
z	side slope of the flow cross-sectional area (horizontal/vertical)
Z_1, Z_2	elevation head at section 1 and 2 (m, ft)
Z_n	section factor (for critical flow, it is denoted as Z_c) ($L^{5/2}$)
α	energy coefficient or Coriolis coefficient
β	momentum coefficient or Boussinesq coefficient
γ	specific weight of water, or unit weight (9.807 kN/m^3, 62.43 lb/ft^3)
θ	slope of the channel (degree)
τ	boundary shear stress (lb/ft^2, N/m^2)
τ_o	maximum shear stress (lb/ft^2, N/m^2)
τ_s	shear stress for the sides of channel (lb/ft^2, kg/m^2)
υ	kinematic viscosity (dynamic viscosity/mass density, m^2/s, ft^2/s)
ϕ	the angle of repose of the material (degree)

8.1 INTRODUCTION

Irrigation systems receive water either through a pipeline or a channel, which in turn receives water from a source. Depending on the size of an irrigation system, which depends on the size of the farm, there may be a network of pipelines or channels. A channel usually has a slope in the direction of flow and the water flows under gravity. The channel can be natural, such as a river, stream, bayou, or brook, or it can be artificial, such as a canal, ditch, tunnel, chute, flume, culvert, or aqueduct. For irrigation, canals or farm channels are constructed in alluvial or other granular material. These channels are erodible unless lined, and their design requires consideration of the stability of their geometry so that flow does not cause unacceptable scouring.

Channels may be lined with concrete, asphalt, exposed or covered membranes, or soil sealants. The primary purpose of lining is to reduce seepage losses through the channel bed and side walls, although lining also reduces maintenance costs and channel size. It is, however, expensive and requires periodic maintenance. Unlined channels are cheaper and easier to construct and relocate.

Channels are a vital part of irrigation systems. They are the link between the source of water and the irrigation field. Channels used in irrigation systems can be either erodible or non-erodible, or earthen or lined. Flow in channels is governed by the principles of hydraulics. This chapter discusses rudimentary aspects of hydraulics and the design of open channels, which depends on the design requirements and characteristics of geometry, flow, and landscape.

8.2 GEOMETRIC ELEMENTS

A channel is called prismatic if its cross-section and slope remain the same throughout its length. In reality, an unlined channel may be prismatic over a certain length but not for its entire length. The cross-section of a channel at any point is the channel section perpendicular to the direction of flow. For example, consider a rectangular channel. Then, at any point along the length, the cross-section area will be the area defined by the product of width and depth of flow at that point in the channel. For defining various geometric elements, a trapezoidal section and longitudinal profile of a channel are sketched, as shown in Figure 8.1. The symbols shown in this figure will be used throughout the chapter.

Depth of flow: This is the vertical distance from the channel bottom to the free surface, denoted d.

Depth of flow section: If the slope of the channel is θ (degrees) then the depth of flow section, denoted y, is equal to $d\cos\theta$ and is the depth perpendicular to the direction of flow. Since the slope is usually small, y is approximately equal to d.

Top width: The top width is the width at the water surface and is denoted by T.

Flow cross-sectional area: This is the cross-sectional area of flow perpendicular to the direction of flow, and is denoted A. It is also sometimes referred to as flow area. For a rectangular cross-section, $A = bd$, where b is the base width.

Wetted perimeter: For a flow cross-section area, it is defined as the length of the channel surface that is in contact with water and is denoted by P. For a rectangular cross-section, $P = b + 2d$.

Hydraulic depth: This is defined as the ratio of flow cross-sectional area and top width, and is denoted D; that is, $D = A/T$.

Hydraulic radius: This is defined by the ratio of flow cross-sectional area and wetted perimeter and is denoted R; that is, $R = A/P$.

Section factor: For uniform flow, this is defined as the product of flow cross-sectional area and two-thirds power of hydraulic radius, and is denoted Z_n; that is, $Z_n = AR^{\frac{2}{3}}$. For critical flow, it is denoted as Z_c and is defined as the product of flow cross-sectional area and the square root of hydraulic depth; that is, $Z_c = A\sqrt{D}$. The dimension of the section factor is $[L^{5/2}]$.

The geometric elements of different channel sections (Chow, 1959) are given in the Supplementary Information Appendix at the end of the chapter.

Example 8.1: Consider a trapezoidal channel with flow depth of 1.5 m, base width of 5 m, and the side slope as 3:1 horizontal to vertical. The bed slope is 10°. Compute the depth of flow section, top width, flow cross-sectional area, wetted perimeter, hydraulic radius, hydraulic depth, and section factor for uniform and critical flows.

Solution: Depth of flow section:

$$y = d\cos\theta = (1.5\text{ m}) \times \cos 10° = 1.48\text{ m}.$$

Top width:

$$T = 5\text{ m} + 2 \times (1.5\text{ m}) \times \frac{3}{1} = 5\text{ m} + 2 \times 4.5\text{ m} = 14\text{ m}.$$

Flow cross-sectional area:

$$A = by = 5\text{ m} \times 1.48\text{ m} = 7.40\text{ m}^2.$$

Wetted perimeter:

$$P = b + 2y = 5\text{ m} + 2 \times (1.48\text{ m}) = 7.96\text{ m}.$$

Hydraulic radius:

$$R = \frac{A}{P} = \frac{7.40\text{ m}^2}{7.96\text{ m}} = 0.93\text{ m}.$$

Hydraulic depth:

$$D = \frac{A}{T} = \frac{7.40\text{ m}^2}{14\text{ m}} = 0.53\text{ m}.$$

Sectional factor for uniform flow:

$$Z_n = AR^{\frac{2}{3}} = 7.40 \times 0.93^{\frac{2}{3}} = 7.05.$$

Sectional factor for critical flow:

$$Z_c = A\sqrt{D} = 7.40 \times \sqrt{0.53} = 5.39.$$

8.3 FLOW CHARACTERISTICS

Flow in an open channel has a number of characteristics that need to be understood before designing the channel, including the type of flow, state of flow, variation in velocity in the flow cross-section, critical flow, mass balance, energy balance, momentum balance, hydraulically most efficient section, and uniform flow. Each of these characteristics is discussed in the following subsections.

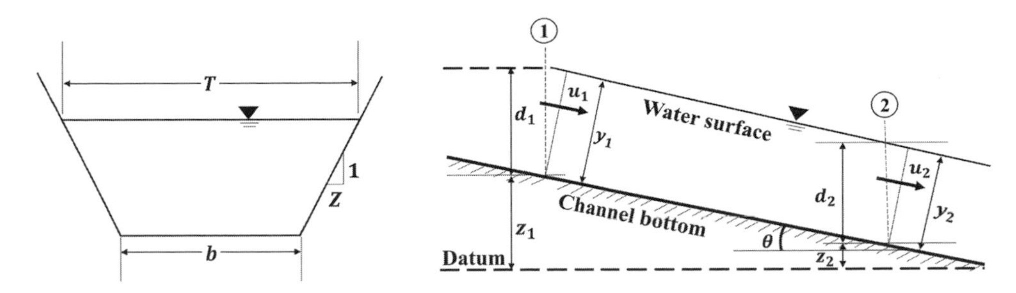

Figure 8.1 Channel section and longitudinal profile in open-channel flow.

8.3.1 Types of Flow

In open channels, flow often varies with space and time. The variation in flow can be characterized by the variation in depth, velocity, or discharge. Since it is easier to visualize or measure the flow depth, flow can be typified by the change in depth. If at a fixed time the flow depth varies along the length, the flow is called nonuniform flow; if it is constant, the flow is called uniform flow. On the other hand, if at a fixed location the flow depth varies in time, the flow is called unsteady flow, and if it remains constant, the flow is called steady flow. If the flow depth varies abruptly over a short distance the flow is called rapidly varied flow. In irrigation channels the flow is generally steady gradually varied, but flow over a weir can be steady rapidly varied. Unsteady gradually varied flow often occurs in rivers, especially during floods, and its treatment is beyond the scope of this chapter.

8.3.2 State of Flow

The state of flow is characterized by the relative effects of inertial force, viscous force, and gravitational force indexed by two dimensionless numbers: the Reynolds number and the Froude number. The Reynolds number (Re) is defined by the ratio of inertial force and viscous force as

$$\mathrm{Re} = \frac{uR}{\upsilon}, \tag{8.1}$$

where u is the cross-sectional mean velocity (m/s, ft/s), R is the hydraulic radius (m, ft), and υ is the kinematic viscosity (dynamic viscosity/mass density; m^2/s, ft^2/s). For irrigation channels, R is often replaced by flow depth. Re is used to classify flow into laminar, transitional, and turbulent flows. If Re is less than 500 the flow is referred to as laminar, in which case the viscous force is dominant. This type of flow occurs in laboratory channels and pavements during the early stages of rainfall. The flow is considered transient if Re is 500–2000, and is turbulent if Re is greater than 2000. In irrigation channels flow is almost always turbulent.

The Froude number (Fr) is defined by the ratio of inertial force and gravitational force:

$$\mathrm{Fr} = \frac{u}{\sqrt{gD}}, \tag{8.2}$$

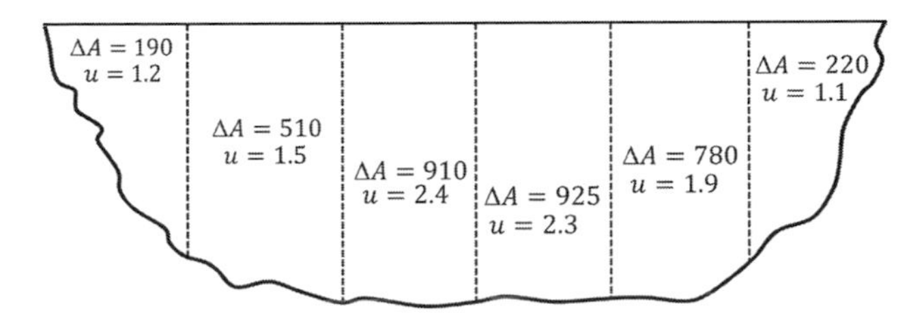

Figure 8.2 Discharge measurement at a channel section (A is in ft^2 and u is in ft/s).

where D is the hydraulic depth (m, ft) and g is acceleration due to gravity (m/s^2, ft/s^2). If Fr is less than 1, the flow is subcritical or tranquil, which is the most dominant flow; in this case, flow velocity is low. If Fr is greater than 1 the flow is called supercritical flow; in this case, velocity is high. This may occur over short distances. If Fr is equal to 1, the flow is called critical, which occurs transitionally, such as during a fall.

It may be interesting to note that the Darcy–Weisbach friction formula for pipes can be applied to uniform subcritical flow. In laminar flow the friction factor (f), expressed by $f = k/\mathrm{Re}$, can be applied to subcritical flow in both rough and smooth channels. Here, k is a factor that depends on the channel shape and roughness. The value of k is greater for rough channels than for smooth channels. In turbulent regions of rough channels, the friction formulas derived for pipe flows are not quite valid because the friction factor depends on channel shape, roughness, and Re. If the flow is supercritical turbulent, the friction factor increases with increasing Fr.

8.3.3 Velocity Distribution in Flow Cross-Section

In irrigation channels, the mean velocity is used to compute the kinetic energy head and the momentum, which are lower than when computed using the actual velocity distribution, which in a flow cross-section varies from wall to wall, as shown in Figure 8.2. The kinetic energy head may therefore need to be corrected as: $\alpha\left(\frac{u^2}{2g}\right)$ and the momentum may be corrected as $\frac{\beta\gamma Q u}{g}$, where α is the energy coefficient or Coriolis coefficient, β is the momentum coefficient or Boussinesq coefficient, γ is the unit weight or specific weight of water, and Q is the discharge. The energy coefficient α can be expressed as

$$\alpha = \frac{\sum_{i=1}^{m} u_i^3 \Delta A_i}{(\overline{u})^3 A}, \tag{8.3}$$

and the momentum coefficient β can be expressed as

$$\beta = \frac{\sum_{i=1}^{m} u_i^2 \Delta A_i}{(\overline{u})^2 A}. \tag{8.4}$$

Depending on the type of channel, the value of α can vary from 1.1 to 2.0 and the value of β can vary from 1.03 to 1.33. For irrigation channels and flumes, α can be from 1.1 to 1.2 and β can be from 1.03 to 1.07.

Example 8.2: Figure 8.2 shows velocity and cross-sectional measurements. Compute the energy and momentum coefficients.

Solution:

$$\text{Mean velocity}(\overline{u}) = \frac{\sum_{i=1}^{6} A_i \times u_i}{\sum_{i=1}^{6} A_i} = \frac{7028.5}{3535} = 1.99\ \text{ft/s}(0.61\ \text{m/s}).$$

The energy coefficient α is

$$\alpha = \frac{\sum_{i=1}^{6} u_i^3 \Delta A_i}{(\overline{u})^3 A} = \frac{31,526.73\,\frac{\text{ft}^3}{\text{s}^3}\cdot \text{ft}^2}{\left(1.99\,\frac{\text{ft}}{\text{s}}\right)^3 (3535\ \text{ft}^2)} = 1.13.$$

The momentum coefficient β is

$$\beta = \frac{\sum_{i=1}^{6} u_i^2 \Delta A_i}{(\overline{u})^2 A} = \frac{14637.95\,\frac{\text{ft}^2}{\text{s}^2}\cdot \text{ft}^2}{\left(1.99\,\frac{\text{ft}}{\text{s}}\right)^2 (3535\ \text{ft}^2)} = 1.05.$$

Section	Area (ΔA) ft^2(m^2)	Velocity (u) ft/s (m/s)	$\Delta A \times u$	$\Delta A \times u^3$	$\Delta A \times u^2$
1	190 (17.7)	1.2 (0.37)	228.0 (6.55)	328.3 (0.90)	273.6 (2.42)
2	510 (47.4)	1.5 (0.46)	765.0 (21.80)	1,721.3 (4.61)	1,147.5 (10.03)
3	910 (84.5)	2.4 (0.73)	2184.0 (61.69)	12,579.8 (32.87)	5,241.6 (45.03)
4	925 (85.9)	2.3 (0.70)	2127.5 (60.13)	11,254.5 (29.46)	4,893.3 (42.09)
5	780 (72.5)	1.9 (0.58)	1482.0 (42.05)	5,350.0 (14.15)	2,815.8 (24.39)
6	220 (20.4)	1.1 (0.34)	242.0 (6.94)	292.8 (0.80)	266.2 (2.36)
Total	3,535 (328.4)	10.4 (3.2)	7,028.5 (199.16)	31,526.7 (82.79)	14,638.0 (126.32)

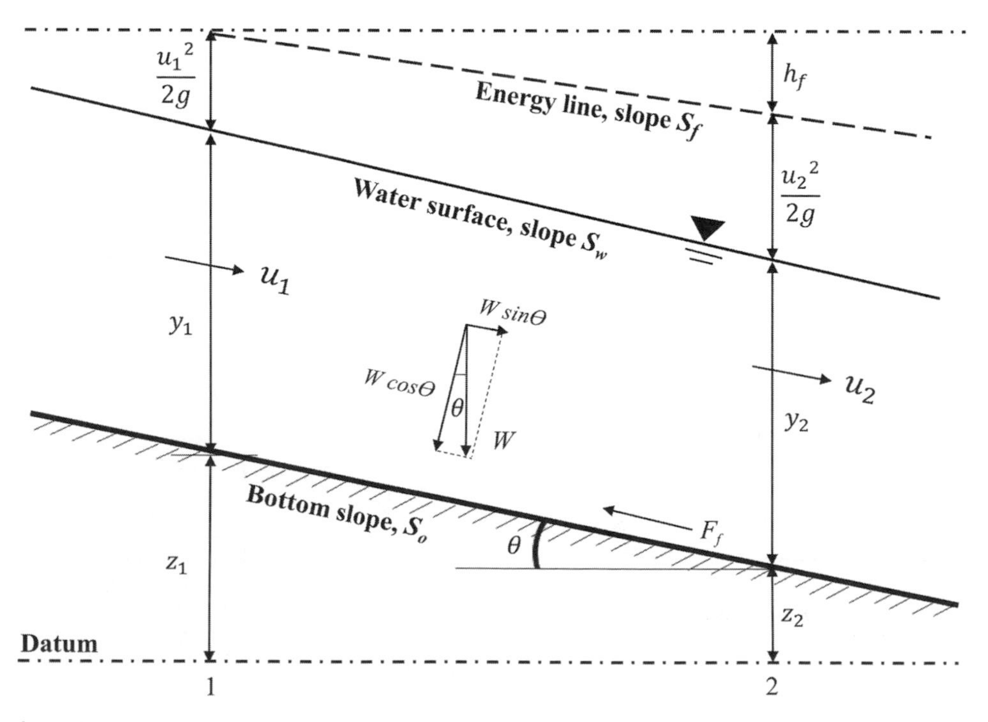

Figure 8.3 Various forms of energy in open-channel flow.

8.3.4 Mass Balance

The mass balance is an expression of the law of conservation of mass. Consider a channel divided into m different reaches. Let the cross-sectional area of an ith reach be denoted as A_i and flow velocity as u_i. Then, the discharge (Q) at the ith section is u_iA_i. If there is no lateral inflow or outflow along the channel, then one can write

$$u_1A_1 = u_2A_2 = \cdots = Q = uA, \tag{8.5}$$

where u is the average velocity and A is the average cross-sectional area.

8.3.5 Energy Balance

In irrigation the total flow energy is expressed by the total energy head, which is the sum of the elevation head (Z), pressure head (y), and kinetic energy head $\left(\frac{u^2}{2g}\right)$. Consider a channel reach whose upstream head is denoted by 1 and the downstream end by 2, as shown in Figure 8.3. Then, from the law of energy conservation, the total energy head at section 1 equals the energy head at section 2 plus the loss of energy during flow from section 1 to section 2 (h_f):

$$Z_1 + y_1 + \frac{u_1^2}{2g} = Z_2 + y_2 + \frac{u_2^2}{2g} + h_f. \tag{8.6}$$

Example 8.3: Compute the discharge per unit width of a broad-crested weir in a rectangular channel, as shown in Figure 8.4. Ignore the losses.

Solution: Since the losses are ignored, then $h_f = 0$ and eq. (8.6) can be written as:

$$Z_1 + y_1 + \frac{u_1^2}{2g} = Z_2 + y_2 + \frac{u_2^2}{2g},$$

$$0 + 12\text{ ft} + \frac{u_1^2}{2g} = 5\text{ ft} + 4\text{ ft} + \frac{u_2^2}{2g},$$

$$\frac{(u_2^2 - u_1^2)}{2 \times \left(32.2\frac{\text{ft}}{\text{s}^2}\right)} = 3\text{ ft},$$

$$3\text{ ft} \times 2 \times \left(32.2\frac{\text{ft}}{\text{s}^2}\right) - \left(u_2^2 - \left(10\frac{\text{ft}}{\text{s}}\right)^2\right),$$

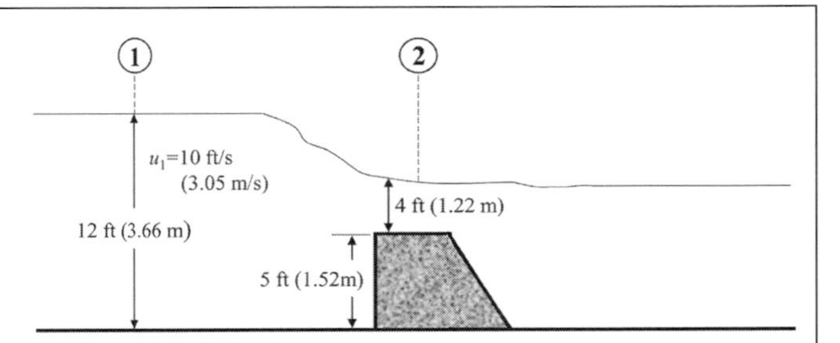

Figure 8.4 Application of the energy principle on a broad-crested weir.

$$u_2 = 17.1\frac{\text{ft}}{\text{s}}(5.2\text{ m/s}).$$

The discharge per unit width of the broad-crested weir in a rectangular channel can be estimated as $Q = A \times u_2 = (4\text{ ft} \times 1\text{ ft}) \times \left(17.1\frac{\text{ft}}{\text{s}}\right) = 68.4\frac{\text{ft}^3}{\text{s}}\,(1.94\text{ m}^3/\text{s})$.

8.3.6 Momentum Balance

Recalling Newton's second law of motion, the resultant external force applied to a body in any direction equals the rate of change of momentum of the body in that direction. When there is a change in momentum, it is transformed into an impulse force (force multiplied by time). Consider two sections, 1 and 2, as shown in Figure 8.5. Then, the rate of change in momentum between the two sections must equal the resultant force between the sections:

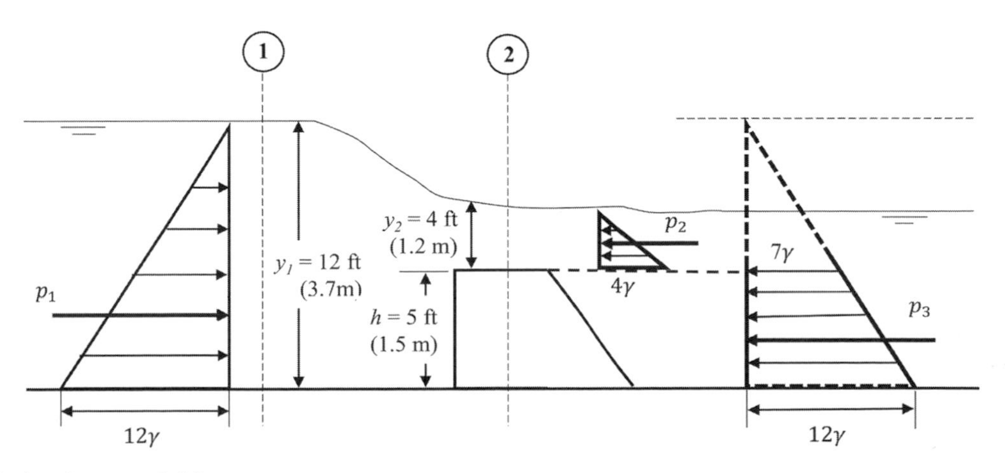

Figure 8.5 Application of moment principle.

$$\rho Q(\beta_2 u_2 - \beta_1 u_1) = \sum F, \tag{8.7}$$

where ΣF is the resultant force equal to the sum of resultant pressure force, weight of water, and friction force:

$$\sum F = P_1 - P_2 + W \sin\theta - F_f, \tag{8.8}$$

in which P_1 is the pressure force at section 1, P_2 is the pressure force at section 2, W is the weight of water between sections 1 and 2, and F_f is the friction force on the surface of contact between the water and the channel, which is sometimes neglected (Figure 8.3).

Example 8.4: Compute the discharge per unit width of a broad-crested weir in a rectangular channel as shown in Figure 8.5 using the momentum principle, given that $F_f = 0, \beta_1 = \beta_2 = 1$.

Solution: Compute the upstream pressure at section 1:

$$P_1 = \frac{1}{2}\gamma b y_1^2 = \frac{1}{2}(\gamma)\,(1\text{ ft})\,(12\text{ ft})^2 = 72\gamma.$$

Compute the pressure at section 2:

$$P_2 = \frac{1}{2}\gamma b y_2^2 = \frac{1}{2}(\gamma)\,(1\text{ ft})\,(4\text{ ft})^2 = 8\gamma.$$

Compute the water pressure on the weir at section 2:

$$P_3 = \frac{1}{2}\gamma h[(y_1 - h) + y_1] = \frac{1}{2}(\gamma)(5\text{ ft})(12 - 5 + 12) = 47.5\gamma.$$

Using the momentum equation:

$$\rho Q(\beta_2 u_2 - \beta_1 u_1) = P_1 - P_2 - P_3 - F_f,$$

$$\frac{\gamma}{g} Q(u_2 - u_1) = P_1 - P_2 - P_3 = (72 - 8 - 47.5)\gamma = 16.5\gamma,$$

$$Q = u_1 \times A_1 = u_2 \times A_2,$$

$$u_1 = \frac{Q}{A_1} = \frac{Q}{12 \times 1},$$

$$u_2 = \frac{Q}{A_2} = \frac{Q}{4 \times 1},$$

$$\frac{1}{g} Q\left(\frac{Q}{4} - \frac{Q}{12}\right) = 16.5,$$

$$Q = 56.5\frac{\text{ft}^3}{\text{s}}\,(1.6\text{ m}^3/\text{s}).$$

8.3.7 Critical Flow

The concept of specific energy is invoked to analyze critical flow.

8.3.7.1 Specific Energy

The specific energy is defined by the sum of pressure energy, which is the static energy and kinetic energy – that is, the energy head is measured from the channel bottom so the elevation head is 0 and the specific energy head is expressed as

$$E = y + \frac{u^2}{2g} = y + \frac{Q^2}{2gA^2}, \tag{8.9}$$

where the energy coefficient is assumed to equal 1. Figure 8.6 plots specific energy, static energy, and kinetic energy against flow depth. When the specific energy is minimum, as indicated by point 0, the flow becomes critical. Let the critical depth be denoted as y_c. The horizontal line passing through this point divides the flow into subcritical and supercritical. Supercritical flow occurs when $y < y_c$ – that is, flow velocity is high and depth is low; subcritical flow occurs when $y > y_c$ (high flow depth and low velocity). It is interesting to note that for a given specific energy there are two alternate depths: one corresponding to subcritical flow and the other corresponding to supercritical flow.

The depth at which critical flow occurs or the specific energy is minimum can be computed by differentiating the specific energy with respect to flow depth and equating the derivative to zero:

$$\frac{dE}{dy} = 1 - \frac{2Q^2}{2gA^3}\frac{dA}{dy} = 0. \tag{8.10}$$

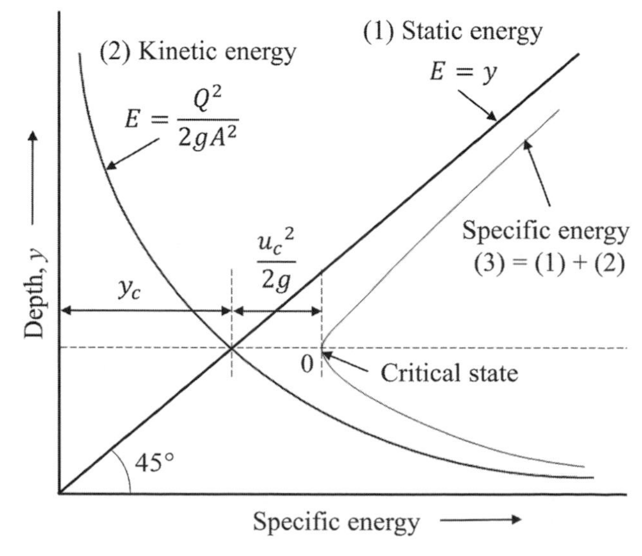

Figure 8.6 Specific energy diagram.

Using $dA/dy = T$ and $A/T = D$ in eq. (8.10), the result is

$$\frac{Q^2}{gA^2}\frac{1}{D} = 1. \tag{8.11}$$

Inserting $Q/A = u$ in eq. (8.11), one obtains

$$\frac{u^2}{gD} = 1 \quad \text{or} \quad \frac{u}{\sqrt{gD}} = 1 \quad \text{or} \quad \text{Fr} = 1. \tag{8.12}$$

Equation (8.12) prescribes the condition for critical flow at which the specific energy is minimum.

8.3.7.2 Computation of Critical Flow

Two cases are considered for critical flow. First, the critical depth is given and discharge is computed. Second, for a given discharge the critical depth is computed. Equation (8.11) can then be expressed as:

$$A\sqrt{D} = \frac{Q}{\sqrt{g}} \tag{8.13}$$

or

$$Z_c = \frac{Q}{\sqrt{g}}, \tag{8.14}$$

in which the energy equation is included. In the first case, the critical section factor is computed and then, using eq. (8.14), Q is computed. In the second case, the critical section factor is computed using eq. (8.14) and then critical flow is computed. Critical flow can be used for a control section. If the depth of flow in the design of a channel is found to be near or equal to critical flow, then the shape or slope of the channel must be changed. The slope at which the computed normal or uniform depth is critical is called the critical slope. Any slope less than the critical slope is called a mild or subcritical slope; any slope greater is called a steep or supercritical slope.

Example 8.5: Consider a rectangular channel 10 m wide having a flow velocity of 1.5 m/s and discharge of 8 m³/s. Compute the specific energy of water, critical depth of water, and critical velocity of flow.

Solution: Compute the specific energy:

$$A = \frac{Q}{u} = \frac{8\,\frac{\text{m}^3}{\text{s}}}{1.5\,\frac{\text{m}}{\text{s}}} = 5.33\ \text{m}^2 = by = (10\ \text{m})(y),$$

$$y = \frac{A}{b} = \frac{5.33\ \text{m}^2}{10\ \text{m}} = 0.533\ \text{m},$$

$$E = y + \frac{u^2}{2g} = 0.533\ \text{m} + \frac{\left(1.5\,\frac{\text{m}}{\text{s}}\right)^2}{2 \times (9.81\ \text{m/s}^2)} = 0.65\ \text{m}$$

Compute the critical depth:

$$Z_c = \frac{Q}{\sqrt{g}} = \frac{8}{\sqrt{9.81}} = 2.55,$$

$$A = b \times y_c = 10y_c,$$

$$D = \frac{A}{T} = \frac{10y_c}{10\ \text{m}} = y_c,$$

$$Z_c = A\sqrt{D} = 10y_c\sqrt{y_c} = 10(y_c)^{\frac{3}{2}} = 2.55\ \text{m},$$

$$y_c = 0.4\ \text{m}.$$

Compute the critical velocity:

$$u_c = \sqrt{gD} = \sqrt{gy_c} = \sqrt{9.81 \times 0.4} = 1.98\,\frac{\text{m}}{\text{s}}.$$

Example 8.6: Consider a trapezoidal channel with a bottom width of 3 m and side slope of 2:1 horizontal to vertical. The discharge in the channel is 20 m³/s. Compute the critical depth, critical velocity, and minimum specific energy.

Solution: Compute the critical depth:

$$Z_c = \frac{Q}{\sqrt{g}} = \frac{20}{\sqrt{9.81}} = 6.39,$$

$$A = b \times y_c + (2y_c)y_c = 3y_c + 2y_c^2,$$

$$T = b + 2 \times (2y_c) = 3 + 4y_c,$$

$$D = \frac{A}{T} = \frac{3y_c + 2y_c^2}{3 + 4y_c},$$

$$Z_c = A\sqrt{D} = \left(3y_c + 2y_c^2\right)\sqrt{\frac{3y_c + 2y_c^2}{3 + 4y_c}} = 6.39\ \text{m}.$$

By trial and error, $y_c = 1.252$ m.
Compute the critical velocity:

$$D = \frac{3 \times 1.252 + 2 \times 1.252^2}{3 + 4 \times 1.252} = 0.861\ \text{m},$$

$$u_c = \sqrt{gD} = \sqrt{g\frac{A}{T}} = \sqrt{9.81 \times 0.861} = 2.91\,\frac{\text{m}}{\text{s}}.$$

Compute the minimum specific energy:

$$E = y_c + \frac{u_c^2}{2g} = 1.252\ \text{m} + \frac{\left(2.91\,\frac{\text{m}}{\text{s}}\right)^2}{2 \times (9.81\ \text{m/s}^2)} = 1.68\ \text{m}.$$

8.3.8 Uniform Flow

The design of irrigation channels is often based on uniform flow, in which flow depth, flow cross-section area, velocity, and discharge do not vary from one section to another. Such a condition may occur when the force in the direction of flow is balanced by the resistance to flow and may be referred to as the kinematic condition, which seldom happens in nature. Uniform flow may occur only over short reaches of a channel.

Uniform flow is described by Chezy's equation:

$$u = C\sqrt{RS} \quad \text{or} \quad Q = CA\sqrt{RS}, \tag{8.15}$$

in which C is Chezy's roughness coefficient and S is the slope of energy line equal to the bed slope for uniform flow. Manning's equation is also used for uniform flow:

$$u = \frac{1}{n}R^{\frac{2}{3}}S^{\frac{1}{2}} \text{ in SI units or } u = \frac{1.486}{n}R^{\frac{2}{3}}S^{\frac{1}{2}} \text{ in British units,} \tag{8.16}$$

where n is Manning's roughness coefficient. Comparing eqs (8.15) and (8.16), it can be seen that

$$C = \frac{R^{1/6}}{n} \text{ (SI units) or } C = \frac{1.486}{n}R^{\frac{1}{6}} \text{ (British units).} \tag{8.17}$$

Equation (8.16) can also be written as

$$Q = \frac{1}{n}AR^{\frac{2}{3}}S^{\frac{1}{2}} \text{ in SI units or } Q = \frac{1.486}{n}AR^{\frac{2}{3}}S^{\frac{1}{2}} \text{ in British units,} \tag{8.18}$$

in which $Z_n = AR^{\frac{2}{3}}$ is the normal section factor for uniform flow.

Sometimes, discharge is expressed in terms of the conveyance factor K:

$$Q = K\sqrt{S}; \quad K = \frac{1}{n}AR^{\frac{2}{3}} \text{ or } K = \frac{1.486}{n}AR^{\frac{2}{3}}. \tag{8.19}$$

Equation (8.18) involves three variables: discharge or velocity, section factor or flow depth, and slope. If two variables are known, then the third variable can be computed. These three cases can be enumerated as follows:

1. Normal depth and slope are given and discharge is to be computed. In this case the section factor is computed and then discharge is computed using eq. (8.18).
2. Discharge and normal depth are known and slope is to be computed. This involves the direct application of eq. (8.18).
3. Discharge and slope are known and the normal depth is computed. Equation (8.18) can be written in terms of section factor as:

$$AR^{\frac{2}{3}} = \frac{Qn}{S^{\frac{1}{2}}} \text{ (SI units) or } AR^{\frac{2}{3}} = \frac{Qn}{1.486S^{\frac{1}{2}}} \text{ (British units).} \tag{8.20}$$

Then, the normal depth can be computed.

Example 8.7: Consider a trapezoidal channel having a normal depth of 1 m, base width of 4 m, and side slope of 2:1 horizontal to vertical. Discharge is 5 m^3/s. Take Manning's n as 0.035. The upstream section of the channel has an elevation of 100 m and the downstream section has an elevation of 90 m. How long is this channel?

Solution:

$$A = b \times y_n + (2y_n)y_n = 4 \times 1 + (2 \times 1) \times 1 = 6\text{ m}^2,$$

$$P = b + 2 \times \sqrt{1^2 + 2^2} = 4 + 2\sqrt{5} = 8.47\text{ m},$$

$$R = \frac{A}{P} = \frac{6\text{ m}^2}{8.47\text{ m}} = 0.708\text{ m},$$

$$AR^{\frac{2}{3}} = \frac{Qn}{S^{\frac{1}{2}}},$$

$$S = \left(\frac{Qn}{AR^{\frac{2}{3}}}\right)^2 = \left(\frac{5\,\frac{\text{m}^3}{\text{s}} \times 0.035}{6\text{ m}^2 \times (0.708\text{ m})^{\frac{2}{3}}}\right)^2 = 0.001348,$$

$$S = \frac{H_1 - H_2}{L} = \frac{100 - 90}{L} = 0.001348,$$

$$L = 7418\text{ m}.$$

Example 8.8: Consider a trapezoidal channel with a bottom width of 3 m and side slope of 2:1 horizontal to vertical. The discharge in the channel is 20 m^3/s. The bottom slope is 0.05% and Manning's n = 0.03. Compute the normal depth, critical slope, and state of flow.

Solution: Compute the normal depth:

$$S = \frac{0.05}{100} = 0.0005,$$

$$Z_n = \frac{Qn}{S^{0.5}} = \frac{20 \times 0.03}{0.0005^{0.5}} = 26.83,$$

$$A = b \times y_n + (2y_n)y_n = 3y_n + 2y_n^2,$$

$$P = b + 2 \times \sqrt{y_n^2 + (2y_n)^2} = 3 + 2\sqrt{5y_n^2} = 3 + 2\sqrt{5}y_n,$$

$$R = \frac{A}{P} = \frac{3y_n + 2y_n^2}{3 + 2\sqrt{5}y_n},$$

$$Z_n = AR^{\frac{2}{3}} = (3y_n + 2y_n^2) \times \left(\frac{3y_n + 2y_n{}^2}{3 + 2\sqrt{5}y_n}\right)^{\frac{2}{3}} = 26.83.$$

By trial and error, $y_n = 2.575$ m.

From Example 8.6, the critical depth is $y_c = 1.252$ m.

Compute the critical slope:

$$A = 3y_c + 2y_c^2 = 6.89\ \text{m}^2,$$

$$P = 3 + 2\sqrt{5}y_c = 8.60\ \text{m},$$

$$R = \frac{A}{P} = \frac{6.89\ \text{m}^2}{8.60\ \text{m}} = 0.801\ \text{m},$$

$$S_c = \left(\frac{Qn}{AR^{\frac{2}{3}}}\right)^2 = \left(\frac{20\,\frac{\text{m}^3}{\text{s}} \times 0.03}{6.89\ \text{m}^2 \times (0.801\ \text{m})^{\frac{2}{3}}}\right)^2 = 0.01.$$

State of flow:

$$S(=0.0005) < S_c(=0.01) \rightarrow \text{Mild slope},$$

$$y_n(=2.575\ \text{m}) > y_c(=1.252\ \text{m}) \rightarrow \text{Subcritical flow}.$$

8.4 DESIGN OF CHANNELS

Design of a channel entails the selection of channel shape, size, alignment, bottom slope, and whether the channel is to be lined. Lining is done to prevent seepage and erosion of bed and sides. However, the cost of lining and maintenance is a major factor. Design depends on whether the channel boundary is erodible or nonerodible. In general, design is done by trial and error, because optimum channel dimensions in most cases cannot be determined directly. Further, economic considerations and specific site characteristics have to be taken into consideration. Design of both rigid boundary channels and erodible channels is considered here.

8.4.1 Design of Rigid Boundary Channels

Channel design involves channel dimensions, slope, and flow depth, and is based on the following considerations.

8.4.1.1 Permissible Flow Velocity

For rigid boundary channels the maximum velocity is not a limitation, but the minimum velocity should be such that it is not conducive to sediment deposition and aquatic growth. In general, a minimum velocity of 2–3 ft/s (0.61–0.91 m/s) is used for open channels as a velocity of 2 ft/s (0.61 m/s) can move 2 mm sand particles.

8.4.1.2 Bottom Longitudinal Slope

A channel is often laid on a predetermined alignment, meaning the slope is fixed and the slope is normally mild. Irrigation channels deliver water at a higher elevation than the head of the field.

8.4.1.3 Channel Side Slope

Side slopes of a channel depend on the material from which the channel is constructed. For lined canals, the US Bureau of Reclamation recommends a side slope of 1:1.5 vertical to horizontal. For sandy soils, a slope of 1:3 vertical to horizontal is often used. The values of suggested side slopes for irrigation channels are summarized here, but these values must be empirically tested (Etcheverry, 1931). For shallow channels (up to 1.2 m or 4 ft) in peat and muck, a vertical side is frequently used. For cuts and fill in stiff (heavy) clay soil, a side slope (horizontal to vertical) of 0.5:1 is often used. For cuts and fill in clay or silt loam soils, a side slope of 1:1 is used. For sandy loam and loose sandy soils, side slopes of 1.5:1 and 2:1 are usually used, respectively. For deep channels (equal to or greater than 1.2 m or 4 ft), the side slopes are flatter than the shallow channels. For example, in peat and muck, a side slope (horizontal to vertical) of 0.25:1 is often used. In stiff (heavy) clay, clay or silt loam, sandy loam, and loose sandy soils, side slopes of 1:1, 1.5:1, 2:1, and 3:1 are used, respectively.

8.4.1.4 Freeboard

The purpose of a freeboard (FB) is to provide extra safety to prevent overtopping of the channel by waves or fluctuating water. Freeboard is the vertical height from the water surface to the top of the channel. The US Bureau of Reclamation suggests

$$FB = \sqrt{cy}, \tag{8.21}$$

where c is a coefficient varying from 1.5 for a capacity of 20 cfs to 2.5 for a capacity of 3000 cfs or more, and y is the flow depth (ft) in the channel.

8.4.1.5 Hydraulic Efficient Sections

For a channel with constant roughness coefficient and slope, discharge increases with the increase of the section factor. If the cross-section is specified then the section factor is the greatest for the least wetted perimeter. The wetter perimeter depends on the shape, and its minimization can yield the relation between flow depth and the best hydraulic section. This leads to the concept of the most efficient hydraulic section or best hydraulic section. This is the section that gives the maximum discharge for a given cross-section area or that leads to the minimum wetted perimeter. The circular section is the most efficient section, because it has the smallest perimeter for a given cross-sectional area, but the circular section is not feasible. In practice, channel dimensions are also determined by cost and other practical considerations, such as scour and erosion, slope, landscape, ease of access, volume of excavation, and transportation of excavated material. From hydraulic considerations only, the base width of a trapezoidal section should not be greater than the

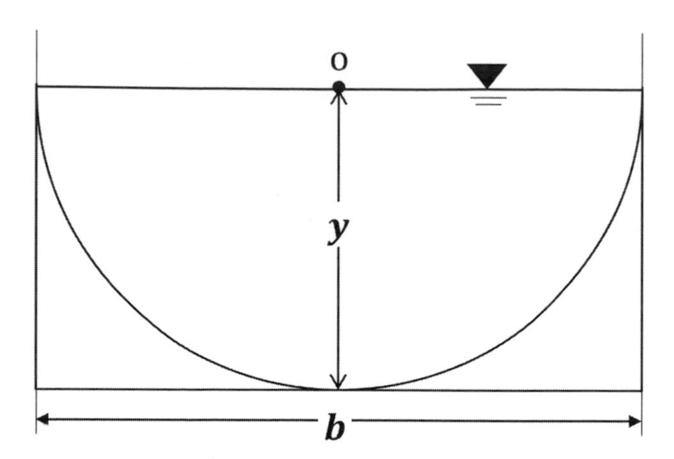

Figure 8.7 A rectangular cross-section.

flow depth, and the width of a rectangular section should be twice the flow depth. Here, the most efficient sections for common geometric shapes are presented.

Rectangular cross-section: For a rectangular cross-section, as shown in Figure 8.7, $A = by$ and $P = b + 2y$. For the most efficient section, P should be minimum for given A. To that end, P can be expressed in terms of A as

$$P = \frac{A}{y} + 2y. \tag{8.22}$$

Differentiating eq. (8.22) with respect to y, keeping in mind that A is constant, and equating the derivative to zero, one gets

$$\frac{dP}{dy} = -\frac{A}{y^2} + 2 = 0. \tag{8.23}$$

Equation (8.23) yields

$$\frac{A}{y^2} = 2. \tag{8.24}$$

Inserting $A = by$ in eq. (8.24),

$$\frac{by}{y^2} = 2 \quad \text{or} \quad y = \frac{b}{2}. \tag{8.25}$$

Then,

$$P = b + 2y = b + \frac{2b}{2} = 2b. \tag{8.26}$$

Thus, a rectangular section is hydraulically most efficient when flow depth is half the channel width or the perimeter is twice the width. Also, the hydraulic radius will be half the flow depth and cross-section will be half of the square of width. If a semicircle with radius y and center O is drawn in the rectangular section then it is seen that the bed and sides are tangential to the semicircle.

Triangular section: For a triangular section, let the side slope be equal on both sides with z:1 horizontal to vertical, as shown in Figure 8.8. Then,

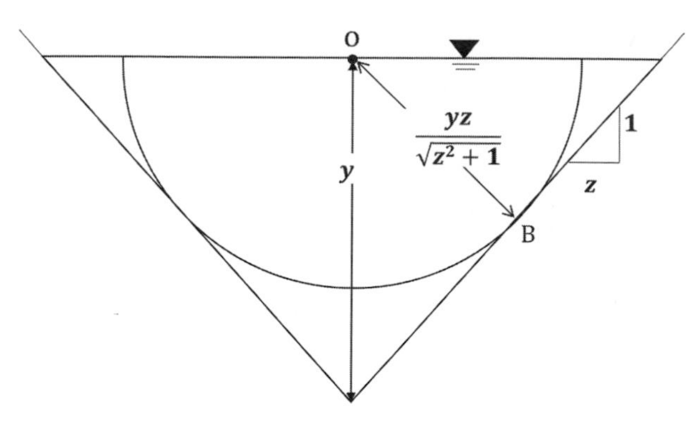

Figure 8.8 A triangular cross-section.

$$A = zy^2, \tag{8.27}$$

$$P = 2\left(\sqrt{1 + z^2}\right)y. \tag{8.28}$$

Equation (8.28) can be expressed in terms of A as

$$P = 2\sqrt{1 + z^2}\left(\frac{A}{z}\right)^{\frac{1}{2}}. \tag{8.29}$$

Equation (8.29) can be expressed by taking the square of both sides as

$$P^2 = 4\left(z + \frac{1}{z}\right)A. \tag{8.30}$$

Differentiating eq. (8.30) with respect to z and equating the derivative to zero, one gets

$$2P\frac{dP}{dz} = 4\left(1 - \frac{1}{z^2}\right)A = 0. \tag{8.31}$$

Equation (8.31) yields $z = 1$. Thus, a triangular section with the sides sloping at 45° or with a central angle of 90° is the most efficient section. Then, the wetted perimeter will be

$$P = 2\sqrt{2}\,y. \tag{8.32}$$

The cross-sectional area will be

$$A = y^2. \tag{8.33}$$

The hydraulic radius will be

$$\frac{z}{2\sqrt{1 + z^2}}y = \frac{y}{2\sqrt{2}}. \tag{8.34}$$

If a semicircle is drawn with radius OB, which is perpendicular to the midpoint on the sloping side to the water surface, then $OB = \frac{yz}{\sqrt{z^2+1}} = y/\sqrt{2}$, with $z = 1$. Then, the semicircle with center at O and radius as $y/\sqrt{2}$ is tangential to the two sides of the triangle.

Trapezoidal section: For a trapezoidal section, as shown in Figure 8.9,

$$A = (b + zy)y, \tag{8.35}$$

$$P = b + 2y\sqrt{1+z^2}. \tag{8.36}$$

Equation (8.36) can be expressed in terms of A as

$$P = \frac{A}{y} - zy + 2y\sqrt{1+z^2}. \tag{8.37}$$

To derive the most efficient section, two possibilities can be considered. First, both A and y can be considered constant and z as variable. Differentiating eq. (8.37) with respect to z and equating the derivative to zero, one gets

$$\frac{dP}{dz} = 0 = -y + y\frac{2}{2\sqrt{1+z^2}}2z \to 2z = \sqrt{1+z^2}. \tag{8.38}$$

Equation (8.38) yields

$$z = \frac{1}{\sqrt{3}} \text{ or } \theta = 60^\circ \tag{8.39}$$

Differentiating eq. (8.37) with respect to y and equating the derivative to zero, one obtains

$$-\frac{A}{y^2} - z + 2\sqrt{1+z^2} = 0 \tag{8.40}$$

or

$$(b + zy) = \left(2\sqrt{1+z^2} - z\right)y \to b = 2\left(\sqrt{1+z^2} - z\right)y. \tag{8.41}$$

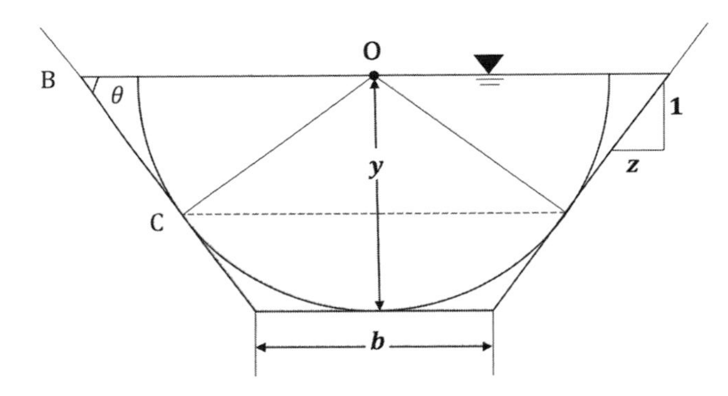

Figure 8.9 A trapezoidal cross-section.

The top water surface width T becomes

$$T = b + 2zy = 2\sqrt{1+z^2}y, \tag{8.42}$$

which is twice the length of the sloping side. The most efficient section is one-half of a hexagon.

If a semicircle is drawn within the trapezoidal section, then the radius of the circle can be expressed as follows. Referring to the triangle OBC in Figure 8.9,

$$OC = OB\sin\theta = \frac{T}{2}\frac{1}{\sqrt{1+z^2}} = \frac{2\sqrt{1+z^2}y}{2}\frac{1}{\sqrt{1+z^2}} = y. \tag{8.43}$$

Equation (8.43) shows that a semicircle with O as the center and y as the radius is tangential to the bed and sides of the best trapezoidal section.

Example 8.9: Compute the width, depth, wetted perimeter, cross-sectional area, and hydraulic radius of the most efficient rectangular section for carrying a discharge of 1.5 m^3/s. The bed slope is 0.005 and Manning's n is 0.025. If the discharge is to be increased to 2.0 m^3/s, keeping the rectangular section the same but increasing the velocity, what should be the bed slope now?

Solution: The most efficient section

$$z = \frac{1}{\sqrt{3}} \text{ or } \theta = 60^\circ,$$

$$b = 2\left(\sqrt{1+z^2} - z\right)y,$$

$$b = 2\left(\sqrt{1+\frac{1}{3}} - \frac{1}{\sqrt{3}}\right)y = \frac{2}{\sqrt{3}}y.$$

Now, plug b and z into A and P equations:

$$A = (b + zy)y = \left(\frac{2}{\sqrt{3}}y + \frac{1}{\sqrt{3}}y\right)y = \frac{3}{\sqrt{3}}y^2,$$

$$P = b + 2y\sqrt{1+z^2} = \frac{2}{\sqrt{3}}y + 2y\left(\frac{2}{\sqrt{3}}\right) = \frac{6}{\sqrt{3}}y,$$

$$R = \frac{A}{P} = \frac{\frac{3}{\sqrt{3}}y^2}{\frac{6}{\sqrt{3}}y} = \frac{1}{2}y.$$

From Manning's equation,

$$Q = \frac{1}{n}AR^{\frac{2}{3}}S^{\frac{1}{2}},$$

$$1.5\frac{\text{m}^3}{\text{s}} = \frac{1}{0.025}\left(\frac{3}{\sqrt{3}}y^2\right)\left(\frac{1}{2}y\right)^{\frac{2}{3}}(0.005)^{\frac{1}{2}},$$

$$\left(\frac{3}{\sqrt{3}}y^2\right)\left(\frac{1}{2}y\right)^{\frac{2}{3}} = \left(1.5\frac{\text{m}^3}{\text{s}}\right)(0.025)\left(\frac{1}{0.005^{\frac{1}{2}}}\right) = 0.53,$$

$$\left(\frac{3}{\sqrt{3}}y^2\right)\left(\frac{1}{2}y\right)^{\frac{2}{3}} = 0.53.$$

Depth:

$$y = 0.764 \text{ m}.$$

Width:

$$b = \frac{2}{\sqrt{3}}y = 0.88 \text{ m}.$$

Wetted perimeter:

$$P=\frac{6}{\sqrt{3}}y=2.65\text{ m}.$$

Cross-sectional area:

$$A=\frac{3}{\sqrt{3}}y^2=1.01\text{ m}^2.$$

Hydraulic radius:

$$R=\frac{1}{2}y=\frac{0.764}{2}=0.38\text{ m}.$$

Now, if the discharge is increased to 2.0 m^3/s by keeping the rectangular section the same:

$$u=\frac{Q}{A}=\frac{2.0}{1.01}=1.98\text{ m/s}.$$

From Manning's equation,

$$u=\frac{1}{n}R^{\frac{2}{3}}S^{\frac{1}{2}}$$

$$1.98=\frac{1}{0.025}(0.38)^{\frac{2}{3}}S^{\frac{1}{2}}$$

Then, $S=0.0089$.

Example 8.10: A trapezoidal section with a bed slope of 0.002 and side slope of 1:2 vertical to horizontal has to carry a discharge of 10 m^3/s. Manning's n is 0.02. Compute the width, top width, cross-sectional area, wetted perimeter, and hydraulic radius of the most efficient section.

Solution: With the most efficient section for a trapezoidal section (when $z=2$),

$$b=2\left(\sqrt{1+z^2}-z\right)y=2\left(\sqrt{5}-2\right)y=0.472y,$$

$$A=(b+zy)y=(0.472y+2y)y=2.472y^2,$$

$$P=b+2y\sqrt{1+z^2}=0.472y+2y\sqrt{5}=4.944y,$$

$$R=\frac{A}{P}=\frac{2.472y^2}{4.944y}=0.5y.$$

From Manning's equation,

$$Q=\frac{1}{n}AR^{\frac{2}{3}}S^{\frac{1}{2}},$$

$$10\frac{\text{m}^3}{\text{s}}=\frac{1}{0.02}\left(2.472y^2\right)(0.5y)^{\frac{2}{3}}(0.002)^{\frac{1}{2}},$$

$$\left(2.472y^2\right)(0.5y)^{\frac{2}{3}}=\left(10\frac{\text{m}^3}{\text{s}}\right)(0.02)\left(\frac{1}{0.002^{\frac{1}{2}}}\right),$$

$$\left(2.472y^2\right)(0.5y)^{\frac{2}{3}}=4.472.$$

Depth:

$$y=1.485\text{ m}.$$

Width:

$$b=0.472y=0.70\text{ m}.$$

Top width:

$$T=b+2zy=0.70+2(2\times 1.485)=6.64\text{ m}.$$

Cross-sectional area:

$$A=2.472y^2=5.45\text{ m}^2.$$

Wetted perimeter:

$$P=4.944y=7.34\text{ m}.$$

Hydraulic radius:

$$R=0.5y=0.74\text{ m}.$$

8.4.1.6 Design Procedure

The following steps are involved in the design of irrigation channels:

1. Select the bed slope S and compute Manning's n.
2. Determine the section factor.
3. For a trapezoidal section, select the side slope, based on the material of construction, and assume b/y. Compute the section factor and solve for the depth.
4. Choose several values of the unknowns, obtain a number of section dimensions, and compare costs.
5. Check for the minimum velocity.
6. Compute the FB and add it to the flow depth.

Example 8.11: Design a rigid boundary trapezoidal earthen channel for carrying a discharge of 1.2 m^3/s. The channel alignment slope is 0.002 and Manning's n is 0.025. Take the side slope as 2:1 horizontal to vertical.

Solution:

1. Based on the channel alignment, $S=0.002$ and $n=0.025$.
2. The section factor for uniform flow can be calculated using eq. (8.20):

$$AR^{\frac{2}{3}} = \frac{Qn}{S^{0.5}} = \frac{1.2 \times 0.025}{0.002^{0.5}} = 0.671.$$

3. For rigid boundary channels, the most efficient section of the trapezoidal has the following b–y relationship based on eq. (8.42): $b = 2\sqrt{1+z^2}y - 2zy$.
   Since $z = 2$, then $b = 2\sqrt{1+2^2}y - 2 \times 2y = 2\sqrt{5}y - 4y = 0.472y$,

$$A = b \times y + (2y)y = 0.472\,y \times y + 2y^2 = 2.472y^2,$$

$$P = b + 2 \times \sqrt{y^2 + (2y)^2} = 0.472\,y + 2\sqrt{5}y = 4.944y,$$

$$R = \frac{A}{P} = \frac{2.472y^2}{4.944y} = 0.5y,$$

$$AR^{\frac{2}{3}} = (2.472y^2)(0.5y)^{\frac{2}{3}} = 0.671,$$

$$y = 0.729 \text{ m}.$$

4. Here, we choose $b/y = 0.472$ based on the most efficient hydraulic section. Also, different ratios of b/y can be chosen to obtain a number of section dimensions and compare costs.
5. Check for velocity:

$$u = \frac{Q}{A} = \frac{1.2\,\frac{\text{m}^3}{\text{s}}}{2.472y^2} = \frac{1.2\,\frac{\text{m}^3}{\text{s}}}{1.31\text{ m}^2}$$

$$= 0.91\frac{\text{m}}{\text{s}} > 0.61\frac{\text{m}}{\text{s}}(= V_{min}) \rightarrow \text{OK}.$$

For FB (since $c = 1.5$ for $20\,\text{ft}^3/\text{s}$, $c = 2.5$ for $3000\,\text{ft}^3/\text{s}$, for $1.2\,\text{m}^3/\text{s} = 42.5\,\text{ft}^3/\text{s}$; assume $c = 1.5$):

$$y = 0.729\text{ m} = 2.39\text{ ft},$$

$$FB = \sqrt{cy} = \sqrt{1.5 \times 2.39}\text{ ft} = 1.89\text{ ft} = 0.576\text{ m}.$$

Total channel depth:

$$\text{Total channel depth} = y + FB = 0.729\text{ m} + 0.576\text{ m} = 1.31\text{ m}.$$

8.4.2 Design of Loose Boundary Channels

Irrigation channels are often loose boundary channels, especially in developing countries. These channels are constructed with alluvial or erodible material and are subject to erosion by water. Table 8.1 shows allowable flow velocity for nonscouring flood control channels (US Army Corps of Engineers, 1991). They must therefore be designed for stability so that bed and banks are not scoured and unacceptable amounts of sediment are not deposited in the channels. Such channels are designed using regime theory or tractive force theory. Regime theory is based on empirical relations and is used for channels that carry sediment, such as irrigation canals, whereas tractive force theory is used for silt and sand channels with clean water.

8.4.2.1 Regime Theory

The regime theory is credited to Lacey (1930) and comprises three basic equations:

1. velocity–depth relation: $u = 1.17\sqrt{f_s R}$; (8.44)
2. velocity–slope relation: $u = 16R^{\frac{2}{3}}S^{\frac{1}{3}}$; and (8.45)
3. width–discharge relation: $P = 2.67Q^{\frac{1}{2}}$, (8.46)

where u is the mean flow velocity (ft/s), R is the hydraulic radius (ft), S is the longitudinal slope, and f_s is the silt factor expressed as

$$f_s = 1.76D_0^{\frac{1}{2}}, \tag{8.47}$$

where D_0 is the mean grain diameter (mm). The mean values of f_s for various materials are given in Table 8.2.

Equations (8.44)–(8.46), when combined, and with Lacey's (1930) own form of Manning's equation, $u = \frac{1.3458}{n}R^{\frac{3}{4}}S^{\frac{1}{2}}$, yield

Table 8.1 *Allowable flow velocity for nonscouring flood control channels*

The material of the channel	Mean channel velocity	
	ft/s	m/s
Fine sand	2.0	0.61
Coarse sand	4.0	1.22
Fine gravel	6.0	1.83
Earth		
Sandy silt	2.0	0.61
Silt clay	3.5	1.07
Clay	6.0	1.83
Grass-lined earth ($S < 0.005$)		
Bermudagrass		
Sandy silt	6.0	1.83
Silt clay	8.0	2.44
Kentucky bluegrass		
Sand silt	5.0	1.52
Silty clay	7.0	2.13
Poor rock (usually sedimentary)	10.0	3.05
Soft sandstone	8.0	2.44
Soft shale	3.5	1.07
Good rock (usually igneous or hard metamorphic)	20.0	6.08

After US Army Corps of Engineers, 1991.

Table 8.2 *Silt factor and permissible unit tractive force*

Material	Size (mm)	Silt factor (f_s)	Average permissible unit tractive force (lb/ft^2) (N/m^2)
Small/medium boulders, cobbles, and shingles	64–256	6.12–9.75	0.92 (44.0)
Coarse gravel	8–64	4.68	0.48 (23.0)
Fine gravel	4–8	2.00	80 × 10^{-3} (3830.4)
Coarse sand	0.5–2	1.44–1.56	50 × 10^{-3} (2394.0)
Medium sand	0.25–0.5	1.31	35 × 10^{-3} (1675.8)
Fine sand	0.06–0.25	1.1–1.3	25 × 10^{-3} (1197.0)
Silt (colloidal)	–	1.00	0.2–0.3 (9.6–14.4)
Fine silt (colloidal)	–	0.4–0.9	0.2–0.3 (9.6–14.4)
Compact clay (colloidal)	–	–	0.3–0.4 (14.4–19.2)
Loose clay (colloidal)	–	–	0.05 (2.4)

$$R = 0.47\left(\frac{Q}{f_s}\right)^{\frac{1}{3}}, \tag{8.48}$$

$$S = \frac{f_s^{\frac{5}{3}}}{1859Q^{\frac{1}{6}}}. \tag{8.49}$$

8.4.2.2 Design Procedure

The procedure for design of loose boundary channels using regime theory includes the following steps:

1. Obtain the slope and design discharge first, which are usually known.
2. Select a value of the silt factor from Table 8.2 (Gupta, 2017) or compute it using eq. (8.47). If the sediment load is greater than 2000 ppm with significant coarse material, then the value of f_s should be greater than the computed f_s. In irrigation channels, the coarse material is normally not significant.
3. Determine P using eq. (8.46).
4. Determine R using eq. (8.48).
5. Compute b and y.
6. Compute S using eq. (8.49) and compare it with the known slope. Usually the computed slope is the minimum slope. If the computed slope is less than the known slope, then it may be necessary to realign the channel. If the computed slope is greater than the known slope, then the channel can be widened.

Example 8.12: Design a loose boundary channel for a maximum discharge of 1000 cfs. The slope is 0.001, the mean sediment diameter is 0.25 mm, and the sediment load is 1000 ppm.

Solution: Compute the silt factor:

$$f_s = 1.76D_0^{\frac{1}{2}} = 1.76(0.25)^{0.5} = 0.88.$$

Compute the width–discharge relation:

$$P = 2.67Q^{\frac{1}{2}} = 2.67(1000)^{0.5} = 84.43 \text{ ft } (25.73 \text{ m}),$$

$$R = 0.47\left(\frac{Q}{f}\right)^{\frac{1}{3}} = 0.47\left(\frac{1000}{0.88}\right)^{\frac{1}{3}} = 4.9 \text{ ft } (1.49 \text{ m}),$$

$$A = PR = 414.11 \text{ ft}^2 \left(38.47 \text{ m}^2\right).$$

Assuming that the side slope is 1:1 horizontal to vertical:

$$A = (b + y)y = 414.11,$$

$$P = b + 2\sqrt{2}y = 84.43.$$

Hence,

$$b = 68.64 \text{ ft } (20.92 \text{ m}),$$

$$y = 5.58 \text{ ft } (1.70 \text{ m}),$$

$$S_{min} = \frac{f_s^{\frac{5}{3}}}{1859Q^{\frac{1}{6}}} = \frac{0.88^{\frac{5}{3}}}{1859 \times (1000)^{\frac{1}{6}}}$$

$$= 0.000137 < S(= 0.001) \rightarrow \text{OK}.$$

Add FB ($Q = 1000$ ft^3/s; assume $c = 2.0$):

$$FB = \sqrt{cy} = \sqrt{2.0 \times 5.58 \text{ ft}} = 3.34 \text{ ft } (1.02 \text{ m}).$$

Total channel depth:

$$\text{Total channel depth} = y + FB = 5.58 + 3.34 = 8.92 \text{ ft } (2.72 \text{ m}).$$

8.4.2.3 Tractive Force Theory

The fundamental hypothesis of this theory is that the force exercised by water on the wetted surface of the channel (called tractive force, F) is balanced by the component of weight of water in the direction of flow; that is, as shown in Figure 8.10,

$$F = W\sin\theta = \gamma ALS, \tag{8.50}$$

where W is the weight, θ is the angle of the bed with the horizontal, A is the cross-sectional area, S is the slope of the energy line, L is the length, and γ is the weight density. The tractive force F equals the contact area multiplied by the boundary shear stress (γALS). The contact area is defined by the product of the wetted perimeter and length (i.e., PL, where P is the wetted perimeter). The boundary shear stress (τ) can be expressed as

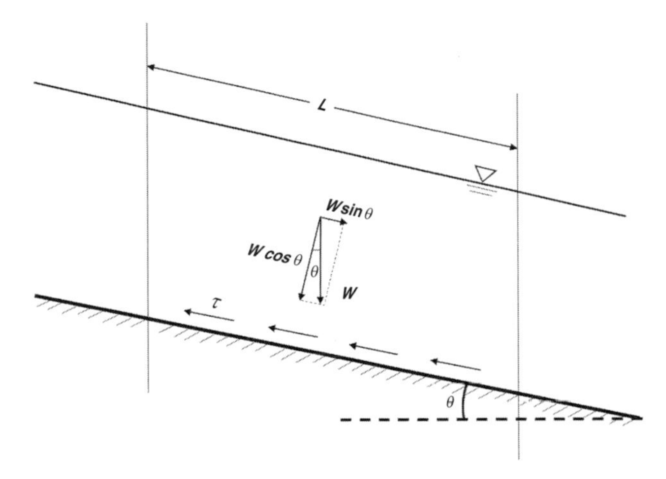

Figure 8.10 Tractive force balanced by the weight of water.

$$\tau = \frac{\gamma ALS}{PL} = \gamma RS. \tag{8.51}$$

Equation (8.51) shows that shear stress is the same as unit tractive force.

For a trapezoidal channel, the maximum shear stress can be written for the bed as

$$\tau_0 = \gamma yS, \tag{8.52}$$

and for the sides as

$$\tau_s = 0.76\,\gamma yS. \tag{8.53}$$

The stress at which the channel material will move is referred to as the critical stress, and the channel will be stable if the maximum stress is below the critical stress. Allowable stresses for noncohesive and cohesive materials have been graphed by the US Department of Agriculture, Natural Resources Conservation Service, and typical values are given in Figures 8.11 and 8.12 (USDA-NRCS, 2007).

Besides the shear stress, the material on the channel sides is also subjected to the gravity force down the slope, causing

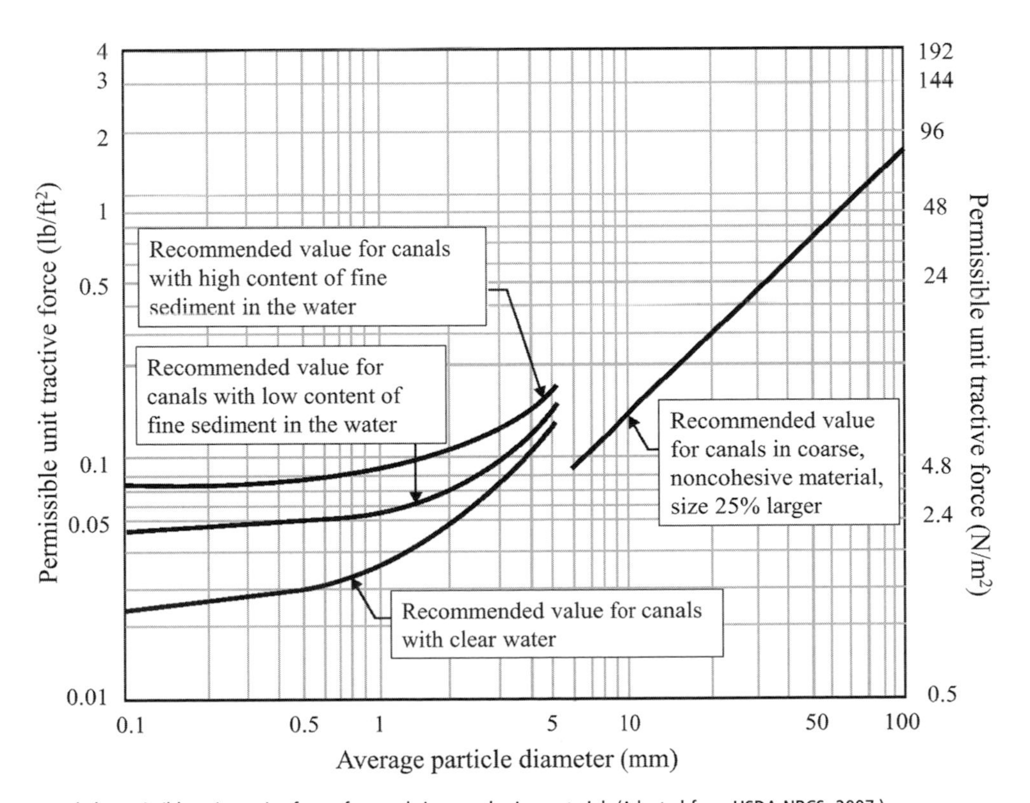

Figure 8.11 Recommended permissible unit tractive forces for canals in noncohesive material. (Adapted from USDA-NRCS, 2007.)

Table 8.3 *Fortier and Scobey's limiting velocities with corresponding tractive force values for straight channels after aging*

Material	Manning's n	For clear water: Velocity (fps) (m/s)	For clear water: Tractive force (psf) (N/m^2)	Water transporting colloidal soils: Velocity (fps) (m/s)	Water transporting colloidal soils: Tractive force (psf) (N/m^2)
Fine sand – noncolloidal	0.02	1.5 (0.46)	0.027 (1.29)	2.50 (0.76)	0.075 (3.59)
Sandy loam – noncolloidal	0.02	1.75 (0.53)	0.037 (1.77)	2.50 (0.76)	0.075 (3.59)
Silt loam – noncolloidal	0.02	2.00 (0.61)	0.048 (2.30)	3.00 (0.91)	0.11 (5.27)
Alluvial silts – noncolloidal	0.02	2.00 (0.61)	0.048 (2.30)	3.50 (1.07)	0.15 (7.18)
Ordinary firm loam	0.02	2.50 (0.76)	0.075 (3.59)	3.50 (1.07)	0.15 (7.18)
Volcanic ash	0.02	2.50 (0.76)	0.075 (3.59)	3.50 (1.07)	0.15 (7.18)
Stiff clay – very colloidal	0.025	3.75 (1.14)	0.26 (12.45)	5.00 (1.52)	0.46 (22.02)
Alluvial silts – colloidal	0.025	3.75 (1.14)	0.26 (12.45)	5.00 (1.52)	0.46 (22.02)
Shales and hardpans	0.025	6.00 (1.83)	0.67 (32.08)	6.00 (1.83)	0.67 (32.08)
Fine gravel	0.02	2.5 (0.76)	0.075 (3.59)	5.00 (1.52)	0.32 (15.32)
Graded loam to cobbles when noncolloidal	0.03	3.75 (1.14)	0.38 (18.19)	5.00 (1.52)	0.66 (31.60)
Graded silts to cobbles when colloidal	0.03	4.0 (1.22)	0.43 (20.59)	5.50 (1.68)	0.80 (38.30)
Coarse gravel – noncolloidal	0.025	4.0 (1.22)	0.30 (14.36)	6.00 (1.83)	0.67 (32.08)
Cobbles and shingles	0.035	5.0 (1.52)	0.91 (43.57)	5.50 (1.68)	1.00 (47.88)

After Fortier and Scobey, 1926.

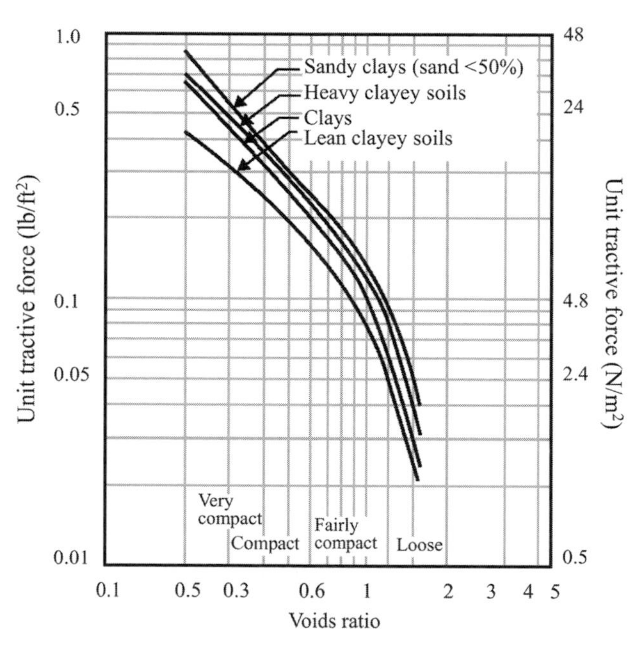

Figure 8.12 Recommended permissible unit tractive forces for canals in cohesive material. (Adapted from USDA-NRCS, 2007.)

a reduction in the shear stress. This reduction is expressed through a reduction factor (K_s) as:

$$K_s = \sqrt{1 - \frac{\sin^2\theta}{\sin^2\phi}}, \tag{8.54}$$

where θ is the slope of the side to the horizontal, and ϕ is the angle of repose of the material.

Lane (1955) computed critical tractive force values for corresponding flow velocities and Manning's roughness coefficient as given in Table 8.3, with the assumption that the depth of flow was 0.9 m, bottom width was 3.1 m, and the side slope was 1.5:1 (Lane, 1955).

8.4.2.4 Design Procedure

The basic premise of design based on the tractive force theory is to determine the channel geometric dimensions considering the maximum stress on the sides and checking for the maximum stress on the bed. The design procedure can be outlined as follows:

1. Choose a value for the permissible (critical) stress for the channel material.
2. Decide on the side slope of the channel.
3. Obtain the angle of repose of the channel material.
4. Compute the shear stress reduction factor K_s.
5. Multiply the chosen permissible stress by K_s. This yields the permissible shear stress for the sides.
6. Equate the computed shear stress for the sides to the theoretical value (i.e., $0.76\gamma yS$) and calculate the value of y.
7. Calculate the width b using Manning's equation.
8. Calculate the theoretical stress on the bed (i.e., γyS), which should be less than the permissible value chosen in step 1.

Example 8.13: A trapezoidal channel is to be designed in a sandy loam landscape where the slope is 0.0025. The mean sediment diameter is 20 mm, Manning's roughness coefficient n is 0.025, and the angle of repose is 30°. The channel is supposed to carry a discharge of 250 cfs. The side slope is considered to be 1:2 vertical to horizontal. Design a trapezoidal channel under these conditions.

Solution: From Figure 8.11, the critical stress for noncohesive material 20 mm in diameter is 0.33 lb/ft². For a 1:2 vertical to horizontal slope, $\theta = 26.57°$:

$$K_s = \sqrt{1 - \frac{\sin^2\theta}{\sin^2\phi}} = \sqrt{1 - \frac{\sin^2 26.57}{\sin^2 30}} = 0.447.$$

Compute the permissible stress on the sides:

$$\tau_s = K_s \times 0.33 \frac{\text{lb}}{\text{ft}^2}$$
$$= 0.447 \times 0.33 \frac{\text{lb}}{\text{ft}^2} = 0.147 \frac{\text{lb}}{\text{ft}^2} = 7.04\ \text{N/m}^2.$$

Compute the water depth:

$$\tau_s = 0.76\,\gamma y S,$$

$$y = \frac{\tau_s}{0.76\,\gamma S} = \frac{0.147 \frac{\text{lb}}{\text{ft}^2}}{0.76 \times 62.4 \frac{\text{lb}}{\text{ft}^3} \times 0.0025} = 1.24\ \text{ft} = 0.38\ \text{m}.$$

From Manning's equation:

$$Q = \frac{1.49}{n}(A)(R)^{\frac{2}{3}}(S)^{0.5},$$

$$250 = \frac{1.49}{0.025}[(b+2y)y]\left[\frac{(b+2y)y}{b+2\sqrt{5}y}\right]^{\frac{2}{3}}(0.0025)^{0.5},$$

$$250 = \frac{1.49}{0.025}[(b+2\times 1.24)\times 1.24] \times \left(\frac{b\times 1.24 + 2\times 1.24\times 1.24}{b+4.47\times 1.24}\right)^{\frac{2}{3}}(0.0025)^{0.5},$$

$$\text{or } 58.62 = \frac{(b+2.48)^{5/3}}{(b+5.5428)^{2/3}}.$$

By trial and error, $b = 58.1$ ft (17.7 m).
Compute the maximum shear stress:

$$\tau_0 = \gamma y S = 62.4 \times 1.24 \times 0.0025$$
$$= 0.19 \frac{\text{lb}}{\text{ft}^2} < 0.33 \frac{\text{lb}}{\text{ft}^2} \text{(from the initial value)} \rightarrow \text{OK}.$$

Add FB ($Q = 250$ ft³/s; assume $c = 1.6$):

$$FB = \sqrt{cy} = \sqrt{1.6 \times 1.24}\ \text{ft} = 1.41\ \text{ft}\ (0.43\ \text{m}).$$

Total channel depth:

$$\text{Total channel depth} = y + FB = 1.24 + 1.41 = 2.65\ \text{ft}\ (0.81\ \text{m}).$$

8.5 GRADUALLY VARIED FLOW

When the flow depth, velocity, and discharge gradually vary along the channel length, the flow is regarded as gradually varied. This occurs when the gravity force causing the water to flow in the downward direction is not balanced by the force resisting the flow. This means that bed slope and energy slope are not the same, or the water surface profile is not parallel to the bed. The governing equation for gradually varied flow can be derived from the energy equation.

8.5.1 Governing Equation

Gradually varied flow over a channel section denoted by points 1 and 2 is sketched in Figure 8.13. From the Bernoulli equation, the total head (H) at point 1 is equal to the total head at point 2 plus the energy head loss (h_f) from point 1 to point 2:

$$H_1 = H_2 + h_f, \quad \text{where } H = Z + y + \frac{u^2}{2g}. \tag{8.55}$$

If the nonuniform distribution of velocity is considered, the term $\frac{u^2}{2g}$ can be replaced by $\alpha\frac{u^2}{2g}$. Here, the velocity across the channel section is assumed to be uniform. That is,

$$Z_1 + y_1 + \frac{u_1^2}{2g} = Z_2 + y_2 + \frac{u_2^2}{2g}. \tag{8.56}$$

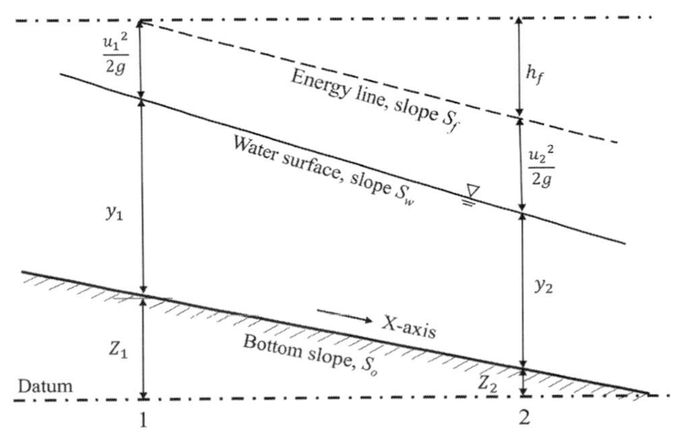

Figure 8.13 Gradually varied flow.

Let Δx be the length of the section. Then, from eq. (8.56) the spatial rate of change in the head can be written as

$$\frac{Z_1 - Z_2}{\Delta x} + \frac{y_1 - y_2}{\Delta x} + \frac{1}{\Delta x}\frac{u_1^2 - u_2^2}{2g} = \frac{h_f}{\Delta x}. \quad (8.57)$$

If Δx tends to a very small value, then eq. (8.57) can be written as

$$\frac{dZ}{dx} + \frac{dy}{dx} + \frac{1}{2g}\frac{du^2}{dx} = \frac{dh_f}{dx}. \quad (8.58)$$

The velocity gradient term can be written as

$$\frac{d}{dx}\left(\frac{u^2}{2g}\right) = \frac{d}{dy}\left(\frac{u^2}{2g}\right)\frac{dy}{dx} \quad (8.59)$$

The term dZ/dx is negative, since the ground elevation is decreasing in the direction of flow, denoted as S_0; and $\frac{dh_f}{dx}$ is negative, denoted as S_f, because $(H_1 - H_2)/\Delta x$ is negative. Therefore, eq. (8.58) can be written as

$$-S_0 + \frac{dy}{dx}\left[1 + \frac{d}{dy}\left(\frac{u^2}{2g}\right)\right] = -S_f. \quad (8.60)$$

Equation (8.60) can be written as

$$\frac{dy}{dx} = \frac{S_0 - S_f}{1 + \frac{d}{dy}\left(\frac{u^2}{2g}\right)}. \quad (8.61)$$

Equation (8.61) is the gradually varied flow equation, which can also be expressed in different ways.

Differentiating the total energy head with respect to x,

$$\frac{dH}{dx} = \frac{dZ}{dx} + \frac{d}{dx}\left(y + \frac{u^2}{2g}\right) = \frac{dZ}{dx} + \frac{dE}{dx}, \quad E = y + u^2/2g. \quad (8.62)$$

Now,

$$\frac{dE}{dx} = \frac{dH}{dx} - \frac{dZ}{dx} = S_0 - S_f,$$

$$\frac{dE}{dy} = 1 + \frac{d}{dy}\left(\frac{Q^2}{2gA^2}\right) = 1 - \frac{Q^2}{gA^3}\frac{dA}{dy} = 1 - \frac{Q^2}{gA^3}T = 1 - \mathrm{Fr}^2, \quad (8.63)$$

where T is the top width. Therefore,

$$\frac{dy}{dx} = \frac{S_0 - S_f}{1 - \mathrm{Fr}^2}. \quad (8.64)$$

For wide sections, the hydraulic radius can be assumed to equal y, and the wetted perimeter equal to the top width T. For Manning's equation, one can write

$$\left(\frac{S_f}{S_0}\right) = \left(\frac{y_0}{y}\right)^{\frac{10}{3}}, \quad (8.65)$$

where y_0 is the normal depth corresponding to bed slope S_0. Then,

$$\frac{dy}{dx} = S_0\left[\frac{1 - \left(\frac{y_0}{y}\right)^{\frac{10}{3}}}{1 - \left(\frac{y_c}{y}\right)^3}\right], \quad (8.66)$$

where y_c is the critical depth, and from eqs 8.13 and 8.14 one can get $y_c^3 = q^2/g$, then

$$Fr^2 = \frac{u^2}{gy} = \frac{q^2}{gy^3} = \left(\frac{y_c}{y}\right)^3, \quad (8.67)$$

where q is the specific discharge, the discharge per unit width. Equation (8.66) is known as the flow equation.

Example 8.14: Consider a wide rectangular channel where the depth of flow is 1.0 m and the bed slope is 0.0015. It carries a specific discharge of 0.5 m^2/s. Manning's n is 0.035. Compute the rate of change in depth. Is the depth increasing or decreasing?

Solution: For a wide channel section,

$$y_c = \left(\frac{q^2}{g}\right)^{\frac{1}{3}} = \left(\frac{0.5^2}{9.81}\right)^{\frac{1}{3}} = 0.294\text{ m},$$

$$y_0 = \left(\frac{qn}{S_0^{0.5}}\right)^{3/5} = \left(\frac{0.5 \times 0.035}{0.0015^{0.5}}\right)^{\frac{3}{5}} = 0.62\text{ m},$$

$$\frac{dy}{dx} = S_0\left[\frac{1 - \left(\frac{y_0}{y}\right)^{\frac{10}{3}}}{1 - \left(\frac{y_c}{y}\right)^3}\right] = 0.0015 \times \left[\frac{1 - \left(\frac{0.62}{1.0}\right)^{\frac{10}{3}}}{1 - \left(\frac{0.294}{1.0}\right)^3}\right] = 0.0012.$$

Since $\frac{dy}{dx}$ is positive, the depth is increasing.

8.5.2 Water Surface Profiles

Equation (8.66) shows that the flow profile depends on (1) the slope of the channel, (2) the depth compared to the critical depth, and (3) the depth compared to the normal depth. A usual convention is that if the water surface is increasing in the direction of flow, then it is known as a backwater curve and is positive; if the water surface is declining in the direction of flow, it is known as a drawdown curve and is negative. In evaluating the profile, it is useful to recall slope–depth relationships:

1. If $y > y_0$, then $S_f < S_0$;
2. If $y < y_0$, then $S_f > S_0$;
3. If $y_0 > y > y_c$, then $S_0 < S_f < S_c$;
4. If $y < y_c$, then $S_f > S_c$ and $\mathrm{Fr} > 1$;
5. If $y > y_c$, then $S_f < S_c$ and $\mathrm{Fr} < 1$,

where S_f is the slope of the total energy line and S_c is the critical slope. The channel slopes can be classified as: slope is mild if $y_0 > y_c$ and Fr < 1; critical if $y_0 = y_c$ and the bed slope is equal to the critical slope and Fr $= 1$; steep if $y_0 < y_c$ and the bed slope is greater than the critical slope and Fr > 1 greater than 1; $S_0 = 0$, the slope is horizontal and y_0 is infinite; and $S_0 < 0$, the slope is adverse and y_0 is nonexistent. These profiles or channels are shown in Figure 8.14. Irrigation channels usually have mild slopes.

Depending on the stage of water surface with respect to normal depth and critical depth, three zones are identified, as shown in Figure 8.15: zone 1, in which y is located above y_0 and y_c – that is, $y > y_0 > y_c$ or $y > y_c > y_0$; zone 2, in which y is located between y_0 and y_c – that is, $y_0 > y > y_c$ or $y_c > y > y_0$; and zone 3, in which y is located below both y_0 and y_c – that is, $y_o > y_c > y$ or $y_c > y_0 > y$.

From the slope of the profile (dy/dx), water surface profiles can be classified as shown in Figure 8.16. (1) If dy/dx is positive, then the water surface is a rising curve; (2) if dy/dx is negative, then the water surface is a falling curve; (3) if $dy/dx = 0$, then flow is uniform and the water surface is parallel to the bed; (4) if dy/dx is $-\infty$, then the water surface forms a right angle with the channel bed – this occurs during the advance phase; (5) if dy/dx is not defined (i.e., equal to ∞/∞), the depth of flow approaches zero – that occurs in irrigation borders and furrows; and (6) if dy/dx equals S_0, the water surface forms a horizontal line.

The water surface profiles can now be classified into 12 types as a profile moves from one stage to another, as shown in Figure 8.17. In zone 1 the flow profile is an M1 curve, and in zones 2 and 3 it is an M2 and M3 curve, respectively. On steep slopes, the profile is an S1 curve in zone 1, S2 curve in zone 2, and S3 curve in zone 3. On critical slopes, the profile is C1 in zone 1 and C3 in zone 3. On horizontal slopes, the profile is H2 in zone 2 and H3 in zone 3. On adverse slopes, the profile is A2 in zone 2 and A3 in zone 3.

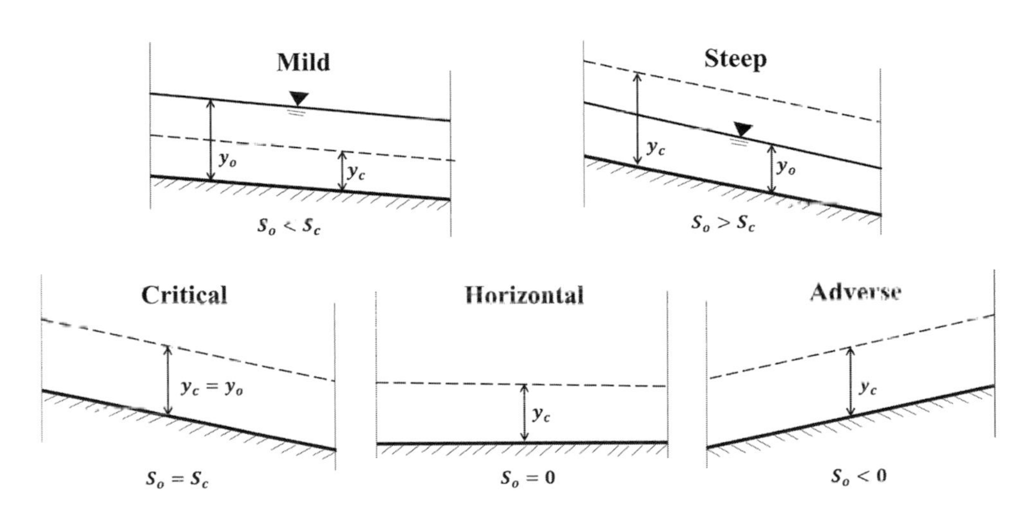

Figure 8.14 Classification of water surface profiles.

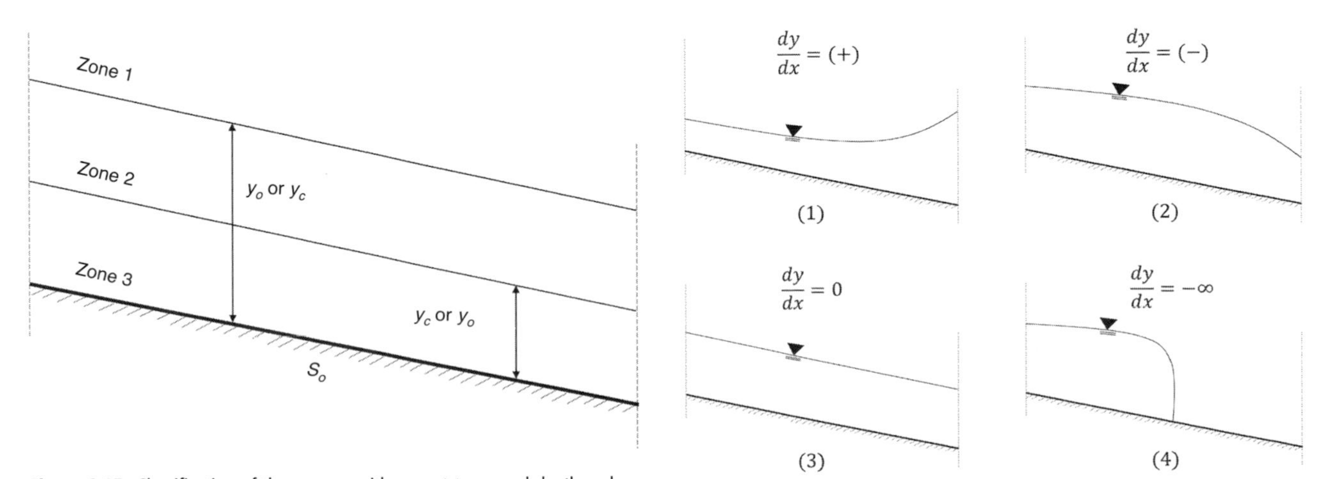

Figure 8.15 Classification of three zones with respect to normal depth and critical depth.

Figure 8.16 Classification of the slope of the water surface profile.

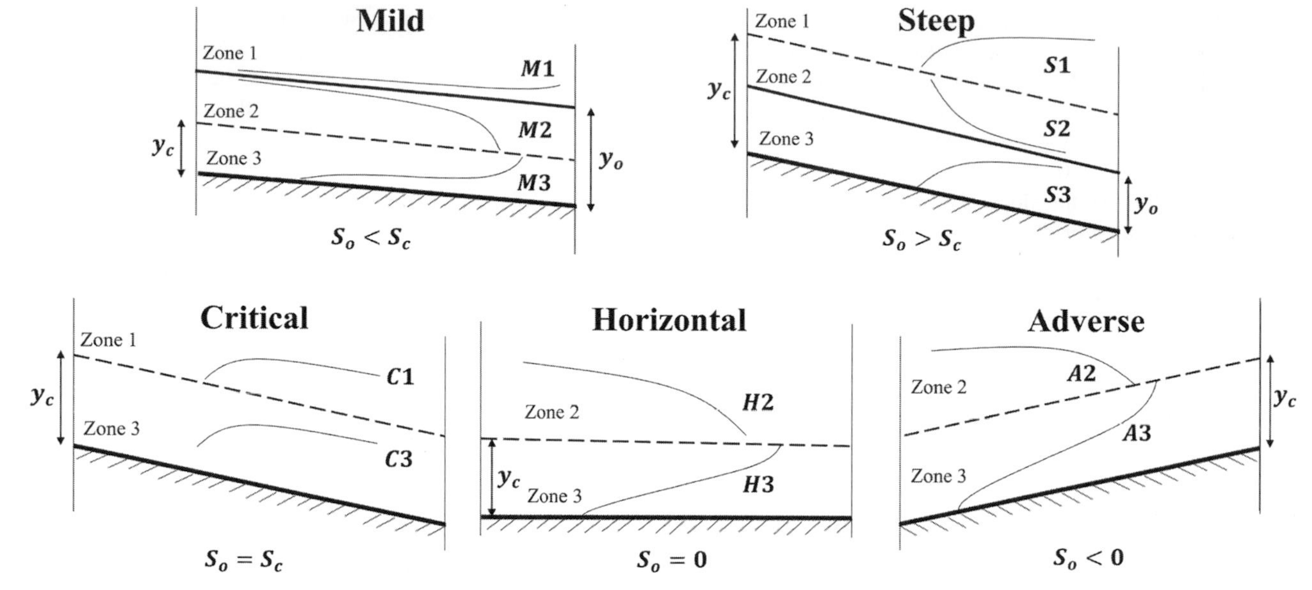

Figure 8.17 Classification of water surface profiles in open channels.

8.5.3 Computation of Flow Profile

Equation (8.66) does not have an analytical solution and is therefore solved numerically. A direct-step method is often used, assuming the channel is prismatic. Applying the Bernoulli equation to two points of a channel section, one can write

$$S_0 \Delta x - S_f \Delta x = E_1 - E_2. \tag{8.68}$$

Hence,

$$\Delta x = \frac{E_1 - E_2}{S_f - S_0}. \tag{8.69}$$

The energy slope between two sections can be taken as the mean of grades at sections 1 and 2. Then, eq. (8.69) can be written as

$$\Delta x = \frac{E_1 - E_2}{\overline{S_f} - S_0}, \tag{8.70}$$

where

$$\overline{S_f} = \frac{S_{f1} + S_{f2}}{2}, \tag{8.71}$$

$$S_f = \frac{u^2 n^2}{R^{\frac{4}{3}}}. \tag{8.72}$$

The computation steps can be outlined as follows:

1. Choose several values of y, beginning with the control point.
2. For a chosen y, compute A, R, $R^{4/3}$, u, and Q.
3. For the chosen y, compute the velocity head, specific energy, and energy slope.
4. For two successive values of y, calculate the difference between the specific energy, ΔE, and average energy slope, $\overline{S_f}$.
5. Compute Δx.

Example 8.15: Consider a trapezoidal channel 3 m wide with side slopes of 2:1 and bed slope of 0.0005. It carries a discharge of 20 m^3/s. Manning's n is given as 0.03. There is an obstruction at the downstream end that backs up the water to a depth of 3.5 m. Compute the backwater profile to a depth 10% greater than the normal depth.

Solution:

1. from Example 8.6: $y_c = 1.252$ m;
2. from Example 8.8: $y_n = 2.575$ m;
3. $y_n(=2.575 \text{ m}) > y_c(=1.252 \text{ m}) \rightarrow$ Mild slope;
4. $y = 3.5$ m is greater than y_n: M1 type profile;
5. section factor:

$$Z_c = \frac{Q}{\sqrt{g}} = \frac{20}{\sqrt{9.81}} = 6.39;$$

6. conveyance factor for uniform flow:

$$K_n = \frac{Q}{\sqrt{S_o}} = \frac{20\,\frac{\text{m}^3}{\text{s}}}{\sqrt{0.0005}} = 894.43.$$

7. At the starting point, the control section depth is 3.5 m. The next calculated point y is selected depending on how many cross-sections are calculated. Normally, increasing the number of cross-sections (i.e., smaller Δy) can reduce the calculation error. The last computed

point is 10% greater than $y_n = 2.575$:
1.10×2.575 m $= 2.83$ m (as shown in Figure 8.18)

8. Subcritical flow is computed from downstream (the obstruction at the downstream as the origin) to proceed upstream. The computations are shown in Table 8.4.

$$T = 3 + 4y,$$

$$A = 3y + 2y^2,$$

$$P = 3 + 2\sqrt{5}y,$$

$$R = \frac{A}{P},$$

$$K = \frac{1}{n}AR^{\frac{2}{3}},$$

$$Z = \sqrt{\frac{A^3}{T}},$$

$$\frac{dx}{dy} = \frac{1}{S_0}\left[\frac{1 - \left(\frac{Z_c}{Z}\right)^2}{1 - \left(\frac{K_n}{K}\right)^2}\right],$$

$$\Delta x = \frac{\left[\frac{dx}{dy_1} + \frac{dx}{dy_2}\right]}{2}(y_2 - y_1).$$

Table 8.4 *Computations for Example 8.15*

y	T	A	P	R	K	Z	dx/dy	Δx	Cumulative x
3.50	17.00	35.00	18.65	1.88	1774.88	50.22	2637.40	–	–
3.30	16.20	31.68	17.76	1.78	1553.31	44.30	2929.85	556.72	556.72
3.10	15.40	28.52	16.86	1.69	1349.46	38.81	3470.34	640.02	1196.74
2.90	14.60	25.52	15.97	1.60	1162.76	33.74	4722.80	819.31	2016.06
2.83	14.32	24.51	15.66	1.57	1101.36	32.06	5640.88	362.73	2378.79

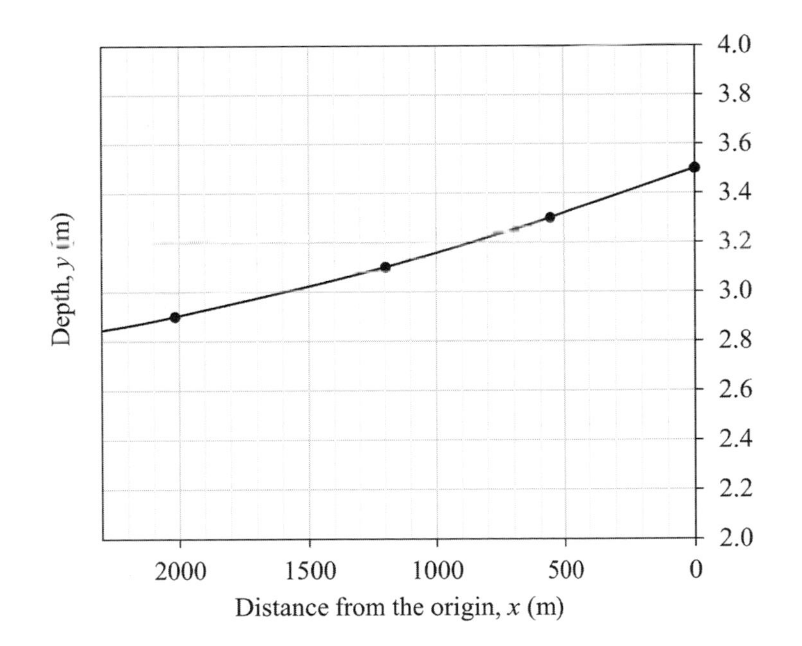

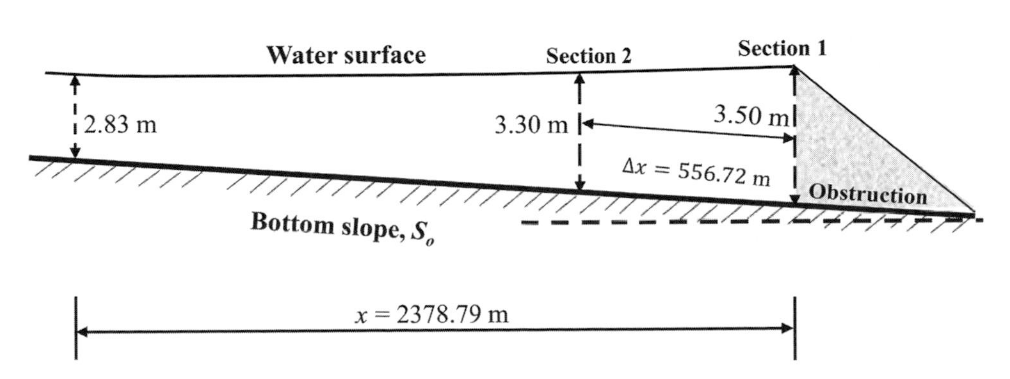

Figure 8.18 Depth profile of Example 8.15.

QUESTIONS

Q.8.1 Flow measurements show the values of velocity and area for different subsections as follows:

Subsection	1	2	3	4	5	6
Area (m^2)	0.5	1.5	2.0	1.75	1.25	0.25
Velocity (m/s)	0.15	0.25	0.40	0.30	0.20	0.10

Calculate the energy and momentum coefficients.

Q.8.2 Consider a 2-m wide rectangular channel reach 100 m long, with a bed slope of 0.001. At section 1 it has 1.5 m of flow depth and flow velocity of 1 m/s. Neglecting energy losses, compute the flow depth and velocity at section 2. Now use the momentum principle to compute flow depth and velocity.

Q.8.3 Consider a rectangular channel reach. At section 1 it has a width of 5 m, and 100 m downstream at section 2 it has a width of 6 m. The channel has a slope of 0.005. At section 1 the flow velocity is 1.5 m/s and the discharge is 5 m^3/s. Calculate the flow depth at sections 1 and 2 and velocity at section 2. Assume there are no energy losses. Also, compute the energy head. Use the momentum principle to solve the same problem.

Q.8.4 A trapezoidal channel carries a discharge of 2 m^3/s and has a bottom width of 1.5 m and a side slope of 2:1. Plot the specific energy curve for this channel. Compute the depth at which the critical flow will occur. Determine the alternative depth to a depth of 0.75 m. Indicate the state of flow at the alternative depth.

Q.8.5 Show for a rectangular channel that the depth of flow equals two-thirds of the minimum specific energy and the velocity head equals one-third of the minimum specific energy.

Q.8.6 Compute the critical depth and critical velocity for a flow of 10 m^3/s carried by a trapezoidal channel with a base width of 2.5 m and a side slope of 1:2.

Q.8.7 A 2-m wide rectangular channel carries a discharge of 2 m^3/s, has a bed slope of 0.001, and has a flow depth of 0.75 m. Calculate the discharge if the flow depth is doubled.

Q.8.8 A 1-m wide rectangular channel carries a discharge of 1.5 m^3/s and has a bed slope of 0.005. Manning's n is 0.035. Compute the normal flow depth, critical flow depth, and state of flow.

Q.8.9 If the bed slope is 0.01, compute the normal flow depth, critical flow depth, and state of flow for the channel in Q.8.8.

Q.8.10 Design a trapezoidal channel for a discharge of 10 m^3/s. The following information is given: Manning's n = 0.025, bed slope = 0.001, side slope = 2:1, and b/y ratio = 0.5.

Q.8.11 Using the regime theory, design a trapezoidal irrigation channel for a discharge of 5 m^3/s on a bed slope of 0.001 and a side slope of 1.5:1. The grain diameter is 0.25 mm and sediment load is 2000 ppm, for which the silt factor is increased by 10%.

Q.8.12 Design the channel in Q.8.11 using the tractive force theory. The angle of repose is 35° and the permissible tractive force is 4 N/m^2.

Q.8.13 A trapezoidal channel has a base width of 2.5 m bed, a slope of 0.002, Manning's n of 0.03, and carries a discharge of 5 m^3/s. Compute the water surface profile using the direct-step method to a depth 10% greater than the normal depth. This channel is a tributary to a river where the water level is 2.5 m above the channel bottom.

Q.8.14 Determine the design capacity of an irrigation canal that has a total depth of 2 m (including freeboard), bottom width of 1.5 m, side slope of 2:1, and a hydraulic gradient of 0.1%. The value of Manning's roughness coefficient is assumed to be 0.025, and the freeboard is 15%.

APPENDIX: SUPPLEMENTARY INFORMATION

Seven channel shapes and their geometric elements are expressed as follows:

1. Rectangular section: The geometric elements of a rectangular channel section can be expressed as follows:

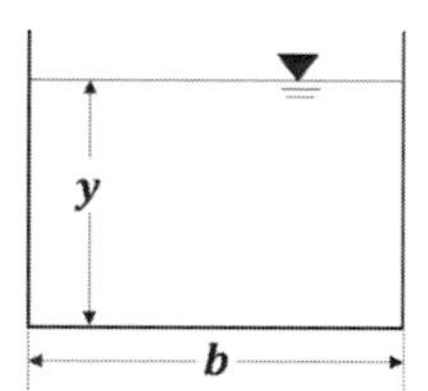

Top width (T): $T = b$.
Flow cross-sectional area (A): $A = by$.
Wetted perimeter (P): $P = b + 2y$.
Hydraulic depth (D): $\frac{\text{Cross-sectional area}}{\text{Top width}} = \frac{by}{b} = D = y$.
Hydraulic radius (R): $\frac{\text{Cross-sectional area}}{\text{Wetted perimeter}} = R = \frac{by}{b+2y}$.

2. Trapezoidal section: The geometric elements of a trapezoidal channel section can be expressed as follows:
Top width (T): $T = b + 2zy$.

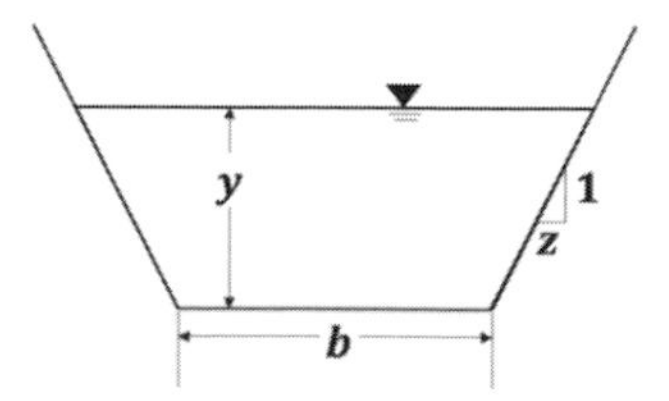

Flow cross-sectional area (A): $A = (b + zy)y$.

Wetted perimeter (P): $P = b + 2y\sqrt{1 + z^2}$.

Hydraulic depth (D): $\frac{\text{Cross-sectional area}}{\text{Top width}} = D = \frac{(b + zy)y}{b + 2zy}$.

Hydraulic radius (R): $\frac{\text{Cross-sectional area}}{\text{Wetted perimeter}} = R = \frac{(b + zy)y}{b + 2y\sqrt{1 + z^2}}$.

3. Triangular section: The geometric elements of a triangular channel section can be expressed as follows:

   Top width (T): $T = 2zy$.

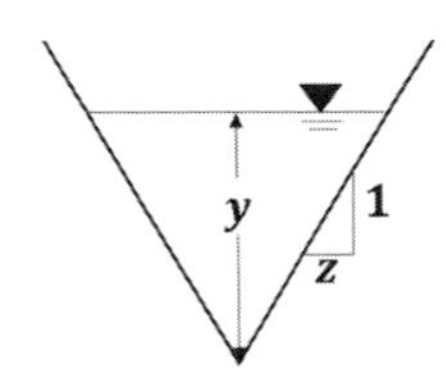

Flow cross-sectional area (A): $A = zy^2$.

Wetted perimeter (P): $P = 2y\sqrt{1 + z^2}$.

Hydraulic depth (D): $\frac{\text{Cross-sectional area}}{\text{Top width}} = \frac{(zy^2)}{2zy} = D = y/2$.

Hydraulic radius (R): $\frac{\text{Cross-sectional area}}{\text{Wetted perimeter}} = \frac{zy^2}{2y\sqrt{1 + z^2}} = R = \frac{zy}{2\sqrt{1 + z^2}}$.

4. Circular section: The geometric elements of a circular channel section can be expressed as follows:

   Top width (T): $T = 2\left(\sin\frac{\theta}{2}\right)r_0$ or $T = \sqrt{y(2r_0 - y)}$.

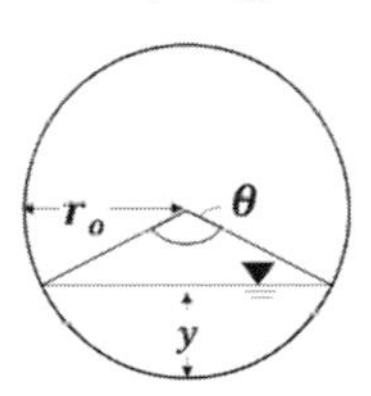

Flow cross-sectional area (A): $A = \frac{(\theta - \sin\theta)r_0^2}{2}$.

Wetted perimeter (P): $P = \theta r_0$.

Hydraulic depth (D): $\frac{\text{Cross-sectional area}}{\text{Top width}} = \frac{\frac{(\theta - \sin\theta)r_0^2}{2}}{2\left(\sin\frac{\theta}{2}\right)r_0} = D = \frac{1}{4}\left(\frac{\theta - \sin\theta}{\sin\frac{\theta}{2}}\right)r_0$.

Hydraulic radius (R): $\frac{\text{Cross-sectional area}}{\text{wetted perimeter}} = \frac{\frac{(\theta - \sin\theta)r_0^2}{2}}{\theta r_0} = R = \left(1 - \frac{\sin\theta}{\theta}\right)\frac{r_0}{2}$

5. Round-cornered rectangular section ($y > r_0$): The geometric elements of a round-cornered rectangular channel section can be expressed as follows:

   Top width (T): $T = b + 2r_0$.

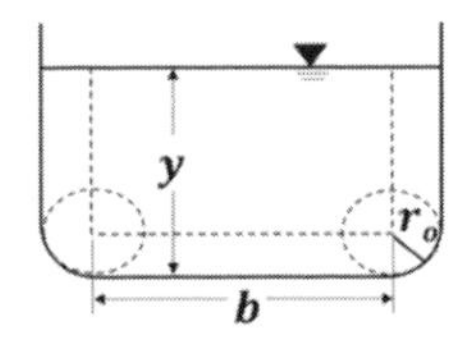

Flow cross-sectional area (A): $A = \left(\frac{\pi}{2} - 2\right)r_0^2 + (b + 2r_0)y$.

Wetted perimeter (P): $P = (\pi - 2)r_0 + b + 2y$.

Hydraulic depth (D): $\frac{\text{Cross-sectional area}}{\text{Top width}} = \frac{\left(\frac{\pi}{2} - 2\right)r_0^2 + (b + 2r_0)y}{b + 2r_0} = D = \frac{\left(\frac{\pi}{2} - 2\right)r_0^2}{b + 2r_0} + y$.

Hydraulic radius (R): $\frac{\text{Cross-sectional area}}{\text{Wetted perimeter}} = R = \frac{\left(\frac{\pi}{2} - 2\right)r_0^2 + (b + 2r_0)y}{(\pi - 2)r_0 + b + 2y}$.

6. Parabolic section: The geometric elements of a parabolic channel section can be expressed as follows:

   Top width (T): $T = \frac{3A}{2y}$.

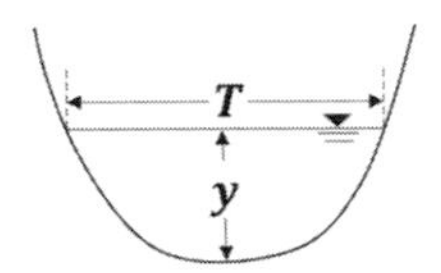

Flow cross-sectional area (A): $A = \frac{2Ty}{3}$.

Wetted perimeter (P): $P = T + \frac{8y^2}{3T}*$.

Hydraulic depth (D): $\frac{\text{Cross-sectional area}}{\text{Top width}} = \frac{\frac{2Ty}{3}}{\frac{3A}{2y}} = D = \frac{2}{3}y$.

Hydraulic radius (R): $\frac{\text{Cross-sectional area}}{\text{Wetted perimeter}} = \frac{\frac{2Ty}{3}}{T + \frac{8y^2}{3T}*} = R = \frac{2T^2y}{3T^2 + 8y^2}*$.

*Satisfactory approximation for the interval $0 < x \leq 1$, where $x = (4y)/T$. When $x > 1$, use the exact expression $P = \frac{T}{2}\left[\sqrt{1 + x^2} + \frac{1}{x}\ln\left(x + \sqrt{1 + x^2}\right)\right]$.

7. Round-cornered triangular section: The geometric elements of a round-cornered triangular channel section can be expressed as follows:

   Top width (T): $T = 2\left[z(y - r_0) + r_0\sqrt{1 + z^2}\right]$.

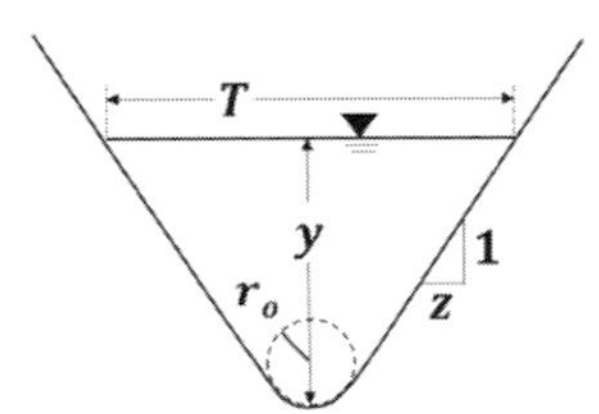

Flow cross-sectional area (A): $A = \frac{T^2}{4z} - \frac{r_0^2}{z}\left(1 - z\cot^{-1}z\right)$.

Wetted perimeter (P): $P = \frac{T}{z}\sqrt{1 + z^2} - \frac{2r_0}{z}\left(1 - z\cot^{-1}z\right)$.

Hydraulic depth (D): $\frac{\text{Cross-sectional area}}{\text{Top width}} = \frac{\frac{T^2}{4z} - \frac{r_0^2}{z}\left(1 - z\cot^{-1}z\right)}{2\left[z(y - r_0) + r_0\sqrt{1 + z^2}\right]} = D = \frac{T^2 - 4r_0^2\left(1 - z\cot^{-1}z\right)}{8z\left[z(y - r_0) + r_0\sqrt{1 + z^2}\right]}$.

Hydraulic radius (R): $\frac{\text{Cross-sectional area}}{\text{Wetted perimeter}} = \frac{\frac{T^2}{4z} - \frac{r_0^2}{z}\left(1 - z\cot^{-1}z\right)}{\frac{T}{z}\sqrt{1 + z^2} - \frac{2r_0}{z}\left(1 - z\cot^{-1}z\right)} = R = \frac{T^2 - 4r_0^2\left(1 - z\cot^{-1}z\right)}{4T\sqrt{1 + z^2} - 8r_0\left(1 - z\cot^{-1}z\right)}$

REFERENCES

Chow, V. T. (1959). *Open-Channel Hydraulics*. New York: McGraw-Hill.

Etcheverry, B. A. (1931). *Land Drainage and Flood Protection*. New York: McGraw-Hill.

Fortier, S., and Scobey, F. C. (1926). Permissible canal velocities. *Transactions of the American Society of Civil Engineers*, 89(1): 940–956.

Gupta, R. S. (2017). *Hydrology and Hydraulic Systems*. Long Grove, IL: Waveland Press.

Lacey, G. (1930). Stable channels in alluvium. *Minutes of the Proceedings of the Institution of Civil Engineers*, 229: 259–292.

Lane, E. W. (1955). Design of stable channels, *Transactions of the ASCE*, 120: 1234–1279.

US Army Corps of Engineers (1991). *Engineering and Design: Hydraulic Design of Flood Control Channels*. Washington, DC: US Army Corps of Engineers.

USDA-NRCS (2007). Stream restoration design. In *National Engineering Handbook*. Washington, DC: USDA.

9 Pipeline Hydraulics

Notation

A	flow cross-sectional area of the pipe (m^2, ft^2)
BP	brake horsepower (hp, kW)
C	Hazen–Williams friction coefficient
d_1, d_2	diameter of pipes (L [length])
D	pipe diameter (m, ft)
D_0	average outside diameter of the pipe (mm, in.)
D_i	average inside diameter of the pipe (mm, ft)
DR	dimension ratio
E_P	pump efficiency (fraction)
f	friction factor obtained from the Moody diagram (ft, mm)
g	acceleration due to gravity (9.81 m/s^2, 32.2 ft/s^2)
h	pressure head (m, ft)
h_f	friction head loss (m, ft)
h_f^*	minor head losses, i.e., head loss due to fitting, valves, etc. (m, ft)
H	total energy head (m, ft)
H_P	head provided by the pumping system (m, ft)
K_f	friction factor representing the Hazen–Williams, Darcy–Weisbach, and Scobey equations
K_F	friction factor, which is used to calculate head loss due to fitting, valves, etc.
K_s	Scobey coefficient
L	length of the pipe (m, ft)
p, p_1, p_2	pressure, the force per unit area exerted by water on the walls of a container (psi, kPa)
PR	pressure rating for outside diameter-based pipe
Q	discharge in the pipe (m^3/s, ft^3/s)
Re	Reynolds number (dimensionless)
S	the slope of energy line, equal to the bed slope for uniform flow (L/L)
S_s	hydrostatic design stress (psi, MPa)
S_{it}	long-term hydrostatic strength (psi, MPa)
t	wall thickness of the pipe (mm, in.)
v	mean flow velocity in the pipe (m/s, ft/s)
z, z_1, z_2	elevation head or gravitational head (m, ft)
γ	specific weight of water (9.807 kN/m^3, 62.43 lb/ft^3)
ε	equivalent roughness of the pipe internal material (ft, mm)

9.1 INTRODUCTION

Irrigation systems may receive water through a pipeline, which in turn receives water from a source. Depending on the size of an irrigation system, which depends on the size of the farm, there may be a network of pipelines. A pipe is a closed conduit, and when it runs full its hydraulics are different from the hydraulics of open channels in which the upper surface of flow is exposed to the atmosphere. Pipes or closed conduits are used in the sprinkler and drip irrigation systems to carry water from the source of water supply to the individual sprinkler or emitter in which water may be received from a reservoir through a pipeline or pumping. These systems are also called pressurized irrigation systems. Because of pressure, water can be made to flow in the desired direction. This chapter reviews the principles of pipeline hydraulics.

9.2 ENERGY EQUATION

Consider a pipe of diameter D resting on a constant slope and carrying a flow discharge as shown in Figure 9.1. At any point on the pipeline, the flow energy is composed of (1) potential energy due to elevation, (2) pressure energy or potential energy due to water pressure, and (3) kinetic energy due to the velocity of flow. Energy has the units

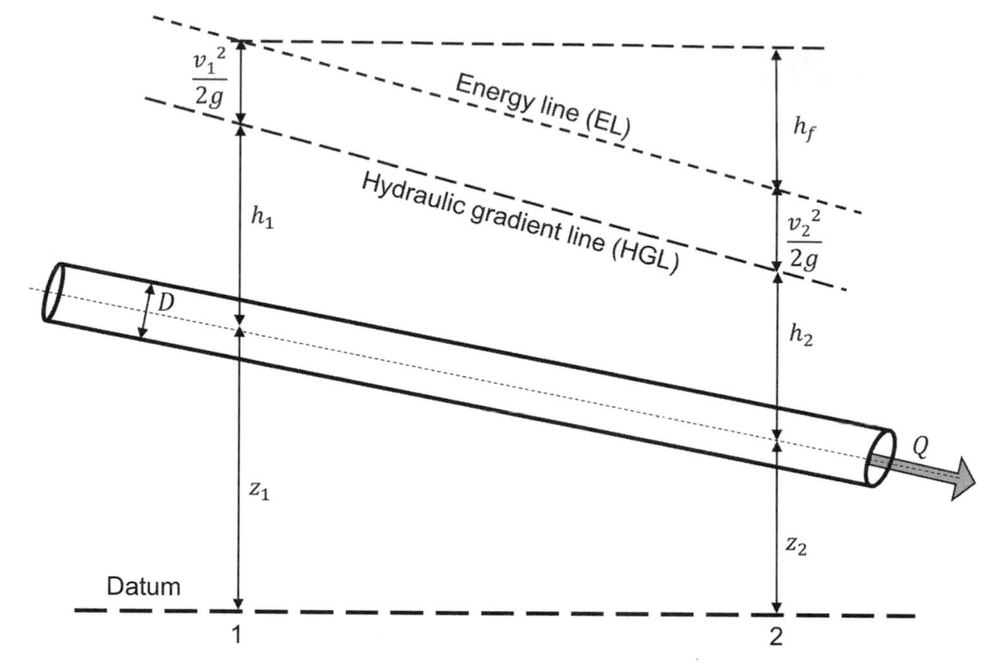

Figure 9.1 Schematic of energy of water flowing in a pipeline.

of force (*F*) times distance (*L*) (*FL*); weight has units of force (*F*). Thus, energy per unit weight has units of *FL*/*F*, or just the dimension of length (*L*), and is often called the energy head, measured in terms of height of water. Hence, the energy of water in an irrigation system includes velocity head, elevation head, and pressure head. The potential energy due to elevation is a result of the location of water relative to an arbitrary reference plane. Water at a higher elevation has more potential energy than water at a lower elevation. The water has the ability to do work as it flows downhill, such as eroding the soil surface, generating power, and transporting material with it. The potential energy of water decreases as it flows downhill. The letter z is often used to represent elevation head, also called gravitational head. The potential energy due to the pressurization of water can be a large component in an irrigation system. Pressure is the force per unit area exerted by water on the walls of a container. The pressure may be expressed as

$$p = \gamma h \text{ or } h = \frac{p}{\gamma}, \tag{9.1}$$

where p is the pressure (e.g., pounds per square inch [psi]), h is the pressure head, and γ is the weight of a unit volume of fluid (specific weight) (e.g., pounds per cubic feet per foot or pounds per square foot).

In general, the maximum recommended average velocity in an enclosed pipeline is 5 ft/s (1.52 m/s). When the velocity in the pipeline exceeds this there is the potential to develop very high-pressure surges, which may damage the pipeline. Pressure surges are due to flow being stopped suddenly while the upstream water has a very large amount of momentum. When the flow is stopped too quickly, the rapid change in momentum results in an impulsive force called water hammer.

Example 9.1: Two columns are filled with water to a height of 10 feet. One column has a cross-sectional area of 1 square inch and the other has a cross-sectional area of 10 square inches. Determine the pressure at the bottom of each column.

Solution: It may be noted that the weight density γ of water equals 62.43 lb/ft^3 or 0.433 psi/ft

$$p = \gamma h = \left(62.43\,\text{lb/ft}^3\right) \times (10\,\text{ft}) \times \left(\text{ft}^2/144\text{ in.}^2\right) = 4.34\text{ lb/in.}^2 = 4.34\text{ psi}.$$

The weight density γ of water equals 9.807 kN/m^3 or 9.778 kPa/m. Given 10 ft = 3.048 m,

$$p = \gamma h = (9.778\text{ kPa/m}) \times (3.048\text{ m}) = 29.8\text{ kPa}.$$

It may be noted that the pressure is independent of the surface area of the column.

Example 9.2: Consider two containers, one having a rectangular cross-section and the other having a trapezoidal cross-section, as shown in Figure 9.2. The height of water above the pressure gage is 5 feet in each container. Determine the pressure at the gage.

Solution: The pressure will be:

$$p = \gamma h = \left(62.43 \frac{\text{lb}}{\text{ft}^3}\right) \times 5 \text{ ft} \times \left(\frac{\text{ft}^2}{144 \text{ in.}^2}\right)$$

$$= 2.17 \frac{\text{lb}}{\text{in.}^2} = 2.17 \text{ psi}$$

Given 5 ft = 1.524 m,

$$p = \gamma h = (9.778 \text{ kPa/m}) \times (1.524 \text{ m}) = 14.9 \text{ kPa}.$$

The pressure will be the same for both containers.

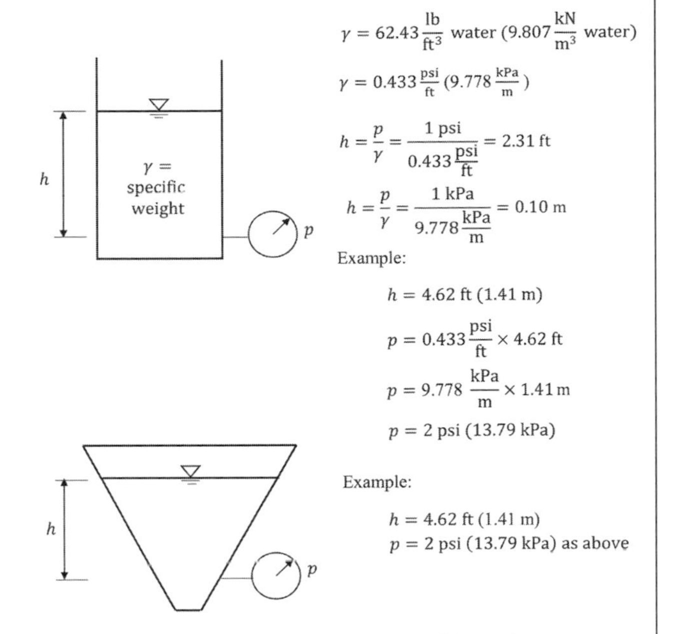

Figure 9.2 A rectangular container and a trapezoidal container.

Kinetic energy is the result of the movement of fluid, and the term *velocity head* or *kinetic energy head* is given by

$$\frac{v^2}{2g}, \tag{9.2}$$

where v is the average velocity at a point in a pipe or channel in ft/s (m/s) and g is the acceleration due to gravity (equal to 32.2 ft/s^2 or 9.81 m/s^2).

When water flows in the pipe from location 1 to location 2, it expends energy. The loss in energy is caused by friction that the flow must overcome. This means that the energy at point 1 is equal to the energy at point 2 plus the loss of energy. The energy at any point can be expressed in terms of head of water as

$$H = z + \frac{p}{\gamma} + \frac{v^2}{2g}, \tag{9.3}$$

where H is the total energy head, z is the elevation of the point with reference to a datum, p is the pressure, γ is the weight density (ρg), ρ is the mass density of water, g is the acceleration due to gravity, p/γ is the pressure head, v is the flow velocity, and $v^2/(2g)$ is the kinetic energy head.

An important law of fluid mechanics is the conservation of energy. Conservation of energy for irrigation systems is described by the Bernoulli equation. Let h_f denote the loss of energy, expressed as head loss. Then, as shown in Figure 9.1, the change in energy between point 1 (upstream) and point 2 (downstream) is given by the energy conservation equation or Bernoulli equation expressed as

$$H_1 = H_2 + h_f \tag{9.4}$$

or

$$z_1 + \frac{p_1}{\gamma} + \frac{v_1{}^2}{2g} = z_2 + \frac{p_2}{\gamma} + \frac{v_2{}^2}{2g} + h_f. \tag{9.5}$$

The head loss from point 1 to point 2 is the sum of friction loss due to the resistance to flow along a pipeline and minor pressure losses of energy through fittings, etc., as shown in Figure 9.3.

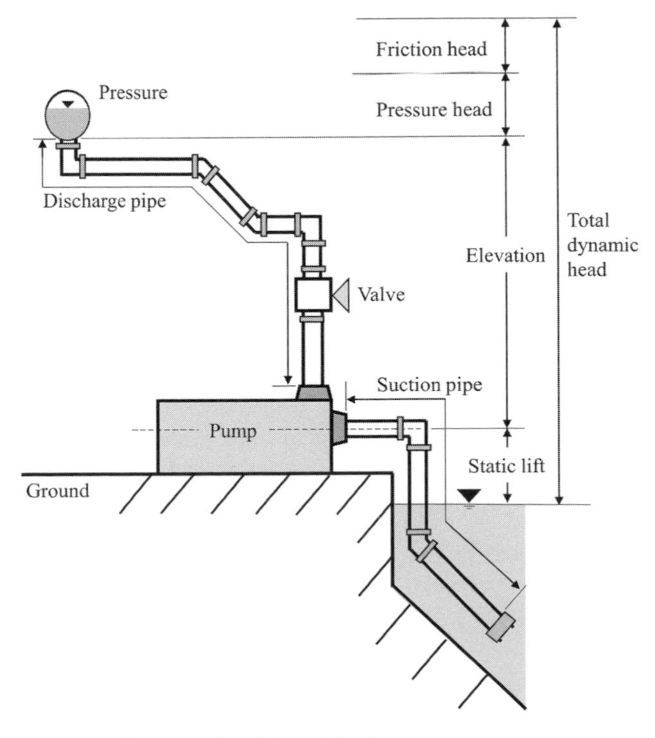

Figure 9.3 Illustration of total dynamic head.

Another important concept of water flow is continuity. In a hydraulic system, the water mass must be conserved. For incompressible fluid such as water, the spatially lumped form of continuity equation is expressed as

$$Q = Av, \tag{9.6}$$

where the flow velocity v is expressed in terms of discharge Q, and A is the cross-sectional area of flow expressed as

$$A = \frac{\pi}{4} D^2, \tag{9.7}$$

where D is the pipe diameter. If flow is constant, that is, $Q_1 = Q_2$, then $v_1 = v_2$, since the flow cross-sectional area remains constant. Equation (9.4) then simplifies to

$$z_1 + \frac{p_1}{\gamma} = z_2 + \frac{p_2}{\gamma} + h_f \tag{9.8}$$

or

$$\frac{p_1}{\gamma} = \frac{p_2}{\gamma} + (z_2 - z_1) + h_f \tag{9.9}$$

or

$$H_1 = H_2 + h_f. \tag{9.10}$$

Equation (9.10) shows that flow in the pipe follows the energy conservation law.

Example 9.3: A 30-cm diameter pipe of finite length is connected to a reservoir near its bottom, as shown in Figure 9.4. The water level in the reservoir from the center of the pipe is 5 m. The pipe is inclined and the elevation difference from the center of the end of the pipe to the water level in the reservoir is 15 m. This shows that the reservoir is at a higher elevation. The datum is the center of the pipe at its end. The discharge carried by pipe is 0.2 m^3/s. What will be the head loss in the pipe?

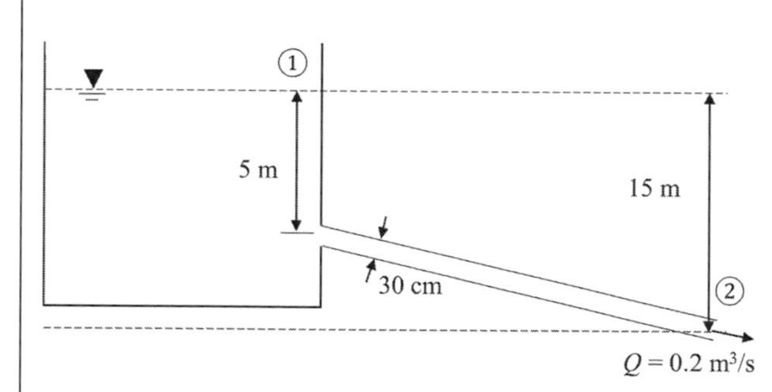

Figure 9.4 A pipe connected to a reservoir.

Solution: The discharge is given as 0.2 m^3/s, so the flow velocity can be computed by computing the pipe cross-sectional area, A, first:

$$A = \frac{\pi}{4}\left(\frac{30}{100}\right)^2 = 0.0707 \text{ m}^2.$$

Therefore, from the continuity equation, $v = \frac{0.2 \text{ m}^3/\text{s}}{0.0707 \text{ m}^2} =$ 2.83 m/s.

The end of the pipe can be considered as a datum, which is designated as point 2. Point 1 can be taken at the reservoir water surface where the pressure is atmospheric. At point 1, the elevation is 15 m, the pressure head is 0, and the velocity head is also 0. At point 2, the pressure head is 0 and the elevation is also 0. Then, applying the Bernoulli equation,

$$z_1 + \frac{p_1}{\gamma} + \frac{v_1^2}{2g} = z_2 + \frac{p_2}{\gamma} + \frac{v_2^2}{2g} + h_f,$$

so

$$15 + 0 + 0 = 0 + 0 + \frac{(2.83 \text{ m/s})^2}{2 \times 9.81 \text{ m/s}^2} + h_f.$$

This gives $h_f = 14.59$ m.

The flow representation, depicted in Figure 9.1 and described by eq. (9.9), is an approximation of the flow that occurs in sprinkler and trickle irrigation systems. In these systems, flow frequently exits the pipe through sprinklers and emitters. The pipe can be on flat ground or upslope or downslope, as shown in Figure 9.5. The distribution of pressure is not uniform and affects the flow from the outlet. The flow from the outlet is not uniform, and it affects the uniformity and efficiency of irrigation. It is not easy to determine the distribution of pressure throughout a sprinkler or trickle irrigation system.

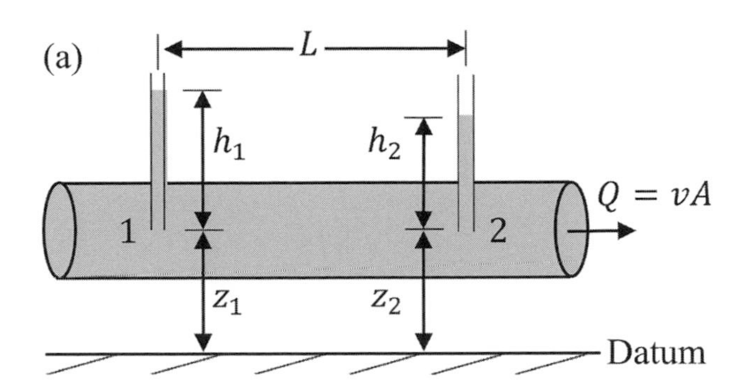

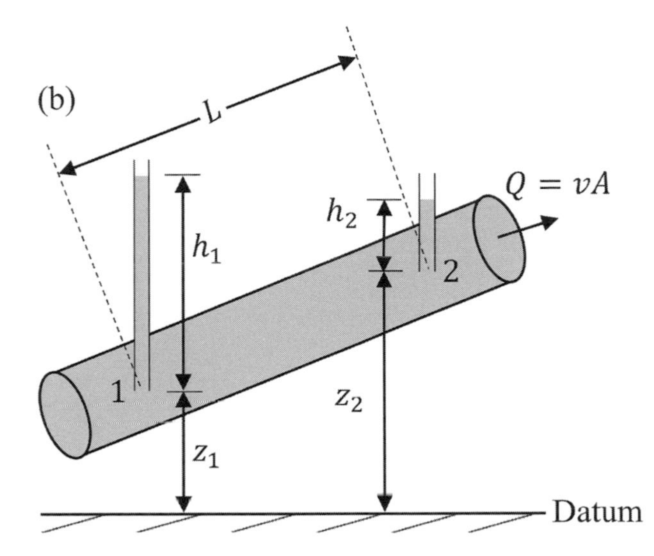

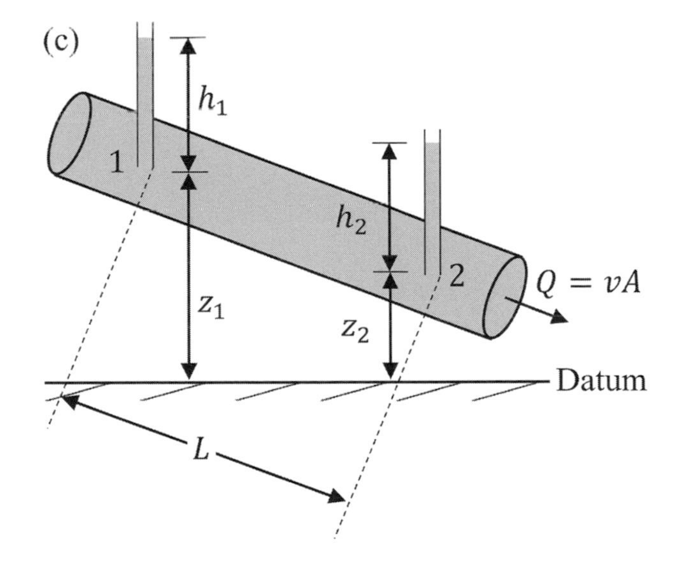

Figure 9.5 Pipe on (a) flat ground, (b) upslope, and (c) downslope.

Example 9.4: For the pipeline system shown in Figure 9.6, determine the head at the inlet into a 5 in. diameter pipeline.

Solution:

$$h = 65\text{ psi} \times \left(2.31\frac{\text{ft}}{\text{psi}}\right) = 150.15\text{ ft }(45.8\text{ m}),$$

$$A = \frac{\pi}{4}(5\text{ in.})^2 = 19.63\text{ in.}^2 \times \left(\frac{1\text{ ft}^2}{144\text{ in.}^2}\right)$$
$$= 0.136\text{ ft}^2\ \left(0.0126\text{ m}^2\right),$$

$$Q = 500\text{ gpm} \times \left(\frac{1\text{ cfs}}{448.83\text{ gpm}}\right) = 1.11\text{ cfs }\left(0.031\text{ m}^3/\text{s}\right),$$

$$v = \frac{Q}{A} = \frac{1.11\,\frac{\text{ft}^3}{\text{s}}}{0.136\ \text{ft}^2} = 8.16\ \frac{\text{ft}}{\text{s}}\ (2.5\ \text{m/s}),$$

$$\text{Velocity head} = \frac{v^2}{2g} = \frac{\left(8.16\,\frac{\text{ft}}{\text{s}}\right)^2}{2 \times \left(32.2\,\frac{\text{ft}}{\text{s}^2}\right)} = 1.03\ \text{ft}\ (0.31\ \text{m}),$$

$$\text{Total head} = 20\,\text{ft} + 150.15\,\text{ft} + 1.03\,\text{ft} = 171.18\,\text{ft}\ (52.2\,\text{m}).$$

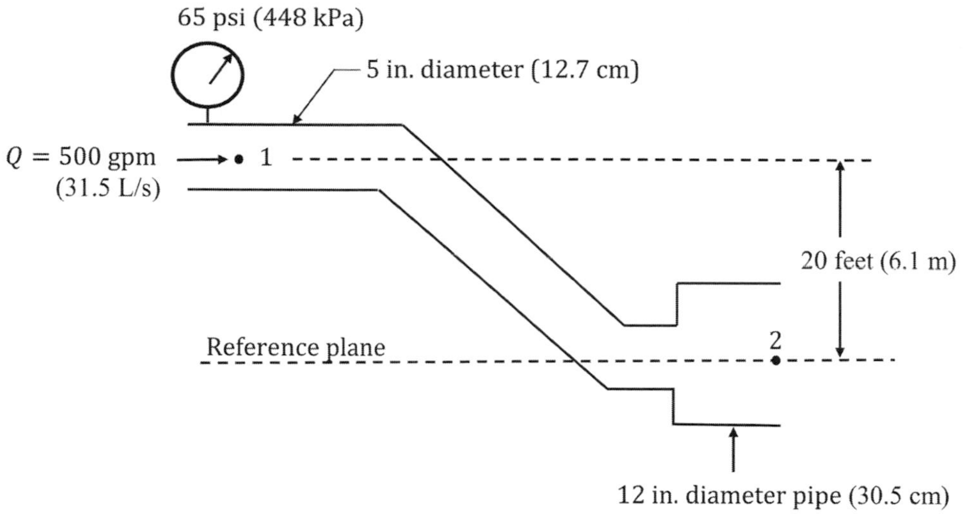

Figure 9.6 The pipeline system in Example 9.4.

Example 9.5: What is the velocity head at point 2 in Example 9.4?

Solution:

$$Q_1 = Q_2 = 500\ \text{gpm}\ (31.5\ \text{L/s}),$$
$$d_2 = 12\ \text{in.}\ (30.5\ \text{cm}),$$

$$A_2 = \frac{\pi}{4}(12\ \text{in.})^2 = 113.10\ \text{in.}^2 \times \left(\frac{1\ \text{ft}^2}{144\ \text{in}^2}\right) = 0.785\ \text{ft}^2\ \left(0.0957\ \text{m}^2\right),$$

$$Q_1 = Q_2 = 500\ \text{gpm} \times \left(\frac{1\ \text{cfs}}{448.83\ \text{gpm}}\right) = 1.11\ \text{cfs}\ \left(0.031\ \text{m}^3/\text{s}\right),$$

$$v_2 = \frac{Q_2}{A_2} = \frac{1.11\,\frac{\text{ft}^3}{\text{s}}}{0.785\ \text{ft}^2} = 1.41\ \frac{\text{ft}}{\text{s}}\ (0.43\ \text{m/s}),$$

$$\text{Velocity head (at point 2)} = \frac{v_2^2}{2g} = \frac{\left(1.41\,\frac{\text{ft}}{\text{s}}\right)^2}{2 \times \left(32.2\,\frac{\text{ft}}{\text{s}^2}\right)} = 0.031\ \text{ft}\ (0.01\ \text{m}).$$

Thus, the velocity head in the 12-inch pipe is only 0.03 times the velocity head in the 5-inch pipe.

In sprinkler and trickle irrigation systems, the main design problem is the selection of pipe sizes in such a manner that energy losses do not exceed the prescribed limits in order to ensure high efficiency and uniformity. From a hydraulic viewpoint, two situations can be considered for the determination of pipe flow: (1) without multiple outlets, and (2) with multiple outlets (laterals and manifolds).

9.3 ENERGY LOSS EQUATIONS

The amount of energy loss incurred by flow depends on the roughness of the interior surface of the pipe, diameter of the pipe, bending of pipes, pipe fittings, viscosity of water, and flow velocity. These factors are combined into what is called friction coefficient, which is determined experimentally. There are other factors that also affect the loss of energy or the friction coefficient, such as debris, sediment deposition, and aging of pipes. These factors occur with time.

9.3.1 Head Loss in Pipes

The head loss in pipelines can be computed in several ways. One of the commonly used equations is the Hazen–Williams equation:

$$h_f = K \frac{L}{D^{1.167}} \frac{v^{1.852}}{C^{1.852}}, \tag{9.11}$$

where h_f is the head loss (L), K is the conversion factor to handle units, C is the Hazen–Williams friction factor, which is a function of pipe material, L is the pipe length (L), D is the internal diameter (L), and v is mean flow velocity (L/T).

For circular pipes in irrigation engineering, the following simplification is often made. The cross-section A is expressed as $A = \left(\frac{\pi}{4}\right)D^2$. The velocity v is expressed in terms of discharge:

$$v = \frac{Q}{\left(\frac{\pi}{4}\right)D^2}. \tag{9.12}$$

Then, eq. (9.11) can be written as

$$h_f = K \frac{L}{D^{4.87}} \left(\frac{Q}{C}\right)^{1.852} \left(\frac{4}{\pi}\right)^{1.852}. \tag{9.13}$$

Equation (9.13) is often expressed as

$$h_f = k\left(\frac{Q}{C}\right)^{1.852} D^{-4.871} L, \tag{9.14}$$

where

$$k = K\left(\frac{4}{\pi}\right)^{1.852}. \tag{9.15}$$

If Q is the discharge in the pipeline (L/s), D is the inside pipe diameter (mm), L is the pipe length (m), h_f is the head loss (m), and C is the friction coefficient for continuous pipe sections, then $k = 1.21 \times 10^{10}$. Coefficient C is often called the Hazen–Williams friction coefficient. The value of C is normally regarded as 120–140 for manifold pipes, and 140–150 for plastic pipes with no discharging outlets.

For computing friction losses using the British system of units, the Hazen–Williams equation is written as

$$h_f = 1047\left(\frac{Q}{C}\right)^{1.852}\left(\frac{1}{D^{4.871}}\right), \tag{9.16}$$

$$p_f = 453\left(\frac{Q}{C}\right)^{1.852}\left(\frac{1}{D^{4.871}}\right), \tag{9.17}$$

where h_f is the friction loss in feet of head per 100 feet of pipe, p_f is the friction loss in psi per 100 feet of pipe, Q is the flow rate in gallons per minute (gpm), D is the internal diameter in inches, and C is the roughness coefficient. The conversion constants for the Hazen–Williams equation for different combinations of units are given in Table 9.1. The values of C are given for different materials in Table 9.2. The values of friction loss for different diameter pipes (4–12 in. or 102–305 mm), discharge values (100–2000 gpm or 6.3–126 L/s), and pipe materials (steel, aluminum, and plastic) are given in Table 9.3. Pipe sizes of different materials (iron and plastic) are given in Table 9.4. Table 9.3 only provides a general friction loss value. A more accurate friction loss value can be obtained if pipe inside diameters and wall thickness are considered.

Table 9.1 *Conversion constants for the Hazen–Williams equation for combinations of units*

h_f	L	Q	D	k
m	m	L/s	mm	1.22×10^{10}
m	m	L/h	mm	3163
m	m	m^3/d	mm	3.162×10^6
ft	ft	ft^3/s	ft	4.73
ft	ft	gpm	ft	10.46

Table 9.2 *Values of* C *in the Hazen–William equation*

Material	C
Aluminum pipe with couplers	120–130
Aluminum pipe with gates	110
Brick sewer	100
Cement asbestos pipe	140
Galvanized steel pipe	140
Galvanized iron	120
Glass	140
Standard steel pipe (new)	130
PVC class 160 irrigation pipe	150
PVC pipe with gates	130

PVC, polyvinyl chloride.

The Hazen–Williams equation is sometimes written directly for computing velocity or discharge in British units as

$$v = 1.318\, CR^{0.63} S^{0.54}, \tag{9.18}$$

$$Q = 0.432\, CD^{2.63} S^{0.54}, \tag{9.19}$$

in which v is the cross-sectional mean velocity (ft/s), Q is the discharge in ft^3/s, C is the Hazen–Williams coefficient of roughness, D is the internal pipe diameter, R is the hydraulic radius (ft), and S is the slope of the energy line (h_f/L), where h_f is the head loss due to friction, and L is the pipe length. In metric units, discharge Q (m^3/s) can be expressed as

$$Q = 0.278\, CD^{2.63} S^{0.54}. \tag{9.20}$$

It may be of interest to note that the Hazen–Williams equation (eq. 9.18) is directly related to the Chezy equation for $R = 1$, $S = 1/1000$, with C being the same as the Chezy coefficient of roughness.

Table 9.3 *Friction loss for different pipe materials (steel, aluminum, and plastic pipe)*

Q in gpm (L/s)	Friction head loss in ft/100 ft (m/100 in.)														
	Diameter														
	4 in. (102 mm)			6 in. (152 mm)			8 in. (203 mm)			10 in. (254 mm)			12 in. (305 mm)		
	Steel	Alum.	Plastic	Steel	Alum.	Plastic	Steel	Alum.	Plastic	Steel	Alum.	Plastic	Steel	Alum.	Plastic
100 (6.3)	1.25	0.81	0.55	0.17	0.11	0.08									
150 (9.5)	3.00	1.73	1.18	0.36	0.23	0.16	0.09	0.06							
200 (12.6)	4.39	3.65	2.01	0.62	0.42	0.28	0.15	0.10	0.07	0.05					
300 (18.9)	9.47	6.35	4.27	1.32	0.92	0.60	0.32	0.21	0.14	0.11	0.07	0.05	0.05	0.05	
350 (22.1)		8.32	5.43	1.73	1.16	0.79	0.43	0.28	0.19	0.14	0.09	0.06	0.06	0.05	
400 (25.2)		10.74	7.39	2.31	1.50	1.02	0.55	0.37	0.25	0.18	0.12	0.09	0.08	0.05	
450 (28.4)			9.24	2.77	1.85	1.27	0.69	0.45	0.32	0.23	0.15	0.11	0.10	0.06	
500 (31.6)			11.55	3.47	2.31	1.55	0.83	0.55	0.39	0.28	0.18	0.13	0.12	0.08	0.05
550 (34.7)				4.11	2.66	1.85	0.99	0.66	0.46	0.33	0.22	0.16	0.13	0.09	0.06
600 (37.9)				4.85	3.19	2.19	1.18	0.79	0.54	0.39	0.25	0.18	0.17	0.11	0.07
650 (41.0)				5.54	3.70	2.54	1.39	0.90	0.63	0.46	0.30	0.21	0.19	0.13	0.09
700 (44.2)				6.47	4.27	2.89	1.62	1.04	0.72	0.53	0.35	0.24	0.22	0.15	0.10
750 (47.3)				7.39	4.85	3.35	1.80	1.16	0.82	0.60	0.39	0.28	0.25	0.16	0.11
800 (50.5)				8.32	5.54	3.70	2.02	1.27	0.89	0.68	0.42	0.31	0.28	0.18	0.13
850 (53.6)				9.24	6.12	4.16	2.31	1.50	1.03	0.76	0.51	0.35	0.32	0.21	0.15
900 (56.8)				10.16	6.93	4.62	2.54	1.67	1.16	0.84	0.55	0.39	0.35	0.23	0.16
950 (59.9)				11.55	7.39	5.20	2.82	1.85	1.35	0.95	0.61	0.43	0.39	0.25	0.18
1000 (63.1)					8.32	5.66	3.07	2.02	1.40	1.06	0.65	0.48	0.43	0.28	0.19
1050 (66.3)					9.01	6.24	3.35	2.25	1.50	1.12	0.74	0.51	0.46	0.31	0.21
1100 (69.4)					9.93	6.93	3.70	2.54	1.65	1.24	0.81	0.56	0.51	0.33	0.23
1200 (75.7)					11.55	8.09	4.39	2.72	1.96	1.46	0.95	0.66	0.60	0.39	0.27
1300 (82.0)						9.24	5.08	3.44	2.28	1.69	1.11	0.76	0.71	0.46	0.31
1400 (88.3)						10.51	5.89	3.81	2.59	1.96	1.25	0.88	0.81	0.52	0.37
1500 (94.7)							6.58	4.39	2.93	2.19	1.47	1.00	0.92	0.60	0.42
1600 (101)							7.39	4.97	3.29	2.54	1.60	1.12	1.04	0.67	0.46
1700 (107)							8.32	5.54	3.70	2.77	1.85	1.27	1.16	0.76	0.52
1800 (114)							9.24	6.12	4.13	3.10	2.08	1.39	1.29	0.84	0.57
1900 (120)							10.16	6.81	4.62	3.47	2.31	1.55	1.46	0.95	0.65
2000 (126)							11.32	7.39	5.08	3.80	2.54	1.70	1.59	1.04	0.69

Adapted from USDA-NRCS, 1997.

Table 9.4 *Pipe sizes of different materials (ANSI/ASAE Standard S376.3)*

Nominal pipe size		Average OD.			50 ft head	SDR 64	SDR 51	SDR 41	SDR 26	SDR 21	SDR 17	Schedule 40[d]	Schedule 80
					Pipe ID[c]	Pipe ID	Pipe ID	Pipe ID	Pipe ID	Pipe ID	Pipe ID	Pipe ID	Pipe ID
mm	in.	mm	in.	Pipe class	(in.)	(in.)	(in.)	(in.)	(in.)	(in.)	(in.)	(in.)	(in.)
4	1/8	10.29	0.405	IPS[a]								0.279	0.255
8	1/4	13.72	0.540	IPS								0.364	0.302
10	3/8	17.14	0.675	IPS								0.493	0.423
15	1/2	21.34	0.840	IPS								0.622	0.546
20	3/4	26.67	1.050	IPS						0.930	0.926	0.834	0.742
25	1	33.40	1.315	IPS					1.195	1.189	1.161	1.094	0.957
32	1 1/4	42.16	1.660	IPS					1.532	1.502	1.464	1.380	1.218
40	1 1/2	48.26	1.900	IPS					1.754	1.720	1.676	1.610	1.500
50	2	60.32	2.375	IPS					2.193	2.149	2.095	2.067	1.939
65	2 1/2	73.02	2.875	IPS					2.655	2.601	2.537	2.469	2.323
80	3	88.90	3.500	IPS				3.330	3.230	3.166	3.088	3.068	2.900
90	3 1/2	101.60	4.000	IPS				3.804	3.692	3.620	3.530	3.548	3.364
100	4	114.30	4.500	IPS		4.36		4.280	4.154	4.072	3.970	4.026	3.826
		114.30	4.500	PIP[b]	4.37		4.338	4.298					
125	5	141.30	5.563	IPS		5.389		5.291	5.135	5.033	4.909	5.047	4.813
150	6	168.23	6.625	IPS		6.417		6.301	6.115	5.993	5.845	6.065	5.761
		155.96	6.140	PIP	6.00		5.90	5.840					
200	8	219.06	8.625	IPS				8.205	7.961	7.799	7.609	7.981	
		207.30	8.160	PIP	8.0	7.986	7.840	7.762					
250	10	273.05	10.750	IPS				10.226	9.924	9.728	9.486	10.02	
		259.08	10.200	PIP	10.0	9.982	9.800	9.720					
300	12	323.85	12.750	IPS				12.128	11.770	11.538	11.250	11.938	
		310.90	12.240	PIP	12.0	11.978	11.760	11.642					
350	14	355.60	14.000	IPS				13.318	12.924	12.664	12.354		
380	15	388.62	15.300	PIP	15.0	14.972	14.70	14.554	14.124	13.844			

Table 9.4 (*cont.*)

Nominal pipe size		Average OD.			50 ft head	SDR 64	SDR 51	SDR 41	SDR 26	SDR 21	SDR 17	Schedule 40[d]	Schedule 80
mm	in.	mm	in.	Pipe class	Pipe ID[c] (in.)	Pipe ID (in.)	Pipe ID (in.)	Pipe ID (in.)	Pipe ID (in.)	Pipe ID (in.)	Pipe ID (in.)	Pipe ID (in.)	Pipe ID (in.)
400	16	405.40	16.000	IPS				15.220	14.770	14.476	14.118		
450	18	457.20	18.000	IPS		17.666		17.122	16.616	16.286	15.882		
		475.00	18.701	PIP		18.301	17.969	17.789	17.263				
500	20	508.00	20.000	IPS		19.564		19.024	18.462	18.096	17.648		
	21	559.99	22.047	PIP		21.571	21.183	20.971	20.351				
600	24	609.60	24.000	IPS		23.486		22.830	22.154	21.714	21.176		
		629.99	24.803	PIP		24.271	23.831	23.593	22.895				
700	27	710.00	27.953	PIP				26.589	25.803				
750	30	762.00	30.000	IPS				28.536	27.692	27.144	26.47		
900	36	914.40	36.000	IPS				34.244	33.230	32.572	31.764		

[a] The pipe OD sizing class designated as IPS (iron pipe size) refers to average OD dimensions for PVC pipe that match these OD dimensions for that standard OD size system known as IPS.

[b] The pipe OD sizing class designated as PIP (plastic irrigation pipe size) refers to average OD dimensions for PVC pipe that match these OD dimensions for plastic irrigation pipe.

[c] Pipe ID = Average OD – 2× minimum wall thickness.

[d] Chlorinated poly(vinyl chloride) (CPVC) plastic pipe, schedule 40 or 80.

Example 9.6: Let $z_1 = 110$ ft (33.5 m), $z_2 = 100$ ft (30.5 m), $p_1 = 50$ psi (345 kPa). The pipe is 6 in. (150 mm) IPS PVC, SDR = 21 (200 psi or 1380 kPa), and length is 1000 ft (304.8 m). Determine p_2 with discharge of 600 and 300 gpm. Friction loss for the pipe under 600 and 300 gpm is given as 2.26 ft/100 ft and 0.62 ft/100 ft.

Solution:

$$h_1 = 50 \text{ psi} \times \left(2.31 \frac{\text{ft}}{\text{psi}}\right) = 115.5 \text{ ft}.$$

Since $v_1 = v_2$, velocity heads cancel from the Bernoulli equation.

$$h_2 + z_2 = h_1 + z_1 - \text{losses},$$

$$h_2 + 100 \text{ ft} = h_1 + 110 \text{ ft} - \text{losses},$$

$$h_2 = 115.5 \text{ ft} + 110 \text{ ft} - 100 \text{ ft} - \text{losses} = 125.5 \text{ ft} - \text{losses},$$

$$p_2 = \frac{h_2 \text{ ft}}{2.31 \frac{\text{ft}}{\text{psi}}}.$$

The friction losses with discharge of 600 and 300 gpm can be calculated as

Q (gpm)	Losses (ft/100 ft)	Losses (per 1000 ft)	h_2 (ft)	p_2 (psi)
600	2.26	22.6	102.9	44.5
300	0.62	6.20	119.3	51.6

Similarly, the results can be expressed in SI units as

Q (L/s)	Losses (m/100 m)	Losses (per 304.8 m)	h_2 (m)	p_2 (kPa)
37.9	2.26	6.89	31.4	306.8
18.9	0.62	1.89	36.4	356.5

Example 9.7: Compute friction losses for $Q = 10$ gpm (0.63 L/s).

Pipe size (in.) (mm)	Pipe material	ID	OD	C
1 in. (25 mm)	SCH 40 PVC	1.049 (26.64)	1.315 (33.40)	150
1 in. (25 mm)	SCH 40 steel	1.049 (26.64)	1.315 (33.40)	130
0.5 in. (15 mm)	SCH 40 PVC	0.622 (15.80)	0.848 (21.54)	150

Solution: For 0.5 in. SCH 40 PVC pipe:

$$h_f = 1047 \times \left(\frac{Q}{C}\right)^{1.852} \times \left(\frac{1}{D}\right)^{4.871}$$
$$= 1047 \times \left(\frac{10 \text{ gpm}}{150}\right)^{1.852} \times \left(\frac{1}{0.622}\right)^{4.871},$$

$$h_f = 70.2 \text{ ft}/100 \text{ ft},$$

$$p_f = 70.2 \text{ ft} \times \frac{\left(\frac{0.433 \text{ psi}}{1 \text{ ft}}\right)}{100 \text{ ft}} = 30.4 \text{ psi}/100 \text{ ft}.$$

For 1-in. SCH 40 PVC pipe:

$$h_f = 1047 \times \left(\frac{Q}{C}\right)^{1.852} \times \left(\frac{1}{D}\right)^{4.871}$$
$$= 1047 \times \left(\frac{10 \text{ gpm}}{150}\right)^{1.852} \times \left(\frac{1}{1.049}\right)^{4.871},$$

$$h_f = 5.5 \text{ ft}/100 \text{ ft},$$

$$p_f = 5.5 \text{ ft} \times \frac{\left(\frac{0.433 \text{ psi}}{1 \text{ ft}}\right)}{100 \text{ ft}} = 2.4 \text{ psi}/100 \text{ ft}.$$

For 1-in. SCH 40 steel pipe:

$$h_f = 1047 \times \left(\frac{Q}{C}\right)^{1.852} \times \left(\frac{1}{D}\right)^{4.871}$$
$$= 1047 \times \left(\frac{10 \text{ gpm}}{130}\right)^{1.852} \times \left(\frac{1}{1.049}\right)^{4.871},$$

$$h_f = 7.17 \text{ ft}/100 \text{ ft}$$

$$p_f = 7.17 \text{ ft} \times \frac{\left(\frac{0.433 \text{ psi}}{1 \text{ ft}}\right)}{100 \text{ ft}} = 3.1 \text{ psi}/100 \text{ ft}.$$

For 15-mm, assuming $L = 100$ m, SCH 40 PVC pipe:

$$h_f = 1.22 \times 10^{10} \times \left(\frac{Q}{C}\right)^{1.852} \times \left(\frac{1}{D}\right)^{4.871} L$$
$$= 1.22 \times 10^{10} \times \left(\frac{0.63}{150}\right)^{1.852} \times \left(\frac{1}{15.80}\right)^{4.871} (100),$$

$$h_f = 70.14 \text{ m or } 70.14/100 \text{ m},$$

$$p_f = 70.14 \text{ m} \times \frac{(9.778 \text{ kPa})}{100 \text{ m}} = 685.8 \text{ kPa}/100 \text{ m}.$$

For 25-mm, assuming $L = 100$ m, SCH 40 PVC pipe:

$$h_f = 1.22 \times 10^{10} \times \left(\frac{Q}{C}\right)^{1.852} \times \left(\frac{1}{D}\right)^{4.871} L$$

$$= 1.22 \times 10^{10} \times \left(\frac{0.63}{150}\right)^{1.852} \times \left(\frac{1}{26.64}\right)^{4.871} (100),$$

$h_f = 5.51$ m or $5.51/100$ m,

$$p_f = 5.51 \text{ m} \times \frac{(9.778 \text{ kPa})}{100 \text{ m}} = 53.8 \text{ kPa}/100 \text{ m}.$$

For 25-mm, assuming $L = 100$ m, SCH 40 steel pipe:

$$h_f = 1.22 \times 10^{10} \times \left(\frac{Q}{C}\right)^{1.852} \times \left(\frac{1}{D}\right)^{4.871} L$$

$$= 1.22 \times 10^{10} \times \left(\frac{0.63}{130}\right)^{1.852} \times \left(\frac{1}{26.64}\right)^{4.871} (100),$$

$h_f = 7.18$ m or $7.18/100$ m,

$$p_f = 7.18 \text{ m} \times \frac{(9.778 \text{ kPa})}{100 \text{ m}} = 70.2 \text{ kPa}/100 \text{ m}.$$

Thus, comparing friction losses,

0.5-in. PVC	30.4 psi/100 ft (685.8 kPa/100 m)
1-in. PVC	2.4 psi/100 ft (53.8 kPa/100 m)
1-in. steel	3.1 psi/100 ft (70.2 kPa/100 m)

Steel has 30% more losses than PVC. Doubling the size means approximately 1/12 of the loss.

Watters and Keller (1978) have found that the Hazen–Williams equation significantly underestimates energy (friction) losses when flow tends to be laminar; that is, the Reynolds number approaches the values in the laminar range. The Darcy–Weisbach equation is regarded as more accurate for determining energy losses and can be written as

$$h_f = f\frac{L}{D}\frac{v^2}{2g}, \tag{9.21}$$

where D is the pipe diameter (m), L is the pipe length (m), v is the average velocity flow velocity (m/s), g is the acceleration due to gravity (9.81 m/s^2), and f is the friction factor. It may be noted that since $v^2/(2g)$ is the kinetic energy head, h_f is expressed as a function of this head. For a pipe $f(L/D)$ is a fixed quantity, where L and D are easily measured, and f needs to be determined.

For circular pipes, the following substitution can be made:

$$h_f = k_1 f L \frac{Q^2}{D^5}, \tag{9.22}$$

where k_1 is the conversion factor. The constants for the Darcy–Weisbach equation for combinations of units are shown in Table 9.5.

Table 9.5 *Conversion constants for the Darcy–Weisbach equation for combinations of units*

h_f	L	Q	D	k_1
m	m	L/s	mm	8.2627×10^7
m	m	L/h	mm	6.3755
m	m	m^3/d	mm	1.10686×10^4
ft	ft	ft^3/s	ft	0.02517
ft	ft	gpm	in	0.03107

The friction factor f is a function of the relative roughness of the pipe internal material (ε/D) with respect to the internal pipe diameter (D) and the Reynolds number (Re), reflecting the quality of flow-laminar: transient or turbulent. The roughness of commercial pipes may not be uniform, so equivalent roughness values of pipes are given as in Table 9.6. The equivalent roughness in millimeters of different pipe materials is shown in Figure 9.7 (Moody, 1944).

It may also be noted that turbulent flow can be further classified into (1) flow in smooth pipe, where the relative roughness (ε/D) is very small, (2) flow in fully rough pipe, and (3) flow in a partially rough pipe where both the relative roughness and the viscosity are significant. Generally, the solution of this equation is often by trial and error as the flow is not known and thus the Reynolds number (and hence

Table 9.6 *Equivalent roughness for different pipe materials*

Pipe material	Equivalent roughness (ε) (ft)	Hazen–Williams coefficient (C)
Brass, copper, aluminum	Smooth	140
PVC, plastic	Smooth	150
Cast iron, new	8.0×10^{-4}	130
Cast iron, old	–	100
Galvanized iron	5.0×10^{-4}	120
Asphalted iron	4.0×10^{-4}	–
Wrought iron	1.5×10^{-4}	–
Commercial and welded steel	1.5×10^{-4}	120
Riveted steel	60.0×10^{-4}	110
Concrete	40.0×10^{-4}	130
Wood stave	20.0×10^{-4}	120

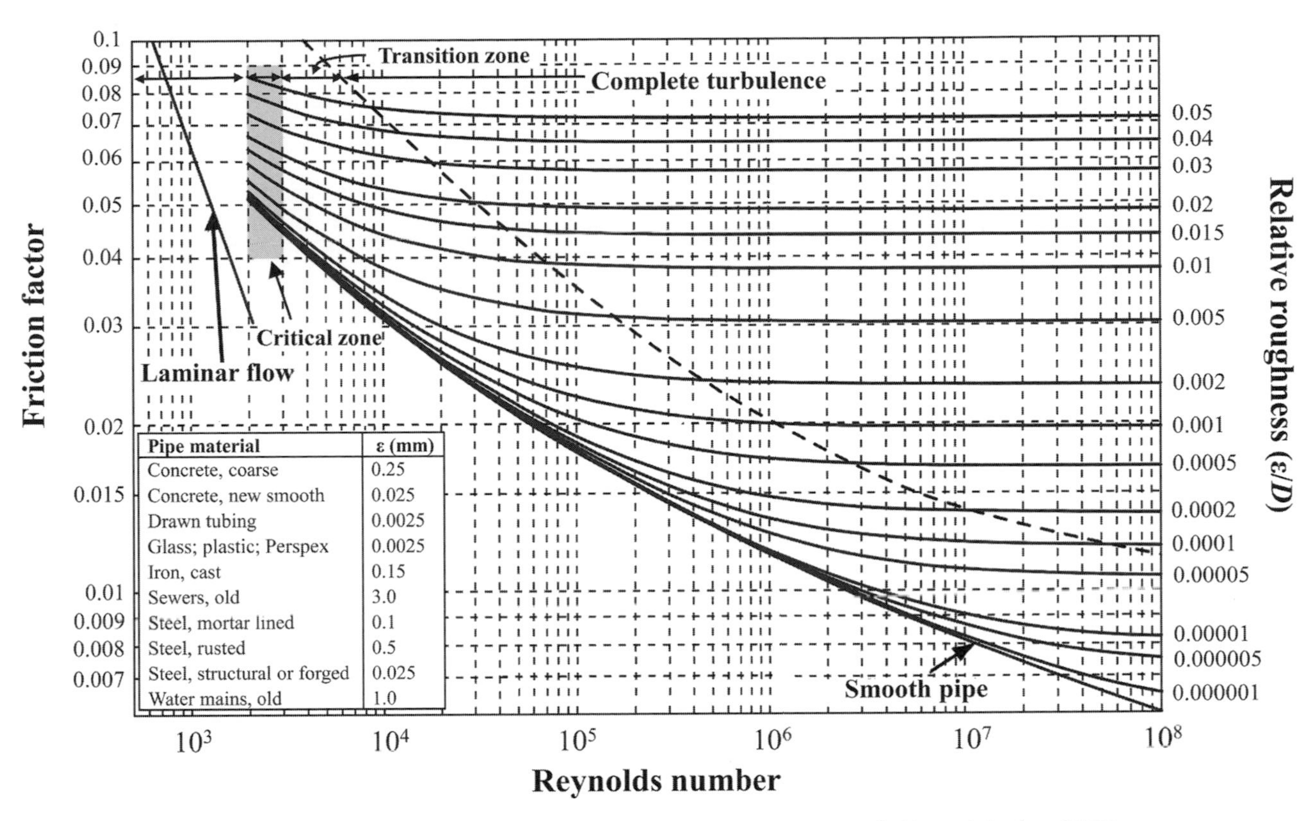

Pipe material	ε (mm)
Concrete, coarse	0.25
Concrete, new smooth	0.025
Drawn tubing	0.0025
Glass; plastic; Perspex	0.0025
Iron, cast	0.15
Sewers, old	3.0
Steel, mortar lined	0.1
Steel, rusted	0.5
Steel, structural or forged	0.025
Water mains, old	1.0

Figure 9.7 Moody diagram. (Adapted from Moody, 1944. Copyright © 1944 by ASME, 66, 671–678. Reprinted with permission from ASME.)

friction factor) is not known. In hydraulics, f is often estimated from the Moody diagram (Figure 9.7), which is a family of curves expressing f as a function of Re for a range of values of roughness height (ε) encompassing smooth to rough surfaces. In agricultural irrigation plastic pipes are common, and hence the curves corresponding to smooth surfaces in the Moody diagram can be utilized. In any case, the first step is to determine the Reynolds number, which can be expressed as

$$\mathrm{Re} = 1.26 \times 10^6 \frac{Q}{D}, \tag{9.23}$$

in which Q is the pipe discharge (L/s) and D is the inside pipe diameter (mm).

The friction factor is estimated as

$$f = \frac{64}{\mathrm{Re}}, \quad \mathrm{Re} \le 2100, \tag{9.24}$$

$$f = 0.04, \quad 2100 \le \mathrm{Re} \le 3000, \tag{9.25}$$

$$f = \frac{0.32}{\mathrm{Re}^{0.25}}, \quad 3000 \le \mathrm{Re} \le 10^5, \tag{9.26}$$

and

$$f = \frac{0.13}{\mathrm{Re}^{0.172}}, 10^5 \le \mathrm{Re} \le 10^7. \tag{9.27}$$

Depending on the value of Re , one of eqs (9.24)–(9.27) is substituted in eq. (9.22) to determine h_f, expressed as

$$h_f = \frac{A_i Q_i^{p_i - 3} L}{D^{p_i}}, \tag{9.28}$$

in which A_i and p_i are given as in Table 9.7, h_f is the head loss (m), L is the pipe length (m), and Q and D are the same as in eq. (9.23) for the calculation of Re.

In sprinkler irrigation using riveted steel pipe (or similar pipe), the Scobey equation is also often employed. This can be expressed as:

$$h_f = \frac{K_s v^{1.9}}{D^{1.1}}, \tag{9.29}$$

where K_s is the retardation factor based entirely on the pipe diameter and characteristics or Scobey's coefficient, v is the velocity in ft/s, D is the inside diameter in ft, and h_f is the head loss per 1000 ft of pipe. Equation (9.29) avoids the computation required for eq. (9.28).

The three equations – Hazen–Williams, Darcy–Weisbach, and Scobey – can be expressed for head loss due to friction as

$$h_f = \frac{K_f c L Q^m}{D^{2m+n}}, \tag{9.30}$$

where c, m, and n are constants; K_f is the friction factor depending on the pipe material; L is the length of the pipe (ft, m); Q is the flow rate (L/s, gpm); and D is the diameter (mm, in.). The friction factor K_f can be expressed as follows. For the Darcy–Weisbach equation:

Table 9.7 *Values of A_i and p_i for corresponding values of* Re

I	A_i	p_i	Re
1	4.1969×10^3	4.00	Re < 2100
2	3.305×10^6	5.00	$2100 < \text{Re} \leq 3000$
3	7.8918×10^5	4.75	$3000 < \text{Re} \leq 10^5$
4	9.5896×10^5	4.828	$10^5 < \text{Re} \leq 10^7$
Hazen–Williams ($C = 140$)	1.283×10^6	4.852	$10^5 < \text{Re} \leq 10^7$

$$K_f = 0.811\left(\frac{f}{g}\right); \tag{9.31}$$

for the Hazen–Williams equation:

$$K_f = (0.285C)^{-1.852}; \tag{9.32}$$

for the Scobey equation:

$$K_f = \frac{K_s}{348}, \tag{9.33}$$

where f is the friction factor obtained from the Moody diagram given in Figure 9.7; g is acceleration due to gravity (9.81 m/s^2, 32.2 ft/s^2); C is the Hazen–Williams coefficient given in Table 9.6; and K_s is the Scobey coefficient. For welded steel pipes the values of K_s are 0.34, 0.33, and 0.32 for OD of 2 and 2.5 in., 3 in., and 4–6 in., respectively. For aluminum tubing without couplers, the K_s value is 0.33. For aluminum tubing with couplers every 20 ft (6 m), 30 ft (9 m), and 40 ft (12 m), the K_s values are 0.43, 0.40, and 0.39, respectively (James, 1988).

The total energy loss is due to friction loss in pipe between upstream and downstream locations, and minor losses due to fittings. The friction loss is multiplied by a factor F, which depends on the number of outlets, such as sprinklers or laterals. For sprinklers the calculation of the value of F is discussed later in the chapter. The values of c, m, and n in eq. (9.30) are different for the Darcy–Weisbach, Hazen–Williams, and Scobey equations. For the Darcy–Weisbach equation, c is 277,778 for SI units and 1.235 for English units, m is 2.0, and n is 1.0. For the Hazen–Williams equation, c is 591,722 for SI units and 1.000 for English units, m is 1.85, and n is 1.0. For the Scobey equation, c is 610,042 for SI units and 1.000 for English units, m is 1.90, and n is 1.10 (James, 1988).

Example 9.8: Consider a 350-m long PVC pipeline with 250 mm of ID laid on a horizontal grid. The pipeline has a steady flow of 50 L/s. The total head available at the inlet is 340 kPa. The temperature of water is 20 °C. Determine the pressure head at a distance of 350 m from the inlet. Use the Hazen–Williams equation for computing the head loss.

Solution: For eq. (9.14), $k = 1.21 \times 10^{10}$, and from Table 9.2 for PVC pipe, $C = 150$. Then,

$$h_f = k \times \left(\frac{Q}{C}\right)^{1.852} \times D^{-4.871} \times L$$

$$h_f = \left(1.21 \times 10^{10}\right) \times \left(\frac{50\,\frac{\text{L}}{\text{s}}}{150}\right)^{1.852} \times (250\text{ mm})^{-4.871} \times (350\text{ m}) = 1.16\text{ m}.$$

For converting to kPa,

$$h_f' = h_f \times \gamma = 1.16\text{ m} \times 9.807\frac{\text{kN}}{\text{m}^3} = 11.38\frac{\text{kN}}{\text{m}^2} = 11.38\text{ kPa}.$$

If the pipeline is laid on a horizontal grid, then

$$H_{1,\,inlet} = H_{2,\,350\text{m}} + h_f',$$

$$340\text{ kPa} = H_{2,\,350\text{m}} + 11.38\text{ kPa},$$

$$H_{2,\,350\text{m}} = 328.62\text{ kPa}.$$

Example 9.9: Solve Example 9.8 using the Darcy–Weisbach equation for head loss.

Solution: Flow velocity is:

$$v = \frac{Q}{A} = \frac{50\,\frac{\text{L}}{\text{s}} \times \left(\frac{1\text{ m}^3}{1000\text{ L}}\right)}{\frac{\pi}{4} \times \left[250\text{ mm} \times \left(\frac{1\text{ m}}{1000\text{ mm}}\right)\right]^2} = 1.02\frac{\text{m}}{\text{s}}$$

The Reynolds number can be calculated using eq. (8.1) or eq. (9.23). Using eq. (8.1), at 20 °C, the dynamic viscosity $\mu = 1.0 \times 10^{-3}\,\frac{\text{kg}}{\text{m/s}}$, the Reynolds number is:

$$\text{Re} = \frac{\rho v D}{\mu} = \frac{\left(1000\,\frac{\text{kg}}{\text{m}^3}\right)\left(1.02\,\frac{\text{m}}{\text{s}}\right)(0.25\text{ m})}{1.0 \times 10^{-3}\,\frac{\text{kg}}{\text{m/s}}} = 2.55 \times 10^5.$$

Using eq. (9.23), the Reynolds number is:

$$\text{Re} = 1.26 \times 10^6\left(\frac{Q}{D}\right)$$

$$= 1.26 \times 10^6\left(\frac{50\,\frac{\text{L}}{\text{s}}}{250\text{ mm}}\right) = 2.52 \times 10^5.$$

The flow is fully turbulent, so f is computed from eq. (9.27):

$$f = \frac{0.13}{(\mathrm{Re})^{0.172}} = \frac{0.13}{(2.52 \times 10^5)^{0.172}} = 0.0153.$$

Then, the head loss is computed using eq. (9.22) or directly using eq. (9.28) without the calculation of f.

Using eq. (9.22):

$$h_f = k_2 \times f \times L \times \frac{Q^2}{D^5}$$

$$= (8.263 \times 10^7) \times (0.0153) \times (350\ \mathrm{m}) \times \frac{(50\,\frac{\mathrm{L}}{\mathrm{s}})^2}{(250\ \mathrm{mm})^5}$$

$$= 1.13\ \mathrm{m}.$$

Using eq. (9.27) and A_i and p_i from Table 9.7 when $I = 4$:

$$h_f = \frac{A_i Q_i^{p_i - 3} L}{D^{p_i}}$$

$$= (9.5896 \times 10^5) \frac{(50\,\frac{\mathrm{L}}{\mathrm{s}})^{4.852-3}}{(250\ \mathrm{mm})^{4.852}} (350\ \mathrm{m}) = 1.09\ \mathrm{m}.$$

Converting to kPa,

$$h_f' = h_f \times \gamma = 1.13\ \mathrm{m} \times 9.807 \frac{\mathrm{kN}}{\mathrm{m}^3}$$

$$= 11.08 \frac{\mathrm{kN}}{\mathrm{m}^2} = 11.08\ \mathrm{kPa}.$$

The total head at 350 m from the inlet is:

$$H_{1,\ inlet} = H_{2,\ 350\mathrm{m}} + h_f',$$

$$340\ \mathrm{kPa} = H_{2,\ 350\mathrm{m}} + 11.08\ \mathrm{kPa},$$

$$H_{2,\ 350\mathrm{m}} = 328.92\ \mathrm{kPa}.$$

The friction factor for turbulent flow in smooth pipes is expressed as

$$\frac{1}{\sqrt{f}} = -2 \log\left(\frac{2.51}{\mathrm{Re}\sqrt{f}}\right), \tag{9.34}$$

and for turbulent flow in fully rough pipes as

$$\frac{1}{\sqrt{f}} = -2 \log\left(\frac{\varepsilon}{3.7D}\right). \tag{9.35}$$

The friction factor for all types of turbulent flow can be expressed as

$$\frac{1}{\sqrt{f}} = -2 \log\left(\frac{\varepsilon}{3.7D} + \frac{2.51}{\mathrm{Re}\sqrt{f}}\right). \tag{9.36}$$

Equation (9.36) reduces to eq. (9.34) if the relative roughness (ε/D) is very small and to eq. (9.35) if the Reynolds number is very high. Jain (1976) proposed an explicit expression for the entire turbulent regime as

$$\frac{1}{\sqrt{f}} = -2 \log\left(\frac{\varepsilon}{3.7D} + \frac{5.72}{\mathrm{Re}^{0.9}}\right). \tag{9.37}$$

Equation (9.37) is an alternative to the Moody diagram for determining the friction factor.

Example 9.10: A 10-cm diameter galvanized iron pipe is being used to convey water at a rate of $0.20\ \mathrm{m^3/s}$ for irrigation. The temperature of the water is 25 °C. What will be the friction factor for flow?

Solution: First, the Reynolds number is computed if the flow is turbulent. To that end, the pipe cross-sectional area is

$$A = \frac{\pi}{4}(10\ \mathrm{cm})^2 = \frac{\pi}{4}(0.1\ \mathrm{m})^2 = 0.007854\ \mathrm{m}^2.$$

Velocity is

$$v = \frac{Q}{A} = \frac{0.20\ \mathrm{m^3/s}}{0.007854\ \mathrm{m}^2} = 25.46\ \mathrm{m/s}.$$

At 25 °C, the dynamic viscosity is

$$\mu = 0.9 \times 10^{-3} \frac{\mathrm{kg}}{\mathrm{m/s}}.$$

The Reynolds number is $\mathrm{Re} = \frac{\rho v D}{\mu} = \frac{(1000\,\frac{\mathrm{kg}}{\mathrm{m}^3})(25.46\,\frac{\mathrm{m}}{\mathrm{s}})(0.1\ \mathrm{m})}{0.9 \times 10^{-3}\,\frac{\mathrm{kg}}{\mathrm{m/s}}} = 2{,}830{,}000.$

Because Re is greater than 4000, flow is turbulent.

The equivalent roughness from Table 9.6 is 5.0×10^{-4} ft. Then, the relative roughness is:

$$\frac{\varepsilon}{D} = \frac{5.0 \times 10^{-4}\ \mathrm{ft}}{0.1\ \mathrm{m} \times \frac{3.28\ \mathrm{ft}}{1\ \mathrm{m}}} = 0.0015.$$

Using eq. (9.37), the friction factor is computed as

$$\frac{1}{\sqrt{f}} = -2 \log\left(\frac{\varepsilon}{3.7D} + \frac{5.72}{\mathrm{Re}^{0.9}}\right)$$

$$= -2 \log\left(\frac{5.0 \times 10^{-4}\ \mathrm{ft}}{3.7\left(0.1\ \mathrm{m} \times \frac{3.28\ \mathrm{ft}}{1\ \mathrm{m}}\right)} + \frac{5.72}{(2.83 \times 10^6)^{0.9}}\right)$$

$$= 6.75,$$

$$\sqrt{f} = \frac{1}{6.75}.$$

Then, $f = 0.022$.

It may be interesting to obtain f from the Moody diagram, which comes out to be:

$$\mathrm{Re} = 2.83 \times 10^6 \rightarrow \frac{\varepsilon}{D} = 0.0015 \rightarrow f = 0.022.$$

9.3.2 Head Loss Due to Other Pipe Characteristics

The flow in a pipeline is also resisted by other pipe characteristics, such as entrances, bends, enlargement, change in cross-sectional area, contractions, and fittings. These characteristics are used in the construction of a pipe system for irrigation. Head losses are due to the friction in the fitting, plus losses resulting from turbulence and changes in the direction of flow. Analogous to the head loss in the pipe, the head losses due to these characteristics (fittings, valves, etc.) are also expressed as a function of velocity head:

$$h_f^* = K_F \frac{v^2}{2g}, \tag{9.38}$$

in which h_f^* is the head loss due to fitting, etc., and K_F is the friction factor that has been reported for various conditions (Pair et al., 1975) as follows:

Entrances:	inward projecting pipe,	$K_F = 0.78$
	square edged inlets,	$K_F = 0.50$
	slightly rounded inlets,	$K_F = 0.23$
	bell-mouthed inlet,	$K_F = 0.04$
Bends:	sharp 90°, $K_F = 1.5$	long 90°, $K_F = 0.25$
	sharp 60°, $K_F = 1.2$	long 60°, $K_F = 0.20$
	sharp 30°, $K_F = 0.9$	

For sudden enlargement (v = velocity of small pipe):

$$K_F = \left[1 - \frac{(d_1)^2}{(d_2)^2}\right]^2, \tag{9.39}$$

where d_1 is the smaller diameter pipe (ID), and d_2 is the larger diameter pipe.

For sudden contraction (v = velocity of smaller pipe):

$$0.10 \leq K_F \leq 0.50. \tag{9.40}$$

Resistance coefficients for various types of fittings and valves are given in Table 9.8.

As a practical guide, the velocity head is often computed as

$$\frac{v^2}{2g} = \frac{2.594 \times 10^{-3}(\text{gpm}^2)}{(\text{ID})^4}.$$

The constant is 8.26×10^4 for Q in L/s and D in mm.

As an example, for $Q = 40$ gpm, ID = 1.5 in., the velocity head is

$$\frac{v^2}{2g} = \frac{2.594 \times 10^{-3} \times 40^2}{(1.5)^4} = 0.82\text{ft}.$$

Table 9.8 *Resistance coefficient K_F for determining head losses*

Fitting or valve	Resistance coefficient
Entrance	
Flush connection	0.5
Projecting connection	0.8
Exit	
Projecting	1.0
Sharp-edged	1.0
Rounded	1.0
Bends	
90° bend and 180° return – threaded	0.8
45° bend – threaded	1.5
90° bend and 180° return – flanged	0.4
45° bend – flanged	0.3
Tee	
Through flow – threaded	0.9
Branched flow – threaded	2.0
Through flow – flanged	0.2
Branched flow – flanged	1.0
Valve	
Gate valve	0.19
Check valve	2.0
Globe valve	10.0
Angle valve	2.0
Butterfly valve	0.3

For $Q = 2.52$ L/s (40 gpm), ID = 38.1 mm (1.5 in.), the velocity head is

$$\frac{v^2}{2g} = \frac{8.26 \times 10^4 \times (2.52)^2}{(38.1)^4} = 0.25 \text{ m } (0.82 \text{ ft}).$$

9.4 HEAD LOSS IN PIPELINES

The head loss equations (Darcy–Weisbach and Hazen–Williams) can be employed in three ways: (1) computation of head loss given a pipe diameter (size) that carries a known discharge or velocity; (2) computation of flow velocity or discharge for a given pipe diameter and head loss; and (3) computation of pipe diameter for a given flow discharge and head loss. These methods are illustrated by the examples that follow.

Example 9.11: A 15-cm cast iron pipe is used to carry a discharge of 0.01 m^3/s over a distance of 400 m having a slope of 1/15 m. What will be the head loss? What will be the pressure difference between the two points? The temperature of the water is 25 °C.

Solution: First, the flow velocity is computed:

$$A = \frac{\pi}{4}(15 \text{ cm})^2 = \frac{\pi}{4}(0.15 \text{ m})^2 = 0.018 \text{ m}^2,$$

$$v = \frac{Q}{A} = \frac{0.01 \frac{\text{m}^3}{\text{s}}}{0.018 \text{ m}^2} = 0.56 \frac{\text{m}}{\text{s}}.$$

At 25 °C, the dynamic viscosity is

$$\mu = 0.9 \times 10^{-3}\ \frac{\text{kg}}{\text{m/s}}.$$

The Reynolds number is

$$\text{Re} = \frac{\rho v D}{\mu} = \frac{\left(1000\ \frac{\text{kg}}{\text{m}^3}\right)\left(0.56\ \frac{\text{m}}{\text{s}}\right)(0.15\ \text{m})}{0.9 \times 10^{-3}\ \frac{\text{kg}}{\text{m/s}}} = 93{,}000.$$

Because Re is greater than 4000, flow is turbulent.

The equivalent roughness from Table 9.6 is 8.0 × 10^{-4} ft for cast iron. Then, the relative roughness is

$$\frac{\varepsilon}{D} = \frac{8.0 \times 10^{-4}\text{ft}}{0.15\ \text{m} \times \frac{3.28\ \text{ft}}{1\ \text{m}}} = 0.0016.$$

From the Moody diagram, $f = 0.024$. Applying the Darcy–Weisbach equation,

$$h_f = f\frac{L}{D}\frac{v^2}{2g} = (0.024) \times \frac{400\ \text{m}}{0.15\ \text{m}} \times \frac{\left(0.56\frac{\text{m}}{\text{s}}\right)^2}{2 \times 9.81\frac{\text{m}}{\text{s}^2}} = 1.0\ \text{m}.$$

For computing the pressure difference between the two points, the Bernoulli equation is applied with respect to point 2 as the datum:

$$z_1 + \frac{p_1}{\gamma} + \frac{v_1^2}{2g} = z_2 + \frac{p_2}{\gamma} + \frac{v_2^2}{2g} + h_f, \quad (v_1 = v_2),$$

$$\frac{400\ \text{m}}{15} + \frac{p_1}{\gamma} + \cancel{\frac{v_1^2}{2g}} = 0 + \frac{p_2}{\gamma} + \cancel{\frac{v_2^2}{2g}} + 1.0\ \text{m},$$

$$\frac{400\ \text{m}}{15} - 1.0\ \text{m} = \frac{p_2 - p_1}{\gamma} = 25.67\ \text{m}.$$

This yields $p_2 - p_1 = (25.67\ \text{m}) \times \gamma = (25.67\ \text{m}) \times \left(9.807\ \frac{\text{kN}}{\text{m}^3}\right) = 252\frac{\text{kN}}{\text{m}^2} = 252\ \text{kPa}.$

Example 9.12: In a large irrigation system, two reservoirs are connected by a 25-cm diameter galvanized iron pipeline 1000 m long, as shown in Figure 9.8. The difference between the water levels of the two reservoirs is maintained as 10 m. What will be the discharge through the pipeline? The head loss due to entrance and exit can be assumed to be negligible. The water temperature is 25 °C.

Solution: Consider two points: one at the water surface of the upstream reservoir and the other at the water surface of the downstream reservoir where the flow velocities can be assumed negligible and pressure is also zero. Take the water surface of the downstream reservoir as the reference datum. Therefore, application of the Bernoulli equation yields

$$z_1 + \frac{p_1}{\gamma} + \frac{v_1^2}{2g} = z_2 + \frac{p_2}{\gamma} + \frac{v_2^2}{2g} + h_f,$$

$$10\ \text{m} + 0 + 0 = 0 + 0 + 0 + h_f \rightarrow h_f = 10\ \text{m}.$$

The roughness of the galvanized iron pipe is $\varepsilon = 5 \times 10^{-4}$ ft. The relative roughness is therefore

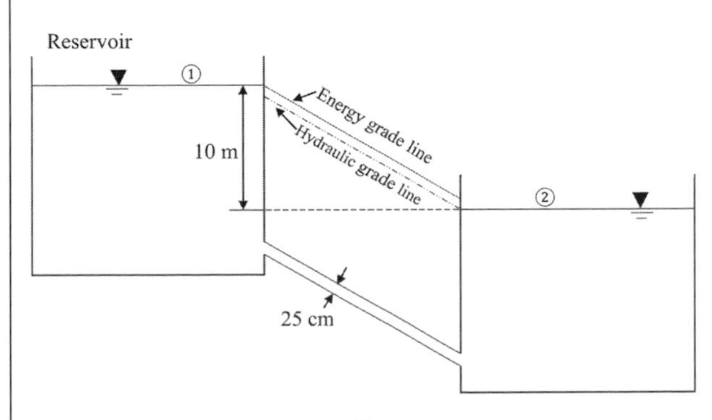

Figure 9.8 Two reservoirs connected by a pipe.

$$\frac{\varepsilon}{D} = \frac{5.0 \times 10^{-4}\text{ft}}{0.25\ \text{m} \times \frac{3.28\ \text{ft}}{1\ \text{m}}} = 0.0006.$$

Method 1: Since the velocity of flow in the pipeline is not known, as a first approximation the flow can be assumed to be rough turbulent where the friction factor is independent of the Reynolds number. From the Moody diagram, $f = 0.018$ for the above relative roughness.

Now the flow velocity can be computed from the Darcy–Weisbach equation as:

$$h_f = \frac{fL}{D}\frac{v^2}{2g} \rightarrow 10\ \text{m} = (0.018) \times \frac{1000\ \text{m}}{0.25\ \text{m}} \times \frac{v^2}{2 \times 9.81\frac{\text{m}}{\text{s}^2}},$$

$$\frac{10\ \text{m} \times \left(2 \times 9.81\frac{\text{m}}{\text{s}^2}\right)}{(0.018) \times \frac{1000\ \text{m}}{0.25\ \text{m}}} = v^2.$$

This yields v as 1.65 m/s.

Now the Reynolds number is computed with dynamic viscosity at 25 °C as

$$\text{Re} = \frac{\rho v D}{\mu} = \frac{\left(1000\ \frac{\text{kg}}{\text{m}^3}\right)\left(1.65\ \frac{\text{m}}{\text{s}}\right)(0.25\ \text{m})}{0.9 \times 10^{-3}\ \frac{\text{kg}}{\text{m/s}}} = 460{,}000.$$

The value of f can be recomputed from the Moody diagram for the above values of Reynolds number and relative roughness as:

$$\text{Re} = 460{,}000 \rightarrow \frac{\varepsilon}{D} = 0.0006 \rightarrow f = 0.019.$$

The velocity can be recomputed from the Darcy–Weisbach equation as:

$$10\ \text{m} = h_f = f\frac{L}{D}\frac{v^2}{2g} = (0.019) \times \frac{1000\ \text{m}}{0.25\ \text{m}} \times \frac{v^2}{2 \times 9.81\frac{\text{m}}{\text{s}^2}},$$

$$\frac{10 \text{ m} \times \left(2 \times 9.81 \frac{\text{m}}{\text{s}^2}\right)}{(0.019) \times \frac{1000 \text{ m}}{0.25 \text{ m}}} = v^2.$$

This velocity is 1.61 m/s.

With this value of velocity, the Reynolds number is recomputed:

$$\text{Re} = \frac{\rho v D}{\mu} = \frac{\left(1000 \frac{\text{kg}}{\text{m}^3}\right)\left(1.61 \frac{\text{m}}{\text{s}}\right)(0.25 \text{ m})}{0.9 \times 10^{-3} \frac{\text{kg}}{\text{m/s}}} = 447{,}000.$$

The friction is recomputed for the new value of the Reynolds number and relative roughness as

$$\text{Re} = 447{,}000 \rightarrow \frac{\varepsilon}{D} = 0.0006 \rightarrow f = 0.019.$$

The friction factor does not change. Therefore, $v = 1.61$ m/s and $Q = Av$,

$$A = \frac{\pi}{4}(0.25 \text{ m})^2 = 0.05 \text{ m}^2,$$

$$Q = vA = \left(1.61 \frac{\text{m}}{\text{s}}\right)\left(0.05 \text{ m}^2\right) = 0.08 \frac{\text{m}^3}{\text{s}}.$$

Method 2: Since Q is unknown, Re and f cannot be determined. The above iteration procedure can be used. However, the velocity can also be computed using the following direct procedure.

Equation (9.21) can be rewritten as

$$f = h_f \frac{D}{L}\frac{2g}{v^2} = (10 \text{ m})\frac{0.25 \text{ m}}{1000 \text{ m}} \times \frac{2 \times 9.81 \frac{\text{m}}{\text{s}^2}}{v^2} = \frac{0.04905}{v^2}.$$

At 25 °C, the dynamic viscosity is

$$\mu = 0.9 \times 10^{-3} \frac{\text{kg}}{\text{m/s}}.$$

The Reynolds number is

$$\text{Re} = \frac{\rho v D}{\mu} = \frac{\left(1000 \frac{\text{kg}}{\text{m}^3}\right) v \, (0.25 \text{ m})}{0.9 \times 10^{-3} \frac{\text{kg}}{\text{m/s}}} = 0.278 v \times 10^6.$$

The roughness for the galvanized iron pipe is $\varepsilon = 5 \times 10^{-4}$ ft. The relative roughness is then

$$\frac{\varepsilon}{D} = \frac{5.0 \times 10^{-4} \text{ft}}{0.25 \text{ m} \times \frac{3.28 \text{ ft}}{1 \text{ m}}} = 0.0006.$$

Using eq. (9.36), the friction factor is computed as

$$\frac{1}{\sqrt{f}} = -2 \log\left(\frac{\varepsilon}{3.7D} + \frac{2.51}{\text{Re}\sqrt{f}}\right),$$

$$\text{Then, } \frac{1}{\sqrt{\frac{0.04905}{v^2}}} = -2 \log\left(\frac{0.0006}{3.7} + \frac{2.51}{\left(0.278 v \times 10^6\right)\sqrt{\frac{0.04905}{v^2}}}\right),$$

$$\frac{v}{\sqrt{0.04905}} = -2 \log\left(\frac{0.00122}{3.7} + \frac{2.51}{\left(0.278 \times 10^6\right)\sqrt{0.04905}}\right),$$

$$v = 1.63 \text{m/s}.$$

The Reynolds number is:

$$\text{Re} = 0.278 \, v \times 10^6 = 0.278 \times 1.63 \times 10^6 = 4.53 \times 10^6$$

As Re $\geq$ 4000, the flow is fully turbulent:

$$Q = Av = \frac{\pi}{4}D^2 v = \frac{\pi}{4} \times (0.25 \text{ m})^2 \times 1.63 \frac{\text{m}}{\text{s}} = 0.08 \frac{\text{m}^3}{\text{s}}.$$

Example 9.13: Two reservoirs are connected by a 300 m cast iron pipeline carrying a discharge of 0.25 m³/s, as shown in Figure 9.9. The difference in the water levels of the two reservoirs is 10 m. Assume a water temperature of 25 °C. What is the diameter of the pipeline?

Solution: First, the head loss is computed considering points 1 and 2, using the water level at point 2 as the reference datum:

$$z_1 + \frac{p_1}{\gamma} + \frac{v_1^2}{2g} = z_2 + \frac{p_2}{\gamma} + \frac{v_2^2}{2g} + h_f + h_m \rightarrow 10 + 0 + 0 = 0 + 0 + 0 + h_f + h_m.$$

This yields $h_f = 10 \text{ m} - h_m$. From the Darcy–Weisbach equation,

$$h_f = \frac{fL}{D}\frac{v^2}{2g} = \frac{fL}{D}\frac{Q^2}{\left[\left(\frac{\pi}{4}\right)D^2\right]^2 \times (2g)} = \frac{fLQ^2}{12.1 \times D^5}.$$

Now minor head losses due to entrance, exit, bend, and valve are computed with K_F for entrance loss as 0.5, for exit loss as 1.0, for two 90° bends as $1.5 \times 2 = 3$, and for the globe valve as 10.0. The total value of K_F comes out to be 14.5. Then, minor losses h_f^* are

$$h_f^* = \sum_{i=1}^{4} K_i \frac{v^2}{2g} = \frac{14.5Q^2}{\left[\left(\frac{\pi}{4}\right)D^2\right]^2 \times 2 \times 9.81} = \frac{14.5Q^2}{12.1 \times D^4}.$$

The total head loss is

$$h_f + h_f^* = 10,$$

$$\frac{fLQ^2}{12.1 \times D^5} + \frac{14.5Q^2}{12.1 \times D^4} = 10,$$

$$\frac{f(300 \text{ m})\left(0.25 \frac{\text{m}^3}{\text{s}}\right)^2}{12.1 \times D^5} + \frac{14.5 \times \left(0.25 \frac{\text{m}^3}{\text{s}}\right)^2}{12.1 \times D^4} = 10,$$

$$\frac{1.55 \times f}{D^5} + \frac{0.075}{D^4} = 10.$$

This gives a five-degree polynomial in D, which can be solved by trial and error.

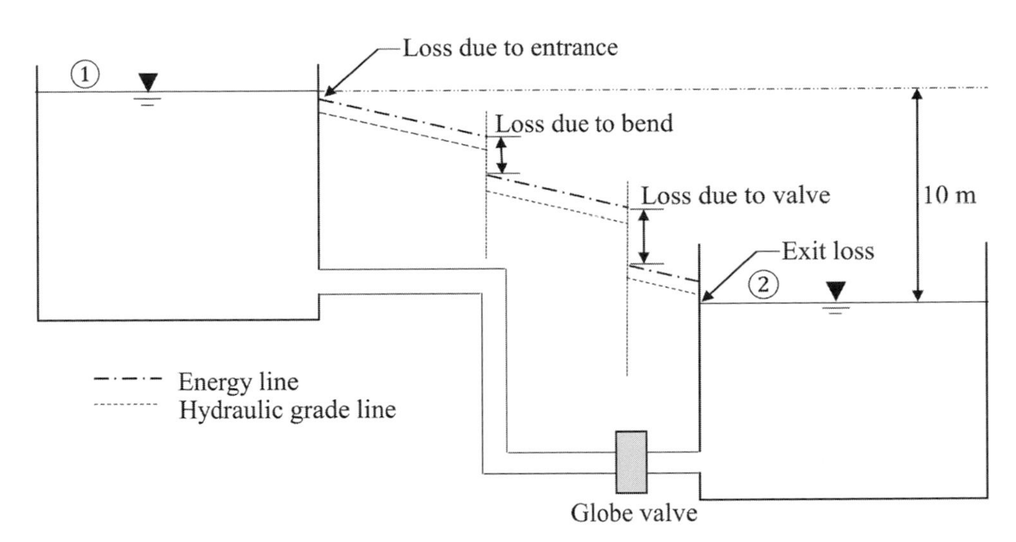

Figure 9.9 Two reservoirs connected by a pipe in Example 9.13.

First trial: Assume $f = 0.015$:

$$\frac{1.55 \times 0.015}{D^5} + \frac{0.075}{D^4} = 10,$$

$$D = 0.345 \text{ m},$$

$$A = \frac{\pi}{4}(0.345 \text{ m})^2 = 0.093 \text{ m}^2,$$

$$v = \frac{Q}{A} = \frac{0.25 \frac{\text{m}^3}{\text{s}}}{0.093 \text{ m}^2} = 2.69 \frac{\text{m}}{\text{s}},$$

$$\text{Re} = \frac{\rho v D}{\mu} = \frac{\left(1000 \frac{\text{kg}}{\text{m}^3}\right)\left(2.69 \frac{\text{m}}{\text{s}}\right)(0.345 \text{ m})}{0.9 \times 10^{-3} \frac{\text{kg}}{\text{m/s}}} = 103{,}000,$$

$$\frac{\varepsilon}{D} = \frac{8.0 \times 10^{-4} \text{ft}}{0.345 \text{ m} \times \frac{3.28 \text{ ft}}{1 \text{ m}}} = 0.0007.$$

From the Moody diagram for the above values of the Reynolds number and relative roughness:

$$\text{Re} = 103{,}000 \rightarrow \frac{\varepsilon}{D} = 0.0007 \rightarrow f = 0.018.$$

Second trial: Assume $f = 0.018$:

$$\frac{1.55 \times 0.018}{D^5} + \frac{0.075}{D^4} = 10,$$

$$D = 0.352 \text{ m},$$

$$A = \frac{\pi}{4}(0.352 \text{ m})^2 = 0.097 \text{ m}^2,$$

$$v = \frac{Q}{A} = \frac{0.25 \frac{\text{m}^3}{\text{s}}}{0.097 \text{ m}^2} = 2.57 \frac{\text{m}}{\text{s}},$$

$$\text{Re} = \frac{\rho v D}{\mu} = \frac{\left(1000 \frac{\text{kg}}{\text{m}^3}\right)\left(2.57 \frac{\text{m}}{\text{s}}\right)(0.352 \text{ m})}{0.9 \times 10^{-3} \frac{\text{kg}}{\text{m/s}}} = 101{,}000,$$

$$\frac{\varepsilon}{D} = \frac{8.0 \times 10^{-4} \text{ft}}{0.352 \text{ m} \times \frac{3.28 \text{ ft}}{1 \text{ m}}} = 0.0007.$$

From the Moody diagram for the above values of the Reynolds number and relative roughness:

$$\text{Re} = 101{,}000 \rightarrow \frac{\varepsilon}{D} = 0.0007 \rightarrow f = 0.018.$$

Since f stabilizes, $D = 0.352$ m.

The pipeline can be designed for different situations, such as a single pipeline with a pump, pipes in series, pipes in parallel, and a pipe network. Each of these cases is briefly illustrated here.

9.4.1 Single Pipeline with a Pump

In irrigation systems pumps are commonly used to lift water, say from a well to a reservoir, or to an irrigation field, or to boost the pressure at an intermediate point. Consider two reservoirs connected by a pipeline with a pump providing added pressure at an intermediate point, as shown in Figure 9.10. From the Bernoulli equation,

$$z_1 + \frac{p_1}{\gamma} + \frac{v_1^2}{2g} + H_p = z_2 + \frac{p_2}{\gamma} + \frac{v_2^2}{2g} + h_f + h_f^*, \tag{9.41}$$

where H_p is the head provided by the pumping system, h_f is the head loss due to friction, and h_f^* is the minor head losses. Here, $v_1 = v_2$. Equation (9.41) can be written as

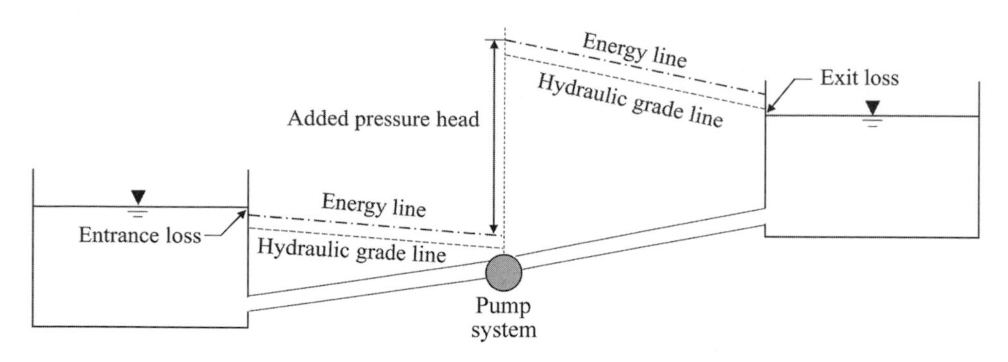

Figure 9.10 Two reservoirs connected by a pipeline with a pump system.

$$H_p = \left(z_2 + \frac{p_2}{\gamma}\right) - \left(z_1 + \frac{p_1}{\gamma}\right) + h_f + h_f^*. \tag{9.42}$$

The brake horsepower of the pump (BP) is related to the energy head H_p (which will be discussed in the next chapter) as

$$BP = \frac{\gamma Q H_p}{550 E_p} \text{ (British units)}, \tag{9.43a}$$

where BP is the horsepower (hp), γ is the specific weight of water in lb/ft^3, Q is the pipe discharge in cfs, H_p is the pump head in ft, and E_p is the pump efficiency; or

$$BP = \frac{\gamma Q H_p}{E_p} \text{ (metric units)} \tag{9.43b}$$

where the BP is in kW, the specific weight of water γ is in kN/m^3, the pipe discharge Q is in m^3/s, and the pump head H_p is in meters.

Example 9.14: Two reservoirs are connected by a 30-cm diameter steel pipeline that is 1000 m long. The upstream reservoir has a water surface elevation of 500 m and the downstream reservoir has a water surface elevation of 550 m. The pipeline carries a discharge of 0.5 m^3/s. Determine the head loss, neglecting the minor losses. Then, compute the head the pump should provide and horsepower the pump should have if its efficiency is 80%. Assume the water temperature is 25 °C.

Solution:

$$v = \frac{Q}{A} = \frac{0.5 \text{ m}^3/\text{s}}{\left[\left(\frac{\pi}{4}\right)(0.3 \text{ m})^2\right]} = 7.07 \ \frac{\text{m}}{\text{s}},$$

$$\text{Re} = \frac{\rho v D}{\mu} = \frac{\left(1000 \ \frac{\text{kg}}{\text{m}^3}\right)\left(7.07 \ \frac{\text{m}}{\text{s}}\right)(0.3 \text{ m})}{0.9 \times 10^{-3} \ \frac{\text{kg}}{\text{m/s}}} = 2{,}400{,}000.$$

Relative roughness (for the steel pipeline) is

$$\frac{\varepsilon}{D} = \frac{1.5 \times 10^{-4} \text{ ft}}{0.3 \text{ m} \times \frac{3.28 \text{ ft}}{1 \text{ m}}} = 0.00015.$$

From the Moody diagram, $f = 0.014$. From the Darcy–Weisbach equation,

$$h_f = \frac{fL}{D}\frac{v^2}{2g} = (0.014) \times \frac{1000 \text{ m}}{0.3 \text{ m}} \times \frac{\left(7.07 \ \frac{\text{m}}{\text{s}}\right)^2}{2 \times 9.81 \ \frac{\text{m}}{\text{s}^2}} = 118.9 \text{ m}.$$

Then,

$$H_p = \left(z_2 + \frac{p_2}{\gamma}\right) - \left(z_1 + \frac{p_1}{\gamma}\right) + h_f + h_f^*, \ (p_1 = p_2 \text{ and } h_f^* = 0)$$

$$H_p = \left(z_2 + \frac{\cancel{p_2}}{\gamma}\right) - \left(z_1 + \frac{\cancel{p_1}}{\gamma}\right) + h_f + \cancel{h_m},$$

$$H_p = (z_2 - z_1) + h_f = 50 \text{ m} + 118.9 \text{ m} = 168.9 \text{ m},$$

$$BP = \frac{\gamma Q H_p}{E_p} = \frac{(9.807 \text{ kN/m}^3)(0.5 \text{ m}^3/\text{s})(168.9 \text{ m})}{(0.8)} = 1035.3 \text{ kW},$$

$1035.3 \text{ kW} = 1388.4 \text{ hp}$.

9.4.2 Pipes in Series

Pipes of different sizes may be connected in series, as shown in Figure 9.11. From the continuity equation, one can write

$$Q = Q_1 = Q_2 = Q_3 = Q_4 + \ldots. \tag{9.44}$$

From the energy equation, one can write

$$h_f = h_{f1} + h_{f2} + h_{f3} + h_{f4} + \ldots. \tag{9.45}$$

For simplicity, the concept of equivalent pipe is employed, which replaces different-sized pipes. The equivalent pipe is of uniform diameter and has a length that will carry Q with the same head loss h_f. This is illustrated by Example 9.15.

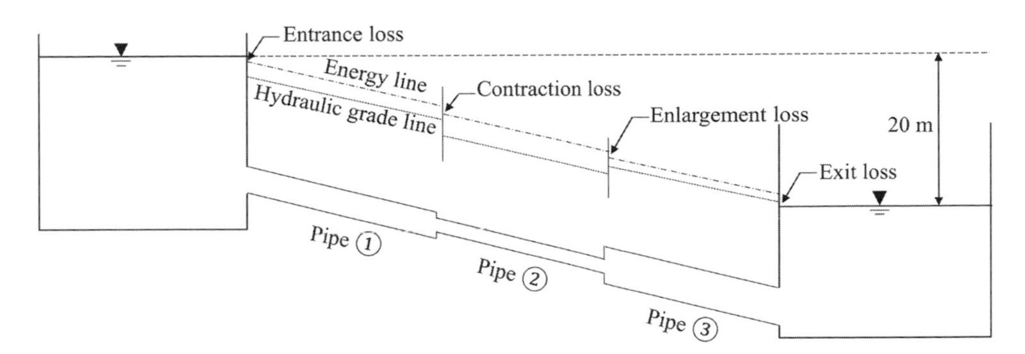

Figure 9.11 Pipes connected in series.

Example 9.15: Three commercial steel pipes are connected in series, as shown in Figure 9.11, and they connect two reservoirs. The difference in the water surface levels of the two reservoirs is 20 m. The pipe connecting the upstream reservoir has a diameter of 20 cm and a length of 400 m; the second pipe connecting to the first pipe has a diameter of 15 cm and a length of 300 m; and the third pipe connecting to the second pipe has a diameter of 25 cm and a length of 500 m. What is the discharge through the pipeline?

Solution: First, assume a discharge of 0.02 m³/s. Compute the friction slope (S) for each pipe using either the Darcy–Weisbach equation or the Hazen–Williams equation. From eq. (9.20),

$$Q = 0.278CD^{2.63}S^{0.54},$$

$$C = 140 \text{ (from Table 9.2).}$$

For pipe 1 ($D_1 = 0.2$ m),

$$0.02\frac{\text{m}^3}{\text{s}} = 0.278(140)(0.2)^{2.63}S_1{}^{0.54},$$

$$S_1 = 0.0021.$$

For pipe 2 ($D_2 = 0.15$ m),

$$0.02\frac{\text{m}^3}{\text{s}} = 0.278(140)(0.15)^{2.63}S_2{}^{0.54},$$

$$S_2 = 0.0084.$$

For pipe 3 ($D_3 = 0.25$ m),

$$0.02\frac{\text{m}^3}{\text{s}} = 0.278(140)(0.25)^{2.63}S_3{}^{0.54},$$

$$S_3 = 0.0007.$$

The head loss for each pipe is obtained by multiplying by the pipe length:

$$h_{f1} = S_1 \times L_1 = 0.0021 \times 400 \text{ m} = 0.84 \text{ m},$$

$$h_{f2} = S_2 \times L_2 = 0.0084 \times 300 \text{ m} = 2.52 \text{ m},$$

$$h_{f3} = S_3 \times L_3 = 0.0007 \times 500 \text{ m} = 0.35 \text{ m}.$$

The total head loss is computed as

$$h_f = h_{f1} + h_{f2} + h_{f3} = 3.71 \text{ m}.$$

A desired size of uniform pipe is selected ($D = 0.2$ m), and with the assumed discharge of 0.02 m³/s the slope is computed as

$$0.02\frac{\text{m}^3}{\text{s}} = 0.278(140)(0.2)^{2.63}(S)^{0.54},$$

$$S = 0.0021.$$

The required length of uniform pipe is computed as

$$L = \frac{h_f}{S} = \frac{3.71 \text{ m}}{0.0021} = 1766.67 \text{ m}.$$

Thus, a uniform pipe of 0.2 m diameter and 1766.67 m length is equivalent to the three pipes connected in series. Given the difference in the water surface levels of the two reservoirs is 20 m, the total head loss $h_f = 20$ m:

$$S = \frac{h_f}{L} = \frac{20 \text{ m}}{1766.67 \text{ m}} = 0.0113.$$

The discharge is computed as

$$Q = 0.278CD^{2.63}S^{0.54} = 0.278(140)(0.2)^{2.63}0.0113^{0.54}$$
$$= 0.05\frac{\text{m}^3}{\text{s}}.$$

9.4.3 Pipes in Parallel

Pipes connected in parallel are also referred to as looping pipes, and are shown in Figure 9.12. The continuity for a parallel arrangement of n pipes can be written as

$$Q = \sum_{i=1}^{n} Q_i, \quad i = 1, 2, \ldots, n. \tag{9.46}$$

The energy equation can be written as

$$h_f = h_{f1} = h_{f2} = \ldots = h_{fn}. \tag{9.47}$$

The concept of looping pipes is illustrated by Example 9.16.

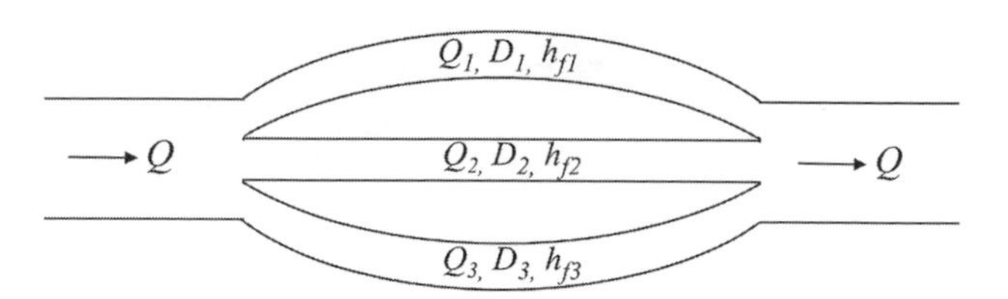

Figures 9.12 Pipes connected in parallel.

Example 9.16: A reservoir supplies water through a standard steel pipeline that is 1.5 km long. The pipeline, as shown in Figure 9.13, has another pipe of the same diameter attached to it in its middle. The diameter of the pipeline is 60 cm and the parallel pipe is 900 m long. What will be the discharge through the pipeline?

Solution: As in the case of pipes in series, the parallel pipes are converted into a single pipe of uniform size. One assumes a head loss through the pipeline as $h_f = 30$ m. For each pipe, the friction slope $S = h_f/L$ is computed as:

$$S_1 = \frac{h_f}{L_1} = \frac{30 \text{ m}}{800 \text{ m}} = 0.0375,$$

$$S_2 = \frac{h_f}{L_2} = \frac{30 \text{ m}}{900 \text{ m}} = 0.0333.$$

The discharge is computed using the Hazen–Williams equation as follows. From eq. (9.20):

$C = 130$ (from Table 9.2),

$$Q_1 = 0.278 C D_1^{2.63} S_1^{0.54}$$

$$= 0.278(130)(0.6)^{2.63}(0.0375)^{0.54} = 1.60 \frac{\text{m}^3}{\text{s}},$$

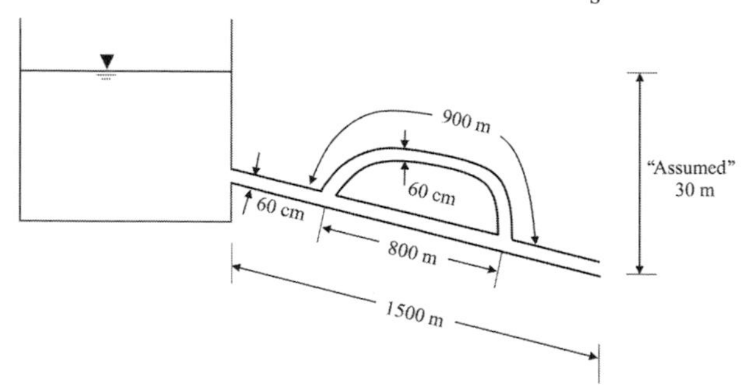

Figure 9.13 A reservoir water supply through the parallel pipe.

$$Q_2 = 0.278 C D_2^{2.63} S_2^{0.54}$$

$$= 0.278(130)(0.6)^{2.63}(0.0333)^{0.54} = 1.50 \frac{\text{m}^3}{\text{s}}.$$

The total discharge is computed as:

$$Q = Q_1 + Q_2 = 3.1 \frac{\text{m}^3}{\text{s}}.$$

A desired diameter of the uniform pipe is selected as $D = 60$ cm. For the selected diameter of the uniform pipe and the computed total discharge, the friction slope is computed as:

$$3.1 \frac{\text{m}^3}{\text{s}} = 0.278(130)(0.6)^{2.63}(S)^{0.54},$$

$$S = 0.1274.$$

The required length of the uniform pipe is computed as:

$$L = \frac{h_f}{S} = \frac{30 \text{ m}}{0.1274} = 235.48 \text{ m}.$$

A uniform pipe of 60 cm diameter and 235.48 m is equal to the pipe in parallel.

The total length of uniform pipe of 60 cm diameter is computed as $800 + 235.48 = 1035.48$ m.

The friction slope is computed as:

$$S = \frac{h_f}{L} = \frac{30 \text{ m}}{1035.48 \text{ m}} = 0.029.$$

The discharge is computed as:

$$Q = 0.278 C D^{2.63} S^{0.54}$$

$$= 0.278(130)(0.6)^{2.63}(0.029)^{0.54} = 1.39 \frac{\text{m}^3}{\text{s}}.$$

9.5 PIPELINES

In farm irrigation, thermoplastic pipes are usually used where common materials are polyvinyl chloride (PVC) and polyethylene (PE). Because of the convenience of fabrication, PVC is more frequently employed. Standards have been established by the American Society of Agricultural Engineers (ASAE, 1987) for thermoplastic pipes, which relate pipeline dimensions and type of material to their strength and ability to resist loads. Based on pipe diameter and design operating pressure, thermoplastic pipes are classified into low-pressure and high-pressure categories as follows:

Low pressure: nominal diameter 114–630 mm (4–24 in.) and internal pressure 550 kPa (80 psi) or less.

High pressure: nominal diameter: 21–710 mm (0.5–27 in.) and internal pressure 550–2170 kPa (80–315 psi), including surge pressure.

9.5.1 Pressure Rating and Hydrostatic Design Stress

Pressure rating (PR) denotes the maximum pressure that is continuously exerted by water in the pipe without causing failure of the pipe with high probability. The hydrostatic design stress (S_s) denotes the maximum tensile stress that is caused by the internal hydrostatic pressure whose continuous application does not cause failure of the pipe with high probability. Thus, both PR and stress relate to long-term operation of the pipeline and are related to a dimensionless parameter, called the dimension ratio (DR), defined as

$$DR = \frac{D}{t}, \tag{9.48}$$

where D is the outside (OD) or inside (ID) pipe diameter (mm), and t is the wall thickness (mm) whose minimum value for thermoplastic pipe is taken as 1.52 mm (0.06 in.).

The PR for OD-based pipe is specified as

$$PR = \frac{2S_s}{DR - 1} \tag{9.49}$$

or

$$PR = \frac{2S_s}{\frac{D_0}{t} - 1}, \tag{9.50}$$

where D_0 is the average OD. The PR for ID-based pipe can be written as

$$PR = \frac{2S_s}{DR + 1} \tag{9.51}$$

or

$$PR = \frac{2S_s}{\frac{D_i}{t} + 1}, \tag{9.52}$$

where D_i is the average ID.

Based on experimental testing, the hydrostatic design stress is related to the long-term hydrostatic strength (S_{it}) as

$$S_s = \frac{S_{it}}{2.0}. \tag{9.53}$$

The values of hydrostatic design stress for different thermoplastic pipe materials and different pipe strengths are reported by the ASABE (2004) as ASAE Standard S376.2, as given in Table 9.9. Table 9.10 shows the PR for different strengths of pipe materials.

Example 9.17: Consider a PVC 2120 material pipe with an ID of 300 mm. Calculate the PR if the wall thickness is 2 mm and pressure is low. Also, consider a wall thickness of 10 mm and high pressure.

Solution: For low pressure,

$$DR_{low} = \frac{D_i}{t} = \frac{300\text{ mm}}{2\text{ mm}} = 150.$$

From Table 9.10, $S = 13.8$ MPa $= 13{,}800$ kPa. The PR is

$$PR_{low} = \frac{2S_s}{DR_{low} + 1} = \frac{2 \times 13{,}800\text{ kPa}}{150 + 1} = 182.8.$$

For high pressure,

$$DR_{high} = \frac{D_i}{t} = \frac{300\text{ mm}}{10\text{ mm}} = 30.$$

The pressure rating is

$$PR_{low} = \frac{2S_s}{DR_{high} + 1} = \frac{2 \times 13{,}800\text{ kPa}}{30 + 1} = 890.3.$$

9.5.2 Underground Pipelines

Underground pipelines are often used to convey water to fields where it is needed. Although more expensive than open channels, they save water by eliminating or reducing losses due to seepage, evaporation, or breaches; they can deliver water where needed irrespective of elevation; they require low maintenance; and their water supply can be fully controlled. However, they require higher head in order to be able to distribute water, and more power than do open channels.

9.5.3 Types of Irrigation Pipeline Systems

Irrigation pipeline systems are (1) completely portable surface pipeline systems, (2) partly surface and partly buried pipeline systems, or (3) fully buried. In the surface pipeline systems, water is conveyed from the source to the field either through gated outlets or from an open end. In the combined system, permanently buried pipelines bring water from the source to risers, which then supply water to surface pipelines. Fully buried pipelines transmit water from the source to the risers directly to borders or furrows or through a channel.

9.5.4 Low-Head Pipelines

Low-head pipelines can be either buried or on the surface, and transmit water from the source to the field, which can be uphill, downhill, or undulating. A pipeline comprises an inlet, one or more outlets, head control devices, surge protection devices, an air relief valve, a flow meter, and sand

and debris removal devices. A pressurized pipeline may also have pressure relief, air release, and vacuum relief valves.

9.5.5 Underground Pipeline Components

A low-head pipeline system requires a pump stand as the inlet or gravity inlet, gate stand, pressure relief valves, outlets, and end plug, as shown in Figure 9.14 (a similar figure can be found in Michael [2010]). In order to transmit water from the source to the pipeline system, an inlet structure is needed that will develop adequate pressure and full flow capacity so that water is distributed at different points on the field. The inlet structure is equipped with a sand trap and trash screen to prevent their entry into the pipeline.

The pump stand, located at the inlet end of the pipeline, is needed to develop sufficient pressure for all pipeline outlets. In order to release entrapped air and dissipate a high-velocity stream, the size of pump stand should be greater than the pipeline diameter, as shown in Figure 9.15 (a similar figure can be found in Michael [2010]).

If the elevation of the surface water source, such as a reservoir or a canal, is high enough to permit gravity flow into the pipeline and provide enough pressure at all outlets, then a gravity inlet, as shown in Figure 9.16, is needed (a similar figure can be found in Michael [2010]).

Gate stands are employed to control flow into branch lines that take off from the main pipeline. For releasing water, the gate stand may be equipped with a sliding gate or gate valve. The gate stand also acts as a surge chamber and prevents high pressure.

Pressure relief valves are employed to relieve any surge that may develop in the pipeline and are set to open at predetermined pressures. They prevent water hammer when rapid changes in flow velocity occur. The valves close when the pressure falls below the set value.

Table 9.9 *Maximum hydrostatic design stress for thermoplastic pipe (ASAE Standard S376.2)*

Compound	Standard code designation	Hydrostatic design stress	
		MPa	psi
PVC	PVC 1120	13.8	2000
PVC	PVC 1220	13.8	2000
PVC	PVC 2120	13.8	2000
PVC	PVC 2116	11.0	1600
PVC	PVC 2112	8.6	1250
PVC	PVC 2110	6.9	1000
PE	PE 3408	5.5	800
PE	PE 3406	4.3	630
PE	PE 3306	4.3	630
PE	PE 2306	4.3	630
PE	PE 2305	3.4	500

Table 9.10 *Pressure rating (PR) for non-threaded thermoplastic pipe: PVC pipe sizes* (ASAE Standard S376.2)*

SDR[a]		PVC 1120 PVC 1220 PVC 2120		PVC 2116		PVC 2112		PVC 2110		PE 3408[b]		PE 3406 PE 3306 PE 2306		PE 2305	
OD-based pipe	ID-based pipe	psi	kPa	psi	kPa	psi	kPa	psi	kPa	psi	kPa	psi	kPa	psi	kPa
	5.3									250	1725	200	1380	160	1105
	7.0									200	1380	160	1105	125	860
11.0	9.0									160	1105	125	860	100	690
13.5	11.5	315	2170	250	1725	200	1380	160	1105						
17.0	15.0	250	1725	200	1380	160	1105	125	860	100	690	80	559	63	435
21.0		200	1380	160	1105	125	860	100	690	80	559	64	440		
26.0		160	1105	125	860	100	690	80	559	64	440	50	345		
32.5		125	860	100	690	80	559	63	435	50	345	40	559		
41.0		100	690	80	550	63	435	50	345	40	275	31	215		
51.0		80	550	63	435	50	345	40	275						
64.0		63	435	50	345	40	275	30	205						
81.0		50	345	40	275	30	205	25	170						
93.5		43	395												
50 ft head (15.24 m)		22	150												

* For water at 23 °C (73.4 F).
[a] Standard dimension ratio (SDR) is the ratio of the average pipe OD to the minimum wall thickness for OD-controlled plastic pipe.
[b] PE refers to polyethylene plastic pipe.

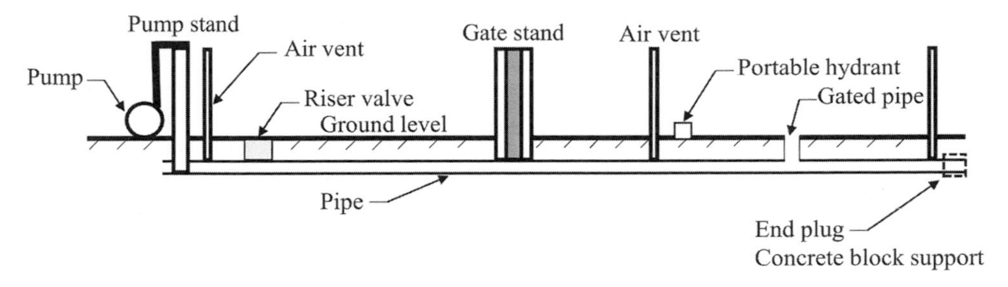

Figure 9.14 Components of an underground pipeline irrigation system.

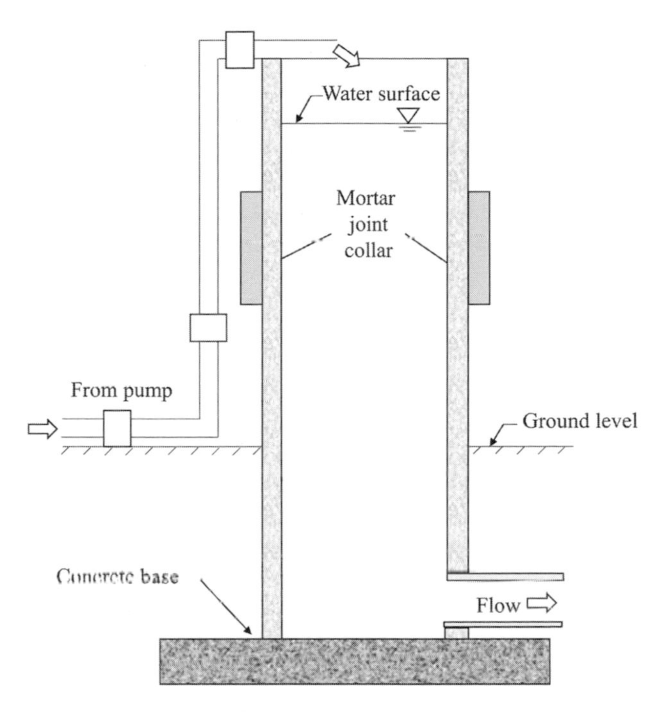

Figure 9.15 Pump stands for underground pipelines.

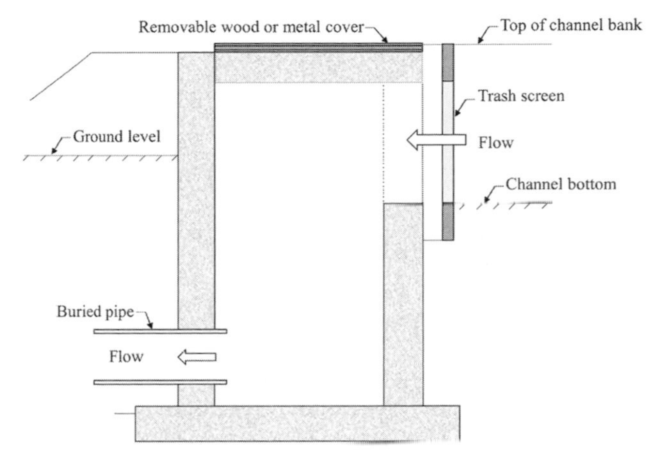

Figure 9.16 A section view of an inlet for taking water from a minor canal into an underground pipeline.

Air release, vacuum relief, inlet valves, or air vents prevent vacuum formation by releasing entrapped air. They are installed at the inlet end near pump stands, sharp bends, high elevation points, and before the ends of pipelines.

Outlet structures, comprising a riser pipe and one or more valves for flow control, release water from pipelines to the location where they are needed. An example of a common outlet consists of a concrete riser pipe and valves. The riser pipe is connected to riser valves, hydrants, and gate pipes which distribute water to a border, furrow, or basin. Hydrants, placed over the riser valve, connect portable gated pipes to the pipeline. They are portable and can also be used to connect the suction hose of a pump to the water supply in the pipeline, helping to develop pressure. The end plug, as the name suggests, closes the pipeline and absorbs the pressure developed at the pipeline. The plug may be supported by a concrete block that is strong enough to withstand unexpected high pressure caused by sudden opening or closing of valves.

9.5.6 Pressure Variation

The pressure in a pipeline varies with location because of the change in elevation of the ground. The pressure variation can be described using the energy equation and head loss equations discussed earlier.

9.5.7 Design of Buried Pipelines

The objective of designing an underground pipeline system is that the system distributes water as uniformly as possible. The first step in the design is to know the land topography, location of the water source, and water discharge. Then, each component of the pipeline system is determined. Beginning with a pump stand, the stand should be high enough to develop sufficient operating head for the pipeline, but not so high as to develop heads greater than needed, which will build up water and excessive pressure in the pipeline. The guideline normally used is that the operating pressure in the pipeline should be less than one-quarter of the internal bursting pressure. Outlet sizes are selected to obtain the needed flow at diversion points.

The height of water in the pump stand is computed as the sum of the reduced level at height and losses in the pipeline. A freeboard of 0.5 m of water head is added to obtain the height of the pump stand. Losses (both major and minor) in the pipeline are computed as discussed earlier. The diameter of the pipeline is computed considering head loss due to friction and discharge. Pumping cost and material of pipe factor in the design for the system cost for large-diameter pipes and pumping cost for small-diameter pipes as increased friction head losses are high.

9.5.8 Laying Out of Pipelines

The first step in laying buried pipes is a contour map that depicts field boundaries, streams, rivers, reservoirs, roads, wells, residential areas, and other features of the area. The pipeline alignment and location of valves should depend on the topographic features. Earthwork is needed to lay the pipes below the ground, depending on the depth, gradient, and location of structures, such as inlet, outlet, water control, and diversion structures, pressure release and air release valves, and end plugs. Pipe leaks should be checked and repaired if there are any.

QUESTIONS

Q.9.1 What will be the friction factor for a 25-cm diameter new cast iron pipe that carries a discharge of 0.25 m^3/s at a temperature of 25 °C?

Q.9.2 Compute the head loss using the Darcy–Weisbach equation for a 20-cm diameter galvanized steel pipe that is 1000 m long and carrying a discharge of 0.25 m^3/s.

Q.9.3 Solve Q.9.2 using the Hazen–Williams equation.

Q.9.4 A reservoir is connected to a pipeline that supplies water at a rate of 0.1 m^3/s, as shown in Figure 9.17. The difference between the water level in the reservoir and the center of the connecting pipe is 10 m. This pipe, 30 cm in diameter, runs horizontal and connects after some distance to a pipe 20 cm in diameter at a 90° downward, which connects after some distance to another pipe 15 cm in diameter at a 90° bend which runs horizontally. The height from the center of the exiting pipe to the center of the first pipe is 15 m. What will be the total loss of energy in the system?

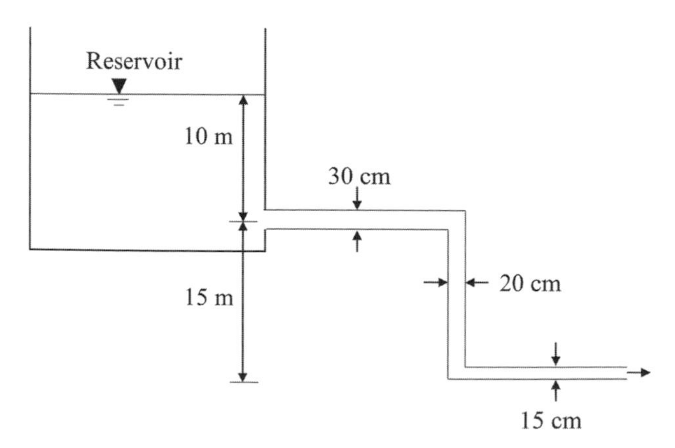

Figure 9.17 A reservoir is connected to a pipeline in Q.9.4.

Q.9.5 Consider a 500-m long cast iron pipe that is 20 cm in diameter. What will be the discharge from the pipe if the pressure drop is limited to 28,000 kg/m^2 at 20 °C? Use the Darcy–Weisbach equation for head loss.

Q.9.6 Solve Q.9.5 using the Hazen–Williams equation.

Q.9.7 Consider a 500-m long, 20-cm diameter PVC pipeline that carries a discharge of 0.1 m^3/s. The inlet of the pipeline is 30 m from the reference datum. The end of the pipeline is 5 m above the datum and the pressure at the outlet is 15,000 kg/m^2. What will be the loss of energy through the pipeline? What will be the piezometric head above the reference datum at a point 100 m away from the outlet? The Hazen–Williams equation can be used to compute the head loss.

Q.9.8 Consider a 10-cm diameter galvanized iron pipe that carries water at a velocity of 1 m/s at a temperature of 25 °C. What will be the friction factor?

Q.9.9 There are two reservoirs, one supplying water to the other. The water is supplied through a cast iron pipeline 2000 m long at a rate of 0.5 m^3/s. The difference between the water levels of the two reservoirs is 25 m. Assuming a water temperature of 25 °C, what is the size of the pipe? The Darcy–Weisbach equation can be used to compute the head loss.

Q.9.10 Solve Q.9.9 using the Hazen–Williams equation.

Q.9.11 Two reservoirs are connected by a 400-m long galvanized iron pipeline that is 30 cm in diameter. The difference between the water levels of the two reservoirs is 30 m. The pipeline has two 90° elbows, a check valve, and an orifice. Assume a water temperature of 25 °C. What will be the flow through the pipeline?

Q.9.12 Two reservoirs are connected by a 5000-m long commercial steel pipeline that is 90 cm in diameter. The difference in the water levels of the two reservoirs is 25 m. The pipeline encounters a hill 20 m above the upper reservoir at a distance of 2000 m and must cross the hill. What will be the discharge through the pipeline? Assume a water temperature of 25 °C. What will be the minimum depth below the summit for laying the pipeline so that the pressure in the pipeline does not fall below the atmospheric pressure? The Darcy– Weisbach equation can be used to compute the head loss.

Q.9.13 A water tank supplies water to another tank through a commercial steel pipeline that is 400 m long and 15 cm in diameter. The supply tank has a 15-m high water level. The height from the bottom of the supply tank to the center of the pipe to the receiving tank is 25 m. The pressure at the delivery end is

45,000 kg/m^2 and discharge there is 0.12 m^3/s. Assume a water temperature of 25 °C. What will be the head needed for pumping and the power required if efficiency is 75%? The entrance and exit losses can be assumed to be negligible.

Q.9.14 Three welded steel pipelines arranged in series are used to supply water. The first pipe is 200 m long and 10 cm in diameter, the second pipe is 800 m long and 15 cm in diameter, and the third pipe is 1500 m long and 20 cm in diameter. The total pressure drop in the pipeline due to friction is 40,000 kg/m^2. What will be the discharge through the pipeline?

Q.9.15 Two reservoirs are connected through two PVC pipes arranged in series. The water surface elevation in the upstream reservoir is 250 m. The pipe connected to it is 250 m long and has a diameter of 20 cm, and is connected to a pipe 1000 m long and 30 cm in diameter. The discharge through the pipeline is 0.3 m^3/s. What is the elevation of the water surface in the downstream reservoir?

Q.9.16 Consider a flow with a discharge of 5 m^3/s that is partitioned into three parallel commercial steel pipes having lengths of 500 m, 250 m, and 400 m, and diameters of 80 cm, 40 cm, and 60 cm, respectively. What will be the head loss and flow through each pipe?

Q.9.17 A reservoir supplies water through a commercial steel pipeline. The elevation of water in the reservoir is 500 m. The pipe connected to the reservoir is 500 m long and 25 cm in diameter; the pipe is connected to two parallel pipes, one of which is 200 m long and 15 cm in diameter and the other of which is 150 m long and 12 cm in diameter. These two pipes are then connected to a single pipe that is 30 cm in diameter and 1000 m in length. The elevation of the exit point is 450 m. What is the discharge in the pipeline?

Q.9.18 Two reservoirs are connected through a commercial steel pipeline that is 4000 m long and 50 cm in diameter. After 2000 m, a second pipe of 60-cm diameter is laid alongside the first pipe for the remainder of the distance of 2000 m. What will be the effect on discharge?

REFERENCES

American Society of Agricultural Engineers (ASAE) (1987). ASAE Standard S376.1. In *Design, Installations and Performance of Underground Thermoplastic Irrigation Pipelines*. St. Joseph, MI: ASAE, pp. 501–511.

American Society of Agricultural and Biological Engineers (ASABE) (2004). ASAE Standard S376.2. In *Design, Installation and Performance of Underground, Thermoplastic Irrigation Pipelines*. St. Joseph, MI: ASABE.

American Society of Agricultural and Biological Engineers (ASABE) (2016). ANSI/ASAE Standard S376.3. In *Design, Installation and Performance of Underground, Thermoplastic Irrigation Pipeline*. St. Joseph, MI: ASABE.

Gupta, R. S. (2017). *Hydrology and Hydraulic Systems*. Long Grove, IL: Waveland Press.

Jain, A. K. (1976). Accurate explicit equation for friction factor. *Journal of Hydraulics Division, ASCE*, 102(HY5): 674–677.

James, L. G. (1988). *Principles of Farm Irrigation System Design*. New York: Wiley.

Michael, A. M. (2010). *Irrigation Theory and Practice*. Noida: Vikas Publishing House PVT Ltd.

Moody, L. (1944). Friction factors for pipe flow. *Transactions of American Society of Mechanical Engineers*, 66: 671–678.

Pair, C. H., Hinz, W. W., Reid, C., and Frost, K. R. (1975). *Sprinkler Irrigation*, 4th ed. Silver Spring, MD: Sprinkler Irrigation Association.

USDA-NRCS (1997). Energy use and conservation. In *National Engineering Handbook*. Washington, DC: USDA.

Watters, G. Z., and Keller, J. (1978). Trickle irrigation tubing hydraulics. ASAE technical paper 782015.

10 Pumps and Pump Selection

Notation

A	flow cross-sectional area of the pipe (m^2, ft^2)	H_{ps}	pressure head on suction side (m, ft)
A_i	irrigated area (ha)	H_{pd}	pressure head on discharge side (m, ft)
BP	brake horsepower (hp, kW)	H_{ss}	static head or lift (m, ft, positive for suction lift and negative for suction head) (m, ft)
D	pipe diameter (m, ft)	H_{sd}	static discharge head (m, ft)
d_n	design daily irrigation requirement (mm/day)	H_v	velocity head (m, ft)
D_L	discharge side lift, is equal to H_{sd} (m, ft)	h_{vap}	vapor pressure of water at the operating temperature (m, ft)
E_i	irrigation efficiency (%)	N	pump speed (rpm)
E_m	motor efficiency for a constant-speed pump (fraction)	$NPSH$	net positive suction head (m, ft)
E_p	pump efficiency, the ratio of water horsepower to brake horsepower (fraction)	$NPSH_a$	net positive suction head available (m, ft)
f_s	safety factor (0.6–1.0 m for pumping water), which is used to consider any inaccurate estimation of pressure drop in the pumping system	$NPSH_r$	net positive suction head required (m, ft)
g	acceleration due to gravity (9.81 m/s^2, 32.2 ft/s^2)	N_s	specific speed of the pump, dimensionless
H	total head, system head, or total dynamic head (TDH) (m, ft)	p_{kw}	power required to operate the pump (kW)
h_{atm}	atmospheric pressure head (m, ft)	Q	discharge of the pump, or pump capacity (L/min, m^3/s, ft^3/s, gal/min)
H_D	head due to drawdown (m, ft)	Q_{sd}	system design capacity (L/min)
h_f	friction head loss in pipes (m, ft) minor head loss due to fitting is h_f^*	Re	Reynolds number (dimensionless)
h_f^*	minor head loss due to fitting (m, ft)	S_L	suction-side lift, is equal to H_{ss} when the pump is above the water surface (m, ft)
h_{fs}	head loss due to friction in the suction line (m, ft)	T	duration of system operation (h/day)
h_{fs}^*	minor head loss due to fitting in the suction line (m, ft)	TDH	total dynamic head (ft, m)
h_{lift}	maximum practical suction lift (m, ft)	v	mean flow velocity in the pipe (m/s, ft/s)
H_o	operating head, i.e., the pressure head required on the suction (H_{ps}) and discharge sides (H_{pd}) (m, ft)	WP	water horsepower (hp, kW)
		Z	elevation head (m, ft)
		γ	specific weight of water (9.807 kN/m^3, 62.43 lb/ft^3)
		μ	dynamic viscosity of water (1.0×10^{-3} kg/m · s at 20 °C)

10.1 INTRODUCTION

Pumps are mechanical devices that are used to move water. These devices provide energy to water and overcome friction during its conveyance. They use mechanical energy generated by a motor to increase the potential and kinetic energy of the water. The motor can be electric, gasoline, or diesel, and may be self-starting or hand-starting. In the olden days, water was lifted either manually or using animal power, or via Persian wheel or water wheel. In farm irrigation systems, pumps are used to lift water from a supply source, such as a well, canal, reservoir, or stream, and convey it through channels or pipes. Pumps are also used to drain excess water from fields for lowering the water table.

Pumps are an integral part of many agricultural irrigation systems. A pump is used to lift groundwater to the ground surface, to raise water from a lower elevation to a higher elevation, to transport water, to overcome friction, or to generate pressure for the operation of sprinkler and trickle irrigation systems. This chapter discusses rudimentary aspects of pumps and their operation and selection.

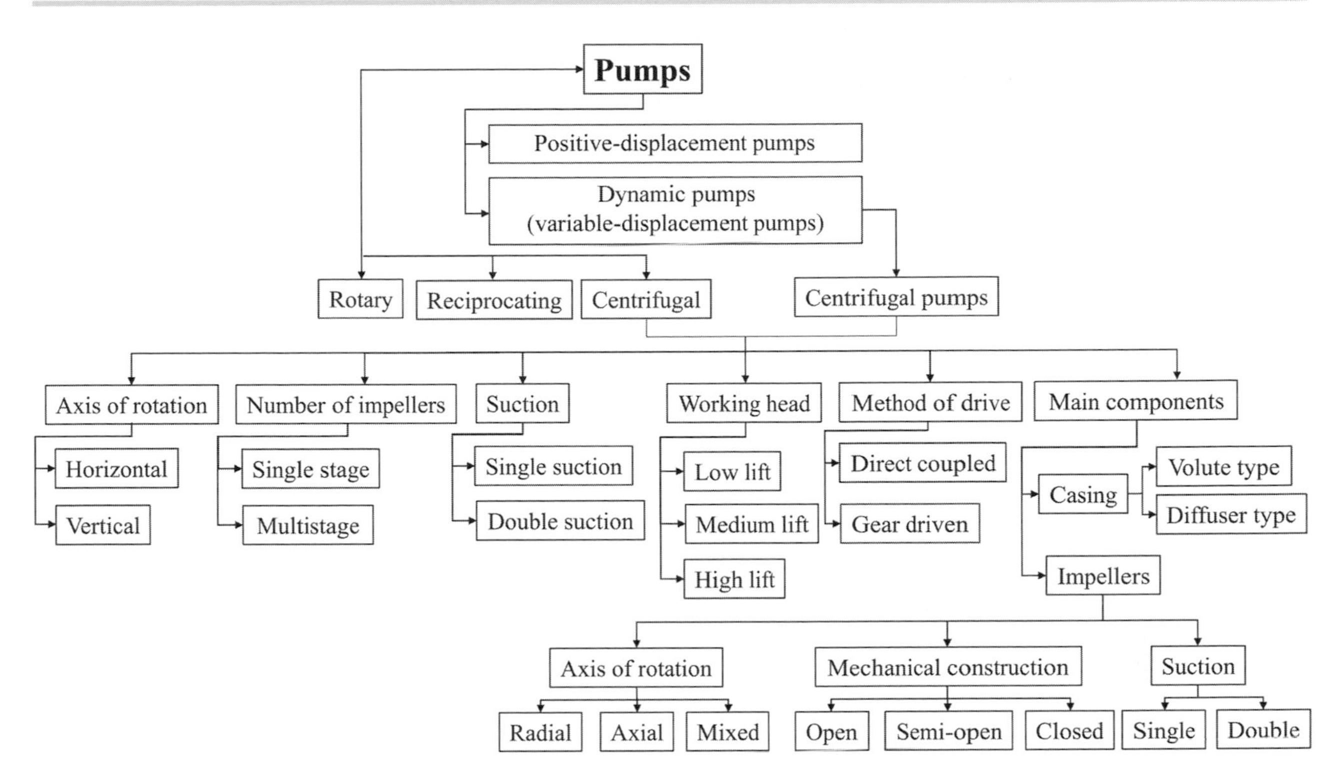

Figure 10.1 Classification of pumps.

10.2 CLASSIFICATION OF PUMPS

Pumps are of different sizes and types, and operate on different principles. They can be classified in different ways, depending on the material of construction, application, spatial orientation, and transfer of energy to water, as shown in Figure 10.1. A common classification of pumps includes rotary, reciprocating, and centrifugal pumps, but a broader classification includes positive displacement and dynamic pumps. A positive displacement pump discharges the same amount of water regardless of head, such as a hand pump commonly used for domestic water supply or home garden irrigation. These pumps are more efficient than centrifugal pumps, but need greater maintenance.

Dynamic pumps, also called variable-displacement pumps, produce discharge which is inversely related to the pressure head; that is, they produce higher discharge at lower head and vice versa and require more energy. The most common example is the centrifugal pump, commonly used in farm irrigation, which will be the focus of this chapter.

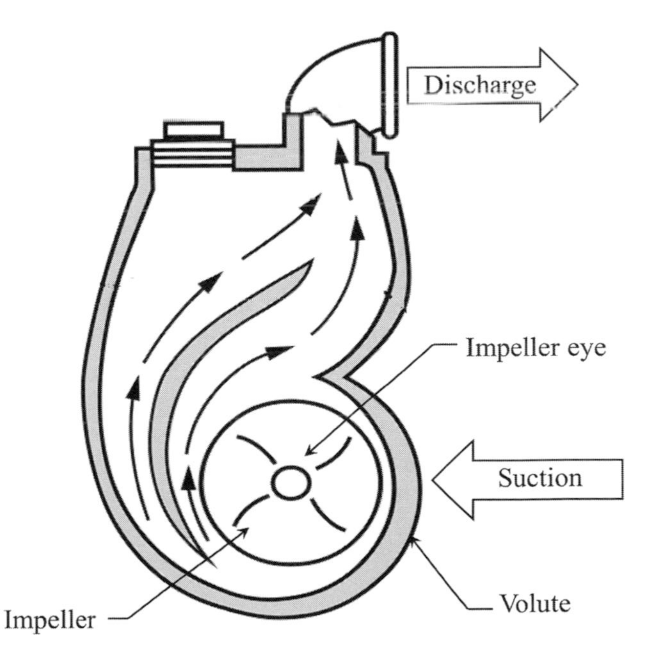

Figure 10.2 Centrifugal pump composition. (Adapted from USDA-NRCS, 2016.)

10.3 CENTRIFUGAL PUMPS

A centrifugal pump comprises a set of vanes, referred to as impellers, that are enclosed within a stationary housing referred to as a casing, as shown in Figure 10.2. The water enters the eye of the impeller by atmospheric or other pressure and is set into rotation by the impeller vanes; it rotates at a very high speed with the help of a prime mover, such as an electric motor or a diesel engine. The impeller rotation generates the centrifugal force that accelerates out the water sitting in the cavities between vanes from the eye of the impeller to its periphery. With water leaving the eye of the impeller, suction is created which causes more water to flow toward the inlet. The water from the periphery of the impeller moves into the casing, where the velocity head (kinetic energy) is reduced because of increased casing cross-sectional area, and converted into pressure or head (potential energy). Then, water is pumped out from the discharge pipe.

Centrifugal pumps are simple in construction, have low to moderate initial cost, exhibit moderate to high efficiency, have a wide range of capacity, have the ability to adapt to several prime movers, are less noisy, do not need internal lubrication, and have low space requirement. However, they have no self-prime ability, have efficiency limited to a narrow range of head and discharge, and are prone to impeller damage by abrasive matter that may exist in the water.

10.3.1 Classification of Centrifugal Pumps Based on the Axis of Rotation

Depending on the orientation of their axis of rotation, centrifugal pumps can be either horizontal or vertical. Horizontal pumps have a horizontal shaft on which vertical impellers are mounted, whereas vertical pumps have a vertical shaft on which horizontal impellers are mounted. Horizontal pumps are used for pumping water from open wells, shallow tubewells, and surface water bodies, such as lakes, streams, and canals. These pumps are installed on the surface and are easy to inspect and maintain, whereas vertical pumps are used below the water table and do not require priming. Horizontal pumps can be classified into four subcategories, depending on the position of the suction inlet: side-suction, end-suction, bottom-suction, and top-suction.

10.3.2 Classification of Centrifugal Pumps Based on Number of Impellers

Based on the number of impellers, centrifugal pumps can be single-stage or multistage. A single-stage pump is essentially a low-lift pump, and the total head is developed by a single impeller, whereas in a multistage pump, a number of impellers are mounted on the same shaft in series and the outflow from one impeller becomes inflow to the second impeller. This increases the pressure as water moves through a series of stages. Multistage pumps are high-lift pumps, and for a given impeller type the increase in head and power requirement are proportional to the number of stages (impellers), but the discharge and efficiency remain the same as for a single-stage pump.

10.3.3 Classification of Centrifugal Pumps Based on Suction

Centrifugal pumps can also be classified as single-suction or double-suction pumps, depending on the number of entrances to the impeller. In a single-suction pump, the water enters the impeller from one side, but in double-suction pumps the water enters from both sides. Single-suction pumps are easy to construct and maintain and have low cost and large waterway. They can be used for pumping water with high concentrations of suspended materials. The double-suction impeller is identical to two single-suction impellers arranged back to back, and is used for pumping large quantities of water. Double-suction pumps have low net positive suction requirements.

10.3.4 Classification of Centrifugal Pumps Based on Working Head

Depending on the working head, pumps can also be classified as low-lift, medium-lift, and high-lift pumps. Low-lift pumps are single-stage horizontal types with volute casing that operate up to a maximum total head of 15 m. Medium-lift pumps are provided with diffuser vanes and may operate up to a maximum head of 40 m. High-lift pumps are multistage turbine pumps with a maximum head of more than 40 m.

10.3.5 Classification of Centrifugal Pumps Based on Method of Drive

Centrifugal pumps can also be classified as direct coupled, gear-driven, or belt-driven. Generally, centrifugal pumps are direct coupled with a prime mover (electric motor) with a common shaft and bearing; the pump and prime mover have the same speed. Gear-driven pumps are used when the pump is run at a speed different from that of the prime mover. They have lower efficiency than do direct coupled pumps, but higher than that of belt-driven pumps. Belt-driven pumps are used when the pump is kept away from the prime mover. As an example, if the source of power is away from the pump, the pump can be powered by a tractor; it is lowered into a well to keep it within the suction limits, but the motor is kept on the ground surface to avoid submergence. In such cases, a belt and pulley arrangement is used to power the pump.

10.4 THE MAIN COMPONENTS

Pumps can also be classified based on their main components – that is, the casing and impeller.

10.4.1 Casing

Casing surrounds the impeller, encloses the pumping water, collects the water discharged from the circumference of the impeller, and conducts it to a suitable discharge point on the pump. It helps convert the kinetic energy of water to potential energy (head). Centrifugal pump casings are either volute or diffuser type (turbine), as shown in Figure 10.3 (a similar figure can be found in Fraenkel [1986]), which differ essentially in construction around the impeller. The turbine pump has fixed diffuser vanes which more efficiently convert velocity (kinetic) energy into pressure energy. The volute pump has a casing whose cross-sectional area gradually increases from the impeller periphery toward the pump outlet. The increase in area

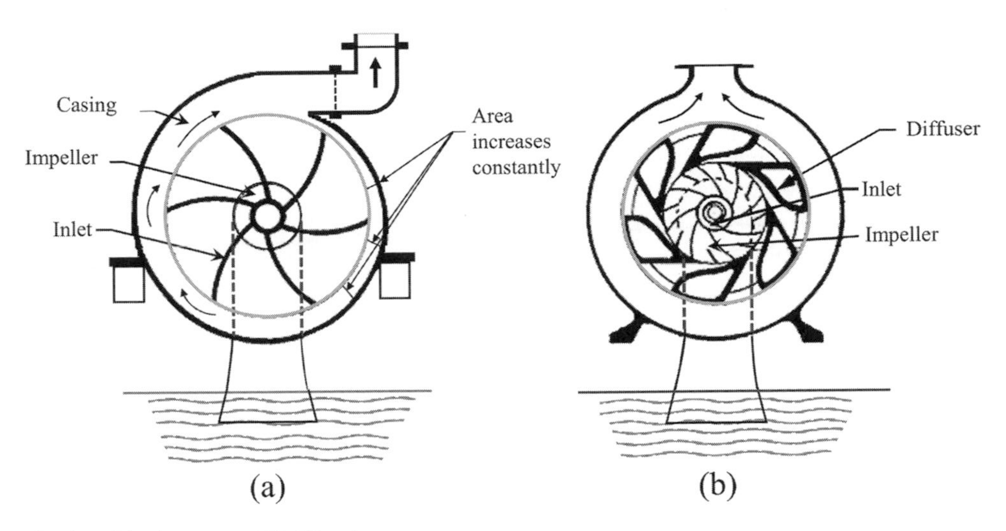

Figure 10.3 Types of casings: (a) volute type and (b) diffuser type.

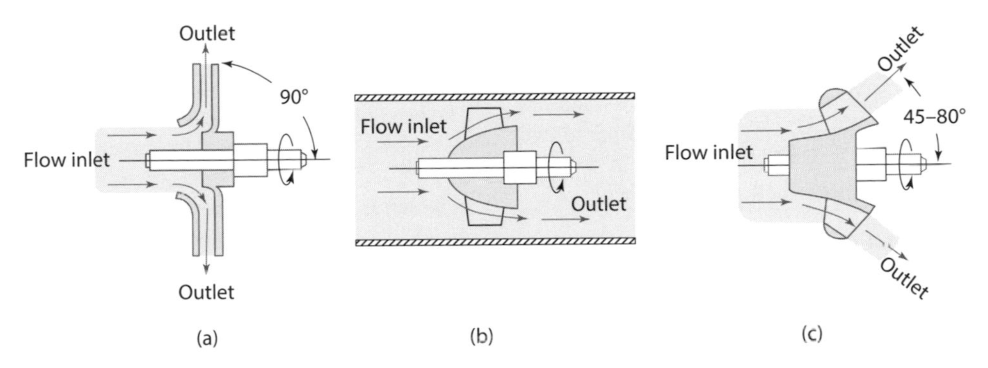

Figure 10.4 Impellers for centrifugal pumps: (a) radial, (b) axial, and (c) mixed-flow impellers.

decreases the flow velocity, which then increases the pressure (head). Thus, a centrifugal pump discharges water that normally has a greater pressure energy than the pressure energy it contained on its entry to the pump. The single volute pump often has nonuniform distribution of radial pressure, which affects the bearing life. Therefore, some centrifugal pumps may have two volutes ending at the same outlet to counterbalance the radial pressure. Most of the horizontal centrifugal pumps used for irrigation are of the volute type.

10.4.2 Impellers and Their Classification

Impellers can be classified based on the axis of rotation, mechanical construction, and suction.

10.4.2.1 Classification Based on Axis of Rotation

Impellers can be classified as radial, axial, or mixed flow based on the direction of flow through the impeller relative to the axis of rotation, as shown in Figure 10.4. In radial-flow impellers the water enters a plane parallel to the axis of rotation and leaves the pump in the perpendicular direction, whereas in the case of axial impellers the water enters and leaves the pump into a plane parallel to the axis of rotation. In the case of mixed-flow impellers the water also enters parallel to the axis of rotation but leaves at some angle. The radial-flow impellers have low specific speed and are suitable for high head and low flows, whereas axial-flow impellers have higher specific speed and are suitable for high discharge at low head.

10.4.2.2 Classification Based on Mechanical Construction

Impellers can be classified as open impellers, semi-open impellers, or closed impellers, based on mechanical construction, as shown in Figure 10.5. Open impellers have vanes (blades) which are open on all sides except where they are attached to the shaft. They are used to pump very viscous fluid or liquid having considerable small solids (sand, silt, and pebble), as found in dredging. Semi-open impellers are shrouded with one side (side wall) only usually at the back and are useful for pumping liquid with suspended solids (paper pulp, sewage water, etc.). Enclosed impellers have side walls on both sides that cover the vanes and are useful for pumping clean water, but are not suitable for pumping water with suspended material. In general, open impellers are more efficient than are semi-open and enclosed impellers for the same specific speed.

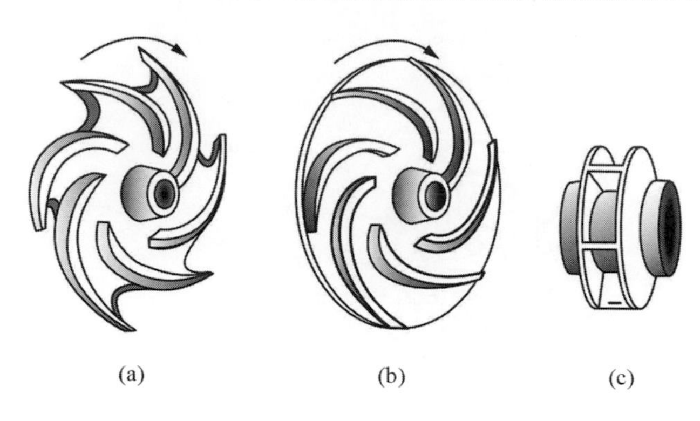

Figure 10.5 Different types of impeller for centrifugal pumps: (a) open, (b) semi-open, and (c) enclosed impeller. (Adapted from USDA-NRCS, 2016.)

10.4.2.3 Classification Based on Suction

Impellers can be single or double suction. In single-suction impellers, water enters through an inlet situated on one side of the impeller and flows symmetrically into both sides of double-sided impellers. Single-suction impellers are used for pumping water with high concentrations of solids, for multi-stage pumps, and for small pumps. They cost less and are easy to maintain. A double-suction impeller is similar to two single-suction impellers arranged back to back in a single casing. It has a low net positive suction requirement.

10.5 PUMP PERFORMANCE

The performance of a pump is described by six parameters: capacity, head, power, efficiency, required net positive suction head, and specific speed. Each of these parameters is discussed here.

10.5.1 Capacity

The capacity of a pump, denoted Q, is defined by the volume of water delivered per unit time and is often expressed in liters per minute (L/min), cubic meters per second (m^3/s), cubic feet per second (cfs, ft^3/s), or gallons per minutes (gal/min). Depending on the size of a pump, an appropriate unit is selected.

10.5.2 Head

The head is defined as the net work done by the pump on a unit weight of water. It expresses the total energy head of water comprising elevation (above a reference point, such as a pump centerline), pressure head, and velocity head, as discussed in Chapter 9 on pipe hydraulics, and its pressure. Depending on the relative position of the source of water and of the centerline of the pump, suction lift and suction head are often used in assessing pump performance. As the name suggests, suction lift is the vertical distance between the centerline of the pump and the free water surface of the source (i.e., the pump is installed above the water level) and suction head then becomes the height from the water surface to the centerline (i.e., the pump is installed below the water level, as shown in

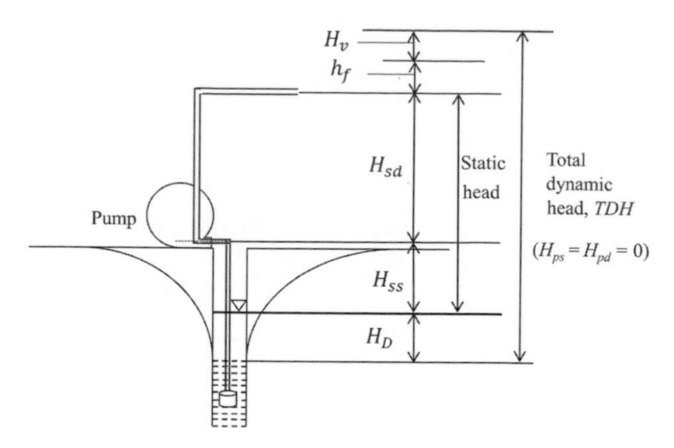

Figure 10.6 A schematic of components of total dynamic head. The pump is above the water level; H_{ss} is the suction lift.

Figure 10.6). The static discharge head is the vertical distance from the pump centerline to the point of free discharge. The sum of static suction head (H_{ss}) and static discharge head (H_{sd}) constitutes the total static head (Figure 10.6). It is clear that a pump has two sides: the suction side and the discharge side. The head of the water (H) can be expressed as

$$H = H_d - H_s, \tag{10.1}$$

in which H_d denotes the head on the discharge side and H_s is the head on the suction side. From the energy equation,

$$H_i = \left[Z + \frac{p}{\gamma} + \frac{v^2}{2g}\right]_i, \quad i = d, s, \tag{10.2}$$

in which subscript d corresponds to the discharge side and subscript s corresponds to the suction side. Z is the elevation head, $\frac{p}{\gamma}$ represents the pressure head, and $\frac{v^2}{2g}$ is the velocity head.

It may be noted that for horizontal pumps, the datum is a horizontal line passing through the centerline of the pump shaft, whereas for vertical pumps the datum is a horizontal plane passing through the eye entrance of the first-stage impeller. When the pump is operating, pressure head, velocity head, and friction head loss become important; when added to the total static head, this constitutes the total dynamic head. Figure 10.6 shows the components of the total dynamic head for a condition when the pressure head is zero at the suction and delivery sides ($H_{pd} = H_{ps}$), and minor head loss is neglected. The total head (H) or total dynamic head (TDH) can be expressed in meters or feet as:

$$TDH = H_{ss} + H_{ps} + H_{sd} + H_D + H_{pd} + h_f + h_f^* + H_v, \tag{10.3}$$

where H_{ss} is the static suction head or lift (m, ft; positive for suction lift and negative for suction head); H_{ps} is the pressure head on the suction side (m, ft); H_{sd} is the static discharge head (m, ft); H_D is the head due to drawdown (m, ft); H_{pd} is the pressure head on the discharge side (m, ft); h_f is the friction head loss in pipes (m, ft); h_f^* is the minor head loss due to fittings (m, ft); and H_v is the velocity head (m, ft). The datum is taken from the centerline of the shaft of the pump for horizontal centrifugal pumps and the entrance eye of the first-stage impeller for vertical pumps.

Example 10.1: Consider a centrifugal pump that pumps groundwater at a rate of 15 L/s, as shown in Figure 10.7. The pump is installed 4 m above the water table and discharges water 6 m above the pump centerline. On both suction and discharge sides, the pipe is 12 cm in diameter and is made of galvanized steel iron. The velocity of flow is 1.33 m/s. On the suction side the vertical pipe is attached to another pipe, 2 m long and 8 cm in diameter, at 90°. Likewise, the discharge side is attached to another pipe which is at 90° and is 12 cm in diameter. After 15 hours of pumping the water table declines by 0.5 m. Calculate the total static and total dynamic head. The temperature of water is 20 °C.

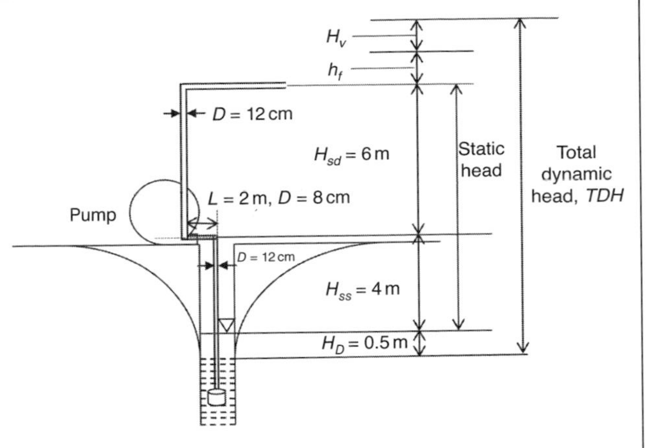

Figure 10.7 Illustration of total dynamic head.

Solution: Given $H_{ss} = 4$ m, $H_{sd} = 6$ m, $H_D = 0.5$ m, $Q = 15\ \text{L/s} = 0.015\ \text{m}^3/\text{s}$, $D = 120$ mm, $H_{pd} = H_{ps} = 0$.

The static water head can be expressed as

$$\textit{Static water head} = H_{ss} + H_{sd} = 4 + 6 = 10\ \text{m}.$$

The TDH can be expressed as

$$\begin{aligned} TDH &= H_{ss} + H_{sd} + H_D + h_f + H_v \\ &= 4 + 6 + 0.5 + h_f + h_f^* + H_v \\ &= 10.5 + h_f + h_f^* + H_v. \end{aligned}$$

The average velocity is

$$v = \frac{Q}{A} = \frac{4Q}{\pi D^2} = \frac{4(0.015\ \text{m}^3/\text{s})}{\pi(0.12\ \text{m})^2} = 1.33\ \text{m/s}.$$

At 20 °C, the dynamic viscosity is

$$\mu = 1.0 \times 10^{-3}\ \frac{\text{kg}}{\text{m/s}}.$$

The Reynolds number is

$$\text{Re} = \frac{\rho v D}{\mu} = \frac{\left(1000\ \frac{\text{kg}}{\text{m}^3}\right)\left(1.33\ \frac{\text{m}}{\text{s}}\right)(0.12\ \text{m})}{1.0 \times 10^{-3}\ \frac{\text{kg}}{\text{m/s}}} = 1.6 \times 10^5.$$

Re is greater than 4000, so flow is turbulent.

The equivalent roughness of galvanized steel iron is 5.0×10^{-4} ft. Therefore, the relative roughness is:

$$\frac{\varepsilon}{D} = \frac{5.0 \times 10^{-4}\text{ft}}{0.1\ \text{m} \times \frac{3.28\ \text{ft}}{1\ \text{m}}} = 0.0015.$$

From the Moody diagram,

$$\text{Re} = 1.6 \times 10^5 \rightarrow \frac{\varepsilon}{D} = 0.0015 \rightarrow f = 0.024.$$

The head loss due to friction in the pipes can be calculated using the Darcy–Weisbach equation.

Head loss due to friction in the suction pipe:

$$h_{fs} = f\frac{L}{D}\frac{v^2}{2g} = 0.024 \times \frac{2\ \text{m}}{0.08\ \text{m}} \times \frac{\left(1.33\frac{\text{m}}{\text{s}}\right)^2}{2 \times 9.81\ \text{m/s}^2} = 0.054\ \text{m}.$$

Head loss due to friction in the delivery pipe:

$$h_{fd} = f\frac{L}{D}\frac{v^2}{2g} = 0.024 \times \frac{(4\ \text{m} + 6\ \text{m})}{0.12\ \text{m}} \times \frac{\left(1.33\frac{\text{m}}{\text{s}}\right)^2}{2 \times 9.81\ \text{m/s}^2} = 0.180\ \text{m}.$$

Therefore, head loss due to friction in the pipe

$$h_f = h_{fs} + h_{fd} = 0.054 + 0.180 = 0.234\ \text{m}.$$

Head loss due to bends is computed with K for two 90° elbows as $1.5 \times 2 = 3$. Therefore, the minor head loss can be computed as:

$$h_f^* = \sum_{i=1}^{2} K_i \frac{v^2}{2g} = 3 \times \frac{\left(1.33\frac{\text{m}}{\text{s}}\right)^2}{2 \times 9.81} = 0.270\ \text{m}.$$

Velocity head, $H_v = 0$, since the pipe diameter on both suction and delivery sides pipes is the same.

The TDH is:

$$\begin{aligned} \text{TDH} &= 10.5 + h_f + h_f^* + H_v \\ &= 10.5 + 0.234 + 0.270 + 0 = 11.00\ \text{m}. \end{aligned}$$

10.5.3 Power

For pumps a distinction is made between brake horsepower (BP), which is defined as the amount of power imparted to the pump shaft, and water horsepower (WP), which is the actual power delivered to the water by the pump. BP is a measure of energy required to operate the pump, part of which is used up by the pump to overcome disc friction, circulation losses, bearing friction, and hydraulic losses. WP (kW, hp) can be expressed as

$$WP = \gamma QH = \frac{QH}{K}, \tag{10.4}$$

where Q is the pump capacity (L/min, m^3/s, gpm, cfs), γ is the weight density (ρg), H is the head (m, ft), and K is the unit constant equal to 6116 for WP in kW and Q in L/min; equal to 0.102 for WP in kW and Q in m^3/s; equal to 3960 for WP in hp and Q in gpm; and equal to 8.81 for WP in hp and Q in cfs.

10.5.4 Efficiency

The pump efficiency is defined by the percentage of power input to the pump shaft (brake power); that is, the ratio of WP to BP. It can be expressed as

$$E_p = \frac{WP}{BP} \times 100. \tag{10.5}$$

Since an electric motor or diesel engine is needed to operate a pump, the electrical energy required for the motor-driven pump can be determined by accounting for the motor efficiency for a constant-speed pump and the wire-to-shaft efficiency of the motor and variable-speed drive for a variable-speed pump, expressed as:

$$P_{kw} = \frac{0.746\ BP}{E_m\ or\ E_{ws}}, \tag{10.6}$$

where P_{kw} is the power required to operate the pump in kW, BP is the brake horsepower in hp, E_m is the motor efficiency for a constant-speed pump as a fraction, and E_{ws} is the wire-to-shaft efficiency of the motor and variable drive (fraction). The value of 0.746 represents the conversion factor for converting horsepower to kW.

Example 10.2: Considering the data from Example 10.1, compute the pump water horsepower and brake horsepower if pump efficiency is 85%. Compute the power required to operate the pump driven by an electric motor having an efficiency of 90%. If the electricity charge is $0.25/kWh, what would be the cost of pumping 15,000 m^3 of water?

Solution: $Q = 15\,L/s = 0.015\,m^3/s$, $H = 11.00\,m$, $E_p = 85\%$, $E_m = 90\%$. The WP can be computed using eq. (10.4) with $K = 1.02$ for WP in kW, Q in m^3/s, and H in m:

$$WP = \frac{QH}{0.102} = \frac{0.015\ \frac{m^3}{s} \times 11.00\ m}{0.102} = 1.62\ kW.$$

Brake horsepower is computed using eq. (10.5):

$$BP = \frac{WP}{E_p} = \frac{1.62\ kW}{85\%} = 1.91\ kW.$$

Since BP is also in kW, eq. (10.6) can be written as:

$$P_{kw} = \frac{BP}{E_m} = \frac{1.91\ kW}{90\%} = 2.12\ kW.$$

The time required to pump 15,000 m^3 of water is

$$\text{Duration (h)} = \frac{15{,}000\ m^3}{0.015\ \frac{m^3}{s} \times 3600\ \frac{s}{h}} = 277.78\ h.$$

The cost of pumping 15,000 m^3 of water is

$$\begin{aligned}\text{Power cost (\$)} &= P_{kw} \times \text{duration} \times \text{per unit electricity charge}\\ &= 2.12\,kW \times 277.78\,h \times 0.25\frac{\$}{kWh}\\ &= \$147.\end{aligned}$$

10.5.5 Net Positive Suction Head

The net positive suction head (NPSH) is the amount of energy needed to prevent the formation of vapor-filled cavities of water within the eye of the single- and first-stage impeller pumps to determine whether the liquid will vaporize at the lowest pressure point in the pump. It is expressed in terms of head and is the difference between the suction head (i.e., the sum of the static head and velocity head) close to the impeller (Figure 10.8a) and the vapor pressure of water at pump suction conditions (Gupta, 1989). Pumps have two values of NPSH: net positive suction head required ($NPSH_r$) and net positive suction head available ($NPSH_a$). $NPSH_r$ denotes the minimum head at the pump entrance that prevents cavitation; it is determined by the manufacturer and is a function of pump design, type, and size, and varies with the speed and capacity of the pump. It increases with increasing flow rate in a given pump and is higher for larger pumps, thus suggesting that cavitation is a problem in large pumps.

The suction head close to the suction inlet is usually inconvenient to measure. In such cases, $NPSH_a$ can be defined as the atmospheric pressure minus all losses on the suction side of the pump, which include vapor pressure head, friction head, and static lift, as shown in Figure 10.8:

$$NPSH_a = h_{atm} - h_{vap} - h_{fs} - h_{fs}^{*} - S_L, \tag{10.7}$$

where $NPSH_a$ is the net positive suction head, h_{atm} is the atmospheric pressure head, h_{vap} is the vapor pressure of water at the operating temperature , h_{fs} is the friction head loss on the suction side of the pump, that is, pipe friction between the pump and suction intake; h_{fs}^{*} is the minor head losses through fittings on the suction side of the pump, and S_L is the static suction lift, all in meters. S_L is equal to H_{ss}

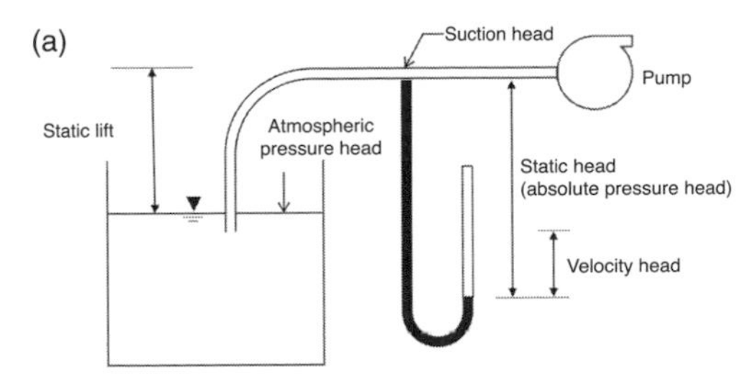

Suction head = static head + velocity head
= atomospheric pressure head – static lift – friction loss
$NPSH_a$ = suction head – vapor pressure head

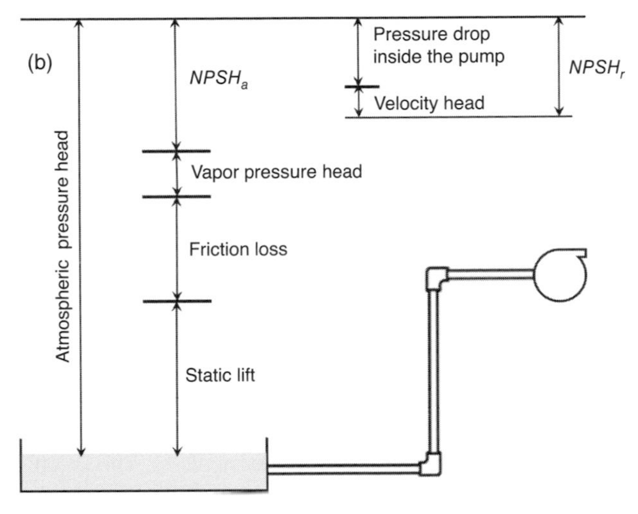

Figure 10.8 (a) Pressure head at the suction inlet and (b) net positive suction head ($NPSH_a$ and $NPSH_r$).

when the pump is above the water surface and is negative when the pump is below the water surface.

It may be noted that the atmospheric pressure at sea level is 14.7 psi (760 mmHg, 101.3 kPa), which corresponds to a head of 10.34 m for clean water at 4 °C. For every 1000 m elevation above mean sea level, a value of 1.2 m head can be subtracted from 10.34 m. To avoid the effect of cavitation on pump performance, $NPSH_a$ should be greater than $NPSH_r$ (Figure 10.8b). Thus, there is a limitation on the elevation at which the pump can be located. Generally, a margin of 1 m of head is considered for pumping water and a margin of 20–30% is taken when water is contaminated.

A centrifugal pump can obtain a theoretical maximum suction lift of 10.34 m with clean cold water at the mean atmospheric pressure at sea level, which corresponds to a pump reducing the absolute pressure to zero at its inlet. In practice, the lift is directly affected by the altitude of the place, temperature, friction losses on the suction side, velocity head, and vapor pressure, and thus would be less than the theoretical lift of 10.34 m. The maximum available practical lift can be expressed as:

$$\begin{aligned} h_{lift} &= NPSH_a - NPSH_r - f_s \\ &= h_{atm} - h_{vap} - h_{fs} - h_{fs}^* - NPSH_r - f_s, \end{aligned} \tag{10.8}$$

where h_{lift} is the maximum practical suction lift in meters and f_s is the safety factor, which may be taken as 0.6–1.0 m for pumping water. The safety factor is used for the consideration of any inaccurate estimation of pressure drop in the pumping system.

Example 10.3: Consider a centrifugal pump with a capacity of 100 L/s operating at an elevation of 100 m above mean sea level. The pump is located at 5 m above the static water level in a reservoir and has a suction PVC pipe 8 m long with an internal diameter of 250 mm. The velocity of flow is 1.6 m/s. The water temperature is 10 °C. The friction head of the pipe is given as 0.07 m, footwall as 0.80 m, and elbow as 0.55 m. Compute the net positive suction pressure available for the pump.

Solution: $S_L = 5$ m, $Z = 100$ m, $Q = 100$ L/s, and $T_{water} = 10$ °C.

Atmospheric pressure at an elevation of 100 m is

$$\begin{aligned} h_{atm} &= 10.33 - 1.17 \times 10^{-3} Z + 5.55 \times 10^{-8} Z^2 \\ &= 10.33 - 1.17 \times 10^{-3} \times 100 + 5.55 \\ &\quad \times 10^{-8} \times (100)^2 \\ &= 10.214 \text{ m.} \end{aligned}$$

Vapor pressure (h_{vap}) in meters can be calculated as follows:

$$\begin{aligned} h_{vap} &= 0.0623 \times \exp\left(\frac{17.27 \times T_{water}}{T_{water} + 237.3}\right) \\ &= 0.0623 \times \exp\left(\frac{17.27 \times 10}{10 + 237.3}\right) \\ &= 0.125 \text{ m.} \end{aligned}$$

$$h_{fs} = 0.07 \text{ m},$$

$$h_{fs}^* = 0.80 \text{ m} + 0.55 \text{ m} = 1.35 \text{ m},$$

$$\begin{aligned} NPSH_a &= h_{atm} - h_{vap} - h_{fs} - h_{fs}^* - S_L \\ &= 10.214 - 0.125 - 0.07 - 1.35 - 5 \\ &= 3.67 \text{ m.} \end{aligned}$$

Example 10.4: Compute the NPSH available for the pump in Example 10.3 if the pump is installed at an elevation of 1200 m above mean sea level.

Solution: The change in elevation would affect the atmospheric head, h_{atm}, which can be estimated as:

$$\begin{aligned} h_{atm} &= 10.33 - 1.17 \times 10^{-3} Z + 5.55 \times 10^{-8} Z^2 \\ &= 10.33 - 1.17 \times 10^{-3} \times 1200 + 5.55 \\ &\quad \times 10^{-8} \times (1200)^2 \\ &= 9.00 \text{ m}, \end{aligned}$$

$$\begin{aligned} NPSH_a &= h_{atm} - h_{vap} - h_{fs} - h_{fs}^* - S_L \\ &= 9.00 - 0.125 - 0.07 - 1.35 - 5 = 2.46 \text{ m}. \end{aligned}$$

10.5.6 Specific Speed

Specific speed (N_s) is a dimensionless index combining pump capacity, head, and speed. It is defined as the speed of a geometrically similar pump when discharging 1 gpm of water against a total head of 1 ft, and can be expressed as:

$$N_s = \frac{NQ^{\frac{1}{2}}}{H^{\frac{3}{4}}}, \tag{10.9}$$

where N_s is the specific speed, N is the pump rotational speed in rpm, Q is the discharge in gpm or pump design capacity (e.g., m³/h, L/s, L/min, m³/min), and H is the total head per stage at the best efficiency point in ft or m. Typical values for N_s for different designs when using Q in gpm and H in ft are:

radial-flow pump: $500 < N_s < 3500$, which is suitable for high head and low flows – typical for centrifugal impeller pumps;

mixed-flow pump: $3500 < N_s < 7500$, which is suitable for flow rates of more than 1000 gpm (227 m³/h) – typical for mixed-impeller single-suction pumps;

axial-flow pump: $7500 < N_s < 15{,}000$, which is used for high discharge of more than 5000 gpm (1136 m³/h) – typical for propellers and axial fans.

If using Q in m³/h and H in m, specific head $N_s(\text{gpm, ft}) = 0.861 N_s(\text{m}^3/\text{h, m})$.

Equation (10.9) can be written in dimensionless form as

$$N_s = \frac{2\pi NQ^{0.5}}{(gH)^{0.75}}, \tag{10.10}$$

where g is the acceleration due to gravity in m/s² and 2π is to convert revolutions to radians (dimensionless). The specific speed leads to a useful limitation in the design of suction condition – that is, for a given total head and suction lift condition, it should remain below the upper allowable shaft speed.

Example 10.5: Select a suitable pump (i.e., radial-flow, mixed- or axial-flow pump) for a sprinkler irrigation system which is designed to deliver 700 gpm at an operating head of 175 ft. The pump should operate at 1500 rpm.

Solution: $N = 1500$ rpm, $Q = 700$ gpm, and $H = 175$ ft.

$$N_s = \frac{NQ^{\frac{1}{2}}}{H^{\frac{3}{4}}} = \frac{(1500 \text{ rpm})(700 \text{ gpm})^{0.5}}{(175 \text{ ft})^{0.75}} = 825.$$

Using $Q = 700 \text{ gpm} = 158.99 \text{ m}^3/\text{h}$ and $H = 175 \text{ ft} = 53.34$ m,

$$N_s = 0.861 \frac{NQ^{\frac{1}{2}}}{H^{\frac{3}{4}}} = \frac{(1500 \text{ rpm})(158.99)^{0.5}}{(53.34)^{0.75}} = 825.$$

Thus, a radial-flow pump would be appropriate for this system.

10.6 PERFORMANCE CHARACTERISTIC CURVES

Selection of a proper pump that operates near its maximum efficiency for a given head and discharge condition is essential for economical operation of a pumping system. Pump characteristic or performance curves can help in selecting a proper pump from a variety of alternatives. These curves show the TDH, BP, efficiency, and NPSH all plotted against the capacity range of the pump. A typical pump characteristic curve for a single-stage centrifugal pump (radial pump) is shown in Figure 10.9. The characteristic curve for a given pump is supplied by the manufacturer.

The head–capacity curve shows how much water a given pump, operating at a particular specific speed, will deliver with a given head. The head related to zero Q (when the pump is operating against a closed valve) is known as the shutoff head. It is useful in designing the discharge side of piping, which must withstand the shutoff head if the delivery-side valve is accidentally closed. However, the efficiency of the pump at this point is zero, because the pump still requires energy to drive it. For all pumps, head decreases as the capacity increases. Figure 10.10 shows characteristic curves for radial-, mixed-, and axial-flow pumps (a similar figure can be found in Walker [1972]). The H–Q curve first decreases slightly with the increase in capacity, and thereafter it decreases rapidly for radial-flow pumps. However, head decreases more rapidly for mixed-flow and axial-flow pumps than for a radial-flow pump. Thus, the shape of the H–Q curve is a function of specific speed. The higher specific speed pump may perform steadily in cases where relatively constant discharge is expected against a fluctuating head.

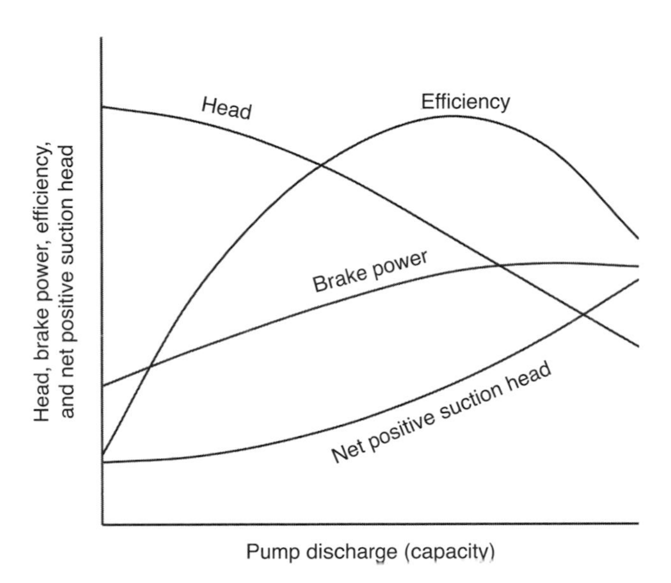

Figure 10.9 Characteristic curve for a single-stage centrifugal pump.

The BP for a radial pump gradually increases with increasing capacity up to a certain maximum value, and thereafter it declines slightly, whereas it increases continuously over the range of capacity in the case of mixed-flow pumps. The BP of the axial-flow pump decreases with increasing discharge, and has a maximum value at zero capacity. Energy input is required to drive the pump even at zero discharge when the pump is operating against the shutoff head.

The efficiency of all pumps increases from zero at zero discharge to a peak value with increasing discharge, and thereafter it declines gradually. The efficiency varies with pump type, size, and manufacturer, and is affected by construction material, finish of casing, quality of bearing, and wear of the casing and impeller. The efficiency of a pump may decline over a period of time due to wear and tear in the impeller and casing, and change in head. Larger pumps generally have higher efficiency. Also, smoother impellers

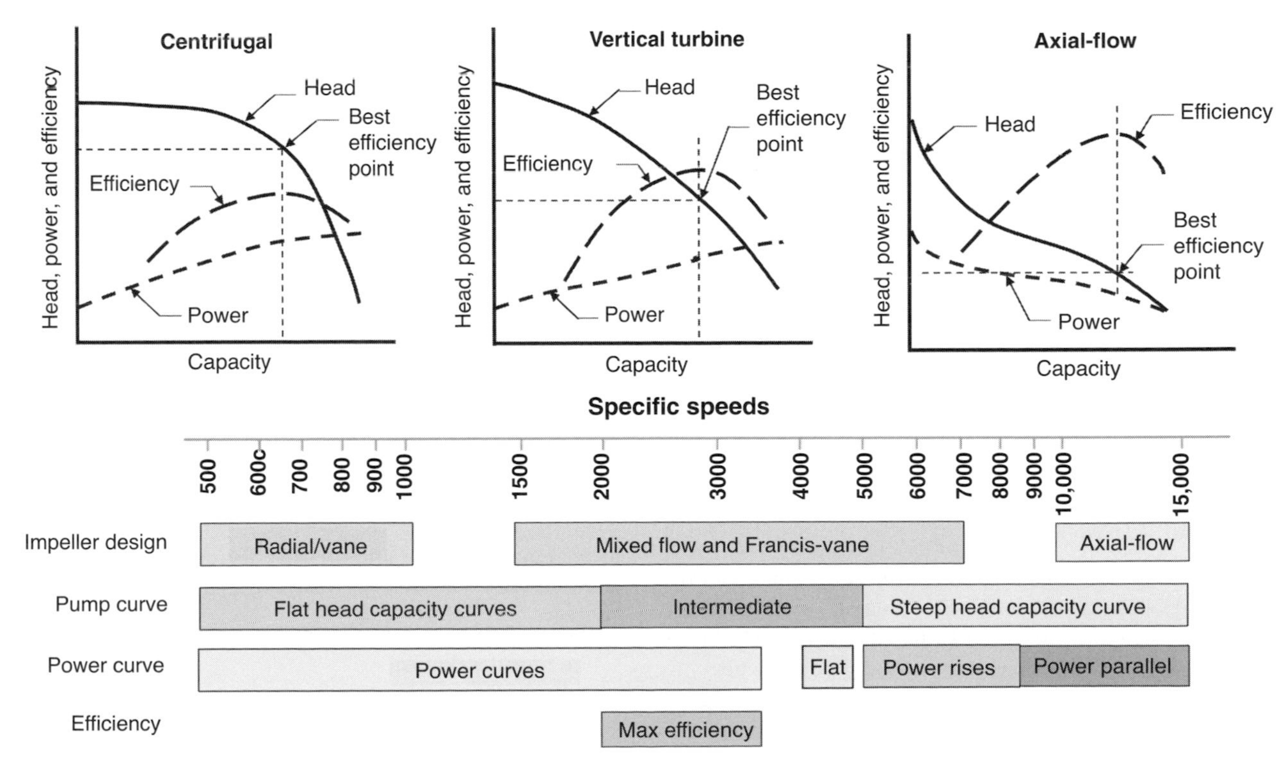

Figure 10.10 H–Q curve as a function of specific speed and impeller design. (Adapted from USDA-NRCS, 2016.)

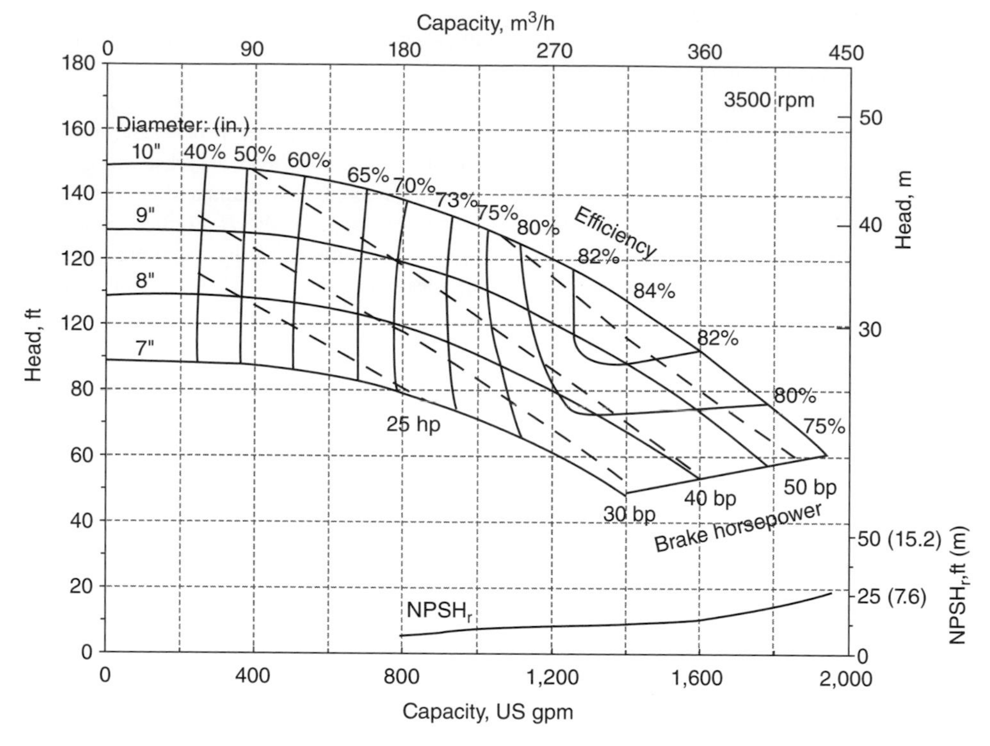

Figure 10.11 Characteristic curves from a pump manufacturer's catalog.

(enameled) will have higher efficiency compared to bronze or steel impellers. Generally, a pump that operates at or near peak efficiency is selected. The NPSH also increases with increasing capacity. The above discussion presents only a generalized description of characteristic curves for radial-, mixed-, and axial-flow impeller pumps. Most turbine pumps would have a characteristic curve between radial- and mixed-flow pumps, depending upon their specific speeds.

Manufacturers generally provide a composite characteristic curve, which represents the characteristic curve related to several possible impeller diameters that can be used with the pump. Constant horsepower, efficiency, and $NPSH_r$ lines are superimposed over different head–capacity curves. These curves are helpful in analyzing a pump behavior for a given speed with various impeller diameters. Figure 10.11 shows typical characteristic curves from a pump manufacturer's catalog.

10.7 AFFINITY LAWS

The performance or characteristic curves of a pump can be changed by changing either impeller diameter or speed. This permits manufacturers to use a single impeller for a variety of head and discharge conditions, and pump owners to change pump performance to match changes in configuration. Since several variables are involved in pump performance, the relationships between them are formulated as affinity laws which express impeller performance from changes in pump speed. The affinity laws are as follows:

1. The pump capacity is directly proportional to the ratio of impeller speeds:

$$Q_2 = Q_1\left(\frac{N_2}{N_1}\right). \tag{10.11}$$

2. The head is directly proportional to the square of the ratio of impeller speeds:

$$H_2 = H_1\left(\frac{N_2}{N_1}\right)^2. \tag{10.12}$$

3. The BP is directly proportional to the cube of the ratio of impeller speeds:

$$BP_2 = BP_1\left(\frac{N_2}{N_1}\right)^3. \tag{10.13}$$

In a similar vein, the affinity laws for changes in impeller diameter with constant speed are formulated as follows:

4. The pump capacity is directly proportional to the ratio of the impeller diameters:

$$Q_2 = Q_1\left(\frac{D_2}{D_1}\right). \tag{10.14}$$

5. The head is directly proportional to the square of the ratio of the impeller diameters:

$$H_2 = H_1\left(\frac{D_2}{D_1}\right)^2. \tag{10.15}$$

6. The BP is directly proportional to the cube of the ratio of the impeller diameters:

$$BP_2 = BP_1\left(\frac{D_2}{D_1}\right)^3, \tag{10.16}$$

where Q is the capacity of the pump, H is the head, BP is the brake horsepower, N is the impeller speed, D is the impeller diameter, and subscripts 1 and 2 represent known and new values of the variables.

Equations (10.14)–(10.16) hold for centrifugal pumps only and are not as accurate as eqs (10.11)–(10.13) related to pump speed. The affinity laws hold within a certain range of speed, impeller diameter, flow rate, and head, and are more accurate in the region of maximum pump efficiency (Merkley and Allen, 2009). These laws are generally used to either change speed or to reduce impeller diameter. However, the impeller diameter should not be reduced by more than 20% of the original diameter.

Example 10.6: A pump operating at 1800 rpm delivers 800 L/s with a head of 150 m. The pump has 25 BP. Compute the change in pump performance for the following conditions: (a) If the impeller speed is changed to 2200 rpm, and (b) if the impeller diameter of 200 mm is reduced by 15%.

Solution:

a. If impeller speed is changed to 2200 rpm:

$$Q_2 = Q_1\left(\frac{N_2}{N_1}\right) = 800\left(\frac{2200}{1800}\right) = 977.8\ \text{L/s},$$

$$H_2 = H_1\left(\frac{N_2}{N_1}\right)^2 = 150\left(\frac{2200}{1800}\right)^2 = 224.1\ \text{m},$$

$$BP_2 = BP_1\left(\frac{N_2}{N_1}\right)^3 = 25\left(\frac{2200}{1800}\right)^3 = 45.6\ \text{hp}.$$

b. If impeller diameter of 200 mm is reduced by 15%:

$$Q_2 = Q_1\left(\frac{D_2}{D_1}\right) = 800\left(\frac{200 \times (1 - 0.15)}{200}\right) = 680\ \text{L/s},$$

$$H_2 = H_1\left(\frac{D_2}{D_1}\right)^2 = 150\left(\frac{200 \times (0.85)}{200}\right)^2 = 108.4\ \text{m},$$

$$BP_2 = BP_1\left(\frac{D_2}{D_1}\right)^3 = 25\left(\frac{200 \times (0.85)}{200}\right)^3 = 15.4\ \text{hp}.$$

10.8 PERFORMANCE CURVES FOR PUMPS IN SERIES

Let there be two pumps, A and B, arranged in series as shown in Figure 10.12; their respective head–capacity curves are shown in Figure 10.13. The combined head–capacity curve is prepared by adding the heads of the two pumps (eq. 10.18) for the same values of discharge (eq. 10.17), as shown in Figure 10.14. Other combined characteristic curves (capacity–BP and capacity–efficiency) can be derived (with subscripts A, B, and s representing pump A, and B and pumps in series) as follows:

$$Q_s = Q_A = Q_B, \tag{10.17}$$

$$H_s = H_A + H_B, \tag{10.18}$$

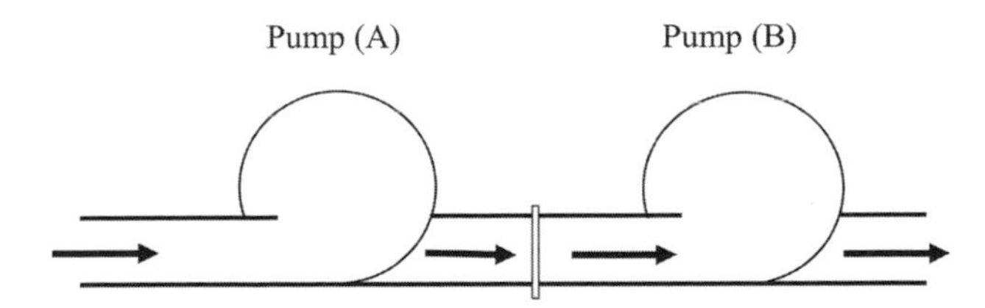

Figure 10.12 Two pumps, A and B, arranged in series.

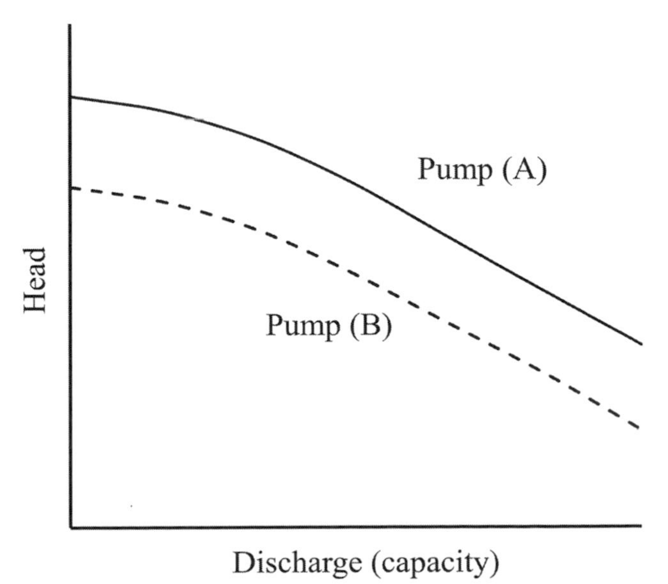

Figure 10.13 H–Q curves for two pumps, A and B.

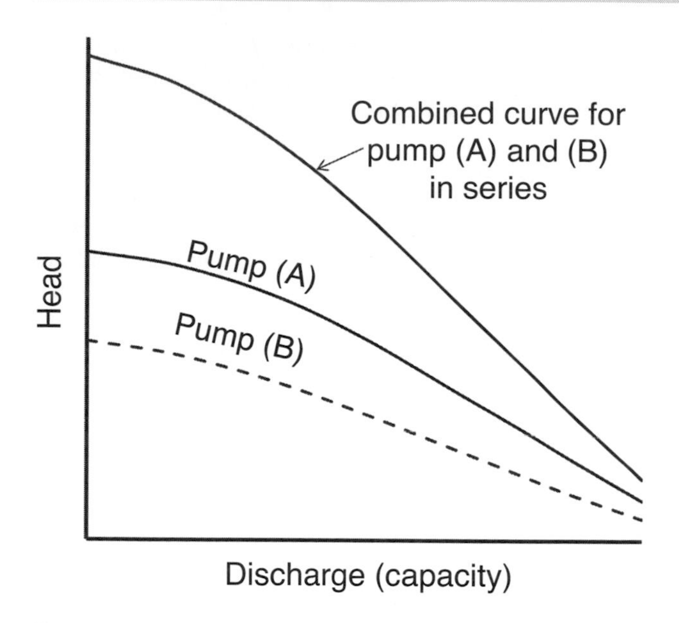

Figure 10.14 Combined *H–Q* curve for two pumps in series.

$$BP_s = BP_A + BP_B, \tag{10.19}$$

$$E_s = \frac{Q_s H_s}{0.102 BP_s}, \tag{10.20}$$

where Q is the capacity in m^3/s, H is the head in m, BP is the BP in kW, and E is the efficiency as a fraction.

The pumps arranged in series must be identical. This arrangement is sometimes needed to achieve the required head or discharge – that is, the discharge from a single pump is sufficient but head is not adequate. As an example, the water is to be pumped from a deep well where the head is more than the suction limit of a single pump. The outflow from the first pump becomes the inflow to the subsequent pump, and so on. The distance between thc two pumps is normally kept to within a few meters. The $NPSH_r$ of the furthest pump upstream needs to be considered when the pumps are in close vicinity. However, the head available at each pump needs to be determined and compared with the respective $NPSH_r$ if the pumps are widely separated. The second pump must be capable of operating at the higher suction pressure provided by the first pump.

Example 10.7: Consider two pumps, A and B, in series, with operating heads of 20 m and 22 m and efficiencies of 75% and 80%, respectively. This arrangement is needed to meet the capacity requirement of 150 L/s. Compute the combined capacity, head, power, and efficiency for the pumping system.

Solution:

$$Q_s = Q_A = Q_B = 150\ \text{L/s} = 0.15\ \text{m}^3/\text{s},$$

$$H_s = H_A + H_B = 20 + 22 = 42\ \text{m}.$$

Combing eqs (10.4) and (10.5) with Q in m^3/s, H in m, BP in kW, the BPs of pump A and B are

$$BP_A = \frac{Q_A H_A}{0.102 \times E_p} = \frac{0.15 \times 20}{0.102 \times 75\%} = 39.22\ \text{kW},$$

$$BP_B = \frac{Q_B H_B}{0.102 \times E_p} = \frac{0.15 \times 22}{0.102 \times 80\%}$$
$$= 40.44\ \text{kW}$$

$$BP_s = BP_A + BP_B = 79.66\ \text{kW},$$

$$E_s = \frac{Q_s H_s}{0.102 BP_s} = \frac{0.15 \times 42}{0.102 \times 79.66} = 0.775.$$

The efficiency for the pumping system is 77.5%.

10.9 CHARACTERISTICS OF PUMPS IN PARALLEL

Pumps can be arranged in parallel, as shown in Figure 10.15, for the purposes of augmenting discharge for the available head. An example of pumps in parallel includes pumping of water from the same source to a common pipeline. The parallel arrangement provides flexibility in controlling discharge and may be useful for irrigation systems where variable flow rate is required during the crop season. Figure 10.16 shows head–capacity curves of pumps A and B as well as the combined head–capacity curve. The combined head–capacity curve is prepared by summing up the discharges of the two pumps (eq. 10.21) for the same values of head (eq. 10.22). Other combined characteristic curves (capacity–BP and capacity–efficiency) can be obtained as follows:

$$Q_{par} = Q_A + Q_B, \tag{10.21}$$

$$H_{par} = H_A = H_B, \tag{10.22}$$

$$BP_{par} = BP_A + BP_B, \tag{10.23}$$

$$E_{par} = \frac{Q_{par} H_{par}}{0.102 BP_{par}}, \tag{10.24}$$

where Q is the capacity in m^3/s, H is the head in m, BP is the BP in kW, E is the efficiency as a fraction, and subscripts A, B, and par represent pump A, pump B, and pumps in parallel, respectively.

Example 10.8 There are two pumps, A and B, that have operating capacities of 50 and 45 L/s and efficiencies of 80% and 85%, respectively. To meet the irrigation system head of 30 m, it is decided to arrange the pumps in parallel. Compute the combined capacity, head, power, and efficiency of the pumping system.

Solution: Given $Q_A = 50\ \text{L/s} = 0.05\ \text{m}^3/\text{s}$, $Q_B = 45\ \text{L/s} = 0.045\ \text{m}^3/\text{s}$.

$$Q_{par} = Q_A + Q_B = 0.095\ \text{m}^3/\text{s},$$
$$H_{par} = H_A = H_B = 30\ \text{m}.$$

Combing eqs (10.4) and (10.5), with Q in m^3/s, H in m, BP in kW, the BP of pumps A and B are

$$BP_A = \frac{Q_A H_A}{0.102 \times E_p} = \frac{0.05 \times 30}{0.102 \times 80\%} = 18.38\ \text{kW},$$
$$BP_B = \frac{Q_B H_B}{0.102 \times E_p} = \frac{0.045 \times 30}{0.102 \times 85\%} = 15.57\ \text{kW},$$
$$BP_{par} = BP_A + BP_B = 33.95\ \text{kW},$$
$$E_{par} = \frac{Q_{par} H_{par}}{0.102 BP_{par}} = \frac{0.095 \times 30}{0.102 \times 33.95} \times 100(\%) = 82.3\%.$$

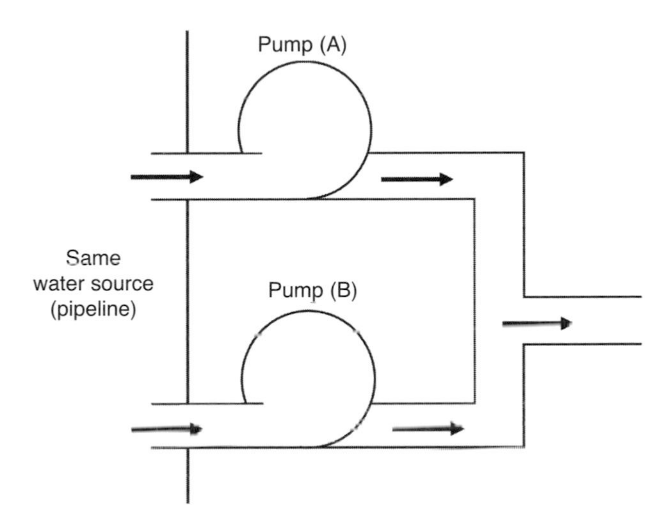

Figure 10.15 Two pumps, A and B, arranged in parallel.

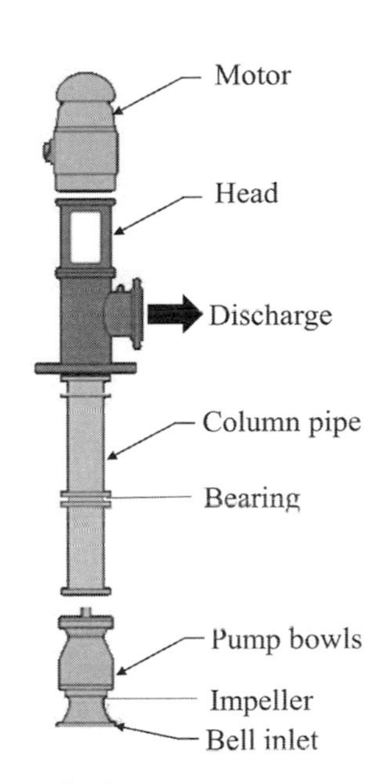

Figure 10.17 A deep-well turbine pump.

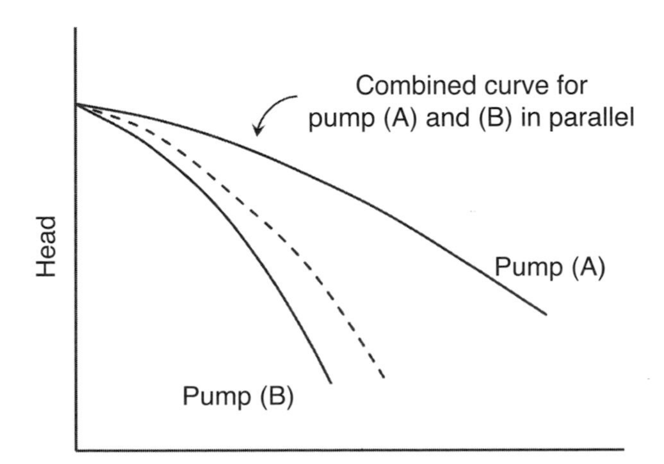

Figure 10.16 Head–capacity curves of pumps A and B with the combined head–capacity curve.

10.10 DEEP-WELL TURBINE PUMP

The deep-well turbine pump, also known as a vertical turbine pump, is a special case of a centrifugal pump having a vertical axis with multiple stages, as shown in Figure 10.17. Deep-well turbine pumps are used to obtain water from small-diameter, machine-constructed wells where the water level is beyond the practical suction limits of a centrifugal pump. They are also used with surface water systems to avoid priming, and at booster stations in the water distribution system to enhance the pressure required for service. In general, vertical turbine pumps are more expensive than centrifugal pumps for the same capacity, but their efficiencies are comparable to or greater than those of centrifugal pumps. They are, however, more difficult to install, inspect, and repair.

10.10.1 Components of Deep-Well Turbine Pumps

The turbine pump has three main parts: the discharge (head) assembly, the shaft and column assembly (discharge column), and the pump element (bowl assembly), as shown in Figure 10.17. The discharge assembly, also known as the head, consists of a base that supports the discharge column, bowl and shaft assemblies, and a discharge head elbow that guides water into the delivery line. It is generally made of cast iron and designed to be installed on a foundation. The head assembly may also support an electric motor, a right-angle gear drive, a belt drive, or a suitable combination of these drives.

The shaft and column assembly connects the bowl assembly to the discharge head assembly, transmits prime mover torque to the impeller shaft, and conveys the discharge from the bowl assembly to the discharge head. The line shaft of a turbine pump may be either open – that is, lubricated by fluid pumped (water lubricated) – or enclosed, with a separate pipe around the shaft for external lubrication (oil lubricated). In an oil-lubricated pump the shaft is protected from abrasive and corrosive elements, which is preferred if the pumped water contains fine sand. A water-lubricated pump must be used to protect water supply for domestic or livestock use from any oil contamination. However, some US states (e.g., Minnesota) also require water-lubricated pumps in all new irrigation wells. In general, water-lubricated pumps are slightly more economical than oil-lubricated pumps.

A pump bowl assembly consists of an impeller, a diffuser, and a bearing. In a deep-well turbine pump, a number of bowls, also known as stages, are arranged in series to overcome the limitations of low head developed by limited-diameter impeller. The impeller diameter is constrained by the diameter of the bowl, which in turn is limited by the diameter of the well. The vertical turbine pump is a special case of pumps in series, where the outflow from one stage becomes the inflow to the subsequent stage, with increase in head. For example, a pump with a three-stage bowl assembly contains three impellers all attached to a common shaft through separate housings or bowls. It will have discharge of a single impeller head, but three times the discharge head of a single-stage pump. Impellers used in turbine pumps may be either semi-open or enclosed. Prior to placing a turbine pump, it is necessary to have information on the water table fluctuations and drawdown so that the bowl can be installed sufficiently below the point of farthest drawdown.

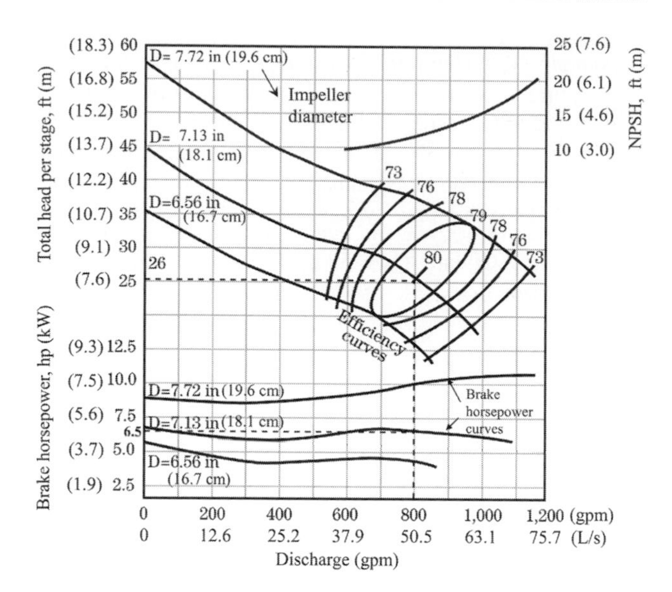

Figure 10.18 Deep-well turbine pump curve. The BP and total head are for one stage. If the pump had five stages, multiply the BP and the total head values by five. The gallons per minute will stay the same no matter how many stages are added. (Adapted from USDA-NRCS, 2016. Original from Scherer 1993.)

10.10.2 Operating Characteristics

The characteristic curves of a deep-well turbine pump depend on bowl design, impeller type, and speed. Vertical turbine pumps are generally designed for particular speed settings. The characteristic curves are similar to those of a centrifugal pump. However, the centrifugal pump gives high efficiency over a wide range of operating speed, whereas the turbine pump cannot give high efficiency over a wide range of speed. Figure 10.18 shows a typical characteristic curve for a vertical turbine pump, which is similar to the centrifugal pump curve, except that instead of curves for various RPMs, the curves are for different diameter impellers. These curves for a turbine pump are generally shown for a single stage and therefore must be multiplied by a number of stages to obtain the TDH and BP. However, the pump capacity will not change with the number of stages.

Example 10.9: What would be the TDH and BP values from the pump curve for a four-stage pump with a 7.13-inch impeller supplying 800 gpm?

Solution: TDH: From Figure 10.18, follow the dashed vertical line from 800 gpm (50.5 L/s) up to where it meets the 7.13-inch (18.1 cm) impeller curve on the upper portion of the chart. Follow the dashed horizontal line left to where it shows 26 ft (7.9 m) of TDH. Multiplying 26 by 4 gives 104 ft (31.7 m) of TDH.

BP: From Figure 10.18, follow the dashed vertical line from 800 gpm (50.5 L/s) up to the 7.13-inch (18.1 cm) impeller BP curve on the lower portion of the chart and then follow the horizontal dashed line left to where it shows 6.5 hp. Multiplying 6.5 hp (4.8 kW) by four stages produces a 26 hp (19.4 kW) requirement for this pump. The pump is operating at its peak efficiency of 80%.

10.11 SUBMERSIBLE PUMPS

Submersible pumps are applied in places that are flooded, at public places that do not have space to install a pump station, and as booster pumps in the suction lines of centrifugal pumps in locations where the depth of water changes significantly during the season and may drop below the level required for the centrifugal volute or diffuser pump. It may be noted that the head at the inlet of the centrifugal pump will change from a suction head to a positive head when using a submersible pump in the suction line. A typical submersible pump is shown in Figure 10.19. The characteristic curves for a submersible pump are similar to those for a vertical turbine pump.

A submersible pump consists of a multistage vertical turbine coupled to an electric motor designed to operate under water. A submersible pump is installed below the water surface in a small-diameter well. This arrangement eliminates the need for a long drive shaft and bearing retainers. It also helps improve the efficiency due to the direct coupling of the motor and its effective cooling by submergence in water. A submersible pump generally uses enclosed impellers. Submersible motors can be either oil lubricated (dry) or water lubricated (wet), and have the same diameter as the pump bowl. However, they are much longer than ordinary motors. Dry motors are hermetically sealed with a light oil to protect the motor from water. Wet motors are open to the well water and both the rotor and the bearings operate in the water. To prevent overheating and burning of the motor they should be constantly submerged.

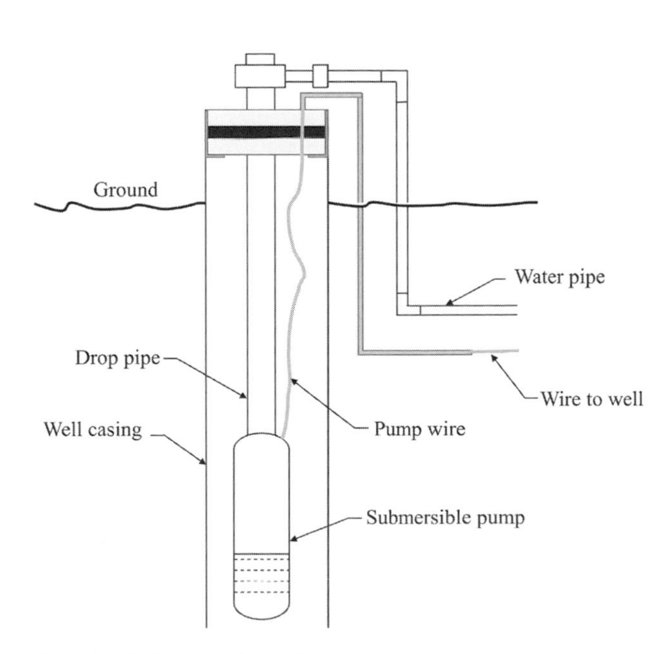

Figure 10.19 A typical submersible pump.

10.12 PUMP OPERATING POINT

A centrifugal pump has a head–discharge (H–Q) characteristic curve and operates at combinations of H and Q. The pump characteristic curve represents the behavior of a pump of a given size and specific speed. The system curve represents the behavior of the piping system. It is a graphical representation of the required head of the piping system under different discharge conditions. The system curve is independent of the pump, except that friction loss occurs in column-type vertical pumps. Both pump characteristic curve and system curve are important in the selection of a pump. Figure 10.20a shows typical system and pump head–capacity curves. The point of intersection of these two curves defines the operating point for the pump. This point gives the combination of head and discharge at which the pump operates. Once this point is determined, the BP, efficiency, and the NPSH can be determined. For constructing the system curve, the head needed for the irrigation system

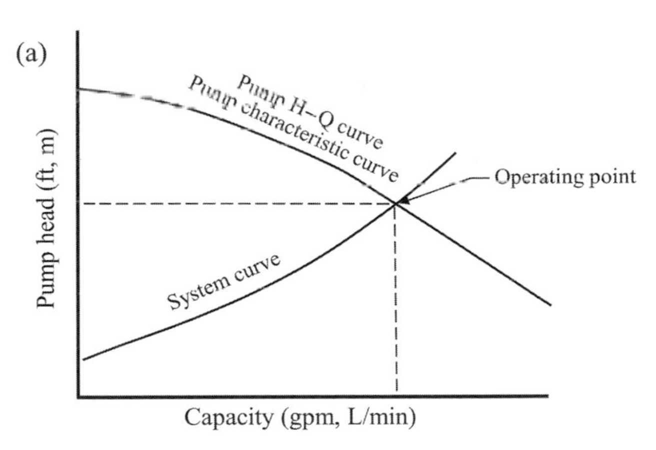

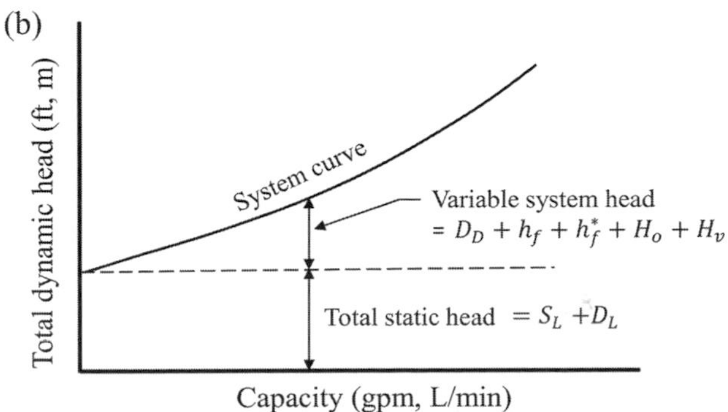

Figure 10.20 (a) Selection of operating point and (b) system curve illustration.

to deliver different volumes of water per unit time is to be computed, which can be done as

$$H = S_L + D_L + D_D + h_f + h_f^* + H_o + H_v, \quad (10.25)$$

where H is the system head head or the total dynamic head (m, ft); S_L is the suction-side lift (m, ft); D_L is the discharge side lift (m, ft); D_D is the water source drawdown (m, ft); h_f is the head loss due to friction; h_f^* defines the minor losses due to fittings (m, ft); H_o is the operating head (m, ft); and H_v is the velocity head (m, ft). It may be noted that eq. (10.25) is similar to eq. (10.3). S_L is the static head or lift in eq. (10.3), which is the vertical distance between the centerline of the horizontal pump and the static surface elevation of the water source, and is positive when the water source is below the pump and negative when it is above. Likewise, D_L is the static discharge head, which is the elevation difference between the point of delivery and the centerline. H_o includes the pressure head required on the suction and discharge sides. For example, a certain head is needed on the delivery side to operate an irrigation system such as a sprinkler system.

As shown in Figure 10.20b, the system curve includes a static component and a dynamic component. The static component is equal to the total static head (i.e., $S_L + D_L$), which remains unaltered and is independent of system flow. The dynamic component varies with change in head. D_D, h_f, h_f^*, H_o, and H_v increase with increasing Q.

To construct the system curve, the velocity head can be computed in terms of Q and D as

$$H_v = \frac{Q^2}{KD^4}, \quad (10.26)$$

where Q is the system discharge (L/min, gpm); D is the diameter of the discharge pipe at the pump (cm, in.); and K is the unit constant equal to 435.7 for H_v in m, Q in L/min, and D in cm; and equal to 385.9 for H_v in ft, Q in gpm, and D in inches.

Example 10.10: Construct the system curve if an irrigation system has 150 sprinklers. The water supply pipeline is a PVC pipe 2500 ft long, 10 inches in diameter, and has a Hazen–Williams coefficient C of 150. The water is lifted 250 ft from a reservoir to the irrigation farm. Assume that minor losses are about 12% of pipe friction loss and there is no water source drawdown. The capacity for each sprinkler is $Q = (1.41 \times P^{0.5})$, and the operating head is $H_o = 2.31P$.

Solution: Given $D = 10$ in., $L = 2500$ ft, assuming the total discharge of the irrigation system is Q in gpm. The head loss in ft due to pipe friction is computed using the Hazen–Williams equation, with $C = 150$:

$$h_f = 10.47\left(\frac{Q}{C}\right)^{1.852} D^{-4.871} L$$

$$= 10.47 \times \left(\frac{Q}{150}\right)^{1.852} (10)^{-4.871} \times 2500$$

$$= \left(3.29 \times 10^{-5}\right) Q^{1.852}.$$

The minor losses due to fittings are

$$h_f^* = 12\% \times h_f = \left(3.95 \times 10^{-6}\right) Q^{1.852}.$$

Since the total discharge is Q, the discharge in each sprinkler is $Q/150$ sprinklers:

$$Q/150 = 1.41 \times P^{0.5}.$$

Then, $P = \left(\frac{Q}{211.5}\right)^2$.

The operating head is

$$H_o = 2.31P = 2.31\left(\frac{Q}{211.5}\right)^2 = 5.16 \times 10^{-5}\, Q^2.$$

The velocity head is computed using eq. (10.26):

$$H_v = \frac{Q^2}{KD^4} = \frac{Q^2}{385.9(10\text{ in.})^4} = 2.59 \times 10^{-7}\, Q^2.$$

Given $S_L + D_L = 250$ ft, $D_D = 0$, the head needed for the irrigation system is

$$H = S_L + D_L + D_D + h_f + h_f^* + H_o + H_v,$$

$$H = 250 + 0 + \left(3.29 \times 10^{-5}\right)Q^{1.852} + \left(3.86 \times 10^{-6}\right) \times Q^{1.852} + \left(5.16 \times 10^{-5}\right)Q^2 + \left(2.59 \times 10^{-7}\right)Q^2,$$

$$H = 250 + \left(3.676 \times 10^{-5}\right)Q^{1.852} + \left(5.19 \times 10^{-5}\right) Q^2.$$

The system curve of the irrigation system is shown in Figure 10.21.

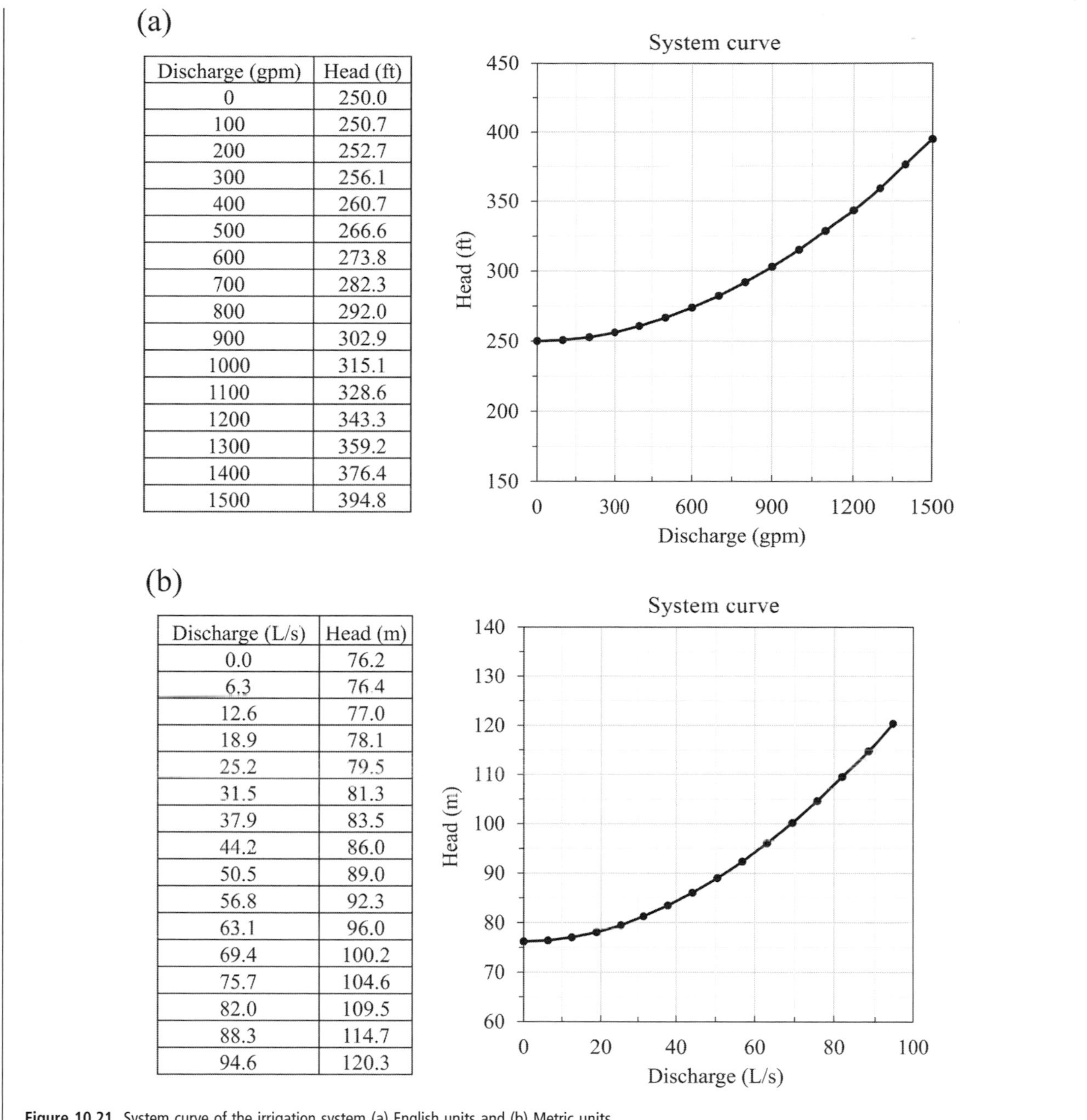

(a)

Discharge (gpm)	Head (ft)
0	250.0
100	250.7
200	252.7
300	256.1
400	260.7
500	266.6
600	273.8
700	282.3
800	292.0
900	302.9
1000	315.1
1100	328.6
1200	343.3
1300	359.2
1400	376.4
1500	394.8

(b)

Discharge (L/s)	Head (m)
0.0	76.2
6.3	76.4
12.6	77.0
18.9	78.1
25.2	79.5
31.5	81.3
37.9	83.5
44.2	86.0
50.5	89.0
56.8	92.3
63.1	96.0
69.4	100.2
75.7	104.6
82.0	109.5
88.3	114.7
94.6	120.3

Figure 10.21 System curve of the irrigation system (a) English units and (b) Metric units.

10.13 PUMP SELECTION

Selection of a pump involves choosing the most suitable pump for a specific irrigation system, and comprises a number of steps, such as specifying the performance requirements of the system, developing the system curve and pump characteristic curve (capacity–head relation), selecting the type of pump (radial, axial, or propeller), identifying alternate pumps that meet the irrigation system requirements, matching the system curve with the pump characteristic curve for all pumps, and finally selecting the pump that meets the requirements of the system and operates most economically at a high level of efficiency.

10.13.1 Performance Requirements

The system head–discharge relation (system curve), which is independent of the pump characteristic curve, must be known or determined. This relation must account for the head needed by the irrigation system to meet different discharge requirements and may change during a growing season due to changes in the groundwater level, cropping area, rainfall, etc. It has a static component and a dynamic component. The static component remains unaltered, but the dynamic component varies with change in head. The system design capacity can be computed as:

$$Q_{sd} = \frac{16667 d_n A_i}{E_i T}, \tag{10.27}$$

where Q_{sd} is the system design capacity (L/s), d_n is the design daily irrigation requirement (mm/day), A_i is the irrigated area (ha), E_i is the irrigation efficiency (%), and T is the duration of system operation (h/day).

10.13.2 Pump Type

Source of water, initial cost, space requirement, specific speed, and system curve are broad guidelines for selecting a pump. Two main pump types for agricultural irrigation systems are horizontal volute and vertical diffuser (turbine) pumps. A centrifugal pump is suitable for a wide range of discharges at high heads, whereas an axial pump is suitable for high discharges at low heads. A horizontal centrifugal pump is less expensive and costs less to install, and is suitable for pumping from surface water sources, whereas a vertical turbine pump is suitable for pumping from groundwater wells. All vertical turbine, submersible, and axial pumps become highly inefficient if design conditions are not met.

The value of $NPSH_a$ can be applied to determine a pump's $NPSH_r$. If $NPSH_a$ is greater than or equal to $NPSH_r$, then the pump will operate without cavitation for all discharges. Similar to eq. (10.7), the $NPSH_a$ can be calculated as

$$NPSH_a = h_{atm} - h_{vap} - h_{fs} - h_{fs}^{*} - H_{vs} - S_L - D_D, \tag{10.28}$$

$$h_{atm} = K_1 - 1.17 \times 10^{-3} Z + K_2 Z^2, \tag{10.29}$$

where h_{atm} is the atmospheric pressure (m or ft of water); h_{vap} is the vapor pressure of water at the operating temperature (m or ft of water); h_{fs} is the head loss due to friction in the suction line (m, ft); h_{fs}^{*} is minor losses in the suction line (m, ft); H_{vs} is the velocity head in the suction line (m, ft); S_L is the suction-side lift (m, ft); D_D is the drawdown (m, ft); Z is the elevation above mean sea level (m, ft); K_1 and K_2 are the unit constants equal, respectively, to 10.33 and 5.55 × 10^{-8} for h_{atm} in m of water and Z in m; and equal to 33.89 and 1.69 × 10^{-8} for P_B in ft of water and Z in ft.

Example 10.11: Compute $NPSH_a$ for the system shown in Figure 10.22. The following information is given: discharge Q equals 500 gpm (31.55 L/s); water level remains unchanged during pumping; elevation is 800 ft (243.84 m) above mean sea level; the suction pipeline is 30 ft (9.14 m) long and 10 in. (25.4 cm) in diameter, and has a Hazen–Williams constant of 140. Consider water temperature of 70 °F (20.11 °C). Assume that there is no water source drawdown.

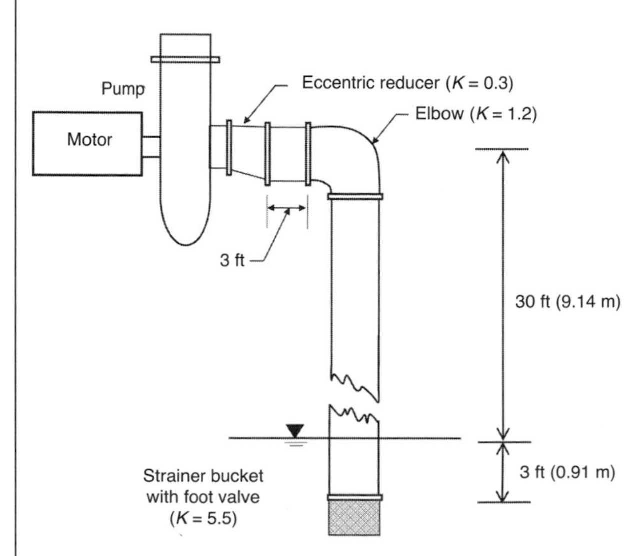

Figure 10.22 Pump system in Example 10.11.

Solution: The atmospheric pressure:

$$h_{atm} = 33.89 - 1.17 \times 10^{-3}(800) + 1.69 \times 10^{-8}(800)^2 = 32.96\,\text{ft},$$

$$h_{atm} = 10.33 - 1.17 \times 10^{-3}(243.84) + 5.55 \times 10^{-8}(243.84)^2 = 10.05\,\text{m}.$$

Given $T = 70\,°F = 20.11\,°C$, vapor pressure (h_{vp}) in meters can be calculated as follows:

$$h_{vap} = 0.0623 \times \exp\left(\frac{17.27 \times T}{T + 237.3}\right) = 0.0623 \times \exp\left(\frac{17.27 \times 20.11}{20.11 + 237.3}\right) = 0.24\,\text{m},$$

$$h_{vap} = 0.24\,\text{m} = 0.79\,\text{ft}.$$

Given $D = 10$ in., $L = 3 + 30 + 3 = 36$ ft, $Q = 500$ gpm. The head loss in feet due to pipe friction is computed using the Hazen–Williams equation, with $C = 140$:

$$h_{fs} = 10.47 \left(\frac{Q}{C}\right)^{1.852} D^{-4.871} L$$
$$= 10.47 \times \left(\frac{500}{140}\right)^{1.852} (10)^{-4.871} \times 36$$
$$= 0.05 \text{ ft}(0.015 \text{ m}).$$

The velocity head is computed by

$$H_{vs} = \frac{v^2}{2g},$$

or directly using eq. (10.26), with $K = 385.9$ for H_{vs} in ft, Q in gpm, and D in inches.

$$H_{vs} = \frac{Q^2}{385.9\, D^4} = \frac{500^2}{385.9\,(10)^4} = 0.065 \text{ ft}(0.020 \text{ m}).$$

The minor losses due to fittings:

$$h_{fs}^* = \sum_{i=1}^{3} K_i \frac{v^2}{2g} = (0.3 + 1.2 + 5.5) \times 0.065 \text{ ft}$$
$$= 0.455 \text{ ft}(0.139 \text{ m}).$$

The suction lift $S_L = 30$ ft, and the drawdown $D_D = 0$. Apply eq. (10.28), the $NPSH_a$ for the system:

$$NPSH_a = h_{atm} - h_{vap} - h_{fs} - h_{fs}^* - H_{vs} - S_L - D_D$$
$$= 32.96 - 0.79 - 0.05 - 0.455 - 0.065 - 30 - 0$$
$$= 1.60 \text{ ft} \ (0.49 \text{ m}).$$

10.13.3 Economic Evaluation

Selecting the most suitable pump or combination of pumps for an irrigation system often hinges on economic considerations. It is therefore necessary to consider the cost of purchase, operation, and maintenance. The total annual costs include operating costs and fixed costs. The total annual operating costs include (a) energy cost and (b) annual maintenance and repair. The total annual fixed costs include (c) annual depreciation, (d) annual interest, and (e) annual taxes and insurance. The calculation of each cost is summarized below:

1. Energy cost calculation. The operating conditions for the pump should be determined under different irrigation conditions. The discharge and head at the operating condition are obtained from the intersection of the head–discharge and the system curves. Then, the efficiency under the operating conditions with the same discharge is obtained. Then, use eq. (10.6) to calculate the power required to operate the pump (P_{kw}) in kW. The energy cost is calculated as:

   Energy cost = P_{kw} × time required to pump the water × per unit electricity charge.

2. Annual maintenance and repair cost. If the initial cost of the pump (*IC*) and the annual maintenance and repair rate are given, the annual maintenance and repair cost is computed by:

   Annual maintenance and repair = initial cost(*IC*) × maintenance and repair rate.

3. Annual depreciation cost. The annual depreciation of the pump is equal to the present worth of the pump (*PW*) multiplied by the depreciation rate (*r*). *r* is the reciprocal of the useful life (*UL*) of the pump.

   Annual depreciation = Present worth(*PW*) × depreciation rate.

   The present worth of the pump is calculated as:

   $$PW = IC - SV\left(\frac{1+r}{1+i}\right)^{AP},$$

   where *SV* is the salvage value, *AP* is the annual period, *r* is the depreciation rate, and *i* is the interest rate.

4. Annual interest cost. Annual interest depends on the interest rate, initial cost, and the AP:

   Annual interest = initial cost(*IC*) ×capital recovery factor(*CRF*),

   where $CRF = \dfrac{i(1+i)^{AP}}{(1+i)^{AP} - 1}$.

5. Taxes and insurance cost. If the annual taxes and insurance rate are given, the annual taxes and insurance is:

   Annual taxes and insurance = initial cost(*IC*) × taxes and insurance rate.

More details of the economic analysis will be discussed in Chapter 29. Here, the steps of the calculation of annual cost are illustrated using a simple example.

Example 10.12: Consider an irrigation system for which the lift varies from 80 to 100 ft and crop water requirement is 25 in. per year. The area of cropland is 120 acres. System curves are given for each lift in Figure 10.23. The pump is operated one-quarter of the time at 80 ft lift, one-half of the time at 90 ft lift, and one-quarter of the time at 100 ft lift. Consider two pumps, A and B, for which head–discharge and efficiency–discharge curves are given in Figure 10.23. Pump A is a three-stage, 8 in. diameter, 1500 rpm vertical turbine pump, and costs $13,000, including a 50-hp electric

Table 10.1 *Solution to Example 10.12, part 1*

		(1)	(2)	(3)	(4)	(5)	(6)	(7)	(8)
	Terms	H	Q	E_p	BP	P_w	T	Energy	Energy cost
Pump	Units	ft	gpm	kW	kW	kW	h	kWh	$
A	80 ft lift	145	970	83	32	36	416	13,376	401
	90 ft lift	148	920	84	31	34	726	22,303	669
	100 ft lift	152	860	85	29	32	323	9,421	283
B	80 ft lift	150	1010	86	33	37	416	13,902	417
	90 ft lift	154	960	87	32	36	726	23,377	701
	100 ft lift	154	940	88	31	35	323	10,075	302

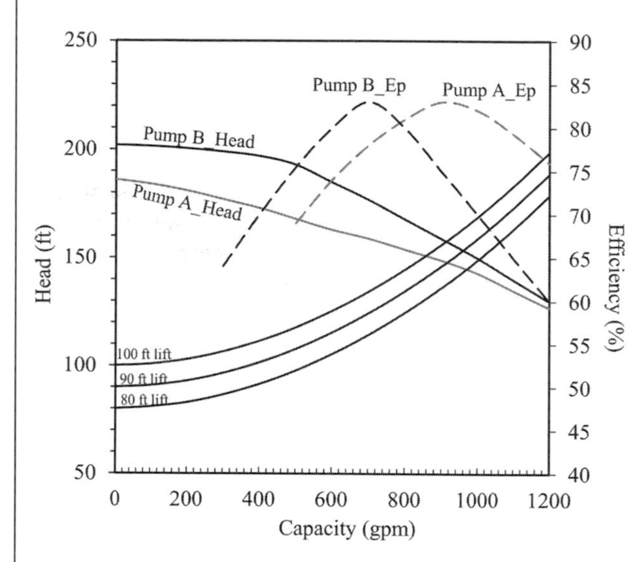

Figure 10.23 System curves for each lift and characteristic curves for pumps A and B in Example 10.12.

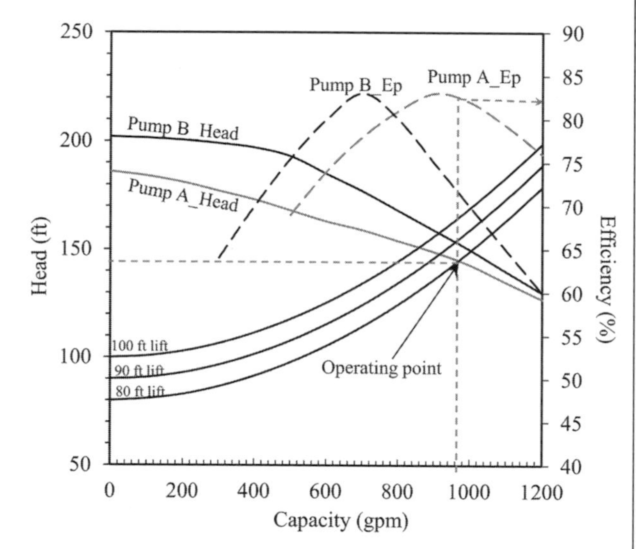

Figure 10.24 Solution to Example 10.12.

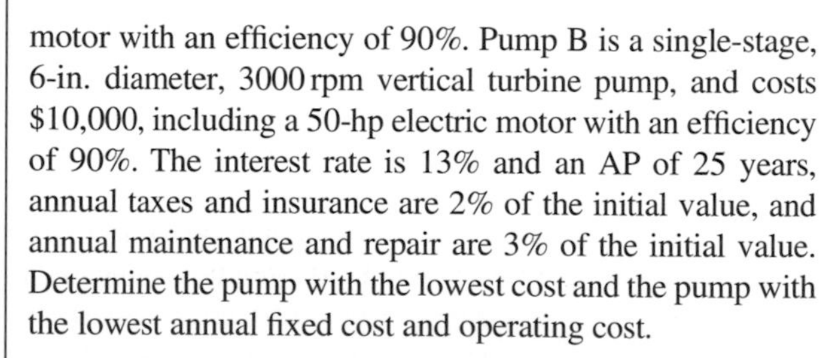

motor with an efficiency of 90%. Pump B is a single-stage, 6-in. diameter, 3000 rpm vertical turbine pump, and costs $10,000, including a 50-hp electric motor with an efficiency of 90%. The interest rate is 13% and an AP of 25 years, annual taxes and insurance are 2% of the initial value, and annual maintenance and repair are 3% of the initial value. Determine the pump with the lowest cost and the pump with the lowest annual fixed cost and operating cost.

Solution: The total annual costs include operating costs and fixed cost. The total annual operating costs include (a) energy cost and (b) annual maintenance and repair. The total annual fixed costs include (c) annual depreciation, (d) annual interest, and (e) annual taxes and insurance.

First, to calculate energy cost, the operating conditions for pumps A and B should be determined under different irrigation conditions. The discharge and head at the operating condition under 80, 90, and 100 ft lift are obtained from the intersection of the head–discharge and the system curves. Then, the efficiency at the operating condition with the same discharge is obtained. Operating conditions including Q, H, and efficiency (E_p) are summarized in Table 10.1 as (1)–(3). Here, Pump A under 80 ft lift irrigation conditions are given as an example. The intersection of the head–discharge and the system curves is $Q = 970$ gpm, $H = 145$ ft. Then, under $Q = 970$ gpm, the efficiency of the pump can be obtained as $E_p = 82.5$, as shown in Figure 10.24.

Given Q in gpm, H in ft, the BP (4) in kW is computed as:

$$BP = \frac{100QH}{E_P 3960} \times (0.746\ \text{kW})/\text{hp}.$$

The electrical energy required (5) in kW using eq. (10.6), with a motor efficiency of 90%, is

$$P_{kw} = \frac{BP}{E_m}.$$

The time required in h is computed by:

$$t = \frac{452.57 \times 120\ \text{acres} \times \text{water required(in.)}}{Q\ (\text{gpm})}.$$

Since the pump is operated one-quarter of the time at 80 ft lift, one-half of the time at 90 ft lift, and one-quarter of the time at 100 ft lift, the time required for each lift condition could be roughly calculated by assuming 80 ft lift supply one-quarter of the total water required, one-half at 90 ft lift, and one-quarter at 100 ft lift. The calculated T (h) are shown as (6).

The total energy (7) is $P_{kw}(5) \times$ time required(6).

The energy cost (8) is Energy cost (\$) $= P_{kw} \times$ time required $\times$ per unit electricity charge.

a. The total energy cost for pump A is \$1353, and for pump B is \$1421. Pump A has a slightly lower energy cost than pump B.

b. Given the maintenance and repair rate at 3%, annual maintenance and repair cost is computed by:

$$\text{Annual maintenance and repair} = \text{initial cost}(IC) \times \text{maintenance rate}.$$

The total annual operating costs = (a) energy cost + (b) annual maintenance and repair cost.

c. Annual depreciation:

$$\text{Annual depreciation} = \text{present worth}(PW) \times \text{depreciation rate},$$

$$\text{where } PW = IC - SV\left(\frac{1+r}{1+i}\right)^{AP}.$$

Assume the salvage value (SV) is 0 and annual period (AP) equals to the useful life (UL) (i.e., $AP = UL = 25$ years), then

$$PW = IC.$$

Given $UL = 25$ years, $L = 25$ years, and depreciation rate $= 1/25 = 4\%$.

d. Given interest rate $i = 13\%$, $AP = 25$ years, the annual interest:

$$\text{Annual interest} = IC \times CRF(\text{capital recovery factor}),$$

$$CRF = \frac{i(1+i)^{AP}}{(1+i)^{AP}-1} = \frac{(0.13)(1+0.13)^{25}}{(1+0.13)^{25}-1} = 0.136.$$

e. Given the annual taxes and insurance rate is 2%, the annual taxes and insurance:

$$\text{Annual taxes and insurance} = IC \times \text{taxes and insurance rate}.$$

All the costs are summarized in Table 10.2.

Table 10.2 *Solution to Example 10.12, part 2*

Operating costs	Pump A	Pump B
(a) Energy	\$1353	\$1421
(b) Annual maintenance and repair	\$390	\$300
Total annual operating costs	\$1743	\$1721
Fixed cost	Pump A	Pump B
(c) Annual depreciation	\$520	\$400
(d) Annual interest	\$1768	\$1360
(e) Annual taxes and insurance	\$260	\$200
Total annual fixed costs	\$2548	\$1960
Total annual costs	\$4291	\$3681

Pump B has lower operating costs, fixed costs, and total costs than pump A.

QUESTIONS

Q.10.1 A reservoir has been constructed for supplying water for irrigation. A farm is located away from the reservoir and a pipeline of 800 m with a diameter of 30 cm has been laid to supply water to a channel that will carry water to the farm. Clearly water will have to be pumped from the pipeline to the channel. The pipeline has four elbows ($k = 1.2$) and a strainer bucket and a foot valve ($k = 10$) on the pump inlet on the suction line. The friction factor of the pipe is 0.025. The water surface in the channel is 10 m higher than the water surface in the reservoir. The static head and static discharge head are assumed to be zero. The pump runs at 4000 rpm and discharges 4500 L/min, has an efficiency of 85%, and has a motor efficiency of 90%. Compute the total head that the pump must provide, the brake power, the water horsepower, and the demand power. Develop a system curve for the conveyance system. If pipeline diameter is reduced to 25 cm it will alter minor and pipe friction losses, so how will it impact the system curve? How will the system curve change if the difference between reservoir water surface elevation and the channel water surface is 15 m?

Q.10.2 A sprinkler system is supplied water by a 20-cm diameter pipe through pumping from a reservoir. The centerline of the pump is 10 m above the reservoir water surface. The system design capacity is 5000 L/min and the pressure immediately downstream of the pump is 150 kPa. Compute the head that the pump provides and the water horsepower (the power that the pump supplies to the water). Determine the specific speed of the pump if the speed is 2000 rpm. How much brake power be needed if the pump efficiency is 80%? How much demand power will be needed if the electric motor has an efficiency of 95%?

Q.10.3 Consider a confined 80-m thick aquifer for which the hydraulic conductivity is 50 m/day. The static water level is 150 m below the ground surface. A well fully penetrating the aquifer is constructed and a vertical turbine pump is used to pump the water. The well is 50 cm in diameter and 300 m deep. The water is to be supplied to a sprinkler system with a design capacity of 7500 L/min, which requires a pressure of 1200 kPa. Minor and pipe friction losses can be assumed to be 145 m. Compute the total head that the pump must provide and the water horsepower.

Q.10.4 Construct characteristic curves, including head–discharge and brake power–discharge for a 25-cm diameter impeller operating at 1500 rpm. Construct the composite head–discharge and brake power–discharge curves if two identical pumps with 25-cm impellers are connected in series. Do the same for parallel connection.

The pump has the following characteristics:

	Pump	
Q (L/min)	Head (m)	Efficiency (%)
0	51.0	
1000	48.5	54
2000	44.3	59
3000	40.5	67
4000	36.3	78
5000	32.0	80
6000	27.4	74

Q.10.5 A pump (seen in Figure 10.8a) is used to pump water from a reservoir to a sprinkler system that has a design capacity of 4500 L/min. The water surface of the reservoir is 100 m above mean sea level. The pipe on the suction side is 30 cm in diameter and 8 m long. It is an aluminum pipe with an elbow ($k = 1.5$) and bucket strainer ($k = 3$), and an eccentric reducer ($k = 6$). The friction factor of the pipe is 0.03. The water temperature is 20 °C. The required NPSH of the pump is 2.5 m. Compute the maximum distance above the reservoir surface at which the pump can be located.

Q.10.6 Consider the pump and data from Q.10.4 for sprinkler irrigation. The pipeline is 800 m long and 30 cm in diameter. The pipeline is a PVC pipe and has a Hazen–Williams coefficient C of 150. The field to be irrigated is located 3 m above the pump. The pump is located on the water surface. What would be the operating point if 120 sprinklers are operating and if 80 sprinklers are operating? Assume that minor losses are about 10% of pipe friction loss and there is no water source drawdown. Let the discharge of an individual sprinkler be described by $Q = 2.31P^{0.5}$, where Q is the sprinkler discharge in L/min and P is the operating pressure in kPa.

Table 10.3 *Data for Q.10.7*

	Pump A		Pump B	
Q (L/min)	Head (m)	Efficiency (%)	Head (m)	Efficiency (%)
0	51.0		37.5	
1000	48.5		39.5	
2000	44.3		40.2	70
3000	40.5	67	39.8	78
4000	36.3	78	36.2	81
5000	32.0	84	32.1	78
6000	27.4	85	23.5	62

Q.10.7 Consider the sprinkler system in Q.10.6 where 120 sprinklers operate 70% of the time and the other 80 sprinklers 30% of the time. The pump operates 800 hours per season. Compute the operating BP of the two pumps under 120- and 80-sprinkler conditions. The pump characters are shown in Table 10.3. Which of the following pumps has the lowest operating costs, the lowest fixed costs, and the lowest total costs? Pump A costs $4000 and pump B costs $3000. These costs include a 40-kW electric motor, which has an efficiency of 90%. The energy cost is $0.05 per kW. The interest rate is 10% and a period of 20 years, annual taxes and insurance is 2% of the initial value, and annual maintenance and repair is 3% of the initial value.

REFERENCES

Fraenkel, P. L. (1986). Water lifting devices. FAO irrigation and drainage paper 43.

Gupta, R. S. (1989). *Hydrology and Hydraulic Systems*. Englewood Cliffs, NJ: Prentice Hall.

Merkley, G. P., and Allen, R. (2009) *Lecture Notes on Pumps and System Curves*. Logan, UT: Utah State University.

Scherer, T. F. (1993). *Irrigation Water Pumps*. North Dakota State: Agricultural Engineer Department.

USDA-NRCS (2016). Irrigation pumping plants. In *National Engineering Handbook*. Washington, DC: USDA.

Walker, R. (1972). *Pump Selection: A Consulting Engineer's Manual*. Ann Arbor, MI: Science Publishers, Inc.

11 Flow Measurement

Notation

A_o, A_1	orifice and upstream flow cross-sectional area of orifice plate (m^2, ft^2)
A_i	cross-sectional area of the *i*th subsection (m^2, ft^2)
b	width of the orifice (m, ft)
C_1, C_2	concentration of tracer solution injected into the stream (g/L)
C_b	background concentration (g/L)
C_c	contraction coefficient
C_d	discharge coefficient
C_f, C_s	coefficient of the head–discharge relation of a Parshall flume
C_v	velocity coefficient
g	acceleration due to gravity (9.81 m/s^2, 32.2 ft/s^2)
h	head differences, water level differences, or the height of the rise of liquid above the pipe (cm, in.)
h_a	velocity head in the approach channel (m, ft)
h_m	head difference in the manometer (m, ft)
H	head of water (m, ft)
H_d	head at the downstream section (m, ft)
H_u	head at the upstream section (m, ft)
K_f, K_s	flume length coefficient
K_{sf}	submergence correction factor
L	length of the weir (m, ft)
n_f, n_s	flow exponent of the head–discharge relation of a Parshall flume
N	number of subsections
N_s	number of revolutions (per second) or rating equation
P_1, P_2	pressures at sections 1 and 2 (N/m^2)
q	rate of flow of the injected tracer solution (m^3/s)
Q	total discharge (L/s, gpm)
Q_i	discharge of the *i*th subsection (L/s, gpm), $i = 0, 1, \ldots, N$
Q_s	submerged-flow discharge (L/s, gpm)
S_m	specific gravity of the liquid in the manometer
S_p	specific gravity of the liquid in the pipeline
S_t	transition submergence
u	velocity of flow (m/s)
u_a	actual velocity of jet at vena contracta (m/s)
u_i	average flow velocity of the *i*th subsection (m/s), $i = 0, 1, \ldots, N$,
V	volume of flow (m^3, acre-feet)
w_i	top width of the *i*th subsection (m, ft), $i = 0, 1, \ldots, N$
W	throat width of flume (ft)
y_i	average flow depth of the *i*th subsection (m, ft), $i = 0, 1, \ldots, N$
Δt	time interval (min)
ρ	mass density (kg/m^3)

11.1 INTRODUCTION

For farm irrigation, water is brought to the field either by open channels or pipes. The volume of water applied and the rate at which it is applied are determined for controlling the application of water, detecting changes in performance due to system malfunctions, determining water allocation, and effectively managing and maintaining irrigation systems. For irrigation management and maintenance, it is necessary to determine the volume of water that is applied to the field and the rate at which this happens. Appropriate measurement devices are therefore included in irrigation systems for measuring both the volume and the flow rate. It is also important to determine the location where these devices should be placed. This chapter discusses different methods and devices that are commonly employed for determining flow rates and volume in agricultural irrigation.

11.2 FLOW VOLUME AND FLOW RATE

Flow volume and flow rates are interrelated. In irrigation, different units are used to measure the volume. In the SI system, these units include liters, cubic meters, hectare-centimeters, and hectare-meters. In the British system of units, these correspondingly are gallons (gal), cubic feet (ft^3), acre-inches (ac-in), and acre-feet (ac-ft). Just to recap, 1 L is the volume equal to one cubic decimeter (1 dm^3) (i.e., 1/1000 m^3); 1 m^3 is equal to the volume of a cube 1 m long, 1 m wide, and 1 m high; one hectare-centimeter

Table 11.1 *Units of flow volume and rates and their conversion*

Volume		Flow rate	
Units	Equivalent units	Units	Equivalent units
1 cubic meter	1000 liters	1 liter/s	15.85 gallons/minute
1 cubic meter	35.31 cubic feet	1 liter/minute	60 L/s
1 cubic meter	264.2 gallons	1 cubic meter/s	1000 L/s
1 liter	1000 cubic centimeter	1 cubic meter/s	35.31 cubic feet/s
1 acre-foot	12 acre-inches	1 gallon/min	3.785 L/min
1 acre-foot	325,851 gallons	1 gallon/h	3.785 L/h
1 acre-foot	43560 cubic feet	1 cubic foot/s	448.8 gallons/min
1 acre-foot	1233.5 cubic meter	1 cubic foot/s	1.9835 acre-feet/day
		1 acre-in/h	452.6 gallons/min
		1 acre-in/h	1.0083 cubic feet/s

(ha-cm) equals the volume that covers an area equal to 1 ha (i.e., 10,000 m^2) at a depth of 1 cm, which translates into 100 m^3 or 100,000 L of water; and one hectare-meter (1 ha-m) equals the volume equal to 1 ha covered to a depth of 1 m – that is, 10,000 m^3 or 10,000,000 L. Likewise, one gallon equals 1/7.48 ft^3; 1 ft^3 equals 7.48 gallons; 1 ac-in equals the volume of 1 acre (i.e., 43,560 ft^2) covered to a depth of 1 inch (i.e., 3,630 ft^3); and 1 ac-ft equals the volume of 1 acre covered to a depth of 1 ft (i.e., 43,560 ft^3).

Flow rate can be expressed in different ways. In the SI system, it is expressed in liters per second (L/s) or cubic meters per second (m^3/s); in the British system of units it is expressed in gallons per minute (gpm) or cubic feet per second (cfs). Conceptually, 1 L/s means 1 L of water passing through a point in 1 s, and 1 m^3/s means flow 1 m wide and 1 m deep occurring at a velocity of 1 m/s. In a similar manner, 1 gpm means 1 gallon of water passing through a point in 1 min, and 1 cfs means flow 1 ft wide and 1 ft deep occurring at a velocity of 1 ft/s. The SI units and the British units can be converted to each other using conversions given in Table 11.1.

The relationship between the volume of flow (V) and the flow rate (Q) can be expressed as

$$V = \frac{Q\Delta t}{K}, \tag{11.1}$$

where Δt is the time interval (min) and K is the unit constant (equal to 16.67 for V in m^3 and Q in L/s, and equal to 325,851 for V in acre-feet and Q in gpm).

Example 11.1: If the volume of water is given as 5000 m^3, then how many acre-inch, acre-feet, gallons, and liters of water will that be?

Solution: V in acre-inches:

$$5000\ \text{m}^3 \times \left(\frac{1\ \text{acre}}{4046.86\ \text{m}^2}\right) \times \left(\frac{1\ \text{in.}}{0.0254\ \text{m}}\right)$$
$$= 48.64\ \text{acre-inches.}$$

V in acre-feet:

$$5000\ \text{m}^3 \times \left(\frac{1\ \text{acre}}{4046.86\ \text{m}^2}\right) \times \left(\frac{1\ \text{ft}}{0.3048\ \text{m}}\right)$$
$$= 4.05\ \text{acre-feet.}$$

V in gallons:

$$5000\ \text{m}^3 \times \left(\frac{1\ \text{gal}}{0.003785\ \text{m}^3}\right) = 1{,}321{,}000\ \text{gal.}$$

V in liters:

$$5000\ \text{m}^3 \times \left(\frac{1\ \text{L}}{0.001\ \text{m}^3}\right) = 5{,}000{,}000\ \text{L.}$$

11.3 FLOW MEASUREMENT IN OPEN CHANNELS

Flow in open channels can be measured using four types of methods: the volumetric method, the velocity–area method, measuring devices, and tracer methods. If the flow rate is known, the volume can be determined using eq. (11.1).

11.3.1 Volumetric Method

The volumetric method involves collecting water in a small container of known volume and measuring the time taken to fill the container. Then, the volumetric rate or discharge is computed by dividing the volume by time. This method can be used to measure small irrigation streams, discharge of individual sprinklers, discharge rates of drip emitters, and flows in furrows. It is recommended that the container is large enough that it takes at least 20 s to fill it for 1% accuracy, 10 s for 2% accuracy, and 4 s for 5% accuracy, provided the time is measured within 0.2 s accuracy.

11.3.2 Velocity–Area Method

The velocity–area method involves the measurement of flow cross-sectional area (A) and velocity of flow (u) at a fixed location. The discharge is then determined by the product of A and u. If the cross-sectional area is not rectangular, which is often the case, it is measured by dividing the area into subsections or segments, measuring the average depth of flow in each subsection, measuring the area of each subsection, and measuring the average velocity in each subsection. Then, the discharge of each subsection is measured and summing the discharges of subsections yields the flow discharge. The question arises as to the number of segments that can be addressed using the following guidelines:

1. The discharge of any subsection should not exceed 10% of the total discharge.
2. The velocities in two adjacent subsections should not differ by more than 20%.
3. The subsection width should not be greater than 1/15 or 1/20 of the total width.

The total discharge with the velocity–area method can be computed as follows. Let there be N subsections; y_i is the average flow depth of the ith subsection, w_i is the top width of the ith subsection, and u_i is the average flow velocity, $i = 0, 1, \ldots, N$, as shown in Figure 11.1. Note the ith subsection is bounded by flow depth y_{i-1} and y_i. Let Q_i be the discharge of the ith subsection. The cross-sectional area of the ith subsection can be computed by multiplying the average flow depth y_i and width and then the discharge by multiplying the cross-sectional area with velocity as

$$A_i = w_i \times \frac{y_{i-1} + y_i}{2}; Q_i = u_i \times w_i \times \frac{y_{i-1} + y_i}{2}. \tag{11.2}$$

The total discharge Q is then computed as

$$Q = \sum_{i=1}^{N} Q_i. \tag{11.3}$$

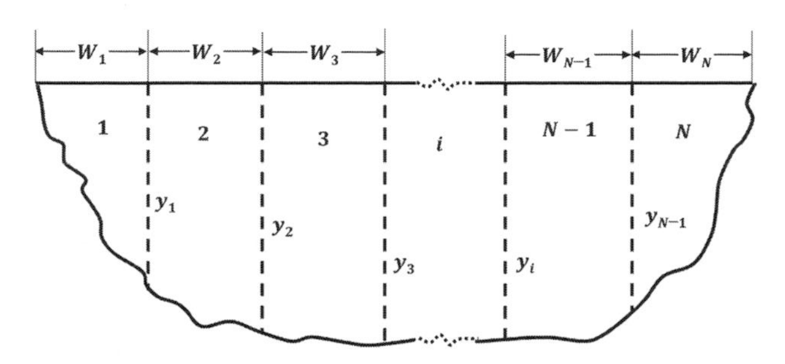

Figure 11.1 Streamflow section for the area–velocity method.

Now the velocity in each subsection needs to be measured, or the average flow velocity can be measured or computed, and then by multiplying it with the total flow cross-sectional area the discharge can be computed. The velocity can be measured using floats, current meters, or tracers.

11.3.2.1 Float Method

This is a simple and inexpensive method for measuring surface velocity; the mean velocity can be obtained by using a correction factor. For this method, a straight section of channel with a uniform cross-section free of surface disturbances and cross currents is needed. This section should be at least three channel widths long and the time of travel should exceed 20 s. A float is placed in the channel a sufficient distance upstream of the section so that it attains the flow velocity before entering the section. Then, the time of travel of the float reaching the midpoint and the end of the section is noted. This measurement is repeated three times and velocity is computed each time at both points and then the average velocity is computed. This average velocity is multiplied by a correction factor varying from 0.66 for the average flow depth of 0.3 m to 0.80 for depth greater than or equal to 6 m.

11.3.2.2 Current Meters

The basic principle of a current meter is that its impeller revolves at a speed proportional to flow velocity. By measuring the time required for a certain number of revolutions, computing the impeller speed, and calibrating the meter, the flow velocity is computed. The flow cross-sectional area is divided into several segments, and in each subsection velocity is measured at several depths and positions. If only two depths (depth greater than 60 cm) are used then velocity is measured at 0.2 times and 0.8 times the depth in each subsection. The velocities measured at these two depths are averaged to represent the velocity in each subsection. If only one depth is used (depth less than 60 cm) then velocity is measured at 0.6 times the depth. It is recommended that 40 s be allowed for each reading and no subsection should have more than 5% of the total discharge. In most of the cases, flow in the open channel has a logarithmic distribution, with the maximum velocity at the surface. The velocity at 0.6 times the depth or mean of 0.2 and 0.8 times the depth is equal to the average velocity of the logarithmic distribution. But the actual flow distribution is not strictly logarithmic. For example, if the 0.8 times depth is higher than the 0.2 times depth velocity or the velocity at 0.2 times depth is twice the value at the 0.8 times depth, it is considered a non-standard condition. Then, the three-point method (i.e., the mean velocity of 0.2 times, 0.6 times, and 0.8 times the depth) should be used.

The relation between flow velocity u (m/s) and number of revolutions (per second) or rating equation is expressed as

$$u = a + bN_s, \tag{11.4}$$

where a (m/s) and b (m/revolutions) are constants of the current meter that are obtained by calibration.

Example 11.2: Compute the discharge of a stream where observations using a current meter are given as follows:

Width measured from the left edge (m)	2	4	6	8	12	14	16	18
Depth (m)	2.0	3.5	5.0	6.0	5.2	4.0	1.8	0
Number of revolutions at 0.6 times depth	30	50	80	100	85	60	25	0
Duration of observations (s)	100	110	120	130	125	120	100	0

For the rating equation, constant $a = 0.1$ m/s and $b = 0.70$ m/revolutions.

Solution:

	Section number (i)	1	2	3	4	5	6	7	8
1	Width measured from the left edge (m)	2	4	6	8	12	14	16	18
2	The top width of the ith subsection (m)	2	2	2	2	4	2	2	2
3	Depth (m)	2.0	3.5	5.0	6.0	5.2	4.0	1.8	0
4	Number of revolutions at 0.6 times depth	30	50	80	100	85	60	25	0
5	Duration of observations (s)	100	110	120	130	125	120	100	0
6	Area $(m^2) = w_i \times \frac{y_{i-1}+y_i}{2}$	2	5.5	8.5	11	22.4	9.2	5.8	1.8
7	Velocity $(m/s) = a + b \times (4)/(5)$	0.31	0.42	0.57	0.64	0.58	0.45	0.28	–
8	Discharge $(m^3/s) = (6) \times (7)$	0.62	2.30	4.82	7.02	12.90	4.14	1.60	–
	Total discharge	33.4 m^3/s							

11.3.2.3 Tracer Method

In this method a tracer is injected into a stream at one point and measured at a point downstream. The rate of injection can be constant or sudden. When a tracer is injected for a finite period of time, its concentration increases at the downstream location, reaches a plateau, and then starts decreasing, as shown in Figure 11.2. For a constant rate of injection in the stream, the tracer injection is continued until the concentration of tracer at the downstream location attains a constant value. If the background concentration is denoted as C_b, the concentration of tracer solution injected into the stream as C_1, the concentration of tracer corresponding to the plateau of the concentration–time curve as C_2, the rate of flow of the injected tracer solution as q, and the discharge of the stream as Q, then the conservation of mass can be expressed as

$$QC_b + qC_1 = (Q+q)C_2. \quad (11.5)$$

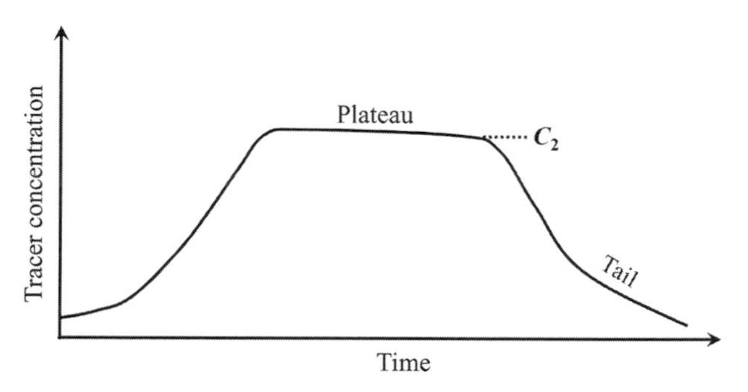

Figure 11.2 Concentration–time curve at downstream sampling site for constant rate injection.

Therefore, Q can be computed as

$$Q = \frac{C_1 - C_2}{C_2 - C_b} \times q. \quad (11.6)$$

Example 11.3: Determine the discharge of a stream whose background tracer concentration is known as 3 ppb (parts per billion). A tracer with 15 g/L of solution is injected into the stream at a constant rate of 20 cm^3/s. At a downstream location sufficiently distant from the location of injection, the tracer is observed to be at the equilibrium concentration of 10 ppb.

Solution: $q = 20\ cm^3/s = 20 \times 10^{-6}\ m^3/s$, $C_b = 3 \times 10^{-9}$ kg/L, $C_1 = 0.015$ kg/L, $C_2 = 10 \times 10^{-9}$ kg/L.

Therefore,

$$Q = \frac{C_1 - C_2}{C_2 - C_b} \times q$$

$$= \frac{0.015 - 10 \times 10^{-9}}{10 \times 10^{-9} - 3 \times 10^{-9}} \times \left(20 \times 10^{-6}\ m^3/s\right)$$

$$= 42.86\ \frac{m^3}{s}.$$

11.3.3 Weirs

A weir is a device installed in an open channel for measuring flow. Weirs are easy to construct and can reasonably accurately measure a wide range of flows. They are portable and adjustable and can lead to accurate discharge rating. However, for reliable measurements, the upstream pool must be clean and free of sediment and a relatively large head should be available for free-flow conditions. The weir crest should be sufficiently high to enable a free fall of water below the weir and should be above the bottom of the approach channel.

There are different types of weirs, some more convenient than others, that can be classified based on their shape of opening or notch, where the shape can be either sharp or broad-crested. Different types of weirs include sharp-crested, broad-crested, suppressed, and contracted. Sharp-crested weirs are classified as rectangular, trapezoidal, and triangular. Figure 11.3 shows weir classification. It may be first useful to define the terms that are commonly used when discussing weirs.

Weir crest is the edge over which water flows. The crest can be sharp or broad. If the crest is thin-edged, the flowing sheet of water has minimum contact with the crest. On the other hand, if the crest is broad the streamlines become parallel to the crest invert, because the crest is nearly horizontal and long.

Pond is the portion of the channel immediately upstream of the weir.

Head is the depth of water in the pond that flows over the weir crest.

Nappe is the sheet of water that overflows a weir.

End contraction is the horizontal distance from the ends of the weir crest to the sides of the weir pond.

Height is the distance from the bottom of the channel to the weir crest.

Weir scale is the scale for measuring the head on the weir crest and is fastened on the sides of the weir. It is located at a distance about four times the head. Its zero scale should be at the same level as the crest.

11.3.3.1 Sharp-Crested Weirs

These weirs are also called thin-plate weirs and comprise a smooth, vertical flat plate which is positioned perpendicular to the flow and across the channel, as shown in Figure 11.4. Due to its positioning, the plate impedes the flow, which causes the water to back up behind the plate and the water to flow over the weir crest. These weirs are generally used for measuring flow on farms. For water to flow free of the crest, a head of at least 6 cm (2.5 in.) and a crest thickness not to exceed 1–2 mm (0.03 in.) is recommended by the US Bureau of Reclamation (1975). Sharp-crested weirs are either rectangular, trapezoidal, or triangular type.

11.3.3.2 Rectangular Weirs

These weirs, as shown in Figure 11.5, have a rectangular notch. The discharge (Q) through the notch depends on the head (H), which is influenced by contraction, velocity of the approaching stream, elevation of the water surface downstream of the weir, and crest condition. These weirs can be partially contracted, fully contracted, or suppressed.

The relation between Q and H can be derived as follows. Let L be the length of the weir and u the velocity of flow, and

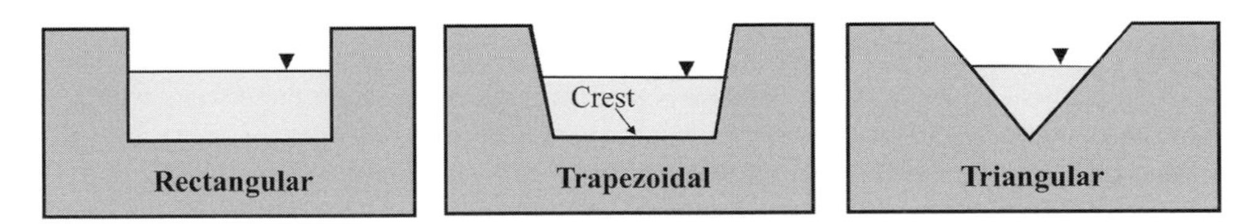

Figure 11.3 Weir classification.

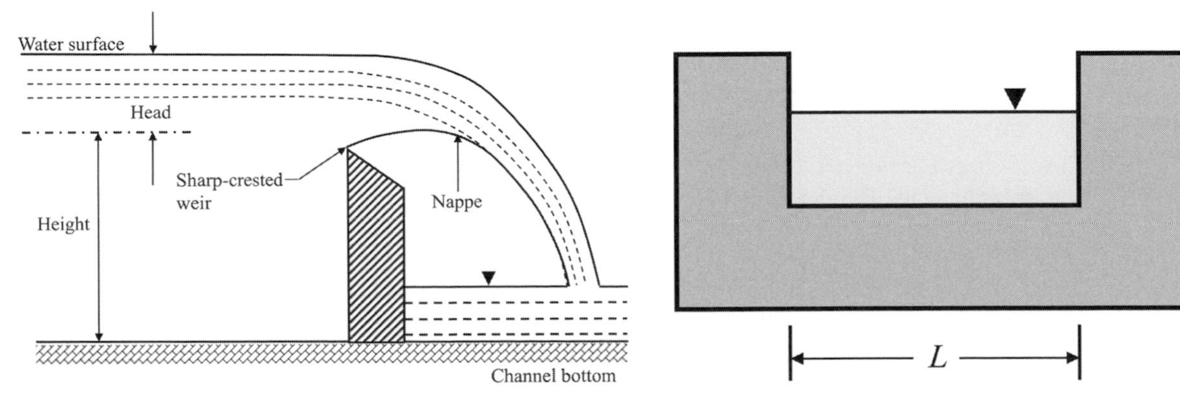

Figure 11.4 Sharp-crested weir.

Figure 11.5 Rectangular weir.

y the flow depth in the weir. Note that y varies from 0 to H. The velocity can be expressed as

$$u = \sqrt{2gy}, \tag{11.7}$$

where g is the acceleration due to gravity. If the incremental depth is denoted as dy, then $dA = Ldy$. Then,

$$dQ = udA = uLdy. \tag{11.8}$$

Integrating eq. (11.8),

$$\int_0^Q dQ = Q = \int_0^H udA = \int_0^H \sqrt{2gy}Ldy = \frac{2}{3}L\sqrt{2g}H^{\frac{3}{2}}. \tag{11.9}$$

If $C = \frac{2}{3}\sqrt{2g}$ denotes the discharge coefficient, then for Q in L/s and L and H in cm, $C = 0.0184$. Thus, the discharge equation for a rectangular weir can be written as

$$Q = 0.0184LH^{\frac{3}{2}}. \tag{11.10}$$

For one-sided contraction, eq. (11.10) becomes

$$Q = 0.0184(L - 0.1H)H^{\frac{3}{2}}. \tag{11.11}$$

For two-sided contraction,

$$Q = 0.0184(L - 0.2H)H^{\frac{3}{2}}. \tag{11.12}$$

Example 11.4: The water is flowing through a two-sided weir that is 100 cm long and has a depth of 20 cm. Compute the flow discharge.

Solution: Applying eq. (11.12) where $H = 20$ cm and $L = 100$ cm,

$$Q = 0.0184(L - 0.2H)H^{\frac{3}{2}} = 0.0184(100 - 0.2 \times 20)20^{\frac{3}{2}}.$$

This yields $Q = 157.99$ L/s.

11.3.3.3 Trapezoidal Weirs

Trapezoidal weirs are also called Cipolletti weirs. As shown in Figure 11.6, the side slope is about 1:4 horizontal to vertical and the length of the notch is at least $3H$, but preferably more than or equal to $4H$. These are fully contracted weirs and their discharge equation can be written as

$$Q = 0.0186LH^{\frac{3}{2}}. \tag{11.13}$$

Example 11.5: Compute the discharge through a trapezoidal weir that has a head of 20 cm and a width of 70 cm at the crest.

Solution: $L = 70$ cm and $H = 20$ cm. Then,

$$Q = 0.0186LH^{\frac{3}{2}} = 0.0186 \times 70 \times (20)^{\frac{3}{2}} = 116.5 \text{ L/s}.$$

11.3.3.4 Triangular Weirs

Triangular weirs are also called "V-notch" weirs (Figure 11.7). The head is the depth of flow above the bottom of the notch and should be measured at a distance of at least $4H$ upstream of the weir. These weirs more accurately measure flow than do rectangular weirs. The discharge equation for these weirs with a 90° notch can be expressed as

$$Q = 0.0138H^{\frac{5}{2}}. \tag{11.14}$$

Example 11.6: Compute the discharge through a 90° V-notch weir if H head is 20 cm.

Solution: $Q = 0.0138H^{\frac{5}{2}} = 0.0138 \times (20)^{\frac{5}{2}} = 24.69$ L/s.

11.3.4 Flumes

A flume is a device that is either constructed or installed in open channels for flow measurement. The device is a specially shaped channel section for developing a stage–discharge relation, which is then used to measure flow. Flumes are not highly sensitive to the velocity of approach, cause small friction losses, and are not prone to the deposition of silt and debris, due to high flow velocities.

A flume has three sections: inlet, throat, and outlet. The inlet section is a converging section that directs the flow in the throat section, which has a level or sloping bed and constricted width and acts as a control. The inlet section serves as a transition between the channel and the throat. The throat opens into the outlet section, which diverges. The water level in the diverging section is kept low so that it does not influence the depth in the inlet section. The throat enables a unique relation between the water level in the inlet section and the discharge through the flume.

Depending on the length of the throat section relative to the upstream head, flumes can be classified into short-

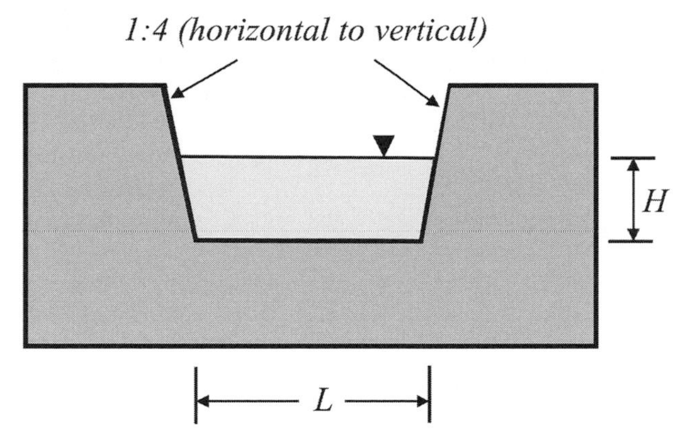

Figure 11.6 Cipoletti or trapezoidal weir.

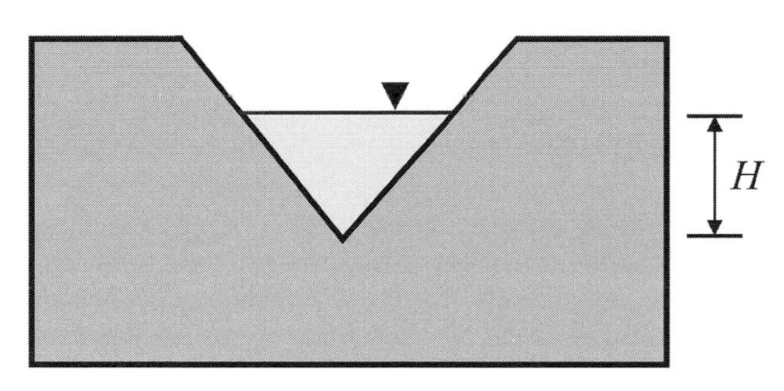

Figure 11.7 V-notch or triangular weir.

throated or long-throated flumes. Examples of short-throated flumes are Parshall, cutthroat, WSC, HS, and various trapezoidal flumes. These flumes are usually designed to meet specific field site requirements, and head–discharge relations are then developed for the sites. Long-throated flumes have throat section length at least twice the upstream head measured above the flume bottom.

11.3.4.1 Parshall Flume

The Parshall flume, shown in Figure 11.8, is extensively used to accurately measure flow in irrigation systems varying from 0.3 to 85,000 L/s (0.01–3000 cfs) (James, 1988). It can be made up of wood, concrete, galvanized sheet metal, or other materials. The US Bureau of Reclamation (1975) has provided dimensions and ranges of flow for various types of standard Parshall flumes. These flumes are classified into three groups, based on the throat width: (1) very small, 25.4–76.2 mm; (2) small, 152.40–2438.4 mm; and (3) large, 3048–15,240 mm. Tables 11.2 and 11.3 provide discharge values for various dimensions of Parshall flumes.

In Parshall flumes discharge can occur under either free or submerged conditions. Free-flow conditions occur when the tailwater depth does not influence flow. If H_u is the head at the upstream section (located at two-thirds of the length of the converging section from the crest, as seen in Figure 11.8) and H_d is the head at the downstream section (located in the

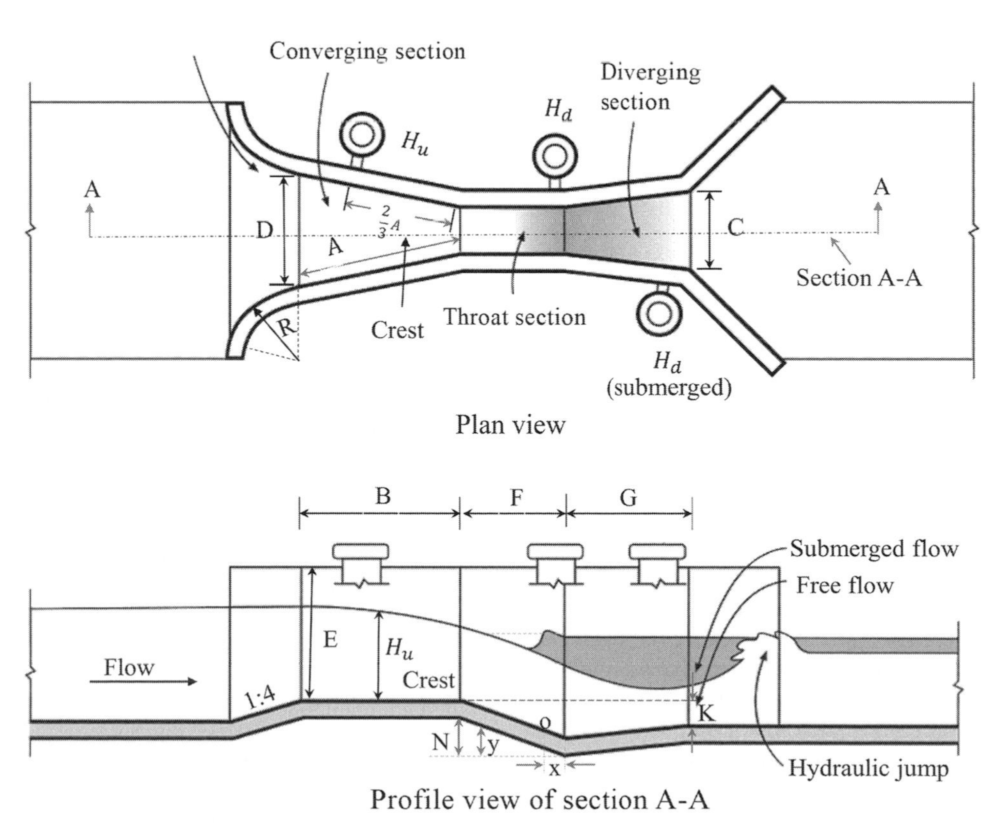

Figure 11.8 Parshall flume. (Adapted from the US Bureau of Reclamation, 1975.)

Table 11.2 *Dimensions and capacities of Parshall flumes of various sizes (see Figure 11.8 for annotation)*

Throat width (cm)	A (cm)	B (cm)	C (cm)	D (cm)	E (cm)	F (cm)	G (cm)	K (cm)	N (cm)	x (cm)	y (cm)	Free-flow capacity Minimum (L/s)	Maximum (L/s)
7.5	31	46	18	26	46	15	30.5	2.5	5.7	2.5	3.8	0.85	28.4
15	41.4	61	39	39.7	61	30.5	61	7.6	12	5.1	7.6	1.4	110.8
23	58.8	86	38	57.5	76	30.5	45.5	7.6	12	5.1	7.6	2.5	253
30	91.5	134	61	84.5	92	61	91.5	7.6	23	5.1	7.6	3.13	456.6

Adapted US Bureau of Reclamation, 1975.

Table 11.3 *Free-flow discharge values (L/s) for Parshall flume*

Head (cm)	Throat width 7.5 cm	15 cm	23 cm	30 cm
3	0.8	1.4	2.6	3.1
4	1.2	2.3	4	4.5
5	1.7	3.3	5.5	7
6	2.3	4.4	7.2	9.6
7	2.7	5.4	8.5	11.4
8	3.4	7.2	11.1	14.4
9	4.3	8.5	13.5	17.7
10	5	10.2	15.9	21.1
11	5.8	11.6	18.1	23.8
12	6.7	13.5	21.1	27.5
13	7.5	15	23.3	31
14	8.5	17.3	26.7	35
15	9.4	19.2	29.5	38.7
16	10.4	21.2	32.5	42.7
17	11.4	23.2	35.6	46.6
18	12.4	25.3	39	51.2
19	13.6	27.8	42.5	55
20	14.3	30	45.8	59.7
21	15.8	32.7	49.3	64.7
22	17.1	35.2	53.3	69.8
23	18.2	37.7	56.8	74
24	19.4	40.1	60.5	79
25	20.7	42.7	64.5	84.1
26	22	45.7	69.3	89
27	23.3	48.7	72.4	94.3
28	24.8	51.5	76.7	100
29	26	54	80.7	105.1
30	27.5	57.3	85.2	111

Adapted US Bureau of Reclamation, 1975.

throat section, as seen in Figure 11.8), then free-flow conditions occur when the ratio $S = H_d/H_u$ is around 0.5, 0.6, and 0.7 for throat width varying from 2.5 to 7.5 cm, 1.5 to 22.5 cm, and 3.0 to 24.0 cm, respectively. When the tailwater depth is high enough to influence flow, the submerged conditions occur. Table 11.4 gives free-flow and submerged-flow coefficient and exponents for Parshall flumes (Skogerboe et al., 1967).

The head–discharge relation of a Parshall flume for free-flow conditions can be expressed as (Michael, 2009)

$$Q = C_f(KH)^{n_f}, \quad S \leq S_t, \tag{11.15}$$

where Q is the discharge (L/s), C_f is the coefficient, H is the head at the upstream section, H_u (m, ft), n_f is the flow exponent, S_t is the transition submergence, and K is the unit constant (equal to 3.28 for H in m, and equal to 1 for H in ft).

For submerged conditions ($S > S_t$), the head–discharge relation can be expressed as

$$Q_s = \frac{C_s[K(H_u - H_d)]^{n_f}}{[-(\log S + C)]^{n_s}}, S > S_t, \tag{11.16}$$

where Q_s is the submerged-flow discharge (L/s), H_u is the upstream head (m, ft), H_d is the downstream head (m, ft), C_s is the coefficient, n_s is the exponent, $C = 0.0044$ for Parshall flumes, and K is the unit constant as before. In the field, reading of H_d in the throat section is usually difficult because of the turbulence of the water surface. To solve this problem, another gage can be used in the diverging section where the water surface is smoother (as seen in Figure 11.8) to relate to H_d.

11.3.4.2 Cutthroat Flumes

Cutthroat flumes are characterized by a rectangular section, a level floor, a converging inlet section, and a diverging outlet section, as shown in Figure 11.9. The throat occurs at the intersection of the converging and diverging sections (James, 1988). These are easier to construct and cheaper than Parshall flumes and can be placed on the channel floor. If L is the total length of the flume, then the length of the converging section (L_1) should be $L/3$; the length of the diverging section should be $2L/3$; the width of the converging and diverging sections should be $W + L/4.5$; the distance to the piezometric tap for measuring H_u (L_a) should be

Table 11.4 *Free-flow and submerged-flow coefficients and exponents for Parshall flumes*

Throat width	C_f for Q in		C_s for Q in		n_f	n_s	S_t
	L/s	cfs	L/s	cfs			
1 in.	9.57	0.338	8.47	0.299	1.55	1	0.56
2	19.14	0.676	17.33	0.612	1.55	1	0.61
3	28.09	0.992	25.91	0.915	1.55	1	0.64
6	58.34	2.06	47.01	1.66	1.58	1.08	0.55
9	86.94	3.07	71.08	2.51	1.53	1.06	0.63
12	113.28	4	88.08	3.11	1.52	1.08	0.62
18	169.92	6	125.17	4.42	1.54	1.115	0.64
24	226.56	8	168.22	5.94	1.55	1.14	0.66
30	283.2	10	204.47	7.22	1.555	1.15	0.67
3 ft	339.84	12	243.55	8.6	1.56	1.16	0.68
4	453.12	16	314.35	11.1	1.57	1.185	0.7
5	566.4	20	383.74	13.55	1.58	1.205	0.72
6	679.68	24	448.87	15.85	1.59	1.23	0.74
7	792.96	28	514.01	18.15	1.6	1.25	0.76
8	906.24	32	577.73	20.4	1.6	1.26	0.78
10	1136.48	40.12	702.05	24.79	1.59	1.275	0.8
12	1345.2	47.5	830.91	29.34	1.59	1.275	0.8
15	1658.42	58.56	1024.33	36.17	1.59	1.275	0.8
20	2180.64	77	1346.9	47.56	1.59	1.275	0.8
25	2702.86	95.44	1669.46	58.95	1.59	1.275	0.8
30	3225.08	113.88	1992.03	70.34	1.59	1.275	0.8
40	4269.24	150.75	2636.88	93.11	1.59	1.275	0.8
50	5313.68	187.63	3282	115.89	1.59	1.275	0.8

After Skogerboe et al., 1967.

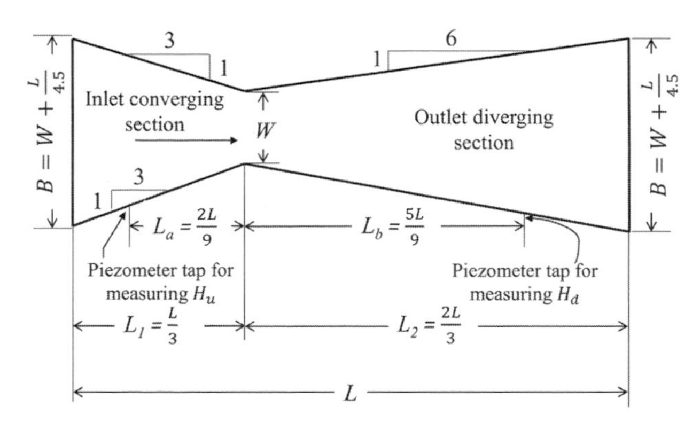

Figure 11.9 A sketch of a cutthroat flume. (Adapted from Skogerboe et al. 1972. Copyright © 1972 by American Society of Civil Engineers, 98(4). Reprinted with permission from ASCE.)

$2L/9$; and the distance to the piezometric tap for measuring $H_d(L_b)$ should be $5L/9$.

The head–discharge relations are the same as in eqs (11.15) and (11.16), but the values of n_f, n_s, and S_t are obtained from Figure 11.10. The value of $C = 0$ and the values of C_f and C_s are determined from:

$$C_f = KK_f W^{1.025}; \qquad C_s = KK_s W^{1.025}, \tag{11.17}$$

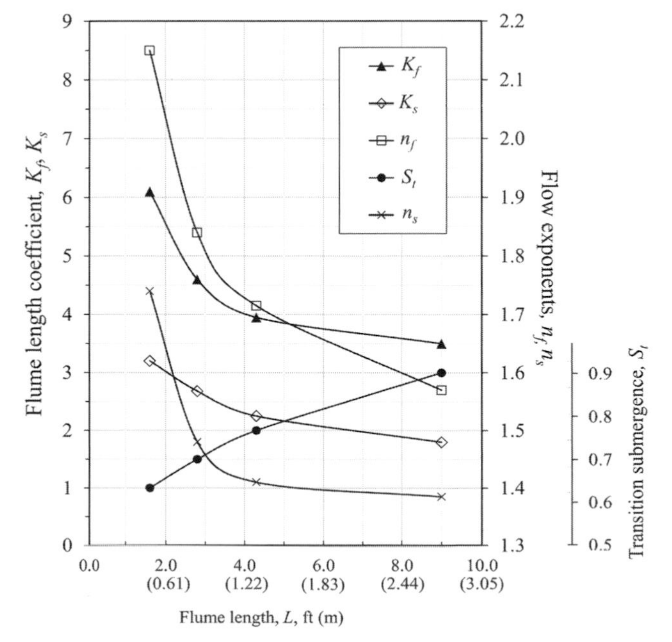

Figure 11.10 Relations among flume length, length coefficient, flow exponents, and transition submergence. (Adapted from Skogerboe et al. 1972. Copyright © 1972 by American Society of Civil Engineers, 98(4). Reprinted with permission from ASCE.)

Table 11.5 *Dimensions of standard calibrated trapezoidal measuring flumes*[a]

Flume	b_1	b_3	Z	H	L_1	L_2	L_3	L_4	L_5	a_1[b]	a_2[b]
						Feet					
1	1.00	0.40	1.00	1.333	1.25	1.422	1.00	1.422	0.50	0.146	0.0625
2	2.00	1.00	1.25	3.00	2.00	3.00	2.50	2.00	1.00	0.50	0.50
3	1.33	0.67	0.58	2.312	1.50	1.326	1.667	1.326	0.50	0.25	0.25
4	0.167	–	0.58	0.562	0.583	0.578	0.583	0.0578	0.25	0.125	–
						Meters					
1	0.305	0.122	0.305	0.406	0.381	0.433	0.305	0.433	0.152	0.045	0.019
2	0.610	0.305	0.381	0.914	0.610	0.914	0.762	0.610	0.305	0.152	0.152
3	0.405	0.204	0.177	0.705	0.457	0.404	0.508	0.508	0.152	0.076	0.076
4	0.051	–	0.177	0.171	0.178	0.176	0.178	0.178	0.076	0.038	–

[a] Dimensions apply to Figure 11.11.
[b] Distance from edge of section to point of depth measurement H_u and H_d.

Table 11.6 *Discharge equations for standardized calibrated trapezoidal flumes*

Flume no.	Equations	H_u range (ft) (m)	Q range (cfs) (L/s)
1	$Q = 3.23H_u^{2.5} + 0.63H_u^{1.5} + 0.05$	0.20–1.20 (0.06–0.37)	0.05–5.96 (1.42–168.8)
2	$Q = 4.27H_u^{2.5} + 1.67H_u^{1.5} + 0.19$	0.30–2.70 (0.09–0.82)	0.54–58.8 (15.3–1665)
3	$Q = 1.46H_u^{2.5} + 2.22H_u^{1.5}$	0.20–2.2 (0.06–0.67)	0.24–17.4 (6.80–493)
4	$Q = 1.55H_u^{2.58}$	0.15–0.50 (0.05–0.15)	0.012–0.26 (0.34–7.36)

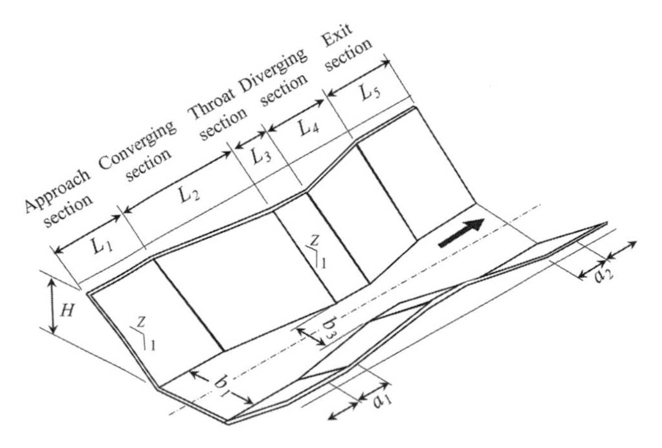

Figure 11.11 Typical standard calibrated trapezoidal flume. Source: adapted from American Society of Agricultural Engineers (ASAE). 1982-83 Agricultural Engineers Yearbook. Copyright © 1982 by ASABE, St Joseph, Michigan. Reprint with permission from ASABE.

where W is the throat width (ft), coefficients K_f and K_s are obtained from Figure 11.10, and K is a unit constant equal to 28.31 for Q and H in L/s and m, respectively, equal to 1.00 for Q and H in cfs and ft, respectively (James, 1988).

11.3.4.3 Trapezoidal Flumes

Trapezoidal flumes have an approach section, converging section, throat, diverging section, and an exit section, as shown in Figure 11.11 (James, 1988). There are several types of trapezoidal flumes, some applicable to unlined irrigation channels and some to concrete lined channels. The ASAE (1982) has reported on the dimensions of these different types, as given in Tables 11.5 and 11.6. The discharge equations for four standard calibrated trapezoidal measuring flumes are given by the ASAE (1982) as in Table 11.7. For submerged flow for flumes 1, 2, and 3, submerged-flow discharge is computed as

$$Q_s = K_{sf} \times Q, \tag{11.18}$$

where K_{sf} is the submergence correction factor given in Table 11.7 and Q is for the free-flow condition.

11.3.4.4 WSC Flume

A trapezoidal flume with a 60° V-notch throat for measuring flow varying from 4 to 100 L/min (1–26 gpm) was designed at Washington State College (WSC) (now Washington State University), as shown in Figure 11.12 (James, 1988; Soil

Table 11.7 *Submergence correction factor for standard calibrated trapezoidal flumes 1, 2, and 3*

Submergence ratio S	Submergence correction factor K_{sf}	
	Flumes 1 and 2	Flume 3
0.70	0.993	1.000
0.75	0.984	1.000
0.80	0.970	0.996
0.85	0.945	0.988
0.90	0.902	0.972
0.92	0.875	0.964
0.94	0.838	0.953
0.95	0.815	0.946

Conservation Service, 1962). The dimensions of the calibrated trapezoidal fume as shown in the figure are given in Table 11.5. The head–discharge relation for a fiberglass version of the WSC flume was given by Trout (pers. comm.: Installation and use of the Powlus furrow flume) as

$$Q = K(H_u - 0.15)^{2.65}, \tag{11.19}$$

in which Q is the discharge (L/min, gpm); H_u is the upstream head measured along the sloping side of the flume inlet section (mm); and K is the unit constant ($K = 5.43 \times 10^{-4}$ for Q in L/min; $K = 1.43 \times 10^{-4}$ for Q in gpm).

11.4 ORIFICES

Orifices can be either freely flowing, partially submerged, or fully submerged. They are either rectangular or circular in

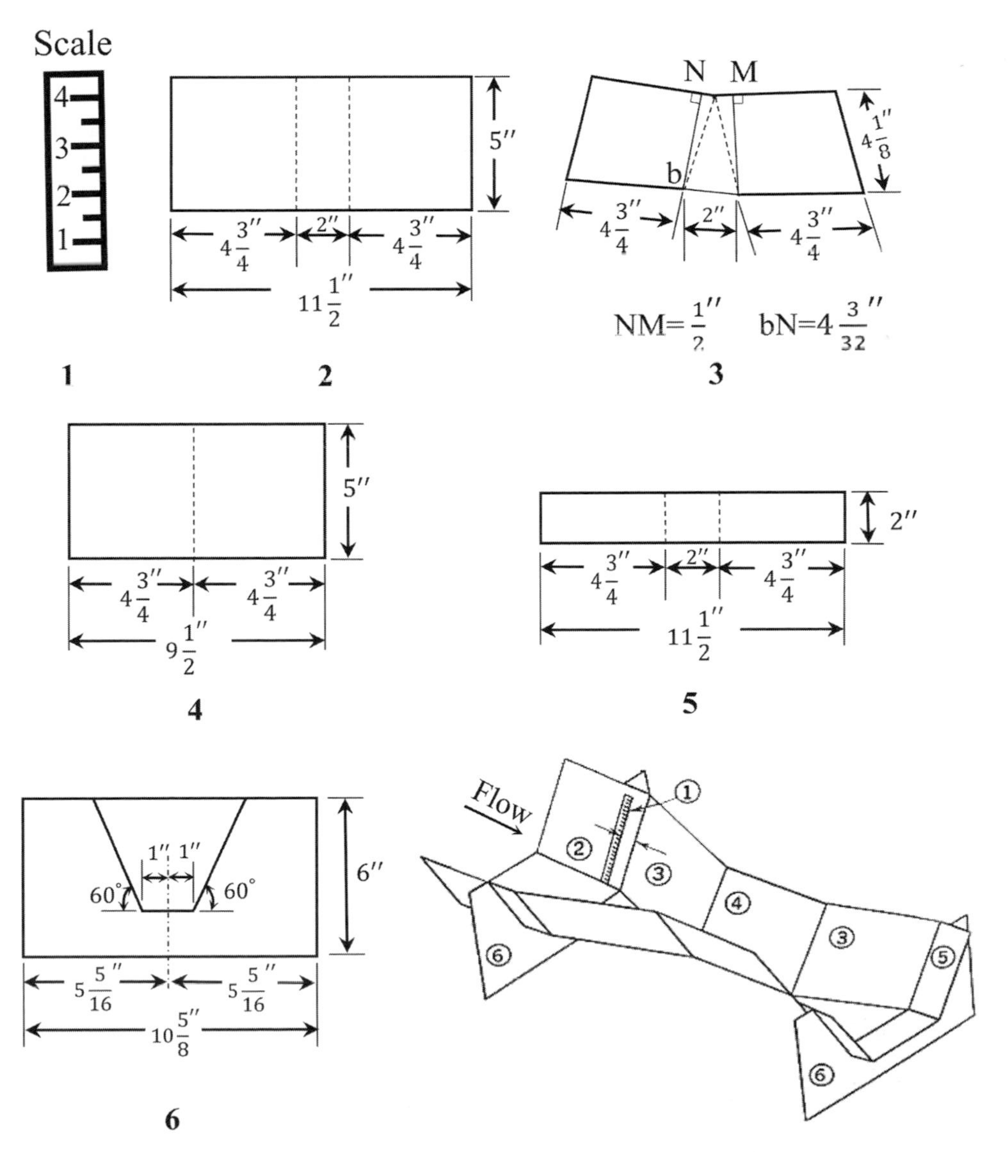

Figure 11.12 Layout and dimension of the WSC V-notch flume (inches). (Adapted from Soil Conservation Service, 1962.)

shape and are placed near the channel bottom on the vertical surface perpendicular to the direction of flow. A short description of each orifice is given.

11.4.1 Freely Flowing Orifice

When there is insufficient fall for a weir and the use of a weir is difficult to justify, orifices can be used. Orifices are generally placed on vertical surfaces near the bottom of the channel perpendicular to the direction of flow. They are generally circular or rectangular in shape. A freely flowing orifice is shown in Figure 11.13, in which the jet of water flows freely. The section where the contraction of the jet is maximum is often referred to as vena contracta. For a circular orifice the vena contracta is about half the diameter of the orifice itself. The velocity of flow through the orifice can be expressed as

$$u = \sqrt{2gh}, \tag{11.20}$$

where h is the head. If A is the cross-sectional area of the orifice, then the discharge of water Q through the orifice can be expressed as

$$Q = C_d A u = C_d \frac{\pi d^2}{4} \sqrt{2gh}, \tag{11.21}$$

where C_d is the discharge coefficient.

For orifices, there are three defined coefficients: velocity coefficient, contraction coefficient, and discharge coefficient. The velocity coefficient, C_v, can be defined as

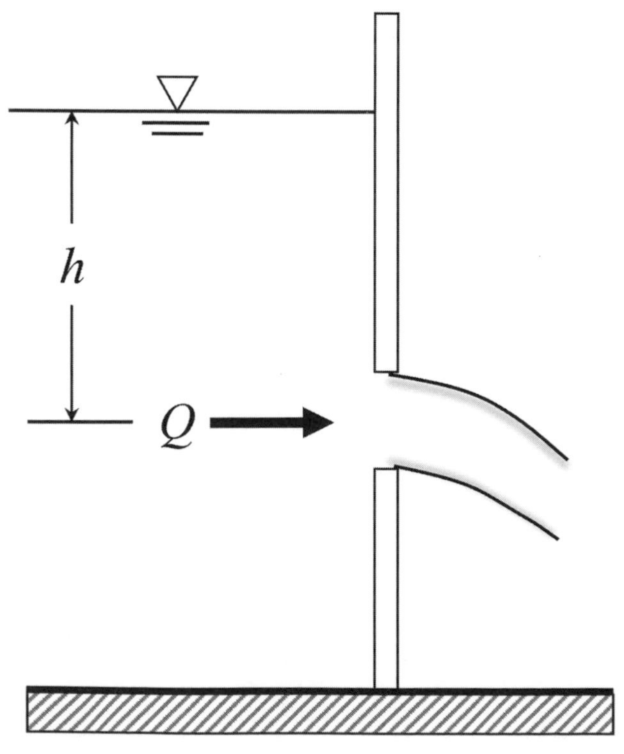

Figure 11.13 A freely flowing orifice.

$$C_v = \frac{\text{Actual velocity of jet at vena contracta}}{\text{Theoretical velocity of jet}} = \frac{u_a}{\sqrt{2gh}}, \tag{11.22}$$

in which u_a is the actual velocity.

The contraction coefficient, C_c, can be defined as

$$C_c = \frac{\text{Area of the jet at vena contracta}}{\text{Actual area of the orifice}}. \tag{11.23}$$

Now the discharge coefficient, C_d, can be defined as

$$C_d = \frac{\text{Actual velocity} \times \text{actual area}}{\text{Theoretical velocity} \times \text{theoretical area}} = \frac{Q}{Q_{theor}} = C_v \times C_c. \tag{11.24}$$

11.4.2 Fully Submerged Orifices

If orifices are placed on vertical surfaces near the bottom of the channel perpendicular to the direction of flow, they are fully submerged, as shown in Figure 11.14. The submergence minimizes the difference between upstream and downstream water levels. The upstream water level must always be above the top of the orifice, and the outlet side is also fully submerged. Fully submerged orifices may be contracted so the opening is far from the walls of the approach channel. For submerged orifices the approach velocity should be negligible, which requires that the water prism 6–9 m (20–30 ft) upstream of the orifice should be at least eight times the area of the orifice (US Bureau of Reclamation, 1975).

The head–discharge relation for fully submerged contracted orifice with negligible approach velocity can be expressed as

$$Q = 0.61 K A H^{\frac{1}{2}}, \tag{11.25}$$

where Q is the discharge (L/s, gpm), A is the area of orifice opening (cm^2, in.^{2}), H is the head (m, ft), and K is the unit constant ($K = 0.433$ for Q in L/s, A in cm^2, and H in m;

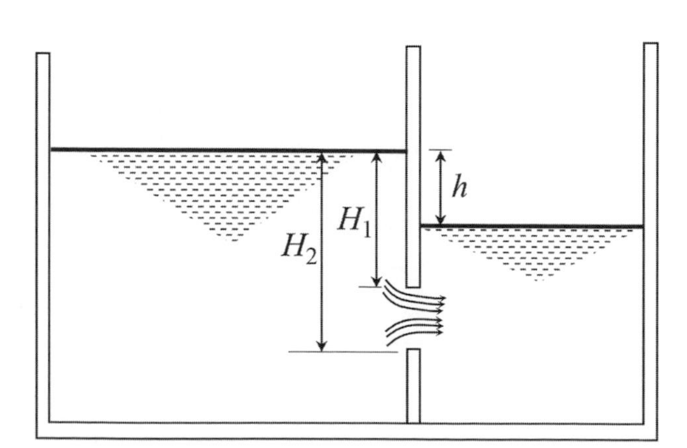

Figure 11.14 Fully submerged orifice.

$K = 25.0$ for Q in gpm, A in in.^{2}, and H in ft). If the approach velocity is not negligible, then eq. (11.25) is modified as

$$Q = 0.61KA(H + h_a)^{\frac{1}{2}}, \tag{11.26}$$

where h_a is the velocity head in the approach channel in m or ft.

The head–discharge relation for a fully submerged orifice can also be expressed as

$$Q = C_d b(H_2 - H_1)\sqrt{2gh}, \tag{11.27}$$

in which H_1 is the head above the top of the orifice on the upstream aside, H_2 is the head above the bottom of the orifice on the upstream side, b is the width of the orifice, and h is the difference in the water levels between upstream and downstream sides.

11.4.3 Partially Submerged Orifices

The orifice has its outlet side partially submerged, as shown in Figure 11.15. The head–discharge relation can be expressed as

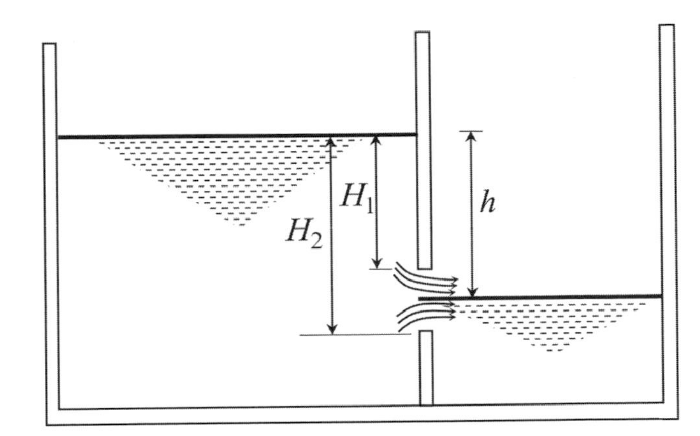

Figure 11.15 Partially submerged orifice.

$$Q = C_d \times b \times (H_2 - h)\sqrt{2gh} + \frac{2}{3}C_d \times b \times \sqrt{2g}\left[h^{\frac{3}{2}} - H_1^{\frac{3}{2}}\right]. \tag{11.28}$$

The two terms on the right-hand side of eq. (11.28) represent, respectively, flow through the submerged portion of the orifice and the flow through the free portion of the orifice.

Example 11.7: A fully submerged orifice having a width of 1 m is used for measuring flow. Its discharge coefficient is 0.61. Measurements show that the head of water above the top of the orifice is 2 m, that on the bottom of the orifice is 2.25 m, and the difference between water levels on the upstream and downstream sides of the orifice is 0.25 m. Determine the discharge of water through the orifice.

Solution: Here, $b = 1$ m, $h = 0.25$, $H_1 = 2$ m, $H_2 = 2.25$ m, and $C_d = 0.61$. Discharge is

$$Q = C_d \times b \times (H_2 - H_1) \times \sqrt{2gh}$$
$$= (0.61) \times 1 \times (2.25 - 2.0) \times \sqrt{2 \times 9.81 \times 0.25}$$
$$= 0.338 \text{ m}^3/\text{s}.$$

Example 11.8: Determine the discharge coefficient, velocity coefficient, and contraction coefficient for a circular orifice 50 cm in diameter. The head of the water over the orifice is 1 m, and coordinates of a point on the jet measured from the vena contracta are 30 cm vertical and 1 m horizontal. The jet falls in a channel 10 m wide and 3 m deep, where the water level rises 0.25 m in 18 s.

Solution: Here, $d = 50$ cm $= 0.5$ m, $h = 1$ m, horizontal distance $= 1$ m, and vertical distance $= 0.5$ m. The cross-section area of the orifice is

$$\frac{\pi}{4} \times (0.5)^2 = 0.196 \text{ m}^2.$$

Theoretical velocity: $u = \sqrt{2gh} = \sqrt{2 \times 9.81 \times 1} =$ 4.429 m/s.

Theoretical discharge: $Q_{theor} = u \times$ area of orifice $=$ $4.429 \frac{\text{m}}{\text{s}} \times 0.196 \text{ m}^2 = 0.87 \frac{\text{m}^3}{\text{s}}$.

The point on the jet from the vena contracta has a horizontal distance $x = 1$ m and vertical distance $y = 30$ cm $= 0.3$ m; the travel time

$$t = \sqrt{\frac{2y}{g}} = 0.2473 \text{ s}.$$

The actual velocity at the vena contracta is

$$u_a = \frac{x}{t} = \frac{1 \text{ m}}{0.2473 \text{ s}} = 4.044 \text{ m/s}.$$

From eq. (11.22), the velocity coefficient is

$$C_v = \frac{u_a}{u} = \frac{4.044 \text{ m/s}}{4.429 \text{ m/s}} = 0.913.$$

The water level rise is 0.25 m in 8 s, so in 1 s the water level rise is

$$\frac{0.25 \text{ m}}{18} = 0.01389.$$

The actual discharge

$$Q = Au_a = 0.01389 \text{ m} \times 10 \text{ m} \times 4.044 \frac{\text{m}}{\text{s}} = 0.562 \frac{\text{m}^3}{\text{s}}.$$

From eq. (11.24), the discharge coefficient is

$$C_d = \frac{Q}{Q_{theor}} = \frac{0.562}{0.870} = 0.646.$$

From eq. (11.24), the contraction coefficient is

$$C_c = \frac{C_d}{C_v} = \frac{0.646}{0.870} = 0.743.$$

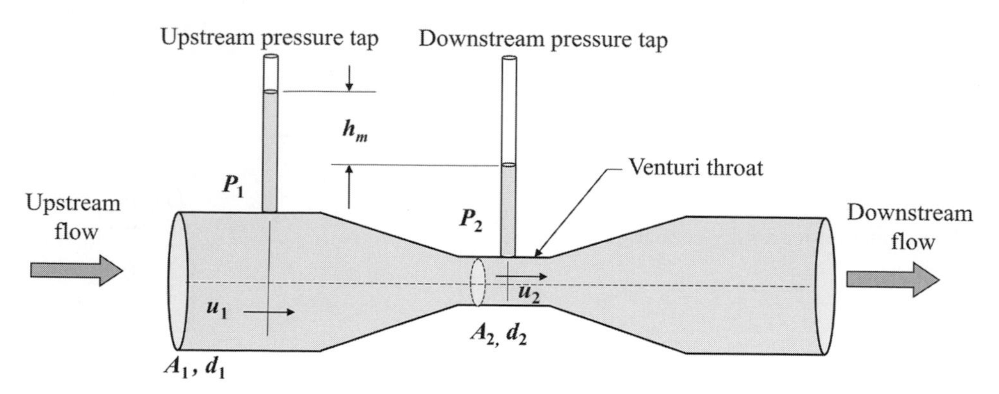

Figure 11.16 Venturimeter and its operations.

11.5 FLOW MEASUREMENT IN PIPES

For irrigation, water is also conveyed by pipelines, especially for sprinkler and drip systems. The volume of flow and volumetric flow rates in pipelines can be measured using several devices, such as differential pressure, rotating mechanical, bypass, ultrasonic, and insertion meters. Some of these devices are briefly discussed here.

11.5.1 Venturimeter

A venturimeter, as shown in Figure 11.16, works on the Bernoulli principle by which an increase in the kinetic energy leads to a reduction in pressure. It comprises a converging part, a throat, and a diverging part. The constriction results in an increase in the kinetic energy due to an increase in flow velocity. As water moves down, it gets accelerated through the converging cone, which has an angle of 15–20°. A pressure difference is created between the upstream part of the cone and the throat where the flow area is minimum, which is an indicator of flow rate. Further down in the diverging part, the water slows down as the kinetic energy is transformed back to pressure energy. It may be noted that there is no vena contracta. The discharge through a venturimeter can be computed as

$$Q = C_d A_2 \sqrt{\frac{2(P_2 - P_1)}{\rho(1 - \beta^4)}}, \tag{11.29}$$

where P_1 and P_2 are the pressures at sections 1 and 2, respectively; $\beta = d_1/d_2$, where d_1 and d_2 are the diameters at point 1 of the pipe and point 2 of the throat, respectively; C_d is the discharge coefficient; A_2 is the cross-sectional area at point 2 of the throat; and ρ is the mass density. Venturimeters are expensive. For a Reynolds number exceeding 20,000, the value of C_d is about 0.98.

If the manometer liquid is heavier than water, then the value of the head difference in the pipe can be computed from the head difference in the manometer as

$$h = h_m \left[\frac{S_m}{S_p} - 1\right]. \tag{11.30}$$

If the manometer liquid is lighter, then the head difference can be computed as

$$h = h_m \left[1 - \frac{S_p}{S_m}\right], \tag{11.31}$$

in which h_m is the head difference in the manometer, S_p is the specific gravity of the liquid in the pipeline, and S_m is the specific gravity of the liquid in the manometer.

Example 11.9: Calculate the discharge through a pipeline if a horizontal venturimeter with an inlet of diameter 30 cm and throat of diameter 15 cm is installed. The differential manometer connected to the inlet and throat reads as 12 cm mercury.

Solution: Here $d_1 = 30$ cm and $d_2 = 15$ cm; assume $C_d = 0.98$ and $P_1 - P_2 = 12$ cmHg.

$$A_2 = \frac{\pi}{4} d_2^2 = \frac{\pi}{4} \times (15\ \text{cm})^2 = 176.7\ \text{cm}^2,$$

$$\beta = \frac{d_2}{d_1} = \frac{15\ \text{cm}}{30\ \text{cm}} = 0.5,$$

$$\frac{2(P_1 - P_2)}{\rho} = 2gh.$$

Using eq. (11.30),

$$h = h_m \left[\frac{S_m}{S_p} - 1\right] = 12\ \text{cm} \times \left[\frac{13.6}{1} - 1\right] = 151.2\ \text{cm of water},$$

$$2gh = 2\left(9.81 \frac{\text{m}}{\text{s}^2}\right)(1.512\ \text{m}) = 29.665 \frac{\text{m}^2}{\text{s}^2},$$

$$Q = C_d \times A_2 \times \sqrt{\frac{2(P_1 - P_2)}{\rho(1 - \beta^4)}} = C_d \times A_2 \times \sqrt{\frac{2gh}{(1 - \beta^4)}},$$

$$Q = 0.98 \times 0.0177\ \text{m}^2 \times \sqrt{\frac{29.665 \frac{\text{m}^2}{\text{s}^2}}{(1 - 0.5^4)}} = 0.10 \frac{\text{m}^3}{\text{s}}.$$

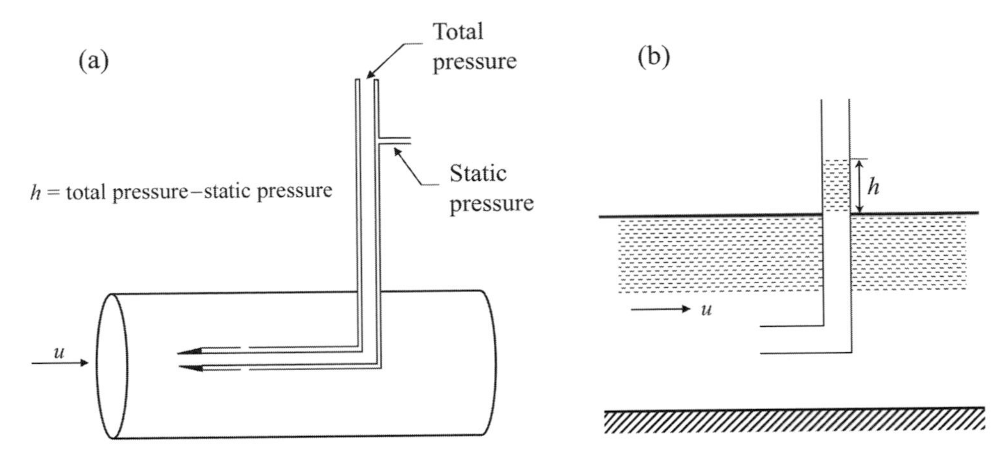

Figure 11.17 Pitot tube: (a) two tubes; (b) single tube.

11.5.2 Pitot Tube

A pitot tube comprises two tubes pointing upstream, as shown in Figure 11.17a. The inner tube measures the kinetic energy and potential energy, while the outer tube measures the potential energy only. The basic working principle is the Bernoulli principle. As the water in the mouth of the inner tube is brought to rest, its kinetic energy is transformed to pressure energy. The difference between the pressure of the outer tube and that of the inner tube should equal the kinetic energy of the flow. Hence, the flow velocity can be computed. For an open channel with a free water surface, a single tube case is enough to determine the velocity, as shown in Figure 11.17b. Pitot tubes cannot be used for measuring turbulent flows, but are simple to install.

The actual flow velocity can be computed as

$$u = C_v\sqrt{2gh}, \tag{11.32}$$

where h is the head difference in the two tube cases or the height of the rise of liquid above the tube, and C_v is the velocity coefficient for the pitot tube. It may be noted that the theoretical velocity can be given by eq. (11.32) with $C_v - 1$.

Example 11.10: Calculate the discharge in a pipeline 20 cm in diameter using a pitot tube. The pressure difference is found to be 100 mm of water. The velocity coefficient is assumed to be 0.98. The mean velocity is assumed to be 0.9 times the central velocity.

Solution:
Pressure difference = 100 mm = 0.1 m of water, diameter of pipeline = 0.20 m, and $C_v = 0.98$. The actual velocity at the center is computed as

$$u = 0.98\sqrt{2gh} = 0.98\sqrt{2 \times 9.81 \times 0.1} = 1.373\ \text{m/s},$$

$$\text{Mean velocity} = 0.90 \times \text{central velocity} = 0.9 \times 1.373\ \text{m/s} = 1.236\ \text{m/s}.$$

Cross-sectional area of the pipe: $\frac{\pi}{4} \times (0.2\,\text{m})^2 = 0.0314\,\text{m}^2$.

$$\text{Discharge} = \text{mean velocity} \times \text{cross-sectional area} = 1.236\ \text{m/s} \times 0.0314\ \text{m}^2 = 0.039\ \text{m}^3/\text{s}.$$

11.5.3 Orifice Plate

An orifice plate or meter can also be used for measuring flow in a pipe. It comprises a thin, flat plate with a circular, sharp-edged hole, called the orifice, which is clamped between flanges in a pipe and is concentric with the pipe, as shown in Figure 11.18. The diameter of the orifice can vary from 0.4 to 0.8 times the diameter of the pipe, although it is generally taken as half of the pipe diameter. The velocity in the orifice increases because of reduced cross-sectional area, which results in a pressure drop between the upstream and downstream sides of the plate. In this case, there is a point of maximum contraction, called the vena contracta. Beyond this point, the velocity and pressure change. The pipe orifice is simple in construction and is cheaper than a venturimeter and pitot tube.

The discharge from an orifice plate can be computed as

$$Q = \frac{C_d A_o A_1 \sqrt{2gh}}{\sqrt{A_1^2 - A_o^2}}, \tag{11.33}$$

where A_o is the orifice cross-sectional area, A_1 is the upstream flow cross-sectional area, h is the differential head, and C_d is the discharge coefficient for the orifice.

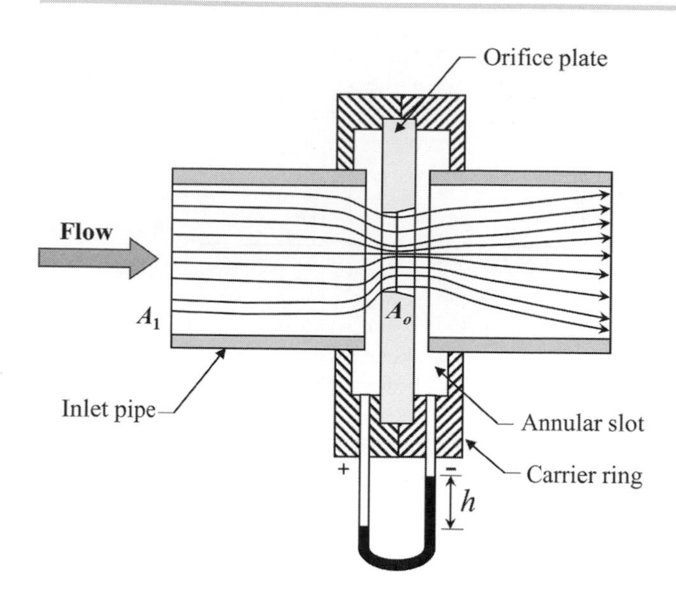

Figure 11.18 Orifice plate.

Example 11.11: Compute the discharge through a pipe 10 cm in dimeter in which an orifice of 5 cm in diameter is installed. The pressure on the upstream side of the orifice is 10 N/cm^2 and that on the downstream side is 8 N/cm^2. Assume the discharge coefficient as 0.61.

Solution: Pipe diameter = 10 cm and orifice diameter = 5 cm.

Orifice area: $A_o = \frac{\pi}{4}(0.05\ \text{m})^2 = 0.002\ \text{m}^2$.

Pipe area: $A_1 = \frac{\pi}{4}(0.1\ \text{m})^2 = 0.008\ \text{m}^2$.

$$P_1 = 10\frac{\text{N}}{\text{cm}^2} = 10 \times 10^4\frac{\text{N}}{\text{m}^2}; \frac{P_1}{\rho g} = \frac{10 \times 10^4}{1000 \times 9.81} = 10.19\ \text{m},$$

$$P_2 = 8\frac{\text{N}}{\text{cm}^2} = 8 \times 10^4\frac{\text{N}}{\text{m}^2}; \frac{P_2}{\rho g} = \frac{8 \times 10^4}{1000 \times 9.81} = 8.15\ \text{m}.$$

Therefore,

$$h = \frac{P_1}{\rho g} - \frac{P_2}{\rho g} = 10.19 - 8.15 = 2.04\ \text{m}.$$

Discharge Q is

$$Q = \frac{C_d A_o A_1 \sqrt{2gh}}{\sqrt{A_1^2 - A_o^2}}$$

$$= \frac{0.61 \times 0.002\ \text{m}^2 \times 0.008\ \text{m}^2 \sqrt{2 \times 9.81\frac{\text{m}}{\text{s}^2} \times 2.04\ \text{m}}}{\sqrt{(0.008\ \text{m})^2 - (0.002\ \text{m})^2}}$$

$$= 0.008\ \frac{\text{m}^3}{\text{s}}.$$

QUESTIONS

Q.11.1 Calculate the volume of irrigation water that is applied to a field at a rate of 250 gpm for a period of 12 hours. Express the volume in gallons, cubic feet, and cubic meters. Also express the rate of application in liters per second, cubic feet per second, and cubic meters per second.

Q.11.2 For the purpose of measuring discharge of a channel, observations have been made using a current meter. The flow velocity is computed by noting the number of revolutions per second, in which parameters are $a = 0.095$ and $b = 0.65$ m/revolutions. The observations are given as follows:

Width measured from the left edge (m)	0.5	1.0	1.5	2.0	2.5	3.0
Depth (m)	0.25	0.75	1.25	1.50	1.0	0.15
Number of revolutions at 0.6 times depth	20	40	60	80	50	10
Duration of observation (s)	60	100	120	120	90	60

Determine the discharge of the channel.

Q.11.3 A tracer is used to measure the discharge of a channel. The background tracer concentration of the channel is found to be 5 ppb. A tracer with 10 g/L of solution is injected into the stream at a constant rate of 10 cm^3/s. The equilibrium concentration of 10 ppb of the tracer is observed at a downstream location sufficiently far away from the location of injection. What is the discharge of the channel?

Q.11.4 Water is flowing through a two-sided weir that is 150 cm long and has a depth of 10 cm. Compute the flow discharge.

Q.11.5 Consider a trapezoidal weir 120 cm long through which water flows. The weir has a depth of 25 cm. What is the flow discharge?

Q.11.6 Consider a contracted rectangular weir 150 cm long having a flow depth of 35 cm. The water enters into a channel 200 cm long (a second weir). What is the depth of water over the second weir? [Hint: The discharge through the second weir should equal the discharge through the first weir.]

Q.11.7 Consider a Cipolletti weir (trapezoidal) that is 80 cm wide, with a head of 40 cm over the crest. What is the discharge through the weir?

Q.11.8 Consider a triangular weir with a 90° V-notch. It has a head of 25 cm. What is the discharge through the weir?

Q.11.9 Consider a fully submerged orifice for measuring flow. It has a width of 0.5 m and its discharge coefficient is 0.55. The head of water above the top of the orifice is 1 m and that on the bottom of the orifice is 1.25 m. The difference between water levels on the upstream and downstream sides of the orifice is 0.20 m. What is the discharge of water through the orifice?

Q.11.10 Consider a circular orifice 30 cm in diameter. The head of the water over the orifice is 0.5 m, and coordinates of a point on the jet measured from the vena contracta are 20 cm vertical and 0.6 m horizontal. The jet falls in a channel 10 m wide and 4 m deep where the water level rises 0.10 m in 15 s. What are the discharge coefficient, velocity coefficient, and contraction coefficient for a circular orifice?

Q.11.11 A horizontal venturimeter with an inlet of 20 cm diameter and throat of 10 cm diameter is installed in a pipeline. The reading of the differential manometer connected to the inlet and throat is 8 cmHg. What is the discharge through the pipeline?

Q.11.12 A pitot tube is used to compute discharge in a pipeline of 10 cm diameter. The pressure difference is observed to be 80 mm of water. The velocity coefficient is assumed to be 1.00. The mean velocity can be taken as 0.8 times the central velocity. What is the discharge in the pipeline?

Q.11.13 An orifice of 6 cm diameter is installed in a pipe of 12 cm diameter that is carrying water. The pressure on the upstream side of the orifice is 15 N/cm^2 and that on the downstream side is 12 N/cm^2. What is the discharge through the pipe, assuming the discharge coefficient as 0.58?

REFERENCES

American Society of Agricultural Engineers (ASAE) (1982). *1982–83 Agricultural Engineers Yearbook*. St Joseph, MI: ASAE.

James, L. G. (1988). *Principles of Farm Irrigation Systems Design*. New York: Wiley.

Michael, A. M. (2009). *Irrigation Theory and Practice*. New Delhi: Vikas Publishing House Pvt. Ltd.

Skogerboe, G. V., Hyatt, M. L., England, J. D., and Johnson, J. R. (1967). Design and calibration of submerged open channel flow measurement structures: Part 2. Parshall flumes. Utah Water Research Laboratory, Utah State University report WG 31-3.

Skogerboe, G. V., Bennett, R. S., and Walker, W. R. (1972). Generalized discharge relations for cutthroat flumes. *Journal of Irrigation and Drainage Engineering*, 98(4): 569–583.

Soil Conservation Service (1962). Measurement of irrigation water. In *SCS National Engineering Handbook*. Washington, DC: USDA.

US Bureau of Reclamation (1975). *Water Measurement Manual*. Washington, DC: US Government Printing Office.

Part III Principles of Hydrology

12 Infiltration

Notation

$i(t)$	infiltration rate at time t (min)
i_0	initial rate of infiltration (cm/h, in./h, mm/min)
i_a	average infiltration rate (cm/h, in./h, mm/min)
i_c	basic infiltration rate (cm/h, in./h, mm/min)
i_f	number of intake family which is equal to the long-term intake rate (in./h)
$I(t)$	amount of water infiltrated or cumulative infiltration at time t (cm, in.)
I_0	cumulative infiltration up to time t_0 (cm, in.)
I_p	potential cumulative infiltration volume (cm, in.)
k	proportionality factor dependent on soil type and initial soil moisture content
K_e	effective hydraulic conductivity (cm/h)
m, n, and β	positive real constants for a given soil–vegetation–land-use complex
M	fillable porosity of the soil (fraction or percentage)
S	maximum soil moisture retention
$S(t)$	potential storage space at any time t
S_0	initial storage (or pore space) space available
S_{av}	average matric suction at the wetting front (cm of water)
t	time (min)
t_0	beginning of the time interval (min)
t_L	time when the long-term intake rate occurs (min)
$W(t)$	amount of water stored (cm)
α	percentage factor indicating reduction in pore space
η	effective porosity of the soil

12.1 INTRODUCTION

The purpose of irrigation is to supply water to plants, and this supply of water is made possible through infiltration into the soil. The infiltrated water is extracted by plant roots. Thus, infiltration plays a fundamental role in the design, management, and operation of irrigation systems. The objective of this chapter is to briefly discuss elementary aspects of infiltration and methods for computing it. Before proceeding further, it is pertinent to provide a number of definitions that are used throughout the chapter.

12.2 SYMBOLS AND DEFINITIONS

The rate of infiltration is defined as the rate at which water enters the soil at the soil surface. If there is no limitation to the supply of water, the water will enter the soil at the highest rate, which is referred to as the potential rate of infiltration. Otherwise, the rate of infiltration will be smaller than the potential rate. This rate is maximum at the beginning and decreases as time progresses, as shown in Figure 12.1. This rate is the initial rate and is denoted as i_0 at time $t = 0$. The rate of decrease depends on a number of factors, including soil type, texture, and structure; vegetation; tillage and land use; and antecedent conditions. Table 12.1 shows the variation of infiltration with soil type, slope, and tilth (USDA-NRCS, 2016). The rate of infiltration is denoted as $i(t)$, where t is time, because i is a function of time. When designing an irrigation system, a key consideration is the rate at which infiltration can be maintained in a given soil.

Cumulative infiltration at any time is defined as the depth of water that has infiltrated up to that time, and is denoted by $I(t)$. Clearly, $I(t)$ is the integral of $i(t)$ expressed as

$$I(t) = \int_0^t i(t)dt. \tag{12.1}$$

By the same token, $i(t)$ can be expressed as the derivative of $I(t)$ as

$$i(t) = \frac{dI(t)}{dt}. \tag{12.2}$$

In general, cumulative infiltration increases more steeply in the beginning, and then the rate of increase starts slowing, as shown in Figure 12.2(a). Between infiltration rate and cumulative infiltration, time is a common parameter, and by eliminating it one can plot infiltration rate as a function

Table 12.1 *Suggested maximum infiltration rates for sprinkler systems (in./h [cm/h])*

	Ground slope			
Soil texture	0–5%	5–8%	8–12%	12–16%
Coarse, deep sandy soil	2.00 (5.08)	1.50 (3.81)	1.00 (2.54)	0.50 (1.27)
Coarse sandy soils over more compact soil	1.50 (3.81)	1.00 (2.54)	0.75 (1.91)	0.40 (1.02)
Light, deep sandy loam soils	1.00 (2.54)	0.80 (2.03)	0.60 (1.52)	0.40 (1.02)
Light sandy loams over more compact soils	0.75 (1.91)	0.50 (1.27)	0.40 (1.02)	0.30 (0.76)
Deep silt loam soils	0.50 (1.27)	0.40 (1.02)	0.30 (0.76)	0.20 (0.51)
Silt loams over more compact soils	0.3 (0.76)	0.25 (0.64)	0.15 (0.38)	0.10 (0.25)
Heavy-textured clays or clay loam soils	0.15 (0.38)	0.10 (0.25)	0.08 (0.02)	0.06 (0.15)

After USDA-NRCS, 2016.

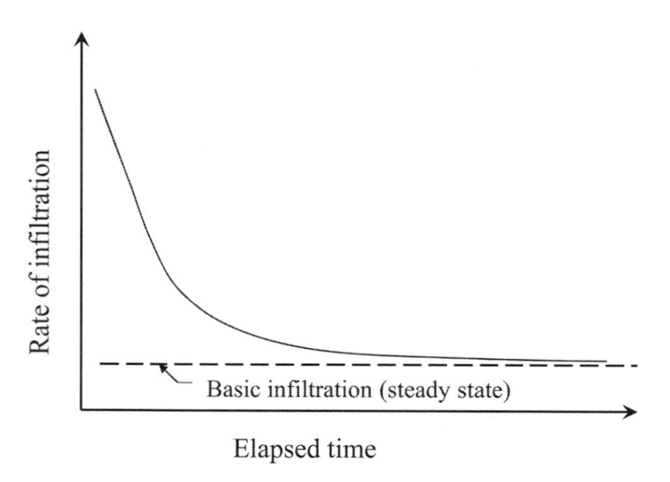

Figure 12.1 Infiltration capacity rate as a function of time.

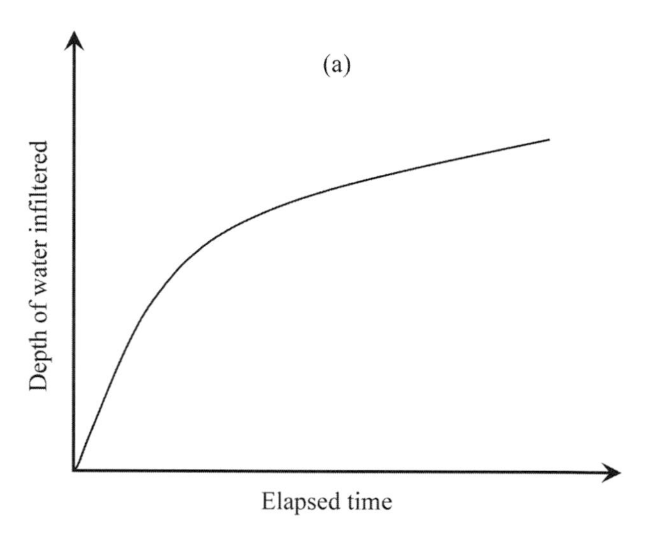

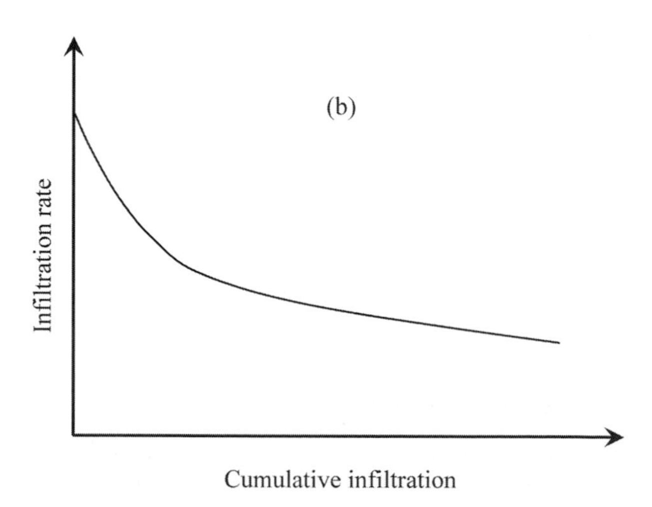

Figure 12.2 Cumulative infiltration as a function of time (a) and infiltration rate and cumulative infiltration (b).

of cumulative infiltration, as shown in Figure 12.2(b). This function can be expressed as

$$i(t) = f[I(t)], \tag{12.3}$$

in which f is a function. The cumulative infiltration curves for different soils, as shown in Figure 12.3, can be used to determine the time for applying irrigation water.

The function f in eq. (12.3) is usually a power-type function, but has also been expressed in exponential form. For simple cases, this function will be examined later. Equation (12.3) can also be regarded as a flux–concentration relation, where $i(t)$ is flux, because it is the volumetric rate of water entering the soil per unit area of the soil, and $I(t)$ is the volume of water per unit area that has entered the soil and is expressed as depth, and can be denoted as concentration.

The average infiltration rate i_a over a time interval $(t - t_0)$ is defined as

$$i_a = \frac{I(t) - I_0}{t - t_0}, \tag{12.4}$$

where t_0 is the beginning of the time interval and t is the end of the interval, and I_0 is the cumulative infiltration up to time t_0. Figure 12.4 shows typical patterns of $i(t)$, $i_a(t)$, and $I(t)$.

After a long time, the infiltration rate becomes quite small and almost approaches a constant value, often referred to as

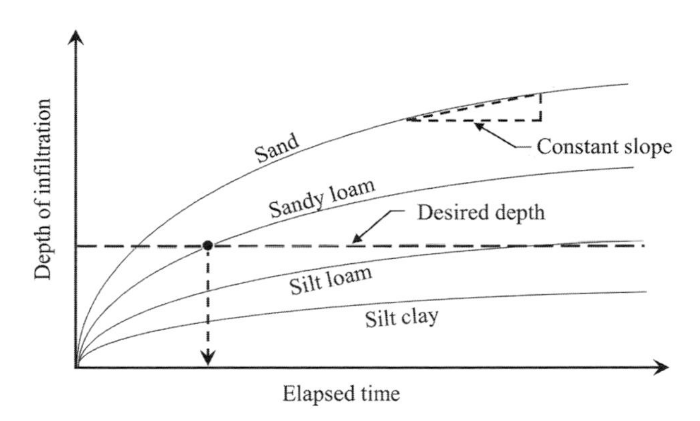

Figure 12.3 Cumulative infiltration curves for different soils.

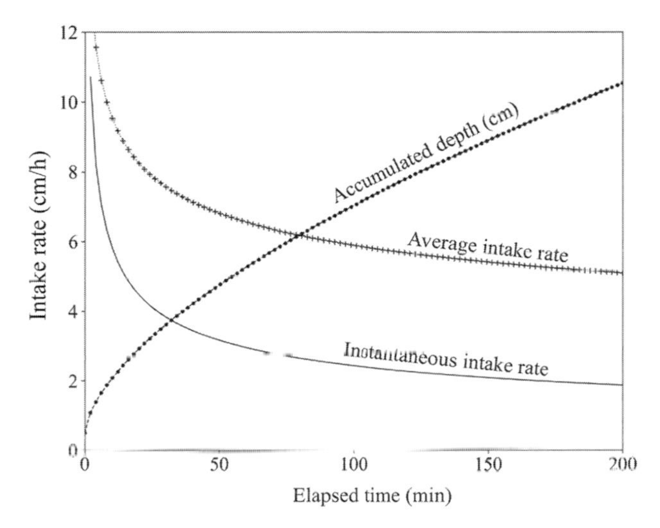

Figure 12.4 Typical patterns for average and instantaneous intake rates and cumulative depth of infiltration.

a steady rate, denoted as i_c, which is the minimum rate. In irrigation, i_c is called the basic infiltration rate. Thus, the rate of infiltration varies between i_0 and i_c. If i_c is regarded as a constant value, then $i(t) - i_c$ is called the excess rate of infiltration – that is, this is the rate in excess of the constant rate of infiltration.

12.3 FACTORS AFFECTING INFILTRATION

The rate of infiltration varies from one type of soil to another type. It varies with the soil structure and texture, root zone, vegetation (or cover) and microbes in the soil, macropores, and antecedent soil moisture. The rate of infiltration for a given soil depends on the initial conditions. Figure 12.5 shows that the rate of infiltration is affected by initial soil moisture (USDA-SCS, 1991). Under the same conditions, cumulative infiltration will depend on the type of soil, as shown in Figure 12.6 (Taylor and Ashcroft, 1972). In practice, infiltration is assumed to depend on the soil type.

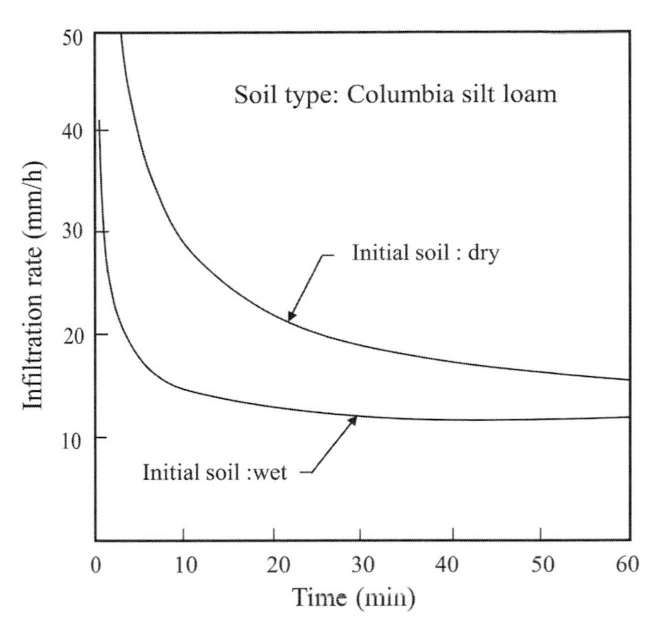

Figure 12.5 Different infiltration rate curves caused by variation in initial water content for Deep Columbia Silt Loam. (Adapted from USDA-SCS, 1991)

12.4 MEASUREMENT OF INFILTRATION

Infiltration should be measured in the field, as laboratory measurements often do not represent field conditions. Methods of field measurement vary with the type of irrigation system. For border or furrow irrigation systems, infiltration is measured with a single- or double-ring infiltrometer, as shown in Figures 12.7 and 12.8. The infiltrometer is filled with water and the depth of water infiltrated at different times is noted. Measurements are commonly taken at 15 minutes and 160 minutes. It is preferable to measure the depth of infiltrated water every 5 or 10 minutes. These measurements are then utilized to estimate parameters of infiltration equations.

For furrow irrigation, infiltration can also be measured using a discharge measuring device, such as a small Parshall flume, placed at two sufficiently distant points in the furrow where discharge is measured. The difference between the discharge values at the two points is considered as the amount of infiltrated water over the length of the points. This, when multiplied by the time of advance, is the volume of water infiltrated over the segment of the furrow. Dividing this volume by the product of furrow spacing and length, the depth of infiltrated water is obtained.

For sprinklers, catch cans are placed at fixed intervals along a line. Then, infiltration rates are measured by a single sprinkler nozzle that is directed to discharge along the line. If no water is observed on the surface at a particular point in the field, then the rate of infiltration at that point is equal to or greater than the rate of application. Then, field conditions are measured and the equivalent depth of water caught in catch cans is measured. From these values, the rate of infiltration is computed.

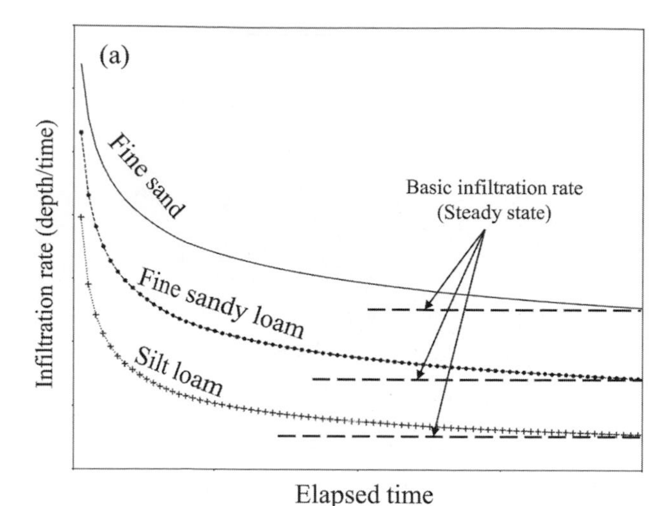

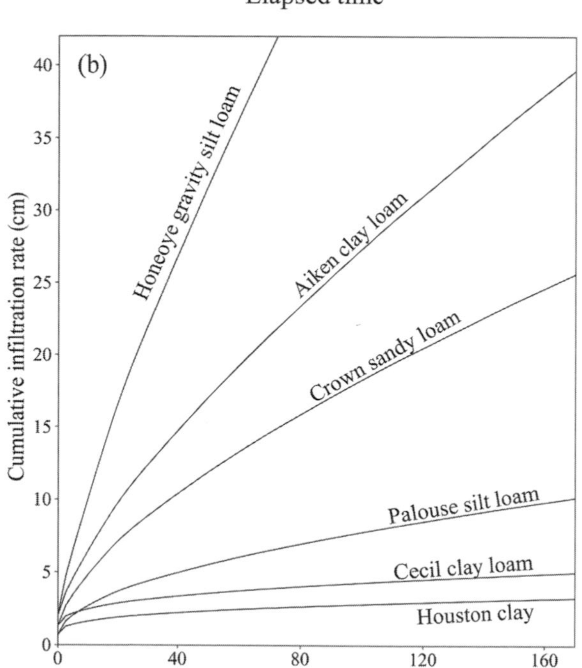

Figure 12.6 Infiltration curves for different soil types (Part (b) data from Taylor and Ashcroft, 1972.)

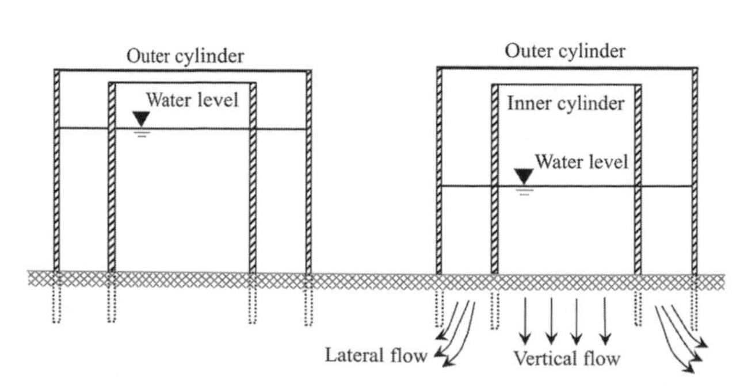

Figure 12.7 Sectional view of a double-ring infiltrometer. View at left is $t = 0$ and view at right is after infiltration has occurred.

12.5 DETERMINATION OF INFILTRATION

There are several equations that can be employed to determine the rate of infiltration for a given soil. Some of these equations are empirical, some quasi-theoretical, and some theoretical. In irrigation engineering, especially in the design of irrigation systems, empirical or quasi-empirical equations are often used. Therefore, only those equations that have been found to be useful in irrigation practice are discussed here. Some of these equations can be derived from the use of mass conservation and a flux law describing the relation between infiltration capacity rate and cumulative infiltration or between flux and concentration.

12.5.1 Mass Conservation

Consider a soil element of unit area as shown in Figure 12.9. This element can be up to the root zone. Let the initial storage space (or pore space) available be denoted as S_0; in other words, this is the space available for infiltrated water to occupy. Let the effective porosity of the soil be denoted as η. If the soil is initially dry, then S_0 will equal the volume of the element multiplied by the effective porosity. In irrigation, the rate at which water is applied is usually greater than the soil infiltration capacity rate, particularly when the method of irrigation is flood irrigation or furrow irrigation. As water is applied to the soil, infiltration starts occurring at a rate $i(t)$. After some time, the soil element becomes saturated and water starts exiting the element at its downstream end. The rate at which water exits is called the seepage rate. This seepage rate can be denoted as $i_c(t)$, but is usually a constant value. Since water is infiltrating the soil, it is occupying pore spaces and is reducing the initially available storage space. Let the potential storage space at any time t be denoted as $S(t)$. In reality, this is equal to the potential cumulative infiltration volume I_p.

The conservation of mass for the soil element in the form of a spatially lumped continuity equation can be expressed as

$$\frac{dS(t)}{dt} = i_c(t) - i(t) \tag{12.5}$$

or

$$S(t) = S_0 + \int_0^t [i_c(t) - i(t)]dt. \tag{12.6}$$

The amount of water stored in the element, $W(t)$, can be expressed as

$$W(t) = S_0 - S(t) = \int_0^t [i(t) - i_c(t)]dt. \tag{12.7}$$

The amount of water stored, $W(t)$, is the same as the amount of water infiltrated, or cumulative infiltration, $I(t)$. Therefore,

$$I(t) = W(t) = S_0 - S(t) = S_0 - I_p(t). \tag{12.8}$$

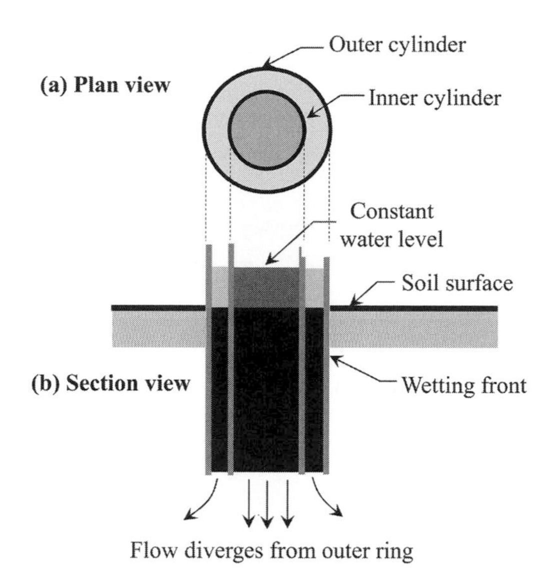

Figure 12.8 Double-ring infiltrometer. A black and white version of this figure will appear in some formats. For the color version, please refer to the plate section.

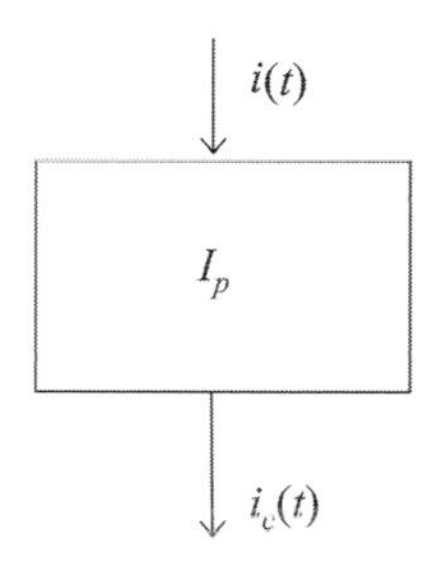

Figure 12.9 A conceptual element representing infiltration.

On farms, part of the soil is occupied by plant roots and therefore the initial space available in the soil for water to occupy is less than what it will be if there were no plant roots in the soil. Consequently, the initial storage space must be multiplied by 1 minus the factor expressed as a fraction (or percentage). Thus, one can write

$$(1-\alpha)S_0 = I(t) + I_p(t) \quad \text{or} \quad \alpha S_0 = I(t) + S(t), \tag{12.9}$$

where α is a percentage factor indicating reduction in pore space; it is about 0.7 for weeds, 0.65 for alfalfa, and 0.3 for crabgrass.

12.5.2 Flux Law

By definition, flux is a volumetric rate per unit area. Thus, for water the volumetric rate of flow per unit area is flux. In the case of infiltration, the capacity rate of infiltration can be regarded as flux. Likewise, one can define energy flux, momentum flux, and volumetric flux. The amount of water infiltrated or cumulative infiltration can be regarded as concentration. It is plausible to formulate a relation between infiltration flux and infiltration concentration. Singh and Yu (1990) proposed a general relation between $i(t)$ and $S(t)$ or $I(t)$ that can be expressed as

$$i(t) - i_c(t) + \frac{\beta[S(t)]^n}{[S_0 - S(t)]^m} - i_c(t) + \frac{\beta\left[I_p(t)\right]^n}{[I(t)]^m}, \tag{12.10}$$

where m, n, and β are positive real constants for a given soil–vegetation land-use complex. From eq. (12.10) it can be seen that $i(t) \to \infty$ as $t \to 0$, and $i(t) \to i_c(t)$ as $t \to \infty$. Different infiltration equations can be derived by using special forms of eq. (12.10).

12.6 INFILTRATION EQUATIONS

There are several infiltration equations developed in soil physics and hydrology, such as the Horton equation, the Philip two-term equation, the Kostiakov equation, the modified Kostiakov equation, the SCS-CN equation, the Green–Ampt equation, the Overton equation, the Holtan equations, and the Singh–Yu generalized equation, to name but a few. Singh and Yu (1990) derived these equations using eq. (12.10) and Singh (2010) derived some of these equations using entropy theory. However, only a few of these have been applied in agricultural irrigation and the more frequently used ones are presented here.

12.6.1 Kostiakov Equation

The Kostiakov (1932) equation can be written as

$$I(t) = at^b \quad \text{or} \quad \log I(t) = \log a + b \log t \tag{12.11}$$

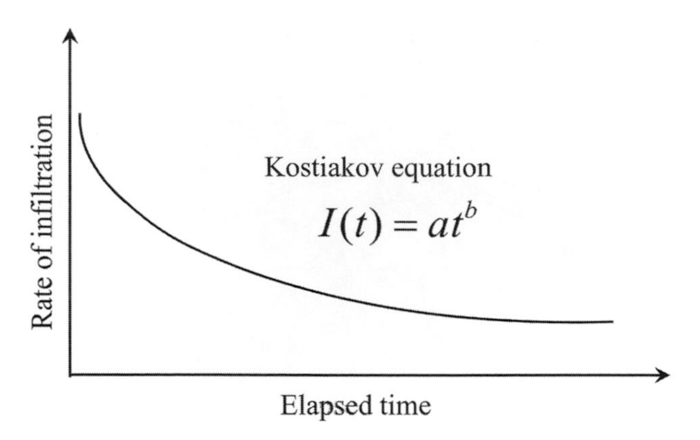

Figure 12.10 Kostiakov equation.

or

$$i(t) = \frac{dI(t)}{dt} = abt^{b-1} \text{ or } \log i(t) = \log(ab) + (b-1)\log t, \quad (12.12)$$

where a is an empirical constant and b is an exponent. The value of parameter a depends on soil characteristics and antecedent soil moisture conditions. Exponent b is about 0.7, but it is better to estimate a and b using observed data. Using entropy theory, Singh (2010) showed that if $b = 0.5$, then $a = \sqrt{2i_c S}$, where S is the maximum soil moisture retention and i_c is the steady-state infiltration rate. Both S and i_c can be obtained for a given soil. Equation (12.11) is an empirical equation and is one of the most frequently used equations in the design of irrigation systems. It seems to represent observed infiltration reasonably well over short periods of time, say a few hours. Figure 12.10 graphs the Kostiakov equation.

Equation (12.11) can be derived using eqs (12.5) and (12.10) with $n = 0$, and $i_c(t) = 0$. Then, eqs (12.5) and (12.10), respectively, become

$$\frac{dS(t)}{dt} = -i(t) \quad (12.13)$$

and

$$i(t) = \frac{\beta}{[S_0 - S(t)]^m} = \frac{\beta}{[I(t)]^m}. \quad (12.14)$$

Equation (12.14) says that the rate of infiltration is inversely proportional to the m power of cumulative infiltration up to that time.

Equation (12.13) can be expressed as

$$\frac{dS(t)}{dt} = \frac{d[S_0 - I(t)]}{dt} = -\frac{dI(t)}{dt} = -i(t). \quad (12.15)$$

Equation (12.14) can be written as

$$\frac{dI(t)}{dt} = \frac{\beta}{[I(t)]^m}. \quad (12.16)$$

Integrating eq. (12.16) with the condition that $I(0) = 0$ at $t = 0$, one gets

$$I(t) = [(m+1)\beta]^{1/(m+1)} t^{1/(m+1)}. \quad (12.17)$$

Differentiating eq. (12.17) and substituting in eq. (12.15), one obtains

$$i(t) = \frac{1}{m+1}[(m+1)\beta]^{1/(m+1)} t^{-m/(m+1)} \quad (12.18a)$$

or

$$i(t) = \beta[(m+1)\beta t]^{-m/(m+1)}. \quad (12.18b)$$

Frequently, the power of t in the Kostiakov equation is taken between 0.5 and 0.7. The power of 0.5 will be obtained if $m = 1$. For this particular case, eq. (12.17) becomes

$$I(t) = \sqrt{2\beta}\, t^{0.5}, \quad (12.19)$$

and eq. (12.18) becomes

$$i(t) = 0.5\,(2\beta)^{0.5}\, t^{-0.5}. \quad (12.20)$$

Parameters a and b are empirical and need to be determined from field measurements for a specific location and time. It is difficult to generalize these parameters for different soils.

Example 12.1: A double-ring infiltrometer test in a soil shows that the amount of cumulative infiltration is 40 mm at the end of one hour and is 60 mm at the end of two hours. If the Kostiakov equation is applied to estimate infiltration for the soil, then what would be the Kostiakov infiltration parameters for this soil? Then, compute the capacity infiltration rate and cumulative infiltration depth at time equal to 90 minutes, 150 minutes, 180 minutes, 210 minutes, and 240 minutes using the Kostiakov equation. Plot the relationship (a) between time and infiltration rate and (b) between time and cumulative infiltration.

Solution: $I(t) = 40$ mm at $t = 60$ min and 60 mm at $t = 120$ min.

From the Kostiakov equation, $I(t) = at^b$ or $\log I(t) = \log a + b \log t$,

$$\log(40) = \log a + b \log(60)$$

and

$$\log(60) = \log a + b \log(120).$$

Solution of these two equations yields: $b = 0.585$ and $a = 3.647$.

Then, the Kostiakov equation for infiltration rate with $i(t)$ in mm/min and t in min is

$$i(t) = abt^{b-1} = 3.647 \times 0.585 \times t^{0.585-1} = 2.133 \times t^{-0.415}.$$

The Kostiakov equation for cumulative infiltration with $I(t)$ in mm and t in min, is

$$I(t) = 3.647 \times t^{0.585}.$$

The infiltration rate and cumulative infiltration depth at time equal to 90 minutes, 150 minutes, 180 minutes, 210 minutes, and 240 minutes are below. The relationships between time and infiltration rate and cumulative infiltration are shown in Figure 12.11.

Time (min)	Infiltration rate (mm/min)	Cumulative infiltration (mm)
90	0.33	50.71
150	0.27	68.37
180	0.25	76.07
210	0.23	83.24
240	0.22	90.00

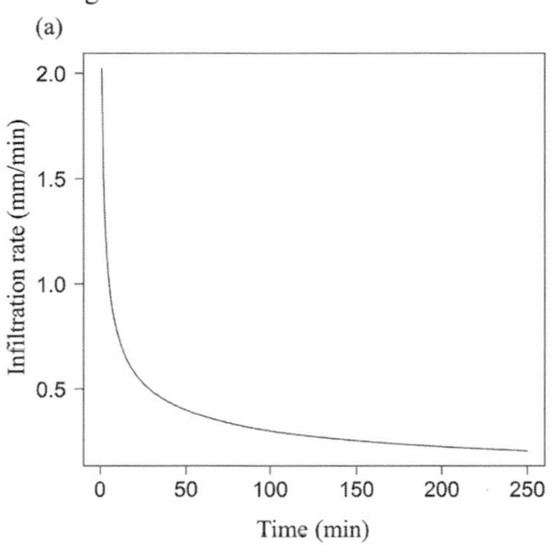

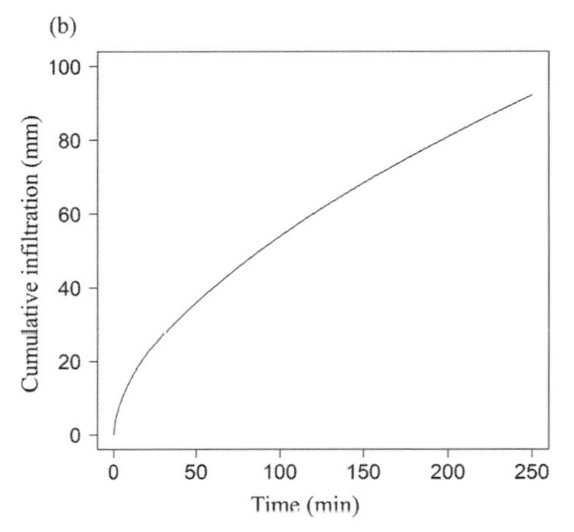

Figure 12.11 Relationships between time and infiltration rate (a) and between time and cumulative infiltration (b).

Example 12.2: A ring infiltrometer test yields the following data:

Time (minutes)	Cumulative infiltration (mm)
1	1.321
15	8.001
30	12.192
60	18.288
150	35.687
200	38.862

Plot the cumulative infiltration data on a log–log paper and fit a straight line. Compute the Kostiakov infiltration parameter. Determine the cumulative infiltration depth at the end of 2, 3, 4, 5, and 6 hours. (Hint: Use "trend line" with the "power" option in Excel.)

Solution: Using "trend line" with the "power" option in Excel, one obtains $a = 1.346$ and $b = 0.6443$, and the equation $I = 1.346t^{0.6443}$ with $R^2 = 0.994$, where t is the time in min and I is the cumulative infiltration in mm. Then, changing the axis to a logarithmic scale, one obtains a straight line, as shown in Figure 12.12.

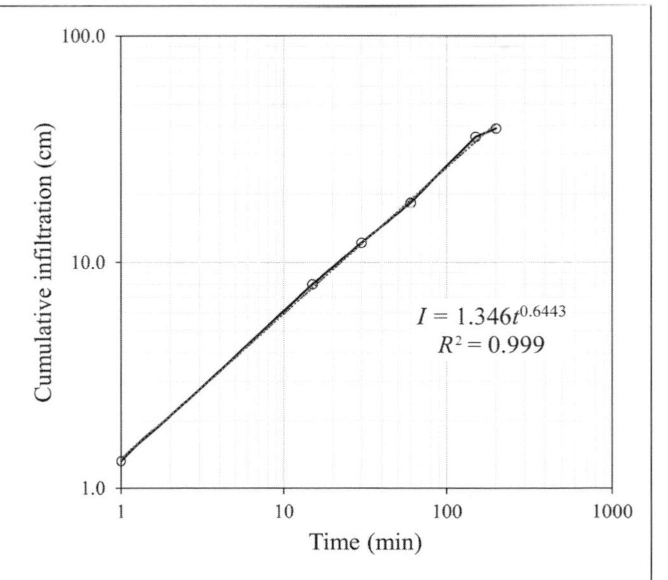

Figure 12.12 Log–log plot of cumulative infiltration as a function of time.

Then, the cumulative infiltration depth at the end of 2, 3, 4, 5, and 6 hours is:

For 120 minutes, $I = 1.346 \times 120^{0.6443} = 29.4$ mm,
For 180 minutes, $I = 1.346 \times 180^{0.6443} = 38.2$ mm,
For 240 minutes, $I = 1.346 \times 240^{0.6443} = 46.0$ mm,
For 300 minutes, $I = 1.346 \times 300^{0.6443} = 53.1$ mm,
For 360 minutes, $I = 1.346 \times 360^{0.6443} = 59.7$ mm.

Example 12.3: For a constant rate of application of 0.25 in./h (0.635 cm/h) for 1.5 h, what will be the depth of water infiltrated using the Kostiakov equation ($i = 0.75t^{-0.35}$; t [min], i [in./h])? Compute the amount of water that may run off the surface.

Solution: First, compute the time when the infiltration rate intersects the application rate of 0.25 in./h:

$$0.25 = 0.75(t_s)^{-0.35}.$$

This yields $t_s = 23.07$ min – 0.385 h.

The total application time is 1.5 h = 90 min, then water may run off the surface when $t > 0.385$ h.

Second, when the infiltration rate ≥ application rate ($t \leq 0.385$ h), all the water is infiltrated and there is no runoff. Then the amount of infiltration during this time is equal to the total application – that is, the area of A_1, as shown in Figure 12.13.

Compute the area A_1 under the application rate (0.25 in./h) from 0 to the time $t_s = 0.385$ h:

$$A_1 = 0.25\frac{\text{in.}}{\text{h}} \times 0.385\ \text{h} = 0.096\ \text{in.}\ (0.244\ \text{cm}).$$

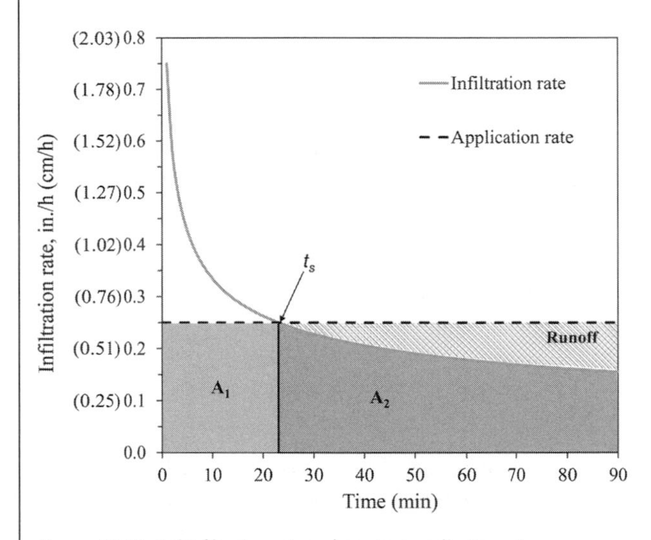

Figure 12.13 Soil infiltration rate and constant application rate.

Third, when the infiltration rate < application rate (0.385 h < $t \leq$ 1.5 h), runoff will be generated and the runoff amount will be equal to the application amount minus the infiltration amount.

The Kostiakov infiltration rate equation is $i = 0.75t^{-0.35}$, with t in min and i in in./h. To obtain the cumulative infiltration in inches, one should change the unit of i into in./min; then, one obtains

$$i = (0.75/60)t^{-0.35} = 0.0125t^{-0.35}.$$

Given $b - 1 = -0.35$ and $a \times b = 0.0125$, this yields $a = 0.01923$, $b = 0.65$. The cumulative infiltration equation $I = at^b$ can be obtained using eqs (12.11) and (12.12):

$$I = at^b = 0.01923t^{0.65}.$$

The cumulative infiltration during 23.07 min < $t \leq$ 90 min, i.e., the area of A_2, is

$$I = 0.01923(90)^{0.65} - 0.01923(23.07)^{0.65}$$
$$= 0.21\ \text{in.}(0.533\ \text{cm}).$$

Another method to compute area A_2 under the infiltration curve from t_s to the time of application ($t_{max} = 1.5$ h) is to integrate the infiltration rate equation. This is given as follows:

$$A_2 = \int_{t_s}^{t_{max}} 0.0125t^{-0.35}dt$$
$$= \frac{1}{-0.35+1} \times 0.0125 \times \left(t_{max}^{-0.35+1} - t_s^{-0.35+1}\right)$$
$$= 0.01923 \times \left(90^{0.65} - 23.07^{0.65}\right)$$
$$= 0.210\ \text{in.}(0.533\ \text{cm}).$$

The total water infiltrated is

$$A_1 + A_2 = 0.096\ \text{in.} + 0.210\ \text{in.} = 0.306\ \text{in}\ (0.777\ \text{cm}).$$

The total water applied is

$$0.25\frac{\text{in.}}{\text{h}} \times 1.5\ \text{h} = 0.375\ \text{in.}\ (0.953\ \text{cm}).$$

The total runoff is

$$0.375\ \text{in.} - 0.306\ \text{in.} = 0.069\ \text{in.}(0.175\ \text{cm}).$$

12.6.2 Soil Conservation Service Equation

The infiltration rate given by the Kostiakov equation approaches zero after a long time. Since zero infiltration rate is often not observed even if water is applied to soils for a prolonged period, the Soil Conservation Service (SCS), now called the Natural Resources Conservation Service (NRCS), of the US Department of Agriculture (USDA) developed an equation that is a slightly modified form of the Kostiakov equation, expressed as

$$I(t) = at^b + c \quad \text{or} \quad \log\left[I(t) - c\right] = \log a + b \log t, \qquad (12.21)$$

$$i(t) = \frac{dI(t)}{dt} = abt^{b-1} \quad \text{or} \quad \log I(t) = \log(ba) + (b-1)\log t, \qquad (12.22)$$

in which parameter a and exponent b are expressed as functions of the intake family, and parameter c has a value of 0.275 if I is in inches and a value of 0.6985 if I is in centimeters. Figure 12.14 plots the SCS infiltration equation.

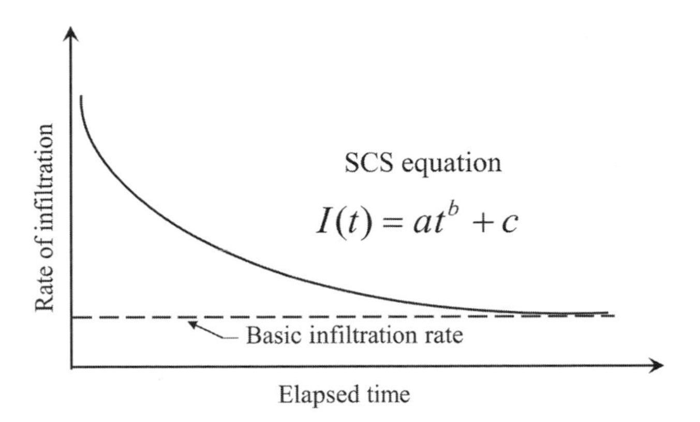

Figure 12.14 SCS infiltration equation.

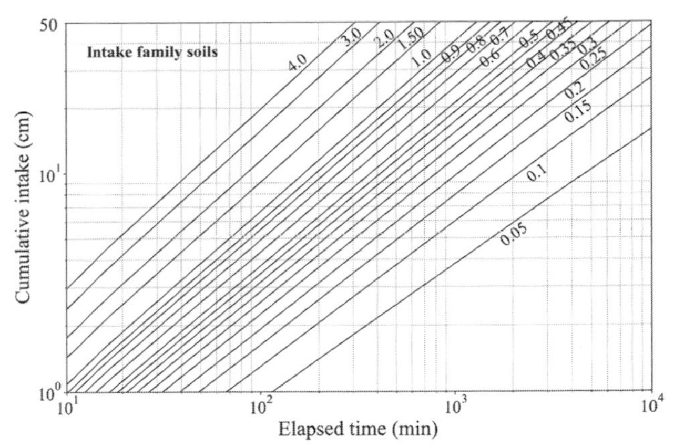

Figure 12.15 Intake families relation between cumulative intake and time on a logarithmic scale (drawn using parameters in Table 12.2).

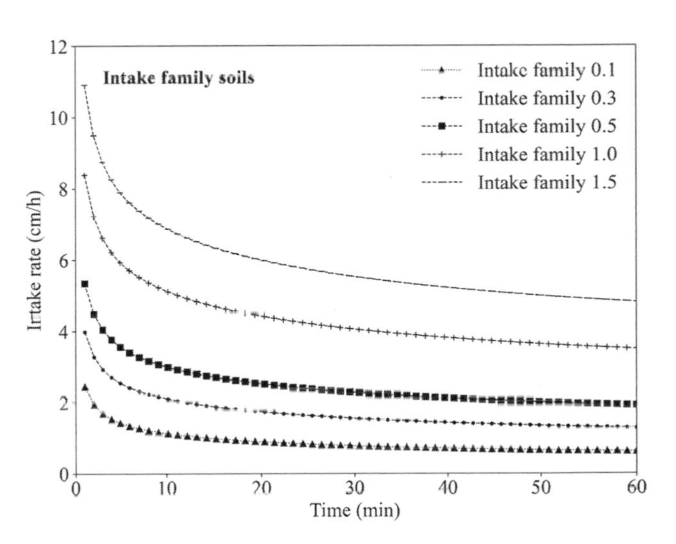

Figure 12.16 Intake families relation between intake rate and time (drawn using parameters in Table 12.2).

The concept of intake family entails the concept of long-term intake rate, which is defined as the point on a Kostiakov-type infiltration curve where the infiltration rate decreases by 5% within 1 h. This can be expressed mathematically as:

$$\frac{di(t)}{dt} = \frac{d^2I(t)}{dt^2} = \frac{0.05}{60\text{ min}}\frac{dI(t)}{dt} = \frac{0.05}{60\text{ min}}i(t). \tag{12.23}$$

Let t_L define the time when the long-term intake rate occurs and i_f define the number of intake family which is equal to the long-term intake rate in the units of inches per hour. Then, i_f can be expressed as

$$i_f = 60 \times i(t) = 60 \times \frac{dI(t)}{dt}. \tag{12.24}$$

Now, t_L can be derived by substituting eq. (12.22) in eq. (12.24) as

$$i_f = 60 \times i(t) = 60 \times ab(t_L)^{b-1}. \tag{12.25}$$

Thus, i_f is equal to the intake rate in inches per hour at time equal to t_L. One can also derive t_L by substituting eq. (12.22) and (12.24) in eq. (12.23) as

$$0.05i_f = 60 \times 60 \times \frac{d^2I}{dt^2} = 60 \times 60 \times ab(b-1)(t_L)^{b-2}. \tag{12.26}$$

The SCS has developed intake families, as shown in Figures 12.15 and 12.16, and has given parameters a and b for each intake family, as shown in Table 12.2. Although each soil has its own intake characteristics, the differences between some soil types are very small and thus can be considered in the same group. The intake family represents the long-term intake rate of different soil groups. For example, the assigned group numbers, such as 0.05, 0.1, and 0.5, represent the long-term intake rates in the unit of inches per hour. It may be noted that the values of parameters a and b in Table 12.2 are based on the assumption that the infiltration begins when about 50% of the soil moisture as the difference between field capacity and crop-extractable water has been depleted. This assumption of 50% moisture depletion is reasonable, because irrigation is typically done when about 50% of the soil moisture has been used up by plants. This assumption is also considered when designing and operating irrigation systems. If the antecedent soil moisture is significantly different from this 50% value, then it is understood that the shape of infiltration curve and the magnitude of infiltration will significantly change. The relationship between long-term intake rate and intake family is useful for designing surface irrigation systems, including border, furrow, or sprinkler. The SCS method is based on a large number of field trials and is useful for designing and operating irrigation systems.

The SCS infiltration equation can be derived using eqs (12.5) and (12.10) as follows. Let $n = 0$ in eq. (12.10).

Table 12.2 *Parameters for calculation of cumulative infiltration using the SCS intake family concept*

Intake family	*a* (cm)	*a* (in.)	*b*
0.05	0.0533	0.0210	0.6180
0.1	0.0620	0.0244	0.6610
0.15	0.0701	0.0276	0.6834
0.2	0.0771	0.0306	0.6988
0.25	0.0853	0.0336	0.7107
0.3	0.0925	0.0364	0.7204
0.35	0.0996	0.0392	0.7285
0.4	0.1064	0.0419	0.7356
0.45	0.1130	0.0445	0.7419
0.5	0.1196	0.0471	0.7475
0.6	0.1321	0.0520	0.7572
0.7	0.1443	0.0568	0.7656
0.8	0.1560	0.0614	0.7728
0.9	0.1674	0.0659	0.7792
1.0	0.1786	0.0703	0.7850
1.5	0.2283	0.0899	0.7990
2.0	0.2753	0.1084	0.8080
3.0	0.3650	0.1437	0.8160
4.0	0.4445	0.1750	0.8230

After USDA-NRCS, 2012.

Further, let the cumulative infiltration be replaced by excess cumulative infiltration. Then,

$$i(t) - i_c(t) = i_e(t) = \frac{dI_e(t)}{dt} = \frac{\beta}{[I_e(t)]^m} = \frac{\beta}{[I(t) - i_c t]^m}. \tag{12.27}$$

Equation (12.27) hypothesizes that the excess rate of infiltration is inversely proportional to some power of excess cumulative infiltration. Integration of eq. (12.27) with the condition, at $t = 0$, $I_e(t) = 0$, yields

$$I_e(t) = [(m+1)\beta]^{1/(m+1)} t^{1/(m+1)} \tag{12.28}$$

or

$$I(t) = i_c t + [(m+1)\beta]^{1/(m+1)} t^{1/(m+1)}. \tag{12.29}$$

Differentiation of eq. (12.29) yields

$$i(t) = i_c + \frac{1}{(m+1)}[(m+1)\beta]^{1/(m+1)} t^{-m/(m+1)}, \tag{12.30}$$

which is the same as eq. (12.21).

Example 12.4: Consider three intake families corresponding to long-term intake rates of 0.3, 0.6, and 1.0 inches per hour. Compute the time to the long-term infiltration rate in minutes and cumulative depth of infiltration in mm.

Solution: From Table 12.2,

for intake family of 0.3 in./h, $a = 0.0364$ and $b = 0.7204$;
for intake family of 0.6 in./h, $a = 0.0520$ and $b = 0.7572$;
for intake family of 1.0 in./h, $a = 0.0703$ and $b = 0.7850$;

and $c = 0.275$ (constant, when the unit is inches).

Given $i_f = 0.3$ in./h, the time to the long-term infiltration rate in minutes can be computed using eq. (12.25):

$$i_f = 60 \times i(t) = 60 \times ab(t_L)^{b-1},$$

$$0.3 = 60 \times 0.0364 \times 0.7204(t_L)^{0.7204-1},$$

which yields $t_L = 375.02$ min.

The cumulative depth of infiltration in mm can be computed using eq. (12.21):

$$I(t) = at^b + c = 0.0364 \times (375.02)^{0.7204} + 0.275$$
$$= 2.88 \text{ in. or } 73.10 \text{ mm.}$$

Similarly, for $i_f = 0.6$ and 1.0 in./h and using the corresponding values of *a* and *b*, the following is obtained:

for $i_f = 0.6$ in./h, $t_L = 282.8$ min, and $I(t) = 4.01$ in. or 101.9 mm;
for, $i_f = 1.0$ in./h, $t_L = 262.1$ min and $I(t) = 5.84$ in. or 148.3 mm.

12.6.3 Philip Equation

From theoretical analysis of one-dimensional vertical infiltration into uniform soils, Philip (1957) derived an equation whose form is similar to the Kostiakov or SCS equation, expressed as

$$I(t) = at + bt^{0.5} \tag{12.31a}$$

or

$$i(t) = 0.5bt^{-0.5} + a, \tag{12.31b}$$

where *b* is the sorptivity and *a* is the conductivity parameter. Both of these parameters have physical meaning. Sorptivity dominates the infiltration process during the early stages of infiltration, whereas conductivity dominates during the later stages. The conductivity parameter is approximated by the saturated hydraulic conductivity – sometimes it is taken as 0.5–0.7 times the saturated hydraulic conductivity. Equation (12.31b) can be derived in the same way as the SCS infiltration shown above, where $m = 1$ in eq. (12.5). Singh (2010) showed that parameter *a* is analogous to steady infiltration and parameter $b = (2aS)^{0.5}$.

12.6.4 Horton Equation

Hypothesizing infiltration as an exhaustion process in which the rate of performing work is proportional to the amount of work remaining to be performed, Horton (1939) derived an infiltration equation. Horton recognized that the infiltration capacity rate $i(t)$ decreased from its initial rate i_0 with time and reached a steady or constant rate denoted as i_c. Hence,

$$\frac{di}{dt} = -k(i - i_c), \tag{12.32}$$

in which k is the proportionality factor dependent on soil type and initial soil moisture content. Using the condition that $i = i_0$ at $t = 0$, integration of eq. (12.32) yields

$$i(t) = i_c + (i_0 - i_c)\exp(-kt). \tag{12.33}$$

The Horton equation can also be expressed in terms of cumulative infiltration $I(t)$ as

$$I(t) = i_c t + \left(\frac{1}{k}\right)(i_0 - i_c)[1 - \exp(-kt)]. \tag{12.34}$$

Equation (12.33) can also be derived from eq. (12.10) by taking $n = 1$, $m = 0$, i_c is replaced with i_0, and $S(t)$ is replaced with $-(I - i_c t)$:

$$i(t) - i_c = \frac{dI(t)}{dt} - i_c = -a(I - i_c t). \tag{12.35}$$

Solution of eq. (12.35) can be expressed as

$$I(t) = i_c t + \left(\frac{1}{a}\right)(i_0 - i_c)[1 - \exp(-at)] \tag{12.36}$$

Differentiation of eq. (12.36) yields

$$i(t) = i_c + (i_0 - i_c)\exp(-at). \tag{12.37}$$

In eqs (12.36) and (12.37), parameter a is equal to parameter k in eq. (12.32). Using entropy theory, Singh (2010) derived a or k as

$$k = a = \frac{(i_0 - i_c)}{S}. \tag{12.38}$$

12.6.5 Green–Ampt Equation

Using the Darcy equation and piston concept, Green and Ampt (1911) derived an infiltration equation that can be expressed in implicit form as

$$t = \frac{1}{i_c}\left[I - \frac{a}{i_c}\left(\ln\left(1 + \frac{Ii}{a}\right)\right)\right]. \tag{12.39}$$

Equation (12.39) can also be derived from eq. (12.10) by taking $n = 0$, $m = 1$, and replacing S^{-1} by I^{-1}:

$$i(t) - i_c = aI^{-1}. \tag{12.40}$$

In eq. (12.39), i_c can be interpreted as almost equal to the saturated hydraulic conductivity. From entropy theory (Singh, 2010), parameter $a = Si_c$ can be obtained from observations.

The Green–Ampt equation can also be expressed in the form

$$I(t) = K_e t + S_{av} M \ln\left[1 + \frac{I(t)}{MS_{av}}\right], \tag{12.41}$$

where K_e is the effective hydraulic conductivity, M is fillable porosity, and S_{av} is the average matric suction at the wetting front. The fillable porosity is the difference between initial and final water contents. The parameters of the Green–Ampt equation can be determined from measured soil properties (Rawls and Brakensiek, 1989).

QUESTIONS

Q.12.1 The following data are obtained from a double-ring infiltrometer test on the sandy soil.

Time, t (min)	Infiltration, I, (inches [cm])
1	0.54 (1.37)
15	3.17 (8.05)
30	4.84 (12.29)
60	7.23 (18.36)
150	13.08 (33.22)
200	15.32 (38.91)

Here, I is cumulative infiltration in inches. Do the following:

a. Determine the observed rate of infiltration and plot it as a function of time.
b. Plot the data on log–log graph paper with the time axis as the abscissa. Determine the approximate best-fit straight line through the data.
c. Use the Kostiakov equation, $I = at^b$ and fit it to the data. Determine the values of a and b for the best-fit line.
d. Determine the rate of infiltration using the Kostiakov equation and compare with the observed rate.
e. Using the computed values of a and b, determine the depth of infiltration after 6 h.
f. Plot the observed rate of infiltration against cumulative infiltration and fit a curve.

Q.12.2 Given a soil having an infiltration rate described by the soil intake family of 0.5, calculate the soil infiltration rate (in./h) and accumulative water infiltrated (in.) for the following time periods: 1 h, 2 h, and 5 h.

Q.12.3 Given a soil having an infiltration rate described by the soil intake family of 1.5, and a constant application rate of 1.5 in./h for 5 h, calculate the following:

a. total depth of water applied;
b. depth of water infiltrated; and
c. depth of water runoff.

Q.12.4 Given a soil having an infiltration rate described by the soil intake family of 0.3 and the center-pivot characteristics pivot length = 1320 feet, radius of sprinklers = 20 feet, depth applied = 1.0 inch, and system flow rate = 900 gpm, calculate the following:

a. depth of water infiltrated (inches),
b. depth of water runoff (inches); and
c. percent of runoff.

Q.12.5 Solve Q.12.4 for the soil intake family of 1.0:

a. depth of water infiltrated (inches);
b. depth of water runoff (inches); and
c. percent of runoff.

Q.12.6 Data on infiltration in field soils have been published in a report by the Agriculture Research Service of the USDA. Two datasets on infiltration in Robertsdale loamy sand in the Georgia Coastal Plain are given in Tables 12.3 and 12.4.

Plot the data on infiltration capacity rate and cumulative infiltration on regular graph paper and plot curves through the data. Plot the data on infiltration capacity rate and cumulative infiltration on log–log graph paper and plot the best-fit straight lines through the data. Express the best-fit straight line equations and show the parameter values. These best-fit straight line equations represent the Kostiakov equation. Determine the square of the coefficient of correlation and comment on the goodness of fit.

Q.12.7 For the data in Tables 12.3 and 12.4, determine the time when the infiltration capacity rate becomes 2 cm/h, 1.5 cm/h, and 1.0 cm/h.

Q.12.8 Plot the infiltration rate data versus the cumulative infiltration from Tables 12.3 and 12.4 and see if there is a relation that can be expressed mathematically.

Q.12.9 Compute the rate of change in infiltration rate as a function of time from Tables 12.3 and 12.4, and plot it on graph paper and see if it follows any relation. Determine the time when the rate of change in infiltration is about 5% within 1 h.

Q.12.10 Assume the intake family long-term infiltration rate is 1.5 in./h. Determine the value of time when the long-term intake rate decreases by 5% on the Kostiakov curve. Compute the cumulative infiltration at this time from Tables 12.3 and 12.4. Now assume the intake family long-term infiltration rate is 1 in./h. Then, determine the value of time when the long-term intake rate decreases by 5% on the Kostiakov curve and compute the cumulative infiltration at this time.

Q.12.11 Consider the cumulative infiltration data at $t = 25$ and $t = 100$ min from Tables 12.3 and 12.4.

Table 12.3 *Observations on infiltration in the Robertsdale loamy sand (ID = 09091D)*

Time from start of water application (min)	Infiltration rate (cm/h)	Accumulated infiltration (cm)
4	12.21	0.81
5	8.55	0.96
10	5.81	1.60
15	4.89	2.03
20	4.78	2.44
25	4.27	2.82
30	3.86	3.16
35	3.51	3.48
40	3.21	3.75
45	3.23	4.03
50	3.10	4.28
55	3.18	4.56
60	3.21	4.82
65	3.36	5.10
70	3.18	5.34
75	3.31	5.62
80	3.16	5.88
85	3.11	6.14
90	2.89	6.38
95	2.51	6.58
100	2.60	6.82
105	2.68	7.04
110	2.31	7.21
115	2.44	7.43
120	2.42	7.61

After Rawls et al., 1976.

Table 12.4 *Observations for Robertsdale loamy sand (ID = 09091W)*

Time from start of water application (min)	Infiltration rate (cm/h)	Accumulated infiltration (cm)
4	8.24	0.55
5	5.49	0.67
10	2.83	0.99
15	2.60	1.18
20	2.22	1.38
25	2.04	1.56
30	1.20	1.73
35	2.02	1.90
40	1.98	2.06
45	2.03	2.24
50	1.93	2.40
55	1.97	2.57
60	2.00	2.73
65	2.04	2.90
70	2.13	3.08
75	2.20	3.26
80	2.21	3.43
85	2.09	3.60
90	2.18	3.78
95	2.43	3.94
100	2.38	4.16
105	2.32	4.34
110	2.47	4.54
115	2.33	4.72
120	2.25	4.90

After Rawls et al., 1976.

Determine the parameters of the Kostiakov equation. Then, compute the rate of infiltration and cumulative infiltration at time equal to 150 minutes.

REFERENCES

Green, W. H., and Ampt, G. A. (1911). Studies on soil physics. *Journal of Agricultural Science*, 4(1): 1–24.

Horton, R. E. (1939). Analysis of runoff plot experiments with varying infiltration capacities. *Transactions of American Geophysical Union*, 20(IV): 693–669.

Kostiakov, A. N. (1932). On the dynamics of coefficient of water percolation in soils and of the necessity of studying it from a dynamic point of view for purposes of amelioration. *Transactions of the Sixth Commission of the International Soil Science Society*, A: 17–29. [In Russian.]

Philip, J. R. (1957). The theory of infiltration: 4. Sorptivity and algebraic equations. *Soil Science*, 84: 257–265.

Rawls, W. J. and Brakensiek, D. I. (1989). Estimation of soil water retention and hydraulic properties. In *Unsaturated Flow in Hydraulic Modeling*, edited by H. J. Morel-Seytous. London: Kluwer Academic Press, pp. 275–300.

Rawls, W., Yates, P. and Asmussen, L. (1976). *Calibration of Selected Infiltration Equations for the Georgia Coastal Plain, Rep. USDA-ARS-S113*. Beltsville, MD: Agricultural Research Service, USDA.

Singh, V. P. (2010). Entropy theory for derivation of infiltration equations. *Water Resources Research*, 46: 3527.

Singh, V. P., and Yu, F. X. (1990). Derivation of infiltration equation using systems approach. *Journal of Irrigation and Drainage Engineering*, 116(6): 837–858.

Taylor, S. A., and Ashcroft, G. L. (1972). *Physical Edaphology*. San Francisco, CA: W.H. Freeman and Company.

USDA-NRCS (2012). Surface irrigation. In *National Engineering Handbook*. Washington, DC: USDA.

USDA-NRCS (2016). Sprinkler irrigation. In *National Engineering Handbook*. Washington, DC: USDA.

USDA-SCS (1991). Soil–plant–water relationships. In *National Engineering Handbook*. Washington, DC: USDA.

13 Soil Water

Notation

AWC	available water capacity or available water holding capacity (in./in. or m/m)
D	capillary diameter (cm)
D_s	thickness of the soil layer (m)
DW_1	depth of soil water content for one meter of soil (mm)
f_d	fraction of available water depleted (fraction)
f_r	fraction of available water remaining (fraction)
F_c	capillary force (N)
F_g	gravitational force (N)
h	capillary rise (cm)
h_s	height of soil water (cm)
K	hydraulic conductivity (in./h, mm/h)
K_s	saturated hydraulic conductivity (in./h, mm/h)
M_a	mass of air (kg)
MAD	management allowed depletion (fraction or percentage)
M_s	mass of dry soil (kg)
M_{sw}	soil mass before oven drying (kg)
M_t	total mass of soil (kg)
M_w	mass of water (kg)
N	porosity of soil (fraction or percentage)
Q	flow in the soil (cfs, m^3/s)
RAW	readily available water (in. or mm)
SWD	soil water deficit, the depth of water required to fill a soil layer to field capacity (in. or mm)
TAW	total available soil water (in. or mm)
u	flow velocity in the soil (m/s, ft/s)
V_a	volume of air (m^3)
V_b	bulk volume of soil (m^3)
V_p	volume of pore spaces (m^3)
V_s	volume of solids (m^3)
V_w	volume of water (m^3)
z	gravitational potential of soil water (m)
γ	specific weight of water (9.807 kN/m^3 or 62.43 lb/ft^3)
θ	water content of soil (fraction or percentage)
θ_{fc}	field capacity of the soil (in./in. or m/m)
θ_m	water content of soil on a mass basis (fraction or percentage)
θ_{ms}	saturated water content on a mass basis (fraction or percentage)
θ_v	water content of soil on a volume basis (fraction or percentage)
θ_{vs}	saturated water content on a volume basis (fraction or percentage)
θ_{wp}	permanent wilting point of soil (in./in. or m/m)
v_r	void ratio of a soil (fraction)
ρ_b	bulk density of soil (kg/m^3)
ρ_s	soil particle density (kg/m^3)
ρ_w	density of water (kg/m^3)
σ	surface tension (bar, pascal, atm, m of water)
Ψ	soil water potential (m of water, kPa, bar, psi, $lb/in.^2$)

13.1 INTRODUCTION

Soil, water, plant, and atmosphere constitute a continuum in which water occupies center stage, for the plant needs water for its survival and growth. The soil stores this water and acts as a reservoir. The water stored in the soil is called soil water. The atmosphere supplies the energy the plant needs in order to withdraw the soil water. The plant transpires the water through the process called transpiration in the atmosphere. Thus, soil, water, plant, and atmosphere are related and their relationship, called the soil–water–plant–atmosphere relationship, is fundamental for agricultural irrigation. Relevant aspects of soil are discussed in Chapter 6 under soils and soil management, and those of atmosphere in Chapter 2 on climate. Some aspects of plants are discussed in Chapter 7 under crops and crop production. Relevant aspects of the connection between soil and plant and the atmosphere are presented in Chapter 14 under evapotranspiration, and in Chapter 15 under crop water use. This chapter focuses on soil water.

Soil water refers to the water in the unsaturated zone, called the vadose zone, or in the soil. It is utilized for crop production. This water is sometimes referred to as soil moisture and can occur in all three phases: gaseous, liquid, and solid. It is dynamic-stored in the soil, and moves around and leaves the soil. When in liquid form it contains dissolved salts, plant nutrients, and other materials. It dissolves and carries nutrients to plant roots. It dissolves air and uses carbon dioxide to form a weak acid that dissolves rocks.

With the decay of plant tissues, it forms organic acids that also dissolve rocks. Characteristics of this water play a fundamental role in the design and operation of agricultural farming and irrigation systems, because they define the availability of water to plants.

Since soil water occurs in the unsaturated zone, it is the only source of water and nutrients for most agricultural crops. Because nutrients, including fertilizers, are dissolved in soil water, it is the only source from which plants can extract them. The objective of irrigation is to maintain enough soil water in the soil. The objective of this chapter is to briefly discuss those characteristics of soil water that are important in irrigation engineering.

13.2 SOIL WATER IN GASEOUS PHASE

The moisture in the gaseous phase in the soil plays an important role in plant development. First, it moves in soil pores and can leave the soil with ease. It is reasonable to assume that the relative humidity (RH) is 100% in the atmosphere within the soil matrix. The soil begins to dry at the top through diffusion because of the difference in humidity between the soil and the air above it that has lower RH. The rate of conversion of moisture into vapor is accelerated by heating from sun and wind.

At night, as the soil at the top begins to cool, some of the vapor may condense into liquid, as dew. Deeper layers of soil are not as cool and it is possible that some of the moisture there may evaporate. If the night is sufficiently cool, the vapor from deeper layers will condense near the surface and will be available for evaporation during the following day. This suggests that soil can lose moisture from layers as deep as 0.7 m by this upward movement, then condensation, and then evaporation.

13.3 SOIL WETNESS

On application, water enters the soil through infiltration and begins to start filling the pores, as shown in Figure 13.1; the soil becomes water-saturated when its pores are filled with water, which is then characterized as the maximum water retention capacity. The matric potential at this capacity is near zero and the volumetric water content is the same as porosity. The soil will remain at this capacity as long as infiltration of water continues, and the excess water will drain as gravitational water under gravity. If the rate of infiltration is greater than gravity drainage, the soil may become supersaturated. Figure 13.2 shows how the wetting of a dry soil progresses with time as water is applied to the soil.

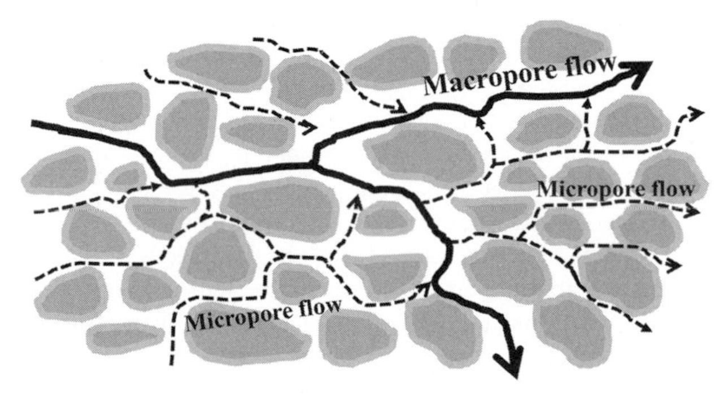

Figure 13.1 Flow of water in soil.

13.4 TYPES OF SOIL WATER

Consider a vertical soil column and assume that the soil is saturated. The soil water in the column, as shown in Figure 13.3, can be identified to be of three types: gravitational water, capillary water, and hygroscopic water. Gravitational water is in excess of field capacity of the soil and is the water in the soil that is rapidly drained under the force of gravity. The water drains from large pores which are then filled with air. The term "gravitational" is derived from the association of this water with the gravitational force. This is also called free water and is not influenced by any forces due to soil particles or pores in the soil matrix. The term "rapid" normally implies the duration of 1–2 days. Gravitational water corresponds to the water that is drained between the states of maximum retentive capacity and field

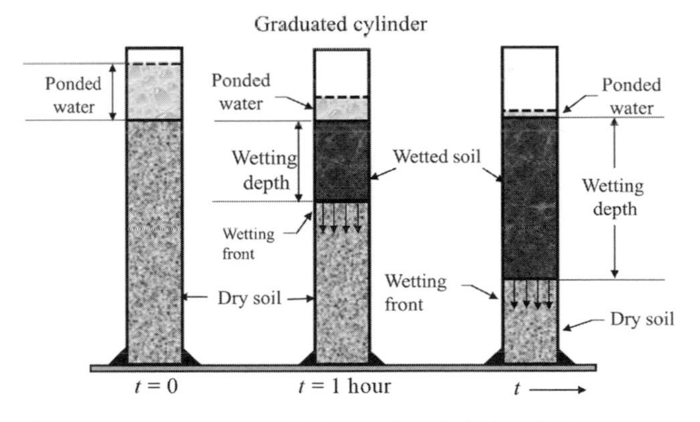

Figure 13.2 Progression of wetting in a graduated cylinder with time.

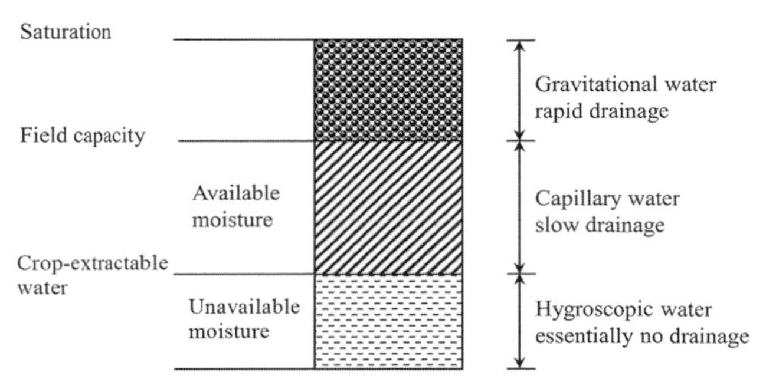

Figure 13.3 Classes of soil, water availability to plants, and characteristics of drainage.

capacity into deeper layers. Most soil leaching occurs as gravitational water and therefore much of this water transports chemicals, such as nutrient ions, pesticides, and organic contaminants that eventually find their way to streams, rivers, and groundwater.

The infiltration of water into the soil is caused by capillarity and gravity. Capillarity is the attraction of water due to matric potential that pulls water into the pores. The capillary force pulls water horizontally and vertically. Gravity is the force that pulls water into and through the soil pores. The force pulls the water vertically through the soil.

Capillary water, also called loosely held water, is the water that is available to plant roots for extraction. The smaller interconnecting pores that hold this water behave like capillaries, hence the name capillary water. The water is stored in the pores and is governed by capillary forces. Capillary water is not free to move, but is also not tightly held, meaning it can be moved by some external force or suction. It remains in the soil micropores after the drainage of macropores by gravity, and drains slowly under capillary forces. The forces under which this water is removed by plant roots are greater than the force of gravity. That is, plants can extract this water from soil by suction, and this water is the water that plants use to meet their needs. This explains why capillary water is also referred to as plant-available water or simply available water. Micropores that hold the capillary water are small enough not to allow the flow under gravity, but large enough to permit capillary flow under matric potential gradients. The water moves under capillary forces, but at a much slower pace.

As the soil loses water by evaporation and extraction by plant roots, it dries and reaches a stage at which the water left forms a thin film around soil particles, which is only a few molecules thick and is tightly held under adhesion. This water is called hygroscopic water or tightly held water – that is, water that adheres to soil particles. The force of adhesion is so great that this water cannot be removed by plant roots. Of course, it can be removed by oven drying. Thus, capillary water falls between gravitational water and hygroscopic water.

These three types of soil water can be illustrated by considering an analogy with a sponge that is submerged in water such that it contains no air bubbles. On lifting the sponge, the excess water drains out quickly, and then the drainage ceases in a short time. This represents gravitational water and gravitational drainage. At this time, large pores in the sponge have emptied and are filled with air. The remainder of the water is held in smaller, interconnecting pores, referred to as capillaries. This water is capillary water. If the sponge is squeezed, this water comes out of the sponge. On further squeezing, no more water comes out, even though the sponge is still wet. The remaining water is the hygroscopic water.

13.5 SOIL MOISTURE AND FIELD CAPACITY

The water content of soil can be defined on a mass basis as well as a volume basis. Let θ denote the water content. On a mass basis, θ can be denoted as θ_m and on a volume basis as θ_v. Then, the water content on the mass basis can be defined as the ratio of mass of water M_w to the mass of dry soil M_s:

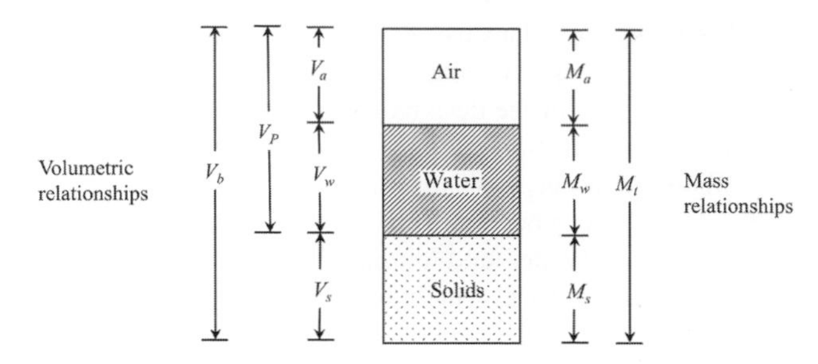

Figure 13.4 Schematic diagram of a soil block as a three-phase system.

$$\theta_m = \frac{M_w}{M_s}. \tag{13.1}$$

The total mass of soil M_t comprises the mass of dry soil M_s, the mass of water M_w, and the mass of air M_a, as shown in Figure 13.4. The mass of air is negligible. Therefore, the total mass of soil M_t is equal to the mass of water and the mass of dry soil.

Similarly, the water content on a volume basis can be defined as the ratio of volume of water V_w to the bulk volume of soil V_b, as shown in Figure 13.4. The bulk volume is composed of the volume of solids, V_s, and the volume of pore spaces, V_p, which, in turn, is composed of the volume of water V_w and the volume of air V_a. Thus, θ_v can be expressed as

$$\theta_v = \frac{V_w}{V_b} = \frac{V_w}{V_s + V_p} = \frac{V_w}{V_s + V_w + V_a}. \tag{13.2}$$

Field capacity defines the wetness of soil after the largest soil pores have drained downward in response to the hydraulic gradient and rapid drainage has become almost negligible – that is, the gravitational water has been drained from the soil under the influence of gravity. Then, the water content of the soil attains a value that is referred to as field capacity. At this stage the soil holds the water on its own. After irrigation or rainfall, it may take several hours to a few days to reach field capacity, depending on the type of soil. At field capacity, matric forces play a larger role in the movement of the remainder of the water. Depending on the soil, it may take about one day in coarser soils to three days in finer soils after irrigation or rainfall for the rapid drainage to cease, as shown in Figure 13.5. Micropores or capillary pores are still filled with water, and are the source of water for plants. The matric potential at field capacity varies slightly with soil but is normally in the range from –10 to –30 kPa. It may be noted that the movement of soil water will continue even below field capacity, but will be at a much slower pace and the rate of change of soil water around the field capacity will be large. Table 13.1 shows typical values of soil water content for different soils.

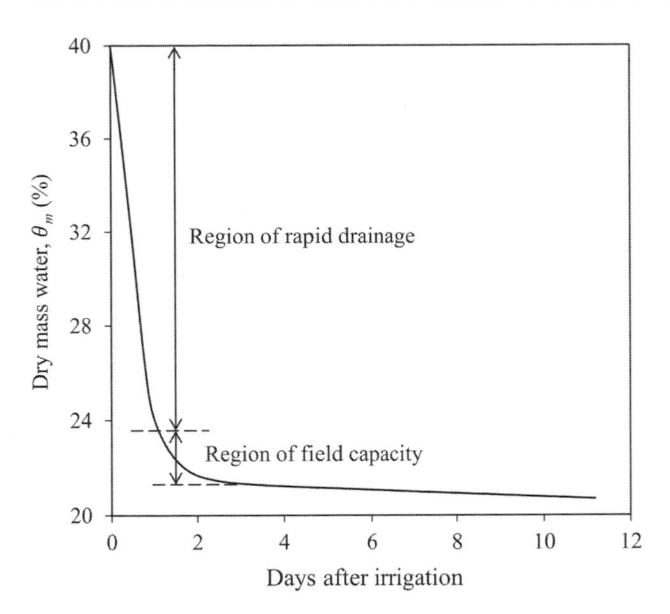

Figure 13.5 Decrease in water content as a function of time following irrigation. (Adapted from Cuenca, 1989. Copyright © by 1989. Prentice Hall, Englewood Cliffs, New Jersey.)

Field capacity is a useful term in agricultural irrigation because it corresponds to an approximate degree of soil wetness which has important implications. First, at field capacity there is sufficient air in pores that permit optimal aeration for most aerobic activity and for plant growth. Second, field capacity corresponds to the optimal wetness for ease of tillage or excavation. This is because at this wetness the soil is near its lower plastic limit. Below this value of wetness, the soil behaves like a crumbly semisolid, and as mud above it. Third, the maximum amount of water that is useful to plants is held by the soil at this capacity. Above it, water is held for a short period, and drainage of this water is essential for optimal plant growth.

13.6 WILTING POINT

As the soil continues to dry – that is, soil water held in large and medium pores has drained – it becomes harder for plants to extract water because the water is now more tightly held in smaller pores. The objective of irrigation is to not allow the soil to reach that stage. When the plants cannot extract enough water to satisfy their demands, they attempt to temporarily adjust to the water shortage by closing down their stomata and slowing down their metabolic activities. The effect of water shortage is normally observed in plant leaves becoming curled, or yellowing. This stage is temporary, because plant health can be restored with irrigation or rainfall. If the soil water continues to decline, plants will suffer permanent physiological damage and will not be saved even if irrigation is applied. This stage is often observed under prolonged drought conditions and is referred to as the permanent wilting point, because plants have permanently wilted. The matric potential at this point is around –15 bar.

Table 13.1 *Typical values of water content held in soil at field capacity and wilting point*

Soil texture	Field capacity (mm/m)	Permanent wilting point (mm/m)	Available water (mm/m)
Fine sand	120	30	90
Sandy loam	160	50	110
Fine sandy loam	210	70	140
Loam	260	100	160
Silt loam	280	120	160
Clay loam	310	150	160
Clay	320	210	110

After Blencowe et al., 1960.

The wilting point depends upon the type of plant, root zone, the rate of water withdrawal, and soil type.

13.7 AVAILABLE WATER

From the standpoint of irrigation, the water that stays in the soil and is available for plants to extract is of primary interest. Only some part of the total available water is accessible to plants for their needs. The remaining water is not easily accessible to plants, and plants have to work extra hard to extract it, which affects their growth. The soil water that is readily available to plants is called readily available water. The point at which irrigation is applied is called the refill point. This means that the readily available water is between field capacity and the refill point. The matric potential is about –10 kPa for field capacity and –60 to –80 kPa for the refill point. This available water is bounded by field capacity on the upper side and wilting point on the lower side. While the gravitational water is draining, a small amount of water above the field capacity is also used by plants. At saturation, there is no tension and it is about 0.3 bar at field capacity but may be closer to 0.1 bar for coarser soils. The tension at the wilting point is about 15 bar.

13.8 SOIL WATER RETENTION

The amount of water a soil can retain depends on its texture and to some extent its structure. Fine-textured soils, such as clay, retain more water in their small pores after the drainage of gravitational water than do coarse soils such as sand. However, water is held more tightly in fine-textured soils due to their smaller pores than in coarse-textured soils. This means less water is available to plants in fine-textured soils. This characteristic of soil plays a fundamental role in determining the type of crop that should be grown on a particular soil.

It may now be appropriate to sketch the volumetric water content as soil water tension, as shown in Figure 13.6.

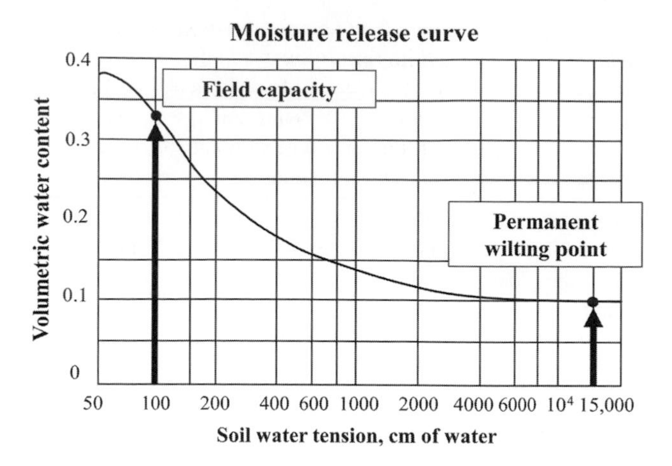

Figure 13.6 Increase in soil water tension with decreased ion volumetric water content.

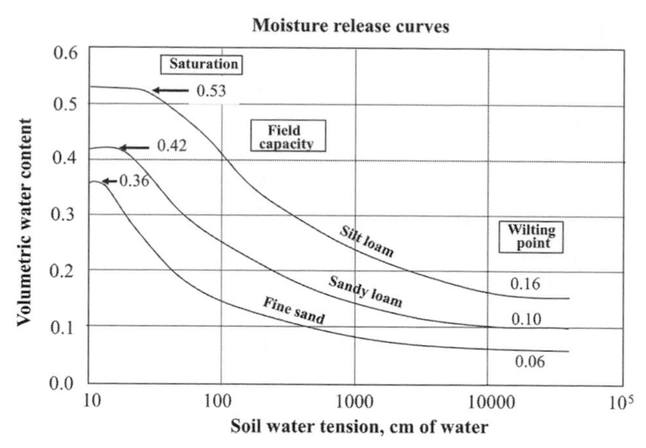

Figure 13.8 Moisture release curves for three different soils.

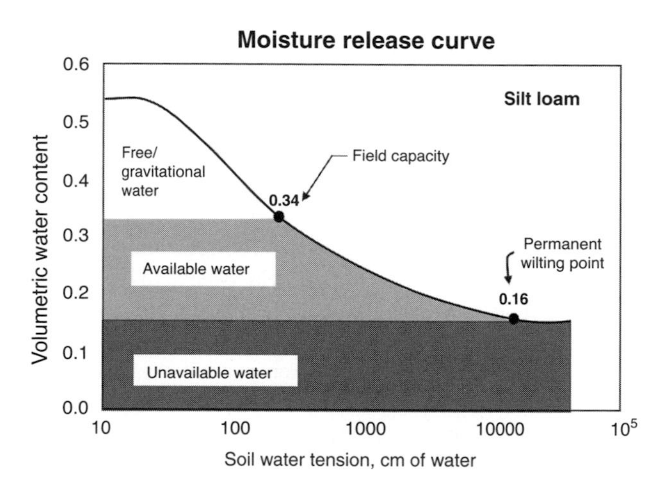

Figure 13.7 Gravitational water, available water for plants, and unavailable water with changes in soil water tension.

This figure shows that as soil becomes dry, that is, its volumetric water content declines, the soil water tension increases as a power function, and this function is referred to as the moisture release curve. At wilting point the soil water tension is two orders of magnitude greater than that at field capacity. Figure 13.7 shows the increase in soil water tension for silt loam with the decrease in the volumetric water content, as well as gravitational water, available water for plants, and unavailable water. Figure 13.8 sketches moisture release curves for three different types of soil.

13.9 SOIL WATER PROPERTIES

There are several properties of soil water that are used in irrigation that are summarized here. It may be recalled that an unsaturated soil comprises soil particles, water, and air. The pore spaces are partly filled with water and partly with air. The soil particles may consist of minerals and organic matter. Normally sand particles are 0.05–2.0 mm in diameter, silt particles are 0.002–0.05 mm in diameter, and clay particles are less than 0.002 mm in diameter.

Saturated water content by volume: Denoted by θ_{vs}, the saturated water content on a volume basis is the ratio of the volume of water V_p ($V_w = V_p$) when the soil is saturated to the volume of bulk soil V_b, expressed as

$$\theta_{vs} = \frac{V_p}{V_b}. \tag{13.3}$$

Saturated water content by mass: Denoted by θ_{ms}, the saturated water content on a mass basis is the ratio of the mass of water $\rho_w V_p$ ($V_w = V_p$) when the soil is saturated to the mass of dry soil M_s, expressed as

$$\theta_{ms} = \frac{\rho_w V_p}{M_s}, \tag{13.4}$$

where ρ_w is the density of water.

Relations between different properties: Soil porosity can be expressed as a function of density. Recalling porosity (*N*),

$$N = \frac{V_p}{V_b} = \frac{V_b - V_s}{V_b}. \tag{13.5}$$

Recalling the definition of density as the ratio of mass to volume, the volume of solids V_s can be written as M_s/ρ_s, where ρ_s is the soil particle density. The bulk volume of soil V_b can be written as M_s/ρ_b, where ρ_b is the bulk density. Bulk density is defined as the weight of dry soil divided by the bulk volume of soil. Then, eq. (13.5) can be expressed in terms of density as

$$N = \frac{V_b - V_s}{V_b} = \frac{(M_s/\rho_b) - (M_s/\rho_s)}{(M_s/\rho_b)} = 1 - \frac{\rho_b}{\rho_s}. \tag{13.6a}$$

Equation (13.6) can be recast in terms of bulk density as a function of porosity as

$$\rho_b = (1 - N)\rho_s. \tag{13.6b}$$

For agricultural soils, the value of ρ_s is normally taken as 2.65 g/cm^3 or 2650 kg/m^3. This is approximately the particle density of silica, sand, granite, and quartz rock which are the parent materials of farm soils. Porosities of agricultural soils are usually in the range of about 50% or 0.5. With this value of porosity, the value of ρ_b from eq. (13.6b) would turn out to be 1.3 g/cm^3, which is the value commonly used.

The void ratio (v_r) of a soil can be defined as the ratio of pore spaces or interstices (V_i) to the volume of soil particles:

$$v_r = \frac{V_i}{V_s} = \frac{V_w}{V_s} = \frac{N}{1-N}. \tag{13.7}$$

Now the volumetric moisture content can be expressed as

$$\theta_v = \frac{V_w}{V_b} = \frac{(M_w/\rho_w)}{(M_s/\rho_b)} = \frac{(M_w/M_s)\rho_b}{\rho_w}. \tag{13.8}$$

Substituting eq. (13.1) in eq. (13.8), the relation between the water content on a volume basis and the water content on a mass basis is obtained as

$$\theta_v = \theta_m \frac{\rho_b}{\rho_w}. \tag{13.9}$$

The quantity (ρ_b/ρ_w) is referred to as the apparent specific gravity of soil. Equation (13.9) is quite useful in irrigation because it is easy to measure θ_m and hence it can be used to determine θ_v, which is used in the design of irrigation systems.

Soil water content: The amount of soil water content present in the soil determines when to irrigate and how much to irrigate. From the above discussion, it can be measured either on a mass basis or volume basis. A soil sample of volume V_b is collected in the field, the soil mass before oven drying is M_{sw} and the mass after oven drying is M_d. On a mass basis it can be defined as

$$\theta_m = \frac{M_{sw} - M_d}{M_d} = \frac{M_w}{M_d}, \tag{13.10}$$

and on a volume basis as

$$\theta_v = \frac{V_w}{V_b} = \frac{(M_{sw} - M_d)/\rho_w}{M_d/\rho_b} = \theta_m \frac{\rho_b}{\rho_w}, \tag{13.11}$$

where M_{sw} is the mass of moist soil – that is, the soil mass before drying – M_w is the mass of water in the soil, M_d is the mass of oven-dry soil, V_w is the volume of water in the soil, V_b is the volume of soil before oven drying, and ρ_b is the bulk density of soil. If the density of water ρ_w is replaced by 1000 kg/m^3 in eq. (13.11), then the relation between θ_m and θ_v is:

$$\theta_v = \frac{\rho_b}{1000}\theta_m. \tag{13.12}$$

The bulk density of soil (ρ_b) is in units of kg/m^3 in eq. (13.12) and can be determined experimentally. In practice, the bulk density is calculated as the weight of soil that has been oven-dried for 16 h or longer at 105 °C, divided by the original volume of soil before oven drying.

Irrigation is often applied on a depth basis. Therefore, it is preferable to express the soil water content in terms of depth of water in the soil profile. The depth of soil water content for 1 m of soil, denoted as DW_1, is related to the volumetric soil water content (%) as

$$DW_1 = 10\theta_v, \tag{13.13}$$

where DW_1 is in mm. The depth of water, DW, in mm available in a soil of given thickness can be computed as

$$DW = 10\theta_v D_s, \tag{13.14}$$

where D_s is the thickness of the soil layer.

If there are more soil layers in the plant root zone, then DW can be computed for each layer and then the DW values can be added to obtain the DW for the soil. It may be noted that the depth of soil water is the amount of soil water, not the depth of penetration of soil water. Not all of this water is available to plants.

The water held between field capacity (θ_{fc}) and permanent wilting point (θ_{wp}) is referred to as the available water, available water capacity (AWC), or available water holding capacity. It is primarily a function of soil texture, as shown in Table 13.2. It should be noted that the values given in the table are subject to significant variability. If the depth of the

Table 13.2 *Values of field capacity, wilting point, and available moisture of different soil textures*

Soil texture	θ_{fc} (in./in. or m/m)	θ_{wp} (in./in. or m/m)	AWC (in./in. or m/m)	AWC (in./ft)
Coarse sand	0.10	0.05	0.05	0.60
Sand	0.15	0.07	0.08	0.96
Loamy sand	0.18	0.07	0.11	1.32
Sandy loam	0.20	0.08	0.12	1.44
Loam	0.25	0.10	0.15	1.80
Silt loam	0.30	0.12	0.18	2.16
Silty clay loam	0.38	0.22	0.16	1.92
Clay loam	0.40	0.25	0.15	1.80
Silty clay	0.40	0.27	0.13	1.56
Clay	0.40	0.28	0.12	1.44

These are examples only. Considerable variation exists from these values within each soil texture.

plant root zone is known, then the total AWC in the plant root zone can simply be computed by multiplying AWC by the depth of the plant root zone. Since plants can extract only a portion of the total available soil water (TAW) before growth and yield are affected, this portion of water is often termed as readily available water (RAW), which for most crops ranges between 40% and 65% of the available water in the root zone. If the irrigation management allowed depletion (MAD; expressed as a fraction), which can be removed, is known, then RAW is the product of AWC and MAD. The values of MAD and rooting depth for different soils are given in Table 13.3.

Example 13.1: For a soil sample, the mass of moist soil is 0.25 kg and the mass of soil after oven drying is 0.21 kg. Assume the bulk density of soil as 1200 kg/m^3. Compute the soil water content if the thickness is 1 m and if it is 0.8 m.

Solution: The soil water content on a mass basis is

$$\theta_m = \frac{0.25 - 0.21}{0.21} \times 100 = 19.05\%.$$

The soil water content on a volumetric basis is

$$\theta_v = \frac{1200}{1000} \times \theta_m = 22.86\%.$$

The depth of water in 1 m of soil is

$$DW_1 = 10 \times \theta_v = 22.86 \times 10 = 228.60 \text{ mm}.$$

For 0.8 m of soil, the depth of soil water content is

$$DW = 10 \times \theta_v \times 0.8 = 22.86 \times 10 \times 0.8 = 182.88 \text{ mm}.$$

Example 13.2: A three-layer soil having the following characteristics is given:

Layer	Depth (in.)	θ_{fc}	θ_v
1	5 (12.7 cm)	0.30	0.25
2	5–15 (12.7–38.1 cm)	0.45	0.30
3	15 + (38.1 + cm)	0.25	0.20

Compute the depth to which 3 in. (7.62 cm) of infiltrated water would penetrate. How much water will be lost if the root zone depth was 25 in. (63.5 cm)?

Solution: First, compute the storage of water.

$$SWD_1 = (\theta_{fc} - \theta_v)L = (0.30 - 0.25) \times 5 = 0.25 \text{ in. } (0.635 \text{ cm}),$$

$$SWD_2 = (0.45 - 0.30) \times 10 = 1.5 \text{ in. } (3.81 \text{ cm}).$$

The sum of depths required to fill the upper two layers is 0.25 + 1.5 = 1.75 in. (4.45 cm). The remaining water is 3.00 – 1.75 = 1.25 in. The depth penetrating the third layer (L_3), is determined as

$$SWD_3 = 1.25 \text{ in.} = (0.25 - 0.20) \times L_3.$$

Therefore, $L_3 = \frac{1.25}{0.05} = 25$ in. Thus, the depth of soil with 3 in. penetration is 5 + 10 + 25 = 40 in. (101.6 cm).

The depth below the root zone is 40 – 25 = 15 in. (38.1 cm). Therefore, water lost = (0.25 – 0.20) × 15 in. = 0.75 in. (1.91 cm).

Example 13.3: Consider peas with a rooting depth of 0.8 m (from Table 13.3). The soil is sandy loam. Compute the RAW.

Solution: From Table 13.2, AWC = 0.12 m/m (for sandy loam).

From Table 13.3, MAD = 0.35 (for peas).

RAW is the product of AWC and MAD = 0.12 × 0.35 = 0.042 m/m.

For the peas with a rooting depth of 0.8 m, RAW = 0.042 m/m × 0.8 m = 0.0336 m.

One can also define the depleted and remaining water as a fraction of available water depleted or fraction of available water remaining. The fraction of available water depleted, f_d, can be expressed as the ratio of the difference between field capacity and available water to the difference between field capacity and wilting point. The fraction of available water remaining, f_r, is the ratio of the difference between available water and wilting point to the difference between field capacity and wilting point. Clearly this fraction is 1 minus the fraction of available water depleted, that is, $f_r = 1 - f_d$.

For irrigation management, it is useful to determine the depth of water required to fill a soil layer to field capacity. Let this depth be denoted as SWD. Then, SWD is equal to the product of AWC, depth of soil layer, and f_d, or the product of soil layer depth and the difference between field capacity and available water.

Table 13.3 *Values of MAD (as a fraction) and rooting depth for different crops*

Crop	Rooting depth	MAD	Crop	Rooting depth	MAD
Alfalfa	1.0–2.0	0.55	Melons	1.0–1.5	0.35
Banana	0.5–0.9	0.35	Olives	1.2–1.7	0.65
Barley	1.0–1.5	0.55	Onions	0.3–0.5	0.25
Beans	0.5–0.7	0.45	Palm trees	0.7–1.1	0.65
Beets	0.6–1.0	0.5	Peas	0.6–1.0	0.35
Cabbage	0.4–0.5	0.45	Peppers	0.5–1.0	0.25
Carrots	0.5–1.0	0.35	Pineapple	0.3–0.6	0.6
Celery	0.3–0.5	0.2	Potatoes	0.4–0.6	0.25
Citrus	1.2–1.5	0.5	Safflower	1.0–2.0	0.6
Clover	0.6–0.9	0.35	Sisal	0.5–1.0	0.8
Cacao	1.0–1.5	0.2	Sorghum	1.0–2.0	0.55
Cotton	1.0–1.7	0.65	Soybeans	0.6–1.3	0.5
Cucumber	0.7–1.2	0.5	Spinach	0.3–0.5	0.2
Dates	1.5–2.5	0.5	Strawberries	0.2–0.3	0.15
Dec. orchards	1.0–2.0	0.5	Sugar beet	0.7–1.2	0.5
Flax	1.0–1.5	0.5	Sugarcane	1.2–2.0	0.65
Grains small	0.9–1.5	0.6	Sunflower	0.8–1.5	0.45
winter	1.5 2.0	0.6	Sweet potatoes	1.0–1.5	0.65
Grapes	1.0–2.0	0.35	Tobacco early	0.5–1.0	0.35
Grass	0.5–1.5	0.5	late		0.65
Groundnuts	0.5–1.0	0.4	Tomatoes	0.7–1.5	0.4
Lettuce	0.3–0.5	0.3	Vegetables	0.3–0.6	0.2
Maize	1.0–1.7	0.6	Wheat	1.0–1.5	0.55
silage		0.5	ripening		0.9

After Doorenbos and Pruitt, 1977.

Example 13.4. Consider a sandy loam soil 80 cm deep that has a volumetric water content of 0.15. Compute f_d, f_r, AWC, and SWD.

Solution: From Table 13.2, for sandy loam soil the field capacity $\theta_{fc} = 0.20$ m/m, the permanent wilting point $\theta_{wp} = 0.08$ m/m,

$$f_d = \frac{\text{field capacity} - \text{available water}}{\text{field capacity} - \text{wilting point}} = \frac{0.2 - 0.15}{0.2 - 0.08} = 0.417,$$

$f_r = 1 - f_d = 1 \quad 0.417 = 0.583.$

The AWC is: $AWC = \theta_{fc} - \theta_{wp} = 0.20$ m/m $-$ 0.08 m/m $= 0.12$ m/m.

For 80 cm = 0.8 m depth of soil, the SWD is:

$$SWD = AWC \times \text{depth of soil layer} \times f_d = 0.12\frac{\text{m}}{\text{m}} \times 0.8\text{ m} \times 0.417 = 0.04\text{ m}.$$

13.10 SOIL WATER PHYSICS

Water is a compound consisting of water molecules, with each molecule consisting of one oxygen atom and two much smaller hydrogen atoms. The water molecules are asymmetric and have polarity because the hydrogen atoms are not symmetrically attached to the oxygen atom. Thus, water molecules are electropositive on the side where hydrogen atoms are located, and electronegative on the opposite side. A hydrogen atom of one water molecule is attracted to the oxygen end of a nearby molecule and forms a low-energy bond between the two molecules. This type of bonding is referred to as hydrogen bonding, and leads to the polymerization of water. Positively charged ions or surfaces are attracted to water molecules through the oxygen (negative) end and negatively charged surfaces are attracted to water molecules through the hydrogen (positive) end. Thus, polarity can explain hydration and dissolution of salt in water. It may be noted that water molecules, when attracted to electrostatically charged ions or clay surfaces, are more closely packed than when they are in pure water. The implication is that water molecules have less freedom of movement and lower energy levels than the molecules in pure water.

The water stored in soil undergoes phase change or moves in all directions. The phase change occurs due to temperature differential and concentration differential. The storage and movement of water occur due to gravitational and capillary forces acting on it, entailing cohesion and adhesion. These forces are now described.

13.10.1 Cohesion and Adhesion

Cohesion is the attraction of water molecules toward each other; adhesion is the attraction of water molecules to solid surfaces. Adhesion is also referred to as adsorption. When water molecules are tightly held at the soil particle surfaces by virtue of adhesion, these water molecules move other molecules away from the surfaces. In this manner, the forces of cohesion and adhesion, resulting from hydrogen bonding, influence the retention and movement of water in the soil.

13.10.2 Surface Tension

At the water–air interface, water molecules have greater attraction to each other than they have to the air above – that is, cohesion is greater than adhesion. The result is that there is an inward force at the surface that pushes the water downward, causing a curved surface. This causes the water to have high surface tension, which is important for capillarity. In comparison with other liquids, water has high surface tension of 72.8 millinewtons/m at 20 °C.

13.10.3 Capillarity

When a clean glass tube of small diameter is placed in water, the water is observed to rise in the tube against the force of gravity, and the rise is higher in a smaller-diameter tube than in a larger-diameter tube. This rise is called the capillary rise. Because of adhesion, water molecules are attracted to the sides of the glass tube. Simultaneously, because of cohesion the water molecules are attracted to each other and cause surface tension, leading to a curved surface at the interface between water and air in the tube. This curved surface is called the meniscus. The pressure under the meniscus is lower and the pressure on the free water is higher. As a result, the free water is pushed up in the tube. The water will continue to rise until its weight balances the pressure difference across the meniscus. The capillary pressure is the difference in pressure between the pressure at the non-wetting fluid and that at the wetting fluid and is related to suction and tension. Thus, capillarity is caused by two forces: adhesion and cohesion. The weight of water lifted up is proportional to the volume of water. If D is the capillary diameter, h is the capillary rise, and γ is the specific weight of water, then the gravitational force, F_g, is

$$F_g = \frac{\pi}{4} D^2 h\gamma. \tag{13.15}$$

The gravitational force is equal to the vertical component of the surface tension times the circumference of the tube. In most cases, the contact angle (as illustrated in Figure 13.10) between the meniscus and the tube is close to 0°, so the vertical component is close to the surface tension. If the surface tension is denoted as σ, then the capillary force, F_c, can be simplified as

$$F_c = \sigma\pi D. \tag{13.16}$$

Since $F_g = F_c$, equating eqs (13.15) and (13.16),

$$\frac{\pi}{4} D^2 h\gamma = \sigma\pi D. \tag{13.17}$$

This yields the capillary rise as

$$h = \frac{4\sigma}{D\gamma}. \tag{13.18}$$

Equation (13.18) says that the capillary rise, h, depends on the capillary tube diameter (or radius, r) and surface tension. However, liquid density, ρ, also influences surface tension. For water at 20 °C, the height of rise can be computed as

$$h = \frac{0.15\ \text{cm}^2}{r}, \tag{13.19}$$

where h and r are in cm.

A point should be noted here. The water that rises comes from some source and this source must not exert a greater opposite pull. The water with the lowest energy level in the soil is a free water surface, such as the water table. The force that moves the water up occurs when there is a water–air interface. If there is no free water surface or lower-energy water source, there will be no upward movement of water.

In soils, small pores can be thought of as capillaries. In unsaturated soils, water moves due to capillary forces and the movement is determined by the pore size distribution. The movement of water can occur in both horizontal and vertical directions. Different soils have different pore size distributions, and as a result the amount and rate of movement of water by capillarity vary. For example, in sandy soils with medium to large pores the initial capillary rise can be rapid, whereas in clay, with a high proportion of fine pores, the capillary rise is slow but ultimately may be higher than in sandy soils.

The cause for water in the soil to move upward from the water table is capillary action. The thickness of the soil layer above the water table moistened by this action is referred to as the capillary fringe. The height of this fringe depends on the capillary diameter, which, in turn, depends on soil texture. For example, in sand, which has coarse texture and large particle sizes, the capillary diameter will be large and hence the capillary fringe will be smaller in height. In clay the opposite will be true. That is, the height of the fringe will be smaller in large-diameter capillaries and greater in smaller-diameter capillaries.

The movement of water in the capillaries depends on the dryness of soil and the net resultant force of gravity push and capillary pull. In large-pore soils the flow path may be broken by large pores and macropores and hence the rise of water may be constrained. On the other hand, in smaller-pore soils, such as silt, there occur many small capillaries that permit continuous flow paths; as a result the water may rise 1 m or more above the water table. In finer soils the rise

Table 13.4 *Equivalent values of soil water tension, pressure, and potential*

Bar	Atmosphere (atm)	Pascal (Pa)	m of water at 4 °C	cm of mercury at 0 °C
1	0.9869	100,000	10.1981	75.06
1.013	1	101,325	10.3322	76.00
0.00001	0.000009869	1	0.00010198	0.000750
0.09807	0.09677	9806.6	1	7.35559
0.013332	0.0131579	1333.22	0.135951	1

may be even greater. In medium-textured soils the rise will be small, say 0.3 m above the water table.

If the capillary fringe is close to the soil surface, the soil surface becomes wet and this may cause a drainage problem. On the other hand, evaporation will be greater from the top soil and may cause a salinity problem. This is because the water will evaporate, leaving the salt behind, and salt will then concentrate on the surface. Further, as more water and salt move up from below to replace the evaporated water, severe salinity problems will be the result. This condition is serious in silts and fine sands, because capillaries are fine enough to exert appreciable tension, say 1 m or so, but not so fine as to slow down the upward movement that a clay soil would have despite its greater capillary pull.

13.10.4 Forces on Soil Water

There are three important forces in soil water. First, there is adhesion, which is the attraction of water to soil particles (matrix) that produces the matric force. This force causes adsorption and capillarity and reduces the energy level of water near particle surfaces. Second, there is osmotic force, caused by the attraction of water molecules to ions and other solutes. This force reduces the energy level of water in the soil solution. The third force is gravity, which pulls water downward. The energy level of soil water at a higher elevation is greater than that at a lower elevation, causing the movement of water downward.

13.10.5 Soil Water Storage and Movement

Different soils, depending on their texture and structure – or the distribution of pore sizes and shapes – have different capacities for storing water and permitting its movement. Pores are large in coarse-textured soils, small in fine-textured soils, and medium in medium-textured soils. It is these pores that constitute paths for the movement of water. The size and shape of pores may change with each soil particle. This means that a soil is composed of tubes of different sizes that are in different directions. These tubes form capillaries when they are partially filled with water. Large capillaries hold large amounts of water and small capillaries hold small amounts of water. There is a correlation between the amount of water held and the size of the pores or capillaries holding it.

When soil is saturated it has no air in it, and the pores are filled with water. At this stage, the water is in storage and is also moving downward. The large pores drain first, with the drained water replaced by air. Then, the capillaries are large and long. With the continued draining of pores, smaller pores empty and are filled with air, replacing the draining water. At field capacity, capillaries become discontinuous and short, suggesting that water does not move very far and ceases to drain down. At this point storage of water becomes dominant over its movement. As the soil gets further dried and tends toward the wilting point – that is, about three-quarters of the pore volume contains air, capillaries as tubes cease to exist.

Thus, the forces acting on soil water are gravitational and capillary. The free downward movement of soil water is primarily due to gravitational force. The soil water is pulled down by the weight of water or pressure above it. That means that soil water will experience more pressure if the depth of water above it is greater; this pressure is proportional to the depth.

The capillary force is a result of surface tension and adhesive and cohesive forces. The capillary force pulls the water in and occurs when the water–air interface is in contact with soil particles. When the soil is unsaturated, the soil matrix exerts tension on the water and this tension is directly proportional to the capillary tension, which is inversely proportional to the diameter of the interconnected pores. This shows that as soil gets drier, the soil matrix exerts more tension on the water and plants have to exert more force to extract water from the soil. This water is loosely held. Tension can be expressed in various equivalent units, as shown in Table 13.4.

With soil drying further, capillaries cease to exist and water exists only as a thin film around the soil particles; at this point the forces acting on the film are cohesive and adhesive. The water is tightly held on the soil particles and plants have difficulty extracting it. There is no movement of water, only storage.

13.10.6 Soil Water Energy

In soil water there are mainly two different kinds of energy: potential energy and kinetic energy. Because soil water moves so slowly, for practical purposes its kinetic energy can be considered negligible. Therefore, potential energy is the energy considered in the retention and movement of water in soils, uptake and translocation of water in plants, and loss of water to the atmosphere. The potential energy is composed of gravitational, pressure, and osmotic energy. For water to move from one location to another there must be a difference in the energy levels. Energy is expressed in different ways and in different units. Two common ways are in terms of potential or head.

13.10.7 Soil Water Potential

Soil water potential expresses the energy possessed by the water in the soil. It can be measured or calculated in different equivalent units, such as bar, height of water, etc., as shown in Table 13.4. Pressure is normally defined with reference to a reference datum, which is the atmospheric pressure. Likewise, the potential of soil water is defined with respect to that of pure water located at a reference elevation at standard pressure and temperature, not affected by soil – that is, potential is defined by the difference in the energy levels. Thus, soil water potential can be defined as the difference between the energy level of the pure water and that of the soil water. The water will move from higher potential to lower potential. It is the difference in potential that causes the movement of soil water from one point to another. For example, water moves from wet soil in the root zone (having higher potential) to the plant roots (having lower potential).

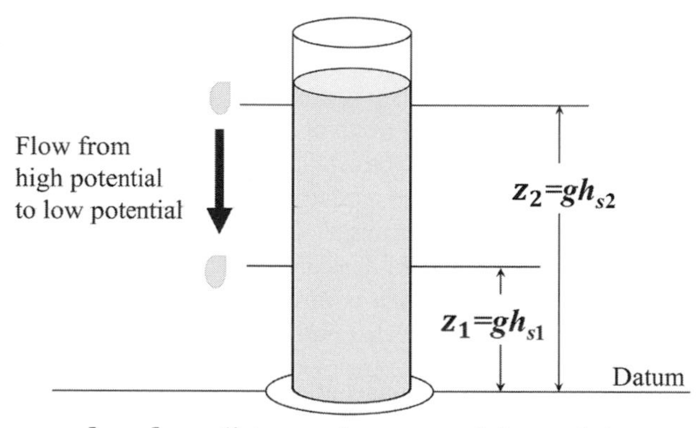

Figure 13.9 Gravitational potential.

The total soil potential is composed of gravitational potential, matric potential, hydrostatic potential, and osmotic potential, each corresponding to differences in energy levels with respect to their forces. The potential may also be caused by soluble material in water. At any point in time, one or more potentials may be acting on the soil water. Each of these potentials is now defined. The gravitational potential of soil water at a particular position, designated z, is the product of acceleration due to gravity, g, and the height of soil water, h_s, as shown in Figure 13.9:

$$z = gh_s. \tag{13.20}$$

The gravitational potential is considered positive above the datum, zero at the datum, and negative below it.

The pressure potential of soil water includes the hydrostatic pressure due to the weight of water in saturated soils and the negative pressure due to the force of attraction between water and soil particles. It is also measured with respect to the atmospheric potential and can be positive or negative. If the pressure potential at a given point in the soil water is above the atmospheric potential, then it is positive. Because the hydrostatic pressure is important in saturated soils, it vanishes in unsaturated soils. There is matric potential due to the attraction between water molecules and soil particles which is always negative, because the energy level is less than that of pure water. If it is less than the atmospheric potential, then it is negative. This is essentially the capillary potential, as shown in Figure 13.10.

The matric potential expresses the energy or force with which the water molecules are held by soil particles, and applies mostly to loosely held water – that is, between field capacity and the wilting point. The water molecules close to the soil particles are held more tightly than those farther away. As distance increases between soil particles and water particles the force becomes weaker, and eventually vanishes at some distance away. This suggests that water molecules

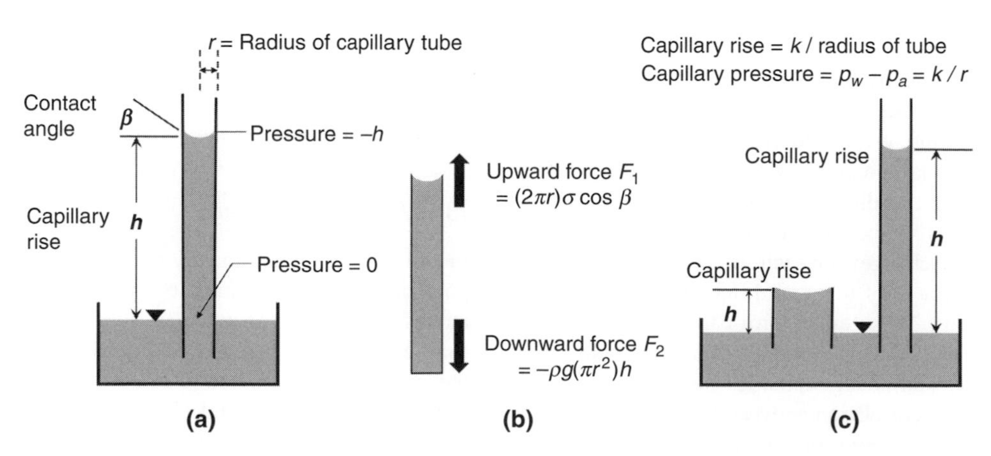

Figure 13.10 (a) Sketch of capillary force in a circular tube, (b) upward and downward force balance, and (c) capillary rise in tubes of different radii.

within the soil–water system are held with different levels of forces, some very tightly and some loosely.

The matric potential does not have a positive value and hence varies from zero to a negative value. It is zero at saturation and negative below saturation. A negative value of the matric potential means that external force or suction will be needed to move the water away from soil particles, because the water itself does not have enough energy to move. The matric potential is also sometimes called soil water suction, because soil particles hold the water tightly with their own suction. The soil suction is regarded as a positive force exerted by soil particles, which means soil suction varies from zero to a positive value.

Soil water has dissolved material in it. When a material is dissolved in water, the water loses part of its energy and is hence left with lower potential. This has implications for irrigation. For example, if irrigation water is salty or soil contains salts, then the salty soil water will have lower potential and hence the movement of water from the soil to plant roots will be slower and will stop earlier, even if there is plenty of water in the soil.

The osmotic potential is the potential due to the concentration of organic and inorganic solutes in the soil–water solution. Since water is attracted to solutes, this attraction reduces the energy level of water molecules, lowering the osmotic potential and limiting the freedom of movement of water. On the other hand, plant roots will also attract water. If the concentration of solutes is high, meaning lower osmotic potential (i.e., greater negative value), then roots will not be able to attract as much water as they want, affecting plant growth. This situation occurs when salt is dissolved in the water and a semi-impermeable membrane is present. The semi-impermeable membrane is a material that permits water, but not salt, to flow through it. Cell walls in roots act as semi-impermeable membranes. Consider a case when the soil water is so salty that its potential is −14.7 bar. A crop growing in this soil will have difficulty extracting water even at field capacity (−0.3 bar) and will experience water shortage almost to the point of the permanent wilting point (−15 bar). This becomes critically important when irrigating crops with saline water.

The total soil water potential can be summarized as the sum of four potentials: gravitational, matric, pressure, and osmotic or solute. The total potential determines the direction and velocity of the flow of soil water from one point to another within the soil system. From the irrigation viewpoint, the matric potential is the most important, and irrigation management aims to control this potential. The value of this potential determines the ease with which plants will be able to extract water from the soil. The availability of soil water to plants is much greater when the potential is zero and becomes lower as the potential becomes negative.

The soil water potential can be expressed as the weight potential or volumetric potential. Note that specific potential is defined as energy per unit mass. The weight potential can be written as

$$\text{weight potential} = \frac{\text{specific potential}}{\text{gravitational constant}}. \tag{13.21}$$

In SI units, eq. (13.21) can be written as

$$\text{weight potential} = \frac{\text{joules/kg}}{\text{m/s}^2} = \frac{\text{newtons (m/kg)}}{\text{m/s}^2} = \text{m}. \tag{13.22}$$

This shows that the weight potential can be expressed as head (m).

Alternatively, the potential can be expressed as

$$\text{volumetric potential} = \text{specific potential} \times \text{density}. \tag{13.23}$$

Then, in SI units,

$$\text{volumetric potential} = (\text{joules/kg}) \times (\text{kg/m}^3) = \frac{\text{newton}}{\text{m}^2} = \text{Pa}. \tag{13.24}$$

The volumetric potential has the same units as does pressure.

Soil water potential is a measure of water availability. Common units of pressure and head and their equivalents are listed in Table 13.5.

Table 13.5 *Common units of pressure and head and their equivalents*

Unit	Equivalent pressure	Water head equivalent
1 atm	101.3 kPa	1034 cmH_2O
	1.013 bar	34 ftH_2O
	1013.25 mb	76 cmHg
	14.7 psi ($lb/in.^2$)	29.9 in.Hg
1 psi ($lb/in.^2$)	6.89 kPa	2.31 ftH_2O

13.10.8 Soil Water Characteristic Curve

The relationship between soil water potential (ψ) and soil moisture content (θ) is referred to as the soil water characteristic curve or release curve. This curve varies with soil type and structure, and is plotted for different soils, as shown in Figure 13.11. The curve is useful for evaluating the availability of soil water for plant use. Fine- and very fine-textured soils have relatively fine uniform pores and hold much water at the wilting point, pointing to the lower availability of water than in moderately fine-textured soils. These soils do not have large pores so there is less gravitational water to drain. Moderately fine-textured soils have fairly uniform distribution of fine particle sizes and lack coarser particles. These soils have large volumes of small pores and low bulk density, and hold large amounts of water at saturation. Also, large amounts of water remain in the soil at the wilting point. The water is held at moderate tensions. The drainage of water is quite slow. It may take up to four days

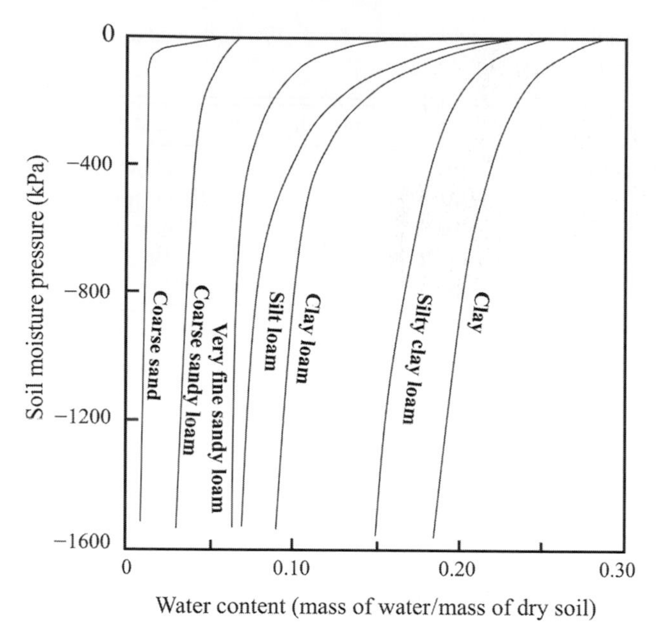

Figure 13.11 Soil water characteristics curves for several soil types. (Adapted from Cuenca, 1989. Copyright © 1989 by Prentice Hall, Englewood Cliffs, New Jersey.)

after irrigation before the soil reaches the field capacity. The tension in capillaries will be about 0.4 bar.

Medium-textured soils contain some large pores and some moderate-sized pores, and hold less water at the field capacity and wilting point than do moderately fine-textured soils. However, because of their larger pore sizes these soils have a larger proportion of water available at lower tensions.

Moderately coarse soils have fewer small pores and more large pores. They have large bulk density, less total pore space to fill at saturation, and less water at field capacity and wilting point, and hence less water available for plant use. However, almost all of the water is held at low tensions and can be easily extracted by plants.

As moisture content increases, the soil water potential decreases. At a given potential, clay holds much more water than does loam or sand. Similarly, at a given moisture content, the water is held more firmly in clay than in sand or loam. Soil structure influences the pore size distribution and hence the soil water holding capacity. For example, compaction causes a reduction in the volume of water the soil can hold, but compacted soil will likely have small to medium pores that will hold water more firmly. Similarly, granulation alters the pore size distribution. A well-granulated soil has more total pore space and hence more water holding capacity than poorly granulated soil. Soil aggregation increases large inter-aggregate pores that hold water more firmly. Thus, soil texture and structure significantly influence the soil water characteristic curve.

As soil is nearly saturated, the pores between the particles are filled with water, as shown in Figure 13.12a – that is, water fills the pores by forming saturated zones around the particles. There is attraction between soil particles and water by virtue of adhesion and cohesion, which constitute the surface tension of water. As soil dries out the matric potential increases and a curved water surface develops between water and soil particles, as shown in Figure 13.12b. To develop soil moisture release curves one can consider three possibilities. First, the soil is saturated with a positive head and drainage occurs, as shown in Figure 13.12c. At equilibrium, the average matric potential is zero and flow stops because of capillary forces, as shown in Figure 13.12d. Under unsaturated conditions, water flows from the soil and soil is unsaturated. The soil water is under tension, and when flow stops the tension in the soil equals the difference in elevation between two points, as shown in Figure 13.12e.

Example 13.5: Compute the matric potential at point 1 for unsaturated conditions shown in Figure 13.12e. The head difference between the two points is 50 cm.

Solution: Considering the datum at point 2, the total water potential $\Psi_2 = 0$. When flow ceases. the total water potential at point 1: $\Psi_1 = \Psi_2 = 0$. The total water potential at point 1 in the soil equals the gravitational potential (z) plus the matric potential (Ψ_{m1}):

$$\Psi_1 = z + \Psi_{m1}.$$

Given that z = elevation head at point 1 = 50 cm and $\Psi_1 = 0$,

The matric potential at point 1: $\Psi_{m1} = \Psi_1 - z = 0 - 50\text{ cm} = -50\text{ cm}$.

Example 13.6: Determine the volumetric water content if the wet weight is 450 g, the dry weight is 350 g, and the bulk volume is 250 cm^3.

Solution: Given the wet weight $M_{sw} = 450$ g, the dry weight $M_d = 350$ g, the bulk volume $V_d = 250\text{ cm}^3$, the water density $\rho_w = 1.0\text{ g/cm}^3$, using eq. (13.11), the volumetric water content is

$$\theta_v = \frac{V_w}{V_d} = \frac{(M_{sw} - M_d)/\rho_w}{V_d} = \frac{(450\text{g} - 350\text{g})\Big/\left(1.0\frac{\text{g}}{\text{cm}}^3\right)}{250\text{ cm}^3} = 0.4$$

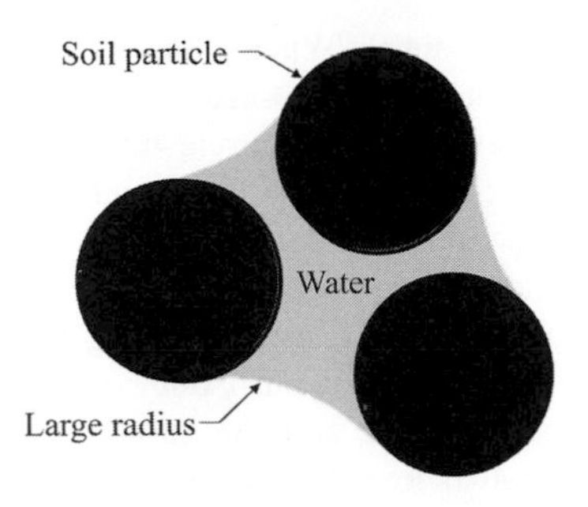

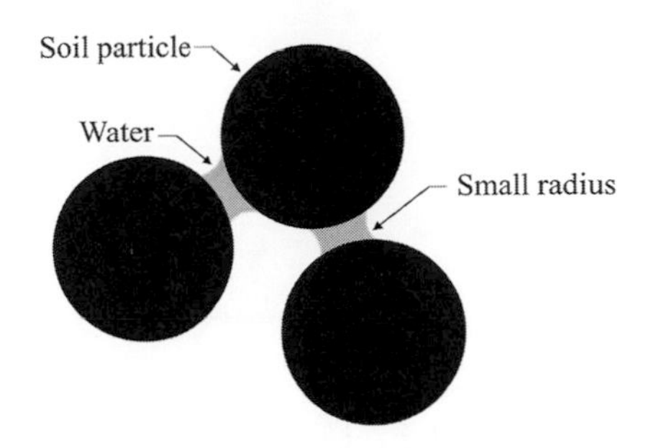

(a) Soil is nearly saturated.
It has small matric potential
so little capillary action

(b) Soil is nearly dry.
It has large matric potential
so large capillary action

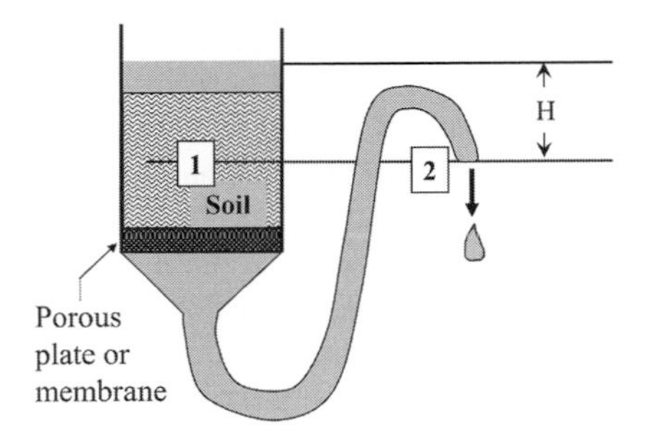

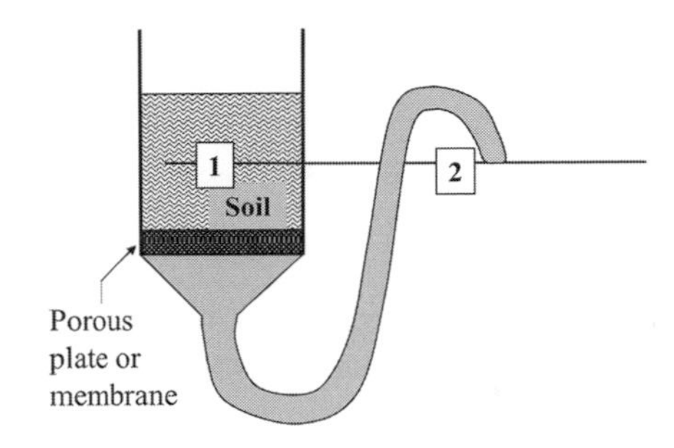

(c) Soil is initially saturated and drainage occurs through the soil because of positive head. The matric potential at point 2 is greater than at point 1.

(d) The flow stops due to the capillary action of the soil and it is a case of equilibrium. The matric potential is zero.

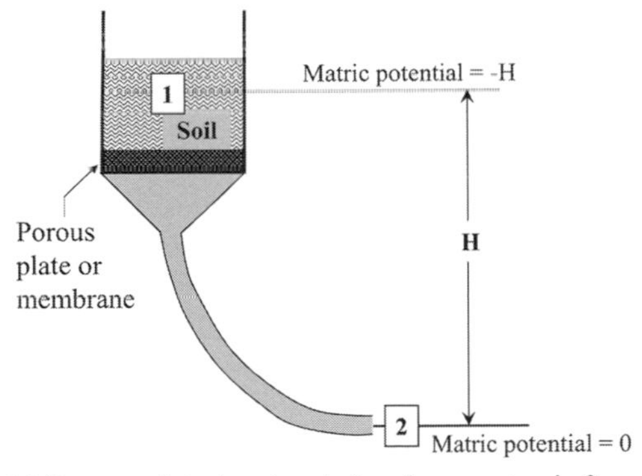

(e) Porous plate is saturated and prevents air from entering the tube. The soil is unsaturated because water flows. The soil water is under tension. When flow ceases the tension in the soil is equal to the difference in elevation between point 1 and 2.

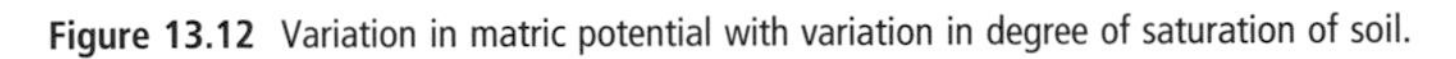

Figure 13.12 Variation in matric potential with variation in degree of saturation of soil.

13.10.9 Hysteresis

The soil characteristic curve when soil is wetting somewhat differs from that when soil is drying, as shown in Figure 13.13. That means there are two curves: one for the wetting phase and one for the drying phase. This phenomenon is called hysteresis. Nonuniformity of soil pores or pore size distribution is among the many factors that affect hysteresis. Consider the wetting phase first. When soil is wetted, water moves to fill larger pores first and some of the smaller pores that have air may not be filled, and the air may get entrapped. The entrapped air resists water penetration. During the drying phase, macropores will lose water first, but micropores may surround macropores, and the entrapped air may prevent the emptying of macropores until

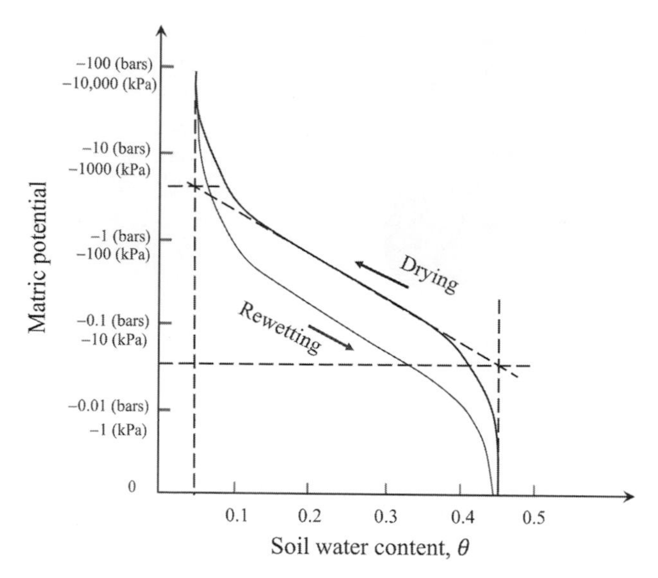

Figure 13.13 Relationship between soil water content and matric potential of a soil upon being dried and then rewetted.

the matric potential is sufficiently low for the surrounding micropores to empty.

13.11 SOIL WATER FLOW

13.11.1 Entry of Water into Soil

Water enters the soil through pore spaces between soil particles. This entry is made possible by the push and pull of water due to the forces acting on it. The rate at which water enters is determined by the pore sizes. For example, sand has large pore spaces so the water, when applied to the surface, soaks in faster than it would in clay, which has smaller pores. Likewise, a compacted soil will absorb water at a much slower pace, because soil particles have been pushed into large pores, and the pores have become smaller and fewer. There is more resistance to the flow of water in small-pore soils. In a capillary tube, friction along the contact surface is greater if the tube is small in diameter and less in the case of a large diameter. The friction per unit length of the tube is proportional to the surface area over which water is moving.

13.11.2 Water Movement within the Soil

The movement of water in the soil occurs from a zone of higher potential to a zone of lower potential until there is equilibrium. Water moves in the soil because it has energy to spend and has more energy than its surroundings. Consider the case where irrigation water is applied to dry soil. At the soil surface there is some ponding of water that generates a gravitational push for water to enter the soil. Below the soil surface, water is attracted by capillary force and moves through capillary tubes. When water reaches the bottom of the tubes it comes into contact with air. The water–air interface defines the wetting front, which exerts a capillary pull on the water just above it. Between the wetting front and the soil surface, the soil is saturated and the tubes are short and full of water. The weight of water in these tubes exerts a gravitational push. Thus, at the wetting front, the water is subjected to gravitational push due to the depth of water above the soil surface and the capillary pull due to the unsaturated soil below. The algebraic sum of these forces moves the water into and through the soil. In saturated soils, the water moves freely because the capillary force is negligible. As the soil starts to become dry due to plant use and evaporation, the water no longer moves freely because water particles held tightly by soil particles cannot move without application of an external force.

The soil can be saturated or unsaturated, and within the unsaturated soil there may be water vapor. The flow of water occurring in saturated soil is referred to as saturated flow, and that in unsaturated soil as unsaturated flow. The movement of vapor is called vapor flow. After irrigation or rainfall, part of the soil may be completely saturated, with all pores – large or small – completely filled with water. For example, when land is irrigated the upper soil gets saturated first. In poorly drained soils the lower horizons are saturated. In well-drained soils the soil above stratified clay layers may be saturated. The flow in saturated soils has its own dynamics and will be discussed in Chapter 25 on drainage. The movement of vapor is beyond the scope of this book. Only unsaturated flow is briefly presented here.

In unsaturated soil, large pores are filled with air and small pores with water. The tenacity with which water is held and the amount of water held determine the rate and direction of water movement. The unsaturated flow occurs due to the difference in matric potential; there is no gravitational potential. The water moves from moist areas with thick moisture films (e.g., matric potential of −1 kPa) to relatively drier areas with thin moisture films (e.g., matric potential of −100 kPa).

Thus, the movement of water in the soil can be either saturated flow or unsaturated flow. In saturated flow the pores are entirely filled with water and there is no air. The force driving the movement of water is gravity, which pulls the water downward. Thus, the gravitational flow occurs in saturated soil and the flow is vertically downward from higher elevation to lower elevation. This occurs soon after irrigation water is applied. Gravitational flow is responsible for groundwater recharge, seepage from a dam, and water flow as springs.

In unsaturated flow there is water and air, and air impedes the flow. The movement of water occurs due to the difference in the matric potential – that is, unsaturated flow occurs from regions with thick water films where water is loosely held to regions with thin water films where water is more tightly held. In order for the unsaturated flow to occur, there should be a continuous film of water. Capillary flow occurs in unsaturated flow and can occur in any direction. It is responsible for the rise of water above the water table. Most of the water reaching plant roots is through unsaturated flow.

13.11.3 Condition for Soil Water Movement

As water enters the soil it must make way for the water that follows. The forces causing the movement are gravitational (downward) and capillary (in all directions). The gravitational force is always there to contribute to the movement. For the capillary force to contribute to the movement, there must be a difference in pull at the two ends. The path between the two ends or points must be continuous, and there must be lower tension at one point that can serve as a source of water and higher tension at the other point to which water will flow. This capillary path is not direct, formed by nonuniform and interconnected pores. The rate of movement along the capillary path will depend on the algebraic sum of forces in the direction of movement. The ease of movement depends on the size of path. The movement is harder along smaller paths.

13.11.4 Effect of Soil Air

When soil is partially saturated, some pores are filled with air; the space occupied by air is not used by water to move, meaning there is less space for the movement of water. As the soil becomes drier there is more air and less space for water movement. At saturation, water moves freely because all pore space in the soil is available for water movement. At the field capacity, up to half of the pore spaces may be filled with air, so the space available for water movement is reduced; in this case the flow is called unsaturated. At field capacity almost all capillary paths become discontinuous. With the drying of soil, paths for flow become more broken and do not permit flow around the surfaces of soil particles.

13.11.5 Zones of Water Movement

The movement of water into and through the soil can be divided into three zones: (1) zone of infiltration or saturation, (2) zone of transmission, and (3) zone of wetting. When water infiltrates the soil surface, the soil gets saturated and there is ponding on the surface. This is the zone of saturation. Vertically downward it exists for a short distance and may consist of the surface crust or just a few top millimeters of a uniform profile. This zone is somewhat disturbed due to human activity. Following this zone is the transmission zone. If the first zone is not restrictive then this zone is nearly saturated, but not entirely saturated because the air gets trapped in the pores when water enters. It may be about 80% saturated. On the other hand, if the first zone is restrictive, this zone will be quite a bit less saturated.

The third zone is the zone of wetting, which is relatively short. The water content declines from the wetter state above to the drier state below. The energy differential between the wetter and drier states is quite large. The front of this zone also defines the interface between water and air and is called the wetting front. The extent of these zones depends on the soil and the antecedent conditions.

13.11.6 Barrier to Flow

The flow in and through the soil depends on the soil structure and texture. Some soils have restrictive layers, and layers have different textures. If the flow path is restricted, the flow may be impeded; this occurs in soils of dissimilar layers. If there is a clay pan below a medium-textured layer, the fine clay layer will exert greater resistance due to its smaller paths. The result will be a slowing of the flow, despite greater pull being exerted by smaller pores. The clay layer may act as a barrier, and water may stack upon it, forming a perched water table. The movement of water through the barrier will depend on the algebraic sum of push and pull forces.

On the other hand, consider a coarse-textured (say, sand) layer underlying a medium-textured layer. On application, the water moves down the upper layer and encounters the coarse-textured layer that exerts less capillary pull in comparison with the overlying layer. The lower face of the fine layer has a stronger pull, so the water stays on the face until a perched water table develops. Now the gravitational push from above and capillary pull from below will tend toward decreasing pull owing to the existence of wetter conditions at the face of the finer layer, and the flow will resume. Even then, there will exist a thin water layer with a capillary fringe above it that will exclude much air. In home gardening a layer of sand is often placed at the bottom of a flower pot to promote drainage and prevent finer particles from washing away through the filtering action. It does the filtering, but does not enhance the drainage much.

Should the bottom, coarse layer rest on some medium-textured layer, this finer layer can pull the water away from the overlying coarse layer. Then, the flow will re-establish through the coarse layer with less resistance than if it were all medium-textured soil. This suggests that layered conditions permit soils to retain a little more water for crop use compared to uniform soils. Hence, the layered condition may be beneficial as long as the perched water table is not detrimental.

13.11.7 Resistance to Soil Water Movement

The soil water flow in capillaries always encounters some resistance, depending on the size, length, and shape of capillaries, which then depend on soil texture and structure. This resistance affects the velocity of flow. The velocity of the wetting front in the first hour may be about 100–150 mm/h in

Table 13.6 *Soil textural classes and related saturated hydraulic conductivity*

Soil texture	Textural class	General	Saturated hydraulic conductivity, *k* (mm/h)	(in./h)
Coarse sand	Coarse	Sandy	>508	>20
Sands	Coarse	Sandy	152–508	6–20
Loamy sands				
Sandy loam	Moderate coarse	Loamy	51–152	2–6
Fine sandy loam				
Very fine sandy loam	Medium	Loamy	15–51	0.6–2
Loam				
Silt loam				
Silt				
Clay loam	Moderate fine	Loamy	5–15	0.2–0.6
Sandy clay loam				
Silty clay loam				
Sandy clay	Fine and very fine	Clayey	1.5–5	0.06–0.2
Silty clay				
Clay				

After USDA-NRCS, 2003.

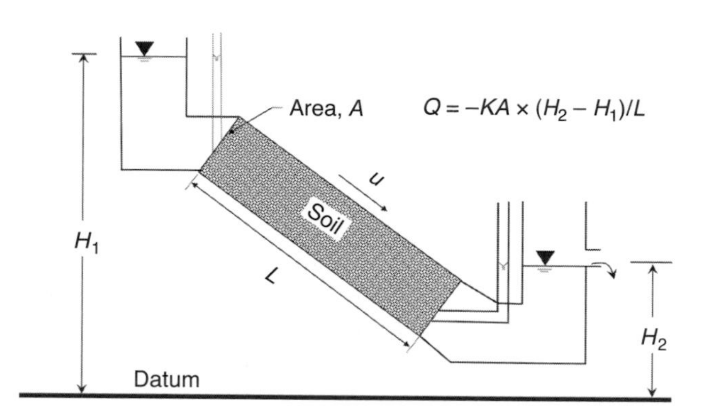

Figure 13.14 Flow through a soil column.

Table 13.7 *Generally used saturated hydraulic conductivity (K_s) of different soils*

Soil type	K_s (in./h)	K_s (mm/h)
Fine sand	10.0	254.0
Sandy loam	4.0	101.6
Loam	0.8	20.3
Silt loam	0.5	12.7
Clay loam	0.3	7.6
Silty clay	0.1	2.5
Clay	0.2	5.1

These are examples only. Considerable variation exists from these values within each soil texture.

a moderately fine-textured soil, about 400 mm/h in moderately coarse soil, and as high as 1500 mm/h in a medium sand. Were there no resistance, the water would just fall as it does in cracks. This shows that the rate of flow depends on the pore size. In field conditions the rate of flow will be greater in sand than in silt, and greater in silt than in clay. Larger pores provide less resistance or friction in the path of flowing water, and the water will spend less energy to flow. Clay swells and shrinks, depending on the air temperature. Swelling may cause pores to close, which may restrict the movement of water; shrinking causes cracks that may facilitate water movement.

13.12 SOIL WATER DYNAMICS

The Darcy equation for movement of water in the soil, as shown in Figure 13.14, can be expressed as

$$u = -K \times \text{hydraulic gradient}$$

or

$$Q = -KA \times \text{hydraulic gradient},$$

where u is the flow velocity (L/T), Q is the flow in the soil (L^3/T), K is the hydraulic conductivity (L/T), A is the flow cross-sectional area (L^2), and the hydraulic gradient is the slope of the water table (L/L), which can be expressed as $(H_1 - H_2)/L$, where H_1 is the hydraulic head at point 1, H_2 is the hydraulic head at point 2, and L is the length of the column. The hydraulic gradient is also expressed as dH/dL, where dH is the change in head over distance dL. The values of saturated hydraulic conductivity of some soils are listed in Tables 13.6 and 13.7.

Example 13.7: As shown in Figure 13.15, the head is 15 cm at the upstream end of a 25-cm long soil column and 10 cm at the downstream end. The diameter of the soil column is 20 cm. The hydraulic conductivity is given as 4 cm/h. Compute the flow and flow velocity.

Solution:

$$u = -K \times \text{hydraulic gradient} = -K \times \frac{\Delta H}{L}$$

$$= -4\frac{\text{cm}}{\text{hour}} \times \frac{(10\text{ cm} - 15\text{ cm})}{25\text{ cm}} = 0.8\frac{\text{cm}}{\text{hour}},$$

$$Q = -KA \times \frac{\Delta H}{L}$$

$$= -4\frac{\text{cm}}{\text{hour}} \times \left(\pi\left(\frac{20\text{ cm}}{2}\right)^2\right) \times \frac{(10\text{ cm} - 15\text{ cm})}{25\text{ cm}}$$

$$= 251.3\frac{\text{cm}^3}{\text{hour}}.$$

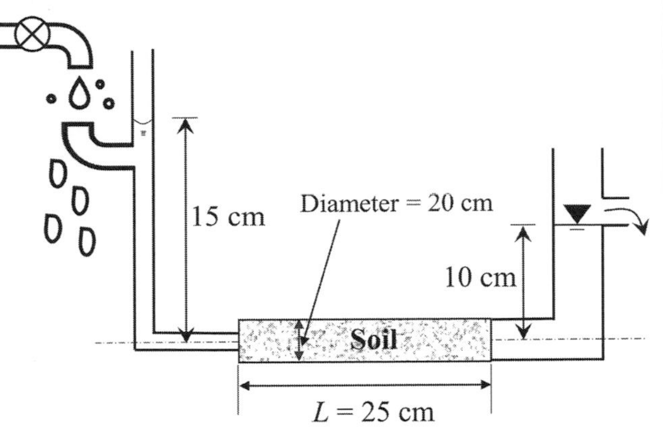

Figure 13.15 Flow through a soil layer.

QUESTIONS

Q.13.1 Consider three types of soil: sandy loam, loam, and clay loam. The following characteristics of these soils are known.

Soil	Field capacity (%)	Crop extractable water (%)	Total available moisture: Volume basis (%)	Total available moisture: Depth basis (mm/m)
Sandy loam	21, range: 15–27	9, range: 6–12	12, range: 9–15	120, range: 90–150
Loam				
Clay loam	31, range: 25–35	14, range: 11–17	17, range: 14–20	170, range: 140–200
	36, range: 31–42	18, range: 15–20	19, range: 16–22	190, range: 160–220

Compute the following: (a) total available moisture for each soil in mm; (b) the available moisture in the root zone if the available water is depleted by 20%, 30%, 40%, and 50%. Take the root zone depth as 1.4 m.

Q.13.2 Consider the capillary rise of water in a tube at 20 °C. Compute the height of the rise if the diameter of the tube is 1 mm, 0.5 mm, and 0.25 mm.

Q.13.3 A tensiometer is connected to a mercury manometer as shown in Figure 13.16. The top of the manometer is open to the atmosphere. Compute the total energy head at point A in the figure and determine if the point A is below, at, or above the water table. The specific gravity of mercury is 13.6.

Q.13.4 A field soil sample (30 cm diameter × 40 cm height) was obtained from a research facility. The core was weighed and placed in an oven to dry, after which it was reweighed. The water loss was 9766 g, the water density is 1.0 g/cm^3, and dry bulk density is 1.3 g/cm^3. Assume a particle density of 2.65 g/cm^3. (a) Determine V_a and V_w. (b) What is the void ratio?

Q.13.5 If the soil sample in Q.13.6 requires 11,500 cm^3 of water to saturate it, what is the porosity of the sample?

Q.13.6 Calculate ρ_b of 1 m^3 of soil. Assume the volume fractions of solid, liquid, and gas of 0.5, 0.34, and 0.16, respectively. Assume $\rho_s = 2.65$ g/cm^3, $\rho_w = 1.0$ g/cm^3, and $\rho_g = 1.3$ kg/m^3.

Q.13.7 Water is applied to a research area to evaluate water movement. If $\rho_b = 1.37$ g/cm^3 and the water content on a mass basis (θ_m) is 0.19, what is the volumetric or volume fraction of water (θ_v) of the research area?

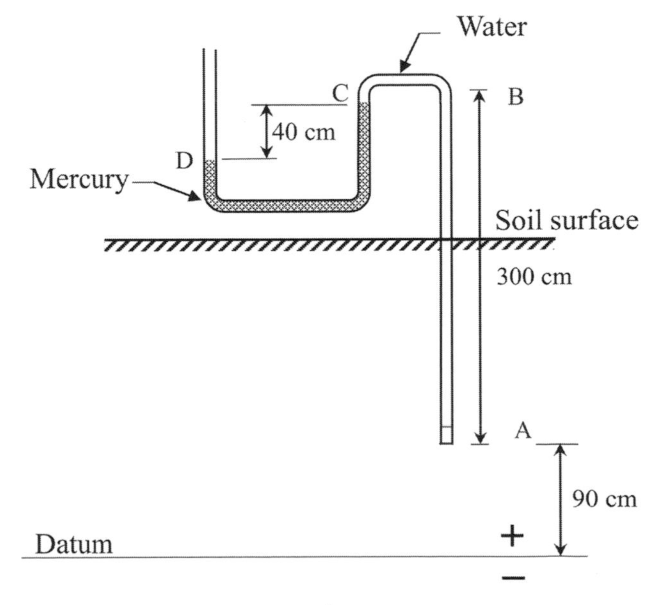

Figure 13.16 A tensiometer connected to a mercury manometer.

Q.13.8 A soil sample collected from the field with a weight of 136 g is placed in an oven to dry at 105 °C for 24 h. After drying it weighs 118 g. Assuming $\rho_b = 1.32\ \text{g/cm}^3$, calculate: (a) water content on a mass basis (θ_m), and (b) on a volume basis (θ_v).

Q.13.9 Using the following data taken from a field sample, compute:

a. average bulk density;
b. percentage moisture, dry and volumetric weight basis for each sample before irrigation;
c. depth of water stored in the 90 cm of soil before irrigation (cm);
d. depth of water stored in the 90 cm of soil after irrigation (cm);
e. depth of water added to the 90 cm of soil.

		Before irrigation		After irrigation
Depth of the sample (cm)	Volume of the sample (cm^3)	Weight of wet sample (kg)	Weight of oven-dried sample (kg)	Dry weight moisture content (%)
0–30	560	0.73	0.62	28.4
30–60	560	0.78	0.64	29.2
60–90	560	0.78	0.67	28.7

Q.13.10 Use the data provided to:

a. compute the soil moisture content (dry weight and volumetric) in each soil layer – assume a bulk density of $1.30\ \text{g/cm}^3$;
b. calculate, if the permanent wilting point of this soil is 12% (dry weight basis) and the soil has an average available water holding capacity of 1.77 mm per centimeter of soil, how much water must be applied to bring the soil to field capacity.

Depth of sample (cm)	Weight of wet sample (g)	Weight of oven-dry sample (g)
0–25	129	108
25–50	131	106
50–75	133	108
75–100	133	109

Q.13.11 Assume a loamy soil with an average apparent specific gravity of 1.5, an available water holding capacity of 12% (volume basis), and a field capacity of 21% (volume basis).

Depth of sample (cm)	Volumetric soil moisture (%)
0–30	13
30–60	16
60–90	18

a. Compute the fraction of soil moisture depletion in each soil depth increment.
b. What is the average soil moisture in the soil profile?
c. What is the approximate soil moisture "tension" in each soil depth?
d. How much water must be added to bring the soil profile to field capacity?

Q.13.12 A farmer wants to establish 20 ha of grass. The root zone of the grass is 40 cm deep. The volumetric water content has to be increased from 14% to 28%. How many cubic meters of water will be needed?

Q.13.13 A tensiometer open to the atmosphere with a porous cup on the end is placed in the field, as shown in Figure 13.17.

a. If H_{Hg} is 12 cm, is point A above or below the water table? What is the potential (cm water)?
b. If H_{Hg} is 2 cm, is point A above or below the water table? What is the potential (cm water)?

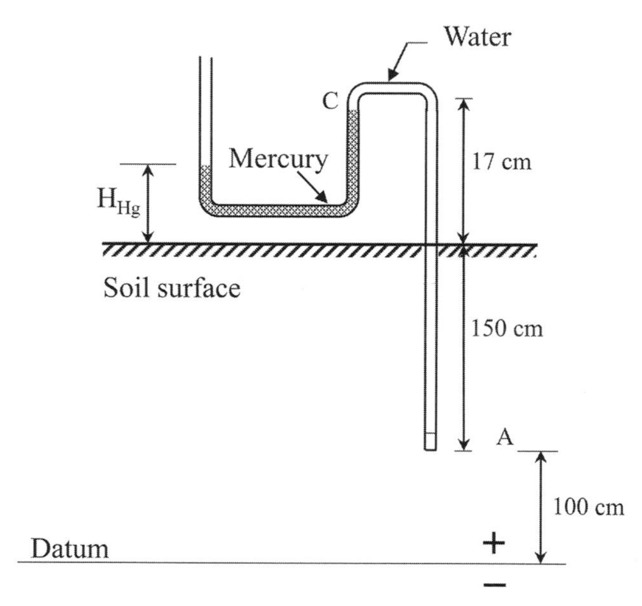

Figure 13.17 A tensiometer connected to a mercury manometer.

Q.13.14 A soil column for measuring the unsaturated hydraulic conductivity is set up in a cylinder, as shown in Figure 13.18. The conductivity is assumed to correspond to the average pressure potential. When the system reaches its steady state, $Q = 125\ \text{cm}^3/\text{h}$. What is the direction of the flow through the sample (up or down), the average pressure potential, and the value of hydraulic conductivity (cm/h)?

Q.13.15 A similar column as in Q.13.14 is set up. For this problem, mercury is used in the manometer to enable reading higher tensions (Figure 13.19). When the system attains its steady state,

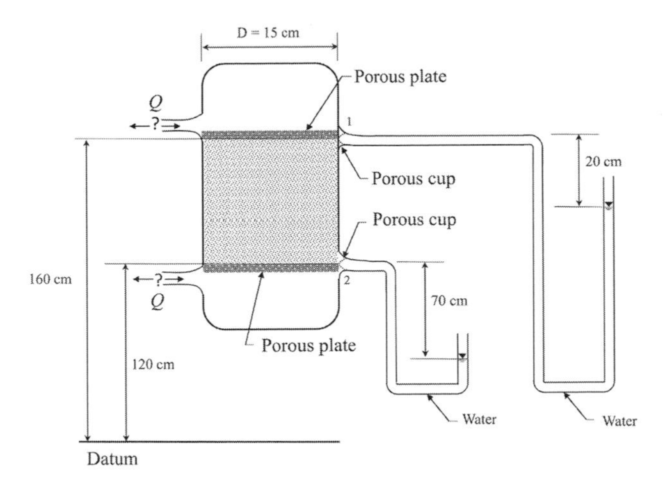

Figure 13.18 A soil column for measuring the unsaturated hydraulic conductivity in a cylinder.

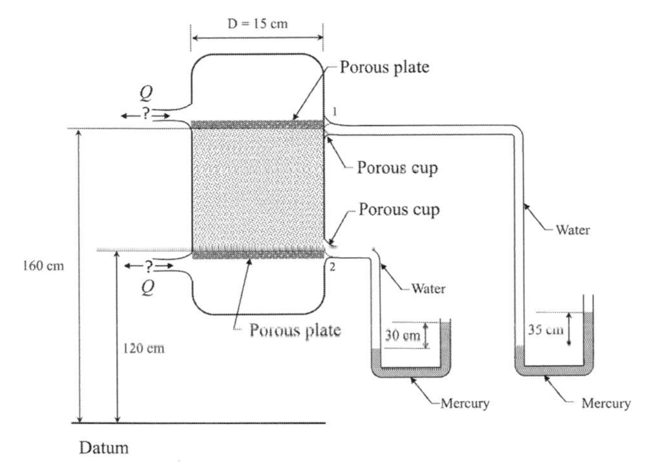

Figure 13.19 A soil column for measuring the unsaturated hydraulic conductivity in a cylinder (mercury manometer).

$Q = 55$ cm^3/h. What is the direction of the flow through the sample (up or down), the average pressure potential, and the value of hydraulic conductivity (cm/h)?

Q.13.16 The following data are given:

Layer, ft (cm)	f_r
0–1 (0–30.5)	0.55
1–2 (30. 5–61.0)	0.65
2–3 (61.0–91.4)	0.60
3–4 (91.4–122.0)	0.80
4–5 (122.0–152.4)	0.80
5–6 (152.4–182.9)	0.80
6–7 (182.9–213.4)	0.80

Fraction of available water depleted

$$= f_d = \frac{\theta_{fc} - \theta_v}{\theta_{fc} - \theta_{wp}},$$

Fraction of available water remaining

$$= f_r = \frac{\theta_v - \theta_{wp}}{\theta_{fc} - \theta_{wp}},$$

$$f_r = 1 - f_d,$$

$$SWD = f_d(AWC)L.$$

a. If the volumetric water contents at field capacity and wilting point are 0.35 and 0.13, respectively, how deep would 4.2 inches (10.7 cm) of infiltrated water penetrate into the soil profile? Give the answer in inches. Assume sandy loam.
b. If the root zone depth is 30 inches (76.2 cm), how many inches of water would percolate below the root zone?

Q.13.17 The following characteristics are given for a soil:

Layer	Depth, in. (cm)	θ_{fc}	θ_v
1	0 ~ 12 (0 ~ 30.5)	0.34	0.2
2	12 ~ 30 (30.5 ~ 76.2)	0.4	0.33
3	30+ (76.2+)	0.3	0.24

Calculate the following:
a. The depth to which 4 inches (10.2 cm) of infiltrated water would penetrate.
b. The depth of water that was lost by deep percolation if the depth of the root zone was 28 inches (71.1 cm).

REFERENCES

Blencowe, J. P. B., Moore, S. D., Young, G. J., et al. (1960). Soil. U.S. Department of Agriculture Bulletin 462.

Cuenca, R. H. (1989). *Irrigation System Design: An Engineering Approach*. Englewood Cliffs, NJ: Prentice Hall.

Doorenbos, J., and Pruitt, W. O. (1977), *Crop Water Requirement*. Rome: FAO.

USDA-NRCS (2003). Saturated hydraulic conductivity: water movement concepts and class history. USDA soil survey technical note 6.

14 Evapotranspiration

Notation

a_w, b_w	constants in the wind function
A	available energy (MJ/m^2·day)
C_p	specific heat at constant pressure of dry air (1.0035 kJ/kg·°C)
C_T	temperature coefficient which is determined as a constant for a given area (°C or °F)
d	zero-plane displacement height, at which the wind speed is considered as zero (m)
D	vapor pressure deficit (kPa)
Dec	declination, i.e., the difference between the magnetic north and the true north (radians)
Dr	relative distance of the sun from the Earth (L)
e_1	saturation vapor pressure of water at the mean monthly minimum air temperature of the warmest month in the year (kPa, mb)
e_2	saturation vapor pressure of water in mb at the mean monthly maximum air temperature of the warmest month in the year (kPa, mb)
e_a	actual vapor pressure or saturation vapor pressure at dew-point (kPa, mb)
e_s	saturated vapor pressure (kPa, mb)
e_{s-dp}	saturation vapor pressure at dew-point or actual vapor pressure (kPa, mb)
E	evaporation (latent heat) flux (MJ/m^2·day, Langley, cal/cm^2·day, mm/day)
E_l	elevation of the field (m)
ET_r	reference evapotranspiration from grass surface (or alfalfa) (mm/day)
$f(u)$	wind function as a function of wind velocity
g	acceleration due to gravity (9.81 m/s^2)
G	soil heat flux (MJ/m^2·day, Langley, cal/cm^2·day, mm/day)
h_c	mean crop height (m)
H	sensible heat flux (MJ/m^2·day, Langley, cal/cm^2·day, mm/day)
I_0	extraterrestrial radiation (MJ/m^2·day, Langley, cal/cm^2·day, mm/day)
J	Julian day (e.g., January 1 = 1)
K	hydraulic conductivity (L/T)
Lat	latitude (radians)
M	number of month (e.g., January, $M = 1$)
Om	sunset hour angle (radians)
p	percentage of daytime hours of a day compared to the entire year (fraction or percentage)
P	atmospheric pressure (mb)
P_0	standard sea-level atmospheric pressure (mb)
Q	rate of flow in the xylem (m^3/s per unit width)
r_a	aerodynamic resistance (s/m)
r_{plant}	resistance to the movement of water into the roots up the xylem and into the leaf (s/m)
r_s	surface resistance of the land cover (s/m)
r_{soil}	resistance to the movement of water in the soil (s/m)
R	universal gas constant for air (287 J/kg·K)
RH	relative humidity (percentage)
RH_{max}	maximum relative humidity (percentage)
RH_{mean}	mean relative humidity (percentage)
RH_{min}	minimum relative humidity (percentage)
R_L	net outgoing longwave radiation (MJ/m^2·day, Langley, cal/cm^2·day, mm/day)
R_{L0}	net outgoing clear-sky longwave radiation (MJ/m^2·day, Langley, cal/cm^2·day, mm/day)
R_n	net radiation flux (MJ/m^2·day, Langley, cal/cm^2·day, mm/day)
R_s	solar radiation (MJ/m^2·day, Langley, cal/cm^2·day, mm/day)
T	average of maximum and minimum daily temperatures (°C)
T_0	standard sea-level temperature (288 K)
T_a	air temperature (°C) or dry-bulb temperature (°C)
TD	average daily temperature range for the period under study (mean daily maximum minus mean daily minimum) (°C)
T_d	dew-point temperature (°C)
T_{dry}	dry-bulb temperature (°C) or air temperature (°C)
T_m	average monthly temperature (°C)
T_{max}	maximum air temperature (°C)
T_{mean}	mean air temperature for the period of interest (°C)
T_{min}	minimum air temperature (°C)

T_p	mean air temperature for the three preceding days (°C)	γ	psychrometric constant (the relationship between vapor pressure deficit and wet-bulb depression) (kPa/°C, mb/°C)
T_w	wet-bulb temperature (°C)	δ	standard lapse rate (K/m)
T_x	intercept of the temperature axis of the linear relationship between ET_0/R_s and air temperature in the JH method (°C)	Δ	slope of the curve of saturation vapor pressure versus temperature at mean temperature (kPa/°C, mb/°C)
u_2	wind velocity at a height of 2 m (km/d, m/s)		
u_{2day}	mean daytime wind velocity at 2 m height (m/s)	ε	effective emissivity
u_{day}/u_{night}	ratio of wind run during the daytime to wind run during the nighttime	ε_w	ratio of the mass of water vapor to mass of dry air (0.62198)
U_z	wind velocity measured at the height of z (m/s)	λ	latent heat of vaporization (MJ/kg, kJ/kg, cal/g)
Z_e	respective heights for humidity measurements (m)	ρ_a	density of air (kg/m^3)
Z_{om}	roughness height governing momentum transfer (m)	σ	Stephan–Boltzmann constant ($\sigma = 4.903 \times 10^{-9}$ MJ/m^2/day/K^4)
Z_{ov}	roughness height governing heat and vapor transfer (m)	ψ_{leaf}	total water potential in the leaf (m of water, kPa, bar, psi, lb/in.^{2})
Z_u	respective heights for wind speed measurements (m)	ψ_{soil}	total water potential in the soil (m of water, kPa, bar, psi, lb/in.^{2})
α	albedo. The value of α is 0.25 for the FAO-modified Penman method and varies with the time of the year to include the effect of sun angle for the Wright modification	ψ_T	turgor pressure in the leaf (m of water, kPa, bar, psi, lb/in.^{2})
		ψ_0	osmotic pressure within the plant (m of water, kPa, bar, psi, lb/in.^{2})

14.1 INTRODUCTION

The objective of irrigation is to supply water to the soil so crops can extract the water they need. The plants then store and transpire the extracted water. The soil also evaporates water. Thus, evaporation and transpiration together constitute the major component of water that is supplied through irrigation. In hydrologic parlance, this is often referred to as loss of water. In actuality it is not loss of water, but a change of phase from liquid to vapor and change of system from soil and vegetation to the atmosphere. The water remains within the hydrologic cycle. The evaporation occurs not only from the soil surface but also from deep layers, and depends on a number of factors, such as amount of water available, radiation and its intensity, wind velocity, temperature, humidity, land use, and cloud cover.

Energy is required for evaporation to occur and the primary source of this energy is heat due to radiation from the sun. Evaporation is higher from wet soils than from dry soils for two reasons. First, when the soil is saturated, the capillary force is nonexistent and it is easier for water to change to vapor. Wet soils tend to be darker and absorb more heat than do light soils; this greater amount of absorbed heat leads to more evaporation. Wind with greater velocity is usually drier than is wind with less velocity, and hence dries the soil, enhancing evaporation. Were there no wind, a thin layer of air close to the soil surface may become quite humid and would decrease evaporation. The drying of soil starts from the surface – that is, the top layer dries first and the process of drying moves down as evaporation continues. When the top layer is drier than the layer below, water moves up from below through capillaries. This occurs rather quickly when the soil below the relatively dry top layer is near or above the field capacity, where paths are well connected, because capillary forces in the top dry layer are relatively large because the water is held tightly but those in the layer below are relatively small where the water is held relatively less tightly. This pressure differential facilitates the upward movement of water. However, as the thickness of the dry layer increases the upward movement slows down, but because of heat the soil water changes to vapor, which leaves the soil easily.

In the night, the soil surface cools down first and the 100% humidity in the soil yields vapor that can condense and become moist in the top layer; the following day it gets evaporated. Thus, after irrigation evaporation occurs in different ways. The top 1 m of soil loses its water primarily through evaporation and transpiration; depending on soil texture, crop type and cover, amount of available water, wind, and cloud cover, this loss of water can be 20–50 mm between two consecutive irrigations.

In cropland, water evaporates from soil and is transpired by plants – that is, water is transported to the atmosphere by evaporation plus transpiration, which together constitute evapotranspiration (ET). Determination of ET is fundamental to determining crop water requirements. Evapotranspiration is an energy-driven process. The energy balance constitutes the basis of quasi-theoretical methods that have been developed for determining evapotranspiration. Empirical methods are also based on data that reflect some components of the energy balance. This chapter discusses the process of evaporation and some of the methods that are used in irrigation engineering to compute it.

14.2 EXTRACTION OF WATER BY PLANTS

When the soil is quite wet, say at field capacity, the water is loosely held and can move easily. The energy potential of soil water is relatively high. Crops transpire water and transpiration creates suction or low water potential in the plants. Because of the energy gradient, water moves from the soil to the plant through roots, and the roots suck up the water. As a result, the water film around soil particles starts to thin and reach lower energy potential. However, the water film away from the root zone is still thick and is at a higher potential; hence, the water moves within the soil toward the roots. The movement of water from the soil to the plants continues until the energy potential of water in the soil is the same as the energy potential in the plants. At this stage, plants will be unable to extract water and will become stressed, and will wilt and eventually die. The purpose of irrigation is to supply water much before this stage is reached.

14.3 ROOT SYSTEM

Different plants, such as orchard crops, vegetables, grain crops, and grasses, have different root systems and different patterns of root development. Seasonal, annual, and permanent crops significantly differ in their ability to access and take up soil water. In the case of permanent plants, the basic root system exists in the same volume of soil for years, but extends a little each year. The extending root zone extracts water and transmits it into the plant. The nature of the root zone has implications for irrigation. In the case of trees, the root system permeates the whole soil mass and after a few years roots of adjacent trees will intersect. In such a scenario, when irrigation covers part of the area, which is not unusual in orchard irrigation, water is applied more frequently. The roots in the intervening areas do not die because they are in the 100% humidity environment, but do not grow new hair roots for extracting moisture.

Annual plants extend their root zones into new areas looking for water and nutrients. Annual crops also extend their root systems into new areas that are moist but will not grow in dry soils. For some crops, such as cotton, the roots may extend 20–30 mm per day. The irrigation water near the plants will be used up first and later farther away and deeper as the root zone extends. This suggests that irrigation water must be applied first to the dry soil near the plant and later farther away, to where the root zone has extended.

For seasonal crops, the root zone permeates the soil mass and comes within the capillary reach of water. The root zone may occupy a large portion of the soil volume, with more roots near the top than below. For example, one rye grass plant in one season can nearly fill a box of size 2 m × 2 m × 2 m, containing several hundred kilometers of roots at a growth rate of 5 km per day. The hair roots extract the moisture from the soil by suction and osmotic force. The extraction of moisture is much larger near the top layer than near the bottom layer, where roots are fewer.

The root development does not occur uniformly in space and time. For example, for a tree the main roots extend first without concentration between them and a year or two later the root zone may become quite uniform. The same may happen for annual plants and crops. For seasonal crops root systems may be erratic. For example, beans and potatoes have root systems where there is intermingling of roots from adjacent plants.

14.4 SOIL WATER EXTRACTION BY ROOTS

The extraction of soil water by roots depends on the way the root zone develops. For annual crops the root system extends both vertically and horizontally in time, but not uniformly. For closely growing crops, the roots essentially expand vertically because the nearby plants also expand their roots laterally and a relatively solid root system develops. The extraction of water occurs more in the upper soil layers, where roots have been for a longer time. The water-absorbing hair roots extend into moist soil, are renewed continually, do not remain active for many weeks, and have limited life span. The hair roots dry the nearby soil first, which may create the movement of moisture within the soil. As the soil gets drier, capillaries break down into a series of disconnected pores; as a result the water cannot move far. The rate at which plants can extract water depends on the root zone development, soil moisture tension, and disconnectivity of capillaries. A clay soil will disconnect slower than will sand.

In general, plants extract water that is more easily available close to their roots. For example, plant roots develop better and more closely in the soil near the surface than in deeper layers. That is where water is extracted more easily. Plants extract water from deeper layers when it is not available in upper layers. The development of plant roots depends on the plant type, soil texture and structure, soil compaction, depth of the water table, salt concentration, and depth of wetting in the soil profile.

The main reason for the movement of water from soil into plants and eventually into leaf tissues is the evaporation of water from stomata and leaf surfaces. This evaporation is referred to as transpiration, and is affected by plants, soil, and the atmospheric conditions, such as radiation, temperature, wind velocity, cloud cover, humidity, and sunshine. The soil water loses its energy as it moves from the soil to the plant leaf to the atmosphere. As long as plants can extract enough moisture to meet their transpiration demands, they will remain turgid and will grow without stress. The purpose of irrigation is to maintain this condition. When plants cannot extract sufficient water to meet their transpiration demand and for photosynthesis, plants lose turgidity and show symptoms of being less than healthy, and eventually wilt. At this stage, plants close their stomata to stop the loss of water. The crop yield is considerably affected if this condition persists for a considerable period of time.

14.5 ENERGY BALANCE

There are four main energy components or fluxes: net radiation flux (R_n), soil heat flux (G), evaporation (latent heat) flux (E) , and sensible heat flux (H). The soil heat flux is positive if soil is warming, evaporation flux is positive if evaporation is occurring, and sensible heat flux is positive if air is warming. The energy balance can be expressed as the net radiation flux equal to the sum of soil heat flux, evaporation flux, and sensible heat flux:

$$R_n = G + E + H. \tag{14.1}$$

As discussed in Chapter 2 on climate, the radiation from the sun is always shortwave and the radiation emanating from the Earth (infrared) is always longwave (Rosenberg et al., 1983). The net radiation is the sum of incoming and outgoing radiation. The radiation reflected from the atmosphere to the Earth is also longwave (infrared).

14.6 METHODS FOR COMPUTING EVAPOTRANSPIRATION

There are different methods for computing ET of crops that are used in irrigation engineering. Some of the methods that are used more commonly are described here.

14.6.1 Blaney–Criddle (BC) Method

Blaney and Criddle (1945, 1950) proposed a method for computing ET which is based on the hypothesis that ET is proportional to the product of mean air temperature and day-length percentage. The monthly constant of proportionality is referred to as the crop growth stage coefficient. Several variants of the BC method have been proposed. A well-known variation in the United States is described in the technical release of the Soil Conservation Service (SCS) (now called the Natural Resources Conservation Service or NRCS) of the US Department of Agriculture (USDA) (SCS, 1970). Another major variant of the BC method was reported by the Food and Agriculture Organization (FAO) of the United Nations (Doorenbos and Pruitt, 1977). This variant first estimates a crop reference ET for grass and is expressed as

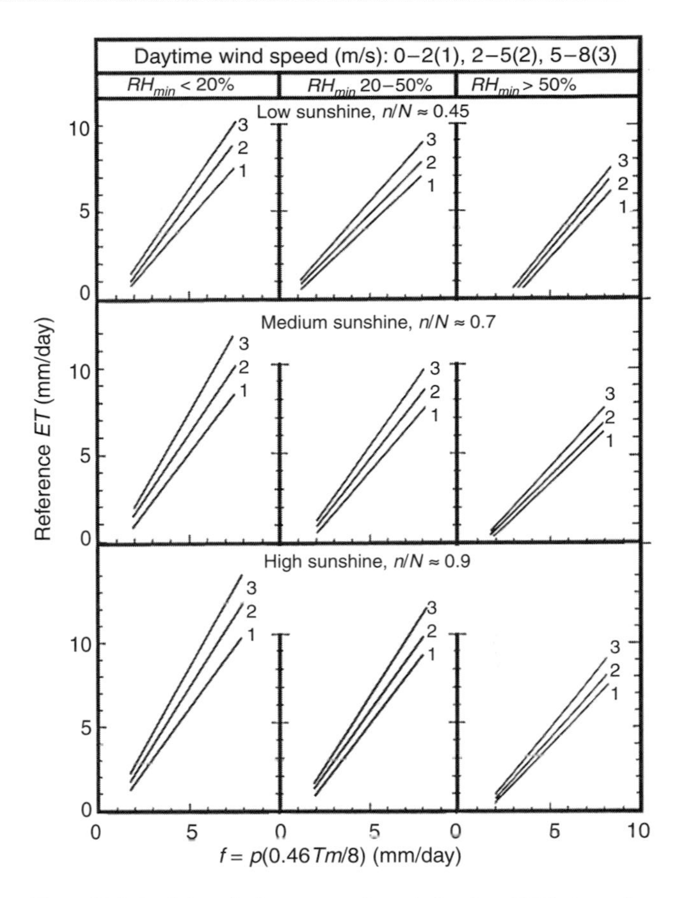

Figure 14.1 Prediction of reference evapotranspiration from BC f factor under different climate conditions. (Adapted from Doorenbos and Pruitt, 1977.)

$$ET_r = a + bf, \quad f = p(0.46T_m + 8), \tag{14.2}$$

where ET_r is the reference ET in mm/day; p is the percentage of daytime hours in a day compared to the entire year; T_m is the average monthly temperature in °C; and a and b are the intercept and the slope of a straight-line relation between ET_r and f (Figure 14.1). The BC method is suitable for monthly ET estimates.

Example 14.1: Estimate ET from a field using the BC method. The following information is given: the percentage (p) of daytime hours in August is 0.31. The field weather data are given as: mean maximum air temperature = 30 °C; mean minimum air temperature = 12 °C; mean air temperature = 19 °C; mean dew-point air temperature = 8.5 °C; mean wind speed = 205 km/day; ratio of wind speed (day/night = 3.5; mean percent sunshine (n/N) = 90%; mean daytime length = 12 h; mean pan evaporation = 8.5 mm/day; mean measured alfalfa ET = 8.0 mm/day; latitude = 42° N; mean solar radiation (R_s) = 600 Langley/day; crop = corn; and site elevation (E_l) = 1150 m.

Solution:

$f = p(0.46T_m + 8),$

$p = 0.31(\text{August}, 42° \text{ N}),$

$f = 0.31(0.46 \times 19 + 8) = 5.19.$

The RH_{min} is required for the BC method:

$$RH_{min} = 100 \times \left[\frac{\text{actual vapor pressure (mean dew-point temperature)}}{\text{saturated vapor pressure (mean maximum daily air temperature)}}\right]$$

$$= 100\% \times \left[\frac{e\ (\text{at } 8.5\ °\text{C})}{e\ (\text{at } 30\ °\text{C})}\right] = 100\% \times \frac{11\ \text{mb}}{42\ \text{mb}} = 26.2\%$$

The wind speed in the BC method is the daytime wind speed:

wind run in km for 24 hour = 205 km;

$$\text{daytime wind run} = 205\ \text{km/day} \times \frac{3.5}{1 + 3.5} = 159.44\ \text{km}$$

$$\text{daytime wind speed} = \frac{159.44\ \text{km}}{12\ \text{hour}} = 13.29\frac{\text{km}}{\text{h}}\ \text{or } 3.69\ \text{m/s};$$

Read f and ET_r from Figure 14.1:

$$f = 5.19 \rightarrow \left(RH_{min} = 26\%, u_{day} = 3.69\frac{\text{m}}{\text{s}}, \text{and } \frac{n}{N} = 0.9\right) \rightarrow ET_r = 6.5\ \text{mm/day}.$$

The result of the BC method is for the reference grass. Given the surface is corn, a crop coefficient could be used to determine crop ET. For corn, one can multiply by a crop coefficient of 1.2: $ET_{corn} = 1.2 \times 6.5\ \text{mm/day} = 7.8\ \text{mm/day}$. More details of the crop coefficients will be discussed in Chapter 15.

14.6.2 Jensen–Haise (JH) Method

Jensen and Haise (1963) proposed a method for estimating ET for a reference crop (such as alfalfa) using temperature and radiation data, which has since been revised (Jensen, 1974). The method can be expressed as

$$ET_r = C_T(T_{mean} - T_x)R_s, \tag{14.3}$$

where R_s is the solar radiation in the same units as ET_r; T_{mean} is the mean air temperature for the period of interest in °C; T_x is the intercept of the temperature axis of the linear relationship between ET_r/R_s and air temperature, which increases as the humidity increases; C_T is the temperature coefficient, which is determined as a constant for a given area, and can be estimated using the following equations if only air temperature data are available:

$$C_T = \frac{1}{C_1 + 7.3C_H}, \tag{14.4}$$

$$C_H = \frac{50\ \text{mb}}{e_2 - e_1}, \tag{14.5}$$

in which the value of 7.3 in eq. (14.4) is for air temperature in °C, and the value is 13 in °F; e_2 is the saturation vapor pressure of water in mb at the mean monthly maximum air temperature of the warmest month in the year (from long-term climatic data), and e_1 is the saturation vapor pressure of water in mb at the mean monthly minimum air temperature of the warmest month in the year; the value of 50 mb is about the maximum value found anywhere of $(e_2 - e_1)$. Hence, the smallest value of C_H is close to 1. C_1 and T_x can be determined by the following equations:

$$C_1 = 38\ °\text{C} - \frac{2\ °\text{C}\ E_l}{305\ \text{m}}, \tag{14.6}$$

$$T_x = -2.5 - 0.14(e_2 - e_1) - \frac{E_l}{550\ \text{m}}, \tag{14.7}$$

where E_l is the elevation of the field in meters. The JH method is recommended for five-day to one-month periods (Jensen, 1974), and local calibration of C_T and T_x is desirable.

Example 14.2: Estimate ET using the Jensen–Haise (JH) method using the field weather data given in Example 14.1.

Solution: The saturated vapor pressure (e_s) can be computed for a given value of temperature using the following equation,

$$e_s = 6.108 \times \exp\left[\frac{17.27T_a}{T_a + 237.3}\right],$$

$e_1 = 14.0$ mb for 12 °C,

$e_2 = 42.4$ mb for 30 °C,

$$C_H = \frac{50\ \text{mb}}{e_2 - e_1} = \frac{50}{42.4 - 14} = 1.76,$$

$$C_1 = 38 - \frac{2E_l}{305} = 38 - \frac{2 \times (1150\ \text{m})}{305} = 30.46,$$

$$C_T = \frac{1}{C_1 + 7.3C_H} = \frac{1}{30.46 + 7.3 \times (1.76)} = 0.0231,$$

$$T_x = -2.5 - 0.14(e_2 - e_1) - \frac{E_l}{550} = -2.5 - 0.14(42.4 - 14.0) - \frac{1150}{550} = -8.57\ °\text{C},$$

$$ET_r = C_T(T_{mean} - T_x)R_s = 0.0231(19 - (-8.57))\left(600\frac{\text{Langley}}{\text{day}}\right) = 382.12\frac{\text{Langley}}{\text{day}}.$$

The standard unit for radiation is MJ/m^2/day or as equivalent evaporation in mm per day (mm/day):

$$1 \text{ mm/day} \cong 2.45 \text{ MJ/m}^2\text{/day}$$

$$1 \text{ Langley} = 1 \text{ cal/cm}^2 = 0.04184 \text{ MJ/m}^2\text{/day}$$

$$1 \text{ Langley} \cong \frac{0.04184}{2.45} \cong 0.017 \text{ mm},$$

$$ET_r = 382.12\frac{\text{Langley}}{\text{day}} \times \frac{0.017 \text{ mm}}{1 \text{ Langley}} = 6.5 \text{ mm/day}.$$

The result of the JH method is for the reference crop. Given the crop surface is corn here, a crop coefficient could be used to determine crop ET. For corn, one can multiply by a crop coefficient of 1.2: $ET_{corn} = 1.2 \times 6.5$ mm/day $= 7.8$ mm/day.

14.6.3 Hargreaves Method

Hargreaves (1975), Hargreaves et al. (1985), and Hargreaves and Samani (1985) proposed a method for estimating *ET* using data on temperature and latitude. The method can be expressed as

$$ET_r = 0.0023 \times I_o \times (T + 17.8) \times TD^{0.50}, \tag{14.8}$$

where T is the mean daily air temperature in °C; I_o is the extraterrestrial radiation in mm of equivalent of water evaporation per day at 20 °C. I_o and ET_r are in the same units; I_o is expressed as

$$I_o = 15.67Dr(Om \times \sin(Lat) \times \sin(Dec) + \cos(Lat) \times \cos(Dec) \times \sin(Om)), \tag{14.9}$$

where *Lat* is the latitude in radians (1 radian = 57.2958 degrees, negative for southern latitudes); *Dec* is the declination – that is, the difference between the magnetic north and the true north, in radians expressed as

$$Dec = 0.40876 \times \cos(0.0172142 \times (J + 192)), \tag{14.10}$$

in which J is the Julian day (e.g., January 1 = 1), *Dr* is the relative distance of the sun from the Earth expressed as

$$Dr = 1.0 + 0.033 \times \cos(0.0172142J), \tag{14.11}$$

and *Om* is the sunset hour angle in radians for latitude <55° expressed as

$$Om = A\cos(-\tan(Lat) \times \tan(Dec)). \tag{14.12}$$

The average I_o for a month can be determined for approximate values of J for the middle of each month as

$$J = 15 + (30.5 \times (M - 1)), \tag{14.13}$$

where M is the number of the month (e.g., January, $M = 1$).

Hargreaves (1994) noted that the average temperature of 10 °C or of 30 °C would produce values of I_o that will differ from those for 20 °C by less than 1.0%. T °C can be computed as the average of maximum and minimum daily temperatures, and *TD* is the average daily temperature range for the period under study (mean daily maximum minus mean daily minimum). The Hargreaves method is recommended for five days or more.

Example 14.3: Estimate *ET* for the month of August using the Hargreaves method for field weather data given as follows: latitude = 42° N; month (M) = August; mean air temperature = 19 °C; mean maximum air temperature = 30 °C; mean minimum air temperature = 12 °C; crop = corn; and site elevation (E_l) = 1150 m.

Solution:

$$Lat = 42° \text{ N} = 42 \times \frac{\pi}{180} = 0.733,$$

$$J = 15 + (30.5 \times (M - 1)) = 15 + (30.5 \times (8 - 1)) = 228.5,$$

$$Dec = 0.40876 \times \cos(0.0172142 \times (J + 192)),$$

$$Dec = 0.40876 \times \cos(0.0172142 \times (228.5 + 192)) = 0.236,$$

$$Dr = 1.0 + 0.033 \times \cos(0.0172142J) = 1.0 + 0.033 \times \cos(0.0172142 \times 228.5) = 0.977,$$

$$Om = A\cos(-\tan(Lat) \times \tan(Dec)) = A\cos(-\tan(0.733) \times \tan(0.236)) = 1.789,$$

$$I_o = 15.67Dr(Om \times \sin(\text{Lat}) \times \sin(Dec) + \cos(Lat) \times \cos(Dec) \times \sin(Om)),$$

$$I_o = 15.67 \times 0.977(1.789 \times \sin(0.733) \times \sin(0.236) + \cos(0.733) \times \cos(0.236) \times \sin(1.789)) = 15.08 \text{ mm/day},$$

$$TD \text{ (temperature range)} = 30\,°C - 12\,°C = 18\,°C,$$

$$T \text{ (Average of maximum and minimum temperatures)} = (30\,°C + 12\,°C)/2 = 21\,°C,$$

$$ET_r = 0.0023 \times RA \times (T + 17.8) \times TD^{0.50} = 0.0023 \times 15.08 \times (21 + 17.8) \times 18^{0.50},$$

$$ET_r = 5.7 \text{ mm/day}.$$

Given the crop surface is corn, one can multiply by a crop coefficient of 1.2: $ET_{corn} = 1.2 \times 5.7$ mm/day = 6.8 mm/day.

14.6.4 Penman Method

Hypothesizing that *ET* is a function of radiation balance and aerodynamics, Penman (1948) developed a method, popularly called the combination method, that combines the energy balance with aerodynamics. The radiation balance accounts for the amount of energy available for evaporation and the aerodynamics accounts for the influence of advection over vegetative canopy on the removal of water vapor. The Penman method has undergone a number of modifications primarily in the way of aerodynamics, consisting of wind function and vapor pressure deficit, but its original hypothesis remains intact.

The original Penman method (with SI units) can be expressed as

$$ET_r = \frac{\Delta}{\Delta+\gamma}(R_n - G) + \frac{\gamma}{\Delta+\gamma}[\Delta e]f(u), \tag{14.14}$$

where ET_r is the reference *ET* from grass in mm/day; Δ is the slope of the curve of saturation vapor pressure versus temperature at mean temperature (T_{mean}); γ is the psychrometric constant (relationship between vapor pressure deficit and wet-bulb depression); $\Delta/(\Delta+\gamma)$ is the weighting function for elevation and temperature; R_n is the net radiation flux in mm/day; *G* is the net soil heat flux in mm/day; Δe is the vapor pressure deficit or the difference between saturation vapor pressure (e_s) at air temperature (T_a) in mb and mean actual vapor pressure (e_a) of air in mb, $\Delta e = e_s - e_a$; and $f(u)$ is the wind function as a function of wind velocity and can be defined as

$$f(u) = m(w_1 + w_2u_2), \tag{14.15}$$

where *m* is a conversion constant, w_1 and w_2 are empirical constants, and u_2 is the wind velocity at a height of 2 m in km/day. It may be noted that

$$\frac{\gamma}{\Delta+\gamma} = 1 - \frac{\Delta}{\Delta+\gamma}. \tag{14.16}$$

Cuenca and Nicholson (1982) have emphasized that when applying a particular form of the Penman method, it is important that the method of calculating the vapor pressure deficit is in agreement with the method of computing the wind function used in that particular form. If radiation is expressed as energy per unit area (J/m^2), then it can be converted to equivalent mm/day by dividing the energy unit by the latent heat of vaporization (kJ/kg) and density of water (kg/m^3), and then converting to mm/day.

14.6.5 Wright's Modification

Wright (1982) modified the Penman method for the arid climate of the western United States. He used alfalfa as a reference. His modified form of the Penman equation can be written as

$$ET_r = \frac{\Delta}{\Delta+\gamma}(R_n - G) + \frac{\gamma}{\Delta+\gamma}\left(\frac{15.36}{0.1\lambda}\right)[e_s - e_a][a_w + b_wu_2], \tag{14.17}$$

where ET_r is the reference *ET* for alfalfa in mm/day; λ is the latent heat of vaporization in cal/g; a_w and b_w are constants in the wind function; and other terms are defined as before. The quantities 15.36 and 0.1 λ in the wind function are needed to convert the results to mm/day. Constants a_w and b_w, which indicate the effect of varying day length on the wind velocity, are computed as

$$a_w = 0.4 + 1.4\exp\left\{-[(J-173)/58]^2\right\}, \tag{14.18}$$

$$b_w = 0.007 + 0.004\exp\left\{-[(J-243)/80]^2\right\}, \tag{14.19}$$

where *J* is Julian day of the year.

14.6.6 FAO Modification

Doorenbos and Pruitt (1975, 1977) developed for the FAO of the United Nations a modification of the Penman method, which can be expressed as

$$ET_r = c\left[\frac{\Delta}{\Delta+\gamma}(R_n - G) + \frac{\gamma}{\Delta+\gamma}[\Delta e]f(u)\right], \tag{14.20}$$

where *c* is the calibration coefficient determined from meteorological data, expressed as a function of maximum relative humidity, solar radiation (R_s, mm/day), daytime wind velocity at a 2 m height (u_2, m/s), and ratio of daytime (between 0700 and 1900 hours) wind velocity (u_{day}) to nighttime wind velocity (u_{night}):

$$c = f\left(RH_{max}, u_{day}/u_{night}, R_s, u_2\right). \tag{14.21}$$

The values of the calibration factor *c* are given in Table 14.1. Equation (14.20) is finally expressed as

$$ET_r = c\left[\frac{\Delta}{\Delta+\gamma}(R_n - G) + \frac{\gamma}{\Delta+\gamma}[e_s - e_a](0.27)(1.0 + 0.01u_2)\right]. \tag{14.22}$$

The FAO modification is based on comparison of lysimeter-measured ET from locations worldwide, with the Penman estimates using meteorological data from the same locations.

Based on regression analysis, Cuenca and Jensen (1987) expressed *c* as

$$\begin{aligned} c = {} & 0.68 + 0.0028RH_{max} + 0.018R_s - 0.068u_{2day} \\ & + 0.013\left(u_{day}/u_{night}\right) + 0.0097u_{2day}\left(u_{day}/u_{night}\right) \\ & + 0.43 \times 10^{-4}RH_{max} \times R_s \times u_{2day}, \end{aligned} \tag{14.23}$$

where u_{2day} is the mean daytime wind velocity at 2 m height in m/s, u_{day}/u_{night} is the ratio of wind run during daytime to wind run during nighttime, RH_{max} is in unit of percent, and R_s is the solar radiation in units of mm/day. Allen (1986)

Table 14.1 *Values of calibration coefficient*

	$RH_{max}=30\%$				$RH_{max}=60\%$				$RH_{max}=90\%$			
R_s, mm/day	3	6	9	12	3	6	9	12	3	6	9	12
u_{day}, m/s	$u_{day}/u_{night}=4.0$											
0	0.86	0.90	1.00	1.00	0.96	0.98	1.05	1.05	1.02	1.06	1.10	1.10
3	0.79	0.84	0.92	0.97	0.92	1.00	1.11	1.19	0.99	1.10	1.27	1.32
6	0.68	0.77	0.87	0.93	0.85	0.96	1.11	1.19	0.94	1.10	1.26	1.33
9	0.55	0.65	0.78	0.90	0.76	0.88	1.02	1.14	0.88	1.01	1.16	1.27
	$u_{day}/u_{night}=3.0$											
0	0.86	0.90	1.00	1.00	0.96	0.98	1.05	1.05	1.02	1.06	1.10	1.10
3	0.76	0.81	0.88	0.94	0.87	0.96	1.06	1.12	0.94	1.04	1.18	1.28
6	0.61	0.68	0.81	0.88	0.77	0.88	1.02	1.10	0.86	1.01	1.15	1.22
9	0.46	0.56	0.72	0.82	0.67	0.79	0.88	1.05	0.78	0.92	1.06	1.18
	$u_{day}/u_{night}=2.0$											
0	0.86	0.90	1.00	1.00	0.96	0.98	1.05	1.05	1.02	1.06	1.10	1.10
3	0.69	0.76	0.85	0.92	0.83	0.91	0.99	1.05	0.89	0.98	1.10	1.14
6	0.53	0.61	0.74	0.84	0.70	0.80	0.94	1.02	0.79	0.92	1.05	1.12
9	0.37	0.48	0.65	0.76	0.59	0.70	0.84	0.95	0.71	0.81	0.96	1.06
	$u_{day}/u_{night}=1.0$											
0	0.86	0.90	1.00	1.00	0.96	0.98	1.05	1.05	1.02	1.06	1.10	1.10
3	0.64	0.71	0.82	0.89	0.78	0.86	0.94	0.99	0.85	0.92	1.01	1.05
6	0.43	0.53	0.68	0.79	0.62	0.70	0.84	0.93	0.72	0.82	0.95	1.00
9	0.27	0.41	0.59	0.70	0.50	0.60	0.75	0.87	0.62	0.72	0.87	0.96

Adapted from Doorenbos and Pruitt, 1977.

found the unadjusted FAO modification (Doorenbos and Pruitt, 1975) without the c calibration coefficient to be more accurate than the adjusted modification (Doorenbos and Pruitt, 1977) for some locations.

14.6.7 Priestley–Taylor Method Modification

Priestley and Taylor (1972) suggested a simpler form of the Penman method by recognizing that the first term in eq. (14.14) is larger than the second term by a factor of about 4. Thus, the general form of the Priestley–Taylor equation expressing the relation between reference crop *ET* (mm/day) and radiation can be written as

$$ET_r = a_{pt}\frac{\Delta}{\Delta+\gamma}(R_n - G), \tag{14.24}$$

in which a_{pt} is a parameter equal to 1.26 for humid climates and equal to 1.74 for arid climates.

14.6.8 Penman–Monteith Method

Monteith (1981) modified the Penman equation by considering the resistance related to stomatal and aerodynamic characteristics of crops, which can be written as

$$ET_r = \frac{1}{\lambda}\frac{\Delta A r_a + \rho_a C_p D}{\Delta r_a + \gamma(r_a + r_s)}, \tag{14.25}$$

in which $D = e_s - e_a$ is the vapor pressure deficit (kPa) at the height for which the aerodynamic resistance r_a is computed in s/m, r_s is the surface resistance of the land cover in s/m, λ is the latent heat of vaporization in MJ/kg, C_p is the specific heat in MJ/kg·°C, Δ is the slope of the vapor pressure curve in kPa/°C, γ is the psychrometric constant in kPa/°C, A is the available energy, i.e., $R_n - G$, in unit of MJ/m^2·day, and ρ_a is the density of air in kg/m^3. Note that the term $\rho_a C_p D$ in eq. (14.25) needs to multiply by 86,400 s/day to make sure the units are the same.

The aerodynamic resistance r_a (s/m) is computed as

$$r_a = \frac{\ln\left[\frac{Z_u - d}{Z_{om}}\right]\ln\left[\frac{Z_e - d}{Z_{ov}}\right]}{(0.41)^2 U_Z}, \tag{14.26}$$

where Z_u and Z_e are the respective heights for wind speed and humidity measurements in m, d is the zero-plane displacement height at which the wind speed is considered as zero, in m, Z_{om} is the roughness height governing momentum transfer in m, Z_{ov} is the roughness height governing heat and vapor transfer, the value 0.41 is the von Kármán constant, and U_Z is the wind speed in m/s. For estimating

r_a, $Z_{om}=0.123\ h_c$, $Z_{ov}=0.0123\ h_c$, and $d=0.67\ h_c$, where h_c is the mean crop height in meters.

To reduce the impact of crop surface, a hypothesis reference grass surface is suggested by FAO for the Penman–Monteith method with an assumed grass height of 0.12 m, an albedo of 0.23, and a fixed surface resistance of ($r_a=70$ s/m) (Allen et al., 1998). The reference level of weather measurements is 2 m height. In this case, aerodynamic resistance r_a in eq. (14.26) can be rewritten as $r_a=208/u_2$, where u_2 is the wind speed at 2 m height.

14.7 COMPUTATION OF INPUT PARAMETERS

The Penman method and its modifications have parameters that can be estimated empirically, following Burman et al. (1983) and Jensen (1974), among others, as follows.

Slope of saturation vapor pressure (Δ in mb/°C): This can be computed as

$$\Delta=2.00\times(0.0073T_{mean}+0.8072)^7-0.00116, \quad (14.27)$$

where T_{mean} is the mean air temperature for the period of interest in °C.

Psychrometric constant (γ, mb/°C): The vapor pressure deficit is related to the wet-bulb depression as

$$e_s-e_a=\gamma(T_a-T_w), \quad (14.28)$$

where e_s is the saturated vapor pressure corresponding to wet-bulb temperature (T_w), and e_a is the actual vapor pressure corresponding to the dry-bulb temperature or air temperature (T_a).

The psychrometric constant can be expressed as

$$\gamma=\frac{c_pP}{\lambda\varepsilon_w}, \quad (14.29)$$

where P is the atmospheric pressure in mb, c_p is the specific heat at constant pressure of dry air = 1.0035 kJ/kg·°C, λ is the latent heat of vaporization in kJ/kg, and ε_w is the ratio of mass of water vapor to mass of dry air = 0.62198. For irrigation purposes, eq. (14.29) can be simplified as

$$\gamma=1.6134\frac{P}{\lambda}. \quad (14.30)$$

The atmospheric pressure varies with elevation. The relation between the US standard atmospheric pressure (mb) and elevation is given as

$$P=1013-0.1055\times\text{elevation}, \quad (14.31)$$

where elevation is given in meters. The expression for any standard atmospheric pressure can be given as

$$P=P_0\left[\frac{T_0-\delta(\text{elevation}-\text{base elevation})}{T_0}\right]^{g/(\delta R)}, \quad (14.32)$$

where P_0 is the standard sea-level atmospheric pressure in mb, T_0 is the standard sea-level temperature in K (K=273.15+°C), g is the acceleration due to gravity = 9.81 m^2/s, R is the universal gas constant for air = 287 J/kg·K, and δ is the standard lapse rate in K/m. For the US standard atmosphere, these values are: the standard sea-level pressure, 1013 mb; the standard lapse rate, 0.00976 K/m; and the standard sea-level temperature, 288 K.

Latent heat of vaporization: The latent heat of vaporization is often taken as a constant at a standard temperature of 20 °C for determining *ET*. Burman et al. (1987) expressed it as a function of dry-bulb temperature as

$$\lambda=2500.78-2.360\times T_a, \quad (14.33)$$

in which λ is in kJ/kg and T_a is in °C. The values of λ have been tabulated as in Table 14.2.

Saturation vapor pressure: The saturation vapor pressure (mb) can be expressed as a function of temperature, as given in eq. 14.34. Algebraically,

$$e_s=33.8639\times[(0.00738T_{mean}+0.8072)^8-0.000019|1.8T_{mean}+48|+0.001316], \quad (14.34)$$

where T_{mean} is in °C.

Net radiation: The net radiation can be expressed as

$$R_n=(1-\alpha)R_s-R_L, \quad (14.35)$$

where α is the albedo, R_s is the incoming shortwave radiation, and R_L is the net outgoing longwave radiation. The value of α is 0.25 for the FAO-modified Penman method and varies with the time of year to include the effect of sun angle for the Wright modification. For clear days when R_s is greater than 70% of clear-sky radiation (R_{so}), it can also be computed as

$$\alpha=0.29+0.06\times\sin[30(M+0.033n+2.25)], \quad (14.36)$$

where M is the number of the month (e.g., $M=12$ for December, $M=1$ for January), n is the day of month, and the sin function is for degrees. For cloudy days, $\alpha=0.3$ when R_s/R_{so} is less than 70%.

The solar radiation, R_s, can be computed as (Fritz and MacDonald, 1949)

$$R_s=\left(0.35+0.61\frac{n}{N}\right)R_{so}, \quad (14.37)$$

where n is the actual sunshine hours, N is the maximum possible sunshine hours, which is a function of the month of the year and latitude, as shown in Table 2.3. The values of R_{so} are given in Table 2.5. The values in Tables 2.3 and 2.5 correspond to the middle of the month, so for other dates the values can be obtained by interpolation.

The net longwave radiation can be computed as

$$R_L=R_{L0}\left[a\frac{R_s}{R_{s0}}+b\right], \quad (14.38)$$

Table 14.2 *Physical properties of liquid water*

Temperature (°C)	Vapor pressure (kPa cm water)		Specific gravity	Specific heat (J/g·°C)	Latent heat of vaporization (kJ/g)	Surface tension (g/s²)	Thermal conductivity (J × 10⁻³/ cm · s · °C)	Viscosity (g × 10⁻²/ cm · s)	Kinematic viscosity (cm²/s)
–10	–		0.99794	4.271	2.526	–	–	–	–
–5	–		0.99918	4.229	2.514	76.4	–	–	–
0	0.611	6.23	0.99987	4.216	2.503	75.6	5.61	1.787	0.0179
4	0.814	8.29	1.00000	4.208	2.493	75.0	5.70	1.567	0.0157
5	0.872	8.89	0.99999	4.204	2.491	74.8	5.74	1.519	0.0152
10	1.228	12.51	0.99973	4.191	2.479	74.2	5.87	1.307	0.0131
15	1.705	17.37	0.99913	4.187	2.467	73.4	5.95	1.139	0.0114
20	2.339	23.81	0.99823	4.183	2.455	72.7	6.03	1.002	0.0101
25	3.170	32.31	0.99708	4.178	2.443	71.9	6.12	0.890	0.0090
30	4.247	43.11	0.99568	4.178	2.432	71.1	6.20	0.798	0.0080
35	5.628	57.05	0.99406	4.178	2.420	70.3	6.28	0.719	0.0073
40	7.385	74.71	0.99225	4.178	2.408	69.5	6.33	0.653	0.0066
45	9.593	96.86	0.99024	4.178	2.396	68.7	6.41	0.596	0.0061
50	12.352	124.42	0.98807	4.183	2.384	67.9	6.45	0.547	0.0056

Table 14.3 *Net radiation coefficients*

Experimental Coefficients for Net Radiation				
Region	a	b	a_1	b_1
Davis, California	1.35	–0.35	0.35	–0.046
Southern Idaho	1.22	–0.18	0.325	–0.044
England	N.A.	N.A.	0.47	–0.065
	N.A.	N.A.	0.44	–0.080
	N.A.	N.A.	0.35	–0.042
Australia	1.2	–0.2	0.39	–0.05
General	1.0	0	–	–

N.A., not available

in which a and b are empirical constants, as given in Table 14.3; and R_{b0} is the net outgoing longwave clear-sky radiation (kJ/m·day), expressed as

$$R_{L0} = \varepsilon\sigma\left[\frac{T_{max}^4 + T_{min}^4}{2}\right], \tag{14.39}$$

where ε is the effective emissivity of the surface and σ is the Stephan–Boltzmann constant, $\sigma = 4.903 \times 10^{-9}$ $\mathrm{MJ/m^2/day/K^4}$. Here, the temperature is in K. However, the outgoing longwave radiation from the Earth's surface is less than that determined by the Stephan–Boltzmann law because of the absorption and downward longwave radiation from the atmosphere and clouds. Greenhouse gases, such as water vapor, carbon dioxide, nitrous oxide, methane, and ozone are absorbers and emitters of longwave radiation, and hence their concentrations are required when assessing the net outgoing longwave radiation. In irrigation applications, however, most of these absorbers are considered at constant concentrations. Only cloudiness and humidity are taken into consideration. The cloudiness is considered in eq. (14.38). With the consideration of humidity, the effective emissivity can be computed as

$$\varepsilon = a_1 + b_1\left[e_{s-dp}\right]^{1/2}, \tag{14.40}$$

where a_1 and b_1 are empirical constants and e_{s-dp} is the saturation vapor pressure at the dew-point or the actual vapor pressure at the current temperature e_a in mb. Wright (1982) used $b_1 = -0.044$ and a_1 as

$$a_1 = 0.26 + 0.1\exp\left\{-[-0.0154 \times (30M + N - 207)]^2\right\}. \tag{14.41}$$

The emissivity can also be computed using mean temperature (in °C; Burman et al., 1983) as

$$\varepsilon = -0.02 + 0.261\exp\left[-7.77 \times 10^{-4}(T_{mean})^2\right]. \tag{14.42}$$

The relative humidity (RH) can be computed using air temperature (T_a, in °C) and dew-point temperature (T_d, in °C) as

$$RH = 100 \times \left[\frac{112 - 0.1T_a + T_d}{112 + 0.9T_a}\right]^8. \tag{14.43}$$

Equation (14.43) has an error of less than 0.6% for the temperature range of –25 to 45 °C. To compute the value of minimum RH (RH_{min}), T_a is replaced by T_{max}; for maximum RH (RH_{max}), T_a is replaced by T_{min} in eq. (14.43).

Equation (14.43) can be rearranged for computing T_d as

$$T_d = \left(\frac{RH}{100}\right)^{1/8}(112 + 0.9T_a) - 112 + 0.1T_a. \quad (14.44)$$

Doorenbos and Pruitt (1977) suggested computing the net outgoing radiation (R_L) from mean temperature (in K) and vapor pressure deficit as

$$R_L = \sigma(T_{mean})^4\left[0.34 - 0.044(e_{s-dp})^{1/2}\right]\left(0.1 + 0.9\frac{n}{N}\right). \quad (14.45)$$

Vapor pressure deficit: Depending on the data available, the vapor pressure deficit can be computed in two ways: the FAO-modified Penman method and a different method for the Wright modification. These methods are given as follows.

1. It is assumed that the values of T_{min}, T_{max}, RH_{min}, and RH_{max} are given. Then, e_s is computed using eq. (14.34) or Table 14.3 with T_{mean}, and e_a is computed as

$$e_a = e_s\frac{RH_{mean}}{100}. \quad (14.46)$$

2. The values of T_{min}, T_{max}, temperature from a dry-bulb thermometer (T_{dry}) or air temperature (T_a), and temperature from a thermometer with a saturated wick surrounding the wick (T_w) are given and the vapor pressure is then computed as

$$e_a = e_{s-wet} - \gamma(T_{dry} - T_w), \quad (14.47)$$

where e_{s-wet} is the saturated vapor pressure from eq. (14.28) with T_w for the temperature. γ is the psychrometric constant from eq. (14.29). Then, e_s is computed from eq. (14.34) with T_{mean}.
3. The values of T_{min}, T_{max}, and T_d are given. Then, e_s is computed from eq. (14.28) with T_{mean} and then e_a is computed from eq. (14.34) with T_d.

In the Wright modification, the vapor pressure deficit is calculated from T_{min}, T_{max}, and T_d, where T_{min} is approximately equal to T_d, provided the RH is near 100%. Then, e_s is computed as

$$e_s = \frac{e_{s-Tmax} + e_{s-Tmin}}{2} \quad (14.48)$$

and

$$e_a = e_{s-dewpoint}. \quad (14.49)$$

In the Wright modification, the soil heat flux G (mm/day) can be computed as

$$G = (T_{mean} - T_p)c_s, \quad (14.50)$$

where T_{mean} is in °C, T_p is the mean air temperature for the three preceding days in °C, and c_s is the specific heat coefficient of the soil in 0.05 mm/day·°C. Jensen et al. (1971) stated that daily heat flux or flux over 10-day periods or soil heat flux during the summer months is negligible for practical purposes.

For long periods (e.g., monthly periods), the heat flux can be computed by making two assumptions. First, the soil temperature to a depth of about 2 m changes with average air temperature. Second, the volumetric heat capacity of soil equals 0.5 cal/cm^3·°C (Jensen et al., 1971). Then,

$$G = \frac{T_{i-1} - T_{i+1}}{\Delta t} = \frac{T_{i+1} - T_{i-1}}{\Delta t}, \quad (14.51)$$

where G is in 100 cal/cm^2·day, T is the mean temperature for period i in °C, and Δt is the time between midpoints of two periods, in days.

The soil heat flux can also be computed as a function of net radiation (Brutsaert, 1982) as

$$G = c_r R_n. \quad (14.52)$$

For bare soil, $c_r = 0.3$. For vegetative cover, the value of G is negligible (Brutsaert, 1982).

Example 14.4: For the purposes of computing the reference *ET* for grass for College Station, Texas, the following data are given:

Latitude and longitude of College Station: 30.60° N and 96.31° W; date: June 1; maximum temperature (T_{max}) is 36 °C; minimum temperature (T_{min}) is 24 °C; maximum RH (RH_{max}) is 80%; minimum RH (RH_{min}) is 50%; wind velocity at 0.50 m height is 10 mph or 16 km/h; and ratio of wind run during the day to wind run during the night (u_{day}/u_{night}) is 3. It shows that three units of wind occurred during the day for each unit of wind that occurred during the night. Daytime is arbitrarily chosen as 0700 to 1900 hours. That means that out of four units, three units are for daytime and one unit is for nighttime. There are 12.5 hours of sunshine. The elevation is 328 ft (99.7 m), and temperatures on May 31, 30, and 29 are, respectively, 73.5, 80, and 77 °F (23.2, 26.7, and 25.0 °C). The wind velocity u_2 at 2 m height can be computed from velocity U_z at any height z (m) as

$$u_2 = U_z\left[\frac{2.0}{z}\right]^{0.2}.$$

For energy, the unit is joules (J) or N·m or kg·m^2/s^2. For force, the unit is N or kg·m/s^2. For pressure, the unit is pascals (Pa) or N/m^2 or kg/m·s^2 or mb. Apply appropriate conversion factors when going from one system of units to another.

Do the following:

1. Compute the mean temperature in Celsius, mean RH, wind velocity in km/day, and daytime wind run in kilometers per 12 hours and daytime wind speed in m/s. Compute the wind function.
2. Compute the slope of the saturation vapor pressure curve. Compute the psychrometric constant γ.
3. Compute the weighting factors in the Penman equation: $\frac{\Delta}{\Delta+\gamma}$ and $\frac{\gamma}{\Delta+\gamma}$.
4. Compute the shortwave radiation in mm/day.

5. Compute effective emissivity and net outgoing solar radiation in mm/day.
6. Compute net radiation with an appropriate value of albedo.
7. Compute vapor pressure deficit.
8. Compute the FAO c factor.
9. Now compute the reference ET for grass.
10. Compute the reference ET for alfalfa using the Wright–Penman method.

Solution:

Step 1: Mean temperature $= \frac{36+24}{2} = 30.0\ °C$.

Similarly, mean RH is $RH_{mean} = \frac{80\% + 50\%}{2} = 65\%$.

At 0.5 m height, wind velocity in km/day: $U_{0.5m} = 16 \times 24\ \text{hour/day} = 384\ \text{km/day}$.

At 2 m height:

$$u_2 = U_z\left(\frac{2}{z}\right)^{0.2} = U_z\left(\frac{2}{0.5}\right)^{0.2} = 506.69\ \text{km/day}.$$

The wind run in km for 24 hours is 506.69 km. With a ratio of U_{day}/U_{night} of 3,

$$\text{daytime wind run} = 506.69\ \text{km} \times \frac{3}{3+1} = 380.02\ \text{km},$$

$$\text{daytime wind speed} = \frac{380.02\ \text{km}}{12\ \text{hour}} = 31.67\frac{\text{km}}{\text{h}}\ \text{or}\ 8.80\ \text{m/s}.$$

Wind function from eqs (14.15) and (14.22):

$$f(u) = 0.27(1.0 + 0.01 \times u_2) = 0.27(1.0 + 0.01 \times 506.69) = 1.638.$$

Step 2: Compute the slope of the saturation vapor pressure curve using eq. (14.27):

$$\Delta = 2.0 \times (0.0073 \times T_{mean} + 0.8072)^7 - 0.00116 = 2.0 \times (0.0073 \times 30 + 0.8072)^7 - 0.00116 = 2.4\frac{\text{mb}}{°\text{C}}.$$

To calculate the psychrometric constant, we need to calculate the atmospheric pressure (P) in mb and latent heat of vaporization (λ) in kJ/kg. Using eq. (14.31), the atmospheric pressure in mb is

$$P = 1013 - 0.1055 \times \text{Elevation} = 1013 - 0.1055 \times 99.77 = 1002.47\ \text{mb}.$$

Using eq. (14.33), latent heat of vaporization (λ) in kJ/kg:

$$\lambda = 2500.78 - 2.360 \times T_a = 2500.78 - 2.360 \times 30 = 2429.98\ \text{kJ/kg}.$$

With $c_p = 1.0035\ \text{kJ/kg}\cdot°\text{C}, \varepsilon_w = 0.62198$, the psychrometric constant is computed using eq. (14.29):

$$\gamma = \frac{c_p \cdot P}{\lambda \cdot \varepsilon_w} = \frac{1.0035 \times 1002.47}{2429.98 \times 0.62198} = 0.666\frac{\text{mb}}{°\text{C}}.$$

For irrigation purposes, eq. (14.30) can also be used:

$$\gamma = 1.6134\frac{P}{\lambda} = \frac{1.6134 \times 1002.47}{2429.98} = 0.666\frac{\text{mb}}{°\text{C}}.$$

Step 3: The weighting factors in the Penman equation are

$$\frac{\Delta}{\Delta+\gamma} = \frac{2.4}{2.4+0.666} = 0.783,$$

$$\frac{\gamma}{\Delta+\gamma} = 1 - \frac{\Delta}{\Delta+\gamma} = 1 - 0.783 = 0.217.$$

Step 4: Shortwave radiation in mm/day:

$$R_s = \left(0.35 + 0.61\frac{n}{N}\right)R_{so},$$

where $n = 12.5$ hours. N is the maximum possible sunshine hours, which is a function of month of year and latitude, calculated from Table 2.3.

For latitude = 30.60° N and the date of June 1, $N = 13.80$ hours (interpolating from the values of May 15 and June 15). Similarly, R_{so} are given in Table 2.5 for particular latitudes and months. From Table 2.5, $R_{so} = 767.92\ \text{cal/cm}^2\cdot\text{day}$ (interpolating from the values of May 15 and June 15).

From step 2, one gets the latent heat of vaporization λ as 2429.98 kJ/kg.

Converting λ into cal/g: $\frac{2429.98}{4.184} = 580.78\ \text{cal/g}$.

$$R_{so} = 767.92\frac{\text{cal}}{\text{cm}^2\cdot\text{day}} \times \frac{10\frac{\text{mm}}{\text{cm}}}{(580.78)\frac{\text{cal}}{\text{g}} \times \frac{1\ \text{g}}{1\ \text{cm}^3}} = 13.22\ \text{mm/day},$$

$$R_s = \left(0.35 + 0.61\frac{12.5}{13.8}\right)(13.22) = 11.93\ \text{mm/day}.$$

Step 5: Effective emissivity,

$$\varepsilon = -0.02 + 0.261\ \exp\left[-7.77 \times 10^{-4}(T_{mean})^2\right] = -0.02 + 0.261\ \exp\left[-7.77 \times 10^{-4}(30)^2\right] = 0.110.$$

For calculation of net outgoing radiation, eq. (14.45) can be used:

$$R_L = \sigma(T_{mean})^4\left[0.34 - 0.044(e_{s-dp})^{1/2}\right]\left(0.1 + 0.9\frac{n}{N}\right).$$

Also, net outgoing radiation in mm/day can be calculated as

$$R_L = R_{L0}\left[a\frac{R_s}{R_{so}} + b\right] = \varepsilon\sigma\left[\frac{T_{max}^4 + T_{min}^4}{2}\right]\left[a\frac{R_s}{R_{so}} + b\right],$$

where a and b are empirical constants, as given in Table 14.3. σ is the Stephan–Boltzmann constant, $\sigma = 4.903 \times 10^{-9}\ \text{MJ/m}^2/\text{day/K}^4$.

$$R_{L0} = \varepsilon\sigma\left[\frac{T_{max}^4 + T_{min}^4}{2}\right] = 0.110 \times (4.903 \times 10^{-9})\left[\frac{(36+273.15)^4 + (24+273.15)^4}{2}\right] = 4.5624\ \text{MJ/m}^2/\text{day}.$$

From step 2, one gets the latent heat of vaporization λ as 2429.98 kJ/kg = 2.42998 MJ/kg. Therefore,

$$R_{L0}=4.5624\frac{\text{MJ}}{\text{m}^2\cdot\text{day}}\times\frac{1000\frac{\text{mm}}{\text{m}}}{(2.42998)\frac{\text{MJ}}{\text{kg}}\times\frac{1000\,\text{kg}}{1\,\text{m}^3}}=1.88\,\text{mm/day},$$

or $1\,\text{mm/day}=2.43\,\text{MJ/m}^2/\text{day}$.

By assuming the experimental coefficients of College Station to be between Davis, California and Southern Idaho and taking the average of both, one gets $a=1.285$ and $b=-0.265$.

$$R_L=(1.88)\left[(1.285)\frac{11.93}{13.22}+(-0.265)\right]=1.68\text{ mm/day}.$$

Step 6: The value of α (albedo) is 0.25 for the FAO-modified Penman method.

The value of α varies with the time of year to include the effect of sun angle for the Wright modification:

$$\frac{R_s}{R_{s0}}=\frac{11.93}{13.22}=0.90>0.7.$$

If $\frac{R_s}{R_{s0}}>0.7$, the following formula is used to calculate albedo, and for June 1, $M=6$, $n=1$:

$$\alpha=0.29+0.06\times\sin[30(M+0.033n+2.25)]$$
$$=0.29+0.06\times\sin[30(6+0.033\times1+2.25)]=0.234.$$

For the FAO-modified Penman method and the Penman method (albedo is 0.25),

$$R_n=(1-\alpha)R_s-R_L=(1-0.25)\times11.93-1.68$$
$$=7.27\text{ mm/day}.$$

For the Wright modification method (albedo is 0.234),

$$R_n=(1-\alpha)R_s-R_L=(1-0.234)\times11.93-1.68$$
$$=7.46\text{ mm/day}.$$

Step 7: Given T_{min}, T_{max}, RH_{min}, and RH_{max}, method (1) can be used.

The saturation vapor pressure e_s is computed using eq. (14.34) with T_{mean}:

$$e_s=33.8639\times[(0.00738T_{mean}+0.8072)^8-0.000019\,|1.8T_{mean}+48|+0.001316]$$
$$=33.8639\times[(0.00738\times30+0.8072)^8-0.000019\,|1.8\times30+48|+0.001316]$$
$$=42.41\text{ mb}.$$

The actual vapor pressure e_a is computed using eq. (14.46):

$$e_a=e_s\frac{RH_{mean}}{100}=42.41\frac{65}{100}=27.57\text{ mb},$$

Vapor pressure deficit $(D)=e_s-e_a=42.41-27.57$
$=14.84$ mb.

Step 8: Given $RH_{max}=80\%$, $u_{day}/u_{night}=3$, the FAO c factor can be calculated using eq. (14.23):

$$c=0.68+0.0028RH_{max}+0.018R_s-0.068u_{2day}+0.013(u_{day}/u_{night})+0.0097u_{2day}(u_{day}/u_{night})+0.43\times10^{-4}RH_{max}\times R_s\times u_{2day}$$
$$=0.68+0.0028\times80+0.018\times11.93-0.068\times8.80+0.013(3)+0.0097\times8.80\times(3)+0.43\times10^{-4}\times80\times11.93\times8.80=1.177$$

Also, using Table 14.1, with $RH_{max}=80\%$, $u_{day}/u_{night}=3$, $R_s=11.93$ mm/day $\approx$ 12 mm/day, and $u_{2\text{day}}=8.8$ m/s $\approx$ 9 m/s, one gets $c=1.14$ (interpolating from the values 1.05 [$RH_{max}=60\%$] and 1.18 [$RH_{max}=90\%$]).

Step 9: Use the Penman and FAO-modified Penman equation.

Soil heat flux (G) is given by

$$G=(T_{mean}-T_p)c_s,$$

where T_p is the mean air temperature for the three preceding days in °C, $c_s=0.05$ (constant, specific heat coefficient of the soil in mm/day·°C).

$$T_p=\frac{23.2+26.7+25.0}{3}=24.97\text{ °C},$$
$$G=(30-24.97)\times0.05=0.25\text{ mm/day}.$$

Penman equation:

$$ET_r=\frac{\Delta}{\Delta+\gamma}(R_n-G)+\frac{\gamma}{\Delta+\gamma}[\Delta e]f(u)$$
$$=(0.783)(7.27-0.25)+(0.217)(14.84)(1.638)$$
$$=10.8\text{ mm/day}.$$

FAO-modified Penman method:

$$ET_r=c\left[\frac{\Delta}{\Delta+\gamma}R_n+\frac{\gamma}{\Delta+\gamma}[\Delta e]f(u)\right]$$
$$=1.177\times[(0.783)(7.27)+(0.217)(14.84)(1.638)]$$
$$=12.9\text{ mm/day}.$$

Step 10: For Wright's modification:

$$a_w=0.4+1.4\exp\left\{-[(J-173)/58]^2\right\}$$

(J = Julian day of the year, in our case = 152),

$$a_w=0.4+1.4\exp\left\{-[(152-173)/58]^2\right\}=1.628,$$
$$b_w=0.007+0.004\exp\left\{-[(152-243)/80]^2\right\}=0.008,$$
$$a_w+b_wu_2=1.628+0.008\times506.9=5.73,$$
$$ET_r=\frac{\Delta}{\Delta+\gamma}(R_n-G)+\frac{\gamma}{\Delta+\gamma}\frac{15.36}{0.1\lambda}[e_s-e_a][a_w+b_wu_2],$$
$$=(0.783)(7.46-0.25)+(0.217)\frac{15.36}{0.1\times580.78}(14.84)(5.73)$$
$$=10.5\text{ mm/day}.$$

Penman equation = 10.8 mm/day;
FAO-modified Penman method = 12.9 mm/day;
Wright's modification = 10.5 mm/day.

Example 14.5: Compute reference *ET* by the Priestley–Taylor method for grass for College Station, Texas, using the data given in Example 14.4, with an assumed albedo of 0.23.

Solution: With albedo = 0.23, and R_s and R_L calculated from Example 14.4,

$$R_n = (1-\alpha)R_s - R_L = (1-0.23)\times 11.93 - 1.68 = 7.51 \text{ mm/day}.$$

From Figure 2.4, the climate in College Station belongs to humid subtropical, where $a_{pt} = 1.26$ is used, with G and $\frac{\Delta}{\Delta+\gamma}$ calculated from Example 14.4:

$$ET_r = a_{pt}\frac{\Delta}{\Delta+\gamma}(R_n - G) = (1.26)(0.783)\left(7.51\frac{\text{mm}}{\text{day}} - 0.25\frac{\text{mm}}{\text{day}}\right) = 7.2\frac{\text{mm}}{\text{day}}.$$

Example 14.6: Compute reference *ET* by the Penman–Monteith method for grass for College Station, Texas, using the data given in Example 14.4, with an assumed albedo of 0.23.

Solution: The Penman–Monteith method is computed using eq. (14.25):

$$ET_r = \frac{1}{\lambda}\frac{\Delta A r_a + \rho_a C_p D}{\Delta r_a + \gamma(r_a + r_s)}.$$

Step 1: With albedo = 0.23, from Example 14.5 one gets $R_n = 7.51$ mm/day. From Example 14.4, $G =$ 0.25 mm/day, then the available energy:

$$A = R_n - G = 7.51 \text{ mm/day} - 0.25 \text{ mm/day} = 7.26 \text{ mm/day}.$$

From Example 14.4, one gets latent heat of vaporization: $\lambda = 2429.98$ kJ/kg = 2.43 MJ/kg, and 1 mm/day = 2.43 MJ/m^2/day. Hence, the available energy in unit of MJ/m^2/day is:

$$A = 7.26 \times 2.43 \text{ MJ/m}^2\text{/day} = 17.64 \text{ MJ/m}^2\text{/day}.$$

Step 2: From Example 14.4, one gets vapor pressure deficit:

$$D = e_s - e_a = 42.41 - 27.57 = 14.84 \text{ mb} = 1.484 \text{ kPa}.$$

Step 3: The aerodynamic resistance r_a is computed from eq. (14.26):

$$r_a = \frac{\ln\left[\frac{Z_u - d}{Z_{om}}\right]\ln\left[\frac{Z_e - d}{Z_{ov}}\right]}{(0.41)^2 U_Z}.$$

Given $Z_u = 2$ m, $Z_e = 2$ m, $Z_{om} = 0.123\,h_c$ and $Z_{ov} = 0.0123\,h_c$, and $d = 0.67\,h_c$, h_c is the mean crop height. In this case, one assumes grass height of 0.12 m: $U_{2m} = u_2 = 506.69$ km/day = 5.86 m/s,

$$r_a = \frac{\ln\left[\frac{2-0.67\times0.12}{0.123\times0.12}\right]\ln\left[\frac{2-0.67\times0.12}{0.0123\times0.12}\right]}{(0.41)^2(5.86)} = 35.41 \text{ s/m}.$$

Step 4: The surface resistance of the land cover r_s is assumed to be 70 s/m for grass.

Step 5: The computation of the slope of the saturation vapor pressure curve $\Delta$is the same as in Example 14.4, but in a different unit:

$$\Delta = 2.4\frac{\text{mb}}{^\circ\text{C}} = 0.24\frac{\text{kPa}}{^\circ\text{C}}.$$

Step 6: The calculation of psychrometric constant γ is the same as in Example 14.3, but in a different unit:

$$\gamma = 0.666\frac{\text{mb}}{^\circ\text{C}} = 0.0666\frac{\text{kPa}}{^\circ\text{C}}.$$

Step 7: Given the density of air is $\rho_a = 1.225$ kg/m^3 and $c_p = 1.0035 \times 10^{-3}$ MJ/kg·°C, the reference ET for grass is:

$$ET_r = \frac{1}{\lambda}\frac{\Delta A r_a + \rho_a c_p D}{\Delta r_a + \gamma(r_a + r_s)}$$

$$= \frac{1}{2.43\frac{\text{MJ}}{\text{kg}}}\frac{0.24\frac{\text{kPa}}{^\circ\text{C}}\times\frac{17.64\frac{\text{MJ}}{\text{m}^2}}{\text{day}}\times 35.41\frac{\text{s}}{\text{m}} + \frac{1.225\text{ kg}}{\text{m}^3}\times 1.0035\times10^{-3}\frac{\text{MJ}^\circ}{\text{kg}}\text{C}\times 1.484\text{ kPa}\times 86{,}400\text{ s/day}}{0.24\frac{\text{kPa}}{^\circ\text{C}}\times 35.41\frac{\text{s}}{\text{m}} + 0.0666\frac{\text{kPa}}{^\circ\text{C}}\left(35.41\frac{\text{s}}{\text{m}} + 70\frac{\text{s}}{\text{m}}\right)}$$

$$= 8.2 \text{ mm/day}$$

Note: the term $\rho_a C_p D$ needs to be multiplied 86400 s/day to make sure the units are the same.

In Examples 14.4–14.6 the same input data are used for the calculation of *ET*. The *ET* calculated by the Penman–Monteith method is 8.2 mm/day, while the calculated *ET*s are 10.8 mm/day (Penman equation), 12.9 mm/day (FAO-modified Penman method), 10.5 mm/day (Wright's modification), and 7.2 mm/day (Priestley–Taylor method) for the others. Compared with the Penman–Monteith method, all three Penman-related methods overestimate ET_r, and the Priestley–Taylor method underestimates it in this case. The main reason is that it is under strong wind conditions – that is, the mean wind speed is 5.86 m/s. The three Penman-form methods tend to overestimate ET_r under moderate to strong winds since a linear wind function is used, which means ET_r linearly increases with increasing wind speed. With the inclusion of r_a (i.e., the surface resistance in both the numerator and denominator), the Penman–Monteith method projects a curvilinear wind function, which is more appropriate in reality. The underestimation of the Priestley–Taylor method is that the aerodynamic term is absent.

14.8 TRANSPIRATION

The movement of water between soil, plant, and atmosphere entails the movement of water through the soil, extraction of water by plant roots, movement of water through the roots, movement of water through the xylem, movement of water into the leaves, conversion of water into water vapors, and movement of vapors into the atmosphere. For the movement to occur, there must exist an energy gradient. For example, there exists a potential difference between soil water and plant roots. Then, there is a potential differential between leaves and xylem, and there is a potential difference between leaves and the atmosphere. Through its roots, the plant extracts water from the soil; this extracted water moves into the roots through the xylem to the leaves, and then converts to water vapor to move through stomata to the atmosphere. The plant absorbs water principally through roots or root hairs, which move through small pore spaces contacting and absorbing water. These root hairs are single cells of the epidermis layer with a large surface area, and each cell contains a large vacuole and thus aids osmosis. As roots grow, they bring in new supplies of water, especially for soil below field capacity. In order to absorb or remove water from the soil, work must be performed through the use of energy, which depends on the status of the soil water and consequently on capillary pressure or soil water tension. As soil moisture decreases, the capillary pressure (negative) increases, meaning that soil water will be more tightly held by soil particles and more work will have to be done and more energy will be used to remove water from the soil. Further, plants renew their roots in established areas and extend their roots to tap additional water stored in soil pore spaces.

Absorption of water by plant cells and intercellular spaces occurs through osmosis, by which water moves from lower salt concentration areas through a semipermeable membrane to higher salt concentration areas. The cells have semipermeable walls, and water moves from the soil through the walls to the inside of the cell. Then the plant transports the water and nutrients through the xylem tissues to the leaves, where the water in the form of water vapor moves through leaf cells or stomata to the atmosphere. The plant has the ability to control the concentration of water molecules inside cells and the movement of water into the cells.

The source of energy for the plant roots is in the form of sugar, which is produced in the leaves by photosynthesis. This sugar is then transported downward through phloem tissues to plant roots. The root cells that are active respire, take in oxygen, and release carbon dioxide. Oxygen intake is fundamental for plant health, and when oxygen is lacking in the root zone, plants will suffocate.

The movement of water through xylem due to the water potential gradient between the soil and leaves can be expressed as

$$Q = \frac{\psi_{leaf} - \psi_{soil}}{r_{plant} + r_{soil}}, \tag{14.53}$$

$$\psi_{leaf} = \psi_T + \psi_o, \tag{14.54}$$

in which Q is the rate of flow, ψ_{leaf} is the total water potential in the leaf, ψ_{soil} is the total water potential in the soil, ψ_T is the turgor pressure in the leaf, ψ_o is the osmotic pressure within the plant, r_{plant} is the resistance to the movement of water into the roots up the xylem and into the leaf, and r_{soil} is the resistance to the movement of water in the soil, which can be expressed as

$$r_{soil} = \frac{1}{K}, \tag{14.55}$$

where K is the hydraulic conductivity. As explained in Chapter 13 on soil water, the water potential in the soil and the hydraulic conductivity increase as moisture content increases. The flow of water into the plant tends to decrease as the moisture potential decreases and the resistance to the movement of water in the soil increases. Should the decrease in the flow of water continue, the result will be dehydration of the plant and the reduction in the turgor pressure in the leaf, and hence the decrease in the water potential in the leaf in order to maintain the transpiration rate. If the water potential in the leaf begins to get low, the leaf will begin to become dehydrated and stomata will start to close, increasing the resistance to the flow of vapors from the leaf to the atmosphere and reducing the rate of transpiration. The stomatal closure will lead to decreased photosynthesis, because carbon dioxide enters the leaf through the same pathway that water vapor escapes. The result will be a reduction in growth, yield, and quality of crop.

Like human beings, plants need water for digestion, photosynthesis, transport of minerals and photosynthates, structural support, growth, and transpiration. Depending on the growth stage, a physiologically active plant has 60–95% water. The plant transpires the water in transpiration, which accounts for almost 99% of the water used by the plant (Wilson et al., 1992). Transpiration can be defined as evaporation from a living surface, and entails conversion of water from the liquid phase to the vapor phase within the leaf and subsequent transport of the vapors through stomata of the leaf into the atmosphere.

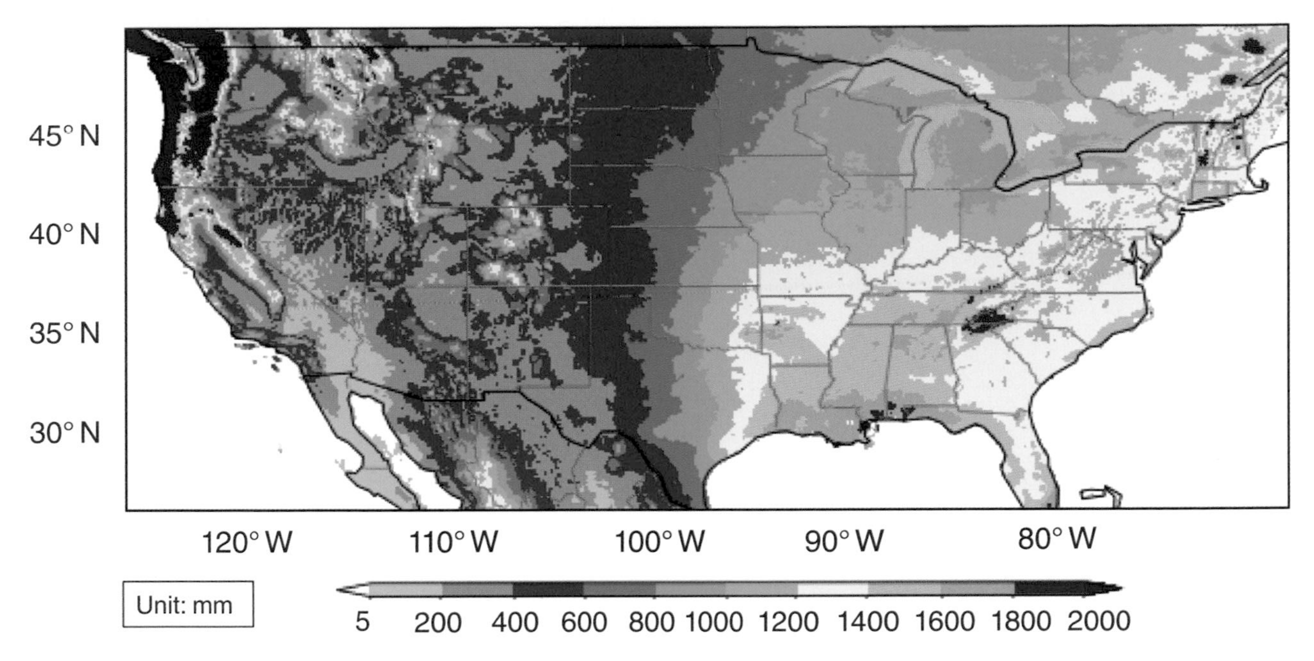

Figure 1.2 Annual annual precipitation in millimeters during 1982–2010 across the contiguous United States. (Data from Daily Surface Weather and Climatological Summaries, DAYMET, https://daymet.ornl.gov.)

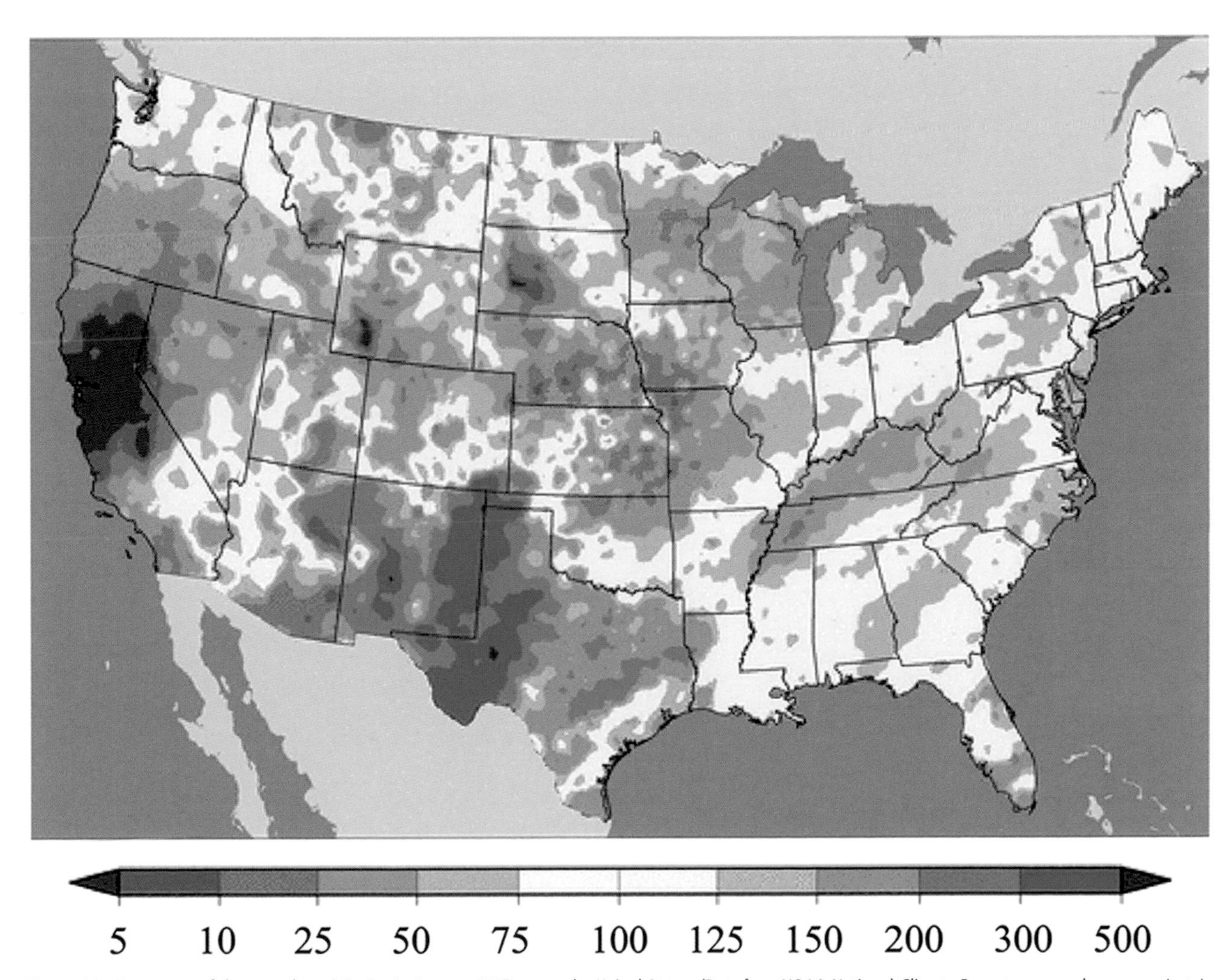

Figure 1.3 Percentage of the normal precipitation in January 2015 across the United States. (Data from NOAA National Climate Report, www.ncdc.noaa.gov/sotc/national/201501.)

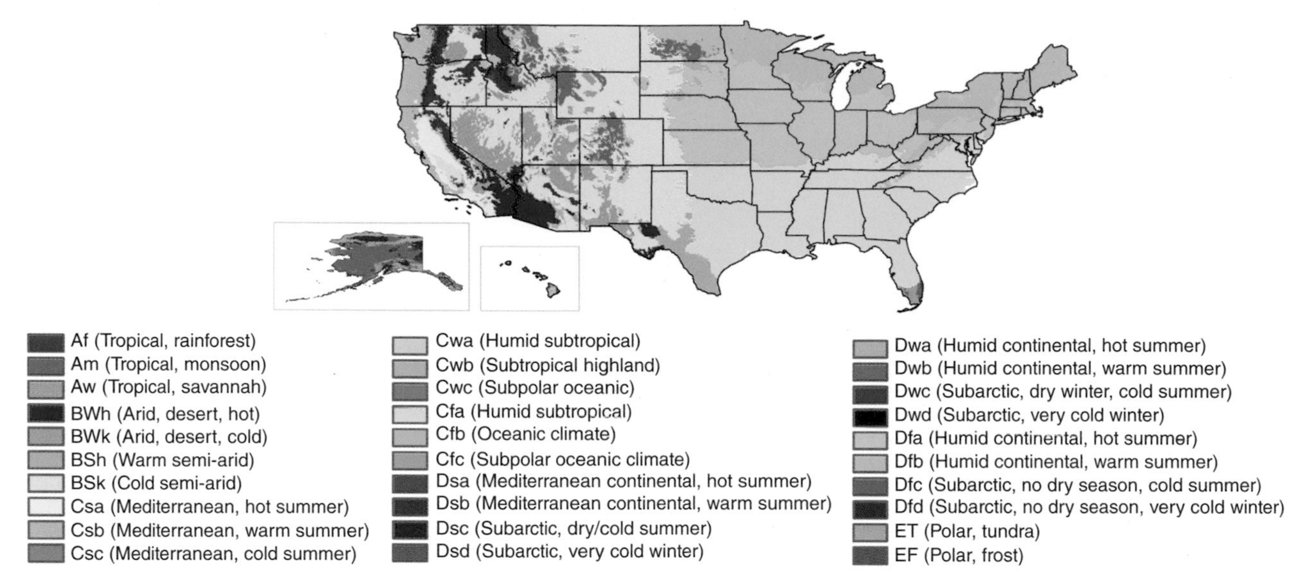

Figure 1.4 Köppen climate classification for the United States. (Data from Beck et al. [2018] using weather data from 1980–2016.)

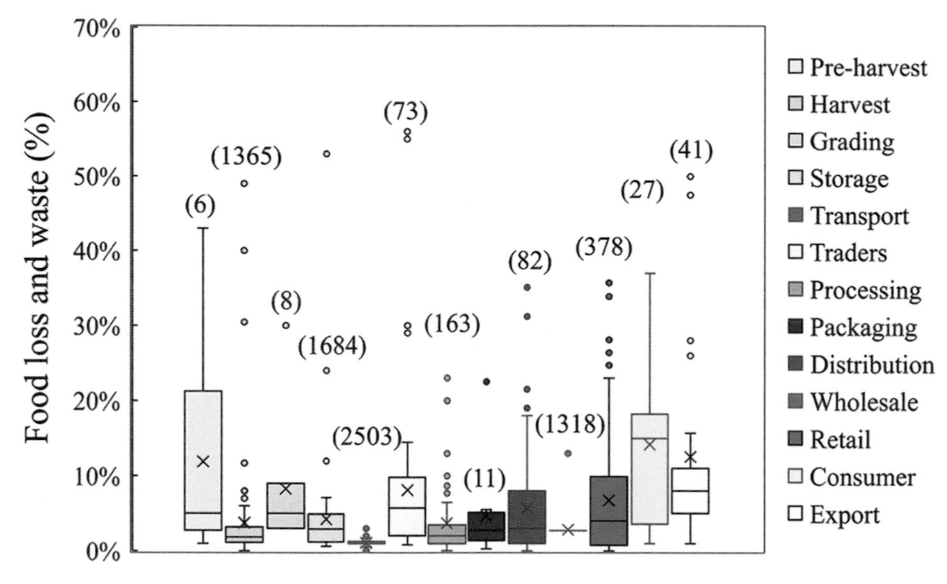

Figure 1.8 Range of reported global food loss and waste percentage during different processes (2001–2017). Number of observations is shown in brackets. (Data from FAO, 2019.)

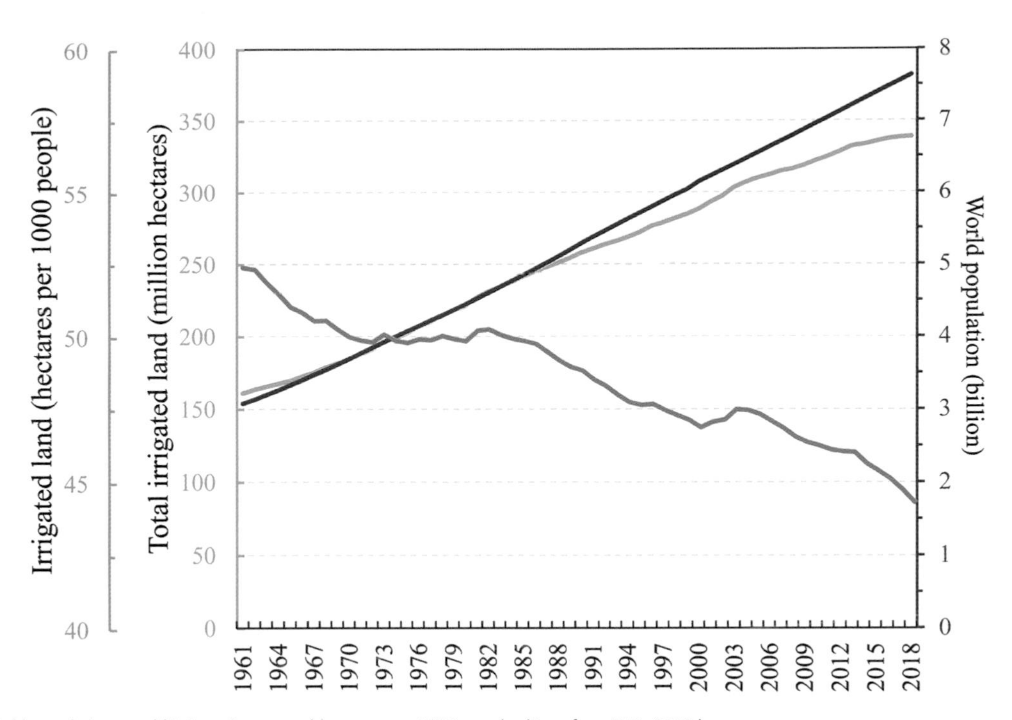

Figure 1.10 World population, world irrigated area, and hectares per 1000 people. (Data from FAO, 2020.)

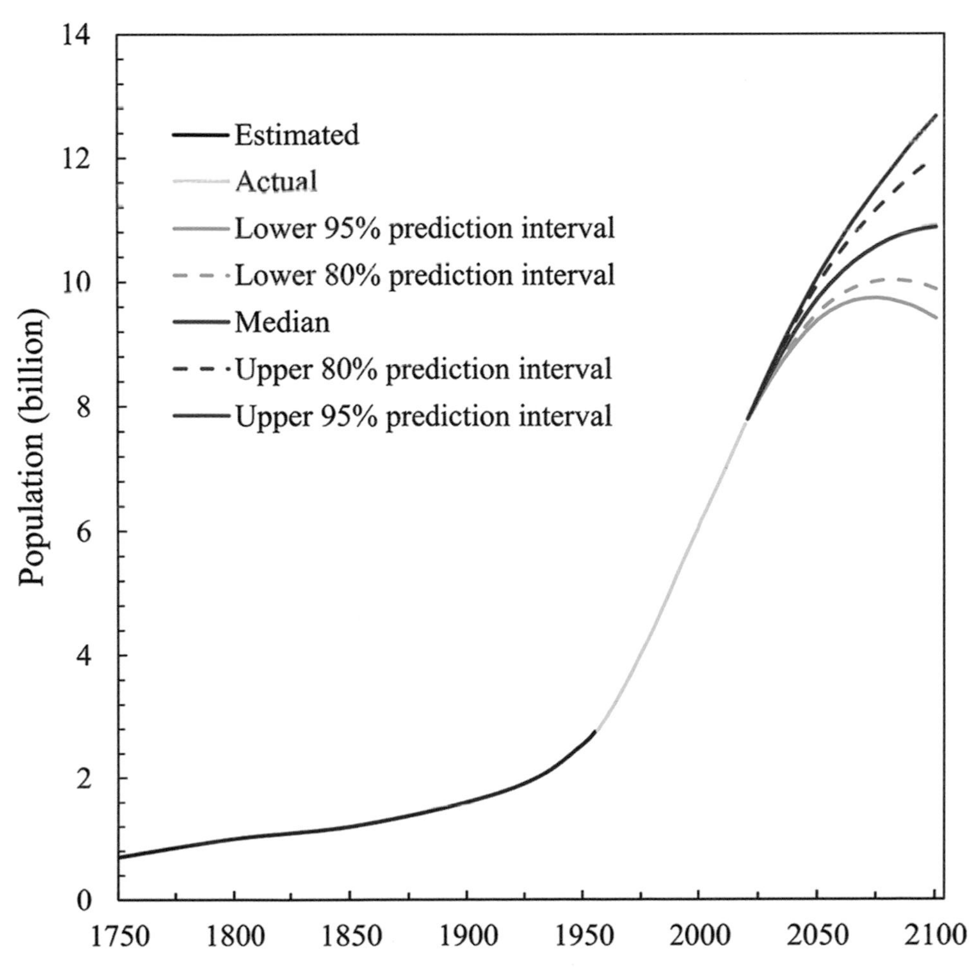

Figure 1.12 Population growth as a function of time. World population estimates from 1750 to 2100, based on the probabilistic median, and the upper and lower 80% and 95% prediction intervals of population projections. (Data from United Nations, DESA, Population Division, World Population Prospects 2019, http://population.un.org/wpp.)

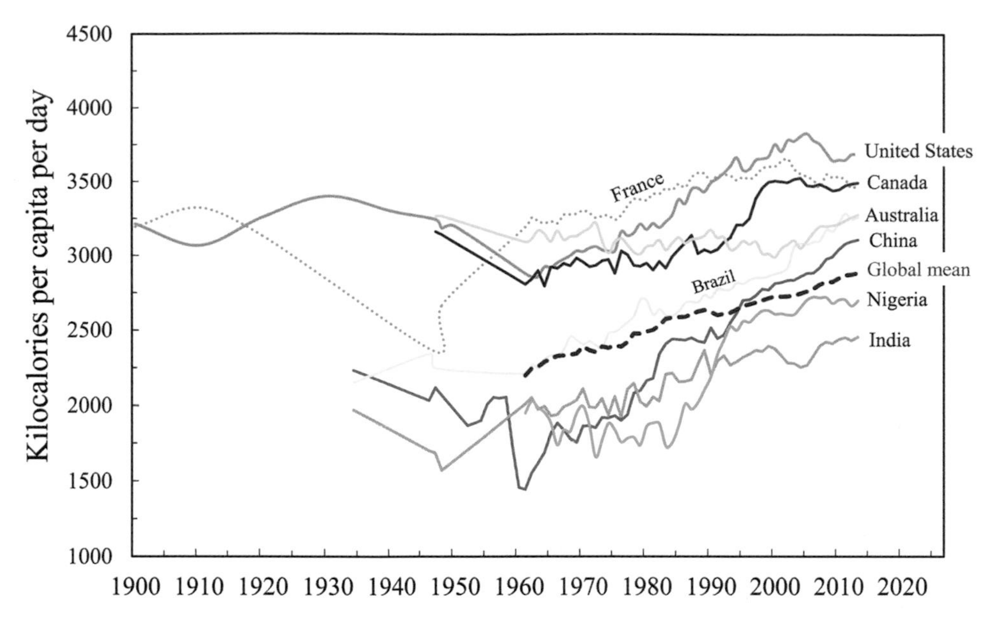

Figure 1.13 Country-level food consumption as a function of time. (Data from FAO, 2020.)

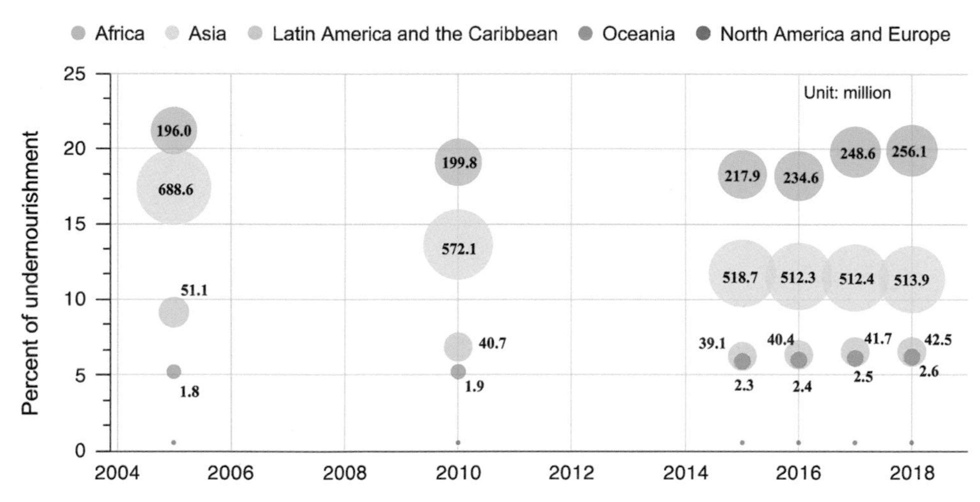

Figure 1.15 The absolute number of undernourished people. The size of the bubble indicates the total number of undernourished people in the unit of millions. (Data from FAO et al., 2019.)

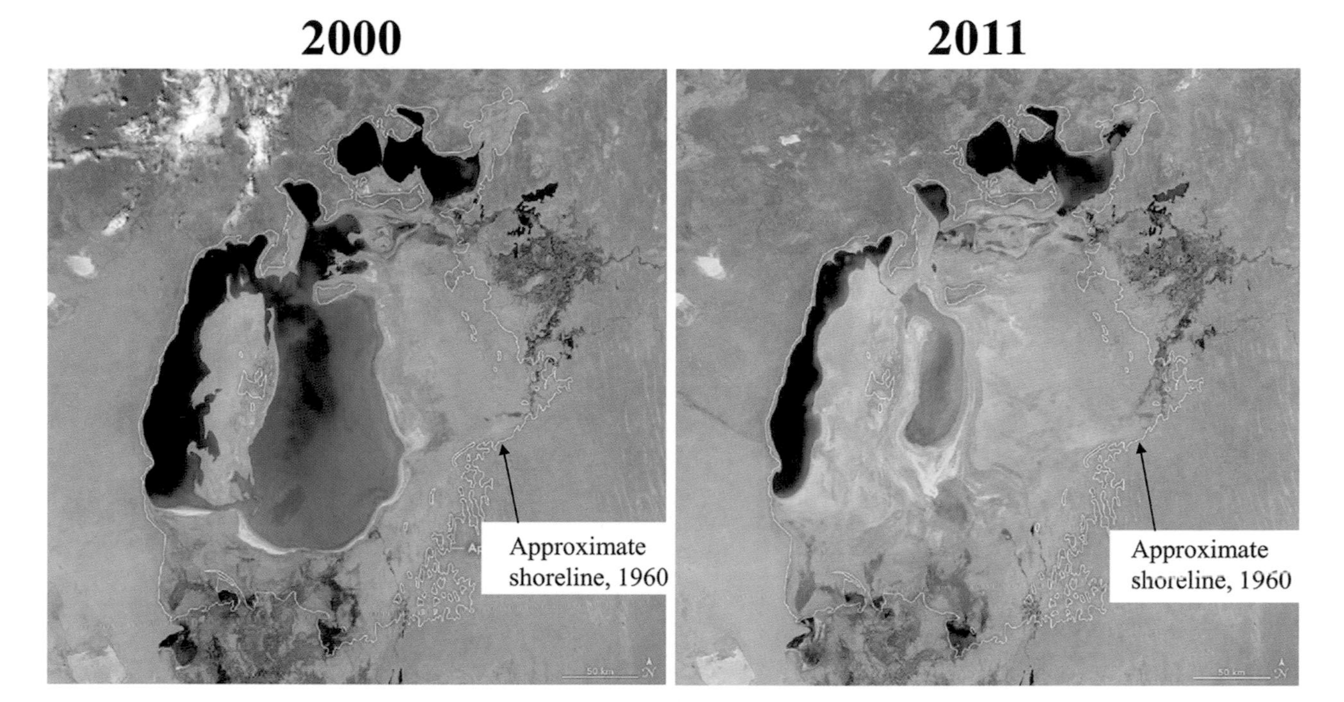

Figure 1.19 Aral Sea in 2000 and 2011. (Adapted from NASA, https://earthobservatory.nasa.gov/world-of-change/AralSea/show-all.)

Figure 1.20 Amudaz River (Amu Darya), 100 miles away from the Aral Sea.

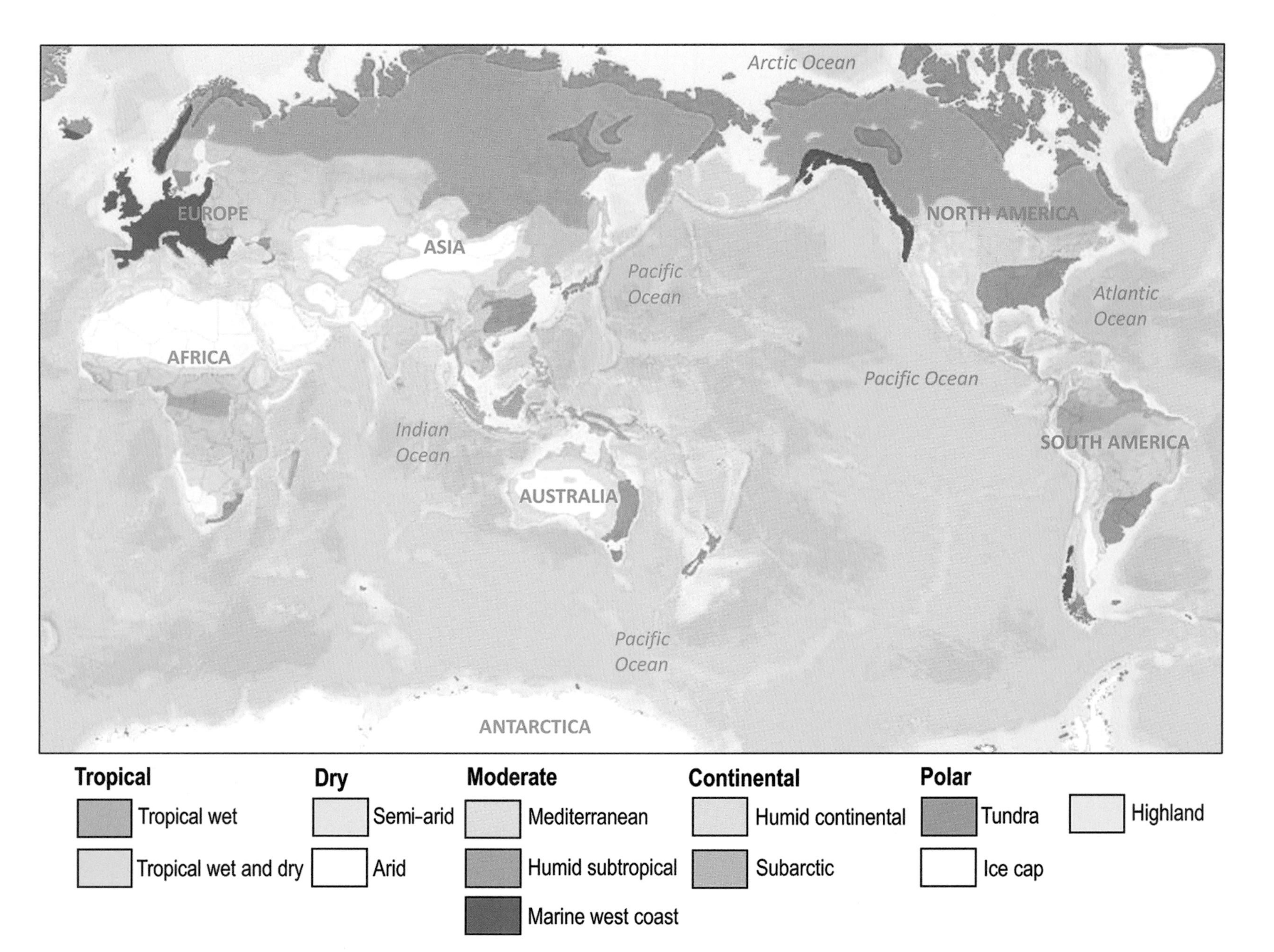

Figure 2.2 Twelve different types of climates on the Earth. (Data from www.nationalgeographic.org/mapmaker-interactive.)

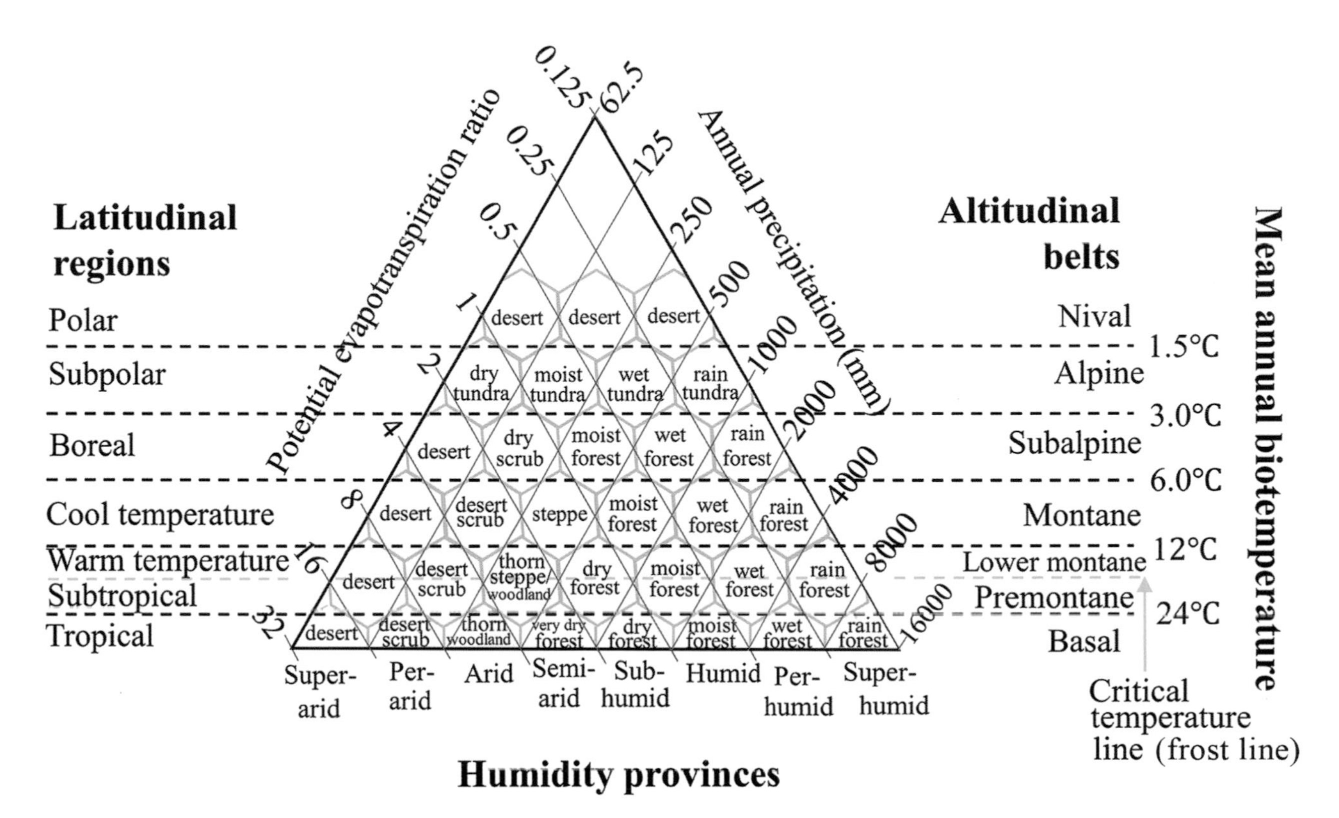

Figure 2.3 Holdridge life zone classification. (Adapted from Lugo et al., 1990.)

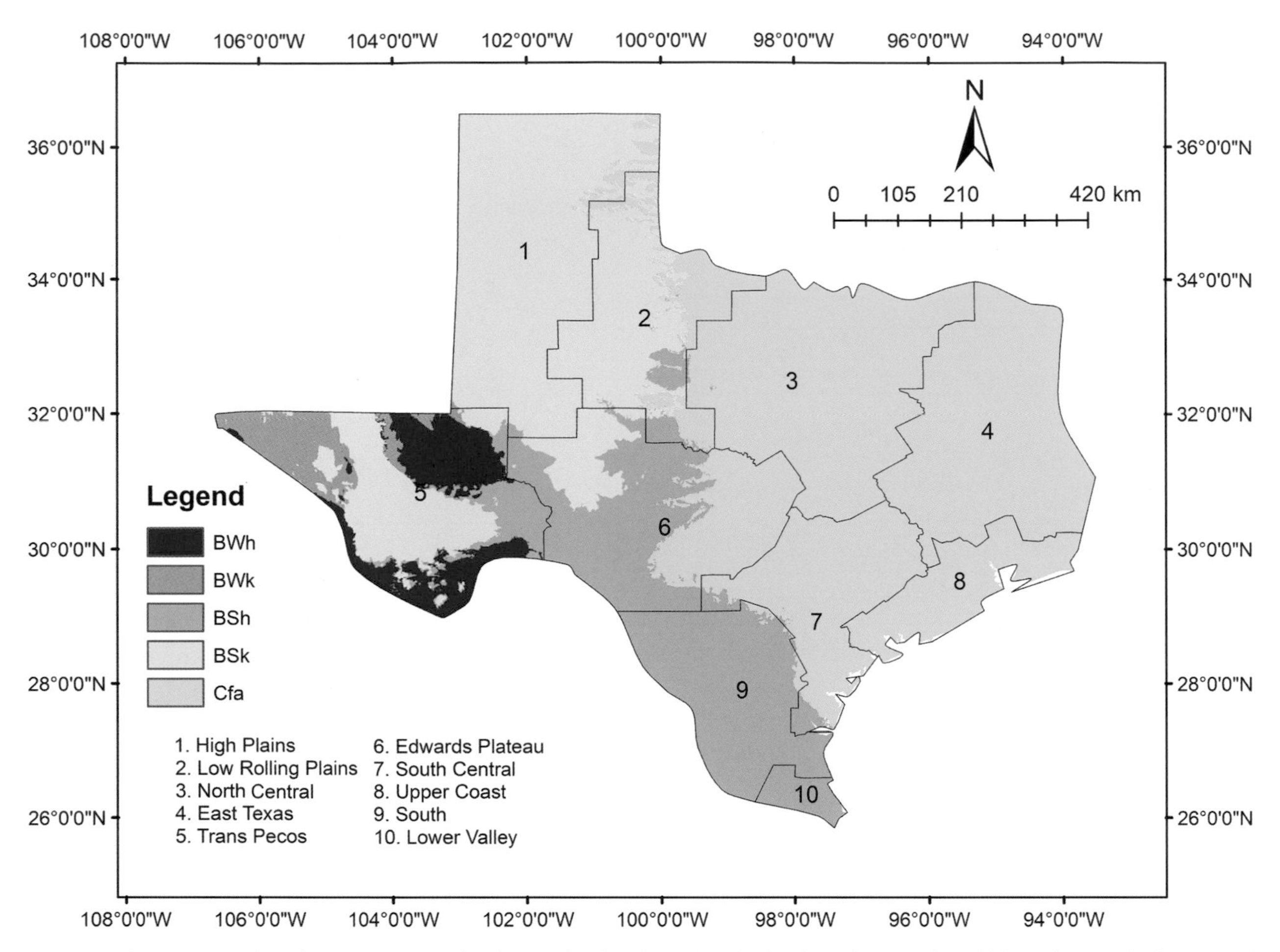

Figure 2.4 Climates in Texas. (Data from Köppen–Geiger classification of Beck et al., 2018). BWh is hot desert climate, BWk is cold desert climate, BSh is hot semi-arid, BSk is cold semi-arid, and Cfa is humid subtropical. Texas is divided into 10 climate regions. Some areas in Region 4 belong to a highland climate and are not shown in the figure. Cfa can be further divided into humid subtropical (Regions 4 and 8, part of Regions 3 and 7) and subhumid subtropical (other regions left in Cfa). The 1980–2016 global monthly datasets are used for the Köppen–Geiger classification.

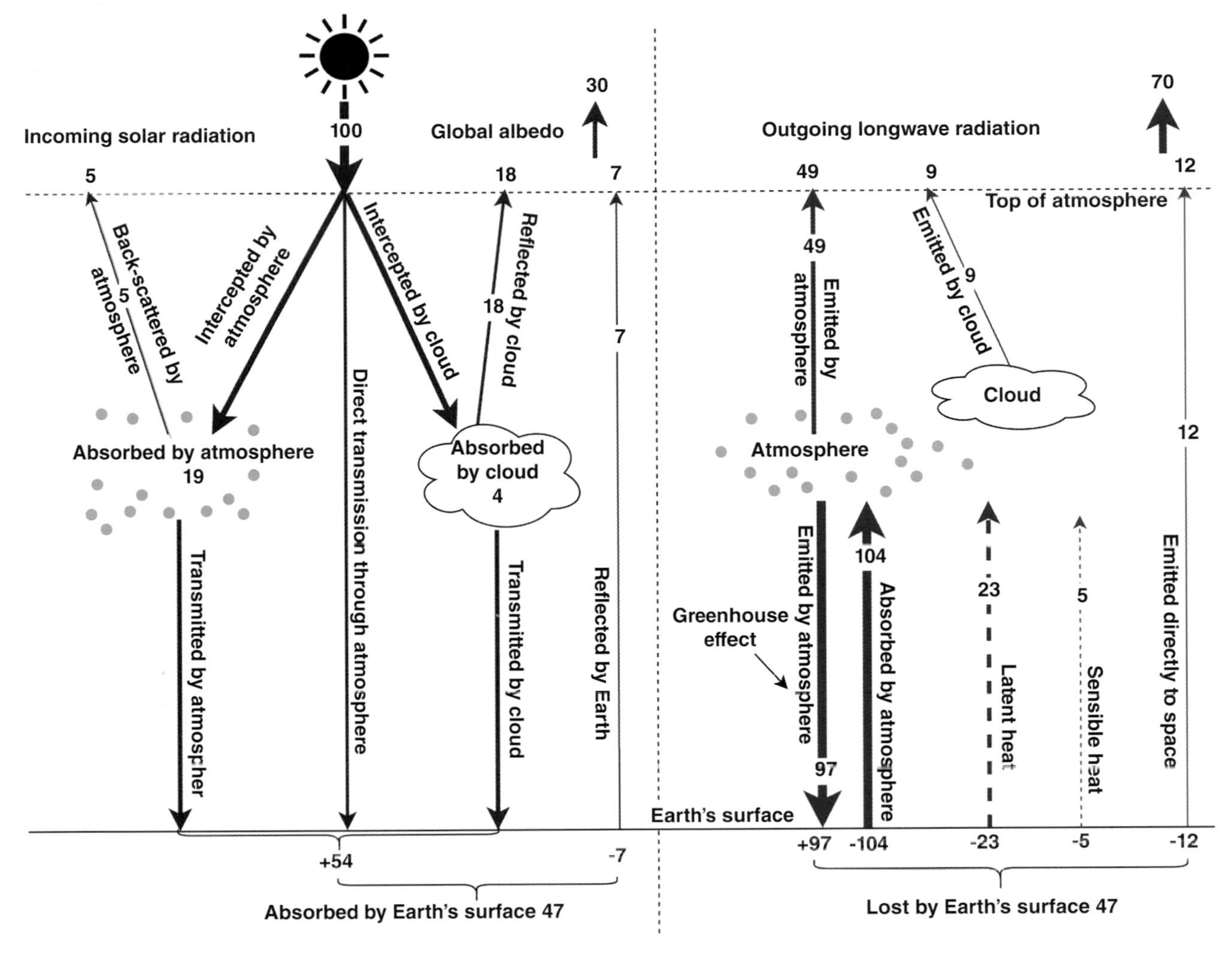

Figure 2.6 Average global energy balance. (Data from Trenberth et al., 2009.)

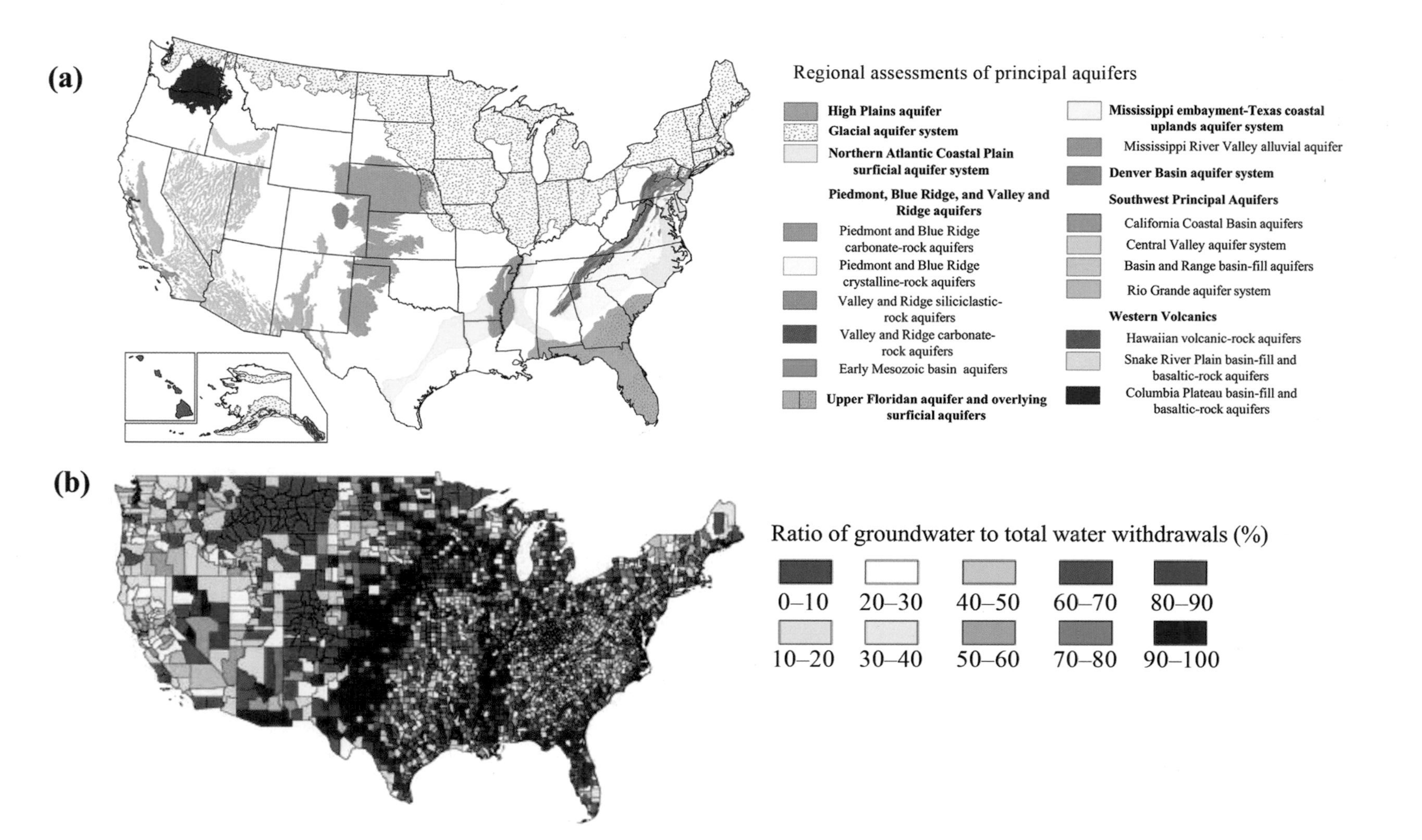

Figure 3.1 (a) Principal aquifers and (b) the percentage of groundwater withdrawals to total water withdrawals in the United States. (Adapted from US Geological Survey, www.usgs.gov/media/images/regional-assessments-principal-aquifers.)

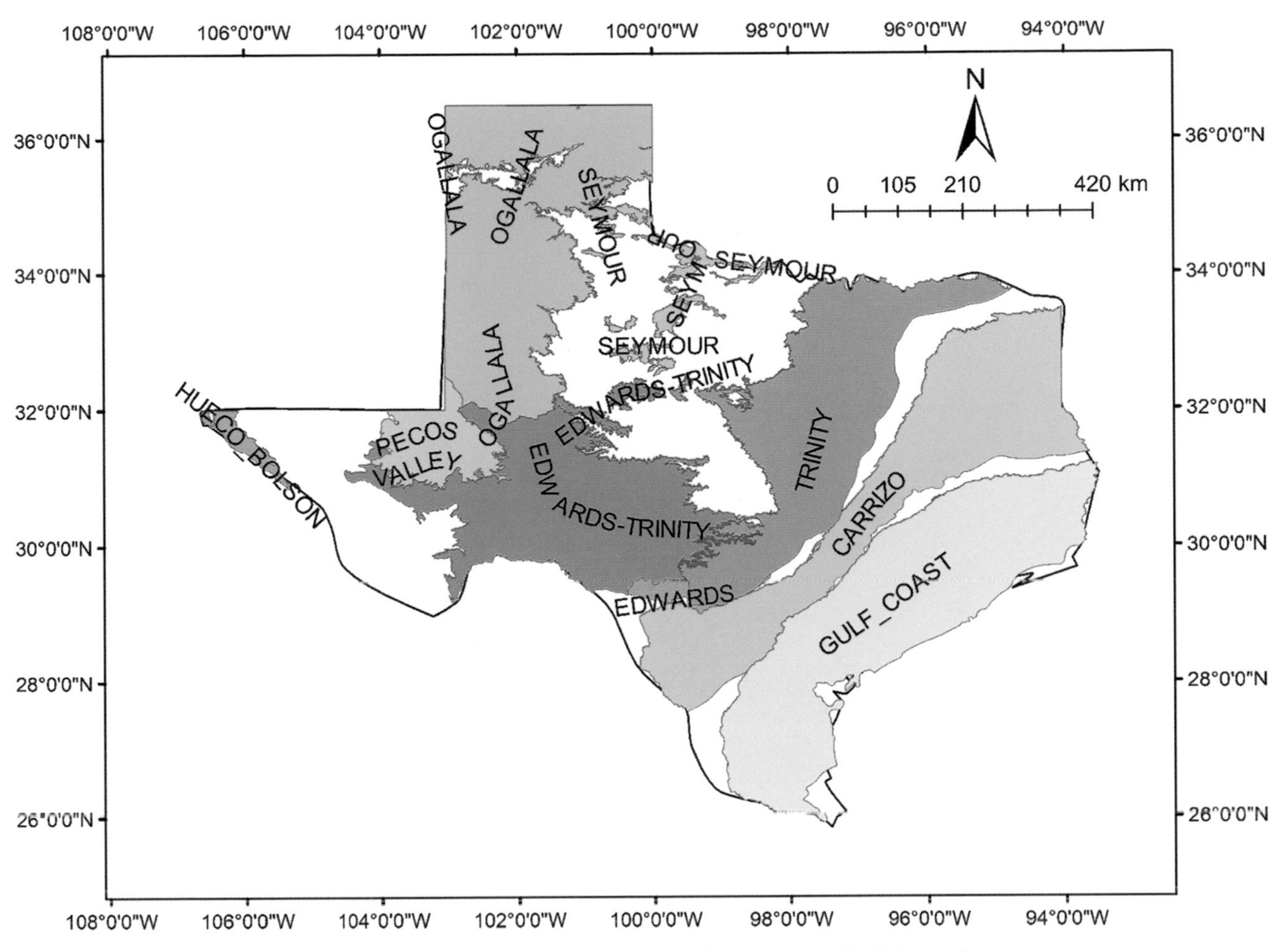

Figure 3.3 Major aquifers in Texas. (Data from Texas Water Development Board, www.twdb.texas.gov/mapping/gisdata.asp.)

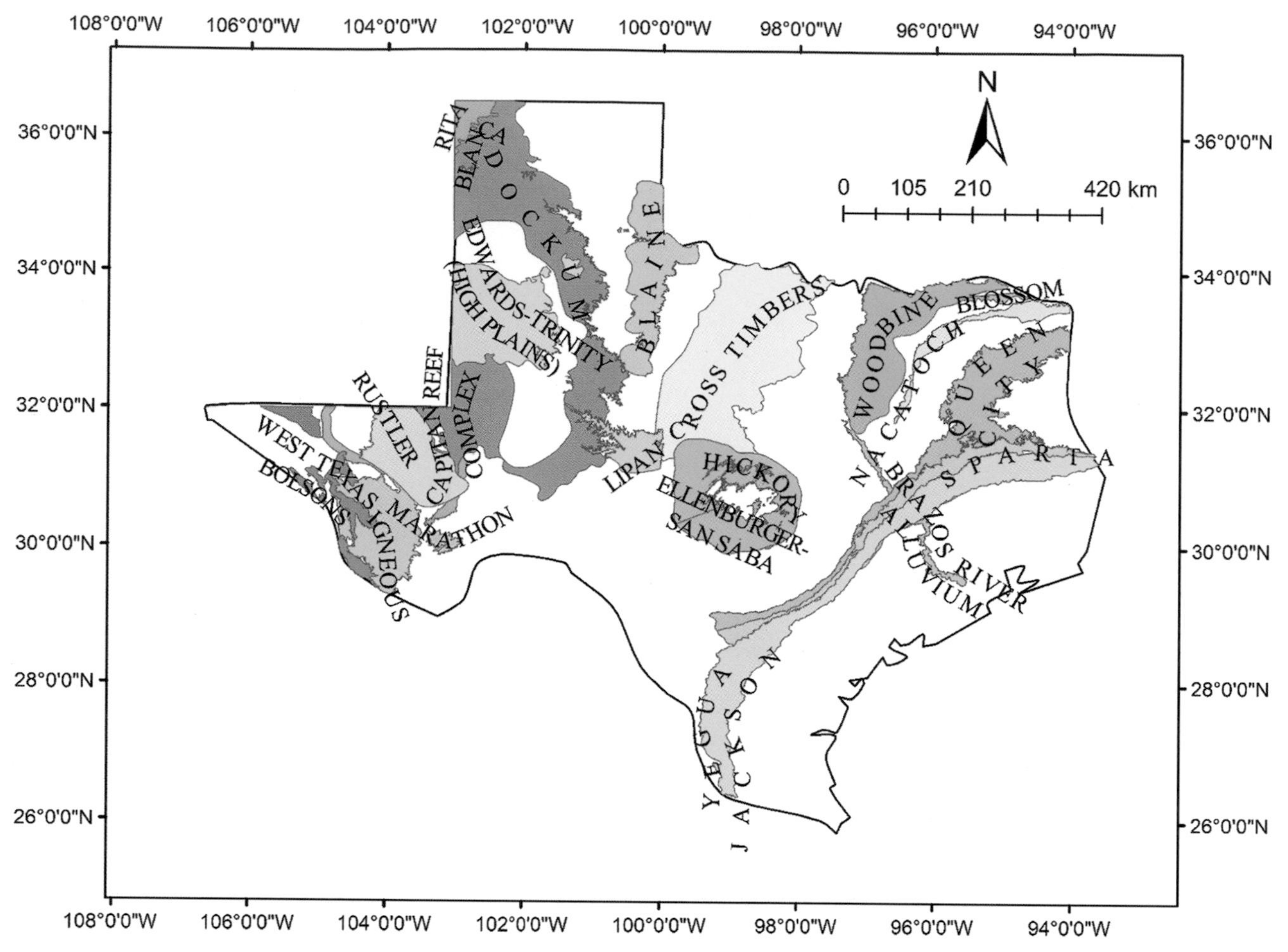

Figure 3.4 Minor aquifers in Texas. (Data from Texas Water Development Board, www.twdb.texas.gov/mapping/gisdata.asp.)

Figure 5.3 Turbidity levels.

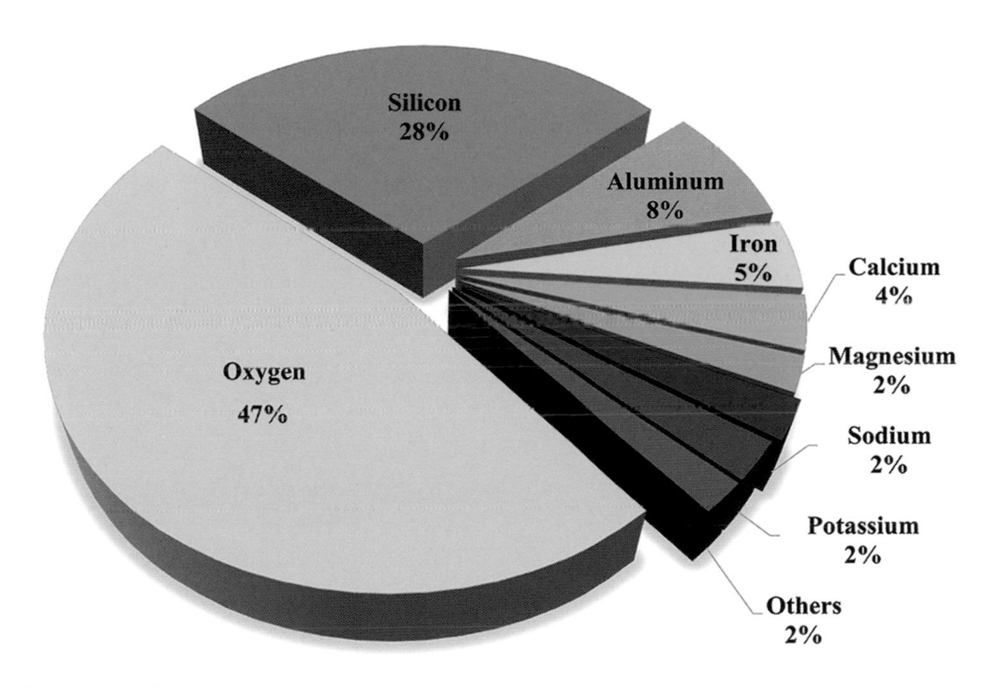

Figure 6.1 Chemical elements in soil.

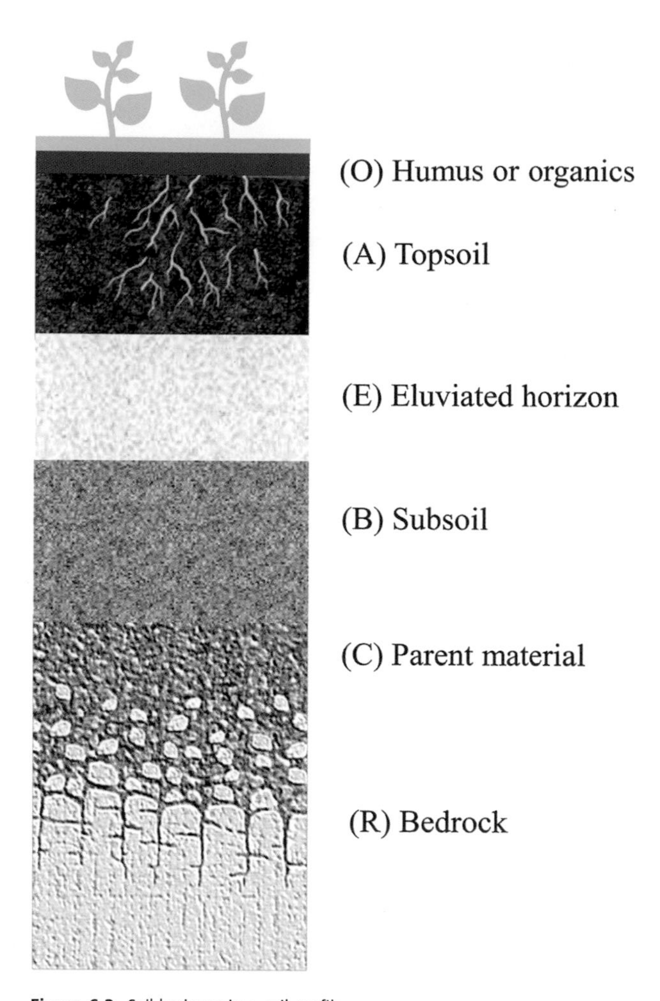

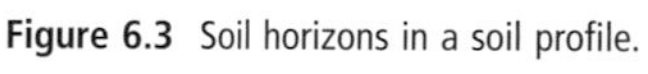

Figure 6.3 Soil horizons in a soil profile.

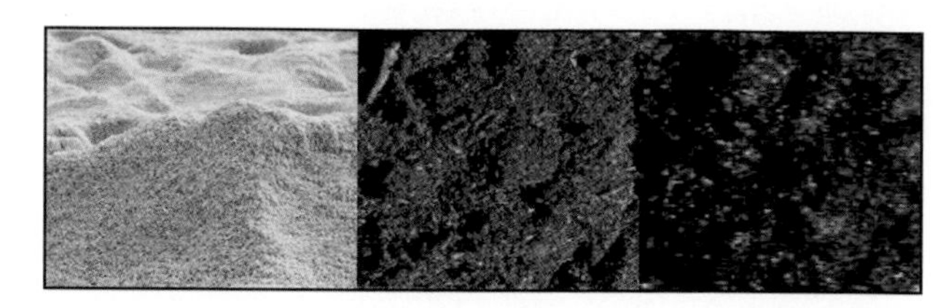

Figure 6.5 Three types of soil.

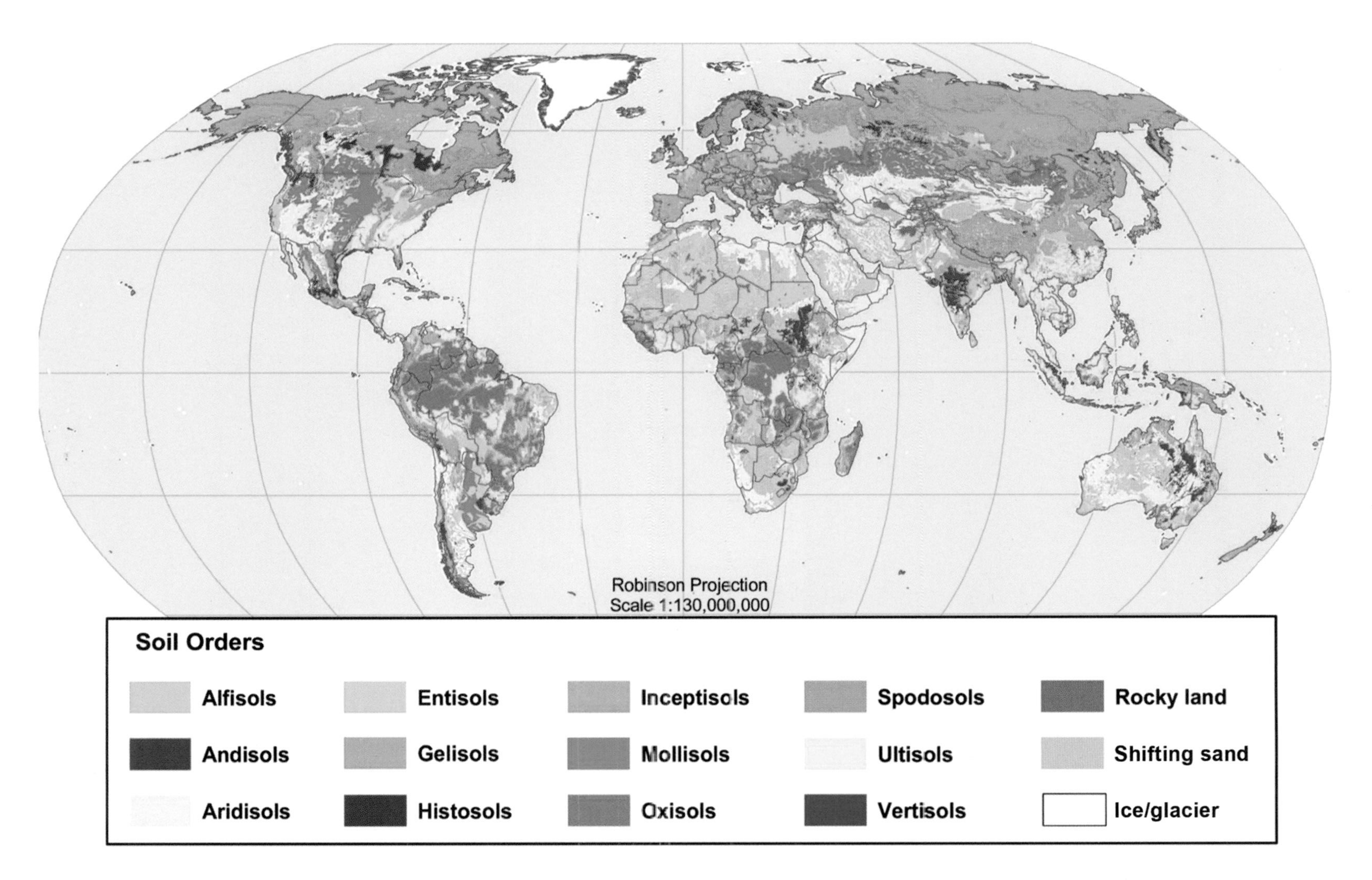

Figure 6.7 Global soil regions. (After USDA-NRCS, 2005.)

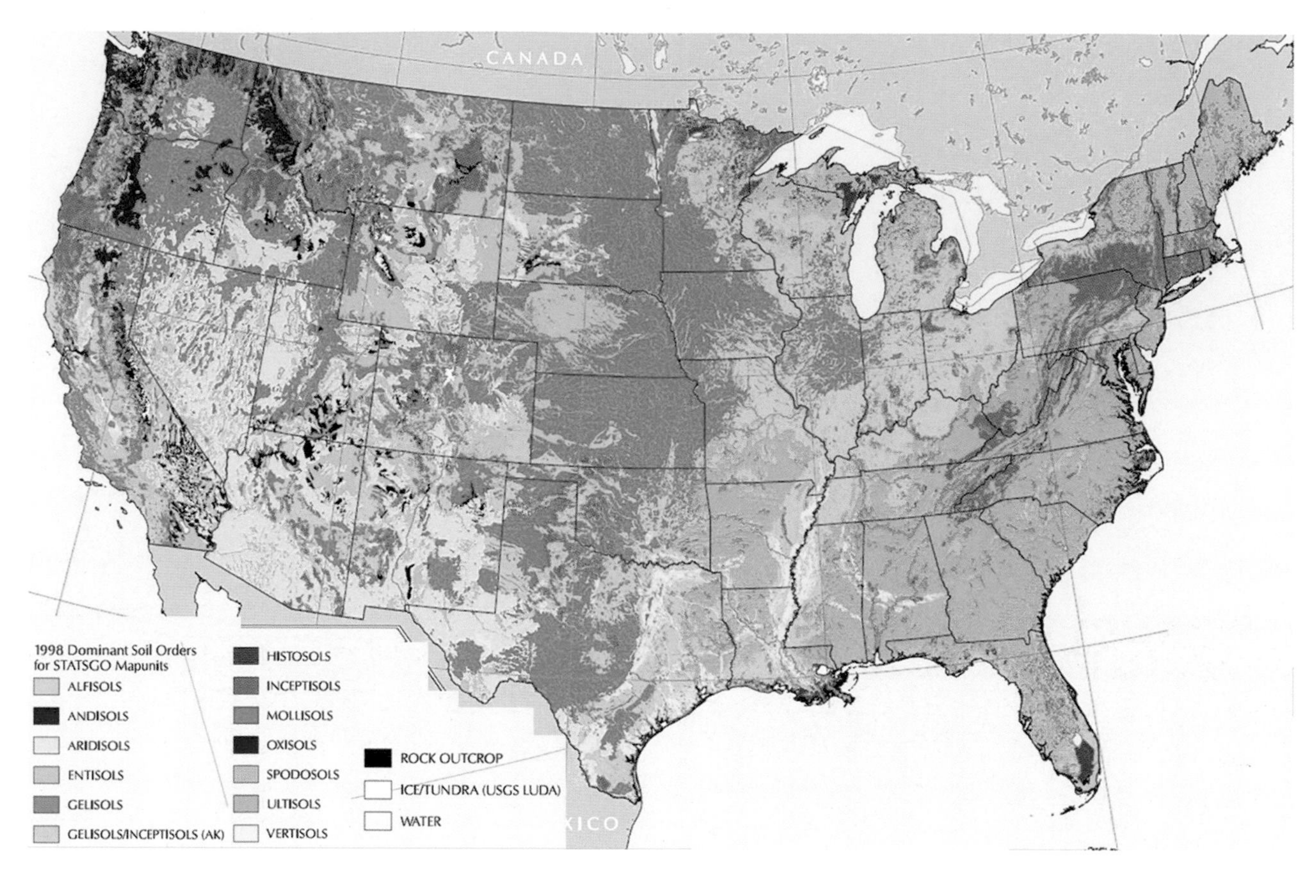

Figure 6.8 Twelve major soil orders in the United States. (After Natural Resources Conservation Service , 2007.)

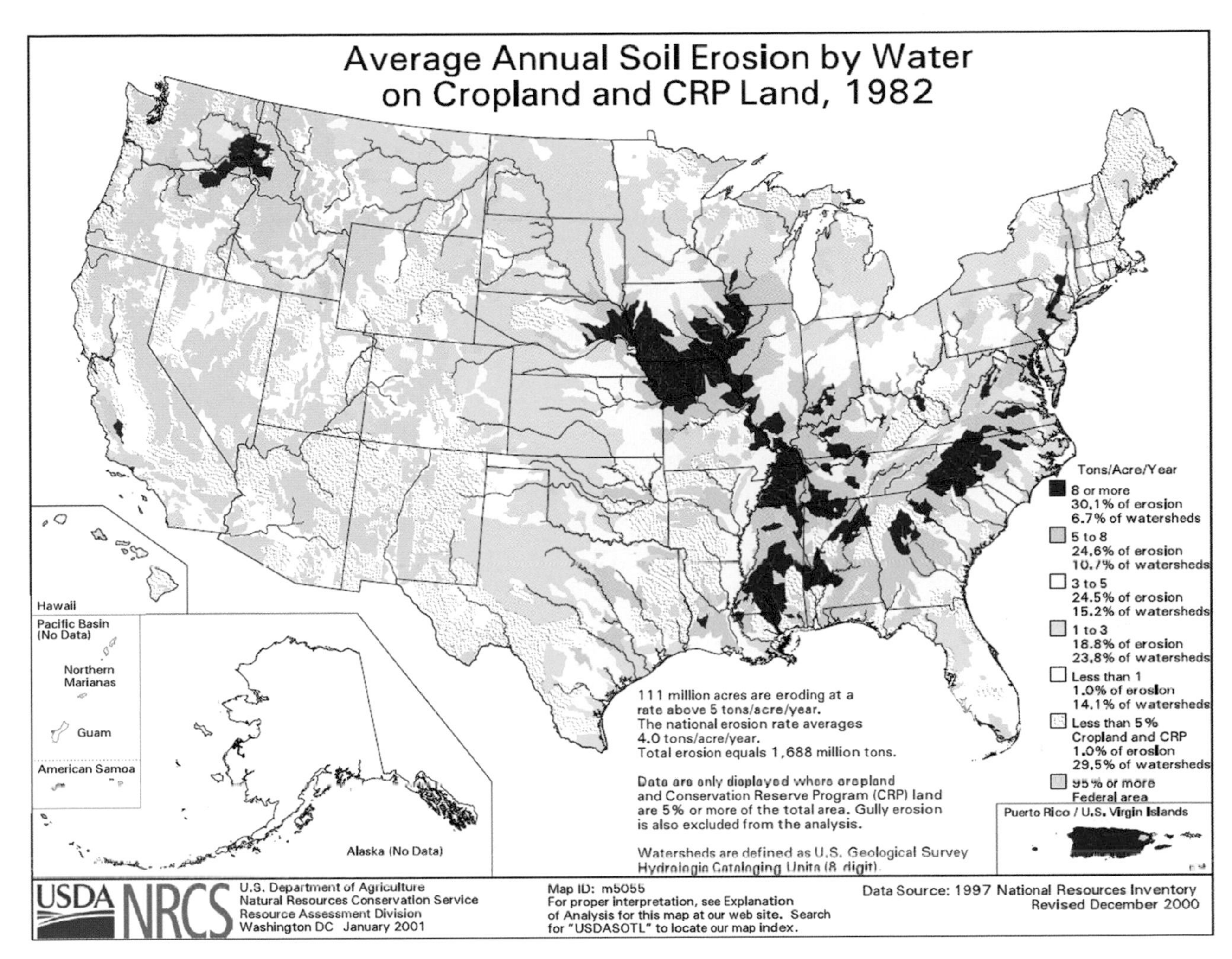

Figure 6.12 Average annual soil loss erosion by water on cropland and Conservation Reserve Program land, 1982. (USDA-NRCS, 2001b.)

Figure 7.1 Cereal crops.

Figure 7.2 Grain crops.

Figure 7.3 Pulses.

Figure 12.8c Double-ring infiltrometer.

Figure 21.2 A portable sprinkler system.

Figure 21.3 Permanent and semi-permanent systems.

Figure 21.4 Solid-set systems: portable and permanent.

Figure 21.5 Set-move systems: hand-move and tow-move.

Figure 21.6 Set-move systems: side-roll and gun-type. (Adapted from USDA-NRCS, 2016.)

Figure 21.7 Spray sprinkler systems: rotating head and perforated.

Figure 21.8 Continuous-move systems. (Traveler systems adapted from USDA-NRCS, 2016.)

Figure 21.11 Different types of sprinkler heads.

Figure 21.12 Types of sprinklers: impact and reaction types.

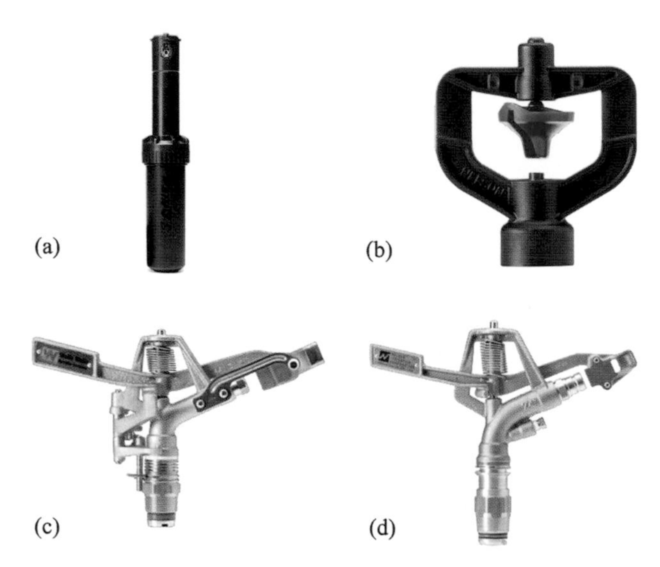

Figure 21.13 (a) Full-circle, impulse-driven, rotating pop-up sprinkler (www.rainbird.com/products/ag5-maxi-paw-pop-impact-sprinklers; Copyright © by Rain Bird. Reprint with permission from Rain Bird) . (b) Full-circle, low application rate agricultural sprinkler (https://nelsonirrigation.com/products/rotator-sprinklers/r10-rotator). (c) Part-circle, medium-sized agricultural sprinkler (https://nelsonirrigation.com/products/impact-sprinklers/part-circle-impacts). (d) Full-circle higher application rate agricultural sprinkler (https://nelsonirrigation.com/products/impact-sprinklers/full-circle-impact-sprinklers). (b–d Copyright © by Nelson Irrigation. Reprint with permission from Nelson Irrigation.)

Figure 21.25 Mid-elevation spray application, low-elevation spray application, low-energy precision application.

Figure 21.27 Big gun sprinkler system. (The left figure is adapted from USDA-NRCS, 2016.)

Figure 22.2 Types of trickle irrigation methods.

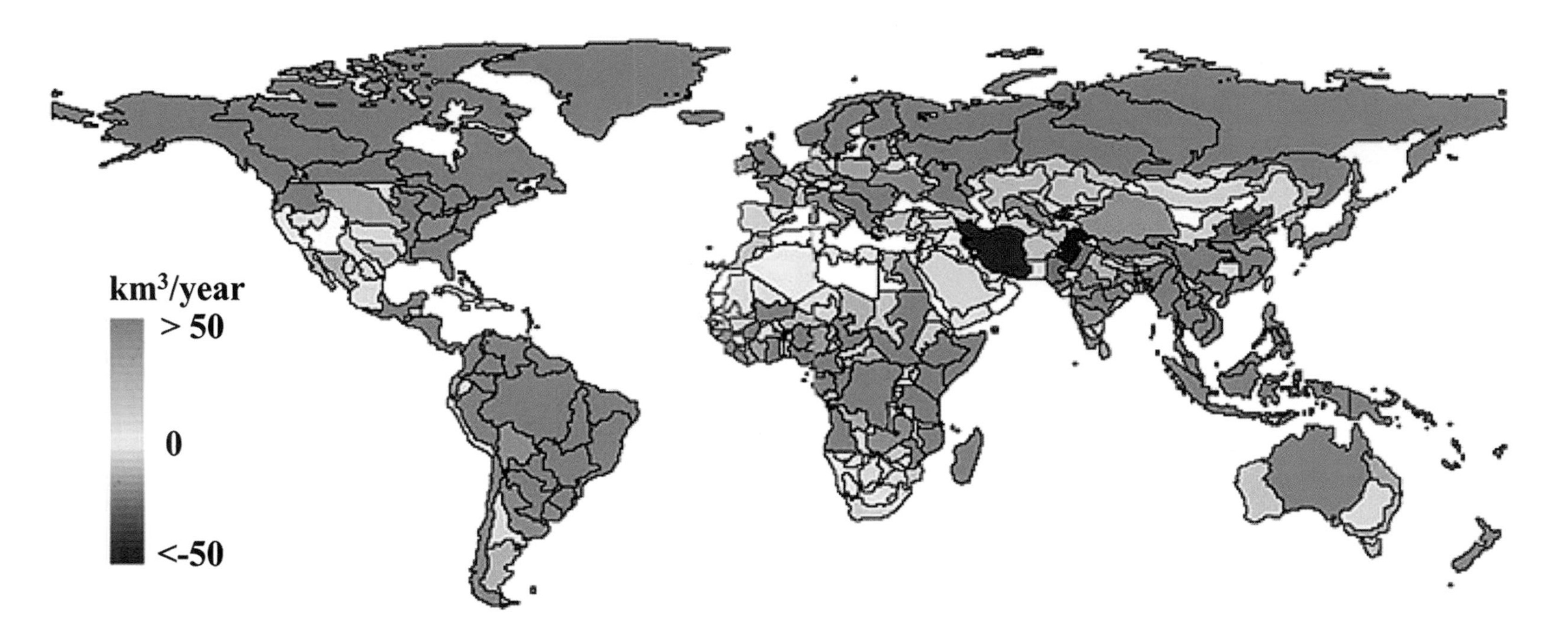

Figure 23.1 The median differences between irrigation demand and available irrigation demand at the end of the century under RCP 8.5 from 10 GHMs and 6 GGCMs. (Adapted from Elliott et al., 2014. Reprinted with permission from PNAS.)

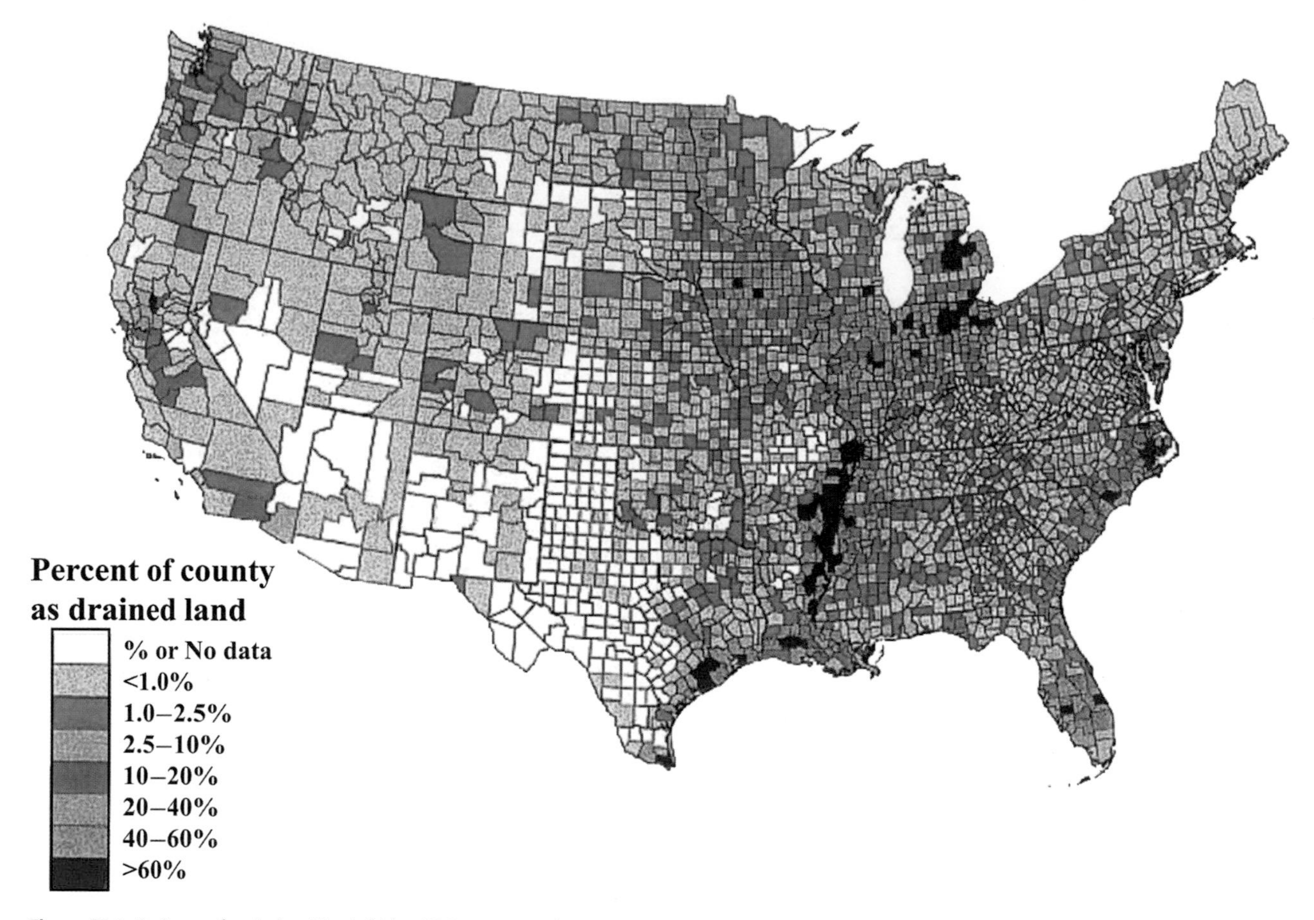

Figure 25.1 Drainage of agricultural land. (After US Department of Commerce, 1981. Graphed by William Battaglin, USGS.)

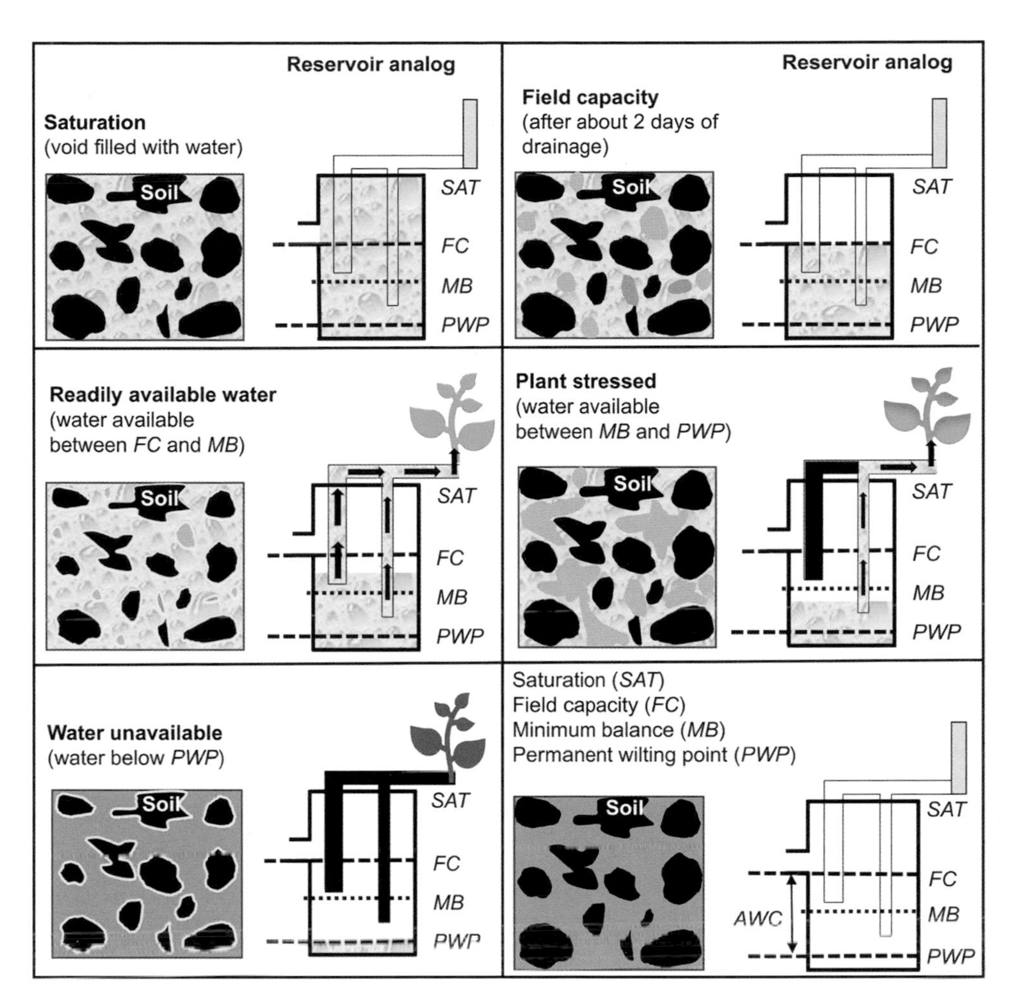

Figure 27.4 Soil water with reservoir analog.

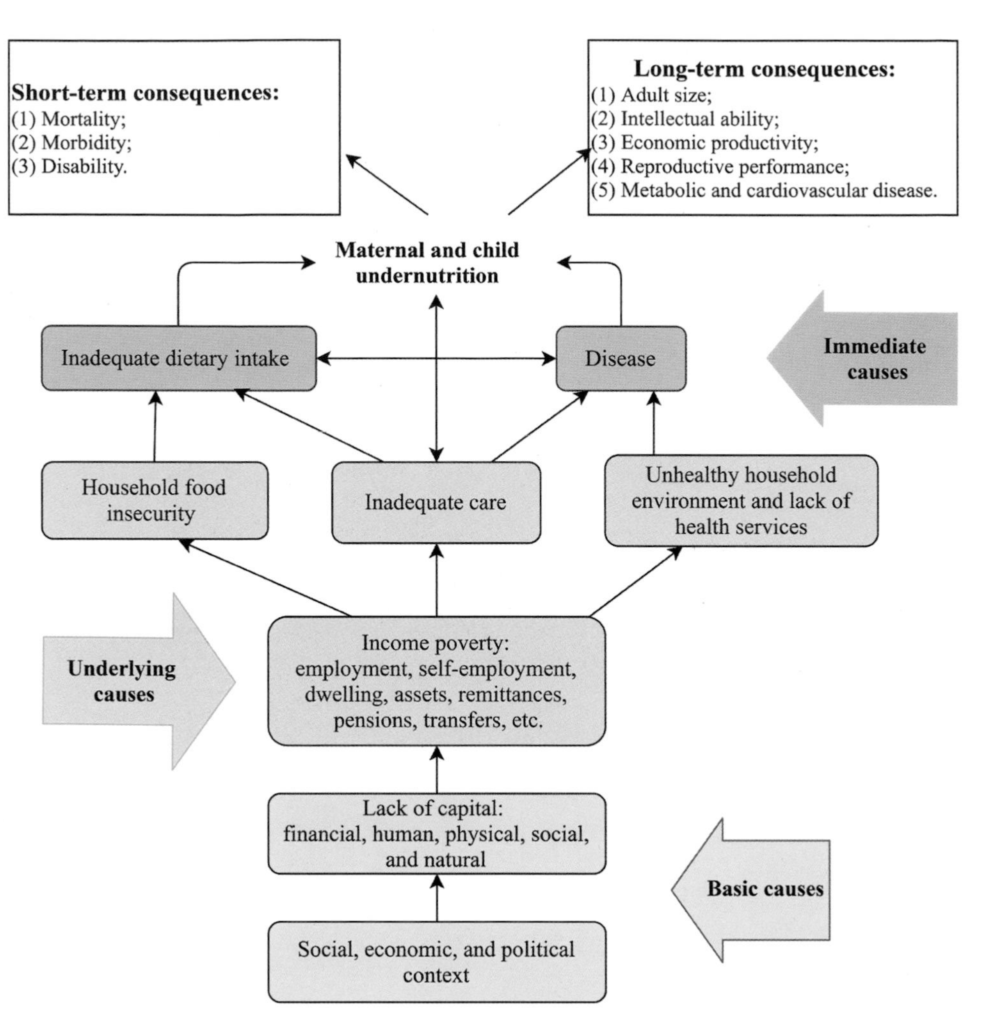

Figure 31.1 Conceptual framework of undernutrition. (After Black et al., 2008. Copyright © 2008. Reprinted with permission from Elsevier.)

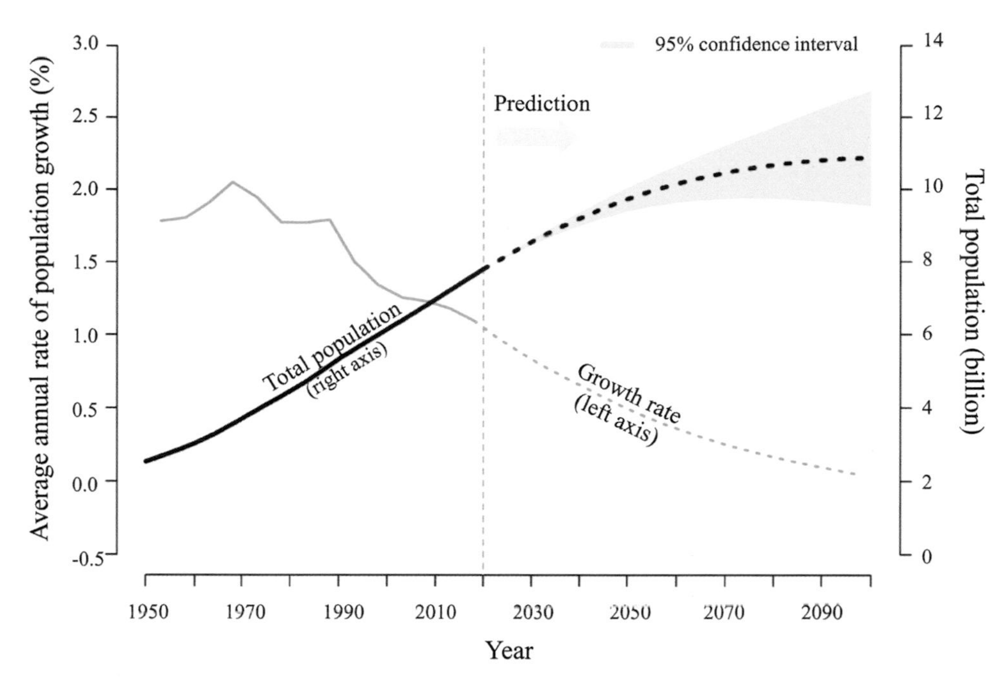

Figure 31.2 Global population and average annual rate of population growth (%) as a function of time. The projections are based on total fertility and life expectancy at birth. The green shaded area represents 95% prediction intervals, 2020–2100. (Data from United Nations, DESA, Population Division, World Population Prospects 2019. http://population.un.org/wpp.)

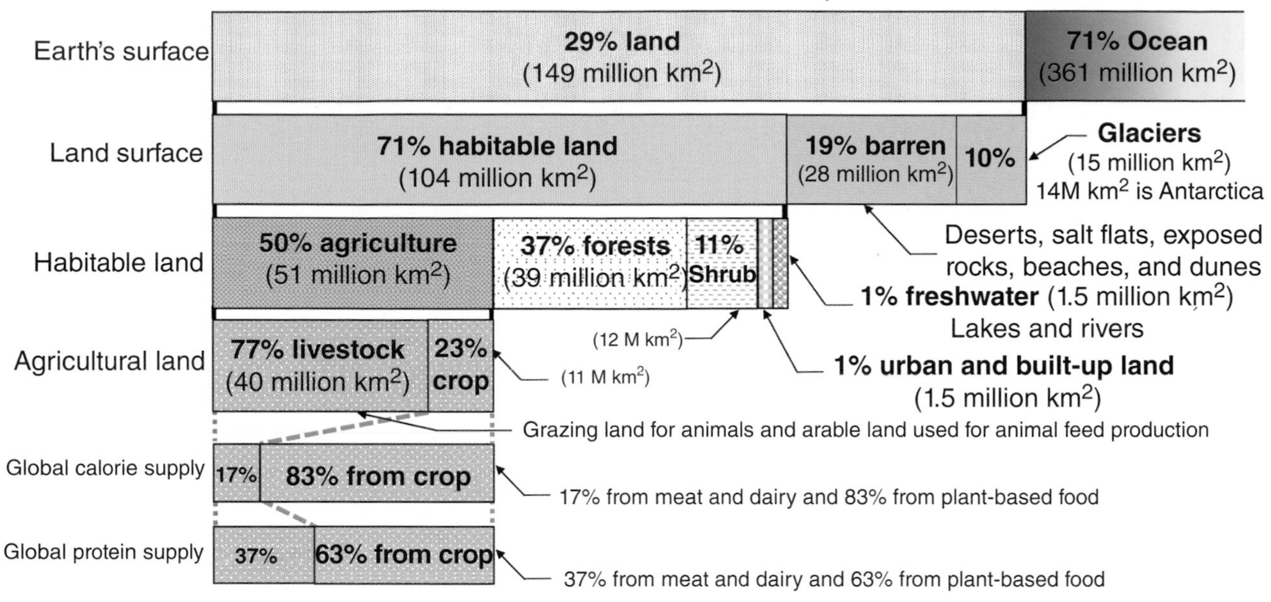

Figure 31.3 Global land use for food production. (Data from FAO, 2016.)

The opening and closing of stomata determine the occurrence and cessation of transpiration. When stomata are open, transpiration occurs, and at this time the vapor pressure in the leaf is generally greater than the vapor pressure of the surrounding air. Since the leaf has some water even when it is wilted, the vapor pressure in the leaf is equal to the saturation vapor pressure. This means there is a vapor pressure gradient when stomata are open and this vapor pressure gradient facilitates the transport of vapors into the atmosphere. However, there is also resistance to vapor flow through stomata, which primarily depends on the movement of air. When stomata close, this resistance increases, indicating that the degree of resistance depends on the degree of stomatal closure. When there is high wind, the resistance decreases because this wind tends to break the boundary layer surrounding the leaf.

QUESTIONS

Q.14.1 Using the BC method, compute *ET* from a field where the percentage (p) of daytime hours in July is given as 0.40, mean maximum air temperature is 32 °C, the mean minimum air temperature is 10 °C, the mean air temperature is 18 °C, the mean dew-point air temperature is 8.0 °C, the mean wind speed is 190 km/day, the ratio of wind speed (day/night) is 2.5, the mean percent sunshine (n/N) is 95%, the mean daytime length is 13 h, the mean pan evaporation is 8.0 mm/day, the mean measured alfalfa *ET* is 7.5 mm/day, latitude is 42° N, the mean solar radiation (R_s) is 550 Langley/day, the crop is wheat, and site elevation (E_l) is 1000 m.

Q.14.2 Estimate *ET* using the JH method, using the field weather data given in Q.14.1.

Q.14.3 Estimate *ET* for the month of July using the Hargreaves method, where the latitude is 40° N, the month (M) is July, the mean air temperature is 25 °C, the mean maximum air temperature is 32 °C, the mean minimum air temperature is 15 °C, the crop is wheat, and site elevation (E_l) is 500 m.

Q.14.4 For the purposes of computing reference *ET* for wheat for Dallas Station, Texas, the following data are given: latitude and longitude of Dallas: 32.85° N and 99.86° W; date: July 1; maximum temperature (T_{max}) is 35.6 °C; minimum temperature (T_{min}) is 26.1 °C; maximum RH (RH_{max}) is 78 %; minimum RH (RH_{min}) is 51%; wind velocity at 0.50 m height is 8.5 km/h; and ratio of wind run during day to wind run during night (u_{day}/u_{night}) is 3/1 It shows that three units of wind occurred during the day for each unit of wind that occurred during the night. Daytime is arbitrarily chosen as 0700 to 1900 hours. The number of sunshine hours is 11.5. The elevation is 439.6 ft (= 134 m); and the temperature for June 30, 29, and 28 is, respectively, 30.3, 29.4, and 29.5 °C. The wind velocity u_2 at 2 m height can be computed from velocity U_z at any height z (m) as

$$u_2 = U_z \left[\frac{2.0}{z}\right]^{0.2}.$$

Note: For energy, the unit is joules (J) or N·m or kg·m²/s². For force, the unit is N or kg·m/s². For pressure, the unit is pascal (Pa) or N/m² or kg/m·s² or mb. Apply appropriate conversion factors when going from one system of units to another.

Do the following calculations:

1. Compute the mean temperature in Celsius, mean *RH*, wind velocity in km/day, and daytime wind run in kilometers per 12 hours and daytime wind speed in m/s. Compute the wind function.
2. Compute the slope of the saturation vapor pressure curve. Compute the psychrometric constant γ.
3. Compute the weighting factors in the Penman equation: $\frac{\Delta}{\Delta+\gamma}$ and $\frac{\gamma}{\Delta+\gamma}$.
4. Compute the shortwave radiation in mm/day.
5. Compute the effective emissivity and net outgoing solar radiation in mm/day.
6. Compute the net radiation with an appropriate value of albedo.
7. Compute the vapor pressure deficit.
8. Compute the FAO c factor.
9. Compute the reference *ET* for grass.
10. Compute the reference *ET* for alfalfa using the Wright–Penman method.

Q.14.5 Compute the reference *ET* by the Taylor Priestley method for grass for Dallas, Texas, using the data given in Q.14.4. Assume albedo is 0.23.

Q.14.6 Compute the reference *ET* by the Penman Monteith method for grass for Dallas, Texas, using the data given in Q.14.4. Assume albedo is 0.23.

Q.14.7 For the purposes of computing reference *ET* for wheat for Beaumont, Texas, the following data are given: latitude and longitude of Beaumont are 29.95° N and 92.42° W; date is July 1; maximum temperature (T_{max}) is 35 °C; minimum temperature (T_{min}) is 24 °C; maximum RH (RH_{max}) is 90%; minimum RH (RH_{min}) is 65%; wind velocity at 0.50 m height is 5.4 km/h; and ratio of wind run during day to wind run during night (u_{day}/u_{night}) is 2/1. The number of sunshine hours is 12.5. The elevation is 16.4 ft (5 m); and temperature on June 30, 29, and 28 is, respectively, 28.9, 26.8, and 27.5 °C. The wind velocity u_2 at 2 m height can be computed from velocity U_z at any height z (m) as

$$u_2 = U_z \left[\frac{2.0}{z}\right]^{0.2}.$$

Note: For energy, the unit is joules (J) or N·m or kg·m²/s². For force, the unit is N or kg·m/s². For pressure, the unit is pascal (Pa) or N/m² or kg/m·s² or mb. Apply appropriate conversion factors when going from one system of units to another.

Do the following calculations:

1. Compute the mean temperature in Celsius, mean RH, wind velocity in km/day, and daytime wind run in kilometers per 12 hours and daytime wind speed in m/s. Compute the wind function.
2. Compute the slope of the saturation vapor pressure curve. Compute the psychrometric constant γ.
3. Compute the weighting factors in the Penman equation: $\frac{\Delta}{\Delta+\gamma}$ and $\frac{\gamma}{\Delta+\gamma}$.
4. Compute the shortwave radiation in mm/day.
5. Compute the effective emissivity and net outgoing solar radiation in mm/day.
6. Compute the net radiation with an appropriate value of albedo.
7. Compute the vapor pressure deficit.
8. Compute the FAO c factor.
9. Compute the reference ET for grass.
10. Compute the reference ET for alfalfa using the Wright–Penman method.

Q.14.8 Compute the reference ET by the Priestley–Taylor method for grass for Beaumont, Texas, using the data given in Q.14.7. Assume albedo is 0.23.

Q.14.9 Compute the reference ET by the Penman–Monteith method for grass for Beaumont, Texas, using the data given in Q.14.7. Assume albedo is 0.23.

REFERENCES

Allen, R. G. (1986). A Penman for all seasons. *Journal of Irrigation and Drainage Division, ASCE*, 112(4): 348–368.

Allen, R. G., Pereira, L. S., Raes, D., and Smith, M. (1998). Crop evapotranspiration: guidelines for computing crop water requirements. FAO irrigation and drainage paper 56.

Blaney, H. F., and Criddle, W. D. (1945). Determining water requirements in irrigated areas from climatological data. Working paper.

Blaney, H. F., and Criddle, W. D. (1950). *Determining Water Requirements in Irrigated Areas from Climatological and Irrigation Data*. Washington, DC: Soil Conservation Service, USDA.

Brutsaert, W. (1982). *Evaporation into the Atmosphere*. Dordrecht: D. Reidel.

Burman, R. D., Cuenca, R. H., and Weiss, A. (1983). Techniques for estimating irrigation water requirements. In *Advances in Irrigation*, Vol. 2, edited by D. Hillel,. New York: Academic Press, pp. 336–394.

Burman, R. D., Jensen, M. E., and Allen, R. G. (1987). Thermodynamic factors in evapotranspiration. In *Proceedings, Irrigation and Drainage Specialty Conference*, edited by L. G. James and M. J. English. Portland, OR: ASCE, pp. 28–30.

Cuenca, R. H., and Jensen, M. E. (1987). Approximating the FAO coefficients: a second look. Unpublished.

Cuenca, R. H., and Nicholson, M. T. (1982). Application of Penman equation wind function. *Journal of Irrigation and Drainage Division, ASCE*, 108(IR 1): 13–24.

Doorenbos, J., and Pruitt, W. O. (1975). Crop water requirements. FAO irrigation and drainage paper 24.

Doorenbos, J., and Pruitt, W. O. (1977). Crop water requirements. FAO irrigation and drainage paper 24, revised.

Fritz, S., and MacDonald, J. H. (1949). Average solar radiation in the United States. *Heating and Ventilating*, 46: 61–64.

Hargreaves, G. H. (1975). Moisture availability and crop production *Transactions of ASAE*, 18(5): 980–985.

Hargreaves, G. H. (1994). *Irrigation Fundamentals*. Logan, UT: Utah State University.

Hargreaves, G. H., and Samani, Z. A. (1985). Reference crop evapotranspiration from temperature. *Applied Engineering in Agriculture*, 1(2): 96–99.

Hargreaves, G. L., Hargreaves, G. H., and Riley, J. P. (1985). Irrigation water requirements for Senegal River basin. *Journal of Irrigation and Drainage Engineering*, 11(3): 265–275.

Jensen, M. E. (1974). *Consumptive Use of Water and Irrigation Water Requirements. Report Prepared by the Technical Committee on Irrigation Water Requirements of the Irrigation and Drainage Division of the American Society of Civil Engineers*. New York: ASCE.

Jensen, M. E., and Haise, H. R. (1963). Estimating evapotranspiration from solar radiation. *Proceedings of the Irrigation and Drainage Division*, 89: 15–41.

Jensen, M. E., Wright, J. L., and Pratt, B. J. (1971). Estimating soil moisture depletion from climate, crop and soil data. *Transactions of ASAE*, 14(5): 954–959.

Keenan, J. H., Keyes, P. G., Hill P. G., and Moore J. G. (1978) *Stream Tables: Thermodynamics Properties of Water Including Vapor, Liquid, and Solid Phase*. New York: Wiley.

Monteith, J. L. (1981). Evaporation and surface temperature. *Quarterly Journal of Royal Meteorological Society*, 107: 1–27.

Penman, H. L. (1948). Natural evaporation for open water, bare soil, and grass. *Proceedings Royal Society of London*, A193: 120–146.

Priestley, C. H. B., and Taylor, R. J. (1972). On the assessment of surface heat flux and evaporation using large scale parameters. *Monthly Weather Review*, 100: 81–92.

Rosenberg, N. J., Blad, B. L., and Verma, S. B. (1983). *Microclimate: The Biological Environment*. New York: Wiley.

Soil Conservation Service (SCS). (1970). Irrigation water requirements. Technical release 21.

Wilson, D. H., Reginato, R. J., and Hollett, K. J. (1992). Evapotranspiration measurements of native vegetation, Owens Valley, California. USGS Water-Resources Investigations Report 91-4159.

Wright, J. L. (1982). New evapotranspiration coefficients. *Journal of Irrigation and Drainage Division, ASCE*, 108(IR 1): 57–74.

Part IV Irrigation Science

Part IV: Irrigation Science

15 Crop Water Use

Notation

AVM	available soil moisture at the time K_c is evaluated (mm)	K_{cf}	crop adjustment factor for wind speed and relative humidity
CEW	limit of extractable water expressed as depth in the root zone (mm)	K_{c-FAO}	crop coefficient based on the grass reference crop
ET_c	crop evapotranspiration (mm/day)	K_{ci}, K'_{cp}, K'_{cm}	basal crop coefficients for initial, crop development, and mid-season period, respectively
ET_r	reference crop evapotranspiration (mm/day)		
$ET_{r-alfalfa}$	reference evapotranspiration for alfalfa (mm/day)	K_{cmax}	maximum crop coefficient for wet soil evaporation
$ET_{r-grass}$	reference evapotranspiration for grass (mm/day)	K_{co}	FAO dual crop coefficient that considers the effect of soil water stress
$f(t)$	wet decay function	K_s	stress factor to account for soil water stress on evapotranspiration
f_c	critical threshold of available water when stress begins (fraction)	K_t	threshold coefficient corresponding to the level of crop-canopy development beyond which soil evaporation is negligible
f_r	fraction of soil water that remains (fraction)		
f_{s1}, f_{s2}, and f_{s3}	fraction of growing season for initial, crop development, and mid-season period, respectively (fraction)	K_w	wet soil evaporation factor
		K_{ws}	coefficient related to evaporation from wet soil surface condition
f_{vc}	vegetation coverage fraction (fraction)		
F_w	fraction of soil surface wetted (fraction)	LAI	leaf area index
		RH_{min}	minimum relative humidity (%)
FC	soil moisture content in the root zone at field capacity (mm)	SMC	soil moisture content in the root zone at the time K_c is evaluated (mm)
h	mean crop height during the mid-season and late season period (m)	t	time since last wetting (day)
		t_d	duration of wet soil surface (day)
I_f	irrigation or rainfall frequency (day)	TAM	total available soil moisture as depth (mm)
K_a	coefficient related to the soil moisture available in the root zone	u_2	mean daily wind speed at 2 m height (m/s)
K_c	crop coefficient		
$K_{c-alfalfa}$	crop coefficient based on the alfalfa reference crop	θ_{fc}	field capacity of the soil (in./in. or m/m)
$K_{c-alfalfa/grass}$	FAO coefficient for alfalfa given in Table 15.4 for peak conditions	θ_v	water content of soil on a volume basis (fraction or percentage)
K_{cb}	basal crop coefficient related to a dry soil surface	θ_{wp}	permanent wilting point of soil (in./in. or m/m)

15.1 INTRODUCTION

Soil water is vital for seeds to germinate, plants to grow, foliage to form, and plant cells to maintain turgidity. Water is absorbed from the soil by plant roots through the water potential gradient between plant leaves and soil. As water evaporates from the plant foliage, plants are cooled. The process of evapotranspiration (ET) satisfies the atmospheric demand for moisture; when the demand exceeds ET, moisture deficit occurs in the plant and stomata close. Then, the movement of water from leaves into the atmosphere ceases and plants begin to wilt. A moisture deficit can be caused by lack of soil moisture, lack of healthy root zone, or damaged plant transport systems, and leads to plant stress. Indications of moisture deficit in plants are darkening of leaf color and lack of turgidity in foliage. Wilting can occur in plants on very hot days despite sufficient moisture in the soil. If wilting is in the early stage, most plants can recover turgidity during the night when the atmospheric demand for water is low. If water is not supplied and the moisture deficit grows, wilting will become severe and plants may not be able to regain turgidity even if enough water is then applied; then, the plants will die. On the other hand, the removal of moisture by ET results in the concentration of the salts remaining in the soil water. High concentration of salts makes it harder for plant root systems to absorb moisture, leading to unsatisfactory plant growth.

Evapotranspiration is either estimated for a reference crop, such as alfalfa or grass, or as a potential *ET* value. The estimated *ET* value is then modified for a specific crop using a crop coefficient, which is what constitutes the crop water use or consumptive use (CU). Estimation of the crop coefficient is therefore fundamental. The objective of this chapter is to discuss the methods for converting the estimated *ET* to a specific crop *ET* (ET_c) which is the same as CU or to compute the crop coefficient.

15.2 CROP WATER USE AND COEFFICIENT

Consumptive use or crop water use includes water used in all of the plant physiological processes, including transpiration, digestion, photosynthesis, structural support, growth, transport of minerals, as well as direct evaporation from plant surfaces and soil. Water use in crop-growing areas occurs through evaporation from the soil surface as well as from wet leaf surfaces and transpiration from plants. Although CU is slightly greater (less than 1%) than *ET*, both are considered equal. Transpiration entails change of water into the vapor phase within the leaf and then transfer through stomata of the leaf into the atmosphere. The process of transpiration accounts for 99% of water used by plants (Wilson and Loomis, 1962). The transfer of water to the atmosphere involves direct evaporation of solid and liquid water from plant surfaces and soil, as well as transpiration. These processes, when combined, constitute *ET* – that is, the sum of evaporation and transpiration is evapotranspiration.

The factors affecting CU are the type of plant grown, leaf canopy, stage of plant growth, climate conditions, soil minerals, and availability of soil water. The rate of transpiration depends on the stage of crop growth even if the atmospheric demand is invariable. Soil and air temperature, radiation, humidity, rainfall, wind, and plant characteristics affect CU. Soil physical factors, including the amount of available soil water, salt concentration, soil temperature, and carbon dioxide concentration, affect the CU. Plants vary in their ability to extract moisture from the soil and their requirement for water. Some plants are more drought-tolerant than others through a variety of physiological characteristics, such as deep and well-developed root systems, wavy leaf surfaces, leaf hairs to reduce airflow past the leaf surface, shiny surfaces to reflect light, and leaves that fold up or drop under stress conditions. On the other hand, too much water can also damage crops due to lack of oxygen, and shallow root systems in cases of frequent irrigation can also harm crops and affect CU.

The CU or crop water use is the actual *ET* of a specific crop, which is normally computed by multiplying either potential *ET* or reference crop *ET* by what is called the crop coefficient. Thus, the crop coefficient plays a fundamental role in computing CU. The crop coefficient, denoted as K_c, is a dimensionless number and varies from crop to crop. For any crop it depends on growth stage, relative humidity, and wind velocity, as well as irrigation and rainfall frequency. It is the ratio of *ET* for a crop for which the water use is to be determined to the *ET* for a reference crop. Doorenbos and Pruitt (1977) defined reference crop *ET* as the "ET from an extensive surface of 8 to 15 cm (3 to 6 in.) tall, green grass cover of uniform height, actively growing, completely shading the ground and not short of water." Wright (1981) defined reference crop *ET* as being "equal to daily alfalfa ET when crop occupies an extensive surface, is actively growing, standing erect and at least 20 cm (8 in.) tall, and is well watered so that soil water availability does not limit ET." Reference *ET* is defined for a specific crop and set of aerodynamic conditions and is therefore preferred to potential *ET*, which varies from crop to crop because of differences in aerodynamic roughness and albedo (surface reflectance) and from place to place because of differences in the amount of sensible and latent heat transfer into the area. Reference *ET* is the potential *ET* for a specific crop, usually taken as grass or alfalfa.

It is clear that crop coefficient pertains to the actual rate of CU and depends on the crop physiology, crop cover, location, and method of computing potential or reference *ET*. The variation of K_c with location and potential *ET* can be reduced by considering reference crop *ET*. There are several methods for determining the crop coefficient that are discussed in what follows.

15.3 FAO CROP COEFFICIENT METHOD

The FAO method can be expressed as

$$ET_c = K_c ET_r, \tag{15.1}$$

where ET_c is the crop *ET*, ET_r is the reference crop *ET*, and K_c is the crop coefficient. It should be noted that the values of K_c in eq. (15.1) must be for the same crop as the values of ET_r. The crop coefficient when plotted against time from the beginning of crop development to the harvest (i.e., growing season) plots as a curve. This curve is called the crop coefficient curve. In the FAO method, this curve is partitioned into four parts, where each part is approximated

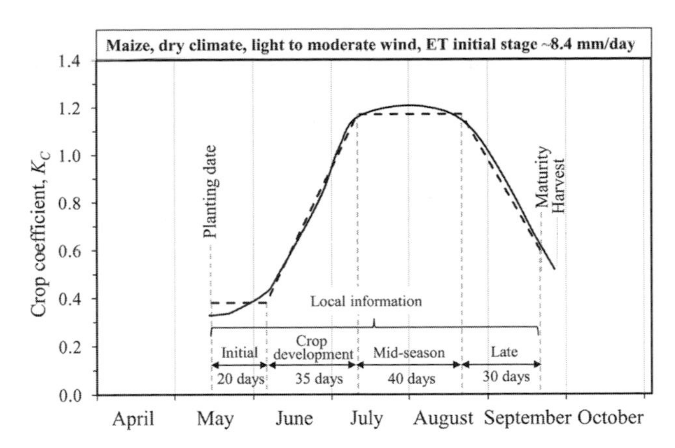

Figure 15.1 Example of a crop coefficient curve. The solid line represents the actual K_c curve. The dashed line represents a linear approximation of the K_c curve using the FAO method. (Adapted from Doorenbos and Pruitt, 1977.)

by a straight line – that is, the curve is approximated by four straight lines. In other words, the growing season is divided into four parts where each part is represented by the corresponding part of the coefficient curve, as shown in Figure 15.1.

The growing season comprises four growth periods or stages: (1) initial period; (2) crop development period; (3) mid-season period; and (4) late season period. Planning and sowing dates can be established using local information. Then, the length of the growing season and the length of the crop development stage can be determined. Table 15.1 provides information on the growing season and crop development stages of selected crops (Doorenbos and Pruitt, 1977). The initial period is defined as the period from the time of planting to the time of 10% ground cover. During the initial period, ET occurs primarily from the bare soil. Thus, K_c for the initial period depends on ET and the soil wetness,

Table 15.1 *Length of growing season and crop development stages of selected field crops (Doorenbos and Pruitt, 1977)*

Crop	Growing season stages
Artichokes	Perennial, replanted every 4–7 years; coastal California planting in April 40/40/250/30 and (360)[a]; subsequent crops with crop growth cutback to ground level in late spring each year at the end of harvest or 20/40/220/30 and (310).
Barley	Also wheat and oats; varies widely with variety; wheat in Central India November planting 15/25/50/30 and (120); early spring sowing, semi-arid, 35–45° latitudes; and November planting in Republic of Korea 20/25/60/30 and (135); wheat sown in July in East African highlands at 2500 m altitude and Republic of Korea 15/30/65/40 and (150).
Beans (green)	February and March planting California desert and Mediterranean 20/30/30/10 and (90); August–September planting California desert, Egypt, Coastal Lebanon 15/25/25/10 and (75).
Beans (dry) Pulses	Continental climates late spring planting 20/30/40/20 and (110); June planting Central California and West Pakistan 15/25/35/20 and (95); longer season varieties 15/25/50/20 and (110).
Beets (table)	Spring planting Mediterranean 15/25/20/10 and (70); early spring planting Mediterranean climates and pre-cool season in desert climates 25/30/25/10 and (90).
Carrots	Warm season of semi-arid to arid climates 20/30/30/20 and (100); for cool season up to 20/30/80/20 and (150); early spring planting Mediterranean 25/35/40/20 and (120); up to 30/40/60/20 and (150) for late winter planting.
Castor beans	Semi-arid and arid climates, spring planting 25/40/65/50 and (180).
Celery	Pre-cool season planting semi-arid 25/40/95/20 and (180); cool season 30/55/105/20 and (210); humid Mediterranean mid-season 25/40/45/15 and (125).
Corn (maize) (sweet)	Philippines, early March planting (late dry season) 20/20/30/10 and (80); late spring planting Mediterranean 20/25/25/10 and (80); late cool season planting desert climates 20/30/30/10 and (90); early cool season planting desert climates 20/30/50/10 and (110).
Corn (maize) (grains)	Spring planting East African highlands 30/50/60/40 and (180); late cool season planting, warm desert climates 25/40/45/30 and (140); June planting sub-humid Nigeria, early October India 20/35/40/30 and (125); early April planting Southern Spain 30/40/50/30 and (150).
Cotton	March planting Egypt, April–May planting Pakistan, September planting South Arabia 30/50/60/55 and (195); spring planting, machine-harvested Texas 30/50/55/45 and (180).
Crucifers	Wide range in length of season due to varietal differences; spring planting Mediterranean and continental climates 20/30/20/10 and (80); late winter planting Mediterranean 25/35/25/10 and (95); autumn planting Coastal Mediterranean 30/35/90/40 and (195).
Cucumber	June planting Egypt, August–October California desert 20/30/40/15 and (105); spring planting semi-arid and cool season arid climates, low desert 25/35/50/20 and (130).
Eggplant	Warm winter desert climates 30/40/40/20 and (130); late spring–early summer planting Mediterranean 30/45/40/25 and (140).
Flax	Spring planting cold winter climates 25/35/50/40 and (150); pre-cool season planting Arizona low desert 30/40/100/50 and (220).

Table 15.1 (*cont.*)

Crop	Growing season stages
Grain, small	Spring planting Mediterranean 20/30/60/40 and (150); October–November planting warm winter climates; Pakistan and low deserts 25/35/65/40 and (165).
Lentil	Spring planting in cold winter climates 20/30/60/40 and (150); pre-cool season planting warm winter climates 25/35/70/40 and (170).
Lettuce	Spring planting Mediterranean climates 20/30/15/10 and (75) and late winter planting 30/40/25/10 and (105); early cool season low desert climates from 25/35/30/10 and (100); late cool season planting, low deserts 35/50/45/10 and (140).
Melons	Late spring planting Mediterranean climates 25/35/40/20 and (120); mid-winter planting in low desert climates 30/45/65/20 and (160).
Millet	June planting Pakistan 15/25/40/25 and (105); central plains United States spring planting 20/30/55/35 and (140).
Oats	See Barley.
Onion (dry)	Spring planting Mediterranean climates 15/25/10/40 and (150); pre-warm winter planting semi-arid and arid desert climates 20/35/110/45 and (210).
(green)	Respectively 25/30/10/5 and (70) and 20/45/20/10 and (95).
Peanuts (groundnuts)	Dry season planting West Africa 25/35/45/25 and (130); late spring planting coastal plains of Lebanon and Israel 35/45/35/25 and (140).
Peas	Cool maritime climates early summer planting 15/25/35/15 and (90); Mediterranean early spring and warm winter desert climates planting 20/25/35/15 and (95); late winter Mediterranean planting 25/30/30/15 and (100).
Peppers	Fresh Mediterranean early spring and continental early summer planting 30/35/40/20 and (125); cool coastal continental climates mid-spring planting 25/35/40/20 and (120); pre-warm winter planting desert climates 30/40/110/30 and (210).
Potato (Irish)	Full planting warm winter desert climates 25/30/30/20 and (105); late winter planting arid and semi-arid climates and late spring–early summer planting continental climate 25/30/45/30 and (130); early–mid-spring planting central Europe 30/35/50/30 and (145); slow emergence may increase length of initial period by 15 days during cold spring.
Radishes	Mediterranean early spring and continental summer planting 5/10/15/5 and (35); coastal Mediterranean late winter and warm winter desert climates planting 10/10/15/5 and (40).
Safflower	Central California early–mid-spring planting 20/35/45/25 and (125) and late winter planting 25/35/55/30 and (145); warm winter desert climates 35/55/60/40 and (190).
Sorghum	Warm season desert climates 20/30/40/30 and (120); mid-June planting Pakistan, May in mid-West United States and Mediterranean 20/35/40/30 and (125); early spring planting warm arid climates 20/35/45/30 and (130).
Soybeans	May planting Central United States 20/35/60/25 and (140); May–June planting California desert 20/30/60/25 and (135); Philippines late December planting, early dry season-dry: 15/15/40/15 and (85); vegetables 15/15/30 and (60); early–mid-June planting in Japan 20/25/75/30 and (150).
Spinach	Spring planting Mediterranean 20/20/15/5 and (60); September–October and late winter planting Mediterranean 20/20/25/15 and (70); warm winter desert climates 20/30/40/10 and (100).
Squash (winter) pumpkin	Late winter planting Mediterranean and warm winter desert climates 20/30/30/15 and (95); August planting California desert 20/35/30/25 and (110); early June planting maritime Europe 25/35/35/25 and (120).
Squash (zucchini) crookneck	Spring planting Mediterranean 25/35/25/15 and (100+); early summer Mediterranean and maritime Europe 20/30/25/15 and (90+); winter planting warm desert 25/35/25/15 and (100).
Sugar beet	Coastal Lebanon, mid-November planting 45/75/80/30 and (230); early summer planting 25/35/50/50 and (160); early spring planting Uruguay 30/45/60/45 and (180); late winter planting warm winter desert 35/60/70/40 and (205).
Sunflower	Spring planting Mediterranean 25/35/45/25 and (130); early summer planting California desert 20/35/45/25 and (125).
Tomato	Warm winter desert climates 30/40/40/25 and (135); and late autumn 35/45/70/30 and (180); spring planting Mediterranean climates 30/40/45/30 and (145).
Wheat	See Barley.

[a] 40/40/250/30 and (360) denote, respectively, initial crop development, mid-season and late season crop development stages in days, and (360) for total growing period from planting to harvest in days.

After Doorenbos and Pruitt, 1977.

Table 15.2 *Values of parameters a and b for computing K_c for growth stage 1 for annual crops given in Table 15.1*

	ET_r (mm/day)		ET_r (in./day)	
Average interval of irrigation or rainfall (days)	*a*	*b*	*a*	*b*
2	1.049	–0.119	0.714	–0.119
4	0.904	–0.216	0.450	–0.216
7	0.743	–0.319	0.264	–0.319
10	0.580	–0.408	0.155	–0.408
20	0.438	–0.455	0.101	–0.455

After Ryan and Cuenca, 1984.

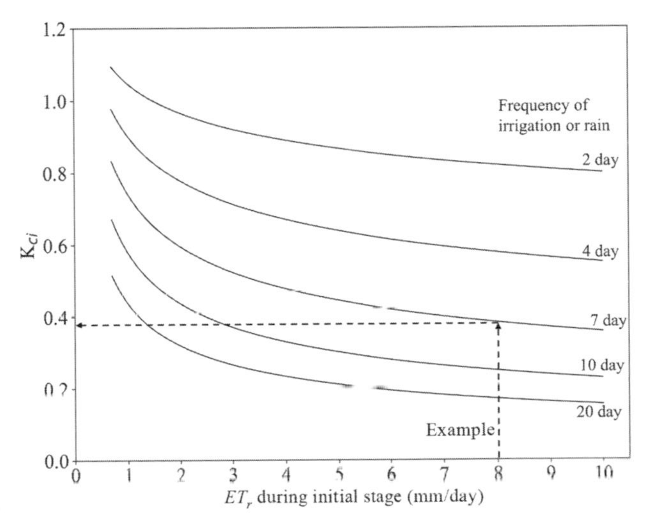

Figure 15.2 Average K_c for the initial stage as a function of average ET_r during initial stage and frequency of irrigation or significant rainfall. Adapted from Doorenbos and Pruitt, 1977.

which is a function of irrigation or rainfall frequency, as shown in Figure 15.2. As shown in the example in Figure 15.2, for ET_r of 8 mm/day and irrigation frequency of 7 days, the K_c for the initial period can be obtained as 0.39. Ryan and Cuenca (1984) developed regression equations for Figure 15.2. For a normal springtime interval between irrigation or significant rainfall, I_f, of less than four days,

$$K_c = (1.286 - 0.27 \ln I_f) \exp[(-0.01 - 0.042 \ln I_f) ET_r]. \tag{15.2}$$

For I_f greater than or equal to four days,

$$K_c = 2(I_f)^{-0.49} \exp[(-0.02 - 0.04 \ln I_f) ET_r]. \tag{15.3}$$

In eqs (15.2) and (15.3), ET_r is the average initial period reference *ET* in mm/day, and I_f is in days.

For growth stage 1 or initial period, Doorenbos and Pruitt (1977) expressed K_c as a function of average daily reference crop ET_r as:

$$K_c = aET_r^b, \tag{15.4}$$

where *a* is a constant and *b* is an exponent given in Table 15.2 with reference to a grass reference crop, which is calculated based on eqs (15.2) and (15.3) (Ryan and Cuenca, 1984).

The crop development period is defined as from the end of the initial period to the beginning of effective full cover, which is characterized by 70–80% ground cover. The mid-season period is defined from the end of the crop development period to the beginning of plant maturity, which may be indicated by change of leaf colors or falling of leaves. The late season period is from the end of the mid-season period to the time of full maturity or harvest. The crop coefficients for the mid-season and the late season have been developed as a function of wind speed and relative humidity for a wide range of crops, as shown in Table 15.3. In this table, the daytime wind speed and minimum relative humidity values are the average values over the specified growth period.

The crop coefficient can be computed for any growth period. It may be noted that the values of FAO K_c stem from coefficients that have average values over different growth periods. Therefore, computation of daily or shorter time *ET* for a crop under consideration may entail a little bit of error. The crop coefficient curve must be used for the computed values of reference *ET*. The weighted values of K_c must be obtained for the time periods for which reference *ET* has been computed. Table 15.3 gives the K_c values for a number of crops based on grass reference *ET* and Table 15.4 gives K_c values for alfalfa, clover, grass-legumes, and pasture (Doorenbos and Pruitt, 1977).

Table 15.3 *FAO crop coefficient (K_c) for field and vegetable crops for different stages of crop growth and climatic conditions*

Crop	Crop stage: initial 1, crop dev. 2, mid-season 3, at harvest or maturity 4	Humidity (RH) RH_{min} >70% Wind speed (m/s) 0–5	RH_{min} >70% 5–8	RH_{min} <20% 0–5	RH_{min} <20% 5–8
Artichokes (perennial-clean cultivated)	3	0.95	0.95	1.0	1.05
	4	0.9	0.9	0.95	1.0
Barley	3	1.05	1.1	1.15	1.2
	4	0.25	0.25	0.2	0.2
Beans (green)	3	0.95	0.95	1.0	1.05
	4	0.85	0.85	0.9	0.9
Beans (dry), pulses	3	1.05	1.1	1.15	1.2
	4	0.3	0.3	0.25	0.25
Beets (table)	3	1.0	1.0	1.05	1.1
	4	0.9	0.9	0.95	1.0
Carrots	3	1.0	1.05	1.1	1.15
	4	0.7	0.75	0.8	0.85
Castor beans	3	1.05	1.1	1.15	1.2
	4	0.5	0.5	0.5	0.5
Celery	3	1.0	1.05	1.1	1.15
	4	0.9	0.95	1.0	1.05
Corn (sweet) (maize)	3	1.05	1.1	1.15	1.2
	4	0.95	1.0	1.05	1.1
Corn (grain) (maize)	3	1.05	1.1	1.15	1.2
	4	0.55	0.55	0.6	0.6
Cotton	3	1.05	1.15	1.2	1.25
	4	0.65	0.65	0.65	0.7
Crucifers (cabbage, cauliflower, broccoli, Brussels sprouts)	3	0.95	1.0	1.05	1.1
	4	0.8	0.85	0.9	0.95
Cucumber	3	0.9	0.9	0.95	1.0
Fresh market	4	0.7	0.7	0.75	0.8
Machine harvest	4	0.85	0.85	0.95	1.0
Eggplant (aubergine)	3	0.95	1.0	1.05	1.1
	4	0.8	0.85	0.85	0.9
Flax	3	1.0	1.05	1.1	1.15
	4	0.25	0.25	0.2	0.2
Grain	3	1.05	1.1	1.15	1.2
	4	0.3	0.3	0.25	0.25
Lentil	3	1.05	1.1	1.15	1.2
	4	0.3	0.3	0.25	0.25
Lettuce	3	0.95	0.95	1.0	1.05
	4	0.9	0.9	0.9	1.0
Melons	3	0.95	0.95	1.0	1.05
	4	0.65	0.65	0.75	0.75
Millet	3	1.0	1.05	1.15	1.15
	4	0.3	0.3	0.25	0.25
Oats	3	1.05	1.1	1.15	1.2
	4	0.25	0.25	0.2	0.2
Onion (dry)	3	0.95	0.95	1.05	1.1
	4	0.75	0.75	0.8	0.85
Onion (green)	3	0.95	0.95	1.0	1.05
	4	0.95	0.95	1.0	1.05
Peanuts (groundnuts)	3	0.95	1.0	1.05	1.1
	4	0.55	0.55	0.6	0.6

Table 15.3 (*cont.*)

Crop	Crop stage: initial 1, crop dev. 2, mid-season 3, at harvest or maturity 4	Humidity (RH)			
		RH_{min} >70%		RH_{min} <20%	
		Wind speed (m/s)			
		0–5	5–8	0–5	5–8
Peas	3	1.05	1.1	1.15	1.2
	4	0.95	1.0	1.05	1.1
Peppers (fresh)	3	0.95	1.0	1.05	1.1
	4	0.8	0.85	0.85	0.9
Potato	3	1.05	1.1	1.15	1.2
	4	0.7	0.7	0.75	0.75
Radishes	3	0.8	0.8	0.85	0.9
	4	0.75	0.75	0.8	0.85
Safflower	3	1.05	1.1	1.15	1.2
	4	0.25	0.25	0.2	0.2
Sorghum	3	1.0	1.05	1.1	1.15
	4	0.5	0.5	0.55	0.55
Soybeans	3	1.0	1.05	1.1	1.15
	4	0.45	0.45	0.45	0.45
Spinach	3	0.95	0.95	1.0	1.05
	4	0.9	0.9	0.95	1.0
Squash	3	0.9	0.9	0.95	1.0
	4	0.7	0.7	0.75	0.8
Sugar beet	3	1.05	1.1	1.15	1.2
	4	0.9	0.95	1.0	1.0
	4 (no irrigation, last month)	0.6	0.6	0.6	0.6
Sunflower	3	1.05	1.1	1.15	1.2
	4	0.4	0.4	0.35	0.35
Tomato	3	1.05	1.1	1.2	1.25
	4	0.6	0.6	0.65	0.65
Wheat	3	1.05	1.1	1.15	1.2
	4	0.25	0.25	0.2	0.2

Many cool season crops cannot grow in dry, hot climates. The values of K_c are given for dry, hot conditions, since they may occur occasionally, and result in the need for higher K_c values, especially for tall rough crops.
After Doorenbos and Pruitt, 1977.

Table 15.4 *K_c values for alfalfa, clover, grass-legumes, and pasture*

Climatic Conditions	Statistic	Alfalfa	Grass for hay	Clover, grass-legumes	Pasture
Humid	Mean	0.85	0.8	1.0	0.95
Light to	Peak	1.05	1.05	1.05	1.05
moderate wind	Low	0.5	0.6	0.55	0.55
Dry	Mean	0.95	0.9	1.05	1.0
Light to	Peak	1.15	1.1	1.15	1.1
moderate wind	Low	0.4	0.55	0.55	0.5
Strong wind	Mean	1.05	1.0	1.1	1.05
	Peak	1.25	1.15	1.2	1.15
	Low	0.3	0.5	0.55	0.5

Light, <2 m/s, moderate, 2–5 m/s; strong wind, 5–8 m/s.
After Doorenbos and Pruitt, 1977.

Example 15.1: Determine the values of K_c for beans (dry) (continental climate, late spring planting) and potato (Irish) (spring–early summer planting, continental climate) if the following is known: average daily level of ET_r during the initial period (stage 1) is 4 mm/day, the interval between irrigations is four days, and the relative humidity is greater than 70% when the wind speed is 5 m/s.

Solution: For beans (sub-humid climate, late spring planting): season of 110 days.

Using eq. (15.4) and parameters from Table 15.2, given $ET_r = 4$ mm/day and the interval between irrigation is four days, the initial stage:

K_c for stage 1: $0.904 \times \left(4\,\frac{\text{mm}}{\text{day}}\right)^{-0.216} = 0.67.$

From Table 15.3, given $RH_{min} > 70\%$ and wind speed is 5 m/s, K_c for stage 3 is 1.05.

K_c for stage 2 varies from 0.67 to 1.05 (K_c between stage 1 and 3).

K_c for stage 4 varies from 1.05 to 0.30.

For potato (Irish) (spring–early summer planting, continental climate): season of 130 days

K_c for stage 1 is the same as for beans above:

$0.904 \times \left(4\,\frac{\text{mm}}{\text{day}}\right)^{-0.216} = 0.67$

K_c for stage 2 varies from 0.67 to 1.05.

K_c for stage 3 is 1.05.

K_c for stage 4 varies from 1.05 to 0.70.

Example 15.2: Compute the crop coefficient K_c for corn in the Brenham (Texas) area and construct the FAO crop coefficient curve. The planting date is May 10 and the lengths of growth periods, based on local information, are as follows:

Initial period: 15 days; crop development period: 30 days; mid-season period: 35 days; and late season period: 25 days.

Use the initial period reference ET as 5.3 mm/day. Take the frequency of irrigation or rainfall as 7 days. The values of relative humidity and wind velocity for the Brenham area from the weather data are as follows:

Average minimum relative humidity (%) is 48.07 for January, 50.82 for February, 49.83 for March, 50.81 for April, 51.33 for May, 52.44 for June, 45.27 for July, 38.86 for August, 42.10 for September, 48.20 for October, 47.75 for November, and 48.49 for December. The average humidity value is 47.83.

Average maximum relative humidity (%) is 69.30 for January, 71.84 for February, 70.48 for March, 72.14 for April, 72.48 for May, 73.83 for June, 69.99 for July, 66.49 for August, 66.47 for September, 71.20 for October, 70.54 for November, and 70.54 for December. The average maximum humidity is 70.4.

Over the course of the year, typical wind speeds vary from 0 mph to 15 mph (calm to moderate breeze), rarely exceeding 21 mph (fresh breeze). The highest average wind speed of 8 mph (gentle breeze) occurs around May 3, at which time the average daily maximum wind speed is 15 mph (moderate breeze). The lowest average wind speed of 4 mph (light breeze) occurs around September 24, at which time the average daily maximum wind speed is 10 mph (gentle breeze).

Solution: The procedure to construct the crop coefficient curve consists of the following:

1. Identifying the four crop growth stages, determining the length of each stage, and selecting the corresponding crop coefficients. Three values for crop coefficients are required to construct the crop coefficient curve: the crop coefficient during the initial period (K_{ci}), the mid-season period ($K_{c\ mid}$), and the late season period ($K_{c\ late}$).
2. Adjusting the K_{ci} for initial period for different frequency of irrigation or significant rainfall and initial period reference evapotranspiration using Figure 15.2.
3. Adjusting the $K_{c\ mid}$ and $K_{c\ late}$ for mid-season and late season periods for different climatic conditions using Table 15.3.
4. Constructing the crop coefficient curve. From Figure 15.2 for $ET_r = 5.3$ mm/day and frequency of irrigation is 7 days, $K_{ci} = 0.45$. Now, we determine $K_{c\ mid}$ and $K_{c\ late}$.

Mid-season starts on June 24 and ends on July 28. The average RH_{min} for July is 45.27%. The average wind speed for July ranges from 4 mph (1.8 m/s) to 8 mph (3.6 m/s). From Table 15.3, for $RH_{min} > 70\%$ and wind speed of 1.8–3.6 m/s, for corn (maize) grain, $K_{c\ mid} = 1.05$. Similarly, for $RH_{min} < 20\%$ and wind speed 1.8–3.6 m/s, $K_{c\ mid} = 1.15$. Interpolating the values for $RH_{min} = 45.27\%$, $K_{c\ mid} = 1.10$.

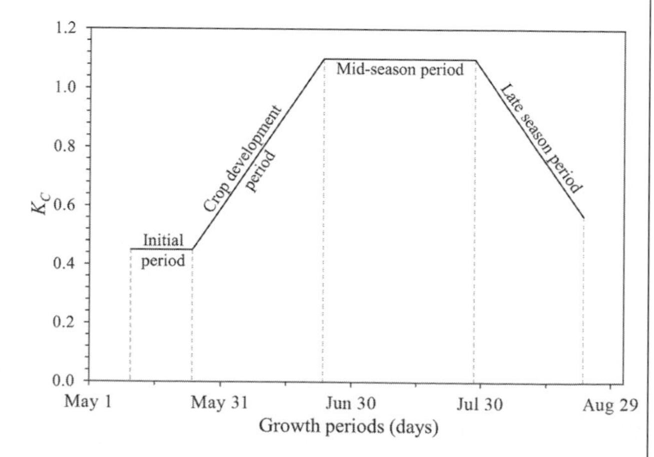

Figure 15.3 Crop coefficient curve for Example 15.2.

5. Late season starts on July 29 and ends on August 22. The average RH_{min} for August is 38.86%. Average wind speed during this period is 1.8–3.6 m/s. From Table 15.3, for $RH_{min} > 70\%$ and wind speed of 1.8–3.6 m/s, for corn (maize) grain, $K_{c\ late} = 0.55$. Similarly, for $RH_{min} < 20\%$ and wind speed of 1.8–3.6 m/s, $K_{c\ late} = 0.6$. Interpolating the values for $RH_{min} = 38.86\%$, $K_{c\ late} = 0.58$. The crop coefficient curve is shown in Figure 15.3.

Example 15.3: Use the K_c values obtained in Example 15.2 with mean monthly ET_r values and growth period data given below:

Location: College Station, Texas
Crop: Corn

For College Station, the average monthly ET_r in in./month is 2.2 for January, 2.71 for February, 4.22 for March, 5.2 for April, 6.25 for May, 6.89 for June, 7.1 for July, 6.85 for August, 5.6 for September, 4.3 for October, 2.8 for November, and 2.2 for December. The total ET_r for all months is 56.32 inches.

Growth period	Number of days	Dates
Initial	15	May 10–May 24
Crop development	30	May 25–June 23
Mid-season	35	June 24–July 28
Late season	25	July 29–August 22

Compute the mean monthly K_c values. Then, determine the monthly and seasonal ET_c.

Solution: We use the K_c values obtained from Example 15.2 with mean monthly ET_r values and growth period data. The total initial period is 15 days (May 10–May 24) and the crop development period is 30 days (May 25–June 23).

For May (22 days): 15-day initial period and 7 days of crop development.

K_c at start of the crop development period is 0.45 and at the end of the crop development period is 1.10.

K_c on May 31: $K_c = 0.45 + \frac{7\times(1.10-0.45)}{30} = 0.602.$

Weighted average K_c(May)

$$= \frac{14 \times 0.45}{22} + \frac{8 \times (0.602 + 0.45)}{22 \times 2}$$
$$= 0.478.$$

For June (30 days): 23 days in crop development and 7 days in mid-season:

K_c on June 1: $K_c = 0.45 + \frac{8\times(1.10-0.45)}{30} = 0.623.$

Weighted average K_c(June)

$$= \frac{23 \times (0.623 + 1.10)}{30 \times 2} + \frac{7 \times 1.10}{30}$$
$$= 0.917.$$

For July (31 days): 28 days in mid-season and 3 days in late season:

K_c at the start of the late season is 1.10 and at the end of the late season is 0.58.

K_c at July 31: $K_c = 1.10 - \frac{3\times(1.10-0.58)}{31} = 1.038.$

Weighted average K_c(July)

$$= \frac{27 \times 1.10}{31} + \frac{4 \times (1.10 + 1.038)}{31 \times 2}$$
$$= 1.096.$$

For August (22 days):

Weighted average K_c(August)

$$= \frac{\left[1.10 - \frac{4 \times (1.10 - 0.58)}{25}\right] + 0.58}{2} = 0.798.$$

Month	ET_r (in./month)	Mean monthly K_c	ET_c (in./month)	ET_c (mm/day)	No. days	Monthly ET_c (mm)
May	6.25	0.478	2.99	2.45	22	53.85
June	6.89	0.917	6.32	5.35	30	160.48
July	7.10	1.096	7.78	6.38	31	197.65
August	6.85	0.798	5.47	4.48	22	98.53
Seasonal total ET_c (mm)						510.52

15.4 WRIGHT COEFFICIENT METHOD

Wright (1982) developed a method that involves alfalfa as a reference crop, and hence the Penman method-based *ET* values are used for this crop. Therefore, the crop coefficients relate to alfalfa, in contrast to the FAO coefficients that are for grass. However, the FAO coefficients can be converted to alfalfa using the alfalfa coefficients given in Table 15.4, or vice versa as follows:

$$K_{c-alfalfa} = \frac{K_{c-FAO}}{K_{c-alfalfa/grass}}, \tag{15.5}$$

in which $K_{c-alfalfa}$ is the crop coefficient based on the alfalfa reference crop, K_{c-FAO} is the crop coefficient based the grass

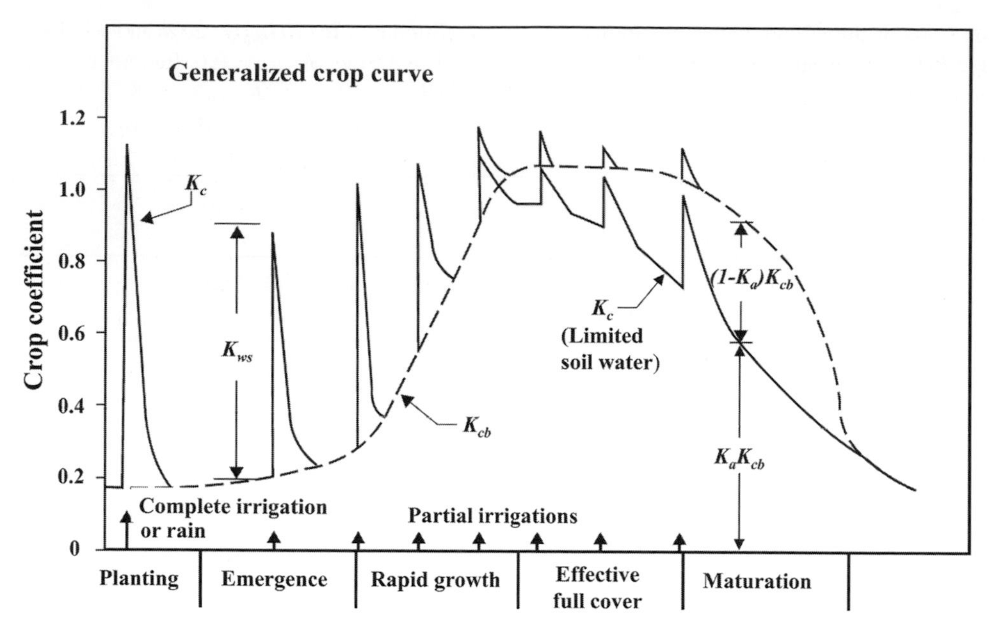

Figure 15.4 Generalized crop coefficient curves showing the effects of growth stage, wet surface soil due to rain or irrigation, and limited available soil water; K_{cb} represents the basal curve; K_c is the adjusted curve based on the surface wetness coefficient, K_{ws}, and available water coefficient, K_a. (Adapted from Wright, 1982. Copyright © 1982 by American Society of Civil Engineers, 108 (1), pp. 57–74. Reprint with permission from ASCE.)

reference crop, and $K_{c-alfaalfa/grass}$ is the FAO coefficient for alfalfa given in Table 15.4 for peak conditions. Equation (15.5) should be used with reference *ET* for alfalfa computed by the Wright–Penman method.

Inversely, the reference crop *ET* by the Wright–Penman method can be transformed to the grass *ET* with the use of the same FAO crop coefficient as follows:

$$ET_{r-grass} = \frac{ET_{r-alfalfa}}{K_{c-alfalfa/grass}}, \tag{15.6}$$

where $ET_{r-grass}$ is the reference *ET* for grass, and $ET_{r-alfalfa}$ is the reference *ET* for alfalfa. Equation (15.6) should be used with the FAO crop coefficients.

Wright (1982) developed crop coefficients that are for reference *ET* for alfalfa computed by the Wright–Penman method and can be used for any alfalfa-based reference *ET*. His method for developing these coefficients is different from that for the FAO coefficients, but his coefficient values have not been developed for a wide range of crops. The Wright method divides crop coefficients into three parts. The first part is the basal coefficient that relates to CU measured when the soil surface is dry. The second part accounts for the soil moisture in the root zone. The third part accounts for the increase in evaporation due to wet soil surface conditions. Wright (1982) derived the coefficient values using lysimeter measurements at Kimberly, Idaho. The crop coefficient K_c can now be expressed as

$$K_c = K_{cb}(K_a) + K_{ws}, \tag{15.7}$$

where K_{cb} is the basal crop coefficient related to a dry soil surface, K_a is the coefficient related to the soil moisture available in the root zone, and K_{ws} is the coefficient related to evaporation from wet soil surface conditions. Wright (1982) graphically presented crop coefficient curves, as shown in Figure 15.4. This formulation of crop coefficient, given by eq. (15.7), permits separating the effects of surface wetness and limited soil moisture in the root zone from the basal crop coefficient. For the Wright coefficients, the growing season is divided into three periods identified by dates for planting, effective full cover, and harvest. Table 15.5 provides the dates and the number of days for these periods at Kimberly, Idaho.

Wright (1982) specified the basal crop coefficient as a function of the percentage of time in 10% increments from planting to effective full cover and then the number of days after full cover, as given in Table 15.6. However, for alfalfa the basal crop coefficient is specified from the time of new growth until harvest as a percentage. In this way, it accounts for reduced water requirements after cutting.

The K_a coefficient in eq. (15.7) is amended for available soil moisture as

$$K_a = \frac{\ln\left[\frac{AVM}{TAM} \times 100 + 1\right]}{\ln(101.0)}, \tag{15.8}$$

in which AVM is the available soil moisture in mm at the time K_c is evaluated:

$$AVM = SMC - CEW, \tag{15.9}$$

where *SMC* is the soil moisture content in mm in the root zone at the time K_c is evaluated, *CEW* is the limit of extractable water, expressed as depth in mm in the root zone, and *TAM* is the total available soil moisture as depth in mm, expressed as

Table 15.5 *Date of various crop growth stages identifiable for crops studied at Kimberly, Idaho, 1968–1979*

	Date of occurrence (month/day)							Days	
Crop	Planting	Emergence	Rapid growth	Full cover	Heading or bloom	Ripening	Harvest	Planting to full cover	Full cover to harvest
Spring grain[a]	4/01	4/15	5/10	6/10	6/10	7/20	8/10	70	61
Peas	4/05	4/25	5/10	6/05	6/15	7/05	7/25	60	50
Sugar beets	4/15	5/10	6/01	7/10	–	–	10/15	85	95
Potatoes	4/25	5/25	6/10	7/10	7/01	9/20	10/10	75	90
Field corn	5/05	5/25	6/10	7/15	7/30	9/10	9/20	72	67
Sweet corn	5/05	5/25	6/10	7/15	7/20	–	8/15	72	30
Beans	5/22	6/05	6/15	7/15	7/05	8/15	8/30	55	45
Winter wheat	(2/15)	(3/01)	3/20	6/05	6/05	7/15	8/10	(110)	60
Alfalfa[b] (first)	4/01		4/20				6/15		76
(second)	6/15		6/25				7/31		46
(third)	7/31		8/10				9/15		46
(fourth)	9/15		10/01				10/30		46

[a] Spring grain includes wheat and barley.
[b] First denotes first harvest; intermediate harvests (second and third) may be one or more depending on length of season; last harvest (fourth) is when crop becomes dormant in cool weather.
Adapted from Wright, 1982. Copyright © 1982 by American Society of Civil Engineers, 108 (1), pp. 57–74. Reprint with permission from ASCE.

$$TAM = FC - CEW, \tag{15.10}$$

where FC is the soil moisture content in mm in the root zone at field capacity. The ratio AVM/TAM varies from 1 when the soil moisture is at field capacity to zero when the soil moisture is non-extractable.

The K_{ws} coefficient in eq. (15.7) is amended for surface evaporation as

$$K_{ws} = (K_t - K_{cb}) \exp(-bt), \tag{15.11}$$

where K_t is the threshold coefficient corresponding to the level of crop-canopy development beyond which soil evaporation is negligible, which can be considered the maximum value of K_{cb} following irrigation or rain, b is the empirical coefficient accounting for the effect of soil moisture and climatic conditions, and t is the time in days after wetting of the soil surface. Burman et al. (1983) and Jensen (1980) reported the values of K_t and b for Kimberly, Idaho. The term $\exp(-bt)$ is a dimensionless evaporation reduction coefficient and Wright (1982) indicated that it is equal to 0.8 for $t = 1$ day; 0.5 for $t = 2$ days; 0.3 for $t = 3$ days; 0 for $t > 3$ days.

15.5 DIVISION OF CROP COEFFICIENT

Two approaches to calculate crop coefficient have been discussed. In Section 15.2, the FAO crop coefficient method uses a "single" crop coefficient with consideration of the average effects of crop transpiration and soil evaporation. The Wright coefficient method in Section 15.3 uses a dual crop coefficient approach. The effects of crop transpiration and soil evaporation are considered separately. The Wright coefficient method uses alfalfa as a reference crop, and the FAO dual crop coefficient method is developed based on the reference grass. The crop coefficient can be further expressed as

$$K_{co} = K_s K_{cb} + K_w, \tag{15.12}$$

where K_{co} is the FAO dual crop coefficient that considers the effect of soil water stress, and K_{cb} is the basal crop coefficient for unstressed crops with a dry soil surface, the value of K_{cb} in Table 15.7 represents K_{cb} for a sub-humid climate and under moderate wind speed. For the consideration of the effects of different humidity and wind speed, K_{cb} can be corrected using Table 15.8 and eq. (15.15). K_s is the stress factor to account for soil water stress on ET, and K_w is the factor to account for increased evaporation from wet soils. The crop coefficients for unstressed, well-managed crops under sub-humid climate ($RH_{min} \approx 45\%$ and wind speed at 2 m height ≈ 2 m/s) have been tabulated as shown in Table 15.7 (Allen et al. 1998), which is used with the FAO Penman–Monteith ET_0 for the calculation of CU ET_c.

Doorenbos and Pruitt (1977) stress that "crop coefficient values relate to evapotranspiration of a disease-free crop grown in large fields under optimum soil water and fertility conditions and achieving full production under the given environment." Crops not meeting these requirements generally use less water unless raised in small fields where the effects of field boundaries can cause evapotranspiration to be significantly higher.

Table 15.6 *Basal crop coefficients, (K_{cb}), for dry surface soil for use with alfalfa reference crop for irrigated crops grown in an arid region with a temperate intermountain climate*

	Basal ET crop coefficients, K_{cb}										
	Percent time from planting to effective full cover (%)										
Crop	0	10	20	30	40	50	60	70	80	90	100
Spring grain[a]	0.15	0.15	0.16	0.20	0.25	0.40	0.52	0.65	0.81	0.96	1.00
Peas	0.15	0.15	0.16	0.18	0.20	0.29	0.38	0.47	0.65	0.80	0.90
Sugar beets	0.15	0.15	0.15	0.15	0.15	0.17	0.20	0.27	0.40	0.70	1.00
Potatoes	0.15	0.15	0.15	0.15	0.15	0.20	0.32	0.47	0.62	0.70	0.75
Corn	0.15	0.15	0.15	0.16	0.17	0.18	0.25	0.38	0.55	0.74	0.93
Beans	0.15	0.15	0.16	0.18	0.22	0.34	0.45	0.60	0.75	0.88	0.92
Winter wheat	0.15	0.15	0.15	0.30	0.55	0.80	0.95	1.00	1.00	1.00	1.00
DT, days after effective cover											
Crop	0	10	20	30	40	50	60	70	80	90	100
Spring grain[a]	1.00	1.00	1.00	1.00	0.90	0.40	0.15	0.07	0.05	–	–
Peas	0.90	0.90	0.72	0.50	0.32	0.15	0.07	0.05	–	–	–
Sugar beets	1.00	1.00	1.00	1.00	0.98	0.91	0.85	0.80	0.75	0.70	0.65
Potatoes	0.75	0.75	0.73	0.70	0.66	0.63	0.59	0.52	0.20	0.10	0.10
Field corn	0.93	0.93	0.93	0.90	0.87	0.83	0.77	0.70	0.30	0.20	0.15
Sweet corn	0.93	0.91	0.90	0.88	0.80	0.70	0.50	0.25	0.15	–	–
Beans	0.92	0.92	0.86	0.65	0.30	0.10	0.05	–	–	–	–
Winter wheat	1.00	1.00	1.00	1.00	0.95	0.50	0.20	0.10	0.05	–	–
Time from new growth or harvest to harvest (%)											
Crop	0	10	20	30	40	50	60	70	80	90	100
Alfalfa[b] (first)	0.40	0.50	0.62	0.80	0.90	0.95	1.00	1.00	0.98	0.96	0.94
(second and third)	0.25	0.30	0.40	0.70	0.90	0.95	1.00	1.00	0.98	0.96	0.94
(fourth)	0.25	0.30	0.40	0.50	0.55	0.50	0.40	0.35	0.30	0.21	0.25

[a] Spring grain includes wheat and barley.
[b] First denotes first harvest; intermediate harvests (second and third) may be one or more depending on length of season; last harvest (fourth) is when the crop becomes dormant in cool weather; see text for further discussion. Cultivar used was ranger.
Adapted from Wright, 1982. Copyright © 1982 by American Society of Civil Engineers, 108 (1), pp. 57–74. Reprint with permission from ASCE.

Table 15.7 *Information on basal crop coefficient for selected crops*

Crop	K_{ci}	K'_{cp}	K'_{cm}	f_{s1}	f_{s2}	f_{s3}	Soil water stress threshold f_c
Alfalfa, first cuttings	0.30	1.15	1.10	0.13	0.53	0.87	0.45
Alfalfa, later cuttings	0.30	1.15	1.10	0.11	0.56	0.78	0.45
Beans, dry	0.15	1.00	0.25	0.25	0.50	0.80	0.55
Beans, greens	0.15	1.00	0.80	0.22	0.56	0.89	0.55
Carrot	0.15	0.95	0.85	0.17	0.42	0.85	0.65
Corn, field	0.15	1.15	0.15*	0.18	0.41	0.71	0.45
Corn, sweet	0.15	1.10	1.00	0.30	0.60	0.90	0.50
Winter wheat	0.15	1.10	0.15	0.11	0.44	0.72	0.45

Note: K_{ci}, K'_{cp}, and K'_{cm} are the basal crop coefficients for initial, crop development, and mid-season period, respectively; f_{s1}, f_{s2}, and f_{s3} are the fraction of growing season for initial, crop development, and mid-season periods, respectively; * denotes that the corn grain is after complete field drying to about 18% moisture.
After Allen et al., 1998.

Table 15.8 *Crop adjustment factor (K_{cf}) for wind speed and relative humidity*

	Average minimum relative humidity (%)						
	20	30	40	50	60	70	80
Wind run (miles/day)	Crop height, 8 feet (2.33 m)						
50 (0.93 m/s)	0.05	0.01	–0.02	–0.06	–0.10	–0.14	–0.17
100 (1.86 m/s)	0.09	0.05	0.01	–0.03	–0.06	–0.10	–0.14
150 (2.79 m/s)	0.12	0.08	0.04	0.01	–0.03	–0.07	–0.11
200 (3.73 m/s)	0.15	0.12	0.08	0.04	0.00	–0.03	–0.07
250 (4.66 m/s)	0.19	0.15	0.11	0.07	0.04	0.00	–0.04
300 (5.59 m/s)	0.22	0.18	0.15	0.11	0.07	0.03	–0.01
350 (6.52 m/s)	0.25	0.22	0.18	0.14	0.10	0.07	0.03

After Allen et al., 1998.

The crop coefficient depends on the prevailing climatic conditions. The ET of tall crops is affected more by wind than that of short crops, such as grass. This effect is enhanced in arid climates. Therefore, Allen et al. (1998) recommended that the basal crop coefficient be adjusted based on wind speed and humidity. The basal crop coefficient is computed from the values in Table 15.7 plus the adjustment factor in Table 15.8. The initial value K_{ci} is not modified; however, the tabulated values of K_{cp} and K_{cm} are adjusted. It may be noted that the climatic data used to adjust the crop coefficient are average values for the time of the year for a specific region. Daily measured climatic conditions are not used to make the adjustment.

$$K_{cp} = K'_{cp} + K_{cf} \tag{15.13}$$

and

$$K_{cm} = K'_{cm} + K_{cf}, \tag{15.14}$$

where K'_{cp} and K'_{cm} are basal crop coefficients and are given in Table 15.7. The value of K_{cf} is given in Table 15.8, with adjustments for seasonal relative humidity and wind speed. Daily measured climatic data are not used to make these adjustments. For the values of K'_{cp} and K'_{cm} higher than 0.45, the adjustments for climates where RH_{min} and wind speed differs from 45% or 2 m/s can be considered using eq. (15.15):

$$K_{cf} = [0.04(u_2 - 2) - 0.004(RH_{min} - 45)]\left(\frac{h}{3}\right)^{0.3}, \tag{15.15}$$

where u_2 is the mean daily wind speed at 2 m height in m/s, RH_{min} is in the range of 20–80%, and h is the mean crop height during the mid-season and late season period in meters. Equation (15.15) can also be used for daily-measure climatic data adjustments.

Example 15.4: Consider field corn. The average maximum daily air temperature during mid-season and at harvest is 90 °F (32 °C) and 50 °F (10 °C), respectively. The average dew-point temperature during mid-season and at harvest is, respectively, 65 °F (18.3 °C) and 40 °F (4.4 °C). The average wind run at mid-season is 200 miles/day (3.73 m/s) and 150 miles/day (2.79 m/s) at harvest. Determine the peak harvest values of the basal crop coefficient for corn at this site.

Solution: The corn will be approximately 8 ft tall. From Table 15.7, we obtain $K'_{cp} = 1.15$ and $K'_{cm} = 0.15$. The values for saturation vapor pressure are determined from Table 2.2.

$$RH_{min} = 100 \times \left[\frac{\text{vapor pressure (average dew-point temperature)}}{\text{vapor pressure (average maximum daily air temperature)}}\right].$$

At mid-season,

$$RH_{min} = 100 \times \left[\frac{e(\text{at } 65\ ^\circ\text{F})}{e(\text{at } 90\ ^\circ\text{F})}\right] = 100 \times \left[\frac{21.1 \text{ mb}}{48.1 \text{ mb}}\right] \approx 44\%.$$

At harvest,

$$RH_{min} = 100 \times \left[\frac{e(\text{at } 40\ ^\circ\text{F})}{e(\text{at } 50\ ^\circ\text{F})}\right] = 100 \times \left[\frac{8.4 \text{ mb}}{12.3 \text{ mb}}\right] \approx 68\%.$$

From Table 15.8, the crop coefficient adjustment factors are:

At mid-season the average relative humidity is 44% and with wind run 200 miles/day. Thus, K_{cf} is approximately 0.06 (interpolating average minimum RH of 40% [0.08] and 50% [0.04]).

At harvest, the average relative humidity is 68% and with wind run 150 miles/day. Thus, K_{cf} is approximately –0.06.

The adjusted basal crop coefficients are

$$K_{cp} = K'_{cp} + K_{cf} = 1.15 + 0.06 = 1.21$$

$$K_{cm} = K'_{cm} + K_{cf} = 0.15 + (-0.06) = 0.09.$$

Example 15.5: Corn was grown using grain planted on May 1 and was harvested on September 30. Using the basal crop coefficients $K_{cp} = 1.15$ and $K_{cm} = 0.15$, determine the basal crop coefficients on May 15, June 15, July 15, August 15, and September 15. Use the elapsed time since planting to describe the canopy development.

Solution: The time from planting to harvest is determined first: 31 days in May, 30 days in June, 31 days in July, 31 days in August, and 30 days in September. Thus, the growing season is 153 days. The fraction of growing season for each date is now computed:

Date	Elapsed time since planting	Fraction of growing season
May 15	15	15/153 = 0.10
June 15	15 + 31 = 46	46/153 = 0.30
July 15	46 + 30 = 76	76/153 = 0.50
August 15	76 + 31 = 107	107/153 = 0.70
September 15	107 + 31 = 138	138/153 = 0.90

From Table 15.7, $f_{s1} = 0.18$, $f_{s2} = 0.41$, and $f_{s3} = 0.71$, as shown in Figure 15.5.

On May 15, $f_s = 0.1$, which is between 0 and f_{s1}, so $K_{cb} = K_{ci} = 0.15$.

On June 15, $f_s = 0.30$, which is between f_{s1} and f_{s2}, so

$$\begin{aligned} K_{cb} &= 0.15 + (K_{cp} - 0.15)(f_s - f_{s1})/(f_{s2} - f_{s1}) \\ &= 0.15 + (1.15 - 0.15)(0.30 - 0.18)/(0.41 - 0.18) \\ &= 0.67. \end{aligned}$$

On July 15, $f_s = 0.50$ which is between f_{s2} and f_{s3}, so $K_{cb} = K_{cp} = 1.15$.

On August 15, $f_s = 0.70$, which is between f_{s2} and f_{s3}, so $K_{cb} = K_{cp} = 1.15$.

On September 15, $f_s = 0.90$, which is between f_{s3} and 1.0, so

$$\begin{aligned} K_{cb} &= K_{cp} - (K_{cp} - K_{cm})(f_s - f_{s3})/(1.0 - f_{s3}) \\ &= 1.15 - (1.15 - 0.15)(0.90 - 0.71)/(1.0 - 0.71) \\ &= 0.49. \end{aligned}$$

The resulting K_{cb} values are:

Date	K_{cb}
May 15	0.15
June 15	0.67
July 15	1.15
August 15	1.15
September 15	0.49

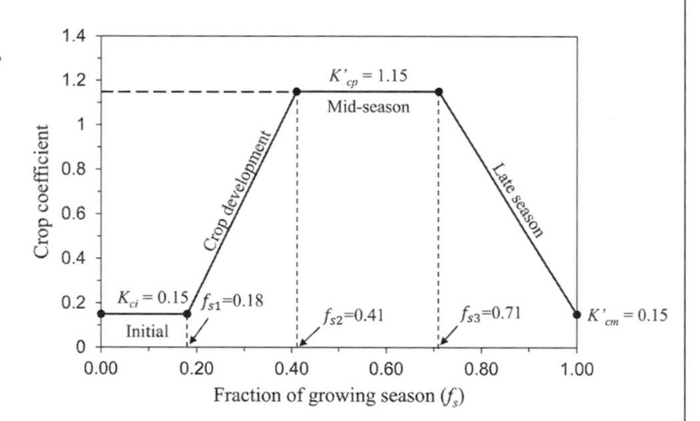

Figure 15.5 Crop coefficient during the growing season in Example 15.5.

15.6 WATER STRESS EFFECT (K_s)

For the purposes of irrigation management, the effect of water stress on ET can be described by using a stress factor K_s, which is based on soil water content and is described in Figure 15.6.

$$\begin{aligned} K_s &= \frac{f_r}{f_c}, \ f_r < f_c \\ &= 1.0, \ f_r \geq f_c, \end{aligned} \qquad (15.16)$$

where K_s is the stress factor, f_r is the fraction of soil water that remains, and f_c is the critical threshold of available water when stress begins (Table 15.7). Note that

$$f_r = \frac{(\theta_v - \theta_{wp})}{(\theta_{fc} - \theta_{wp})}, \qquad (15.17)$$

where θ_v is the water content of soil on a volume basis (fraction or percentage), θ_{wp} is the permanent wilting point of soil (in./in. or m/m), and θ_{fc} is the field capacity of soil (in./in. or m/m).

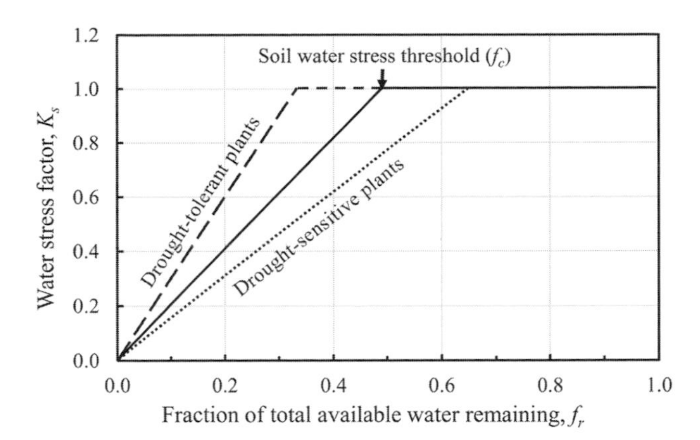

Figure 15.6 Stress factor and fraction of total available water remaining.

Example 15.6: Consider a sandy loam soil that has current volumetric water content as 0.15, volumetric water content at field capacity as 0.24, and volumetric water content at the wilting point as 0.12. Also given are: $ET_r = 0.30$ in./day, $K_{cb} = 1.20$, and critical water content (f_c) is 0.45. Determine the value of stress factor and the crop ET for the day.

Solution: $f_r = (0.15 - 0.12)/(0.24 - 0.12) = 0.25$
Since $f_r < f_c$, that is, $(0.25 < 0.45)$,

$$K_s = \frac{f_r}{f_c} = \frac{0.25}{0.45} = 0.56,$$

$$ET_c = K_s K_{cb} ET_r = (0.56)(1.20)(0.30) = 0.20 \text{ in./day}.$$

15.7 WET SOIL EVAPORATION FACTOR (K_w)

As shown in Figure 15.7, evaporation from bare soil can be assumed to take place in two phases: an energy limiting phase and a soil limiting phase. Immediately following rain or irrigation, K_w has the maximum value (K_{cmax}) and evaporation occurs at the maximum rate which is determined by available energy at the soil surface. When the soil surface dries out, less water is available for evaporation and the evaporation rate will reduce in proportion to the amount of water remaining in the surface soil layer. K_w is small or even zero, and there is no water near the surface for evaporation. As illustrated in eq. (15.18), the calculation of wet soil evaporation factor includes the determination of (1) the maximum crop coefficient for wet soil evaporation (K_{cmax}), which can be calculated with energy balance equations and will be discussed later; (2) the fraction of soil surface wetted (F_w), which is influenced by the types of irrigation systems and the amount of canopy development; and (3) the wet soil decay $f(t)$, which is determined by the hydraulic properties of the soil (Table 15.9).

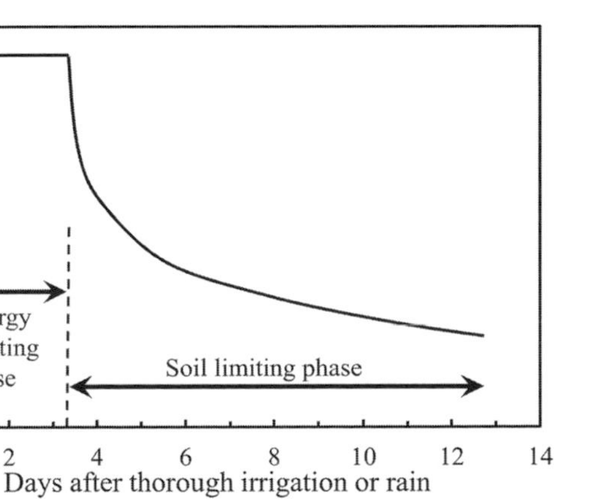

Figure 15.7 Example of evaporation from bare soil.

Table 15.9 *Duration of wet soil evaporation (t_d) for selected soil textures and values of the wet soil decay function $f(t)$ for time since wetting (t)*

Time since wetting (t) days	Clay	Clay loam	Silt loam	Sandy loam	Loamy sand	Sand
	Duration of wet soil evaporation (t_d) days					
	10	7	5	4	3	2
0	1.00	1.00	1.00	1.00	1.00	1.00
1	0.68	0.62	0.55	0.50	0.42	0.29
2	0.55	0.47	0.37	0.29	0.18	0.00
3	0.45	0.35	0.23	0.13	0.00	
4	0.37	0.24	0.11	0.00		
5	0.29	0.15	0.00			
6	0.23	0.07				
7	0.16	0.00				
8	0.11					
9	0.05					
10	0.00					

Adapted from Hill et al., 1983.

Table 15.10 *Fraction of the soil surface wetted for various types of irrigation systems*

Method	F_w
Rain	1.0
Above-canopy sprinklers	1.0
LEPA system (alternate furrows wetted)	0.5
Borders and basin irrigation	1.0
Furrow irrigation (large)	1.0
Furrow irrigation (small)	0.5
Furrow irrigation (alternate)	0.5
Trickle/drip irrigation	0.25

LEPA, low energy precise application.
After Allen et al., 1998.

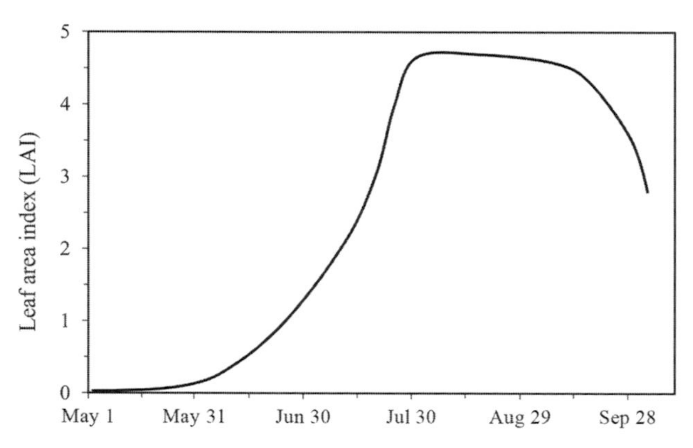

Figure 15.8 Example of *LAI* variations during crop growth.

$$K_w = F_w(K_{cmax} - K_{cb})f(t), \qquad (15.18)$$

$$f(t) = 1 - \sqrt{t/t_d}, \qquad (15.19)$$

where K_w is the wet soil evaporation factor, K_{cmax} is the maximum crop coefficient for wet soil evaporation, K_{cb} is the basal crop coefficient with the consideration of humidity and wind speed, F_w is the fraction of soil surface wetted, t is the time since last wetting in days, t_d is the duration of wet soil surface in days, and $f(t)$ is the wet decay function.

Table 15.11 *Values of vegetation coverage fraction (f_{vc}) and sunlight exposure fraction ($1 - f_{vc}$)*

Method	f_{vc}	$1 - f_{vc}$
Initial period	0.0–0.1	1.0–0.9
Crop development period	0.1–0.8	0.9–0.2
Mid-season period	0.8–1.0	0.2–0.0
Late season period	0.8–0.2	0.2–0.8

After Allen et al., 1998.

K_{cmax} represents an upper limit wet soil evaporation from any cropped surface and can be determined by:

$$K_{cmax} = \max\{1.2 + K_{cf}, k_{co} + 0.05\}, \qquad (15.20)$$

where K_{cf} is the crop adjustment factor for wind speed and RH, which can be calculated using eq. (15.15), or using the values in Table 15.8. The coefficient of 1.2 in eq. (15.20) reflects the increased aerodynamic roughness of the crop during crop development periods, the impact of reduced albedo due to wet soil, and the heat stored in dry soil before the wetting event. If irrigation or rain events are daily or every two days, then less heat is stored in the soil, and the coefficient of 1.2 can be reduced to 1.1. k_{cb} is the basal crop coefficient with the consideration of humidity and wind speed. Equation (15.20) ensures that the value of K_{cmax} is always larger than or equal to $k_{cb} + 0.05$.

As shown in Figure 15.8, the leaf area index (*LAI*) varies greatly during the crop growth stages, and *LAI* = 3 can be considered full coverage (Wright, 1982). In cropland with incomplete coverage by the crop leaf, evaporation from the soil is not the same over the entire surface. The evaporation is greater between crops with a lower *LAI*, where more vapor could be transferred from the soil surface to the canopy, especially when only part of the soil surface is wetted through irrigation. Table 15.10 lists typical values of F_w for various types of irrigation systems. To consider the impacts of exposure to sunlight, the values of vegetation coverage fraction (f_{vc}) during different crop growth periods could be used. If ($1 - f_{vc}$) is lower than F_w, then ($1 - f_{vc}$) should be used instead of F_w for various types of irrigation systems. Table 15.11 lists the values of vegetation coverage fraction ranges during different crop growth stages.

Example 15.7: A silt loam soil was irrigated two days earlier. A corn (field) crop is grown in a dry (30% minimum RH) and windy climate (wind speed at 2 m height is 300 miles/day, or 5.588 m/s). The crop is about 2 ft tall (0.61 m) and is expected to have a basal crop coefficient of 0.6 before adjustments are made for wind and humidity. The reference surface *ET* is 0.35 inches/day (8.9 mm/day). The field is irrigated with a LEPA system. Determine the water use on this day. Since the field is irrigated, soil water stress effects can be neglected. The given coefficient related to evaporation from wet soil surface condition (K_s) is 1.0, the time since last wetting (t) is 2 days, and the duration of wet soil surface (t_d) is 5 days.

Solution:

1. Determination of K_{cmax}:
   Using eq. (15.15), given $h = 2$ ft $= 0.61$ m,
   $u_2 = 300$ miles/day $= 5.588$ m/s, $RH_{min} = 30\%$:

$$K_{cf} = [0.04(U_2 - 2) - 0.004(RH_{min} - 45)]\left(\frac{h}{3}\right)^{0.3} = 0.126.$$

The basal crop coefficient with wind and humidity adjustment is

$$K_{cb} = 0.6 + K_{cf} = 0.6 + 0.126 = 0.726.$$

Given time since last wetting (t) is 2 days, the coefficient of 1.1 is used in eq. (15,20):

$$K_{cmax} = 1.1 + 0.126 = 1.226.$$

2. Determination of F_w:
   From Table 15.10, the fraction of the soil surface wetted for the LEPA system is $F_w = 0.5$.
   Given the crop is corn and the height is 2 ft and it is during the crop development period, then the sunlight exposure fraction ($1 - f_{vc}$) can be selected as 0.6 from Table 15.11. Since $1 - f_{vc} > F_w$, $F_w = 0.5$ is used.
3. Determination of $f(t)$:
   For silt loam soil, with the time since last wetting (t) as 2 days and the duration of wet soil surface (t_d) as 5 days, one gets $f(t) = 0.37$ from Table 15.9.
   Or, using eq. (15.19),

$$f(t) = 1 - \sqrt{t/t_d} = 1 - \sqrt{2/5} = 1 - 0.63 = 0.37.$$

4. Calculation of K_w and ET_c:

$$K_w = F_w(K_{cmax} - K_{cb}) \times f(t) = 0.5 \times (1.226 - 0.726) \times 0.37 = 0.09$$

$$ET_c = K_{co}ET_r = (K_s K_{cb} + K_w)ET_r = (1.0 \times 0.726 + 0.09) \times 0.35 = 0.29 \text{ in./day (7.4 mm/day)}.$$

Example 15.8: The volumetric water contents at field capacity and the permanent wilting point are 25% and 10%, respectively. Soil water was measured in two fields with the following results:

Field A: Available water in the root zone is 5 in. (127 mm).
Field B: Available water in the root zone is 2 in. (50.8 mm).

The crop root zone is 4 ft (1.22 m) deep in both fields. The reference crop ET rate is 0.30 in./day (7.6 mm/day), and the basal crop coefficient is 1.1 at this time of year. The crop grown at the sites has a critical soil water threshold value of $(f_c) = 0.35$. Compute ET for each field.

Solution: The total available water in the root zone is

48 inches × (25 – 10)/100 = 7.2 inches.

The fraction of soil water that remains in each field is

Field A: $f_r = (5/7.2) = 0.69$,
Field B: $f_r = (2/7.2) = 0.28$.

Crop evapotranspiration in Field A:

Since $f_r > 0.35, K_s > 1.0$, the ET rate is:
$ET_c = K_{cb}K_s ET_r = 1.1 \times 1.0 \times 0.3 = 0.33$ in./day (8.4 mm/day).

Crop evapotranspiration in Field B:

$f_r = 0.28$, which is less than 0.35, so

$K_s = f_r/f_c = 0.28/0.35 = 0.80$,

$$ET_c = K_{cb}K_s ET_r = 1.1 \times 0.80 \times 0.3 = 0.26 \text{ in./day (6.7 mm/day)}.$$

QUESTIONS

Q.15.1 Determine the values of K_c for beans (dry) (subhumid climate, late spring planting) and potato (Irish) (spring–early summer planting, continental climate) if the following are known: average daily level of ET_r during the initial period (stage 1) is 5 mm/day, the interval between irrigations is 5 days, and the relative humidity is greater than 60% when the wind speed is 4 m/s.

Q.15.2 Compute the crop coefficient K_c for corn in the Houston (Texas) area and construct the FAO crop coefficient curve. The planting date is April 20 and the lengths of growth periods, based on local information, are as follows:

Initial period: 15 days; crop development period: 30 days; mid-season period: 35 days; and late season period: 25 days.

Use the initial period reference ET as 4.5 mm/day. Take the frequency of irrigation or rainfall as 5 days. The values of RH and wind velocity for the Houston area from the weather data are: Average minimum RH (%) is 46.85 for January, 45.39 for February, 46.62 for March, 48.28 for April, 51.69 for May, 53.01 for June, 51.66 for July, 51.28 for August, 50.81 for September, 46.10 for October, 45.77 for November, and 45.74 for December. The average humidity value is 48.60.

Average maximum RH (%) is 61.80 for January, 66.00 for February, 72.70 for March, 79.10 for

April, 85.30 for May, 90.83 for June, 93.71 for July, 93.51 for August, 89.00 for September, 81.40 for October, 71.67 for November, and 64.89 for December. The average maximum humidity is 79.16.

Over the course of the year, typical wind speeds vary from 0 mph to 15 mph (calm to moderate breeze), rarely exceeding 21 mph (fresh breeze). The highest average wind speed of 9.4 mph (gentle breeze) occurs around April 3, at which time the average daily maximum wind speed is 14.5 mph (moderate breeze). The lowest average wind speed of 6.4 mph (light breeze) occurs around August 16, at which time the average daily maximum wind speed is 9.8 mph (gentle breeze).

Q.15.3 Use the K_c values obtained in Q.15.2 with mean monthly ET_r values and growth period data given below:

Location: Houston, Texas
Crop: Corn

For Houston, the average monthly ET_r (inches/month) is 2.36 for January, 2.83 for February, 4.32 for March, 5.01 for April, 6.11 for May, 6.57 for June, 6.52 for July, 6.08 for August, 5.57 for September, 4.28 for October, 2.90 for November, and 2.35 for December. The total ET for all months is 54.90 inches.

Growth period	Number of days	Dates
Initial	15	April 20–May 4
Crop development	30	May 5–June 3
Mid-season	35	June 4–July 8
Late season	25	July 9–August 2

Compute the mean monthly K_c values. Then, determine monthly and seasonal ET_c.

Q.15.4 Consider a wheat field. The average maximum daily air temperature during mid-season and at harvest is 85 °F and 55 °F, respectively. The average dew-point temperature during mid-season and at harvest is, respectively, 70 °F and 45 °F. The average wind run at mid-season is 150 miles/day and 100 miles/day at harvest. Determine the peak harvest values of the basal crop coefficient for wheat at this site.

Q.15.5 Wheat was grown for grain planted on March 1 and was harvested on July 31. Use the basal crop coefficients $K_{cp} = 1.10$ and $K_{cm} = 1.15$. Determine the basal crop coefficients on March 15, April 15, May 15, June 15, and July 15. Use the elapsed time since planting to describe the canopy development.

Q.15.6 Consider a sandy loam soil that has current volumetric water content as 0.20, volumetric water content at field capacity as 0.30, and volumetric water content at the wilting point as 0.10. Also given are: $ET_r = 0.40$ in./day (10.16 mm/day), the basal crop coefficient at this time of year is $K_{cb} = 1.15$, and critical water content $(f_c) = 0.50$. Determine the value of stress factor and the crop ET for the day.

Q.15.7 A sandy loam soil was irrigated two days earlier. A wheat (field) crop is grown in a dry (20% minimum RH) and windy (150 miles/day; 2.79 m/s) zone. The crop is about 0.75 ft tall and is expected to have a basal crop coefficient of 0.4 before adjustments are made for wind and humidity. The reference surface ET is 0.25 in./day (6.35 mm/day). The field is irrigated with a LEPA system. Determine the water use on this day. Since the field is irrigated, soil water stress effects can be neglected. The coefficient related to evaporation from the wet soil surface condition (K_s) is given as 0.75, the time since last wetting (t) is 4 days, and the duration of wet soil surface (t_d) is 6 days.

Q.15.8 The volumetric water contents at field capacity and the permanent wilting point are 40% and 15%, respectively. Soil water was measured in two fields with the following results:

Field A: Available water in the root zone is 3.5 in. (88.9 mm).
Field B: Available water in the root zone is 1.5 in. (38.1 mm).

The crop root zone is 1.5 ft (0.46 m) deep in both fields. The reference crop ET rate is 0.25 in./day (6.35 mm/day) and the basal crop coefficient is 0.8 at this time of year. The crop grown at the sites has a critical soil water threshold value $(f_c) = 0.45$. Compute the evapotranspiration for each field.

REFERENCES

Allen, R. G., Pereira, L. S., Raes, D., and Smith, M. (1998). Crop evapotranspiration (guidelines for computing crop water requirements). FAO irrigation and drainage paper 56.

Burman, R. D., Cuenca, R. H., and Weiss, A. (1983). Techniques for estimating irrigation water requirements. In *Advances in Irrigation*, Vol. 2, edited by D. Hillel. New York: Academic Press, pp. 336–394.

Doorenbos, J., and Pruitt, W. O. (1977). Crop water requirements. FAO irrigation and drainage paper 24, revised.

Hill, R. W., Johns, E. L., and Frevert. D. K. (1983). *Comparison of Equations Used for Estimating Agricultural Crop Evapotranspiration with Field Research*. Washington, DC: Bureau of Reclamation.

Jensen, M. E. (ed.) (1980). *Design and Operation of Farm Irrigation Systems*. St. Joseph, MI: American Society of Agricultural Engineers.

Ryan, P. K., and Cuenca, R. H. (1984). Feasibility and significance of revising consumptive use and net irrigation requirements for Oregon. Project completion report, Department of Agricultural Engineering, Oregon State University.

Wilson, C. L., and Loomis, W. E. (1962). *Botany*. New York: Holt, Rinehart and Wilson.

Wright, J. L. (1981). Crop coefficients for estimates of daily crop evapotranspiration. In *Proceedings of the American Society of Agricultural Engineers Irrigation Scheduling Conference*. St. Joseph, MI: American Society of Agricultural Engineers, pp. 18–26.

Wright, J. L. (1982). New evapotranspiration crop coefficients. *Journal of Irrigation and Drainage Division, ASCE*, 108(1R): 57–74.

16 Irrigation Efficiency

Notation

a	average depth of the 25% observations having the least water depth or y_a (cm or in.)
$AELQ$	application efficiency of the lower quarter (%)
C_u	uniformity coefficient (%) or Christiansen uniformity coefficient (%)
d	average depth of water stored (cm or in.)
$\overline{d}$	mean depth of water of the catch in all cans (cm or in.)
d_a	depth applied from the original source (cm or in.)
d_e	effective depth of irrigation (cm or in.)
d_{ev}	effective depth of evaporation and drift (cm or in.)
d_g	gross irrigation depth (cm or in.)
d_i	measured depth of water in equally spaced catch cans at the ith grid (cm or in.)
d_{LQ}	low quarter depth of irrigation (cm or in.)
d_m	mean depth infiltrated (or average depth over the entire field) (cm or in.)
d_r	effective depth of runoff (cm or in.)
DP	deep percolation (m^3)
DU	distribution uniformity (fraction or percentage)
e_a	water application efficiency (%)
e_c	water conveyance efficiency (%)
e_i	overall irrigation system performance (%)
e_r	reservoir storage efficiency (%)
e_s	soil water storage efficiency (%)
e_u	water use efficiency (%)
I	irrigation requirement (m^3)
L	leaching requirement (m^3)
N	number of catch cans
O	operational losses (m^3)
PE	Soil Conservation Service pattern efficiency (%)
P_r	deep percolation ratio (fraction)
RO	runoff (m^3)
S	standard deviation of the measured depths of water in equally spaced catch cans (cm or in.)
SWD	soil water depletion (cm or in.)
TW_r	tailwater ratio (fraction)
UCH	Hart uniformity coefficient (%)
$UCWS$	Wilcox–Swailes uniformity coefficient (%)
V_e	volume of water evaporation from the reservoir (m^3)
V_i	volume of inflow (L/min)
V_o	volume of outflow (L/min)
V_s	volume of seepage water (m^3)
W_{ac}	volume of water before irrigation or antecedent soil water (m^3)
W_c	volume of water in the root zone at field capacity minus the antecedent water or volume of water in the root zone before irrigation (m^3)
W_d	volume of water delivered to the area being irrigated (m^3)
W_{fc}	volume of water in the root zone at field capacity (m^3)
W_i	volume of water diverted from the source (m^3)
W_{ro}	volume of surface runoff or tailwater (m^3)
W_s	amount of water stored in the root zone (m^3)
W_{sw}	volume of water stored in the root zone due to irrigation (m^3)
W_u	amount of water beneficially used (m^3)
y	average of the absolute values of deviations of the depth of water stored from average depth of water stored (cm or in.)
y_a	average depth of infiltrated water over the one-quarter of the field with the least infiltration (cm or in.)
ΔS	change in the reservoir storage, which can be negative or positive (m^3)

16.1 INTRODUCTION

Only part of the water applied to an irrigation field is used by plants, called consumptive use (CU); the remainder is for non-CU. Likewise, part of the water applied is for beneficial use. The CU includes evapotranspiration (ET) by crops, evaporation for cooling, and evaporation for frost protection; whereas non-CU comprises water for leaching. Both of these uses are beneficial for crop production. On the other hand, the CU comprising ET by weeds and phreatophytes, evaporation from soil, evaporation from channels and reservoirs, and spray evaporation, is not beneficial. Similarly, non-CU consisting of deep percolation, surface runoff, and operational spill is not for any beneficial use. Other factors, such as sabotage, misallocation, water rights, irrigation at nighttime, social and cultural behavior, climate, and management, affect the use of water. This discussion suggests that irrigation efficiency is impacted by the amounts of water used for both CU and non-CU, and their beneficial and non-beneficial forms.

When an agricultural field is irrigated, the common thought is that it should be irrigated efficiently so that crop yield is maximum. Does this mean that irrigation efficiency should be maximum? Is maximum efficiency desirable under all circumstances? How does irrigation efficiency relate to the type and design of the irrigation system? Should irrigation efficiency be entirely based on the amount of water consumed by plants? Before addressing these and related questions, we must first address the question: What is meant by efficient irrigation or irrigation efficiency? This chapter discusses the concept of irrigation efficiency and related aspects.

16.2 FACTORS AFFECTING IRRIGATION EFFICIENCY

There is a multitude of factors affecting irrigation efficiency. These include irrigation system type and design, operation and management of the irrigation system, field size and topography, climate, soil, and stage of plant growth. Irrigation efficiency is often evaluated for an individual irrigation event, but it is not equal to the average irrigation efficiency for the whole growing season. Likewise, irrigation efficiency of a field is not equal to the irrigation efficiency of the entire farm or watershed. This suggests that irrigation efficiency varies with space and time scales.

16.3 IRRIGATION EFFICIENCIES

In many irrigation systems there is a source from which water is extracted, there is a conveyance system that carries this water, and there is the field that is to be irrigated with this water. The source of water can be a tubewell, a stream, a reservoir, or a lake. The conveyance system may comprise open channels – earthen or lined – or closed conduits such as pipes or tubes. The channels have transmission losses through seepage and evaporation and spills. Even pipes and tubes have leakages through joints and cracks. There is also some loss due to evaporation. This means that the water delivered to the field is less than the water received from the source. Thus, when measured in terms of water received and water delivered, the efficiency is usually less than 100%, as shown in Figure 16.1. Of course, one always wants to reduce these losses.

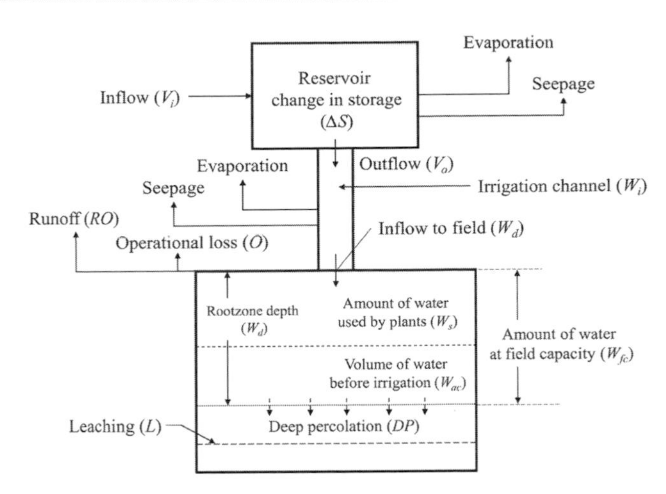

Figure 16.1 Partitioning of irrigation water supply.

Irrigation efficiency is employed for designing irrigation systems and evaluating their performance. Many definitions of irrigation efficiency have been reported in the literature. Each definition has a particular objective and should be used to meet that objective. More commonly used definitions are discussed here. For additional discussion, reference is made to Israelson (1950), Jensen et al. (1967), ASCE (1978), Painter and Carran (1978), Bos (1979), Hansen et al. (1980), Jensen et al. (1983), Walker and Skogerboe (1987), and Burt et al. (1997).

16.3.1 Reservoir Storage Efficiency

Reservoirs are used for storing water and then supplying it whenever needed. Part of the water in the reservoir is lost to evaporation and seepage. Thus, reservoir storage efficiency (e_r) can be defined as the ratio of outflow from the reservoir plus the change in storage due to evaporation and seepage to the inflow to the reservoir, expressed as

$$e_r = \frac{V_o + \Delta S}{V_i} \times 100 = \left(1 - \frac{V_s + V_e}{V_i}\right) \times 100, \qquad (16.1)$$

where V_o is the volume of outflow, V_i is the volume of inflow, V_s is the volume of seepage water, V_e is the volume of water evaporation from the reservoir, and ΔS is the change in the reservoir storage, which can be negative or positive.

Example 16.1: A reservoir receives an inflow of 5000 L/min for 24 hours and the change in its storage within 24 hours is 250 m^3, resulting from the removal of water used to restore the initial water level. The amount of water being withdrawn from the reservoir for irrigation is 3000 L/min. What is the reservoir efficiency for the 24-hour period?

Solution: The volume of inflow is

$$V_i = 5000\left(\frac{\text{L}}{\text{min}}\right) \times 60\frac{\text{min}}{\text{h}} \times 24\text{ h} \times \left(\frac{1\text{ m}^3}{1000\text{ L}}\right)$$
$$= 7200\text{ m}^3.$$

The volume of outflow is

$$V_o = 3000\left(\frac{\text{L}}{\text{min}}\right) \times 60\frac{\text{min}}{\text{h}} \times 24\text{ h} \times \left(\frac{1\text{ m}^3}{1000\text{ L}}\right)$$
$$= 4320\text{ m}^3.$$

Therefore,

$$e_r = \frac{V_o + \Delta S}{V_i} \times 100 = \frac{4320 + 250}{7200} \times 100 = 63.5\%.$$

16.3.2 Water Conveyance Efficiency

Because of the losses in the conveyance system, the volume of water delivered to the field is often less than the water received from the source. The water conveyance efficiency (e_c) can be defined as

$$e_c = \frac{W_d}{W_i} \times 100, \tag{16.2}$$

where e_c is the water conveyance efficiency (%), W_d is the volume of water delivered to the area being irrigated (inflow), and W_i is the volume of water diverted from the source (outflow). The difference between W_d and W_i is the volume of water lost during conveyance. Increasing W_d or decreasing W_i will increase e_c. However, the increase will come at a cost and may not always be economically justified. Some losses, such as spills, may be reduced without much cost. Proper maintenance always pays off in the long run.

Example 16.2: An agricultural field has 100 border strips for irrigation. The channel supplying water to the field supplies water at the rate of 20 L/min to 20 border strips and 30 L/min to 80 border strips. The channel is receiving water from the source at a rate of 3200 liters per minute. Compute the conveyance efficiency of the irrigation system.

Solution:

$$e_c = \frac{W_d}{W_i} \times 100$$
$$= \frac{20\text{ L/min} \times 20 + 30\text{ L/min} \times 80}{3200\text{ L/min}} \times 100 = 87.5\%.$$

16.3.3 Water Application Efficiency

Not all of the water that is applied to the field is used by plants. Some of the water is lost to deep percolation. In the case of sprinkler irrigation, some of the water drifts away from the field, some water is evaporated before reaching the ground, and some may run off the field. However, the spray water may reduce the potential *ET* and may not reduce the application efficiency. But the water intercepted by the vegetative canopy is evaporated at a higher rate than the rate of transpiration by well-watered plants, and hence it may reduce application efficiency. Thus, water application efficiency (e_a) can be defined as

$$e_a = \frac{W_s}{W_d} \times 100, \tag{16.3}$$

where e_a is the water application efficiency (%) and W_s is the amount of water stored in the root zone. W_d is the volume of water delivered to the area being irrigated (inflow).

Example 16.3: A 1 ha area having 100 border strips is irrigated once per week. Wheat is grown on 60 border strips and barley on 40 border strips. The readily available water (RAW) is 12 cm for wheat and 10 cm for barley; that is, RAW is the crop irrigation requirement. Wheat is irrigated at the rate of 25 L/min for each strip and barley is irrigated at the rate of 20 L/min. Compute the application efficiency for the wheat field, barley field, and the farm.

Solution: For the wheat field, the total planting area is $1\text{ ha} \times \frac{60}{60+40} = 0.6\text{ ha}$, the volume of water required is $W_s = 0.6\text{ ha} \times 12\text{ cm} \times \frac{1\text{ m}}{100\text{ cm}} \times \frac{10{,}000\text{ m}^2}{\text{ha}} = 720\text{ m}^3$, and the volume of water supplied or delivered is

$$W_d = 60\text{ border strips} \times 25\frac{\text{L}}{\text{min}} \times 60\frac{\text{min}}{\text{h}} \times 24\text{ h} \times \frac{1\text{ m}^3}{1000\text{ L}}$$
$$= 2160\text{ m}^3.$$

The application efficiency for the wheat field is $e_a = \frac{720}{2160} \times 100 = 33.3\%$.

For the barley field, the total planting area is $1\text{ ha} \times \frac{40}{60+40} = 0.4\text{ ha}$.

The volume of water required is

$$W_s = 0.4\text{ ha} \times 10\text{ cm} \times \frac{1\text{ m}}{100\text{ cm}} \times \frac{10{,}000\text{ m}^2}{\text{ha}} = 400\text{ m}^3,$$

and the volume of water supplied or delivered is

$$W_d = 40\text{ border strips} \times 20\frac{\text{L}}{\text{min}} \times 60\frac{\text{min}}{\text{h}} \times 24\text{ h}$$
$$\times \frac{1\text{ m}^3}{1000\text{ L}} = 1152\text{ m}^3.$$

The application efficiency for the barley field is $e_a = \frac{400}{1152} \times 100 = 34.7\%$

For the farm, the average application efficiency is

$$e_a = \frac{720 + 400}{2160 + 1152} \times 100 = 33.8\%.$$

16.3.4 Water Use Efficiency

For a farm irrigation system the efficiency, water use efficiency, or irrigation efficiency can be defined as the ratio of the amount of water supplied to the farm that is beneficially used to the amount of water that is supplied. The beneficial use may include water used for leaching (L) required. The water that is supplied to the farm is partly lost to deep percolation (DP), runoff (RO), accidental spills, leakage from channels, and other operational losses (O). Not all water that is applied to the field is beneficially used.

It may be noted that the definitions of water application efficiency and water use do not represent a similar concept. In the case of water use efficiency the leaching water is considered as beneficial use, whereas in the case of water application efficiency the leaching water is considered as loss of water.

The water use efficiency (e_u) or irrigation efficiency can be defined as (ASCE, 1978):

$$e_u = \frac{W_u}{W_d} \times 100. \tag{16.4}$$

The beneficial use can be expressed as $I + L = W_d - DP - RO - O$. Therefore,

$$\begin{aligned} e_u &= \frac{\text{Beneficial use of water}}{\text{Supply}} \times 100 \\ &= \frac{I+L}{S} \times 100 = \frac{W_d - DP - RO - O}{W_d} \times 100, \end{aligned} \tag{16.5}$$

in which e_u is the water use efficiency (%), W_u is the amount of water beneficially used (W_u can also be interpreted as the volume of water beneficially used by the plants, and in that sense it may be equal to the sum of irrigation requirement and leaching requirement), W_d is the amount of water delivered, I is the irrigation requirement, and L is the leaching requirement.

Example 16.4: Consider a 10-ha farm on which corn is grown on 4 ha and wheat is grown on 6 ha. Irrigation is done once per week. The RAW for corn is 10 cm and that for wheat is 12 cm; that is, RAW equals the amount of water received for the crop. The farm receives a supply of water at a rate of 15,000 L/min for a period of 24 hours. Assume that water is applied uniformly over the farm. Compute the irrigation efficiency of the farm.

Solution: The amount of water supplied to the farm is:

$$\begin{aligned} W_d &= 15{,}000\left(\frac{\text{L}}{\text{min}}\right) \times 60\,\frac{\text{min}}{\text{h}} \times 24\text{ h} \times \frac{1\text{ m}^3}{1000\text{ L}} \\ &= 21{,}600\text{ m}^3. \end{aligned}$$

Assume that there is no leaching (i.e., $L = 0$). The irrigation requirement I is computed as

$$\begin{aligned} I &= (10\text{ cm} \times 4\text{ ha} + 12\text{ cm} \times 6\text{ ha}) \times \frac{1\text{ m}}{100\text{ cm}} \\ &\quad \times \frac{10{,}000\text{ m}^2}{\text{ha}} = 11{,}200\text{ m}^3, \end{aligned}$$

$$e_u = \frac{11{,}200}{21{,}600} \times 100 = 52\%.$$

This shows that 52% of water is used for irrigation and the remainder is lost to deep percolation, runoff, evaporation, and spillage. It may be noted that the actual irrigation requirement of any crop may not equal the amount of evaporation and the application of water will not be uniform, meaning losses will be impacted by increased percolation in over-irrigated areas and decreased percolation in under-irrigated areas.

16.3.5 Overall System Efficiency

When evaluating the overall irrigation system performance, it may often be desirable to evaluate the efficiency of each component – storage, conveyance, and application – which will help identify which component is performing well and which is not. The overall irrigation system performance, then, is the product of individual component efficiencies, expressed as

$$e_i = \left(\frac{e_r}{100}\right) \times \left(\frac{e_c}{100}\right) \times \left(\frac{e_a}{100}\right). \tag{16.6}$$

Example 16.5: For an irrigation system, the reservoir efficiency is 80%, the conveyance efficiency is 60%, and the application efficiency is 70%. What is the overall system or irrigation efficiency?

Solution:

$$e_i = \left(\frac{80}{100} \times \frac{60}{100} \times \frac{70}{100}\right) \times 100 = 33.6\%.$$

16.3.6 Soil Water Storage Efficiency

When irrigation is applied, it increases the soil water in the root zone to field capacity, but not beyond. At the time of application, the soil already has some water, called antecedent water. That means that the amount of water to be applied by irrigation should be the difference between the field capacity and antecedent water. In practice, it is difficult to know the antecedent water because of its time-varying characteristics, unless one is simulating the hydrology of the irrigated field on a continuing basis. Since that is seldom done, the soil water storage efficiency is difficult to apply. The soil water storage efficiency (e_s) can be defined as

$$e_s = \frac{W_{sw}}{W_c} \times 100, \tag{16.7}$$

where e_s is the soil water storage efficiency (%), W_{sw} is the volume of water stored in the root zone due to irrigation, and W_c is the volume of water in the root zone at field capacity minus the antecedent water or volume of water in the root zone before irrigation. Thus,

$$W_c = W_{fc} - W_{ac}, \tag{16.8}$$

where W_{fc} is the volume of water in the root zone at field capacity and W_{ac} is the volume of water before irrigation or antecedent soil water.

Example 16.6: Consider an irrigated field 1.5 ha in area. The effective depth of the root zone is 1.5 m. The available moisture-holding capacity is 20 cm/m depth of the soil which was supplied by irrigation. The water at field capacity is 40 cm/m. The moisture level prior to irrigation was 20% of the field capacity. Compute the water storage efficiency.

Solution: The volume of water stored in the root zone due to irrigation is $W_{sw} = 20\ \text{cm/m} \times 1.5\ \text{m} = 30\ \text{cm} = 0.3\ \text{m}$, so the volume of water stored in the root zone due to irrigation of the 1.5-ha area is $0.3\ \text{m} \times 15{,}000\ \text{m}^2 = 4500\ \text{m}^3$.

The volume of water in the root zone at field capacity is $W_{fc} = 40\ \text{cm/m} \times 1.5\ \text{m} = 60\ \text{cm} = 0.6\ \text{m}$.

The volume of water in the root zone before irrigation or antecedent soil water is $W_{ac} = 0.6\ \text{m} \times 20\% = 0.12\ \text{m}$. Therefore, $W_c = W_{fc} - W_{ac} = 0.6\ \text{m} - 0.12\ \text{m} = 0.48\ \text{m}$ or W_c of 1.5-ha area is $0.48\ \text{m} \times 15{,}000\ \text{m}^2 = 7200\ \text{m}^3$.

The soil water storage efficiency is computed using eq. (16.7):

$$e_s = \frac{W_{sw}}{W_c} \times 100 = \frac{0.3\ \text{m}}{0.48\ \text{m}} \times 100\% = 62.5\%$$

or

$$e_s = \frac{W_{sw}}{W_c} \times 100 = \frac{4500\ \text{m}^3}{7200\ \text{m}^3} \times 100\% = 62.5\%.$$

16.3.7 Deep Percolation Ratio

The deep percolation ratio (P_r) is defined as the ratio of the volume of water percolated below the root zone to the volume of water delivered, expressed as

$$P_r = \frac{W_{dp}}{W_d}, \tag{16.9}$$

where P_r is the deep percolation ratio defined as a fraction, W_{dp} is the volume of water percolated below the root zone, and W_d is the volume of water delivered to the field. Deep percolation results in a rise of the water table and subsurface return flow to streams. If the quality of water return flow to streams is poor, streams may degrade.

16.3.8 Tailwater Ratio

In fields that are not bunded or diked at the downstream end, the water is lost from the lower end. This lost water is called tailwater and is a loss unless there is a way to recirculate it by pumping to the upstream end of the field or direct it to another field. In many cases, tailwater is allowed to flow to a nearby stream, which may affect the stream health. The tailwater ratio is defined as

$$TW_r = \frac{W_{ro}}{W_d}, \tag{16.10}$$

where TW_r is the tailwater ratio and W_{ro} is the volume of surface runoff or tailwater.

16.3.9 Irrigation Uniformity

An important concept not considered in the definitions of irrigation efficiencies discussed above is the uniformity or nonuniformity of water applied to the field. For the same volume of water applied to a field, the water application may not be uniform – that is, on some parts of the field there may be excess water, and at other places there may be water shortages. This often happens when fields are not prepared or leveled well. In either case, plants are harmed. Excess water may cause surface runoff or deep percolation. When the soil is saturated for many days, plants may suffer due to the lack of oxygen in the root zone. Deep percolation may raise the water table or cause a perched water table, depending on the substratum conditions. Of course, percolation is needed to leach salts that accumulate in the root zone.

The question arises: How do we measure the uniformity of irrigation? Many definitions of uniformity have been advanced. One way to measure the uniformity is the uniformity coefficient, which describes the spatial distribution of water applied by irrigation. It indicates the degree to which water has penetrated to a uniform depth throughout the field and is defined as

$$C_u = \left(1 - \frac{y}{d}\right) \times 100, \tag{16.11}$$

where C_u is the uniformity coefficient (%), y is the average of the absolute values of deviations of the depth of water stored from the average depth of water stored, and d

is the average depth of water stored. When the deviation from the average depth is zero, the uniformity coefficient is 100%.

A variation of eq. (16.11) is also employed to define the distribution uniformity (*DU*) as

$$DU = \frac{y_a}{d} \times 100, \qquad (16.12)$$

where y_a is the average depth of infiltrated water over the quarter of the field with the least infiltration (or average depth over the low-quarter depth of water received), and d is the average depth over the entire field (or mean depth infiltrated). The distribution uniformity is used for microirrigation and sprinkler irrigation systems.

Christiansen uniformity coefficient (C_u): Of the many definitions describing the irrigation uniformity, the uniformity coefficient defined by Christiansen (1942) for evaluating sprinkler systems is most commonly used. It is defined as

$$C_u = 100 \times \left[1 - \frac{\sum_{i=1}^{N}|(d_i - d_m)|}{N d_m}\right], \qquad (16.13)$$

where C_u is the Christiansen uniformity coefficient, N is the number of catch cans, d_i is the measured depth of water in equally spaced catch cans at the ith grid, and d_m is the mean depth of water of the catch in all N cans. The implication here is that depths represent equal areas, which may not be true for center-pivot systems. In center-pivots, catch cans are equally spaced along a radial line from the pivot to the outer end, so each measurement must be weighted based on the area it represents.

Wilcox–Swailes uniformity coefficient (*UCWS*): Wilcox and Swailes (1947) defined a uniformity coefficient by replacing the absolute value of the mean deviation by the standard deviation as:

$$UCWS = 100 \times \left[1 - \frac{S}{\bar{d}}\right], \qquad (16.14)$$

where S is the standard deviation of the measured depths of water in equally spaced catch cans (where d_i is the depth at the ith grid, $i = 1, 2, 3, \ldots, N$), and $\bar{d}$ is the mean depth of water of the catch in all cans.

Soil Conservation Service pattern efficiency (*PE*): The Soil Conservation Service of the US Department of Agriculture (USDA) proposed a concept of pattern efficiency (Criddle et al., 1956) as

$$PE = 100\left(\frac{a}{\bar{d}}\right), \qquad (16.15)$$

where a is the average depth of the 25% observations having the least water depth.

Hart uniformity coefficient (*UCH*): Assuming the water distribution by sprinklers to be normal, Hart (1961) and Hart and Reynolds (1965) expressed a uniformity coefficient as

$$UCH = 100\left[1 - \frac{0.798\,S}{\bar{d}}\right]. \qquad (16.16)$$

If the water from sprinklers is assumed to be normal distributed, then the term in eq. (16.13) $\left(\frac{\sum_{i=1}^{N}|(d_i - d_m)|}{N}\right)$, which is the mean of the absolute value of the deviation from the mean, is equal to $0.798S$. Therefore, for a truly normal distribution, $UCH = C_u$.

Different uniformity coefficients have been related. Hart and Heermann (1976) stated:

$$C_u = 0.03 + 0.958\ UCH. \qquad (16.17)$$

Seniwongse et al. (1972) stated:

$$UCWS = -11.287 + 0.92\,(C_u) + 0.0020\,(C_u)^2 \qquad (16.18)$$

Dabbous (1962) showed:

$$UCWS = -0.25 + 1.25\,C_u \qquad (16.19)$$

and

$$PE = -0.45 + 1.45\,C_u. \qquad (16.20)$$

These relationships allow the comparison of different uniformity coefficients.

Example 16.7: Compute the uniformity coefficient using the measured depth of water from 20 cans, given in Table 16.1.

Table 16.1 *Depth measured by 20 cans ranked from the lowest depth to the highest depth*

Rank	Depth (cm)	Deviation (cm)	Absolute deviation (cm)
1	3.30	–1.666	1.666
2	3.81	–1.156	1.156
3	4.32	–0.646	0.646
4	4.32	–0.646	0.646
5	4.32	–0.646	0.646
6	4.57	–0.396	0.396
7	4.57	–0.396	0.396
8	4.83	–0.136	0.136
9	4.83	–0.136	0.136
10	4.83	–0.136	0.136
11	5.08	0.114	0.114
12	5.08	0.114	0.114
13	5.33	0.364	0.364
14	5.33	0.364	0.364
15	5.33	0.364	0.364
16	5.59	0.624	0.624
17	5.84	0.874	0.874
18	5.84	0.874	0.874
19	6.10	1.134	1.134
20	6.10	1.134	1.134
	Mean (d_m) = 4.966		Sum = 11.92

Solution: As shown in Table 16.1, the mean depth is $d_m = 4.966$ cm.

The standard deviation of all data is $s = \sqrt{\frac{\sum_{i=1}^{N}(d_i - d_m)^2}{N}} = 0.749$ cm.

Method 1: Using the C_u method,

$$C_u = 100 \times \left[1 - \frac{\sum_{i=1}^{N}|(d_i - d_m)|}{Nd_m}\right]$$

$$= 100 \times \left(1 - \frac{11.92}{20 \times 4.966}\right) = 88.0\%.$$

Method 2: Using the UCWS method,

$$UCWS = 100 \times \left[1 - \frac{S}{\bar{d}}\right] = 100 \times \left(1 - \frac{0.749}{4.966}\right) = 84.9\%.$$

16.3.10 Adequacy of Irrigation

As the term "adequacy" suggests, adequacy of irrigation implies the percentage of the farm that receives a sufficient amount of water to achieve profitable crop production. Profitability includes both the quantity and quality of crop yield and will depend on crop, soil, climate, and market conditions. Such a definition becomes unwieldy in practice. Therefore, a simpler definition of irrigation is defined in terms of the percentage of the farm that receives the desired amount of water or more. Thus, the adequacy of irrigation can be determined by plotting the cumulative percentage area against the amount of water infiltrated, as shown in Figure 16.2. This is analogous to a cumulative probability distribution of a random variable, if the depth of soil moisture can be considered a random variable. Figure 16.2 shows that 50% of the area receives the desired amount, meaning the adequacy of irrigation is 50%. In order to construct the cumulative distribution graph, the values of infiltrated depth for different areas should be known. This is illustrated in Example 16.8.

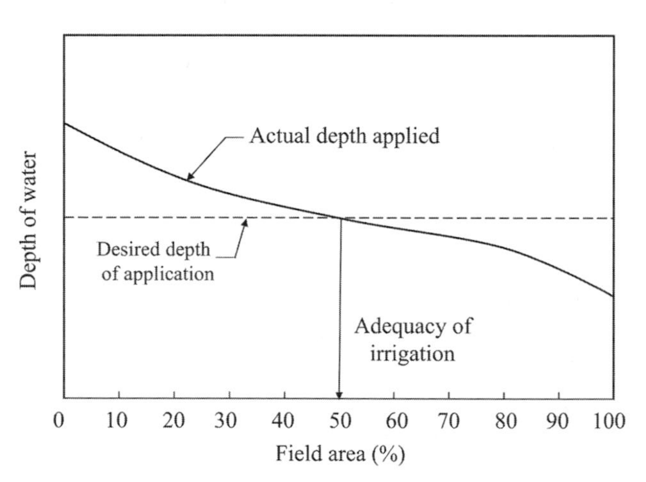

Figure 16.2 Cumulative distribution for determining the adequacy of irrigation.

Example 16.8: The amount of infiltration (cm) observed at 25 different parts of equal size of a farm are:

5.0	3.5	3.7	5.3	3.2
4.5	3.9	4.3	5.6	3.3
3.8	4.1	4.6	3.6	4.0
4.2	4.7	4.8	4.4	5.9
5.2	5.1	5.4	4.9	5.8

Determine the adequacy of irrigation if the desired depth of irrigation is 4.5 cm.

Solution: For ease of understanding, the following steps are used to solve the problem.

Step 1: Arrange the depth values in descending order, as shown in Table 16.2.

Step 2: Indicate the percent of area represented by each depth of infiltration value. In this case the percent area is the same, as shown in Table 16.2.

Step 3: Determine the cumulative percent area for each depth value, as shown in Table 16.2.

Table 16.2 *Cumulative percent area and depth*

Depth of infiltration (cm)	Percent area of field	Cumulative percent area of field
5.9	4	4
5.8	4	8
5.6	4	12
5.5	4	16
5.3	4	20
5.2	4	24
5.1	4	28
5.0	4	32
4.9	4	36
4.8	4	40
4.7	4	44
4.6	4	48
4.5	4	52
4.4	4	56
4.3	4	60
4.2	4	64

Table 16.2 (*cont.*)

Depth of infiltration (cm)	Percent area of field	Cumulative percent area of field
4.1	4	68
4.0	4	72
3.9	4	76
3.8	4	80
3.7	4	84
3.6	4	88
3.5	4	92
3.3	4	96
3.2	4	100

Step 4: Plot the cumulative area against depth, as shown in Figure 16.3.

Step 5: Determine the adequacy of irrigation from the plot, as shown in Figure 16.3.

From Figure 16.3, the adequacy of irrigation is 52%.

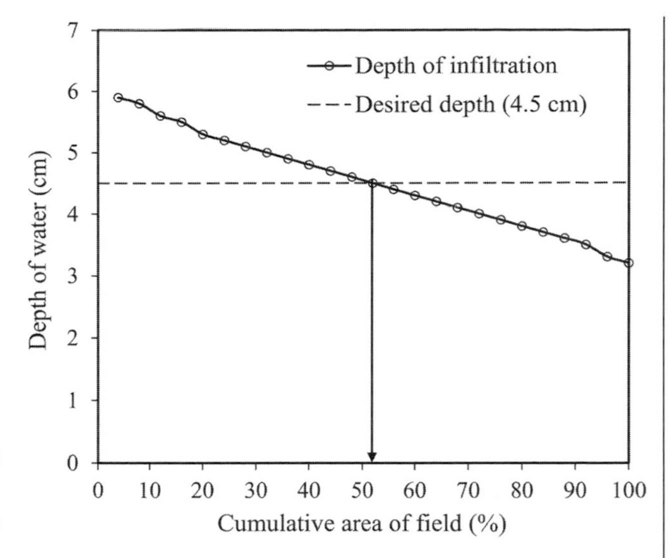

Figure 16.3 Percent of the area against the depth of infiltrated water with respect to the desired depth.

Example 16.9: Compute the Christiansen uniformity coefficient and distribution uniformity for data given in Table 16.2.

Solution:

As shown in the table, the mean depth of infiltration is $d_m = 4.516$ cm, the sum of the absolute deviations of depth is 16.416 cm.

Depth of infiltration (cm)	Deviation (cm)	Absolute deviation (cm)
5.9	1.384	1.384
5.8	1.284	1.284
5.6	1.084	1.084
5.5	0.984	0.984
5.3	0.784	0.784
5.2	0.684	0.684
5.1	0.584	0.584
5	0.484	0.484
4.9	0.384	0.384
4.8	0.284	0.284
4.7	0.184	0.184
4.6	0.084	0.084
4.5	–0.016	0.016
4.4	–0.116	0.116
4.3	–0.216	0.216
4.2	–0.316	0.316
4.1	–0.416	0.416
4	–0.516	0.516
3.9	–0.616	0.616
3.8	–0.716	0.716
3.7	–0.816	0.816
3.6	–0.916	0.916
3.5	–1.016	1.016
3.3	–1.216	1.216
3.2	–1.316	1.316
Mean		**Sum**
4.516		**16.416**

The C_u is computed using eq. (16.13):

$$C_u = 100 \times \left[1 - \frac{\sum_{i=1}^{N} |(d_i - d_m)|}{N d_m}\right]$$

$$= 100 \times \left(1 - \frac{16.416}{25 \times 4.516}\right) = 85.5\%.$$

The average depth of infiltrated water over one-quarter ($25/4 \approx 6$) of the field with the least infiltration is $y_a = (3.8+3.7+3.6+3.5+3.3+3.2)/6 = 3.517$ cm. Using eq. (16.12), DU is:

$$DU = \frac{y_a}{d} \times 100 = \frac{3.517}{4.516} \times 100 = 77.9\%.$$

16.3.11 Effectiveness of Irrigation

In irrigation engineering, effectiveness of irrigation is often evaluated by the desired combination of application efficiency, uniformity, and adequacy of irrigation. The implication is that the desired combination will maximize the net profit. However, the maximum net profit is not necessarily generated by the maximum application efficiency, uniformity, and adequacy, but by a certain combination. Therefore, it is desirable to determine the relationship between application efficiency, uniformity, and adequacy to identify strategies to maximize the net profit.

Relation between application efficiency and uniformity: For an irrigation system the percentage of field area against soil moisture depth or infiltrated depth due to irrigation is shown in Figure 16.4. Knowing the desired depth of application or depth of full irrigation, it shows the deep percolation and water stress for a percentage of field area. The intersection of soil moisture profile and the line of desired depth yields the adequacy of irrigation. It also reflects the adequacy and efficiency of irrigation. If deep percolation and crop water stress are reduced, application efficiency would increase and the water would be more uniformly applied. However, as the depth of water applied increases, leaching increases, stress decreases, and adequacy increases, but the application efficiency may not necessarily increase.

Figure 16.4 shows the dimensionless cumulative area versus dimensionless depth of application of water, and shows the areas of water stored, water lost, and water deficit. It thus reflects relative uniformity of irrigation. Figure 16.5 illustrates application efficiency (E_a) defined in terms of the ratio of area of water stored to the area of water applied (i.e., $E_a = (\text{Area A})/(\text{Area A} + \text{Area B})$), or the depth of water stored area to the depth of water applied (i.e., $E_a = (\text{Water stored})/(\text{Water applied})$). Figure 16.5 shows that 78% of the area receives the desired amount, meaning the adequacy of irrigation is 78%.

Relation between application efficiency and adequacy: Figure 16.6 illustrates the change in adequacy with change in infiltrated depth for the same value of C_u. Different soil moisture profiles lead to different values of adequacy, and the application efficiency would be different. Figure 16.7

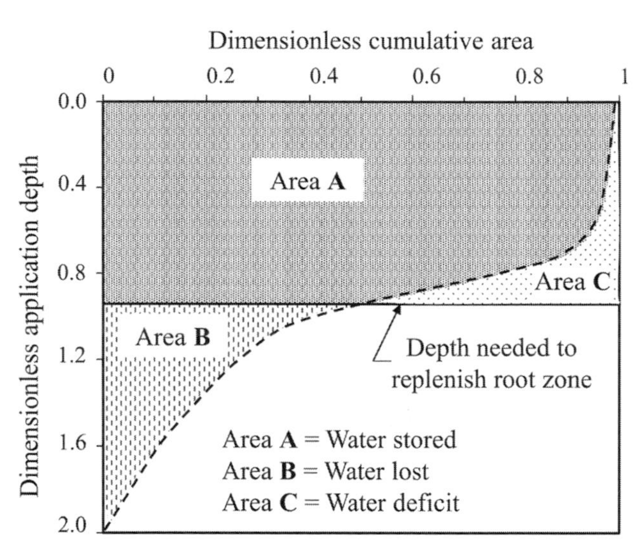

Figure 16.4 Dimensionless application depths.

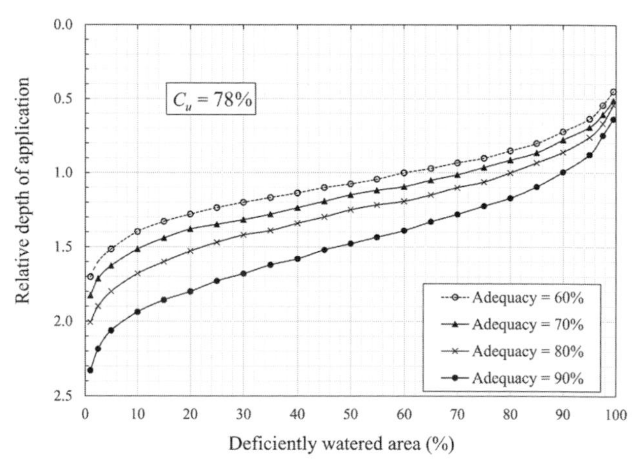

Figure 16.6 Variation of adequacy with different soil moisture profiles. (Data from USDA-NRCS, 2016.)

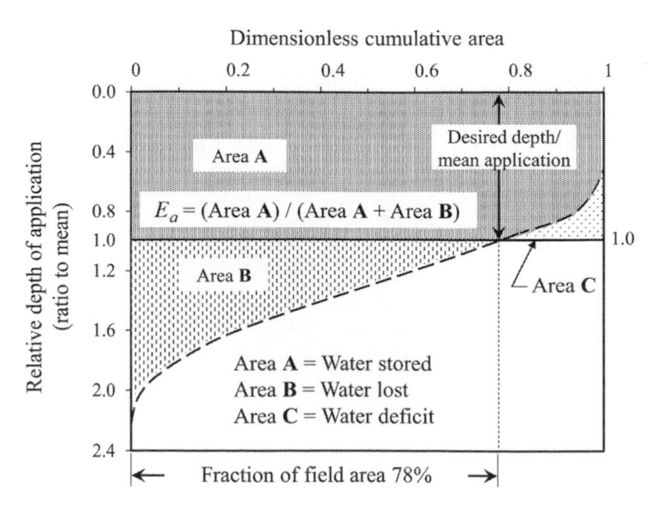

Figure 16.5 Definition of application efficiency.

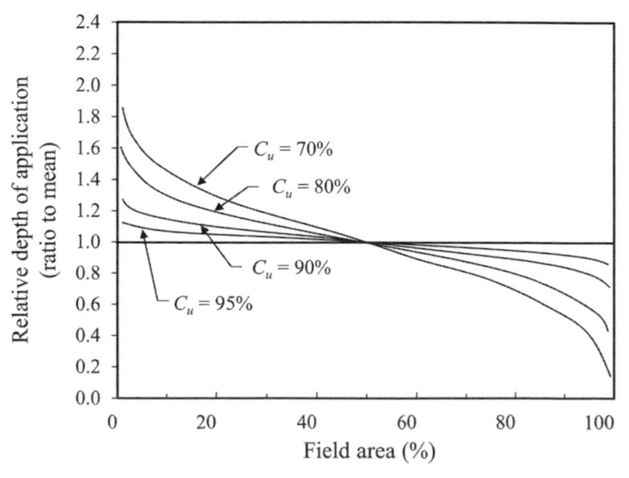

Figure 16.7 Variation in C_u with different soil moisture profiles. (Data from USDA-NRCS, 2016.)

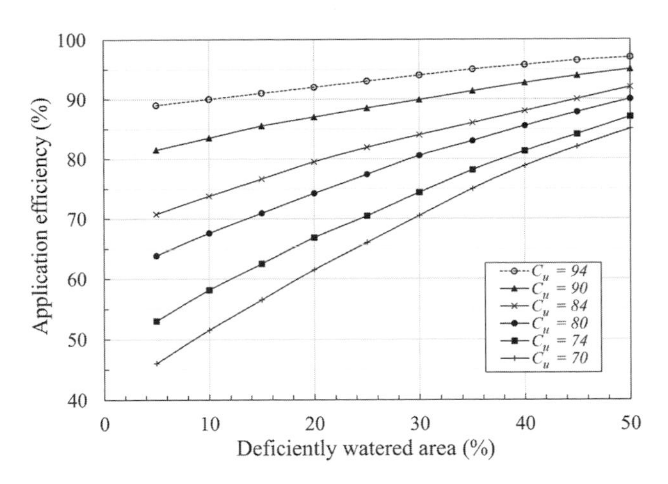

Figure 16.8 Use of application efficiency. (Data from USDA-NRCS, 2016.)

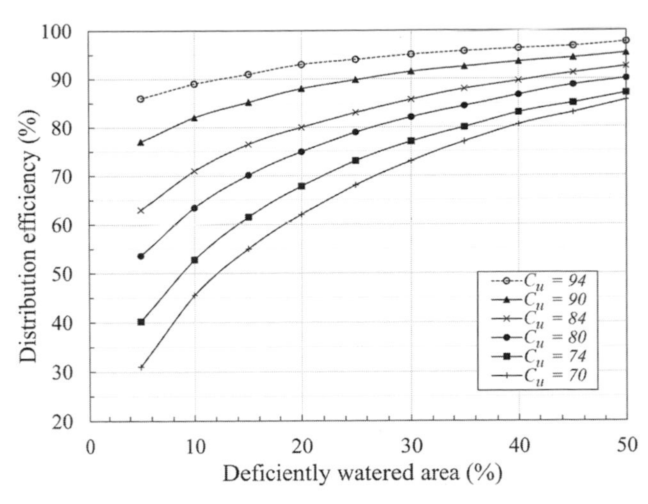

Figure 16.9 Use of distribution efficiency. (Data from USDA-NRCS, 2016.)

shows the variation of depth profile with the variation of C_u for the same value of adequacy. Here also, the application efficiency will be different for different profiles. The variation of application efficiency with water deficit area for varying C_u is illustrated in Figure 16.8. The variation of distribution efficiency with water deficit area for varying C_u is illustrated in Figure 16.9.

Example 16.10: Given $C_u = 80\%$, SWD (soil water depletion) is 1.2 inches (3.05 cm), and adequacy is 50%. Determine irrigation efficiency in percentage and gross irrigation depth (d_g) in inches (cm).

Solution: Given $C_u = 80\%$ and adequacy = 50%, the deficiently watered area is 100% − adequacy = 100% − 50% = 50%. From Figure 16.10, the irrigation application efficiency $E_a = 90\%$.

Given $SWD = 1.2$ inches (3.05 cm), the gross irrigation depth

$$d_g = SWD/E_a = 1.20 \text{ inches}/0.9$$
$$= 1.33 \text{ inches (3.4 cm)}.$$

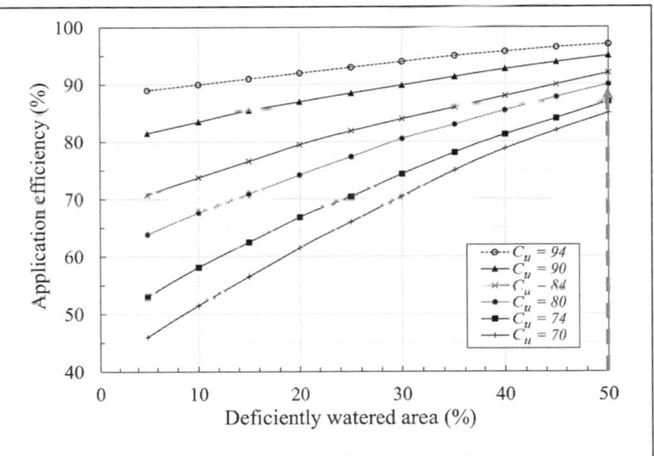

Figure 16.10 Application efficiency for different values of C_u.

Example 16.11: Given $C_u = 80\%$, $SWD = 1.2$ inches (3.05 cm), and adequacy = 80%, determine the irrigation application efficiency (E_a) in percent and gross irrigation depth (d_g).

Solution: Given $C_u = 80\%$ and adequacy = 80%, the deficiently watered area is 100% − adequacy = 100% − 80% = 20%. From Figure 16.10, the irrigation application efficiency $E_a = 74\%$.

Given $SWD = 1.2$ inches (3.05 cm), the gross irrigation depth

$$d_g = SWD/E_a = 1.20 \text{ inches}/0.74$$
$$= 1.62 \text{ inches (4.1 cm)}.$$

16.3.12 Application Efficiency of Low Quarter

Unification of uniformity and efficiency is

$$d_g = d_m + d_r + d_{ev}, \tag{16.21}$$

where d_g is the gross irrigation application depth, d_m is the mean depth infiltrated (or average depth over the entire field), d_{ev} is the effective depth of evaporation and drift, and d_r is the effective depth of runoff. Rearranging eq. (16.21), one gets

$$d_m = d_g - d_r - d_{ev}. \quad (16.22)$$

Also,

$$d_{LQ} = DU \times d_m, \quad (16.23)$$

where d_{LQ} is the low quarter depth of irrigation and DU is the distribution uniformity.

If the water were uniformly applied across the irrigated area, the lower quarter depth of irrigation (d_{LQ}) is equal to the mean depth infiltrated over the entire field (d_m), $d_{LQ} = d_m$, as shown in Figures 16.11 and 16.12. Under perfect irrigation conditions (Figure 16.11), the actual application is equal to the desired application and there is no deep percolation and runoff. When there is excess water irrigated (Figure 16.12), deep percolation happens and $d_m(= d_{LQ})$ is higher than SWD. If the water were nonuniformly applied with no excess (Figure 16.13), part of the field is deficiently irrigated and d_{LQ} is smaller than SWD. If the water were nonuniformly applied with excess irrigation (Figure 16.14), this means that all the field is well irrigated and d_{LQ} is higher than SWD.

Application efficiency of the lower quarter ($AELQ$) can be summarized as:

$$\text{if } d_{LQ} \leq SWD, \text{then } d_e = d_{LQ}, \quad (16.24)$$

$$\text{if } d_{LQ} > SWD, \text{then } d_e = SWD, \quad (16.25)$$

$$AELQ = \frac{d_e}{d_a} \times 100, \quad (16.26)$$

where $AELQ$ is in %, d_e is the effective depth of irrigation and d_a is the depth applied from the original source.

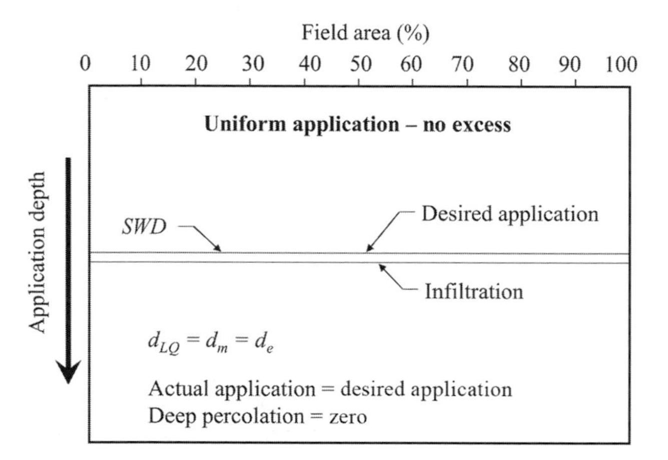

Figure 16.11 Unification of uniformity and efficiency (perfect irrigation-impossible).

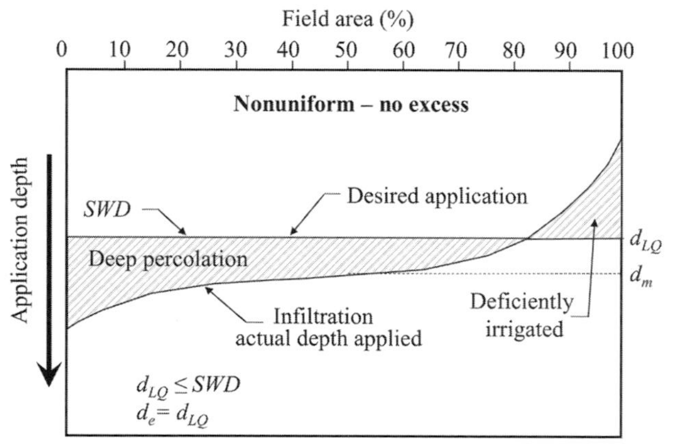

Figure 16.13 Unification of uniformity and efficiency (nonuniform-no excess).

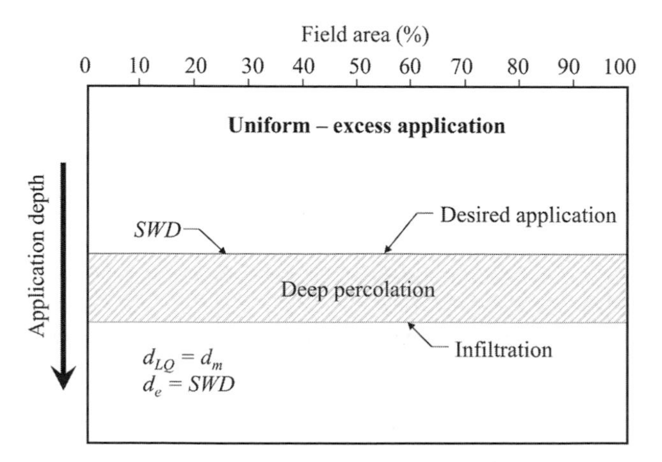

Figure 16.12 Unification of uniformity and efficiency (perfect uniform-too much water).

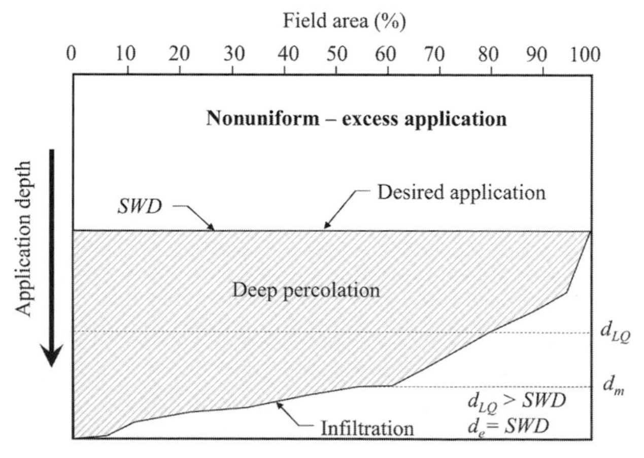

Figure 16.14 Unification of uniformity and efficiency (nonuniform-excess application).

Example 16.12: Given $DU = 0.75$, $d_m = 2.0$ in. (5.1 cm), $d_r = 0.0$ in., $d_g = 2.2$ in. (5.6 cm), and $SWD = 1.6$ in. (4.1 cm), determine d_{ev} and $AELQ$.

Solution:

$d_{ev} = d_g - d_m - d_r = 2.2 - 2.0 - 0.0 = 0.2$ in. (0.5 cm),

$d_{LQ} = DU \times d_m = 0.75 \times 2.0 = 1.5$ in. (3.8 cm),

$d_{LQ} \leq SWD$, then $d_e = d_{LQ} = 1.5$ in. (3.8 cm).

Since $d_r = 0.0$ in., $d_a = d_g = 2.2$ in. (5.6 cm). Thus,

$$AELQ = \frac{d_e}{d_a} \times 100 = \frac{1.5}{2.2} \times 100 = 68\%.$$

Example 16.13: Consider the data from Example 16.12, but $SWD = 1.2$ in. (3.0 cm); that is, $DU = 0.75$, $d_m = 2.0$ in. (5.1 cm), $d_r = 0.0$ in., $d_g = 2.2$ in.(5.6 cm). Determine d_{ev} and $AELQ$.

Solution:

$d_{ev} = d_g - d_m - d_r = 2.2 - 2.0 - 0.0 = 0.2$ in. (0.5 cm),

$d_{LQ} = DU \times d_m = 0.75 \times 2.0 = 1.5$ in. (3.8 cm).

In this case, $d_{LQ} > SWD$, then $d_e = SWD = 1.2$ in (3.0 cm).

Since $d_r = 0.0$ in., $d_a = d_g = 2.2$ in. (5.6 cm). Thus,

$$AELQ = \frac{d_e}{d_a} \times 100 = \frac{1.2}{2.2} \times 100 = 55\%.$$

QUESTIONS

Q.16.1 Determine the reservoir storage efficiency if the reservoir inflow rate is 5000 L/min and the water diverted from the reservoir is 4000 L/min, both for a period of 12 h, and the change in storage in the reservoir is 200 m^3.

Q.16.2 Consider a field 2 ha in area that is irrigated by water received from a channel fed by a canal. The period of irrigation is 10 h. The discharge in the channel is 150 L/s and 120 L/s is supplied to the field. The field is not diked at the end so when irrigated there is surface runoff at a rate of 10 L/s. The depth of soil moisture due to irrigation varies from 1.8 m at the head of the field to 1.4 m at the tail end. The effective root zone depth is considered to be 1.6 m. It is known that the soil has a moisture-holding capacity of 20 cm per meter of depth and the moisture extraction level prior to irrigation is 50% of the available moisture. Determine the water conveyance efficiency, water application efficiency, water storage efficiency, and water distribution uniformity.

Q.16.3 A farm 25 ha in area receives irrigation water at the rate of 10 m^3/s for a period of 5 h. It is estimated that 10% of the water runs off the farm. It is found that 0.25 m of water is stored in the root zone. Determine the application efficiency.

Q.16.4 Compute the water application efficiency and uniformity coefficient if a field receives 0.1 m^3/s for 4 h. The tailwater runoff was 5% of inflow to the field. The root zone depth is 1.45 m and the depth of moisture due to irrigation varies linearly from 1.45 m at the head to 1.25 m at the tail end.

Q.16.5 A farm receives 1.5 m^3/s from a pumping system which produces 1.75 m^3/s for a period of 3 h. The surface runoff from the farm is 0.1 m^3/s. The remainder 0.15 m^3/s recharges the groundwater. Determine the water conveyance efficiency and water application efficiency.

Q.16.6 On farm the dry density of soil is 1250 kg/m^3. A 10-cm irrigation increased the average soil moisture content of 0.75 m of the top soil from 15% to 20% (on a dry weight basis). Compute the water application efficiency.

Q.16.7 Consider a farm that receives 2000 L/min for 20 h for irrigating 0.5 ha of wheat and 1.0 ha of potato. Readily available water is 5 cm for wheat and 10 cm for potato. Assuming that water is uniformly applied to the field, determine the irrigation efficiency.

Q.16.8 What will be the reservoir storage efficiency if 4000 L/min of water is supplied to the reservoir and 3200 L/min is withdrawn for irrigation over 20 h? The change in storage of the reservoir is 400 m^3 to restore the initial water level.

Q.16.9 A canal carries water at the rate of 2600 L/min and supplies water to a farm for irrigation. The farm has 100 furrows, 65 of which receive 25 L/min each and 35 receive 15 L/min each. Compute the conveyance efficiency.

Q.16.10 Compute the application efficiency for a farm that has 0.5 ha of corn and 0.5 ha of alfalfa. Both corn and alfalfa have 25 furrows each. The RAW is 6 cm for corn and 12 cm for alfalfa. Each furrow discharges 1500 L/min for corn and 2000 L/min for alfalfa. It is assumed that water is applied uniformly over the field.

Q.16.11 A 1-ha wheat field is irrigated by sprinklers. The readily available water-holding capacity of the soil is 8 cm. If the water is applied at the rate of 1500 L/min for a period of 15 h, what will be the irrigation efficiency?

Q.16.12 A 50-ha wheat farm is irrigated by water from a canal with a discharge of 10,000 L/min. The irrigation is done for 10 days. If the readily available water-holding capacity of the soil is 12 cm, determine the irrigation efficiency.

Q.16.13 A canal supplies water at a rate of 0.5 m^3/s to a reservoir from which the seepage and evaporation losses are 50 L/s. The reservoir supplies water to a 500-m channel to irrigate a 20-ha farm, where the seepage and evaporation losses from the channel are 0.75 L/min/m. The average daily irrigation requirement is 12 cm/day and irrigation is done for a period of 24 h. Determine the reservoir storage efficiency, conveyance efficiency, application efficiency, and overall system efficiency.

Q.16.14 Compute the uniformity of application, distribution uniformity, and storage efficiency if the soil moisture data for the top 80 cm of soil on a 80-m grid are available before and after irrigation. The water content at field capacity is 40% by volume.
Soil water content prior to irrigation (A):

15.2	17.1	18.0	17.6	16.5	15.9	14.2
15.6	16.2	17.5	17.4	14.6	15.4	14.4
14.5	15.5	16.3	17.3	14.7	16.1	14.3
13.2	14.8	15.8	16.4	14.9	16.6	15.6
15.1	16.8	16.3	16.7	13.9	15.7	17.2

Soil water content after irrigation (B):

30.1	32.2	31.5	33.4	32.6	30.4	28.4
30.5	32.5	32.0	33.6	28.5	30.9	28.2
28.6	28.2	33.8	33.4	28.9	32.6	28.6
29.1	28.9	32.0	32.3	28.7	32.1	30.3
31.2	31.9	32.4	32.7	27.9	30.7	34.2

Q.16.15 What is the adequacy of irrigation in Q.16.14 if the desired depth of irrigation was 11 cm, 12 cm, and 13 cm.

Q.16.16 Determine the application efficiency for a desired depth of 12 cm using the cumulative distribution curve from Q.16.15.

Q.16.17 If the cumulative distribution curve from Q.16.15 is used, what would be the depth of application for achieving 90% adequacy? What will be the application efficiency?

Q.16.18 Given $DU = 0.80$, $d_m = 2.2$ in. (5.59 cm), $d_r =$ 0.1 in. (0.25 cm), $d_g = 2.60$ in. (6.60 cm), and $SWD = 1.77$ in. (4.50 cm), determine d_{ev} and $AELQ$.

REFERENCES

ASCE (1978). Describing irrigation efficiency and uniformity. *Journal of Irrigation and Drainage Division, ASCE*, 104(IR1): 35–41.

Bos, M. G. (1979). Standards for irrigation efficiencies of ICID. *Journal of Irrigation and Drainage Division, ASCE*, 105(1): 37–43.

Burt, C. M., Clemmens, A. J., Strelkoff, T. S., et al. (1997). Irrigation performance measures: efficiency and uniformity. *Journal of Irrigation and Drainage Engineering*, 123(6): 423–442.

Christiansen, J. E. (1942). Irrigation by sprinkling. University of California Agricultural Experiment Station bulletin 670.

Criddle W. D., Davis, S., Pair, C. H., et al. (1956). *Methods for Evaluating Irrigation Systems*. Washington, DC: US Government Printing Office.

Dabbous, B. (1962). A study of sprinkler uniformity evaluation methods. MS Thesis, Utah State University.

Hansen, V. E., Israelson, O. W., and Stringham, G. E. (1980). *Irrigation Principles and Practices*, 4th ed. New York: Wiley.

Hart, W. E. (1961). Overhead irrigation pattern parameters. *Agricultural Engineering*, 42(7): 354–355.

Hart, W. E., and Heermann, D. F. (1976). Evaluating water distributions of sprinkler irrigation systems. Colorado State University Experiment Station bulletin 128.

Hart, W. E., and Reynolds, W. N. (1965). Analytical design of sprinkler system. *Transactions of the ASCE*, 8(1): 83–85, 89.

Israelson, O. W. 1950. *Irrigation Principles and Practices*, 2nd ed. New York: Wiley.

Jensen, M. E., Swamer, L. R., and Phelan, J. T. (1967). Improving irrigation efficiencies. In *Irrigation of Agricultural Land*, edited by R. M. Hagan, H. Haise, T. Edminster, et al. Madison, WI: American Society of Agronomy, pp. 1120–1142.

Jensen, M. E., Harrison, D. S., Korven, H. C., and Robinson, F. E. (1983). The role of irrigation in food and fiber production. In *Design and Operation of Farm Irrigation Systems*, edited by G. Hoffman, R. G. Evans, M. Jensen, et al. Washington, DC: ASCE, pp. 15–41.

Painter, D., and Carran, P. (1978). What is irrigation efficiency? *Soil and Water*, 14: 15–22.

Seniwongse, C., Wu, I. P., and Reynolds, W. N. (1972). Skewness and kurtosis influence on uniformity coefficient and application to sprinkler irrigation design. *Transaction of the ASAE*, 15(2): 266–271.

USDA-NRCS (2016). Sprinkler irrigation. In *National Engineering Handbook*. Washington, DC: USDA.

Walker, W. R., and Skogerboe, G. V. (1987). *Surface Irrigation, Theory and Practice*. Englewood Cliffs, NJ: Prentice Hall.

Wilcox, J. C., and Swailes, G. E. (1947). Uniformity of water distribution by some undertree orchard sprinklers. *Scientific Agriculture*, 27(11): 565–583.

17 Surface Irrigation Preliminaries

Notation

Symbol	Description	Symbol	Description
$\overline{A}$	average flow cross-sectional area (L^3/T or $L^3/T/L$)	S_0	channel bottom slope (L/L)
A	cross-sectional area of flow (L^2)	S_f	channel friction slope or the slope of the energy grade line (L/L)
A_0	inlet area (L^2)	t	time since the beginning of irrigation (min)
d	average depth of water stored (cm or in.)	t_s	time when water front reaches the distance (T)
f_0	basic infiltration rate (m/min, L/min/m or gpm/ft)	t_x	advance time to distance X since the beginning of irrigation (min)
g	gravitational acceleration (L/T^2)	T	duration of irrigation (min)
h, y	flow depth (L)	T_a	advance time (min)
h_0, y_0	flow depth at the head or inlet flow depth (L)	V	average velocity in the flow cross-section (L/T)
I	volume rate of infiltration per unit length of the channel or border (L/T)	V_y	volume of surface storage at time t (m^3)
k, a	parameters in the Kostiakov and the modified Kostiakov (Kostiakov–Lewis) equations	V_z	volume of infiltrated water at time t (m^3)
		x	distance in the direction of flow (L)
L	length of the field (m)	X	advance distance (m)
n	Manning's roughness coefficient	Z	cumulative infiltration volume (m^3/unit area)
p, r	fitted parameters in power advance volume balance model	Z_0	infiltrated volume per unit area at the field inlet
q	unit flow rate in border irrigation ($L^3/T/L$)	α	kinematic roughness parameter α ($L^{1/3}/T$ where T is in seconds)
Q	flow rate (L^3/T or $L^3/T/L$ in border irrigation)	ρ_1, ρ_2	empirical shape factors
Q_0	steady inflow rate (m^3/min or m^3/min/unit width)	σ_y	surface shape factor (varying between 0.70 and 0.80, but often taken as 0.77)
Q_r	runoff (outflow) rates (m^3/min)	σ_z	subsurface shape factor
R	hydraulic radius (L)	τ	intake opportunity time (min)

17.1 INTRODUCTION

Surface irrigation is the oldest method used for irrigation in the world. It is commonly practiced around the world, more so in developing countries. Over 90% of the irrigated land in the world, which is about 16% of the total cultivable land, is irrigated by surface irrigation. In the United States, about 40% of all the irrigated land is irrigated by surface irrigation, whereas in India it is over 90%. In California, nearly 57% of all irrigated land is irrigated by gravity methods (Dillon et al., 1999), and in Texas 78% of all irrigated land is watered by surface irrigation systems (Bloodworth and Gillett, 2010). The advantages of surface irrigation methods are low expenditure in energy, minimum capital investment, and simple equipment. Properly designed and managed surface irrigation systems can have application efficiencies comparable to other irrigation systems (e.g., pressurized irrigation systems). The disadvantages include large labor input, large stream size, land leveling, and often low efficiency.

Surface irrigation entails a broad class of irrigation methods in which water is applied at the head of the field and flows freely as overland flow over the land surface under gravity. The purpose is to allow water to infiltrate as much as possible to refill the crop root zone. There are different methods by which the water is applied to the field. These methods can be broadly classified into five types: (1) border irrigation, (2) basin irrigation, (3) water spreading or wild flooding, (4) furrow irrigation, and (5) contour ditch irrigation. One can also add sub-irrigation and subsurface irrigation methods. Before discussing these methods, it is pertinent to briefly discuss the total irrigation system, often called the physical system, of which farm irrigation is one component. The system supplies water for irrigation. Selection of a particular method depends on a number of factors, including climate, soil, crop, water availability, landscape, availability of labor, energy, costs and benefits, and traditions. This chapter discusses the preliminaries of the entire irrigation system.

17.2 PHYSICAL SYSTEM

The physical system comprises four subsystems: the water supply, water delivery, water use, and water removal and recycle subsystems, as shown in Figures 17.1 and 17.2. The system may also have measuring devices and turnouts.

17.2.1 Water Supply Subsystem

The source of water is either surface or subsurface. The surface systems are mainly reservoirs, lakes, ponds, or river diversions. A surface system can be artificial or natural. The subsurface source is groundwater, which is pumped for irrigation.

17.2.2 Water Delivery Subsystem

The delivery system brings water from the source to the field. The delivery system is also called the conveyance system. If the water is brought from the reservoir, then the system may involve a main canal, minor canals, orifices or outlets, and field channels. If the water is taken out of groundwater reservoirs, then pumps and channels are involved. If the water is taken out of a river, then either pumping or a diversion and then channels may be involved. Depending on the type of delivery system, the loss of water through seepage may be significant.

17.2.3 Water Use Subsystem

The water is received from the water delivery system and is applied to the field, which is the objective of farm irrigation. The application of water may be through surface or

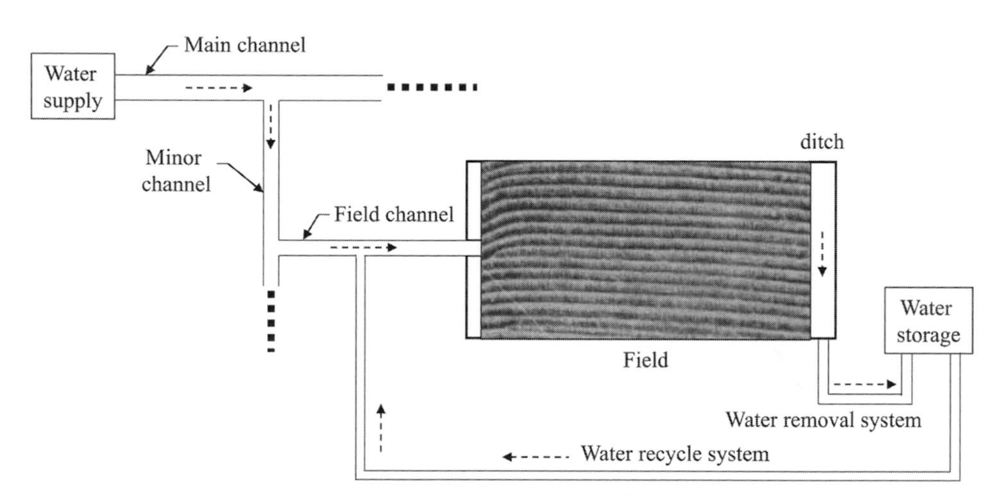

Figure 17.1 Components of a typical irrigation system.

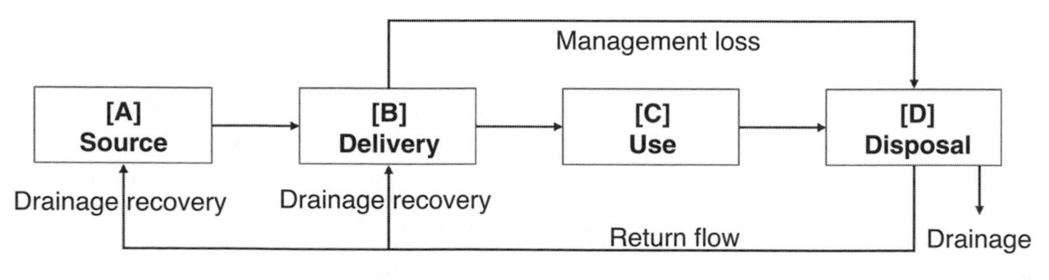

Figure 17.2 Different subsystems and their interconnections.

pressurized methods. The objective of either method is to apply water to the field in the desired amount as uniformly as possible for optimal crop production and with minimum environmental damage.

17.2.4 Water Removal and Reuse Subsystem

The water removal subsystem entails removal and disposal of surface and subsurface water from the field to facilitate agricultural operations and promote agricultural productivity. Excess water may be caused by irrigation water or rainwater, and it is important to dispose of it. Removal of excess water is essential to lower the groundwater table and provide proper root aeration and maintain an appropriate salinity level in the soil.

The excess water may be collected in a reservoir or ditch and can be reused. The reuse subsystem has several advantages. First, it allows higher rates of application that increase the uniformity of application. Second, it saves excess water, which can be reused. Third, fertilizers, pesticides, and salts that wash from the field can be intercepted and prevented from entering the receiving stream. Measurement devices are used to measure the amount of water applied for irrigation. Turnouts are used to release water at the head of the border or furrow.

17.3 IRRIGATION METHODS

There are four basic methods of water application by irrigation: (1) surface, (2) sprinkler, (3) trickle, and (4) below-surface. A short preview of each method is given here and a more detailed discussion of the first three methods will follow separately in succeeding chapters.

As the name suggests, in surface methods water is applied on the ground and flows under gravity. Border, basin, and furrow methods (Figure 17.3) are the main surface methods for water application. Furrows are normally used for row crops or crops on beds, such as potatoes, vegetables, and fruit crops. Borders are used for closely growing crops, such as peas, sugarcane, wheat, maize, and mustard. Surface methods are used for flat or gently sloping fields. Basin irrigation is like border irrigation, but the fields are level with perimeter dikes. This is a very common method of irrigation in developing countries, especially for rice cultivation.

Sprinklers spray water through the air like rain and can therefore be used for any landscape. Since sprinklers are expensive, they are not as popular in developing countries as in the developed world. In trickle irrigation, water is applied through emitters to individual plants and involves an extensive network of pipes, as shown in Figure 17.4. This is also expensive and is usually used for fruit and vegetable crops.

Below-surface irrigation is distinguished as sub-irrigation and subsurface irrigation. The sub-irrigation method supplies water to the root zone by artificially regulating the groundwater table elevation. Check dams and gates are used to maintain the water level in the soil from open ditches, as shown in Figure 17.5, or from jointed or perforated pipes, as

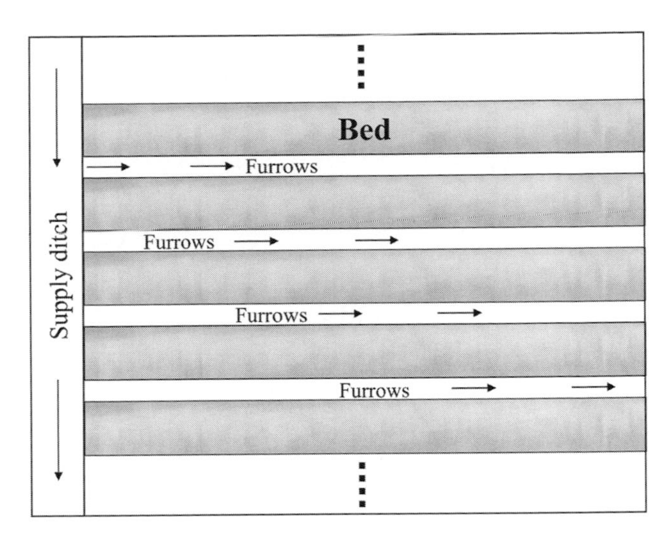

Figure 17.3 Furrow irrigation.

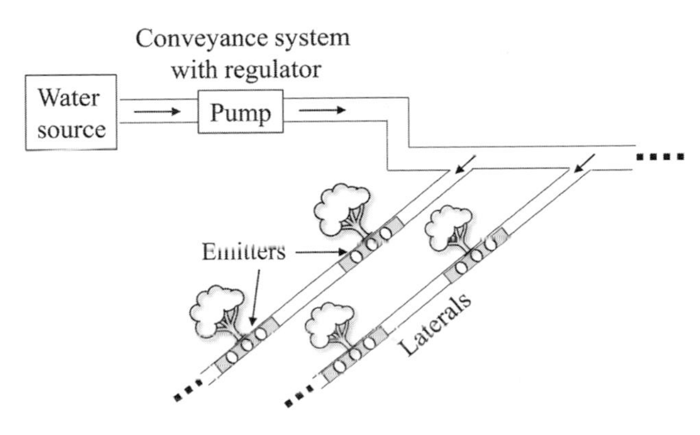

Figure 17.4 Trickle Irrigation.

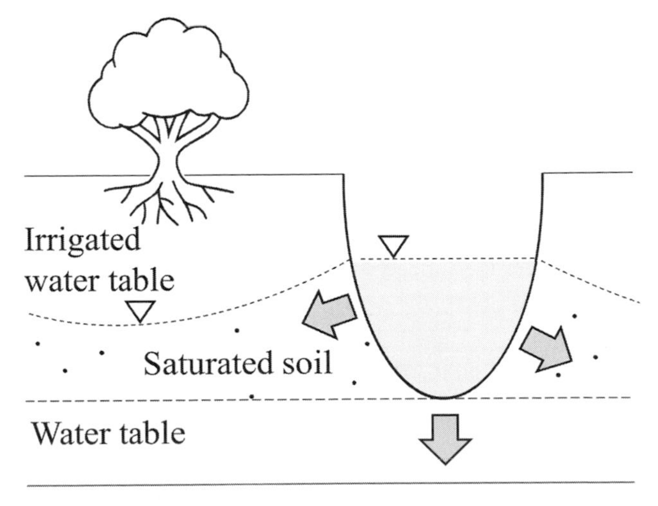

Figure 17.5 Sub-irrigation by the open-ditch method.

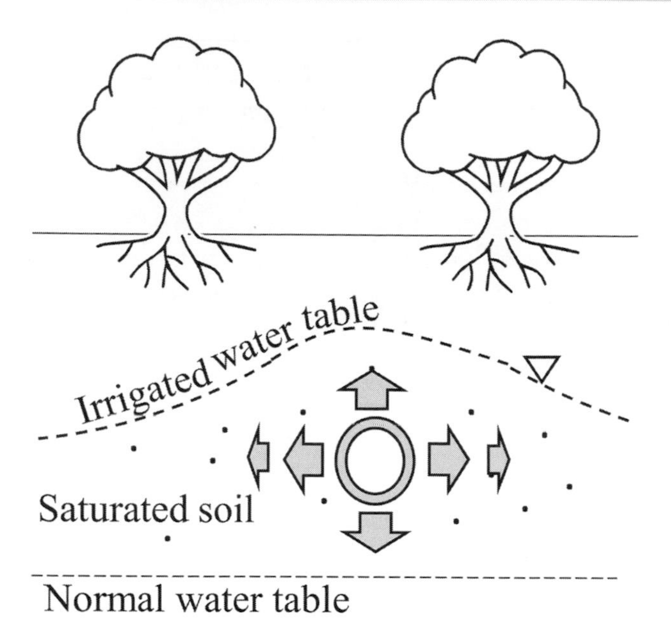

Figure 17.6 Subsurface irrigation by the underground perforated pipe method.

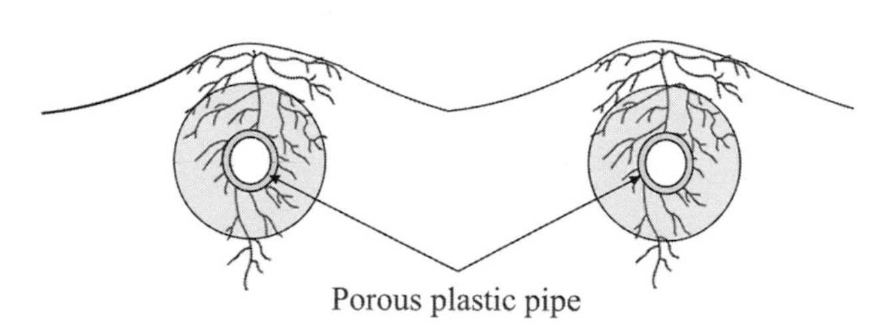

Figure 17.7 Subsurface irrigation with water directed into the plant root zone by a small perforated pipe.

shown in Figure 17.6, below the root zone (30–91 cm or 12–36 in.). In this method, the water level can be lowered or raised, depending on the crop root zone.

The subsurface method applies water below the surface by perforated or porous plastic pipes, as shown in Figure 17.7. It is similar to the trickle irrigation method, but the pipe is placed under the surface.

17.4 SURFACE IRRIGATION METHODS

Some irrigation methods can be classified based on land slope, because some methods are designed for level land and some for land with some slope. Level land is considered as land having a less than 0.1% slope (10 cm per 100 m or 1.2 in. per 100 ft), whereas sloping land is considered as land with a slope of 0.1–15% (10 cm to 15 m per 100 m; 1.2–18 in. per 100 ft). Thus, surface irrigation methods can be considered as level systems or graded systems.

17.4.1 Level Systems

Level systems can be classified as level border or basin, contour levee, or level furrow. A level border or basin is made up of a level area enclosed by earthen ridges, dikes, or levees. This type of system is very popular in developing countries like India, where land holdings are very small. A predetermined amount of water is released into the strip and water is allowed to stand until it is absorbed. The rate at which water is released into the strip is at least twice as much as the soil can absorb. A general rule of thumb is that at least 1.7 m^3/min (1 ft^3/s) is available for each 8 ha (20 acres) strip, but 6.7 m^3/min (4 ft^3/s) may be better. Another consideration for constructing a strip is that the total elevation drop from the head to the tail end is not more than half of the depth of water to be applied. This will increase the speed of water on the strip.

The difference between a contour levee and a level border is that the contour levee is adjusted to the sloping land and hence the field is enclosed by the levee on the contour. The strips are leveled and are rectangular fields.

The level furrow system is similar to the level border system, except that in a field there are several level furrows. The size and shape of a furrow depends on the crop to be grown, equipment to be used, and the spacing between furrows. About 15% more water is applied than is actually needed to get enough water on the strip.

17.4.2 Graded Systems

Graded systems can be classified as graded border, contour ditch, graded furrow, corrugation, and contour furrow. A graded border has some slope in the direction of slope and can be laid out on contours on sloping fields by bench leveling.

A contour ditch is constructed on sloping land after the crop is planted. The spacing between contour ditches depends on the slope, intake rate, and the amount of water to be applied, and varies from 24 to 914 m (80–3000 ft). In a contour ditch system, irrigation is done from the first strip to the next.

The graded furrow system comprises small channels between rows and beds which uniformly slope in the direction of flow. These are constructed straight down the field and are similar to level furrows except for the grade. The corrugation system is a small graded furrow method used for sodded or sown fields. The contour furrow system is similar to the graded furrow system, except that furrows are laid out across the slope on a uniform slope. The Natural Resources Conservation Service of the US Department of Agriculture (USDA-NRCS, 2012) has provided a general comparison of surface irrigation methods, as shown in Table 17.1.

Table 17.1 *General comparison of surface irrigation methods*

Selection criteria	Furrow irrigation	Border irrigation	Basin irrigation
Necessary development costs	Low	Moderate to high	High
Most appropriate field geometry	Rectangular	Rectangular	Variable
Amount and skill of labor required	High labor and high skill required	Moderate labor and high skill required	Low labor and moderate skill required
Land leveling and smoothing	Minimal required but needed for high efficiency Smoothing needed regularly	Moderate initial investment and regular smoothing is critical	Extensive land leveling required initially, but smoothing is less critical if done periodically
Soils	Coarse- to moderate-textured soils	Moderate- to fine-textured soils	Moderate- to fine-textured soils
Crops	Row crops	Row/solid-stand crops	Solid-stand crops
Water supply	Low discharge, long duration, frequent supply	Moderately high discharge, short duration, infrequent supply	High discharge, short duration, infrequent supply
Climate	All, but better in low rainfall	All, but better in low to moderate rainfall	All
Principal risk	Erosion	Scalding	Scalding
Efficiency and uniformity	Relatively low to moderate	High with blocked ends	High

Adapted from USDA-NRCS, 2012.

17.5 CHOICE OF AN IRRIGATION METHOD

The choice of a particular method of irrigation depends on the specific needs of a farmer and the methods that best suit the local conditions. Each method has its advantages and disadvantages. However, there are broad guidelines that can be utilized to select the method that is best for the farmer. The suitability of a particular method depends on a number of factors, including natural conditions (including climate, soil characteristics, field slope and slope variability, water quality, supply and availability – amount, duration, and timing of delivery); crop type and cropping pattern; cultural factors; field geometry – sizes, shapes, and land leveling; type of technology; historical practices; required labor input; and costs of system and its appurtenances and net benefit. The cultural factor is often the deciding factor. The crop to be irrigated may determine the method of irrigation. For example, basin irrigation may be the logical choice for paddy rice. Sometimes the ground slope may determine the method of irrigation. For example, for wheat on a highly undulating terrain, sprinkler irrigation may be the preferred method. For grapes, drip irrigation may be the preferred method.

17.5.1 Natural Factors

If the climate is very windy, then sprinkler irrigation may not be suitable; instead drip or surface irrigation methods may be preferred. For supplementary irrigation, sprinkler or drip irrigation may be selected because they are more flexible and can easily be adjusted for varying irrigation demands.

Soil texture and structure determine the storage capacity and infiltration rate, which in turn point to a method of irrigation that will be suitable for the particular soil. For example, surface irrigation is ideal for clay soils, which have low infiltration rates (less than 13 mm/h (0.5 in./h). Surface, drip, and sprinkler irrigation methods are suitable for loam and clay soils (13–76 mm/h, or 0.5–3.0 in./h of infiltration). For soils with high intake rates (>76 mm/h, or 3.0 in./h) the choice of method may be limited to drip, sprinkler, or subsurface. If the soil type widely varies on a farm, then a drip or sprinkler irrigation method is preferable because that will lead to a uniform distribution of water.

The water-holding capacity of the soil influences the method of irrigation. In general, coarse-textured soils hold more water than fine-textured soils. This means that for fine-textured soils water must be applied more frequently.

The ground slope is an important determinant for selecting an irrigation method. If the ground is undulating or steeply sloping then surface irrigation is not viable and drip or sprinkler irrigation is preferred. On steep slopes, terraces can be employed and rice can be grown. Land leveling can be used to change the land surface, but it is always expensive, so when choosing an irrigation method it is important to consider whether leveling is needed.

Field geometry and topography are often the determining factors. In developing countries, field sizes are very small and staple crops are grown for meeting daily needs. Topography is kept as is and is divided into small plots for cultivation, allowing for greater uniformity.

The availability and quality of water may determine the method of irrigation. If the amount of water available is limited, then drip or sprinkler irrigation may be the only option that uses water efficiently. If water is sufficiently available, then border, basin, or furrow irrigation may be used. However, the efficiency of surface irrigation can also be enhanced. If the water has too much sediment, then surface irrigation is preferable because sediment will clog drips or sprinklers. If the water contains dissolved salts, then drip irrigation is more suitable because it uses less water and will hence add less salt to the soil. Sprinkler irrigation leaches out salts more efficiently than do surface irrigation methods. The uniformity of water application is another consideration. Surface irrigation methods are not as efficient as sprinkler or drip irrigation, but with modern technologies they can be made more efficient.

Other important factors are inflow rate and efficiency, both of which are heavily influenced by soil characteristics. For example, low inflow rate will lead to poor uniformity and low efficiency due to slow advance and more time for infiltration. The inflow rate for a particular crop also depends on the crop growth stage. In order to achieve high efficiency and uniformity, inflow rates have to be adjusted. When to terminate the inflow becomes important in deciding the uniformity and efficiency.

These days, mechanization is becoming popular even in developing countries. Farmers are shifting to sprinkler or drip irrigation. Automation is difficult to achieve on smallholdings. However, for larger farms it is a viable option. Automation of delivery systems has been done in many developing countries.

17.5.2 Crop Type

Barring a few exceptions, all irrigation methods can be theoretically used for all crops but cost becomes a determining factor. Drip irrigation can be used for irrigating row crops, such as vegetables and grapes, but is not suitable for close-growing crops, such as wheat, maize, and rice. Sprinkler and drip irrigation methods require large investments so they are often used for high-value cash crops, such as vegetables and fruit trees. However, with declining costs and growing scarcity of water, sprinkler irrigation is now being used for low-value staple crops.

The water tolerance of crops is also an important factor. Different crops have different tolerance levels. For example, potatoes do not tolerate water standing around their roots. Surface irrigation methods may not be suitable. Some crops, such as string beans, develop fungi or disease under high moisture conditions. Sprinkler irrigation causes the presence of water around the leaves, which tends to promote disease and may hence not be recommended. Weed growth and seeding of crops may be impacted by the surface irrigation method.

17.5.3 Type of Technology

Surface irrigation methods are simple and need less sophisticated equipment that is easy to construct and maintain, except for pumps. Sprinkler and drip irrigation systems are more complicated and cost more. The maintenance of the equipment needed requires technical know-how. A farmer's choice may therefore be dictated by the type of technology and its cost.

17.5.4 Irrigation Practice

Sometimes the tradition of irrigation practice within a region or on a farm influences the choice of irrigation method. The farmer may feel at home using what has been used before and may be reluctant to use an unknown method. He has to consider the cost and life of equipment, maintenance cost, technical know-how needed, and net benefit.

17.5.5 Labor

Different irrigation methods require different labor inputs. For example, surface irrigation methods require more labor for their construction, operation, and maintenance, but low technical know-how. They may, however, require land leveling, which is done by expensive machinery and is hence expensive. On the other hand, drip and sprinkler irrigation methods cost more but require little leveling, and their operation and maintenance are less labor-intensive.

17.5.6 Cost and Benefit

Cost and benefit are key to choosing an irrigation method. Costs involve the cost of equipment, installation cost, and maintenance and operation costs. Benefits include crop yield, fodder, soil health, environmental quality, reduced loss of productive soil, etc. Unless there is a higher net benefit, it will be difficult for a farmer to justify the choice of a particular method.

An important factor is the operational costs, including costs of water, labor, and maintenance. Thus, the total cost of the system must be considered in choosing a method on a long-term basis. Sprinklers require a high expenditure in the beginning, but operation and maintenance costs may not be high. The same applies to drip irrigation. On the other hand, border and furrow irrigation methods are not capital intensive but require yearly or seasonal maintenance.

The factors that affect the selection of an irrigation method are listed in Table 17.2 (USDA-NRCS, 1997). The Food and Agriculture Organization of the United Nations has given guidelines for selecting an irrigation method based on the depth of net irrigation application, as shown in Table 17.3 (Brouwer et al., 1985).

17.5.7 Selection of a Surface Irrigation Method

There are a number of factors that affect the selection of a surface irrigation method, such as ground slope, soil intake rate, field shape, soil conditions, and crop type. The distribution of water on the field by a surface irrigation method is greatly determined by the method. The USDA

Table 17.2 *Factors affecting the selection of an irrigation method*

	Factors affecting selection			
Irrigation method	Crop	Land and Soil	Water supply	Climate
Sprinkler	Adaptable for use on most crops	Adaptable to both level and sloping ground surfaces Adaptable to any soil intake rate	Can apply water at rates of less than 0.1 in./h (2.5 mm/h) to more than 2 in./h (50.8 mm/h)	Wind drift and evaporation may affect application efficiency
Trickle	Adaptable to many specialty fruits, vegetables, and most crops	Adaptable to both level and sloping ground surfaces Adaptable to any soil intake rate	A low, continuous flow rate is preferred	No effect
Sub-irrigation	Adaptable to most crops	Land must be nearly level and smooth Adaptable to soils with low available water-holding capacity and high intake rates Soil must have either a natural high water table or impermeable layer in the substratum	A low, continuous flow rate is preferred	No effect
Surface irrigation	Adaptable to most crops May be harmful to crops that cannot tolerate prolonged standing water and root crops	Land leveling is generally required to obtain the proper soil slope for uniform water distribution	Flow rate affects application uniformity and efficiency	No effect High winds may affect application efficiency on bare soil

After USDA-NRCS, 1997.

Table 17.3 *Selection of an irrigation method based on the depth of net irrigation application*

Soil type	Rooting depth of the crop	Net irrigation depth per application (mm)	Irrigation method
Sand	Shallow	20–30	Short furrows
	Medium	30–40	Medium furrows, short borders
	Deep	40–50	Long furrows, medium borders, small basins
Loam	Shallow	30–40	Medium furrows, short borders
	Medium	40–50	Long furrows, medium borders, small basins
	Deep	50–60	Long borders, medium basins
Clay	Shallow	40–50	Long furrows, medium borders, small basins
	Medium	50–60	Long borders, medium basins
	Deep	60–70	Large basins

After Brouwer et al., 1985.

(USDA-NRCS, 1997) has enumerated the factors that affect the selection of a surface irrigation system, as shown in Table 17.4.

17.6 SURFACE IRRIGATION DECISION VARIABLES

The objective of irrigation is to apply the required irrigation depth to replenish the soil moisture depletion in the crop root zone uniformly throughout the field while minimizing the loss of water due to deep percolation and runoff. However, it is very difficult to uniformly apply water with high application efficiency, because a number of factors (parameters and variables) affect the performance of surface irrigation. Irrigation parameters are those factors that do not change during the irrigation event or during the season, whereas variables can change during an irrigation event and also during the season. These factors and variables can be grouped into three broad categories: field geometry (length, width, slope, and shape of furrow); field conditions

Table 17.4 *Factors affecting the selection of surface irrigation systems*

	Level		Graded			Contour		
Item	Border/basin	Furrow	Border	Furrow	Corrugation	Levee	Furrow	Ditch
Crop								
Field – close growing	N	N	N	–	N	N	–	N
Field – row	N	N	–	+	–	N	–	N
Vegetable – fresh	–	N	–	+	–	–	N	–
Vegetable – seed	–	N	–	+	–	–	N	–
Orchards, berries, grapes	N	N	N	N	–	–	N	–
Alfalfa hay	N	–	N	–	N	N	–	N
Corn	–	N	–	+	–	–	N	–
Cotton	–	N	–	+	–	–	N	–
Potatoes, sugar beets	–	N	–	+	–	–	N	–
Land and soil								
Low available water capacity	N	N	N	N	N	N	N	–
Low infiltration rate	+	+	N	N	N	+	N	N
Moderate infiltration rate	N	N	N	N	N	N	N	N
High infiltration rate	–	–	–	+	–	–	–	–
Variable infiltration rate	–	–	–	N	–	–	–	–
High salinity or sodicity	+	+	N	+	–	–	–	N
Highly erodible	–	–	–	–	–	–	–	–
Undulating topography	–	–	–	–	–	–	–	–
Steep topography	–	–	–	–	–	–	–	–
Odd-shaped fields	+	+	–	–	–	N	N	N
Obstructions	–	–	–	–	–	–	–	–
Stony, cobbly	–	–	–	–	–	–	–	–
Water supply								
Low, continuous flow rate	–	–	–	N	N	–	N	N
High, intermittent flow rate	+	+	+	–	–	–	–	N
High salinity	+	+	N	N	N	N	–	N
High sediment content	N	N	N	N	N	N	N	N
Delivery schedule								
continuous	–	–	N	N	N	N	N	N
rotation	+	+	N	N	N	N	N	N
arranged, flexible	+	+	N	N	N	N	N	N
demand	N	N	N	N	N	N	N	N
Climate								
Humid and sub-humid	–	–	–	–	–	–	–	–
Arid and semi-arid	N	N	N	N	N	N	N	N
Windy	N	N	N	N	N	N	N	N
High temperature – humid	N	N	N	N	N	N	N	N
High temperature – arid	N	N	N	N	N	N	N	N
Social/institutional								
Easy to manage	N	+	–	–	–	–	–	–
Automation potential	+	+	N	+	–	N	N	–

\+ indicates positive effects or preferred selection; – indicates negative effects or another method or system should be considered; N indicates neutral effect or no influence on selection.
After USDA-NRCS, 1997.

(infiltration and roughness); and management variables (inflow rate, cutoff time, and irrigation requirement). The combined effect of these parameters and variables is assessed using irrigation performance measures (efficiency, adequacy, and uniformity).

17.6.1 Field Geometry

Field geometry (boundary) is often fixed, which limits the length of furrows, basin, or borders. However, in some cases the length can be considered as a design variable if the available inflow supply is inadequate to complete advance or reduce the irrigation set time. The width of basins and borders is often dictated by the machinery width. Furrow spacing depends on agronomic considerations and furrow shape depends on the available farm equipment and local practices.

Field slope significantly affects the performance of surface irrigation. The field slope can be modified by land leveling, but it is expensive. In the design of graded borders and furrows, too high field slopes should not be used in order to prevent soil erosion. Likewise, too low slopes should not be used either, to avoid slow water advance. Basins are level and thus have no slope either in the longitudinal or transverse direction. The longitudinal slope in borders usually ranges between 0.5% and 1%, but it may go up to 4% if sod crops are grown (Jurriens et al. 2001). The furrow irrigation system works well with longitudinal slopes of 0.05–3%, and transverse slopes of 0.5–1.5%. However, the flow velocity in graded furrows should not exceed the maximum non-erosive velocity (V_{max}), which ranges from 8 m/min for erosive silt soils to about 13 m/min for the more stable clay and sandy soils (Walker and Skogerboe, 1987).

17.6.2 Field Conditions

Flow resistance is generally considered using Manning's roughness coefficient (n). The roughness characteristics not only vary in space but also vary during the season. Furthermore, roughness is influenced by the growth of vegetation, surface sealing, tillage operation, and flow geometry. The surface roughness affects flow velocity and thereby water advance, infiltrated depth, and overall irrigation performance. For design of furrow irrigation, an n (Manning's roughness coefficient) value of 0.04 is generally used. Recommended roughness values for the design of surface irrigation systems are given in Table 17.5.

Table 17.5 *Recommended Manning's n values for design of surface irrigation systems*

n value	Field conditions	Irrigation methods
0.03–0.05	Freshly constructed furrows; 0.04 is used for first irrigation and 0.02 is used for later irrigations	Furrow
0.04	Bare soil	Basin and border
0.10	Small grains (drilled lengthwise)	Basin and border
0.15	Alfalfa, mint, broadcast small grains	Basin and border
0.20	Dense alfalfa or alfalfa on long fields	Basin and border
0.25	Dense crops or small grain drilled crosswise	Basin and border

Adapted from USDA-NRCS, 2012.

Infiltration controls advance, percolation, runoff, and recession, and thus affects the performance of surface irrigation and is a basic design variable. For a given soil, however, infiltration characteristics are not constant and vary in both space and time. Therefore, irrigation systems should be designed with field representative infiltration characteristics. The Soil Conservation Service (USDA-SCS, 1974) has developed intake families and provided guidelines for choosing infiltration characteristics based on soil texture. Several infiltration equations have been developed, as discussed in Chapter 12. Among these infiltration equations, the Kostiakov and the modified Kostiakov (Kostiakov–Lewis) equations are most commonly used for the design of surface irrigation systems:

$$Z = k\tau^a, \tag{17.1}$$

$$Z = k\tau^a + f_0\tau, \tag{17.2}$$

where Z is the cumulative infiltration volume (m^3/unit area), k is a parameter (m/min^a), and a is a fitted parameter, τ is the intake opportunity time (min), and f_0 is the basic infiltration rate (m/min).

The SCS intake family infiltration function is similar to eq. (17.2), but the second term is represented by a constant. USDA-NRCS (2012) provided values of parameters (k, a, and f_0) of the modified Kostiakov infiltration equation corresponding to the SCS intake families for the first and later irrigations. These values are presented in Tables 19.7 and 19.8.

17.6.3 Management Variables

The required irrigation depth, inflow rate, and cutoff time are management decision variables, along with field dimensions to a limited extent. The required irrigation depth can be determined using irrigation scheduling. Once the decision to irrigate a field is made or irrigation requirement is known, the main task is to irrigate the field by choosing a suitable combination of the inflow rate and cutoff time to obtain better irrigation performance within the existing constraints. However, among all irrigation variables the inflow rate and time of cutoff offer the most flexibility to a decision-maker. This flexibility in inflow rate and cutoff time is related to the delivery system. For example, in a warabandi system each piece of the land receives water for a fixed amount of time at almost a fixed rate on a weekly rotation basis. Thus, there is almost no choice in either selecting the flow rate or cutoff time. On the other hand, a demand system (also groundwater pumping) gives complete choice to the irrigator to choose inflow rate and cutoff time.

The cutoff time is measured from the beginning of irrigation to the termination of inflow, and its ideal value is when the infiltrated depth equals the irrigation requirement in the least-watered portion of the field. Cutoff influences deep percolation, surface runoff, and the amount of losses and hence irrigation efficiency and adequacy.

Inflow affects the rate of advance, uniformity, efficiency, and adequacy of irrigation and is hence a key variable. If it is too high it will cause erosion; if it is too low the water will not advance to the downstream end. Thus, it affects the net result of an irrigation event.

The amount of water to be applied should be the amount of the water that needs to be stored in the crop root zone for normal crop growth. The crop root zone depends on the type of crop, stage of growth, depth to the water table, and soil type. The amount of water that can be stored per unit depth of soil depends on the soil type. These factors together with climatic conditions determine the amount of water that should be applied per irrigation.

17.7 HYDRAULICS OF SURFACE IRRIGATION

During surface irrigation, the hydraulics of flow changes with time and space. As a result, different flow regimes develop at different times and spaces. However, the dominating flow regime that prevails over a majority of time can be characterized by gradually varied unsteady free surface flow. A typical change in water surface profile over a small time interval during the water advance is shown in Figure 17.8. The flow is unsteady because the flow rate and depth at each point increase with time due to the time-dependent intake characteristics of soil, and it is nonuniform because both flow rate and depth decrease gradually down the field.

The state of flow is likely to be transitional or laminar, with Reynolds number well below 1000. Further, the Froude number is generally well below unity and thus flow remains subcritical. In addition, other flow regimes also occur, particularly just behind the wetting front during advance, just ahead of the drying front during recession, and near certain boundaries. The flow of water over the soil surface during surface irrigation is spatially varied unsteady flow and can be characterized by the well-known St. Venant equations of continuity and momentum (Ram et al., 1986a,b). These equations can be expressed as:

$$\frac{\partial Q}{\partial x} + \frac{\partial A}{\partial t} + I = 0, \tag{17.3}$$

$$\frac{\partial V}{\partial t} + V\frac{\partial V}{\partial x} + g\frac{\partial y}{\partial x} = g(S_0 - S_f) + \frac{VI}{2A}, \tag{17.4}$$

where Q is the flow rate (L^3/T or $L^3/T/L$) in border irrigation, A is the cross-sectional area of flow (L^2), I is the volume rate of infiltration per unit length of the channel or border (L/T), V is the average velocity in the flow cross-section (L/T), g is the gravitational acceleration (or ratio of weight to mass) (L/T^2), y is the flow depth (L), S_0 is the channel bottom slope, S_f is the channel friction slope, x is the distance in the direction of flow (L), and t is time (T).

In eq. (17.4), $g\partial y/\partial x$ is the unbalanced hydrostatic pressure force on the surface water, gS_0 is the component of the gravitational force in the direction of flow, S_f is the slope of the energy grade line, $\partial V/\partial t$ is the local acceleration (a measure of unsteadiness), $V\partial V/\partial x$ is the convective acceleration (a measure of nonuniformity), and $(VI)/(2A)$ is the net acceleration stemming from the removal of zero-velocity components of the surface stream at the bed by infiltration (Strelkoff, 1969; Bassett et al., 1980; Sherman and Singh, 1978, 1982). Equations (17.3) and (17.4) are based on the following assumptions: (1) the fluid is incompressible, i.e., the density of water is constant; (2) the flow is one-dimensional; (3) pressure is hydrostatic; (4) the streamline curvature is small; and (5) the bottom slope of the channel is small.

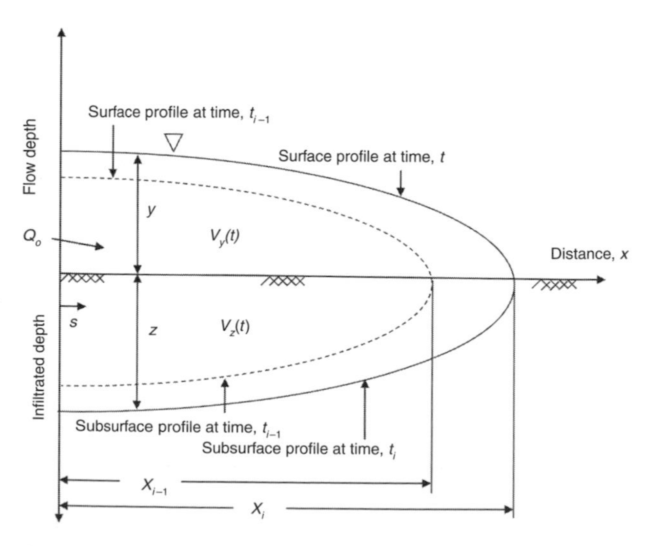

Figure 17.8 Surface and subsurface profiles at a short time increment during irrigation.

17.7.1 Hydraulic Surface Irrigation Models

Surface irrigation models, based on eqs (17.3) and (17.4), are known as full hydrodynamic models. These equations can only be solved numerically. Several investigators modeled surface irrigation by using the full hydrodynamic equations. Although these full hydrodynamic models are expected to be more accurate, they require more information that may not be normally available.

Over the years, a number of surface irrigation models have been developed based on simplified forms of the St. Venant equations, such as the zero-inertia or kinematic-wave models. The simplification is actually done by simplifying eq. (17.4). As the name suggests, the zero-inertia simplification neglects the inertial and acceleration terms in eq. (17.4), resulting in:

$$\frac{\partial y}{\partial x} = S_0 - S_f. \tag{17.5}$$

This simplification is reasonable, because flow velocities during most surface irrigation events are low and hence accelerations are small and can be neglected. Furthermore, local and convective acceleration terms are of the same order of magnitude but of opposite sign. Surface irrigation models based on eqs (17.3) and (17.5) are called zero-inertia or diffusion-wave models. The zero-inertia models have been found to yield results as accurate as full hydrodynamic models (Zerihun and Junna, 1996).

A further simplification is obtained by neglecting the water depth gradient term in eq. (17.5) that reduces the momentum equation to

$$S_0 = S_f. \tag{17.6}$$

This simplification is called kinematic-wave approximation and is reasonable if the bottom slope is sufficiently steep. When S_0 is large enough, the water wave is sufficiently long and flat and thus $\partial y/\partial x$, $V\partial V/\partial x$, and $\partial V/\partial t$ are small compared to S_0. Equations (17.3) and (17.6) constitute the basis of kinematic-wave models. Equation (17.6) can be expressed as a relationship between depth and discharge, such as Manning's or Chezy's equation (Seyedzadeh et al., 2019) . The kinematic-wave approximation is also known as normal depth approximation if the relationship between depth and discharge is based on the normal depth. The kinematic-wave models have limited application to sloping and free-draining conditions. Thus, kinematic-wave solutions may not be suitable for borders with zero or small slope or blocked borders. Over the years, a number of kinematic-wave models have been developed for border and furrow irrigation (Sherman and Singh 1978, 1982; Ram et al., 1983; Singh and Prasad, 1983; Singh and Ram, 1983a–c, 1984; Singh and Sherman, 1983; Ram and Singh, 1985; Jain and Singh, 1989; Singh et al., 1990). These studies have shown that kinematic-wave models accurately simulate surface irrigation processes for steep slopes.

As shown in Figure 17.9, the general surface irrigation process includes four phases: advance phase, storage phase, depletion phase, and recession. As the water is applied upstream of the field, it begins to advance downward (Figure 17.9a). As the water advances, the field is divided into a dry part and a wet part at any point in time. The dividing line is a moving one and is called an advance front. The rate of the advance of the water front depends on the inflow rate, soil infiltration rate, surface roughness, and slope. To complete the advance phase, the inflow rate must be greater than the soil infiltration rate. The time between the start of irrigation and the time when the advance front reaches the end of the field is called the advance time. The advance time at any location is the time elapsed between the beginning of irrigation and the arrival of the advance front at that point. When adequate irrigation depth at the downstream end is achieved, inflow is cutoff.

The storage phase, also called the ponding or soaking phase, begins with the time elapsed between the arrival of the water front at the downstream end and the cutoff of inflow at the upstream end. Surface storage continues to contribute to both infiltration and runoff. With the elapsed time, the depth of storage decreases, and a time comes when the water front starts disappearing from the soil surface. The depletion phase, or lag time phase, begins with the time when inflow is cutoff and the time when the depth of water at the upstream end reaches zero.

The recession phase begins with the appearance of the first bare soil that was under water. The recession normally starts at the upstream end and progresses downward. The phase ends when the soil at the downstream end also has zero depth. In reality, the recession phase may start at the upstream end,

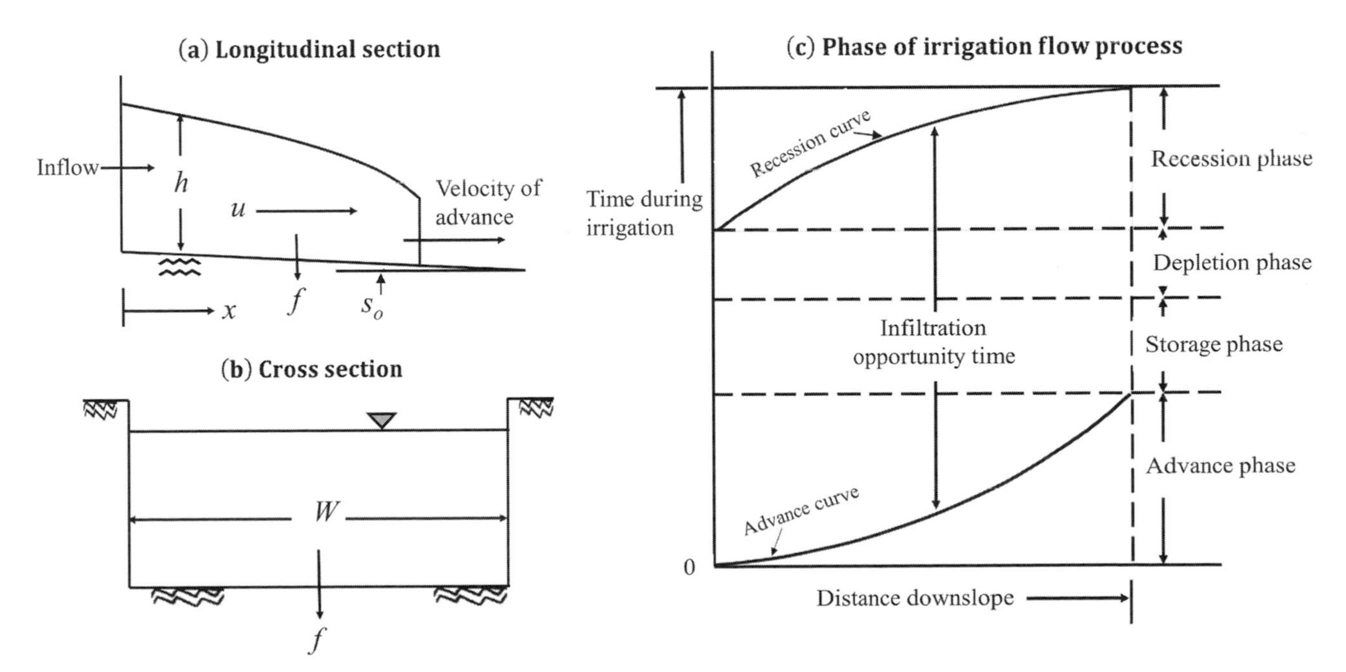

Figure 17.9 Surface irrigation process. (After Sherman and Singh, 1978.)

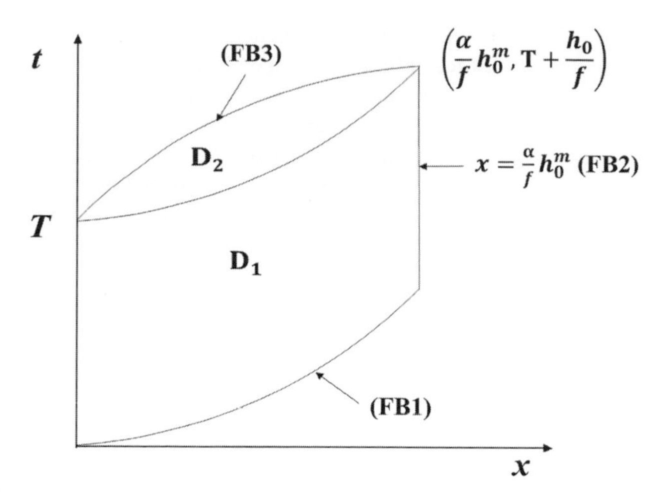

Figure 17.10 Solution domain for the case when f = const. > 0, $h_0(t)$ = const., $t > 0$, $0 \le t \le T$, and α is constant. (After Sherman and Singh, 1978.)

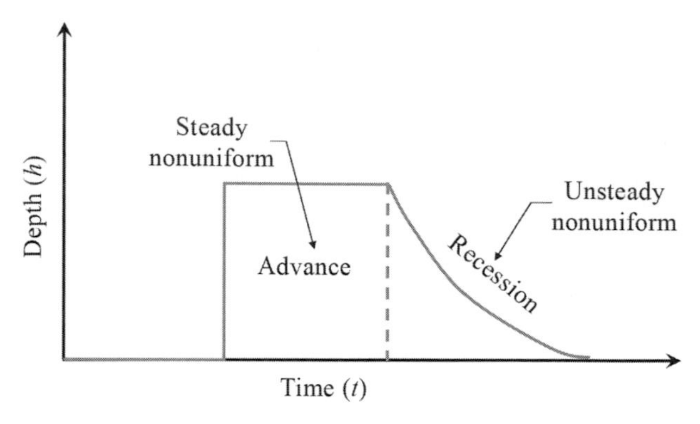

Figure 17.11 Flow depth change during advance and recession function at certain position (x).

downstream end, or both, or anywhere along the field. The recession phase completes when there is no visible standing water on the soil surface. The recession time at any point in the field is the time when the water front completely recedes and bare soil becomes visible. A graphical representation between recession distance and time is known as the recession curve, as shown in Figure 17.9c.

These four phases may occur sequentially, but if the inflow stream is stopped before the advance phase is complete, the storage or even the depletion phase could be eliminated, and the recession phase could occur concurrently with the advance phase. In Figure 17.9c, the infiltration opportunity time is the difference between the advance and recession curves. At any point in the field, the infiltration opportunity time represents the period during which water was available for infiltration.

Sherman and Singh (1978) derived analytical expressions for surface irrigation, sketched in Figure 17.9. Assuming a constant inflow of water at the head ($x=0$) of duration T entering a border or furrow with a constant rate of infiltration f(m/min) and kinematic-wave roughness parameter α and exponent m, the solution domain can be sketched as in Figure 17.10. The advance function is given as:

$$x=\frac{\alpha}{f}\left[h_0^m-\left(h_0-\frac{ft}{m}\right)^m\right], \tag{17.7}$$

where h_0 is the flow depth at the head. By defining the friction slope as parallel to the bed slope (i.e., a uniform flow assumption), eq. (17.6) can be expressed as a relationship between depth and discharge (i.e., $q=\alpha h^m$), where q is the rate of flow in units of $L^3/T/L$ in border irrigation. With $S_0=S_f$, Manning's equation can be written as $q=\frac{1}{n}R^{\frac{5}{3}}S_0^{\frac{1}{2}}$, where R is the hydraulic radius and n is Manning's roughness coefficient. In border irrigation the hydraulic radius (R) can be roughly considered equal to the water depth (h) (i.e., $R=h$), considering that the width of flow is usually more than five times the depth (Seyedzadeh et al., 2019). Combing these two equations, the kinematic roughness parameter α (dimension: $L^{1/3}/T$ where T is in seconds) and the exponent m can be expressed as

$$\alpha=\frac{S_0^{\frac{1}{2}}}{n};m=5/3. \tag{17.8}$$

The depth at any position x on the border to the left of the advance curve is given as

$$h(x,t)=\left(h_0^m-\frac{fx}{\alpha}\right)^{\frac{1}{m}}. \tag{17.9}$$

The recession curve is given as

$$x=\alpha f^{m-1}(t-T)^m. \tag{17.10}$$

The depth of flow to the right of the recession curve is given as

$$x=\frac{\alpha}{f}\{[h+f(t-T)]^m-h^m\}. \tag{17.11}$$

The location of the front wall is given by

$$x=\alpha\frac{h_0^m}{f}. \tag{17.12}$$

The flow depth in terms of times (t) with fixed position (x) is shown in Figure 17.11. During the advance phases, the flow is steady and nonuniform. The flow depth is 0 m from $t=0$ to the advance time of the fixed position (x). Then the depth suddenly reaches the given flow depth at the head. It remains the same depth during the irrigation time. Then, after the irrigation time, the flow depth begins to decrease along the t-axis and the flow is unsteady and nonuniform.

Example 17.1: Let the flow depth at the head of a 300-m long field be 0.25 m, the average infiltration rate be 0.25 cm/min, and the duration of irrigation be 5 h. Manning's roughness coefficient (n) is 0.1 and slope is 0.002. Compute the advance function, recession function, and flow depth as functions of space and time.

Solution:
Step 1: The kinematic-wave roughness parameter is computed using Manning's equation as:

$$\alpha = \frac{S_0^{0.5}}{n} = \frac{(0.002)^{0.5}}{0.1}.$$

It may be noted that α has dimensions of $L^{1/3}/T$, where the length dimension (L) is in meters and the time dimension (T) is in seconds. Since infiltration is given per minute, it may be convenient to convert the time dimension of α to minutes and convert infiltration to meters per minute. Therefore, one can write as

$$\alpha = \frac{(0.002)^{0.5}}{0.1} \times 60 \frac{\text{s}}{\text{min}} = 26.83,$$

$$f = 0.25 \frac{\text{cm}}{\text{min}} = 0.0025 \frac{\text{m}}{\text{min}}.$$

If there is no barrier or bund at the downstream end of the field the advance front will extend over the distance as:

$$x = \alpha \frac{h_0^{\frac{5}{3}}}{0.0025 \frac{\text{m}}{\text{min}}} = (26.83) \frac{(0.25\ \text{m})^{\frac{5}{3}}}{0.0025 \frac{\text{m}}{\text{min}}} = 1065\ \text{m}.$$

Since the field length is given as 300 m, calculations will be done for the length of 300 m only.

Step 2: The advance function is now given as (t in min)

$$x = \frac{\alpha}{f}\left[h_0^m - \left(h_0 - \frac{ft}{m}\right)^m\right]$$
$$= \frac{26.83}{0.0025 \frac{\text{m}}{\text{min}}}\left[0.25^{\frac{5}{3}} - \left(0.25 - \frac{0.0025 \frac{\text{m}}{\text{min}} \times t}{(5/3)}\right)^{\frac{5}{3}}\right].$$

For different values of time, the advance distance is as given in Table 17.6 and is plotted in Figure 17.12. When $t = 10$ min, using eq. (17.7), the advance distance is:

$$\text{x} = \frac{26.83}{0.0025 \frac{\text{m}}{\text{min}}}\left[0.25^{\frac{5}{3}} - \left(0.25 - \frac{0.0025 \frac{\text{m}}{\text{min}} \times 10\ \text{min}}{(5/3)}\right)^{\frac{5}{3}}\right]$$
$$= 104.3\ \text{m}.$$

Now calculate the advance time (t) for $x = 300$ m. It takes 30.01 min to reach the distance of 300 m.

$$300\ \text{m} = \frac{26.83}{0.0025 \frac{\text{m}}{\text{min}}}\left[0.25^{\frac{5}{3}} - \left(0.25 - \frac{0.0025 \frac{\text{m}}{\text{min}} \times t}{(5/3)}\right)^{\frac{5}{3}}\right]$$
$$\rightarrow t = 30.01\ \text{min}.$$

The advance function is used when (or before 30.01 min) the distance reaches 300 m. In this example, we only look at the field domain until 300 m.

Step 3: In domain D_1, that is, to the left of the advance function, the flow depth is given as

$$h(x,t) = \left(h_0^m - \frac{fx}{\alpha}\right)^{\frac{1}{m}} = \left(0.25^{\frac{5}{3}} - \frac{0.0025 \frac{\text{m}}{\text{min}} \times x}{26.83}\right)^{\frac{3}{5}}.$$

For different values of x the flow depth h is given in Table 17.7 and is plotted in Figure 17.13.

Table 17.6 *Calculation for advance function (data shown in Figure 17.12)*

t (min)	x (m)
0.00	0.0
5.00	52.7
10.00	104.3
15.00	154.9
20.00	204.3
25.00	252.7
30.00	299.9
30.01	300.0

Table 17.7 *Calculation for Figure 17.13 (flow depth in domain D_1)*

x (m)	h (m)	x (m)	h (m)
0	0.250	160	0.227
20	0.247	180	0.224
40	0.244	200	0.221
60	0.241	220	0.218
80	0.239	240	0.214
100	0.236	260	0.211
120	0.233	280	0.208
140	0.230	300	0.205

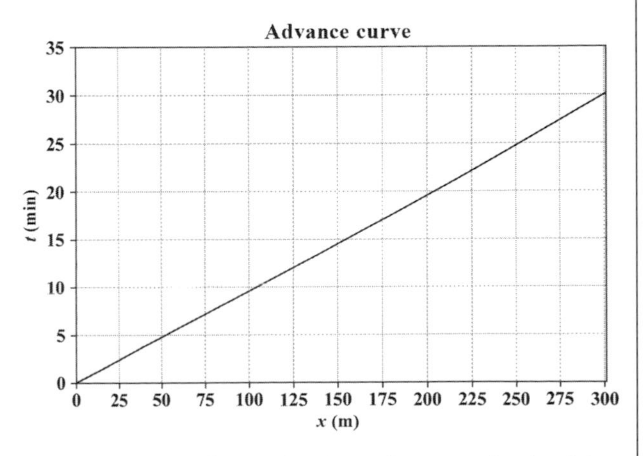

Figure 17.12 Advance function: the advance distance as a function of time.

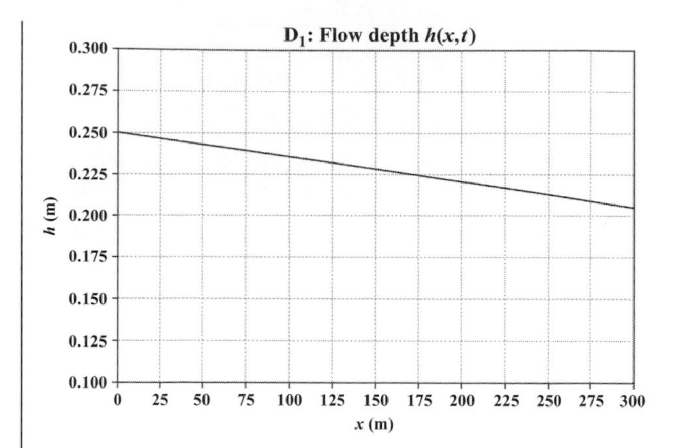

Figure 17.13 Flow depth as function of distance x, $h(x)$.

When $x = 100$ m, the flow depth $h = 0.236$ m using eq. (17.9). As time (t) increases, the flow depth (h) decreases. Likewise, the flow depth at 300 m is 0.205 m.

$$h = \left(0.25^{\frac{5}{3}} - \frac{0.0025\,\frac{\text{m}}{\text{min}} \times 100\ \text{m}}{26.83}\right)^{\frac{3}{5}} = 0.236\ \text{m}.$$

Step 4: The recession function is given as

$$x = \alpha f^{m-1}(t - T)^{m}$$
$$= 26.83 \times \left(0.0025\frac{\text{m}}{\text{min}}\right)^{\frac{5}{3}-1} \times \left(t - 5\ \text{h} \times \frac{60\ \text{min}}{\text{h}}\right)^{\frac{5}{3}}.$$

For different values of time, the recession distance is given in Table 17.8 and is plotted in Figure 17.14. The recession function is used when (or after) time reaches the duration of irrigation (5 h). When $t = 315$ min, the distance is calculated using eq. (17.10):

$$x = 26.83 \times \left(0.0025\frac{\text{m}}{\text{min}}\right)^{\frac{5}{3}-1} \times (315\ \text{min} - 300\ \text{min})^{\frac{5}{3}}$$
$$= 22.94\ \text{m}.$$

Now we determine the duration (t) when recession reaches the distance of 300 m using eq. (17.11):

$$300\ \text{m} = 26.83 \times \left(0.0025\frac{\text{m}}{\text{min}}\right)^{\frac{5}{3}-1} \times (t - 300\ \text{min})^{\frac{5}{3}}.$$

This yields $t = 346.76$ min.

Step 5a: The depth in domain D_2, that is, to the right of the recession function, is given as

$$x = \frac{\alpha}{f}\left([h + f(t - T)]^{m} - h^{m}\right)$$
$$= \frac{26.83}{0.0025\,\frac{\text{m}}{\text{min}}}\left[\{h + 0.0025(t - T \times 60)\}^{\frac{5}{3}} - h^{\frac{5}{3}}\right].$$

For different values of time and space, the flow depth is given in Table 17.9 and is plotted in Figures 17.15 and 17.16.

Table 17.8 *Calculation for Figure 17.14 (recession function)*

x (m)	t (min)
0.00	300.00
7.23	305.00
22.94	310.00
45.09	315.00
72.84	320.00
105.65	325.00
143.16	330.00
185.10	335.00
231.24	340.00
281.39	345.00
300.00	346.76

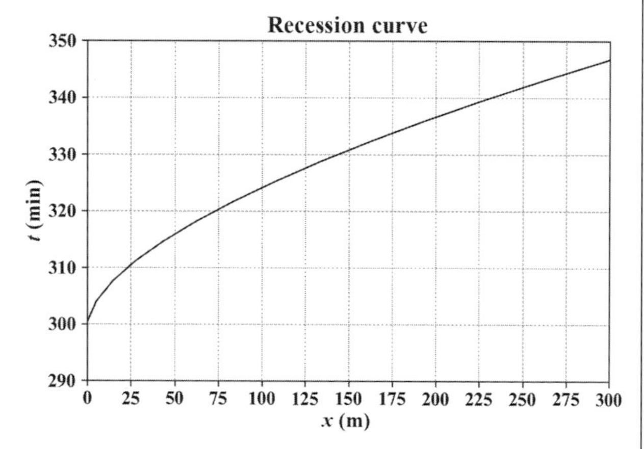

Figure 17.14 Recession function: recession distance as a function of time.

Now we determine how the depth (h) will change in terms of time (t) with space (x). We take $x = 50$ m as an example, and repeat the same steps for $x = 100$, 150, 200, 250, and 300 m. From the advance function (eq. 17.7), the advance time duration is calculated as

$$50\ \text{m} = \frac{26.83}{0.0025\,\frac{\text{m}}{\text{min}}}\left[0.25^{\frac{5}{3}} - \left(0.25 - \frac{0.0025\,\frac{\text{m}}{\text{min}} \times \text{t}}{(5/3)}\right)^{\frac{5}{3}}\right].$$

This yields $t = 4.74$ min.

The depth (h) is 0 m from $t = 0$ min to $t = 4.74$ min. Then, the depth suddenly reaches 0.243 m, which is the flow depth in domain D_1. It remains the same depth (h) for 5 h (300 min). Then, we determine the recession time at $x = 50$ m using eq. (17.10):

$$50\ \text{m} = 26.83 \times \left(0.0025\frac{\text{m}}{\text{min}}\right)^{\frac{5}{3}-1} \times (t - 300\ \text{min})^{\frac{5}{3}}.$$

This results in $t = 315.96$ min.

The flow depth begins to decrease along the t-axis from $t = 300$ min, the depth reaches $h = 0$ m at $t = 315.96$ min. To see the gradual depth changes, we

Table 17.9 *Calculation for Figure 17.15 (fixed space)*

x (m)	t (min)	h (m)	x (m)	t (min)	h (m)	x (m)	t (min)	h (m)
	0	0		0	0		0	0
	4.74	0		**9.55**	0		**14.50**	0
	4.74	**0.243**		**9.55**	**0.236**		**14.50**	**0.228**
50 m	300	0.243	**100 m**	300	0.236	**150 m**	300	0.228
	303	0.224		306	0.220		309	0.216
	307	0.055		312	0.066		315	0.087
	312	0.014		318	0.022		325	0.019
	315.96	0		**324.19**	0		**330.85**	0
	0	0		0	0		0	0
	19.60	0		**24.70**	0		**30.01**	0
200 m	**19.60**	**0.221**	**250 m**	**24.70**	**0.213**	**300 m**	**30.01**	**0.205**
	300	0.221		300	0.213		300	0.205
	312	0.213		315	0.209		318	0.205
	320	0.081		325	0.075		330	0.069
	330	0.021		335	0.022		340	0.021
	336.66	0		**341.92**	0		**346.76**	0

add more time steps between 300 min and 315.96 min. For example, the flow depth (h) is calculated as a function of t for a given distance (x) using eq. (17.11). Determine the flow depth (h) at $t=303, 307,$ and 312 min. For example, when $t=303$ min, $x=50$ m, the depth (h) is

$$50\,\text{m}=\frac{26.83}{0.0025\frac{\text{m}}{\text{min}}}\left\{[h+0.0025(303\text{ min}-300\text{ min})]^{\frac{5}{3}}-h^{\frac{5}{3}}\right\}$$
$$\rightarrow h=0.224\text{ m}.$$

Likewise, when $t=307$ min and $x=50$ m, the depth (h) is

$$50\,\text{m}=\frac{26.83}{0.0025\frac{\text{m}}{\text{min}}}\left\{[h+0.0025(307\text{ min}-300\text{ min})]^{\frac{5}{3}}-h^{\frac{5}{3}}\right\}$$
$$\rightarrow h=0.055\text{ m}.$$

The same steps are applied for $x=100, 150, 200, 250,$ and 300 m.

For fixed space, the flow depth at different times is given in Table 17.9 and is plotted in Figure 17.15.

Step 5b: Determine how the depth (h) will change in terms of space (x) with time (t).

During 5 Hours of Irrigation

We already know from Step 2 the advance time (t) until $x=300$ m; it takes 30.01 min to reach the distance of 300 m. Therefore, the change of depth only depends on the advance function until $t=30.01$ min. We take $t=5$ min as an example and repeat the same steps for $t=15$ and 25 min. From the advance function (eq. 17.7), the distance (x) is calculated as:

$$x=\frac{26.83}{0.0025\frac{\text{m}}{\text{min}}}\left[0.25^{\frac{5}{3}}-\left(0.25-\frac{0.0025\frac{\text{m}}{\text{min}}\times 5\text{ min}}{(5/3)}\right)^{\frac{5}{3}}\right]$$
$$=52.7\text{ m}.$$

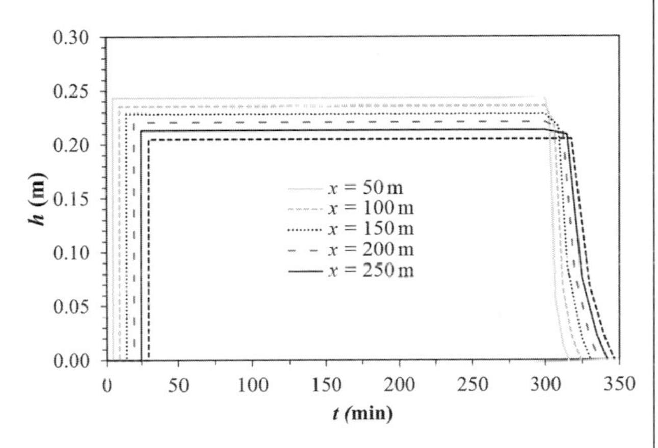

Figure 17.15 Flow depth in terms of time (t) with fixed space (x).

When $t=15$, the depth (h) decreases from 0.25 m as distance (x) increases. Using eq. (17.9), the flow depth (h) is calculated in terms of distance (x). Determine how the flow depth decreases at $x=10, 20, 30, 40,$ and 50 m as:

$$h=\left(0.25^{\frac{5}{3}}-\frac{0.0025\frac{\text{m}}{\text{min}}\times 10\text{ m}}{26.83}\right)^{\frac{3}{5}}=0.249\text{ m }(x=10).$$

From $x=52.7$ m to $x=300$ m, the flow depth (h) remains 0 m. The same steps are applied for $t=15$ and 25 min.

After 5 Hours of Irrigation

After 5 h (300 min) of irrigation, the change of the depth only depends on the recession function from $t=300$ min. We take $t=320$ min as an example and repeat the same steps for $t=330$ min. From the recession function (eq. 17.10), the recession distance ($x=72.84$ m) is calculated when $t=320$ min.

$$x=26.83\times\left(0.0025\frac{\text{m}}{\text{min}}\right)^{\frac{5}{3}-1}\times(320\text{ min}-300\text{ min})^{\frac{5}{3}}$$
$$=72.84\text{ m}.$$

Table 17.10 *Calculation for Figure 17.16 (fixed time)*

t (min)	x (m)	h (m)	t (min)	x (m)	h (m)	t (min)	x (m)	h (m)
t = 5 min	0	0.250	t = 15 min	0	0.25	t = 25 min	0	0.25
	10	0.249		20	0.247		60	0.241
	20	0.247		40	0.244		120	0.233
	30	0.246		60	0.241		170	0.225
	40	0.244		80	0.239		220	0.218
	50	0.243		100	0.236		250	0.213
	52.7	0.243		**154.9**	0.227		**252.7**	0.212
	52.7	0.000		**154.9**	0		**252.7**	0
	300	0.000		300	0		300	0
t = 320 min	0	0	t = 330 min	0	0			
	72.84	**0**		**143.16**	**0**			
	100	0.013		150	0.002			
	150	0.044		175	0.011			
	200	0.081		200	0.021			
	225	0.101		225	0.032			
	250	0.123		250	0.044			
	275	0.146		275	0.056			
	300	0.169		300	0.069			

When $t = 320$ min, the depth (h) increases from 0 m as distance (x) increases (when $x \geq 72.84$ m). Using eq. (17.10), the flow depth (h) is calculated in terms of distance (x). We determine the flow depth (h) at $x = 100, 150, 200, 225, 250, 275,$ and 300 m. For example, when $x = 100$ m, $t = 320$ min, the depth (h) is

$$100\text{ m} = \frac{26.83}{0.0025\frac{\text{m}}{\text{min}}}\left\{[h + 0.0025 \times (320\text{ min} - 300\text{ min})]^{\frac{5}{3}} - h^{\frac{5}{3}}\right\}$$

$$\rightarrow h = 0.013\text{ m}.$$

When $x = 300$ m, $t = 320$ min, the flow depth (h) is

$$300\text{ m} = \frac{26.83}{0.0025\frac{\text{m}}{\text{min}}}\left\{[h + 0.0025 \times (320\text{ min} - 300\text{ min})]^{\frac{5}{3}} - h^{\frac{5}{3}}\right\}$$

$$\rightarrow h = 0.169\text{ m}.$$

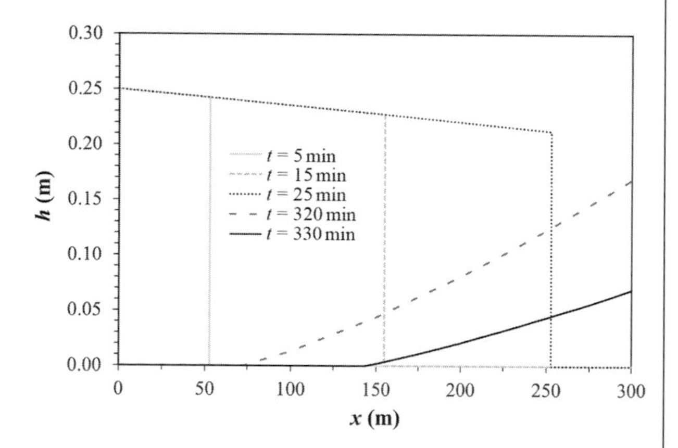

Figure 17.16 Flow depth in terms of distance (x) with fixed time (t).

The same steps are applied for $t = 330$ min.

For fixed time, the flow depth at different distances is given in Table 17.10 and is plotted in Figure 17.16.

17.7.2 Volume Balance Irrigation Models

Another approach for surface irrigation modeling is one that is entirely based on the volume balance or mass conservation only. This approach is called the volume balance approach, and it mainly differs from other approaches in which the volume balance equation is applied to the entire flow profile over a distance x at irrigation time t rather than to a differential element (Ram and Singh, 1982a, b; Singh and Yu, 1987a–c, 1988a,b, 1989a,b; Singh and He, 1988; Singh et al., 1988; Yu and Singh, 1990a,b). The volume of water applied during any time t equals the sum of volume of water stored over the surface and volume of water infiltrated, as illustrated in Figure 17.8. The volume balance equation for any time (t) can be expressed as:

$$Q_0 t = V_y(t) + V_z(t) \qquad t < T_a, \tag{17.13}$$

where Q_0 is the steady inflow rate (m^3/min), t is the time since the beginning of irrigation (min), T_a is the advance time (min), V_y is the volume of surface storage at time t (m^3), and V_z is the volume of infiltrated water at time t (m^3).

The volume of surface storage at any time t over the advance distance can be determined by integrating the flow area as:

$$V_y(t) = \int_0^x A(s,t)\,ds = \overline{A}x = \sigma_y A_0 x, \tag{17.14}$$

where A is the cross-sectional area of flow; $\overline{A}$ is the average flow cross-sectional area; σ_y is the surface shape factor (varying between 0.70 and 0.80, but often taken as 0.77); s is the variable of integration; x is the advance front distance; and A_0 is the inlet area related to the normal depth corresponding to the inflow rate, roughness, field slope, and hydraulic radius at the field inlet, which is expressed as:

$$A_0 = \left(\frac{Q_0^2 n^2}{3600 \rho_1 S_0}\right)^{1/\rho_2}, \tag{17.15}$$

where Q_0 is the inflow rate (m^3/min/unit width), n is Manning's roughness coefficient, S_0 is the field slope, and ρ_1 and ρ_2 are the empirical shape factors.

For basin and borders, $\rho_1 = 1$ and $\rho_2 = 3.33$ (Walker and Skogerboe, 1987). If there exists a level slope condition (e.g., basin), the friction slope in Manning's equation is assumed to equal the inlet flow depth (y_0) divided by the advance front distance (x). Thus, eq. (17.15) becomes:

$$A_0 = y_0 = \left(\frac{Q_0^2 n^2 x}{3600}\right)^{0.24}. \tag{17.16}$$

The surface area varies from A_0 at the field inlet to zero at the advancing tip. The volume balance approach neglects the space–time variation of A and assumes a constant average area. The average area is related to the inlet area by the surface shape factor, which is the ratio of the average area to the inlet area.

The infiltrated volume over the advance distance at any time t can be determined as follows:

$$V_z(t) = \int_0^x Z(s,t)\,ds = \int_0^x Z(t - t_s)\,ds = \sigma_z Z_0 x, \tag{17.17}$$

where Z is the infiltrated volume per unit area, $t - t_s$ is the intake opportunity time, t_s is the time when the water front reaches the distance s, Z_0 is the infiltrated volume per unit area at the field inlet, and σ_z is the subsurface shape factor (the ratio of infiltrated volume over the distance s to infiltrated volume at the field inlet). In eq. (17.17), it is assumed that $Z(s, t)$ is not a function of water surface depth but is dependent on the intake opportunity time.

Substituting the volume of surface storage (eq. 17.15) and the volume of infiltration (eq. 17.17) terms into eq. (17.13), one obtains:

$$Q_0 t = \overline{A}x + \int_0^x Z(t - t_s)\,ds = \sigma_y A_0 x + \sigma_z Z_0 x, \tag{17.18}$$

which is known as the Lewis-Milne (1938) volume balance equation.

17.7.3 Power Advance Volume Balance Model

The solution techniques used for solving eq. (17.18) can be grouped into four categories: the numerical or recursive approach (Hall, 1956), the kernel function approach (Hart et al., 1968), Laplace transformation (Philip and Farrell, 1964), and the power advance approach (Christiansen et al., 1966; Elliott and Walker, 1982; Singh 1979–1980). The integral term in eq. (17.18) is a distance integral of a time-dependent function. Unlike other approaches, the power advance approach assumes the following functional relationship between the advance rate and time a priori:

$$X = p t_x^r, \tag{17.19}$$

where X is the advance distance for the basin, border, or furrow (m), t_x is the advance time to distance X since the beginning of irrigation (min), and p and r are the fitted parameters.

Using the power advance given by eq. (17.19) along with the modified Kostiakov infiltration equation (eq. 17.2), Elliott and Walker (1982) presented the following solution to the volume balance eq. (17.18):

$$Q_0 t_x = \sigma_y A_0 X + \sigma_z X k t_x^a + \sigma_z' f_0 t_x X, \tag{17.20}$$

where Q_0 is the inflow rate (m^3/min), A_0 is the cross-sectional area of flow at the inlet (m^2), X is the advance distance (m), t_x is the advance time to distance X since the beginning of irrigation (min), k and a are the coefficients of the modified Kostiakov equation, f_0 is the basic infiltration rate (m^3/m/min), σ_y is the surface storage factor, and σ_z and σ_z' are defined as:

$$\sigma_z = \frac{a + r(1-a) + 1}{(1+a)(1+r)}, \tag{17.21a}$$

$$\sigma_z' = \frac{1}{1+r}, \tag{17.21b}$$

where r is the empirical parameters of the advance curve in eq. (17.9).

Equation (17.20) can be used to either determine the modified Kostiakov infiltration parameters using the advance distance and time information at two points along the field or advance to the end of the field knowing the infiltration parameters. This method is also known as the "two-point method." A convenient form of eq. (17.20) is obtained by substituting the advance time given by eq. (17.12) as:

$$\frac{Q_0 T_a}{K} - 0.77 A_0 L - \sigma_Z L k T_a^a - \sigma_Z' \frac{f_0 T_a L}{K} = 0, \tag{17.22}$$

where K is a unit constant equal to 1000 for Q_0 in L/min, A_0 in m^2, and f_0 is L/min/m; K is equal to 7.48 for Q_0 in gpm, A_0 in ft^2, and f_0 is gpm/ft.

17.7.4 Evaluation of Infiltration Parameters

The two points most often correspond to half of the field length ($L/2$) and the field length (L) and their respective advance times, and constitute time–distance pairs. Substituting these pairs in eq. (17.19), one obtains two volume balance equations corresponding to a half field length and full field length. These equations can be solved to determine the infiltration parameters k and a as follows:

$$Q_0 t_{L/2} = \sigma_y A_0 \frac{L}{2} + \sigma_z k t_{L/2}^a \frac{L}{2} + \sigma_z' \frac{f_0 t_{L/2} L}{2}, \tag{17.23a}$$

$$Q_0 t_L = \sigma_y A_0 L + \sigma_z k t_L^a L + \sigma_z' f_0 t_L L. \tag{17.23b}$$

For convenience of computation, eqs (17.23a) and (17.23b) can be substituted with the advance time given by eq. (17.12) as:

$$\frac{Q_0 t_{\left(\frac{L}{2}\right)}}{K} = \sigma_y A_0 \frac{L}{2} + 0.5\sigma_z L k T_a^a + \sigma_z{}' \frac{f_0 T_a L/2}{K}, \tag{17.24a}$$

$$\frac{Q_0 t}{K} = \sigma_y A_0 L + \sigma_z L k T_a^a + \sigma_z{}' \frac{f_0 T_a L}{K}. \tag{17.24b}$$

Multiplying eq. (17.23a) by $L/2$ throughout, one obtains

$$V_{L/2} = \frac{2Q_0 t_{L/2}}{L} - \sigma_y A_0 - \sigma_z' f_0 t_{L/2}. \tag{17.25a}$$

Thus, eq. (17.23a) can be expressed as

$$\sigma_z k t_{L/2}{}^a = V_{L/2} \text{ or } a \ln\left(t_{L/2}\right) + \ln \sigma_z + \ln k = \ln\left(V_{L/2}\right). \tag{17.25b}$$

Likewise, dividing eq. (17.23b) by L throughout, one obtains

$$V_L = \frac{Q_0 t_L}{L} - \sigma_y A_0 - \sigma_z' f_0 t_L. \tag{17.25c}$$

Thus, eq. (17.23b) can be expressed as

$$\sigma_z k t_L{}^a = V_L \text{ or } a \ln(t_L) + \ln k + \ln \sigma_z = \ln(V_L) \tag{17.26}$$

From eqs (17.25a) and (17.25c), constant a can be obtained as:

$$a = \frac{\ln\left(V_L/V_{L/2}\right)}{\ln\left(t_L/t_{L/2}\right)} \tag{17.27}$$

and k as:

$$k = \frac{V_L}{\sigma_z t_L^a}, \tag{17.28a}$$

or eq. (17.28a) can be written as follows by substituting σ_z with eq. (17.21a):

$$k = \frac{(1+a)(1+r)\ V_L}{[a + r(1-a) + 1]\, t_L^a}. \tag{17.28b}$$

The two-point method does not estimate the basic infiltration rate (f_0), which is generally determined using the information on inflow and outflow as:

$$f_0 = \frac{Q_0 - Q_r}{L}, \tag{17.29}$$

where Q_0 and Q_r are the inflow and runoff (outflow) rates, respectively (m^3/min), and L is the length of the field (m).

Equation (17.28) can be used to determine the time of advance across the field, but r depends on the stream size. Therefore, eqs (17.18) and (17.20) have to be solved simultaneously.

Example 17.2: Consider a surface irrigation system in which the border is 10 m wide and 50 m long. The flow cross-sectional area is $0.5\ m^2$, the stream size is 2500 L/min, the basic infiltration rate is 1.5 L/min/m, and the advance time for full field length is 20 min and for half field length is 8 min. K (unit constant) is 1000. Compute the value of parameters r, a, k, the volume of cumulative infiltration after 150 min (m^3/m), and the average depth of infiltration (cm).

Solution: Given the stream size $Q_0 = 2500$ L/min or $2.5\ m^3/min$, the length, width, and cross-section area of the field are $L = 50$ m, $W = 10$ m, and $A_0 = 0.5\ m^2$, respectively, the advance time for length $T_L = 20$ min, for half field length $T_{L/2} = 8$ min, and the basic infiltration rate is 1.5 L/min/m or $0.0015\ m^3/min/m$.

From eq. (17.19), one gets

$$50 = p t_L^r,$$

$$25 = p t_{L/2}^r.$$

Then, combing these equations one obtains

$$r = \frac{\ln 2}{\ln t_L - \ln t_{L/2}} = \frac{\ln 2}{\ln 20 - \ln 8} = 0.756.$$

Then,

$$\sigma_z' = \frac{1}{1+r} = \frac{1}{1+0.756} = 0.569.$$

Assuming surface shape factor $\sigma_y = 0.77$, $V_{L/2}$ and V_L are calculated using eq. (17.25):

$$\begin{aligned} V_{L/2} &= \frac{2Q_0 t_{L/2}}{L} - \sigma_y A_0 - \sigma_z' f_0 t_{\frac{L}{2}} \\ &= \frac{2 \times 2.5 \times 8}{50} - 0.77 \times 0.5 - 0.569 \times 0.0015 \times 8 \\ &= 0.408\ m^3/m, \end{aligned}$$

$$\begin{aligned} V_L &= \frac{Q_0 t_L}{L} - \sigma_y A_0 - \sigma_z' f_0 t_L \\ &= \frac{2.5 \times 20}{50} - 0.77 \times 0.5 - 0.569 \times 0.0015 \times 20 \\ &= 0.598\ m^3/m. \end{aligned}$$

a can be obtained using eq. (17.27):

$$a = \frac{\ln V_L - \ln V_{0.5L}}{\ln t_L - \ln t_{0.5l}} = \frac{0.598 - 0.408}{\ln 20 - \ln 8} = 0.417.$$

σ_z can be obtained using eq. (17.21a):

$$\sigma_z = \frac{a + r(1-a) + 1}{(1+a)(1+r)} = \frac{0.417 + 0.756 \times (1 - 0.417) + 1}{(1 + 0.417) \times (1 + 0.756)}$$

$$= 0.747.$$

k can be calculated using eq. (17.28):

$$k = \frac{V_L}{\sigma_z t_L^a} = \frac{0.598}{0.747 \times 20^{0.417}} = 0.230.$$

Based on the modified Kostiakov equation (eq. 17.2), the infiltration volume at the inlet after 150 min can be calculated:

$$Z_0 = k \times t^a + f_0 \times t$$

$$= 0.23 \times (150)^{0.417} + 0.0015 \times (150) = 2.083\ \frac{\text{m}^3}{\text{m}}.$$

Z_0 in the unit of depth is:

$$Z_0 = \frac{2.083\ \frac{\text{m}^3}{\text{m}}}{10\ \text{m}} \times \left(\frac{100\ \text{cm}}{1\ \text{m}}\right) = 20.83\ \text{cm}.$$

Example 17.3: Compute the advance time for the full field length (T_a), half field length ($T_{0.5a}$), advance coefficients p and r for the data in Example 17.2, that is, the border is 10 m wide and 50 m long, the stream size is 2500 L/min, and the basic infiltration rate is 1.5 L/min/m. Given that the flow cross-sectional area is 0.5 m^2, a is 0.42 and k is 0.23.

Steps to Find a Solution

Step 1: Guess an r value.
Step 2: Compute σ_z.
Step 3: Compute σ.
Step 4: Initially guess T_a.
Step 5: Solve the following equation with T_a (until the equation result is equal to or less than 0.005):

$$\frac{(Q)(T_a)}{K} - 0.77(A_o)(L) - \sigma_z(L)(k)(T_a)^a - \sigma\frac{(f_o)(T_a)(L)}{K}.$$

Step 6: Initial guess $T_{0.5a}$: Set the initial guess of $T_{0.5a}$ equal to T_a obtained in Step 5, and use the $T_{0.5a}$ to calculate the left parts in eq. (17.22) (until the equation result is equal to or less than 0.005).

Solution:

Step 1: Guess r value as 0.7.
Step 2: Compute σ_z:

$$\sigma_z = \frac{a + r(1-a) + 1}{(1+a)(1+r)} = \frac{0.42 + 0.7 \times (1 - 0.42) + 1}{(1 + 0.42) \times (1 + 0.7)}$$

$$= 0.756.$$

Step 3: Compute σ:

$$\sigma = \frac{1}{1+r} = \frac{1}{1+0.7} = 0.588.$$

Step 4: Initial guess T_a:

$$T_a = \frac{L}{V} = \left(\frac{A_o}{Q}\right) \times L = \left(\frac{0.5\ \text{m}^2}{2500\frac{\text{L}}{\text{min}}}\right) \times 50\ \text{m} \times \left(\frac{1000\ \text{L}}{1\ \text{m}^3}\right)$$

$$= 10\ \text{min}.$$

The initial value of T_a should be greater than or equal to the actual T_a. An approximation is required here. The above $T_a - 10$ min is multiplied by 4 to exceeds the actual T_a. Now we have $T_a = 10 \times 4\ \text{min} = 40$ min.

Step 5: Calculate the following,

$$\frac{(Q)(T_a)}{K} - 0.77(A_o)(L) - \sigma_z(L)(k)(T_a)^a - \sigma\frac{(f_o)(T_a)(L)}{K}.$$

When $T_a = 40$ min,

$$\frac{(2500)(40)}{1000} - 0.77(0.5)(50) - 0.756(50)(0.23) \times (40)^{0.42} - 0.588\frac{(1.5)(40)(50)}{1000}$$

$$= 38.1 > 0.005.$$

When $T_a = 30$ min,

$$\frac{(2500)(30)}{1000} - 0.77(0.5)(50) - 0.756(50)(0.23) \times (30)^{0.42} - 0.588\frac{(1.5)(30)(50)}{1000}$$

$$= 18.2 > 0.005.$$

When $T_a = 22$ min,

$$\frac{(2500)(22)}{1000} - 0.77(0.5)(50) - 0.756(50)(0.23) \times (22)^{0.42} - 0.588\frac{(1.5)(22)(50)}{1000}$$

$$= 3.4 > 0.005.$$

When $T_a = 20.4$ min,

$$\frac{(2500)(20.4)}{1000} - 0.77(0.5)(50) - 0.756(50)(0.23) \times (20.16)^{0.42} - 0.588\frac{(1.5)(20.4)(50)}{1000} \approx 0.0.$$

Step 6: Initial guess $T_{0.5a}$: Set the initial guess of $T_{0.5a}$ equal to the final T_a in Step 5:

$$\frac{(Q)(T_{0.5a})}{K} - 0.77(A_o)\left(\frac{L}{2}\right) - \sigma_z\left(\frac{L}{2}\right)(k)(T_{0.5a})^a - \sigma\frac{(f_o)(T_{0.5a})\left(\frac{L}{2}\right)}{K}.$$

When $T_{0.5a} = 20.4$ min,

$$\frac{(2500)(20.4)}{1000} - 0.77(0.5)\left(\frac{50}{2}\right) - 0.756\left(\frac{50}{2}\right)(0.23) \times (20.4)^{0.42} - 0.588\frac{(1.5)(20.4)\left(\frac{50}{2}\right)}{1000} = 25.5 > 0.005.$$

When $T_{0.5a} = 10$ min,

$$\frac{(2500)(10)}{1000} - 0.77(0.5)\left(\frac{50}{2}\right) - 0.756\left(\frac{50}{2}\right)(0.23) \times (10)^{0.42} - 0.588\frac{(1.5)(10)\left(\frac{50}{2}\right)}{1000} = 3.7 > 0.005.$$

When $T_{0.5a} = 8.11$ min,

$$\frac{(2500)(8.11)}{1000} - 0.77(0.5)\left(\frac{50}{2}\right) - 0.756\left(\frac{50}{2}\right)(0.23) \times (8.11)^{0.42} - 0.588\frac{(1.5)(8.11)\left(\frac{50}{2}\right)}{1000} \approx 0.0.$$

Step 7: Compute r':
From Example 17.2, one gets

$$r' = \frac{\ln\left(\frac{a}{0.5a}\right)}{\ln T_a - \ln T_{0.5a}} = \frac{\ln 2}{\ln 20.4 - \ln 8.11} = 0.751,$$

Since the absolute value of $r - r' = |0.7 - 0.751| = 0.051 > 0.001$, then

*Repeat Steps 2–7 with $r = r' = 0.751$:

$$\sigma_z = \frac{a + r(1-a) + 1}{(1+a)(1+r)} = \frac{0.42 + 0.751 \times (1 - 0.42) + 1}{(1 + 0.42) \times (1 + 0.751)} = 0.746,$$

$$\sigma = \frac{1}{1+r} = \frac{1}{1+0.751} = 0.571.$$

When $T_a = 20.4$ min,

$$\frac{(2500)(20.4)}{1000} - 0.77(0.5)(50) - 0.746(50)(0.23) \times (20.4)^{0.42} - 0.571\frac{(1.5)(20.4)(50)}{1000} = 0.43 > 0.005.$$

When $T_a = 20.16$ min,

$$\frac{(2500)(20.16)}{1000} - 0.77(0.5)(50) - 0.746(50)(0.23) \times (20.16)^{0.42} - 0.571\frac{(1.5)(20.16)(50)}{1000} \approx 0.0.$$

When $T_{0.5a} = 20.16$ min,

$$\frac{(2500)(20.16)}{1000} - 0.77(0.5)\left(\frac{50}{2}\right) - 0.746\left(\frac{50}{2}\right)(0.23) \times (20.16)^{0.42} - 0.571\frac{(1.5)(20.16)\left(\frac{50}{2}\right)}{1000} = 25.2 > 0.005.$$

When $T_{0.5a} = 8.04$ min,

$$\frac{(2500)(8.04)}{1000} - 0.77(0.5)\left(\frac{50}{2}\right) - 0.746\left(\frac{50}{2}\right)(0.23) \times (8.04)^{0.42} - 0.571\frac{(1.5)(8.04)\left(\frac{50}{2}\right)}{1000} \approx 0.0.$$

$$r' = \frac{\ln 2}{\ln T_a - \ln T_{0.5a}} = \frac{\ln 2}{\ln 20.16 - \ln 8.04} = 0.754,$$

$$|r - r'| = |0.751 - 0.754| = |0.003| > 0.001.$$

Repeat Steps 2–7 with $r = r' = 0.754$:
Then, one gets $T_a = 20.16$ min and $T_{0.5a} = 8.04$ min:

$$r' = \frac{\ln 2}{\ln T_a - \ln T_{0.5a}} = \frac{\ln 2}{\ln 20.16 - \ln 8.04} = 0.7540,$$

Step 8: Compute p using eq. (17.19):

$$p = \frac{L}{(T_a)^r} = \frac{50\text{ m}}{(20.16\text{ min})^{0.754}} = 5.19.$$

QUESTIONS

Q.17.1 Compute the advance distance for a field at the head of which flow enters with a depth of 0.25 m for a period of 6 h. The average depth rate of infiltration of the soil is 10 cm/h, the ground slope is 0.005, and the value of Manning's n is 0.10.

Q.17.2 Compute the advance function by taking values of time as 1, 2, 3, 5, and 6 h for the data in Q.17.1.

Q.17.3 What will be the maximum advance distance for the data in Q.17.1?

Q.17.4 Compute the flow depth at each advance distance corresponding to the time in Q.17.2.

Q.17.5 Compute the recession function when the duration of irrigation is 6 h for the data in Q.17.1.

Q.17.6 Compute the depth of flow to the right of the recession curve for the data in Q.17.1.

Q.17.7 Consider a surface irrigation system in which the border is 15 m wide and 40 m long. The flow cross-sectional area is 0.45 m^2, the stream size is 3000 L/min, the basic infiltration rate is 1.6 L/min/m, and the advance time for full field length is 25 min and for half field length is 10 min. K (unit constant) is 1000. Compute the value of parameters r, a, k, depth of cumulative infiltration after 140 minutes, and the average depth of infiltration (cm).

Q.17.8 Compute the advance time for the full field length (T_a), half field length ($T_{0.5a}$), advance coefficients p and r for the data in Q.17.7; that is, the border is 15 m wide and 40 m long, the stream size is 3000 L/min, and the basic infiltration rate is 1.6 L/min/m. The flow cross-sectional area is 0.45 m^2. a is 0.30 and k is 0.72.

Q.17.9 Let the flow depth at the head of a 300 m field be 0.3 m, the average infiltration rate be 0.2 cm/min, the duration of irrigation is 3 h, Manning's coefficient (n) be 0.15, and slope be 0.002. Compute the advance function, recession function, and flow depth as functions of space and time.

REFERENCES

Bassett, D. L., Fangmeier, D. D., and Strelkoff, T. (1980). Hydraulics of surface irrigation. In *Design and Management of Farm Irrigation Systems*. St Joseph, MI: ASAE.

Bloodworth, M. E., and Gillett, P. T. (2010) Irrigation. In *Handbook of Texas Online*. Texas State Historical Association. www.tshaonline.org/handbook/online/articles/ahi01.

Brouwer, C., Prins, K., Kay, M., and Heibloem, M. (1985). *Irrigation Water Management: Irrigation Methods*. Rome: FAO.

Christiansen, J. E., Bishop, A. A., Kiefer, F. W., and Fok, Y. S. (1966). Evaluation of intake rate constants as related to advance of water in surface irrigation. *Transactions of the ASAE*, 9(5): 671–674.

Dillon, J., Edinger-Marshall, S., and Letey, J. (1999). Farmers adopt new irrigation and fertilizer techniques: changes could help growers maintain yields, protect water quality. *California Agriculture*, 53(1): 24–31.

Elliott, R. L., and Walker, W. R. (1982). Field evaluation of furrow infiltration and advance functions. *Transactions of the ASAE*, 25(2): 396–400.

Hall, W. A. (1956). Estimating border irrigation flow. *Agricultural Engineering*, 37: 263–265.

Hart, W. E., Bassett, D. L., and Strelkoff, T. (1968). Surface irrigation hydraulics-kinematics. *Journal of Irrigation and Drainage Division, ASCE*, 94(IR4): 419–440.

Jain, S. K., and Singh, V. P. (1989). A numerical kinematic wave model for border irrigation. *Irrigation Science*, 10: 253–263.

Jurriens, M., Zerihun, D., Boonstra, J., and Feyen, J. (2001). *SURDEV: Surface Irrigation Software; Design, Operation, and Evaluation of Basin, Border, and Furrow Irrigation*. Wageningen: International Institute for Land Reclamation and Improvement.

Lewis, M. R., and Milne, W. E. (1938). Analysis of border irrigation. *Agricultural Engineering*, 19: 267–272.

Philip, J. R., and Farrell, D. A. (1964). General solution of the infiltration advance problem in irrigation hydraulics. *Journal of Geophysical Research*, 69: 621–631.

Ram, R. S., and Singh, V. P. (1982a). A design procedure for closed end irrigation borders. *Agricultural Water Management*, 5: 1–14.

Ram, R. S., and Singh, V. P. (1982b). Evaluation of models of border irrigation recession. *Journal of Agricultural Engineering Research*, 27: 235–252.

Ram, R. S., and Singh, V. P. (1985). Application of kinematic wave equations to border irrigation design. *Journal of Agricultural Engineering Research*, 32: 57–71.

Ram, R. S., Singh, V. P., and Prasad, S. N. (1983). *Mathematical Modeling of Surface Irrigation*. Baton Rouge, LA: Department of Civil Engineering, Louisiana State University.

Ram, R. S., Singh, V. P., and Prasad, S. N. (1986a). A quasi-steady state integral model for closed-end border irrigation. *Agricultural Water Management*, 11: 39–57.

Ram, R. S., Singh, V. P., and Prasad, S. N. (1986b). A quasi-steady state integral model for border irrigation. *Irrigation Science*, 7: 113–141.

Seyedzadeh, A., Panahi, A., Maroufpoor, E., and Singh, V. P. (2019). Development of an analytical method for estimating Manning's coefficient of roughness for border irrigation. *Irrigation Science*, 37: 523–531.

Sherman, B., and Singh, V. P. (1978). A kinematic model of surface irrigation. *Water Resources Research*, 14(2): 357–364.

Sherman, B., and Singh, V. P. (1982). A kinematic model of surface irrigation: an extension. *Water Resources Research*, 18(3): 659–667.

Singh, V. P. (1979–1980). Derivation of shape factors for border irrigation advance. *Agricultural Water Management*, 2: 271–288.

Singh, V. P., and He, Y. C. (1988). A Muskingum model for furrow irrigation. *Journal of Irrigation and Drainage Engineering, ASCE*, 114(1): 89–103.

Singh, V. P., and Prasad, S. N. (1983). Derivation of mean depth in Lewis–Milne equation for border irrigation advance. In *Proceedings of the ASCE Specialty Conference on Advances in Irrigation and Drainage: Surviving External Pressures,* July 20–22, Jackson, OH, pp. 234–241.

Singh, V. P., and Ram, R. S. (1983a). A kinematic model of surface irrigation: verification by experimental data. *Water Resources Research*, 19(6): 1599–1612.

Singh, V. P., and Ram, R. S. (1983b). A semi-analytical approach to kinematic wave equations for design of border irrigation. In *Proceedings of the ASCE Specialty Conference on Advances in Irrigation and Drainage: Surviving External Pressures*, July 20–22, Jackson, OH, pp. 242–249.

Singh, V. P., and Ram, R. S. (1983c). *Some Aspects of the Hydraulics of Border Irrigation*. Baton Rouge, LA: Department of Civil Engineering, Louisiana State University.

Singh, V. P., and Ram, R. S. (1984). Mathematical modeling of farm irrigation. In *Proceedings of the IAHR/UNESCO International Seminar on Water Resources Management*, Zaria, Nigeria, pp. 317–344.

Singh, V. P., and Sherman, B. (1983). *A Kinematic Study of Surface Irrigation: Mathematical Solutions*. Baton Rouge, LA: Department of Civil Engineering, Louisiana State University.

Singh, V. P., and Yu, F. X. (1987a). A mathematical model for border irrigation: I. Advance and storage phases. *Irrigation Science*, 8: 151–174.

Singh, V. P., and Yu, F. X. (1987b). A mathematical model for border irrigation: II. Vertical and horizontal recession phases. *Irrigation Science*, 8: 175–190.

Singh, V. P., and Yu, F. X. (1987c). A mathematical model for border irrigation: III. Evaluation of border irrigation models. *Irrigation Science*, 8: 191–213.

Singh, V. P., and Yu, F. X. (1988a). A farm irrigation system (FIS) model. *Water Resources Management*, 2, 173–181.

Singh, V. P., and Yu, F. X. (1988b). A model for simulating closed border irrigation. *Journal of the Institution of Engineers, Agricultural Engineering Division*, 69: 34–41.

Singh, V. P., and Yu, F. X. (1989a). An analytical closed border irrigation model: 1. Theory. *Agricultural Water Management*, 15: 223–241.

Singh, V. P., and Yu, F. X. (1989b). An analytical closed border irrigation model: 2. Experimental verification. *Agricultural Water Management*, 15: 243–252.

Singh, V. P., Scarlatos, P. D., and Raudales, S. A. (1988). A Muskingum model for border irrigation. *Journal of Irrigation and Drainage Engineering, ASCE*, 114(2): 266–280.

Singh, V. P., Scarlatos, P. D., and Prasad, S. N. (1990). An improved Lewis–Milne equation for the advance phase of border equation. *Irrigation Science*, 11(1): 1–6.

Strelkoff, T. (1969). One-dimensional equations of open channel flow. *Journal of the Hydraulics Division*, 95(HY3): 861–876.

USDA-NRCS (1997). Selecting an irrigation method. In *National Engineering Handbook*. Washington, DC: USDA.

USDA-NRCS (2012). Surface irrigation. *National Engineering Handbook*. Washington, DC: USDA.

USDA-SCS (1974). Surface irrigation. In *National Engineering Handbook*. Washington, DC: USDA.

Walker, W. R. and Skogerboe, G. V. (1987). *Surface Irrigation: Theory and Practice*. Englewood Cliffs, NJ: Prentice Hall.

Yu, F. X., and Singh, V. P. (1990a). Analytical model for furrow irrigation. *Journal of Irrigation and Drainage Engineering*, 116(2): 154–171.

Yu, F. X., and Singh, V. P. (1990b). Analytical model for border irrigation. *Journal of Irrigation and Drainage Engineering*, 116(6): 982–999.

Zerihun, D., and Junna, M. R. (1996). Sensitivity analysis of furrow-irrigation performance parameters. *Journal of Irrigation and Drainage Engineering*, 122(1): 49–57.

Part V Methods of Irrigation

18 Basin Irrigation

Notation

A_0	cross-sectional area at the upstream end (m^2, ft^2)
A_b	basin area (m^2, ft^2)
D_{IR}	design daily irrigation requirement (mm/day, in./day)
e, f, g	parameters in the SCS cumulative infiltration equation
E_a	application efficiency (percentage)
f_0	basic infiltration rate (L/min/m, gpm/ft)
F	cumulative infiltration or depth of infiltration (mm, in.)
F_n, Z_{req}	desired net application depth or infiltration depth (mm, in.)
k, a	coefficients of the modified Kostiakov equation
L	basin length (m, ft)
n	Manning's roughness coefficient
N_b	number of basins irrigated per set
N_T	total number of basins being irrigated
Q	stream size (L/min, m^3/s, ft^3/s)
Q_0	unit inflow rate (m^3/s/m, m^2/s)
r	exponent in the advance function
R	efficiency advance ratio
t	time since infiltration began or opportunity time (min)
$T_{0.5}$	advance time for half the length of the basin (min)
T_a	advance time for the full length of the basin (min)
T_i	inflow time (min)
T_n	net opportunity time or time required (min)
X	distance that water advanced across the basin (m, ft)

18.1 INTRODUCTION

Basin irrigation is a common method for surface irrigation, especially in developing countries, where water is applied by flooding. A basin is an agricultural field with zero to little slope and is diked from all sides. A basin can be a rectangular, square, or circular plot of an agricultural field. Normally it is level or has a very small slope. The size of the plot can widely vary, ranging from 4 to 4000 m^2. The plot is bounded by dikes, ridges, or bunds on all sides. Fields that are irrigated by the basin method are divided into plots and water is supplied at one or more points of the plot until the desired volume of water has been applied. This is irrigation by flooding. The water is supplied to the basin either by flooding through one basin and then into another, or directly from a ditch to each basin. The flow rate is normally large so that the flow covers the entire basin in about 60–75% of the time the soil needs to absorb the desired amount of water. Since the rate of application of water is larger than the rate of infiltration, the water ponds until it is infiltrated. Because the land is nearly level, the depth of water is more uniform.

The basin method of irrigation is more suitable for fine-textured soils with moderate to low intake rates (50 mm/h) and percolation rates that allow water to remain on the surface and in the root zone for a longer period. The land slope should be smooth, gentle, and uniform. If the land has steep slopes or is undulating it can be prepared for basin irrigation by land grading if the soils are deep.

Basin irrigation can be applied for most crops, but is more widely used for close-growing crops, such as rice, small grains, legumes, grasses, alfalfa, groundnut, and vegetable crops, as well as for grapes, berries, and orchards or tree crops. It can also be applied to row crops, such as sorghum, corn, finger millet, cotton, and sugar beets, but is not suitable for crops that cannot tolerate waterlogging for more than a day, such as potato, cassava beet, and carrot. In orchards, each tree can have one basin or 2–5 trees that may be included in each basin. Basins can be employed for leaching of salts below the root zone by deep percolation. A good discussion of basin irrigation is presented by Hart et al. (1981), and this chapter draws from their work on the basin method and its design.

18.2 ADVANTAGES AND DISADVANTAGES

Basin irrigation has advantages as well as limitations. Its advantages include the following: (1) water can be applied uniformly; (2) it has high application efficiency; (3) it does not require skilled labor; (4) small streams can also be used for irrigating crops; (5) it can be easily automated; (6) simple and cheap equipment is used for constructing bunds; (7) the basin size is flexible and can be as large as 16 ha if appropriately leveled; (8) many crops can be grown in sequence without major changes in design, layout, or operation; (9) there is no runoff; (10) rainfall can be maximally utilized; (11) deep percolation can be controlled; and (12) leaching can be done without altering either the layout or the method of operation.

The disadvantages of basin irrigation include the following: (1) land must be leveled, which can be expensive; (2) since a field is divided into diked basins a significant amount of land (up to 30%) is covered by ditches and bunds; (3) bunds interfere with intercultural operations; (4) it is difficult to maintain ridge heights on sandy soils or fine-textured soils that crust or crack when dry; (5) if the basin method is not properly managed then prolonged waterlogging and crop scalding can occur; (6) provision must be made for the drainage of excess water; (7) more labor is needed for field layout and irrigation; (8) since relatively large inflow rates are needed for basin irrigation, provision, such as special structures, must be made to prevent erosion; and (9) on steep slopes that require benching, flows need to be controlled through pipelines, lined ditches, or drop structures.

18.3 TYPES OF BASINS

Basins can be mainly distinguished as check basins and ring basins. Check basins are either square or rectangular.

18.3.1 Check Basins

Check basins, or checks, can be classified into rectangular and contour types, as shown in Figure 18.1. The size of a check basin depends on the purpose, crop, soil, water supply, and topography. For example, for vegetables and intensive crops the size can be as small as 1 m^2, and for crops like rice the size can be 1–2 ha. The irrigation is by flooding

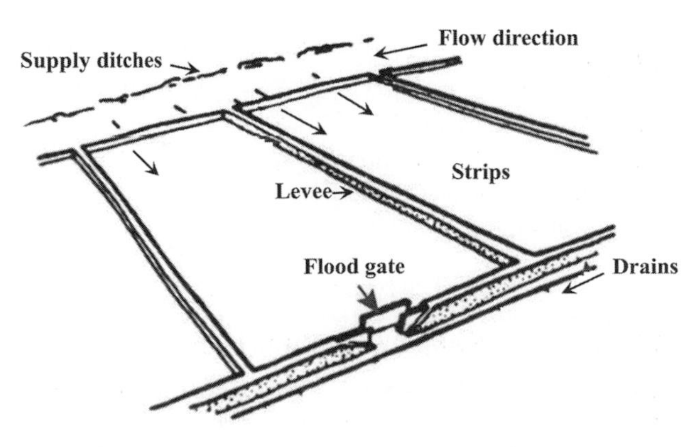

Figure 18.1 Layout of border irrigation system. (Adapted from USDA-SCS, 1974.)

and is well suited to very permeable soils where excessive deep percolation loss near the ditch is controlled by quick coverage of the soil with water. As mentioned before, it is also suitable for heavy soils with low intake rates.

Contour checks are also prepared by building dikes along contours with 0.2–0.4 ft (6–12 cm) intervals and connecting them with cross (perpendicular) levees at appropriate places. Such checks are useful where large streams are available or when streams are flowing full during torrential rains, so water can be applied without loss.

Levees should be 6–8 ft (1.8–2.4 m) wide at the base and 10–12 in. (25–30 cm) high. These dimensions allow growing crops on the levees and the use of machinery without obstruction. However, where land holdings are small and farmers do not use machinery, the levee size is much smaller, with a width of no more a foot or two (0.3–0.6 m).

18.3.2 Ring Basins

Ring basins are used for sparsely grown tree orchards; as the name suggests, a circular bund or ring is constructed around each tree, constituting a basin, and in some cases more trees are included in each basin, which is connected to a water-supply ditch. One basin may be connected to another basin and finally to the supply ditch.

18.4 DESIGN OF BASINS

The design of level basin systems involves optimizing system layout, inflow rate, inflow time, basin dimensions, and the number of basins per set to be irrigated for achieving the desired level of efficiency, uniformity, adequacy, convenience of operation, and cost. Each design step is now discussed.

18.4.1 Layout

The system layout depends on the location of water source, topography, and basin size. To maximize the spacing between supply ditches/pipelines, the long axis of basins should be perpendicular to them. Sometimes it may be necessary to provide drains for removing excess water due to overirrigation or intense rainfall, as shown in Figure 18.1. At other times, basins are laid out as a cascade, as shown in Figure 18.2.

18.4.2 Location of Water Source

The key considerations are twofold. First, the water source should be located so that all basins can be irrigated under gravity, or pumping may be needed. Second, to minimize the size of supply channels or pipelines, the water source should be near the center of the irrigated area. This consideration is

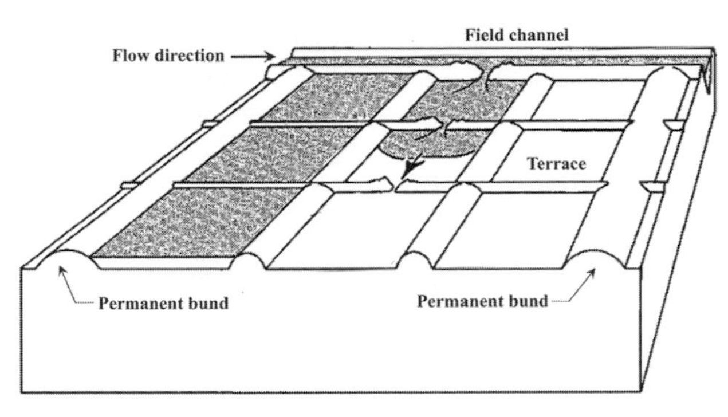

Figure 18.2 Layout of border irrigation system (cascade irrigation). (Adapted from Walker, 1989).

often not in the farmer's hands, especially if the source of water is a canal.

18.4.3 Topography

The basin shape depends on the terrain. If the terrain is steeply sloping then contour basins can be constructed, or terracing can be used to obtain level basins. If the topography is undulating, then basin shapes are irregular. On level or uniformly sloping lands, rectangular basins are appropriate.

18.4.4 Basin Sizes

The basin size depends on two factors: (1) stream size and (2) soil infiltration characteristics. If the infiltration rate is high, as in sandy soils, then basins should be small even if the available stream size is large. This will ensure better uniformity of irrigation. On fine-textured soils, the basin size can be small or large, depending on the stream size. The size of a basin can also be influenced by the flow depth. If the required flow depth is large, the basin can be large. Similarly, if the required depth is small, then the basin size should be small to get good water distribution.

Booher (1974) developed an empirical method for the design of basins. He suggested maximum basin areas for different soil types and rates of flow, as shown in Table 18.1. If soils have high infiltration rates, then basin size should be small even if the inflow rate is high, otherwise uniformity of flow will not be high. For low intake rates, both small and large basin sizes are fine. The basin size should be such that the entire area is irrigated in a reasonable length of time with a high degree of uniformity over the basin.

18.4.5 Basin Width

The basin width primarily depends on the topographic slope, depth of fertile soil in the case of undulating terrain, method of basin construction, and agricultural practices. For steep slopes, the width should be small. Otherwise, too much earth will be removed to obtain level basins. If land grading is to be done, then the soil should be deep, otherwise less fertile soil will be exposed. For machine operations, the basin width should be larger, but for hand operations the width can be narrow. Table 18.2 provides a guideline for basin widths, depending on different slopes.

Example 18.1: What will be the dimension of the basin if the soil is deep sandy loam and the slope is 0.8%? The basin construction is mechanized and thus the basin width should be maximized. The stream size is 60 L/s. Use the Booher empirical method.

Solution: From Table 18.1, the maximum basin size for a sandy loam soil with a stream size of 60 L/s is 012 ha = 1200 m^2. From Table 18.2, the maximum basin width for a sandy loam soil is 30 m (range: 15–30 m). If the total area is 1200 m^2 and the width is 30 m, the basin length is 1200/30 = 40 m.

18.4.6 Land Smoothing

In order to improve the uniformity and efficiency of irrigation, reduce labor requirements, permit construction of rectangular basins, and facilitate the layout of channels, roadways, and drainageways, land leveling is done to remove high and low areas within basins.

18.4.7 Stream Size

The stream size is determined by three factors. First, stream size should be as large as possible in order to maximize uniformity and application efficiency of irrigation. Second, it should not cause undue erosion. Third, it should not overtop the basin ridges. Large stream sizes reduce differences in infiltration opportunity times and deep percolation losses.

18.4.8 Irrigation Time

Irrigation time is defined by the infiltration opportunity time needed to infiltrate the desired depth of irrigation water. In Chapter 17, the Kostiakov equation (eq 17.1) and the modified Kostiakov (Kostiakov–Lewis) equation (eq. 17.2) were introduced for the design of surface irrigation systems. The infiltration opportunity time can be determined as follows by changing the second term of the modified Kostiakov

Table 18.1 *Suggested maximum basin areas for different soil types and rates of flow*

Area (ha)					
Flow rate		Soil type			
Liters/s	Cubic meters/h	Sand	Sandy loam	Clay loam	Clay
30	108	0.02	0.06	0.12	0.2
60	216	0.04	0.12	0.24	0.4
90	324	0.06	0.18	0.36	0.6
120	432	0.08	0.24	0.48	0.8
150	540	0.10	0.30	0.60	1.0
180	648	0.12	0.36	0.72	1.2
210	756	0.14	0.42	0.84	1.4
240	864	0.16	0.48	0.96	1.6
270	972	0.18	0.54	1.08	1.8
300	1080	0.20	0.60	1.20	2.0
Area in acres					
Flow rate		Soil type			
ft^3/s	Gallons/min (US)	Sand	Sandy loam	Clay loam	Clay
1	450	0.05	0.15	0.3	0.5
2	900	0.10	0.30	0.6	1.0
3	1350	0.15	0.45	0.9	1.5
4	1800	0.20	0.60	1.2	2.0
5	2250	0.25	0.75	1.5	2.5
6	2700	0.30	0.90	1.8	3.0
7	3150	0.35	1.05	2.1	3.5
8	3600	0.40	1.20	2.4	4.0
9	4000	0.45	1.35	2.7	4.5
10	4500	0.50	1.50	3.0	5.0

After Booher, 1974.

Table 18.2 *Approximate value for the maximum basin widths for different slopes*

	Maximum width (m)	
Slope (%)	Average	Range
0.2	45	35–55
0.3	37	30–45
0.4	32	25–40
0.5	28	20–35
0.6	25	20–30
0.8	22	15–30
1.0	20	15–20
1.2	17	10–20
1.5	13	10–20
2.0	10	5–15
3.0	7	5–10
4.0	5	3–8

After Brouwer et al., 1984.

Table 18.3 *Values of e, f, and g for eq. (18.1) for basins*

	e			*g*	
Intake family	(mm)	(in.)	*f*	(mm)	(in.)
0.1	0.6198	0.0244	0.661	6.985	0.275
0.3	0.9347	0.0368	0.721	6.985	0.275
0.5	1.1862	0.0467	0.756	6.985	0.275
1.0	1.7805	0.0701	0.785	6.985	0.275
1.5	2.2835	0.0899	0.799	6.985	0.275
2.0	2.7534	0.1084	0.808	6.985	0.275
3.0	3.6500	0.1437	0.816	6.985	0.275
4.0	4.4450	0.1750	0.823	6.985	0.275

After USDA-SCS, 1974.

equation into a constant. Let infiltration be determined by the USDA-SCS (1974) method, which defines the cumulative infiltration F or depth of infiltration as

$$F = et^f + g, \tag{18.1}$$

in which t is the time since infiltration began or opportunity time in min, and $e, f,$ and g are parameters to be determined. The USDA-SCS (1974) classified major soils into groups, referred to as intake families, and specified these parameters for each family, as discussed in Chapter 12. Table 18.3 gives values of these parameters for some intake families. Substituting the desired depth in eq. (18.1) yields the opportunity time.

18.4.9 Inflow Time

The inflow time is defined by the time that water flows into the basin. This time is determined by permitting the desired depth of irrigation or infiltration to take place at the far end of the basin. Since there is no runoff for basins, this time equals the sum of the advance time and the time needed to provide the volume of water to enable the desired depth of irrigation. Thus, the inflow time (min) can be expressed as

$$T_i = T_a + \frac{Z_{req} A_b}{KQ}, \tag{18.2}$$

in which T_i is the inflow time (min), T_a is the advance time across the basin (min), Z_{req} is the desired depth of irrigation (mm, in.), A_b is the basin area (m^2, ft^2), Q is the stream size (L/min, m^3/s, ft^3/s), and K is the unit constant ($K = 10$ for Z_{req} in mm, A_b in m^2, and Q in L/min; $K = 60{,}000$ for Z_{req} in mm, A_b in m^2, and Q in m^3/s; and $K = 720$ for Z_{req} in inches, A_b in ft^2, and Q in ft^3/s).

The cross-sectional area at the upstream end (A_0) of the level basin can be determined by using Manning's equation, with slope assumed to equal the upstream flow depth divided by the basin length. The advance time T_a can be determined (Walker and Skogerboe, 1987) as

$$\frac{QT_a}{K} - 0.77A_0L - \sigma_z LkT_a^a - \sigma\frac{f_0T_aL}{K} = 0, \tag{18.3}$$

in which T_a is the advance time for the full length of the basin (min), Q is the stream size (L/min, gpm), K is a unit constant equal to 1000 for Q_0 in L/min, A_0 is the cross-sectional area flow at the inlet (m^2, ft^2), L is the basin length (m, ft), f_0 is the basic infiltration rate (L/min/m, gpm/ft), k and a are the coefficients of the modified Kostiakov equation, and k can be expressed as:

$$k = \frac{V_a}{\sigma_z T_a^a}, \tag{18.4}$$

where

$$V_a = \frac{QT_a}{KL} - 0.77A_0 - \frac{f_0T_a}{K(r+1)}, \tag{18.5}$$

$$\sigma_z = \frac{a + r(1-a) + 1}{(1+a)(1+r)}, \tag{18.6}$$

where r is the exponent in the advance function,

$$X = pt^r, \tag{18.7}$$

defined as

$$r = \frac{\ln 2}{\ln T_a - \ln T_{0.5}}, \tag{18.8}$$

$$\sigma = \frac{1}{1+r}. \tag{18.9}$$

X is the distance that water advanced across the basin, p is the fitted parameter, $T_{0.5}$ is the time to advance half the length of the basin:

$$V_{0.5} = \frac{2QT_{0.5}}{KL} - 0.77A_0 - \frac{f_0T_{0.5}}{K(r+1)}, \tag{18.10}$$

$$F = kt^a + \frac{f_0t}{K}, \tag{18.11}$$

$$a = \frac{\ln V_a - \ln V_{0.5}}{\ln T_a - \ln T_{0.5}}, \tag{18.12}$$

where K is the unit constant ($K = 1000$ for Q in L/min, A_0 in m^2, and f_0 in L/min/m; $K = 7.48$ for Q in gpm, A_0 in ft^2, and f_0 in gpm/ft).

18.4.10 Application Efficiency

A criterion for evaluating the basin design is application efficiency, which can be computed as

$$E_a = \frac{Z_{req}A_b}{KQT_i} \times 100, \tag{18.13}$$

where E_a is the application efficiency in percent and K is the unit constant, as for eq. (18.2). The basin size and inlets may have to be adjusted to achieve the desired efficiency.

18.4.11 Ridge Dimensions

Ridges or bunds constitute the boundaries of the basin. The height of a bund should be equal to or higher than the maximum flow depth plus 5 cm (2 in.). At the base temporary bunds have a width of 60–120 cm and a height of 15–30 cm above the ground surface, including a freeboard of 10 cm. This means the depth of irrigation can be 5–20 cm. Temporary bunds are obliterated each season and reconstructed when crops are grown again. Permanent bunds are 130–160 cm wide at their base and are 60–90 cm high, but their settled height reduces to 40–50 cm.

18.4.12 Number of Basins Irrigated per Set

The number of basins irrigated per set can be computed as

$$N_b = \frac{N_T T_i D_{IR} E_a}{144{,}000 \times Z_{req}}, \tag{18.14}$$

where N_b is the number of basins irrigated per set, N_T is the total number of basins being irrigated, T_i is the inflow time (min), D_{IR} is the design daily irrigation requirement (mm/day, in./day), Z_{req} is the desired depth of irrigation (mm, in.), and E_a is the application efficiency (percent).

18.4.13 Delivery System

After the irrigation time, stream size, inflow time, basin dimensions, and number of basins irrigated per set have been determined, the delivery system should be designed. The delivery system may be open channels, discussed in Chapter 8, or pipelines, presented in Chapter 9. Appropriate structures, such as checks, check-drops, valves, or gates, should be provided in the delivery system for the control and regulation of water flow.

18.4.14 Supply Ditches

Supply ditches should be constructed with 0.1% grade or less and be able to convey the design inflow rate of each basin or multiples of design inflow rates if more than one basin is irrigated at the same time. The water surface in the ditch should be 0.15–0.3 m above the ground surface in the basin. If the water surface is below the ground surface then a low-head pump will be needed.

18.4.15 Supply Ditch Outlets

The water from the ditch is supplied through outlets to the basin. Outlets are of different types, such as gated rectangular or trapezoidal and gated orifice. If the entire ditch flow discharges into a basin, then gated rectangular or trapezoidal outlets are commonly used. If more than one basin is irrigated at the same time, then gated orifice outlets are appropriate. If the invert of the outlet is at or above the water surface in the ditch, then outlet gates are not needed.

18.4.16 Pipeline Outlets

Sometimes pipelines are used to supply water to the field. If the pipeline is underground then a vertical rise is attached to bring water to the basin and a gate or valve is installed in the riser to regulate flow or discharge. If the pipeline is on the ground surface and a riser is not required, then the valve or gate is attached to the pipeline. Alfalfa valves are in common use.

18.4.17 Erosion

In order to avoid erosion or scour adjacent to the turnout, the flow velocity is kept within 1 m/s. To limit the flow velocity, turnout structures with energy dissipation capability are emplaced.

18.4.18 Outlet Location

Depending on the flow rate and the width, a basin may have one or more outlets. If the basin width is 60 m or less and the flow rate is less than 0.4 m/s, then one outlet with energy dissipation capability is usually adequate. Although more outlets and spacing them properly may provide more uniform distribution of water over the basin, fewer outlets reduce labor and facilitate automation.

18.5 HYDRAULICS OF BASIN IRRIGATION

The hydraulics of basin irrigation comprises four phases: (1) initial spread of the entrance stream, (2) advance, (3) storage, and (4) recession. Upon entering the basin the stream spreads along the width till the entire width is covered with water. It also advances longitudinally along the slope.

The water advances longitudinally, depending on the slope, roughness, stream size, infiltration characteristics, and soil characteristics. The advance function of curve defines the lower boundary of the opportunity time and defines the advance phase. Water starts to get ponded once it has reached the downstream end of the basin and defines the storage phase. The volume of water stored in the storage phase is the volume of water admitted to the basin minus the volume of water infiltrated. Once the depth of ponding at the upstream end becomes zero, recession starts and continues until the depth at the downstream end goes to zero. This defines the recession phase.

18.6 USDA-SCS DESIGN METHOD

The USDA-SCS (1974) (now called the Natural Resources Conservation Service [NRCS]) developed design equations for basin irrigation, which are discussed here. These equations are based on the premise that the volume of water applied to a unit width of basin strip during the time of water advance from the head to the tail is equal to the volume of intake plus the volume in temporary storage (surface and subsurface) during the same period. Further, one must know the cumulative infiltration characteristics of the soil, Manning's roughness for the crops to be irrigated, and application depth to be used for design.

18.6.1 Opportunity Time

The opportunity time needed for the infiltration of the net irrigation depth is determined from the SCS cumulative infiltration equation as

$$T_n = \left[\frac{F_n - g}{e}\right]^{\frac{1}{f}}, \tag{18.15}$$

where T_n is the net opportunity time or time required, F_n is the desired net application depth or infiltration depth, and e, f, and g are infiltration parameters as defined for eq. (18.1).

18.6.2 Advance Time

The advance time is the time required for the unit flow rate to reach the downstream end of the basin. The required advance time for the desired water application efficiency is obtained by the product of the net opportunity time and the efficiency advance ratio (R) whose values are given in Table 18.4. The quantity R is defined by the ratio of advance time (T_a) to the net opportunity time (T_n). The water application efficiency (in percent) is the ratio of average net application to gross application. Figure 18.3 graphs efficiency versus the efficiency advance ratio.

Table 18.4 *Efficiency as a function of the efficiency advance ratio*

Efficiency E_a	Efficiency advance ratio ($R = T_a/T_n$)
95	0.16
90	0.28
85	0.40
80	0.58
75	0.80
70	1.08
65	1.46
60	1.90
55	2.45
50	3.20

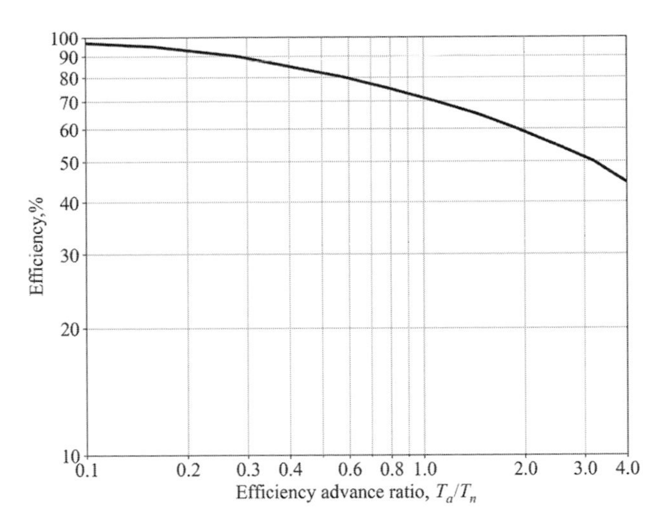

Figure 18.3 Efficiency versus the efficiency advance ratio for border irrigation. (After USDA-SCS, 1974).

18.6.3 Basin Length and Inflow Rate

The basin length is expressed as a function of unit inflow rate (Q_0) and advance time (T_a):

$$L = \frac{6 \times 10^4 Q_0 T_a}{\frac{\sigma T_a^f}{1+f} + 7.0 + 1798 n^{\frac{3}{8}} Q_0^{\frac{9}{10}} T_a^{\frac{3}{16}}}, \tag{18.16}$$

where L is the basin length (m), Q_0 is the unit inflow rate ($m^3/s/m$ or m^2/s), T_a is the required advance time for the desired efficiency (min), n is Manning's roughness coefficient, and e, f, and $g = 6.985 \approx 7.0$ are constants in the cumulative intake function. Manning's n, commonly used in basin and border irrigation, is given in Table 17.5. A freshly tilled basin or border has an n value which is the same as for furrows: 0.03–0.05. Compared to furrows, Manning's n for basin and border are more affected by crop growth and the geometry of its crown. Before substantial crop growth, the n value is about 0.15–0.20, but can be as high as 0.80 for dense crops such as alfalfa–grass mix (USDA-NRCS, 2012).

18.6.4 Inflow Time

The time required to apply the gross application rate to the basin, called inflow time, is computed as

$$T_i = \frac{F_n L}{600 Q_0 E_a}, \tag{18.17}$$

where T_i is the inflow time (min) for the unit inflow rate Q_0 ($m^3/s/m$), F_n is the net depth of application (mm), and E_a is the efficiency (percent).

18.6.5 Maximum Flow Depth

The maximum flow depth is computed as

$$d = 2250 n^{\frac{3}{8}} Q_0^{\frac{9}{16}} T_i^{\frac{3}{16}}, \tag{18.18}$$

where d is the flow depth at the inlet (mm). If T_a is larger than T_i, then T_i is replaced by T_a in eq. (18.18).

18.6.6 Design Charts

The USDA-SCS (1974) prepared charts for designing basins, as shown in Figure 18.4. A separate chart can be prepared for any combination of Manning's roughness coefficient, cumulative intake function, and net application depth. This chart should entail the relationship between length, time of inflow, inflow rate, and depth of flow for a given efficiency.

Example 18.2: A basin with the following characteristics is being irrigated. Inflow rate is 0.01 $m^3/s/m$, the maximum allowable depth of flow is 200 mm, Manning's n is 0.20, the desired depth of application is 150 mm, and intake family is 0.3. Assuming the desired efficiency of 85%, what will be the basin length, required inflow time, required opportunity time, and the maximum depth of flow?

Solution:

1. For intake family 0.3, the parameters are:

$$F = 0.9347 t^{0.721} + 6.985.$$

2. Given the desired depth of application $F_n = 150$ mm, the opportunity time (T_n) can be calculated using eq. (18.15):

$$T_n = \left(\frac{F_n - 6.985}{0.9347}\right)^{1/0.721}$$

$$= \left(\frac{150 - 6.985}{0.9347}\right)^{1/0.721} = 1071.8 \text{ min}.$$

3. For the advance efficiency ratio, from Table 18.4, given the desired efficiency is 85%, $T_a/T_n = 0.4$.
4. Advance time: $T_a = 0.4 T_n = 0.4 \times 1071.8$ min $=$ 428.7 min.
5. Basin length: Given the inflow rate $Q_0 = 0.01$ $m^3/s/m$, advance time $T_a = 428.7$ min, $n = 0.20$, the basin length is computed using eq. (18.16):

$$L = \frac{6 \times 10^4 Q_0 T_a}{\frac{eT_a^f}{1+f} + 7.0 + 1798 n^{\frac{3}{8}} Q_0^{\frac{9}{10}} T_a^{\frac{3}{16}}}$$

$$= \frac{6 \times 10^4 (0.01)(428.7)}{\frac{0.9347(428.7)^{0.721}}{1+0.721} + 7.0 + 1798(0.2)^{\frac{3}{8}}(0.01)^{\frac{9}{10}}(428.7)^{\frac{3}{16}}}$$

$$= 2612.2 \text{ m}.$$

6. Inflow time: Given the net application rate is $F_n = 150$ mm, $L = 2612.2$ m, the efficiency is 85%, the inflow rate $Q_0 = 0.01$ m^3/s, the inflow time is computed using eq. (18.16):

$$T_i = \frac{F_n L}{600 Q_0 E_a} = \frac{150 \times 2612.2}{600 \times 0.01 \times 85} = 768.3 \text{ min}.$$

7. Maximum depth of flow: Given that T_a is smaller than T_i, the maximum flow depth is computed using eq. (18.18):

$$d = 2250 n^{\frac{3}{8}} Q_0^{\frac{9}{16}} T_l^{\frac{3}{16}} = 2250(0.2)^{\frac{3}{8}}(0.01)^{\frac{9}{16}}(768.3)^{\frac{3}{16}}$$
$$= 287.5 \text{ mm}.$$

Since the calculated maximum depth of flow is larger than the maximum allowable flow depth, the flow rate needs to be reduced.

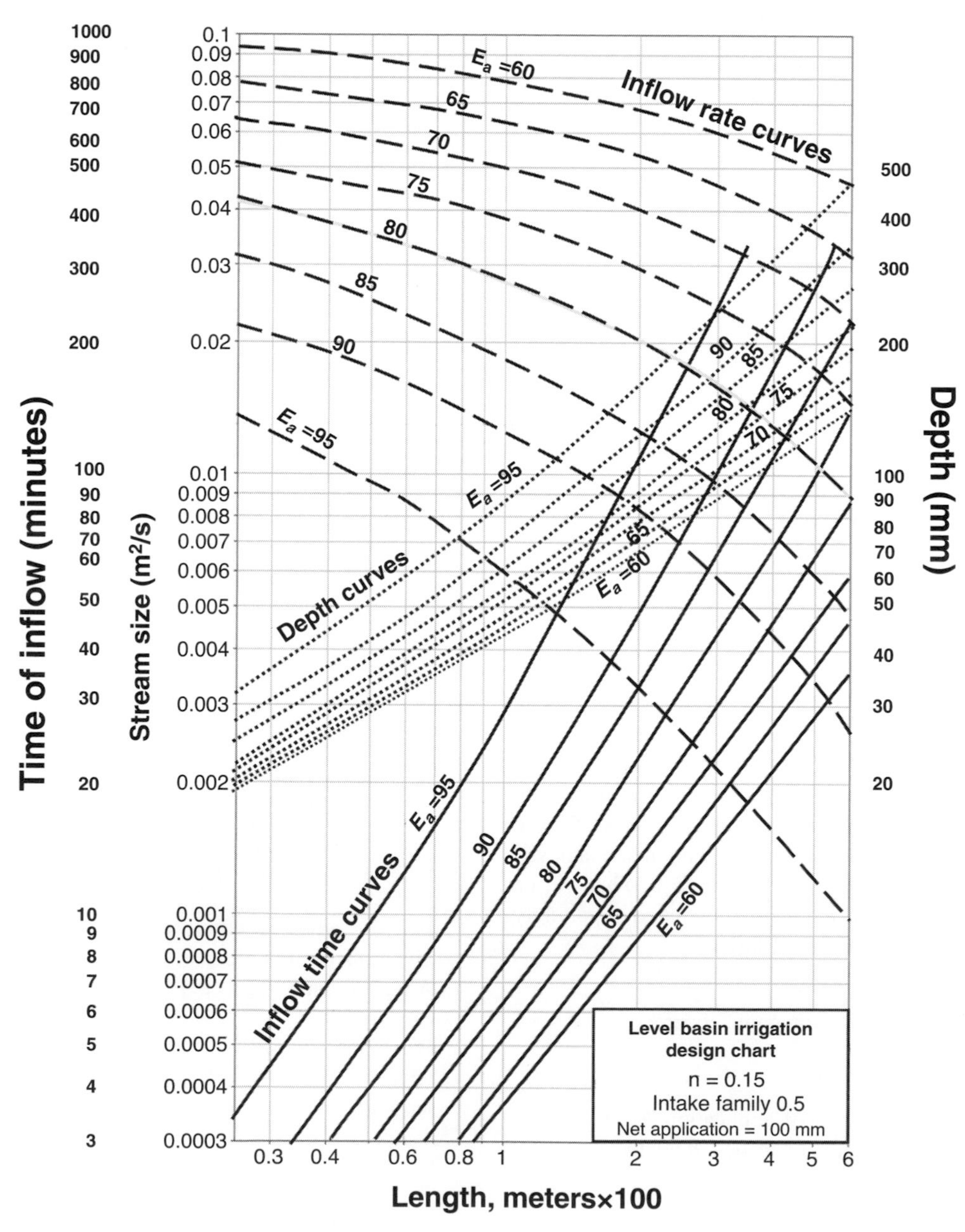

Figure 18.4 Basin irrigation chart sample. (After USDA-SCS, 1974.).

QUESTIONS

Q.18.1 What will be the dimension of the basin if the soil is clay loam and the slope is 0.6%. The basin construction is mechanized and thus the basin width should be maximized. The stream size is 90 L/s. Use the Booher empirical method.

Q.18.2 Compute the basin areas for sand, sandy loam, clay loam, and clay for inflow rates varying from 50 L/s to 500 L/s at intervals of 50 L/s. Tabulate the areas and plot area versus inflow rate for these four soil types.

Q.18.3 Compute the ratios of areas computed in Q.18.2, taking clay loam as the reference soil. In other words, the ratio should be computed as the area of a soil divided by the area of clay loam for a given inflow rate. Tabulate the ratios and then plot them against the inflow rate for different soils.

Q.18.4 What is the efficiency advance ratio if the application efficiency is 80%, 70%, and 60%?

Q.18.5 What will be the net opportunity time if the desired net application depth is 150 mm, 175 mm, and 200 mm? Take the soil as belonging to the intake family of 0.5.

Q.18.6 Based on the net opportunity time calculated in Q.18.5, what will be the required advance time if the application efficiency is 80%, 70%, and 60%, as in Q.18.4?

Q.18.7 What is the application efficiency if the efficiency advance ratio is 0.40, 0.80, and 1.45?

Q.18.8 Compute the basin length if Manning's n is 0.04, the inflow rate 0.005 $m^3/s/m$, the required advance time is 200 min, and the intake family is 0.5.

Q.18.9 Compute the inflow time if the application efficiency is 75%, the basin length is 400 m, the net application depth is 180 mm, and the inflow rate is 0.008 $m^3/s/m$.

Q.18.10 What will be the maximum depth of flow at the inlet end of the basin if the inflow rate is 0.01 $m^3/s/m$, the inflow time is 100 min, and Manning's n is 0.035?

REFERENCES

Booher, L. J. (1974). Surface irrigation. FAO agricultural development paper 95, pp. 80–90.

Brouwer, C., Prins, K., Kay, M., and Heibloem, M. (1984). *Irrigation Water Management: Irrigation Methods*. Rome: FAO.

Hart, W. E., Woodward, G., and Humphreys, A. S. (1981). Design and operation of gravity or surface systems. In *Design and Operation of Farm Irrigation Systems*, edited by M. E. Jensen. St Joseph, MI: ASAE.

USDA-NRCS (2012). Surface irrigation. In *National Engineering Handbook*. Washington, DC: USDA.

USDA SCS (1974). Border irrigation. In *National Engineering Handbook*. Washington, DC: USDA.

Walker, W. R. (1989). Guidelines for designing and evaluating surface irrigation systems. FAO irrigation and drainage paper 45.

Walker, W. R., and Skogerboe, G. V. (1987). *Surface Irrigation: Theory and Practice*. Englewood Cliffs, NJ: Prentice Hall.

19 Border Irrigation

Notation

A	cross-sectional area per unit width (m^2)	t_d	time at the end of depletion (min)
d	normal depth of flow at the upstream end of the border (mm, in.)	T_a	advance time (min)
d_n	irrigation depth to be applied (mm)	T_I	infiltration opportunity time for the desired depth of application, or net infiltration time (min)
E_a	application efficiency (percentage)	T_i	inflow time or cutoff time (T_{co}) (min)
E_d	distribution efficiency (percentage)	T_L	lag time (min)
f_0	basic infiltration rate (m/min)	T_r	recession time, time of recession from the beginning of irrigation (min)
I	average infiltration rate over the length (m/s)	W	field width (m)
k, a	coefficients of the modified Kostiakov equation	W_b	border set width that contains even number of borders of satisfactory width (m)
L	length of the field (m)	WP	wetted perimeter (m)
n	Manning's roughness coefficient	y_L	depth at the downstream end (m)
N_b	integer number of sets	y_0	inflow depth at the inlet (m)
Q	total water supply (m^3/min)	y_{max}	maximum flow depth at the border inlet (m)
Q_0	unit inflow rate or stream size (m^3/min/m)	Z_o	infiltrated depth at the border inlet (m)
Q_{max}	maximum permissible nonerosive inflow rate or stream size (m^3/min/m)	Z_L	infiltrated depth in the case of deficit irrigation (m)
Q_{min}	minimum inflow rate (m^3/min/m)	Z_{req}	required irrigation depth, or the design irrigation requirement (m)
Q_r	runoff (m^3/min/m)	τ_{req}	time to satisfy the irrigation requirement using eq. (19.2), which is the same as T_I (min)
R	hydraulic radius equal to A/WP (m)		
S_0	slope of the border (L [length]/L)		
S_y	slope of the depth of flow, with flow assumed to change with distance at a uniform rate over the entire length of border (L/L)		

19.1 INTRODUCTION

Border irrigation is one of the popular methods of surface irrigation, especially in developing countries, largely because there is little energy required to irrigate agricultural fields as water flows under gravity, the cost involved is low, and the skill needed to construct borders is minimal. The border method of irrigation is suitable for close-growing crops such as wheat, barley, legumes, and fodder crops, but for rice, which requires standing water, the border is bunded all around. Also, it is not suitable for crops that are sensitive to wet soil conditions. The border irrigation method is most suited for soils with moderate low to moderate high intake rates without crusting. It is not suitable for soils with very low and very high intake rates.

The source of water for border irrigation is usually a ditch or channel flowing perpendicular to the borders or low-pressure pipelines or tubes. Thus, water is introduced at the upstream end of the field and is allowed to flow into the border as a sheet of flow which covers the entire width of the border, and flows along the longitudinal direction. In this manner, the entire border is wetted. The inflow to the border is allowed for a time such that the volume of water infiltrated into the soil is almost equal to the desired depth. After the flow into the border is cut off, the water front recedes from the upstream end to the downstream end.

Border irrigation design aims to achieve an equal opportunity time everywhere along the border time by balancing advance and recession curves. This can happen if the volume of water supplied to the border is adequate to cover it with an average depth equal to the gross application, or the intake opportunity time at the head of the border is equal to the time needed for the soil to infiltrate the net application. Fundamental to design is the intake rate of the soil and

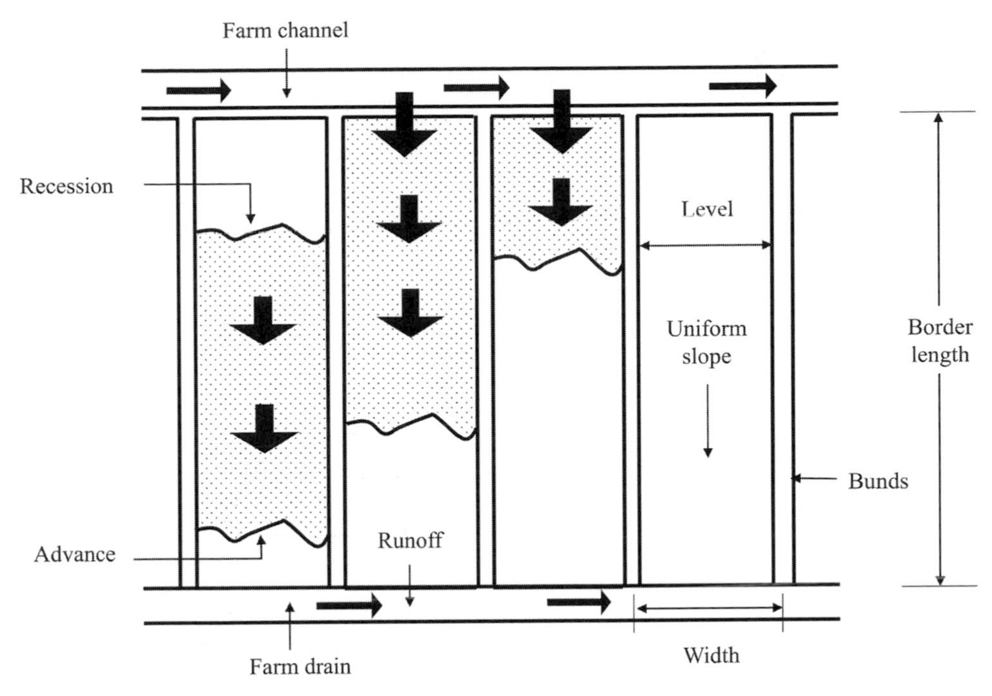

Figure 19.1 Typical layout for graded border irrigation system. (Adapted from Jurriëns et al., 2001.)

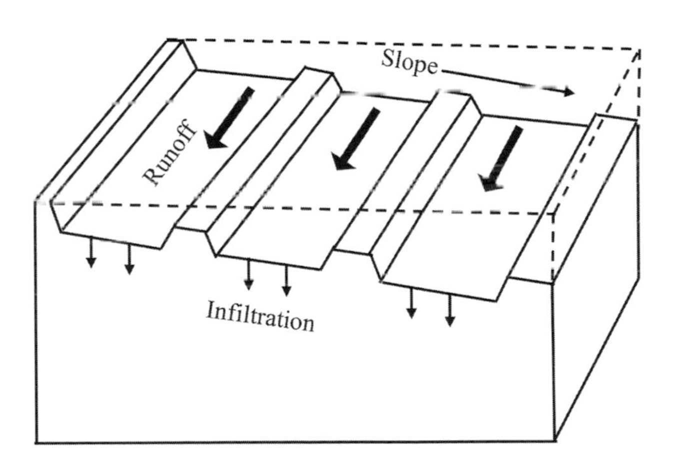

Figure 19.2 Contour border irrigation.

roughness coefficient. This chapter discusses the method and design of border irrigation.

19.2 BORDERS

A border is a strip bounded by low ridges. Thus, for border irrigation a field is divided into strips by constructing parallel ridges along the slope. The border strips have no cross-slope, but they may be level (level border) or graded (graded border) in the longitudinal direction. The border length can vary from 3 m to 800 m. Graded borders can be either open or closed (diked) at the downstream end. If the downstream end is open, the water will freely drain from the border. Diking is therefore done to prevent runoff of irrigation water, or a drainway across the lower end of adjacent borders may be needed to carry away runoff water. Level borders are often diked at the downstream end, which leads to higher efficiency and uniformity. Borders are mostly rectangular planes.

In general, borders are constructed on lands that have a low grade or are level, or that can be leveled economically without reducing field productivity. On land that has an excessive slope and undulating topography, leveling may not be economical. On such lands, borders are also laid across the longitudinal slope following contours; these borders are called contour borders. Contour borders have a uniform longitudinal slope and no cross-slope. The field to be irrigated using borders is divided into graded strips bounded by parallel dikes or ridges, and can be open or closed.

19.3 LAYOUT OF BORDERS

A typical border layout consists of a series of strips that are parallel to the field boundary, with the source of water on the upstream end and drainage ditch at the downstream end, as shown in Figure 19.1. On the other hand, contour borders divide the field into a series of strips following contours where each strip is leveled cross-wise, forming a series of benches, as shown in Figure 19.2. The elevation difference between consecutive benches should be limited to 30 cm and should not exceed 60 cm in any case (Michael, 1978).

The border layout also depends on the field size, location of water source, and length and width of borders. If a field is small and the soil infiltration capacity is low, then borders extend over the full length of the field. If the field is large and its soil infiltration capacity is high, then two or more borders across the field may be needed. It is desirable to have the source of water supply located such that it allows all borders to be irrigated by gravity. If the source of water supply is a canal, then the farm channel will convey water at the head of the border.

Table 19.1 *Typical border lengths for different soils*

	Typical border length	
Soil	m	ft
Clay	180–350	600–1150
Clay loam	90–300	300–1000
Sandy loam	90–250	300–800
Loamy sand	75–150	250–500
Sand	60–90	200–300

After Booher, 1974.

19.3.1 Border Length

The length of a border depends on field size, soil characteristics, topographic slope, stream size, and depth of irrigation required. Booher (1974) has given typical border lengths for different soils for different combinations of soil texture, slope, and irrigation requirement, based on experience in the southwestern United States, as shown in Table 19.1. He reported that acceptable irrigation performance can be achieved with border lengths up to 800 m on low-intake soils, whereas a length of less than 100 m may be required on high-intake soils. Existing field size and shape are often practical limits to the size of borders. Savva and Frenken (2002) have presented typical border lengths, as shown in Tables 19.2 and 19.3, for large and small farm sizes, respectively. The values in the tables are guidelines and actual border length depends on field conditions, ease and economy of field operations, and irrigation performance. Border length can be longer on heavy soils than on light soils, and can be longer on the same soil with larger stream size. The border can be longer on steeper slopes, but precautions should be taken against erosion. Short basin length may provide better irrigation performance but require more labor and system cost.

Table 19.2 *Typical border strip dimensions in meters as related to soil type, slope, irrigation depth, and stream size*

Soil type	Slope (%)	Depth applied (mm)	Flow (L/s)	Strip width (m)	Strip length (m)
Coarse	0.25	50	240	15	150
		100	210	15	250
		150	180	15	400
	1.00	50	80	12	100
		100	70	12	150
		150	70	12	250
	2.00	50	35	10	60
		100	30	10	100
		150	30	10	200
Medium	0.25	50	210	15	250
		100	180	15	400
		150	100	15	400
	1.00	50	70	12	150
		100	70	12	300
		150	70	12	400
	2.00	50	30	10	100
		100	30	10	200
		150	30	10	300
Fine	0.25	50	120	15	400
		100	70	15	400
		150	40	15	400
	1.00	50	70	12	400
		100	35	12	400
		150	20	12	400
	2.00	50	30	10	320
		100	30	10	400
		150	20	10	400

After Withers and Vipond, 1974. Copyright © 1974 by Batsford, London. Reprint with permission from Batsford.

Table 19.3 *Suggested maximum border widths and length for smallholder irrigation schemes; flow is given per meter width of the border and the total flow into a border is equal to the unit flow multiplied by the border width in meters*

Soil type	Border strip slope (%)	Unit flow per meter width (L/s)	Border strip width (m)	Border strip length (m)
Sand (infiltration rate greater than 25 mm/h)	0.2–0.4	10–15	12–30	60–90
	0.4–0.6	8–10	9–12	80–90
	0.6–1.0	5–8	6–9	75
Loam (infiltration rate of 10–25 mm/h)	0.2–0.4	5–7	12–30	90–250
	0.4–0.6	4–6	9–12	90–180
	0.6–1.0	2–4	6	90
Clay (infiltration rate less than 10 mm/h)	0.2–0.4	3–4	12–30	180–300
	0.4–0.6	2–3	6–12	90–180
	0.6–1.0	1–2	6	90

After Savva and Frenken, 2002.

Table 19.4 *Recommended maximum border widths for different slopes in the direction of flow*

	Maximum border width	
Slope (%)	m	ft
Level	60	200
0.0–0.1	35	120
0.1–0.5	20	60
0.5–1.0	15	50
1.0–2.0	12	40
2.0–4.0	9	30
4.0–6.0	6	20

After USDA-SCS, 1974.

19.3.2 Border Width

Booher (1974) has recommended that the difference in elevations of ridges of a border should not exceed 3 cm (1 in.), or the difference between ground surface elevations of uphill and downhill sides of a ridge should not exceed 6 cm (2 in.). Thus, the border width should not exceed 9 m (30 ft.) on a 1.0% cross-slope (9 cm/0.01 = 9 m). Alternatively, the entire field can be graded to a uniform 1% slope and the maximum width will then be 3 cm/0.01 = 3 m. For zero field slope and 1% cross-slope, the width can be 6 cm/0.01 = 6 m (20 ft). The width of a border also depends on the stream size and land declivity, and can be 3–30 m. For small stream sizes, the width can be less but should not be less than 3 m in any case. The width should not be greater than 9 m on 1% cross-slopes (James, 1988). If machinery is used for cultivation, then the width should be sufficient to allow one pass. However, the width should be sufficient to allow an even number of passes.

If the cross-border slope is zero between adjacent ridges, then the maximum border width depends on how well the water spreads laterally from the inlet of the border to the ridges. This spread depends on the slope in the direction of flow. The values of maximum border width, depending on the slope, have been given by the USDA-SCS (1974), as shown in Table 19.4.

19.3.3 Slope

Three issues should be considered in the border slope. First, borders should have enough slope to permit water to flow downstream over the surface. Second, it should allow some water to infiltrate into the soil but prevent deep percolation at the upstream end. Third, flow velocity should not be so large as to cause significant soil erosion. The maximum slope depends on the potential for soil erosion. Savva and Frenken (2002) suggest a minimum slope of 0.05–0.1% needed for water to flow downstream over the border. Soil slope is generally greater on coarse-textured soils (0.25–0.6% for sand) than on fine-textured soils (0.05–0.20% for clay, 0.20–0.40% for loams) (Michael, 1978). A larger slope can be used on sod conditions than on non-sod cover conditions. In order to improve the application uniformity of border irrigation, land smoothing may be required to remove furrows or depressions that concentrate the flow. It may be necessary to adjust slopes to improve irrigation efficiency. Minimum slopes of 0.2–0.3% and maximum slopes of 2% for sandy loams and up to 7% for pastured clay soils with water-stable aggregates have been recommended by Booher (1974).

19.3.4 Ridges

The top width and height of the border ridge must be about the same. The height should be sufficient to accommodate the maximum depth of flow. The height of ridges should be at least 3 cm (1 in.) higher than the maximum flow depth. Also, a freeboard of about 2.5 cm must be provided. It is difficult to fully wet ridges wider than 30 cm, and in salinity-prone areas salts may accumulate on ridge crests. The side slope should not be greater than 2.5:1 for cohesive soils and

3:1 for non-cohesive soils (USDA-SCS, 1984), and must be stable when wet. The base widths of ridges constructed in clay soils may be 60 cm (2 ft) and may be up to 2.4 m (8 ft) in sandy soils (Booher, 1974).

The maximum depth of flow can be determined using Manning's equation as:

$$y_{max} = \left(\frac{Q_{max} n}{60 S_0^{0.5}}\right)^{3/5}, \tag{19.1}$$

where y_{max} is the maximum flow depth at the border inlet (m), Q_{max} is the maximum permissible nonerosive inflow rate or stream size (m^3/min/m), and S_0 is the slope (fraction). Equation (19.1) yields the inflow depth for a given inflow. The depth of flow, d, at the upstream end of a border can also be obtained as

$$d = K_1 T_L^{\frac{3}{16}} Q_0^{\frac{9}{16}} n^{\frac{3}{8}} \text{ for } S_0 \leq 0.4\%, \tag{19.2}$$

$$d = K_2 Q_0^{0.6} n^{0.6} S_0^{-0.3} \text{ for } S_0 > 0.4\%, \tag{19.3}$$

in which d is the normal depth of flow at the upstream end of the border (mm, in.); T_L is the lag time (min); Q_o is the stream size or unit inflow rate (m^3/s/m, ft^3/s/ft); n is Manning's roughness coefficient; S_0 is the slope of the border (m/m, ft/ft); K_1 is the unit constant equal to 2454 for Q_0 in m^3/s/m and d in mm, and equal to 25.4 for Q_0 in ft^3/s/ft and d in inches; and K_2 is the unit constant equal to 1000 for Q_0 in m^3/s/m and d in mm, and equal to 9.46 for Q_0 in ft^3/s/ft and d in inches.

19.3.5 Number of Borders and Number of Sets

The number of borders and sets can be determined by selecting the unit inflow rate (flow rate per unit width of the border), which is in the range of the maximum and minimum inflow rates. The set width that contains an even number of borders of satisfactory width for ease of other farming operations can be computed as:

$$W_b = \frac{Q}{Q_0}, \tag{19.4}$$

where Q is the total water supply (m^3/min), Q_0 is the unit inflow rate (m^3/min/m), and W_b is the border set width (m) that contains an even number of borders of satisfactory width. The integer number of sets (N_b) can be obtained as

$$N_b = \frac{W}{W_b}. \tag{19.5}$$

where W is the field width (m).

Example 19.1: If a field has a width of 200 m and the border set width is 50 m, what should be the number of border sets?

Solution: Using eq. (19.5),

$$N_b = \frac{200 \text{ m}}{50 \text{ m}} = 4.$$

Example 19.2: If a channel has a flow of 2 m^3/min and the inflow to a border set is 0.20 m^3/min/m, what would be the border set width?

Solution: Using eq. (19.4),

$$W_b = \frac{Q}{Q_0} = \frac{2 \frac{\text{m}^3}{\text{min}}}{0.2 \frac{\text{m}^3}{\text{min} \times \text{m}}} = 10 \text{ m}.$$

19.3.6 Water Delivery

Water is delivered from the source of supply to the field either through an open channel or a low-pressure pipeline. For a farm, the delivery system should have sufficient capacity to meet the irrigation demand everywhere in the farm. The water delivery system is designed with knowledge of the inflow rate, border length, border slope, and number of borders per set (set width). The principles for the design of open channels and pipelines are discussed in Chapters 8 and 9.

19.3.7 Cutoff Time (Inflow Time)

A border irrigation event is defined by the duration within which the desired depth of irrigation occurs at the end of the border. A substantial amount of surface storage remains in the border at the time when inflow is cut off, and this water infiltrates into the soil as well as runs off the border during the depletion and recession phases. Thus, the recession time (T_r) at the end of the border equals the sum of the advance time (T_a) and intake opportunity time (T_I) – that is, $T_r = T_a + T_I$, as shown in Figure 19.3. Knowing the recession time, the time of depletion and cutoff time must be determined for each chosen inflow rate. Assuming that advance and recession curves are parallel (i.e., infiltration opportunity time is considered the same along the whole border length), the inflow time can be computed as

$$T_i = T_I - T_L, \tag{19.6}$$

where T_i is the inflow time or cutoff time (T_{co}) (min); T_I is the infiltration opportunity time for the desired depth of application (min); and T_L is the lag time (min).

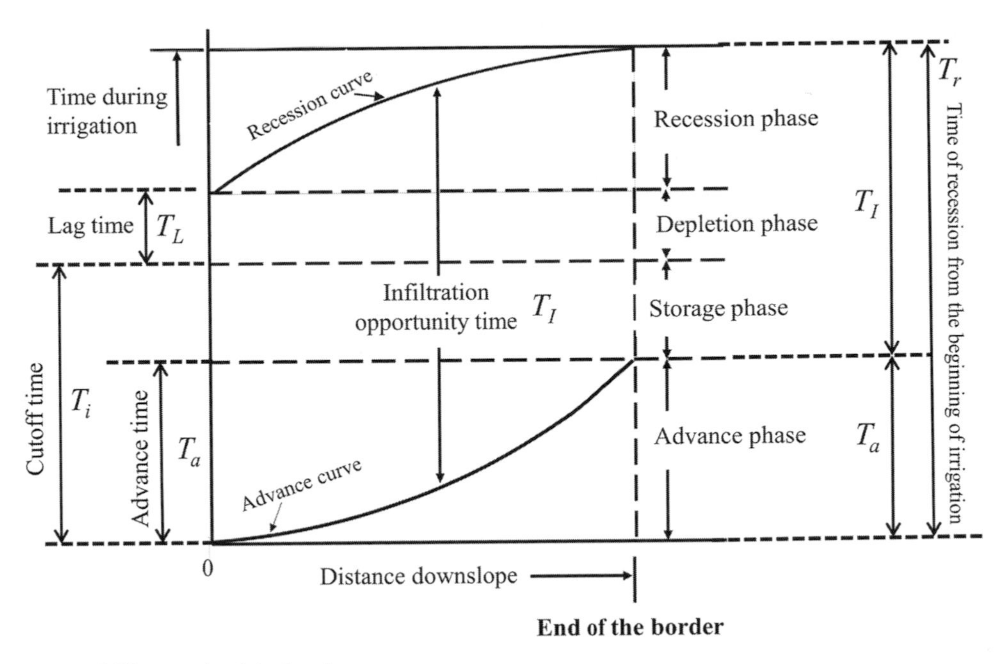

Figure 19.3 An illustration of different surface irrigation phases.

Example 19.3: For how long should a farmer irrigate a field if the irrigation opportunity is estimated to be 90 min and the depletion time is 20 min?

Solution: Using eq. (19.6), the inflow time can be computed as

$$T_i = T_I - T_L = 90\,\text{min} - 20\,\text{min} = 70\,\text{min}.$$

19.3.8 Irrigation Stream Size

Inflow rate and time of cut off are key design variables in border irrigation, and they offer maximum flexibility in the design process. The stream size must be such that it is nonerosive and water adequately spreads across the width of the border and must reach the end of the border. Also, the resulting advance and recession times are about the same. The maximum nonerosive stream size, Q_{max}, for non-sod-forming crops, such as grains and alfalfa, can be obtained as

$$Q_{max} = KS_0^{-0.75}, \qquad (19.7)$$

where K is a unit constant equal to 1.765×10^{-4} for Q_{max} in $m^3/s/m$ and equal to 1.899×10^{-3} for Q_{max} in $ft^3/s/ft$; and S_0 is the slope of the border in m/m. The value of Q_{max} computed using eq. (19.7) should be doubled for dense well-established sod crops.

The USDA-SCS (1974) provided the following guidelines for selecting the maximum nonerosive inflow rate (Q_{max}) and minimum inflow rate (Q_{min}). Q_{max} for non-sod-forming crops (alfalfa and small grain) can be obtained as:

$$Q_{max} = 0.01059S_0^{-0.75}, \qquad (19.8)$$

where Q_{max} is in $m^3/min/m$ and S_0 in m/m. Eqs (19.7) and (19.8) are the same, if they were changed to the same unit. For dense sod-forming crops, the value of Q_{max} can be twice that obtained from eq. (19.7). The value of Q_{min} can be obtained as:

$$Q_{min} = 0.000357\frac{L}{n}S_0^{0.5}, \qquad (19.9)$$

where Q_{min} is in $m^3/min/m$ and S_0 in m/m, the constant in eq. (19.9) is 5.95×10^{-6} for Q_{min} in $m^3/s/m$ and 6.4×10^{-5} for Q_{min} in $ft^3/s/ft$; L is the border length in m; and n is Manning's roughness coefficient. The stream size must be within the range of minimum and maximum inflow rates that results in the maximum application efficiency.

Example 19.4: Compute the maximum border inflow rate in $m^3/min/m$ if the field slope is given as 0.002.

Solution: Using eq. (19.7),

$$Q_{max} = KS_0^{-0.75} = \left(1.765 \times 10^{-4}\right) \times (0.002)^{-0.75}$$

$$= 0.019\frac{m^3}{s \times m}\frac{60\,s}{min} = 1.12\frac{m^2}{min}.$$

Now, using eq. (19.8),

$$Q_{max} = 0.01059S_0^{-0.75} = (0.01059) \times (0.002)^{-0.75}$$

$$= 1.12\frac{m^2}{min}.$$

Example 19.5: What will be the maximum depth at the inlet for the border in Example 19.4 if Manning's roughness coefficient is 0.08?

Solution: Using eq. (19.1), the maximum depth of flow can be determined as:

$$y_{max} = \left(\frac{Q_{\max} \times n}{60 \times S_0^{0.5}}\right)^{\frac{3}{5}} = \left(\frac{1.12 \times 0.08}{60 \times (0.002)^{0.5}}\right)^{\frac{3}{5}} = 0.130 \text{ m}.$$

The stream size should also depend on the advance, storage, depletion, and recession phases. The duration of the depletion phase, also called lag time (T_L), as shown in Figure 19.3, can be considered as the time required to drain the water above the upstream end. For borders this can be determined as

$$T_L = \frac{n^{1.2} Q_0^{0.2}}{120 \times K^{0.2} S_0^{1.6}} \quad \text{for } S_0 > 0.4\%, \tag{19.10}$$

$$T_L = \frac{n^{1.2} Q_0^{0.2}}{120 \times K^{0.2} A^{1.6}} \quad \text{for } S_0 \leq 0.4\%, \tag{19.11}$$

where

$$A = S_0 + \frac{0.0094 n Q_0^{0.175}}{K^{0.175} T_I^{0.88} S_0^{0.5}}, \tag{19.12}$$

in which T_L is the lag time in minutes; n is Manning's roughness coefficient; Q_0 is stream size (m³/s/m, ft³/s/ft); S_0 is the border slope in m/m or ft/ft; T_I is the infiltration opportunity time for desired application depth (min); and K is a unit constant equal to 1 for Q_0 in m³/s/m and equal to 10.76 for Q_0 in ft³/s/ft. For flow at normal depth (i.e., when $S_0 > 0.4\%$), eq. (19.10) can be obtained by the depletion volume ($d^2/2S_0$) dividing by the stream size (Q_0), where the flow depth at the upstream end of the border (d) is replaced with eq. (19.3).

Example 19.6: What will be the depletion time (or lag time) for a border if the ground slope is 0.002, inflow discharge is 0.003 m²/s, the intake opportunity time is 100 min, and Manning's roughness coefficient is 0.15? Compute the depth at the inlet. What will be the lag time if the border slope is 0.01? What will be the depth at the inlet?

Solution: Given $S_0 = 0.002 < 0.4\%$, the inflow rate $Q_0 = 0.003 \text{ m}^2/\text{s}$, the intake opportunity time $T_I = 100$ min, Manning's roughness coefficient $n = 0.15$, the lag time can be computed using eqs (19.11) and (19.12):

$$A = S_0 + \frac{0.0094 n Q_0^{0.175}}{K^{0.175} T_I^{0.88} S_0^{0.5}}$$

$$= 0.002 + \frac{0.0094 \times 0.15 \times \left(0.003 \frac{\text{m}^2}{\text{s}}\right)^{0.175}}{1^{0.175} (100 \text{ min})^{0.88} (0.002)^{0.5}}$$

$$= 0.0022,$$

$$T_L = \frac{n^{1.2} Q_0^{0.2}}{120 \times K^{0.2} A^{1.6}} = \frac{0.15^{1.2} \times (0.003)^{0.2}}{120 \times 1^{0.2} \times (0.0021)^{1.6}}$$

$$= 4.78 \text{ min}.$$

The inlet depth is computed using eq. (19.2) as

$$d = K_1 T_L^{\frac{3}{16}} Q_0^{\frac{9}{16}} n^{\frac{3}{8}} = 2454 \times (4.78)^{\frac{3}{16}} \times (0.003)^{\frac{9}{16}} \times (0.15)^{\frac{3}{8}}$$

$$= 61.5 \text{ mm}.$$

Given $S_0 = 0.01 > 0.4\%$, the lag time can be computed using eq. (19.10):

$$T_L = \frac{n^{1.2} Q_0^{0.2}}{120 \times K^{0.2} S_0^{1.6}} = \frac{0.15^{1.2} \times (0.003)^{0.2}}{120 \times 1^{0.2} \times (0.01)^{1.6}}$$

$$= 0.42 \text{ min}.$$

The depth of flow at the inlet is computed using eq. (19.3):

$$d = K_2 Q_0{}^{0.6} n^{0.6} S_0^{-0.3}$$

$$= 1000 \times 0.003^{0.6} \times 0.15^{0.6} \times 0.01^{-0.3}$$

$$= 39.1 \text{ mm}.$$

The procedure for the calculation of net infiltration time or the infiltration opportunity time (T_I), recession lag time (T_L), unit inflow rate (Q_0), cutoff time (T_{CO}) or inflow time (T_i), and depth of flow (d) for a border irrigation system is illustrated as follows shown in Example 19.7.

Example 19.7: Calculate the net infiltration time, recession time, unit inflow rate, time to cutoff, and depth flow for a border irrigation system with 100% application efficiency.

The information given is as follows: Intake family is 0.4, irrigation depth to be applied is 80 mm, ground slope is 0.002, Manning's roughness coefficient is 0.25, allowable flow depth is 120 mm, distribution pattern efficiency is 75%, border length is 200 m, and the crop is wheat.

Solution: Given the irrigation depth to be applied $d_n = 80$ mm, the net infiltration time T_I can be calculated using the modified form of the Kostiakov equation developed by the USDA-SCS, as shown in eq. (12.21):

$$T_I = \left(\frac{d_n - c}{a}\right)^{1/b} = \left(\frac{80 - 7.0}{1.064}\right)^{\frac{1}{0.736}} = 312.7 \text{ min},$$

in which the values of a, b, and c are extracted from the table for an intake family of 0.4 ($a = 1.064$, $b = 0.736$, and $c = 7.0$ if T_I is in mm). These parameters are shown in Table 19.5.

Table 19.5 *Parameters for the calculation of infiltration rate using the SCS intake family concept for depth of infiltration in mm and time in minutes*

Intake family	a	b	c
0.1	0.6198	0.661	7.0
0.2	0.7772	0.699	7.0
0.4	1.064	0.736	7.0
0.5	1.196	0.748	7.0
1.0	1.786	0.785	7.0
1.5	2.284	0.799	7.0
2.0	2.753	0.808	7.0

After USDA-SCS, 1974.

The recession lag time for gently sloping borders is shown in Table 19.6.

The recession lag time T_L for $S_0 = 0.002$, $n = 0.25$, and $T_I = 312.7$ must be within the following limits:

$$4.8 \text{ min} < T_L < 13.8 \text{ min}.$$

Given the border length $L = 200$ m, the distribution efficiency $E_d = 75\%$, the unit inflow rate Q_0 (m³/s/m) is computed by solving the following equations by trial and error as:

$$Q_0 = \frac{0.00167 \times d_n \times L}{(T_I - T_L)E_d},$$

where 0.00167 is obtained by changing the unit of d_n from mm to m, T_I and T_L from min to s, and E_d is a percentage.

Initial trial: $T_L = 10$ min:

$$Q_0 = \frac{0.00167 \times d_n \times L}{(T_I - T_L)E_d} = \frac{0.00167 \times 80 \times 200}{(312.7 - 10)(75)}$$
$$= 0.001177 \frac{\text{m}^2}{\text{s}}.$$

Referring to Table 19.6, T_L is revised to 7.8 min for the second trial.

Table 19.6 *Recessional lag times (min) in gently sloping borders*

	Border slope, S_0 (m/m)							
	0.001				0.002			
	Inflow rate, Q_0 (m²/s)							
	0.0001	0.001	0.01	0.02	0.0001	0.001	0.01	0.02
Opportunity time T_n (min)	Manning's roughness coefficient $n = 0.15$							
10	2.5	2.7	2.7	2.7	1.6	2.1	2.5	2.6
25	4.4	5.4	6.2	6.4	2.2	3.0	4.1	4.4
50	5.7	7.7	9.8	10.4	2.4	3.6	5.1	5.7
100	6.8	9.7	13.4	14.6	2.6	3.9	5.9	6.6
200	7.5	11.2	16.3	18.1	2.7	4.2	6.4	7.3
500	8.1	12.4	18.9	21.4	2.8	4.3	6.8	7.7
1000	8.3	12.9	20.0	22.8	2.8	4.4	6.9	7.9
2000	8.4	13.2	20.7	23.6	2.8	4.4	7.0	8.0
	Manning's roughness coefficient $n = 0.25$							
10	2.8	2.8	2.7	2.6	2.2	2.7	2.9	3.0
25	5.8	6.6	7.0	7.1	3.4	4.4	5.6	6.0
50	8.5	10.6	12.5	12.9	4.1	5.8	7.9	8.5
100	10.9	14.9	19.1	20.4	4.5	6.7	9.7	10.8
200	12.8	18.5	25.6	28.1	4.8	7.4	11.1	12.4
500	14.3	21.7	32.2	36.0	5.0	7.8	12.1	13.8
1000	15.0	23.1	35.3	39.9	5.1	8.0	12.5	14.3
2000	15.3	24.0	37.2	42.4	5.2	8.1	12.8	14.6

After USDA-SCS, 1974.

Second trial: $T_L = 7.8\,\text{min}$:

$$Q_0 = \frac{0.00167 \times d_n \times L}{(T_I - T_L)E_d} = \frac{0.00167 \times 80 \times 200}{(312.7 - 7.8)(75)}$$
$$= 0.001169\frac{\text{m}^2}{\text{s}}.$$

Since $S_0 < 0.4\%$, check the calculated T_L using eqs. (19.11) and (19.12):

$$T_L = \frac{Q_0^{0.2} \times n^{1.2}}{120K^{0.2}\left(S_0 + \frac{0.0094 \times n \times Q_0^{0.175}}{T_I^{0.88} S_0^{0.5}}\right)^{1.60}}$$
$$= \frac{0.001169^{0.2} \times 0.25^{1.2}}{120 \times 1^{0.2}\left(0.002 + \frac{0.0094 \times 0.25 \times 0.001169^{0.175}}{312.7^{0.88} \times 0.002^{0.5}}\right)^{1.60}}$$
$$= 7.86\,\text{min}.$$

Check if the calculated recession lag time is equal to the assumed lag time:

$$T_L = 7.86\ \text{min} \approx 7.8\,\text{min}.$$

Thus, the unit inflow rate is:

$$Q_0 = 0.001169\frac{\text{m}^2}{\text{s}}.$$

Since the application efficiency $E_a = 100\%$, the final unit inflow rate is 0.001169 $\frac{\text{m}^2}{\text{s}}$.

The cutoff time is calculated using eq. (19.6):

$$T_i = T_I - T_L = 312.7\ \text{min} - 7.8\ \text{min} = 304.9\ \text{min}.$$

The flow depth is computed using eq. (19.2):

$$d_n = K_1 T_L^{\frac{3}{16}} Q_0^{\frac{9}{16}} n^{\frac{3}{8}}$$
$$= 2454 \times (7.8)^{\frac{3}{16}} \times (0.001169)^{\frac{9}{16}} \times (0.25)^{\frac{3}{8}}$$
$$= 48.1\ \text{mm}.$$

Now check the adequacy of the design.

The crop is wheat. For non-sod-forming crops (alfalfa and small grain), the maximum inflow rate is computed using eq. (19.7):

$$Q_{max} = 1.765 \times 10^{-4} \times S_0^{-0.75}$$
$$= 1.765 \times 10^{-4} \times (0.002)^{-0.75}$$
$$= 0.0187\frac{\text{m}^2}{\text{s}} > 0.001169\frac{\text{m}^2}{\text{s}}.$$

The minimum inflow rate is computed using eq. (19.9).

To change the unit of Q_{min} to m^3/s/m, eq. (19.9) can be expressed as:

$$Q_{min} = 5.95 \times 10^{-6}\frac{L}{n}S_0^{0.5}$$
$$= 5.95 \times 10^{-6} \times \frac{200}{0.25} \times 0.002^{0.5}$$
$$= 0.00021\frac{\text{m}^2}{\text{s}} < 0.001169\frac{\text{m}^2}{\text{s}}.$$

The maximum slope can be computed based on the criteria for the minimum depth of flow. Combining the minimum flow eq. (19.9) and the unit inflow rate Q_0 (m^3/s/m) equation, one gets

$$S_{max} = \left[\frac{nQ_{min}}{5.95 \times 10^{-6}L}\right]^2$$
$$= \left[\frac{n}{5.95 \times 10^{-6}L}\frac{0.00167 \times d_n \times L}{(T_I - T_L)E_d E_a}\right]^2$$
$$= \left[\frac{n}{0.00356}\frac{d_n}{(T_I - T_L)E_d E_a}\right]^2,$$

$$S_{max} = \left[\frac{0.25}{0.00356} \times \frac{80}{(312.7 - 7.8)(75)(100\%)}\right]^2$$
$$= 0.06 > 0.002.$$

The maximum length is limited by the maximum inflow rate, i.e., $Q_{max} = 0.0187\,\frac{\text{m}^2}{\text{s}}$. For $S_0 < 0.4\%$, the lag time is calculated using eqs. (19.11) and (19.22):

$$T_{L,max} = \frac{0.0187^{0.2} \times 0.25^{1.2}}{120 \times 1^{0.2}\left(0.002 + \frac{0.0094 \times 0.25 \times 0.0187^{0.175}}{1^{0.175}312.7^{0.88} \times 0.002^{0.5}}\right)^{1.60}}$$
$$= 13.04\ \text{min}.$$

The maximum length is computed as:

$$L_{max} = \frac{Q_{max} \times E_d \times (T_I - T_{L,max})}{0.00167 \times d_n}$$
$$= \frac{(0.0187) \times 75 \times (312.7 - 13.04)}{0.00167 \times 80}$$
$$= 3146\ \text{m}.$$

$L_{max} = 3146\ \text{m} > 200\ \text{m}$. For some soils with a low intake rate on small slopes, the theoretical maximum length of the border can be several kilometers. However, as mentioned in Section 19.1.3, border lengths over 800 m are seldom designed.

In addition, the calculated depth (48.1 mm) is smaller than the allowable flow depth (120 mm).

It is concluded that the computed values for unit inflow rate, slope, border length, and depth of flow are within design standards.

19.3.9 Simplified Border Irrigation Design

Following Walker and Skogerboe (1987), it is assumed that the surface water profile at the time of cutoff of inflow (T_{CO}) or the inflow time (T_i) and that at the end of depletion (t_d), which is also the beginning of recession (T_r), are straight lines with endpoints corresponding to uniform flow conditions, as shown in Figure 19.4. Further, the depth at the downstream end (y_L) remains constant and runoff (Q_r) occurs at a constant rate during the depletion phase. During the depletion and recession phases, the sum of

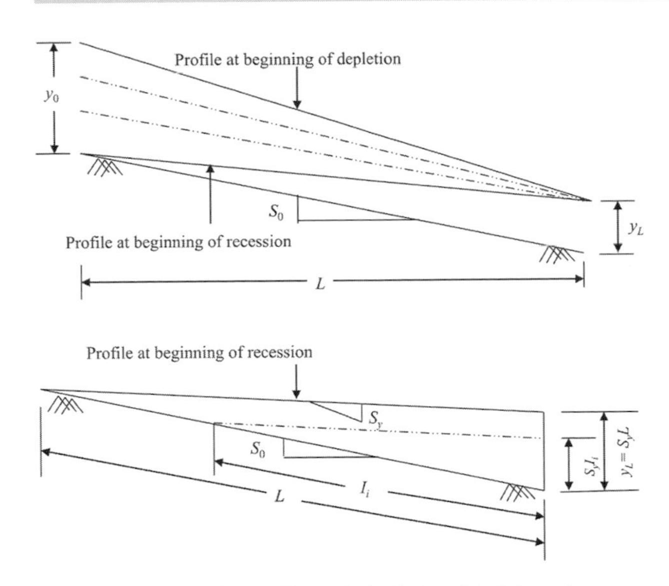

Figure 19.4 Water surface profiles at the beginning of depletion and recession phases.

infiltration (I) and runoff (Q_r) is equal to the pre-cutoff unit inflow rate (Q_0).

From Figure 19.4, it is seen that the time required from the cutoff of inflow to the end of depletion phase (i.e., when the upstream depth becomes zero) is equal to the time required to remove the volume of water defined by the triangle of length L and height y_0 at a constant rate of Q_0 through both infiltration and runoff. This can be expressed as:

$$t_d = T_{CO} + \frac{y_0 L}{2Q_0}. \tag{19.13}$$

At the beginning of recession, the depth of flow is assumed to change with distance at a uniform rate over the entire length of the border, and its slope can be expressed as:

$$S_y = \frac{y_L(t_d)}{L}, \tag{19.14}$$

where y_L is a function of Q_r at time t_d, which can be determined as:

$$Q_r(t_d) = Q_0 - I \times L = A\frac{R^{\frac{2}{3}}S_0^{\frac{1}{2}}}{n}, \tag{19.15}$$

where A is the cross-sectional area per unit width and R is the hydraulic radius equal to A/WP, where WP is the wetted perimeter, Q_0 is in m^3/s/m, and I is the average infiltration rate (m/s) over the length, L. For borders, $A = y$ and $WP = 1$, and hence $R = y$ or S_yL. Therefore, with the use of eq. (19.15), eq. (19.14) becomes

$$S_y = \frac{1}{L}\left[\frac{(Q_0 - I \times L) \times n}{60 \times S_0^{\frac{1}{2}}}\right]^{\frac{3}{5}}, \tag{19.16}$$

where Q_0 is in m^3/min/m, I is in m/min, and L is in m; I can be assumed to be the average value of infiltration rate at the upstream end [$I(t_d)$] and at the downstream end $I(t_d - T_a)$:

Table 19.7 *Parameters of the modified Kostiakov infiltration equation corresponding to the SCS intake families (continuous flow border/basin irrigation) for the first irrigation*

Intake family	Soil name	a	k (m/min)	f_0 (m/min)
0.02	Heavy clay	0.192	0.00042	0.00002
0.05	Clay	0.247	0.00075	0.00004
0.10	Clay	0.303	0.00103	0.00005
0.15	Silty clay	0.348	0.00127	0.00007
0.20	Silty/sand clay	0.385	0.00147	0.00008
0.25	Sandy clay	0.416	0.00165	0.00010
0.30	Sandy clay	0.442	0.00182	0.00011
0.35	Silty clay loam	0.464	0.00197	0.00013
0.40	Silty clay loam	0.483	0.00211	0.00014
0.45	Clay loam	0.499	0.00224	0.00016
0.50	Clay loam	0.514	0.00236	0.00017
0.60	Sandy clay loam	0.537	0.00259	0.00019
0.70	Sandy clay loam	0.556	0.00279	0.00021
0.80	Silt loam	0.572	0.00298	0.00023
0.90	Silt	0.585	0.00315	0.00025
1.00	Loam	0.597	0.00331	0.00027
1.50	Sandy loam	0.638	0.00397	0.00034
2.00	Loamy sand	0.6666	0.00450	0.00038
4.00	Sand	0.751	0.00614	0.00046

After USDA-NRCS, 2012.

$$I = \frac{ak}{2}\left[t_d^{a-1} + (t_d - T_a)^{a-1}\right] + f_0, \tag{19.17}$$

where T_a is the advance time, f_0 is the basic infiltration rate and a and k are parameters of the modified Kostiakov infiltration equation. The USDA-NRCS (2012) provided values of these parameters (k, a, and f_0) of the modified Kostiakov infiltration equation corresponding to the SCS intake families for the first and later irrigations. These values are presented in Tables 19.7 and 19.8.

The recession time can be determined by the equation given by Walker and Skogerboe (1987) as:

$$T_r = t_d + \frac{0.095 n^{0.47565} S_y^{0.20735} L^{0.6829}}{I^{0.52435} S_0^{0.237825}}. \tag{19.18}$$

Now, a step-by-step design procedure for free-drained borders can be outlined as follows (Walker and Skogerboe, 1987):

1. Obtain information on field characteristics, soil, crop, and water supply.
2. Determine the maximum (Q_{max}) and minimum (Q_{min}) values of unit inflow rate Q_0 (m^3/min/m) using eqs (19.8) and (19.9), respectively. The flow should be limited to within the nonerosive velocity, with sufficient depth to spread laterally.
3. Select the unit flow rate (Q_0) between Q_{max} and Q_{min} that results in a set width comprising an even number of

Table 19.8 *Parameters of the modified Kostiakov infiltration equation corresponding to the SCS intake families (continuous flow border/basin irrigation) for the later irrigation*

Intake family	Soil name	a	k (m/min^a)	f_0 (m/min)
0.02	Heavy clay	0.153	0.00036	0.00002
0.05	Clay	0.197	0.00063	0.00003
0.10	Clay	0.242	0.00088	0.00004
0.15	Silty clay	0.278	0.00108	0.00006
0.20	Silty/sand clay	0.308	0.00125	0.00007
0.25	Sandy clay	0.333	0.00141	0.00008
0.30	Sandy clay	0.354	0.00154	0.00009
0.35	Silty clay loam	0.371	0.00168	0.00010
0.40	Silty clay loam	0.386	0.00179	0.00011
0.45	Clay loam	0.399	0.00190	0.00012
0.50	Clay loam	0.411	0.00201	0.00014
0.60	Sandy clay loam	0.43	0.00220	0.00015
0.70	Sandy clay loam	0.445	0.00237	0.00017
0.80	Silt loam	0.458	0.00253	0.00019
0.90	Silt	0.468	0.00268	0.00020
1.00	Loam	0.478	0.00282	0.00022
1.50	Sandy loam	0.1	0.00337	0.00027
2.00	Loamy sand	0.533	0.00383	0.00030
4.00	Sand	0.601	0.00522	0.00037

After USDA-NRCS, 2012.

borders of satisfactory width and an integer number of sets using eqs (19.4) and (19.5), respectively.

4. Determine the inflow depth at the inlet y_0 (m) using eq. (19.1).
5. Compute the time required (τ_{req}) in minutes to satisfy the irrigation requirement using eq. (17.2) or eq. (19.21).
6. Compute the time of advance to the end of the border, T_a (minutes). The advance time can be calculated using the power advance approach of eq. (17.19). The advance time can be obtained by solving T_a in eq. (17.19) as $T_a = \left(\frac{L}{p}\right)^{1/r}$, where p and r are the fitted parameters. The two-point method illustrated in Example 17.3 can be used to define the two empirical parameters, p and r. The value of r typically ranges between 0.1 and 0.9.
7. Compute the time of recession (T_r) in minutes from the beginning of irrigation, assuming that the design will meet irrigation requirements at the end of the border:

$$T_r = \tau_{req} + T_a. \tag{19.19}$$

8. Compute the depletion time, t_d (min) numerically, say using the Newton–Raphson method, as follows:
   a. Assume an initial guess of t_d as $t_d^i = T_r$.
   b. Determine the average infiltration (I) by substituting $t_d = t_d^i$ in eq. (19.17).
   c. Determine S_y using eq. (19.16).
   d. Determine a new value of t_d as t_d^{i+1} using eq. (19.18) as follows:

$$t_d^{i+1} = T_r - \frac{0.095 n^{0.47565} S_y^{0.20735} L^{0.6829}}{I^{0.52435} S_0^{0.237825}}. \tag{19.20}$$

   e. Compare the initial guess (t_d^i) with the new computed value $\left(t_d^{i+1}\right)$. If the values are equal, then t_d is the right value, so continue with step 9. Otherwise, set $t_d^i = t_d^{i+1}$ and repeat steps b–e.
9. Determine the infiltrated depth at the border inlet (Z_0) and compare it with Z_{req} to determine the status of irrigation (complete irrigation: $Z_0 \geq Z_{req}$; deficit irrigation $Z_0 < Z_{req}$):

$$Z_0 = k t_d^a + f_0 t_d. \tag{19.21}$$

The values of the parameters k, a, and f_0 for border irrigation under a continuous inflow rate can be found in Tables 19.7 and 19.8.

10. If irrigation is complete, then determine T_{CO} and E_a as follows:

$$T_{CO} = t_d - \frac{y_0 L}{2Q_0}, \tag{19.22}$$

$$E_a = \frac{Z_{req} L}{Q_0 T_{CO}}. \tag{19.23}$$

11. In the case of deficit irrigation, increase the cutoff time and compute the new t_r value as follows:
    a. Calculate the new T_{CO} by substituting τ_{req} in place of t_d in eq. (19.13).
    b. Calculate the average infiltration (I) by substituting $t_d = \tau_{req}$ in eq. (19.17).
    c. Calculate S_y using eq. (19.16).
    d. Calculate T_r by substituting $t_d = \tau_{req}$ in eq. (19.17).
    e. Calculate Z_L:

$$Z_L = k(T_r - T_a)^a + f_0(T_r - T_a). \tag{19.24}$$

    f. Compute E_a using eq. (19.23).
12. Check whether the water availability is satisfied and repeat steps 4–12 for other unit inflow rates. Choose the design which gives the maximum E_a value.

Example 19.8: Design a border irrigation system based on the following requirements: field length $L = 400$ m; field width $W = 600$ m; longitudinal slope, $S_0 = 0.005$; Manning's roughness coefficient $n = 0.04$; soil texture is silt; design irrigation requirement (Z_{req}) = 5 cm. Infiltration function parameters are: $k = 0.0028$, $a = 0.356$, and $f_0 = 0.00017$ (i.e., $Z_{req} = 0.0028\tau_{req}^{0.356} + 0.00017\tau_{req}$, where Z_{req} is in m); available supply rate, $Q = 50$ (m^3/min), and supply duration = 12 h. Soils appear to be relatively non-erosive and have been tested and found to have the following advance coefficients: $p = 18.5$ and $r = 0.72$.

Solution: The following steps are involved in solving this problem:

Step 1: Determine the maximum (Q_{max}) and minimum (Q_{min}) limits for unit inflow rate Q_0 using eqs (19.8) and (19.9), respectively:

$$Q_{max} = 0.01059 S_0^{-0.75} = 0.01059 \times 0.005^{-0.75}$$
$$= 0.563 \frac{\text{m}^2}{\text{min}},$$

$$Q_{min} = 0.000357 \frac{L}{n} S_0^{0.5} = 0.000357 \times \frac{400}{0.04} \times 0.005^{0.5}$$
$$= 0.252 \frac{\text{m}^2}{\text{min}}.$$

Step 2: Select Q_0 within the range of Q_{max} and Q_{min}. The flow rate is 0.252–0.563 m^3/min/m. Considering the unit flow rate as Q_{max} (0.563 m^3/min/m) results in a border width (W_b) of 88.8 m, with 6.8 as the number of borders (N_b).

$$W_b = \frac{Q}{Q_0} = \frac{50 \frac{\text{m}^3}{\text{min}}}{0.563 \frac{\text{m}^3}{\text{min} \times \text{m}}} = 88.8 \text{ m}, \quad N_b = \frac{600 \text{ m}}{88.8 \text{ m}} = 6.8.$$

It is impractical since the number of borders should be an integer number. The border width and the unit inflow rate are adjusted, and possible combinations are as follows:

Number of borders (N_b)	Border width, W_b (m)	Unit inflow rate, Q_0 (m^3/min/m)
6	100	0.5000
5	120	0.4167
4	150	0.3333
3	200	0.2500

Taking $Q_0 = 0.50$ m^3/min/m as an example:

Step 3: Calculate the inflow depth at inlet y_0 (m) using Manning's equation as:

$$y_0 = \left(\frac{Q_o \times n}{60 \cdot S_0^{0.5}}\right)^{\frac{3}{5}} = \left(\frac{0.5 \times 0.04}{60 \times (0.005)^{0.5}}\right)^{\frac{3}{5}} = 0.0402 \text{ m};$$

the y_0 value should be less than the ridge height.

Step 4: Calculate τ_{req} (min) to satisfy the irrigation requirement:

$$Z_{req} = k \times \tau_{req}^a + f_o \times \tau_{req},$$

$$0.05 \text{ m} = 0.0028 \tau_{req}^{0.356} + 0.00017 \tau_{req},$$

$$\tau_{req} = 187.89 \text{ min}.$$

Step 5: Calculate the time of advance to the end of border T_a (min):

$$T_a = \left(\frac{L}{p}\right)^{1/r} = \left(\frac{400}{18.5}\right)^{1/0.72} = 71.45 \text{ min}.$$

Step 6: Calculate the time of recession (T_r in minutes since the beginning of irrigation), assuming that the design will meet the irrigation requirement at the end of the border:

$$T_r = \tau_{req} + T_a = 187.89 + 71.45 = 259.34 \text{ min}.$$

Step 7: Calculate the depletion time, t_d (min) using the Newton–Raphson method:

a. Assume an initial guess of t_d as $t_d^i = T_r = 259.34$ min.
b. Calculate the average infiltration (I) by substituting $t_d = t_d^i$ in eq. (19.17):

$$I = \frac{ak}{2}\left[\left(t_d^i\right)^{a-1} + \left(t_d^i - T_a\right)^{a-1}\right] + f_0$$
$$= \frac{0.356 \times 0.0028}{2} \times \left[(259.34)^{0.356-1} + (259.34 - 71.45)^{0.356-1}\right] + 0.00017$$
$$= 0.000201 \frac{\text{m}^2}{\text{min}}.$$

c. Calculate S_y using eq. (19.16):

$$S_y = \frac{1}{L}\left[\frac{(Q_0 - I \times L) \times n}{60 \times S_0^{\frac{1}{2}}}\right]^{\frac{3}{5}}$$
$$= \frac{1}{400}\left[\frac{(0.5 - 0.000201 \times 400) \times 0.04}{60 \times (0.005)^{0.5}}\right]^{\frac{3}{5}}$$
$$= 0.0000904.$$

d. Calculate the new value of t_d as t_d^{i+1} using eq. (19.20):

$$t_d^{i+1} = T_r - \frac{0.095 n^{0.47565} S_y^{0.20735} L^{0.6829}}{I^{0.52435} S_0^{0.237825}}$$
$$= 259.34 - \frac{0.095(0.04)^{0.47565}(0.0000904)^{0.20735}(400)^{0.6829}}{(0.000201)^{0.52435}(0.005)^{0.237825}}$$
$$= 204.78 \text{ min}.$$

e. The initial guess ($t_d^i = 259.34$ min) is not close to the new computed value ($t_d^{i+1} = 204.78$ min). Therefore, set ($t_d^i = 204.78$ min) and repeat steps b–e:

$$I = \frac{0.356 \times 0.0028}{2} \times \left[(204.78)^{0.356-1} + (204.78 - 71.45)^{0.356-1}\right] + 0.00017$$
$$= 0.0002075 \frac{\text{m}^2}{\text{min}},$$

$$S_y = \frac{1}{L}\left[\frac{(Q_0 - I \times L) \times n}{60 \times S_0^{\frac{1}{2}}}\right]^{\frac{3}{5}}$$

$$= \frac{1}{400}\left[\frac{(0.5 - 0.0002075 \times 400) \times 0.04}{60 \times (0.005)^{0.5}}\right]^{\frac{3}{5}}$$

$$= 0.0000901,$$

$$t_d^{i+1} = 259.34 \text{ min}$$

$$- \frac{0.095(0.04)^{0.47565}(0.0000901)^{0.20735}(400)^{0.6829}}{(0.0002075)^{0.52435}(0.005)^{0.237825}}$$

$$= 205.72 \text{ min}.$$

The second guess (204.78 min) is close to the new computed value of 205.72 min.

Step 8: The correct value of t_d is 205.72 min.

Step 9: Calculate the infiltrated depth at the border inlet (Z_0) and compare it with Z_{req}:

$$Z_0 = kt_d^a + f_0 t_d$$

$$= 0.0028 \times (205.72)^{0.356} + 0.00017 \times 205.72$$

$$= 0.054 \text{ m}.$$

Step 10: Since the infiltrated depth at the end (Z_0 = 0.054 m) is higher than Z_{req} (0.05 m), it is a case of complete irrigation. The cutoff time is calculated as

$$T_{CO} = t_d - \frac{y_0 L}{2Q_0} = 205.72 - \frac{0.0402 \times 400}{2 \times 0.5} = 189.6 \text{ min}.$$

Calculate E_a using eq. (19.25):

$$E_a = \frac{Z_{req}L}{Q_0 T_{CO}} = \frac{0.05 \times 400}{0.5 \times 189.6} \times 100 = 21.1\%.$$

Step 11: Check the water availability constraint and repeat steps 4–10 for other unit inflow rates. Choose the design that gives the maximum E_a value.

Number of borders (N_b)	Border width, W_b (m)	Unit inflow rate, Q_0 (m^3/min/m)	T_{CO} (min)	E_a (%)	Total irrigation time (h)
6	100	0.5000	189.6	21.1	19.0
5	120	0.4167	189.9	25.3	15.9
4	150	0.3333	190.0	31.6	12.7
3	200	0.2500	190.2	42.1	9.5
2	300	0.1667*	190.3	63.0	6.3

* Flow less than Q_{min}.

Choose the design that gives the maximum E_a value. The unit inflow rate of 0.25 m^3/min/m, with a border width of 200 m and three borders gives the maximum E_a value of 42.1%. The number of borders is 3, the total duration of the irrigation is 3 × 190.2/60 = 9.5 h. Since the water supply duration is 12 h, the water supply can stratify the irrigation requirement. However, an even number of borders with satisfactory width is usually used for ease of other farming operations.

Example 19.9: A border irrigation system is to be designed for a field 250 m long and 150 m wide. The land slope is 0.1% along the length and 0.075% along the width. Also given is: Manning roughness n = 0.04; soil texture is clay loam; design irrigation requirement = 6.5 cm; available supply rate, Q = 15 (m^3/min), and supply duration = 15 h. The advance coefficients are p = 9.45 and r = 0.68. The following infiltration equations have been found to be adequate:

First irrigation: $Z_{req} = 0.005\,\tau_{req}^{0.42} + 0.00003\,\tau_{req}$.
Second irrigation: $Z_{req} = 0.003\,\tau_{req}^{0.40} + 0.00003\,\tau_{req}$.

Solution: The solution is computed step by step as follows:

1. Determine the maximum inflow (Q_{max}) per unit width for the first irrigation along the field length of 250 m (S_0 = 0.1%), where erosion is likely under a high inflow rate:

$$Q_{max} = 0.01059 S_0^{-0.75} = 0.01059 \times 0.001^{-0.75}$$

$$= 1.88 \frac{\text{m}^2}{\text{min}}.$$

Compute the maximum depth:

$$y_{max} = \left(\frac{Q_{max} \times n}{60 \times S_0^{0.5}}\right)^{\frac{3}{5}} = \left(\frac{1.88 \times 0.04}{60 \times (0.001)^{0.5}}\right)^{\frac{3}{5}}$$

$$= 0.144 \text{ m}.$$

The inlet depth should not exceed the field dikes.

2. Compute the maximum inflow along the width direction (S_0 = 0.075%).

$$Q_{max} = 0.01059 S_0^{-0.75} = 0.01059 \times 0.00075^{-0.75}$$

$$= 2.34 \frac{\text{m}^2}{\text{min}}.$$

Compute the maximum depth:

$$y_{max} = \left(\frac{Q_{max} \times n}{60 \times S_0^{0.5}}\right)^{\frac{3}{5}} = \left(\frac{2.34 \times 0.04}{60 \times (0.00075)^{0.5}}\right)^{\frac{3}{5}}$$

$$= 0.179 \text{ m}.$$

3. Compute the minimum flow for 250 m length where water spreading may not be uniform:

$$Q_{min} = 0.000357\frac{L}{n}S_0^{0.5}$$
$$= 0.000357 \times \frac{250}{0.04} \times 0.001^{0.5}$$
$$= 0.071\frac{\text{m}^2}{\text{min}}.$$

4. Compute the minimum flow along the width (150 m):

$$Q_{min} = 0.000357\frac{L}{n}S_0^{0.5}$$
$$= 0.000357 \times \frac{150}{0.04} \times 0.00075^{0.5}$$
$$= 0.037\frac{\text{m}^2}{\text{min}}.$$

First Irrigation: $L = 250$ m (Length)

5. Select the flow and adjust it for possible combinations of border width and number of borders:

Select Q_0 within the range of Q_{max} and Q_{min}. The flow rate is 0.071–1.88 m³/min/m. The unit flow rate of 1.88 m³/min/m is too high to be considered realistic, thus a lower value of 0.5 m³/min/m is used as Q_{max}.

$$W_b = \frac{Q}{Q_0} = \frac{15\,\frac{\text{m}^3}{\text{min}}}{0.5\,\frac{\text{m}^3}{\text{min} \times \text{m}}} = 30\text{ m}, \qquad N_b = \frac{150\text{ m}}{30\text{ m}} = 5.$$

Other possible border width and the unit inflow rate combinations are as follows:

Number of borders (N_b)	Border width, W_b (m)	Unit inflow rate, Q_0 (m³/min/m)
5	30.0	0.50
4	37.5	0.40
3	50.0	0.30
2	75.0	0.20
1	150	0.10

Taking $Q_0 = 0.50$ m³/min/m as an example:

6. Calculate the inflow depth at inlet y_0 (m) using Manning's equation:

$$y_0 = \left(\frac{Q_0 \times n}{60 \times S_0^{0.5}}\right)^{\frac{3}{5}} = \left(\frac{0.5 \times 0.04}{60 \times (0.001)^{0.5}}\right)^{\frac{3}{5}} = 0.0651\text{ m}.$$

This y_0 value should be less than the ridge height.

7. Compute the intake opportunity time in minutes to satisfy the irrigation requirement for the first irrigation:

$$Z_{req} = 0.065\text{ m} = 0.005\tau_{req}^{0.42} + 0.00003\tau_{req},$$
$$\tau_{req} = 310.62\text{ min}.$$

8. Calculate the time of advance to the end of border T_L (min) and calculate the time of recession (T_r in minutes since the beginning of irrigation), assuming that the design will meet the irrigation requirement at the end of the border: $\tau_{req} = T_r - T_a$:

$$T_a = \left(\frac{L}{p}\right)^{1/r} = \left(\frac{250}{9.45}\right)^{1/0.68} = 123.57\text{ min},$$

$$T_r = \tau_{req} + T_a = 310.62\text{ min} + 123.57\text{ min}$$
$$= 434.19\text{ min}.$$

9. Calculate the depletion time, t_d (min) using the Newton–Raphson method as follows:
   a. Assume an initial guess of t_d as $t_d^1 = T_r = 434.19$ min.
   b. Calculate the average infiltration (I) by substituting $t_d = t_d^i$ in eq. (19.17):

$$I = \frac{ak}{2}\left[\left(t_d^i\right)^{a-1} + \left(t_d^i - T_a\right)^{a-1}\right] + f_0$$
$$= \frac{0.42 \times 0.005}{2} \times \left[(434.19)^{0.42-1} + (434.19 - 123.57)^{0.42-1}\right] + 0.00003$$
$$= 0.0000986\frac{\text{m}^2}{\text{min}}.$$

   c. Calculate S_y using eq. (19.16):

$$S_y = \frac{1}{L}\left[\frac{(Q_0 - I \times L) \times n}{60 \times S_0^{\frac{1}{2}}}\right]^{\frac{3}{5}}$$
$$= \frac{1}{250}\left[\frac{(0.5 - 0.00010 \times 250) \times 0.04}{60 \times (0.001)^{0.5}}\right]^{\frac{3}{5}}$$
$$= 0.0002527.$$

   d. Calculate the new value of t_d as t_d^2 using eq. (19.20):

$$t_d^{i+1} = T_r - \frac{0.095n^{0.47565}S_y^{0.20735}L^{0.6829}}{I^{0.52435}S_0^{0.237825}},$$

$$t_d^2 = 434.19 - \frac{0.095(0.04)^{0.47565}(0.0002527)^{0.20735}(250)^{0.6829}}{(0.0000986)^{0.52435}(0.001)^{0.237825}}$$
$$= 329.84\text{ min}.$$

   e. The initial guess ($t_d^1 = 434.19$ min) is not close to the new computed value ($t_d^2 = 329.84$ min). Therefore, set ($t_d^2 = 329.84$ min) and repeat steps b–e.

$$I = \frac{0.42 \times 0.005}{2} \times \left[(329.84)^{0.42-1} + (329.84 - 123.57)^{0.42-1}\right] + 0.00003$$
$$= 0.0001141\frac{\text{m}^2}{\text{min}},$$

$$S_y = \frac{1}{L}\left[\frac{(Q_0 - I \times L) \times n}{60 \times S_0^{\frac{1}{2}}}\right]^{\frac{3}{5}}$$
$$= \frac{1}{250}\left[\frac{(0.5 - 0.0001141 \times 250) \times 0.04}{60 \times (0.001)^{0.5}}\right]^{\frac{3}{5}}$$
$$= 0.0002515,$$

$$t_d^3 = 434.19 - \frac{0.095(0.04)^{0.47565}(0.0002515)^{0.20735}(250)^{0.6829}}{(0.0001141)^{0.52435}(0.001)^{0.237825}}$$

$$= 337.62 \text{ min.}$$

The second guess ($t_d^2 = 329.84$ min) is not close to the new computed value ($t_d^3 = 337.62$ min). Therefore, set ($t_d^3 = 337.62$ min) and repeat steps b–e. One gets $t_d^4 = 336.94$ min.

10. The correct value of $t_d = 336.94$ min.
11. Calculate the infiltrated depth at the border inlet (Z_0) and compare it with Z_{req}:

$$Z_0 = kt_d^a + f_0 t_d$$

$$= 0.005 \times (336.94)^{0.42} + 0.00003 \times 336.94$$

$$= 0.068 \text{ m.}$$

Since the infiltrated depth at the end ($Z_0 = 0.068$ m) is higher than Z_{req} (0.065 m), it is a case of complete irrigation. The cutoff time is calculated as

$$T_{CO} = t_d - \frac{y_0 L}{2Q_0} = 336.94 - \frac{0.0651 \times 250}{2 \times 0.5}$$

$$= 320.7 \text{ min.}$$

12. Calculate E_a using eq. (19.25):

$$E_a = \frac{Z_{req}L}{Q_0 T_{CO}} = \frac{0.065 \text{ m} \times 250 \text{ m}}{0.5 \frac{\text{m}^2}{\text{min}} \times 320.7 \text{ min}} \times 100(\%)$$

$$= 10.1\%.$$

13. Check the water availability constraint and repeat steps 5–12 for other unit inflow rates. Choose the design that gives the maximum E_a value.

First Irrigation $L = 250$ m (Length)

Number of borders (N_b)	Border width, W_b (m)	Unit inflow rate, Q_0 (m^3/min/m)	T_{CO} (min)	E_a (%)	Total irrigation time (h)
5	30	0.50	320.7	10.1	26.7
4	37.5	0.40	321.8	12.6	21.5
3	50	0.30	323.0	16.8	16.1
2	75	0.20	324.1	25.1	10.8
1	150	0.10	325.3	50.0	5.4

14. Repeat steps 5–13 for later irrigation with $L = 250$ m (along the length), first irrigation with $L = 150$ m (along the width), and later irrigation with $L = 150$ m (along the width). T_{CO} and E_a will be updated along the length and width slopes.

Later Irrigation $L = 250$ m (Length)

Number of borders (N_b)	Border width, W_b (m)	Unit inflow rate, Q_0 (m^3/min/m)	T_{CO} (min)	E_a (%)	Total irrigation time (h)
5	30	0.50*	736.0	4.4	61.4
4	37.5	0.40*	734.5	5.5	49.0
3	50	0.30*	732.3	7.4	36.6
2	75	0.20*	728.8	11.1	24.3
1	150	0.10	728.8	22.3	12.2

* Flow results initially incomplete irrigation.

First Irrigation $L = 150$ m (Width)

Number of borders (N_b)	Border width, W_b (m)	Unit inflow rate, Q_0 (m^3/min/m)	T_{CO} (min)	E_a (%)	Total irrigation time (h)
8	31.25	0.48*	299.8	6.8	40.0
7	35.71	0.42*	299.2	7.6	34.9
6	41.67	0.36*	298.5	9.1	29.9
5	50.00	0.30*	297.6	10.9	24.8
4	62.50	0.24*	296.3	13.7	19.8
3	83.33	0.18*	294.6	18.4	14.7
2	125.0	0.12*	291.8	27.8	9.7
1	250.0	0.06*	285.8	56.9	4.8

* Flow results initially incomplete irrigation.

Later Irrigation $L = 150$ m (Width)

Number of borders (N_b)	Border width, W_b (m)	Unit inflow rate, Q_0 (m^3/min/m)	T_{CO} (min)	E_a (%)	Total irrigation time (h)
8	31.25	0.48*	741.5	2.7	98.9
7	35.71	0.42*	740.9	3.1	86.5
6	41.67	0.36*	740.2	3.7	74.0
5	50.00	0.30*	739.2	4.4	61.6
4	62.50	0.24*	738.0	5.5	49.2
3	83.33	0.18*	736.3	7.4	36.8
2	125.0	0.12*	733.5	11.1	24.5
1	250.0	0.06*	727.4	22.3	11.5

* Flow results initially incomplete irrigation.

These tables give a summary of the selected options for the first and later irrigations. Irrigation along the width slope with border width of 250 m and one set border give the maximum E_a value of 56.9% and 22.3% for the first and later irrigations, respectively. However, the differences of the irrigation along the length and width are small. An even number of borders with satisfactory width is usually used for ease of other farming operations, e.g., two set borders along the width can be used.

19.3.10 Singh–Yu Design Method

Using a volume balance approach, Singh et al. (1990) and Singh and Yu (1987a–c) developed a mathematical model for border irrigation design. The model gives analytical equations for advance, recession, and vertical and horizontal phases. The intake rate is represented by the Kostiakov equation, and flow by Manning's equation. The model was verified using data from nine borders (Roth, 1971) and was particularly accurate for Reynold's number less than 2500. Yu and Singh (1989) extended the model using the modified Kostiakov equation, which is the same as the USDA-SCS infiltration method. Singh et al. (1987) presented one-, two-, and three-dimensional infiltration equations for surface irrigation. Singh and Yu (1989a,b) extended the design method to closed borders. Ram and Singh (1982) evaluated five models of border irrigation recession, including the Wu model (1972), the Sherman and Singh model (1978, 1982), the Strelkoff model (1977), the Singh and McCann model (1979), and the Ram and Lal model (1971), and found the Sherman and Singh model to be the most accurate. Singh et al. (1988) developed the Muskingum model for border irrigation design. Singh and Yu (1988) developed a farm irrigation system design by simulating the entire irrigation cycle for open or closed borders or furrows. This system is based on volume balance and simple flow profile curves (Singh, 1979–1980).

19.3.11 Selection of Pump and Power Unit

A pump is normally needed if the source of water is groundwater. In some cases, water may also be pumped directly from rivers, streams, canals, or other water bodies. In that case, an appropriate size of pump should be selected by matching the system curve with the pump characteristics curve. Based on the selected pump, a suitable power unit can be chosen. The details of the selection of pumps and power unit were discussed in Chapter 10.

QUESTIONS

Q.19.1 Compute the border inflow rate in m^3/min/m if the field slope is given as 0.001. What will be the maximum depth at the inlet for the border if Manning's roughness coefficient is 0.07?

Q.19.2 Calculate the net infiltration time, recession time, unit inflow rate, time to cutoff, and depth flow for a border irrigation system with 100% application efficiency. The information given is as follows: Intake family is 0.1, irrigation depth to be applied is 65 mm, ground slope is 0.001, Manning's roughness coefficient is 0.25, allowable flow depth is 120 mm, distribution pattern efficiency is 82%, border length is 175 m, and the crop is wheat.

Q.19.3 Design a border irrigation system for the following requirements: field length $L = 350$ m; field width $W = 580$ m; longitudinal slope $S_0 = 0.0055$; Manning's roughness coefficient $n = 0.04$; soil texture is silt; design irrigation requirement $(Z_{req}) = 5.5$ cm; available supply rate, $Q = 50$ m^3/min, and supply duration = 24 h. The advance coefficients are $p = 14.5$ and $r = 0.68$. The soil appears to be relatively nonerosive and has been tested to the following infiltration function (Z_{req} is required irrigation in m):

First irrigation: $Z_{req} = 0.00492 \times \tau_{req}^{0.412} + 0.00007 \times \tau_{req}$,

Second irrigation $Z_{req} = 0.00552 \times \tau_{req}^{0.325} + 0.00004 \times \tau_{req}$.

Q.19.4 What will be the depletion time (or lag time) for a border if the ground slope is 0.001, inflow discharge is 0.004 m^2/s, the intake opportunity time is 120 minutes, and Manning's roughness coefficient is 0.10? Compute the depth at the inlet. What will be the lag time if the border slope is 0.02? What will be the depth at the inlet?

Q.19.5 Calculate the net infiltration time, recession time, unit inflow rate, time to cutoff, and depth flow for a border irrigation system with 90% application efficiency. The information given is as follows: Intake family is 0.2, irrigation depth to be applied is 70 mm, ground slope is 0.002, Manning's roughness coefficient is 0.15, allowable flow depth is 120 mm, distribution pattern efficiency is 78%, border length is 180 m, and the crop is wheat.

Q.19.6 Design a border irrigation system for the following requirements: field length $L = 375$ m; field width $W = 550$ m; longitudinal slope $S_0 = 0.0045$; Manning's roughness coefficient $n = 0.04$; soil texture is silt; design irrigation requirement $(Z_{req}) = 6$ cm; the advance coefficients are $p = 15.5$ and $r = 0.71$; available supply rate $Q = 55$ m^3/min, and supply duration = 24 h. The soil appears to be relatively nonerosive and has been tested to the following infiltration function (Z_{req} is required irrigation in m):

First irrigation: $Z_{req} = 0.0051 \times \tau_{req}^{0.388} + 0.00006 \times \tau_{req}$,

Second irrigation: $Z_{req} = 0.0054 \times \tau_{req}^{0.327} + 0.00004 \times \tau_{req}$.

Q.19.7 Compute the border inflow rate in m^3/min/m if the field slope is given as 0.002. What will be the maximum depth at the inlet for the border if Manning's roughness coefficient is 0.06?

Q.19.8 Calculate the net infiltration time, recession time, unit inflow rate, time to cutoff, and depth flow for a border irrigation system with 80% application efficiency. The information given is as follows: Intake family is 0.5, irrigation depth to be applied is 75 mm, ground slope is 0.001, Manning's roughness coefficient is 0.25, allowable flow depth is 120 mm, distribution pattern efficiency is 80%, border length is 195 m, and the crop is wheat.

Q.19.9 Design a border irrigation system for the following requirements: field length $L = 410$ m; field width $W = 550$ m; longitudinal slope $S_0 = 0.0055$; Manning's roughness coefficient $n = 0.04$; soil texture is silt; design irrigation requirement $(Z_{req}) = 6$ cm; the advance coefficients are $p = 15.5$ and $r = 0.71$; available supply rate $Q = 60\ \text{m}^3$/min, and supply duration = 20 h. The soil appears to be relatively nonerosive and has been tested to the following infiltration function (Z_{req} is required irrigation in m):

First irrigation: $Z_{req} = 0.0054 \times \tau_{req}^{0.395} + 0.00007 \times \tau_{req}$,
Second irrigation: $Z_{req} = 0.0051 \times \tau_{req}^{0.352} + 0.00005 \times \tau_{req}$.

Q.19.10 What will be the depletion time (or lag time) for a border if the ground slope is 0.003, inflow discharge is $0.0035\ \text{m}^2$/s, the intake opportunity time is 150 min, and Manning's roughness coefficient is 0.12? Compute the depth at the inlet. What will be the lag time if the border slope is 0.03? What will be the depth at the inlet?

Q.19.11 Calculate the net infiltration time, recession time, unit inflow rate, time to cutoff, and depth flow for a border irrigation system with 70% application efficiency. The information given is as follows: Intake family is 1.0, irrigation depth to be applied is 85 mm, ground slope is 0.002, Manning's roughness coefficient is 0.15, allowable flow depth is 120 mm, distribution pattern efficiency is 75%, border length is 210 m, and the crop is wheat.

Q.19.12 Design a border irrigation system for the following requirements: field length $L = 415$ m; field width $W = 500$ m; longitudinal slope $S_0 = 0.004$; Manning's roughness coefficient $n = 0.04$; soil texture is silt; design irrigation requirement $(Z_{req}) = 6.5$ cm; infiltration function parameters: the advance coefficients are $p = 15.5$ and $r = 0.71$; available supply rate $Q = 65\ \text{m}^3$/min, and supply duration = 15 h. The soil appears to be relatively nonerosive and has been tested to the following infiltration function (Z_{req} is required irrigation in m):

First irrigation: $Z_{req} = 0.0048 \times \tau_{req}^{0.412} + 0.00007 \times \tau_{req}$,
Second irrigation: $Z_{req} = 0.0056 \times \tau_{req}^{0.335} + 0.00005 \times \tau_{req}$.

Q.19.13 For border irrigation the cumulative infiltration (m^3/m) is given as:

$$Z_{req} = 0.0075\tau_{req}^{0.25} + 0.00025\tau_{req},$$

and the opportunity time is given in minutes. The advance time for the field was 100 min and then the water drained off the field before the inflow was cut off. Once the flow was cut off, it took 20 min for the surface water to recede from the field. Assume that 8 cm is the depth of water to be applied at the lower end of the field. Compute the time for the inflow to be cut off.

Q.19.14 Design a border irrigation system for a field 400 m by 200 m with a slope in the longitudinal direction at 0.15% and as 0.03% across the width. Manning's roughness coefficient $n = 0.04$. The soil is silty loam. The depth of water to be applied is 5 cm per irrigation, available supply rate $Q = 10\ \text{m}^3$/min, and duration of water supply is 20 h. The Kostiakov infiltration parameters are: $k = 0.003$, $a = 0.35$, and $f_0 = 0.0001$. The advance coefficients are $p = 10.5$ and $r = 0.71$.

Q.19.15 Design a freely draining border irrigation system if the following information is given: The soil is sandy loam, the slope is 0.004, and there is no cross-slope; Manning's roughness coefficient $n = 0.05$, $L = 350$ m, $W = 250$ m, water supply discharge = $0.30\ \text{m}^3$/s on a two-day basis; and amount of irrigation to be applied is 8 cm for each irrigation. The Kostiakov cumulative infiltration (m^3/m) parameters are: $k = 0.0033$, $a = 0.568$, and $f_0 = 0.000155$. The advance coefficients are $p = 12.5$ and $r = 0.70$.

Q.19.16 The following information is given for designing a draining border irrigation system: The soil is silt loam, the slope is 0.0008, and there is no cross-slope; Manning's roughness coefficient $n = 0.07$, $L = 250$ m, $W = 150$ m, water supply discharge = $0.50\ \text{m}^3$/s on a three-day basis; and amount of irrigation to be applied is 6.5 cm for each irrigation. The Kostiakov cumulative infiltration (m^3/m) parameters are: $k = 0.008$, $a = 0.25$, and $f_0 = 0.0002$. The advance coefficients are $p = 11.5$ and $r = 0.69$.

REFERENCES

Booher, L. J. (1974). *Surface Irrigation*. Rome: FAO.

James, L. G. (1988). *Principles of Farm Irrigation System Design*. New York: Wiley.

Jurriëns, M., Zerihun, D., Boonstra, J., and Feyen, J. (2001). *SURDEV: Surface Irrigation Software; Design, Operation, and Evaluation of Basin, Border, and Furrow Irrigation*. Wageningen: International Institute for Land Reclamation and Improvement.

Michael, A. M. (1978). *Irrigation Theory and Practice*. New Delhi: Vikas Publishing House Pvt Ltd.

Ram, R. S., and Lal, R. (1971). Recession flow in border irrigation. *Journal of Agricultural Engineering*, 8(3): 62–70.

Ram, R. S., and Singh, V. P. (1982). Evaluation of models of border irrigation recession. *Journal of Agricultural Engineering Research*, 27: 235–252.

Roth, R. L. (1971). Roughness during border irrigation. MS Thesis, University of Arizona, Tucson, AZ.

Savva, A. P., and Frenken, K. (2002). *Irrigation Manual: Planning, Development, Monitoring and Evaluation of Irrigated Agriculture with Farmers' Participation*. Rome: FAO.

Sherman, B., and Singh, V. P. (1978). A kinematic model for surface irrigation. *Water Resources Research*, 14(2): 357–363.

Sherman, B., and Singh, V. P. (1982). Kinematic model of surface irrigation: an extension. *Water Resources Research*, 18(3): 659–667.

Singh, V. P. (1979–1980). Derivation of shape factors for border irrigation advance. *Agriculture Water Management* 2: 271–288.

Singh, V. P., and McCann, R. C. (1979). Mathematical modeling of hydraulics of irrigation recession. In *Proceedings of Second International Conference on Mathematical Modeling*, St. Louis, Missouri, July 11–13.

Singh, V. P., and Yu, F. X. (1987a). A mathematical model for border irrigation: I. Advance and storage phases. *Irrigation Science*, 8: 151–174.

Singh, V. P., and Yu, F. X. (1987b). A mathematical model for border irrigation: II. Vertical and horizontal recession phases. *Irrigation Science*, 8: 175–190.

Singh, V. P., and Yu, F. X. (1987c). A mathematical model for border irrigation: III. Evaluation of border irrigation models. *Irrigation Science*, 8: 191–213.

Singh, V. P., and Yu, F. X. (1988). A farm irrigation system (FIS) model. *Water Resources Management*, 2: 173–181.

Singh, V. P., and Yu, F. X. (1989a). An analytical closed border irrigation model: 1. Theory. *Agricultural Water Management*, 15: 223–241.

Singh, V. P., and Yu, F. X. (1989b). An analytical closed border irrigation model: 2. Experimental verification. *Agricultural Water Management*, 15: 243–252.

Singh, V. P., He, Y. C., and Yu, F. X. (1987). 1-D, 2-D and 3-D infiltration for irrigation. *Journal of Irrigation and Drainage Engineering, ASCE*, 113(2): 266–278.

Singh, V. P., Scarlatos, P. D., and Raudales, S. A. (1988). A Muskingum model for border irrigation. *Journal of Irrigation and Drainage Engineering, ASCE*, 114(2): 266–280.

Singh, V. P., Scarlatos, P. D., and Prasad, S. N. (1990). An improved Lewis–Milne equation for the advance phase of border equation. *Irrigation Science*, 11(1): 1–6.

Strelkoff, T. (1977). Algebraic computation of flow in border irrigation. *Journal of Irrigation and Drainage Division, ASCE*, 103 (IR3): 325–342.

USDA-NRCS (2012). Surface irrigation. In *National Engineering Handbook*. Washington, DC: USDA.

USDA-SCS (1974). Border irrigation. In *National Engineering Handbook*. Washington, DC: USDA.

USDA-SCS (1984). Engineering surveys. In *Engineering Field Handbook*. Washington, DC: USDA.

Walker, W. R., and Skogerboe, G. V. (1987). *Theory and Practice of Surface Irrigation*. Englewood Cliffs, NJ: Prentice-Hall.

Withers, B., and Vipond, S. 1974. *Irrigation: Design and Practice*. London: Batsford.

Wu, I. P. (1972). Recession flow in surface irrigation. *Journal of Irrigation and Drainage Division, ASCE*, 103(IR3): 357–377.

Yu, F. X., and Singh, V. P. (1989). An analytical model for border irrigation. *Journal of Irrigation and Drainage Engineering, ASCE*, 115(6): 982–999.

20 Furrow Irrigation

Notation

A	cross-sectional area in the furrow (m^2, ft^2)	S_0	furrow slope (m/m, fraction, or percentage)
A_0	cross-sectional area at the field inlet (m^2, ft^2)	t	advance time since the start (min)
b, c	parameters in the power advance function	t_L, $t_{L/2}$	time necessary for flow to advance the full (L) and half the furrow ($L/2$) distance (min)
d	depth of the wetting front (mm, in.)	T	top width of flow (m)
d_{av}	average depth of application (mm)	$T_{0.5a}$	advance time to half of the length (min)
d_g	gross depth of water application (mm)	T_a	advance time, which is the time to advance to the end of the field (min)
d_n	required net depth of irrigation, or Z_n, Z_{req} (mm)	T_{cb}	cutback time, which equals the advance time (min)
d_p	depth of percolation (mm)	T_{CO}	cutoff time (min)
d_r	depth of surface runoff from furrow (mm)	T_i	inflow time or cutoff time (T_{CO}) (min)
D_{IR}	design daily irrigation requirement (mm/day, in./day)	T_I, τ_{req}, T_n	infiltration opportunity time for the desired depth of application (min)
E_a	application efficiency (percentage)	T_r	recession time, time of recession from the beginning of irrigation (min)
E_d	distribution efficiency (percentage)	v_{max}	maximum permissible velocity (m/min)
f_0	basic or final infiltration rate (L/min/m, gpm/ft)	$V_{0.5a}$	volume of water during half length (m^3/m)
FAR	fractional advance ratio	V_a	volume of water on the full length during the advance time (m^3/m)
F_s	furrow spacing (m)	V_{in}	inflow volume (L)
I_f	intake family	V_{out}	outflow volume (L)
k, a	coefficients of the Kostiakov and modified Kostiakov equations	V_s	volume of water in storage (L)
L	distance between inflow and outflow measurements (m)	W	field width (m)
mx, md, i	maximum, middle, and individual values of depth and top width, respectively	x	advance distance (m, ft)
n	Manning's roughness coefficient	y	flow depth in the furrow geometry (m)
N_f	number of furrows in a given field	Z_a	average intake or infiltration depth (mm)
N_{fs}	number of furrows per set	Z_{av}	average infiltration depth under cutback conditions (mm)
N_s	number of sets	Z_n, Z_{req}	desired depth of irrigation or d_n (mm, in.)
P	wetted perimeter (m)	α_1, α_2, σ_1, σ_2, γ_1, γ_2	fitted parameters of the furrow geometry
Q	total inflow rate or the stream size (m^3/min, L/s)	ρ_a, ρ_b	advance coefficients
Q_0	inflow rate per furrow (m^3/min)	σ_z	subsurface shape factor
Q_{max}	maximum permissible nonerosive inflow rate or stream size (L/min, gpm, m^3/min)		
R	hydraulic radius (m)		

20.1 INTRODUCTION

Furrow irrigation is a surface irrigation method, but it differs from other surface irrigation methods because in this method only one-fifth to one-half of the ground surface is covered with water. The partial coverage of the field by water leads to less evaporation and results in improved soil structure, and makes it possible to cultivate the soil sooner after irrigation than for other surface irrigation methods.

A furrow irrigation system is suitable for row crops, vegetable crops, and orchards. For row crops, furrows are usually parabolic in cross-section or have flat bottoms and about 2:1 side slope. Furrow irrigation is well suited for crops that are subject to injury from ponded surface water or susceptible to fungal root rot. It can be used on most soils, except coarse-textured sand and loamy sand with very high vertical infiltration that results in poor lateral distribution of water between furrows and high deep percolation losses. Land grading is essential to provide a uniform slope for obtaining uniform water application.

Water is applied to furrows so it moves in well-defined channels or furrows and infiltrates from the bottom and the sides of the furrows in vertical and lateral directions. Water is diverted into furrows either from an open ditch (a field canal, a tertiary canal, etc.) by means of siphons placed over the side of the ditch or pipes. In general, portable gated pipes or single and multiple outlet risers connected to a hydrant fitted to low-pressure buried pipelines are used for water distribution in the field. The objective of this chapter is to discuss the basic aspects and design of furrow irrigation.

20.2 CLASSIFICATION OF FURROWS

A furrow is a small channel that runs either along or across the slope of the field. A field has a system of furrows and hence a furrow irrigation system comprises ridges and furrows. Furrows can be classified by their size and spacing into deep furrows and corrugations. Corrugations are essentially small, shallow furrows or channels of V or U shape, about 6–10 cm deep, 1:1 side slope, and 40–75 cm apart. The hydraulics of flow is essentially the same for furrows and corrugations. Corrugation irrigation is used for non-cultivated close-growing crops, such as small grains and pastures on steep slopes. Corrugations form after the seeding of crops, but in perennial crops they are usually reshaped to maintain the desired cross-section.

Furrows can be classified on the basis of alignment, size, and spacing. Based on alignment, furrows can be distinguished as straight furrows and contour furrows. Straight furrows may be graded or level. Level furrows do not have a longitudinal slope and are diked at the end to prevent or minimize runoff. Such furrows are suitable for fine-textured soils with low infiltration rates. Graded furrows are usually straight channels constructed on the prevailing land slope. These furrows have a nearly uniform slope in the direction of irrigation. Contour furrows are curved to fit the topography and carry water across rather than along the field. The furrows are provided with a gentle slope along the length. The contour furrow system can be adopted for slopes up to 5% in the case of light soils, whereas the furrow irrigation method can be used to irrigate stable soils for orchards with slopes up to 10% (Michael, 1978). The graded contour furrow system is used with uneven topography where it is not practical to use straight furrows within permissible grade limitations. The furrow system is sometimes used in combination with the bench terrace system in naturally steep areas.

Based on their size and spacing, furrows can be classified as deep furrow and corrugation. Corrugations are small, shallow furrows that are suitable for irrigating close-growing crops, such as small grains and meadow, on moderately steep slopes. Corrugations are often used in conjunction with border irrigation in poorly leveled border strips.

20.3 LAYOUT OF A FURROW IRRIGATION SYSTEM

The layout of a furrow irrigation system depends on the location of the water source, terrain characteristics, soil characteristics, infiltration capacity, furrow length, and crop type. A typical layout of a graded furrow system is shown in Figure 20.1. It is seen that furrows are laid parallel to the field boundary, with the delivery channel or pipeline at their upstream end. The drainage channel or runoff recovery system is placed at the lower end of the field to collect runoff, particularly from free-draining furrows. If a field is small or has a low infiltration capacity soil, then furrows can extend the full length of the field. If the field is long or has high infiltration capacity, then it is divided into smaller portions and furrow lengths are equal to an even fraction of the total field length. If the terrain is steep, then bench terracing can be used to achieve mild areas and furrows can be constructed. If the terrain is irregular, then graded furrows following field contours can be used. Vegetable crops grown on slopes up to 8–10% can be irrigated with contour furrows.

20.3.1 Source of Water

The water source should be located near the center of the irrigation field, which would reduce the size of delivery channels and pipelines. Also, the location should be such that the entire field can be irrigated by gravity.

20.3.2 Furrow Length

Furrow length depends on field size, machinery to be used, type of soil, type of crop, water supply, frequency of

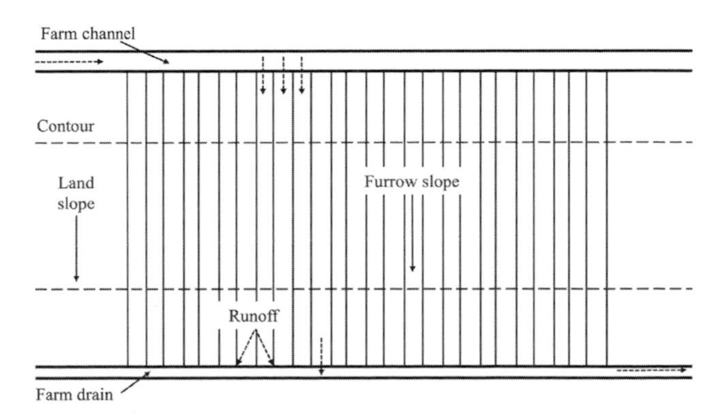

Figure 20.1 Graded furrow system. (Adapted from Jurriëns et al., 2001.)

Table 20.1 *Furrow lengths in meters as related to soil type, slope, stream size, and irrigation depth*

		Soil type								
		Fine			Medium			Coarse		
		Average irrigation depth (mm)*								
Furrow slope (%)	Stream size (L/s)	50	100	150	50	100	150	50	100	150
0.25	3.00	300	450	530	250	375	420	150	210	260
0.50	1.50	220	310	380	170	240	290	120	150	180
1.00	0.75	170	250	280	130	180	220	70	110	120
1.50	0.50	130	190	250	100	140	170	60	90	120
2.00	0.37	120	160	200	90	120	150	50	70	90
3.00	0.25	90	130	160	70	100	120	25	60	70

* Indicates including leaching water

After Withers and Vipond, 1974. Copyright © 1974 by Batsford, London.

Table 20.2 *Practical values of maximum furrow lengths in meters depending on soil type, slope, stream size, and irrigation depth for small-scale irrigation*

	Soil type	Clay		Loam		Sand	
		Net irrigation requirement (mm)					
Furrow slope (%)	Maximum stream size per furrow (L/s)	50	75	50	75	50	57
0.0	3.0	100	150	60	90	30	45
0.1	3.0	120	170	90	125	40	60
0.2	2.5	130	180	110	150	60	95
0.3	2.0	150	200	130	170	75	110
0.5	1.2	150	200	130	170	75	110

After Brouwer et al., 1984.

irrigation, degree of mechanization, and landscape. The length should be such that there exists a reasonable balance between efficiency of farming operations, labor requirement, system cost, application efficiency, and uniformity of applied water. If a field is short, then furrow length can be constrained by the field length (or field boundary). On the other hand, in long, mechanized fields it is desired to obtain a reasonable efficiency of the farming operation and high application efficiency and uniformity. Application efficiency and uniformity normally increase as the furrow length decreases. Therefore, the furrow length may be set equal to an even fraction of the total field length.

Short furrows may be suitable in high-intake soils, while and low and long furrows are more appropriate for low-intake soils. Short furrows may be most suitable in areas with low water-holding-capacity soils, small and frequent application depth, inexpensive labor, and limited water supply. However, the system layout cost is greater for short furrows than for large furrows. Tables 20.1 and 20.2 give some guidelines for selecting furrow length based on soil type, available flow, land slope, and application depths for large mechanized and small farms, respectively. The values given in Tables 20.1 and 20.2 should be used only as a guide in the initial planning. However, the furrow length should be determined based on local conditions such as soil, crop, water availability, frequency of irrigation, degree of mechanization, and more.

20.3.3 Furrow Slope

Furrow slope should normally be below 1% to minimize the risk of rain-induced erosion, except in arid areas, where it can be up to 3% (Hart et al., 1980). In humid areas, the furrow slope should not be more than 0.3–0.5% (James, 1988). The maximum slope should not exceed 0.5% for smallholdings. The maximum furrow slope for erodible soils should be limited to 60 mm/(P30), where P30 is the 30-minute rainfall in mm for a two-year frequency. For less erodible soils (sandy and clayey soils), this limit can be exceeded by 25% (USDA, 1984 taken from Clemmens et al., 2001). A minimum furrow slope of 0.03–0.05% is needed in humid areas for proper surface drainage.

20.3.4 Furrow Shape

The shape of a furrow depends on the soil type and stream size; it can be either V-shaped or U-shaped in cross-section. Furrows are 10–20 cm deep and 20–30 cm wide at the top (James, 1988), and their shape changes during the season due to the flow of water. Shallow furrows are suited to uniformly graded fields and shallow-rooted crops. In low-intake soils, shallow, wide, U-shaped furrows are used to reduce the velocity of flow and to increase the wetted perimeter to enhance infiltration. On the other hand, V-shaped furrows are used to reduce the wetted perimeter to reduce infiltration in sandy soils. In any case, furrows should be large enough to carry the desired inflow rate without overtopping. The wetted perimeter (P, in m) can be expressed as (USDA-SCS, 1983)

$$P = 0.265 \times \left(\frac{nQ_0}{S_0^{0.5}}\right)^{0.425} + 0.227, \tag{20.1}$$

where Q_0 is the volumetric inflow rate per furrow (L/s), n is Manning's roughness coefficient, and S_0 is the furrow slope (m/m). This relationship was developed based on the calculation of wetted perimeter from a series of trapezoidal shapes ranging from 60 mm bottom width with 1:1 side slopes to 150 mm bottom width with 2:1 side slopes. Note that the constant in eq. (20.1) accounts for both vertical and lateral infiltration.

Example 20.1: Compute the wetted perimeter of a furrow if Manning's roughness coefficient is 0.04, inflow discharge is 0.5 L/s, and the slope is 0.05.

Solution: Using eq. (20.1), the wetted perimeter (P, in m) can be computed as

$$P = 0.265 \times \left(\frac{nQ_0}{S_0^{0.5}}\right)^{0.425} + 0.227$$

$$= 0.265\left[\frac{0.04 \times 0.5}{0.05^{0.5}}\right]^{0.425} + 0.227 = 0.322 \text{ m}.$$

20.3.5 Furrow Spacing

The spacing of furrows depends on soil texture, agronomic practices, and type of machines used in planting and cultivation. Furrows should be spaced such that water spreads laterally between adjacent furrows and adequately wets the entire crop root zone. Figure 20.2 shows the moisture distribution pattern in homogeneous clay, loam, and sandy soils, whereas moisture distribution under a nonhomogeneous soil profile is shown in Figure 20.3. The sandy soil has more vertical movement of water due to the predominance of gravitational forces than do clay and loam soils, which have a predominance of capillary forces; thus, closer spacing is required in sandy soils. Further, soils with nonuniform profiles have greater lateral movement of water than soils with

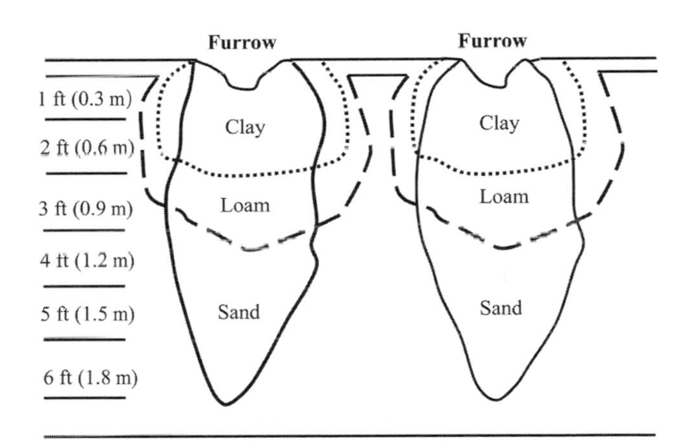

Figure 20.2 Typical pattern of moisture penetration by the same amount of water from furrows in different homogeneous soils. (Adapted from USDA-SCS, 1983.)

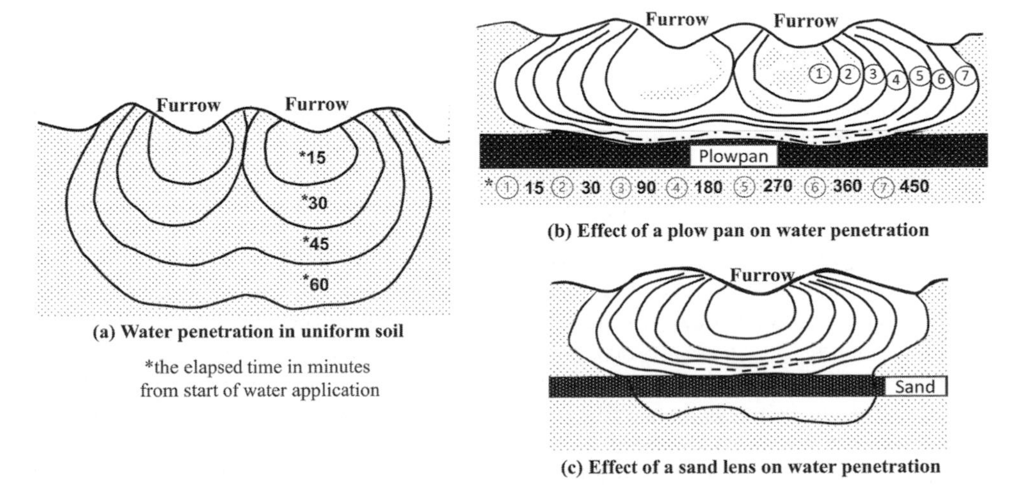

Figure 20.3 Example of water penetration from an irrigation furrow. (Adapted from USDA-SCS, 1983.)

uniform profiles (Figure 20.3). In most cultivated soils, the presence of sand lens or plow pan retards the vertical movement of water, which helps in better lateral movement. In general, furrow spacing of 0.3 m, 0.6 m, and up to 1.2 m is recommended for coarse soils, fine soils, and heavy clay soils, respectively.

If the total number of furrows in the field is N_f and the number of furrows irrigated per set is N_{fs}, then the number of furrows per set can be computed as

$$N_{fs} = \frac{N_f T_i D_{IR} E_a}{144{,}000 Z_{req}}, \tag{20.2}$$

where T_i is the inflow time (min), D_{IR} is the design daily irrigation requirement (mm/day, in./day), E_a is the application efficiency (%), and Z_{req} is the desired depth of irrigation (mm, in.).

Example 20.2: Compute the number of furrows per set in a field for which the total number of furrow sets to be irrigated is 100, the inflow time is 120 min, the design daily irrigation requirement is 60 mm/day, the application efficiency is 90%, and the desired depth of irrigation is 150 mm.

Solution: Using eq. (20.2), the number of furrows per set is computed as

$$N_{fs} = \frac{N_f T_i D_{IR} E_a}{144{,}000 Z_{req}} = \frac{100 \times 120 \times 60 \times 90}{144{,}000 \times 150} = 3 \text{ (furrows per set)}.$$

20.3.6 Cutoff Time (Inflow Time)

The purpose of irrigation is to replace the soil moisture deficit in the root zone. To ensure the minimum infiltrated depth is equal to the desired depth of irrigation at the downstream end, the cutoff time must include the time to advance to the end of the field and the time to infiltrate the desired irrigation depth. Assuming that depletion and recession times are negligible, the cutoff time for free-draining graded furrows can then be expressed as:

$$T_{CO} = \tau_{req} + T_a, \tag{20.3}$$

where T_{CO} is the time of cutoff or inflow time (min), τ_{req} is the intake opportunity time required to infiltrate the desired irrigation (design) depth (min), Z_{req}; and T_a is the time to advance to the end of the field or advance time (min). The cumulative infiltration can be obtained from the Kostiakov equation,

$$Z_{req} = k\tau_{req}{}^a, \tag{20.4}$$

or the modified Kostiakov equation,

$$Z_{req} = k\tau_{req}{}^a + f_0\tau_{req}. \tag{20.5}$$

It should be noted that when the irrigation time is relatively short, as occurs in some basin and border irrigation systems, eq. (20.4) will not greatly underestimate irrigation at the final stage. However, when the irrigation opportunity is longer than 3–4 h, which commonly occurs in furrow irrigation and large border and basin irrigation systems, eq. (20.4) may significantly underestimate infiltration. Hence, eq. (20.5) is usually suggested in such conditions.

The time τ_{req} for achieving Z_{req} can be obtained analytically from eq. (20.4) or numerically from eq. (20.5) using the Newton–Raphson method:

1. Assume an initial value of τ_{req} as τ_{req}^i (i is an iteration number).
2. Compute a new value ([$i + 1$] iteration) as follows:

$$\tau_{req}^{i+1} = \tau_{req}^i - \frac{f(\tau_{req})}{f'(\tau_{req})}, \tag{20.6}$$

$$f(\tau_{req}) = Z_{req} - k\left(\tau_{req}^i\right)^a - f_0\tau_{req}^i, \tag{20.7}$$

$$f'(\tau_{req}) = -ak\left(\tau_{req}^i\right)^{a-1} - f_0. \tag{20.8}$$

3. Compare the values of τ_{req}^i and τ_{req}^{i+1}, and if the two values are within a specified tolerance limit, then τ_{req}^{i+1} so obtained is the value of τ_{req}. Otherwise, repeat steps 2 and 3, considering a new value (τ_{req}^{i+1}) as the initial guess (τ_{req}^i).

From eq. (20.4), the time of infiltration can be expressed as

$$\tau_{req} = \left(\frac{Z_{req}}{k}\right)^{\frac{1}{a}}. \tag{20.9}$$

Equation (20.9) can be interpreted to yield the time required to apply irrigation of depth Z_{req}. If the soil is homogeneous, with uniform initial water content, then the depth of the wetting front, d, measured from the soil surface, can be written as

$$d = cZ_{req}, \tag{20.10}$$

in which c is the proportionality constant. Inserting eq. (20.10) into eq. (20.4), one gets

$$d = ck\tau_{req}{}^a. \tag{21.11}$$

If the time required to infiltrate the desired depth of water is T_I, then the advance time, the time required for the water advance to reach the end of the field, T_a, assuming the

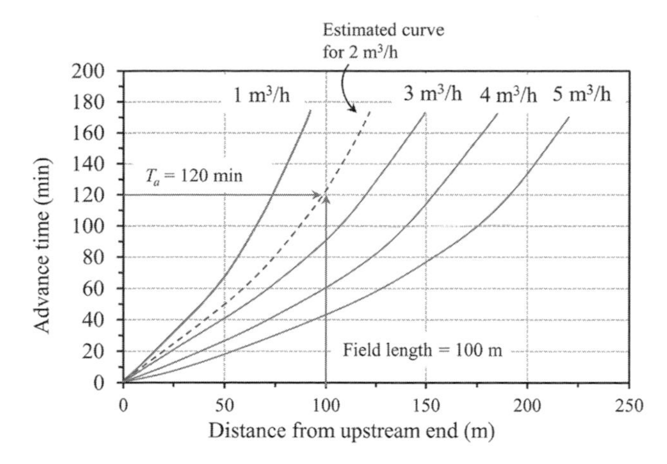

Figure 20.4 A graphical procedure for the design stream size. The measured advance data for different stream sizes are obtained from James (1988).

one-fourth rule, is $T_a = T_I/4$. In the field, the wetting front advances as water moves from the head down the field, as shown in Figure 20.4. Figure 20.4 can be plotted based on measured advance data for different streams, and then it can be used to determine the stream size. For example, based on the design depth of irrigation, we can calculate the time required to infiltrate the desired depth of water T_I. Using the one-fourth rule, T_a can be obtained. An example is shown in Figure 20.4, in which we can estimate the design stream size as $2\,m^3/h$ with $T_a = 120\,min$, and the field length is 100 m. Clearly, the depth of infiltrated water will be larger at the head than at the downstream end. At different times the wetting front will be different. Likewise, the depth of water percolating below the root zone will also be different at different locations in the field. The wetting front can now be calculated as follows.

For the convenience of calculation and reasonable accuracy, the total infiltration time can be divided into four equal parts:

$$\Delta\tau_1 = \Delta\tau_2 = \Delta\tau_3 = \Delta\tau_4 = \frac{T_I}{4}. \tag{20.12}$$

The depth of water infiltrated during time $\Delta\tau_1$ can be expressed as

$$\Delta Z_1 = Z_1 - Z_0. \tag{20.13}$$

To start with, if $Z_0 = 0$,

$$\Delta Z_1 = Z_1. \tag{20.14}$$

From the Kostiakov equation,

$$\Delta Z_1 = Z_1 = k\tau_1^a. \tag{20.15}$$

The depth of wetted soil during $\Delta\tau_1$ can be given as

$$\Delta d_1 = ck\tau_1^a - 0 = ck\tau_1^a. \tag{20.16}$$

In a similar manner, for time interval $\Delta\tau_2$, the wetting soil depth can be written as

$$\Delta Z_2 = Z_2 - Z_1. \tag{20.17}$$

Then,

$$\Delta Z_2 = Z_2 - \Delta Z_1 = k\tau_2^a - k\tau_1^a. \tag{20.18}$$

The depth of wetted soil during $\Delta\tau_2$ can be given as

$$\Delta d_2 = ck\tau_2^a - \Delta d_1 = ck\tau_2^a - ck\tau_1^a. \tag{20.19}$$

It is noted that $\tau_2 = 2\tau_1$. Then, eq. (20.19) can be written as

$$\Delta d_2 = ck(2\tau_1)^a - ck(\tau_1)^a. \tag{20.20}$$

This shows that

$$\Delta d_2 = \Delta d_1[(2)^a - 1]. \tag{20.21}$$

In the same way, one can write

$$\Delta d_3 = \Delta d_1[(3)^a - (2)^a], \tag{20.22}$$

$$\Delta d_4 = \Delta d_1[(4)^a - (3)^a], \tag{20.23}$$

$$\Delta d_5 = \Delta d_1[(5)^a - (4)^a]. \tag{20.24}$$

At the downstream end of the field, the infiltration opportunity time is $\tau_4 = 4\tau_1$. The wetted depth is

$$d_4 = ck(4\tau_1)^a = \Delta d_1(4)^a. \tag{20.25}$$

When the opportunity time at the end of the field is T_I, then using the one-fourth rule, the opportunity time at the head of the field can be written as

$$\tau_5 = T_I + T_a = T_I + \frac{T_I}{4} = 5\tau_1. \tag{20.26}$$

The wetted depth at the head can then be written as

$$d_5 = \Delta d_1(5)^a. \tag{20.27}$$

One can approximately represent the wetting front as a linear function of distance. Then, the average depth of soil wetted by deep percolation can be written as

$$\frac{\Delta d_5}{2} = \frac{d_5 - d_4}{2} = \frac{\Delta d_1[(5)^a - (4)^a]}{2}. \tag{20.28}$$

The average depth of wetted soil can be determined as

$$\frac{d_5 + d_4}{2} = \frac{\Delta d_1[(5)^a + (4)^a]}{2}. \tag{20.29}$$

The ratio of the average depth of percolation to the average depth of wetted soil can be computed as

$$\frac{(\Delta d_5)/2}{(d_5 + d_4)/2} = \frac{[(5)^a - (4)^a]}{[(4)^a + (5)^a]}. \tag{20.30}$$

The fractional advance ratio (FAR) can be expressed by the ratio of advance time to the net time of irrigation at the end of the field:

$$FAR = \frac{T_a}{T_I}. \tag{20.31}$$

Table 20.3 gives the expected percentages of percolation losses (r_a) for different FARs subject to different values of a.

Table 20.3 *Percentages of percolation losses (r_a) for fractional advances ratios, subject to various values of a*

	a								
r_a	0.1	0.2	0.3	0.4	0.5	0.6	0.7	0.8	0.9
$FAR = 1$	3.5	6.9	10.4	13.8	17.2	20.5	23.8	27.0	30.2
$FAR = 1/2$	2.0	4.1	6.1	8.1	10.1	12.1	14.1	16.1	18.0
$FAR = 1/3$	1.4	2.9	4.3	5.7	7.2	8.6	10.1	11.5	12.9
$FAR = 1/4$	1.1	2.2	3.3	4.5	5.6	6.7	7.8	8.9	10.0
$FAR = 1/5$	0.9	1.8	2.7	3.6	4.6	5.5	6.4	7.3	8.1
$FAR = 1/10$	0.5	1.0	1.4	1.9	2.4	2.9	3.3	3.8	4.3

Example 20.3: Consider that the advance time is one-quarter of the net time of irrigation. Assume the Kostiakov infiltration exponent *a* is 0.5. Derive the percentage of percolation loss and compare this loss with the value that can be obtained from Table 20.3.

Solution: Given $FAR = 1/4$, the percentage loss from Table 20.3 is 5.6%.

The ratio of the average depth of percolation to the average depth of wetted soil can be computed using eq. (20.30) as:

$$\frac{(\Delta d_5)/2}{(d_5+d_4)/2} = \frac{[(5)^a-(4)^a]}{[(4)^a+(5)^a]} = \frac{[(5)^{0.5}-(4)^{0.5}]}{[(4)^{0.5}+(5)^{0.5}]}$$

$$= \frac{0.236}{4.236} \times 100\% = 5.57\% \approx 5.6\%.$$

This value is similar to the value obtained from Table 20.3.

Example 20.4: Determine the net time of irrigation and the advance time that the water takes to reach the downstream end of the field if the net irrigation required is 10 cm. It is assumed that the Kostiakov infiltration parameters for depth in cm and time in min are $a = 0.75$ and $k = 0.25$, and the percolation loss is 8.5%.

Solution: Given that $a = 0.75$, the percolation loss is 8.5%, the *FAR* is about 0.258 using interpolation from Table 20.3:

$$FAR = \frac{T_a}{T_I} = 0.255.$$

From the Kostiakov equation, the net irrigation T_I can be computed as

$$T_I = \left(\frac{10}{0.25}\right)^{\frac{1}{0.75}} = 136.8 \text{ min}.$$

Therefore,

$$T_a = 0.255 \times 136.8 \text{ min} = 35 \text{ min}.$$

20.3.7 Inflow Rate (Stream Size)

The inflow rate and cutoff time are critically important, but are also flexible design variables. The cutoff time is related to the stream size (inflow rate). A large inflow rate would clearly take less time to reach the downstream end of the field and would result in better uniformity compared with a small inflow rate. However, inflow rate also affects irrigation performance (efficiency). For example, a large inflow rate may result in more runoff from freely draining graded furrows and a small inflow rate may result in more deep percolation losses at the head end of the field, resulting in reduced application efficiency. Therefore, the goal of furrow irrigation design is to determine inflow rates that would lead to maximum application efficiency, while irrigation is completed subject to available water supply constraints. The minimum inflow rate must be such that the advance is completed within the stipulated time. Likewise, the maximum flow rate should correspond to the maximum nonerosive velocity (8 m/min in erosive silt soils to about 13 m/min in stable clay and sandy soils). The furrow stream size that does not exceed the maximum nonerosive stream size can be obtained as

$$Q_{max} = \frac{K}{S_0}, \tag{20.32}$$

where Q_{max} is the maximum stream size (L/min, gpm), S_0 is the furrow slope in the flow direction (%), and K is the unit constant equal to 40 for Q_{max} in L/min and equal to 10 for Q_{max} in gpm.

The maximum permissible value of inflow rate can also be obtained by considering the maximum nonerosive velocity for a given soil as (Walker and Skogerboe, 1987):

$$Q_{max} = \left[v_{max}^{a_2} \frac{n^2}{3600 a_1 S_0} \right]^{1/(a_2 - 2)}, \tag{20.33}$$

where Q_{max} is the maximum permissible inflow rate (m^3/min), v_{max} is the maximum permissible velocity (m/min), S_0 is the slope as a fraction, n is Manning's roughness coefficient, and a_1 and a_2 are the empirical furrow geometry parameters.

Sometimes the inflow rate is selected based on the one-fourth rule, which states that the advance time to the end of the furrow should be one-quarter of the time required to infiltrate the desired depth of water (T_I or τ_{req}). To obtain a higher efficiency of a furrow irrigation system, either inflow rate is cut back after the completion of advance phase or a reuse system is placed to recirculate runoff. Efficiency can also be improved by changing the furrow length and slope.

Example 20.5: Compute the stream size using the one-fourth rule – that is, the advance time is equal to one-quarter of the net time of infiltration for a field 150 m long, and the net infiltration time is 300 min. Advance time for different stream sizes is given in Figure 20.4.

Solution: The advance time T_a is computed using the one-quarter rule as $T_a = 300/4 = 75$ min. Then, from Figure 20.4, the value of stream size is found as: 5 m^3/h.

20.3.8 Number of Furrows per Set and Number of Sets

The number of sets and number of furrows per set depends on field width, furrow spacing, total discharge available, and the chosen inflow rate. The number of furrows per set and the number of sets have to be integer values and may require some adjustment in either the chosen inflow rate or total water supply to get these integer values. The number of furrows in a given field (N_f), number of sets (N_s), and number of furrows per set (N_{fs}) can be obtained as:

$$N_f = \frac{W}{F_s}, \tag{20.34}$$

$$N_s = \frac{N_f Q_0}{Q}, \tag{20.35}$$

$$N_{fs} = \frac{N_f}{N_s}, \tag{20.36}$$

where Q_0 is the selected inflow rate per furrow (m^3/min), Q is the total inflow rate (m^3/min), F_s is the furrow spacing (m), and W is the field width (m).

20.3.9 Delivery System

A delivery system that delivers water from the supply source to a field is either an open channel or low-pressure pipeline. The design of the delivery system is carried out once inflow rate, furrow length, furrow slope, and the number of furrows per set are finalized. The delivery system should have sufficient capacity to meet the demand everywhere in the farm that it is needed. The principles for the design of open channels and pipelines are discussed in Chapters 8 and 9.

20.3.10 Two-Point Method for Computing Infiltration Parameters

This method assumes a power function relationship between advance time and distance across a furrow as:

$$x = bt^c, \tag{20.37}$$

where b and c are parameters, x is the advance distance (m, ft), and t is the advance time since the start (min). The advance time for the full length of the furrow is denoted as T_a, and $T_{0.5a}$ is the advance time to half of the length. The exponent c can be expressed as

$$c = \frac{\ln 2}{\ln T_a - \ln T_{0.5a}}. \tag{20.38}$$

Rewriting the modified Kostiakov equation for the cumulative infiltration (m^3/m, ft^3/ft) again,

$$Z = kt^a + \frac{f_0 t}{K}, \tag{20.39}$$

where

$$a = \frac{\ln V_a - \ln V_{0.5a}}{\ln T_a - \ln T_{0.5a}}, \tag{20.40}$$

where V_a is the volume of water on the full length during the advance time and $V_{0.5a}$ is the volume during half length, expressed, respectively, as

$$V_a = \frac{QT_a}{KL} - 0.77A_0 - \frac{f_0 T_a}{K(c+1)}, \quad (20.41)$$

$$V_{0.5a} = \frac{2QT_{0.5a}}{KL} - 0.77A_0 - \frac{f_0 T_{0.5a}}{K(c+1)}, \quad (20.42)$$

$$k = \frac{V_a}{\sigma_z T_a^a}, \quad (20.43)$$

$$\sigma_z = \frac{a + c(1-a) + 1}{(1+a)(1+c)}, \quad (20.44)$$

where Q is the stream size (L/min, gpm); f_0 is the basic or final infiltration rate (L/min/m, gpm/ft); L is the length of furrow (m, ft); A_0 is the cross-sectional area at the field inlet (m^2, ft^2); K is a unit constant (equal to 1000 for Q in L/min, A_0 in m^2, σ_z is the subsurface shape factor, and f_0 is in L/min/m; or 7.48 for Q in gpm, A_0 in ft^2, and f_0 in gpm/ft).

Example 20.6: Consider a furrow system in which the furrow cross-sectional area is 0.2 m^2, the stream size is 2500 L/min, the final infiltration rate is 1.5 L/min, and the field length and width are 60 m and 10 m, respectively, with advance time for full field length as 18 min and for half field length as 8 min. Compute the values of parameters k, a, and c, and depth cumulative infiltration after 150 min.

Solution: Given $T_a = 18$ min and $T_{0.5a} = 8$ min,

$$c = \frac{\ln 2}{\ln T_a - \ln T_{0.5a}} = \frac{\ln 2}{\ln 18 - \ln 8} = 0.855.$$

Given the furrow cross-sectional area $A_0 = 0.2\ m^2$, the stream size $Q = 2500$ L/ min, $K = 1000$, $f_0 = 1.5$ L/ min, V_a and $V_{0.5a}$ can be computed using eq. (20.41) and (20.42), respectively:

$$V_a = \frac{QT_a}{KL} - 0.77A_0 - \frac{f_0 T_a}{K(c+1)} = \frac{2500 \times 18}{1000 \times 60} - 0.77 \times 0.2 - \frac{1.5 \times 18}{1000 \times (0.855+1)} = 0.581 \frac{m^3}{m},$$

$$V_{0.5a} = \frac{2QT_{0.5a}}{KL} - 0.77A_0 - \frac{f_0 T_{0.5a}}{K(c+1)} = \frac{2 \times 2500 \times 8}{1000 \times 60} - 0.77 \times 0.2 - \frac{1.5 \times 8}{1000 \times (0.855+1)} = 0.506 \frac{m^3}{m}.$$

a can be obtained using eq. (20.40):

$$a = \frac{\ln V_a + \ln V_{0.5a}}{\ln T_a - \ln T_{0.5a}} = \frac{\ln 0.581 - \ln 0.506}{\ln 18 - \ln 8} = 0.17.$$

σ_z can be obtained using eq. (20.44):

$$\sigma_z = \frac{a + c(1-a) + 1}{(1+a)(1+c)} = \frac{0.17 + 0.855(1 - 0.17) + 1}{(1+0.17)(1+0.855)} = 0.866.$$

k can be calculated using eq. (20.43):

$$k = \frac{V_a}{\sigma_z T_a^a} = \frac{0.581}{(0.866)(18)^{0.17}} = 0.41.$$

The value of cumulative infiltration for 150 min can be calculated using eq. (20.39):

$$Z = kt^a + \frac{f_0 t}{K} = 0.41 \times (150)^{0.17} + \frac{1.5 \times 150}{1000} = 1.19 \frac{m^3}{m}.$$

Z in the unit of depth is:

$$Z = \frac{1.19 \frac{m^3}{m}}{10\ m} \times \left(\frac{100\ cm}{1\ m}\right) = 11.9\ cm.$$

20.3.11 Design Relationships

The average intake Z_a (mm) over the furrow length can be computed as

$$Z_a = \frac{1}{LP}(V_{in} - V_{out} - V_s), \quad (20.45)$$

where L is the distance between inflow and outflow measurements (m), P is the adjusted wetted perimeter (m), V_{in} is the inflow volume (L), V_{out} is the outflow volume (L), and V_s is the volume of water in storage (L).

The volume of furrow storage V_s can be expressed as

$$V_s = \frac{L}{0.305}\left[2.947\left(\frac{Q_0 n}{S_0^{0.5}}\right)^{0.735} - 0.0217\right]. \quad (20.46)$$

The required infiltration depth for a furrow system must be written as equivalent length over the entire field. The infiltration depth given by the modified Kostiakov equation must be multiplied by the ratio of adjusted wetted perimeter to the furrow spacing (F_s, in m) as

Table 20.4 *Intake family and advance coefficients for depth of infiltration in mm, time in min, and length in m*

Intake family	k	a	f_0	ρ_a	ρ_b
0.05	0.5334	0.618	7.0	7.16	1.088×10^{-4}
0.10	0.6198	0.661	7.0	7.25	1.251×10^{-4}
0.15	0.7110	0.683	7.0	7.34	1.414×10^{-4}
0.20	0.7772	0.699	7.0	7.43	1.578×10^{-4}
0.25	0.8534	0.711	7.0	7.52	1.741×10^{-4}
0.30	0.9246	0.720	7.0	7.61	1.904×10^{-4}
0.35	0.9957	0.729	7.0	7.70	2.067×10^{-4}
0.40	1.064	0.736	7.0	7.79	2.230×10^{-4}
0.45	1.130	0.742	7.0	7.88	2.393×10^{-4}
0.50	1.196	0.748	7.0	7.97	2.556×10^{-4}
0.60	1.321	0.757	7.0	8.15	2.883×10^{-4}
0.70	1.443	0.766	7.0	8.33	3.209×10^{-4}
0.80	1.560	0.773	7.0	8.50	3.535×10^{-4}
0.90	1.674	0.779	7.0	8.68	3.862×10^{-4}
1.00	1.786	0.785	7.0	8.86	4.188×10^{-4}
1.50	2.284	0.799	7.0	9.76	5.819×10^{-4}
2.00	2.753	0.808	7.0	10.65	7.451×10^{-4}

After USDA-SCS, 1974, 1983.

$$Z_a = (kt^a + f_0)\frac{P}{F_s}. \tag{20.47}$$

The advance time for a stream of size per furrow Q_0 (L/s) can be written as

$$T_a = \frac{x}{\rho_a}\exp\left[\frac{\rho_b x}{Q_0 S_0^{0.5}}\right], \tag{20.48}$$

in which ρ_a and ρ_b are advance coefficients, as shown in Table 20.4, and x is the advance front distance.

The infiltration opportunity time (T_I) is the sum of cutoff time (T_{CO}) and recession time (T_r) minus advance time (T_a):

$$T_I = T_{co} - T_a + T_r. \tag{20.49}$$

The cutoff time is normally regarded as the advance time plus the required net infiltration time minus the recession time. If Z_n is the desired net infiltration depth, then the net infiltration time can be determined from eq. (20.47) as

$$T_n = \left[\frac{Z_n\left(\frac{F_s}{P}\right) - f_0}{k}\right]^{\frac{1}{a}}. \tag{20.50}$$

For graded furrows it is not unrealistic to assume that the recession time is zero. Therefore, eq. (20.49) becomes

$$T_I = T_n = T_{CO} - T_a. \tag{20.51}$$

The cutoff time can be expressed as

$$T_{CO} = T_a + T_n. \tag{20.52}$$

The average infiltration opportunity time over distance x of the furrow can be expressed as

$$T_I(x) = T_{CO} - \frac{0.0929}{\rho_a x\left[\frac{0.305\beta}{x}\right]^2}[(\beta - 1)\exp(\beta) + 1]; \beta = \frac{\rho_b x}{Q_0 S_0^{0.5}}. \tag{20.53}$$

If x in eq. (20.53) is replaced by L, then the average infiltration opportunity time for the entire furrow length will be obtained. When the infiltration depth given by eq. (20.47) is divided by this opportunity time, the result is the average depth of infiltration over the entire furrow.

Now the gross depth of water application, d_g, can be determined. This can be defined as the required net depth of irrigation (d_n) divided by the product of application and distribution efficiencies (E_d):

$$d_g = \frac{d_n}{\frac{E_a}{100} \times \frac{E_d}{100}}. \tag{20.54}$$

In furrow irrigation evaporation losses are negligible, so application efficiency can be assumed to be close to 100%. Therefore, eq. (20.56) reduces to

$$d_g = \frac{d_n}{\frac{E_d}{100}}. \tag{20.55}$$

The depth of application can be expressed as a function of inflow rate and field geometry as:

$$d_g = \frac{60QT_{CO}}{WL} = \frac{60Q_0T_{CO}}{F_sL}, \tag{20.56}$$

in which Q is in L/s, W is in m, L is in m, and d_a is in mm.

The depth of surface runoff from furrow (d_r) can be regarded as the difference between gross and average depths of application:

$$d_r = d_g - d_{av}. \tag{20.57}$$

The depth of percolation (d_p) can now be computed as the difference between the average depth of application (d_{av}) and the net depth of application (d_n):

$$d_p = d_{av} - d_n. \tag{20.58}$$

Example 20.7: Compute the time of cutoff, depth of surface runoff, depth of deep percolation, and distribution efficiency for a furrow system with furrow length of 300 m, slope of 0.005 m/m, furrow spacing of $F_s = 0.5$ m, Manning's roughness coefficient of 0.04, inflow rate per furrow of 0.5 L/s, and net irrigation depth of 80 mm. The intake family I_f is 0.35.

Solution: First, compute the advance time T_a:

From Table 20.4, for intake family of 0.35, $\rho_b = 2.067 \times 10^{-4}$, $\rho_a = 7.7$,

$$\beta = \frac{\rho_b L}{Q_0 S_0^{0.5}} = \frac{0.000207 \times 300}{0.5 \times (0.005)^{0.5}} = 1.754,$$

$$T_a = \frac{L}{\rho_a} \exp\left[\frac{\rho_b L}{Q_0 S_0^{0.5}}\right] = \frac{300}{7.7} \exp[1.754] = 225 \text{ min}.$$

Compute the wetted perimeter using eq. (20.1):

$$P = 0.265 \times \left(\frac{Q_0 n}{S_0^{0.5}}\right)^{0.425} + 0.227$$

$$= 0.265 \times \left[\frac{0.5 \times 0.04}{0.005^{0.5}}\right] + 0.227 = 0.382 \text{ m}.$$

From Table 20.4, for the intake family of 0.35, $k = 0.9957$, $a = 0.729$, and $f_0 = 7$, compute the net infiltration time using eq. (20.50):

$$T_n = \left[\frac{Z_n\left(\frac{F_s}{P}\right) - f_0}{k}\right]^{\frac{1}{a}} = \left[\frac{80\left(\frac{0.5}{0.382}\right) - 7}{0.9957}\right]^{\frac{1}{0.729}}$$

$$= 540 \text{ min}.$$

Compute the cutoff time using eq. (20.52):

$$T_{CO} = T_a + T_n = 225 + 540 = 765 \text{ min}.$$

Compute the gross application depth using eq. (20.56):

$$d_g = \frac{60QT_{CO}}{WL} = \frac{60Q_0T_{CO}}{F_sL} = \frac{60 \times 0.5 \times 765}{0.5 \times 300} = 153 \text{ mm}.$$

Compute the average infiltration time using eq. (20.53):

$$T_I(L) = T_{CO} - \frac{0.0929}{\rho_a L\left[\frac{0.305\beta}{L}\right]^2}[(\beta - 1)\exp(\beta) + 1]$$

$$= 765 - \frac{0.0929}{(7.7) \times (300)\left[\frac{0.305 \times 1.754}{300}\right]^2}$$
$$[(1.754 - 1) \times \exp(1.754) + 1]$$
$$= 697 \text{ min}.$$

Compute the average infiltration depth using eq. (20.47):

$$Z_a = (kt^a + f_0)\frac{P}{F_s} = \left(0.9957 \times 697^{0.729} + 7\right) \times \frac{0.382}{0.5}$$
$$= 95.3 \text{ mm} = d_{av}.$$

Compute the surface runoff depth using eq. (20.57):

$$d_r = d_g - d_{av} = 153 - 95.3 = 57.7 \text{ mm}.$$

Compute the percolation depth using eq. (20.58):

$$d_p = d_{av} - d_n = d_{av} - Z_n = 95.3 - 80 = 15.3 \text{ mm}.$$

Compute the distribution efficiency using eq. (20.55):

$$E_d = \frac{d_n}{d_g} \times 100 = \frac{Z_n}{d_g} \times 100 = \frac{80}{153} \times 100 = 52.3\%.$$

20.3.12 Cutback Systems

When the FAR is smaller, the inflow rate is larger and the amount of deep percolation is smaller. Keeping this in mind, the inflow at the head of the furrow is reduced once the water reaches the end of the furrow. This consideration gives rise to the concept of the cutback system in which the cutback time equals the advance time and final inflow rate is half the initial inflow rate. For simplicity, subscript 1 is used to denote the initial condition and subscript 2 to denote cutback conditions. The following modifications are then made for cutback systems.

The wetted perimeter under cutback conditions is computed using eq. (20.1), with Q replaced by Q_2. The required net infiltration time is computed using eq. (20.50), with P replaced by P_2. The average opportunity time for infiltration during the advance phase is given by eq. (20.53), with x replaced by L as

$$T_{av} = \frac{0.0929}{\rho_a L\left[\frac{0.305\beta}{L}\right]^2}[(\beta - 1)\exp(\beta) + 1]; \beta = \frac{\rho_b L}{Q_0 S_0^{0.5}}. \tag{20.59}$$

The average infiltration under cutback conditions can be expressed as

$$Z_{av} = \{k[T_{co} - T_{av}]^a + f_0\}\frac{p_2}{F_s} + \{k[T_{av}]^a + f_0\}\frac{(p_1 - p_2)}{F_s}. \tag{20.60}$$

The gross depth of application can be computed as

$$d_g = \frac{60}{F_s L}[Q_1 T_a + Q_2 T_n]. \tag{20.61}$$

Example 20.8: Compute the cutoff time, depth of surface runoff, depth of percolation, and distribution efficiency for a furrow system given in Example 20.7 with a cutback system where the stream size is reduced by half.

Solution: The cutback time is equal to the advance time, which is equal to:

$$T_{cb} = T_a = 225\ \text{min}.$$

The initial wetted perimeter is $P_1 = 0.382$ m. The perimeter during reduced flow is computed eq. (20.1):

$$P_2 = 0.265 \times \left(\frac{\frac{Q_0}{2} \times n}{S_0^{0.5}}\right)^{0.425} + 0.227$$

$$= 0.265 \times \left[\frac{0.25 \times 0.04}{0.005^{0.5}}\right]^{0.425} + 0.227$$

$$= 0.342\ \text{m}.$$

Now the net infiltration time (T_n) is computed under reduced flow conditions using eq. (20.50):

$$T_n = \left[\frac{Z_n\left(\frac{F_s}{P_2}\right) - f_0}{k}\right]^{\frac{1}{a}} = \left[\frac{80\left(\frac{0.5}{0.342}\right) - 7}{0.9957}\right]^{\frac{1}{0.729}}$$

$$= 634\ \text{min}.$$

The time of cutoff, which is the sum of T_a and T_n, is computed as

$$T_{CO} = T_a + T_n = 225 + 634 = 859\ \text{min}.$$

The average infiltration time for infiltration during the advance phase is computed using eq. (20.59):

$$T_{av} = 859 - \frac{0.0929}{(7.7) \times (300)\left[\frac{0.305 \times 1.754}{300}\right]^2} [(1.754 - 1) \times \exp(1.754) + 1]$$

$$= 791\ \text{min}.$$

The average infiltration depth under cutback conditions is computed using eq. (20.60):

$$Z_{av} = \{k[T_{co} - T_{av}]^a + f_0\}\frac{P_2}{F_s} + \{k[T_{av}]^a + f_0\}\frac{(P_1 - P_2)}{F_s}$$

$$= \left[0.9957(859 - 68)^{0.729} + 7\right]\frac{0.342}{0.5} + \left[0.9957(68)^{0.729} + 7\right] \times \frac{(0.382 - 0.342)}{0.5}$$

$$= 95.5\ \text{mm}.$$

The gross application depth is computed using eq. (20.61):

$$d_g = \frac{60}{F_s L}[Q_1 T_a + Q_2 T_n] = \frac{60}{0.5 \times 300}[0.5 \times 225 + 0.25 \times 634]$$

$$= 108.4\ \text{mm}.$$

The surface runoff depth is computed using eq. (20.57):

$$d_r = d_g - d_{av} = d_g - Z_{av} = 108.4 - 95.5 = 12.9\ \text{mm}.$$

The percolation depth is computed using eq. (20.56):

$$d_p = d_{av} - d_n = d_{av} - Z_n = 95.5 - 80 = 15.5\ \text{mm}.$$

The distribution efficiency is computed using eq. (20.55):

$$E_d = \frac{d_n}{d_g} \times 100 = \frac{Z_n}{d_g} \times 100 = \frac{80}{108.4} \times 100 = 73.8\%.$$

20.3.13 Selection of Pump and Power Unit

Most furrow irrigation systems do not require a pump. A pump is required if the source of water is a well and groundwater is pumped. In some cases, water is also pumped directly from a river, a stream, or another water body. In these cases, an appropriate size of pump can be selected by matching the system curve with the pump characteristics curve (Chapter 10). Based on the selected pump, a suitable power unit can be chosen.

20.4 DESIGN OF FREE-DRAINING GRADED FURROW SYSTEMS

The furrow system design procedure presented here is based on the volume-balance principle. The design procedure uses the two-point method as proposed by Walker and Skogerboe (1987). The design steps are as follows:

1. Collect the following required field data:
   a. field characteristics: field length (L), width (W), area, topography (slope, S_0), and roughness (Manning's roughness coefficient, n, representative of surface or cover condition);
   b. soil characteristics: texture, water-holding capacity, field representative infiltration parameters for the first and later irrigations (k, a, and f_0);
   c. crop characteristics: type, design irrigation requirement (Z_{req}), irrigation schedule, sensitive crop stages;
   d. water supply: source, flow rate (Q), and duration;
   e. furrow characteristics: spacing (F_s), shape and geometry (depth and width data for computing furrow shape factors, ρ_1 and ρ_2).
2. Determine the furrow geometry parameters:
   The flow cross-sectional area (A), wetted perimeter (P), and hydraulic radius (R) affect the furrow irrigation performance and are functions of flow depth (y) and

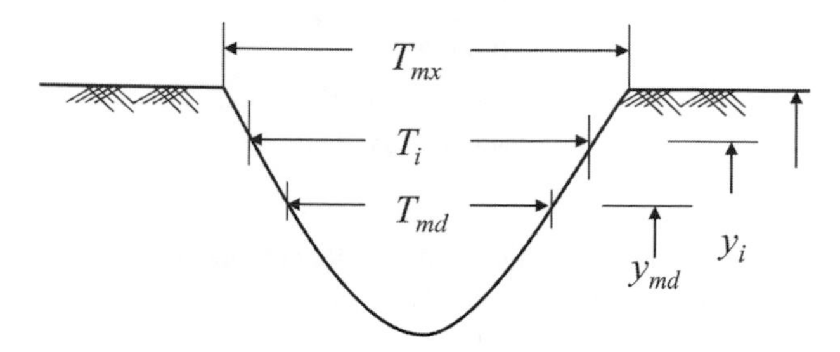

Figure 20.5 A schematic of a typical furrow cross-section.

geometry of the section. Although the shape of the cross-section may vary during the season, it can be assumed that the final stable shape follows a power law, which relates top width of flow (T), flow area (A), and wetted perimeter (P) to flow depth (y) as follows:

$$T = \alpha_1 y^{\alpha_2}, \tag{20.62}$$

$$A = \sigma_1 y^{\sigma_2}, \tag{20.63}$$

$$P = \gamma_1 y^{\gamma_2}, \tag{20.64}$$

where α_1, α_2, σ_1, σ_z', γ_1, and γ_2 are the fitted parameters.

Figure 20.5 shows a typical cross-section of a furrow for which these parameters can be determined based on pairs of furrow depth and top width measurements as follows:

$$\alpha_2 = \frac{\log(T_{mx}/T_{md})}{\log(y_{mx}/y_{md})}, \tag{20.65}$$

$$\alpha_1 = \frac{T_{mx}}{y_{mx}^{\alpha_2}}, \tag{20.66}$$

$$\sigma_1 = \frac{\alpha_1}{\alpha_2 + 1}, \tag{20.67}$$

$$\sigma_2 = \alpha_2 + 1, \tag{20.68}$$

$$\gamma_2 = \frac{\log\left(P|_{y_{mx}}/P|_{y_{md}}\right)}{\log(y_{mx}/y_{md})}, \tag{20.69}$$

$$\gamma_1 = \frac{P|_{y_{mx}}}{y_{mx}^{\gamma_2}}, \tag{20.70}$$

$$P|_{y_{md}} = \sum_{i=0}^{n/2} \left\{ 2\left[(y_i - y_{i-1})^2 + [0.5(T_i - T_{i-1})]^2 \right]^{0.5} \right\}, \tag{20.71}$$

$$P|_{y_{mx}} = \sum_{i=0}^{n} \left\{ 2\left[(y_i - y_{i-1})^2 + [0.5(T_i - T_{i-1})]^2 \right]^{0.5} \right\}, \tag{20.72}$$

where y is the depth (m), T is the top width (m), n is the number of depth and top width pairs, and subscripts mx, md, and i represent the maximum, middle, and individual values of depth and top width, respectively. The values of α_1, γ_1, and σ_1, are the unit width of the flow. ρ_1 is the square unit width. The values of α_1, α_2, γ_1, γ_2, σ_1, and σ_2 for borders and basins are 1.0, 0.0, 1.0, 0.0, 1.0, and 1.0, respectively.

The hydraulic section can be computed by combining eqs (20.63) and (20.64):

$$A^2 R^{4/3} = \rho_1 A^{\rho_2}, \tag{20.73}$$

where

$$\rho_2 = \frac{10}{3} - \frac{4\gamma_2}{3\sigma_2}, \tag{20.74}$$

$$\rho_1 = \frac{\sigma_1^{\frac{10}{3} - \rho_2}}{\gamma_1^{4/3}}. \tag{20.75}$$

Using eq. (20.75), the cross-sectional flow area at the field inlet can be estimated as a function of the flow at the normal depth, using the furrow shape factors (ρ_1 and ρ_2) and Manning's equation as:

$$A_0 = \left(\frac{Q_0^2 n^2}{3600 \rho_1 S_0} \right)^{1/\rho_2}. \tag{20.76}$$

For borders and basins, ρ_1 and ρ_2 are 1.0 and 3.33, respectively. For most furrow systems, ρ_2 ranges from 0.65 to 0.75, but the furrow hydraulics are not very sensitive to the changes in ρ_2, and $\rho_2 = 0.675$ is usually adequate. The value of ρ_1 depends on the furrow size and shape, ranging between 0.09 and 0.49.

3. Compute the intake opportunity time (τ_{req}) at the end of the field that exactly satisfies the irrigation requirement (Z_{req}):
   a. Assume an initial value.
   b. Use the Newton–Raphson method (eq. 20.5) and compute the new value.
   c. Repeat step b by setting the computed value as the initial guess until the initial guess and the computed value are within a specified tolerance limit.
4. Compute the maximum furrow discharge (Q_{max}) considering the maximum nonerosive velocity for a given soil texture (8 m/min in erosive silt soils to 13 m/min in less erosive clay and sandy soils) using eq. (20.33).
5. Select a furrow discharge (Q_0, m^3/min) near Q_{max} which results in an integer number of furrow sets (N_s [eq. 20.35]) and the number of furrows per set (N_{fs} [eq. 20.36]).
6. Calculate the cross-sectional flow area at the field inlet (A_0) using eq. (20.63) and check the maximum flow velocity and depth of channel:

$$V_{max} = \frac{Q_0}{A_0}. \tag{20.77}$$

If V_{max} is greater than the maximum erosive velocity, or flow area is insufficient, reduce Q_0 and repeat Step 6.

7. Assume an initial value of advance rate exponent (r) equal to 0.5, $r^j = 0.5$.

8. Compute σ_z and σ_z' by substituting $r = r^j$ in eqs (20.78) and (20.79) as follows:

$$\sigma_z = \frac{a + r^j(1-a) + 1}{(1+a)(1+r^j)}, \tag{20.78}$$

$$\sigma_z' = \frac{1}{1+r^j}. \tag{20.79}$$

9. Compute the time necessary for flow to advance half the furrow (L/2) distance, $t_{L/2}$:
   a.

$$t_{L/2}^i = \frac{2.5A_0L}{Q_0}, \tag{20.80}$$

   b.

$$t_{L/2}^{i+1} = t_{L/2}^i - \frac{Q_0t_{L/2}^i - \sigma_yA_0(0.5L) - \sigma_zk\left(t_{L/2}^i\right)^a(0.5L) - \sigma_z'f_0\left(t_{L/2}^i\right)(0.5L)}{Q_0 - \sigma_zak\left(t_{L/2}^i\right)^{a-1}(0.5L) - \sigma_z'f_0(0.5L)} \tag{20.81}$$

   c. Is $t_{L/2}^i = t_{L/2}^{i+1}$ or within the permissible tolerance limit? If this condition is not satisfied, then repeat Step b by setting $t_L^i = t_L^{i+1}$. Otherwise, $t_L = t_L^{i+1}$ and proceed to Step 10.
10. Compute the time necessary for flow to advance L distance, t_L:
   a.

$$t_L^i = \frac{5A_0L}{Q_0}, \tag{20.82}$$

   b.

$$t_L^{i+1} = t_L^i - \frac{Q_0t_L^i - \sigma_yA_0(L) - \sigma_zk(t_L^i)^a(L) - \sigma_z'f_0(t_L^i)(L)}{Q_0 - \sigma_zak(t_L^i)^{a-1}(L) - \sigma_z'f_0(L)}. \tag{20.83}$$

   c. Is $t_L^i = t_L^{i+1}$ or within the permissible tolerance limit? If this condition is not satisfied then repeat Step b by setting $t_L^i = t_L^{i+1}$. Otherwise, $t_L = t_L^{i+1}$ and proceed to Step 11.

   Note that there will be no convergence if the chosen Q_0 is insufficient to complete the advance. In this case, either increase Q_0 or decrease furrow (field) length.
11. Compute a revised estimate of the advance rate exponent (r^{j+1}) as follows:

$$r^{j+1} = \frac{\ln(2)}{\ln\left(\frac{t_L}{t_{L/2}}\right)}. \tag{20.84}$$

12. Compare the initial guess value of r^j with the computed value r^{j+1} (eq. 20.84). If both values are equal (or within the permissible range), continue with Step 13, otherwise set $r^j = r^{j+1}$ and repeat Steps 8–12.
13. Compute the time of cutoff, T_{CO} (min) by substituting the values of τ_{req} and t_L as obtained in Steps 3 and 10, respectively, as:

$$T_{CO} = \tau_{req} + t_L. \tag{20.85}$$

   Note, a simultaneous recession is assumed.
14. Compute the application efficiency E_a as follows:

$$E_a = \frac{Z_{req}L}{Q_0T_{co}}. \tag{20.86}$$

15. Application efficiency should be maximized considering different values of furrow inflow rate, subject to the limitations on erosion velocity, integer number of sets, the availability and the total discharge of water supply, and other farming practices. At least three Q_0 values should be tried in order to identify a value that maximizes E_a.
16. The same procedure needs to be repeated considering the representative values of infiltration parameters for later irrigations during the season. It is generally observed that infiltration rate decreases for later irrigations.

20.5 YU–SINGH METHOD FOR FURROW IRRIGATION DESIGN

Yu and Singh (1990) developed an analytical method for the design of furrow irrigation. The model transforms any furrow shape to a discharge-equivalent semicircular shape, thus accommodating any shape of furrow. The advance phase is modeled using the continuity equation and assuming parabolic shapes for surface and subsurface profiles (Singh, 1979–1980), with their coefficients being determined from the conditions from the gradually varied flow region. Infiltration is simulated in three dimensions rather than in one or two dimensions (Singh et al., 1987). The storage phase is modeled by balancing inflow and outflow (both surface and subsurface). Both the vertical recession and the horizontal recession are modeled by modifying the Strelkoff model (Strelkoff, 1977) with the incorporation of the time-varying rate of infiltration. The Yu–Singh model was verified using data from six furrows. Singh and He (1988) developed a Muskingum model for design of furrow irrigation, and Singh and Yu (1988) developed a farm irrigation system model for design of surface irrigation, including furrow irrigation.

QUESTIONS

Q.20.1 Compute the maximum furrow length for a farm located in a semi-arid area having sandy loam soil. The irrigation requirement is 75 mm per irrigation, the nonerosive inflow rate is 2.5 L/s, and the ground slope is 0.01%.

Q.20.2 Compute the gross depth of application for a furrow system with furrow length of 300 m, furrow spacing of 1 m, ground slope of 0.02%, Manning's roughness coefficient of 0.04, net irrigation requirement of 90 mm, inflow rate to furrow of 0.60 L/s, and furrow intake family of 0.50.

Q.20.3 Compute the time of cutoff, distribution efficiency, and field slope for a furrow system for a field of length 120 m, furrow spacing of 0.5 m, wetted perimeter of 0.5 m, inflow rate of 0.25 L/s for each furrow, the net time of infiltration of 360 min, the advance time of 120 min, and intake family of 0.5.

Q.20.4 Compute the volumetric inflow rate for each furrow in L/s for a furrow irrigation system with field slope of 0.02%, net irrigation requirement of 80 mm, and net infiltration time of 300 min, where spacing between furrows is 0.5 m. Assume the intake family of 0.4 and Manning's roughness coefficient 0.04.

Q.20.5 Compute the depth infiltration in mm over the surface area of a field at the end of 4 h of irrigation for a furrow system with a slope of 0.01%, inflow rate of 0.3 L/s, length of furrow of 180 m, and runoff is 0.04 L/s. Assume Manning's roughness coefficient is 0.04.

Q.20.6 Compute Manning's roughness coefficient for a furrow system if the furrow length is 250 m, time of cutoff is 1025 min, furrow spacing is 0.5 m, slope is 0.002, net irrigation requirement is 100 mm, and inflow rate per furrow is 0.5 L/s. Assume the intake family is 0.3.

Q.20.7 Compute the desired time of cutoff, the equivalent depths of surface runoff and deep percolation, and distribution efficiency for furrow irrigation for which the following information is given: inflow rate per furrow is 0.75 L/s, net irrigation depth is 85 mm, Manning's roughness coefficient is 0.045, the intake family is 0.4, furrow length is 300 m, furrow spacing is 0.5 m, and furrow slope is 0.005 m/m.

Q.20.8 Solve Q.20.7 if the inflow rate is reduced by 20%.

Q.20.9 What will be the average depth of water applied to a furrow system consisting of 100 m long furrows that are 50 cm apart, having a slope of 0.02% if a nonerosive stream of 0.5 L/s is applied for each furrow for a period of 30 min?

Q.20.10 Compute the cutoff time, depth of surface runoff, depth of percolation, and distribution efficiency for the furrow system given in Q.20.7 with a cutback system where the stream size is reduced by half.

Q.20.11 Consider that the advance time is one-quarter of the net time of irrigation. Assume the Kostiakov infiltration exponent a is 0.7. Derive the percentage of percolation loss and compare this loss with the value that can be obtained from Table 20.3.

Q.20.12 Determine the net time of irrigation and the advance time the water takes to reach the downstream end of the field if the net irrigation required is 8 cm. It is assumed that the Kostiakov infiltration parameters for depth in cm and time in min are $a = 0.6$ and $k = 0.31$ (i.e., $Z_{req} = 0.31\left(\tau_{req}\right)^{0.6}$), and the percolation loss is 5.5%.

REFERENCES

Brouwer, C., Prins, K., Kay, M., and Heibloem, M. (1984). *Irrigation Water Management: Irrigation Methods*. Rome: FAO.

Clemmens, A. J., Wahl, T. L., Bos, M. G., and Replogle, J. A. (2001). *Water Measurement with Flumes and Weirs*. Wageningen: International Institute for Land Reclamation and Improvement.

Hart, W. E., Collins, H. G., Woodward, G., and Humphreys, A. S. (1980). Design and operation of gravity or surface irrigation systems. In *Design and Operation of Farm Irrigation Systems*, edited by M. S. Jensen. St. Joseph, MI: ASAE, pp. 501–580.

James, L. J. (1988). *Principles of Farm Irrigation System Design*. New York: Wiley.

Jurriëns, M., Zerihun, D., Boonstra, J., and Feyen, J. (2001). *SURDEV: Surface Irrigation Software; Design, Operation, and Evaluation of Basin, Border, and Furrow Irrigation*. Wageningen: International Institute for Land Reclamation and Improvement.

Michael, A. M. (1978). *Irrigation Theory and Practice*. New Delhi: Vikas Publishing House Pvt. Ltd.

Singh, V. P. (1979–1980). Derivation of shape factors for border irrigation advance. *Agricultural Water Management*, 2: 271–285.

Singh, V. P. and He, Y.-C. (1988). Muskingum model for furrow irrigation. *Journal of Irrigation and Drainage Engineering*, 114 (1): 89–103.

Singh, V. P., and Yu, F. X. (1988). A farm irrigation system (FIS) model. *Water Resources Management*, 2: 173–181.

Singh, V. P., He, Y.-C., and Yu, F. X. (1987). 1-D, 2-D, and 3-D infiltration for irrigation. *Journal of Irrigation and Drainage Engineering*, 113(2): 266–278.

Strelkoff, T. (1977). Algebraic computation of flow in border irrigation. *Journal of Irrigation and Drainage Engineering*, 103 (3): 16–20.

USDA-SCS (1974). Border irrigation. In *National Engineering Handbook*. Washington, DC: USDA.

USDA-SCS (1983). Furrow irrigation. In *National Engineering Handbook*. Washington, DC: USDA.

Walker, W. R., and Skogerboe, G. V. (1987). *Surface Irrigation*. Englewood Cliffs, NJ: Prentice Hall.

Withers, B., and Vipond, S. (1974). *Irrigation: Design and Practice*. London: Batsford.

Yu, F. X., and Singh, V. P. (1990). Analytical model for furrow irrigation. *Journal of Irrigation and Drainage Engineering*, 116(2): 154–171.

21 Sprinkler Irrigation

Notation

a	cross-sectional area of the opening of the nozzle (m^2, ft^2, $in.^2$)
A	area to be irrigated (m^2, ha, acre)
AD	soil water depletion allowed between irrigations (mm, in.)
A_i	cross-sectional area of the opening in the *i*th nozzle (m^2, ft^2)
C	Hazen–Williams friction coefficient
C_i	coefficient depending on the shape and roughness of the *i*th nozzle
C_0	sprinkler coefficient
d_1, d_2	diameters of nozzles (in., mm)
d_a	net application rate (cm/h)
$d_{a\ max}$	maximum application rate (mm/h)
d_g	gross application rate (cm/h)
d_n	design daily irrigation requirement rate (mm/day, in./day)
D	inside diameter of the pipeline (in., mm)
D_a	net depth of application (mm)
D_g	gross irrigation requirement depth (mm)
D_{g-max}	maximum amount of application depth (mm)
D_n	net depth of irrigation requirement or the desired depth of application (mm)
D_r	depth of active root zone (m)
D_w	diameter of the wetted area (m, ft)
E_a	application efficiency (fraction or percentage)
E_m	motor efficiency for constant-speed pump (fraction)
E_p	efficiency of the pump (fraction)
ET_c	average crop evapotranspiration rate during the peak use period (mm/day)
ET_m	peak monthly evapotranspiration (mm, in.)
ET_{seas}	seasonal crop water requirement (mm)
F	friction loss factor, which depends on the value of *m* and the number of sprinklers
FC	field capacity (mm/m, or m/m)
g	acceleration due to gravity (9.81 m/s^2)
h	hour of operation per day (h/day)
h_f	head loss due to friction (m)
$h_{f\text{-}pi}$	head loss due to friction from the pump to point *i* (m)
$h_{f\text{-}s}$	head loss due to friction on the suction side of the pump (m)
h_T	total head loss in the entire lateral (m)
H	total dynamic head required of the pump (m, $lb/in.^2$)
H_a	nozzle design pressure head (m)
H_e	increase in elevation between two sprinklers in the flow direction; will be negative for downhill-sloping laterals (m)
H_{e1-2}	increase in elevation head between points 1 and 2 (m)
H_{e-si}	increase in the elevation head from level of water source to point *i* (m)
H_{f1-2}	head loss due to friction between points 1 and 2 (m)
H_i	pressure head at point *i* (m)
H_L	maximum allowable head loss due to friction (m/m)
H_{L-ac}	actual head loss due to friction (m/m)
H_{L-p}	equivalent head loss due to friction in the through-flow pipe (m/m)
H_m	pressure required at the mainline entrance (m)
HP	power needed for pumping water for sprinkler systems (horsepower, kW)
H_r	height of sprinkler riser (m)
H_{v1-2}	increase in the velocity head between points 1 and 2 (m)
i	minimum intake rate of soil (in./h, mm/h)
I	optimum application rate (in./h, mm/h)
I_a	average application rate (mm/h)
K_l, K_s	constant depending on the sprinkler spacing pattern and wind
l_n	distance between the first and the last sprinklers (m)
L_t	length of the lateral or pipe (m or ft)
m, n	exponents in the friction loss equation
MAD	management allowed depletion (fraction)
N	number of sprinklers
N_l	number of laterals operating concurrently
P or p	operating pressure in the nozzle (kPa, $lb/in.^2$)
P_f	percentage of field irrigated when the sprinkler system is operating (fraction)
P_i	operating pressure in the *i*th nozzle (kPa)
P_{kW}	power required to operate the pump (kW)

P_p	total pumping head (lift and friction losses) (lb/in.^{2})	s_r	sprinkler spacing at distance r (m)
q	sprinkler discharge (L/s, L/min, gpm)	s_t	towpath spacing (m)
q_r	sprinkler discharge at distance r from the pivot point (L/s)	T	irrigation interval for a given set (h)
Q	system discharge or capacity (L/min, gpm)	T_a	time of application (h)
Q_a	discharge rate for water for total lateral arm in the linear-move sprinkler system (L/s)	TAM	total available moisture (mm/m, or m/m)
Q_t	total flow rate into the upstream end of the lateral (L/min, gpm)	T_i	maximum irrigation interval (d)
r	distance from the sprinkler to the pivot point (m)	T_{max}	maximum permissible interval (d)
r_{max}	maximum radius of the irrigated area (m)	T_r	time per revolution of pivot arm (h)
R	radius of a circular area in center pivot system (m)	T_{seas}	seasonal operational time (h)
s_l, S	lateral spacing or spacing between sprinklers along the lateral (m)	v	mean velocity in the pipeline (m/min)
s_m, L	mainline spacing or distance between laterals (m)	v_t	traveler speed (m/min)
		WP	limit of crop-extractable water (mm/m, or m/m)
		x_i	exponent in the discharge equation for the ith nozzle; the value of x is about 0.5
		Δ	maximum allowable pressure difference (fraction)
		Φ	portion of circle wetted by sprinklers (degrees)

21.1 INTRODUCTION

Sprinkler irrigation is now becoming a standard irrigation method and is no longer confined to developed countries. This is primarily because water is becoming scarce, and agriculture, being the largest user of water, is now being compelled to compete with other water uses, such as industrial, energy generation, waste disposal, and domestic, and is being forced to be more efficient. As a result, irrigation is becoming more efficient and sprinkler irrigation use is increasing each year. For example, in the United States, in 2010 about 51% of the area was irrigated by sprinkler systems, and the withdrawal of water declined by 9% between 2005 and 2010 despite a 2% increase in total irrigated area and the area irrigated by sprinkler irrigation increasing by 3% (Maupin et al., 2014).

Besides increased irrigation efficiency, sprinklers have other applications. For example, they can be used for disposal of waste (municipal, domestic, and agricultural) (Pettygrove and Asano, 1986), improvement of germination by lightly wetting the soil surface after seeding, application of fertilizers, weedicides, and pesticides, frost protection under adverse weather conditions, and dust suppression. For agricultural irrigation, sprinklers have advantages and disadvantages which are briefly stated below. The objective of this chapter is to discuss different aspects of sprinkler irrigation.

21.2 ADVANTAGES AND DISADVANTAGES

21.2.1 Advantages of Sprinkler Irrigation

Sprinkler irrigation has a number of advantages:

1. Sprinkler irrigation systems are flexible and can be used for irrigating any kind of soil, topography, and crop.
2. Sprinkler irrigation can be used for high-permeability or low water-holding capacity soils. Likewise, it can be used for dense soils with low permeability.
3. Because of its flexibility, dry land and wastelands can be irrigated by sprinkler irrigation. The result is thousands of hectares of additional land now being under sprinkler irrigation and producing crops, as can be observed in Kansas, Western Nebraska, Eastern Colorado, and California.
4. Sprinkler irrigation minimizes the wastage of water.
5. Sprinklers can yield approximately uniform applications of water and achieve the desired irrigation uniformity.
6. Sprinklers have more efficient control over water application. That is, they can apply water at a rate that is compatible with the intake rate of the soil and hence avoid ponding and runoff.
7. On saline soils, sprinkler irrigation can lead to better leaching and crop germination.
8. Where labor and water costs are high, sprinkler irrigation is perhaps the most economical.
9. For vegetable crops and fruits where color and quality are important, sprinkler irrigation increases yield.
10. Sprinkler systems, bought for agricultural irrigation, can also be used for crop cooling, frost control, and application of fertilizers, pesticides, and herbicides. This helps reduce the cost of sprinkler irrigation.
11. Sprinkler irrigation is easily done on farms that are large and where modern equipment is used.

21.2.2 Disadvantages of Sprinkler Irrigation

Sprinkler irrigation is not without pitfalls:

1. It has been found that poor-quality water applied by sprinklers to the foliage reduces the yield of some crops.
2. Poor-quality water applied by sprinklers can lead to undesirable deposits or coloring on the leaves of fruits of crops.

3. Sprinkler irrigation can increase the incidence of certain crop diseases, such as blight in pears, fungi, or foliar bacteria.
4. Sprinklers, especially soil-set systems, are relatively expensive and not all farmers can afford them.
5. Sprinklers may need large amounts of energy to operate when the gravity head is not enough to supply the needed pressure. In many developing countries, electricity is not available on farms and sprinkler systems therefore cannot be operated.

21.3 CLASSIFICATION OF SPRINKLER SYSTEMS

Different types of sprinkler systems have been developed to meet different needs arising from topographic conditions, availability of water and land resources, application requirements, labor situations, and economic conditions. A sprinkler system consists of a pump, a mainline, laterals, and sprinklers. A fully portable system comprises a portable or stationary pump and portable mainlines, laterals, and sprinklers. Sprinkler systems can be classified in different ways (Figure 21.1), depending on portability, spray pattern, field layout, equipment to be used, and type of system to be irrigated, such as agricultural irrigation, revegetation of mined areas, irrigation of fruit crops, irrigation of golf courses, and irrigation of lawns and playgrounds. Based on spray pattern, sprinkler systems can be classified into rotating head, perforated type, and center-pivot systems. These systems are now briefly discussed.

21.3.1 Portable Sprinkler Systems

Based on portability, sprinkler systems can be classified as portable, semi-portable, semi-permanent, permanent, solid-set, and set-move systems.

A sprinkler system is fully portable when all of the system components can be moved from one area to another (Figure 21.2). That is, mainlines, submains, laterals, and pumping units can be moved. In semi-portable systems, all components can be moved, except the pumping unit and the main pipeline.

Based on the method of moving the laterals, portable and semi-portable systems can be further classified, and there is a large variety of ways in which lateral pipelines are moved from set to set. These systems can also be called move–stop systems. The laterals can be from hand-moved to tractor-towed to motor-moved laterals, giving rise to names such as hand-moved, hose-pull, end tow, boom-type, self-propelled, and side-roll. Perforated pipe systems can also be included in this classification.

In a hand-moved system, laterals are moved by hand to the next lateral position, whereas in a side-roll lateral system the lateral is placed on wheels and an engine is placed at its center that enables moving it to the next position. The hose-pull system has buried mains and submains and a large number of hoses with sprinklers mounted on. This system is popular in parks, golf courses, and lawns. A boom-type sprinkler system has one boom sprinkler on each lateral. A boom is a slowly rotating nozzled pipeline suspended from a portable tower which is moved to the next position by a tractor. A self-propelled system comprises a radial pipeline that is supported on towers mounted on wheels and is rotated around the pivot point in the center of the field. A perforated pipe system comprises lightweight tubing with a small hole on the top side of the pipe.

The portable or semi-portable systems have been designed primarily to reduce equipment needs and to reduce interference with farming operations. They are less expensive to buy and install, but require more labor and maintenance. Since they are moved from wet areas to dry areas, the movement may damage crops and hence reduce yield.

21.3.2 Permanent and Semi-permanent Systems

In permanent systems the main pipeline, submains, and laterals are buried in the ground and the pumping unit is fixed in the irrigation area, whereas in semi-permanent systems, the mainlines, submains, pumping source, and water source are fixed but laterals are portable

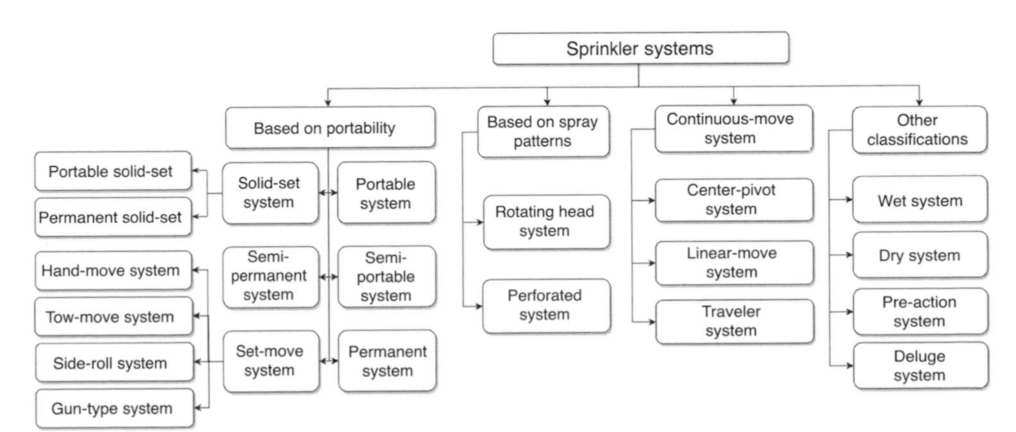

Figure 21.1 Classification of sprinkler systems.

Figure 21.2 A portable sprinkler system. A black and white version of this figure will appear in some formats. For the color version, please refer to the plate section.

Figure 21.3 Permanent and semi-permanent systems. A black and white version of this figure will appear in some formats. For the color version, please refer to the plate section.

(Figure 21.3). Mainlines and submains are normally buried and risers are provided for nozzle connections. Stationary pipelines but portable sprinklers are often used in golf courses and parks.

21.3.3 Solid-Set Systems

A solid-set system (Figure 21.4) has two basic characteristics. First, the sprinkler system remains in a fixed location during the irrigation season. Second, it has a fixed network of pipes that can be arranged in many different ways, but they are of two basic types: aluminum and plastic. Aluminum pipes are generally laid on the ground surface at the start of the irrigation season and collected seasonally or annually to make room for cultivation practices. Plastic (usually PVC), coated aluminum, coated steel, or asbestos cement pipes are generally buried permanently in order to avoid deterioration due to sunlight, with only the sprinklers and a portion of the risers above the ground.

Solid-set systems are more expensive because they irrigate the entire field with single sets of components, whereas other sprinkler systems employ various components, such as laterals, for multiple uses. The labor and maintenance costs are minimal, but cultural operations, such as cultivation, spraying, planning, harvesting, etc., may be restricted. Solid-set systems are suitable for perennial or annually harvested crops, turf areas, and revegetation sites.

21.3.4 Set-Move Systems

Here, a field is irrigated in a sequence of steps. First, the system is set at one position for one irrigation; when the irrigation is completed, it is moved either by hand or mechanically to another position. During irrigation, the set-move system remains stationary. When irrigation is done, the water is shut off; once the laterals have drained the system is moved to another position. This cycle is repeated until the entire field is irrigated.

Figure 21.4 Solid-set systems: portable and permanent. A black and white version of this figure will appear in some formats. For the color version, please refer to the plate section.

Figure 21.5 Set-move systems: hand-move and tow-move. A black and white version of this figure will appear in some formats. For the color version, please refer to the plate section.

Figure 21.6 Set-move systems: side-roll and gun-type. (Adapted from USDA-NRCS, 2016.) A black and white version of this figure will appear in some formats. For the color version, please refer to the plate section.

In set-move systems there is a single mainline laid through the center of the field, with one or more laterals on each side of the mainline. Multiple laterals are equally spaced and the entire field is irrigated when a lateral reaches the starting position of the lateral ahead of it. The set-move systems are of different kinds, such as hand-move, tow-move (Figure 21.5), side-roll, and gun-type (Figure 21.6).

Hand-move: Laterals are moved by hand from one position to the next. This involves uncoupling lateral sections and carrying and reassembling them. Lateral sections are usually 6, 9, or 12 m long and have a diameter of 50 to 100 mm.

Tow-move: As the name suggests, the mainline is hooked to a tractor and laterals are attached to the mainline, which is

Figure 21.7 Spray sprinkler systems: rotating head and perforated. A black and white version of this figure will appear in some formats. For the color version, please refer to the plate section.

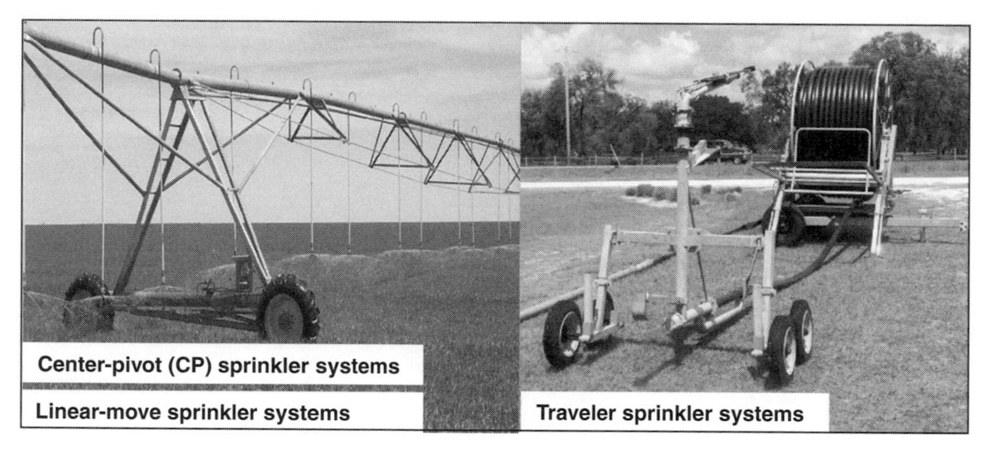

Figure 21.8 Continuous-move systems. (Traveler systems adapted from USDA-NRCS, 2016.) A black and white version of this figure will appear in some formats. For the color version, please refer to the plate section.

dragged. Each section of a lateral has skids or wheels so the entire lateral can be moved to the next position.

Side-roll: In a side-roll movement, each section of a lateral has a wheel, with the pipe serving as the axle of the wheel. An engine is used to supply the power to roll the lateral. A large volume (big gun) sprinkler with a capacity up to 4700 L/min (about 1250 gpm) and operating pressure ranging from 480 to 896 kPa (70–130 psi) is mounted on a wheeled cart or trailer for moving from one position to another.

Gun-type: This type of system has a large volume (big gun) sprinkler on a wheeled cart or trailer, which is moved from one set to another with a tractor or even by hand. These systems may have operating pressures of 480–896 kPa, wetted diameters of up to 180 m (about 600 ft), and discharge capacities of 4700 L/min (about 1250 gpm).

21.3.5 Spray Sprinkler Systems

Based on spray patterns, sprinkler systems can be classified as rotating head or revolving systems and perforated pipe systems (Figure 21.7).

Rotating head: These systems have a pump, mainline, submain, and laterals. Laterals are pipes laid on the ground. These pipes have fixed risers mounted with nozzles that have small spray size. The risers are high enough to clear the crop canopy. The sprinkler heads or nozzles rotate 90–360° to irrigate a rectangular strip. Depending on the nozzle size, the discharge rate is 4.0–20.0 mm/h and the operating pressure is 2.0–4.0 kg/cm^2.

There are different types of rotating sprinkler systems, which can be classified into the same systems as portable systems.

Perforated: These systems involve pipes having holes or nozzles along their length which spray water under pressure. These systems are commonly used in lawns, parks, and house yards. The operating pressure is usually low, such as 1 kg/cm^2, and the rate of discharge is 1.25–5 cm/h.

21.3.6 Continuous-Move Systems

The continuous-move systems (Figure 21.8) are the most widely used sprinkler systems for agricultural irrigation. Center-pivot, linear-move, and big gun systems are typical types of continuous-move systems. A continuous-move system covers the irrigated area by continuously moving, and the equipment is automated and mobile. The major advantage of a continuous system is its labor savings.

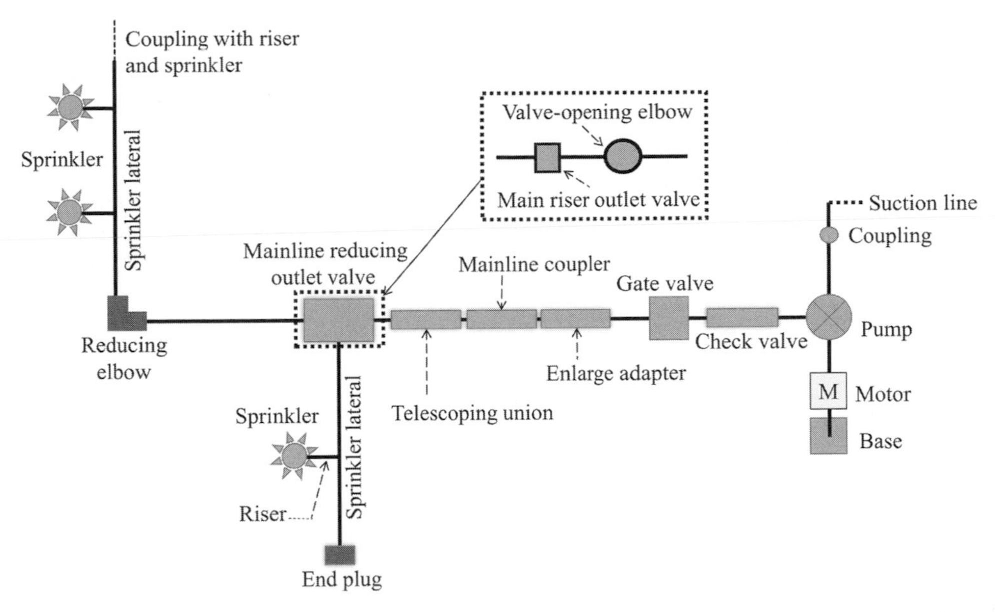

Figure 21.9 Components of a portable sprinkler system. (Adapted from Schwab et al., 1993. Copyright © 1993 by Wiley: New York, USA.)

The irrigation uniformity is improved, but sometimes such a system leads to high precipitation rates and field runoff.

Center-pivot (CP): A CP system has a lateral supported by towers spaced every 24–76 m (80–250 ft) along the lateral and trusses and cable, which rotates in a circle around a fixed pivot structure. The supply of water to the lateral is at the pivot point. Mounted on wheeled towers, they are driven by electric motors or by hydraulic oil or water-driven motors. Laterals can be 60–790 m (200–2600 ft) but are generally 365–400 m (1200–1300 ft) long. The operating sprinkler pressure can range from 140 kPa (about 20 psi) to more than 790 kPa (about 100 psi). These systems are the most suitable for coarse-textured soils having high infiltration capacities and where frequent but light irrigations are needed.

Linear-move: Like CP systems, these systems have towers and alignment systems to control the movement of the towers, but the laterals move linearly across the field. A flexible hose is hooked to a mainline and supplies water to it. There are other ways to supply water, such as pumps and risers. There are several options for irrigation by linear-move systems, entailing the movement of the machine, such as from the left edge of the field to its center, across the field, back and forth across the field, starting at one end and moving to the middle of the field, and from one end of the mainline to the other.

Traveler: In these systems the mainline is either buried or potable and a high-capacity gun is mounted on a cart, while a hose supplies water to the sprinkler. The cart is moved across the field by a cable or the hose itself. Travelers can be soft- or hard-hose systems.

21.3.7 Another Classification

Sprinkler systems can also be classified as wet, dry, pre-action, and deluge systems. Wet sprinkler systems constantly have water, which allows for quick response upon start-up; these are the most common. Usually these are cost-effective and require low maintenance. Dry sprinkler systems, as the name suggests, do not retain water upon the conclusion of operation. To start they need pressurized air and water, and they take a few minutes to start spraying. These systems are most suited to cold climates where water freezes. Pre-action systems are filled with air and prevent water from sprouting if there is a false alarm or mechanical failure. These are best suited for within buildings. Deluge systems respond quickly and flood the area with water. They have open sprinkler heads and are used in hazardous areas, such as aircraft hangers, where flammable liquids are stored.

21.4 SPRINKLER COMPONENTS

A multi-sprinkler irrigation system consists of a number of components, including the head, distribution pipelines or laterals, flow controllers, sprinklers, filters, injectors, pump, and power source, as shown in Figure 21.9.

21.4.1 Pumps

For most sprinkler systems, the source of water supply is by pumping. The pump supplies the energy or pressure to the sprinkler system, which forces the water to spray. The power to operate pumps comes from electricity or some kind of fuel. Pumps are either centrifugal or deep-well turbines. The selection of a pump depends on the system discharge and operating pressure. A high-speed centrifugal or turbine pump is used for sprinkler irrigation in individual fields. A centrifugal pump is used when the lift of water, that is the distance from the water surface to the pump inlet, is less than 8 m. For greater lifts a submersible pump can be used.

21.4.2 Controllers

Controllers are designed to coordinate the water supply (usually pumps) and delivery network. Following a predetermined schedule, various sections of the field can be irrigated differently and at different times. Controllers also safeguard the system in case of failure.

21.4.3 Filters and Injectors

Filters are used to filter the water for debris and sediment near the source of supply.

21.4.4 Fertilization Application Unit

Injectors are used to apply fertilizers and pesticides through the sprinkler system. In this case, an applicator, consisting of a tank and necessary tubings and connections, is attached to the mainline.

21.4.5 Distribution Pipelines

Components of a sprinkler system are shown in Figure 21.10 (a similar figure can be found in Cuenca, 1989). The sprinkler head is connected to the lateral with a riser pipe and then to the water supply through the manifold, auxiliary, submain, and main pipes. The length of pipe is 20–30 ft (6–9 m). Pipes are available in various grades to accommodate different processes and for convenience during system movement.

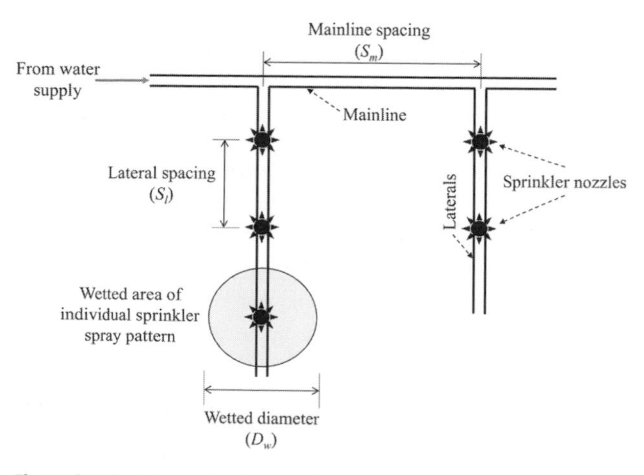

Figure 21.10 Components of a sprinkler system.

21.4.6 Sprinkler Head

The flow distribution device is referred to as the sprinkler head and is a basic element of the system. There are different kinds of distribution devices, such as impact head, gear-driven rotating head, wobbler, and fixed-position spray head, as shown in Figure 21.11. The choice of sprinkler head depends on the crop to be irrigated and the intake rate of the soil. The sprinkler head determines the spacing of laterals, manifolds, and other pipe sections, since the wetted diameter and other overlapping criteria influence the distance between sprinklers.

The choice of sprinkler head determines the operational cost and system efficiency. High-discharge heads require high pressures that, in turn, require large energy expenditures. High pressures enable adequate uniformity, and since wetted perimeters are large, application rates can be small for tight soils. Small application rates minimize runoff, waterlogging, and water quality problems.

21.4.7 Flow Regulators

Pressure and discharge have to be managed in sprinkler systems. There are different types of devices, called couplers and fitting accessories. Some of the devices, commonly referred to as valves, are available to regulate flow. There are pressure regulators to manage pressure variation. Pressure regulation can be applied at the sprinkler head to maintain constant discharge at major pipe junctions to accommodate large topographical variations and at the water source. The flow can be regulated with valves that open and close manually, hydraulically, or electronically. Some of the flow regulators are used to ensure the safety of the system. For example, backflow-prevention valves, pressure-release

Figure 21.11 Different types of sprinkler heads. A black and white version of this figure will appear in some formats. For the color version, please refer to the plate section.

Figure 21.12 Types of sprinklers: impact, gear-driven, and reaction types. A black and white version of this figure will appear in some formats. For the color version, please refer to the plate section.

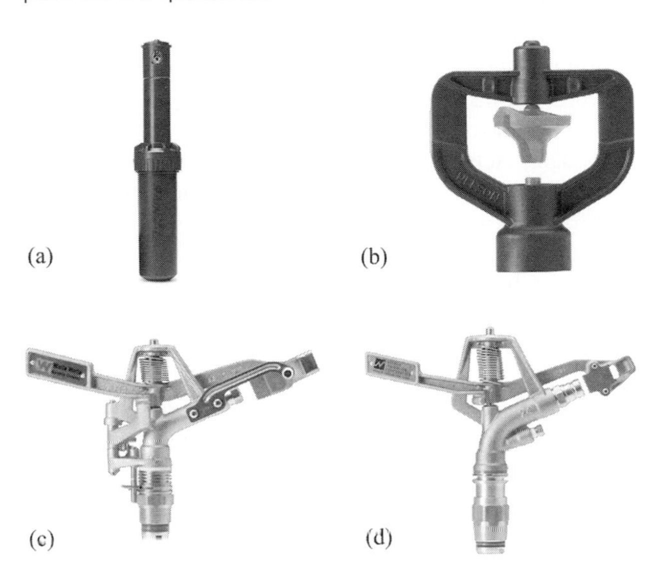

Figure 21.13 (a) Full-circle, impulse-driven, rotating pop-up sprinkler (www.rainbird.com/products/ag5-maxi-paw-pop-impact-sprinklers; Copyright © by Rain Bird. Reprint with permission from Rain Bird). (b) Full-circle, low application rate agricultural sprinkler (https://nelsonirrigation.com/products/rotator-sprinklers/r10-rotator). (c) Part-circle, medium-sized agricultural sprinkler (https://nelsonirrigation.com/products/impact-sprinklers/part-circle-impacts). (d) Full-circle higher application rate agricultural sprinkler (https://nelsonirrigation.com/products/impact-sprinklers/full-circle-impact-sprinklers). (b–d Copyright © by Nelson Irrigation. Reprint with permission from Nelson Irrigation.) A black and white version of this figure will appear in some formats. For the color version, please refer to the plate section.

valves, and vacuum-release valves function during shutdown or set changes.

Couplers are employed to connect pipes and should be such that they are easy to use and uncouple, flexible, light, noncorrosive and durable, and do not leaking at the joints. Water meters are used to measure the volume of water delivered to the system. Flange, couplings, and nipples are needed for proper connection to the pump, suction, and delivery. Bends, tees, reducers, elbows, hydrants, and plugs are other components of a sprinkler system.

21.5 TYPES OF SPRINKLERS

There are different types of sprinklers, but most of them are rotating or fixed-head sprinklers. For agricultural irrigation most of them are the slowly rotating type with one or two nozzles, which rotate around a vertical axis. Rotating sprinklers can be classified as impact, gear-driven, and reaction type, whereas fixed-head sprinklers are generally spray type (Figure 21.12).

Impact sprinklers: These have one or more nozzles discharging jets of water into the air. Jets rotate in a start–stop manner. The nozzles for impact sprinklers are constant diameter, constant discharge, and diffuse-jet nozzles.

Gear-driven sprinklers: There is a small water turbine located at the base of the sprinkler that drives the rotating sprinklers. The high speed of the turbine is educed by a series of gears, hence the name.

Reaction sprinklers: The sprinklers are rotated by the torque, which is generated by the water leaving the sprinkler.

Most sprinklers rotate around a vertical axis, as shown in Figure 21.13. The nozzles are from about 1.5 mm (1/16 in.) to 5 cm (2 in) in diameter, discharge water at less than 4 L/min (1 gal/min) to more than 4000 L/min (1000 gal/min), and cover a circular area with a diameter of 6–180 m (20–600 ft). In common applications, sprinklers are of small to medium size and discharge at 7.5–75 L/min (2–20 gal/min), covering a circular area with a diameter of 6–40 m (18–120 ft) and operating at pressures of 1.4–4.5 kg/cm^2 (20–60 lb/in.^{2}). Figure 21.13 shows four types of sprinklers. The first type is a full-circle, impulse-driven, rotating pop-up sprinkler that is often used for irrigating turf. The second type is a full-circle low application rate agricultural sprinkler. It discharges water at a rate of 0.3–0.7 gal/min and operates at a pressure of 25–50 lb/in.^{2}. The third type is a part-circle medium-sized agricultural sprinkler. It discharges water at a rate of 7–34 gal/min and operates at a pressure of 30–80 lb/in.^{2}. The fourth type is a full-circle higher-rate agricultural sprinkler. It discharges water at a rate of 21–110 gal/min and operates at a pressure of 40–100 lb/in.^{2}. Larger sprinklers discharge at a rate of 200–600 gal/min and operate at a pressure of about 100 lb/in.^{2}. Boom sprinklers, which have arms of 12 m (35 ft) or longer, cover areas with a diameter of 90–120 m (300–400 ft) and have application rates of 6–25 mm/h (0.5–1 in./h). Fixed-head sprinklers are used for lawns and gardens.

Fixed-head sprinklers: These sprinklers produce full or nearly full-circle sprays or several small streamlets that are discharged around the sprinkler circumference.

21.6 CONSIDERATIONS FOR DESIGN OF SPRINKLER SYSTEMS

Before a sprinkler system is designed, certain considerations must be kept in mind. The sprinkler system should be designed so that it irrigates with satisfactory uniformity of water distribution at minimum operating cost, including depreciation, power, and labor costs. In other words, the design should be the most economical. Also, long-term changes in costs and cropping patterns should be taken into account. The following considerations should be kept in mind in the design procedure (Christiansen and Davis, 1967): (1) The source of water supply should be as close to the center of the area to be irrigated as possible. This will lead to minimum cost, minimum friction losses in laterals, smaller pipe sizes, and more uniform distribution of water. (2) On sloping fields, mainlines should be placed on the slope and laterals on the contour or slightly downhill to minimize pressure variations in the pipeline. (3) The system should be able to irrigate a variety of soils and crops. (4) It should be possible to expand the system to cover a greater irrigation area. (5) The system should have minimum interference with farm operations. (6) The number of sprinklers operating at any time should conform to the irregularity of the area. (7) The system should have desirable supplementary controls, such as safety valves, air-relief valves, pressure-control valves, pressure gages, and automatic controls. (8) Booster pumps or reservoirs may be considered if they lead to reduction in overall costs.

21.6.1 Field Layout

The layout of a sprinkler system depends on the topography and the type of sprinkler system. The key point is to minimize costs and achieve uniform distribution of water by balancing the pressure distribution along the laterals. Costs of mainlines are significantly higher than the costs of laterals, which means the mainline length should be minimized. The positioning of the mainline should account for topography and the source of water supply. Laterals run perpendicular to the mainline in the direction of the minimum slope, which will lead to the highest pressure and distribution uniformity. Different types of layouts are illustrated in Figure 21.14.

The field layout of a sprinkler system varies with the type of system. A typical layout of a CP system is shown in Figure 21.15 (a similar figure can be found in Walker, 1980), where water enters the pivot and moves to the lateral for individual sprinklers to discharge, which increases along the lateral. In this system there is only one lateral, whose one end is fixed on the pivot structure and the other end moves continuously around the pivot.

A continuously moving linear system has a field layout, as shown in Figure 21.16 (a similar figure can be found in Walker, 1980), where water is pumped from a ditch. The system travels to the lower end of the field and shifts to the

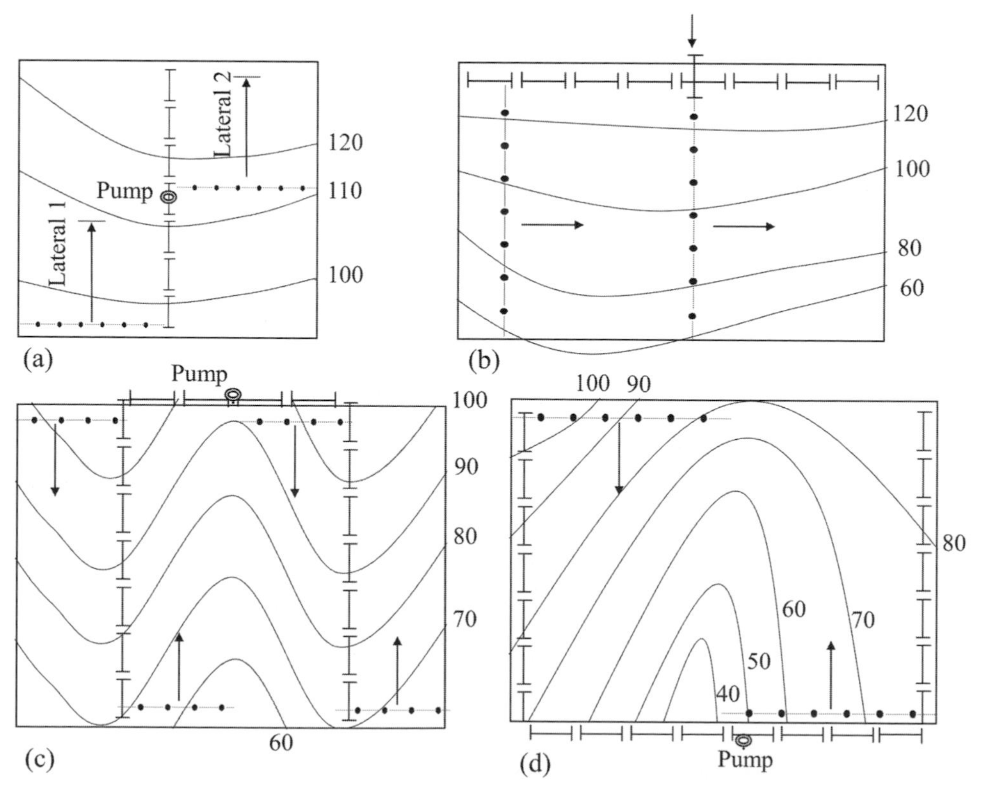

Figure 21.14 Layouts for set-move sprinkler systems. (a) Layout on moderate, uniform slopes with water supply at the center. (b) Layout with gravity pressure where pressure gain approximates friction loss and allows running laterals downhill. (c) Layout with two sub-mainlines on ridges to avoid running laterals uphill. (d) Layout with two sub-mainlines on the sides of the area to avoid running laterals uphill. (Adapted from USDA-SCS, 1983.)

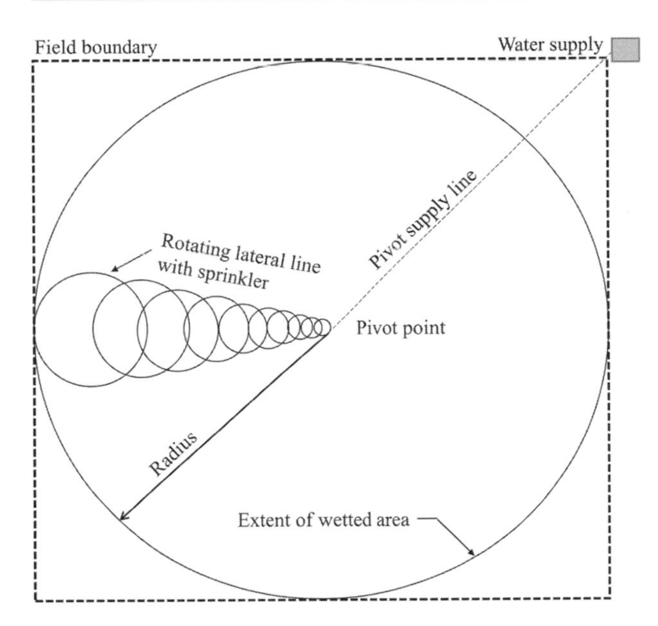

Figure 21.15 A typical layout of a CP system.

other half of the field through inward or outward pivots. The system can be towed to the other end.

21.6.2 Sprinkler System Selection

To select a sprinkler system, the following considerations should be kept in mind. (1) The system should have the capacity to meet the crop requirements. (2) It should discharge water at a rate that does not exceed the minimum intake rate of the soil. (3) The water application should satisfy the minimum economic uniformity. (4) The annual cost of irrigation, including capital, labor, and power, should be as low as possible. (5) The system should be selected for crops that justify the use of the system.

21.6.3 Distribution of Water

The distribution of water by sprinklers in the area being irrigated is fundamental to the design of sprinkler systems. The water applied to the field is measured in terms of depth of precipitation that occurs during the period of irrigation, and this depth varies from point to point. Hart (1975) discussed three factors that affect the distribution of water: (1) atmospheric conditions, (2) sprinkler characteristics, and (3) operational practices. Atmospheric conditions include wind speed and direction and evaporation potential. Wind causes the drifting of droplets and the distortion of wetting patterns. Evaporation of droplets causes loss of water, thus reducing water reaching the ground and infiltration. Rotation of sprinkler heads, the spacing of heads along the lateral, the spacing between laterals, the height of sprinkler heads above the crop canopy, and turbulence of the flow entering the head

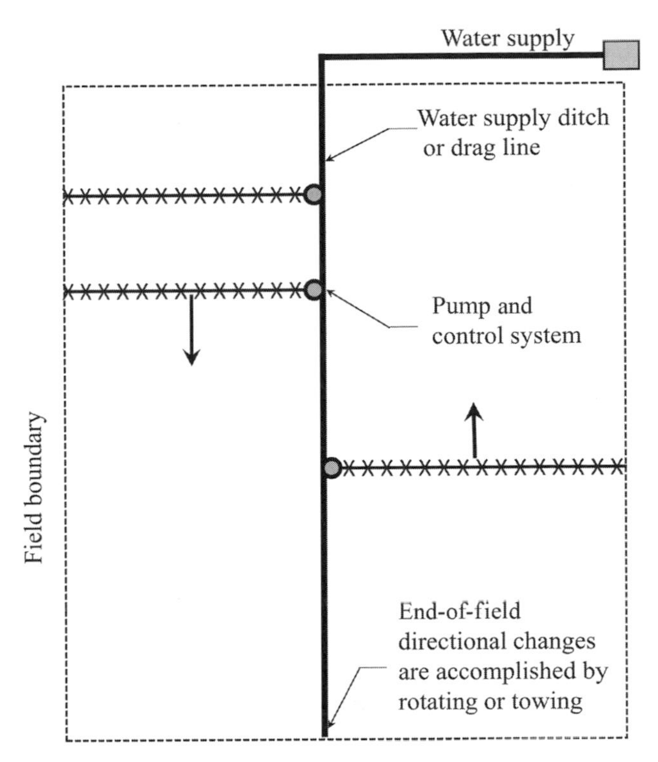

Figure 21.16 A typical layout of a linear configuration move system: the lateral line may be replaced by a giant gun traveling system.

affect the distribution of water. Operating pressure influences the formation of droplet size, uniformity of precipitation, drift, and evaporation. Atmospheric factors cannot be controlled, but their impact can be minimized with proper operational practices. Other factors can be controlled by proper design and management.

The distribution of water is a measure of the performance of sprinkler systems. Several indices have been defined that provide an indication of the uniformity of water application and efficiency of sprinkler systems. During irrigation by sprinklers, some water is evaporated from the field, some runs off the field, and some infiltrates the field. Part of the infiltrated water may reach the groundwater or drainage system. Because the depth of precipitation on the ground varies widely, each of these terms will also vary. The efficiency of the system can be defined in two ways. The first is the application efficiency, which is the ratio of water stored in the root zone or consumed by evapotranspiration to the water supplied by the sprinkler system. The second is the water requirement efficiency, defined as the ratio of soil moisture stored in the root zone to the root zone deficit.

21.6.4 Uniformity Coefficient

There are different methods for quantifying the uniformity of application of water, some of which were discussed in Chapter 16 and will not be repeated here.

21.6.5 Sprinkler Performance

The operation of sprinklers is primarily controlled by operating pressure and nozzle geometry characterized by nozzle opening size, shape, and angle. The sprinkler performance is measured by its discharge, distance of throw, distribution patterns, application rate, and droplet size. A short discussion of each of these factors is now given.

Sprinkler discharge: The sprinkler discharge (q) is the volume of water exiting the sprinkler per unit time and is measured either in L/s or gpm. It is a function of operating pressure and nozzle geometry and can be expressed as

$$Q = \sum_{i=1}^{n} q_i = \sum_{i=1}^{n} KC_i A_i P_i^{x_i}, \tag{21.1}$$

where A_i is the cross-sectional area of the opening in the ith nozzle, P_i is the operating pressure in the ith nozzle, x_i is the exponent for the ith nozzle, K is the constant depending on the unit used, and C_i is the coefficient depending on the shape and roughness of the ith nozzle. For a multi-nozzle sprinkler, the discharge will be the sum of nozzle discharges (Q). The value of x is about 0.5.

Distance of throw: The distance a sprinkler throws water depends on the operating pressure and the size, shape, and nozzle opening. This distance of throw increases and then decreases as the nozzle angle rises above the horizontal. It increases with the increase of operating pressure and also increases as the nozzle size increases. Nozzles emitting larger droplets wet a larger area than those emitting smaller droplets.

Distribution pattern: The volume and rate of water application varies with the distance from the sprinkler; the pattern of this variation is referred to as the distribution pattern, which is about the same for a given operating pressure, nozzle geometry, and wind. Nozzle shape and opening size do not normally influence the distribution pattern. Figure 21.17 shows distribution patterns for different operating pressures. At low pressures, nozzles emit almost uniform-size droplets and lead to donut-shaped distribution patterns; at high pressures, the distribution pattern with a wide range of droplet sizes is triangular; and at extremely high pressures the patterns with smaller droplets is bell-shaped but with a flatter top.

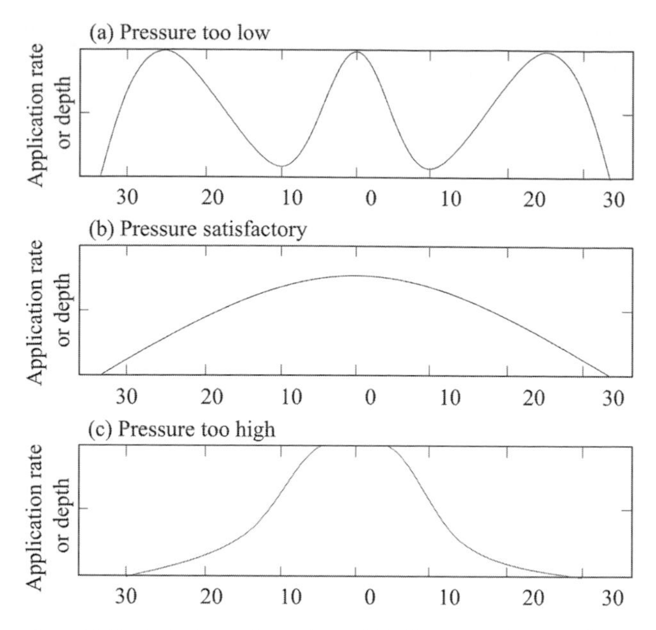

Figure 21.17 Application patterns of an individual impact sprinkler for different operating pressures.

Application rate: Application rate is important for matching sprinklers to soil, crop, and terrain types. The average application rate of an individual sprinkler depends on the nozzle geometry: opening size and shape. The average application rate (I_a) of an individual sprinkler can be computed as

$$I_a = K\frac{q}{D_w}, \tag{21.2}$$

where q is the sprinkler discharge, D_w is the wetted area, and $K = 60$ for I_a in mm/h, q in L/min, and D_w in m^2), and $K = 96.3$ for I_a in in./h, q in gpm, and D_w in ft^2. If there are several identical sprinklers spaced in an L (spacing between laterals) by S (spacing between sprinklers along the lateral) grid, then the average application rate can be computed as

$$I_a = \frac{Kq}{LS}. \tag{21.3}$$

The average application rate of a lateral of sprinklers can be determined as

$$I_a = \frac{KQ_t}{L_t L}, \tag{21.4}$$

where Q_t is the total flow rate into the upstream end of the lateral, and L_t is the length of the lateral.

Droplet size: Droplet sizes are important for sprinkler performance because they affect the formation of sealing on the bare soil and distribution pattern. The nozzle opening shape controls the droplet size, but the nozzle angle does not.

21.6.6 Irrigation Pipelines

Irrigation pipelines comprise many materials. Currently, the most common materials used for aboveground sprinkler systems and gated-pipe surface irrigation systems are aluminum and ultraviolet-protected PVC (polyvinyl chloride plastic) pipes. Center-pivot and linear-move systems commonly use galvanized steel as the pipeline material.

For pipelines that are buried below the ground, the most common material in agricultural applications is PVC, and in landscape and turf applications it is either PVC or PE (polyethylene plastic). PE is also commonly used for aboveground microirrigation systems.

Table 21.1 *Pipeline flow rates (velocity at 5 ft/s [1.5 m/s])*

Nominal pipe diameter (in.)	Flow rate (gpm)
6	441
8	783
10	1224
12	1763
14	2399
15	2754

After USDA-NRCS (1997).

Sizing mainlines is usually based on a maximum of 5–6 ft/s (1.5–1.8 m/s) average velocity. The typical flow ranges for aluminum pipe and class 160 PVC pipe at various nominal sizes and reasonable flow velocities are shown in Table 21.1.

21.6.7 Hydraulics of Sprinkler Systems

Basic knowledge of hydraulics is fundamental to the planning and design of sprinkler irrigation systems. The discharge from a sprinkler can be expressed as

$$q = C_0 a (2gH_a)^{\frac{1}{2}}, \tag{21.5}$$

where q is the sprinkler discharge, C_0 is the sprinkler coefficient, a is the cross-sectional area of the opening of the nozzle, g is the acceleration due to gravity, and H_a is the pressure head. Expressing discharge q in gpm, A in in.^{2}, and H_a in lb/in.^{2}, eq. (21.5) can be written as

$$q = 38 C_0 a P^{\frac{1}{2}} \tag{21.6}$$

or

$$q = 30 C_0 (d_1^2 + d_2^2) P^{\frac{1}{2}}, \tag{21.7}$$

where P is the pressure in lb/in.^{2}, a is the area of nozzles in in.^{2}, and d_1 and d_2 are the diameters of nozzles in inches. The constant in eq. (21.7) is 0.00111 for q in L/s, d_1 and d_2 are in mm, and P is in kPa. The value of C depends on the sprinkler and nozzle design and varies from less than 0.80 to 0.95.

The head loss h_f due to friction that has to be overcome by flow in pipes and other conduits can be expressed in a general form as

$$h_f = \frac{K L_t v^m}{D^n}, \tag{21.8}$$

where L_t is the length of the pipe; K is the friction factor depending on the relative roughness of the walls; D is the inside diameter of the pipeline; m is an exponent between 1.85 and 2.0, depending on the roughness; n is an exponent between 1.0 and 1.25; and v is the mean velocity in the pipeline. Sometimes $n = 3.0 - m$.

The head loss can also be expressed in terms of Q as

$$h_f = \frac{K L_t Q^m}{D^{2m+n}}. \tag{21.9}$$

The values of friction loss (in feet of head or in lb/in.^{2}) for a given length of pipe have been tabulated. For lines with multiple uniformly spaced outlets, the total friction loss can be estimated following Christiansen (1942) as

$$h_f = F\left(\frac{K L_t Q^m}{D^{2m+n}}\right), \tag{21.10}$$

where F depends on the value of m and the number of sprinklers and can be evaluated as

$$F = \frac{1}{m+1} + \frac{1}{2N} + \frac{(m-1)^{\frac{1}{2}}}{6N^2}, \tag{21.11}$$

where N is the number of sprinklers. The value of m is 1.85 for the Hazen–Williams equation, 1.9 for Scobey's equation, and 2.0 for the Darcy–Weisbach equation. If the first sprinkler is located at a distance of one-half the sprinkler interval from the mainline, then

$$F = \frac{2N}{2N-1}\left[\frac{1}{m+1} + \frac{(m-1)^{\frac{1}{2}}}{6N^2}\right]. \tag{21.12}$$

The power (HP) needed for pumping water for sprinkler systems can be expressed as

$$HP = \frac{Q P_p}{1715 E_p}, \tag{21.13}$$

where HP is the horsepower, Q is the total discharge of the pump (gal/min), P_p is the total pumping head (lift and friction losses) (lb/in.^{2}), and E_p is the efficiency of the pump. The constant in eq. (21.13) is 6116 for HP in kW, Q in L/min, and P_p in m. For any system, HP is proportional to 1.5 power of the total pressure:

$$HP = K P_p^{1.5}. \tag{21.14}$$

21.7 DESIGN OF SPRINKLER SYSTEMS

The design of sprinkler systems involves several steps, as shown in Figure 21.18, and several of these steps require iterations to get acceptable design alternatives. An investigation of the given field should be performed before the design of the sprinkler system, including topography, water quality and quantity, soils, crops, source of power, and farm operation schedules. The first five steps of the procedure are usually referred to as the preliminary design factors. Other steps are referred to as the final design steps. The design of a sprinkler irrigation system may include the following steps (USDA-NRCS, 2016):

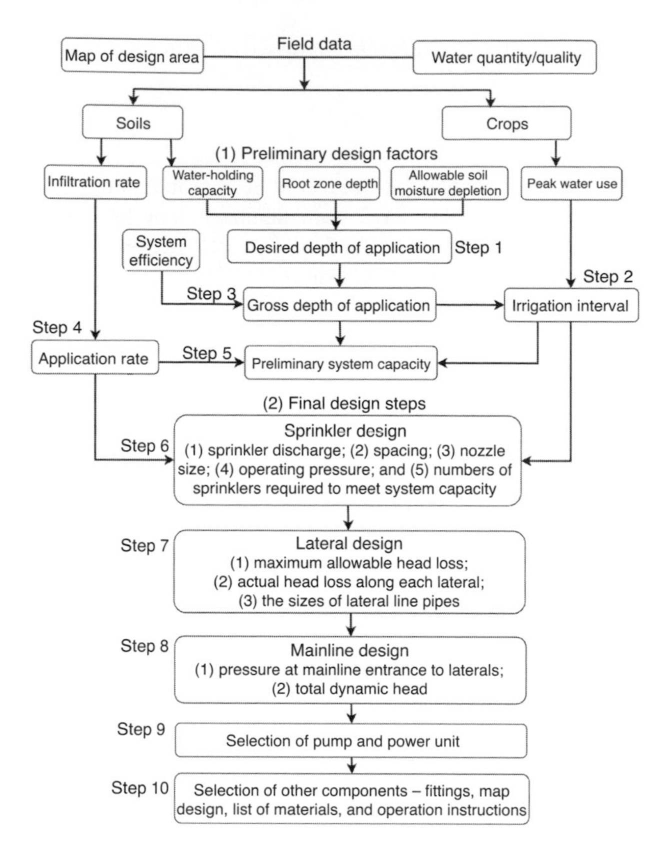

Figure 21.18 Design steps of a solid-set or set-move sprinkler system.

Step 1: Determination of the net depth of water application. The soil water-holding capacity, root zone depth, and allowable soil moisture depletion are used to calculate this depth (more details are discussed in Chapter 13).

Step 2: Calculation of the irrigation interval at peak demand. The peak demand is based on the peak crop consumptive use rate, which is detailed in Chapters 14 and 15.

Step 3: Calculation of the gross depth of application based on the system efficiency. The methods for quantifying the system efficiency are discussed in Chapter 16.

Step 4: Determination of the optimum application rate. The maximum water application rate should neither cause runoff nor damage the crop under the prevailing wind condition. The basic infiltration rate of the soil (detailed in Chapter 12) is used as a guide to select a sprinkler with an irrigation rate lower than the infiltration rate. The optimum application rate can be adjusted later, once the preliminary design factors are obtained.

Step 5: Determination of the capacity requirements of the system.

For solid-set or set-move systems, the procedure includes:

Step 6: Sprinkler design, including the determination of sprinkler discharge, spacing, nozzle sizes, and operating pressure for the optimum water application rate, and determination of the number of sprinklers required to meet the system capacity.

Step 7: Lateral design, including (1) the calculation of maximum allowable head loss between two sprinklers; (2) actual head loss along each lateral, based on the maximum allowable head loss; and (3) determination of the sizes of lateral pipes.

Step 8: Mainline design, including (1) the determination of pressure at the mainline entrance to laterals; (2) determination of the total dynamic head; and (3) determination of the sizes of mainline pipes.

Step 9: Selection of a pump and power unit for maximum operating efficiency.

Step 10: Preparation of a map of design and selection of other components – fittings, list of materials, and operation instructions.

For continuous-move sprinkler systems, Steps 6–10 are different and will be discussed in Section 21.8.

21.7.1 Net Depth of Application (Step 1)

The net depth of irrigation requirement or the desired depth of application (D_n, mm) can be defined as the product of total available moisture (*TAM*, mm/m), management allowed depletion (*MAD*, fraction), and depth of the active root zone (D_r, m):

$$D_n = TAM \times MAD \times D_r, \tag{21.15}$$

where *TAM* is the total available moisture (mm/m), which is the difference between field capacity (*FC*) and the limit of crop-extractable water (*WP*) (i.e., $TAM = FC - WP$). The values of *TAM* of different soil textures can be found in Table 13.2. The values of *MAD* and D_r for different crops can be found in Table 13.3.

Example 21.1: Consider an agricultural farm that has sandy loam soil with a total depth of 2 m. Maize is grown on the farm and is to be irrigated by the sprinkler system. The active rooting depth of cotton is 1.0 m. Calculate the net depth of application.

Solution: From Table 13.2, the value of *TAM* for sandy loam soil is 120 mm/m. The value of *MAD* from Table 13.3 is 0.65. Given $D_r = 1.0$ m, the net depth of application is

$$D_n = TAM \times MAD \times D_r = 120\ \text{mm/m} \times 0.5 \times 1.0\ \text{m}$$
$$= 60\ \text{mm}.$$

If the area of the field is A (ha), the net depth of irrigation requirement can be expressed in terms of volume,

$$\text{The volume of water to be applied } (m^3) = 10 \times A \times D_n. \quad (21.16)$$

21.7.2 Irrigation Interval (Step 2)

The irrigation interval is defined as the time between successive irrigations and is equal to the ratio of the net depth of irrigation requirement (D_n) to the evapotranspiration rate (ET_c):

$$T_i = \frac{D_n}{ET_c}, \quad (21.17)$$

where T_i is the irrigation interval in days and ET_c is the average crop evapotranspiration rate during the peak use period in mm/day. If the system has different crops to be irrigated, the crop with the highest peak evapotranspiration will be used. More details are given in Chapter 25.

Example 21.2: Based on the information in Example 21.1, the crop evapotranspiration rate of maize during the peak period is 8 mm/day. Determine the irrigation interval.

Solution:

$$T_i = \frac{D_n}{ET_c} = \frac{60 \text{ mm}}{8 \text{ mm/day}} = 7.5 \text{ days} \approx 7 \text{ days}.$$

Thus, the irrigation interval can be taken as 7 days.

The range of the irrigation interval for agricultural crops is usually 0.25–8 days. The calculated T_i can be rounded down to the nearest integer number of days. In some cases, it is also acceptable to round up if the value is very close to the next highest integer number. T_i can be fractional if an automated sprinkler system is used.

Maximum permissible irrigation interval: Now the maximum permissible interval (T_{max}) between irrigations is to be determined. This will depend on the allowable loss of water from the soil. It is assumed that soil moisture is maintained above the midpoint between the wilting point (WP) and FC for a root zone depth (D_r, m), and that soil moisture is restored to field capacity after two days of application. In this case, $MAD = 0.5$. Then, the irrigation interval T_{max} in days should not exceed:

$$\begin{aligned} T_{max} &= \frac{TAM \times MAD \times D_r}{ET_c} + 2 + \frac{t}{24} \\ &= \frac{TAM \times D_r}{2 \times ET_c} + 2 + \frac{t}{24}, \end{aligned} \quad (21.18)$$

where t denotes the hours of operation per application. The value of T_{max} should be reduced by 1 or 2 days for very coarse and shallow-rooted crops.

Example 21.3: Determine the maximum irrigation interval in days $TAM = 120$ mm, $ET_c = 8$ mm/day, $D_r = 1.0$ m, and $t = 22$ h.

Solution: The value of T_{max} is determined as:

$$T_{max} = \frac{TAM \times D_r}{2 \times ET_c} + 2 + \frac{t}{24} = \frac{(120) \times (1.0)}{2 \times 8} + 2 + \frac{22}{24}$$

$$= 10.4 \text{ days} \approx 10 \text{ days}.$$

21.7.3 Gross Depth of Application (Step 3)

The gross depth of application is defined as the net depth of application divided by the irrigation efficiency:

$$D_g = \frac{D_n}{E_a}, \quad (21.19)$$

where D_g is the gross application depth (mm) and E_a is the application efficiency as a fraction, accounted for both evaporation, wind drift, and deep percolation.

Example 21.4: Based on the data in Example 21.2, the irrigation efficiency is given as 80%. Calculate the gross depth of application.

Solution:

$$D_g = \frac{D_n}{E_a} = \frac{60 \text{ mm}}{80\%} = 75 \text{ mm}.$$

The amount of water to be delivered by the system in one irrigation should not exceed

$$D_{g-max} = \frac{ET_c \times T_{max}}{E_a}. \quad (21.20)$$

Example 21.5: Compute the maximum amount of delivery of water considering the data in Examples 21.2 and 21.3 if $E_a = 80\%$.

Solution: The maximum amount of application depth is computed as

$$D_{g-max} = \frac{ET_c \times T_{max}}{E_a} = \frac{8 \text{ mm/day} \times 10 \text{ days}}{80\%} = 100 \text{ mm}.$$

21.7.4 Application Rate (Step 4)

This is the amount of water that should be applied to the field within a minimum period of time, which will depend on the minimum intake rate of soil (i) after several hours of wetting. This time (t) can be estimated as

$$t = \frac{D_g}{i}. \tag{21.21}$$

Example 21.6: Compute the value of t for the data from Example 21.4 if $i = 0.25$ in./h (6.35 mm/h).

Solution: The value of t is computed as

$$t = \frac{D_g}{i} = \frac{75 \text{ mm}}{6.35 \text{ mm/h}} = 11.8 \text{ h}.$$

The application rate less the intake rate of the soil is the upper bound, but the lower bound should consider evaporation and drift of water from the nozzle and the discharge rate of the nozzle should be high enough to allow adequate infiltration of water into the root zone. Considering the lateral and mainline geometry, the application rate is defined by the nozzle discharge converted to the amount of water applied to the field, expressed as

$$d_g = \frac{q}{2.778 \times s_l \times s_m}, \tag{21.22}$$

where d_g is the gross application rate (cm/h), q is the sprinkler nozzle discharge (L/s), s_l is the lateral spacing or distance between sprinklers on lateral (S) (m), and s_m is the mainline spacing or distance between lateral lines (L) (m). Part of d_g will be used up by evaporation and drift. Thus, the net application rate that will be applied to the soil can be expressed as

$$d_a = d_g(1 - L_s), \tag{21.23}$$

where d_a is the net application rate (cm/h) and L_s is the loss due to evaporation and drift (as a fraction). The Soil Conservation Service (SCS) of the US Department of Agriculture (USDA) (1983) has recommended minimum gross application rates for different climatic zones, as given in Table 21.2. The maximum net application rates for sprinklers for average soil slope and tilth are given in Table 21.3.

21.7.5 Sprinkler System Capacity (Step 5)

The capacity of a system is influenced by the area to be irrigated, crops to be grown, maximum evapotranspiration during an irrigation interval, gross depth of water to be applied at each irrigation, soil moisture storage capacity of the soil between the wilting point and field capacity, accounting for the root zone, operational schedule, and source of water supply. The required minimum capacity of a sprinkler system depends on the area to be irrigated, the gross depth of water application, and the operating time to apply this depth. Based on the volume balance, the product of capacity and application duration is equal to the product

Table 21.2 *Minimum gross application rates recommended by USDA-SCS (1983)*

	Minimum gross application rate (cm/h)		Minimum gross application rate (in./h)	
Climate zone	From	to	From	to
Cold maritime	0.25	0.40	0.10	0.15
Warm maritime	0.40	0.50	0.15	0.20
Cool dry continental	0.40	0.50	0.15	0.20
Warm dry continental	0.50	0.75	0.20	0.30
Cool desert	0.75	1.25	0.30	0.50
Hot desert	1.25	1.90	0.50	0.75

After USDA-SCS, 1983.

Table 21.3 *Maximum net application rates for sprinklers for average soil, slope, and tilth suggested by USDA-SCS (1983)*

	0–5% slope		5–8% slope		8–12% slope		12–16% slope	
Soil texture	in./h	cm/h	in./h	cm/h	in./h	cm/h	in./h	cm/h
Coarse sandy soil to 2 m (6 ft)	2.00	5.00	1.50	3.80	1.00	2.50	0.50	1.30
Coarse sandy soils over more compact soils	1.50	3.80	1.00	2.50	0.75	1.90	0.40	1.00
Light sandy loams to 2 m (6 ft)	1.00	2.50	0.80	2.00	0.60	1.50	0.40	1.00
Light sandy loams over more compact soils	0.75	1.90	0.50	1.30	0.40	1.00	0.30	0.80
Silt loams to 2 m (6 ft)	0.50	1.30	0.40	1.00	0.30	0.80	0.20	0.50
Silt loams over more compact soils	0.30	0.80	0.25	0.60	0.15	0.40	0.10	0.30
Heavy textured clays or clay loams	0.15	0.40	0.10	0.25	0.08	0.20	0.06	0.15

After USDA-SCS, 1983.

of the application depth and area. Then, the system capacity can be estimated as

$$Q = K\frac{D_g \times A}{T_i \times h} = K\frac{D_n \times A}{E_a \times T_i \times h}, \tag{21.24}$$

where Q is the required minimum system capacity (L/s or gpm), A is the area to be irrigated (ha or acres), h is the hours of operation per day (h/day); T_i is the irrigation interval (days); D_g and D_n are the gross and net irrigation application (mm or in.); E_a is the application efficiency (fraction); $K = 2.778$ if Q is in L/s, A is in ha, and D_g and D_n are in mm; and $K = 453$ if Q is in gpm, A is in acres, and D_g and D_n are in inches.

From eq. (21.17), D_n/T_i can be replaced by ET_c, then eq. (21.24) can be expressed as

$$Q = K\frac{ET_c \times A}{E_a \times h}, \tag{21.25}$$

where ET_c is the average evapotranspiration rate during the peak use period (mm/day or in./day).

In the design of a sprinkler system to be discussed later, the number of sprinklers operating at the same time will be determined. We can also calculate the system capacity by multiplying the numbers of sprinklers and individual sprinkler discharge. Here, the calculation of a single sprinkler discharge is given. Explicitly considering the distance between laterals (or spacing) (L in m, ft), spacing between sprinklers (S in m, ft), time interval between beginnings of successive irrigations of a given set (T, h), depth to be applied for irrigation (mm, in.), and downtime for moving set-move systems and/or maintenance (T_m, h), the sprinkler discharge q (L/min, gpm) or capacity can also be expressed for application efficiency E_a (in percent) as

$$q = \frac{K \times D_n \times L \times S}{(T - T_m)E_a}, \tag{21.26}$$

where $K = 1.67$ if q is in L/min, d is in mm, and L and S are in m; and $K = 1.04$ if q is in gpm, d is in inches, and L and S are in ft. For the desired depth of irrigation (D_n, mm, in.), design daily irrigation requirement (d_n, mm/day, in./day), and the percentage of the field irrigated when the sprinkler system is operating (P_f), the interval T (h) can be estimated as

$$T \leq \frac{0.24 P_f D_n}{d_n}. \tag{21.27}$$

The value of D_n must be the same as that employed for computing d_n and hence it can normally be regarded as equal to readily available water (RAW). Following the SCS, the value of d_n (mm/day, in./day) can be computed from peak monthly evapotranspiration (ET_m, mm, in.) for various values of soil water depletion allowed between irrigations (AD, mm, in.) as

$$d_n = 0.034\frac{ET_m^{1.09}}{AD^{0.09}}. \tag{21.28}$$

With the value of T selected, which should be less than or equal to the value computed using eq. (21.27), D_n can be computed as

$$D_n = \frac{T d_n}{0.24 P_f}. \tag{21.29}$$

For the length of lateral (L_t, m, ft), number of laterals operating concurrently (N_l), and total field area (A, ha, acres), the value of P_f can be determined as

$$P_f = \frac{L_t L N_l}{KA}, \tag{21.30}$$

where $K = 100$ when L and L_t are in m and A is in ha; and $K = 435.6$ when L and L_t are in ft and A is in acres.

For solid-set, set-move, and traveler sprinkler systems, the value of spacing between laterals can be determined as

$$L \leq K_l D_w, \tag{21.31}$$

and spacing between sprinklers can be determined as

$$S \leq K_s D_w, \tag{21.32}$$

where K_l and K_s are constant, depending on the sprinkler spacing pattern and wind, and D_w is the diameter of the wetted area (m, ft). Sprinkler spacing patterns for set-move and solid-set systems are rectangular, square, and triangular,

Table 21.4 *Values of K_l and K_s (a denotes $K_l = 0.86\ K_s$)*

Wind velocity		Rectangular		Square		Triangular	
(m/s)	(mph)	K_l	K_s	K_l	K_s	K_l	K_s
0–1.3	(0–3)	0.60	0.50	0.55	0.55	a	0.60
1.8–2.1	(4–7)	0.60	0.45	0.50	0.50	a	0.55
3.6–5.4	(8–12)	0.60	0.40	0.45	0.45	a	0.50

After USDA-NRCS, 2016.

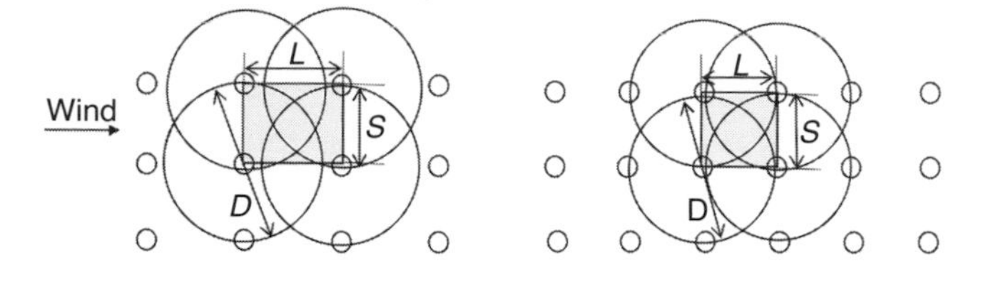

Figure 21.19 Sprinkler spacing patterns.

as shown in Figure 21.19. The values of K_l and K_s for different sprinkler patterns are given in Table 21.4.

The spacing between laterals equals the distance that a linear-move system moves per irrigation, which is normally the length of the field. For a CP system, L can be defined as

$$L = 2\pi r, \tag{21.33}$$

where r is the distance from the pivot to the sprinkler along the lateral. The value of L will be different for each sprinkler along the lateral.

Example 21.7: Determine the system capacity in gal/min for a 50-acre (20.23 ha) field with a maximum evapotranspiration rate of 0.3 in./day (7.62 mm/day), application efficiency of 80%, and 22 h/day of operating time.

Solution: The required capacity is determined as

$$Q = 453 \times \frac{ET_c \times A}{E_a \times h} = 453 \times \frac{\left(0.30 \frac{\text{in.}}{\text{day}} \times 50 \text{ acres}\right)}{\left(0.80 \times 22 \frac{\text{h}}{\text{day}}\right)}$$

$$= 386.1 \frac{\text{gal}}{\text{min}},$$

$$Q = 2.778 \times \frac{ET_c \times A}{E_a \times h} = 2.778 \times \frac{\left(7.62 \frac{\text{mm}}{\text{day}} \times 20.23 \text{ ha}\right)}{\left(0.80 \times 22 \frac{\text{h}}{\text{day}}\right)}$$

$$= 24.3 \text{ L/s}.$$

Example 21.8: Determine the system capacity for a 50-acre (20.23 ha) corn field where the depth moisture to be replaced at each irrigation is 2.5 inches (63.5 mm). The irrigation period (N) is 10 days in a 16-day interval. The irrigation application efficiency is 75%.

Solution: The irrigation period is 10 days in a 16-day interval, so the system is to be operated 15 hours a day. The system capacity is

$$Q = \frac{453 \times A \times D_g}{Nh} = \frac{453 \times 50 \text{ acre} \times 2.5 \text{ in.}}{10 \text{ day} \times 15 \frac{\text{h}}{\text{day}}}$$

$$= 377.5 \frac{\text{gal}}{\text{min}},$$

$$Q = \frac{2.778 \times A \times D_g}{Nh} = \frac{2.778 \times 20.23 \text{ ha} \times 63.5 \text{ mm}}{10 \text{ day} \times 15 \frac{\text{h}}{\text{day}}}$$

$$= 23.8 \text{ L/s}.$$

If more than one crop is to be irrigated by the system, then the capacity for each area is computed and then capacities are summed up to obtain the system capacity. This discussion can also be extended to crops in rotation.

Example 21.9: Consider a set-move sprinkler system that is 80% efficient. The system has six laterals, each 1350 ft (411.5 m) long, spaced 70 ft (21.3 m) apart. The sprinklers along the laterals are spaced 30 ft (9.1 m) apart. The total field area for irrigation is 120 acres (48.6 ha), the desired depth of irrigation is 2.25 in. (57.15 mm), and the design daily irrigation requirement is 0.35 in./day (8.89 mm/day). The downtime for moving the set-move system is 1 h. Compute the sprinkler capacity for the system.

Solution: The percentage of total field irrigated is computed first as

$$P_f = \frac{L_t \times L \times N_l}{K \times A} = \frac{1350 \times 70 \times 6}{435.6 \times 120} = \frac{567{,}000}{52{,}296} = 10.85\%.$$

Now the time interval between the beginnings of successive irrigations is computed as

$$T \leq \frac{0.24 P_f D_n}{d_n} = \frac{0.24 \times 10.85 \times 2.25}{0.35} = \frac{5.85}{0.35} = 16.74 \text{ h}.$$

One can use 14 hours (less than 16.74 hours), allowing for moving the system approximately twice per day. The depth applied can be determined as

$$D_n = \frac{T \times d_n}{0.24 \times P_f} = \frac{14 \times 0.35}{0.24 \times 10.84} = 1.88 \text{ in. } (47.8 \text{ mm}).$$

The sprinkler capacity is determined as

$$q = \frac{K \times D_n \times L \times S}{(T - T_m)E_a} = \frac{1.04 \times 1.88 \times 70 \times 30}{(14 - 1) \times 80} = 3.95 \frac{\text{gal}}{\text{min}} (0.25 \text{ L/s}).$$

21.7.6 Sprinkler Design (Step 6)

Sprinkler design entails determination of sprinkler discharge, operating pressure, spacing, number, and selection.

Sprinkler discharge: The sprinkler nozzle discharge depends on the nozzle diameter, type, and operating pressure, and can be expressed as

$$q = K p^{0.5}, \tag{21.34}$$

where q is the nozzle discharge (L/s), p is the nozzle pressure (kPa), and K is the proportionality constant. The value of K for different nozzle diameters is shown in Table 21.5. The values of sprinkler diameter have been tabulated for different values of discharge, wetted diameter, and operating pressure, as shown in Table 21.6 (Rain Bird, 2017).

The required discharge of an individual sprinkler can also be expressed as (USDA-SCS, 1968):

$$q = \frac{S_l \times S_m \times I}{96.3}, \tag{21.35}$$

where q is the required discharge (gal/min), S_l is the spacing of sprinklers along laterals (ft), S_m is the spacing of laterals along the mainline (ft), and I is the optimum application rate (in./h). The constant in eq. (21.35) is 3600 if q is in L/s, S_l and S_m are in m, and I is in mm/h. For low-, moderate-, and intermediate-pressure sprinklers, S_l should be within 50% of the wetted diameter, and S_m should be within 65% of the wetted diameter. For high-pressure sprinklers the maximum diagonal distance between two sprinklers on adjacent laterals should be within two-thirds of the wetted diameter. The diagonal distance (S_d) can be computed for square or rectangular spacing as

$$S_d = \sqrt{S_l^2 + S_m^2}, \tag{21.36}$$

and for sprinklers on alternate laterals as

$$S_d = \sqrt{\left(\frac{1}{2S_l}\right)^2 + S_m^2}. \tag{21.37}$$

This value of diameter is compared with two-thirds of the wetted diameter specified by the manufacturer.

Sprinkler spacing: Sprinkler spacing involves mainline spacing and lateral spacing. Mainline spacing essentially means the spacing of valves through which water is supplied to laterals. More than one lateral may receive water from one valve, as shown in Figure 21.20.

Typically, mainline spacing is larger than lateral spacing. Thus, the spacing of laterals depends on wind, flow rate, pressure, wetted diameter, and uniformity coefficient. Rain Bird (2017) provided manufacturer sprinkler specifications, as given in Table 21.6, from which the wetted diameter D_w can be obtained under no-wind conditions. Table 21.7 gives recommended sprinkler spacing (s_l) for laterals and mains (s_m) as a fraction of D_w for given wind speeds. This table can be used to compute D_w if information on spacings is known, say for manufacturers. Because the uniformity changes with wind speed, spacing (lateral spacing followed by mainline spacing) and operating pressure, as shown in Figure 21.21, for a double-nozzle sprinkler is set on various rectangular lateral spacings.

Table 21.5 *Values of coefficient K for sprinkler discharge*

Nozzle diameter (mm)	K (q in L/s, p in kPa)	Nozzle diameter (in.)	K (q in gpm, p in psi)
4	0.0172	1/6	0.81
5	0.0278	0.2	1.16
10	0.1113	0.4	4.63
15	0.2553	0.6	10.63
20	0.4451	0.8	18.53
25	0.6991	1.0	29.10
30	1.021	1.2	42.51
36	1.382	1.4	57.52
41	1.812	1.6	75.39

After USDA-NRCS, 2016.

Table 21.6 *Sprinkler specifications from the manufacturer*

Pressure (kPa)	Diameter (mm)	D-wet (m)	Flow rate (L/s)	Diameter (in.)	D-wet (ft)	Flow rate (gpm)
172 kPa (25 psi)	1.57	–	–	0.062	–	–
	1.70	9.8	0.139	0.067	32	2.2
	1.83	10.7	0.177	0.072	35	2.8
	2.03	11.6	0.265	0.080	38	4.2
	2.21	11.9	0.347	0.087	39	5.5
241 kPa (35 psi)	1.57	11.3	0.126	0.062	37	2.0
	1.70	11.3	0.170	0.067	37	2.7
	1.83	11.6	0.208	0.072	38	3.3
	2.03	12.5	0.303	0.080	41	4.8
	2.21	12.8	0.398	0.087	42	6.3
310 kPa (45 psi)	1.57	11.6	0.145	0.062	38	2.3
	1.70	11.9	0.189	0.067	39	3.0
	1.83	12.2	0.233	0.072	40	3.7
	2.03	12.8	0.341	0.080	42	5.4
	2.21	13.4	0.448	0.087	44	7.1
379 kPa (55 psi)	1.57	11.6	0.158	0.062	38	2.5
	1.70	12.5	0.208	0.067	41	3.3
	1.83	12.5	0.259	0.072	41	4.1
	2.03	13.1	0.379	0.080	43	6.0
	2.21	13.7	0.498	0.087	45	7.9
414 kPa (60 psi)	1.57	11.6	0.164	0.062	38	2.6
	1.70	12.5	0.221	0.067	41	3.5
	1.83	12.8	0.265	0.072	42	4.2
	2.03	13.4	0.404	0.080	44	6.4
	2.21	13.7	0.530	0.087	45	8.4

After Rain Bird, 2017.

The figure shows that, depending on wind speed, high-value crops require closer spacing of 30 × 40 ft, 40 × 40 ft, or 30 × 50 ft for given uniformity values. For moderate-value crops, the spacing can be higher, at 40 × 50 ft, 50 × 50 ft, or 40 × 60 ft. Equations (21.26) and (21.35) show that for a given nozzle the gross and net application rates vary with sprinkler spacing. This suggests that sprinkler spacing must be such that these rates are within the maximum and minimum specified bounds.

Table 21.7 *Recommended sprinkler spacing along laterals and mains under different wind conditions*

Wind speed (km/h)	s_l/D_w	s_m/D_w
0–8	0.60	0.65
8–16	0.50	0.50
>16	0.35	0.50

Adapted from Cuenca, 1989. Copyright © 1989 by Prentice Hall, Englewood Cliffs, New Jersey.

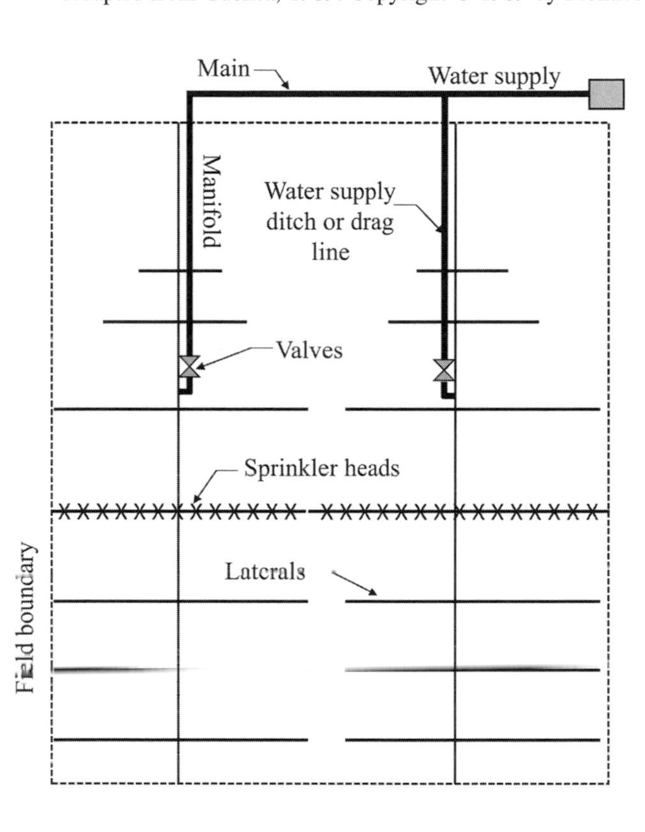

Figure 21.20 Main, manifold, lateral, and valve layout.

Example 21.10: Compute the design wetted diameter for a sprinkler system with manufacturer's specification of lateral spacing of 15 m (50 ft) and mainline spacing of 18 m (60 ft) if the wind speed is 20 km/h.

Solution: Given wind speed is 20 km/h, the ratio of $s_l/D_w = 0.35$, from Table 21.7. With the value of lateral spacing:

$$D_w = \frac{s_l}{0.35} = \frac{15 \text{ m}}{0.35} = 42.9 \text{ m}.$$

With the value of mainline spacing:

$$D_w = \frac{s_m}{0.5} = \frac{18 \text{ m}}{0.5} = 36 \text{ m}.$$

For better uniformity, a higher value is chosen for sprinkler nozzle selection, so

$D_w = 42.9$ m.

Example 21.11: Consider a solid-state sprinkler system that operates during the day only when the wind speed is 10 mph (16 km/h), the minimum relative humidity is 50%, and the maximum temperature is 80 °F (26.7 °C). The nozzle has a diameter of 1/6 in. (4 mm), operating pressure of 55 psi (379 kPa), and wetted diameter of 75 ft (22.9 m). The design soil intake rate is 0.2 in./h (50.8 mm/h). Compute the spacing between sprinklers along the lateral, spacing between laterals, and nozzle discharge.

Solution: Given wind speed is 10 mph = 16 km/h, the ratio of $s_l/D_w = 0.50$, and $s_m/D_w = 0.50$ from Table 21.7, where S_l is the spacing of sprinklers along laterals (ft), S_m is the spacing of laterals along the mainline (ft), and I is the optimum application rate (in./h).

The lateral spacing is computed as:

$$s_l = 0.5 \times D_w = 0.5 \times 75 \text{ ft} = 37.5 \text{ ft} \quad (11.4 \text{ m}).$$

The mainline spacing is computed as:

$$s_m = 0.5 \times D_w = 0.5 \times 75 \text{ ft} = 37.5 \text{ ft} \quad (11.4 \text{ m}).$$

The value of K from Table 21.5 for a 1/6-in. diameter nozzle is 0.81. The discharge q is given by eq. (21.34) as:

$$q = 0.81 \times (55 \text{ psi})^{0.5} = 6.0 \text{ gpm } (0.4 \text{ L/s}).$$

Given the designed soil intake rate is 0.2 in./h, the required discharge of an individual sprinkler can also be computed as:

$$q = \frac{S_l \times S_m \times I}{96.3} = \frac{37.5 \times 37.5 \times 0.2}{96.3} = 2.9 \text{ gpm}.$$

For the optimum performance, nozzle discharge is 2.9 gpm (0.2 L/s).

Nozzle selection: Nozzle selection depends on nozzle diameter, operating pressure, sprinkler spacing, and the physical requirements of the irrigation system. Balancing these factors involves trial and error. Shearer et al. (1965) have provided guidelines as an aid in the nozzle selection process.

Number of sprinklers: The number of sprinklers (N) can be computed if the design discharge (q) of the nozzle is known, as:

$$N = \frac{Q}{q}, \tag{21.38}$$

where Q is the total system capacity.

21.7.7 Lateral Design (Step 7)

The pressure along a lateral varies, and as a result the operating pressure for every nozzle is not the same. However, by design the pressure variation should be within a specified limit so the variation in nozzle discharge is within an acceptable range, which normally is less than 10%. The variation in operating nozzle pressure is therefore limited to 20%; the implication here is that the pressure difference between two critical sprinklers (that have the maximum pressure difference) should be limited to 20%. Where these two sprinklers are located depends on the ground slope. For example, these may be the first and last sprinklers if the lateral is laid on level ground, a constant, or a constant moderate downslope. If the ground has an upslope or a depression in the middle, the head loss due to friction may be outweighed by the gain in potential energy head due to the elevation difference, and a critical sprinkler may be located near the middle of the lateral. The design of laterals depends on the length of the lateral, head loss due to friction in the lateral, and area elevation difference.

Head loss in sprinkler laterals: A pipeline with outlets has a lower friction loss than a conveyance pipe because the velocity decreases with distance along the pipe. This is because of the reduction of flow each time an outlet is passed. To correct for the effects of outlets, a multiple outlet factor, F, proposed by Christiansen (1942), is used. The value of F is 1 for pipelines without outlets. The derivation of the equation for F is described as follows. In general, the friction loss in pipes with no outlets can be expressed as

$$h_f = K \frac{L_t}{D^n} v^m, \tag{21.39}$$

where m is the velocity exponent and n is the diameter exponent. Often the Scobey formula, a special case of eq. (21.39), is used:

$$h_f = \frac{K_s \times L_t \times v^{1.9}}{1000 \times D^{1.1}}, \tag{21.40}$$

where h_f is the friction loss (ft), K_s is the coefficient of retardation depending only on the pipe diameter and pipe material condition, L_t is the length of the line (ft), v is the velocity of flow (ft/s), and D is the diameter (ft).

Equation (21.39) can be simplified as follows:

$$v = \frac{Q}{A} = \frac{Q}{\left(\frac{\pi}{4} D^2\right)}. \tag{21.41}$$

Thus,

$$h_f = K \frac{L_t}{D^n} v^m = K \frac{L_t}{D^n} \frac{Q^m}{\left(\frac{\pi}{4} D^2\right)^m}. \tag{21.42}$$

Thus, in general,

$$h_f = k_3 \frac{L_t}{D^{2m+n}} Q^m. \tag{21.43}$$

Assume that the spacing (S) between sprinklers is equal and the total number of sprinklers is N. If the total flow into the lateral is Q, then each sprinkler flow (q) is equal to Q/N. Further, the sprinkler lateral length (L_t) is given as SN, as shown in Figure 21.22. The first sprinkler is located at distance S from the inlet. To compute the head loss in the lateral, the calculation starts from the last sprinkler outlet and works back to the first sprinkler outlet. The distance between the last two sprinklers is defined as section 1, as shown in Figure 21.22. The flow rate in section 1 is q and the head loss in section 1 is calculated from eq. (21.43):

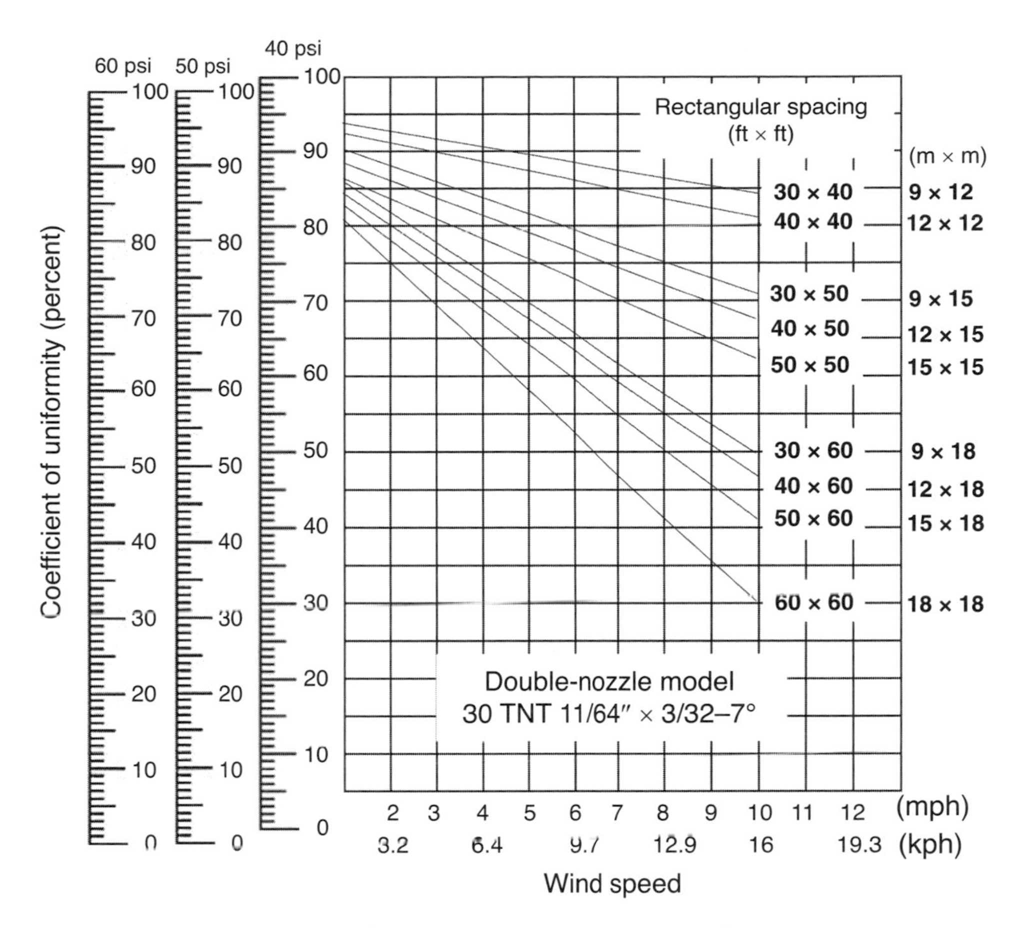

Figure 21.21 Change in uniformity coefficient with wind speed, spacing, and operating pressure for a particular double-nozzle sprinkler (US Standard 11/64″ × 3/32-7°). (Adapted from Cuenca, 1989. Copyright © 1989 by Prentice Hall, Englewood Cliffs, New Jersey.)

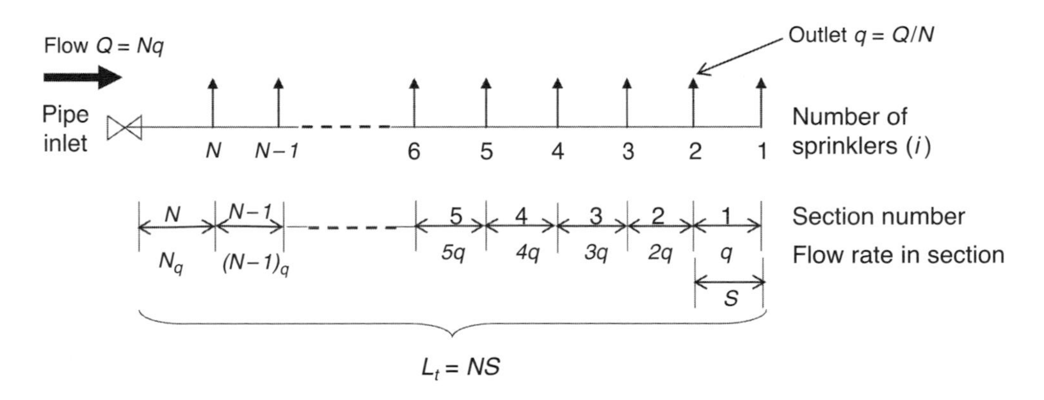

Figure 21.22 Computing head loss in a sprinkler lateral with an equal flow rate and spacing $S = L/N$.

$$h_{f1} = k_3 \frac{S}{D^{2m+n}} q^m. \tag{21.44}$$

Likewise, the head losses in section 2 and the next sections are calculated as

$$h_{f2} = k_3 \frac{S}{D^{2m+n}} (2q)^m, \tag{21.45}$$

$$h_{f3} = k_3 \frac{S}{D^{2m+n}} (3q)^m, \tag{21.46}$$

$$h_{fi} = k_3 \frac{S}{D^{2m+n}} (iq)^m. \tag{21.47}$$

The total head loss in the entire lateral is now the sum of head losses of all sections of the lateral, or

Table 21.8 *Multiple outlet factors for laterals with equally spread outlets of the same discharge** *(based on eq. 21.51)*

	Values of F						
No. of outlets	$m=1.85$	$m=1.90$	$m=2.00$	No. outlets	$m=1.85$	$m=1.90$	$m=2.00$
1	1.0	1.0	1.0	16	0.382	0.377	0.365
2	0.639	0.634	0.635	17	0.380	0.376	0.363
3	0.535	0.528	0.518	18	0.379	0.373	0.361
4	0.486	0.480	0.469	19	0.377	0.372	0.360
5	0.457	0.451	0.440	20	0.376	0.370	0.359
6	0.435	0.433	0.421	22	0.374	0.368	0.357
7	0.425	0.419	0.408	24	0.372	0.366	0.355
8	0.415	0.410	0.398	26	0.370	0.364	0.353
9	0.409	0.402	0.391	28	0.369	0.363	0.351
10	0.402	0.396	0.385	30	0.368	0.362	0.350
11	0.397	0.392	0.380	35	0.365	0.359	0.347
12	0.394	0.388	0.376	40	0.364	0.357	0.345
13	0.391	0.384	0.373	50	0.361	0.355	0.343
14	0.387	0.381	0.370	100	0.356	0.350	0.338
15	0.384	0.379	0.366	>100	0.351	0.345	0.337

* $F=0.54$ for CPs without end guns; $F=0.56$ for CPs with an end gun.

$$h_T=\sum_{i=1}^{n}h_{fi}=\sum_{i=1}^{n}\left[k_3\frac{S}{D^{2m+n}}(iq)^m\right], \tag{21.48}$$

$$h_T=\sum_{i=1}^{n}h_{fi}=\sum_{i=1}^{n}\left[k_3\frac{\frac{L}{N}}{D^{2m+n}}(iq)^m\right]=\frac{k_3\left(\frac{L}{N}\right)}{D^{2m+n}}\sum_{i=1}^{n}(iq)^m. \tag{21.49}$$

Recall that $Q=Nq$, and for the entire lateral, if it were a mainline, the friction loss h_f is given by eq. (21.43); then, eq. (21.49) can be expressed as:

$$h_T=\frac{k_3}{D^{2m+n}}\frac{L}{N^{m+1}}Q^m\sum_{i=1}^{N}(i^m)$$
$$=h_f\frac{1}{N^{m+1}}\sum_{i=1}^{N}(i^m)=h_f\frac{\sum_{i=1}^{N}(i^m)}{N^{m+1}}. \tag{21.50}$$

As shown in eq. (21.50), the head loss in a pipe with multiple outlets can be determined by first estimating the head loss in a mainline with no outlets (h_f), and then multiplying this head loss by a factor F, which is given as:

$$F=\frac{\sum_{i=1}^{N}i^m}{N^{m+1}}=\frac{1}{m+1}+\frac{1}{2N}+\frac{\sqrt{m-1}}{6N^2}. \tag{21.51}$$

The values of F based on eq. (21.51) are given in Table 21.8. m is the exponent in different head loss equations discussed in Chapter 9, with $m=1.85$ for the Hazen–Williams equation, $m=1.90$ for the Scobey equation, and $m=2.00$ for the Darcy–Weisbach equation (more details can be found in Table 9.11). Figure 21.23 graphs F as a function of the number of sprinklers in the lateral with different values of m. If the first sprinkler outlet is located half-spacing from the inlet to the lateral, then the factor F can be approximated as

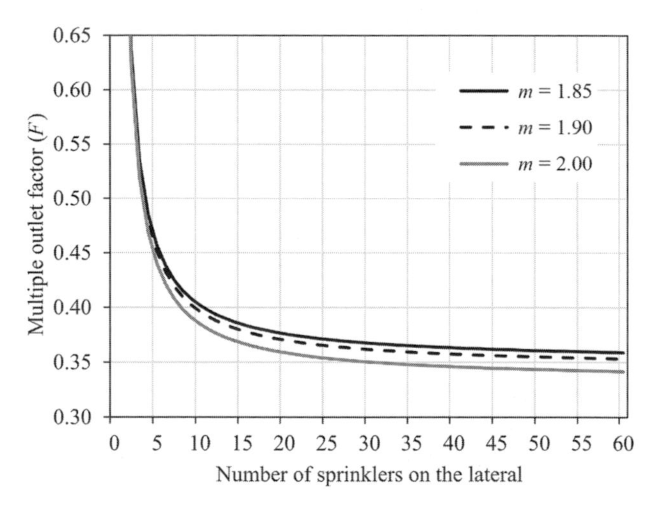

Figure 21.23 Factor F as a function of the number of sprinklers in the lateral.

$$F=\frac{1}{2N-1}+\frac{2}{(2N-1)N^m}\left(\sum_{i=1}^{N-1}(N-i)^m\right)$$
$$=\frac{2N}{2N-1}\left[\frac{1}{m+1}+\frac{\sqrt{m-1}}{6N^2}\right]. \tag{21.52}$$

For lateral pipelines with constant-spaced outlets and nearly the same discharge per outlet, one can use Table 21.8. With CPs, the sprinkler discharge increases with distance from the pivot point. These factors are given at the bottom. When the distance to the first sprinkler is half the sprinkler head spacing, Table 21.9 can be used.

Table 21.9 *Multiple outlet factors for laterals when the distance to the first sprinkler is equal to one-half of the sprinkler head spacing with outlets of the same discharge (based on eq. 21.52)*

	Values of F						
No. of outlets	$m=1.85$	$m=1.90$	$m=2.00$	No. outlets	$m=1.85$	$m=1.90$	$m=2.00$
1	1.00	1.00	1.00	16	0.363	0.357	0.345
2	0.518	0.512	0.500	17	0.362	0.356	0.344
3	0.441	0.434	0.422	18	0.361	0.355	0.343
4	0.412	0.405	0.393	19	0.361	0.355	0.343
5	0.397	0.390	0.378	20	0.360	0.354	0.342
6	0.387	0.381	0.369	22	0.359	0.353	0.341
7	0.381	0.375	0.363	24	0.359	0.352	0.341
8	0.377	0.370	0.358	26	0.358	0.351	0.340
9	0.374	0.367	0.355	28	0.357	0.351	0.340
10	0.371	0.365	0.353	30	0.357	0.350	0.339
11	0.369	0.363	0.351	35	0.356	0.350	0.338
12	0.367	0.361	0.349	40	0.355	0.349	0.338
13	0.366	0.360	0.348	50	0.354	0.348	0.337
14	0.365	0.358	0.347	100	0.353	0.347	0.335
15	0.364	0.357	0.346				

Maximum allowable head loss: The maximum allowable head loss due to friction (H_L, m/m) between two sprinklers can be expressed as

$$H_L = \frac{\Delta \times H_a - H_e}{l_n}, \tag{21.53}$$

where H_a is the nozzle design pressure head (m), H_e is the increase in elevation between two sprinklers in the flow direction, which will be negative for downhill-sloping laterals, Δ is the maximum allowable pressure difference (fraction), and l_n is the distance between the first and the last sprinklers (m). For laterals uphill, the USDA-SCS (1983) recommends that the allowable pressure loss due to friction can be considered as about 20% of the average design operating pressure minus the pressure needed to overcome the elevation in the line. On the other hand, for laterals downhill the allowable pressure loss due to friction can be regarded as 20% of the average design operating pressure plus the pressure gained by the difference in elevation in line.

Example 21.12: Consider a sprinkler lateral laid on ground with a slope of 0.001 m/m, with a design nozzle operating pressure of 300 kPa. The distance between the first and last nozzles is 350 m. What is the maximum allowable head loss due to friction (m/m)?

Solution: The two sprinklers on the lateral are critical. The nozzle operating pressure is converted to head (1 m of head = 9.807 kPa):

$$H_a = \frac{300{,}000\ \text{Pa}}{1000\,\frac{\text{kg}}{\text{m}^3} \times 9.807\,\frac{\text{m}}{\text{s}^2}} \text{ or } \frac{300{,}000\ \text{Pa}}{9807\ \frac{\text{Pa}}{\text{m}}} = 30.6\ \text{m}$$

The elevation difference between the two sprinklers equals 350 m × 0.001 m/m = 0.35 m. For downhill-sloping laterals, $H_e = -0.35$ m.

The allowable pressure difference is 20%. Therefore,

$$H_L = \frac{\Delta \times H_a - H_e}{l_n} = \frac{0.2 \times 30.6\ \text{m} - (-0.35\ \text{m})}{350\ \text{m}}$$
$$= 0.018\ \text{m/m}.$$

Actual head loss: Along each lateral, the volumetric flow rate decreases because each nozzle removes water. The actual head loss in the lateral depends on the volumetric flow rate and pipe diameter and must be compared with the allowable head loss. To that end, it is assumed for simplicity that sprinklers are spaced at equal distance s_l, discharge is the same for each nozzle, and the total flow into the lateral is discharged through the lateral.

It is assumed that the volumetric flow at the inflow rate of the lateral is constant. The actual head loss along the lateral can be computed using the head loss in an equivalent through-flow pipe having the same diameter and friction coefficient as

$$H_{L-ac} = F \times H_{L-p}, \tag{21.54}$$

where H_{L-ac} is the actual head loss due to friction (m/m), H_{L-p} is the equivalent head loss due to friction in the through-flow pipe (m/m), and F is the friction factor accounting for the decrease in flow along the lateral (fraction). The actual head loss is a function of pipe material, volumetric flow rate, and pipe diameter and length. The actual head loss should be less than the allowable head loss. To ensure this condition, pipe diameter can be adjusted and, depending on circumstances, sometimes pipe length can be adjusted, which also implies that the volumetric flow rate can be adjusted.

The equivalent head loss can be computed using the Hazen–Williams equation or another of the equations described earlier. The head loss for through-flow pipe can be expressed for per unit length as:

$$H_{L-p} = \frac{h_f}{L_t}, \tag{21.55}$$

where h_f is the head loss due to friction given by the Hazen–Williams equation or any other formula. The friction factor is given in Table 9.2.

Example 21.13: Consider Example 21.12 in which a sprinkler lateral is laid on ground with a slope of 0.001 m/m and the design nozzle operating pressure is 300 kPa. The distance between the first and last nozzles is 350 m. Assume the sprinkler spacing s_1 as 10 m and the first sprinkler is located at a distance of s_1 from the mainline. The design nozzle discharge is 0.30 L/s. The minimum inside pipe diameter is 50.8 mm. Determine the pipe diameter needed to ensure that the actual head loss is within the allowable limit.

Solution: The number of sprinklers on the lateral is: $N = 350\ \mathrm{m}/10\ \mathrm{m} = 35$.

The volumetric flow rate $Q = N \times q = 35 \times 0.30\ \mathrm{L/s} = 10.5\ \mathrm{L/s}$.

Take the pipe diameter of 50.8 mm (2 in.) as an example:

Given $Q = 10.5$ L/s, $D = 50.8$ mm, and an assumed value of friction coefficient $C = 135$, the equivalent head H_{L-p} can be computed per unit length using the Hazen–Williams equation (eq. 9.16) with conversion constants in Table 9.1 as

$$\begin{aligned} H_{L-p} &= 1.21 \times 10^{10} \times \left(\frac{Q}{C}\right)^{1.852} \times \frac{1}{D^{4.87}} \\ &= 1.21 \times 10^{10} \times \left(\frac{10.5}{135}\right)^{1.852} \times \frac{1}{(50.8)^{4.87}} \\ &= 0.526\ \mathrm{m/m}. \end{aligned}$$

The Christiansen factor F when the first sprinkler is at a distance s_1 from the mainline can be computed as

$$\begin{aligned} F &= \frac{1}{m+1} + \frac{1}{2N} + \frac{(m-1)^{0.5}}{6N^2} \\ &= \frac{1}{1.852+1} + \frac{1}{2 \times 35} + \frac{(1.852-1)^{0.5}}{6 \times 35^2} = 0.365. \end{aligned}$$

The actual head loss is computed using eq. (21.54) as:

$$H_{L-ac} = F \times H_{L-p} = 0.365 \times 0.526 = 0.192\ \mathrm{m/m}.$$

H_{L-p} can be computed for different values of pipe diameter. Assume an inside pipe diameter of 2.0–5.0 in. (50.8–127 mm) with an interval of 0.5 in. Then, the actual head loss is computed using eq. (21.54) for each value of equivalent through-flow pipe head loss. These values are:

Diameter (inch)	Diameter (mm)	H_{L-p} (m/m)	H_{L-ac} (m/m)
2	50.8	0.526	0.192
2.5	63.5	0.177	0.065
3	76.2	0.073	0.027
3.5	88.9	0.034	0.012
4	101.6	0.018	0.007
4.5	114.3	0.010	0.004
5	127	0.006	0.002

From Example 21.12, the maximum allowable head loss is 0.018 m/m. Allowing for the pressure variation of less than 20%, the minimum required pipe diameter is 3.5 in. or 88.9 mm. One should also check the maximum velocity criterion (v_{max}, less than or equal to 5 ft/s [1.5 m/s]) to examine the flow uniformity, especially at the entrance to the lateral and sprinkler riser.

When $D = 88.9$ mm, the velocity at the entrance can be computed as

$$\begin{aligned} v &= \frac{Q}{A} = \frac{10.5\ \mathrm{L/s}}{\pi \times (0.0889\ \mathrm{m})^2/4} \times 10^{-3}\frac{\mathrm{m}^3}{\mathrm{L}} \\ &= 1.69\frac{\mathrm{m}}{\mathrm{s}} > 1.5\ \mathrm{m/s}. \end{aligned}$$

The velocity criterion cannot be satisfied.

When $D = 101.6$ mm, the velocity at the entrance can be computed as

$$\begin{aligned} v &= \frac{Q}{A} = \frac{10.5\ \mathrm{L/s}}{\pi \times (0.1016\ \mathrm{m})^2/4} \times 10^{-3}\frac{\mathrm{m}^3}{\mathrm{L}} \\ &= 1.30\frac{\mathrm{m}}{\mathrm{s}} < 1.5\ \mathrm{m/s}. \end{aligned}$$

The velocity criterion is satisfied. The minimum required pipe diameter is 4 in. or 101.6 mm.

21.7.8 Mainline Design (Step 8)

Pressure at mainline entrance to laterals: The pressure in the mainline must be adequate so that the pressure at the entrance to the lateral and the nozzle operating pressure are as per design, accounting for elevation changes and the height of the connecting riser between the lateral and the sprinkler nozzle. The pressure required at the mainline entrance (H_m, kPa) to the lateral can be computed as

$$H_m = H_a + 9.807\frac{\text{kPa}}{\text{m}} \times \left[0.75\left(h_f + H_e\right) + H_r\right], \qquad (21.56)$$

where H_a is the design nozzle operating pressure (kPa), h_f is the total head loss due to friction in the lateral (m), H_e is the increase in elevation of the lateral from the inlet to the location of the critical sprinkler (minimum operating pressure) (m), H_r is the height of the sprinkler riser (m), and 0.75 is the factor necessary to produce the average operating pressure near the midpoint of the lateral. The minimum sprinkler riser height as a function of riser diameter is given in Table 21.10.

Table 21.10 *Minimum sprinkler riser height as a function of riser diameter*

Riser diameter		Minimum height	
cm	in.	cm	in.
1.27	0.50	7.60	3.00
1.90	0.75	15.00	6.00
2.54	1.00	30.00	12.00
7.60	3.00 (big gun)	90.00	36.00

Adapted from Cuenca, 1989. Copyright © 1989 by Prentice Hall, Englewood Cliffs, New Jersey.

Example 21.14: Consider Examples 21.12 and 21.13, in which a sprinkler lateral is laid on ground with a slope of 0.001 m/m and the design nozzle operating pressure is 300 kPa. The distance between the first and last nozzles is 350 m. Assume the sprinkler spacing s_1 as 10 m, and the first sprinkler is located at a distance of s_1 from the mainline. The design nozzle discharge is 0.30 L/s. It is assumed that the riser height is 1 m. Determine the entrance pressure at the main line.

Solution: From Example 21.13, the lateral pipe diameter needed to ensure that the actual head loss is within the allowable limit is 101.6 mm and the value of H_{L-ac} is 0.007 m/m $(= F \times H_{L-P} = 0.37 \times 0.018)$. The value of h_f is $H_{L-ac} \times L = 0.007$ m/m $\times$ 350 m $=$ 2.45 m.

The increase in the elevation head is –0.35 m. Using eq. (21.56),

$$H_m = 300\text{ kPa} + 9.807\frac{\text{kPa}}{\text{m}} \times [0.75(2.45 - 0.35) + 1\text{ m}]$$
$$= 325.3\text{ kPa}.$$

Critical pressure requirement on mainline: The pressure at any point (denoted as 1) on the mainline is the sum of pressure at the next point (denoted as 2) on the mainline, head loss due to friction between these two points, increase in elevation between these two points, and increase in velocity head between these two points, expressed as

$$H_1 = H_2 + h_{f_{1-2}} + H_{e_{1-2}} + H_{v_{1-2}}, \qquad (21.57)$$

where H_1 is the pressure head required at point 1 (m), H_2 is the pressure head required at point 2 (m), $H_{f_{1-2}}$ is the head loss due to friction between points 1 and 2 (m), $H_{e_{1-2}}$ is the increase in elevation head between points 1 and 2, and $H_{v_{1-2}}$ is the increase in the velocity head between points 1 and 2.

In general, the velocity head is small and can be neglected. Because the calculation of the pressure head required at any point involves the calculation of pressure head at the next point, one starts with pressure calculations at the end of the line and works backward. The critical point on the mainline is defined as the point that requires the highest pressure, considering all pressure losses beginning from the pump.

The pressure head required at the pump is the sum of the pressure head required at the critical point in the mainline, total head loss due to friction from the pump to the critical point in the mainline, elevation head from the source of water to the critical point in the mainline, head loss due to friction from the pumping water level to the centerline of the pump, and velocity head at the critical point. The total pressure head required is the total dynamic head, which must be calculated at each point in the mainline; where the head requirement is the highest is the critical point. The total dynamic head (TDH_i) at any point i can be expressed as

$$TDH_i = H_i + h_{f-pi} + H_{e-si} + h_{f-s} + \frac{v_i^2}{2g}, \qquad (21.58)$$

where H_i is the pressure head at point i (m), h_{f-pi} is the head loss due to friction from the pump to point i (m), H_{e-si} is the increase in the elevation head from the level of water source to point i (m), and h_{f-s} is the head loss due to friction on the suction side of the pump (m).

Example 21.15: Consider a sprinkler system that has four laterals that are operated simultaneously. Each lateral has 25 nozzles, each discharging at the rate of 0.40 L/s. The pressure required on the mainline at the inflow into each lateral is 400 kPa. The mainline is an aluminum pipe with an inside diameter of 200 mm, and its Hazen–Williams friction coefficient can be assumed to be 140. The head loss due to friction from the water surface to the centerline of the pump is 0.35 m. Calculate the critical pressure point in the mainline and the total dynamic head requirement of the pump. The elevation of each point on the pipe line and the distance between points are given in Table 21.11.

Table 21.11 *Elevation and distance between points*

Point	Elevation (m)	Distance between points (m)
Pump	20.50	
		35.00
1	19.80	20.00
2	20.75	20.00
3	20.20	20.00
4	20.10	

Solution: Calculations start at the end of the mainline and proceed toward the pump. The discharge required at each lateral is $Q = 25 \times 0.40\ \text{L/s} = 10.00\ \text{L/s} = 0.1\ \text{m}^3/\text{s}$, then $Q_{3-4} = 10.00\ \text{L/s}$, $Q_{2-3} = 20.00\ \text{L/s}$, $Q_{1-2} = 30.00\ \text{L/s}$, and $Q_{0-1} = 40.00\ \text{L/s}$.

The velocity head at each point is

$$H_{v-4} = \frac{\left(\frac{Q}{A}\right)^2}{2g} = \frac{\left(\frac{0.01\ \frac{\text{m}^3}{\text{s}}}{\frac{1}{4}\pi(0.2\ \text{m})^2}\right)^2}{2 \times 9.81\frac{\text{m}}{\text{s}^2}} = 0.005\ \text{m},$$

$$H_{v-3} = \frac{\left(\frac{Q}{A}\right)^2}{2g} = \frac{\left(\frac{0.02\ \frac{\text{m}^3}{\text{s}}}{\frac{1}{4}\pi(0.2\ \text{m})^2}\right)^2}{2 \times 9.81\frac{\text{m}}{\text{s}^2}} = 0.02\ \text{m},$$

$$H_{v-2} = \frac{\left(\frac{Q}{A}\right)^2}{2g} = \frac{\left(\frac{0.03\ \frac{\text{m}^3}{\text{s}}}{\frac{1}{4}\pi(0.2\ \text{m})^2}\right)^2}{2 \times 9.81\frac{\text{m}}{\text{s}^2}} = 0.045\ \text{m},$$

$$H_{v-1} = \frac{\left(\frac{Q}{A}\right)^2}{2g} = \frac{\left(\frac{0.04\ \frac{\text{m}^3}{\text{s}}}{\frac{1}{4}\pi(0.2\ \text{m})^2}\right)^2}{2 \times 9.81\frac{\text{m}}{\text{s}^2}} = 0.08\ \text{m}.$$

1. The total head required at point 4 is the sum of H_{m-4} and the velocity head H_{v-4}.

The pressure required at point 4, which is the last point, is the pressure at the entrance of the lateral (H_m), which can be computed as

$$H_{m-4} = \frac{400{,}000\ \text{Pa}}{1000\frac{\text{kg}}{\text{m}^3} \times 9.807\frac{\text{m}}{\text{s}^2}} = 40.78\ \text{m}.$$

The total head required at point 4 is

$$H_4 = H_{m-4} + H_{v-4} = 40.78 + 0.005 = 40.785\ \text{m}.$$

2. Calculation of the total head required at point 3:
The head loss due to friction during the section of points 3–4 is

$$\begin{aligned} H_{f_{3-4}} &= 1.21 \times 10^{10} \times \left(\frac{Q}{C}\right)^{1.852} \times \frac{L}{D^{4.87}} \\ &= 1.21 \times 10^{10} \times \left(\frac{10}{140}\right)^{1.852} \times \frac{20}{(200)^{4.87}} \\ &= 0.011\ \text{m}. \end{aligned}$$

The increase in elevation head during the section of points 3–4 is

$$H_{e_{3-4}} = H_{e_3} - H_{e_4} = 20.20 - 20.10 = 0.1\ \text{m}.$$

The velocity head increase between points 3 and 4 is

$$H_{v_{3-4}} = H_{v-3} - H_{v-4} = 0.02\ \text{m} - 0.005\ \text{m} = 0.015\ \text{m}.$$

The total head at point 3 is

$$\begin{aligned} H_3 &= H_4 + H_{f_{3-4}} + H_{e_{3-4}} + H_{v_{3-4}} \\ &= 40.785 + 0.011 + 0.1 + 0.015 = 40.911\ \text{m}. \end{aligned}$$

3. Calculation of the total head required at point 2:
The head loss due to friction during the section of points 2–3 is

$$\begin{aligned} H_{f_{2-3}} &= 1.21 \times 10^{10} \times \left(\frac{Q}{C}\right)^{1.852} \times \frac{L}{D^{4.87}} \\ &= 1.21 \times 10^{10} \times \left(\frac{20}{140}\right)^{1.852} \times \frac{20}{(200)^{4.87}} \\ &= 0.041\ \text{m}. \end{aligned}$$

The increase in elevation head during the section of points 2–3 is

$$H_{e_{2-3}} = H_{e_2} - H_{e_3} = 20.75 - 20.20 = 0.55\ \text{m}.$$

The velocity head increase between points 2 and 3 is

$$H_{v_{2-3}} = H_{v-2} - H_{v-3} = 0.045 - 0.02 = 0.025\ \text{m}.$$

The total head at point 2 is

$$\begin{aligned} H_2 &= H_3 + H_{f_{2-3}} + H_{e_{2-3}} + H_{v_{2-3}} \\ &= 40.911 + 0.041 + 0.55 + 0.025 = 41.527\ \text{m}. \end{aligned}$$

4. Calculation of the total head required at point 1:
The head loss due to friction during the section of points 1–2 is

$$\begin{aligned} H_{f_{1-2}} &= 1.21 \times 10^{10} \times \left(\frac{Q}{C}\right)^{1.852} \times \frac{L}{D^{4.87}} \\ &= 1.21 \times 10^{10} \times \left(\frac{30}{140}\right)^{1.852} \times \frac{20}{(200)^{4.87}} \\ &= 0.087\ \text{m}. \end{aligned}$$

The increase in elevation head during the section of points 1–2 is

$$H_{e_{1-2}} = H_{e_1} - H_{e_2} = 19.80 - 20.75 = -0.95 \text{ m}.$$

The velocity head increase between points 1 and 2 is

$$H_{v_{1-2}} = H_{v-1} - H_{v-2} = 0.08 - 0.045 = 0.035 \text{ m}.$$

The total head at point 2 is

$$H_1 = H_2 + H_{f_{1-2}} + H_{e_{1-2}} + H_{v_{1-2}}$$
$$= 41.527 + 0.087 - 0.95 + 0.035 = 40.699 \text{ m}.$$

5. Calculation of the total head required at the pump: The head loss due to friction between point 1 and the pump is

$$H_{f_{0-1}} = 1.21 \times 10^{10} \times \left(\frac{Q}{C}\right)^{1.852} \times \frac{L}{D^{4.87}}$$
$$= 1.21 \times 10^{10} \times \left(\frac{40}{140}\right)^{1.852} \times \frac{35}{(200)^{4.87}}$$
$$= 0.259 \text{ m}.$$

The head loss due to friction between point 2 and the pump is:

$$H_{f_{0-2}} = H_{f_{1-2}} + H_{f_{0-1}} = 0.087 + 0.259 = 0.346 \text{ m}.$$

The increase in elevation head between point 2 and the pump is:

$$H_{e_{0-2}} = H_{e_0} - H_{e_2} = 20.50 - 20.75 = -0.25 \text{ m}.$$

The total head at the pump is:

$$H = H_2 + H_{f_{0-2}} + H_{e_{0-2}} + H_{f-s} + H_{v_2}$$
$$= 41.527 + 0.346 - 0.25 + 0.35 + 0.045 = 42.02 \text{ m}.$$

The total dynamic head requirement of the pump is 42.02, and the critical pressure point in the mainline is point 2.

21.7.9 Selection of Pump and Power (Step 9)

Given the total dynamic head is H, eq. (21.13) can also be expressed as

$$HP = \frac{QH}{6116E_p}, \tag{21.59}$$

where HP is the power needed for pumping water for sprinkler systems (kW), Q is the discharge (L/min), E_p is the pump efficiency (fraction), and H is the dynamic head of the irrigation system (m).

Since an electric motor or diesel engine is needed to operate a pump, the electrical energy required for the motor-driven pump can be determined by accounting for the motor efficiency for a constant-speed pump and wire-to-shaft efficiency of the motor and variable-speed drive for a variable-speed pump, expressed as

$$P_{kW} = \frac{HP}{E_m}, \tag{21.60}$$

where P_{kW} is the power required to operate the pump (kW), HP is the power need for pumping water (kW), and E_m is the motor efficiency for a constant-speed pump (fraction).

Example 21.16: Considering a sprinkler system has a discharge of 0.015 m^3/s, with a total dynamic head of 11.01 m, compute the pump water horsepower and brake horsepower if pump efficiency is 85%. Compute the power required to operate the pump driven by an electric motor having an efficiency of 90%. If the electricity charge is \$0.25/kWh, what would be the cost of pumping 15,000 m^3 of water?

Solution: Given $Q = 0.015 \text{ m}^3/\text{s} = 15 \text{ L/s} = 15 \times 60 \text{ L/min} = 900 \text{ L/min}$, $H = 11.01$ m, $E_p = 85\% = 0.85$, $E_m = 90\% = 0.90$, the power needed for pumping water for the sprinkler system in kW can be expressed as

$$HP = \frac{QH}{6116E_p} = \frac{(900 \text{ L/min})(11.01 \text{ m})}{6116\ (0.85)} = 1.9 \text{ kW}.$$

The power required to operate the pump in kW is

$$P_{kW} = \frac{HP}{E_m} = \frac{1.9 \text{ kW}}{0.90} = 2.11 \text{ kW}.$$

The time required to pump 15,000 m^3 water is

$$\text{Duration(h)} = \frac{15{,}000 \text{ m}^3}{0.015 \frac{\text{m}^3}{\text{s}} \times 3600 \frac{\text{s}}{\text{h}}} = 277.78 \text{ h}.$$

The cost of pumping 15,000 m^3 of water is

$$\text{Power cost(\$)} = P_{kW} \times \text{Duration} \times \text{per unit electricity charge}$$
$$= 2.11 \text{ kW} \times 277.78 \text{ h} \times 0.25 \frac{\$}{\text{kWh}}$$
$$= \$147.$$

21.8 CENTER-PIVOT SYSTEMS

Center-pivot systems require low labor and maintenance, perform well, and are easy to operate, flexible, and convenient. They also conserve water, energy, time, and money. Their energy and water requirements are reduced through low-energy precision application (LEPA) and low-elevation spray application (LESA).

21.8.1 Types

In CP systems the source of water is located at the center of the irrigated field and the laterals pivot around the CP, with the result that nozzles close to the pivot move at a much lower speed than those toward the end of the lateral, causing uneven distribution of water. Hence, to produce an equal depth of applied water along the arm, two types of systems have been designed: constant and variable spacing, as shown in Figure 21.24.

In the constant-spacing case, nozzles are placed at equal distance along the lateral, but a differential application rate is achieved along the lateral by combining the pressure distribution in the lateral and selecting nozzles of different diameters. Large-diameter nozzles will require greater pressure.

In the variable-spacing case, nozzles are of diameters with approximately the same discharge, but the distances between nozzles are varied in order to attain differential application rates. The sprinkler size, coverage, and operating pressure are less than in the constant-spacing case.

Center-pivot systems sometimes require pressure regulators on sprinklers in order to maintain proper pressure distribution along the lateral. A booster pump may be needed at the end of the pivot arm to develop sufficient pressure to operate a large-diameter end gun. Towers, which are vertical supports on wheels to elevate the pivot arm, must be driven in unison to ensure the pivot arm rotates around the pivot point at the design speed. Center-pivot systems are cost-effective where 50–67 ha (125–165 acres) can be irrigated by a single pivot.

There are several types of spray applicators, including impact sprinklers, low-pressure applicators, mid-elevation spray application, low-elevation spray application, and LEPA, as shown in Figure 21.25. High-pressure impact

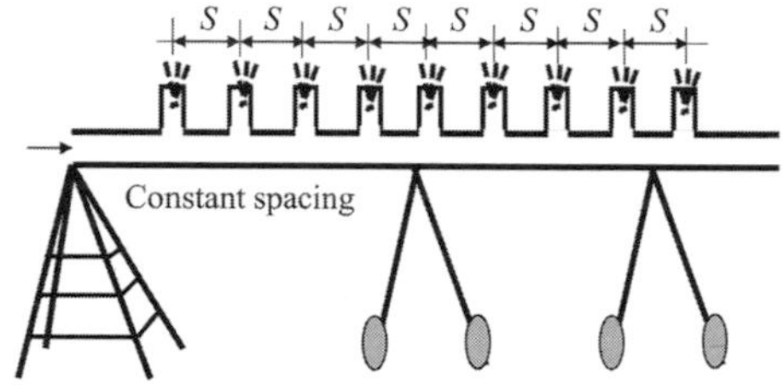

Figure 21.24 Center-pivot systems with constant and variable spacing of nozzles.

Figure 21.25 Mid-elevation spray application, low-elevation spray application, low-energy precision application. A black and white version of this figure will appear in some formats. For the color version, please refer to the plate section.

sprinklers are mounted on the CP mainline and release water at about 15–27°. This type of sprinkler is not used much. Low-pressure applicators require less energy, operate at low water pressure, and have high application efficiency when placed 16–18 inches above the ground. Mid-level spray applicators are placed about midway between the mainline and ground level, and apply water above the crop canopy. If applicators are positioned 60–80 in. apart, the operating pressure can be reduced to 6 psi. Low-elevation spray applicators have water application efficiency of 60–90% and are usually placed 60–80 in. apart, corresponding to two-row crops. Each applicator is connected to a flexible drop hose, which is attached to a gooseneck on the mainline. Low-energy precision applicators are positioned about 12–18 in. above the ground and 60–80 in. apart, about twice the row spacing. They release water through drag socks or hoses in a bubble pattern between alternate crop rows. LEPA has an application efficiency of 95–98%.

21.8.2 Operation Parameters

In the CP systems, the sprinkler arm makes one complete revolution within the irrigation interval, as given by eq. (21.61) for the peak period. This means that the net irrigation requirement is fulfilled during each revolution. Normally the minimum time for one revolution is 24 h, to reduce wear on the equipment. The seasonal operational time (T_{seas}) can then be computed as

$$T_{seas} = \frac{ET_{seas}}{D_n} \times T_r, \tag{21.61}$$

where ET_{seas} is the seasonal crop water requirement (mm), D_n is the net depth of irrigation (mm), and T_r is the time per revolution of the pivot arm (h).

Since the time of application varies along the pivot arm, being minimum at the far end, it can be computed at any radial distance r (m) from the pivot point by dividing the wetted diameter of the sprinkler nozzle at distance r (D_w, m) by the speed of the lateral:

$$T_a = \frac{D_w T_r}{2\pi r}, \tag{21.62}$$

where T_a is the time of application (h). The speed of the lateral is computed by dividing the circumference at distance r from the pivot point by the time per revolution. Equation (21.62) shows that as r increases, T_a decreases. For a selected wetted diameter, r is maximum at the outer edges of the irrigation cycle and T_a is minimum.

The net depth of application (D_a, mm) can be computed as the gross depth (D_g, mm) minus the loss due to wind and drift (L_s, fraction):

$$D_a = D_g(1 - L_s). \tag{21.63}$$

The maximum application rate (d_{a-max}, mm/h) at any point along the pivot arm can be calculated by assuming an

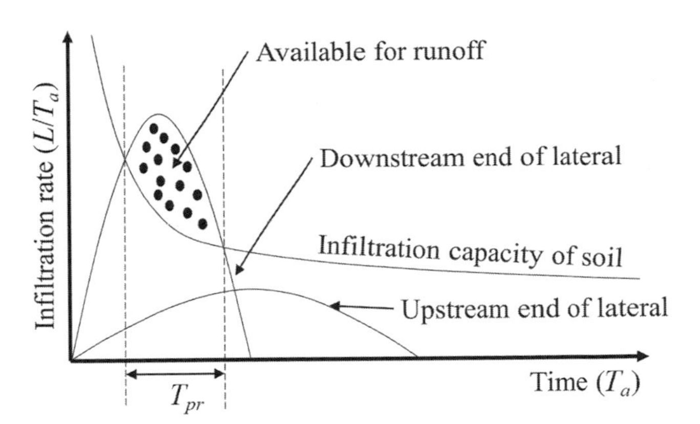

Figure 21.26 Typical relationship between soil infiltration capacity and application rates beneath CP laterals.

elliptical water application rate pattern as (USDA-SCS, 1983):

$$d_{a-max} = \frac{4}{\pi}\frac{D_a}{T_a}. \tag{21.64}$$

The maximum application rate is obtained by equating the elliptical area (e.g., the upstream/downstream end of the lateral as shown in Figure 21.26), to the depth of application. Recall that the average application rate equals $(\pi/4)d_{a-max}$.

Now the application rate is contrasted with soil intake rate. At the end of the pivot arm the application rate is high, which should be constrained by the soil intake rate. The pattern of water application by a CP system can be characterized by half an ellipse. When an infiltration rate curve is superimposed on an elliptical pattern of water application over time T_a, the application rate will exceed the rate of infiltration at a certain time, designated as T_{pr}, when runoff will begin, as shown in Figure 21.26.

Reducing the net depth of application will reduce the time of application, that is, the system will provide light but frequent irrigation. However, the peak application rate would remain the same, because the reduction in d_a will be compensated by the reduction in T_a.

The discharge needed at the pivot can be computed as

$$Q = K\frac{AD_g}{T_r}, \tag{21.65}$$

where Q is the discharge required at the pivot (L/s), A is the total area irrigated by the pivot system (ha), and K is the conversion factor (2.78 for SI units and 453 for British units: Q in gpm, A in acres, and D_g in inches).

The sprinkler discharge required can be calculated as

$$q_r = rs_r\frac{2Q}{(r_{max})^2}, \tag{21.66}$$

where q_r is the sprinkler discharge at distance r from the pivot point (L/s), and s_r is the sprinkler spacing at distance r (m), which is the sum of half the distance to the next upstream nozzle and half the distance to the next downstream nozzle. r_{max} (m) is the maximum radius of the irrigated area.

The sprinkler capacity, assuming 100% application efficiency, can be determined as follows:

$$Q = \frac{D_n A}{t} = \frac{D_n \pi R^2}{t}, \tag{21.67}$$

where Q is the system capacity (L^3/T), D_n is the desired depth of application (L), A is the area to be irrigated (L^2), R is the radius of a circular area in the CP system (L), and t is the total time of operation per revolution (T). The area covered by one revolution of the lateral by a sprinkler is $2\pi r dr$, where r is the distance from the sprinkler to the pivot point. The sprinkler capacity for this area is given as

$$q = D_n(2\pi r)\frac{dr}{t}. \tag{21.68}$$

Integration of this equation from r to R yields the flow rate in the lateral at any point r (Chu and Moe, 1972):

$$q_r = \frac{\pi R^2 D_n}{t}\left(1 - \frac{r^2}{R^2}\right). \tag{21.69}$$

The discharge needed for the end gun at the outer end of the lateral at length L is given by eq. (21.69), with r replaced by L. The discharge of sprinklers along the lateral is given by the difference between eqs (21.67) and (21.69).

Example 21.17: Determine the irrigation interval, seasonal operating time, discharge required at the pivot, and maximum net application rate for a CP sprinkler system with the maximum possible radius of pivot of 350 m. Assume that evaporation and wind drift losses are 10%, the combined distribution and application efficiency is 80%, the net irrigation requirement is 30 mm, the peak period evaporation is 8 mm/day, and the seasonal crop water requirement is 400 mm. Compute the maximum net application rate for constant nozzle spacing with a wetted diameter of 50 m and variable spacing with a wet diameter of 30 m.

Solution: Given the net irrigation requirement $D_n = 30$ mm, the distribution and application efficiency $E_a = 80\% = 0.8$, the gross depth of irrigation is calculated as

$$D_g = D_n/E_a = 30/0.80 = 37.5 \text{ mm}.$$

Given the evaporation and wind drift losses $L_s = 10\% = 0.1$, the net depth of application D_a is

$$D_a = D_g(1 - L_s) = 37.5 \times (1 - 0.10) = 33.75 \text{ mm}.$$

Given the crop peak period evaporation $ET_c = 8$ mm/day, the potential irrigation interval is

$$T_i = D_n/ET_c = 30 \text{ mm}/(8 \text{ mm/day}) = 3.75 \text{ days or } 90 \text{ h} (\sim T_r).$$

Assume the initial time of revolution is equal to the computed irrigation interval. Using eq. (21.61), the seasonal operating time is calculated as

$$T_{seas} = (ET_{seas}/D_n) \times T_r = (400 \text{ mm}/30 \text{ mm}) \times (3.75 \text{ days}) \times (24 \text{ h/day}) = 1200 \text{ h}.$$

Given the maximum possible radius of the pivot, $r_{max} = 350$ m, the total irrigated area is

$$A = \pi (r_{max})^2 = (3.14)(350 \text{ m})(350 \text{ m})(1 \text{ ha}/10{,}000 \text{ m}^2) = 38.5 \text{ ha}.$$

The discharge Q at the pivot is computed using eq. (21.65):

$$Q = 2.78\frac{AD_g}{T_r} = 2.78 \times \frac{(38.5 \text{ ha}) \times (37.5 \text{ mm})}{(90 \text{ h})} = \frac{44.6 \text{ L}}{\text{s}}.$$

Given the wetted diameter of constant spacing $D_w = 50$ m, the time of application at the end of the pivot is computed using eq. (21.62) as:

$$T_a = \frac{D_w T_r}{2\pi r} = \frac{(50 \text{ m})(90 \text{ h})}{2\pi\,(350 \text{ m})} = 2.05 \text{ h}.$$

Then, the maximum application rate (d_{a-max}, mm/h) for the constant-spacing case is computed using eq. (21.64):

$$d_{a-max} = \frac{4}{\pi}\frac{D_a}{T_a} = \frac{4}{\pi}\frac{(33.75 \text{ mm})}{(2.05 \text{ h})} = 21 \text{ mm/h}.$$

Given the wetted diameter of variable spacing $D_w = 30$ m, the time of application at the end of the pivot is computed using eq. (21.62) as:

$$T_a = \frac{D_w T_r}{2\pi r} = \frac{(30 \text{ m})(90 \text{ h})}{2\pi\,(350 \text{ m})} = 1.23 \text{ h}.$$

Then, the maximum application rate (d_{a-max}, mm/h) for the variable-spacing case is computed using eq. (21.64):

$$d_{a-max} = \frac{4}{\pi}\frac{D_a}{T_a} = \frac{4}{\pi}\frac{(33.75 \text{ mm})}{(1.23 \text{ h})} = 35 \text{ mm/h}.$$

21.9 LINEAR-MOVE SPRINKLER SYSTEMS

These systems operate on rectangular fields and irrigate the entire field. The source of water is a pipeline or open channel along one side or down the center of the field. However, the lateral arm should be moved back to the starting position before starting the next irrigation cycle. These systems are known to have the highest uniformity of application.

The sprinkler spacing and nozzle discharge are constant along the lateral. The speed and movement of the linear-move system are such that the gross depth of application is similar to that of a standard sprinkler system. The maximum application rate d_{a-max} (mm/h) can be obtained by revising eq. (21.64) as

$$d_{a-\max} = \frac{4}{\pi} 3600 \frac{Q_a}{D_w L_t}, \tag{21.70}$$

where Q_a is the discharge rate for water for total lateral arm (L/s), D_w is the wetted diameter of the sprinkler nozzle (m), and L_t is the total length of the lateral arm (m). Equation (21.70) can be cast relative to the spacing between nozzles along the lateral as

$$d_{a-max} = \frac{4}{\pi} 3600 \frac{q}{D_w s_l}, \tag{21.71}$$

in which q is the discharge rate for a single nozzle (L/s), and s_l is the lateral spacing.

21.10 BIG GUN AND BOOM SPRINKLER SYSTEMS

Big gun sprinkler systems, as shown in Figure 21.27, have large-diameter nozzles and discharge at high volumetric flow rates in a circular flow pattern, and require operating pressure.

Often the nozzle does not make a full circle and leaves a portion dry in front of the vehicle. It is connected by a flexible hose to the mainline, which is normally positioned in the center of the field. Sprinklers are able to irrigate paths up to 180 m in width.

Stationary boom sprinklers are on a lateral arm at about the same height as a CP arm, which makes a full rotation every 1–5 min. The lateral arm is about 35–75 m long and is able to irrigate a circular area up to 90 m in diameter.

The actual application rate for big gun sprinklers is given by

$$d_a = 360 \frac{4q}{\pi (0.9 D_w)^2} \frac{360}{\Phi}, \tag{21.72}$$

where d_a is the application rate (cm/h), q is the nozzle discharge rate (L/s), and Φ is the portion of the circle wetted by sprinklers (degrees). The wetted area is based on 90% of the wetted diameter.

The value of K in the nozzle discharge equation is given in Table 21.5, and typical wetted diameters for big gun sprinklers with 24° trajectory angle and tapered nozzles operating under no-wind conditions are given in Table 21.12.

Table 21.12 *Typical wetted diameters for big gun sprinklers with 24° trajectory angle and tapered nozzles operating under no-wind conditions*

	Nozzle diameter (mm)				
	20	25	30	36	41
Nozzle pressure (kPa)	Wetted diameter (m)				
414	87	99	111	–	–
483	91	104	116	133	–
552	94	108	120	139	146
621	98	111	125	143	151
689	101	114	128	146	155
758	104	114	131	149	158
827	107	120	134	152	163
	Nozzle diameter (in.)				
Nozzle pressure (psi)	0.8	1.0	1.2	1.4	1.6
	Wetted diameter (ft)				
60	285	325	365	–	–
70	300	340	380	435	–
80	310	355	395	456	480
90	320	365	410	470	495
100	330	375	420	480	510
110	340	385	430	490	520
120	350	395	440	500	535

Adapted from USDA-NRCS, 2016.

Figure 21.27 Big gun sprinkler system. (The left figure is adapted from USDA-NRCS, 2016.) A black and white version of this figure will appear in some formats. For the color version, please refer to the plate section.

Example 21.18: Consider a big gun sprinkler with a 25° trajectory and a 25-mm diameter nozzle. The operating pressure at the nozzle is 630 kPa. It is assumed that loss by wind drift and evaporation is 8%. Determine the application rate for the big gun sprinkler. Compute the application if the tapered nozzle is replaced by an orifice-ring nozzle.

Solution: The value of K from Table 21.11 for a 25-mm diameter nozzle is 0.6991. The discharge q is given by eq. (21.34) as $q = 0.6991 \times (630\ \text{kPa})^{0.5} = 17.55\ \text{L/s}$.

Given that loss by wind drift and evaporation $L_s = 8\%$, the actual discharge rate q_a is

$$q_a = q(1 - L_s) = 17.55 \times (1 - 0.08) = 16.14\ \text{L/s}.$$

The wetted perimeter for the tapered nozzle of 25 mm diameter from Table 21.12 by linear interpolation at a pressure of 630 kPa is given as:

$$D_w = 111 + [(630 - 621)/(689 - 621)] \times (114 - 111) = 111.4\ \text{m}.$$

Degree of circle irrigated: $\Phi = 360° - 25° = 335°$.

Application rate for tapered nozzle from eq. (21.70) is given as

$$d_a = 360 \frac{4q_a}{\pi(0.9D_w)^2} \frac{360}{\Phi} = 360 \frac{4 \times 16.14}{\pi(0.9 \times 111.4)^2} \frac{360}{335} = 0.79 \frac{\text{cm}}{\text{h}}.$$

For a ring-orifice nozzle, D_w is reduced by 5%, so $D_w = 0.95 \times 111.4\ \text{m} = 105.83\ \text{m}$

Therefore,

$$d_a = 360 \frac{4q_a}{\pi(0.9D_w)^2} \frac{360}{\Phi} = 360 \frac{4 \times 16.14}{\pi(0.9 \times 105.83)^2} \frac{360}{335} = 0.88\ \text{cm/h}.$$

The depth of application depends on the traveler speed, nozzle discharge, and towpath spacing:

$$D_g = \frac{6.0\,q}{s_t v_t}, \tag{21.73}$$

where D_g is the gross depth of application (cm), q is the nozzle discharge (L/s), s_t is the towpath spacing (m), and v_t is the traveler speed (m/min). The towpath spacings for sprinklers with orifice-ring nozzle and tapered nozzle under different wind conditions are listed in Table 21.13.

Table 21.13 *Towpath spacing for traveling sprinklers with orifice-ring nozzle (lower value) and tapered nozzle (higher value) under different wind conditions (ft)*

	Percentage of wetted diameter						
	50	55	60	65	70	75	80
Sprinkler wetted diameter	Wind over 10 mph		Wind up to 10 mph		Wind up to 5 mph		No wind
200	100	110	120	130	140	150	160
250	125	137	150	162	175	187	200
300	150	165	180	195	210	225	240
350	175	192	210	227	245	262	280
400	200	220	240	260	280	300	320
450	226	248	270	292	315	338	360
500	280	275	300	325	350	375	400
550	275	302	330	358	385	412	440
600	300	330	360	390	420	–	–

After USDA-SCS, 1983.

Example 21.19: Consider a big gun system where the traveler speed is 0.25 m/min, the average wind speed is 5 km/h, the nozzle operating pressure is 621 kPa, and the orifice-ring nozzle diameter is 30 mm. Determine the gross irrigation depth.

Solution: The value of K from Table 21.5 for a 30-mm diameter nozzle is 1.021. The discharge q is given by eq. (21.34) as $q = 1.021 \times (621\ \text{kPa})^{0.5} = 25.4\ \text{L/s}$.

The wetted perimeter for the tapered nozzle of 30 mm diameter and at a pressure of 621 kPa from Table 21.12 is $D_w = 125$ m (410 ft).

For a ring-orifice nozzle, D_w is reduced by 5%, so $D_w = 0.95 \times 410\ \text{ft} = 389.5$ ft.

Given wind speed is 5 km/h, and the wetted diameter is 389.5 ft, the towspacing (S_t) is 70% of the wetted diameter: $0.7 \times 0.95 \times 125\ \text{m} = 83.125$ m.

Then, the gross depth of irrigation is:

$$D_g = \frac{6.0\,q}{S_t v_t} = \frac{6.0 \times 25.4 \frac{\text{L}}{\text{s}}}{83.125\ \text{m} \times 0.25 \frac{\text{m}}{\text{min}}} = 7.33\ \text{cm}$$

QUESTIONS

Q.21.1 Compute the required design uniformity to operate a sprinkler system which is designed considering an evaporation loss of 10%, application efficiency of 70%, and adequacy equal to 80%.

Q.21.2 Compute the design operating pressure for the sprinkler nozzle if the pressure needed at the mainline is 450 kPa, the lateral is on a ground slope of 0.002 m/m, the lateral length is 280 m, the friction loss in the lateral is 0.008 m/m, and a rise of 1.5 m is required for the crop.

Q.21.3 Compute the number of days of operation of the sprinkler system for a farm during the irrigation interval if available water-holding capacity is 100 mm/m, the root zone depth is 2 m at mid-season, depletion of available water during irrigation is 25%, and the mid-season peak water consumption is 10 mm/day. During each irrigation interval, the sprinkler system is off for a day.

Q.21.4 Consider a solid-state sprinkler system that operates during the day only when the wind speed is 12 mph (19.31 km/h), the minimum relative humidity is 50%, and the maximum temperature is 80 °F (26.7 °C). The nozzle has a diameter of 0.2 in. (50.8 mm), operating pressure of 50 psi (345 kPa), and wetted diameter of 80 ft (24.4 m). The design soil intake rate is 0.25 in./h (6.35 mm/h). Compute the spacing between sprinklers along the lateral, spacing between laterals, and nozzle discharge.

Q.21.5 Compute the required sprinkler system capacity to apply water at a rate of 15 mm/h. The system has two sprinkler laterals each with 20 sprinklers. The spacing between sprinklers is 10 m and the spacing between laterals is 20 m. Compute the number of hours required to apply 60 mm of water to a square 20 ha farm, if one hour is allowed for moving each lateral. Compute the number of 12-h days to apply 60 mm of water.

Q.21.6 Compute the irrigation frequency in days for a farm if the available moisture is 10 cm/m, the management allowed depletion is 50%, the root zone depth at the peak period is 2 m, and the peak period evapotranspiration is 10 mm/day.

Q.21.7 Compute the required entrance pressure to the lateral at the mainline if the length of the lateral is 180 m, the nozzle operating pressure is 450 kPa, the friction head loss in the lateral is 0.025 m/m, the ground slope from the mainline is –0.005 m/m, and the height of the sprinkler above the ground is 1 m.

Q.21.8 Design a sprinkler system to irrigate a square 20 ha farm within a 10-day period. A well 25 m deep is located at the center of the field that has the following discharge–drawdown characteristics: 15 L/s – 10 m; 18 L/s – 12 m, and 20 L/s – 15 m. The maximum elevation in the farm is 1.5 m above the well site. The depth of water application must be 8 mm/day. The maximum operating time is 22 h/day. The design average pressure at the nozzle is 280 kPa and 1.2 m risers are needed at the sprinklers. Assume a pump efficiency of 70% and the engine will provide 80% of its rated output for continuous operation. Determine the rated output for the combustion engine.

Q.21.9 Compute the discharge rate of one sprinkler operating at 280 kPa pressure and having two nozzles of 4.0 mm and 3.0 mm diameters.

Q.21.10 A sprinkler system has a uniformity coefficient of 80%, assuming a normally distributed application pattern, and deep percolation of 25%. Determine the percentage of the total that is adequately irrigated, the percentage of the total area to be adequately irrigated if the percolation was reduced to 15%, and the combined application and distribution efficiency if the evaporation and spray losses were 10%.

Q.21.11 Consider a sprinkler system that is laid out with 25 × 15 m rectangular spacing with a nozzle having a discharge of 0.60 L/s, and evaporation and wind losses are 8%. The farm to be irrigated has an average slope of 5%. Discuss the type of soil such a system will be suitable for. Why?

Q.21.12 A sprinkler system is laid out on level ground with an average operating pressure of 280 kPa at the nozzle, a friction loss of 5 m in the mainline and 3.5 m in the lateral, a drawdown of 5 m in the well at the required discharge of 30 L/s, a rise height of 1.0 m, and friction loss of 2 m in all valves. Compute the total pumping head for the sprinkler system and the power requirement for the pump with an operating efficiency of 70%.

Q.21.13 Determine the depth rate of application of a sprinkler head discharging at 0.9 L/s where sprinkler spacing is 15 × 20 m.

Q.21.14 Compute the number of laterals required to irrigate a farm of 20 ha that has 20 cm of available water, the management allowed depletion is 30%, the crop ET during the peak period is 10 mm/day, and the hand-move system has 1200 m of mainline with valves at both ends and every 15 m. Two sets are operated each day and during the peak period the system is not operated for one day for each irrigation interval.

Q.21.15 Compute the rate of water application at the end of the lateral if the gross depth of application is 10 mm and spray and drift losses are 10%, and the wetted diameter of the sprinkler at the end of the lateral is 30 m. The CP system with variable spacing has a lateral length of 420 m and makes one revolution in 24 h.

Q.21.16 A sprinkler system is designed to irrigate a 20-ha farm at a maximum rate of 12 mm/h. Consider a root zone depth of 1 m, the available soil moisture capacity of the soil is 15 cm/m, the water application efficiency is 80%, the peak rate of moisture use is 10 mm/day, and irrigation is done when the available moisture capacity is depleted by 50%. Compute the net depth of irrigation per irrigation, depth of water to be pumped, number of days to cover the field, and the area to be irrigated per day.

Q.21.17 Compute the length of the lateral of a sprinkler system if the slope of the land from the inlet along the lateral is 0.01 m/m, the sprinkler spacing along the lateral is 10 m, with the first sprinkler being 5 m from the mainline, the operating pressure of nozzles is 212 kPa, the maximum allowable difference in pressure between two sprinklers along the lateral is 15%, and the equivalent head loss given by the Hazen–Williams equation as 0.05 m/m for an equal-diameter pipe of the same length as the lateral with all flow passing through the pipe.

Q.21.18 Compute the operating pressure of the nozzles if a sprinkler system is operated at the design limit, the ground slope from the inlet along the lateral is 0.001 m/m, the lateral has 20 sprinkler nozzles, the equivalent head loss computed by the Hazen–Williams equation is 0.05 m/m, sprinkler spacing along the lateral is 10 m, and the first sprinkler from the mainline is 5 m. The maximum allowable pressure difference between two sprinklers along the lateral is 15%.

Q.21.19 Consider a big gun system where the traveler speed is 0.20 m/min, the average wind speed is 10 km/h, the nozzle operating pressure is 552 kPa, and the tapered nozzle diameter is 25 mm. Determine the gross irrigation depth.

Q.21.20 Consider a sprinkler lateral laid on ground with a slope of 0.002 m/m and with a design nozzle operating pressure of 460 kPa. The lateral length is 345 m. What is the maximum allowable head loss due to friction (m/m)? Assume the sprinkler spacing as 10 m and the first sprinkler is located at a distance of 5 m from the mainline. The design nozzle discharge is 0.4 L/s. The minimum inside pipe diameter is 50.8 mm. Determine the pipe diameter needed to ensure that the actual head loss is within the allowable limit. The friction coefficient $C = 135$.

Q.21.21 Consider a sprinkler system that has four laterals that are operated simultaneously. Each lateral has 30 nozzles, each discharging at the rate of 0.35 L/s. The pressure required on the mainline at the inflow into each lateral is 400 kPa. The mainline is an aluminum pipe with an inside diameter of 200 mm and its Hazen–Williams friction coefficient can be assumed to be 140. The head loss due to friction from the water surface to the centerline of the pump is 0.45 m. Calculate the critical pressure point in the mainline and the total dynamic head requirement of the pump. The elevation of each point on the pipeline and the distance between points are given in Table 21.14.

Table 21.14 *Elevation and distance between points*

Point	Elevation (m)	Distance between points (m)
Pump	15.50	
		35.00
1	15.30	
		25.00
2	15.50	
		25.00
3	15.40	
		25.00
4	15.30	

Q.21.22 Determine the irrigation interval, discharge required at the pivot, and maximum net application rate for a CP sprinkler system with the maximum possible radius of pivot as 300 m. Assume that evaporation and wind drift losses are 10%, the combined distribution and application efficiency is 80%, the net irrigation requirement is 25 mm, and the peak period evaporation is 8.5 mm/day. Compute the maximum net application rate for constant nozzle spacing with a wetted diameter of 30 m.

Q.21.23 Consider an agricultural farm that has an area of 10 ha. The soil is clay loam with a total depth of 1.5 m. The clay loam soil has a field capacity of 0.40 m/m and a wilting point of 0.25 m/m. Soybean is growing on the farm and is to be irrigated by a set-move sprinkler system. The active rooting depth of soybean is 0.8 m. Irrigation is applied when the total available moisture is depleted by 50%. The crop evapotranspiration rate of soybean during the peak period is 6 mm/day. (1) Compute the desired depth of application (mm). (2) Compute the desired value of application in terms of volume (m^3). (3) The set-move sprinkler system has four laterals, and each is 100 m long; they are spaced 25 m apart. The sprinklers along the laterals are spaced 10 m apart. The downtime for moving the set-move system is 1 h. The application efficiency of the sprinkler system is 80%. Compute the percentage of total field irrigated when the sprinkler system is operating. (4) Compute the discharge of an individual sprinkler – that is, the sprinkler capacity. (5) If the nozzle diameter of the sprinkler is 4 mm, what is the operating pressure?

Q.21.24 In a sprinkler system, the slope of the land from the inlet along the lateral is 0.001 m/m (downhill slope), the sprinkler spacing along the lateral is 10 m, with the first sprinkler being 5 m from the mainline, the operating pressure of the nozzles is 212 kPa, the maximum allowable difference in pressure between two sprinklers along the lateral is 15%, and the equivalent head loss given by the Hazen–Williams equation is 0.05 m/m for an equal-diameter pipe of the same length as the lateral with all flow passing through the pipe. (1) Convert the nozzle operating pressure (kPa) to water head (m). (2) Compute the Christiansen factor F. (3) Compute the length of the lateral. (4) Compute the number of emitters.

REFERENCES

Christiansen, J. E. (1942). Irrigation by sprinkling. California Agriculture Experimental Station bulletin 670.

Christiansen, J. E., and Davis, J. R. (1967). Sprinkler irrigation systems. In *Irrigation of Agricultural Lands*, edited by R. M. Hagan, H. R. Haise, and T. W. Edminster. Madison, WI: American Society of Agronomy.

Chu, S. T., and Moe, D. L. (1972). Hydraulics of a center pivot system. *Transactions of the ASAE*, 15(5): 894–896.

Cuenca, R. H. (1989). *Irrigation System Design: An Engineering Approach*. Englewood Cliffs, NJ: Prentice Hall.

Hart, W. E. (1975). *Irrigation System Design*. Fort Collins, CO: Department of Agricultural Engineering, Colorado State University.

Maupin, M. A., Kenny, J. F., Hutson, S. S., et al. (2014). Estimated use of water in the United States in 2010: US Geological Survey Circular 1405. https://pubs.usgs.gov/circ/1405.

Pettygrove, G. S., and Asano, T. (1986). *Irrigation with Reclaimed Municipal Wastewater: A Guidance Manual*. Boca Raton, FL: CRC Press.

Rain Bird (2017). Irrigation catalog. www.rainbird.com/sites/default/files/media/documents/201805/RainBirdCPCatalog2017.pdf.

Schwab, G. O., Fangmeier, D. D., Elliot, W. J., and Fervert, R. K. (1993). *Soil and Water Conservation Engineering*, 4th ed. New York: Wiley.

Shearer, M., Hagood, M., Larsen, D., and Wolfe, J. (1965). *Sprinkler Irrigation in the Pacific Northwest – A Troubleshooter's Guide*. Corvallis, OR: Pacific Northwest Cooperative Extension.

USDA-NRCS (1997). Farm distribution components. In *National Engineering Handbook*. Washington, DC: USDA.

USDA-NRCS (2016). Sprinkler irrigation. In *National Engineering Handbook*. Washington, DC: USDA.

USDA-SCS (1968). Sprinkler irrigation. In *National Engineering Handbook*. Washington, DC: USDA.

USDA-SCS (1983). Sprinkler irrigation. In *National Engineering Handbook*. Washington, DC: USDA.

Walker, W. R. (1980). *Sprinkler and Trickle Irrigation*. Fort Collins, CO: Colorado State University, Engineering Renewable and Growth.

22 Trickle Irrigation

Notation

Symbol	Description
a	cross-sectional area of the opening of the nozzle (mm^2, $in.^2$)
A	cross-sectional flow area in the lateral (L^2 [length])
A_i	area irrigated by the emission device (m^2, ft^2)
C_0	orifice coefficient
C	Hazen–Williams coefficient
C_L	emitter–connection loss (m, ft)
$Cons$	constant of integration in eq. (22.193)
C_u	uniformity coefficient (fraction or percentage)
CV	coefficient of variation
CV_H	coefficient of variation of pressure head expressed in percent
CV_q	coefficient of variation of discharge expressed in percent
d	diameter of the emitter (mm, in.)
D	diameter of the lateral (L)
D_a	depth of water applied (mm, in.)
DEV_l	deviation between H_l and $\overline{H}$ at distance l (L)
D_{IR}	design daily irrigation requirement (mm/day, in./day)
D_s	submain diameter (m, ft)
D_t	distance of throw (m, ft)
E_a	application efficiency (fraction)
f	Darcy–Weisbach friction factor
f_{n+1}	Darcy–Weisbach friction factor for the pipe reach between emitters n and $n+1$
F	pressure forces (F)
g	acceleration due to gravity (9.81 m/s^2)
h_{var}	pressure head variation (%)
h_i	pressure head at any location i on the lateral (L)
$h_{j(s)}$	pressure head at the inlet of the jth lateral (L)
H	pressure head (L)
$\overline{H}, \overline{h}, H_{av}$	average pressure head along the lateral (L)
H_{in}	pressure head at the inlet (L)
H_d	pressure head at the downstream end of the lateral (L)
H_f	head loss due to friction (L)
H_{f0}	total friction loss of a similar pipe conveying the entire flow over its length (L)
H_{fn+1}	friction head loss between emitters n and $n+1$ (L)
H_l	pressure head at length l (L)
H_{max}	maximum pressure head at the inlet (L)
H_n	pressure head at the nth emitter in the lateral (L)
H_s	required design pressure head (L)
H_T	total energy head at any point in the lateral (L)
H_x	energy head at distance x from the inlet (L)
k	proportionality factor which characterizes the physical dimensions of the emitter flow path
l_0	distance from the inlet of the lateral where the maximum pressure difference is located (L)
l	distance from the inlet of the lateral in the direction of flow (m)
L_f	length of the flow path in the emitter or emitter length (cm, ft)
L	length of the lateral (m, ft)
L_0	ideal length design (m, ft)
L_e	equivalent lateral length (m, ft)
n, N	number of emitters
N_d	ratio of the number of emission devices operating per irrigation to the total number of emission devices multiplied by 100 (percentage)
N_e	number of emission devices at each emission point
P	operating pressure (kPa, psi)
P_i	percentage of crop area being irrigated (%)
q	discharge from an emitter (L/min, gpm)
$q(x)$	emitter outflow per unit length (L^3/T/L)
q_0	emitter flow at the inlet computed using the operating pressure (L^3/T)
q_{0j}	flow from the first emitter of the jth lateral (L^3/T)
q_i	emitter flow at any section i of the lateral (L^3/T)
q_{ij}	emitter flow from the jth lateral at the ith location (L^3/T)
$\overline{q}$	mean emitter flow (L^3)
q_n	outflow from the nth emitter (L^3/T)
Q	total discharge in the lateral (L)
$Q(x)$	total outflow rate passing through the lateral section located at distance x from the downstream end (L^3/T)
Q_C	emission device capacity (L/h, gph)

Q_{in}	total flow rate at the inlet of the lateral (L^3/T)
Q_l	flow rate at distance l from the inlet of the lateral(L^3/T)
Q_m	mean discharge in the lateral (L^3/T)
Q_n	flow in the lateral at the nth emitter (L^3/T)
Q_s	specific discharge for a given lateral ($L^3/T/L$)
Q_x	total discharge in the line at a given length x from the inlet (L^3/T)
Re	Reynolds number (dimensionless)
R_i	energy head drop ratio along the lateral (L/L)
R_j	ratio of energy head drop due to friction along the submain
R_j^e	energy gain ratio along the subdomain due to downslope with respect to the length ratio j
s	spacing between emitters (m, ft)
S_0	slope of the lateral (L/L)
s_e	spacing between emission devices of an emission point or the spacing between emitters (m, ft)
S_f	friction slope (positive since x is measured from the downstream end) (L/L)
S_l	spacing between laterals (m, ft)
S_p	spacing between emission points (m, ft)
S_r	spacing between adjacent rows of plants (m, ft)
T	time used to apply D_a hours of irrigation (h)
T_0	time when irrigation is not done due to maintenance (normally 0.5 h/day is taken)
v	cross-sectional average velocity of flow in the lateral in the x-direction (L/T)
V	dimensionless velocity, i.e., $V = v/v_0$
V_n	velocity between emitters $n - 1$ and n (L/T)
V_{n+1}	velocity between emitters n and $n + 1$ (L/T)
x	distance from the inlet of the lateral in the direction of flow (m)
X	dimensionless distance, i.e., $X = x/x_0$
x_0	characteristic length (m)
z	elevation head (L)
γ	specific weight of water (9.807 kN/m^3, 62.43 lb/ft^3)
μ	dynamic viscosity of water ($N\text{-}s/m^2$, $lb\text{-}s/ft^2$, kg/m·s, 1.0×10^{-3} kg/m·s at 20 °C)
ρ	fluid density (M/L^3)
υ	kinematic viscosity (m^2/s, ft^2/s)
ΔH	total energy head drop at the end of the lateral (L)
ΔH_0	total head drop due to friction at the end of the first lateral (L)
ΔH^e	total energy gain at the end of the lateral due to downslope (L)
ΔH_f	head drop at distance l due to friction (L)
ΔH_i	head drop over any given length (L)
ΔH_i^e	gain in energy head for downsloping lateral at position i (L)
ΔH_j	total energy head drop of the jth lateral due to friction (L)
ΔH_l	total head drop at the end of the lateral (L)
ΔH_l^e	total energy gain or loss due to uniform slope at the end of the lateral (L)
ΔH_m	head drop with the mean discharge (L)
Δh_{max}	maximum pressure difference along the lateral (L)
ΔH_s	total head drop at the end of the submain (L)
ΔH_s^e	total energy gain or loss at the end of the submain due to uniform slope (L)
ΔH_{sj}	energy head drop at a given length ratio j (L)
ΔL	length between emitters (L)
$\overline{\Delta q}$	mean deviation of emitter flow along the lateral (L^3)

22.1 INTRODUCTION

Trickle irrigation is becoming popular almost throughout the world largely because of the shortage of water available for agriculture. Hence, its efficient use is now a necessity. Since agriculture, on a global scale, is the largest user of water (about 70% of all freshwater used), considerable water saving can be achieved by reducing the share of water used by agriculture even by a small proportion, say 10%. Trickle irrigation is an effective way to reduce the agricultural use of water. In contrast with surface and sprinkler irrigation methods, in trickle irrigation water can be applied to the root zone of individual plants. For crops, water may be applied under low pressure and water use efficiency can be as high as 90% or more, and the ground slope is no longer a serious limitation.

Trickle irrigation is the direct application of water either into the root zone or onto the surface slowly and frequently, as shown in Figure 22.1. This means that only a portion of the land is irrigated, and a near-optimum soil moisture content is maintained in the root zone. Irrigation of a limited

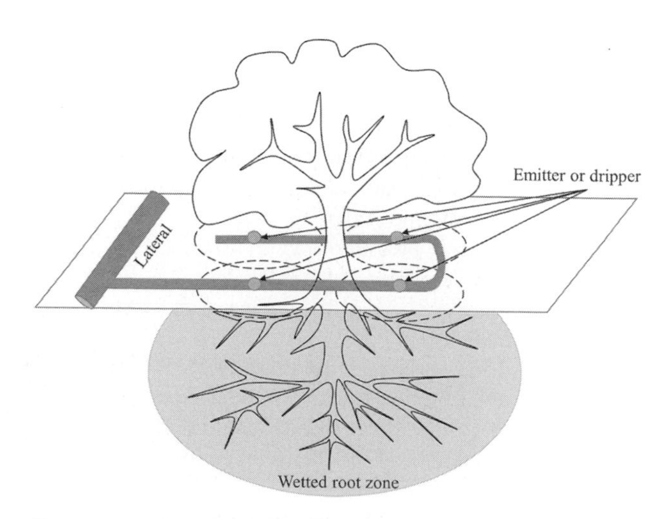

Figure 22.1 Trickle irrigation into the root zone.

portion of land means limited evaporative surface, resulting in limiting evaporation, rendering the unirrigated portion of land resistant to weed growth, and reducing interruption of cultural operations. Trickle irrigation systems can operate daily for the whole period, if needed, and can be designed either analytically or numerically under the assumption of constant discharge or varying discharge. This chapter presents basic concepts of trickle irrigation and different design approaches, and illustrates them with examples.

22.2 ADVANTAGES AND DISADVANTAGES

The advantages of trickle irrigation can be summarized as follows: (1) water is applied very efficiently; (2) evaporation losses are minimized; (3) deep percolation losses are greatly reduced; (4) high average temporal soil water content can be maintained; (5) fertilizer can be applied through trickle irrigation, with minimum loss through leaching; (6) the potential for weed growth is significantly reduced because the wetted surface is only a fraction of the total surface; (7) by maintaining high moisture content, formation of a hard surface crust can be avoided; (8) the problem of root penetration in soils at low soil water content can be avoided by maintaining high water content; (9) in theory, the rate of water application can equal the rate of water use by plants; (10) trickle irrigation reduces fungus and other pests that require a moist environment; (11) it reduces stress under saline conditions, allowing for conventional systems; (12) for some crops the period of maturation is reduced; (13) labor cost is low, because trickle systems are generally permanent; and (14) there is no loss of water at the edges of the field through drift.

However, trickle irrigation also has some disadvantages, such as (1) emitters have small openings, allowing for small flow that can cause frequent clogging; (2) point or strip wetting is not always advantageous; (3) salinity tends to accumulate a short distance away from the emitter and can move to the root zone during a heavy rainfall event; (4) the root zone tends to be smaller and more densely distributed, and can lead to aeration problems for some crops; (5) in windy areas, dry regions between emitters can cause dust problems; and (6) trickle irrigation may mean greater demand for pesticide.

22.3 TRICKLE IRRIGATION METHODS

There are several methods of trickle irrigation, including drip, bubbler, spray, and subsurface, which are briefly discussed here (Figure 22.2).

Figure 22.2 Types of trickle irrigation methods. A black and white version of this figure will appear in some formats. For the color version, please refer to the plate section.

22.3.1 Drip Irrigation

Drip irrigation is often also called trickle irrigation or micro-irrigation. In drip irrigation water is applied to the root zone of individual plants under low pressure. Drip irrigation involves the application of water through emitters as discrete drops, slowly and almost continuously either at a single point on the land surface or as a line source. An emitter applies the water from the distribution system to the soil. The single-point application covers a small wetted area, whereas a line source has either closely spaced emitters or tubes with closely spaced or continuous openings. The discharge rate is usually less than 12 L/h (3 gph) for point-source emitters and less than 12 L/h per meter (1 gph per foot) for lateral or line-source emitters. The wetted area depends on the type of soil and emitter discharge, as shown in Figure 22.3.

An added advantage is that fertilizer can also be mixed with water for directly reaching the root zone. Hence, water use efficiency, defined as crop yield per unit of applied water, can be increased by 50% or more (Hiler and Howell, 1972). Wierenga (1977) reported an increase of cotton yield by more than 8% while using 24% less water than surface irrigation. Drip irrigation reduces water requirements and minimizes return flow.

Current drip systems are expensive to install, at least in the near term, and small-scale farmers, especially in developing countries, cannot afford them. Generally, drip irrigation is practiced for high-value crops, such as fruits, vegetables, and vine crops. Since emitters are provided for each plant, it is more suitable for crops, trees, and other plants that are grown in rows. Other advantages of drip irrigation are that it can be applied to agricultural land of any slope (upslope, downslope, or flat) or varying elevation changes. It can also be applied to almost any soil, but, of course, emitter discharge will have to vary with the soil type in terms of its infiltration rate. For example, for sandy soils, greater emitter discharge will be required to enable adequate lateral wetting of the soil. For clay soils, on the other hand, low emitter discharge will be required to prevent surface water ponding and runoff.

In drip irrigation, emitters have small openings of 0.2–2 mm diameter. It is therefore important that water is free of contaminants in order to avoid blocking these openings. The contaminants may include sediment, algae, and fertilizer deposits, as well as dissolved chemicals that may precipitate as calcium and iron. However, drip irrigation is quite suitable for poor-quality water, such as saline water.

22.3.2 Bubbler Irrigation

Bubbler irrigation involves the application of water to the land surface as a small stream. Laterals are buried and tubes that apply water to the point of interest are attached to the lateral. Tubes may have a 10-mm diameter or more and the discharge rate is usually less than 225 L/h (1 gpm). The flow rate can be computed as

$$q = K\left(\frac{P}{L_f}\right)^{0.57} d^{2.74}, \tag{22.1}$$

where q is the discharge from a bubbler tube (L/min, gpm); P is the operating pressure (kPa, psi); L_f is the length of flow path in the bubbler tube (cm, ft); d is the diameter of the bubbler tube (mm, in.); and K is the unit constant (equal to 9.02×10^{-2} for q in L/min, P in kPa, L_f in cm, and d in mm; and equal to 71.5 for q in gpm, P in psi, L_f in ft, and d in inches).

22.3.3 Spray Irrigation

Spray irrigation is like micro-sprinkler irrigation, spraying water as a mist on the surface with discharge rate of less than 115 L/h (0.5 gpm). With closely spaced micro-sprinklers, the entire land surface can be covered. The normally required operating pressure is 35–300 kPa (5–40 psi) and the wetted diameter is in the range 2–9 m (6–30 ft).

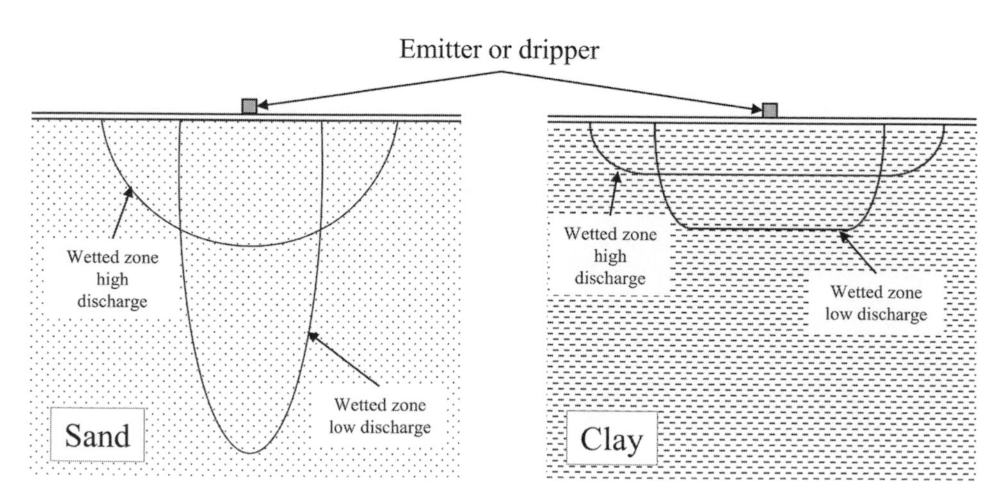

Figure 22.3 Wetting patterns for sand and clay soils with high and low discharge rates.

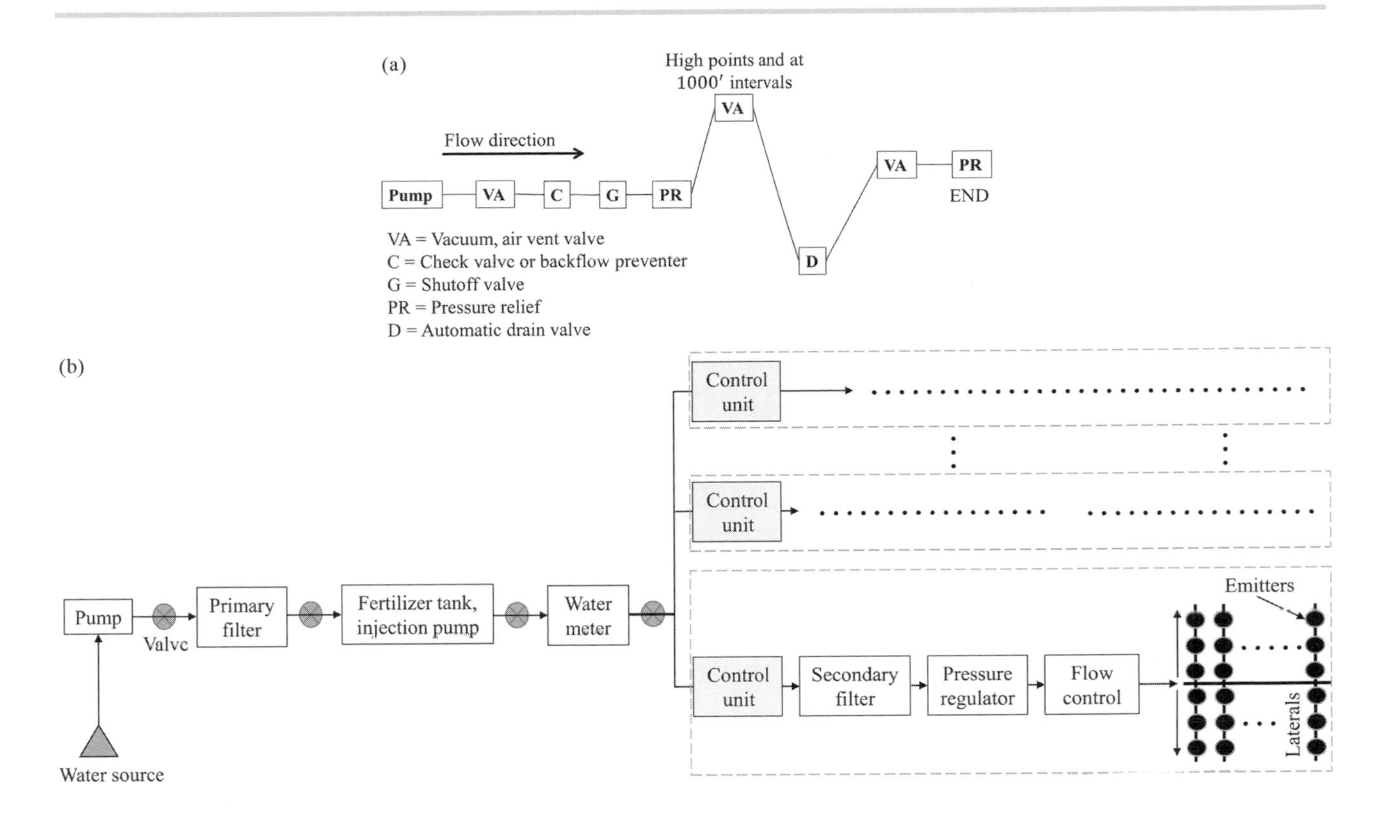

Figure 22.4 (a) A pipeline for trickle irrigation. (b) Layout of a drip irrigation system.

22.3.4 Subsurface Irrigation

Subsurface irrigation involves the application of water below the soil surface through point or line-source emitters, with discharge rates as in drip irrigation.

22.4 DRIP IRRIGATION SYSTEM LAYOUT

A typical drip irrigation system, as shown in Figures 22.4a and 22.4b, comprises a pump unit, control head, mainline, submain or manifold, laterals, and emitters or drippers. The pump unit pumps the water from the source, such as a groundwater well, stream, or reservoir, and delivers it with the right pressure to the pipe system or distribution system.

The control head consists of valves and may have filters, a flow meter, pressure gages, a fertilizer or nutrient injector, a pressure regulator, and a controller. The valves control the discharge and pressure in the entire system. The flow meter and fertilizer injector are not necessary, but are desirable. A controller is needed only if the system is to be automated. Anti-surge valves are often needed on large systems. Filters clean the water and are either screen filters or graded sand filters which remove fine sediment suspended in water. The purpose of the nutrient head is to add fertilizer, in a measured way, to the water during irrigation. The distribution network comprises piping, pipe fittings, emitters, and circuit valves. For an automatic system, valves are actuated electronically or hydraulically.

Mainlines, submains (or manifolds), and laterals constitute the main components of the drip irrigation system, and transport water from the control head to the field through emitters. These components are made from PVC or polyethylene hose. Because they degrade easily if exposed to direct solar radiation they are buried below the ground. Lateral pipes are normally 13–32 mm diameter.

Emitters or drippers, as the name suggests, control the discharge of water and emit it to the soil near the plants. There can be one or more emitters for a single plant, and the spacing between emitters is usually more than 1 m. However, for row crops the emitters may be more closely spaced. For large trees sub-lateral loops may be used. A drip irrigation system is installed for the long term and is considered permanent. Hence, it is conducive to automation and leads to saving of labor, which is desirable, but automation calls for specialist skills.

22.5 TYPES OF EMITTERS

There are two main types of emitters, point source and line source, that are applied. Examples of different models of emitters are shown in Figure 22.5. From an operational point of view, emitters can be classified as long-path emitters, short-orifice emitters, vortex emitters, pressure-compensating emitters, and porous type or tube emitters.

Point-source emitters are used for widely spaced crops, such as vineyards and orchards, whereas line-source emitters

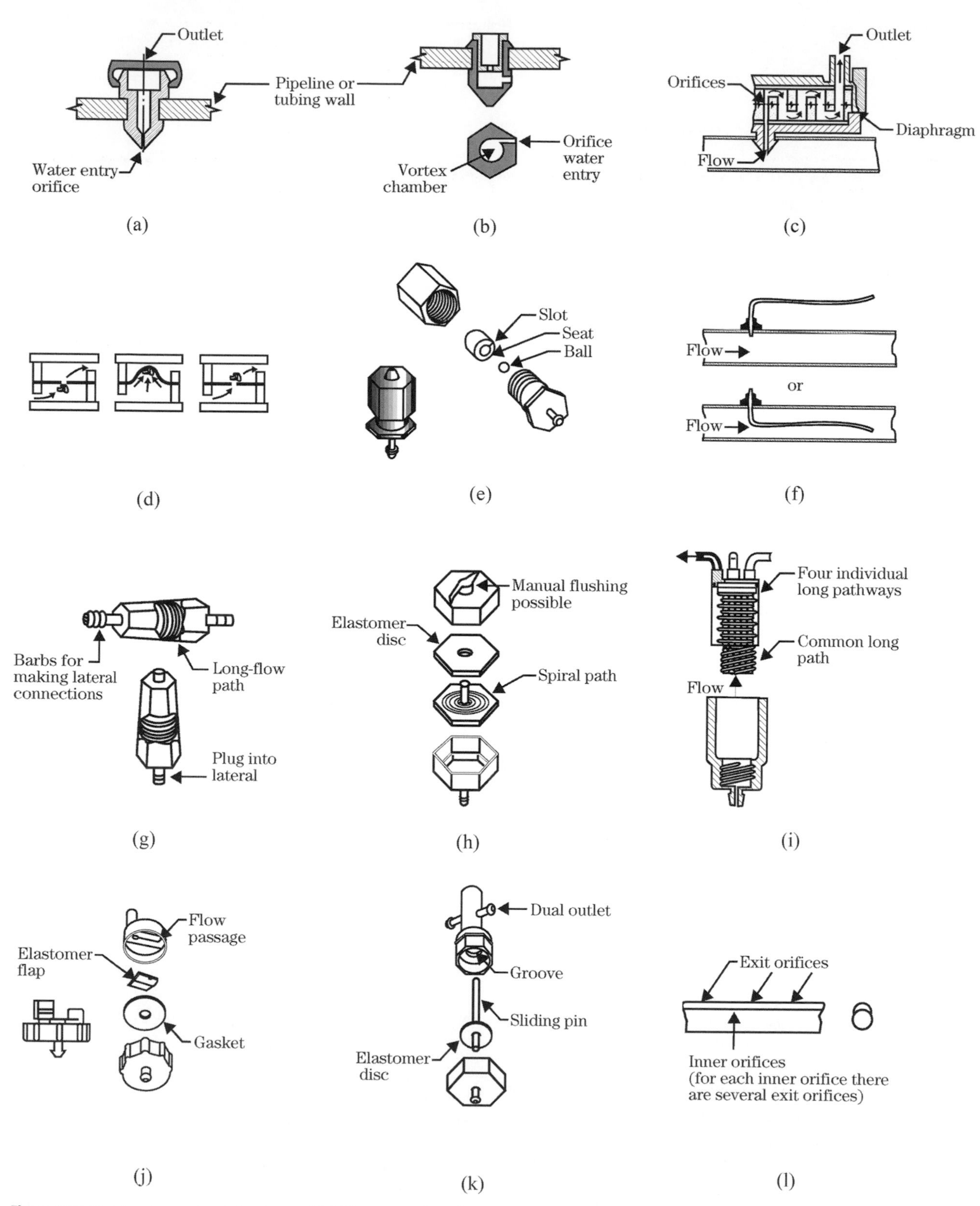

Figure 22.5 Sketches of several emission devices. (a) Orifice emitter. (b) Orifice–vortex emitter. (c) Emitter using flexible orifices in series. (d) Continuous flow principle for multiple flexible orifices. (e) Ball and slotted seat emitter. (f) Long-path emitter small tube. (g) Long-path emitter. (h) Compensating long-path emitter. (i) Long-path multiple-outlet emitter. (j) Groove and flap short-path emitter. (k) Groove and disc short-path emitter. (l) Twin-wall emitter lateral. (Adapted from USDA-NRCS, 2013).

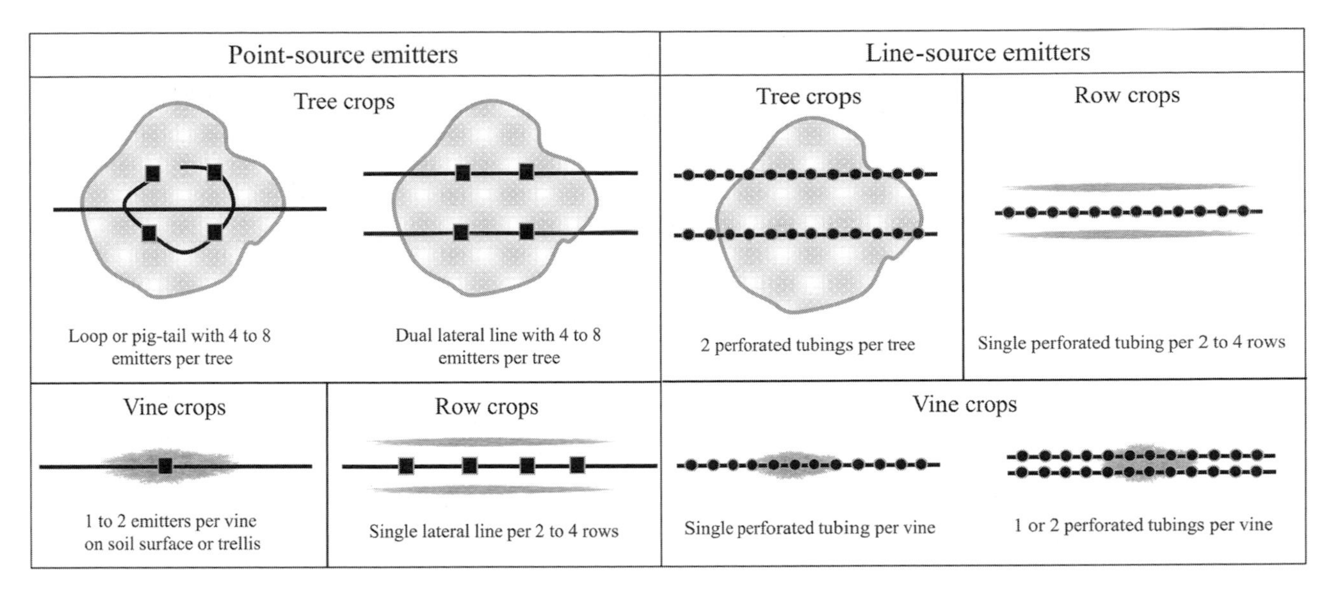

Figure 22.6 Installation of point- and line-source emitters in various cropping situations.

are used for more closely spaced crops, such as agricultural crops. Point-source emitters can be long-path, orifice, or pressure-compensating emitters. The discharge–pressure relation for a point-source emitter can be written as

$$q = kP^y, \tag{22.2}$$

where q is the discharge (L^3/T), P is the operating pressure (F/L^2), y is the exponent that characterizes the flow regime of the emitter, and k is a proportionality factor that characterizes the physical dimensions of the emitter flow path. In general, the values of k and y can be obtained from the manufacturer. They can also be computed by plotting q versus P on a log-log scale. The exponent and proportionality constant are different for these emitters; for example, $y = 1$ for a long path for a laminar flow type emitter, $y = 1$ for an orifice-type source emitter, and y is positive and nearly zero for a pressure-compensating emitter. James (1988) gives theoretical values of k and y. For laminar flow (Re, Reynold number <2000), $k = \frac{\rho d^2 g a}{32\mu L_f}$ and $y = 1.00$. For turbulent flow in a smooth pipe ($3000 < \text{Re} < 10^5$), $k = 2.87a\left(\frac{g}{L_f}\right)^{0.57} d^{0.14}\left(\frac{\rho}{\mu}\right)^{0.14}$ and $y = 0.57$. For the fully turbulent case ($\text{Re} > 10^5$), $k = a\left(\frac{2gd}{fL_f}\right)^{0.50}$ and $y = 0.50$. μ is the dynamic viscosity of fluid (FT/L^2); L_f is the length of the emitter flow path (L); a is the cross-sectional flow area in the emitter (L^2); ρ is the fluid density (M/L^3); d is the characteristic diameter of the emitter flow path (L); g is the acceleration due to gravity (L/T^2); and f is the friction factor in the Darcy–Weisbach equation. Equation (22.2) shows the sensitivity of discharge to pressure and hence of an emitter. The effect of pressure on discharge decreases with decreasing y. For example, fully pressure-compensating emitters have y near zero, then the flow rate would be relatively constant with different pressure ranges. Under this condition, the uniformity of the system is theoretically perfect.

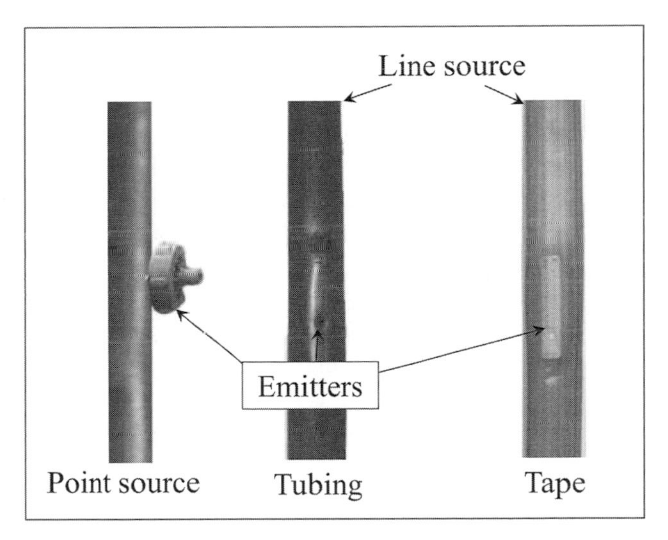

Figure 22.7 Examples of point- and line-source emitters from drip lines.

Line-source emitters are porous pipes, perforated pipes, or laterals with closely spaced emitters that discharge water along their entire length. The bubblers discharging into furrows are a kind of line-source emitter. Equation (22.2) can be employed for determining discharge from line-source emitters with the value of y for laminar, turbulent, and fully turbulent flow regimes. Figure 22.6 shows the application of point- and line-source emitters in a mixture of cropping situations. Figure 22.7 shows the examples of point-source and line-source emitters from drip lines.

The key features desired for emitters are that they do not degrade due to sunlight and temperature variations, are

Table 22.1 *Performance data for emission devices*

Emission device	Figure 22.5 part	exp. y	Coefficient of variation (CV)	Diameters
Orifice–vortex–orifice	a	0.42	0.07	0.6
Multiple flexible orifices (I)	a	0.70	0.05	–
Multiple flexible orifices (II)	c, d	0.70	0.07	–
Ball and slotted set – non-compensating	e	0.50	0.27	0.3^b
Ball and slotted set – pressure compensating (I)	e	0.25	0.09	0.3^b
Ball and slotted set – pressure compensating (II)	e	0.15	0.35	0.3
Long-path type:				
Small tube (I)	f	0.70	0.05	1.0
Small tube (II)	f	0.80	0.05	1.0
Spiral long path – non-flushing	g	0.65	0.02	0.7
Spiral long path – manual flushing	g	0.75	0.06	0.8
Long path – pressure compensating (I)	h	0.40	0.05	0.75^b
Long path – pressure compensating (II)	h	0.20	0.06	0.75^b
Tortuous long path	i	0.65	0.02	1.0^b
Short-path type:				
Groove and flap short path	j	0.33	0.02	0.3
Slot and disk short path (I)	k	0.11	0.10	0.3
Slot and disk short path (II)	k	0.11	0.08	0.3
Line-source devices				
Porous pipe	–	1.00	0.40	–
Twin-wall lateral	l	0.61	0.17	0.4^b

I and II indicate different devices of the same type; superscript ^{b} indicates Solomon's probable estimates; CV is the emitter coefficient of variation, which is defined as the standard deviation of flow rate divided by the average flow rate from a sample of emitters.

available in small increments of discharge of 1 L/h, are able to adjust the flow in response to pressure in response to operating pressure, have a longer life, and do not clog easily. Emitter types include long-path single exit, long-path multi-exit, turbulent vortex flow, and pressure compensating, among others. These models have different flow characteristics. Manufacturer's coefficient of variation (CV) is defined as the standard deviation of flow rate divided by the average flow rate from a sample of emitters, which reflects quality control during the manufacturing process. Table 22.1 shows the manufacturer variation and the exponent y in eq. (22.2) of different types of emitters in Figure 22.5.

22.6 DESIGN CONSIDERATIONS

An important design objective of a drip irrigation system is to achieve a uniform distribution of water delivered through the emitters. To that end, there are three components that relate to the characteristics of ground, drip system, and flow. The ground characteristics are normally described by slope, which can be uniformly flat, down, or up, although soil characteristics also impact the flow distribution. However, the ground can also be undulating, wherein the slope is up and down. The drip system characteristics include type of emitters, number of emitters, spacing of emitters, length of laterals, diameter of laterals, smoothness of pipe, and arrangement of laterals. Flow characteristics include type of flow, including laminar, turbulent, and fully turbulent; head loss due to friction; inlet pressure; spatial variation of pressure; spatial variation of discharge; spatial variation of flow velocity; and desired uniformity coefficient. Computation of flow distribution entails knowledge of these characteristics. These characteristics determine the drip irrigation design. Some of the drip characteristics, such as lateral diameter, smoothness of pipe, and spacing, may be fixed by the manufacturer, and the designer may not have full control on fixing the values of these characteristics. In a like manner, ground characteristics are fixed for the most part, although some, such as surface undulations, can be altered with significant investment. Thus, in practical design, the designer has limited options. Most of the time, for a given uniformity coefficient, the designer determines the length of laterals or inflow to the drip system. For design of a drip irrigation system, basic concepts of hydraulics are needed, which are discussed now.

22.7 BASIC HYDRAULICS: GOVERNING EQUATIONS

In a drip irrigation system water is delivered by the mainline to the submain and by the submain to the laterals. Emitters are attached to the laterals and distribute water for irrigation. From a hydraulic consideration, Keller and Karmeli (1974,

1975) classified emitter characteristics into flow regime, pressure dissipation, lateral connection, water distribution, flow cross-section, cleaning characteristics, pressure compensation, and construction material.

Flow in the submain and laterals can be regarded as steady nonuniform (spatially varied) flow with lateral outflows (Wu and Gitlin, 1973). The flow from the submain line to the laterals or the outflow from emitters is controlled by the distribution of pressure along the submain and laterals. The pressure distribution along the submain or a lateral is controlled by the head loss or head drop (or energy drop) due to friction and energy gain due to slope down or energy loss due to slope up.

By determining the pressure distribution, uniform irrigation can be made possible in three ways: (1) adjusting the size of emitters (Myers and Bucks, 1972); (2) adjusting the length and size of the microtube (a special type of emitter [Kenworthy, 1972]); and (3) slightly adjusting the spacing of emitters (Wu and Gitlin, 1973). Since the pressure distribution may not be uniform, the variation of discharge from emitters along a lateral will be a function of the total flow rate, total length, inlet pressure, and emitter spacing. Thus, the design involves determining the right combination of length and pressure in order to achieve an acceptable degree of irrigation uniformity. Then, the key to the design of a drip irrigation system is the determination of head losses and irrigation uniformity. The uniformity is primarily affected by hydraulic design, manufacturer's variation, the effect of temperature, and plugging.

22.7.1 Emitter Discharge–Head Relation

Consider a lateral pipeline with multiple emitters, as shown in Figure 22.8, where emitters are identical and are equally spaced. As mentioned earlier, flow in a lateral pipe is considered steady and spatially varied (nonuniform), with decreasing outflow in the downstream direction. As discharge decreases along the lateral, the energy gradient line decreases.

The flow discharge from an nth emitter can be expressed as a power function of the head in the lateral pipe at the emitter (Howell and Hiler, 1974; Keller and Karmeli, 1974):

$$q_n = cH_n^y, \tag{22.3}$$

where q_n is the outflow from the nth emitter, H_n is the piezometric (or pressure) head in the lateral pipe at the nth emitter, c is the emitter coefficient that takes into account the effect of area and discharge, and y is the exponent, which varies with the flow regime, whether laminar or turbulent, and emitter type. It is assumed that the cross-sectional area (A), emitter spacing (s), and emitter constant (c) are constant for a lateral. Also, the velocity head is negligible. The value of y should vary from 0 for a pressure-compensating emitter to 1 for an emitter in a laminar flow regime, and should be close to 0.5 for emitters in a turbulent flow regime. The higher the value of y, the greater the degree of care needed to maintain the proper pressure distribution along the lateral (Cuenca, 1989).

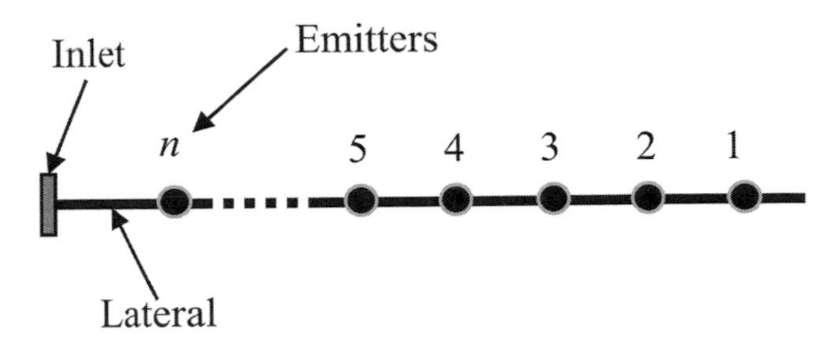

Figure 22.8 A lateral with multiple emitters.

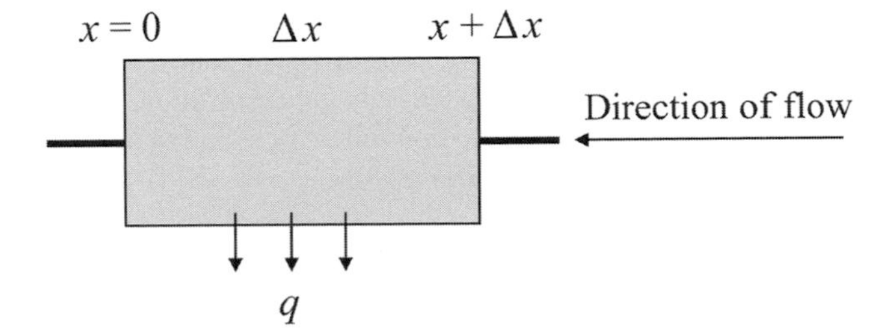

Figure 22.9 Elemental control volume.

If it is assumed that the number of emitters is sufficiently large, with spacing between emitters denoted by s, and outflow varies continuously along the lateral, then the outflow per unit length, denoted as q, can be expressed as

$$q = \left(\frac{c}{s}\right)H^y. \tag{22.4}$$

Equation (22.4) assumes that q and H are continuous functions of coordinate x along the direction of flow.

22.7.2 Conservation of Mass: Continuity Equation

Now consider an elemental control volume of length Δx, as shown in Figure 22.9. The conservation of mass of water along the lateral can be expressed by the continuity equation as

$$\frac{dQ}{dx} + \frac{dA}{dt} = -q, \tag{22.5}$$

where Q is the lateral discharge, A is the cross-sectional area of the lateral, and x and t are the space and time coordinates. Because the flow in the lateral is assumed steady and the cross-sectional area A is constant, the term dA/dt becomes 0. Noting that $Q = Av$, where v is the cross-sectional average velocity of flow in the lateral in the x-direction, eq. (22.5) reduces to

$$A\frac{dv}{dx} = -q. \tag{22.6}$$

Referring to Figure 22.8, the continuity eq. (22.4) can be expressed in discrete form for a pipe reach between two successive emitters, n and $n+1$, as

$$Q_{n+1} = Q_n - q_n, \tag{22.7}$$

where Q_n is the flow in the lateral at the nth emitter, Q_{n+1} is the flow at the $(n+1)$ emitter, which is the value of Q_n reduced by the outflow from the nth emitter. Substituting eq. (22.4) in eq. (22.6), the result is

$$H = \left[-\left(\frac{As}{c}\right)\left(\frac{dv}{dx}\right) \right]^{1/y}. \tag{22.8}$$

Equation (22.8) shows that the pressure head at any position depends on the change in the velocity at that position, which will be negative because velocity will decline due to the loss of energy head in the direction of flow.

22.7.3 Conservation of Momentum

In the lateral pipe, the discharge of flow decreases in the downstream direction; hence, there will be a change in the momentum. From the conservation of momentum for flow in a lateral, one can write

$$g\frac{dH}{dx} = -v\frac{dv}{dx}, \tag{22.9}$$

where g is the acceleration due to gravity. Equation (22.9) can be written in terms of flow discharge Q as

$$\frac{dH}{dx} = -\frac{Q}{gA^2}\frac{dQ}{dx}, \tag{22.10}$$

or in discrete form as

$$\Delta H = -\frac{Q}{gA^2}\Delta Q. \tag{22.11}$$

One can write the momentum conservation equation in discrete form for the pipe reach between two consecutive (n and $n+1$) emitters as

$$\Sigma F = \rho(Q_{n+1}V_{n+1} - Q_n V_n), \tag{22.12}$$

where ΣF is the change of pressure forces, V_n is the velocity between emitters $n-1$ and n, and V_{n+1} is the velocity between emitters n and $(n+1)$. The left side of eq. (22.12) can be converted to the change of pressure head by dividing by (γA, where γ is the specific weight of water $= g\rho$) and the velocity terms on the right side can be converted to discharge as

$$\frac{\Sigma F}{\gamma A} = \Delta H_{n+1}\frac{\rho}{\gamma A}\left(\frac{Q_{n+1}\times Q_{n+1}}{A} - \frac{Q_n \times Q_n}{A}\right) = \frac{(Q_{n+1}^2 - Q_n^2)}{gA^2}, \tag{22.13}$$

where g is the acceleration due to gravity. Equation (22.13) expresses the change in pressure head in the lateral pipe reach between two consecutive emitters n and $n+1$ due to the momentum change. It is seen that eq. (22.13) is a discrete form of eq. (22.11).

22.7.4 Conservation of Energy

Recalling the Bernoulli equation, the total energy head at any point in the lateral can be expressed as

$$H_\gamma = z + \frac{p}{\gamma} + \frac{v^2}{2g}, \tag{22.14}$$

where z is the elevation head, p is the pressure, γ is the specific weight, and ρ is the mass density. The sum of elevation head and pressure head is defined as piezometric head, H:

$$H = z + \frac{p}{\gamma}. \tag{22.15}$$

If the ground slope is 0 and the datum is the same as the ground or lateral, then the piezometric head is the same as the pressure head. Also, the flow velocity is normally quite low so the velocity head may be small or even negligible. In that case the total energy head is the same as the piezometric head, $H_T = H$.

Now the equation of energy conservation can be written for the pipe reach between two successive emitters n and $n+1$ as

$$H_n + \frac{v_n^2}{2g} + z_n = H_{n+1} + \frac{v_{n+1}^2}{2g} + z_{n+1} + H_{fn+1} + \Delta H_{n+1}, \tag{22.16}$$

where H_{fn+1} is the friction head loss between emitters n and $n+1$, H_n is the pressure head at emitter n, H_{n+1} is the pressure head at emitter $n+1$, z_n is the elevation of emitter n above a datum, z_{n+1} is the elevation of emitter $n+1$, ΔH_{n+1} is the loss or drop in energy head due to other factors (and is relatively small), $v_n^2/2g$ is the velocity head at emitter n, and $v_{n+1}^2/(2g)$ is the velocity head at emitter $(n+1)$. The head loss due to friction needs to be determined.

22.7.5 Head Loss Equation

The head loss equation for pipe flow can be expressed by either the Darcy–Weisbach equation, the Hazen–Williams equation, or a power relation.

Darcy–Weisbach Equation

The Darcy–Weisbach equation can be written as

$$H_f = f\left(\frac{L}{D}\right)\left(\frac{v^2}{2g}\right) = \frac{8f}{\pi^2}\frac{L}{D^5}\frac{Q^2}{g}, \quad v = \frac{Q}{A} = \frac{4Q}{\pi D^2}, \tag{22.17}$$

where f is the Darcy–Weisbach friction factor, which is obtained from the tabulated values, D is the diameter of the lateral (m), v is the average velocity flow velocity (m/s), g is the acceleration due to gravity (9.81 m/s^2), and L is the length over which the head loss is computed (m). Now we need to express the friction loss in laterals. For small-diameter smooth lateral pipes, eq. (22.17) can be used to compute the friction head loss as

$$H_{f_{n+1}} = f_{n+1} \frac{8L}{D^5} \frac{Q_{n+1}^2}{\pi^2 g}, \quad (22.18)$$

where Hf_{n+1} is the head loss between emitters n and $n+1$, f_{n+1} is the Darcy–Weisbach friction factor for the pipe reach between emitters n and $n+1$, and D is the internal diameter of the lateral pipe.

For smooth lateral pipes, the friction factor f depends on the Reynolds number (Re) expressed as

$$\text{Re} = \frac{vD}{\upsilon} = \frac{\rho VD}{K\mu}, \quad (22.19)$$

where υ is the kinematic viscosity equal to the dynamic viscosity (μ) divided by the density of the fluid (ρ). Here, K is a unit constant, equal to 10 for ρ in g/cm^3, D in cm, v in cm/s, and μ in N-s/m^2; and equal to 12 for ρ in slug/ft^3, D in inches, v in ft/s, and μ in lb-s/ft^2; and equal to 1 for ρ in kg/m^3, V in m/s, D in m, and μ in kg/m·s.

Warrick and Yitayew (1987) expressed friction factor f as a function of velocity as

$$f = f_0 v^{m-2}, \quad (22.20)$$

where m is the exponent and f_0 depends on the flow velocity and lateral diameter. If Re is less than 2000, that is the flow is laminar, then

$$f = \frac{64}{\text{Re}} = \frac{64\upsilon}{vD} \quad (22.21)$$

or

$$f_0 = \frac{64\upsilon}{D} \quad (m = 1 \text{ for laminar Re} < 2000). \quad (22.22)$$

If Re is between 2000 and 10^5 (i.e., turbulent flow), then f can be given by the Blasius equation:

$$f = \frac{0.316}{\text{Re}^{0.25}} = 0.316\left(\frac{\upsilon}{vD}\right)^{0.25} \quad (22.23)$$

or

$$f_0 = 0.316\left(\frac{\upsilon}{D}\right)^{0.25} \quad (m = 1.75 \text{ smooth}). \quad (22.24)$$

If Re is between 10^5 and 10^7, that is for fully turbulent flow, then f can be given by (Watters and Keller, 1978):

$$f = \frac{0.130}{\text{Re}^{0.172}} = 0.130\left(\frac{\upsilon}{vD}\right)^{0.172} \quad (22.25)$$

or

$$f_0 = 0.130\left(\frac{\upsilon}{D}\right)^{0.172} \quad (m = 1.828). \quad (22.26)$$

The Hazen–Williams Equation

The head loss due to friction is more frequently computed by the Hazen–Williams equation for pipe flow, which can be expressed as

$$v = 1.318C\left(\frac{D}{4}\right)^{0.63}\left(\frac{H_f}{L}\right)^{0.54} \quad (22.27)$$

or

$$Q = 1.318\pi CD^{2.63}\left(\frac{1}{4}\right)^{1.63}\left(\frac{H_f}{L}\right)^{0.54}, \quad (22.28)$$

where C is the Hazen–Williams coefficient and (H_f/L) is the slope of the energy line, also denoted as S_f. Here, the units are in the (British) system: Q is in gpm, D is in inches, L is in feet, and H_f is in feet. The head loss or drop can be computed from eq. (22.28) as

$$H_f = KL\frac{Q^{1.852}}{D^{4.871}}, \quad (22.29)$$

in which K is a constant and is equal to 9.76×10^{-4} for British units and 1.13×10^6 for metric units (Q in L/s, L in m, D in mm, and H_f in m).

Power Function Form

Sometimes the head loss due to friction is also expressed as

$$H_f = av^m S_f, \quad (22.30)$$

where a is given by the Hazen–Williams equation when Re > 2300 and $m = 1.852$ as

$$a = \frac{K}{C^m A^{0.585}}, \quad (22.31)$$

in which C is the Hazen–Williams coefficient, K is the coefficient, and m is the exponent describing the flow regime. When Re < 2300 and $m = 1$, the value of a is given as

$$a = \frac{32\mu}{gD^2}. \quad (22.32)$$

Discharge Estimation for Different Emitter Types

The relation between emitter discharge and operating pressure depends on the type of emitter and flow regime. For an orifice emitter, most of the head loss occurs when water flows through the opening of the emitter. The flow is fully turbulent, and the emitter discharge q (L/h) can be expressed as

$$q = 3.6 \times a \times C_0 \times (2gH)^{0.5}, \quad (22.33)$$

where a is the emitter flow cross-sectional area (mm^2), C_0 is the orifice coefficient, which depends on the nozzle characteristics and ranges from 0.6 to 1.0. $g = 9.81$ m/s^2, and H is the orifice operating pressure head in m. From eq. (22.33) it is possible to determine the orifice diameter with given head loss and discharge.

For a long-path emitter, most of the head loss occurs in the long flow path section. The discharge can be given by the Darcy–Weisbach equation. The discharge from a long-path emitter under laminar flow can be expressed using eq. (22.17) as

$$q = 0.1138a\left[2g\left(\frac{Hd}{fL_f}\right)\right]^{0.5}, \quad (22.34a)$$

where L_f is the length of the flow path in the emitter (m), d is the diameter of the emitter (mm, in.), and f is the friction factor. As shown in eq. (22.34a), with given discharge the change in emitter diameter can significantly influence the length of the flow path and head loss in the long-path emitter. From eq. (22.3), the discharge q is proportional to $H^{0.5}$ for turbulent flow, then the discharge from a long-path emitter under turbulent flow can be expressed as

$$q = 0.1138a\left[2g\left(\frac{H^{0.5}d}{fL_f}\right)\right]^{0.5}. \quad (22.34b)$$

Head Loss in Trickle Irrigation

In trickle irrigation, head loss is also caused by trickle emitters themselves. The trickle emitters are of two types: (1) barbed pop-on and (2) in-line emitters. For computing head loss due to trickle emitters, the equivalent pipe length method is used. The equivalent lateral length L_e is defined as

$$L_e = L\left[\frac{s + C_L}{s}\right], \quad (22.35)$$

where L_e is the equivalent lateral length in m, s is the emitter spacing along the lateral in m, and C_L is the emitter–connection loss values, which is the equivalent length of the pipe that would cause the same head loss as the trickle emitter in m or ft. Urbina (1976) and Watters and Keller (1978) prescribed values of C_L as shown in Figure 22.10.

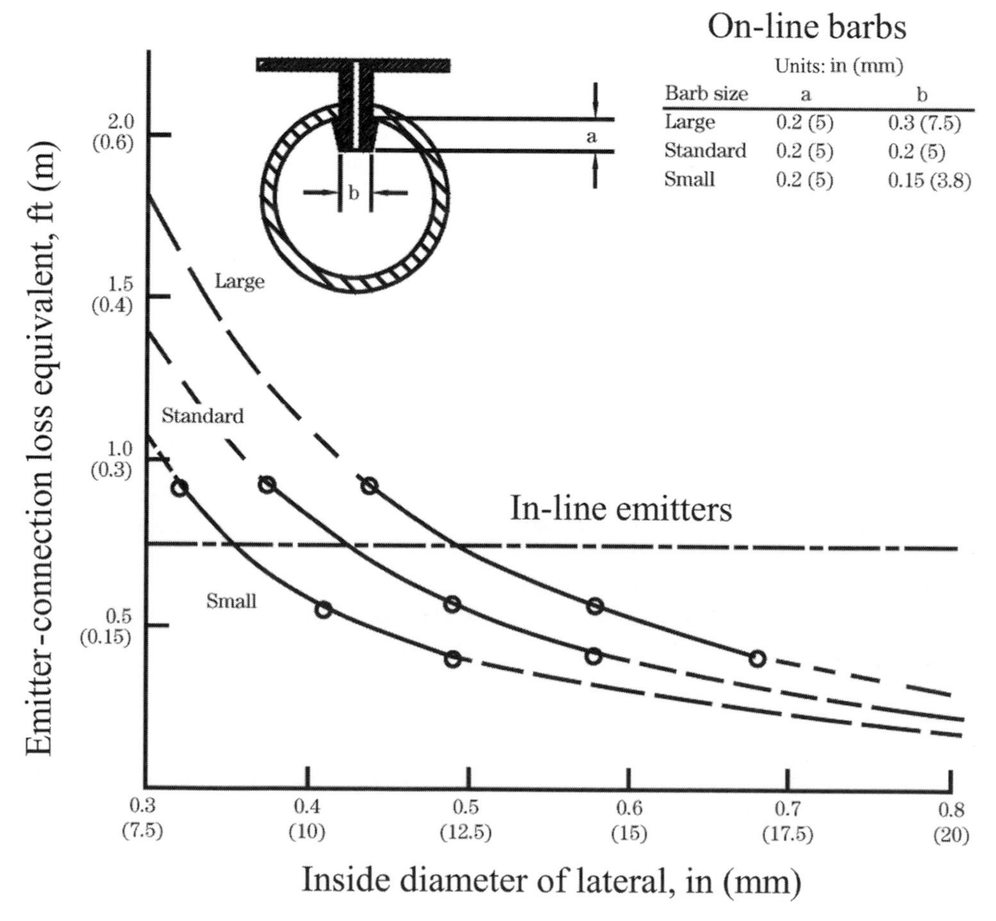

Figure 22.10 Emitter–connection loss (C_L) values for various sizes of barbs and inside diameters of laterals. (After USDA-NRCS, 2013.)

Example 22.1: Determine the head loss due to friction in a drip lateral with the following data: the internal diameter of the lateral is 20 mm; the length of the lateral is 100 m with "standard" emitters 1 m apart; normal water temperature is 20 °C; and the design discharge of each emitter as 1.25 L/h. (Given K (for Re) = 10, $\rho = 0.997\ \text{g/cm}^3$, $\mu = 0.001\ \text{N-s/m}^2$, and $F = 0.335$.)

Solution: First, the total discharge from the lateral is computed as

$$Q = 1.25\left(\frac{\text{L}}{\text{h}}\right) \times 100 \times \left(\frac{\text{h}}{60\ \text{min}}\right) = 2.08\frac{\text{L}}{\text{min}}.$$

The cross-sectional area is

$$A = \frac{\pi}{4}(2.0\ \text{cm})^2 = 3.14\ \text{cm}^2.$$

The velocity of flow is

$$v = \frac{Q}{A} = \frac{2.08\frac{\text{L}}{\text{min}}}{3.14\ \text{cm}^2} = \frac{2080\frac{\text{cm}^3}{\text{min}} \times \frac{1\ \text{min}}{60\ \text{s}}}{3.14\ \text{cm}^2} = 11.04\frac{\text{cm}}{\text{s}}.$$

The Reynolds number is

$$\text{Re} = \frac{\rho D v}{K\mu} = \frac{0.997\frac{\text{g}}{\text{cm}^3} \times 2\ \text{cm} \times 11.04\frac{\text{cm}}{\text{s}}}{10 \times 0.001\frac{\text{N}\cdot\text{s}}{\text{m}^2}} = 2201$$

(Turbulent flow; $2000 < \text{Re} < 100{,}000$).

The friction factor f, depending on the value of Re, in eq. (22.23) is

$$f = \frac{0.316}{\text{Re}^{0.25}} = \frac{0.316}{(2201)^{0.25}} = 0.046.$$

The length L is corrected for barb losses. The value of emitter–connection loss is from Figure 22.10.

$$C_L = 20\ \text{mm} \rightarrow \text{(graph read)} \rightarrow 0.8\ \text{in.(standard emitter)} = 0.076\ \text{m},$$

$$L_e = L\left[\frac{s + C_L}{s}\right] = (100)\left[\frac{1 + 0.076}{(1)}\right] = 107.6\ \text{m}.$$

Given the length $L_e = 107.6$ m, the pipe diameter $D = 20\ \text{mm} = 0.02$ m, and the velocity of flow is $11.04\ \text{cm/s} = 0.1104\ \text{m/s}$, the head loss in the pipe is computed using eq. (22.17) as

$$H_f = f\left(\frac{L}{D}\right)\left(\frac{v^2}{2g}\right) = 0.046\left(\frac{107.6\ \text{m}}{0.02\ \text{m}}\right)\left[\frac{\left(0.1104\frac{\text{m}}{\text{s}}\right)^2}{2\left(9.81\frac{\text{m}^2}{\text{s}}\right)}\right] = 0.154\ \text{m}.$$

Given the factor $F = 0.335$, the actual head loss in the drip lateral is

$$H_f = 0.335 \times 0.154\ \text{m} = 0.052\ \text{m}.$$

Example 22.2: What would be the diameter of an orifice emitter having a design discharge of 12 L/h under turbulent flow and an operating pressure head of 8 m? The orifice coefficient (C_o) is 0.6.

Solution: Using eq. (22.33), the flow cross-sectional area can be computed as

$$a = \frac{q}{(3.6)C_o(2gH)^{0.5}} = \frac{12}{3.6 \times 0.6 \times [2 \times 9.81 \times 8]^{0.5}} = 0.44\ \text{mm}^2,$$

$$a = \frac{\pi}{4}d^2, d = \sqrt{\frac{4a}{\pi}} = \sqrt{0.5605} = 0.75\ \text{mm}.$$

Hence, $d = 0.75$ mm.

Example 22.3: Consider a trickle irrigation system with plastic microtubing with an inside diameter of 1.5 mm. What will be the required length of a long-path emitter with a design discharge of 5 L/h and operating pressure head of 8 m?

Solution: The velocity of flow in the tube can be computed as

$$v = \frac{q}{a} = \frac{5\left(\frac{\text{L}}{\text{h}}\right)\left(\frac{1\ \text{m}^3}{1000\ \text{L}}\right)\left(\frac{1\ \text{h}}{3600\ \text{s}}\right)}{\left(\frac{\pi}{4}\right)\left(1.5\ \text{mm} \times \frac{1\ \text{m}}{1000\ \text{mm}}\right)^2} = 0.786\frac{\text{m}}{\text{s}}$$

At 20 °C, $\mu = 1.0 \times 10^{-3}$ kg/m-s, $\rho = 1.0 \times 10^3\ \text{kg/m}^3$, with $d = 1.5\ \text{mm} = 1.5 \times 10^{-3}$ m, $v = 0.786$ m/s, $K = 1$, Reynolds number is

$$\text{Re} = \frac{\left(1000\frac{\text{kg}}{\text{m}^3}\right)\left(1.5 \times 10^{-3}\ \text{m}\right)\left(0.786\frac{\text{m}}{\text{s}}\right)}{1.0 \times 1.0 \times 10^{-3}\frac{\text{kg}}{\text{m}\cdot\text{s}}} = 1179.$$

The flow is laminar (Re < 2000). The friction factor is computed using eq. (22.21).

$$f = \frac{64}{R_e} = \frac{64}{1179} = 0.054.$$

The long-path emitter discharge can be computed by eq. (22.34a) as

$$5 = 0.1138\left[\frac{\pi}{4} \times (1.5)^2\right]\left[2(9.81)\frac{(8)(1.5)}{(0.054)L_f}\right]^{0.5}.$$

Then, $L_f = 7.05$ m.

Example 22.4: For a trickle irrigation system with plastic microtubing with an inside diameter of 1.5 mm, what will be the required length of a long-path emitter with a design discharge of 25 L/h and operating pressure head of 12 m?

Solution: The flow velocity in the microtube can be computed as:

$$v = \frac{q}{a} = \frac{25\left(\frac{L}{h}\right)\left(\frac{1\ m^3}{1000\ L}\right)\left(\frac{1\ h}{3600\ s}\right)}{\left(\frac{\pi}{4}\right)\left(1.5\ mm \times \frac{1\ m}{1000\ mm}\right)^2} = 3.93\frac{m}{s}.$$

At 20 °C, $\mu = 1.0 \times 10^{-3}$ kg/m·s, $\rho = 1.0 \times 10^3$ kg/m³, with $D = 1.5$ mm $= 1.5 \times 10^{-3}$ m, $v = 3.93$ m/s, $K = 1$, Reynolds number is

$$Re = \frac{\rho D v}{K\mu} = \frac{\left(1000\frac{kg}{m^3}\right)\left(1.5 \times 10^{-3}\ m\right)\left(3.93\frac{m}{s}\right)}{1.0 \times 1.0 \times 10^{-3}\frac{kg}{m \cdot s}} = 5895.$$

The flow is turbulent flow ($2000 < Re < 100{,}000$). The friction factor is computed using eq. (22.23):

$$f = \frac{0.316}{Re^{0.25}} = \frac{0.316}{(5895)^{0.25}} = 0.036.$$

The long-path emitter discharge can be computed by eq. (22.34b) as

$$25 = 0.1138\left[\frac{\pi}{4} \times (1.5\ mm)^2\right]\left[2(9.81)\frac{(12)^{0.5}(1.5)}{(0.036)L_f}\right]^{0.5}.$$

Then, $L_f = 0.18$ m.

Example 22.5: Consider a trickle irrigation system for a vineyard on a level field that has a lateral 12 mm in diameter with 25 single-orifice emitters 1.25 m apart, where the first emitter is at the full space from the lateral entrance from the submain (friction factor, $F = 0.355$). The pressure head at the entrance is 12 m and the emitter design discharge is 15 L/h. What will be the discharge from the final emitter (type B) on the lateral?

Solution: First, the total discharge from the lateral is computed as

$$Q = Nq = 25 \times 15\frac{L}{h} = 375\frac{L}{h}\left(\frac{1\ h}{3600\ s}\right) = 0.104\frac{L}{s}.$$

The flow velocity in the lateral is

$$v = \frac{Q}{A} = \frac{375\frac{L}{h}\left(\frac{1\ m^3}{1000\ L}\right)\left(\frac{1\ h}{3600\ s}\right)}{\left(\frac{\pi}{4}\right)\left(12\ mm \times \frac{1\ m}{1000\ mm}\right)^2} = 0.921\frac{m}{s}.$$

At 20 °C, $\mu = 1.0 \times 10^{-3}$ kg/(m·s), $\rho = 1.0 \times 10^{-3}$ kg/m³, with $D = 12$ mm $= 0.012$ m, $v = 0.921$ m/s, $K = 1$, Reynolds number is

$$Re = \frac{\rho D v}{K\mu} = \frac{\left(1000\frac{kg}{m^3}\right)(0.012\ m)\left(0.921\frac{m}{s}\right)}{1.0 \times 1.0 \times 10^{-3}\frac{kg}{m \cdot s}} = 11{,}052.$$

The flow is turbulent ($2000 < Re < 100{,}000$). The friction factor is computed using eq. (22.23) as

$$f = \frac{0.316}{Re^{0.25}} = \frac{0.316}{(11{,}052)^{0.25}} = 0.031.$$

The head loss due to friction in the lateral (eq. 22.29) is

$$H_f = f\left(\frac{L}{D}\right)\left(\frac{v^2}{2g}\right) = 0.031\left(\frac{25 \times 1.25\ m}{0.012\ m}\right)\left[\frac{\left(0.921\frac{m}{s}\right)^2}{2\left(9.81\frac{m^2}{s}\right)}\right]$$

$= 3.49$ m.

Considering the discharge through the emitters, the actual head loss due to friction can be computed as

$H_{ac} = F \times H_f = 0.355 \times 3.49\ m = 1.24$ m.

The actual head loss is equal to the total head loss as the field is level:

$H_{final} = 12\ m - H_{ac} = 12\ m - 1.24\ m = 10.76$ m.

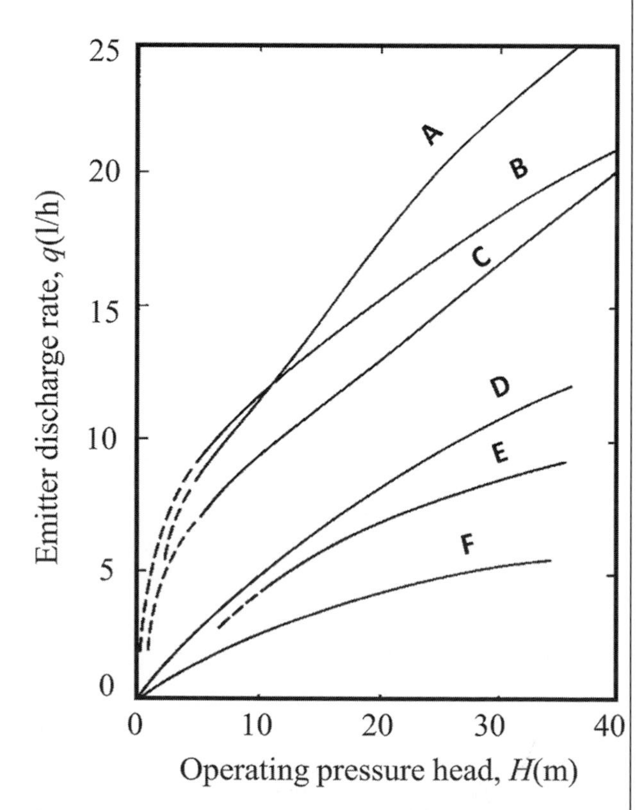

Figure 22.11 Discharge rates for various emitter designs as a function of operating head. (A) Large long path; (B) orifice; (C) long path; (D) medium long path; (E) small long path; (F) double tube. (Adapted from Keller and Karmeli, 1975. Reprint with permission from Rain Bird.)

The operating pressure at the last emitter is 10.76 m. From Figure 22.11, the orifice emitter (type B) discharge for operating pressure of 10.76 is 12 L/h. The variation in discharge along the lateral is computed as

$$\Delta q = \left(\frac{15\frac{\text{L}}{\text{h}} - 12\frac{\text{L}}{\text{h}}}{15\frac{\text{L}}{\text{h}}} \right) \times 100\% = 20\%$$

22.7.6 Energy Slope

The change in energy with respect to the length of the lateral in the direction of flow can be expressed from eq. (22.15) as

$$\frac{dH_T}{dx} = \frac{dz}{dx} + \frac{dH}{dx}, \tag{22.36}$$

where dH_T/dx is the energy slope $= -S_f$, dz/dx is the slope of the lateral $= -S_0$, and dH/dx is the slope of the pressure head or pressure variation along the lateral. Therefore, one can write

$$\frac{dH}{dx} = S_0 - S_f. \tag{22.37}$$

Equation (22.37) shows that if the energy gradient line is known, then the pressure variation along the lateral can be determined.

22.7.7 Pressure Profiles along a Lateral

The pressure head profile along a lateral or submain, as shown in Figure 22.12, can be determined from the inlet pressure H, head loss due to friction ΔH_f, and change in energy head due to slope ΔH_l^e, which can be positive for downslope and negative for upslope, as

$$H_l = H - \Delta H_f + \Delta H_l^e, \tag{22.38}$$

where H_l is the pressure head at location l. The head drop due to friction at distance l can be expressed as (to be shown later):

$$\Delta H_f = \left[1 - \left(1 - \frac{l}{L} \right)^{2.852} \right] \Delta H \tag{22.39}$$

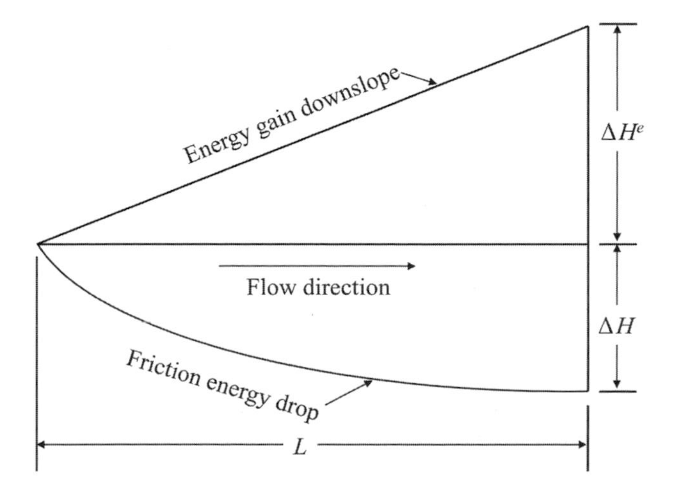

Figure 22.12 Friction energy drop and slope gain (or loss) along a drop lateral.

Substituting eq. (22.39) in eq. (22.38), one gets

$$H_l = H - \left[1 - \left(1 - \frac{l}{L} \right)^{2.852} \right] \Delta H + \Delta H_l^e. \tag{22.40}$$

Depending on the ratio of total energy gain due to uniform downslope or loss due to uniform upslope at the end of the lateral and the total energy head drop at the end of the lateral due to friction, different types of pressure profiles may occur. Following Wu et al. (1983), the profiles are of three main types: type I, type II (subtypes II-a, II-b, and II-c), and type III. As shown in Figure 22.12, the energy gain due to uniform downslope is a straight line, but the pressure head drop is a curve. By superimposing the pressure head drop curve on the energy gain or loss along the lateral, the pressure profile can be determined. These profiles, as shown in Figure 22.13, are now enumerated.

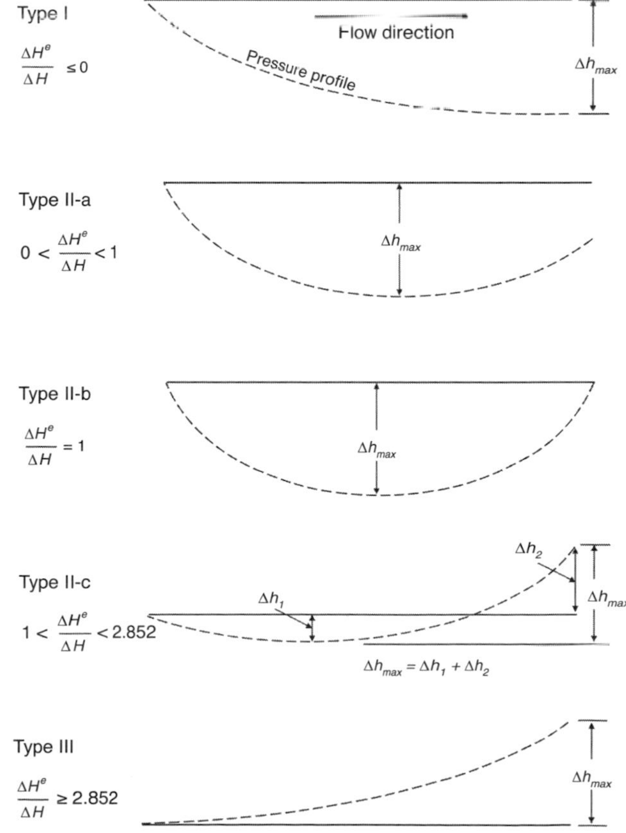

Figure 22.13 Pressure profile along a lateral line for different $\Delta H^e / \Delta H$ values.

Table 22.2 *General guidelines for trickle irrigation system needs*

Type of crop	Row spacing (m)	Plants (per ha)	Emitters (per ha)	Lateral length (m/ha)
Ordinary orchards	6.1	250	500–1,500	1,900
Dwarf orchards and vineyards	3.7	1,000	2,000	3,000
Berries and wide-spaced row crops	1.5	>15,000	>7,500	6,800
Greenhouse & close-spaced row crops	1.0	>25,000	>10,000	10,600

Adapted from Hanks and Keller, 1972.

Pressure profile type I: The pressure head decreases along the lateral and the maximum pressure is at the inlet and minimum pressure is at the downstream end. The dimensionless ratio $\Delta H^e/\Delta H \leq 0$. The energy head loss is due to friction. This type of profile occurs on flat or uphill slopes.

Pressure profile type II: The pressure head decreases along the lateral length, reaches a minimum, and then increases along the length. This profile is classified into three subtypes, depending on the value of $\frac{\Delta H^e}{\Delta H}$.

Profile type II-a: The pressure at the end of the lateral is less than the operating pressure. The lateral is on the downslope. The gain in the energy head is greater than the head loss due to friction, but the pressure at the end of the lateral is still less than the inlet pressure. The maximum pressure is at the inlet and the minimum pressure is somewhere along the lateral. This profile occurs when $\frac{\Delta H^e}{\Delta H}$ is larger than zero but less than unity.

Profile type II-b: The pressure at the end of the lateral is equal to the operating pressure. This profile occurs when $\frac{\Delta H^e}{\Delta H} = 1$. The maximum pressure is at the inlet and at the end of the lateral. The minimum pressure is located near the middle of the lateral.

Profile type II-c: The pressure at the end of the lateral is larger than the operating pressure because the slope is steeper. This profile occurs when $\frac{\Delta H^e}{\Delta H}$ is larger than 1 but less than 2.852. The maximum pressure is at the downstream end and the minimum pressure is somewhere along the lateral.

Pressure profile type III: The pressure increases along the lateral length such that $\frac{\Delta H^e}{\Delta H}$ is equal to or greater than 2.852. Here, the energy gain due to uniform steep downslope is larger than the pressure head drop due to friction along the lateral. The maximum pressure is at the downstream end and the minimum pressure is at the inlet.

Wu and Gitlin (1980) have reported that profile II-b for which $\frac{\Delta H^e}{\Delta H}$ equals 1 corresponds to the minimum pressure difference for a given lateral and operating pressure and is therefore the ideal profile.

22.8 UNIFORMITY OF EMITTER FLOW

The uniformity of emitter flow along a lateral constitutes the main criterion for drip irrigation design by achieving as little variation in pressure as practically possible. The variation of emitter flow is controlled by the pressure variation. There are several uniformity measures that are employed in drip system design. Some of these are discussed in Chapter 18 and will not be repeated here.

22.9 EMISSION DEVICE SELECTION AND CAPACITY

The selection of an emission device depends on the type of crop, soil infiltration, frost protection for a crop, cost, and operator preference, and is to be made from point-source emitters, line-source emitters, bubblers, and micro-sprinklers. Micro-sprinklers are preferred for frost protection, and erosion, pest, and disease control. For soil with high infiltration, bubblers and micro-sprinklers may be considered. For row crops, line-source emitters or closely spaced point-source emitters may be desirable. Hanks and Keller (1972) provided general guidelines for equipment needs in trickle irrigation systems for different types of crops, as shown in Table 22.2.

Example 22.6: Consider an orchard on a rectangular field 500 m long and 300 m wide. A trickle irrigation system is planned and each tree is served by five emitters. Determine the number of emitters required, required emitter discharge, and length of lateral, if, for peak period requirements at full tree maturity, the peak period water requirement is 4 mm/day, the distribution pattern efficiency is 90%, and the operating time is 20 h/day.

Solution: First, the number of emitters needs to be determined. Table 22.2 gives a plant density of 250 trees per hectare for an ordinary orchard. Therefore, the number of emitters needed can be calculated as

$$N = (500\text{ m} \times 300\text{ m}) \times \left(\frac{1\text{ ha}}{10{,}000\text{ m}^2}\right) \times \left(\frac{250\text{ trees}}{\text{ha}}\right) \times \left(\frac{5\text{ emitters}}{\text{tree}}\right) = 18{,}750.$$

The water application per hour, d_a, is now computed as

$$d_a = \frac{4\frac{\text{mm}}{\text{day}}}{0.9 \times 20\frac{\text{h}}{\text{day}}} = 0.222\frac{\text{mm}}{\text{h}}.$$

The required emitter discharge q is determined as

$$q = \left(0.222\frac{\text{mm}}{\text{h}}\right) \times \left(\frac{1\ \text{m}}{1000\ \text{mm}}\right) \times (500\ \text{m} \times 300\ \text{m}) \times \left(\frac{1000\ \text{L}}{1\ \text{m}^3}\right) \times \frac{1}{18{,}750} = 1.78\frac{\text{L}}{\text{h}}.$$

The required length of the lateral, using 1900 m/ha from Table 22.2, can be computed as

$$L = \left(1900\frac{\text{m}}{\text{ha}}\right) \times (500\ \text{m} \times 300\ \text{m}) \times \left(\frac{1\ \text{ha}}{10{,}000\ \text{m}^2}\right) = 28{,}500\ \text{m}$$

The capacity of an emission device can be computed as

$$Q_C = \frac{KD_aA_i}{(T - T_0)E_a}, \tag{22.41}$$

where Q_C is the emission device capacity (L/h, gph); D_a is the depth of water applied (mm, in.); A_i is the area irrigated by the emission device (m^2, ft^2); T is the time used to apply D_a hours of irrigation; T_0 is the time in hours when irrigation is not done (for maintenance, normally taken as 0.5 h/day); E_a is the application efficiency (%); and K is a unit constant equal to 100 for Q_C in L/h, D_a in mm, A_i in m^2, T in hours, and T_0 in hours; and equal to 62.33 for Q_C in gph, D_a in inches, A_i in ft^2, T in hours, and T_0 in hours.

The depth of water applied per irrigation can be computed as

$$D_a = \frac{T \times D_{IR}}{0.24 \times N_d}, \tag{22.42}$$

where D_{IR} is the design daily irrigation requirement (mm/day, in./day) corresponding to the desired depth of irrigation (mm or in.); and N_d is the ratio in percentage of the number of emission devices operating per irrigation to the total number of emission devices multiplied by 100. The value of T is computed based on operator preference, but must satisfy

$$T \leq \frac{0.24 \times N_d \times D_a}{D_{IR}}. \tag{22.43}$$

The area irrigated by an emission device can be determined as

$$A_i = \frac{S_rS_pP_i}{100 \times N_e}, \tag{22.44}$$

where A_i is the area irrigated (m^2, ft^2); S_r is the spacing between adjacent rows (m, ft); S_p is the spacing between emission points (m, ft); P_i is the percentage of crop area being irrigated; and N_e is the number of emission devices at each emission point. N_e can be computed as the total area to be wetted divided by the area wetted per emission device. Figure 22.14 shows the layout of single and double laterals. For single laterals with equally spaced emission points, N_e can be determined as

$$N_e = \frac{K \times P_i \times S_p \times S_r}{D_w \times s_e}, \tag{22.45}$$

where N_e is the number of emission devices per emission point; P_i is the percentage S_p times S_r irrigated; D_w is the maximum diameter of the wetted circle formed by a single-point emission device (cm, in.); s_e is the spacing between emission devices of an emission point or the spacing between emitters (cm, in.); and K is a unit constant equal to 100 for S_p and S_r in m and D_w and s_e in cm; and equal to 1.44 for S_p and S_r in ft and D_w and s_e in inches. For a single lateral, the spacing between laterals is usually set as the spacing between adjacent rows of plants (i.e., $S_r = S_l$). For a single lateral where the emitter distance s_e is greater than the optimum emitter spacing ($s_e = s'_e = 0.8D_w$), s_e in eq. (22.45) should be replaced by $0.8D_w$.

For double laterals, zigzag, pigtail, and multi-exit layouts, N_e can be computed as

$$N_e = \frac{2 \times K \times P_i \times S_p \times S_r}{s'_e(D_w \times s_e)}, \tag{22.46}$$

where s'_e is the optimum emitter spacing ($s'_e = 0.8D_w$). In order to have the largest A_i and no significant dry areas between the double laterals, the spacing between them should equal D_w. If the maximum wetting is not achieved and $s_e < 0.8D_w$, then s'_e in eq. (22.46) should be replaced by s_e.

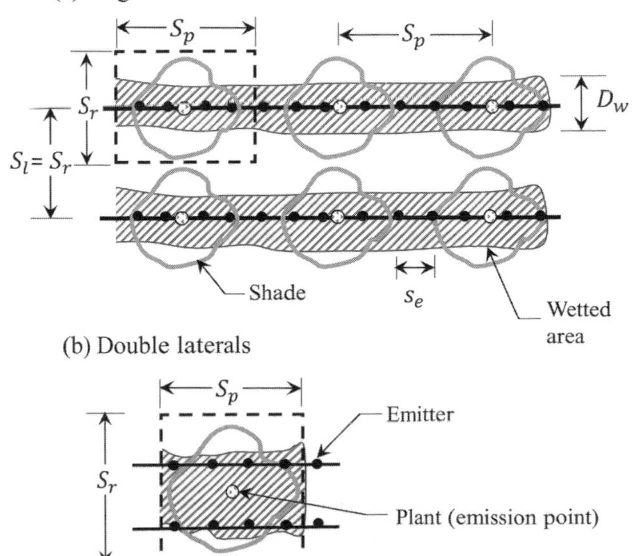

Figure 22.14 Single and double laterals layouts.

For micro-sprinkler irrigation, the number of micro-sprinklers per emission point can be determined as

$$N_e = \frac{P_i \times S_p \times S_r}{100 \times (A_s + \frac{D_w P_i S_p}{2K})}, \quad (22.47)$$

$$s_e = D_t + \frac{D_w}{2K}, \quad (22.48)$$

where A_s is the area wetted by a micro-sprinkler (m^2, ft^2); $P_i \times S_p$ is the perimeter of the area wetted by the micro-sprinkler (m, ft); D_t is the distance of throw (m, ft); and K is a unit constant equal to 100 for s_e and D_t in m and D_w in cm; and equal to 12 for s_e and D_t in ft and D_w in inches.

Example 22.7: Compute the number of emission devices per plant and emission device capacity for a drip system if the readily available water is 4.00 in., D_{IR} is 0.25 in./day, D_w is 48 in., P_i is 25%, the spacing of trees is 16 × 16 ft, N_d is 15%, and E_a is 90%.

Solution: Given the spacing of trees is $S_r = 16$ ft $= 192$ in. and $P_i = 25\%$, assume s_e is equal to the optimum drip emitter spacing, then

$$s_e = s_e' = 0.8D_w = 0.8 \times 48 \text{ in.} = 38.4 \text{ in.}$$

For one lateral, the number of emitters is $N_e = S_r/s_e = 192 \text{ in.}/38.4 = 5$.

From eq. (22.45), P_i is the percentage of crop area being irrigated:

$$P_i = \frac{N_e \times D_w \times s_e}{2 \times K \times S_p \times S_r} = \frac{(5)(48)(38.4)}{(2)(1.44)(16)(16)} = 12.5.$$

Therefore, one lateral cannot satisfy the requirement.

For a double lateral, the number of emission devices per plant is

$$N_e = \frac{2 \times K \times P_i \times S_p \times S_r}{s_e'(D_w \times s_e)} = \frac{2 \times 1.44 \times 25 \times (16 \times 16)}{38.4 \times (38.4 + 48)} = 6.$$

The area irrigated by an emission device can be determined as

$$A_i = \frac{(S_p \times S_r) \times P}{100 \times N_e} = \frac{(16 \text{ ft} \times 16 \text{ ft}) \times 25}{100 \times 6} = 10.67 \text{ ft}^2.$$

Given the desired depth of irrigation $D_a = 4.00$ in., N_d is 15%, $D_{IR} = 0.25$ in./day, the irrigation hours are computed as

$$T \leq \frac{0.24 \times N_d \times D_a}{D_{IR}} = \frac{0.24 \times 15 \times 4.00}{0.25} = 57.6 \text{ h}.$$

Assume the time in hours when irrigation is not done due to the need for maintenance is $T_0 = 0.5 \frac{\text{h}}{\text{day}}$, then the total time during the irrigation hours is

$$T_0 = 0.5 \frac{\text{h}}{\text{day}} \times 57.6 \text{ h} \times \frac{1 \text{ day}}{24 \text{ h}} = 1.2 \text{ h}.$$

The emission device capacity for the drip system is computed as

$$Q_C = \frac{K D_a A_i}{(T - T_0)E_a} = \frac{62.33 \times 4 \times 10.67}{(57.6 - 1.2) \times 90} = 0.52 \text{ gph } (2.0 \text{ L/h}).$$

22.10 METHODS OF DRIP IRRIGATION DESIGN

In drip irrigation there are four basic equations: the emitter discharge relation (eq. 22.3), the continuity equation (eq. 22.5), the energy conservation equation coupled with momentum conservation (eq. 22.16), and the head loss equation (Darcy–Weisbach [eq. 22.17] or Hazen–Williams [eq. 22.27]), plus the uniformity equation (one of eqs [18.11]–[18.20]). There are four unknown variables: Q_{n+1}, q_{n+1}, H_{n+1}, and H_{fn+1} at any location of the lateral ($n + 1$) with known values of Q_n, q_n, H_n, and H_{fn}, as well as of other parameters z_n, z_{n+1}, f_{n+1}, D, s, c, and y. There are many methods to solve for these unknowns and hence for designing drip irrigation systems. Some of the methods are analytical and some are numerical. The focus of this chapter is on simple analytical and numerical methods.

Figure 22.15 shows typical drip system design procedures. The preliminary design factors (Steps 1–5) and Steps 8–10 were discussed in Chapter 21 and will not be introduced here. In the previous section, emission device selection and capacity (Step 6) are also discussed. Here, the focus is the lateral design of drip irrigation (Step 7). In drip irrigation calculations, we may often calculate discharge variation along the lateral, coefficient of variation of discharge, mean discharge, pressure variation, maximum pressure difference, design length of the lateral and ideal design length, average head loss, and ratio of energy gain or loss due to elevation to total energy.

22.10.1 Energy Gradient Approach: Zero Slope

This approach, in which flow in a lateral is assumed steady but spatially varied with decreasing discharge in the line, was proposed by Wu and Gitlin (1973, 1975). It is assumed here that the lateral is on a flat slope (S_0 is zero); hence, the energy head is the same as the pressure head and the head loss is due to friction. It is also assumed that all emitter flows are constant or are subject to small variation, which, strictly speaking, is not true. The energy head (H) or pressure head (h) gradient at any position x along the lateral can be expressed as

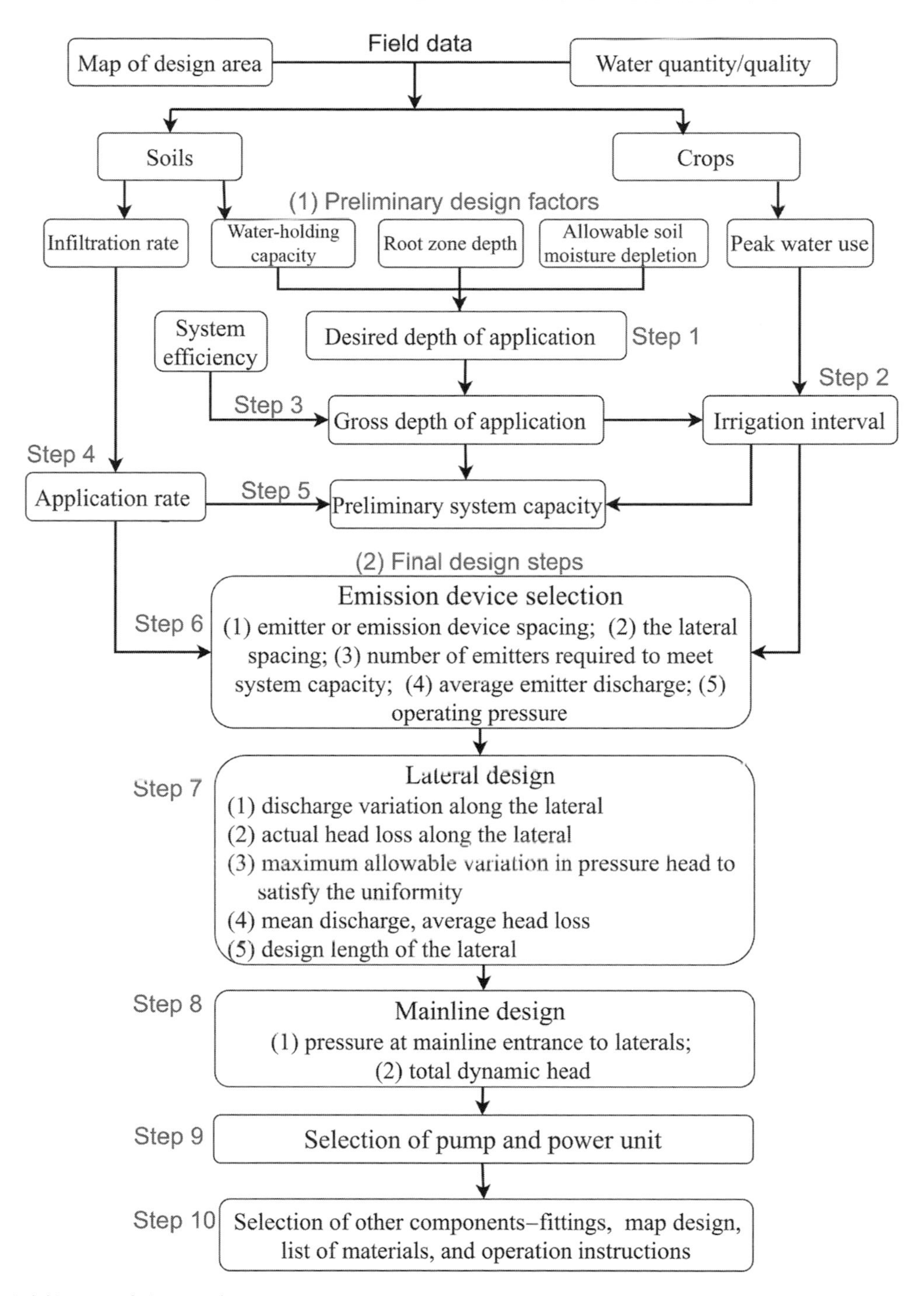

Figure 22.15 Typical drip system design procedures.

$$\frac{dh}{dx} = -KQ_x^m, \tag{22.49}$$

where K is a constant and m is an exponent which equals 1 for laminar flow, 1.75 for turbulent flow, and 2 for fully turbulent flow. K can be obtained from the Blasius equation (eq. 22.23) for smooth pipes and the Hazen–Williams equation (eq. 22.27), that is

$$K = \frac{2.53\nu^{0.25}A^{0.25}}{g\pi^2 D^{5.25}}. \tag{22.50}$$

Let the emitters be equally spaced with a spacing of s along the lateral. If there are n emitters, as shown in Figure 22.16, then the total discharge Q in the lateral and the discharge in any section j can be expressed as

$$Q = \sum\nolimits_{i=1}^{n} q_i; \;\; Q_j = \sum\nolimits_{j=1}^{n} q_i^{\,j} \tag{22.51}$$

where q_i is the outflow from the ith emitter. Since the emitter outflow is assumed constant, $q_i = q$, $i = 1, 2, \ldots, n$. Then,

$$Q_j = (n - j + 1)q. \tag{22.52}$$

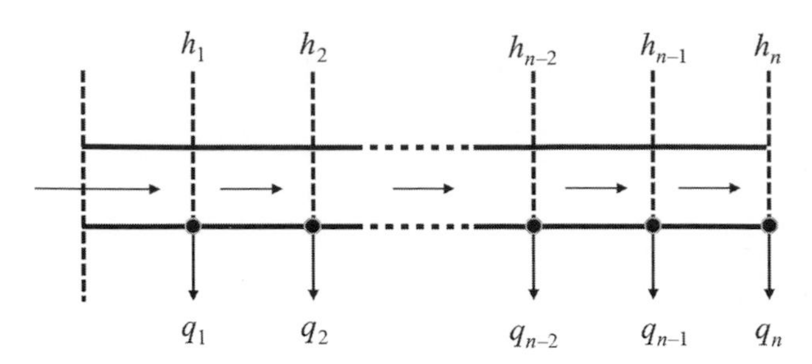

Figure 22.16 Pressure and discharge along the laterals.

The pressure head drop over any length ΔL can be computed using eq. (22.50). If ΔL is fixed, then over any section j the head drop Δh_j can be written as

$$\Delta h_j = -KQ_j^{1.75}\Delta L. \tag{22.53}$$

The total head drop ΔH is the sum of head drops over all the sections as

$$\Delta H = \sum_{i=1}^{n} \Delta h_i = -K\left(\sum_{i=1}^{n} Q_i^{1.75}\right)\Delta L. \tag{22.54}$$

The head drop over any section from eq. (22.52) can be expressed in terms of emitter flow as

$$\Delta h_j = -K[(n-j+1)q]^{1.75}\Delta L. \tag{22.55}$$

The total head drop over the lateral can be written as

$$\Delta H = -Kq^{1.75}\sum_{i=1}^{n}[n-i+1]^{1.75}\Delta L. \tag{22.56}$$

The head drop can also be computed as an approximation using the mean discharge in the lateral. The total discharge in section 1 is nq. Hence, the mean discharge Q_m can be expressed as $(n+1)q/2$. Then, the head drop ΔH_m with the mean discharge can be written as

$$\Delta H_m = -Kq^{1.75}\left[\frac{n+1}{2}\right]^{1.75} L. \tag{22.57}$$

Wu and Gitlin (1973) showed that the error due to mean discharge can be reduced significantly by taking the mean discharge over several sections rather than using the entire lateral length.

The uniform emitter flow can be obtained by adjusting the size and spacing of emitters and the length of the lateral. The discharge from each emitter depends on the emitter size and the pressure head at the emitter. The emitter diameter can be determined as follows.

The emitter discharge q can be expressed as

$$q = av = \frac{\pi}{4}d^2 v, \tag{22.58}$$

where a is the cross-sectional area of the opening of the nozzle and v is average velocity. Also,

$$q = aC_0\sqrt{2gh} = \frac{\pi}{4}C_0\sqrt{2g}d^2h^{1/2}, \tag{22.59}$$

where C_0 is the discharge coefficient. Since q is constant, differentiating eq. (22.59) yields

$$\frac{1}{2}d^2h^{-1/2}dh + 2h^{1/2}d\,dd = 0. \tag{22.60}$$

Equation (22.60) can be arranged as

$$\frac{dh}{h} = -4\frac{dd}{d}. \tag{22.61}$$

Let the initial condition be $d = d_0$ at $h = h_0$. Then, integration of eq. (22.61) yields

$$\frac{h}{h_0} = \left(\frac{d}{d_0}\right)^4. \tag{22.62}$$

Equation (22.62) shows that emitter diameter d can be determined if h_0 and d_0, which actually control the design uniform discharge, are known.

Now the length of the lateral can be determined. From the Darcy–Weisbach equation, one can write

$$h_f = f\frac{L}{d}\frac{v^2}{2g} = f\frac{L}{d}\frac{q^2}{a^2 2g}. \tag{22.63}$$

If the diameter is known, then eq. (22.63) can be written as

$$h_f = C_L L, \quad C_L = \frac{fq^2}{2gda^2}. \tag{22.64}$$

If an initial condition h_0 at L_0 is known, then eq. (22.64) can be written as

$$\frac{h}{h_0} = \frac{L}{L_0}, \quad \text{or } L = \frac{h}{h_0}L_0. \tag{22.65}$$

Now the spacing between emitters can be determined. The spacing can be adjusted such that the discharge per unit length is constant. Expressed algebraically,

$$\frac{q}{\Delta L} = \frac{q_1}{\Delta L_1} = \frac{q_2}{\Delta L_2} = \cdots = \frac{q_n}{\Delta L_n}. \tag{22.66}$$

The spacing between emitters can be determined as

$$\Delta L_i = \left(\frac{q_i}{q}\right)\Delta L. \tag{22.67}$$

Example 22.8: Consider a 0.75 in. (19 mm) diameter lateral (plastic pipe) 250 ft (76 m) long with emitters 2 ft (0.6 m) apart and emitter discharge of 1.5 gph (5.7 L/h). The pressure distribution is given as shown in Figure 22.17a. Determine the design of uniform drip irrigation. Wu and Gitlin (1973) have solved this example, which was given by Myers and Bucks (1972).

Solution: Here, $L = 250$ ft (76 m), $\Delta L = 2$ ft (0.6 m), and $q = 1.5$ gph (5.7 L/h). The pressure variation line is divided into several segments and the average pressure is computed for each segment. If the pressure and emitter diameter in the first segment are considered to define the initial condition, then the initial condition is: $h_0 = h_1 = 4.5$ ft (1.4 m) and $d_0 = d_1 = 0.029$ in ID (0.74 mm).

The diameter of the emitter for the second segment is computed as

$$d_2 = \left(\frac{h_0}{h_1}\right)^{0.25} d_0 = \left(\frac{4.5\text{ ft}}{4\text{ ft}}\right)^{0.25} \times 0.029\text{ in.}$$
$$= 0.0299\text{ in. } (0.76\text{ mm}).$$

In a similar manner, the diameters of other segments can be computed: $d_3 = 0.0308$ in. (0.78 mm); $d_4 = 0.031$ in. (0.8 mm); and $d_5 = 0.0326$ in. (0.83 mm). The graphical solution (Figure 22.17b) of Myers and Bucks (1972) yields: $d_1 = 0.029$ in. (0.74 mm); $d_2 = 0.030$ in. (0.76 mm); $d_3 = 0.031$ in. (0.78 mm); $d_4 = 0.032$ in. (0.81 mm); and $d_5 = 0.033$ in. (0.84 mm). These values are close to the above values calculated analytically.

Now the design length of the lateral is computed. The length of each segment is calculated using eq. (22.65). Segment 1 is the initial condition: $h_1 = h_0 = 4.5$ ft (1.4 m) and $L_1 = L_0 = 2.8$ in. (71 mm). The length of the second segment is $L_2 = (4/4.5) \times 2.8 = 2.5$ in. (64 mm). Likewise, lengths of other segments are: $L_3 = 2.2$ in. (56 mm), $L_4 = 2.0$ in. (51 mm), and $L_5 = 1.75$ in. (45 mm).

The spacing between emitters is now computed using the head–discharge relation shown in Figure 22.17b. Following Myers and Buck (1972), emitter 2 is selected (0.030 in. ID) as an example and shown in Figure 22.17b. As shown in Figure 22.17, with $h_1 = 4.5$ ft, the discharge

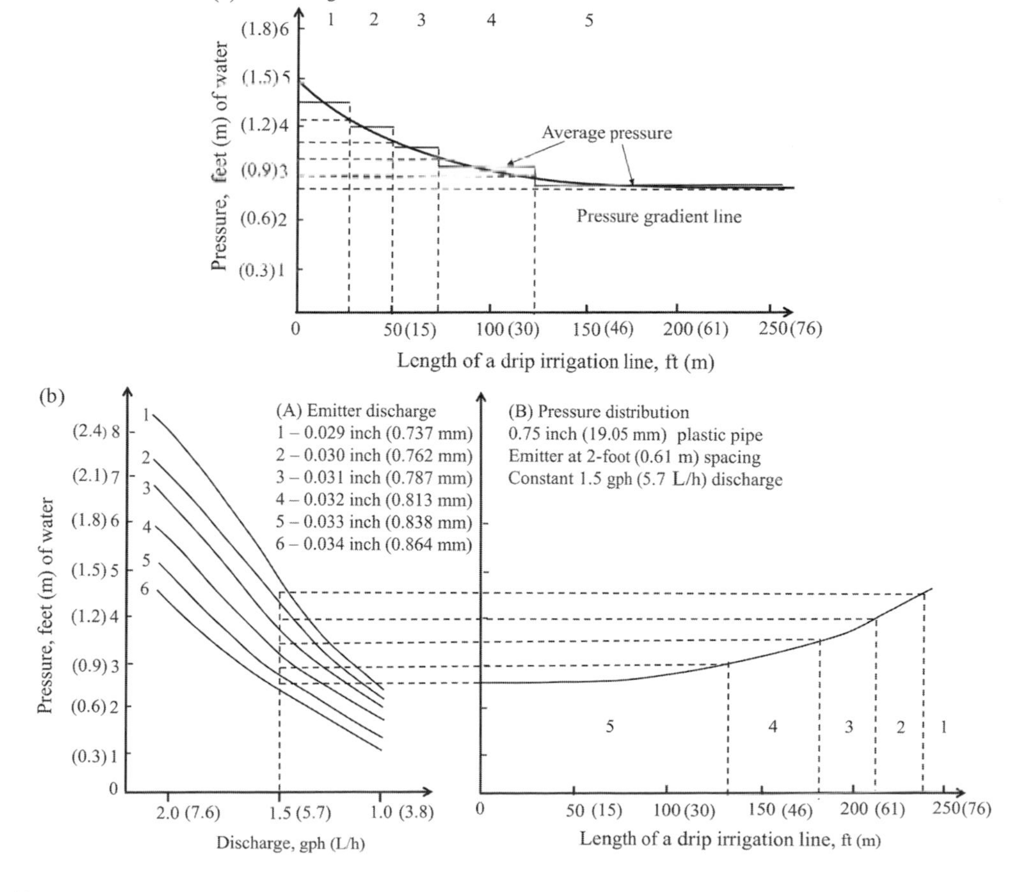

Figure 22.17 (a) Pressure gradient line and average pressure for several sections along a drip irrigation line. (b) Graphic solution of emitter size distribution. (Adapted from Myers and Bucks, 1972. Copyright © 1972 by American Society of Civil Engineering. Reprinted with permission from ASCE.)

of emitter 1 (0.029 in. ID) is $q_1 = 1.56$ gph. Given $q = 1.5$ gph and $\Delta L = 2\,\text{ft} = 24\,\text{in.}$, the emitter spacing in the segment $L_1 = (1.56/1.5) \times 24 = 25\,\text{in.}$ (640 mm). As shown in Figure 22.18, with $h_2 = 4$ ft (1.2 m), $q_2 = 1.45$ gph. The emitter spacing in segment 2 is: $\Delta L_2 = (1.56/1.5) \times 24 = 25\,\text{in.}$ (580 mm). Likewise, $\Delta L_3 = 22\,\text{in.}$ (560 mm), $\Delta L_4 = 21\,\text{in.}$ (530 mm), and $\Delta L_5 = 19\,\text{in.}$ (480 mm).

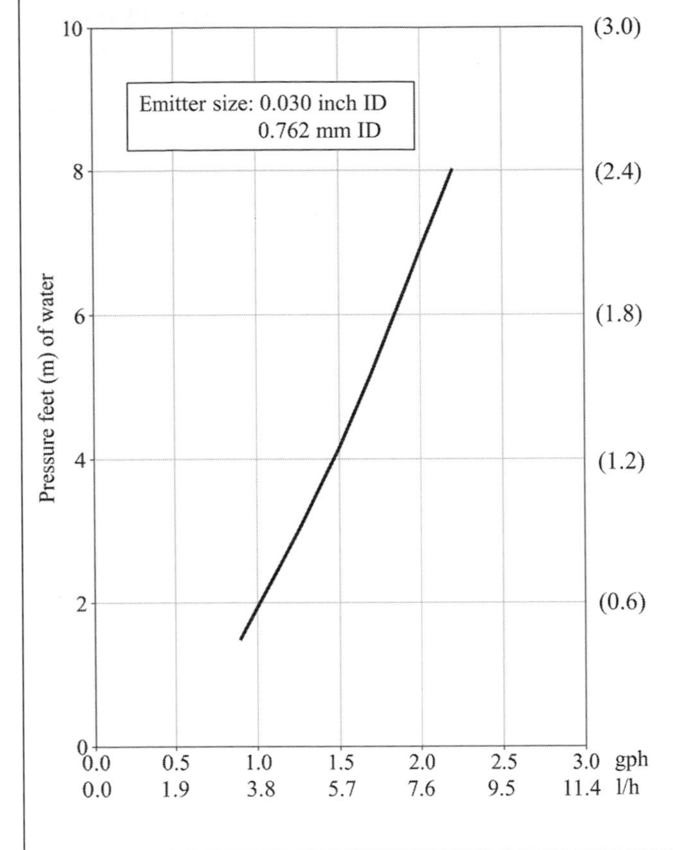

Figure 22.18 Discharge–head (pressure) relation for emitter 2. (Adapted from Myers and Bucks, 1972. Copyright © 1972 by American Society of Civil Engineering. Reprinted with permission from ASCE.)

The above energy gradient approach can be generalized by considering the head variation and the consequent discharge variation as continuous functions of space. The total discharge Q_x in the line at a given length x from the inlet is a linear function of length x and can be expressed as

$$Q_x = \left(N - \frac{x}{s}\right)q, \tag{22.68}$$

where N is the total number of emitters in the lateral. Combining eqs (22.49) and (22.68),

$$\frac{\Delta H}{dx} = -aq^m\left(N - \frac{x}{s}\right)^m. \tag{22.69}$$

If the length of the lateral is L, then $L = N_s$. Equation (22.69) can be written as

$$\frac{dH}{dx} = -\frac{aq^m}{s^m}(L - x)^m. \tag{22.70}$$

At $x = 0$, the energy head is H_0 and at distance x it is denoted as H_x. Then, the energy head drop over the distance or length is $H_0 - H_x = \Delta H_x$. The total energy drop at the length x, ΔH_x, can be written as

$$H_x - H_0 = -\Delta H_x = \frac{-aq^m}{s^m}\int_0^x (L - x)dx. \tag{22.71}$$

The initial condition for eq. (22.71) can be written as follows: At $l = 0$, $\Delta H_l = 0$. Solution of eq. (22.71) yields the total energy head drop over length x as:

$$\Delta H_x = -\frac{aq^m}{s^m}\frac{(L - x)^{m+1}}{m + 1} - \frac{aq^m}{s^m(m + 1)}L^{m+1} \tag{22.72}$$

or

$$\Delta H_x = \frac{aq^m}{s^m(m + 1)}\left[(L - x)^{m+1} - L^{m+1}\right]. \tag{22.73}$$

The total energy head drop due to friction (also called friction drop) ΔH at the downstream end of the lateral (i.e., $x = L$), can be written as

$$\Delta H = \frac{aq^m}{s^m(m+1)} L^{m+1}. \tag{22.74}$$

Now the ratio of energy head drop R_x can be written as

$$R_x = \frac{\Delta H_x}{\Delta H} = -\left[\frac{(L-x)^{m+1} - L^{m+1}}{L^{m+1}}\right] = 1 - \left(\frac{L-x}{L}\right)^{m+1}. \tag{22.75}$$

Denoting $i = x/L$, the energy head drop ratio can be written as

$$R_i = 1 - (1-i)^{m+1}. \tag{22.76}$$

The friction drop ratio for the three types of flow (Figure 22.19) can be written as follows:

$$\text{Laminar flow}: \quad R_i = 1 - (1-i)^2. \tag{22.77}$$

Turbulent flow in smooth pipes with the Blasius equation:

$$R_i = 1 - (1-i)^{2.75}. \tag{22.78}$$

Fully turbulent flow with friction coefficient

$$f \text{ constant}: \quad R_i = 1 - (1-i)^3. \tag{22.79}$$

If the Hazen–Williams equation is used, then the energy head drop ratio can be written as

$$R_i = 1 - (1-i)^{2.852}. \tag{22.80}$$

This equation lies between the equations of turbulent flow in smooth pipe and fully turbulent flow.

Now the total energy head drop along the lateral can be computed from the dimensionless energy gradient line, provided the total energy head drop at the end of the lateral is known. From eq. (22.74), with $L = Ns$, the total energy head drop can be expressed as a function of total discharge $Q_t = Nq$ as

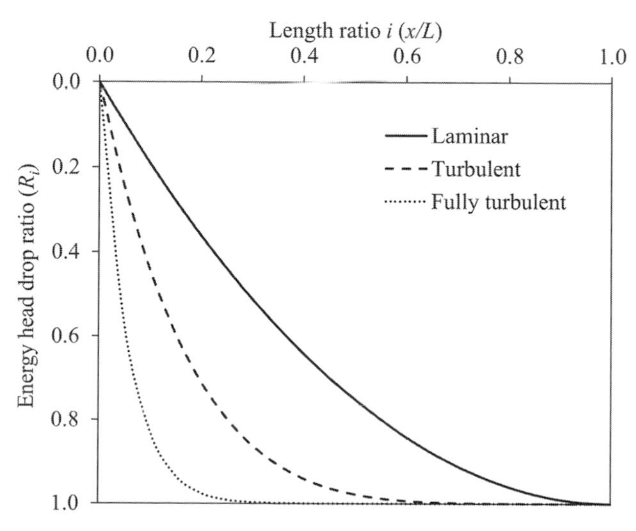

Figure 22.19 The energy head drop ratio as a function of length ratio for the three types of flow.

$$\Delta H = \frac{aQ_t^m}{m+1} L. \tag{22.81}$$

The total energy head drop can also be expressed in terms of average discharge Q_m as

$$\Delta H = -\frac{(2Q_m)^m}{m+1} L. \tag{22.82}$$

If the Hazen–Williams equation is used then the total energy head drop at the end of the lateral can be expressed as (Wu and Gitlin, 1975):

$$\Delta H = \frac{K}{2.852} \frac{Q^{1.852}}{D^{4.871}} L, \tag{22.83}$$

where K is a constant for a given friction coefficient C in the Hazen–Williams equation for pipe flow with constant discharge.

22.10.2 Energy Gradient Approach: Nonzero Slope

Consider the case when the lateral is on sloping ground (Wu, 1992). The pressure head variation along a lateral can be computed directly from the energy gradient line (EGL). The pressure head h_i at any location i on the lateral can be expressed as

$$h_i = H - \Delta H_i + \Delta H_i^e, \tag{22.84}$$

where H is the total energy at the inlet, given as the operating pressure, ΔH_i is the head drop due to friction, and ΔH_i^e is the gain in energy head for the downsloping lateral at position i. Dividing eq. (22.84) by H, one gets

$$\frac{h_i}{H} = 1 - R_i \frac{\Delta H}{H} + R_i^e \frac{\Delta H^e}{H}, \tag{22.85}$$

where R_i is the energy head (friction) drop ratio given by eq. (22.76), ΔH^e is the total energy gain at the end of the lateral due to downslope, ΔH is the total energy head drop at the end of the lateral, and $R_i^e = \frac{\Delta H_i^e}{\Delta H^e}$ has the same value for i as for uniform slope. Equation (22.85) shows that the pressure head along the lateral can be computed by knowing the operating pressure, total energy head drop, and total energy gain at the end of the lateral.

Now the emitter flows can be computed using eq. (22.85) as

$$q_i = q_0 \left(1 - R_i \frac{\Delta H}{H} + R_i^e \frac{\Delta H^e}{H}\right)^y, \tag{22.86}$$

where q_i is the emitter flow at the length ratio i and q_0 is the emitter flow at the inlet, computed using the operating pressure.

The emitter flows in a submain can also be computed by the EGL along the submain using a similar approach. If the number of outlets along the submain is more than five, Wu and Gitlin (1975) defined

$$R_j = 1 - (1 - j)^{m+1}, \tag{22.87}$$

in which j is the length ratio along the submain; $R_i = \Delta H_{sj}/\Delta H_s$ is the ratio of energy head drop due to friction along the submain, ΔH_{sj} is the energy head drop at a given length ratio j, and ΔH_s is the total head drop at the end of the submain.

The emitter flows along any lateral that connects the submain at the length ratio j can be expressed as

$$q_{ij} = q_{0j}\left(1 - R_i\frac{\Delta H_j}{h_{j(s)}} + R_i^e\frac{\Delta H^e}{h_{j(s)}}\right)^y, \tag{22.88}$$

where q_{ij} is the emitter flow from the jth lateral at the ith location; q_{0j} is the flow from the first emitter of the jth lateral; ΔH_j is the total energy head drop of the jth lateral due to friction; ΔH^e is the total energy gain at the end of the lateral due to downslope; and $h_{j(s)}$ is the pressure head at the inlet of the jth lateral.

A general expression for all emitter flows in a submain was derived by Wu and Irudayaraj (1989):

$$q_{ij} = q_{00}\left[\left(1 - R_j\frac{\Delta H_s}{H} + R_j^e\frac{\Delta H_s^e}{H}\right) - R_i\frac{\Delta H_0}{H}\right.$$
$$\left.\times\left(1 - R_j\frac{\Delta H_s}{H} + R_j^e\frac{\Delta H_s^e}{H}\right)^{\frac{m}{2}} + R_i^e\frac{\Delta H^e}{H}\right]^y, \tag{22.89}$$

where q_{00} is the flow from the operating pressure H and is regarded as flow from the first emitter of the first lateral; ΔH_0 is the total head drop due to friction at the end of the first lateral; ΔH_s is the head drop due to friction at the end of the submain; ΔH_s^e is the total energy gain by the submain due to downslope; and R_j^e is the energy gain ratio along the subdomain due to downslope with respect to the length ratio j. It may be noted that ΔH_s is computed from the total discharge Q_0, and ΔH_0 is computed with the total discharge Q_0 for the first lateral, nQ_0/N, where N is the total number of emitters in the submain unit and n is the number of emitters along a lateral. Equation (22.89) can be used to compute flows from all emitters in a submain.

Mean discharge approximation: The total discharge of a lateral can be calculated as the mean emitter flow and the number of emitters. Anyoji and Wu (1987) expressed mean emitter flow $\overline{q}$ if the coefficient of variation of the pressure head was less than 20% as

$$\overline{q} = k\overline{h}^y, \tag{22.90}$$

where $\overline{h}$ is the mean pressure head, k is the proportionality constant, and y is the emitter exponent. Both k and y are constant for a given emitter. The mean pressure head along the lateral can be expressed as (Anyoji and Wu, 1987):

$$\overline{h} = H - \frac{m+1}{m+2}\Delta H + \frac{1}{2}\Delta H^e. \tag{22.91}$$

The total discharge Q of the lateral can be written as the sum of all emitter flows:

$$Q = n\overline{q}, \tag{22.92}$$

where n is the total number of emitters along the lateral. The total discharge Q_0 due to the operating pressure can be expressed as

$$Q_0 = nq_0, \tag{22.93}$$

where q_0 is the emitter flow due to the operating pressure. The discharge ratio can be expressed as

$$\frac{Q}{Q_0} = \frac{\overline{q}}{q_0} = \left(\frac{\overline{h}^y}{H}\right). \tag{22.94}$$

Inserting eq. (22.91) in eq. (22.94), one obtains

$$\frac{Q}{Q_0} = \left(1 - \frac{m+1}{m+2}\frac{\Delta H}{H} + \frac{1}{2}\frac{\Delta H^e}{H}\right)^y. \tag{22.95}$$

If the total energy head drop is computed using the Hazen–Williams equation with $m = 1.852$, then eq. (22.95) can be expressed as

$$\frac{Q}{Q_0} = \left(1 - \frac{m+1}{m+2}\frac{\Delta H}{H} + \frac{1}{2}\frac{\Delta H^e}{H}\right)^y, \tag{22.96}$$

where K_1 is the coefficient in the Hazen–Williams equation for determining the total energy head drop and equals $K/2.852$, where K is given in eq. (22.83).

The mean pressure in a submain unit can be computed by assuming that the mean pressure is located at a lateral that is connected to the submain at the location of mean submain pressure. The mean submain pressure $\overline{h}$ can be expressed as (Anyoji and Wu, 1987):

$$\overline{h} = H - \frac{m+1}{m+2}(\Delta H_s + \Delta H_l) + 0.5\left(\Delta H_s^e + \Delta H_l^e\right), \tag{22.97}$$

where H is the operating pressure at the inlet of the submain; ΔH_s is the total head drop at the end of the submain due to friction computed from the total discharge equal to $N\overline{q}$; N is the total number of emitters in the submain unit; $\overline{q}$ is the mean emitter flow; ΔH_l is the total head drop at the end of the lateral connecting the submain at its mean pressure location computed from the total discharge equal to $n\overline{q}$; ΔH_s^e is the total energy gain or loss at the end of the submain due to the uniform slope; and ΔH_l^e is the total energy gain or loss due to the uniform slope at the end of the lateral connecting the submain at its mean pressure.

Now the discharge ratio for the submain can be expressed as

$$\frac{Q}{Q_0} = \left[1 - 0.7404\left(K_1\frac{Q^{1.852}}{D_s^{4.871}} + K_1\frac{\left(\frac{nQ}{N}\right)^{1.852}}{D^{4.871}}\right)\right.$$
$$\left.+\frac{1}{2}\left(\Delta H_s^e + \Delta H_l^e\right)^y\right], \tag{22.98}$$

where K_1 is the constant in the Hazen–Williams equation for the lateral and submain; D_s is the submain diameter; and D is the lateral diameter.

Computation of maximum pressure difference: A lateral has several emitters (orifices) and flow in each section between two emitters can be regarded as a pipe flow whose friction drop can be computed using the Hazen–Williams equation. Then, the friction drop along the lateral can be determined by summing up friction drops of all the sections.

The total friction drop at length l can be written as

$$\Delta H_l = \frac{K_1}{2.852}\frac{LQ^{1.852}}{D^{4.871}}\left[1-\left(1-\frac{l}{L}\right)^{2.852}\right]. \tag{22.99}$$

The total friction drop at the end of the lateral can be written as

$$\Delta H = \frac{K_1}{2.852}\frac{LQ^{1.852}}{D^{4.871}}. \tag{22.100}$$

Dividing eq. (22.99) by eq. (22.100), we obtain

$$\frac{\Delta H_l}{\Delta H} = \left[1-\left(1-\frac{l}{L}\right)^{2.852}\right]. \tag{22.101}$$

On a uniform slope the energy gain downslope at any length can be expressed as

$$\frac{\Delta H_l^e}{\Delta H^e} = \frac{l}{L} \quad \text{then } \Delta H_l^e = \frac{\Delta H_l^e}{\Delta H^e}\Delta H^e = \left(\frac{l}{L}\frac{\Delta H^e}{\Delta H}\right)\Delta H. \tag{22.102}$$

The pressure difference along the lateral for pressure profiles type I (zero slope), type II a, and II-b can be obtained from eqs (22.101) and (22.102) as

$$\frac{\Delta h}{\Delta H} = \left[1-\left(1-\frac{l}{L}\right)^{2.852} - \left(\frac{l}{L}\frac{\Delta H^e}{\Delta H}\right)\right]. \tag{22.103}$$

If the slope is upward, then eq. (22.103) becomes

$$\frac{\Delta h}{\Delta H} = \left[1-\left(1-\frac{l}{L}\right)^{2.852} + \left(\frac{l}{L}\frac{\Delta H^e}{\Delta H}\right)\right]. \tag{22.104}$$

The maximum pressure difference Δh_{max} for upslope and flat slope will be at the end of the lateral. For upslope it can be written as

$$\frac{\Delta h_{max}}{\Delta H} = \left(1+\frac{\Delta H^e}{\Delta H}\right)\Delta H. \tag{22.105}$$

It should be noted that the absolute values of $\Delta H^e/\Delta H$ are used in eqs (22.104) and (22.105). For flat slopes, eq. (22.105) reduces to

$$\frac{\Delta h_{max}}{\Delta H} = 1. \tag{22.106}$$

Now the location of the maximum pressure difference for profiles II-a and II-b can be determined by differentiating eq. (22.103) with respect to l/L and equating the derivative to zero as

$$2.852\left(1-\frac{l_0}{L}\right)^{1.852} = \frac{\Delta H^e}{\Delta H}, \tag{22.107}$$

in which l_0/L denotes the section where the maximum pressure difference is located. By substituting back into eq. (22.103), the maximum pressure difference can be determined.

Now we compute the maximum pressure difference for profile type II-c, where the maximum pressure difference involves the pressure drop Δh_1 from the operating pressure along the lateral and the pressure difference Δh_2 higher than the operating pressure at the end of the lateral. From eq. (22.103), Δh_1 can be determined as

$$\frac{\Delta h_1}{\Delta H} = \left[1-\left(1-\frac{l_0}{L}\right)^{2.852} - \left(\frac{l_0}{L}\frac{\Delta H^e}{\Delta H}\right)\right], \tag{22.108}$$

where l_0/L is obtained from eq. (22.107). The maximum pressure gain at the end of the lateral is computed as

$$\frac{\Delta h_2}{\Delta H} = \left(\frac{\Delta H^e}{\Delta H}-1\right). \tag{22.109}$$

From eqs (22.108) and (22.109), the maximum pressure difference for profile II-c can be obtained as

$$\frac{\Delta h_{max}}{\Delta H} = \left[1-\left(1-\frac{l_0}{L}\right)^{2.852} - \left(\frac{l_0}{L}\frac{\Delta H^e}{\Delta H}\right)\right] + \left(\frac{\Delta H^e}{\Delta H}-1\right). \tag{22.110}$$

For downslope, $\Delta H^e/\Delta H$ is larger than 2.852. The maximum pressure difference Δh_{max} is located at the end and is equal to

$$\frac{\Delta h_{max}}{\Delta H} = \left(\frac{\Delta H^e}{\Delta H}-1\right). \tag{22.111}$$

Dimensionless design parameter $\Delta H^e/\Delta H$: For uniform downslope, the ratio $\Delta H^e/\Delta H$ can be employed to calculate the shape of the pressure profile and the maximum pressure difference Δh_{max}. For different values of this ratio, the values of maximum pressure difference can be computed and $\Delta h_{max}/\Delta H$ versus $\Delta H^e/\Delta H$ can be graphed as shown in Figure 22.20. For ideal profile II-b, the pressure difference is minimum (only $0.37\Delta H$).

For design, the maximum pressure difference Δh_{max}, if $\Delta H^e/\Delta H$ is known, and the value of $\Delta h_{max}/\Delta H$ is determined, which entails the determination of the total friction drop ΔH. To that end, one can write

$$\frac{\Delta h_{max}}{\Delta H} = \frac{\Delta H^e}{\Delta H}\frac{\Delta h_{max}}{\Delta H^e}, \tag{22.112}$$

plotted and superimposed on Figure 22.20 to show the values of $\Delta h_{max}/\Delta H^e$ for given values of $\Delta H^e/\Delta H$ for downslope or upslope. Hence, once $\Delta H^e/\Delta H$ is known, Δh_{max} can be determined. Thus, $\Delta H^e/\Delta H$ can be considered as a design parameter for the design of laterals on uniform slopes. The values of ΔH^e can be computed from slope S_0 and length L of the lateral as

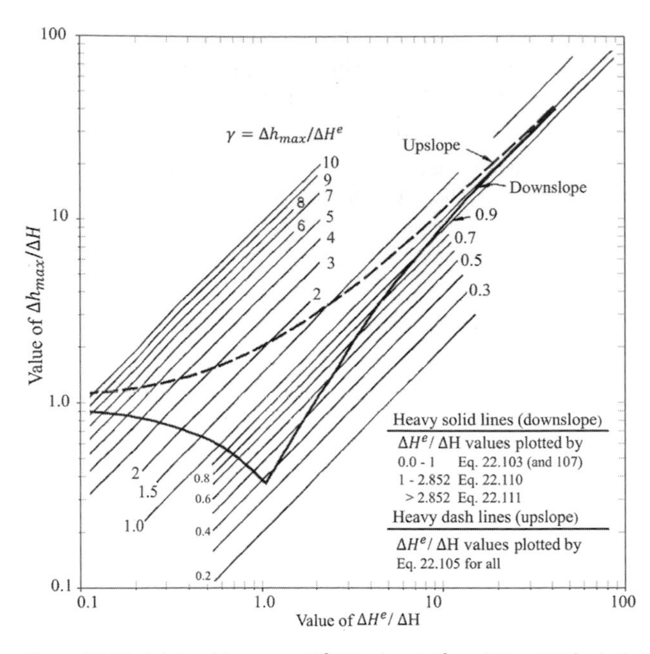

Figure 22.20 Relationship among $\Delta H^e/\Delta H$, $\Delta h_{max}/\Delta H^e$, and $\Delta h_{max}/\Delta H$ for both up- and downslope situations. (After Wu et al., 1983. Copyright © 1983 by Springer.)

$$\Delta H^e = S_0 L. \tag{22.113}$$

Using eq. (22.100), with Q replaced by specific discharge and length ($Q_s L$), we can write

$$\frac{\Delta H^e}{\Delta H} = \frac{2.852 S_0 D^{4.871}}{K_1 Q_s^{1.852} L^{1.852}}. \tag{22.114}$$

If $\Delta H^e/\Delta H$ is the ideal design length, L_0 can be determined from eq. (22.114) as

$$L_0 = \frac{(2.852)^{0.54} S_0^{0.54} D^{2.63}}{K_1^{0.54} Q_s}, \tag{22.115}$$

where K_1 is a constant equal to 9.76×10^{-4} for British units and 1.13×10^6 for metric units.

The maximum pressure at the ideal profile can be computed as

$$(\Delta h_{max})_0 = 0.37 S_0 L \text{ for downslope; } (\Delta h_{max})_1 = 2 S_0 L \text{ for upslope.} \tag{22.116}$$

Relation between ideal design and design length for a given design criterion: For a given design criterion, the design length and the maximum pressure difference corresponding to the ideal profile are the design parameters. There are two possibilities. First, the design length can be decreased if the maximum pressure difference is too large. Second, the length can be increased if the maximum pressure difference is too small. If the lateral diameter D, slope S_0, and specific discharge Q_s are fixed for a given lateral, then the length needs to be determined. The dimensionless length ratio can be obtained from eqs (22.114) and (22.115):

$$\frac{\left(\frac{\Delta H^e}{\Delta H}\right)}{\left(\frac{\Delta H^e}{\Delta H}\right)_0} = \left(\frac{L}{L_0}\right)^{-1.852}, \tag{22.117}$$

where subscript 0 denotes the ideal profile. For a given value of $\Delta H^e/\Delta H$, $\Delta h_{max}/\Delta H^e$ can be determined (see Figure 22.20). For simplicity, let $\Delta h_{max}/\Delta H^e$ be denoted as γ and $(\Delta h_{max}/\Delta H^e)_0$ for the ideal pressure profile be denoted as γ_0. Then, the ratio of $\Delta h_{max}/\Delta H^e$ and $(\Delta h_{max}/\Delta H^e)_0$ can be written for downslope corresponding to the ideal profile as

$$\frac{\left(\frac{\Delta h_{max}}{\Delta H^e}\right)}{\frac{(\Delta h_{max})_0}{(\Delta H^e)_0}} = \frac{\gamma}{\gamma_0} = \frac{\gamma}{0.37}. \tag{22.118}$$

For the ideal profile, eq. (22.118) can be simplified as

$$\frac{\Delta h_{max}}{(\Delta h_{max})_0} = \frac{\Delta H^e}{(\Delta H^e)_0}\left(\frac{\gamma}{0.37}\right) \tag{22.119}$$

or

$$\frac{\Delta h_{max}}{(\Delta h_{max})_0} = \left(\frac{\gamma}{0.37}\right)\left(\frac{L}{L_0}\right). \tag{22.120}$$

The relationship between L/L_0 and $\Delta h_{max}/(\Delta h_{max})_0$ can be determined for downslope from eqs (22.117) and (22.120). It may be noted that for pressure profile II-c, the maximum pressure difference Δh_{max} is constant, and beyond the optimum point ($\Delta H^e/\Delta H = 1$) the maximum pressure increases steeply when the lateral length increases.

The relationship between L/L_0 and $\Delta h_{max}/(\Delta h_{max})_0$ can also be determined for upslope for which there is no ideal solution. However, the design length corresponding to the dimensionless design parameter $\Delta H^e/\Delta H = 1$ can be determined. The dimensionless length ratio can be determined from eq. (22.117) as

$$\frac{\left(\frac{\Delta H^e}{\Delta H}\right)}{\left(\frac{\Delta H^e}{\Delta H}\right)_1} = \left(\frac{L}{L_1}\right)^{-1.852}, \tag{22.121}$$

where subscript 1 denotes the upslope situation. Similar to eq. (22.118), we can write

$$\frac{\left(\frac{\Delta h_{max}}{\Delta H^e}\right)}{\frac{(\Delta h_{max})_1}{(\Delta H^e)_1}} = \frac{\gamma}{\gamma_1} = \frac{\gamma}{2}. \tag{22.122}$$

Equation (22.122) can be simplified for upslope as

$$\frac{\Delta h_{max}}{(\Delta h_{max})_1} = \frac{\Delta H^e}{(\Delta H^e)_1}\left(\frac{\gamma}{2}\right) \tag{22.123}$$

or

$$\frac{\Delta h_{max}}{(\Delta h_{max})_1} = \left(\frac{\gamma}{2}\right)\left(\frac{L}{L_1}\right). \tag{22.124}$$

The relation between L/L_1 and $\Delta h_{max}/(\Delta h_{max})_1$ for upslope can be calculated.

Lateral design: The design procedure consists of the following steps: (1) Determine the ideal length using eq. (22.115). (2) Calculate $(\Delta h_{max})_0$ using the relation $(\Delta h_{max}/\Delta H^e)_0 = 0.37$. (3) Determine the pressure variation H_{var} and determine whether the design is acceptable. (4) If the design is not acceptable, then compute the allowable maximum pressure difference Δh_{max}, based on the design criterion for the pressure profile. (5) Calculate the design length L from L/L_0 versus $\Delta h_{max}/(\Delta h_{max})_0$. The same design procedure can also be used for upslope.

The pressure variation can be computed as

$$h_{var} = \frac{(\Delta h_{max})_0}{H + \Delta H^e - \Delta H} \times 100. \tag{22.125}$$

Now the design length can be computed.

Lateral design on zero slope: Using the Hazen–Williams equation for an allowable maximum pressure difference Δh_{max} at the end, we can write

$$\Delta h_{max} = \frac{K_1}{2.852} \frac{Q_s^{1.852}}{D^{4.871}} L^{2.852} \tag{22.126}$$

or

$$L = \frac{(2.852)^{0.35}}{\left(K_1^{0.35}\right)} \frac{h_{max}{}^{0.35}}{Q_s^{0.65}} D^{1.71}. \tag{22.127}$$

Example 22.9: Determine the length of a lateral whose inside diameter (D) is 15 mm, downslope (S_0) is 1%, and specific discharge (Q_s) is 0.001 lps/m. The operating pressure (H) head is assumed as 10 m and the allowable head variation is 15%.

Solution:

1. Determine the optimal length L_0 using eq. (22.115), which is:

$$L_0 = \frac{(2.852)^{0.54} S_0^{0.54} D^{2.63}}{K_1^{0.54} Q_s} = \frac{2.852^{0.54} \times 0.01^{0.54} \times (15)^{2.63}}{\left(1.13 \times 10^6\right)^{0.54} \times 0.001} = 98 \text{ m}.$$

2. Determine the maximum pressure difference for the ideal case, $(\Delta h_{max})_0$, using eq. (22.116) for downslope, which is:

$$(\Delta h_{max})_0 = 0.37 S_0 L_0 = 0.37 \times 0.01 \times 98 \text{ m} = 0.36 \text{ m}.$$

3. Determine the pressure variation for the ideal case:

$$h_{var} = \frac{(\Delta h_{max})_0}{H} = \frac{0.36 \text{ m}}{10 \text{ m}} = 3.6\% < 15\% \quad \text{(accepted)}.$$

Determine the pressure profile from Figure 22.21, which is type II-a.

4. The allowable maximum pressure difference for profile II-a can be calculated as:

$$\Delta h_{max} = 0.15 \times 10 \text{ m} = 1.5 \text{ m}.$$

5. The dimensionless ratio of maximum pressure drop is

$$\frac{\Delta h_{max}}{(\Delta h_{max})_0} = \frac{1.5 \text{ m}}{0.36 \text{ m}} = 4.17$$

6. The dimensionless lateral length can be computed from Figure (22.21):

$$\frac{\Delta h_{max}}{(\Delta h_{max})_0} = 4.17 \quad \rightarrow \quad \frac{L}{L_0} = 1.6.$$

7. The design length to satisfy the design criterion of pressure variation of 15% is computed as:

$$L = 1.6 \times L_0 = 1.6 \times 98 \text{ m} = 156.8 \text{ m}.$$

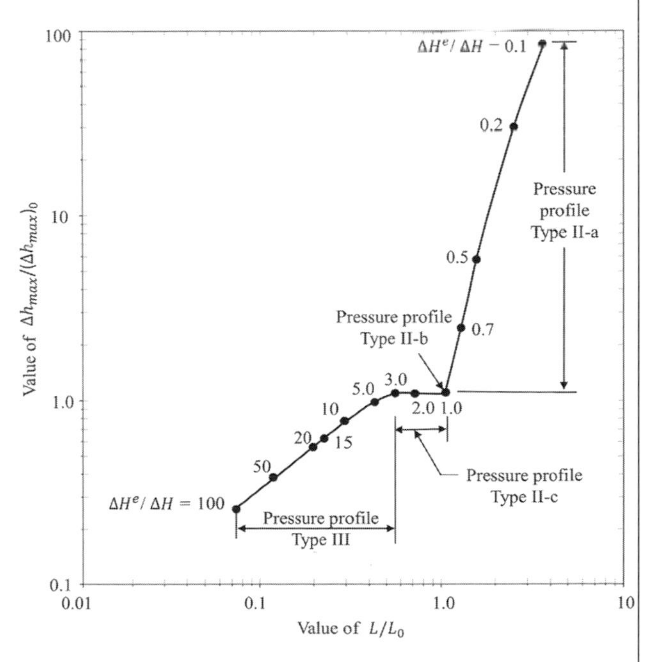

Figure 22.21 Relationship between $\Delta h_{max}/(\Delta h_{max})_0$ and L/L_0 for downslope situations. (After Wu et al., 1983. Copyright © 1983 by Springer.)

Example 22.10: Determine the length of a lateral whose inside diameter (D) is 15 mm, downslope (S_0) is 4%, and specific discharge (Q_s) is 0.001 lps/m. The operating pressure (H) is assumed as 10 m and the allowable head variation is 15%.

Solution:

1. Determine the optimal length L_0 using eq. (22.115), which is:

$$L_0 = \frac{(2.852)^{0.54} S_0^{0.54} D^{2.63}}{K_1^{0.54} Q_s} = \frac{2.852^{0.54} \times 0.04^{0.54} \times (15)^{2.63}}{\left(1.13 \times 10^6\right)^{0.54} \times 0.001} = 207 \text{ m}.$$

2. Determine the maximum pressure difference for the ideal case, $(\Delta h_{max})_0$, using eq. (22.116) for downslope, which is:

$$(\Delta h_{max})_0 = 0.37 S_0 L_0 = 0.37 \times 0.04 \times 207 \text{ m} = 3.06 \text{ m}.$$

3. Determine the pressure variation for the ideal case:

$$h_{var} = \frac{(\Delta h_{max})_0}{H} = \frac{3.06 \text{ m}}{10 \text{ m}} = 31\% > 15\% \text{ (not accepted)}.$$

The ideal length $L_0 = 207$ m is not accepted. To satisfy the design criterion of $h_{var} = 15\%$, the lateral length should be decreased. Determine the pressure profile from Figure 22.21, which is type II-c or type III.

4. The allowable maximum pressure difference for assumed profile III can be calculated as:

$$h_{var} = \frac{\Delta h_{max}}{10 \text{ m} + \Delta h_{max}} = 0.15, \quad \Delta h_{max} = 1.76 \text{ m}.$$

5. The dimensionless ratio of maximum pressure drop is

$$\frac{\Delta h_{max}}{(\Delta h_{max})_0} = \frac{1.76 \text{ m}}{3.06 \text{ m}} = 0.58.$$

Since $\Delta h_{max}/(\Delta h_{max})_0 < 1$, the assumption of a type III profile is correct.

6. The dimensionless lateral length can be computed from Figure (22.21):

$$\frac{\Delta h_{max}}{(\Delta h_{max})_0} = 0.58 \quad \rightarrow \quad \frac{L}{L_0} = 0.20.$$

7. The design length to satisfy the design criterion of pressure variation of 15% is computed as:

$$L = 0.20 \times L_0 = 0.20 \times 207 \text{ m} = 41.4 \text{ m}.$$

Example 22.11: Determine the length of a lateral whose inside diameter (D) is 18 mm, downslope (S_0) is 2%, and specific discharge (Q_s) is 0.001 lps/m. The operating pressure (H) is assumed as 10 m and the allowable head variation is 15%.

Solution:

1. Determine the optimal length L_0 using eq. (22.115), which is:

$$L_0 = \frac{(2.852)^{0.54} S_0^{0.54} D^{2.63}}{K_1^{0.54} Q_s} = \frac{2.852^{0.54} \times 0.02^{0.54} \times (18)^{2.63}}{\left(1.13 \times 10^6\right)^{0.54} \times 0.001} = 229.6 \text{ m}.$$

2. Determine the maximum pressure difference for the ideal case, $(\Delta h_{max})_0$, using eq. (22.116), which is:

$$(\Delta h_{max})_0 = 0.37 S_0 L_0 = 0.37 \times 0.02 \times 229.6 \text{ m} = 1.70 \text{ m}.$$

3. Determine the pressure variation for the ideal case:

$$H_{var} = \frac{(\Delta h_{max})_0}{H} = \frac{1.70 \text{ m}}{10 \text{ m}} = 17\% > 15\%$$

(not accepted).

The ideal length $L_0 = 229.6$ m is not accepted.

To satisfy the design criterion of $H_{var} = 15\%$, the lateral length should be decreased. Determine the pressure profile from Figure 22.21, which is type II-c or type III.

4. The allowable maximum pressure difference for *assumed* profile III can be calculated as:

$$H_{var} = \frac{\Delta h_{max}}{10 \text{ m} + \Delta h_{max}} = 0.15, \quad \Delta h_{max} = 1.76 \text{ m}.$$

5. If the pressure profile is type II-c, then the design length can be:

$$\frac{\Delta h_{max}}{(\Delta h_{max})_0} = \frac{1.76 \text{ m}}{1.70 \text{ m}} = 1.04.$$

Since $\Delta h_{max}/(\Delta h_{max})_0 > 1$, the assumption of type III profile is not correct. The type II-c is correct.

6. The graphical solution (Figure 22.21) is not used; the design length can be calculated from eq. (22.127):

$$L = \frac{(2.852)^{0.35}}{\left(K_1^{0.35}\right)} \frac{\Delta h_{max}^{\ \ 0.35}}{Q_s^{0.65}} D^{1.71} = \frac{(2.852)^{0.35}}{\left(1.13 \times 10^6\right)^{0.35}} \frac{(1.76)^{0.35}}{(0.001)^{0.65}} (18)^{1.71} = 167.2 \text{ m}.$$

Example 22.12: Determine the length of a lateral whose inside diameter (D) is 14 mm, upslope (S_0) is 3%, and specific discharge (Q_s) is 0.001 lps/m. The operating pressure head (H) is assumed as 10 m and the allowable head variation is 15%.

Solution:

1. Determine the optimal length L_1 for ($\Delta H^e/\Delta H = 1$) using eq. (22.115), which is:

$$L_1 = \frac{(2.852)^{0.54} S_0^{0.54} D^{2.63}}{K_1^{0.54} Q_s} = \frac{2.852^{0.54} \times 0.03^{0.54} \times (14)^{2.63}}{\left(1.13 \times 10^6\right)^{0.54} \times 0.001} = 147.6 \text{ m}.$$

2. Determine the maximum pressure difference at the end of the lateral, $(\Delta h_{max})_1$, using eq. (22.116), which is:

$$(\Delta h_{max})_1 = 2S_0L_1 = 2 \times 0.03 \times 147.6 \text{ m} = 8.86 \text{ m}.$$

3. Determine the pressure variation H_{var}:

$$H_{var} = \frac{(\Delta h_{max})_1}{H} = \frac{8.86\text{m}}{10\text{ m}} = 88.6\% > 15\% \quad (\text{not accepted}).$$

The design length $L_1 = 147.6$ m is not accepted.

4. The allowable maximum pressure difference for assumed type-I can be calculated as:

$$\Delta h_{max} = 0.15 \times 10 \text{ m} = 1.5 \text{ m}.$$

5. The dimensionless ratio of maximum pressure drop is

$$\frac{\Delta h_{max}}{(\Delta h_{max})_1} = \frac{1.5 \text{ m}}{8.86 \text{ m}} = 0.17.$$

Since $\Delta h_{max}/(\Delta h_{max})_1 < 1$, the assumption of type I profile is correct.

6. The dimensionless lateral length can be computed from Figure (22.22):

$$\frac{\Delta h_{max}}{(\Delta h_{max})_1} = 0.17 \quad \rightarrow \quad \frac{L}{L_1} = 0.3.$$

7. The design length to satisfy the design criterion of pressure variation of 15% is computed as:

$$L = 0.3 \times L_1 = 0.3 \times 147.6 \text{ m} = 44.3 \text{ m}.$$

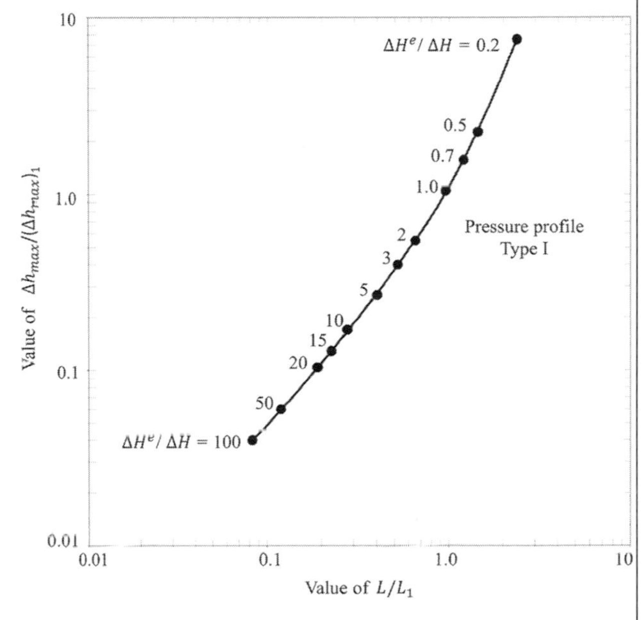

Figure 22.22 Relationship between $\Delta h_{max}/(\Delta h_{max})_1$ and L/L_1 for upslope situations. (After Wu et al., 1983. Copyright © 1983 by Springer.)

Variation of discharge in drip laterals on zero slope: Following Wu and Gitlin (1975), the flow in a lateral is spatially varying with decreasing discharge in the flow direction and pressure head gradient line is of exponential type:

$$\frac{dH}{dl} = \frac{7.89 \times 10^5 \times Q_l^m}{D^{m+3}}, \tag{22.128}$$

where H is the pressure head (m), l is the distance from the inlet of the lateral in the direction of flow (m), Q_l is the flow rate (lps) at distance l, D is the diameter of the lateral (mm), and m is the pipe flow exponent ($m = 1.75$) for the Blasius equation with the Darcy–Weisbach equation ($m = 1.851$ for the Hazen–Williams equation).

Average head loss: Let ΔH_l be the head loss due to friction over length l measured from the inlet. Let dA be the area of the strip:

$$dA = \Delta H_l dl. \tag{22.129}$$

Wu and Gitlin (1975) expressed ΔH_l as

$$\Delta H_l = \frac{7.89 \times 10^5 Q^m L}{D^{m+3}(m+1)}\left[1 - \left(1 - \frac{l}{L}\right)^{m+1}\right]. \tag{22.130}$$

The average head loss due to friction $\overline{\Delta H}$ can be written as

$$\overline{\Delta H} = \frac{1}{L}\int_0^L \Delta H_l dl. \tag{22.131}$$

Inserting eq. (22.130) in eq. (22.131) and integrating, we obtain

$$\overline{\Delta H} = \frac{7.89 \times 10^5 Q^m L(m+1)}{D^{m+3}(m+1)(m+2)}. \tag{22.132}$$

Average pressure head: The total head loss at the end of the lateral due to friction can be written as

$$\Delta H = \frac{7.89 \times 10^5 Q^m L}{D^{m+3}(m+1)}. \tag{22.133}$$

From eqs (22.132) and (22.133),

$$\overline{\Delta H} = \Delta H\left(\frac{m+1}{m+2}\right). \tag{22.134}$$

Let S_0 be the slope of the lateral. The pressure head at length l can be written as

$$H_l = H_0 - \Delta H_l + blS_0 \quad (b = 1 \text{ for downslope}, b = -1 \text{ for upslope}). \tag{22.135}$$

The average pressure head can be expressed as

$$\overline{H} = H_0 - \overline{\Delta H} + b\frac{LS_0}{2}. \tag{22.136}$$

Coefficient of variation of discharge: Let the deviation between H_l and $\overline{H}$ at distance l be denoted as DEV_l. Then,

$$DEV_l^2 = \left(H_l - \overline{H}\right)^2. \tag{22.137}$$

Inserting eqs (22.135) and (22.136) in eq. (22.137), the result is

$$DEV_l^2 = \left[-\Delta H_l + blS_0 + \overline{\Delta H} - b\frac{LS_0}{2}\right]^2. \tag{22.138}$$

Now head loss at l can be written as

$$\Delta H_l = \Delta H\left[1 - \left(1 - \frac{l}{L}\right)^{m+1}\right]. \tag{22.139}$$

Let CV_q be the coefficient of variation of discharge expressed in percent. Bralts and Edwards (1986) expressed the relation between the coefficient of variation of pressure head (CV_H) and CV_q as

$$CV_q = yCV_H, \tag{22.140}$$

where y is the pressure and discharge exponent. Hence,

$$CV_q = y\frac{100}{H}\left[\frac{1}{L}\int_0^L DEV_l^2 dl\right]. \tag{22.141}$$

Equation (22.141) can be written as

$$CV_q = \frac{100y}{H}\left\{\left[\Delta H^2\left(1 + \frac{1}{2m+3} - \frac{2}{m+2}\right) + \frac{S_0^2L^2}{3} + \left(\overline{\Delta H} - b\overline{\Delta S_0}\right) - \left[bS_0\Delta HL\left(1 - \frac{2}{(m+2)(m+3)}\right)\right] + b\left(\overline{\Delta H} - b\overline{\Delta S_0}\right)S_0L - 2\left(\overline{\Delta H} - b\overline{\Delta S_0}\right)\Delta H\left(\frac{m+1}{m+2}\right)\right\}^{0.5}. \tag{22.142}$$

If $m = 1.75$, then

$$CV_q = \frac{y}{H}\left[927\Delta H^2 + 833(LS_0)^2 - 1540b\Delta HLS_0\right]^{0.5}. \tag{22.143}$$

Often the size of the lateral needs to be selected so that the coefficient of variation is below a desired level, say 3–5% (Wu, 1997). The coefficient of variation for different emitter types can be found in Table 22.1. One can compute the diameter of a lateral for any coefficient of variation if all other parameters are known. Rearranging eq. (22.143),

$$\left(\frac{1}{D^{4.75}}\right)^2 a + \left(\frac{1}{D^{4.75}}\right)b + c = 0, \tag{22.144}$$

where

$$a = \left(KQ^{1.75}L\right)^2 + 109.355y^2 - 0.071CV_q^2, \tag{22.145}$$

$$b = KQ^{1.75}L\left[LS_0\left(0.267bCV_q^2 - 560by^2\right) + 0.533CV_q^2H_l\right], \tag{22.146}$$

$$c = 833(LS_0y)^2 - CV_q^2\left(H_l^2 + \left(\frac{LS_0}{2}\right)^2 + bH_lLS_0\right). \tag{22.147}$$

22.10.3 Uniformity-Based Approach

The distribution of discharge along the lateral can be computed from the dimensionless EGL, and then a desirable combination of lateral length and inlet pressure of a drip irrigation system can be computed. From the Darcy–Weisbach equation, the head drop between emitters can be computed as

$$\Delta h_l = f\frac{\Delta L}{D}\frac{v^2}{2g}, \tag{22.148}$$

where f is the Darcy–Weisbach friction factor, D is the lateral diameter, v is the velocity, g is the acceleration due to gravity, and ΔL is the length between emitters. Using the Blasius equation for turbulent flow in a smooth pipe, the head drop can be written as

$$\Delta h_l = KQ^{1.75}\Delta L, \tag{22.149}$$

where Q is the discharge in the lateral and K is the constant. For small discharge variations (less than 10%) and equally spaced emitters with uniform emitter flow, the head drop at the jth section can be written as

$$\Delta h_j = K[(n - j + 1)q]^m\Delta L, \tag{22.150}$$

in which n is the number of emitters and q is the emitter flow. The total head drop at the jth section can be expressed as

$$\Delta H_j = Kq^m\left[\sum_{j=1}^{j}(n - j + 1)^m\Delta L\right]. \tag{22.151}$$

Likewise, the head drop at the end of the lateral can be written as

$$\Delta H = Kq^m\left[\sum_{j=1}^{n}(n - j + 1)^m\Delta L\right]. \tag{22.152}$$

The head drop ratio can now be written as

$$\frac{\Delta H_i}{\Delta H} = \frac{\sum_{j=1}^{j}(n-j+1)^m}{\sum_{j=1}^{n}(n-j+1)^m}. \quad (22.153)$$

The emitter flow q_i at any section i of the lateral can be expressed as

$$q_i = C_1\sqrt{h_i}, \quad (22.154)$$

where C_1 is a constant and h_i is the pressure head at the ith section. The pressure head can be determined from the total inlet pressure head H, the head drop over any given length ΔH_i, and the head drop at the end of the lateral due to slope ΔH_i^e. The discharge can then be written for downslope as

$$q_i = C_1\sqrt{H - \Delta H_i + \Delta H_i^e}. \quad (22.155)$$

The ratio of emitter discharge at any location to the emitter discharge at the inlet can be expressed as

$$\frac{q_i}{q} = \sqrt{\frac{H - \Delta H_i + \Delta H_i^e}{H}} = \sqrt{1 - \frac{\Delta H_i}{\Delta H} + \frac{\Delta H_i^e}{H}}. \quad (22.156)$$

The ratio of ΔH_i and ΔH can be computed for turbulent flow in a smooth pipe:

$$R_i = \frac{\Delta H_i}{\Delta H}. \quad (22.157)$$

The gain or loss of pressure due to slope can also be written as

$$R_i^e = \frac{\Delta H_i^e}{\Delta H^e}. \quad (22.158)$$

The emitter flow ratio can now be expressed as

$$\frac{q_i}{q} = \sqrt{1 - R_i\frac{\Delta H}{H} + R_i^e\frac{\Delta H^e}{H}}. \quad (22.159)$$

The uniformity coefficient can now be determined if the discharge distribution is known:

$$C_u = 1 - \frac{\overline{\Delta q}}{\overline{q}}, \quad (22.160)$$

where $\overline{q}$ is the mean discharge and $\overline{\Delta q}$ is the mean deviation.

For drip line design, the following steps can be employed. (1) For different types of flow (laminar, turbulent, and complex) plot the pressure head drop versus length ratio. (2) Compute total inlet pressure head H, the pressure head drop ΔH_i at any given length i, maximum friction drop at the end of line ΔH, and the pressure gain or loss due to slope ΔH_i^e at position i and the maximum pressure gain or loss due to slope ΔH^e. (3) Compute the pressure drop ratios R_i and R_i^e. (4) Compute the discharge ratio for each head drop ratio. (5) Compute the mean discharge and mean of deviation of discharge from mean discharge. (6) Compute the uniformity coefficient. (7) Compute the ratio of L/H and check whether the uniformity coefficient is verified.

If emitter discharge and spacing are given, one can select a set of H (pressure) and L (length) values which represent field conditions and lead to the desired degree of uniformity.

Example 22.13: The pressure at the inlet is $H = 6$ psi or 13.84 feet of water. The lateral has a length L of 275 feet, inside diameter of 0.5 in., and slope down (positive) of 2%. The total inlet discharge is 1.95 gpm. Compute the uniformity coefficient. Now consider the upslope of 2%.

Solution: For slope down (positive) of 2%:

1. Calculate $\frac{L}{H} = \frac{275}{13.84} = 19.9$.
2. From the design chart of a 0.5 in. lateral line (down slope) (Figure 22.23), draw a vertical line in quadrant II corresponding to $\frac{L}{H} = 19.9$ and discharge of 1.95 gpm at point P_1.

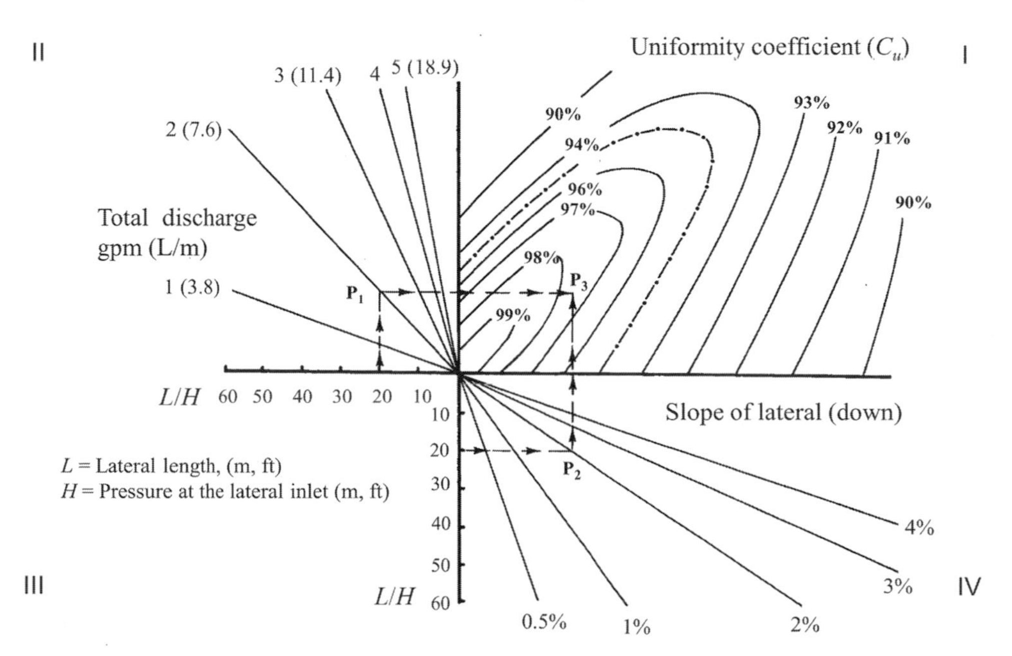

Figure 22.23 Design chart of a 0.5-in. (12.7 mm) lateral line (downslope). (Adapted from Wu and Gitlin, 1974.)

3. Compute the point P_2 corresponding to $\frac{L}{H} = 19.9$ and slope of 2% in quadrant IV.
4. Corresponding to the two points in Steps 2 and 3, P_3 is found and then $C_u = 97.5\%$. The result shows the design is acceptable.

For upslope of 2%, the design chart of a 0.5-inch lateral line (upslope) (Figure 22.24) is used. Repeat Steps 2 and 3, and find the new P_1 and P_2, as shown in Figure 22.24. Corresponding to the two points, P_3 is found and then $C_u = 87\%$.

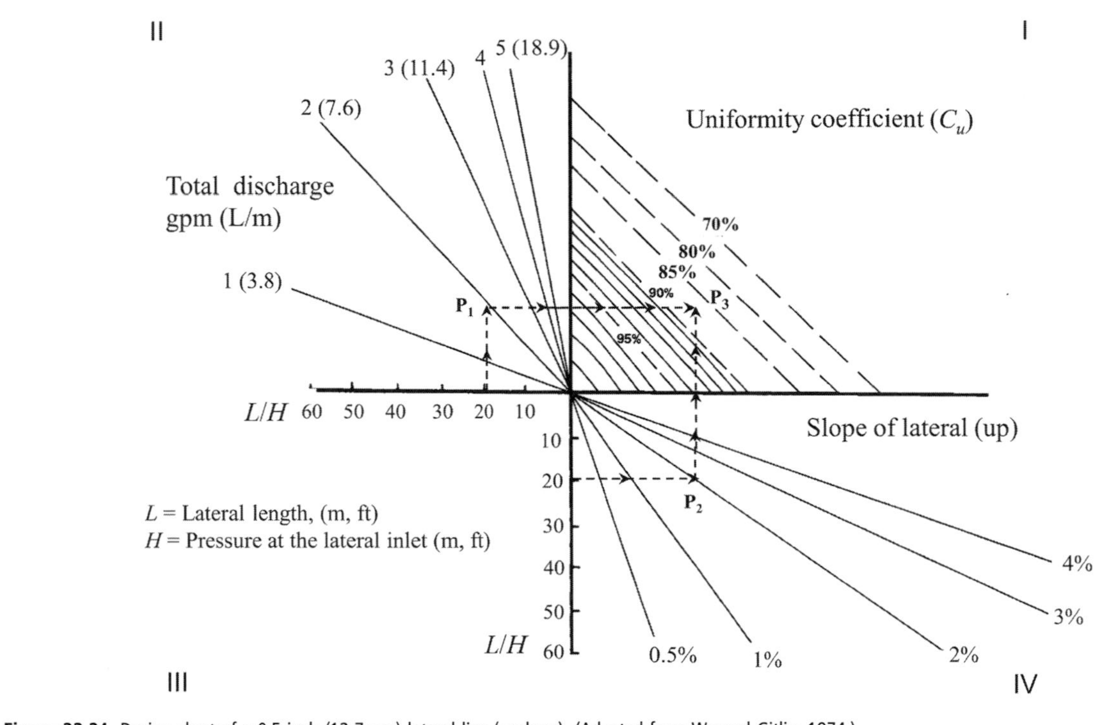

Figure 22.24 Design chart of a 0.5-inch (12.7 mm) lateral line (upslope). (Adapted from Wu and Gitlin, 1974.)

22.10.4 Analytical Approach: Constant Discharge

In the above approaches it is assumed that emitter outflow along the lateral is constant. In this approach, this assumption is relaxed, so that emitter outflow varies along the lateral (Valiantzas, 1998).

Emitter outflow: Assuming the number of emitters is sufficiently large, the emitter outflow can be assumed to vary continuously in space along the lateral. Hence, the emitter outflow per unit length can be expressed as

$$q(x) = \frac{cH^y(x)}{s}, \tag{22.161}$$

where s is the emitter spacing, x is the distance measured from the downstream end of the lateral, $H(x)$ is the piezometric head at distance x, and $q(x)$ is the emitter outflow per unit length.

The friction drop is computed assuming constant emitter outflow along the lateral as:

$$q(x) = \frac{\overline{q}}{s} = \frac{Q_{in}}{L}, \tag{22.162}$$

in which Q_{in} is the total flow rate at the inlet of the lateral, $\overline{q}$ is the average emitter flow, and L is the lateral length. One can write from eq. (22.162):

$$Q(x) = Q_{in}\left(\frac{x}{L}\right), \tag{22.163}$$

where $Q(x)$ is the total outflow rate passing through the lateral section located at distance x from the downstream end.

Energy gradient: The friction slope of the EGL can be determined from the Darcy–Weisbach equation as

$$\frac{dH_f}{dx} = S_f(x) = K\frac{Q^m(x)}{D^{m+1}}, \tag{22.164}$$

in which $H_f(x)$ is the friction head at distance x, S_f is the friction slope (positive since x is measured from the downstream end), K is the roughness coefficient,

$$K = \frac{a_1 \upsilon^{2-m}}{2g}\left(\frac{4}{\pi}\right)^m, \tag{22.165}$$

where υ is the kinematic viscosity, D is the inside diameter, m is empirical constant, and a_1 is constant. For laminar flow

(Re < 2000), $m = 1$ and $a_1 = 64$. For turbulent flow in a smooth pipe ($3000 < \text{Re} < 10^5$), $m = 1.75$ and $a_1 = 0.316$. For the fully turbulent case (Re $> 10^5$), $m = 1.828$ and $a_1 = 0.13$.

Integration of eq. (22.165) between limits 0 and x yields

$$H_f(x) = \frac{H_{f0}}{(m+1)}\left(\frac{x}{L}\right)^{m+1}, \tag{22.166}$$

where

$$H_{f0} = \frac{KLQ_{in}^m}{D^{m+3}} \tag{22.167}$$

is the total friction loss of a similar pipe conveying the entire flow over its length.

In the energy equation, the velocity head is small in the lateral. The pressure head at any point x of the lateral can be given as

$$H(x) = H_d + H_f(x) - S_0(x), \tag{22.168}$$

in which H_d is the pressure head at the downstream end of the lateral and S_0 is the ground slope assumed to be uniform. The downslope lateral will have positive slope values, since x is measured from the downstream end. The friction losses in eq. (22.168) are computed assuming constant emitter outflow. The average pressure H_{av} can be calculated by using eqs (22.168) and (22.166) as

$$H_{av} = H_d + \frac{H_{f0}}{(m+1)(m+2)} - \frac{S_0 L}{2}. \tag{22.169}$$

In order to consider the spatial variation of emitter outflow, the total outflow rate can be expressed by a power function as

$$Q(x) = Q_{in}\left(\frac{x}{L}\right)^{\propto}, \tag{22.170}$$

in which α is an exponent. Differentiating eq. (22.170), one obtains

$$q(x) = \frac{\propto Q_{av}}{s}\left(\frac{x}{L}\right)^{\propto - 1}. \tag{22.171}$$

Estimation of α: To estimate the value of α, the lower half of the lateral ($L/2$) is considered. We then compute the average pressure head over the lower half. Subtracting eq. (22.169) from eq. (22.168), we get

$$H(x) = H_{av} + \frac{H_{f0}}{(m+1)}\left[\left(\frac{x}{L}\right)^{m+1} - \frac{1}{m+2}\right] + S_0\left(\frac{L}{2} - x\right). \tag{22.172}$$

The average pressure can be evaluated by integrating eq. (22.172) between 0 and $L/2$ and dividing by $L/2$ as

$$H_{0.5} = H_{av} + \frac{H_{f0}}{(m+1)(m+2)}\left[0.5^{m+1} - 1\right] + \frac{S_0 L}{4}, \tag{22.173}$$

where $H_{0.5}$ is the average pressure over the distance $L/2$. Then, the average emitter flow $q_{0.5}$ over the lower half can be computed from average pressure head $H_{0.5}$ as

$$q_{0.5} = cH_{0.5}^y. \tag{22.174}$$

Equation (22.173) is sufficiently accurate for pressure head variation of less than 20% (Anyoji and Wu, 1987). The total discharge passing through the lower half of the lateral at $x_0 = L/2$ is equal to the sum of outflow of all emitters between 0 and $L/2$, and therefore

$$Q_{0.50} = \left(\frac{N}{2}\right) q_{0.5} \tag{22.175}$$

in which N is the total number of emitters. From eqs (22.173)–(22.175),

$$Q_{0.5} = \left(\frac{N}{2}\right) c\left[H_{av} + \frac{H_{f0}\left(0.5^{m+1} - 1\right)}{(m+1)(m+2)} + \frac{S_0 L}{4}\right]^y. \tag{22.176}$$

Exponent α can be evaluated from eq. (22.170) as

$$\alpha = \log\left(\frac{Q_{0.5}}{Q_{in}}\right)\frac{1}{\log(0.5)}. \tag{22.177}$$

The total discharge at the inlet of the lateral can be computed as

$$Q_{in} = Nq_{av} = NcH_s^y, \tag{22.178}$$

where H_s is the required design pressure head.

Then, from eqs (22.176)–(22.178), α can be computed as

$$\alpha = 1 + y\log\left[1 + \frac{H_{f0}\left(0.5^{m+1} - 1\right)}{(m+1)(m+2)H_s} + \frac{S_0 L}{4H_s}\right]\frac{1}{\log(0.5)}. \tag{22.179}$$

Inlet pressure head: From eqs (22.170) and (22.172), the pressure head profile along the lateral can be expressed as

$$H(x) = H_{av} + \frac{H_{f0}}{(\alpha m+1)}\left[\left(\frac{x}{L}\right)^{\alpha m+1} - \frac{1}{\alpha m+2}\right] + S_0\left(\frac{L}{2} - x\right). \tag{22.180}$$

The inlet pressure can be obtained as

$$H_{in} = H_{av} + \frac{H_{f0}}{\alpha m+2} - \frac{S_0 L}{2}. \tag{22.181}$$

H_{av} can be approximated by H_s. More accurately,

$$H_{av} = H_s\left[1 + \frac{1}{2}\frac{1}{y}\left(\frac{1}{y} - 1\right)CV^2\right], \tag{22.182}$$

where CV is the coefficient of variation. For normal distributions of emitter discharge in the lateral, CV is related to the uniformity coefficient C_u (Warrick and Yitayew, 1988) as $CV = \frac{1-C_u}{0.798}$.

Determination of uniformity: The coefficient of uniformity C_u can be approximately expressed as

$$C_u = 1 - \frac{0.798y}{H_s}\left[\frac{H_{f0}^2}{(2\alpha m+3)(\alpha m+2)^2} + \frac{(S_0 L)^2}{12} + \frac{H_{f0} S_0 L}{(\alpha m+2)(\alpha m+3)}\right]^{0.5}. \tag{22.183}$$

For design applications, the steps are: (1) Compute exponent α using eq. (22.179). (2) Compute C_u from eq. (22.183) and L. (3) Compute CV from the relationship with C_u for normal distributions of emitter discharge. (4) Compute the inlet pressure using eqs (22.181) and (22.182).

Example 22.14: Using the constant discharge method, determine the length of the lateral and inlet pressure head (H_0) given the following information: pressure–discharge relationship coefficient $c = 3.60 \times 10^{-7}$; emitter exponent $y = 0.5$; required average emitter flow $q_{av} = 5$ L/h or 1.4×10^{-3} lps; the spacing between emitters $s = 1$ m; the insider diameter $D = 8.4$ mm; and the lateral slope $S_0 = 0$. Assume the flow as turbulent in a smooth pipe with $m = 1.75$. The required design pressure head is $H_s = (q_{av}/c)^{1/y} = 15.30$ and the design uniformity coefficient C_u is 0.9 ($K = 1.13 \times 10^6$). This example follows Valiantzas (1998).

Solution:

Step 1: When the constant discharge method is used, exponent α can be considered as 1.

Step 2: Compute C_u and L. For $S_0 = 0$, eq. (22.183) for C_u can be simplified as

$$C_u = 1 - \frac{0.798y}{H_s}\left[\frac{H^2{}_{f0}}{(2\alpha m+3)(\alpha m+2)^2} + 0 + 0\right]^{0.5}.$$

Then, $H_{f0} = \frac{(1-C_u)(2\alpha m+3)^{0.5}(\alpha m+2)H_s}{0.798y}$.

Replace $M = \alpha m$ and $H_{f0} = WL^{m+1}$; then we obtain a direct solution for L as

$$L = \left[\frac{(1-C_u)(2M+3)^{0.5}(M+2)H_s}{0.798yW}\right]^{\frac{1}{(1+M)}}$$

in which,

$$W = \frac{H_{f0}}{L^{m+1}} = \frac{KLQ_{in}^m}{D^{m+3}L^{m+1}} = \frac{KQ_{in}^m}{D^{m+3}L^m} = \frac{Kq_{av}^m}{D^{m+3}}\cdot\frac{N^m}{L^m} = \frac{Kq_{av}^m}{D^{m+3}}\cdot\frac{1}{s^m}$$

$$= \frac{(1.13\times10^6)(1.4\times10^{-3})^{1.75}}{(8.4)^{1.75+3}}\cdot\frac{1}{(1)^{1.75}} = 4.6\times10^{-4}$$

Given $\alpha = 1$, then $M = m = 1.75$. L can be calculated as

$$L = \left[\frac{(1-0.9)(2\times1.75+3)^{0.5}(1.75+2)(15.3)}{0.798\times0.5\times(4.6\times10^{-4})}\right]^{\frac{1}{(1+1.75)}}$$

$$= 60.6 \text{ m}$$

Step 3: Compute CV. Given $C_u = 0.9$, CV is computed as

$$CV = \frac{1-C_u}{0.798} = \frac{1-0.9}{0.798} = 0.125.$$

Step 4: Compute H_{av} and H_{in}.

The average pressure head H_{av} is computed using eq. (22.182):

$$H_{av} = H_s\left[1 + \frac{1}{2}\frac{1}{y}\left(\frac{1}{y}-1\right)CV^2\right]$$

$$= 15.3\times\left[1+\frac{1}{2}\frac{1}{0.5}\left(\frac{1}{0.5}-1\right)(0.125)^2\right].$$

$$= 15.5 \text{ m}$$

The inlet pressure is calculated using eq. (22.181):

$$H_{in} = H_{av} + \frac{H_{f0}}{\alpha m+2} - \frac{S_0 L}{2} = H_{av} + \frac{WL^{m+1}}{m+2}$$

$$= 15.5 + \frac{(4.6\times10^{-4})(60.6)^{1.75+1}}{1.75+2} = 25.3 \text{ m}.$$

Example 22.15: Solve Example 22.14 with the variable discharge method.

Solution: First, an initial guess of L as L_0 is made with $\alpha = 1$ and $M = \alpha m = 1.75$. Then, using the result $L_0 = 60.6$ m obtained from Example 22.14, eq. (22.179) is used to compute the value of α as

$$\alpha = 1 + y\log\left[1 + \frac{H_{f0}(0.5^{m+1}-1)}{(m+1)(m+2)H_s} + \frac{S_0 L}{4H_s}\right]\frac{1}{\log(0.5)}.$$

With $S_0 = 0$,

$$\alpha = 1 + y\log\left[1 + \frac{WL_0{}^{m+1}(0.5^{m+1}-1)}{(m+1)(m+2)H_s}\right]\times\frac{1}{\log(0.5)}$$

$$= 1 + 0.5\log\left[1+\frac{(4.6\times10^{-4})(60.6^{2.75})(0.5^{2.75}-1)}{(2.75)(3.75)(15.3)}\right]$$

$$\times\frac{1}{\log(0.5)} = 1.16$$

This yields $\alpha = 1.16$. Then, $M = 1.16$ m, and L is computed using the equation from Example 22.14:

$$\left[\frac{(1-0.9)(2\times(1.16\times1.75)+3)^{0.5}((1.16\times1.75)+2)(15.3)}{0.798\times0.5\times(4.6\times10^{-4})}\right]^{\frac{1}{(1+(1.16\times1.75))}}$$

$$= 43.1 \text{ m}.$$

The average pressure head H_{av} is computed using eq. (22.182):

$$H_{av} = H_s\left[1 + \frac{1}{2}\frac{1}{y}\left(\frac{1}{y} - 1\right)CV^2\right]$$
$$= 15.3 \times \left[1 + \frac{1}{2}\frac{1}{0.5}\left(\frac{1}{0.5} - 1\right)(0.125)^2\right] = 15.5 \text{ m.}$$

The inlet pressure is calculated using eq. (22.181):

$$H_0 = H_{av} + \frac{H_{f0}}{am+2} - \frac{S_0 L}{2} = H_{av} + \frac{WL^{m+1}}{am+2}$$
$$= 15.5 + \frac{(4.6 \times 10^{-4})(43.1)^{1.75+1}}{1.16 \times 1.75 + 2} = 19.1 \text{ m.}$$

22.10.5 Analytical Solution: Variable Discharge

For small-diameter smooth pipes used in drip systems, Watters and Keller (1978) stated that the Darcy–Weisbach equation provides accurate estimates of head loss due to friction. Therefore, the change in the total head loss along the lateral can be equated to the head loss. Differentiating the energy equation and using the Darcy–Weisbach equation (eq. 22.164), one obtains

$$\frac{d}{dx}\left(H + \frac{v^2}{2g}\right) + \frac{fv^2}{2gD} + S_0 = 0. \tag{22.184}$$

Substituting for H from eq. (22.8), eq. (22.184) can be expressed as

$$\frac{d}{dx}\left[\left(-\frac{As}{c}\right)\left(\frac{dv}{dx}\right)\right]^{\frac{1}{y}} + \left(\frac{v}{g}\right)\frac{dv}{dx} + \frac{fv^2}{2gD} + S_0 = 0. \tag{22.185}$$

The friction factor f can be expressed by eq. (22.20).

The solution described here is from Warrick and Yitayew (1988). Let the dimensionless velocity V and dimensionless distance X be expressed as

$$V = \frac{v}{v_0}; X = \frac{x}{x_0}, \tag{22.186}$$

where v_0 is v at $x = 0$, and x_0 is the characteristic length. Equation (22.185) can be expressed as

$$\frac{d}{dX}\left(-\frac{dV}{dX}\right)^{1/y} + aV\left(\frac{dV}{dX}\right) + V^m + S_0 = 0, \tag{22.187}$$

in which

$$x_0^{1+1/y} = \frac{2^{1-2y}\pi^{1/y} g D^{1+2/y} s^{1/y}}{c^{1/y} f_0 v_0^{m-1/y}}, \tag{22.188}$$

$$a = \left(\frac{2D}{f_0 x_0}\right) v_0^{2-m}, \tag{22.189}$$

$$S_0 = (2gDS_0)/(f_0 v_0^m). \tag{22.190}$$

Equation (22.187) is the governing equation, which is a nonlinear second-order, ordinary differential equation. It should be subject to the boundary conditions $V = 1$ at $X = 0$ and $V = 0$ at the end of the lateral $x = L$ (or $X = L/x_0 = X_0$). For an analytical solution, let

$$p = \left(-\frac{dV}{dX}\right)^{1/y}. \tag{22.191}$$

Since $dV/dX < 0$, p is positive and real for all X. With the use of eq. (22.191), eq. (22.187) can be written with $a = 0$ as

$$-p^y\left(\frac{dp}{dV}\right) + V^m + S_0 = 0. \tag{22.192}$$

Integration of eq. (22.192) yields

$$\frac{p^{1+y}}{1+y} = \frac{V^{m+1}}{m+1} + S_0 V + Cons \tag{22.193}$$

where $Cons$ is the constant of integration. Combining eqs. (22.192) and (22.193), the result is

$$\frac{dV}{dX} = -\frac{1}{F(V)}, F(V) = \left[(1+y)\left(\frac{V^{m+1}}{m+1} + S_0 V\right) + Cons\right]^{-y/(1+y)}. \tag{22.194}$$

From eq. (22.194),

$$X = \int_V^1 F(V)dV. \tag{22.195}$$

The upper limit of integration is defined such that $X = 0$ at $V = 1$. The second boundary condition that $V = 0$ at $X = X_0$ is met by choosing $Cons$ such that

$$X_0 = \int_0^1 F(V)dV. \tag{22.196}$$

With $Cons$ known, X can be specified as a function of V by eq. (22.195). Now, emitter flow per unit length of the lateral can be given as

$$q = -q_{av} X_0 \left(\frac{dV}{dX}\right), \tag{22.197}$$

or by eq. (22.194) as

$$q = \frac{q_{av} X_0}{F(V)}, \tag{22.198}$$

where q_{av} is the average outflow per unit length ($v_0 A/L$).

Example 22.16: Compute the velocity and discharge along a lateral 15 mm in diameter and 200 m long resting on flat ground (slope = 0). Assume $m = 1.85$, spacing of outlet = 1 m, $y = 0.5$, kinematic viscosity = 1.05×10^{-6} m²/s, and $c = 1.96 \times 10^{-7}$. The outlet discharge per unit length is 1.85 L/h/m, and the operating pressure is 6.9 m.

Solution: Given $m = 1.85$, the fully turbulent case (eq. 22.26) can be used to calculate f_0. Then, with kinematic viscosity $v = 1.05 \times 10^{-6}$ m²/s, and diameter $D = 15$ mm = 0.015 m,

$$f_0 = 0.130\left(\frac{v}{D}\right)^{0.172} = 0.130\left(\frac{1.05 \times 10^{-6}}{0.015}\right)^{0.172} = 0.025.$$

Given $q_{av} = 1.85$ L/h/m = 5.15×10^{-7} m²/s, $L = 200$ m, $D = 0.015$ m, the average velocity is

$$V_{av} = \frac{q_{av}L}{\pi\left(\frac{D}{2}\right)^2} = \frac{5.15 \times 10^{-7} \times 200}{\pi\left(\frac{0.015}{2}\right)^2} = 0.58 \text{ m/s}.$$

Given slope $S_0 = 0$, $y = 0.5$, $m = 1.85$, eq. (22.194) can be written as

$$F(V) = \left[(1+y)\left(\frac{V^{m+1}}{m+1} + S_0 V\right) + Cons\right]^{-y/(1+y)}$$
$$= \left[(1+0.5)\left(\frac{V^{1.85+1}}{1.85+1}\right) + Cons\right]^{-\frac{0.5}{1+0.5}}$$
$$= \left(0.526V^{2.85} + Cons\right)^{-1/3}.$$

Then, $X_0 = \int_0^1 \left(0.526V^{2.85} + Cons\right)^{-1/3} dV$.

Choose an initial guess of $Cons > 0$, e.g., $Cons = 1$. Check if the integral of eq. (22.196) is sufficiently close to X_0, if so the $Cons$ can be defined. Otherwise, if the integral is too large (small), a larger (smaller) $Cons$ is selected. In this example, a $Cons$ value of 1.5 is obtained when $X_0 = 0.85$.

Specify the values of V and calculate the corresponding X using eq. (22.195). The relative velocity V as a function of the lateral position is shown in Figure 22.25a. The velocity can be computed by multiplying the average velocity with V, and is shown in Figure 22.25b. The relative discharge (q/q_{av}) and discharge can be computed using eq. (22.198), and is shown in Figure 22.25c and d, respectively.

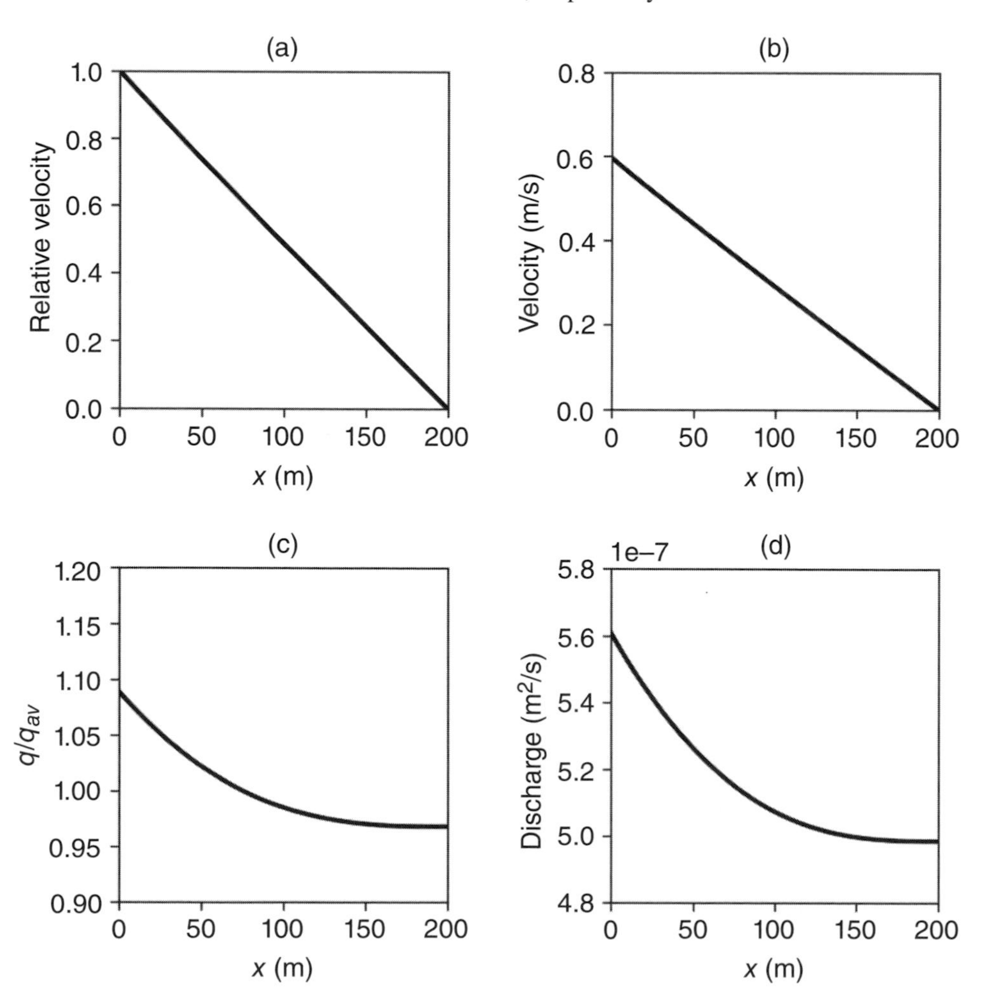

Figure 22.25 Variation of relative velocity (a), velocity (b), relative discharge (c), and discharge (d) along the lateral.

22.10.6 Energy Method: Simple Numerical Approximation

It is assumed that the ground slope is flat. Referring to Figure 22.9 for an elemental control volume, the continuity equation is given by eq. (22.6), which with the use of eq. (22.3) can be expressed as

$$\frac{dV}{dx} = -\frac{c}{A\Delta x} H^y. \tag{22.199}$$

The Bernoulli energy equation can be expressed as

$$H_x + \frac{V_x^2}{2g} = H_{x+dx} + \frac{V_{x+dx}^2}{2g} + H_f, \tag{22.200}$$

where H_f is the head loss due to friction between x and $x + \Delta x$. Expanding the pressure head and velocity terms on the right side, eq. (22.200) can be written as

$$H_x + \frac{V_x^2}{2g} = H_x + \frac{dH_x}{dx} dx + \frac{1}{2g}\left(V_x^2 + 2V_x \frac{dV_x}{dx} dx + \left(\frac{dV_x}{dx} dx\right)^2\right) + H_f. \tag{22.201}$$

In eq. (22.201), $\left(\frac{dV_x}{dx} dx\right)^2$ is negligibly small. Therefore, eq. (22.201) reduces to

$$\frac{dH_x}{dx} dx + \frac{V}{g}\frac{dV_x}{dx} dx + H_f = 0. \tag{22.202}$$

With the use of eq. (22.199) and the head loss

$$H_f = aV^m dx, \tag{22.203}$$

eq. (22.202) can be written as

$$\frac{dH_x}{dx} = -aV^m + \frac{V}{g}\frac{c}{A\Delta x} H^y, \tag{22.204}$$

where a for Re > 2300 and m = 1.852 can be expressed as

$$a = \frac{K}{C^m A^{0.5853}}, \tag{22.205}$$

in which K is constant, C is the Hazen–Williams coefficient, and m is the flow regime exponent. For Re < 2300, $m = 1$ and a can be given as

$$a = \frac{32\mu}{gD^2}. \tag{22.206}$$

Equations (22.119) and (22.204) can be solved for V and H. Zella and Kettab (2002) used the fourth-order Runga–Kutta method by assuming the velocity as 0 ($V_{x=L} = 0$) at the end of the lateral and the pressure head $H_{x=0} = H_{min}$ as known.

It is more convenient to measure the distance from the end of the lateral. Hence, let $X = L - x$. Then, eqs (22.119) and (22.204) can be recast as

$$\frac{dV}{dX} = \frac{c}{A\Delta x} H^y, \tag{22.207}$$

$$\frac{dH}{dX} = aV^m - \frac{V}{g}\frac{c}{A\Delta x} H^y. \tag{22.208}$$

The initial conditions are then $V_{X=0} = 0$ and $H_{X=0} = H_{min}$. Equations (22.207) and (22.208) can be solved simultaneously. For the integration of eqs (22.207) and (22.208), two estimates of the pressure head at the downstream end of the lateral ($X = 0$) need to be given, denoted as H^0_{min} and H^1_{min}.

Example 22.17: Consider a lateral line 12 mm in diameter and 200 m long resting on a flat ground (zero slope). The material of the line is in black polyethylene matter. The lateral has 40 equally placed emitters. For emitters, assume the values of $c = 9.5 \times 10^{-7}$, $y = 0.5$, $m = 1.852$, $K = 5.88$, $g = 9.81$ m/s², and $\mu = 10^{-6}$ m²/s. H_{max} is given as 35 m. Determine the velocity, flow, pressure distribution, and percentage variation of discharge along the lateral.

Solution: Given $m = 1.852$, $K = 5.88$, lateral diameter $D = 12$ mm $= 0.012$, and Hazen–Williams coefficient $C = 150$ for black polyethylene matter, a can be calculated using eq. (22.205):

$$a = \frac{K}{C^m A^{0.5853}} = \frac{K}{C^m\left[\pi\left(\frac{D}{2}\right)^2\right]^{0.5853}} = \frac{5.88}{150^{1.852}\left[\pi\left(\frac{0.012}{2}\right)^2\right]^{0.5853}} = 0.1120.$$

Given $L = 200$ m, and the number of emitters $N = 40$, one can select $\Delta x = L/N = 200/40 = 5$ m:

$$\frac{dV}{dX} = \frac{c}{A\Delta x} H^y = \frac{9.5 \times 10^{-7}}{\pi\left(\frac{0.012}{2}\right)^2(5)} H^{0.5} = 0.001680 H^{0.5},$$

$$\frac{dH}{dX} = aV^m - \frac{V}{g}\frac{c}{A\Delta x} H^y = 0.1120 V^{1.852} - 0.0001713 V H^{0.5}.$$

Equations (22.207) and (22.208) can be solved simultaneously using the fourth-order Runga–Kutta method. Below is the equation used to compute the value of H_{n+1} based on the previous value H_n, Δx is used as the step height with $X_{n+1} = X_n + \Delta x$. The values of n are 0, 1, 2, . . ., 39.

$$k_1 = \Delta X\left[aV_n^m - \frac{V_n}{g}\frac{c}{A\Delta x} H_n^y\right] = 0.5599 V_n^m - 0.000856 V_n H_n^y,$$

$$k_2 = \Delta X\left[aV_n^m - \frac{V_n}{g}\frac{c}{A\Delta x}\left(H_n + \frac{k_1}{2}\right)^{0.5}\right] = 0.5599 V_n^m - 0.000856 V_n \left(H_n + \frac{k_1}{2}\right)^{0.5},$$

$$k_3 = \Delta X\left[aV_n^m - \frac{V_n}{g}\frac{c}{A\Delta X}\left(H_n + \frac{k_2}{2}\right)^{0.5}\right]$$
$$= 0.5599V_n^m - 0.000856V_n\left(H_n + \frac{k_2}{2}\right)^{0.5},$$
$$k_4 = \Delta X\left[aV_n^m - \frac{V_n}{g}\frac{c}{A\Delta X}(H_n + k_3)^{0.5}\right]$$
$$= 0.5599V_n^m - 0.000856V_n(H_n + k_3)^{0.5},$$
$$H_{n+1} = H_n + \frac{k_1}{6} + \frac{k_2}{3} + \frac{k_3}{3} + \frac{k_4}{6}.$$

Similarly, the value of V_{n+1} based on the previous value V_n can be expressed as

$$k_1 = k_2 = k_3 = k_4 = \Delta x\frac{c}{A\Delta x}H^y = 0.008H^{0.5},$$
$$V_{n+1} = V_n + \frac{k_1}{6} + \frac{k_2}{3} + \frac{k_3}{3} + \frac{k_4}{6} = V_n + 0.0084H^{0.5}.$$

The initial conditions are $V_{X=0} = 0$ and $H_{X=0} = H_{min}$. Two estimates of the pressure head at the downstream end of the lateral ($X = 0$) need to be given, denoted as H^0_{min} and H^1_{min}. Using the above equations, the maximum pressure at the inlet can be calculated and denoted as H^0_{max} and H^1_{max}. Then, a new H_{min} can be calculated by interpolating the Lagrange polynomial of degree one, which is described as

$$H_{min} = \frac{H_{max} - H^1_{max}}{H^0_{max} - H^1_{max}}H^0_{min} + \frac{H_{max} - H^0_{max}}{H^1_{max} - H^0_{max}}H^1_{min},$$

with $H_{max} = 35$ m. The process is continued until $\frac{\text{The new calculated } H - \text{Previous } H}{\text{The new calculated } H} < 10^{-5}$. After the interaction process, the velocity, discharge, pressure distribution, and percentage variation of discharge along the lateral (m) are as shown in Figure 22.26. The discharge is computed using eq. (22.3).

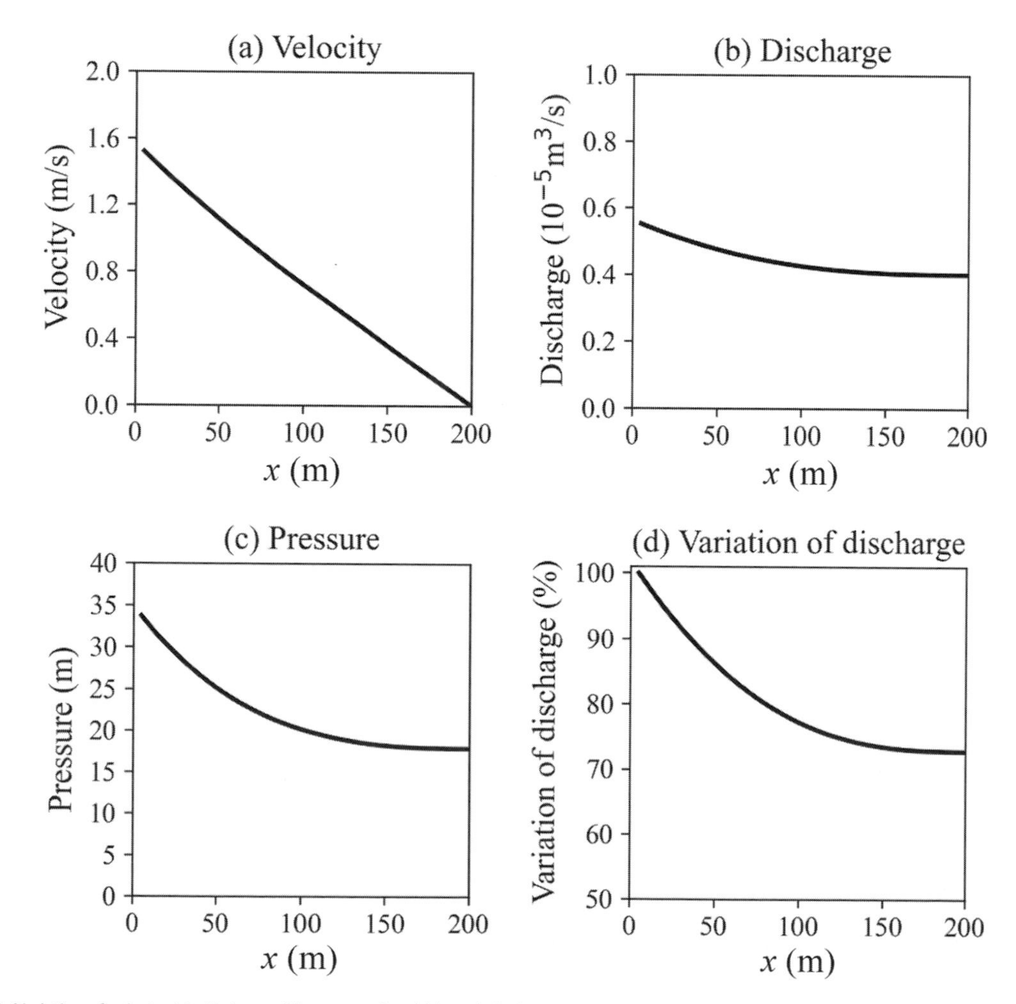

Figure 22.26 Variation of velocity (a), discharge (b), pressure head (c), and discharge (%) (d) along the lateral.

QUESTIONS

Q.22.1 What will be the operating pressure for a bubbler tube 2.5 mm in diameter and 4 m long to deliver water at the rate of 1.25 L/min?

Q.22.2 What will be the capacity of emission devices for a garden with trees on a 4 × 4 m grid? The soil has a readily available water-holding capacity of 15 cm and design daily irrigation requirement is 6 mm/day. Two emission devices are used for each tree and the cropped area of 40% is to be irrigated daily. The efficiency of the devices is 90% and 10% of the number of emission devices are operating per irrigation, i.e., $N_d = 10\%$.

Q.22.3 What will be the length of a bubbler tube 2.5 mm in diameter required to deliver water at the rate of 1.25 L/min at both the upstream and downstream ends of a lateral where the pressures, respectively, are 140 kPa and 125 kPa?

Q.22.4 What will be the length of a bubbler tube 3 mm in diameter for delivering water at the rate of 1.5 L/min at the operating pressure of 160 kPa?

Q.22.5 A polyethylene lateral 10 mm in diameter and 120 m in length is placed in a flat field. It has point-source emitters with 1.5 m spacing and has a pressure of 140 kPa at its upstream end. What would be the emission uniformity along the lateral? Assume the water temperature as 25 °C. The emitter discharge Q (L/h) can be computed by $Q = 0.21P^{0.6}$ with P = operating pressure (kPa).

Q.22.6 What would be the emission uniformity for the data in considering Q.22.5 if the field had an uphill slope of 2%?

Q.22.7 A polyethylene lateral 10 mm in diameter and 105 m in length is placed in a flat field. It has point-source emitters with 1.5 m spacing and has a pressure of 130 kPa at its upstream end. What would be the emission uniformity along the lateral? Assume the water temperature as 25 °C. The emitter discharge Q (L/h) can be computed by $Q = 0.21P^{0.6}$ with P = operating pressure (kPa).

Q.22.8 What would be the emission uniformity for the data in Q.22.7 if the on-line point-source device has standardized barbs? The discharge equation remains the same.

Q.22.9 Determine the length of a lateral whose inside diameter (D) is 15 mm, downslope (S_0) is 3%, and specific discharge (Q_s) is 0.001 lps/m. The operating pressure (H) head is assumed as 12 m and the allowable head variation is 15%.

Q.22.10 What will be the length of a lateral if upslope (S_0) is 3% and other parameters are the same as in Q.22.9.

Q.22.11 What will be the Reynolds number of flow in a fully turbulent emitter with a diameter of 2 mm in a drip irrigation system with a head of 10 m? The orifice coefficient (C_o) is 0.6.

Q.22.12 What will be the limits of emitter discharge for irrigating an area of 1.5 ha with laterals spaced every 2.5 m and emitters along the lateral spaced at 70 cm? Assume the total time of irrigation for each eight-day period is 16 h. The distribution efficiency is 90% and irrigation requirement is 5 mm/day.

Q.22.13 What will be the required operating pressure for a long-path emitter 2.5 mm in diameter operating in the laminar regime for a circular conduit flow area where the design discharge is 10 L/h and emitter length is 5 m?

Q.22.14 What will be the operating pressure head for a long-path emitter with a diameter of 3.0 mm and design discharge of 12 L/h and an equivalent length of 4 m?

Q.22.15 What will be the pressure at the last emitter on the line, the average pressure and its location, and the average, maximum, and minimum discharge in a trickle irrigation system having a multiple-outlet lateral with 40 emitters, inlet pressure head of 10 m, and 10% of average pressure as pressure head loss? The emitter discharge is given by $q = 0.95H^{0.695}$, where q is in L/h and H in m.

Q.22.16 What will be the velocity, pressure, and discharge distribution of emitters along a lateral, and percentage variation of discharge if the lateral has a length of 80 m and an inside diameter of 12 mm, and emitters have a design discharge of 4.5 L/h and minimum operating pressure of 12 m? The discharge of the emitter is assumed to be a function of the pressure head raised to the power of one-half, i.e., $y = 0.5$. For emitters, assume the value of $c = 3.61 \times 10^{-7}$, $m = 1.853$, $K = 5.88$, $g = 9.81$ m/s^2, and $\mu = 10^{-6}$ m^2/s.

Q.22.17 The pressure at the inlet is $H = 6.5$ psi or 15.0 feet of water. The lateral has a length L of 250 ft, inside diameter of 0.5 in., and upslope of 3%. The total inlet discharge is 3 gpm. Compute the uniformity coefficient. Then do the same for the downslope of 3%.

Q.22.18 Determine head loss due to friction in a drip lateral with the following data: internal diameter of the lateral as 15 mm, length of the lateral as 80 m with "standard" emitters 1 m apart, normal water temperature as 20 °C, and design discharge of each emitter as 1.20 L/h. (K [for Re] = 10, $\rho = 0.997$ g/cm^3, $\mu = 0.001$ N-s/m^2, and $F = 0.351$.)

Q.22.19 Consider a trickle irrigation system for a vineyard on a level field that has a lateral 15 mm in diameter with 20 single-orifice emitters 1.25 m apart where the first emitter is at the full space from the lateral entrance from the submain (friction factor $F = 0.356$). The pressure head at the entrance is 14 m and the emitter design discharge is 16 L/h. What will be the discharge from the final emitter (type B) on the lateral (Figure 22.11)? What will be the discharge variation (%) between the first and last emitter?

Q.22.20 Using the analytical approach constant discharge method, determine the length of the lateral and inlet pressure head (H_{in}), given the following information: pressure–discharge relationship coefficient $c = 1.96 \times 10^{-7}$, emitter exponent $y = 0.5$, required average emitter flow $q_{av} = 2.5$ lph or 6.94×10^{-4} L/s, spacing between emitters $s = 1$ m, inside diameter $D = 12$ mm, and lateral slope $S_0 = 0$. Assume the flow as fully turbulent with $m = 1.828$. The required design pressure head is $H_s = (q_{av}/c)^{1/y} = 12.55$ m, and the design uniformity coefficient is 0.95. ($K = 1.13 \times 10^6$.)

Q.22.21 Solve Q.22.20 with the analytical approach variable discharge method.

Q.22.22 A drip lateral has the following data: internal diameter of the lateral is 15 mm; length of the lateral is 60 m with "small" emitters 0.8 m apart; normal water temperature is 20 °C. Each emitter has an inside diameter of 0.4 mm, and the operating pressure head is 6 m. (K (for Re) $= 10$, $\rho = 0.997$ g/cm^3, $\mu = 0.001$ N-s/m^2, and $F = 0.357$.)

Q.22.23 A trickle irrigation system has a multiple-outlet lateral with 50 emitters, the inlet pressure head (H_{in}) is 12 m, and the head loss in the lateral is 10% of the average pressure (H_{av}). The emitter pressure–discharge relation is $q = 0.92H^{0.69}$, where q is in L/h and H is in m. Assume the flow is turbulent in a smooth pipe, with $m = 1.75$. The lateral slope $S_0 = 0$. Using the analytical approach constant discharge method,

1. compute the average pressure head (H_{av});
2. compute the pressure at the last emitter in the lateral (H_{last});
3. determine the average, maximum, and minimum emitter discharge in the lateral; and
4. compute the total discharge at the inlet of the lateral.

REFERENCES

Anyoji, H., and Wu, I. P. (1987). Statistical approach for drip lateral design. *Transactions of the American Society of Agricultural Engineering*, 30: 187–192.

Bralts, V. F., and Edwards, D. M. (1986). Field evaluation of drip irrigation submain units. *Transactions of the American Society of Agricultural and Biological Engineers*, 29(6): 1659–1664.

Cuenca, R. H. (1989). *Irrigation System Design: An Engineering Approach*. Englewood Cliffs, NJ: Prentice Hall.

Hanks, R. J., and Keller, J. (1972). New irrigation method saves water but it's expensive. *Utah Science*, 33: 79–82.

Hiler, E. A., and Howell, T. A. (1972). Crop response to trickle and subsurface irrigation. Presented at the Annual Meeting American Society of Agricultural Engineers. Chicago, IL.

Howell, T. A., and Hiler, E. A. (1974). Designing trickle irrigation laterals for uniformity. *Journal of Irrigation & Drainage Engineering*, 100(IR4): 443–453.

James, L. G. (1988). *Principles of Farm Irrigation System Design*. New York: Wiley.

Keller, J., and Karmeli, D. (1974). Trickle irrigation design parameters. *Transactions of the ASAE*, 17(4): 678–684.

Keller, J., and Karmeli, D. (1975). *Trickle Irrigation Design*. Glendora, CA: Rain Bird Sprinkler Management Corporation.

Kenworthy, A. L. (1972). *Trickle Irrigation: The Concept and Guidelines for Use*. East Lansing, MI: Michigan Agricultural Experiment Station.

Myers, L. E., and Bucks, D. A. (1972). Uniform irrigation with low pressure trickle systems. *Journal of Irrigation and Drainage Division, ASCE*, 98(IR3): 341–346.

Solomon, K. (1979). Manufacturing variation of trickle emitters. *Transactions American Society of Agricultural Engineers*, 22 (5): 1034–1038.

Urbina, J. L. (1976). Head loss characteristic of trickle irrigation hose with emitters. MS Thesis, Utah State University.

USDA-NRCS (2013). Microirrigation. In *National Engineering Handbook*. Washington, DC: USDA.

Valiantzas, J. D. (1998). Analytical approach for direct drip lateral hydraulic calculation. *Journal of Irrigation & Drainage Engineering*, 124(6): 300305.

Warrick, A. W., and Yitayew, M. (1987). An analytical solution for flow in a manifold. *Advances in Water Resources*, 10: 59–63.

Warrick, A. W., and Yitayew, M. (1988). Trickle lateral hydraulics: I. Analytical solutions. *Journal of Irrigation & Drainage Engineering*, 114(2): 281–288.

Watters, G. Z. and Keller, J. (1978). Trickle irrigation tubing hydraulics. ASAE technical paper 78-2015.

Wierenga, P. J. (1977). *Influence of Trickle and Surface Irrigation on Return Flow Quality*. Environmental Protection Agency, Report EPA-600/2-77-093. Ada, OK: Office of Research and Development.

Wu, I. P. (1992). Energy gradient line approach for direct hydraulic calculation in drip irrigation design. *Irrigation Science*, 13: 21–29.

Wu, I. P. (1997). An assessment of hydraulic design of micro-irrigation systems. *Agricultural Water Management*,32: 275–284.

Wu, I. P., and Gitlin, H. M. (1973). Hydraulics and uniformity for drip irrigation. *Journal of the Irrigation and Drainage Division*, 99(2): 157–168.

Wu, I. P., and Gitlin, H. M. (1974). Design of drip irrigation lines. College of Tropical Agriculture and Human Resources, University of Hawaii at Manoa, technical bulletin 96. https://scholarspace.manoa.hawaii.edu/bitstream/10125/31040/driplines.pdf.

Wu, I. P., and Gitlin, H. M. (1975). Energy gradient line drip irrigation laterals. *Journal of Irrigation & Drainage Engineering*, 101(IR4): 323–236.

Wu, I. P., and Gitlin, H. M. (1980). Preliminary concept of a drip irrigation network design. Paper presented at the Annual Meeting of the American Society of Agricultural Engineers, Hilo, HI.

Wu, I. P., and Irudayaraj, J. M. (1989). Sample size determination for evaluating drip irrigation systems. *Transactions of the American Society of Agricultural Engineers*, 32(6): 1961–1965.

Wu, I. P., Saruwatari, C. A., and Gitlin, H. M. (1983). Design of drip irrigation lateral length on uniform slopes. *Irrigation Science*, 4: 117–135.

Zella, L., and Kettab, A. (2002). Numerical methods of microirrigation lateral design. *Biotechnology, Agronomy and Society and Environment*, 6(4), 231–235.

Part VI Design

23 Irrigation Planning

Notation

BCR	benefit/cost (BC) ratio (fraction)	IRR	internal rate of return (fraction)
C	total consumer spending (dollars)	n	period (years or months)
C_n	cash flow in the nth period (dollars)	N	total number of periods
G	total government spending (dollars)	NPV	net present value (dollars)
GDP	gross domestic product (dollars)	NX	total net exports or the difference between exports and imports (dollars)
i	discount rate or interest rate (fraction)	r	internal rate of return (fraction)
I	total business spending on capital (dollars)	R_t	net cash flow at time t (dollars)

23.1 INTRODUCTION

Irrigation planning begins with an assessment of water resources availability and irrigation potential. Then, planning of an irrigation system depends on the size of the system. Small systems may be owned by individual farmers, and farmers plan these systems on their own, with limited outside help. On the other hand, large systems are owned by governments or groups of farmers, and their planning is quite technical. This chapter discusses the rudimentary aspects of irrigation planning.

23.2 AVAILABILITY OF WATER RESOURCES

Irrigation planning for an area is subject to the availability of water resources, which greatly vary throughout the world. It will be interesting to look at water resources at the global level. Table 23.1 shows the annual average long-term precipitation and renewable water resources in different regions (FAO, 2013). The annual average long-term global renewable water resource is 43,022 m^3, of which Americas have the largest share (45%), followed by Asia (29%), Europe (15%), Africa (9%), and Oceania (2%). Current global irrigation water withdrawal only accounts for 6% of the total renewable water resources, but the values vary across different regions. In Northern Africa, Western Asia, Central Asia, and South Asia, more than 50% of the renewable water resources are used for irrigation. By contrast, in Southern America, Eastern Europe, and Russia, irrigation only accounts for 1% of the water resources.

Country-wise availability of water resources for irrigation shows even more considerable variability. During the period 2005–2007, water use for irrigation in Libya (Northern Africa), Saudi Arabia (Western Asia), Yemen (Western Asia), and Egypt (Northern Africa) exceeded their annual renewable water resources. The FAO report on the state of the world's water resources for food and agriculture showed that 11 countries used more than 40% of their available water resources for irrigation, this being the threshold considered as critical water use pressure (FAO, 2013). Another eight countries used 20% of their water resources for irrigation, indicating impending water scarcity. Large spatial variations are also found in several countries. For example, northern China faces severe water shortages, but southern China has sufficient water resources. Groundwater mining is found in certain parts of Central America, South and East Asia, and the Caribbean, although water balance may still be positive at the national level.

23.3 IRRIGATION POTENTIAL

Irrigation potential depends not only on the availability and quality of water resources, but also on climate, soil, and availability of energy. Elliott et al. (2014) analyzed global irrigation potential and constraints at the end of the century based on the projections from 10 global hydrological models (GHMs) and 6 global gridded crop models (GGCMs). The projected differences between future irrigation water demand and available renewable water at the end of the century in terms of m^3/year are shown in Figure 23.1. The agricultural production in a currently irrigated region is considered irrigation constrained if the projected future irrigation water demand is greater than the available renewable water (i.e., the areas with negative values in Figure 23.1).

Table 23.1 *Annual average renewable water resources and irrigation water withdrawal in different regions*

Regions	Precipitation (mm)	Renewable water resources[a] (km^3)	Irrigation withdrawal (km^3)	Pressure on water resources due to irrigation[b] (%)
Africa	**678**	**3,931**	**184**	**5**
Northern Africa	96	47	80	170
Sub-Saharan Africa	815	3,884	105	3
Americas	**1,091**	**19,238**	**385**	**2**
Northern America	636	6,077	258	4
Central America and Caribbean	2,011	781	15	2
Southern America	1,604	12,380	112	1
Asia	**827**	**12,413**	**2,012**	**16**
Western Asia	217	484	227	47
Central Asia	273	263	150	57
South Asia	1,602	1,766	914	52
East Asia	634	3,410	434	13
Southeast Asia	2,400	6,490	287	4
Europe	**540**	**6,548**	**109**	**2**
Western and Central Europe	811	2,098	75	2
Eastern Europe and the Russian Federation	467	4,449	35	1
Oceania	**586**	**892**	**19**	**2**
Australia and New Zealand	574	819	19	2
Pacific Islands	2,062	73	0.05	0.1
World	**809**	**43,022**	**2,710**	**6**

[a] Refers to internal renewable water resources, which excludes incoming flows at the regional level; [b] refers to renewable water resources divided by irrigation withdrawal.
Adapted from FAO, 2013.

Figure 23.1 The median differences between irrigation demand and available irrigation demand at the end of the century under RCP 8.5 from 10 GHMs and 6 GGCMs. (Adapted from Elliott et al., 2014. Reprinted with permission from PNAS.) A black and white version of this figure will appear in some formats. For the color version, please refer to the plate section.

Areas with positive values indicate they have irrigation adaptation potential. At the end of the century, some currently heavily irrigated regions in Western and Central Asia, Northern Africa, northern China, Mexico, and the western United States are projected to have no irrigation potential, resulting in a loss of 600–2900 Pcal of food production, with the change from irrigated to rainfed cropland due to insufficient water resources. Other regions such as Europe, northern/eastern United States, South Asia, and parts of South America have high irrigation potential, which could support a net increase in crop production with the change from rainfed to irrigated cropland. However, the availability of energy in these regions should also be considered if substantial irrigation infrastructure investments are required.

23.4 LAND OWNERSHIP

The ownership of farmland can vary from individual farmers to government. In democratic countries, it is the farmers who own and may cultivate or lease the land. In non-democratic countries, it may be the government that owns the land. On the other hand, there are certain countries in Africa where land ownership is not clearly defined. Ownership, size of the agricultural land, climate, soil, source and quality of water, crops, and market conditions determine the conception and planning of an irrigation system.

In many countries, such as India, Pakistan, Bangladesh, China, and Sri Lanka, farm holdings are small, and large irrigation systems involving canals are conceived, planned, designed, and implemented by government. In contrast, in countries such as Australia, Canada, and the United States, some farmers own large tracts of land – up to thousands of acres – and irrigation systems may be planned by individual farmers or groups of farmers or irrigation districts. For large irrigation systems, the government may play a limited role.

Thus, planning of an irrigation system can vary from an individual farmer to a community of farmers to a government agency. In irrigation planning and implementation, both farmers and government are involved in one way or another. In principle, irrigation is meant for farmers who can increase crop yield and grow more crops. Therefore, farmers' participation should be a key element in the entire planning process. However, this is not always the case, such as when the government is the owner of the irrigation system. This chapter discusses elements of planning and implementation of an irrigation system.

23.5 ENERGY REQUIREMENT FOR IRRIGATION TECHNOLOGIES

Agricultural irrigation is the primary consumer of energy on farms (Naylor, 1996), so any change to the irrigation method used is expected to change on-farm energy consumption. Energy consumption and irrigation in agrarian economies are intertwined (Plappally and Lienhard, 2012), and this relationship is called the energy–irrigation nexus, the electricity–irrigation nexus, or the diesel–irrigation nexus. The energy expended to irrigate a field is dependent on the amount of water pumped, area of the field, soil characteristics of the location, geology, slope, crop varieties or cropping patterns, precipitation or climate of the location, temperature, type of irrigation, irrigation scheduling, application effectiveness, pumping system type, pressure requirement at the point of use, and energy cost (Martin et al., 2011). The energy required for pumping depends on crop water requirement, total dynamic head, flow rate, and system efficiency (Lal, 2004). Direct energy inputs are primarily the fuel sources used to operate farm machinery and pumps, while indirect energy inputs refer to energy used to produce equipment and other goods and services that are used on farms (Pimentel, 1992). Between 23% and 48% of direct energy used for crop production is used for on-farm pumping (Hodges et al., 1994). If a gravity-fed irrigation method is used in conjunction with a surface water source, the energy required to transport and apply water to the field is negligible (Stout, 1990). However, where pressurized groundwater extraction is used, energy is required for pumping and delivery to the field. Crops with a higher water requirement result in a larger amount of water being pumped and increase energy consumption: This means that summer crops will typically consume more energy than winter crops. Pressurized microirrigation systems can decrease energy consumption if operating pressures (and therefore total dynamic head) and pumping volumes are reduced (Hodges et al., 1994).

Surface irrigation by flooding is less energy intensive when surface water is used instead of groundwater. In a study on the effects of drought on California electricity supply and demand (Figure 23.2), Benenson et al. (1977) evaluated the energy use per cubic meter of water supplied by surface irrigation at four different locations in the California Central Valley. The lift to pump water varied considerably: at the Sacramento River, 18.6 m; at the Sacramento–San Joaquin Delta, 29.5 m; at San Joaquin, 38.7 m; and at Tulare Lake, 57.3 m.

Griffiths-Sattenspiel and Wilson (2009) reported that surface irrigation from a surface source may be assumed to consume no energy, while 3 m groundwater source was found to consume 0.024 kW h/m^3 of energy. Surface or furrow irrigation by pumping water from a river at a pressure of approximately 1 bar consumed 0.045 kW h/m^3, and pumping water from a bore well at a pressure of 4.4 bar consumed 0.2 kW h/m^3 in Tasmania, Australia (Richards and Smith, 2003). Surface irrigation using either surface or groundwater pumping consumed energy between a low of 0.138 kW h/m^3 and a high of 0.9 kW h/m^3 in Spain (Muñoz et al., 2010).

High-, intermediate-, and low-pressure sprinkler irrigation systems operate at 4.8–9.6 bar, 2.4–4.8 bar, and 1.4–2.4 bar, respectively. Table 23.2 provides reported values of energy

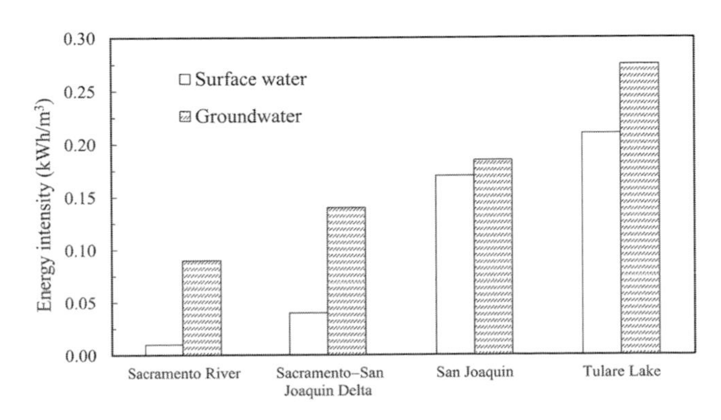

Figure 23.2 Surface irrigation energy intensity (energy use per cubic meter of water supplied) at four different locations at the Central Valley, California. (Data from Benenson et al., 1977.)

Table 23.2 *Energy requirements and water use in different types of irrigation systems*

Type	Irrigation system	Energy consumption (kWh/ha/yr)	Water irrigated (m^3/ha/yr)
Mobile			
	Sprinkler	800	12,000
	Nozzle carrier	410	9,000
Stationary			
	Central-pivot nozzle	340	9,000
	Linear lateral nozzle	340	9,000
	Side-roll irrigation tube	650	12,000
Row sprinkler			
	Movable hose	650	12,000
Microirrigation			
	Drip system	135	8,400

After Chiaramonti et al., 2000.

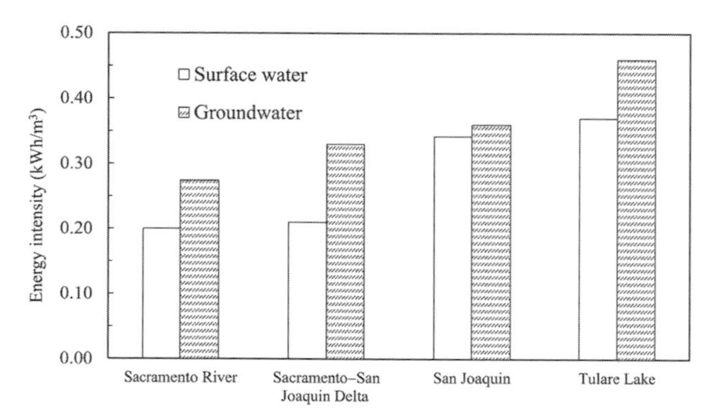

Figure 23.3 Energy intensities for irrigating alfalfa, cotton, and vegetables at different locations in Central Valley, CA using sprinkler irrigation. (Data from Benenson et al., 1977.)

for irrigating a farm using different irrigation systems. Energy consumption in sprinkler irrigation is comparatively higher than in surface irrigation; Figures 23.3 and 23.4 provide a comparison. Considerable differences are observable in energy consumption for pressurized systems such as sprinkler irrigation versus surface irrigation.

From Figure 23.4 it can be observed that in sprinkler irrigation systems there is an increase in energy use per unit volume with decreasing amounts of water. A similar behavior is not seen in drip irrigation systems, which provide water to the root zone of the plant and have their water outlets near the roots. Energy consumption for drip irrigation ranged from 0.32 to 1.1 kW h/m^3, while for sprinkler irrigation it is 0.6–1.3 kW h/m^3 (Moreno et al., 2010).

Studies on traveler irrigation systems from Australia operating at approximately 8.3 bar using river (surface water) pumping, 8.8 bar using bore well pumping, and 11.8 bar discharge pressure have operational energy consumption of 0.38 kW h/m^3, 0.4 kW h/m^3, and 0.54 kW h/m^3, respectively (Richards and Smith, 2003). Figure 23.5 illustrates energy intensities of different irrigation technologies in use for irrigation in Australia.

23.6 SYSTEM PLANNING

The key elements in planning an irrigation system include: (1) ownership; (2) level of financing; (3) time horizon for completion; (4) irrigation command area; (5) source and quality of water; (6) climate; (7) soils; (8) crops; (9) irrigation methods; (10) source of energy; (11) preparation of system proposal; (12) alternative proposals; (13) economic analysis; and (14) political, social, cultural, and environmental considerations.

Planning begins with the ownership of the system. If a farmer owns the system, he has a lot of flexibility at each step of the planning process, but if the government is involved then there are usually a lot of procedural issues involved that have to be followed. The time horizon for completing the planning directly hinges on the ownership and size. Furthermore, the financing of the system is also directly tied to the ownership and size. An individual farmer or a group of farmers usually does not have the wherewithal to finance the system and seeks an investor – be that a bank or a financial institution – whereas the government does not have this limitation. The level of finance can determine the size of the system, which is directly connected with the irrigation command area for the system. For a large system, thousands of acres of land to be irrigated under the system is not uncommon.

Other considerations for planning an irrigation system are the source, quantity, and quality of water, as well as the energy available for irrigation. Not all areas have local sources of enough good-quality water. This means that water will be brought from a distance, which will greatly impact the cost. Energy determines the system operation and its cost. Climate and soils determine the crops that can be grown. The increase in crop yield due to irrigation, the number of crops that can be grown, and the quality of produce play a key role in determining the financial viability of the system. The methods to be envisaged for irrigating crops are important for system planning, as they involve different costs and labor.

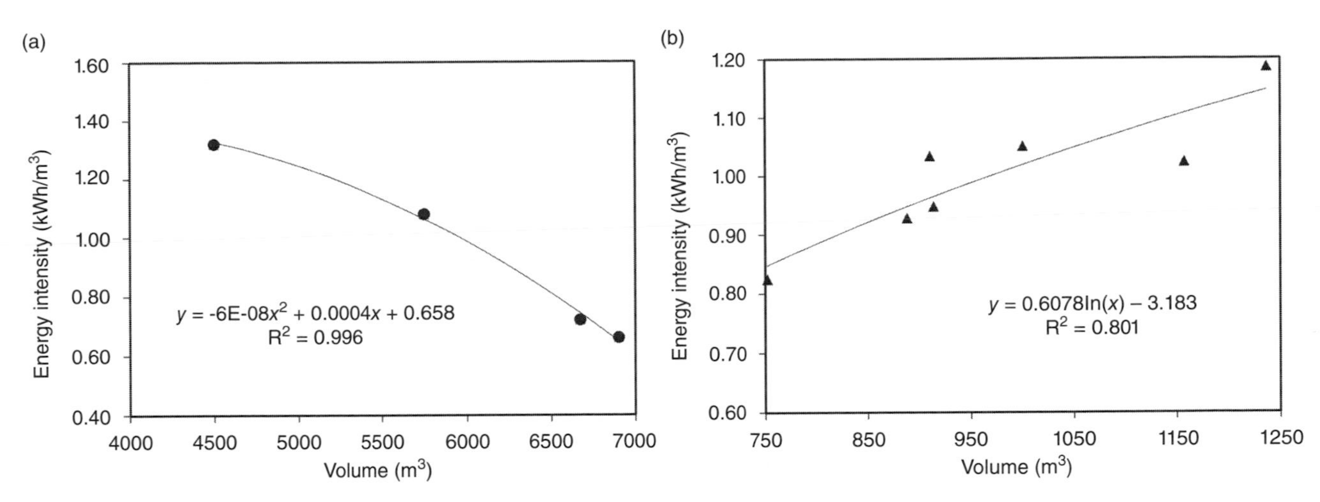

Figure 23.4 Energy use and water application in sprinkler irrigation systems (a) and drip irrigation systems (b) in Spain. (Data from Moreno et al., 2010.)

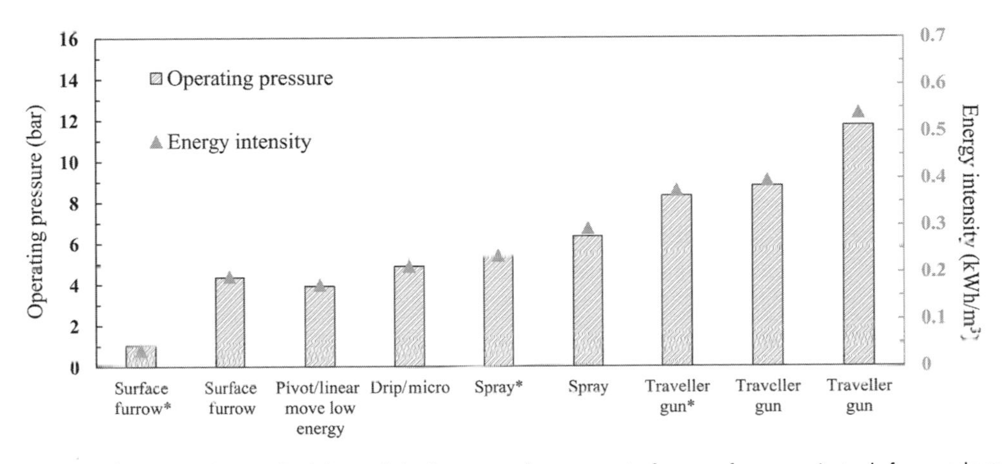

Figure 23.5 Energy intensities for irrigation in Australia. * denotes irrigation systems that pump water from a surface source instead of a groundwater source. All other systems pumped water from groundwater. (Data from Richards and Smith, 2003.)

Once the above considerations have been determined, the preparation of a system proposal begins. The proposal includes a feasibility study, defining system objectives, alternative strategies for achieving the objectives, appraisal, implementation, monitoring and evaluation, economic analysis, and meeting political, social, and environmental constraints. Each of these items is now briefly discussed.

23.6.1 Feasibility Study

A feasibility study to determine whether the system will be viable needs to be undertaken. This study determines whether a more advanced study or the full proposal preparation is desirable. In order to accomplish the feasibility study, it is important to collect as much information on the system as possible so that a preliminary assessment can be made for the viability of achieving objectives and likely alternative strategies for achieving the objectives.

23.6.2 Proposal Preparation

When the feasibility study signals the preparation of a full proposal, a complete system plan is prepared. This includes all relevant technical as well as non-technical details. This is essentially a blueprint of the system. It should clearly detail system implementation, monitoring and evaluation, and financial analysis.

23.6.3 Review of System Plan(s)

The system plan(s) should be carefully evaluated from both technical and non-technical aspects. The evaluation should be done by experienced experts in concerned areas. It is important to keep in mind that the people doing the feasibility study and proposal preparation should not be involved in proposal evaluation, otherwise there will be bias in the review. Both the time frame for implementation and financial viability should be carefully assessed.

23.6.4 System Implementation

Once a system plan is approved, implementation is the next step. This consists of developing a plan of action, collection of labor, acquisition of material, obtaining the needed finance, preparing the budget, involving stakeholders and assigning responsibilities, initiation of field work, contracting, supervising, managing, and contingency plans.

23.6.5 Monitoring and Evaluation

Monitoring is an integral part of system implementation, which means it starts from day one. Close monitoring saves time and money and helps keep track of progress on a regular and timely basis and ensures completion of the system in time. Likewise, evaluation helps ensure the accomplishment of system objectives, implementation of planned activities, anticipation of unforeseen difficulties, and gain of anticipated benefits. Both monitoring and evaluation apply to ongoing as well as completed systems.

23.6.6 Financial Analysis

During system implementation, it is important to keep track of finances and make sure that the budget allocated for each activity is not exceeded. In government-supported systems it is not uncommon that the final system budget is double the amount initially allocated during the planning stage. This happens not only in developing countries but also in countries like the United States. For completed systems, the cost is already incurred and operational and maintenance costs are recurring. It is important to analyze benefits and examine whether these are accruing as proposed in the planning process, and whether over the life of the system benefits will exceed costs.

Benefits accrue primarily from increased crop production, flexibility in cropping patterns, and increased frequency of crops per year. Other indirect benefits are reclaimed wasteland, creation of employment, and tax revenue by selling water to other water users. Costs include the cost of production, cost of labor, cost of equipment, construction, operation, monitoring, and management.

23.6.7 Economic Vocabulary

There are a number of terms that are used in doing economic analysis of an irrigation system. It is therefore useful to define those terms here.

Economic analysis: This is the systematic analysis of the costs of resources and social and environmental costs, as well as benefits, of a system to individual farmers, community, or government.

Economic value: This refers to the difference in national income caused by the return or revenue from a system.

National income: The national income refers to the money or economic value of all goods and services produced within a period of time, say a year.

Economic rate of return (ERR): This refers to the rate at which costs and benefits of a system, discounted over its life, are equal. It accounts for the effect of price controls, subsidies, and tax breaks.

Financial rate of return: This refers to the internal rate of return computed using market value.

Gross domestic product (GDP): This refers to the monetary value of finished goods and services within a country within a period, say a year. There are three ways to calculate GDP, including the production approach (or output/added value approach), the income approach, and the expenditure approach. The expenditure approach is to sum the final uses of all goods and services, including private and public consumption, government outlays, investments, and exports minus imports, produced. It can be expressed as

$$\mathrm{GDP} = C + G + I + NX, \tag{23.1}$$

where C is the total consumer spending, G is the total government spending, I is the total business spending on capital, and NX is the total net exports or the difference between exports and imports.

Net benefit: This denotes benefits minus costs due to an irrigation system, calculated in terms of the present value.

Gross benefit: This denotes the incremental value of system output.

Incremental net benefit: This refers to the increase in net benefit from the system, computed as cash value.

Indirect benefit: This refers to the secondary benefits of a system.

Opportunity cost: This denotes the benefit lost due to the investment of capital in a system, and is utilized to determine the discount rate needed for computing the benefit–cost ratio and net present worth of the investment.

Discount rate: This denotes the interest rate for computing the present worth of a future value.

Efficiency price: This refers to the economic value or the opportunity cost or value of a good or service in a system. For computation, the price should account for market fluctuations.

Inflation: This refers to the rate at which the general level of prices increases and the purchasing power falls.

Shadow price: This refers to the value of cost or benefit in a system when the market price does not give a realistic estimate of the economic value.

Current ratio: This refers to the ability to pay short-term obligations, and can be computed as

$$\text{Current ratio} = \frac{\text{Current assets}}{\text{Current liabilities}}. \tag{23.2}$$

The current ratio is also called the liquidity ratio or cash asset ratio.

Current price: This refers to the price or value of a commodity or service that accounts for inflation.

Farm gate price: This refers to the price of a product sold at the farm, which is usually less than the retail consumer price.

Cutoff rate: This refers to the minimum acceptable rate of return and is taken as the opportunity cost.

Transfer payment: This refers to the payment made without receiving goods or services in return.

Economic life: This refers to the period an asset is capable of providing services for.

Grace period: This refers to the period during which no payment of the principal is required.

Grant: This denotes a payment made to promote a specific activity by a government or an agency to an individual or community without expectation of services or goods in return.

Work day: This denotes the time in a day a person devotes to an activity. It is usually eight hours.

Equity: This refers to the residual amount remaining after subtracting liabilities from assets.

23.6.8 Investment Analysis

To finance the system money is borrowed and is invested in the system. In order for the system to remain viable, this money must be paid back from the income generated from crop production and tax collection. The investment cost should also include the cost of land, cost of water supply, energy cost, repair and maintenance cost, cost of replacement of certain items, and other costs. The investment has to be carried out over the system period, meaning costs and benefits are spread over time, keeping in mind that future revenue should be converted to the present worth.

23.6.9 System Worth

The system worth considering the timing of benefits and costs can be evaluated using the net present value (NPV), benefit–cost (BC) ratio, or internal rate of return (IRR), which are briefly presented here.

23.6.9.1 Net Present Value

The NPV is used to compute the time value of money, that is, the present value of future income or the future value of present income. In an irrigation system income will be generated year after year, and likewise costs will be incurred year after year. Costs like equipment are incurred initially but payment will occur over a certain length of time. Therefore, it is important to compute the present value (PV) of an individual cash flow. The NPV can be computed as

$$\mathrm{NPV} = \frac{R_t}{(1+i)^t}, \tag{23.3}$$

where R_t is the net cash flow (i.e., inflow minus outflow) at time t, and i is the discount rate or interest rate. Equation (23.3) can be used to calculate the PV of each cash flow.

23.6.9.2 Benefit–Cost Ratio

The BC ratio (or BCR) is an important metric in BC analysis. As the name suggests, the BCR is the ratio of benefits of a system to costs, both expressed in monetary terms. All benefits and costs should be expressed in terms of discounted PVs. For all systems, the BCR should be greater than 1, preferably much higher.

23.6.9.3 Internal Rate of Return

The IRR of an investment can be defined as the discount rate at which the NPV of costs (negative cash flows) is equal to the NPV of benefits (positive cash flows). It is the annualized effective compounded rate of return or rate of return that renders the NPV equal to zero. The NPV includes all cash flows – positive as well as negative – and can be computed with the use of eq. (23.4). The higher the IRR value, the more desirable is the system. The NPV can be calculated as

$$\mathrm{NPV} = \sum_{n=0}^{N} \frac{C_n}{(1+r)^n}, \tag{23.4}$$

where N is the total number of periods, n is the period, r is the IRR, and C_n is the cash flow in the nth period. The period is usually in years but can be another period like months.

23.7 SYSTEM COSTS

Irrigation costs are both fixed and variable, which are outlined below.

23.7.1 Fixed Costs

Fixed costs usually include investment or initial costs, which may be the costs of: land purchase, water supply, evacuation and rehabilitation of people, planning and design, storage reservoir, distribution system, hydraulic structures, head regulator, canals/open channels, command area development and surveying, land development operations, on-farm conveyance and control, surface/subsurface drainage system development, road construction, water application equipment, electric power system (connection, metering, and recording), pump house, wells (construction, pumps, electric motors, pumping plant and accessories), and automation equipment. Annual fixed costs include the interest on the total initial investment.

23.7.2 Variable Costs

Variable costs are recurring annual costs, including the costs of: operating manpower (salaries, social benefits, housing insurance, medical treatment, travel, and transportation), maintenance of machinery, maintenance of structures, maintenance of water distribution networks, electricity, fuel, repair, lubricants, painting, maintenance of drainage systems (desilting, weed control, and repair), and layout of fields for surface irrigation (renewal of borders, ridges, and field channels). The annual interest cost is calculated as

$$\text{Annual interest rate} = \frac{(\text{value of installation} - \text{salvage value}) \times \text{interest rate}}{2}. \quad (23.5)$$

23.7.3 Depreciation

Depreciation is the rate at which the value of an asset, such as equipment or structure, reduces owing to its physical use, wear and tear, or obsolescence. It is calculated as

$$\text{Annual depreciation} = \frac{\text{Original cost} - \text{Salvage value}}{\text{Useful life in years}}. \quad (23.6)$$

It can be used to allocate a certain portion of the original cost of a fixed asset to the accounting period of each. In this manner, the value of the asset can be written off during its life span.

23.7.4 Computation of Variable Costs

The cost of power is one of the most important variable costs if pumping systems are employed. Power includes both electricity expressed in kilowatt-hours per hour and fuel measured in liters per hour. The energy consumption of an electric motor can be calculated as

$$\text{Energy consumption} = \frac{\text{Brake horsepower}}{\text{Motor efficiency}} \times 0.746, \quad (23.7)$$

where energy consumption is in kW, brake horsepower is in hp, and the value 0.746 represents the conversion factor for converting horsepower to kW. Motor efficiencies usually range from 75% to 90%, but manufacturers supply performance data. More details were introduced in Chapter 10.

The total annual energy consumption can be obtained by the product of the demand of electrical power for hourly operation and the annual hours of operation. Then, the annual cost of power is the annual energy demand multiplied by the cost per unit of electrical energy.

The cost of fuel can be calculated as

$$\text{Fuel cost} = \text{BHP} \times \text{specific fuel consumption} \times \text{cost of fuel per liter}, \quad (23.8)$$

where BHP is the brake horsepower. The consumption of diesel in irrigation pumping is usually 0.2 L/bhp-hour.

The lubrication oil consumption is about 4.5 liters per 1000 bhp-hour. The annual cost of lubricants can be computed from the cost of lubricants per hour of operation.

Repair and maintenance costs can be approximated either from field studies or local experience. Table 23.3 provides guidelines on these costs.

23.7.5 Service Period

If groundwater is the source of irrigation, then wells, pumps, motors, and pipes are used. These items have a limited service life that needs to be specified, often in hours of operation. Then, their service life in years can be calculated by dividing the service life by the average annual hours of operation.

Example 23.1: Compute the discount rate for the first year, the second year, third year, fourth year, and fifth year, if the opportunity cost of the capital is 8% per year.

Solution: The discount factor is given as

$$\text{Discount factor} = \frac{1}{(1+i)^n},$$

where n is the number of years. The discount rate for the first year is: $\frac{1}{(1+0.08)^1} = 0.926$.

For the second year: $\frac{1}{(1+0.08)^2} = 0.857$.

For the third year: $\frac{1}{(1+0.08)^3} = 0.794$.

For the fourth year: $\frac{1}{(1+0.08)^4} = 0.735$.

For the fifth year: $\frac{1}{(1+0.08)^5} = 0.681$.

Table 23.3 *Guidelines for estimation of service life and annual maintenance and repair costs as a percentage for irrigation system components*

Systems and components	Expected economic life in years	Annual maintenance as percentage of initial investment
Sprinkler systems	10–15	2–6
Hand-move	10–15	2
Side or wheel roll	15+	2
End tow	15+	3
Side move	10+	4
Stationary gun type	15+	2
Center pivot – standard	15+	5
Linear move	15+	6
Cable tow	10+	6
Hose pull	15+	6
Traveling gun type	10+	6
Fixed or solid set		
Permanent	20+	1
Portable	15+	2
Sprinkler gear-driven, impact and spray heads	5–10	6
Valves	10–25	3
Microsystems*	1–20	2–10
Drip	5–10	3
Spray	5–10	3
Bubbler	15+	2
Semi-rigid, buried	10–20	2
Semi-rigid, surface	10	2
Flexible, thin wall, buried	10	2
Flexible, thin wall, surface	1–5	10
Emitters & heads	5–10	6
Filters, injectors, valves	10+	7
Surface and subsurface systems	15	5
Related components		
Pipelines		
buried thermoplastic	25+	1
buried steel	25	1
surface aluminum	20+	2
surface thermoplastic	5+	4
buried nonreinforced concrete	25+	1
buried galvanized steel	25+	1
buried corrugated metal	25+	1
gated pipe, rigid surface	10+	2
surge valves	10+	6
Pumps		
pump only	15+	3
electric motors	10+	3
internal combustion engine	10+	6
Wells	25+	1
Linings		
nonreinforced concrete	15+	5
flexible membrane	10+	5
reinforced concrete	20+	1
Land grading, leveling	Indefinite with adequate maintenance	
Reservoirs	Indefinite with adequate maintenance of structures, watershed	

* Indicates microsystems with no disturbance from tillage and harvest equipment.
After USDA-NRCS, 1997.

Example 23.2: A farmer wants to create a rainy-day fund that he can use in case of emergency. He deposits $5000 each year which will earn an interest of 5%. How much money will there be in the rainy-day fund after 15 years, assuming that the farmer has not withdrawn any money?

Solution: The annuity is $5000. The number of annuity payments is 15. The interest rate is 5% = 0.05. According to the equation of the sum of geometric series,

$$\text{Rainy-day fund} = \$5000 \times \left[\frac{(1+0.05)^{15} - 1}{0.05}\right] = \$107{,}893.$$

Example 23.3: What is the present value of a 30-year annuity when discounted at 3%? The annual payment to be made is $2000.

Solution: The annual payment is $2000. The number of payments is 30, and the discount rate is 3%. According to the equation of the sum of geometric series, the present worth of the annuity can be computed as

$$\text{Present worth} = \frac{\$2000 \times \left[(1+0.03)^{30} - 1\right]}{0.03 \times (1+0.03)^{30}} = \$39{,}201.$$

Example 23.4: A sum of $50,000 is deposited in a bank that gives an interest rate of 2% per year. What is the worth of this amount at the end of the 15-year period in the future?

Solution:

$$\text{Future worth} = \text{Present amount} \times (1+0.02)^{15} = \$50{,}000 \times (1+0.02)^{15} = \$67{,}293.$$

QUESTIONS

Q.23.1 The opportunity cost of the capital is 9% per year. Compute the discount rate for the first year, the second year, third year, fourth year, and fifth year.

Q.23.2 A farmer wants to create a reserve fund by depositing $10,000 each year which will earn interest of 4%. How much money will there be in the fund after 10 years, assuming the farmer has not withdrawn any money?

Q.23.3 A 10-year annual payment of $8000 is made. Calculate the present value when discounted at 4%?

Q.23.4 Calculate the future value at the end of a 20-year period if a sum of $20,000 is deposited in a bank that gives an interest rate of 2.5% per year.

REFERENCES

Benenson, P., Greene, B., Kahn, E., et al. (1977). Effects of drought on California electricity supply and demand. Energy Analysis Program, Energy & Environment Division, Lawrence Berkeley Laboratory.

Chiaramonti, D., Grimm, H., Bassam, N. E., and Cendagorta, M. (2000). Energy crops and bioenergy for rescuing deserting coastal area by desalination: feasibility study. *Bioresource Technology*, 72: 131–146.

Elliott, J., Deryng, D., Müller, C., et al. (2014). Constraints and potentials of future irrigation water availability on agricultural production under climate change. *Proceedings of the National Academy of Sciences*, 111(9): 3239–3244.

FAO (Food and Agriculture Organization) (2013). *The State of the World's Land, and Water Resources for Food and Agriculture: Managing Systems at Risk.* London: Routledge.

Griffiths-Sattenspiel B., and Wilson, W. (2009). *The Carbon Foot Print of Water.* Boulder, CO: River Network.

Hodges, A. W., Lynne, G. D., and Rahmani, M. F. (1994). *Adoption of Energy and Water-Conserving Irrigation Technologies in Florida.* Gainesville, FL: University of Florida.

Lal, R. (2004). Carbon emission from farm operations. *Environment International*, 30: 981–990.

Lotze-Campen, H. (2011). Climate change, population growth, and crop production: an overview. In *Crop Adaptation to Climate Change*, edited by S. S. Yadav, R. J. Reden, J. L. Hatfield, et al. Chichester: Wiley, pp. 1–11.

Martin, D. L, Dorn, T. W., Melvin, S. R, Corr, A. J., and Kranz, W. L. (2011). Evaluating energy use for pumping irrigation water. In: *23rd Annual Central Plains Irrigation Conference*, pp. 104–116.

Moreno, M. A., Ortega, J. F., Córcoles, J. I., and Tarjuelo, J. M. (2010). Energy analysis of irrigation delivery systems: monitoring and evaluation of proposed measures for improving energy efficiency. *Irrigation Science*, 28: 445–460.

Muñoz, I., Mila-i-Canals, L., and Fernandez-Alba, A. R. (2010). Life cycle assessment of water supply plans in Mediterranean Spain: the Ebro River transfer versus the AGUA Programme. *Journal of Industrial Ecology*, 14: 902–918.

Naylor, R. L. (1996). Energy and resource constraints on intensive agricultural production. *Annual Review of Energy and the Environment*, 21: 99–123.

Pimentel, D. (1992). Energy inputs in production agriculture. In *Energy in Farm Production*, edited by R. C. Fluck. Amsterdam: Elsevier.

Plappally, A. K., and Lienhard, V. (2012). Energy requirements for water production, treatment, end use, reclamation, and disposal. *Renewable and Sustainable Energy Reviews*, 16(7): 4818–4848.

Richards, A., and Smith, P. (2003). *How Much Does It Cost to Pump?* Orange, NSW: AGFACT NSW Agriculture.

Stout, B. A. (1990). *Handbook of Energy for World Agriculture*. New York: Elsevier.

USDA-NRCS (1997). Irrigation water management. In *National Engineering Handbook*. Washington, DC: USDA.

24 Land Leveling

Notation

a	slope of the best-fit line through the average x-direction elevation
A_1	area of cut (or fill) C_1 (m^2, ft^2)
A_2	area of cut (or fill) C_2 (m^2, ft^2)
A_{ij}	area represented by the stake (m^2, ft^2)
A_s	area represented by the standard grid dimensions (m^2, ft^2)
b	slope of the best-fit line in the y-direction
c	elevation of origin ($x = 0, y = 0$)
e_i	average elevation of the ith row (m, ft)
e_{ij}	elevation of the (i, j) coordinate obtained from observations (m, ft)
e_j	average elevation of the jth cross-slope row (m, ft)
E, e	elevation (m, ft)
H	difference between C_1 and C_2 (m, ft)
H_c	sum of cuts on four corners of a square grid (m, ft)
H_f	sum of fills on four corners of a square grid (m, ft)
(i, j)	a grid point or stake (the ith advance-slope row and jth cross-slope stake row)
L	grid spacing (m, ft)
N	number of stakes in the row-slope direction
S_x	slope of the plane in the x-direction (% or in feet per station)
S_y	slope of plane in the y-direction (%, or in feet per station)
V	volume of cut (or fill) between areas (m^3, ft^3)
V_c	volume of cut (m^3, ft^3, yd^3)
V_f	volume of fill (m^3, ft^3, yd^3)
w_{ij}	weighting factor for the grid point
x	distance of the centroid from the origin in the x-direction (m, ft)
x, y	longitudinal and transverse coordinates
x_i	distance from the origin to the ith stake row position (m, ft)
x_j	distance from the origin to the jth stake row position (m, ft)
y	distance of the centroid from the origin in the y-direction (m, ft)

24.1 INTRODUCTION

Land surface characteristics, especially slope and highs and lows or irregularities, substantially affect the efficiency of the irrigation system. Indeed, the type of irrigation system to be employed is determined by the land surface itself. Ideally the land surface should be such that the irrigation water moves as uniformly as possible, but the natural landscape is not always so. Therefore, the natural landscape or topography is altered, entailing the movement of earth from one place (high) to another (low). Land leveling or smoothing is one of the most important surface irrigation management and design practices. The objective of this chapter is to discuss the methodology for altering the landscape and various aspects thereof.

24.2 PRELIMINARY CONSIDERATIONS

"Land leveling" is a general term, encompassing grading or leveling that may involve major earth movement and land planning, smoothing, or floating that involves smoothing of small irregularities and roughness (USDA-SCS, 1970). However, it is different from land planning, land smoothing, or land floating, which eliminate minor irregularities using special equipment and do not alter the general topography of the land surface. Rough leveling entails removing knolls, mounds, or ridges, and filling pockets or swales in a field not intended for a planned grade. In this case, no stakes are used and visual observation is employed to achieve the desired field surface. However, rough leveling is not suitable for surface irrigation. An excellent discussion of land leveling is given by the Soil Conservation Service (USDA-SCS, 1970), and this chapter draws from this discussion.

The movement of earth is expensive, involving skilled labor, equipment, and energy. Hence, the topographical modification should be as small as possible without significantly impacting the irrigation efficiency and hence crop yield. It should be noted that the topographic slope and the method of irrigation should be compatible. For example, for border irrigation, the transverse slope should be zero and the slope in the direction of flow is about 0.05–0.1%. Furrow irrigation is amenable to steep slopes of 0.5–3% and cross-slopes of 0.5–1.5%. Further, leveling should be done keeping crop and cropping pattern in mind. Large-scale leveling may be justified for high-value crops, while low-level leveling may be suitable for low-value crops. It is important that returns from increased crop yield are greater than the expenses incurred in leveling.

Prior to leveling, soil profile conditions and the maximum cut without impacting agricultural productivity must be determined. The depth of excavation is strictly determined by the type of soil. For example, shallow soils limit this depth and further limit it on undulating topography or steep slopes.

Leveling makes irrigation more efficient and allows the field to be irrigated more easily. These days, sophisticated equipment is available that allows leveling to be done more precisely. There are other aspects of leveling that border on leveling being negative. Earth movement involves the removal of fertile topsoil from one area and deposition on other areas. Unless caution is exercised, removal of topsoil may result in the reduction of crop yield. It is therefore important that cut areas are enriched with soil amendments and fertilizers. Repeated movement of leveling equipment compacts or pulverizes the soil, reducing the penetration of water and consequently crop yield. Climate should factor in any leveling scheme. If rainfall storms are of short duration and high intensity, then the field should have limited slope or else erosion will occur. Likewise, if the annual rainfall is high, any leveling that might impede drainage should be avoided. Leveling alters infiltration and roughness characteristics.

Since the method of irrigation is determined by the crops to be grown, these crops indicate the amount of leveling needed. For example, vegetable crops may require high leveling, whereas a hay crop may require low leveling. On the same field, several methods of irrigation, depending on the crop and season, may be used and each method has its own leveling requirements. Therefore, leveling should be done for the method having the strictest requirement, such as level border or flooding for cross-slope restriction. It is also important to take into account what the farmer wants, considering future conditions – that is, what the farmer wants today may be different from what they may want next year.

For agricultural production, a plane surface on a nearly level grade is most desirable. Table 24.1 provides a classification of surface relief with regard to its influence on surface irrigation. Class C is normally ranked as the lowest satisfactory for conservation irrigation, and sprinkler or trickle irrigation may the best. Class A_1 may be the only one usable in some areas, whereas in shallow soils and steep slopes Class E may be the best option.

There are four aspects of land leveling that need to be considered in surface irrigation: (1) field design, (2) field preparation, (3) land leveling, and (4) field shape changes. In the context of land leveling, the purpose of field design is to determine the grade of the field surface that best suits local conditions and farming enterprises. This entails measuring the existing condition and computing the new condition.

24.3 SMALL-SCALE LEVELING

As the name suggests, small-scale leveling involves removing soil from high spots and depositing it on low spots. The removal of soil may be no more than a couple of inches. In developing countries, where animals are still used for farming, it may be done with equipment pulled by animal power. Over a period of time the fields are smoothened, permitting relatively uniform irrigation. Localized ponding guides the farmer to where the filling should be done. Trial and error is often used, even though laser-guided systems are available these days. This type of leveling is relatively inexpensive.

24.4 LARGE-SCALE LEVELING

Large-scale leveling is normally done mechanically on large farms that are commonplace in the developed world, especially in Australia, Canada, and the United States. It involves substantial earth movement and is quite expensive. It is done with a laser-guided system and is computerized.

24.5 PREPARATORY STEPS

Since land leveling is an expensive practice, it is important to undertake certain steps before leveling is done. To initiate a land-leveling program, Hart (1975) outlined five general steps:

1. Vegetative material and large debris that hinder equipment operation should be removed. The soil should be conditioned, depending on the equipment to be used. For moving the soil from cut areas to fill areas it is preferable not to have a loose soil surface. For smooth operation, loose soil is, however, better.
2. The farm should be divided into areas with the same slope and soil characteristics. Leveling should be done on a field-by-field basis. There should be a topographic survey of the area to establish horizontal and vertical control on the ground. The topography should be determined either by staking the field in a uniform grid or by a laser-controlled system.
3. The desired field topography should be determined by computing longitudinal and lateral slopes.
4. Cut and fill volumes should be determined and grid stakes should be marked with cut and fill depths – that is, determining the excavation and embankment.
5. The leveling operation is done and the finished grade is checked in order to ensure that the leveling objectives have been achieved. The irrigation system must be planned in order to determine the location of field boundaries, irrigation water supply, and field roads.

24.6 LAND SURVEYING AND MAPPING

Surveying is done to map the land topography. A topographic survey can be undertaken using transit survey, field cross-section, plane table, or grid system. For a grid system, first a

Table 24.1 *Relief classes for surface irrigated land*

Class	Irrigation slope[a]	Cross-slope	Possible irrigation water efficiencies	Irrigation operation labor requirement	Method limitations	Leveling requirement
A_1	Uniform but not more than 0.05%	None	High	Very low	None	None
A_2	Uniform				Length of level borders is restricted	
B_1		Uniform but not more than 0.3%			Length of level borders restricted. Border widths are restricted	Leveling desirable to increase length or width of level borders
B_2		Variable but not more than 0.3%		Low		Leveling desirable
B_3		Either uniform or variable and more than 0.3% but not more than 0.5%		Moderately low	Border widths are very restricted. Level borders not permissible. Shallow furrows not permissible on coarse- or very coarse-textured soils. Corrugations must have downslope of at least four times cross-slope	
C_1	Fairly uniform (when slopes are over 0.5%, convex slopes have maximum grades not over twice minimum; concave slopes have maximum grade not over 1.5 times minimum. Undulation slopes not permissible)	Uniform or variable but not more than 0.3%	Good		Level borders not permissible. Border widths are restricted	Leveling desirable to reduce labor requirement and improve irrigation efficiencies
C_2		Uniform or variable but not more than 0.5%				
C_3	Either uniform or fairly uniform as defined above	Either uniform or variable and more than 0.5%		Moderately high	Applicable only for contour ditches or to cross-slope or contour furrow irrigation within special limitations of furrow depth and soil texture	
D_1	Variable but without level reaches or reverse grade	Variable but not more than 0.3%	Poor	Moderate	Border widths are very restricted. Level borders or corrugations not permissible. Shallow borders not permissible on coarse- or very coarse-textured soils	Leveling required for conservation irrigation
D_2		Variable but not more than 0.5%		High		
E		Variable and more than 0.5%	Very poor	Very high	Applicable only for contour ditches or furrows within special limitations of furrow depth and soil erosion	

[a] Maximum and minimum downslope grades are limited by (1) requirements for drainage, (2) protection from erosion by storm runoff, and (3) the criteria for irrigation.
After USDA-SCS, 1970.

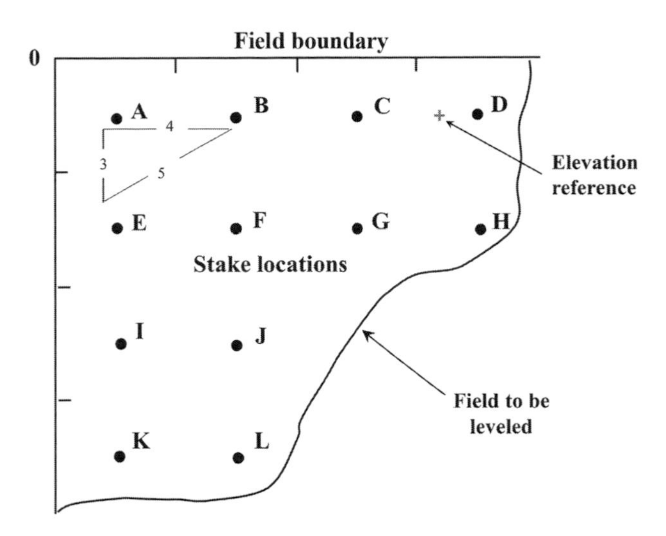

Figure 24.1 Land surveying and mapping.

uniform grid is set up on the field. One corner of the field is selected as a reference point and a stake is located one half-spacing from the adjacent boundary, as shown by stake A in Figure 24.1. Then, a row of stakes is set up along a line parallel to the boundary. This is done by going to the opposite boundary and placing a stake one-half of the grid spacing from the edge. Then, the first row of stakes (say, A, B, C, and D) is put in place by using the level for alignment.

Next, at a 90° angle with respect to the first row, stakes are located in the same way as the first row (say, A, E, I, and K). In a similar manner, other rows are set up with stakes, and elevations of grid points are determined.

In order to determine elevations of grid points, it is a good idea to set up a benchmark near the field and reference the readings as elevations (Figure 24.1).

We level or smooth land so it will irrigate more uniformly and efficiently. An existing field will have a new topography at the end of leveling based on cost and type of surface irrigation system. The cost depends on how much soil has to be moved to new places on the field. As a rule of thumb, we should move as little soil as possible. This means that we want to find the best plane through the existing elevations.

24.7 SELECTION OF FIELD SLOPES

Fundamental to leveling is the selection of field slopes, which depends on a number of factors discussed earlier, such as soil characteristics, land topography, crops, farming practices, irrigation method, climate, cost, labor, and skill. A formal method, called the "plane method" is presented here. This method entails fitting a two-dimensional plane to field elevations obtained by surveying.

Let X and Y be the longitudinal and transverse coordinates and E be elevation. Let x be a particular value of X and y a particular value of Y, and e a particular value of E. Considering a two-dimensional plane, elevation e of any position denoted by (x, y) on the plane can be expressed in terms of x and y as

$$e(x,y) = ax + by + c, \tag{24.1}$$

where a and b are regression coefficients and c is the elevation of origin $(x = 0, y = 0)$. The values of a, b, and c can be determined by the least-squares method.

The areas represented by different stakes or grids are not the same. For example, the boundary grid points usually represent smaller or larger areas due to irregular field shapes and sizes. Therefore, weighted-average elevations of each stake row in both directions need to be determined using weighting factors. Let a grid point or stake be defined by (i, j) (the ith advance-slope row and jth cross-slope stake row) and the area represented by the stake by A_{ij}. Let the area represented by the standard grid dimensions be denoted by A_s. Then, the weighting factor for the grid point can be defined as w_{ij}:

$$w_{ij} = \frac{A_{ij}}{A_s}. \tag{24.2}$$

The average elevation of the ith row, e_i, can be expressed using eq. (24.2) as

$$e_i = \frac{\sum_{j=1}^{M} w_{ij} e_{ij}}{\sum_{j=1}^{M} w_{ij}}, \tag{24.3}$$

where M denotes the number of cross-slope rows and e_{ij} is the elevation of the (i, j) coordinate obtained from observations.

Likewise, the average elevation of the jth cross-slope row, e_j, can be expressed as

$$e_j = \frac{\sum_{i=1}^{N} w_{ij} e_{ij}}{\sum_{i=1}^{N} w_{ij}}, \tag{24.4}$$

where N is the number of stakes in the row-slope direction.

Next, the centroid of the field with the grid system needs to be determined. To that end, an origin is to be established one grid spacing in each direction from the first stake. The distance x of the centroid from the origin in the x-direction can be determined as

$$x = \frac{\sum_{j=1}^{M} w_j x_j}{\sum_{j=1}^{M} w_j}, \tag{24.5}$$

where x_j is the distance from the origin to the jth stake row position, and

$$w_j = \sum_{i=1}^{N} w_{ij}. \tag{24.6}$$

Likewise, the distance y of the centroid from the origin in the y-direction can be determined as

$$y = \frac{\sum_{i=1}^{N} w_i x_i}{\sum_{i=1}^{N} w_i}, \tag{24.7}$$

where x_i is the distance from the origin to the ith stake row position, and

$$w_i = \sum_{j=1}^{M} w_{ij}. \tag{24.8}$$

Now the least-square fit line through the average row elevation in both directions is determined. The slope of the best-fit line through the average x-direction elevation (e_j) is obtained as

$$a = \frac{\sum_{j=1}^{M} x_j e_j - \frac{\left(\sum_{j=1}^{M} x_j\right)\left(\sum_{j=1}^{M} e_j\right)}{M}}{\sum_{j=1}^{M} x_j^2 - \frac{\left(\sum_{j=1}^{M} x_j\right)^2}{M}}. \tag{24.9}$$

The slope of the best-fit line in the y-direction can be written as

$$b = \frac{\sum_{i=1}^{N} y_i e_i - \frac{\left(\sum_{i=1}^{N} y_i\right)\left(\sum_{i=1}^{N} e_i\right)}{N}}{\sum_{i=1}^{N} y_i^2 - \frac{\left(\sum_{i=1}^{N} y_i\right)^2}{N}}. \tag{24.10}$$

Now the parameter needs to be determined as

$$c = e_f - ax - by, \tag{24.11}$$

where e_f is the average field (centroid) elevation. This elevation can be determined by summing the values of e_i or e_j and dividing by the corresponding number of grid rows.

Elevation at each grid point can now be computed and compared with the observed value. The differences between observed and computed elevations point to the cut or fill areas. The computations involved, as above, are handled by various software available these days. For example, the Utah State University's LandLeveler is a software package written to perform a number of land-leveling computations. Safe limits of longitudinal land slope for efficient irrigation for different types of soils are: 0.05–0.20% for heavy (clay) soils, 0.2–0.4% for medium (loamy) soils, and 0.25–0.65% for light (sandy) soils (Michael, 2010). Likewise recommended cross-slope limits for different furrow grades are: 0.3% for 0.1–0.3% furrow grade, 0.4% for grade of 0.4%, and 0.5% for furrow grade of 0.5% (University of Missouri, n.d.).

Example 24.1: A typical grid of surveying stakes on a rectangular field of size 40 × 40 m is shown in Figure 24.2 (elevation in meters). The first stake is located at the upper left corner and was placed one half-spacing from both sides of the field to start the staking. Furrow irrigation and the least disturbing cut–fill plane are to be selected. It is irrigated from bottom to top. Compute the average field elevation (e_f).

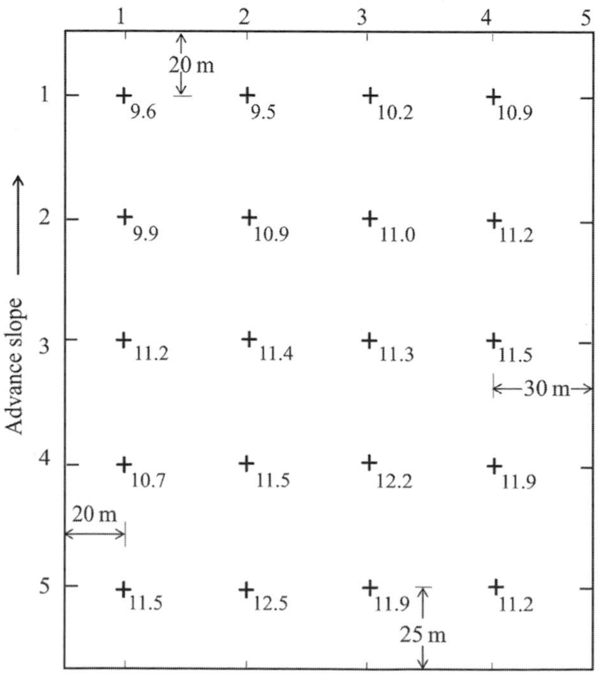

Figure 24.2 Elevation grid of a field.

Solution: Calculate the average row (i advance-slope and j cross-slope) elevations using eqs (24.3) and (24.4). The weighting factors are given as follows:

Weighting factor	$j = 1$	$j = 2$	$j = 3$	$j = 4$
$i = 1$	1.0	1.0	1.0	1.25
$i = 2$	1.0	1.0	1.0	1.25
$i = 3$	1.0	1.0	1.0	1.25
$i = 4$	1.0	1.0	1.0	1.25
$i = 5$	1.125	1.125	1.125	1.406

Take $e_i(i = 1$, advance slope) and $e_j(j = 1$, cross slope) as examples

$e_i(i = 1$, advance slope)

$$= \frac{(9.6 \times 1.0) + (9.5 \times 1.0) + (10.2 \times 1.0) + (10.9 \times 1.25)}{(1.0 + 1.0 + 1.0 + 1.125)}$$

$= 10.10\,\text{m}$

$e_j(j = 1$, cross slope)

$$= \frac{(9.6 \times 1.0) + (9.9 \times 1.0) + (11.2 \times 1.0) + (10.7 \times 1.0) + (11.5 \times 1.125)}{(1.0 + 1.0 + 1.0 + 1.0 + 1.125)}$$

$= 10.60\,\text{m}$

Likewise,

	$j = 1$	$j = 2$	$j = 3$	$j = 4$	$e_i \downarrow$
$i = 1$	9.6	9.5	10.2	10.9	10.10
$i = 2$	9.9	10.9	11	11.2	10.78
$i = 3$	11.2	11.4	11.3	11.5	11.36
$i = 4$	10.7	11.5	12.2	11.9	11.59
$i = 5$	11.5	12.5	11.9	11.2	11.74
$e_j \rightarrow$	10.60	11.19	11.33	11.34	**11.12**

The average field elevation is 11.12 m.

Example 24.2: From the data and results from Example 24.1, compute the elevation of the origin (c) when the distance from the origin to the centroid in the x-coordinate is 75.4 m, and y-coordinate is 92.7 m. The best-fit slope in the x-direction is 0.0057 and y-direction is 0.022.

Solution: Using eq. (24.11),

$$c = e_f - ax - by$$

$$c = 11.12\ \text{m} - (75.4\ \text{m} \times 0.0057) - (92.7\ \text{m} \times 0.022)$$
$$= 8.65\ \text{m}.$$

The elevation of the origin is 8.65 m.

24.8 METHODS FOR LAND-LEVELING DESIGN

Commonly used methods for designing land leveling can be classified into the plane method, profile method, plan-interaction method, and contour-adjustment method. Each method has advantages and disadvantages.

24.8.1 Plane Method

The plane method is used for securing class A_1, A_2, and B_1 surface relief and leads to the land surface with uniform down-field slope and uniform cross-slope. Land leveling by the plane method involves the following steps:

1. Divide the field to be leveled into subareas, as shown in Figure 24.3, because some fields cannot be leveled into single planes. When doing the subdivision, it is important to keep in mind the location of the water supply or proposed ditch.
2. Level each subarea to a plane surface. Subdivision boundaries can be located based on the topographic map.
3. Consider each subarea as a separate plane. When doing leveling, common boundaries of subareas are considered in the leveling of adjacent areas. Compute the area of each subarea.
4. Compute the centroid of the field.
5. If the field is rectangular then the centroid is located at the intersection of its diagonals.
6. If the field is triangular then its centroid is located at the intersection of lines drawn from the corners to the mid-points of the opposite sides.
7. If the field is irregular then it should be divided into rectangles or triangles alone, and lines of reference should be established. Compute the centroid of each subarea and its distance from the line of reference. The centroid of the field from the reference line is computed as the ratio of the sum of products of subareas and distances of their centroids from the reference line to the area of the entire field. The exact point of the centroid can be computed by computing the distance to the centroid from the lines of reference perpendicular to each other (see Figure 24.4).
8. Compute the average elevation of the field. This can be done by summing the elevations of all grid points divided by the total number of points.
9. Compute the volume of cut and fill by passing a plane at the average elevation of the field through the centroid. The volume of cut will equal the volume of fill. In the field it is usually necessary to have more cut than fill to obtain a balance due to the shrinkage of soil, which is detailed in Section 24.10. To attain this, the whole field can be lowered a few hundredths of a foot (e.g., 0.04 ft). As a result, the amount of excavation is increased, and the fill required is reduced.
10. Compute the slope of the best-fit plane. If the grid points are located at 100-ft centers then the slope of the best-fit plane on a rectangular field can be expressed as

$$S_x = \frac{\Sigma(d_x H_y) - C_a H}{C_b}, \tag{24.12}$$

$$S_y = \frac{\Sigma(d_y H_x) - C_a H}{C_b}, \tag{24.13}$$

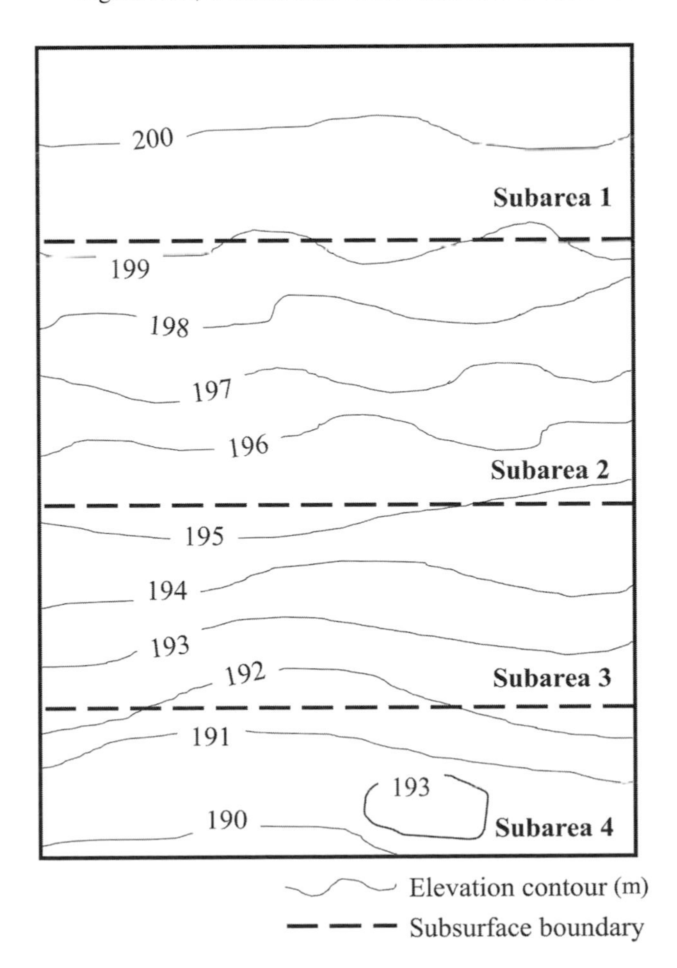

Figure 24.3 Field subdivision for land leveling.

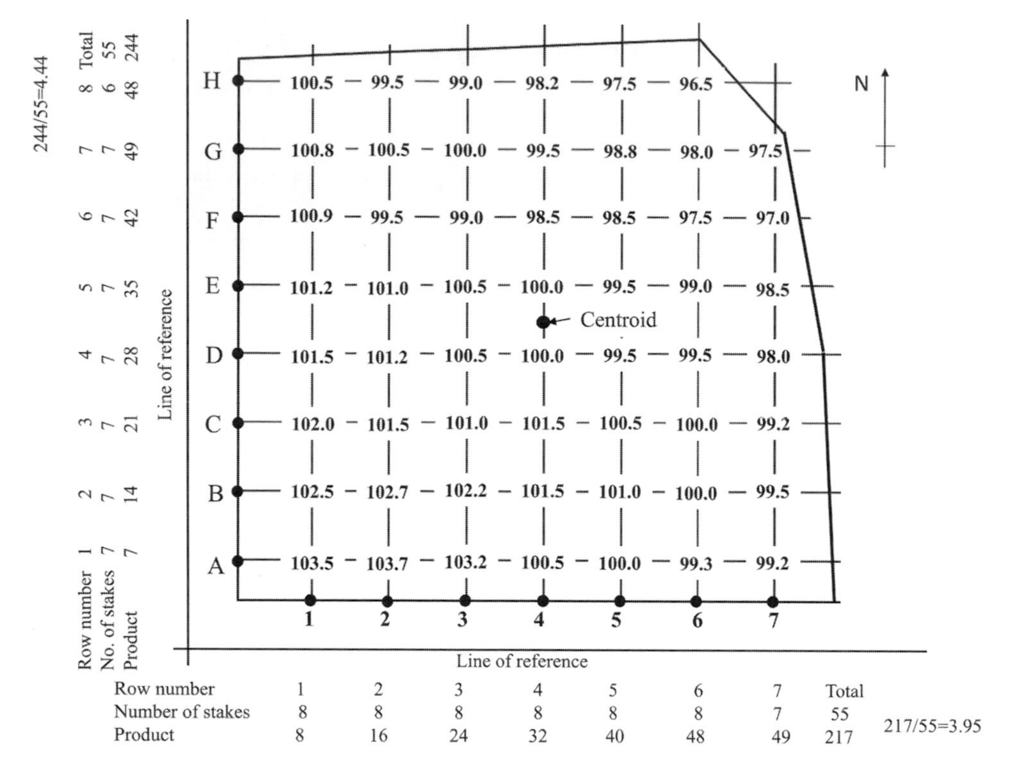

Figure 24.4 Division in irregular fields to find the location of the centroid of a field.

where S_x is the slope of the plane in the x-direction in percent or in feet per station (feet/100 feet); S_y is the slope of the plane in the y-direction in percent or in feet per station; d_x is the distance in number of stations from the y-axis; d_y is the distance in the number of stations from the x-axis; H is the sum of elevations of all grid points; H_x is the sum of elevations of grid points along the line in the x-direction; H_y is the sum of elevations of grid points along the line in the y-direction; and C_a and C_b are the constants from Table 24.2. Note that the units of distance d_x and d_y are in the number of stations, with the assumption of a grid size of 100 feet; the results of S_x and S_y are in the unit of feet per station (feet/100 feet) or percent.

If the spacing of grid points is not equal to 100 feet, the slopes S_x and S_y computed from eqs (24.12) and (24.13), respectively, can be corrected as

$$\text{Corrected } S_x = \frac{\text{Spacing of grid points in the } x\text{-direction}}{100} \times S_x, \tag{24.14}$$

$$\text{Corrected } S_y = \frac{\text{Spacing of grid points in the } y\text{-direction}}{100} \times S_y. \tag{24.15}$$

Example 24.3: Compute the centroid of the field shown in Figure 24.4.

Solution: Assuming a line of reference 100 ft south of line A, the number of stakes in each line multiplied by the row number is:

$$(1 \times 8 + 2 \times 8 + 3 \times 8 + 4 \times 8 + 5 \times 8 + 6 \times 8 + 7 \times 7)$$
$$/(8 + 8 + 8 + 8 + 8 + 7) = 217/55 = 3.95 \text{ stations}$$
$$\text{or } 100 \text{ ft} \times 3.95 = 395 \text{ ft}.$$

The centroid is between lines C and D, and 395 ft from the reference line, or 95 ft from line C.

Another line of reference is assumed to be 100 ft west of line 1. The number of stakes in each line multiplied by the row number is:

$$(1 \times 7 + 2 \times 7 + 3 \times 7 + 4 \times 7 + 5 \times 7 + 6 \times 7 + 7 \times 6)/$$
$$(7 + 7 + 7 + 7 + 7 + 6) = 244/55 = 4.44 \text{ stations}$$
$$\text{or } 100 \text{ ft} \times 4.44 = 444 \text{ ft}.$$

The centroid is between lines 4 and 5, and 444 ft from the reference line, or 44 ft from line 4.

Hence, with the two dimensions, the centroid is located at (line C + 95 ft) and (line 4 + 44 ft).

Table 24.2 *Constants for the determination of plane of the best fit*

Number of stations in the direction slope of the best fit is being determined														
		2	3	4	5	6	7	8	9	10	11	12	13	14
	Value of C_a													
		1.5	2.0	2.5	3.0	3.5	4.0	4.5	5.0	5.5	6.0	6.5	7.0	7.5
	Value of C_b													
Number of stations transverse to the direction slope of the best fit	1	0.5	2	5	10	17.5	28	42	60	82.5	110	143	182	227.5
	2	1.0	4	10	20	35	56	84	120	165	220	286	364	455
	3	1.5	6	15	30	52.5	84	126	180	247.5	330	429	546	682.5
	4	2.0	8	20	40	70	112	168	240	330	440	572	728	910
	5	2.5	10	25	50	87.5	140	210	300	412.5	550	715	910	1,137.5
	6	3.0	12	30	60	105	168	252	360	495	660	858	1,092	1,365
	7	3.5	14	35	70	122.5	196	294	420	577.5	770	1,001	1,274	1,592.5
	8	4.0	16	40	80	140	224	336	480	660	880	1,144	1,456	1,820
	9	4.5	18	45	90	157.5	252	378	540	742.5	990	1,287	1,638	2,047.5
	10	5.0	20	50	100	175	280	420	600	825	1,100	1,430	1,820	2,275
	11	5.5	22	55	110	192.5	308	462	660	907.5	1,210	1,573	2,002	2,502.5
	12	6.0	24	60	120	210	336	504	720	990	1,320	1,716	2,184	2,730
	13	6.5	26	65	130	227.5	364	546	780	1,072.5	1,430	1,859	2,366	2,957.5
	14	7.0	28	70	140	245	392	588	840	1,155	1,540	2,002	2,548	3,185
	15	7.5	30	75	150	262.5	420	630	900	1,237.5	1,650	2,145	2,730	3,412.5
	16	8.0	32	80	160	280	448	672	960	1,320	1,760	2,288	2,912	3,640
	17	8.5	34	85	170	297.5	476	714	1,020	1,402.5	1,870	2,431	3,094	3,867.5
	18	9.0	36	90	180	315	504	756	1,080	1,485	1,980	2,574	3,276	4,095
	19	9.5	38	95	190	332.5	532	798	1,140	1,567.5	2,090	2,717	3,458	4,322.5
	20	10.0	40	100	200	350	560	840	1,200	1,650	2,200	2,860	3,640	4,550
	21	10.5	42	105	210	367.5	588	882	1,260	1,732.5	2,310	3,003	3,822	4,777.5
	22	11.0	44	110	220	385	616	924	1,320	1,815	2,420	3,146	4,004	5,005
	23	11.5	46	115	230	402.5	644	966	1,380	1,897.5	2,530	3,289	4,186	5,232.5
	24	12.0	48	120	240	420	672	1,008	1,440	1,980	2,640	3,432	4,368	5,460
	25	12.5	50	125	250	437.5	700	1,050	1,500	2,062.5	2,750	3,575	4,550	5,687.5
	26	13.0	52	130	260	455	728	1,092	1,560	2,145	2,860	3,718	4,732	5,915

Table 24.2 (*cont.*)

	Number of stations in the direction slope of the best fit is being determined											
	15	16	17	18	19	20	21	22	23	24	25	26
Value of C_a												
	8.0	8.5	9.0	9.5	10	10.5	11	11.5	12	12.5	13	13.5
Value of C_b												
Number of stations transverse to the direction slope of the best fit: 1	280	340	408	484.5	570	665	770	885.5	1,012	1,150	1,300	1,462.5
2	560	680	816	969	1,140	1,330	1,540	1,771	2,024	2,300	2,600	2,925
3	740	1,020	1,224	1,453.5	1,710	1,995	2,310	2,656.5	3,036	3,450	3,900	4,387.5
4	920	1,360	1,632	1,938	2,280	2,660	3,080	3,542	4,048	4,600	5,200	5,850
5	1,100	1,700	2,040	2,422.5	2,850	3,325	3,850	4,427.5	5,060	5,750	6,500	7,312.5
6	1,280	2,040	2,448	2,907	3,420	3,990	4,620	5,313	6,072	6,900	7,800	8,775
7	1,460	2,380	2,856	3,391.5	3,990	4,655	5,390	6,198.5	7,084	8,050	9,100	10,237.5
8	1,640	2,720	3,264	3,876	4,560	5,320	6,160	7,084	8,096	9,200	10,400	11,700
9	1,820	3,060	3,672	4,360.5	5,130	5,985	6,930	7,969.5	9,108	10,350	11,700	13,162.5
10	2,000	3,400	4,080	4,845	5,700	6,650	7,700	8,855	10,120	11,500	13,000	14,625
11	2,180	3,740	4,488	5,329.5	6,270	7,315	8,470	9,740.5	11,132	12,650	14,300	16,087.5
12	2,360	4,080	4,896	5,814	6,840	7,980	9,240	10,626	12,144	13,800	15,600	17,550
13	2,540	4,420	5,304	6,298.5	7,410	8,645	10,010	11,511.5	13,156	14,950	16,900	19,012.5
14	2,720	4,760	5,712	6,783	7,980	9,310	10,780	12,397	14,168	16,100	18,200	20,475
15	2,900	5,100	6,120	7,267.5	8,550	9,975	11,550	13,282.5	15,180	17,250	19,500	21,937.5
16	3,080	5,440	6,528	7,752	9,120	10,640	12,320	14,168	16,192	18,400	20,800	23,400
17	3,260	5,780	6,936	8,236.5	9,690	11,305	13,090	15,053.5	17,204	19,550	22,100	24,862.5
18	3,440	6,120	7,344	8,721	10,260	11,970	13,860	15,939	18,216	20,700	23,400	26,325
19	3,620	6,460	7,752	9,205.5	10,830	12,635	14,630	16,824.5	19,228	21,850	24,700	27,787.5
20	3,800	6,800	8,160	9,690	11,400	13,300	15,400	17,710	20,240	23,000	26,000	29,250
21	3,980	7,140	8,568	10,174.5	11,970	13,965	16,170	18,595.5	21,252	24,150	27,300	30,712.5
22	4,160	7,480	8,976	10,659	12,540	14,630	16,940	19,481	22,264	25,300	28,600	32,175
23	4,340	7,820	9,384	11,143.5	13,110	15,295	17,710	20,366.5	23,276	26,450	29,900	33,637.5
24	4,520	8,160	9,792	11,628	13,680	15,960	18,480	21,252	24,288	27,600	31,200	35,100
25	4,700	8,500	10,200	12,112.5	14,250	16,625	19,250	22,137.5	25,300	28,750	32,500	36,562.5
26	4,880	8,840	10,608	12,597	14,820	17,290	20,020	23,023	26,312	29,900	33,800	38,025

After USDA-SCS, 1970.

Example 24.4: The field shown in Figure 24.5 is to be leveled using the plane method. The grid is 100 × 100 ft (30.5 × 30.5 m) and the elevation given at each grid point is in feet. Compute the centroid of the field, the elevation of the centroid, and the average slope in both horizontal and transverse directions. Calculate the sum of cuts and fills in feet and also show in meters.

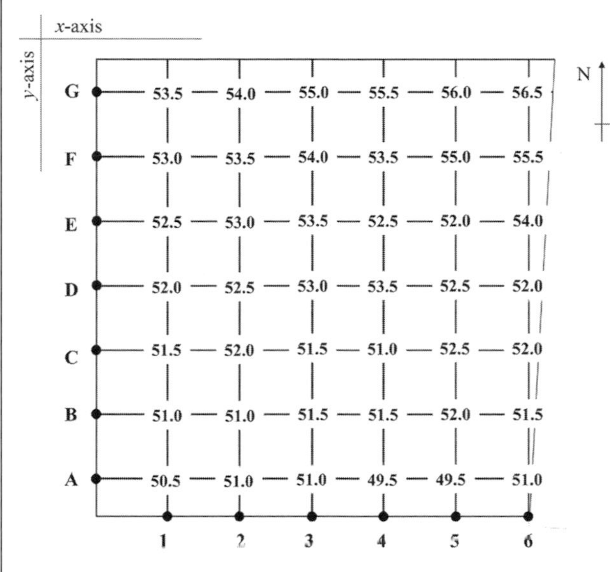

Figure 24.5 A gridded field for the plane method.

Solution:

Step 1: The field is nearly rectangular, and thus the centroid is located at the diagonals or at the coordinate (line 3 + 50 ft, line D + 00 ft), as shown in Figure 24.6.

As shown in Figure 24.5, taking line G and line 1 as examples:

For line G: $H_x = 53.5 + 54.0 + 55.0 + 56.0 + 56.5 = 330.5$ ft.

For line 1: $H_y = 53.5 + 53.0 + 52.5 + 52.0 + 51.5 + 51.0 + 50.5 = 364$ ft.

Step 2: The average elevation of the field is $2210/42 = 52.62$ ft (16.04 m).

Step 3: Given the sum of elevations of all grid points $H = 2210$, for the slope of the plane in the x-direction the number of stations along the x-axis is 6, and this value is found on the horizontal column in Table 24.2; then, $C_a = 3.5$ for 6 stations. The number of stations along the y-axis is 7, then with the same number along the x-axis $C_b = 122.5$; the slope of the plane in the x-direction is

$$S_x = \frac{\Sigma(d_x H_y) - C_a H}{C_b} = \frac{7758 - 3.5 \times 2210}{122.5} = 0.1878\%.$$

Step 4: For the slope of the plane in the y-direction, the number of stations along the y-axis is 7, and this value is found in the horizontal column in the table, then $C_a = 4.0$ for 7 stations. The number of stations along the x-axis is 6,

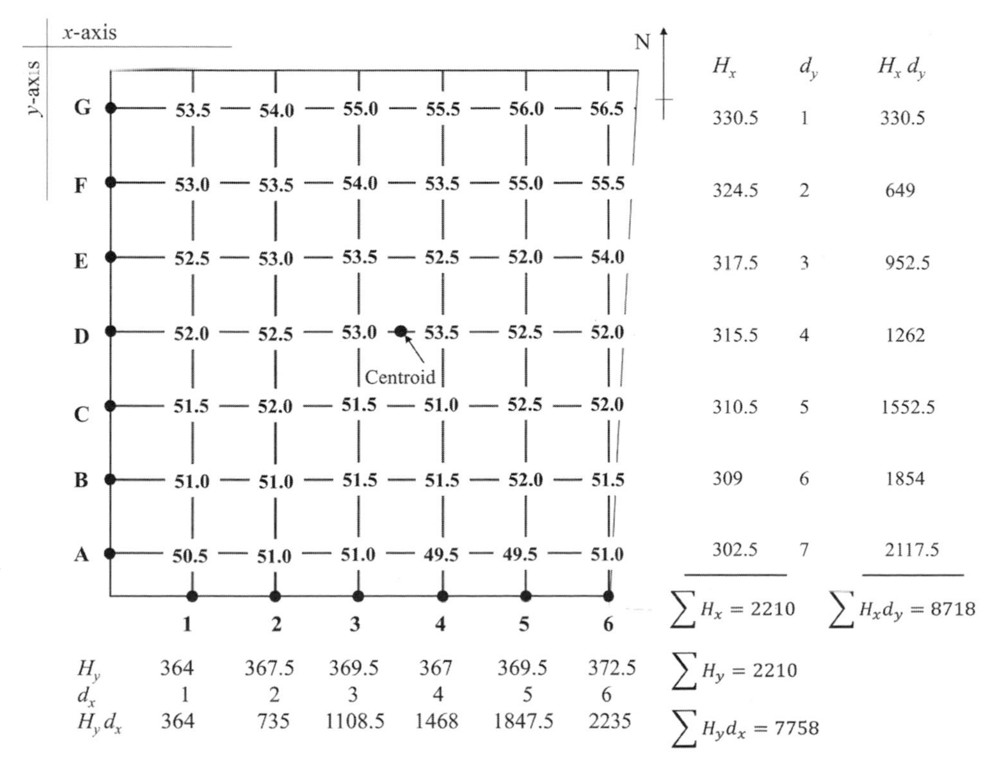

Figure 24.6 Illustration of the determination of the best-fit plane.

then with the same number along the y-axis $C_b = 168$; the slope of the plane in the x-direction is

$$S_y = \frac{\Sigma(d_y H_x) - C_a H}{C_b} = \frac{8781 - 4.0 \times 2210}{168} = -0.7262\%.$$

Note that S_y is negative, which means that the elevation of the plane decreases as the distance from the x-axis increases.

The calculations of $\Sigma(d_x H_y)$ and $\Sigma(d_y H_x)$ are illustrated in Figure 24.6.

Step 5: It is usually necessary to have more cut than fill to obtain a balance due to the shrinkage of soil. Here, we can lower the whole field a few hundredths of a foot (e.g., 0.04 ft). As a result, the amount of excavation is increased and the fill required is reduced. The elevation of the centroid now is $52.62 - 0.04 = 52.58$ ft (16.03 m).

The following is known:
location of the centroid: (line 3 + 50, line D + 00)
elevation of the centroid: 52.58 ft (16.03 m)
the slope of the plane in the x-direction: 0.188%.

The slope of the plane in the y-direction is –0.726%. Thus, the elevation of point (line 4 + 00, line D + 00) = $52.68 + 1/2 \times 0.188 = 52.67$ ft (16.05 m). The elevations of all the other points on line D can be calculated by adding (subtracting) 0.188 to the right (left), respectively. The elevation of line E can be obtained by adding $0.7262\% \times 100 = 0.726$ ft to the proposed elevations on line D. Similarly, all the proposed elevations of each grid are determined and shown as follows:

	Elevation (feet)					
Lines	Line 1	Line 2	Line 3	Line 4	Line 5	Line 6
Line G	54.29	54.48	54.66	54.85	55.04	55.23
Line F	53.56	53.75	53.94	54.13	54.31	54.50
Line E	52.84	53.02	53.21	53.40	53.59	53.78
Line D	52.11	52.30	52.49	52.67	52.86	53.05
Line C	51.38	51.57	51.76	51.95	52.14	52.32
Line B	50.66	50.85	51.03	51.22	51.41	51.60
Line A	49.93	50.12	50.31	50.50	50.68	50.87

The differences between the proposed and original elevations are calculated as follows, in which the positive values indicate fills and negative values indicate cuts.

	Elevation (feet)							
Lines	Line 1	Line 2	Line 3	Line 4	Line 5	Line 6	Sum of cuts	Sum of fills
Line G	+0.79	+0.48	–0.34	–0.65	–0.96	–1.27	–3.22	1.27
Line F	+0.56	+0.25	–0.06	+0.63	–0.69	–1.00	–1.75	1.44
Line E	+0.34	+0.02	–0.29	+0.90	+1.59	–0.22	–0.51	2.85
Line D	+0.11	–0.20	–0.51	–0.83	+0.36	+1.05	–1.54	1.52
Line C	–0.12	–0.43	+0.26	+0.95	–0.36	+0.32	–0.91	1.53
Line B	–0.34	–0.65	–0.47	–0.28	–0.59	+0.10	–2.33	0.10
Line A	–0.57	–0.88	–0.69	+1.00	+1.18	–0.13	–2.27	2.18
Total							–12.53	10.89

To check the calculation, one can calculate the variation per station, which should be less than 0.005 ft since the proposed elevations are rounded to hundredths of a foot. The variation per station is calculated as follows:

Sum of the cuts $= |-12.53| = 12.53$ ft (3.82 m).

Sum of the fills = 10.89 ft (3.32 m).

Difference $= 12.53 - 10.89 = 1.64$ ft (0.50 m).

The number of stations (stakes) $= 6 \times 7 = 42$.

Total shrinkage adjustment $= 42 \times 0.04 = 1.68$ ft (0.51 m).

Total variation $= 1.68 - 1.64 = 0.04$ ft (0.012 m).

Variation per station $= 0.04/42 = 0.001$ ft (0.0003 m).

Since the variation per station is less than 0.001 ft, the calculation is correct.

The calculation of the volume of cuts and fills will be introduced in Section 24.9.

24.8.2 Profile Method

This method is employed to level flat lands or lands with undulating topography for which a surface relief of class B_2, C_1, and C_2 is desired. It entails grid profiles, not elevations. One way to do leveling by this method is to select a profile that leads to a balance between cut and fill with reasonable short-haul distance and meeting irrigation criteria. The method involves the following steps:

1. Divide the field into a rectangular grid, as shown in Figure 24.7. The horizontal grid lines can be regarded as datum lines. The distance between the datum lines D and E is the same as the distance between line 1 and line 2.

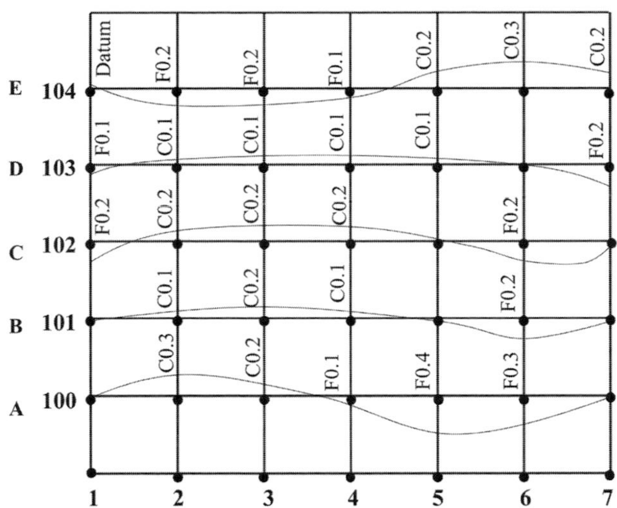

Figure 24.7 The profile method of land leveling.

2. Draw the profiles in one direction, mostly horizontal. Sometimes profiles are also drawn across the slope. Likewise, one can also do a two-way plot.
3. Depending on the irrigation plan and leveling criteria, establish trial gradelines.
4. Place trial gradelines on control lines, say A, E, and C in Figure 24.7.
5. Compute the volume of cut and fill and the ratio between the area of cut and the area of fill. If the volume of cut is approximately equal to the volume of fill and the ratio of their areas is about 1, then the trial gradelines are satisfactory. Otherwise establish another set of trial gradelines.
6. Examine the gradelines between profiles by computing elevations of one relative to the other. Compute the volume of cut and fill and determine if the trial gradelines are satisfactory. Otherwise, move the trial gradelines up or down, noting that the total amount the trial gradelines are moved upward must be the same as the total amount the other lines are moved downward.
7. Compute the balance between the volume of cut and that of fill using the summation method (see Section 24.9). If the balance is not adequate, adjust the profiles until proper balance is achieved.

Example 24.5: Compute the sums of cut and fill for the field shown in Figure 24.7. The grid size is 100 × 100 ft and the unit of elevation is in feet.

Solution: The differences between the proposed and original elevations are shown as follows, in which the positive values indicate fills and negative values indicate cuts.

The sum of the fills = 2.2 ft (0.67 m).

The sum of the cuts = |–2.6| = 2.6 ft (0.79 m).

The ratio of the cuts and fills = 2.6/2.2 = 1.18.

The calculation of the volume of cuts and fills is introduced in Section 24.9.

	Elevation (feet)								
Lines	Line 1	Line 2	Line 3	Line 4	Line 5	Line 6	Line 7	Sum of cuts	Sum of fills
Line E	0	+0.2	+0.2	+0.1	–0.2	–0.3	–0.2	–0.7	+0.5
Line D	+0.1	–0.1	–0.1	–0.1	–0.1	0	+0.2	–0.4	+0.4
Line C	+0.2	–0.2	–0.2	–0.2	0	+0.2	0	–0.6	+0.4
Line B	0	–0.1	–0.2	–0.1	0	+0.2	0	–0.4	+0.2
Line A	0	–0.3	–0.2	+0.1	+0.4	+0.3	0	–0.5	+0.8

24.8.3 Plan-Inspection Method

This is a trial and error method involving considerable judgment, but is similar to the profile method and is employed for leveling lands other than class A or B surface relief. This method involves the following steps:

1. Survey the field and plot the grid.
2. Establish the down-field and cross-slope limitations.
3. Select the elevation that will meet these limitations, considering down-field slope, cross-slope, earthwork balance, and haul distance.
4. Draw the control lines and determine the elevations for intermediate points.
5. Compute the volume of cut and fill and compute the earthwork balance. If the balance is adequate, the leveling is accepted.

Example 24.6: Compute the sums of cut and fill for the field shown in Figure 24.8. The grid size is 100 × 100 ft, and elevations are given in feet.

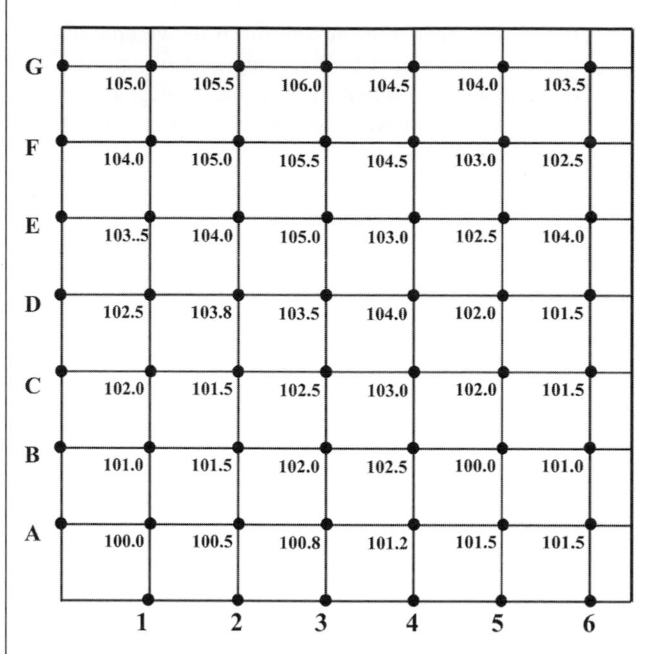

Figure 24.8 Field grid for the plan-inspection method.

Solution:

1. The field is divided as shown in Figure 24.9.
2. Establish the down-field and cross-slope limitations. The maximum cross-slope is 0.005 and could be changed – that is, the elevation difference of each grid along the cross-slope is less than 0.5 ft (0.15 m). Maximum down-field slope is 0.01 and down-field slope can be changed – that is, the elevation difference of each grid along the down-field slope is less than 1 ft (0.30 m), but on convex slopes the maximum grade cannot exceed twice the minimum grade; on concave slopes, maximum grades cannot exceed 1.5 times the minimum grade. The ratio of cuts and fills should be within 1.5 ± 0.05.
3. Elevation is first tentatively selected for the G line so that a ditch can run in both directions. Note that the elevation of the water surface should also be considered to make sure the highest point is not too high. In practice, a minimum head of 0.5 ft (0.15 m) is required for irrigation from ditches. With the G and A lines as the control lines, elevations for other stations are determined considering the design criteria.
4. G and A are the control lines. By trial and error, the intermediate points are shown in Figure 24.9.

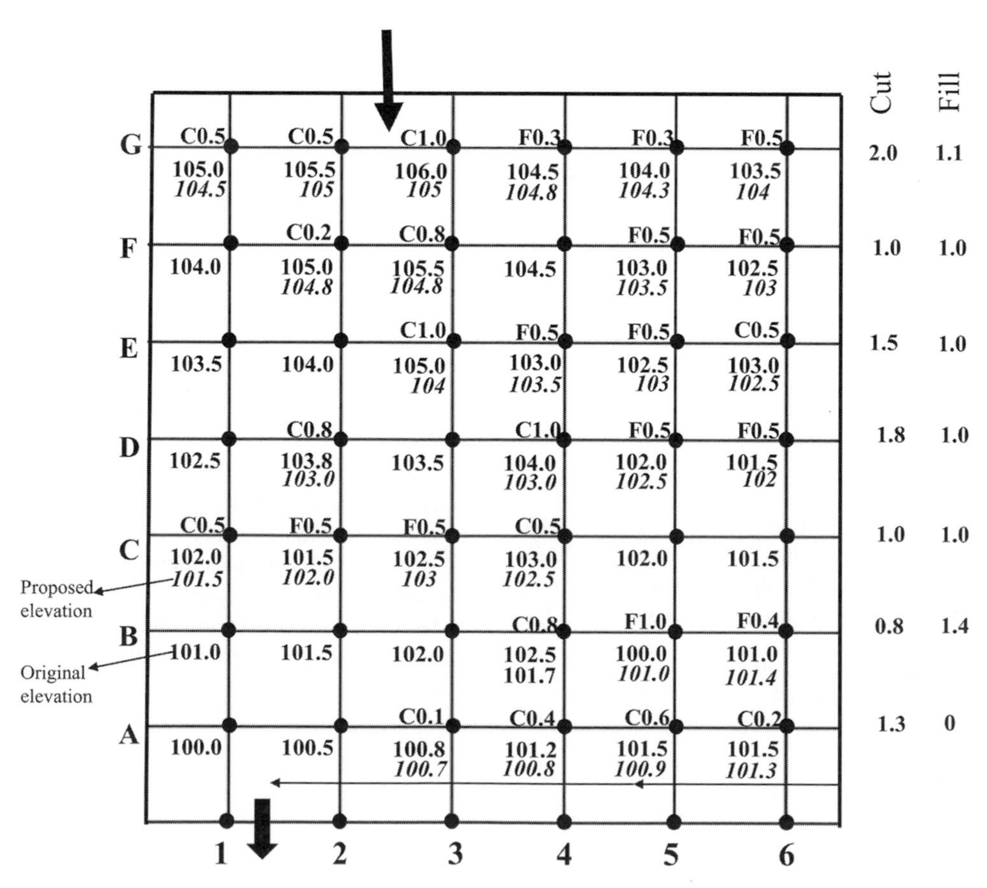

Figure 24.9 The plan-inspection method for land leveling.

5. The total cut = 2.0 + 1.0 + 1.5 + 1.8 + 1.0 + 0.8+ 1.3 = 9.4 ft (2.9 m).

The total fill = 1.0 + 1.0 + 1.0 + 1.0 + 1.0 + 1.4 = 6.5 ft (2.0 m).

The ratio of cuts and fills = 9.4/6.5 = 1.45, which is within 1.5 ± 0.05.

This plan satisfies all the criteria.

24.8.4 Contour-Adjustment Method

This method is used to level lands that are irrigated by contour methods, and fields where the cross-slope can be made across the field. It involves a trial and error adjustment of contour lines on a plan map and hence requires considerable judgment to keep earthwork and haul distance to a minimum. It requires a contour map and involves the following steps:

1. Establish the trial contours.
2. Draw the lines of equal cut or fill through the intersections of proposed and actual contours.
3. Compute the volume of earthwork through the horizontal-plane method.
4. Adjust the trial contours if the volume of cut is not in balance with the volume of fill.

Example 24.7: Compute the ratio of cut and fill for the field as shown in Figure 24.10. Elevations are given in feet, and the field is irrigated from ditches.

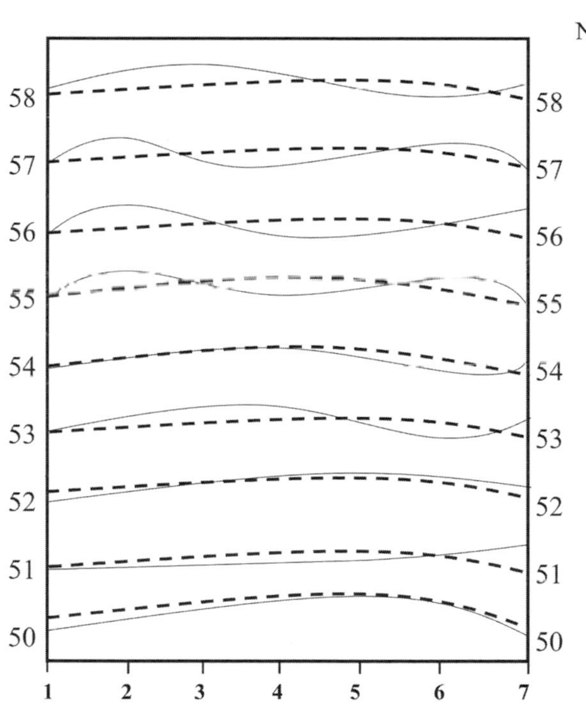

Figure 24.10 Contour map of the field.

Solution: The following limitations are considered:

1. The maximum cross-slope is less than 0.3 ft (0.09 m) per station.
2. Down-field slopes are to be "fairly uniform," as described in Table 24.1.

As shown in Figure 24.11, the direction of irrigation can be determined from the contour map, which is from the north edge of the field to the west.

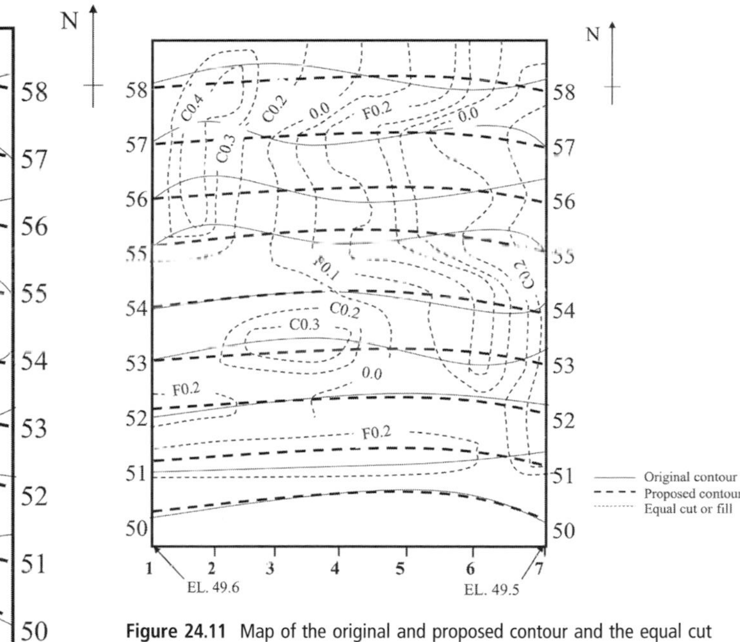

Figure 24.11 Map of the original and proposed contour and the equal cut or fill.

The area between the proposed and actual contours represent excavation and fill, and the excavation should be slightly higher than the area of excavation. Make sure the distance of each contour is less than 100 ft (30 m).

The volume of earthwork is calculated by the horizontal-plane method and is introduced in Section 24.9.

24.9 CALCULATION OF EARTHWORK

Earthwork can be computed using several methods, which are discussed here.

24.9.1 Prismoidal Formula

The volume of earthwork can be calculated using the prismoidal formula, expressed as

$$V = \frac{L(A_1 + 4A_m + A_2)}{6}, \tag{24.16}$$

where V is the volume (ft^3 or m^3), L is the perpendicular distance between end planes (ft or m), A_1 is the area of one end plane (ft^2 or m^2), A_2 is the area of the other end plane (ft^2 or m^2), and A_m is the area in the middle section parallel to the end planes (ft^2 or m^2).

24.9.2 Four-Point Method

Let V_c be the volume of cut (m^3 or ft^3), V_f be the volume of fill (m^3 or ft^3), L be the grid spacing (m or ft), H_c be the sum of cuts on four corners of a square grid (m or ft), and H_f be the sum of fills on four corners of a square grid (m or ft). Then, the four-point method states that

$$V_c = \frac{L^2}{4}\left(\frac{H_c^2}{H_c + H_f}\right), \tag{24.17}$$

$$V_f = \frac{L^2}{4}\left(\frac{H_f^2}{H_c + H_f}\right). \tag{24.18}$$

The volume of cut and that of fill can be determined using eqs (24.17) and (24.18) for each grid square and by summing the total volume of cuts and fills. The volume of cut and fill can also be expressed in yd^3 if eqs (24.17) and (24.18) are divided by 27, with L, H_c, and H_f in ft.

Example 24.8: Consider a square grid (100 × 100 ft) shown in Figure 24.12. The cut and fill are in feet. Compute the volume of cut and fill.

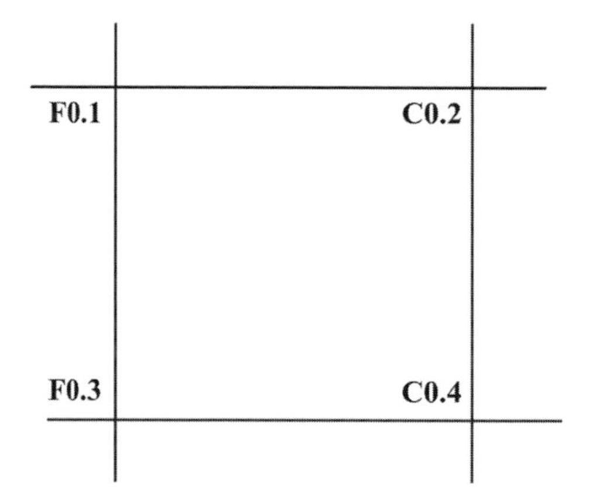

Figure 24.12 Cut and fill in a square grid.

Solution: Given the sum of cuts $H_c = 0.2 + 0.4 = 0.6$ ft (0.18 m), the sum of fills $H_f = 0.1 + 0.3 = 0.4$ ft (0.12 m), the grid spacing $L = 100$ ft (30.48 m), the volume of cut (V_c) and fill (V_f) are calculated using eqs (24.17) and (24.18), respectively.

$$V_c = \frac{L^2}{4}\left(\frac{H_c^2}{H_c + H_f}\right) = \frac{100^2}{4}\left(\frac{0.6^2}{0.6+0.4}\right) = 900\,\text{ft}^3\,(25.5\,\text{m}^3),$$

$$V_f = \frac{L^2}{4}\left(\frac{H_f^2}{H_c + H_f}\right) = \frac{100^2}{4}\left(\frac{0.4^2}{0.6+0.4}\right) = 400\,\text{ft}^3\,(11.3\,\text{m}^3).$$

The volume of cut and fill can also be expressed in $V_c = 33.33$ yd^3 and $V_f = 14.81$ yd^3.

Example 24.9: Using the field data in Example 24.4, compute the volume of cut and fill (m^3) using the horizontal-plane method.

Solution: The fills (+) and cuts (–) in meters are summarized as:

	Elevation (m)							
Lines	Line 1	Line 2	Line 3	Line 4	Line 5	Line 6	Sum of cuts	Sum of fills
Line G	+0.24	+0.15	–0.10	–0.20	–0.29	–0.39	–0.98	0.39
Line F	+0.17	+0.08	–0.02	+0.19	–0.21	–0.30	–0.53	0.44
Line E	+0.10	+0.01	–0.09	+0.27	+0.48	–0.07	–0.16	0.86
Line D	+0.03	–0.06	–0.16	–0.25	+0.11	+0.32	–0.47	0.46
Line C	–0.04	–0.13	+0.08	+0.29	–0.11	+0.10	–0.28	0.47
Line B	–0.10	–0.20	–0.14	–0.09	–0.18	+0.03	–0.71	0.03
Line A	–0.17	–0.27	–0.21	+0.30	+0.36	–0.04	–0.69	0.66
Total							–3.82	3.31

Using the four-point method, with the grid spacing $L = 100$ ft $= 30.48$ m, the volume of cuts for each grid is summarized as follows:

	Volume (m^3)					
Lines	Line 1	Line 2	Line 3	Line 4	Line 5	Line 6
Line G						
Line F		0.00	10.02	46.46	127.81	277.51
Line E		0.00	13.99	4.61	8.85	73.79
Line D		4.23	69.40	74.34	13.25	1.06
Line C		46.30	65.72	49.85	40.10	4.39
Line B		109.02	93.97	20.32	49.13	46.64
Line A		172.73	190.43	60.16	17.57	18.35

The total volume of cuts is 1710.0 m^3.

The volume of fills for each grid is summarized as follows:

	Volume (yd^3)					
Lines	Line 1	Line 2	Line 3	Line 4	Line 5	Line 6
Line G						
Line F		147.25	33.39	16.72	9.59	0.00
Line E		82.83	8.32	88.15	180.87	51.13
Line D		23.34	0.03	22.66	156.25	197.87
Line C		1.00	3.42	40.65	48.59	101.38
Line B		0.00	2.64	52.88	29.31	9.12
Line A		0.00	0.00	29.01	110.31	57.99

The total volume of cuts is 1504.7 m^3.

24.9.3 End-Area Method

For the end-area method, let V be the volume of cut (or fill) (m^3 or ft^3), A_1 be the area of cut (or fill) at one end (m^2 or ft^2), A_2 be the area of cut (or fill) at the other end (m^2 or ft^2), and L be the distance between end areas (m or ft). The end-area method states that

$$V = \frac{L(A_1 + A_2)}{2}. \tag{24.19}$$

For each line in one direction, the total areas (end area) of cut and of fill are calculated from the profile and then the volumes are computed. The volume of cut and fill can also be expressed in yd^3 if eq. (24.19) is divided by 27, with L, A_1, and A_2 in ft.

Example 24.10: Using the field data in Example 24.4. Compute the volume of cut and fill using the end-area method.

Solution: The total end area of cut and fill for each line in one direction, i.e., A, B, C, etc., are summarized as:

		Cuts			Fills		
Line	L	End area (A)	$A_1 + A_2$	Product	End area (A)	$A_1 + A_2$	Product
Line G	100	322			127		
Line F	100	175	497	49,700	144	270	27,100
Line E	100	51	226	22,600	285	429	42,900
Line D	100	154	205	20,500	152	437	43,700
Line C	100	91	245	24,500	153	305	30,500
Line B	100	233	324	32,400	10	163	16,300
Line A	100	227	460	46,000	218	228	22,800
Total				195,700			183,300

Using eq. (24.19),

$$\text{The volume of cuts} = \frac{195{,}700}{2} = 97{,}850 \text{ ft}^3 \left(2771 \text{ m}^3 \text{ or } 3624 \text{ yd}^3\right),$$

$$\text{The volume of fills} = \frac{183{,}300}{2} = 91{,}650 \text{ ft}^3 \left(2595 \text{ m}^3 \text{ or } 3394 \text{ yd}^3\right).$$

The ratio of cuts and fills $3624/3394 = 1.07$.

24.9.4 Horizontal-Plane Method

This method is employed when the contour-adjustment method is applied for leveling and where heavy cuts and fills are involved. It is the end-area method applied to land areas in place of areas from cross-sections. The volume of cut or fill can be computed by this method as

$$V = \frac{H(A_1 + A_2)}{2}, \tag{24.20}$$

where V is the volume of cut (or fill) between areas (m^3 or ft^3), A_1 is the area of cut (or fill) C_1 (m^2 or ft^2), A_2 is the area of cut (or fill) C_2 (m^2 or ft^2), and H is the difference between C_1 and C_2 (m or ft). The value of H should be 0.1–0.2 ft. The volume of cut and fill can also be expressed in yd^3 if eq. (24.20) is divided by 27, with L, A_1, and A_2 in ft. It should be noted that the accuracy of this method depends on the H value. The smaller the value of H, the more accurate the result. The H value as high as 0.5 feet is only for obtaining an approximate value.

Example 24.11: Using the field data in Example 24.7, compute the volume of cut and fill using the horizontal-plane method.

Solution: As shown in Figure 24.11, the areas between the lines of equal cut and equal fill are calculated and summarized as follows:

L	H (ft)	Area (ft^2)	Sum area	Product	Volume (ft^3 [m^3])	Volume (yd^3)
0		38,000				
0.2	0.2	51,000	89,000	17,800	8,900 (252.2)	329.63
0.3	0.1	25,000	76,000	7,600	3,800 (107.6)	141.74
0.4	0.1	8,000	33,000	3,300	1,650 (46.7)	61.11
Total cuts						531.48
L	H (ft)	Area (ft^2)	Sum area	Product	Volume (ft^3 [m^3])	Volume (yd^3)
0		60,000				
0.1	0.1	47,000	107,000	10,700	5,350 (151.5)	198.15
0.2	0.1	45,000	92,000	9,200	4,600 (130.3)	170.37
Total fills						368.52

The ratio of cuts and fills = 531.48/368.52 = 1.44.

24.9.5 Summation Method

This method assumes that a given cut or fill at a stake represents the area midway to the next stake on a 100 × 100 ft grid (10,000 ft^2). Thus, every foot of cut would correspond to a volume of

$$\frac{1}{27}(100 \times 100 \times 1.0) = 370 \text{ cubic yards.} \tag{24.21}$$

Then, the sum of all the cuts or fills is calculated, and the total is multiplied by 370 to get the yardage. If the tenths of cut or fill are added, then the total will be multiplied by 37.

Example 24.12 Consider the field in Example 24.4, shown in Figure 24.5. Compute the volume of cut and fill using the summation method.

Solution: From Example 24.4, the sum of cuts = 12.53 ft and the sum of fills = 10.89 ft. Then:
volume of cuts = 12.53 × 370 = 4636.1 yd^3,
volume of the fills = 10.89 × 370 = 4029.3 yd^3.

24.10 EARTHWORK BALANCE

The earthwork balance involves a balance between excavation and fill without borrow or waste. Two terms that are commonly used in earthwork are *swell* and *shrinkage*. When the field surface is loosened by excavation, its volume increases with respect to the original condition; this increase is referred to as swell. If this loosened volume is placed in a fill and compacted, then its volume decreases. This decrease is referred to as shrinkage. The swell and shrinkage will depend on the type of soil. For land leveling, a high-shrinkage factor may be required because the topsoil moved during leveling normally has high organic content and low density, and the cut areas are compacted by earth-moving equipment. To illustrate the increase in shrinkage, consider that 1 yd^3 of excavated material will make 0.85 cubic hard of fill. Then, the cut to fill ratio will be 1.0/0.85 = 1.18. The shrinkage factor then becomes (1 – 0.85)/0.85 = 0.18 or 18%, and the percentage cut to percentage fill is 1.0/1.85 = 0.54 to 0.85/1.85 = 0.46, or 54% to 46%.

A common approach to balancing earthwork is to achieve an acceptable value cut to fill ratio. The shrinkage factor normally varies from 10% for heavy leveling on firm field surfaces to 100% for leveling with shallow cuts and fills, but is usually 15–60%. In the plane method of leveling, the lowering of field surface can be assumed to vary from 0.02 to 0.03 feet for very compact soils to as high as 0.10 feet for loose soils.

24.11 CONSTRUCTION

Depending on the method of leveling, cut and fill areas are marked either by stakes or marked on the grid stakes. To achieve the design leveling involves (1) leveling equipment, (2) time required for completion of the work, (3) operator skill, (4) haul distance, (5) construction procedure, (6) preservation of topsoil, (7) construction tolerance, (8) checking, (9) finishing, and (10) maintenance. These items are briefly discussed.

The leveling equipment is mostly a tractor-driven scraper. There is a variety of scrapers to choose from. The choice of a scraper depends on the nature of leveling and field conditions. It is important to estimate the time needed to accomplish leveling, which must fit into the cropping sequence. This will impact the choice of leveling equipment. To operate the equipment, a skilled operator is needed – land leveling is a sophisticated job and requires an experienced operator. The excavated material is to be moved from the cut areas to the fill areas and the distance of haul depends on the landscape. This also affects the choice of equipment and operator.

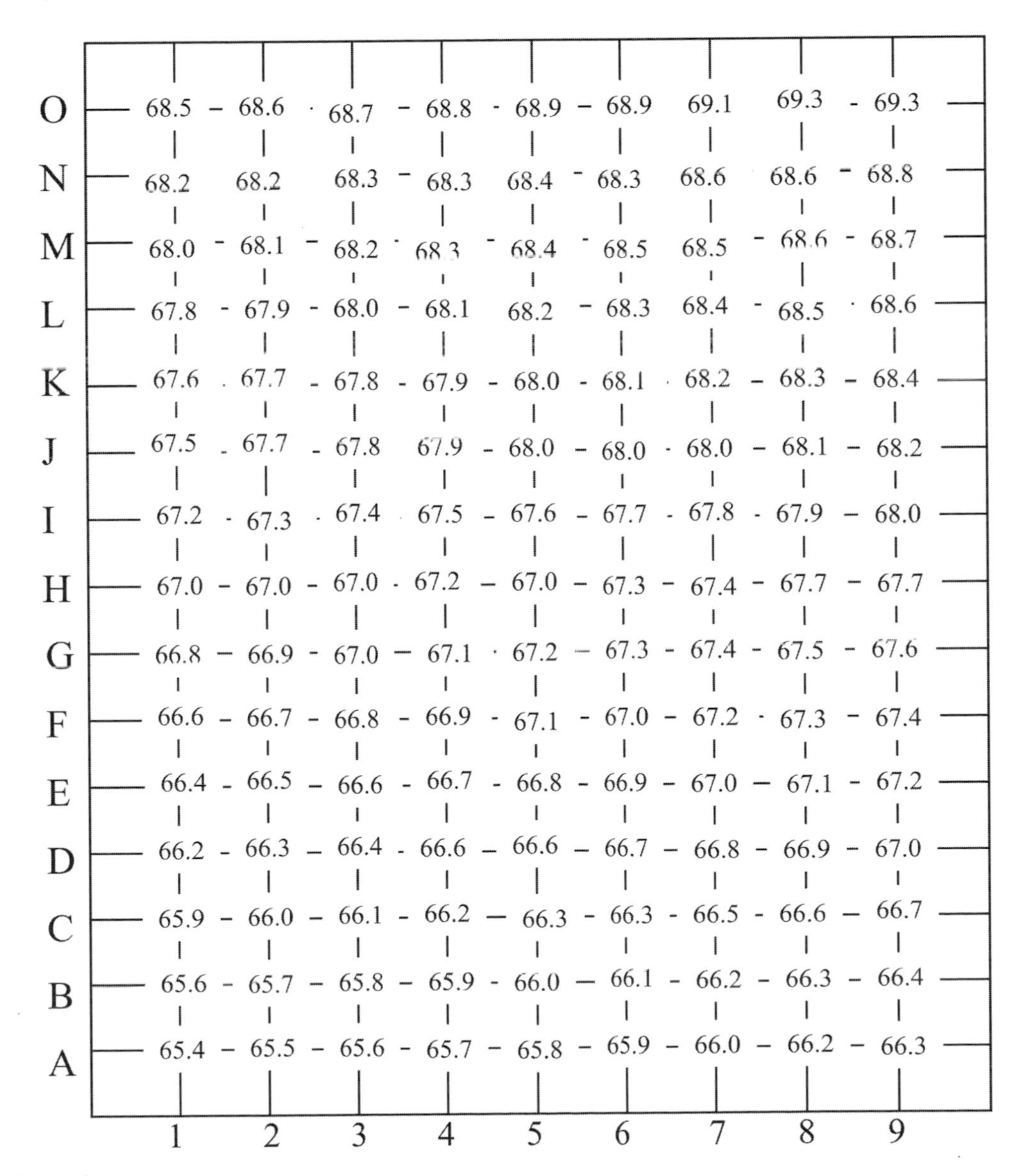

Figure 24.13 Field grid for Q.24.1.

The procedure for land leveling requires first a map showing cut areas and fill areas. Since the haul distance should be as short as possible, it is desirable to move the soil to the fill areas from the nearest cut areas. The starting point can be the north side. The work can be done along the line by moving the soil from east or west, and it can progress until the south side is completed. The earthwork must be balanced and grid stakes should not be disturbed. It is common to provide extra fill height to account for settlement.

Since topsoil (usually 6 inches) is nutrient rich, often it is scraped and stored. Once excavation and filling are completed, the topsoil is brought back. This means extra earthmoving work and expense, so it must be justified. The permissible tolerance for scraping is usually 0.1 foot. For irrigation, the grade should be in the direction of flow. Leveling should be checked before grid stakes are removed. After leveling is done, there may be settlement following irrigation. The settled areas should be filled. Perennial crops are not recommended until settlement has occurred. Leveled fields require maintenance because of erosion by wind and water, and improper use of farming equipment may change the surface characteristics, which should be corrected from time to time.

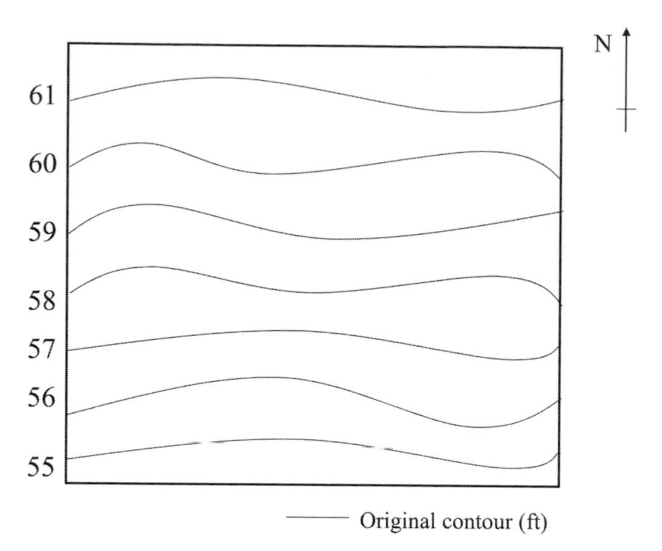

Figure 24.14 Field contour map.

QUESTIONS

Q.24.1 A 10-ha field, as shown in Figure 24.13, is to be leveled. A grid survey with a 25 × 25 m spacing is also shown. The elevation is in meters. Demarcate the location of the centroid of the field. Compute the average elevation of the field and average slope in both horizontal and transverse directions.

Q.24.2 Compute the volume of cut and fill for Q.24.1.

Q.24.3 Using the field in Figure 24.9, compute the volume of cut and fill using the four-point method, end-area method, and summation method.

Q.24.4 Figure 24.14 shows a profile map of a field that is to be leveled. The field size is 700 × 700 ft, and the elevation varies by 55–61 ft. Use the profile method for land leveling. Select the trial gradelines and compute the volume of cut and fill.

Q.24.5 Compute the balance between cut and fill in Q.24.4.

Q.24.6 Select another set of trial gradelines and compute the volumes of cut and fill for Figure 24.14.

Q.24.7 The field shown in Figure 24.5 is to be leveled using the plan-inspection method. Compute the volumes of cut and fill and earthwork.

Q.24.8 The field shown in Figure 24.11 is to be leveled using the contour-adjustment method. Compute the volumes of cut and fill and earthwork.

REFERENCES

Hart, W. E. (1975). *Irrigation System Design*. Fort Collins, CO: Colorado State University.

Michael, A. M. (2010). *Irrigation Theory and Practice*. New Delhi: Vikas Publishing House PVT Ltd.

University of Missouri (n.d.). Land grading for irrigation: design and construction. https://extension.missouri.edu/g1641.

USDA-SCS (1970). Land leveling. In *National Engineering Handbook*,. Washington, DC: USDA.

25 Drainage

Notation

A	cross-sectional area in the pipe (m^2)
A_f	area of the field or area covered by one lateral (m^2, ha)
b	depth of the impermeable layer to the maximum height of the water table (m, ft)
B_c	outside diameter of the conduit (L)
B_d	width of the ditch at the top of the conduit or pipe (L)
C_c	load coefficient for projecting conduit
C_d	load coefficient for ditch conduit
d	depth of drain above the impermeable layer (m, ft)
d_e	effective depth of the distance between drains and the impermeable layer (m, ft)
D	inside diameter of the drainage tubing or tile (m, ft)
D_a	average depth of flow (m, ft)
h_d	depth from the soil surface to the center of the drain (m, ft)
h_w	level of water in the drain (m, ft)
H	maximum height of the water table above the drain (open ditches) (L)
i, v, D_c	rate of replenishment of water, i is irrigation or rainfall, v is velocity of flow in the drain, and D_c is the drainage coefficient (mm/day)
K	constant hydraulic conductivity (in./h, cm/h, m/day)
K_a	saturated hydraulic conductivity of the layer above the drain (m/day)
K_b	saturated hydraulic conductivity of the layer below the drain (m, ft)
L	length of the tile drain (m, ft)
m	maximum height of the water table above the drain (drain tube) (m)
m_0	initial height of the water table above the center of the drain at the midpoint (m)
m_t	height of water table above the center of the drain at the midpoint after time t (m)
n	Manning's roughness coefficient
p	depth of ponding (m)
q	flow rate into the drain from both sides per unit length of soil (L^2/T)
q_x	flow rate across a vertical plain at any point $P(x, y)$
Q	flow rate in the drain (m^3/s)
r	outside radius of the drain (effective radius of the gravel envelope) (m, ft)
R	hydraulic radius (m)
S	spacing between drains (m, ft)
S_y, f	specific yield (S_y) or drainable porosity (f)
S_0	slope of the energy grade line (for uniform flow it can be taken as the slope of the drain) (L/L)
t	time for water table to fall from m_0 to m (days)
v, V	average velocity of flow in the drain (m/s)
w	unit weight of fill (FL^{-3})
W_c	total load on the pipe per unit length (FL^{-1})

25.1 INTRODUCTION

Drainage is the orderly removal of excess water from the soil surface, as well as from the soil profile or root zone. It directs the removal in a manner such that it does not erode the soil and damage crops. By so doing it provides a suitable environment for the maximization of plant growth, keeping in mind financial constraints.

Excess water over the soil surface is often generated by rainfall, snowmelt, excess irrigation, rainfall over irrigation, runoff and seepage from adjoining areas or canal, outflow from streams, poor outlet conditions, and inadequate land leveling. Excess water in the root zone may be caused by excess irrigation, shallow water table, perched water table, and lateral ground water flow to the agricultural land. The objective of this chapter is to present the rudiments of agricultural drainage.

25.2 EFFECT OF EXCESS WATER

Excess water has a detrimental effect on agricultural productivity. First, consider the effect on soil. Soil structure is adversely affected. When soil gets saturated water may pond on the surface, which impedes air circulation in the soil and prevents bacterial activities that are essential for the health of plants and soil. Excess water excludes air from the root zone, thus providing an unhealthy environment for root development. It prevents chemical action from making nutrients available to plants. Excess water also prevents bacterial activities that convert nitrogen and other nutrients to forms that can be utilized by plants. If the soil has salts and alkalinity, they would tend to be concentrated on the surface or in the soil profile. If the field is not properly leveled and has depressions, then water may deposit in the depressions and would impact farm operations. Excess water delays tillage, cultivation, and harvesting operations, erodes fertile soil, and worsens the ill effects of droughts.

Now consider the effect on plant growth. Excess water raises the water table, which then impedes root zone development, particularly root penetration. Wet soil requires more heat to warm up and thus shortens the crop-growing season. Evapotranspiration takes heat from the soil and lowers its temperature. Excess water promotes certain plant diseases and parasites and increases winter deaths and frost heaving by plants.

25.3 INDICATORS OF POOR DRAINAGE

Poor drainage results in poor soil conditions and plant conditions. Soil color is an indicator of soil drainage. For example, if the soil color is brown, then it can be regarded as a well-drained soil. If the soil color is dark gray, blue, or black, then it reflects waterlogging. If the soil is rusty and mottled, then water logging has occurred at certain times of the year. Poorly drained soils develop salt crusting. If there are swamp areas, wet spots, or areas with standing water on the surface, then drainage is poor. Plant conditions are also indicative of poor drainage. If the plant leaves start turning yellow, then that is an indication of poor drainage. Stunted plants, a patchy crop, poor quality crop, and poor yield all point to poor drainage.

25.4 ADVANTAGES OF DRAINAGE

Drainage is vital for crop productivity and soil management. It improves soil structure, controls erosion, improves trafficability, removes toxic salts from the soil, and reduces health hazards. It allows the soil to warm early in spring, resulting in a longer growing season. It provides increased aeration, deepening the root zone. It increases the water retention capacity of the soil and reduces runoff and flood damage. It increases land value, reduces maintenance of farm machinery, and reduces production costs.

25.5 REQUIREMENTS FOR DRAINAGE

Drainage requirements are determined by the type of crop, type of soil, climatic conditions, soil and crop management practices, availability of nutrients, and biological activities. More specifically, how much drainage is needed depends on the maximum duration and the extent of surface ponding, the maximum depth of the water table, and the maximum rate at which the water table can be lowered. Crop growth and the water table are related. The depth of the water table affects the supply of nitrogen and the penetration of plant roots. The quality of the water table also affects plant growth.

25.6 DRAINAGE COEFFICIENT

The drainage coefficient (D_c) is defined as the depth of water to be removed from the land in 24 hours. In general, D_c is expressed in mm/day. In humid areas, common values of the drainage coefficient are 3/8–3/4 in./day (9.5–19.1 mm/day). However, in arid areas the values of D_c depend on the design of irrigation systems. Generally, the values of D_c are between 0.012 in./h and 0.24 in./day. The US Natural Resources Conservation Service (USDA-NRCS, 2001) has given values of D_c for humid areas as in Table 25.1. Drainage of soils in humid and arid areas is compared in Table 25.2. Figure 25.1 shows a map of agricultural land drainage in the United States.

25.7 CLASSIFICATION OF DRAINAGE

Drainage can be classified as surface drainage and subsurface drainage, as shown in Figure 25.2. Each type is now discussed.

25.7.1 Surface Drainage

Surface drainage entails collection, transportation, and removal of water from the soil surface. The removal is done by gravity flow through overland surface and through channels. To that end, the soil surface is reshaped to induce gravity flow.

Surface drainage applies to flat or gently sloping lands where water gets ponded because of a number of factors. Flat lands with slow infiltration rate, low hydraulic conductivity, or shallow impermeable boundary or rock are prone to generating excess water. The ponding of water on these lands can be caused when the land surface is uneven, with pockets or ridges that retard natural runoff. Waterlogging may be caused by excess water from surrounding areas. If

Table 25.1 *Drainage coefficients (D_c) for pipe drains in humid areas*

Crops and degrees of surface drainage	Drainage coefficient (in./day)	
	Mineral soil[a]	Organic soil
Field crops	$\frac{3}{8}$ to $\frac{1}{2}$	$\frac{1}{2}$ to $\frac{3}{4}$
Truck crops	$\frac{1}{2}$ to $\frac{3}{4}$	$\frac{3}{4}$ to $1\frac{1}{2}$

[a] These values may vary depending on special soil and crop conditions. Where available, local recommendations should be followed.

After USDA-NRCS, 2001.

Table 25.2 *Comparison of drainage of soils in humid and arid areas*

	Humid areas	Arid areas
Excess water sources	Precipitation	Excess irrigation and groundwater from surrounding areas
Requirement	Required	Part of the irrigation system
Water table control	Control to avoid interfering with plant growth –	Control to avoid interfering with plant growth Control salinity and alkalinity in the soil and groundwater
Drain depth	About 0.6–1.5 m	Deeper than in humid areas (about 1.5–3.0 m). The minimum effective depth is 1.8 m for medium-textured soils and up to 2.1 m for fine-textured soil
Drain spacing	The greater the depth of the drain, the wider the spacing. Drain spacing is about 10–50 m or even 90 m	Drain spacing is about 50–200 m
Depth to the water table	Depth to the water table from the ground surface is about 60–75 cm	It is necessary to maintain the groundwater at a depth of 1.2–1.5 m midway between the drains. Shallow water table over longer periods of time can cause the accumulation of salts in the root zone

After USDA-NRCS, 2001.

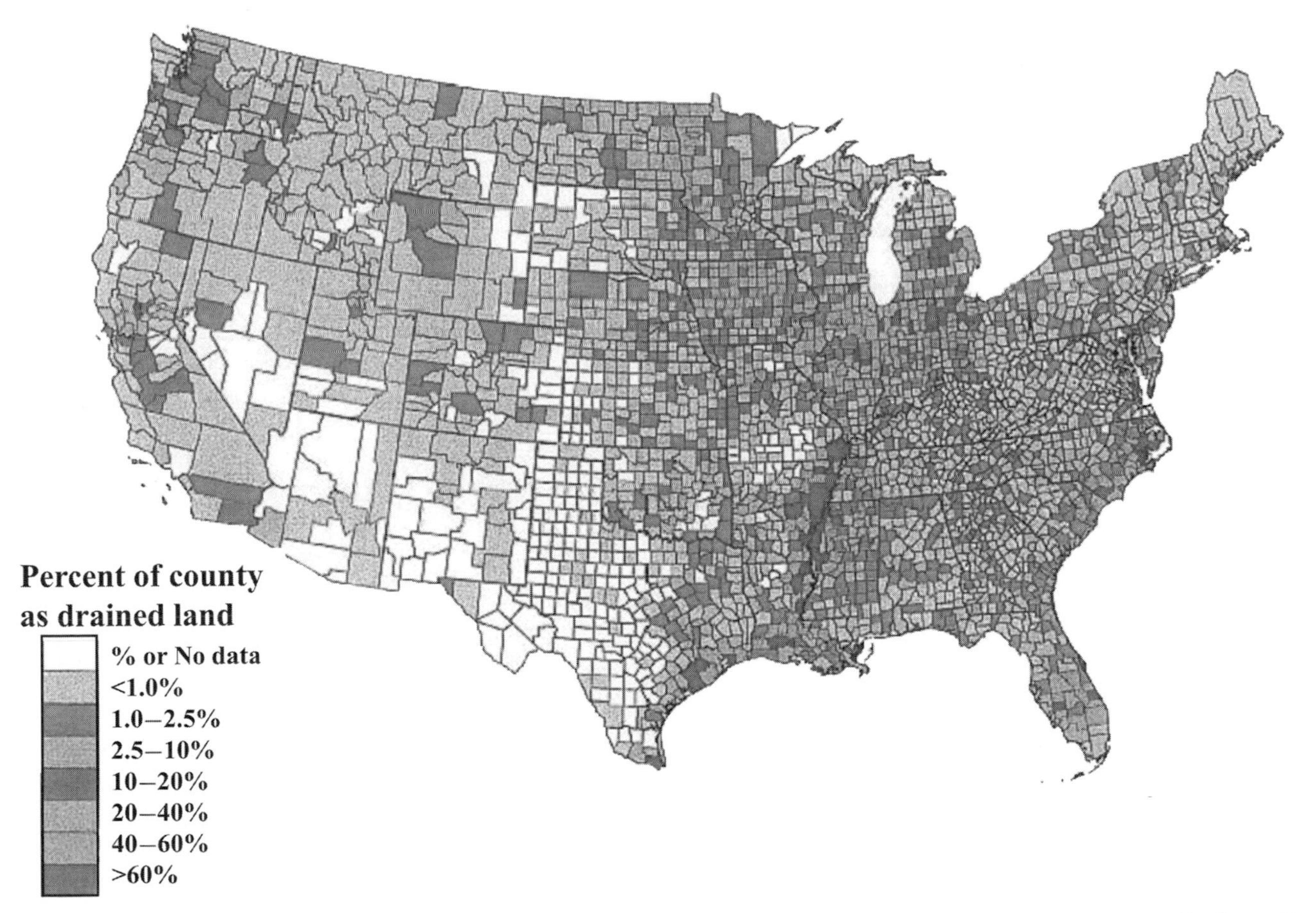

Figure 25.1 Drainage of agricultural land. (After US Department of Commerce, 1981. Graphed by William Battaglin, USGS.) A black and white version of this figure will appear in some formats. For the color version, please refer to the plate section.

the water disposal system has low capacity that removes water slowly, then ponding may be caused. If the outlet conditions are such that water is held on the soil surface for longer periods of time, the result is ponding.

25.7.2 Subsurface Drainage

As the name suggests, subsurface drainage occurs below the soil surface. Thus, subsurface drainage removes excess

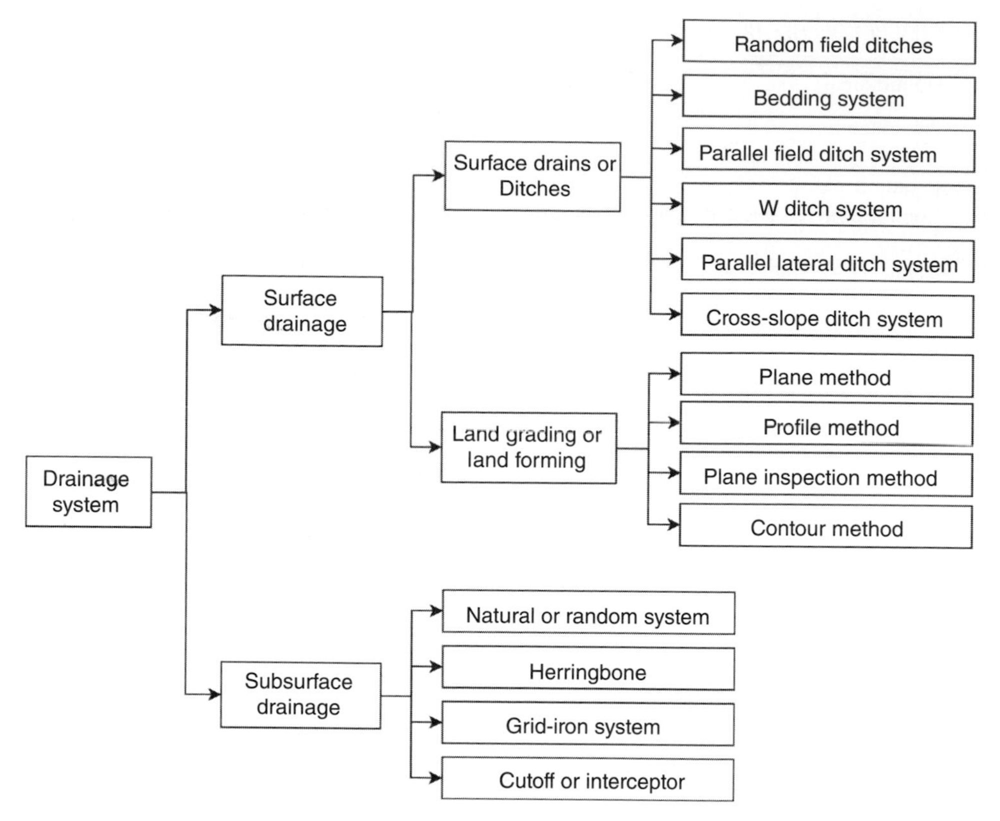

Figure 25.2 Classification of different drainage systems.

water from the root zone or soil profile. It applies to soils with low hydraulic conductivity, high groundwater table, poor outlet conditions, or low hydraulic conductivity at moderate depth below the root zone.

25.8 TYPE OF SURFACE DRAINAGE SYSTEMS

The excess water from the soil surface is removed by land leveling, land grading, bedding, and ditching, as well as by diverting or excluding water from land by diversion ditches. Surface drainage entails collection of water, a disposal system, and outlet (natural or pumped) conditions. Two types of surface drainage systems are regarded: (1) surface drains or ditches and (2) land grading or land forming. Surface drains or ditches include random field ditches, bedding or humps and hollows, the parallel field ditch system, the parallel lateral ditch system, and the cross-slope ditch system.

25.8.1 Random Field Ditches

Random field ditches, as shown in Figure 25.3, are suitable for drainage of scattered depressions where the depth of cut is not greater than 1 m and several potholes are drained with a single drain. For draining an area greater than 2 ha, a 10-year recurrence interval is used to determine the

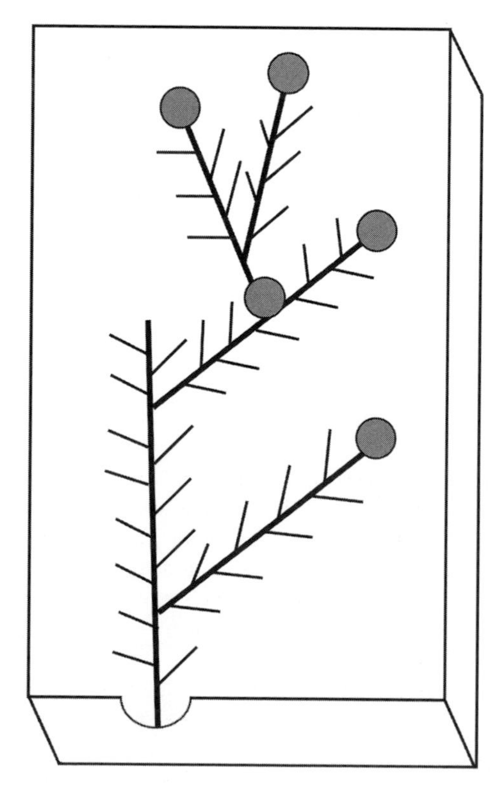

Figure 25.3 Random field drains.

Table 25.3 *Channel dimensions for drainage*

Farming operations perpendicular to the channel	Farming operations parallel to the channel
$Z > 8{:}1$ for $d < 30\,\text{cm}$ $Z > 10{:}1$ for $d > 60\,\text{cm}$	Minimum $Z = 4{:}1$

Z is the side slope horizontal to vertical; and *d* is the depth of the channel.

After USDA-NRCS, 2001.

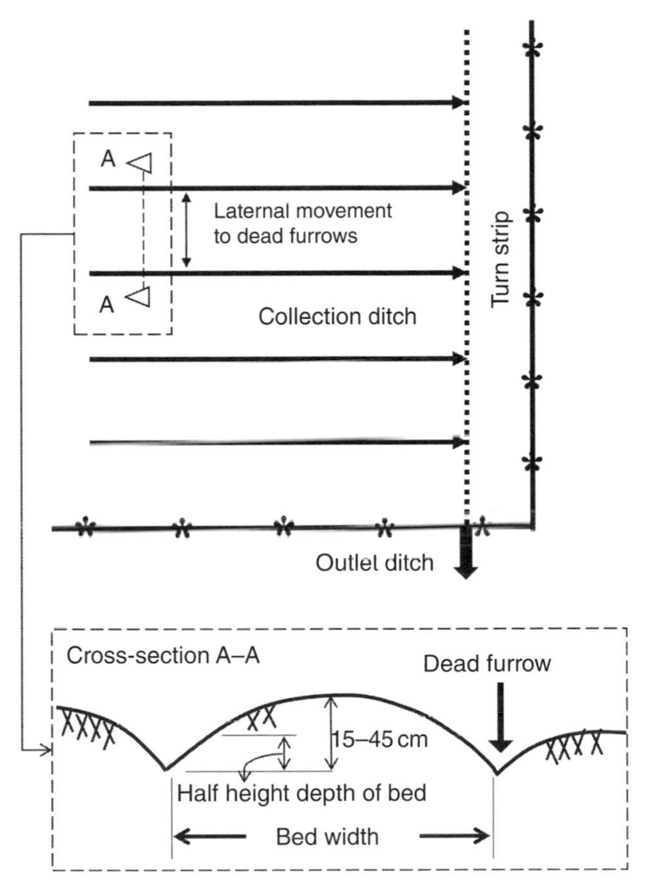

Figure 25.4 Bedding system.

capacity. For an area less than 2 ha, the capacity is not considered. The channel should follow the route of minimum cut and least interference with farming operations. Usually, the channel shape is regarded as V-shaped or parabolic. For dimensioning channels, the depth of channel and side slopes depends upon the nature of farming operations, as shown in Table 25.3.

The depth of the ditch is affected by the topography of the area, the outlet elevation, and area to be drained (capacity of the channel). The velocity of flow should be such that erosion and sedimentation are controlled. It depends on soil conditions, vegetative cover, and flow depth. If the flow

Table 25.4 *Bed width for drainage*

Soil	Bed width (m)
Very slow internal drainage	7–11
Slow internal drainage	13–16
Fair internal drainage	18–28

depth is less than 100 cm, then a velocity of 0.3–0.6 m/s is satisfactory.

The channel grade depends on soil characteristics. For sandy soils the approximate grade is 0.2% and for clay soil it is 0.5%. The roughness coefficient depends on the nature of the vegetation. A value of 0.04 for Manning's *n* can be used if good estimates are not available.

25.8.2 Bedding

The bedding system, as shown in Figure 25.4, is suitable for draining flat lands with slope less than 1.5% where the soils are slowly permeable and pipe drainage is not economical. The important design parameters are width and length of the bed. The width of the bed depends on the land slope, drainage characteristics of the soil, and cropping pattern. Typical values of bed width are given in Table 25.4.

The bed is 90–300 m long and is V-shaped or parabolic. Farming practices may be parallel or normal to the dead furrow. Tillage operations parallel to a dead furrow have the tendency to retard flow to the dead furrow. Plowing is always parallel to the dead furrow.

25.8.3 Parallel Field Ditch System

The parallel field ditch system, as shown in Figure 25.5, is similar to the bedding system except that channels are spaced farther apart and the capacity may be greater than that of bedding. It is suitable for poorly drained flat land. Its design is very similar to that of the bedding system, except that beds are not equally spaced and the water enters the bed from one side. The maximum row spacing is about 180 m, the maximum top width is about 360 m, and for highly erodible soils the length of the bed should be less than 90 m. The capacity depends on the soil type, grade, and drainage area. The minimum depth is 0.2 m, the minimum cross-sectional area is 0.5 m^2 and the side slope is 8:1 (minimum *Z*). The shape is trapezoidal, V-shaped, or parabolic. Farm operations, such as plowing, are parallel to the drain and planting and cultivation are normal to the direction of dead furrow.

25.8.4 W Ditch System

The W ditch system, as shown in Figure 25.6, is a parallel field ditch system with narrow spacing (5–15 m) and all spoil

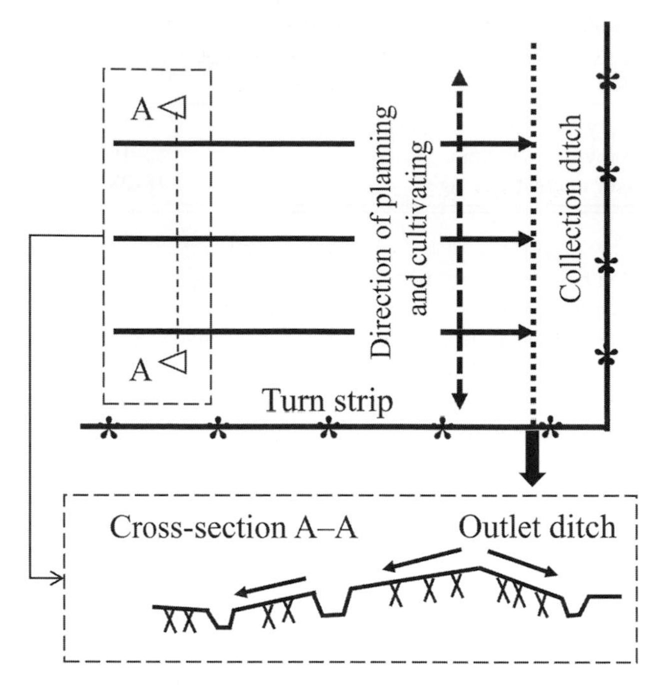

Figure 25.5 Parallel field ditch system.

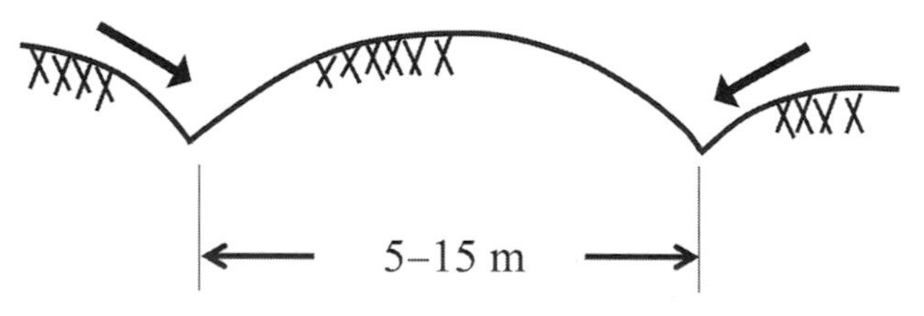

Figure 25.6 W ditch system.

is placed between channels. The advantages of this system include the following. It can be constructed or maintained with ordinary farm equipment, it may serve as a field road, it allows better row drainage, it may be seeded with grass or other row crops, and it may be used as a turn row. Its disadvantages are that a large area is occupied by drains, it involves more earth work, and the spoil is not available for filling depressions.

25.8.5 Parallel Lateral Ditch System

The parallel lateral ditch system, as shown in Figure 25.7, is similar to the parallel field ditch system, except that the ditches are deeper and cannot be crossed by machinery. The depth is greater than 60 cm and side slope is <4:1, and may have more capacity than has bedding. It is suitable for deep, permeable soils underlain with an impermeable layer. It is used to provide surface drainage and initial subsidence of organic soil prior to subsurface drainage, and

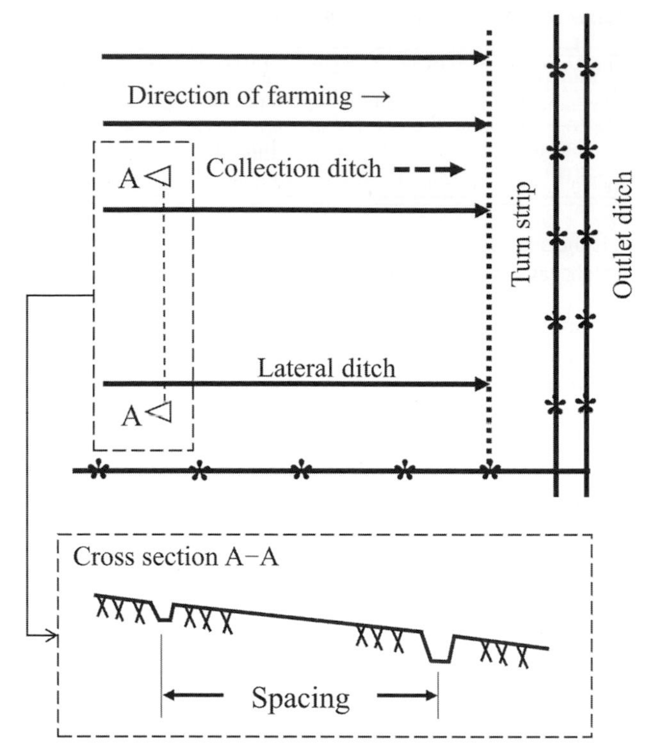

Figure 25.7 Parallel lateral ditch system.

controls the water table. The control of the water table depends on the spoil type, seasonal conditions, topography, and climatic conditions. In organic soils the water table control is desirable to provide water for plant growth, control subsidence, reduce fire, and reduce soil erosion. The water table is maintained at 45–120 cm below the surface. Its design may entail the following:

Spacing: The minimum spacing is 200 m for sandy soils, 100 m for other mineral soils, and 60 m for organic soils.
Bottom width: The minimum bottom width is 1.2 m for sandy soils, 0.3 m for other organic soils, and 0.3 m for organic soils.
Depth: The minimum depth is 1.2 m for sandy soils, 0.8 m for other mineral soils, and 0.9 m for organic soils.

For farm operations, machinery cannot cross because the ditches are too deep, plowing is parallel to the channel, and planting and cultivation are parallel to the ditch.

25.8.6 Cross-Slope Ditch System

The cross-slope ditch system is suitable for drainage of sloping land where soils have poor internal drainage, subsurface drainage is not economical, and soil is very steep for bedding and ditches. It is used for both drainage and erosion control. For drainage it is called a diversion ditch and for erosion control it is called a terrace.

25.8.7 Land Grading

The purpose of land grading is to provide proper grade to allow adequate drainage. There are several ways to do land grading, such as the plane method, profile method, plan inspection method, and contour method (Schwab et al., 1993). Land grading was more fully described in Chapter 24.

25.9 TYPES OF SUBSURFACE DRAINAGE SYSTEMS

Subsurface drainage involves collection systems, disposal systems, and outlet (natural or pumped) conditions. Ditches or buried tubes are installed in the soil profile at an appropriate depth to collect and convey excess water in the root zone to a gravity and pumped outlet. The objective here is to prevent long saturation during the growing season, improve soil moisture conditions that will allow timeliness of tillage, higher storage capacity of the soil, and deeper root zone, increased aeration, and warmer soil. There are several types of subsurface drainage systems (Schwab et al., 1993): natural or random, herringbone, grid-iron, and cutoff or interceptor.

25.9.1 Natural or Random System

The natural or random system, as shown in Figure 25.8, is good for the drainage of isolated areas, and is adopted where the field does not require complete drainage with equally spaced drains.

25.9.2 Herringbone

The herringbone system, as shown in Figure 25.9, is suitable for areas where laterals are long and waterways require thorough drainage. It is adopted where the area has a concave surface, and the main is laid near normal to the slope. Some areas have double drainage.

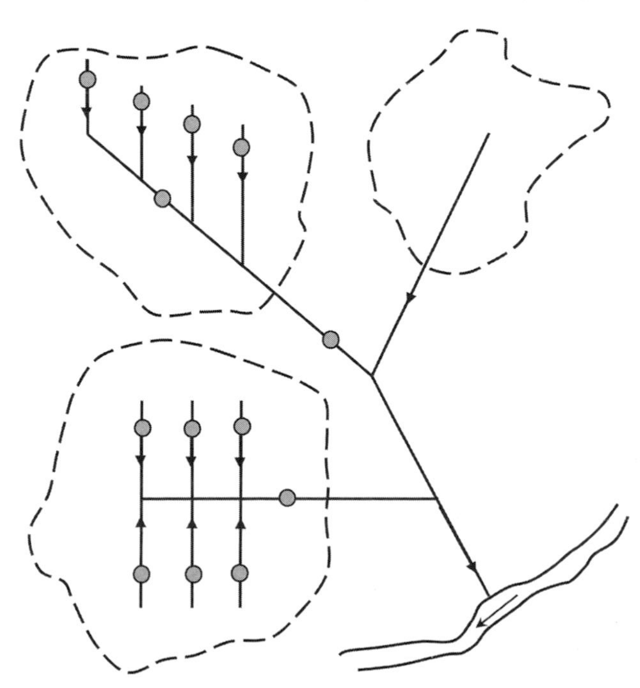

Figure 25.8 Natural or random system.

25.9.3 Grid-Iron System

The grid-iron system, as shown in Figure 25.10, is most common and is similar to the herringbone system, except that the laterals enter the main from one side. It is more economical than the herringbone, because the number of junctions and double drainage is reduced, and it is more suitable where the waterway is of considerable width and it is difficult for the drain line to cross the waterway due to erosion. In such cases the grid-iron system can be laid on each side of the waterway.

25.9.4 Cutoff or Interceptor

Interceptors, as shown in Figure 25.11, are generally located near the upper edge of the wet area and are used to intercept groundwater entering the field from the surrounding areas. These conditions are more common due to the shallow impermeable layer and seepage from the waterway.

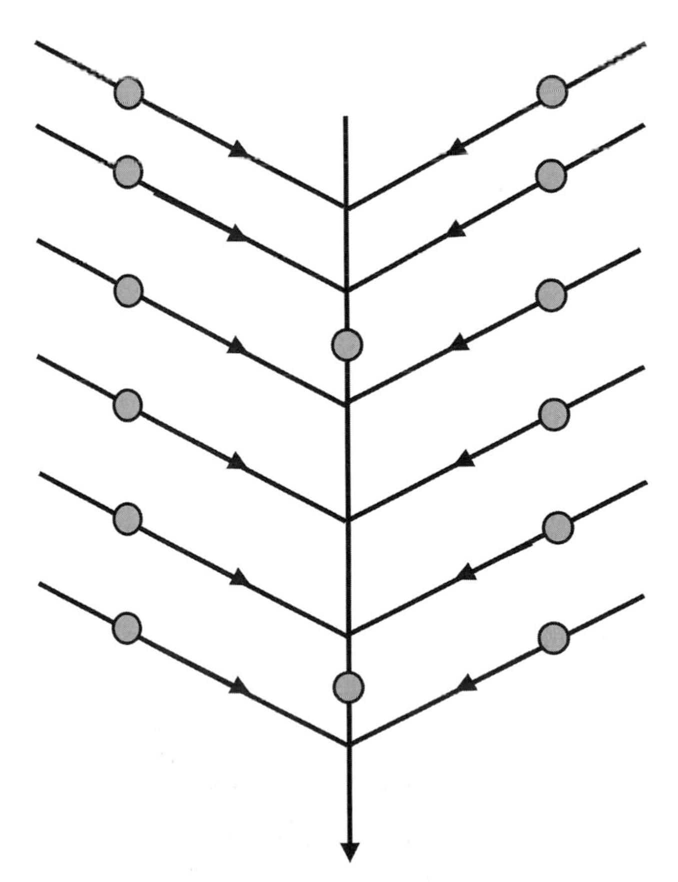

Figure 25.9 Herringbone system.

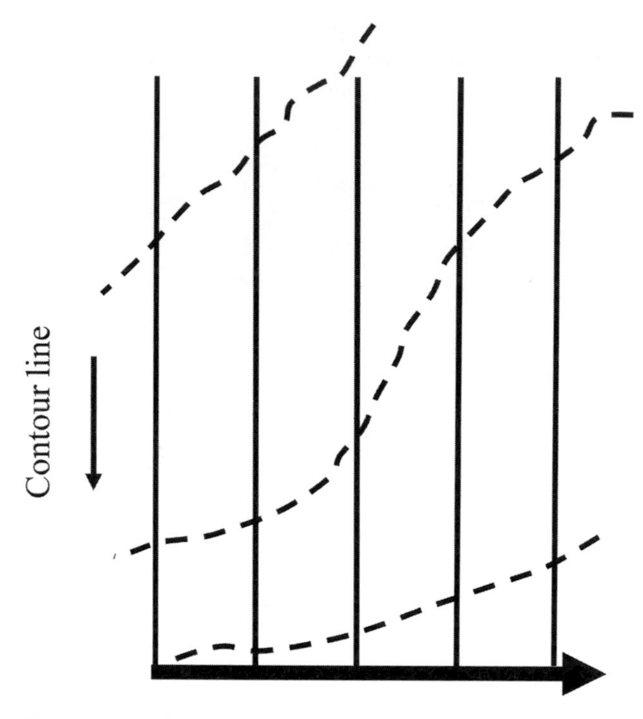

Figure 25.10 Grid-iron system.

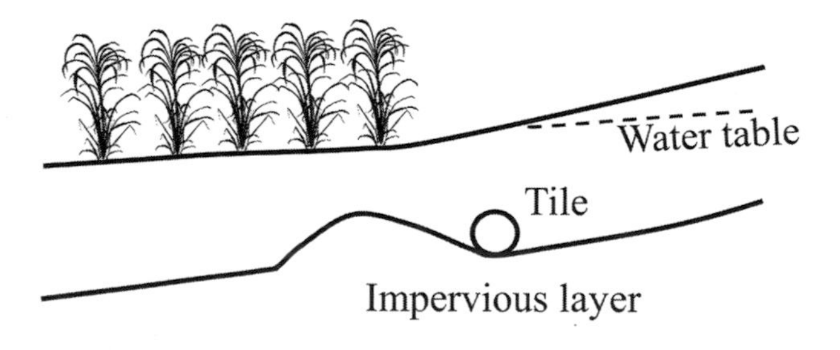

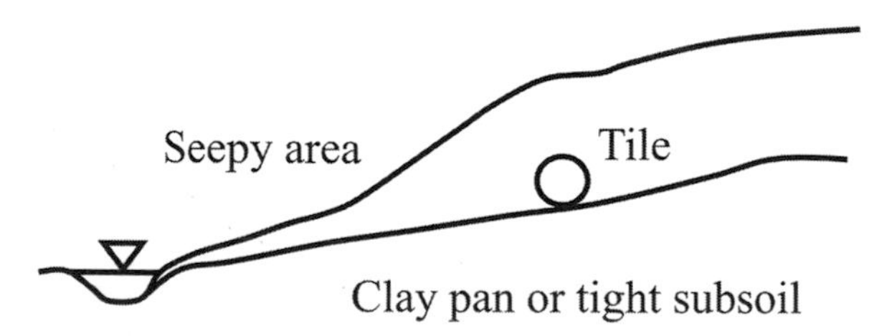

Figure 25.11 Cutoff or interceptor.

25.10 DESIGN OF SUBSURFACE DRAINAGE SYSTEMS

There are three issues that need to be addressed for design of subsurface systems: (1) depth of drains, (2) spacing of drains, and (3) size of drains. In general, depth and spacing of drains vary with the type of crop, soil hydraulic conductivity, crop and soil management practices, and the extent of drainage. The depth of drains, measured to the bottom of drains, is affected by the depth of the impermeable layer, spacing of laterals, elevation of the outlet, and the limitations of the trenching equipment. Spacing of drains is affected by climatic conditions, as shown in Table 25.5 and Table 25.6 (Schwab et al., 1993).

Table 25.5 *Subsurface drainage requirement for humid areas*

Crop	Depth and rate of drop of water table	Location
Mineral crops		
Field crops	30 cm in 24 hours and 50 cm in 48 hours	Illinois
Field crops	20 cm per day	Virginia
Field crops	Initial depth: 15 cm; 30 cm/day for next 15 minutes	Minnesota
Grass	20 cm/day for next 15 minutes	
Arable	Constant depth 50 cm or less Constant depth 0.9 to >1.3 m	The Netherlands
Organic soils		
Grass	Maximum depth = 50 cm	United States
Vegetables (shallow-rooted)	Maximum depth = 60 cm	United States
Field crops (deep-rooted)	Maximum depth = 80 cm	United States
Cereals/short grass/sugar beet	Optimum depth = 80–90 cm	England
Truck crop/grass/ sugarcane	Optimum depth = 30–60 cm	Florida

After Schwab et al., 1993. Copyright © 1993 by John Wiley & Sons, New York, United States.

25.10.1 Deep Percolation and Buildup

When computing drain spacing, deep percolation and buildup in the water table due to irrigation or rainfall should be estimated or measured. If measurements are not available, the amount of expected deep percolation from each irrigation must be estimated. The buildup can be determined by dividing the deep percolation by the specific yield of the material in the rootzone. The US Bureau of Reclamation (1984) has given approximate values of deep percolation from net surface irrigation input as a percentage for various soil textures and infiltration rates of upper root zone soils, as shown in Table 25.7 (US Bureau of Reclamation, 1984). The specific yield corresponding to various hydraulic conductivity can be found in Figure 25.12.

Table 25.6 *Average depth and spacing of tile drains*

Soil	Hydrologic Class	Hydraulic conductivity (cm/h)	Spacing (m)	Depth (cm)
Clay	Very slow	0.1	9–15	90–110
Clay loam	Slow	0.1–0.5	12–21	90–110
Loam	Moderately slow	0.5–2.0	18–30	110–120
Fine sandy loam	Moderate	2.0–6.5	30–37	120–140
Sandy loam	Moderately rapid	6.5–13	30–60	120–150
Peat and muck	Rapid	13–25	30–90	120–150
Irrigated soils	Variable	2.5–25	45–180	150–300

After Schwab et al., 1993.

Table 25.7 *Approximate deep percolation from surface irrigation (percentage of net input)*

By texture			
Texture	Percentage	Texture	Percentage
LS (loamy sand)	30	CL (Clay loam)	10
SL (Sandy loam)	26	SiCL (Silt clay loam)	6
L (loam)	22	SC (Sandy clay)	6
SiL (Silt loam)	18	C (Clay)	6
SCL (Sandy clay loam)	14		
By infiltration rate			
Infiltration rate (in./h [cm/h])	Deep percolation (%)	Infiltration rate (in./h [cm/h])	Deep percolation (%)
0.05 (0.13)	3	1.00 (2.54)	20
0.10 (0.25)	5	1.25 (3.18)	22
0.20 (0.51)	8	1.50 (3.81)	24
0.30 (0.76)	10	2.00 (5.08)	28
0.40 (1.02)	12	2.50 (6.35)	31
0.50 (1.27)	14	3.00 (7.62)	32
0.60 (1.52)	16	4.00 (10.16)	37
0.80 (2.03)	18		

After US Bureau of Reclamation, 1984.

Example 25.1: Compute the buildup of the water table per irrigation if the soil is sandy loam with an infiltration rate of 1.5 in./h and irrigation is 5 in. each time. About 5% of irrigation water runs off the irrigation field.

Solution: The net inflow of water into the soil per irrigation is (100 – 5) percent of 5 in. = 4.75 in. (12.07 cm). The deep percolation for the infiltration rate of 1.5 in./h from Table 25.7 is 24%, which amounts to 4.75 × 24% = 1.14 in. (2.90 cm). It is assumed that the hydraulic conductivity in the soil between the root zone and the drain depth is 1.5 in./h (0.91 m/day). The specific yield corresponding to this hydraulic conductivity value from Figure 25.12 is about 13%. The buildup of the water table for a given irrigation can be computed as the deep percolation divided by the specific yield:

$$\text{Buildup} = \frac{1.14}{0.13} = 8.77 \text{ in. } (22.3 \text{ cm})$$

Example 25.2: Compute the buildup of the water table if the total readily available moisture is 4.5 in. (allowable consumptive use between irrigations) and the rate of infiltration is 1.25 in./h with corresponding deep percolation of 25%.

Solution: The net inflow into the soil per irrigation 4.5 in./(100% – 25%) = 6 in. (15.2 cm). The deep percolation is 6 in. – 4.5 in. = 1.5 in. (3.8 cm). The buildup of the water table per irrigation is equal to the deep percolation divided by the specific yield in the root zone between the drain and the maximum allowable water table is 1.5 in./0.115 = 13.04 in. (33.1 cm). The value of 0.115 is obtained from Figure 25.12 for the hydraulic conductivity of 1.25 in./h (0.76 m/day).

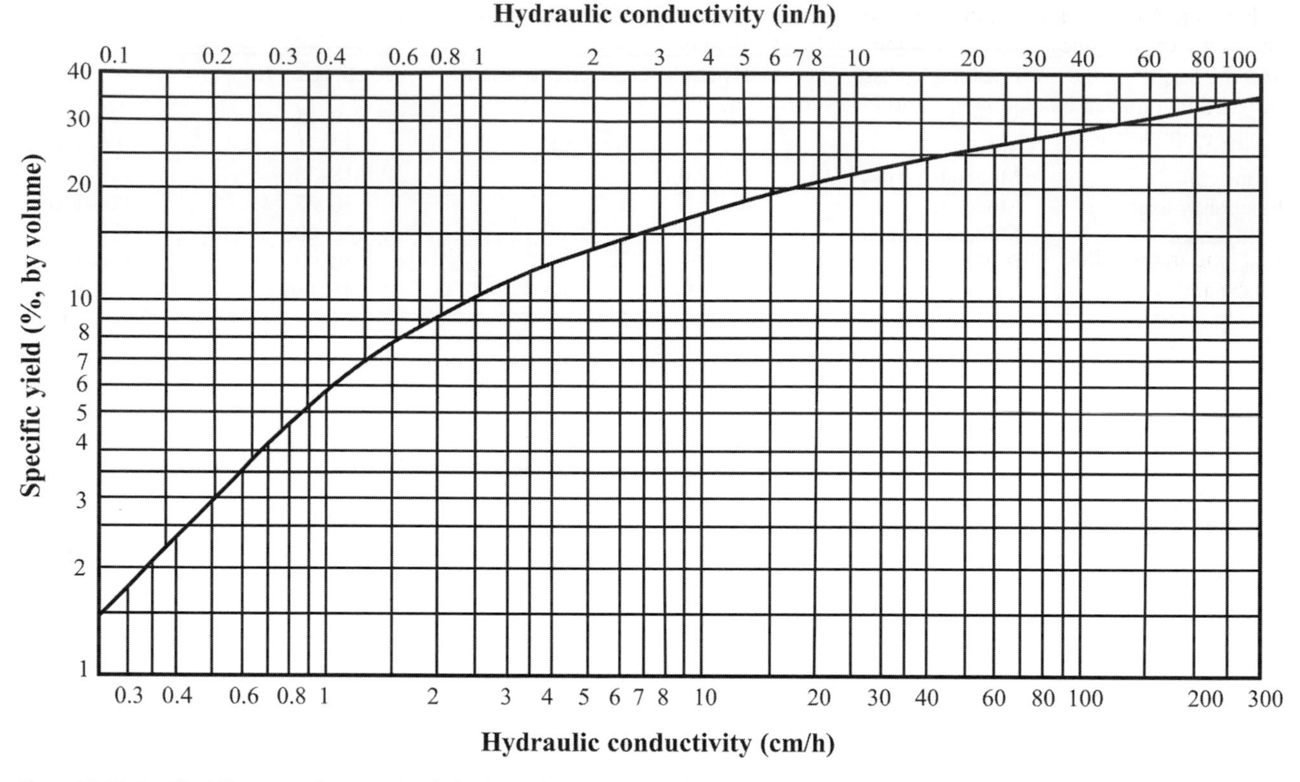

Figure 25.12 Specific yield corresponding to various hydraulic conductivity values. (After US Bureau of Reclamation, 1993.)

25.10.2 Computation of Drain Spacing under the Steady-State Condition

Drain spacing is computed under two conditions: (1) steady state and (2) unsteady state. For the steady-state condition, it is assumed that rainfall intensity or irrigation rate is constant, drainage is constant, the rainfall intensity is equal to the drainage rate, and the water table depth does not change with time. Further, it is assumed that soil is homogeneous, with a constant hydraulic conductivity K, drains are equally spaced at a distance S apart, Darcy's law is valid through the soil, the impermeable layer is a depth d from the drain, and the rate of replenishment of water (i, v, D_c) is constant. It is assumed that the Dupuit–Forchheimer (DF) assumptions hold – that is, the hydraulic gradient at any point is equal to the slope of the water table at that point and the flow to the drain is horizontal. For the purposes of analysis, the origin of the coordinate system is below the drain, as shown in Figure 25.13.

Consider a point P on the curve in Figure 25.13a. For constant vertical addition through point P, the volume of water that must be drained from the region denoted by $[(S/2) - x]$ can be expressed as a function of x as

$$q_x = \left[\frac{\frac{S}{2} - x}{\frac{S}{2}}\right]\frac{q}{2}, \tag{25.1}$$

where q_x is the flow rate across a vertical plain at any point $P(x, y)$, S is the spacing between drains (L), and q is the total flow rate into the drain from both sides (L/T) per unit length of soil.

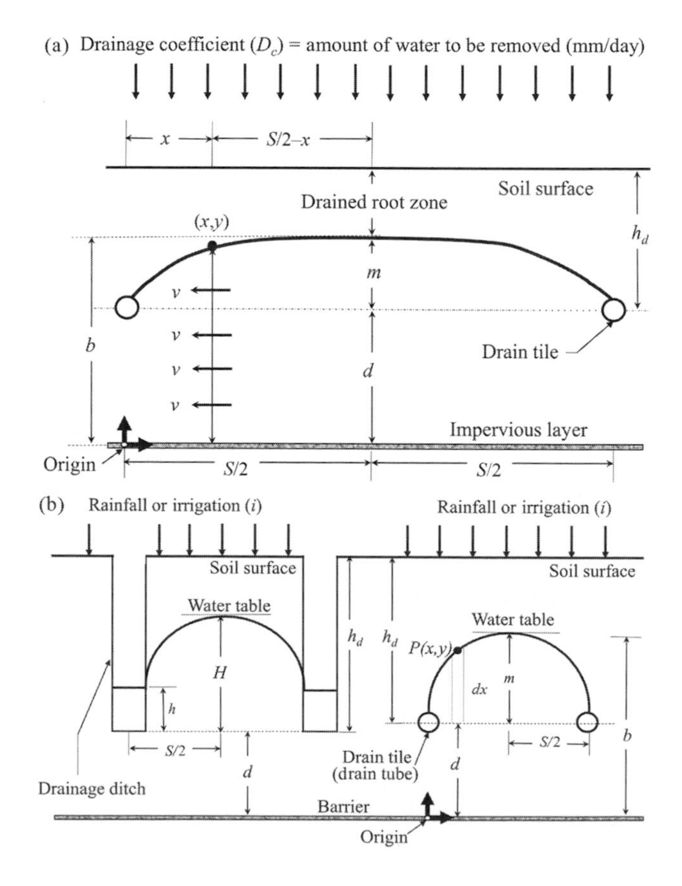

Figure 25.13 (a) Steady-state condition. (b) Steady-state subsurface drainage.

Now the flow per unit length of the drain line that passes across the vertical line at x is equal to the product of velocity and the height of the water table at point (x, y). This volumetric flow rate can be computed from Darcy's law and the DF assumptions as

$$q_x = Ky\frac{dy}{dx}. \tag{25.2}$$

Equating the two flow rates given by eqs (25.1) and (25.2), we obtain:

$$q_x = \left[\frac{\frac{S}{2}-x}{\frac{S}{2}}\right]\frac{q}{2} = Ky\frac{dy}{dx}. \tag{25.3}$$

Equation (25.3) can be expressed as

$$ydy = \frac{q}{KS}\left[\frac{S}{2}-x\right]dx. \tag{25.4}$$

Integrating eq. (25.4),

$$\int_d^b ydy = \frac{q}{KS}\int_0^{S/2}\left[\frac{S}{2}-x\right]dx. \tag{25.5}$$

Solution of eq. (25.4) follows:

$$S = \frac{4K\left(b^2-d^2\right)}{q}. \tag{25.6}$$

Recall that

$$q = Si = Sv = SD_c. \tag{25.7}$$

Substituting eq. (25.7) in eq. (25.6), one gets

$$S^2 = \frac{4K\left(b^2-d^2\right)}{D_c}. \tag{25.8a}$$

Equation (25.8a) is known as the ellipse equation.

One can also express eq. (25.8a) in terms of d and m as:

$$S^2 = \frac{4K\left(2dm+m^2\right)}{D_c}, \tag{25.8b}$$

where $b = d + m$, and hence $b^2 = d^2 + 2dm + m^2$. m is the maximum height of the water table above the drain (in m); d is the depth of the drain above the impermeable layer (in m).

Example 25.3: It is given that the depth to the center of the drain is 2 m and the minimum depth to the water table is 1.75 m (as shown in Figure 25.14). The impervious layer is a depth of 8 m and the hydraulic conductivity is 0.4 m/day. The excess rate of irrigation amounts to a drainage coefficient of 1.3 mm/day. Compute the drain spacing.

Solution: Given that the depth to the center of the drain is 2 m and the minimum depth to the water table is 1.75 m, the maximum height of the water table above the drain $m = 2\text{ m} - 1.75\text{ m} = 0.25\text{ m}$. The impervious layer is a depth of 8 m, the depth of drain above the impermeable layer $d = 8\text{ m} - 2\text{ m} = 6\text{ m}$, as shown in Figure 25.14.

Using eq. 25.8b, with $K = 0.4\text{ m/day}$, $D_c = 1.3\text{ mm/day} = 0.0013\text{ m/day}$, one gets

$$S^2 = \frac{4K\left(2dm+m^2\right)}{D_c} = \frac{4\times 0.4\left(2\times 6\times 0.25+0.25^2\right)}{0.0013},$$

$S = 61.4$ m.

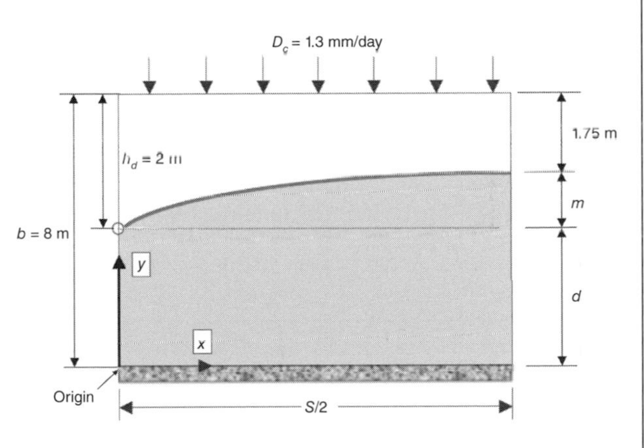

Figure 25.14 Illustration for Example 25.3.

25.10.2.1 Open Ditches

For open ditches,

$$S^2 = \frac{4K\left(H^2-h^2+2dH-2dh\right)}{v}, \tag{25.9}$$

where

$$v = i = D_c. \tag{25.10}$$

Equation (25.9) is called Hooghoudt's equation, where H is the maximum height of the water table above the drain (in m), h is the depth of water in the drain (m), d is the height of drain above the impermeable layer, as shown in the left panel in Figure 25.13b.

Assuming that the ditch is empty in the beginning (i.e., $h = 0$) and recalling $H = b - d$, eq. (25.9) can be written as

$$S^2 = \frac{4K\left[(b-d)^2+2d(b-d)\right]}{D_c}. \tag{25.11}$$

Equation (25.11) can be written as

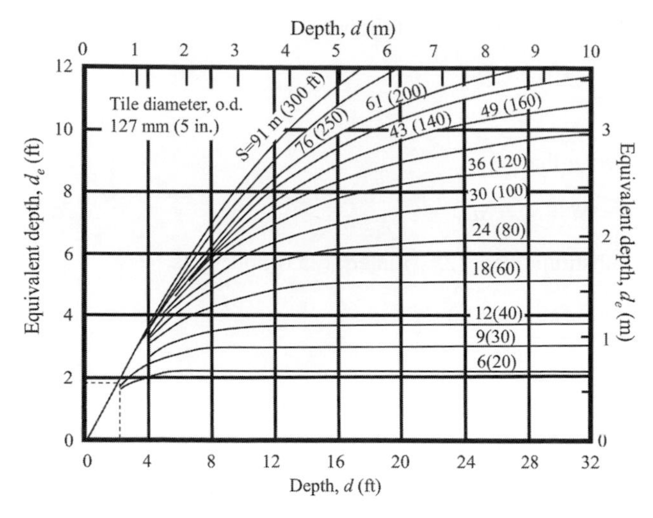

Figure 25.15 Equivalent depth for the water-conducting layer below the drain. (After van Schilfgaarde, 1963. Original data is from Hooghoudt, 1940. Reprinted with permission from ASCE.)

$$S^2 = \frac{4K\left[b^2 + d^2 - 2bd + 2bd - 2d^2\right]}{D_c}. \quad (25.12)$$

Equation (25.12) simplifies to

$$S^2 = \frac{4K\left[b^2 - d^2\right]}{D_c}. \quad (25.13)$$

In this derivation, the size of opening in the tile drain has negligible effect. The convergence of flow lines near the drain is ignored. The assumption of horizontal flow is acceptable if $S >>> d$. In order to correct this problem, d is replaced by equivalent depth d_e in eq. (25.13) (Figure 25.15) as

$$S^2 = \frac{4K\left[b^2 - d_e^{\,2}\right]}{D_c}. \quad (25.14)$$

25.10.2.2 Open Drains

$$S^2 = \frac{4}{D_c}\left(K_a H^2\right) + \frac{8}{D_c} d_e H, \quad K_a > K_b, \quad (25.15)$$

where K_a is the saturated hydraulic conductivity of the layer above the drain, and K_b is the saturated hydraulic conductivity of the layer below the drain.

25.10.2.3 Tile Drains

The assumption of horizontal flow is not valid near the drain, because streamlines have considerable curvature, especially when flow enters the drain. Equation (25.8a) needs to be corrected using a correction for the effective depth (d_e) of the distance between drains and impermeable layer. Therefore,

$$S^2 = \frac{4}{D_c}\left[K_a(b - d_e)^2\right] + \frac{8}{D_c}(b - d_e)d_e, \; K_a > K_b \quad (25.16a)$$

or

$$S^2 = \frac{4K[2d_e m + m^2]}{D_c}, \quad (25.16b)$$

where m is the maximum height of the water table above the drain (i.e., $m = b - d_e$).

Example 25.4: It is given that the maximum height above the center of the tile is 0.6 m, the drainage coefficient is 0.005 m/day, the soil hydraulic conductivity is 1.5 m/day, and the depth of the drain above the impermeable layer is 2.5 m. The drains are standard 5-in. nominal diameter tile drains. Determine the equivalent depth (d_e) and calculate the drain spacing.

Solution: Given the maximum height above the center of the tile is $m = 0.6$ m, the height of the drain above the impermeable layer $d = 2.5$ m, $K = 1.5\ \text{m/day}$, $D_c = 0.005\ \text{m/day}$.

Make an initial guess of $S = 61$ m and the equivalent depth (d_e) from Figure 25.15 is 1.9 m, with $d = 2.5$ m.

Using eq. 25.16b, the calculated drain spacing is:

$$S^2 = \frac{4K[2d_e m + m^2]}{D_c} = \frac{4 \times 1.5 \times \left[2 \times 1.9 \times 0.6 + 0.6^2\right]}{0.005}.$$

Then, $S = 56.3$ m.

The new d_e with $S = 56.3$ m and $d = 2.5$ m is 1.85 m. From Figure 25.15, the calculated drain spacing is:

$$S^2 = \frac{4K[2d_e m + m^2]}{D_c} = \frac{4 \times 1.5 \times \left[2 \times 1.85 \times 0.6 + 0.6^2\right]}{0.005}.$$

Then, $S = 55.6$ m.

The change in d_e is less than 5% (1.85–1.90 m), so the final drain spacing is 55.6 m.

25.10.3 Computation of Drain Spacing under the Unsteady-State Condition

The steady-state condition describes a constant relationship between the water table and the drainage. However, in reality, the groundwater recharge varies with time, and groundwater flow toward the drains is unsteady. When the water table drops because of drainage, only the water exceeding the field capacity – that is, between the saturated water content (porosity) and field capacity – will drain.

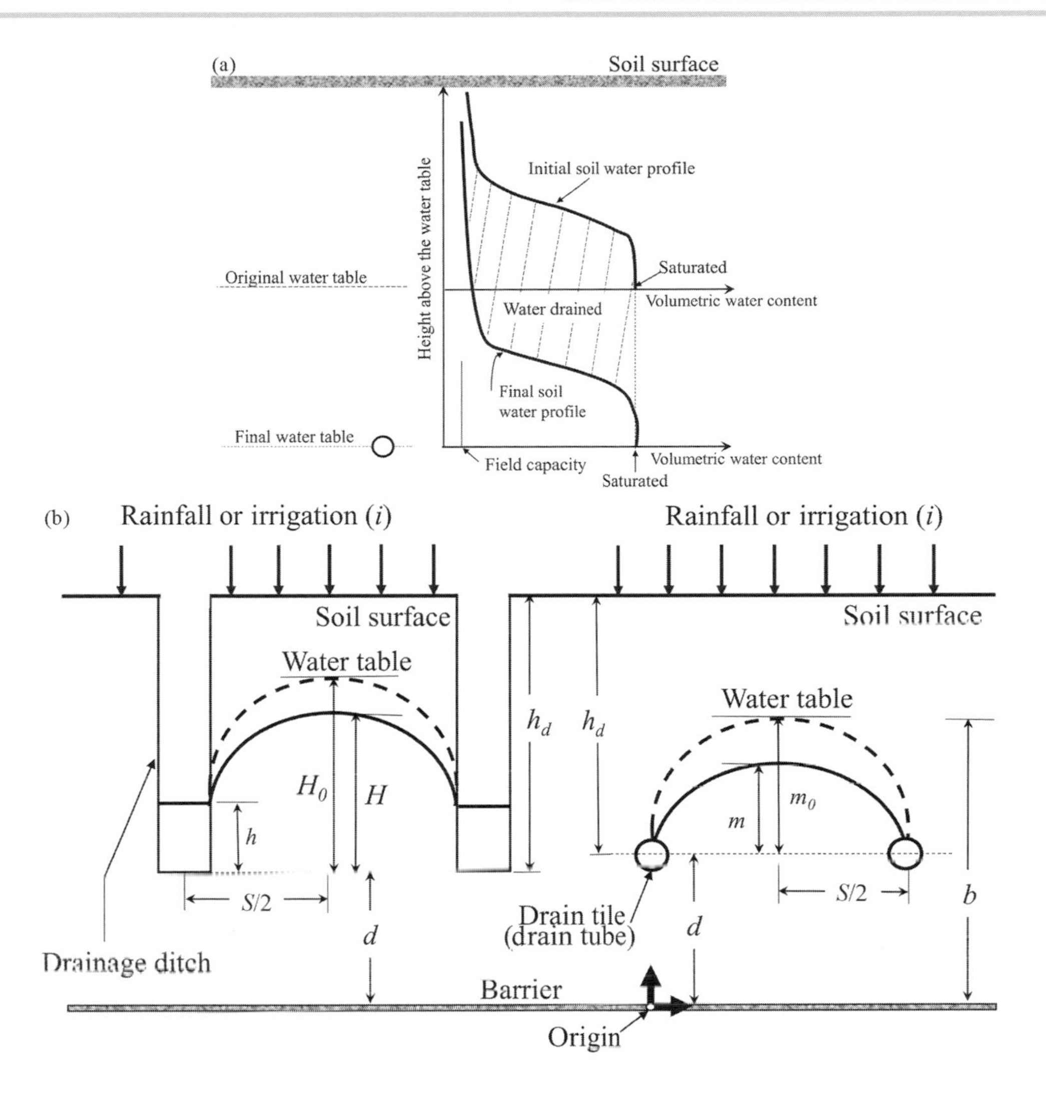

Figure 25.16 (a) Unsteady-state condition. (b) Unsteady-state subsurface drainage.

However, because of capillary fringe above the water table, the soil is wetter than field capacity, as shown in Figure 25.16. It is assumed that the DF assumption is valid – that is, the hydraulic gradient at any point is equal to the slope of the water table above that point and the flow of water to the drain is horizontal:

$$\frac{\partial^2 y}{\partial x^2} = \frac{f}{K\left(d + \frac{m_0}{2}\right)} \frac{\partial y}{\partial t}. \tag{25.17}$$

The solution of eq. (25.17) is

$$S = \pi \frac{Kt\left(d + \frac{m_0}{2}\right)}{f \ln\left(\frac{4}{\pi}\right)\left(\frac{m_0}{m}\right)}. \tag{25.18}$$

Equation (25.18) is known as the Glover equation. This equation ignores the convergence of flow lines near the drain and is based on the average depth of the water-conducting zone.

Considering the above limitations, eq. (25.18) was revised by van Schilfgaarde (1974) as

$$S^2 = \frac{9Ktd_e}{f\{\ln\left[m_0\left(2d_e + m\right)\right] - \ln\left[m(2d_e + m_0)\right]\}}, \tag{25.19}$$

where S is the drain spacing (m), K is the saturated hydraulic conductivity of soil (m/day), d_e is the equivalent depth (m), m_0 is the initial height of the water table above the center of the drain at the midpoint (m), m is the height of the water table above the center of the drain at the midpoint after time t (m), t is the time for the water table to fall from m_0 to m (days), and f is the drainable porosity (voids drained at 60 cm matric potential [fraction]), which is also called specific yield (S_y).

Example 25.5: The water table is falling and the data given are as follows: The hydraulic conductivity is 0.5 m/day, the drainable porosity $f = 0.25$, initial height of the water table m_0 is 1 m, the height of the table above the center of the drain m is 0.5 m, the drain depth to the center line is 1 m, the depth of the impervious layer below the drain d is 0.5 m, and the time for the water to drop from the soil surface to 0.25 m below is 1 day. Compute the drain spacing.

Solution: Given $K = 0.5\ \text{m/day}$, $f = 0.25$, the initial height of the water table above the center of the drain is $m_0 = 0.5$ m, the time for the water table to fall from $m_0 = 0.5$ m to $m = 0.5 - 0.25 = 0.25$ m is $t = 1$ day.

For $d = 0.5$ m, the equivalent depth (d_e) with different drain spacing is all close to 0.5 m, as shown in Figure 25.15. Then it is not necessary to assume drain spacing here.

Using eq. (25.19), the calculated drain spacing is:

$$S^2 = \frac{9Ktd_e}{f\{\ln[m_0(2d_e + m)] - \ln[m(2d_e + m)]\}}$$

$$= \frac{9 \times 0.5 \times 1 \times 0.5}{0.25\{\ln[0.5(2 \times 0.5 + 0.25)] - [\ln[0.25(2 \times 0.5 + 0.5)]\}}.$$

Then, $S = 4.2$ m.

25.10.4 Determination of Drain Spacing under the Steady-State Condition Using the Visser Nomograph

The Visser nomograph is illustrated in Figure 25.17. The step-by-step procedure to use the Visser nomograph is as follows.

Step 1:

1. Determine K and $v(D_c)$ and compute the ratio K/v.
2. Select the appropriate nomograph for the computed K/v ratio, as shown in Figure 25.17.
3. Locate points for K and $v(D_c)$ on the K and v lines.
4. Join these points by straight lines in order to locate the pivot point on the K/v line.

Step 2:

1. Determine the values of d and H (m).
2. Compute $d/H(d/m)$ and locate this point on the $d/H(d/m)$ line.
3. Locate a pivot point on the $S/H(S/m)$ line by joining points on the $d/H(d/m)$ line and the pivot point on the $K/v(K/D_c)$ line.

Step 3:

1. Locate the point for H (m) on the H (m) line.
2. Join the point on the H (m) line and the pivot point on the $S/H(S/m)$ line, and extend this line to the S line.
3. The point of intersection on the S line yields the drain spacing.

The nomograph for the determination of drainage spacing with K/v greater than or equal to 100 and less than 100 are shown in Figure 25.18.

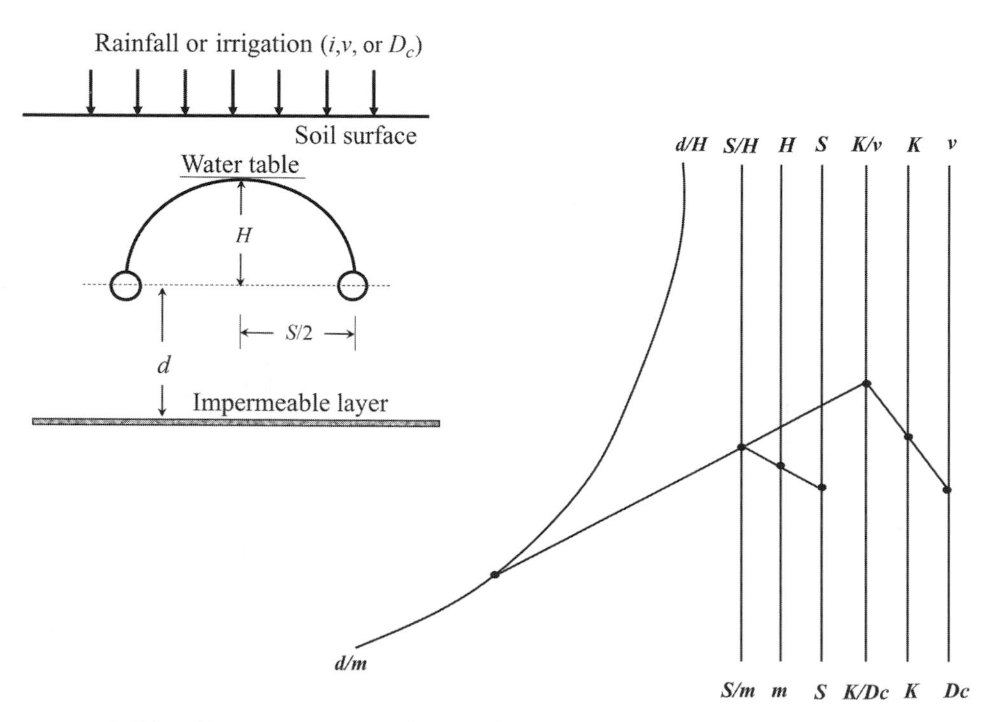

Figure 25.17 Visser nomograph. (Adapted from Visser, 1954. Copyright © 1954 by Elsevier.)

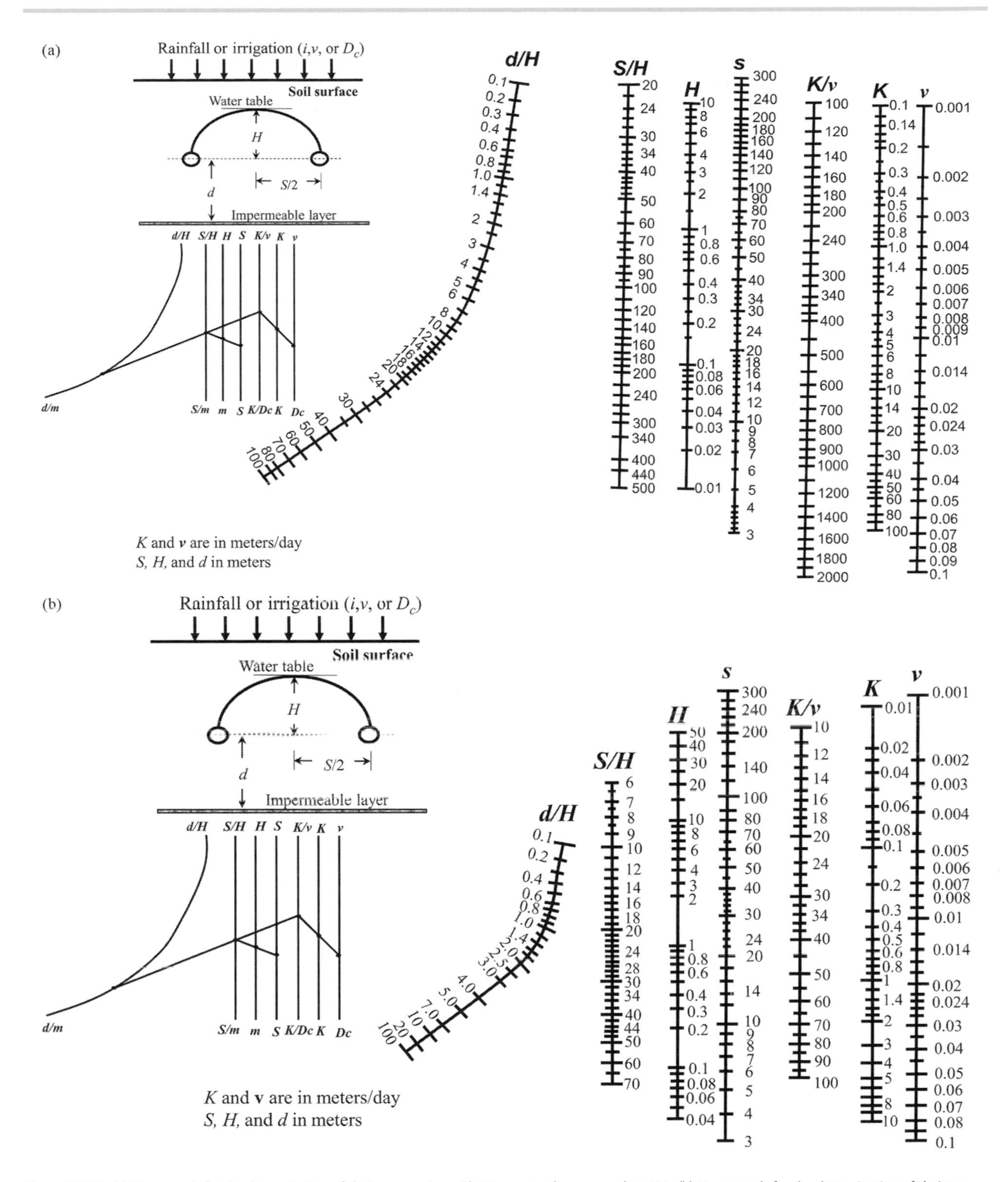

Figure 25.18 (a) Nomograph for the determination of drainage spacing with K/v greater than or equal to 100. (b) Nomograph for the determination of drainage spacing with K/v less than 100. (Adapted from Visser, 1954. Copyright © 1954 by Royal Netherlands Society for Agricultural Sciences (KLV)).

Example 25.6: Compute the drain spacing using the Visser nomograph, if the hydraulic conductivity is 0.5 m/day, the drainage coefficient is 1.5 mm/day, the depth of the drain is 2.0 m, the depth of the water table from the ground surface is 1.75 m, and the depth of the impervious layer from the ground surface is 6.0 m.

Solution:

1. $K = 0.5$ m/day, $D_c = 1.5$ mm/day $= 0.0015$ m/day and the ratio $K/v(D_c) = 0.5/0.0015 = 333$. Since the ratio is greater than 100, Figure 25.18a is used. Locate points for K and v (D_c) on the K and v lines and join these points by straight lines to locate the pivot point on the K/v line, as seen in the light gray dashed line in Figure 25.19.
2. The depth of drain is 2.0 m, the depth of the water table from the ground surface is 1.75 m, and the depth of the impervious layer from the ground surface is 6.0 m. The height of the drain above the impermeable layer $d = 6.0\text{ m} - 2.0\text{ m} = 4.0$ m, and $H = 2.0\text{ m} - 1.75\text{ m} = 0.25$ m. Then, $d/H = 4.0/0.25 = 16$. Locate a pivot point on the $S/H(S/m)$ line by joining points on the d/H (d/m) line and the pivot point on the K/v (K/D_c) line, as shown in the light gray line.
3. Locate the point for $H(m) = 0.25$ on the H (m) line. Join the point on the H (m) line and the pivot point on the $S/H(S/m)$ line and extend this line to the S line, as shown in the dark gray dashed line. The point of intersection on the S line yields the drain spacing $S = 34$ m.

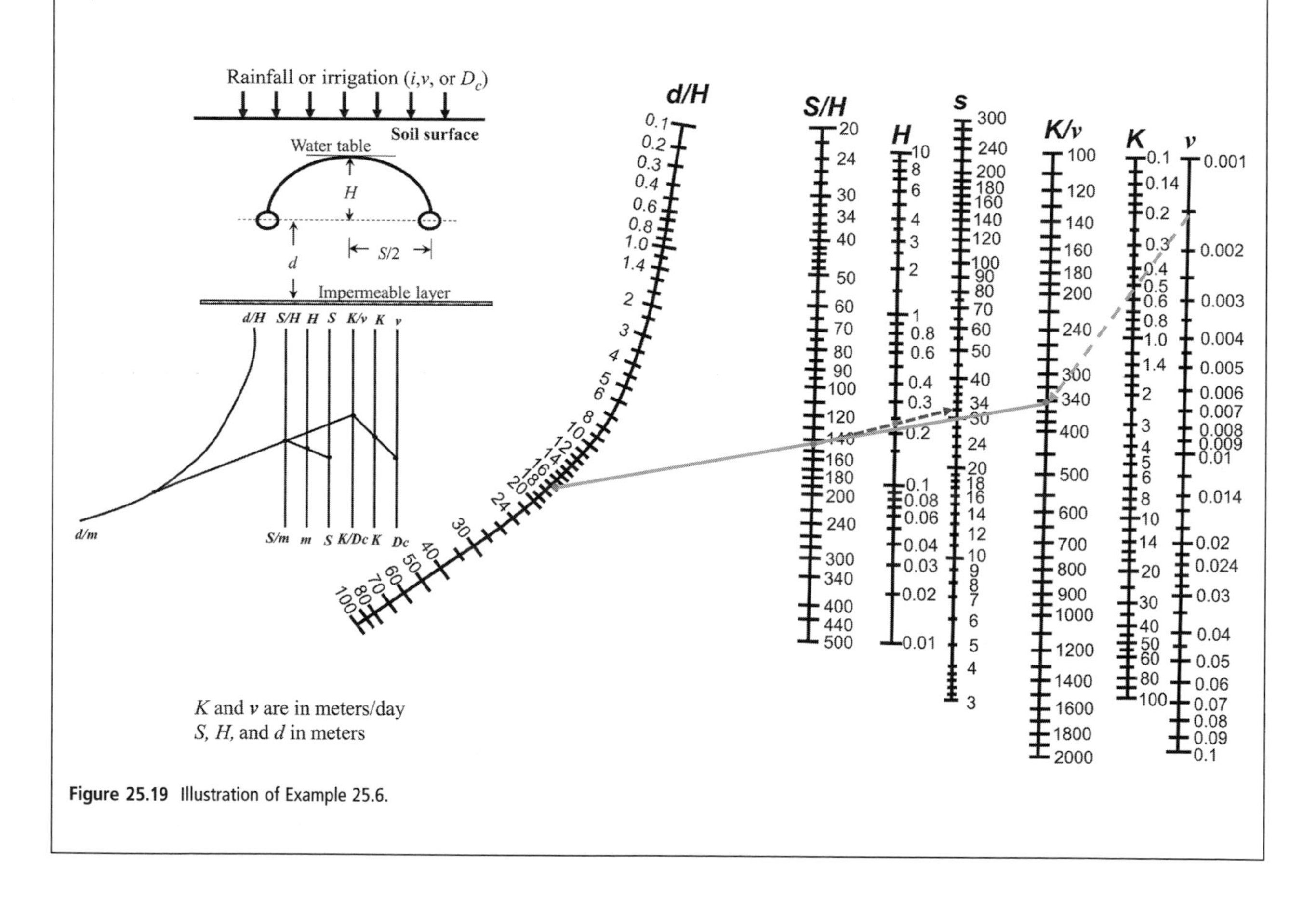

Figure 25.19 Illustration of Example 25.6.

25.10.5 Determination of Drain Spacing under the Steady-State Condition Using the Graphical Solution of Kirkham

The step-by-step procedure for the graphical solution of Kirkham, as shown in Figure 25.20, is as follows:

1. Determine the values of $K, i(D_c), H(m)$, and $h(d)$, and compute the value of L
2. Locate the point for L on the y-axis.
3. Determine the size of drain (r) and compute the $h/(2r)(d/(2r))$ ratio.
4. Select the $h/(2r)(d/2r)$ curve based on the value determined in Step 2.
5. Determine the value of S/h or S/d (x-axis) from L and $h/2r$ using the graph shown in Figure 25.20.
6. Determine the pipe drain spacing S from $S/h(S/d)$.

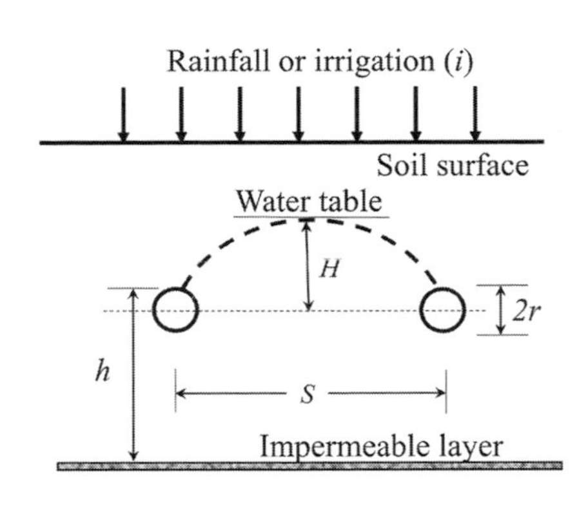

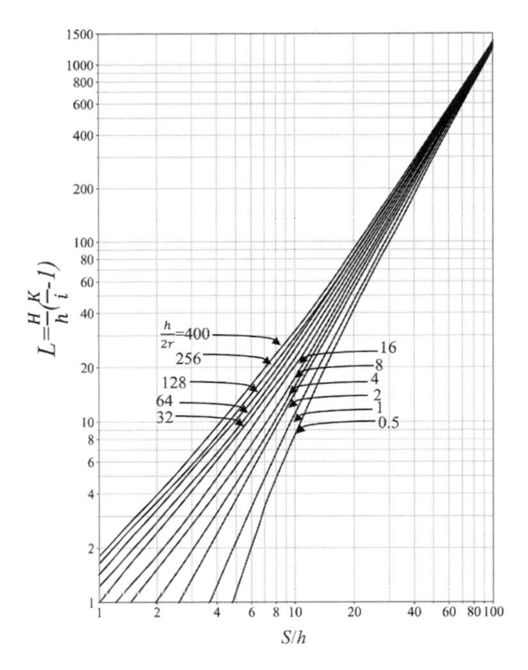

Figure 25.20 Graphical solution of Kirkham for the steady-state case. Adapted from Toksöz and Kirkham, 1961.)

Example 25.7: Determine the drain spacing if the maximum height of water above the pipe drain is 0.5 m, the saturated hydraulic conductivity of the soil is 1.35 m/day, the radius of the pipe drain is 0.06 m, the rate of rainfall is 0.002 m/day, and the distance from the impervious layer to the water table immediately over the pipe drain is 5.0 m.

Solution:

1. Given that $H = 0.5$ m, $K = 1.35$ m/day, the rate of rainfall $i = 0.002$ m/day, $h = 5.0$ m, then the value of L is:

$$L = \frac{H}{h}\left(\frac{K}{i} - 1\right) = \frac{0.5}{5.0}\left(\frac{1.35}{0.002} - 1\right) = 67.4.$$

2. Locate the point for L on the y axis.
3. Given the size of drain $r = 0.06$ m and $h/(2r) = 5.0/0.06 = 83.3$.
4. Select the $h/(2r)$ curve between 64 and 128.
5. Determine the value of S/h (x-axis) from L and $h/2r$. One gets $S/h = 17$.
6. Given $h = 5.0$ m, the pipe drain spacing is: $S = 17 \times 5.0 = 85$ m.

25.10.6 Determination of Drain Spacing under the Unsteady-State Condition Using the USBR Graphical Solution When the Drain Is Not at the Impermeable Layer

Figure 25.21 shows the drain is above the impermeable layer, and a graphical solution under this situation is shown in Figure 25.22.

Referring to Figure 25.22, the procedure consists of the following steps:

1. Determine the hydraulic conductivity (K), specific yield (S_y or f), height of the water table above the drain at the midpoint in the beginning (y_0 or m_0) and at time t (y or m) and the depth of the impermeable layer (barrier) from the center of the drain (d or h).
2. Assume S.
3. Compute d_e from S and d using the graph in Figure 25.15.
4. Compute the value of average flow depth D_a, $D_a = d_e + y_0/2$ or $(d_e + m_0/2)$.
5. Compute the ratio $y/y_0(m/m_0)$.
6. Read the value of $KD_at/(fS^2)$ or $KD_at/(S_yS^2)$ corresponding to $y/y_0(m/m_0)$.
7. Rearrange and compute the spacing S.
8. If the spacing assumed in Step 2 is not equal to the spacing computed in Step 7, go to Step 2.

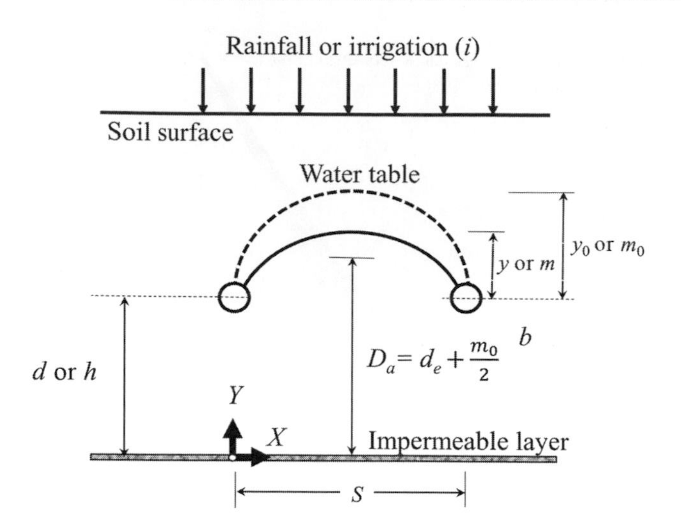

Figure 25.21 The drain not at the impermeable layer. (Adapted from US Bureau of Reclamation, 1984.)

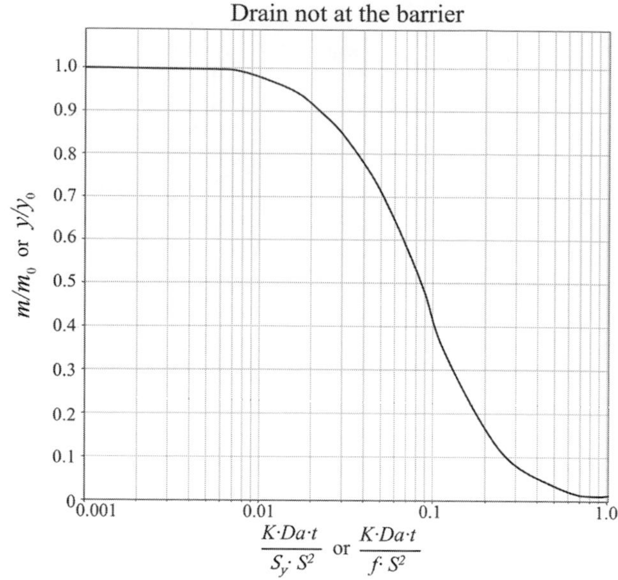

Figure 25.22 Graphical solution. (Adapted from US Bureau of Reclamation, 1984.)

Example 25.8: Determine the drain spacing for a falling water table if the hydraulic conductivity is 0.25 m/day, the value of f is 0.025, the drain depth to the center is 1.0 m, the value of m_0 is 1.0, the value of m is 0.75 m, the time for the water table to drop 0.25 m from the ground surface is 1 day, and the depth of the impermeable layer from the center of the drain is 0.5 m.

Solution:

1. Given the hydraulic conductivity $K = 0.25$ m/day, $f = 0.025$, $m_0 = 10$ m, $m = 0.75$ m, the time required from the depth of m_0 to m is 1 day, the depth of the impermeable layer (barrier) from the center of the drain $d = 0.5$ m.
2. Assume $S = 50$ m.
3. Using the graph in Figure 25.15, $d_e = 0.5$ m, with $S = 50$ m and $d = 0.5$ m.
4. $D_a = d_e + m_0/2 = 0.5 + (1.0/2) = 1$.
5. $m/m_0 = 0.75/1.0 = 0.75$.
6. From Figure 25.22, $KD_at/(fS^2) = 0.043$.
7. Rearrange and compute $S = 15.2$ m.
8. Since the computed S is not equal to the assumed one, use the new computed $S = 15.2$ m, and go to Step 2. Using the graph in Figure 25.15, $d_e = 0.5$ m. The computed $S = 15.2$ m, which is equal to the assumed S. Then, $S = 15.2$ m.

25.10.7 Determination of Drain Spacing under the Unsteady-State Condition Using the USBR Graphical Solution When the Drain Is at the Barrier

Referring to Figure 25.23, the procedure consists of the following steps:

1. Determine the hydraulic conductivity (K), specific yield (S_y or f), and the height of the water table from the bottom of the drain in the beginning (y_0 or m_0) and at time t (y or m).
2. Compute the ratio y/y_0 (m/m_0).
3. Read the value of $km_0t/(S_yS^2)$ or $km_0t/(fS^2)$ corresponding to known y/y_0 (m/m_0) from the graph, as shown in Figure 25.24.
4. Rearrange and compute spacing S.

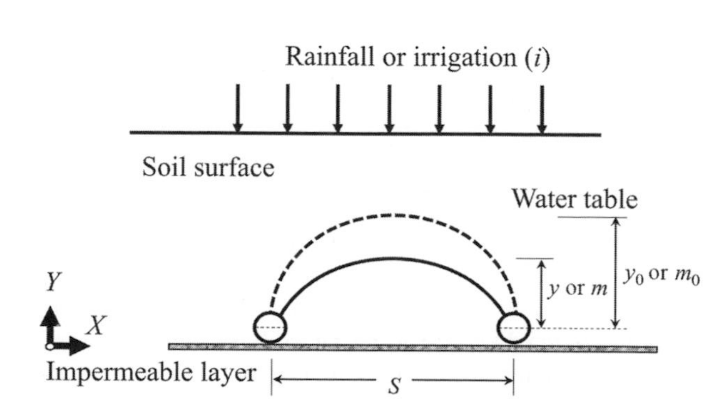

Figure 25.23 Drain at the barrier. (Adapted from US Bureau of Reclamation, 1984.)

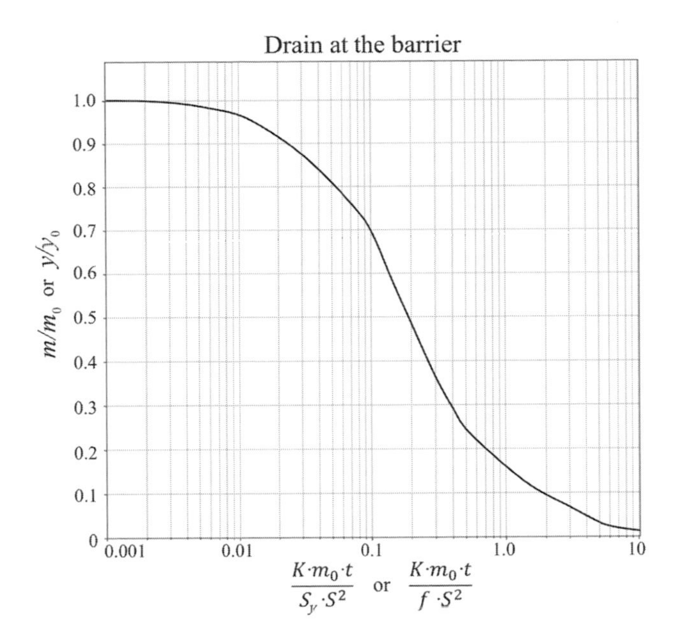

Figure 25.24 Graphical solution for drain at the barrier. (Adapted from US Bureau of Reclamation, 1984.)

Example 25.9: Determine the drain spacing for a falling water table if the hydraulic conductivity is 0.25 m/day, the value of f is 0.025, the drain depth to the center is 1.0 m, the value of m_0 is 1.0 m, the value of m is 0.75 m, the time for the water table to drop 0.25 m from the ground surface is 1 day, and the drain lies at the impermeable layer.

Solution:

1. Given the hydraulic conductivity $K = 0.25$ m/day, $f = 0.025$, $m_0 = 1.0$ m, $m = 0.75$ m, the time required from the depth of m_0 to m is 1 day.
2. $m/m_0 = 0.75/1.0 = 0.75$
3. Read the value of $km_0t(fS^2) = 0.07$ corresponding to known $m/m_0 = 0.75$ from Figure 25.24.
4. The drain spacing $S = 12$ m.

25.10.8 Selection of Drain Size under the Steady-State Condition

The subsurface drain size depends on the grade of the tile line, internal pipe roughness, and drainage characteristics (the area of the contributing watershed area rather than the drained pipe area). The size should be such that it can accommodate drainage flow for a known drainage rate. The tile drain size can be computed under three conditions: steady state, unsteady state, and surface ponding conditions.

The water table is assumed to be a constant depth below the ground. The drainage coefficient approach is used. The flow is characterized as gravity flow, for which Manning's equation can be used. From the continuity equation,

$$Q = VA, \tag{25.20}$$

where Q is the flow rate in a pipe in m^3/s, V is the average velocity of flow in the pipe in m/s, and A is the cross-sectional area in m^2.

From Manning's equation,

$$V = \frac{1}{n} R^{2/3} S_0{}^{1/2}, \tag{25.21}$$

in which V is the average velocity (m/s), R is the hydraulic radius (m), s_0 is the slope of the energy grade line (for uniform flow it can be taken as the slope of the drain), and n is Manning's roughness coefficient ($n = 0.0115$ for concrete tile, $n = 0.0115$ for clay tiles, and $n = 0.015$ for plastic tubing).

The drainage rate for steady-state conditions can be expressed as

$$Q = D_c A_f, \tag{25.22}$$

where Q is the drainage rate from a field (m^3/day), D_c is the drainage coefficient (m/day), and A_f is the area of the field or area covered by one lateral (m^2).

Assuming the tile is running full, the flow cross-sectional area can be expressed as

$$A = \frac{\pi}{4} D^2, \tag{25.23}$$

where D is the inside diameter of the drainage tubing or tile (m). From eqs (25.20)–(25.23) we obtain

$$D_c A_f = \left[\frac{1}{n}\left(\frac{D}{4}\right)^{2/3} S_0{}^{1/2}\right]\left[\frac{\pi}{4}D^2\right][3600 \times 24]. \quad (25.24)$$

In commonly used units, eq. (25.24) simplifies to

$$D = 51.7\left(D_c A_f n\right)^{0.375} S_0{}^{-0.1875}, \quad (25.25)$$

in which D is in mm, D_c is in mm/day, A_f is in ha, and S_0 is in m/m.

For concrete or clay tiles, eq. (25.25) can be simplified as

$$D = 9.7\left(D_c A_f\right)^{0.375} S_0{}^{-0.1875}. \quad (25.26)$$

For plastic tubing, eq. (25.25) can be simplified as

$$D = 10.7\left(D_c A_f\right)^{0.375} S_0{}^{-0.1875}. \quad (25.27)$$

It should be noted that if surface water is allowed to enter the subsurface drains then an additional appropriate drainage coefficient should be selected for surface inlets.

Example 25.10: What will be the diameter of corrugated plastic tubing and flow rate if the slope is 0.24%, the drainage area is 10 ha, and the drainage coefficient is 10 mm/day?

Solution: Given $S_0 = 0.0024$, $A_f = 10\text{ ha} = 10{,}000\text{ m}^2$, $D_c = 10\text{ mm/day} = 0.01\text{ m/day}$, the diameter of the corrugated plastic tubing is computed using eq. (25.27):

$$D = 10.7\left(D_c A_f\right)^{0.375} S_0{}^{-0.1875}$$
$$= 10.7(10 \times 10)^{0.375} 0.0024^{-0.1875} = 186.5\text{ mm}.$$

The flow rate is computed using eq. (25.22):

$$Q = D_c A_f = 0.01\frac{\text{m}}{\text{day}} \times 10{,}000\text{ m}^2 = 100\text{ m}^3/\text{day}.$$

25.10.9 Selection of Drain Size under the Unsteady-State Condition

The tile drain size for a falling water table can be determined by computing the drain discharge using the USBR equations and Manning's equation for flow in pipes for two cases: drain on the barrier and drain not on the barrier.

25.10.9.1 Drain on the Barrier

The drain discharge can be computed as

$$Q = \frac{4Km_0{}^2 L}{S}, \quad (25.28)$$

where Q is the drain discharge (m^3/day), m_0 is the height of the water table above the drain at the midpoint at time $t = 0$ (m), K is the saturated hydraulic conductivity of the soil (m/day), L is the length of the tile drain (m), and S is the drain spacing.

Using Manning's equation, the drain discharge can be estimated as

$$Q = VA \times (3600 \times 24). \quad (25.29)$$

Therefore,

$$D_c A_f = \left[\frac{1}{n}\left(\frac{D}{4}\right)^{2/3} S_0{}^{1/2}\right]\left[\frac{\pi}{4}D^2\right][3600 \times 24]. \quad (25.30)$$

Simplification of eq. (25.30) yields

$$D = 1551(Qn)^{0.375} S_0{}^{-0.1875}, \quad (25.31a)$$

$$D = 122.6(Qn)^{0.375} S_0{}^{-0.1875}, \quad (25.31b)$$

where D is the inside diameter of the drainage tubing or tile (mm), Q is the drain discharge in eq. (25.31a) (m^3/s) and in eq. (25.31b) (m^3/day). It may be noted that for surface inlets an appropriate drainage rate should be used to estimate drainage discharge.

Example 25.11: Determine the diameter of concrete tile and flow rate for a falling water table if the hydraulic conductivity is 0.25 m/day, the length of the tile drain is 100 m, and the drain spacing is 20 m. The drain slope is 0.22%. The height of the water table above the drain at the midpoint at time $t = 0$ is 1 m and the drain lies at the impermeable layer.

Solution: Given the saturated hydraulic conductivity of the soil $K = 0.25$ m/day, the initial height of water table above the drain $m_0 = 1$ m, the length of the tile drains $L = 100$ m, the drain spacing $S = 20$ m, the drain discharge is computed using eq. (25.28) as:

$$Q = \frac{4Km_0{}^2 L}{S} = \frac{4(0.25)(1)^2(100)}{20} = 5\text{ m}^3/\text{day}.$$

Given Manning's roughness coefficient $n = 0.0115$ for concrete tile, the drain slope $S_0 = 0.0022$, the diameter of the concrete tile is computed using eq. (25.31b):

$$D = 122.6(Qn)^{0.375} S_0{}^{-0.1875}$$
$$= 122.6(5 \times 0.0115)^{0.375}(0.0022)^{-0.1875} = 132.3\text{ mm}.$$

25.10.9.2 Drain Not on the Barrier

The drain discharge can be computed as

$$Q = \frac{2\pi K m D_a L}{S}, \tag{25.32}$$

where Q is the drain discharge (m^3/day), m is the height of the water table above the drain at the midpoint at time t, K is the saturated hydraulic conductivity of the soil (m/day), D_a is the average depth of flow $(d_e + m_0/2)$ (m), d_e is the equivalent depth from the drain to the barrier, and m_0 is the height of the water table above the drain at the midpoint at time $t = 0$. L is the length of the tile drain and S is the drain spacing. The inside diameter of the drainage tubing or tile can be computed with eq. (25.31) using the same method as for the case of the drain on the barrier.

Example 25.12: Determine the diameter of concrete tile and flow rate for a falling water table if the hydraulic conductivity is 0.25 m/day, the length of the tile drain is 100 m, and the drain spacing is 20 m. The drain slope is 0.22%. The height of the water table above the drain at the midpoint at time $t = 0$ is $m_0 = 1$ m and at time $t = 1$ day is $m = 0.75$ m. The drain is 3 m above the impermeable layer.

Solution: The drain lies above the impermeable layer at 3 m; the equivalent depth is $d_e = 1.6$ m, with the drain spacing at 20 m using Figure 25.15.

The average depth of flow $D_a = d_e + m_0/2 = 1.6$ m $+ 1$ m$/2 = 2.1$ m.

Given the saturated hydraulic conductivity of the soil $K = 0.25$ m/day, the height of the water table above the drain at time t is $m = 0.75$ m, the length of the tile drains $L = 100$ m, the drain spacing $S = 20$ m, the drain discharge is computed using eq. (25.28) as:

$$Q = \frac{2\pi K m D_a L}{S} = \frac{2\pi(0.25)(0.75)(2.1)(100)}{20}$$
$$= 12.37 \text{ m}^3/\text{day}.$$

Given Manning's roughness coefficient $n = 0.0115$ for concrete tile, the drain slope $S_0 = 0.0022$, the diameter of the concrete tile is computed using eq. (25.31b):

$$D = 122.6(Qn)^{0.375} S_0{}^{-0.1875}$$
$$= 122.6(12.37 \times 0.0115)^{0.375}(0.0022)^{-0.1875}$$
$$= 185.9 \text{ mm}.$$

25.10.9.3 Surface Ponding Condition

Under this condition, the drain discharge can be computed as

$$Q = \frac{2\pi K(p + h_d - h_w)L}{\ln\frac{2h_d}{r}}, \tag{25.33}$$

where Q is the drain discharge (L^3/T), K is the saturated hydraulic conductivity of the soil (L/T), p is the depth of ponding (L), h_d is the depth from the soil surface to the center of the drain (L), h_w is the level of water in the drain (L), r is the outside radius of the drain (effective radius of the gravel envelope; L), and L is the length of the tile drain (m).

Assuming the tile is running full, then

$$Q = \frac{2\pi K(p + h_d - 2r)L}{\ln\frac{2h_d}{r}}. \tag{25.34}$$

The drain size is estimated by Manning's equation, as described under unsteady-state conditions. The effect of inlets can be included by selecting an appropriate drainage coefficient for surface water.

Example 25.13: Determine the flow rate and diameter of concrete tile with surface ponding of 0.05 m, the saturated hydraulic conductivity is 0.2 m/day, the depth from the soil surface to the center of the drain is 0.5 m, the outside radius of the drain is 0.25 m, assuming the tile is running full. The drain slope is 0.002 and the length of the drain is 120 m.

Solution: Given the saturated hydraulic conductivity of the soil $K = 0.2$ m/day, surface ponding $p = 0.05$ m, the depth from the soil surface to the center of the drain $h_d = 0.5$ m, the outside radius of the drain is $r = 0.25$ m, and the tile is running full, the drain discharge is computed using eq. (25.34) as:

$$Q = \frac{2\pi K(p + h_d - 2r)L}{\ln\frac{2h_d}{r}}$$
$$= \frac{2\pi(0.2)(0.05 + 0.5 - 2 \times 0.25)(120)}{\ln\left(2 \times \frac{0.5}{0.25}\right)}$$
$$= 5.44 \text{ m}^3/\text{day}.$$

Given Manning's roughness coefficient $n = 0.0115$ for concrete tile, the drain slope $S_0 = 0.002$, the diameter of the concrete tile is computed using eq. (25.31b):

$$D = 122.6(Qn)^{0.375} S_0{}^{-0.1875}$$
$$= 122.6(5.44 \times 0.0115)^{0.375}(0.002)^{-0.1875}$$
$$= 139.0 \text{ mm}.$$

25.10.9.4 Entry of Water into the Tile Drain

Drain inflow is directly proportional to the hydraulic conductivity of the soil. The tile diameter is not a major factor in determining tile inflow in the usual range of lateral size. If the depth of the impermeable layer is not encountered, the inflow varies directly with the depth of tile. About 60% of the inflow to the tile line enters from the surface from 60 cm on either side of the drain. Tile should be laid directly under the area where water accumulates. About 50% of the equipotential lines lie within two times the diameter of the drain. This indicates that the tile effectiveness can be increased by increasing the saturated hydraulic conductivity of the soil surrounding the tile. This can be achieved by providing gravel pack or a gravel envelope. If the soil particle sizes are 0.5–1 mm, an envelope should be used to prevent soil entry into the drain. Otherwise a gravel envelope can be used to increase the hydraulic effectiveness of the drains.

25.10.10 Drain Opening

Drain opening depends on the type of tiles to be used.

25.10.10.1 Clay Tiles

For 0.3-m long tiles the practice is to have an opening of 2–6 mm. An opening equal to about 1% of the outside diameter of the tile is possible without serious restrictions to drainage. Doubling the crack (opening) will increase the inflow by about 10%.

25.10.10.2 Corrugated Plastic Tubing

For 100 mm tubing the slot size is 88 (1.66 mm times 25 mm) slots per meter length. Doubling the slot area or number of slots will increase the inflow rate by about 20%. The shape of the slot is not important if they are uniform.

25.10.11 Filter Design Criteria

In the subsurface drainage system gravel envelopes can be used to improve hydraulic effectiveness of the tile drain and reduce sediment entry into the tile line. The data required to select a suitable envelope material include size distribution of the base soil and size distribution of the gravel. Information required include D_{15} and D_{85} of the base soil and D_{85} of the finer material and maximum size of the opening in the drain tube or tile. Selection of filter material should satisfy the following conditions (Schwab et al., 1993):

$$\frac{D_{15}(\text{filter})}{D_{15}(\text{base})} = 5 \text{ to } 40,$$

provided the filter does not contain more than 5% finer than 0.074 mm; and

$$\frac{D_{15}(\text{filter})}{D_{85}(\text{base})} = 5 \text{ or less},$$

$$\frac{D_{15}(\text{filter})}{\text{Maximum size of opening in the tube or tile}} = 5 \text{ or more}.$$

25.10.12 Tile Line Grade

Tile grade refers to the slope of the tile line. The tile line is installed on stable ground. Grades are limiting where pipes (tiles) are designed for near full capacity, and pipes are embedded in unstable ground (Table 25.8). Under such conditions, special care is required to provide joints that fit snugly against each other. Desirable minimum working grade is 0.2%. Where sufficient drop is not available, grade may be reduced following Schwab et al. (1993).

Under certain conditions, higher grade may be required to prevent sedimentation. As far as possible minimum velocity at full flow should be 0.5 m/s. The grade should not be less than the minimum unless recommended. Special precautions are required during the construction period to establish the specified grades.

Table 25.8 *Minimum pipe and tile line grade*

	Fine sand or silt not present[a]		Fine sand or silt enters the drain[b]	
	Pipe	Tile	Pipe	Tile
Tile size (inside diameter; mm)	Grade (%)			
75	0.10	0.08	0.81	0.60
100	0.07	0.05	0.55	0.41
125	0.05	0.04	0.41	0.30
150	0.04	0.03	0.32	0.24

[a] Minimum cleaning velocity = 0.15 m/s; [b] minimum cleaning velocity = 0.42 m/s.
After Schwab et al., 1993.

25.10.13 Tile Drain Material

The characteristics of a good drain tile include the following: The tile material should be resistant to weathering and deterioration. The tile should have sufficient strength to support static and impact loads under conditions for which they are designed. The tile material should have low water absorption characteristics (i.e., it should have high density). The tile material should be resistant to freezing and thawing. The tile should be relatively free from defects (i.e., no cracks and ragged ends). The tile should have uniform shape and wall thickness.

25.10.14 Commonly Used Materials for Drainage Tiles

Commonly used drainage tiles are clay, concrete, or corrugated plastic tubing.

25.10.14.1 Clay Tiles

Clay tiles are available in standard, extra, and heavy-duty quality. They can be damaged by frost action where the tile depth is less than 0.5 m and are not recommended for alternate freezing and thawing conditions. They are generally not affected by soil chemicals (acid or alkaline). If tile depth is less than 0.7 m extra quality tile should be used. Well-burned tiles should be used.

25.10.14.2 Concrete Tiles

Concrete tiles are available in standard and extra quality. They are generally resistant to freezing and thawing conditions, but are affected by soil chemicals (acid or alkaline). Properly cured tiles (28 days at 22 °C or 30 hours at 60 °C or steam cure) should be used.

25.10.14.3 Corrugated Plastic Tubing

Plastic tubing is light in weight and easy to handle, and is available in standard and extra quality. It is generally resistant to soil chemicals. Tubing of uniform color and density and free from all visible defects should be used.

$$\text{Stiffness} = 1380 \text{ kPa at } 6.5\% \text{ deflection}$$
$$= 1055 \text{ kPa at } 10\% \text{ deflection}$$

It should be noted that all tile material should meet the American Society for Testing and Materials (ASTM) standards.

25.10.15 Accessories for Subsurface Drainage Systems

Schwab et al. (1993) have described accessories needed for subsurface drainage systems.

25.10.15.1 Inlets

Surface or open inlets are shown in Figure 25.25 and Figure 25.26. Blind inlets or French inlets are shown in Figure 25.27.

Surface or open inlets: Surface or open inlets are used for the removal of surface water from potholes, road ditches, other depressions, and farmsteads. Whenever practicable, surface water should be removed by surface rather than surface inlets with subsurface systems. Surface inlets should be used where a shallow surface drain is not practical and there is a need for rapid removal of water from the soil surface. Surface inlets should be properly located and constructed. They should be located at the lowest point along the fence rows or on land that is under permanent vegetation. In cultivated fields the area immediately around the surface inlets should be grassed. Generally surface inlets are

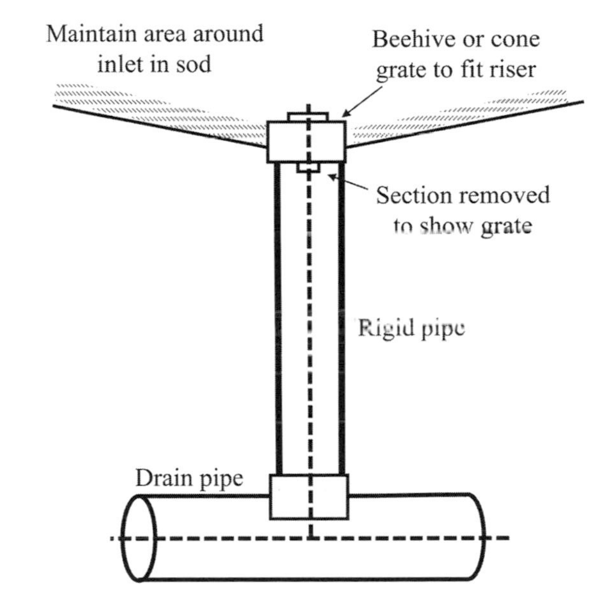

Figure 25.25 Surface inlet.

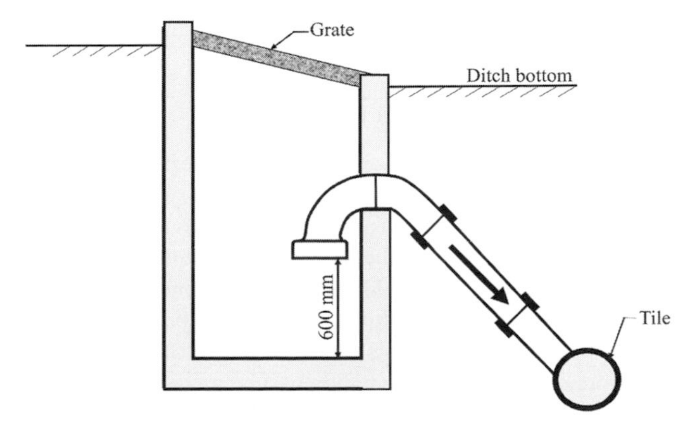

Figure 25.26 Ditch inlet.

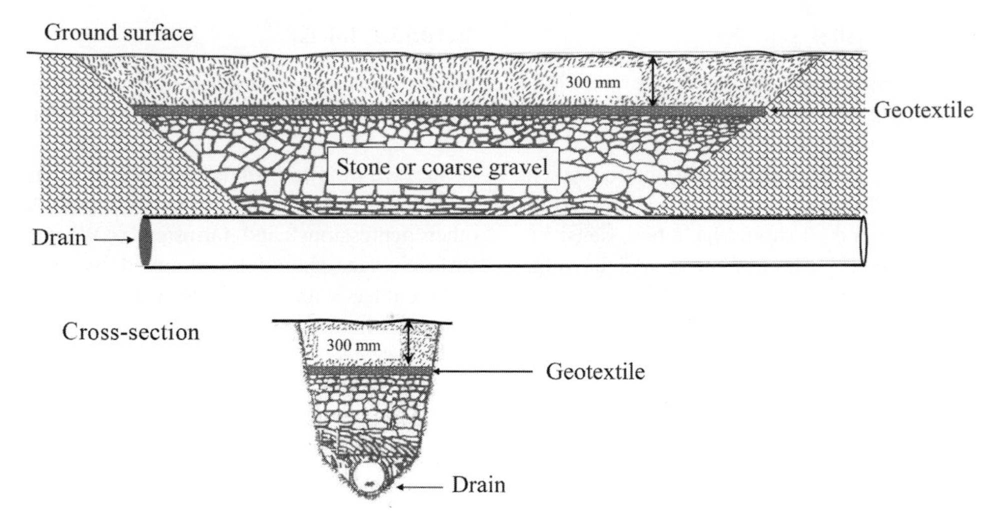

Figure 25.27 Blind inlets or French inlets.

constructed from bell and spigot tiles, but galvanized steel can also be used. Manholes constructed with bricks or concrete are also satisfactory. On the top end of the riser, a beehive cover or other suitable grate is necessary to prevent trash from entering the tile line. A surface inlet is used if the water is relatively free from sediment. Where water contains sediments, either sediment traps should be provided or blind inlets should be used. The cost of maintenance is low but they interfere with farm operations.

Blind inlet or French drain: Blind inlets are used where surface ponding occurs, but the quantity of surface water to be removed is small and the level of sediments in the surface water is too high to permit the installation of surface drains. Blind inlets are constructed by backfilling the tile trench with various gradations of material, with coarser material placed immediately over the tile line. The particle size is gradually reduced toward the ground surface. The soil surface over the blind inlet should be kept under the grass or permanent vegetation to reduce or prevent surface sealing.

Blind inlets have an inlet length of 3 m, an inlet width of 0.5 m, and size of refill of 9–15 mm. Blind inlets improve drainage temporarily and have limited (a few years) life span, but this depends on maintenance. Blind inlets do not interfere with farm operations.

25.10.15.2 Sedimentation Basins

Sedimentation basins are used where the slope downstream is low and there is a good probability of sedimentation (Figure 25.28). In these basins several laterals join the main and surface water enters the subsurface system. If possible, sedimentation basins should be located in fenced areas and non-cultivated fields. If cultivated area has to be used, the top of the sedimentation basin should be at least 300 mm below the ground surface. Sedimentation basins can be circular, with a diameter of 750 mm, rectangular with a minimum dimension of 600 mm, or square with a minimum length of 600 mm.

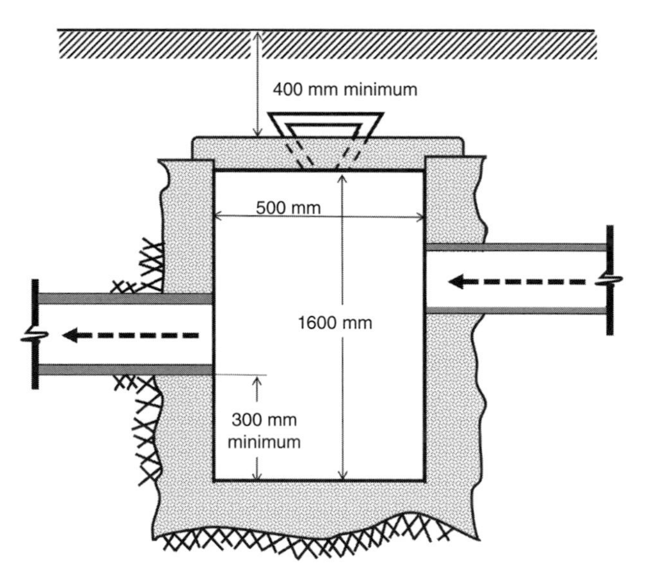

Figure 25.28 Sedimentation basin.

25.10.15.3 Outlets for Subsurface Drains

The outlet is an important component of a subsurface drainage system. Drainage systems have a high percentage of failure because of faulty outlets. The drainage outlet can be either a gravity outlet or a pumped outlet.

Gravity outlet: The gravity outlet is the most common type of outlet. A good gravity outlet is required to provide a free outlet with minimum maintenance. It discharges without serious erosion or damage to the pipe and keeps out rodents and other small animals, protects the end of the outlet pipe against damage by livestock and climate (freeze–thaw), and prevents the entrance of flood waters where the outlet pipe is submerged for several hours. The inside diameter of the outlet pipe should not exceed the outside diameter of the drain pipe by more than 50 mm, and the gate opening should not exceed 25 mm. The outlet pipe should have a

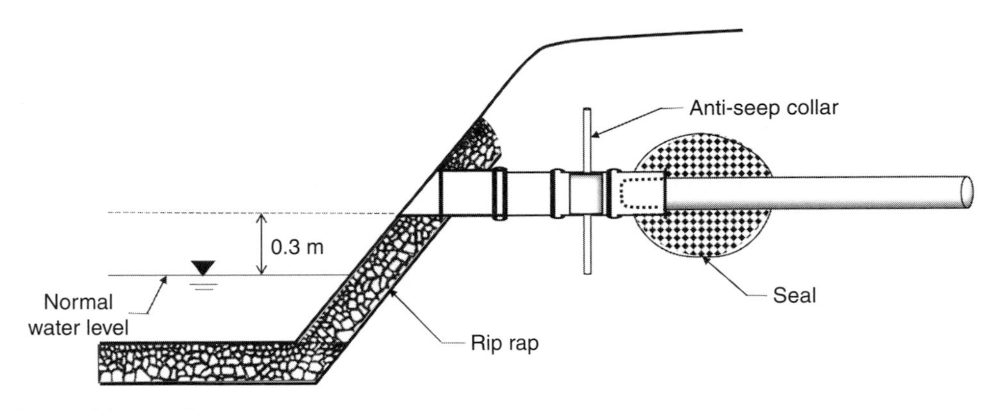

Figure 25.29 Flush-mounted drain outfall.

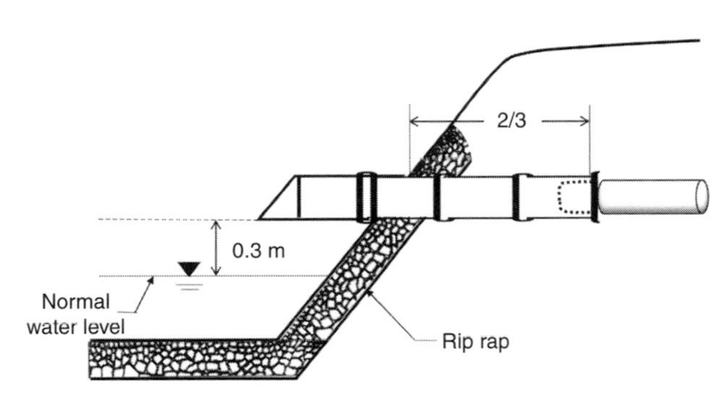

Figure 25.30 Cantilever-style drain outfall.

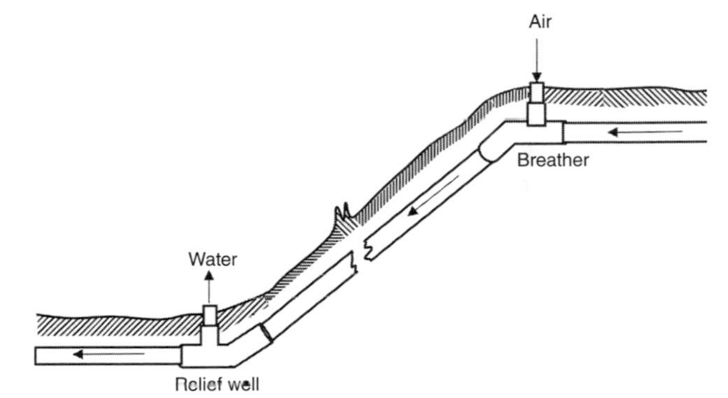

Figure 25.31 Breathers and relief well.

minimum freeboard of 300 mm above the designed normal water level. To protect the exposed end of the pipe from high velocities in the ditch and failure from snow loads, the exposed end of the pipe should not extend more than one-third of its total length beyond the bank.

Pumped outlet: A pumped outlet, shown in Figures 25.29 and 25.30, may be suitable where a long and costly drain or deep outlet ditch is required. Depending on the outflow pipe, it includes flush-mounted and cantilever styles. Its major components are the pump and sump. The pump may also be used to drain isolated areas that cannot be drained by gravity. The pump is usually a self-priming submersible sump pump, which has the advantages of high capacity and low cost. Automatic operation is desirable (i.e., the motor can be started and stopped by a float-operated switch or electrodes located at controls at the selected water level). This can be easily achieved by an electronic motor. Compared to the flush-mounted option, cantilever-style drain outfall is more susceptible to ice and floating debris damage.

25.10.15.4 Other accessories

The other accessories include breathers and relief wells, as shown in Figure 25.31.

25.10.16 Load on Drain Pipe of the Tile Drain

Schwab et al. (1993) have given loads on tile drains, which include the load of soil or fill material and impact load of animals, machinery, etc. The load on a drain pipe depends on the depth of the drain pipe, the size of the trench used to place the pipe, and the type of bedding of the drain pipe.

25.10.16.1 Classification of Trenches Used for the Installation of Tile Drains

Drainage trenches are classified as narrow trenches or ditch-type trenches, and wide trenches or projecting-type trenches.

25.10.16.2 Narrow Trench or Ditch-Type Trench: $B_d < 2B_c$

The fill material for these trenches has a lower bulk density. As the material settles in the trench there is a friction between the fill material and trench walls. The load on the pipe can be expressed as

$$W_c = C_d w B_d^2, \tag{25.36}$$

where W_c is the total load on the pipe per unit length (FL^{-1}), C_d is the load coefficient for the ditch conduit, w is the unit weight of fill (FL^{-3}), and B_d is the width of the ditch at the top of the conduit or pipe (L), as shown in Figure 25.32.

Table 25.9 *Wheel load (percent) transmitted on rigid conduit*

Depth of backfill over top of tile (m)	Trench width at top of drain (m)				
	0.3	0.6	0.9	1.2	1.8
0.3	17.0	26.0	28.6	29.7	30.2
0.6	8.3	14.2	18.3	20.7	22.7
0.9	4.3	8.3	11.3	13.5	15.8
1.2	2.5	5.2	7.2	9.0	11.5
1.5	1.7	3.3	5.0	6.3	8.3
1.8	1.0	2.3	3.7	4.7	6.2

This includes live load and impact load transmitted to a 0.3 m length of tile.

After Schwab et al., 1993. Copyright © 1993 by John Wiley & Sons, New York, United States.

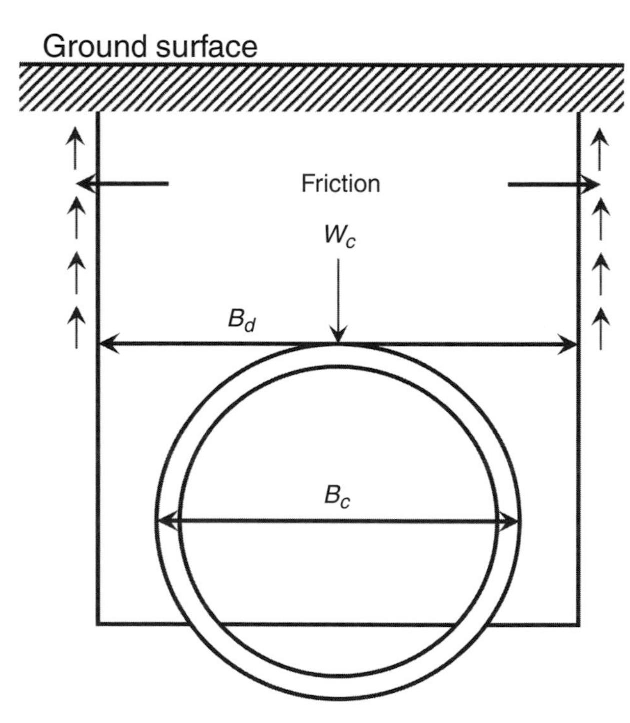

Figure 25.32 Narrow trench or ditch-type trench.

25.10.16.3 Wide Trench or Projecting-Type Trench: $B_d > 2B_c$

The friction between the fill material and the walls of the trench is unimportant. The material directly above the pipe will settle less than the material on either side of the pipe. The load on the pipe increases due to settling of the material to the sides:

$$W_c = C_c w B_c^2, \tag{25.37}$$

where C_c is the load coefficient for projecting conduit and B_c is the outside diameter of the conduit (L), as shown in Figure 25.33.

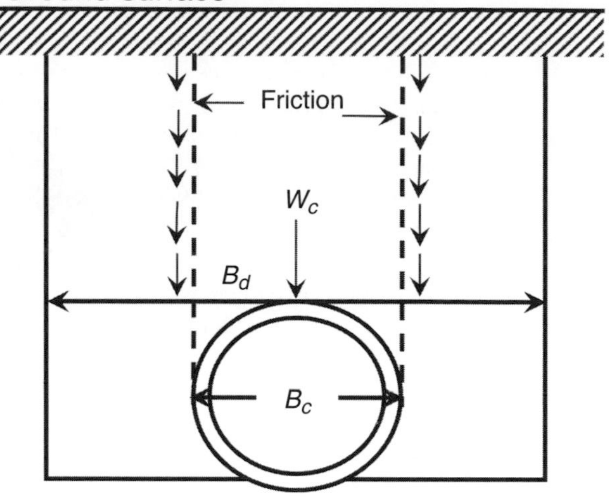

Figure 25.33 Wide trench or projecting-type trench.

25.10.17 Classification of Bedding

The allowable load on a pipe equals the crushing strength times the load factor. The load factor (i.e., strength under given bedding condition/strength determined by the three-edge bearing test) varies with the type of bedding: concrete credible bedding, first class bedding, ordinary bedding, and impermeable bedding (Schwab et al., 1993). Table 25.9 shows wheel load (percent) transmitted on rigid conduit under different trench widths and various depths of backfill over the top of the drain (Schwab et al., 1993).

QUESTIONS

Q.25.1 Compute the drain spacing if the maximum height above the center of the drain is 0.8 m, the drainage coefficient is 0.015 m/day, the soil hydraulic conductivity is 0.4 m/day, and the depth of the drain above the impermeable layer is 1.5 m.

Q.25.2 Compute the buildup of the water table if 4 in. of irrigation is applied each time, the rate of infiltration is 0.95 in./h, and 5% of irrigation water runs off the field.

Q.25.3 Compute the buildup of the water table if the rate of infiltration of soil is 1.25 in./h, the deep percolation is 22%, and the total readily available moisture is 5 in.

Q.25.4 Determine the time needed to lower the water table 4 ft below the ground surface if the hydraulic conductivity is 1.25 ft/day, the specific yield is 12%, the distance from the barrier to the drain is 18 ft, the depth to the drain is 10 ft, the water table is at the ground at time $t = 0$, and the drains are 250 ft apart. Here, the drain is above the barrier.

Q.25.5 Using the data from Q.25.4, determine the drain spacing needed to drop the water table 4 ft below the ground surface in 25 days.

Q.25.6 Considering transient flow for drains located above the barrier, calculate drain spacing if deep percolation is 1.25 in., the hydraulic conductivity of the soil is 5 ft/day, the number of days between irrigations during the peak season is 12 days, and the drop of the water table during this period is 3 ft. The distance from the barrier to the drain is 18 ft and the maximum water table height above the drain is 5 ft.

Q.25.7 Using the data from Q.25.6, compute the drain spacing if the drain is on the barrier.

Q.25.8 Compute the drain spacing under the steady-state condition using the Visser nomograph for an irrigated area if the hydraulic conductivity of the soil is 1 m/day, the drainage coefficient (excess irrigation) is 1.5 mm/day, the impervious layer is 5 m below the ground surface, the depth to the center of the drain is 1.5 m, and the minimum depth to the water table is 1.25 m.

Q.25.9 Compute the drain spacing for a falling water table if the depth of the impervious layer below the drain is 1 m, the hydraulic conductivity is 0.5 m/day, the value of drainable porosity *f* is 0.03, the drain depth to the center line is 1 m, and the initial height of the water table above the center of the drain is 1.0 m. The time for the water table to drop from the initial height to 0.3 m below is 1 day.

Q.25.10 What will be the diameter of concrete tiles and flow rate if the slope is 0.002, the drainage area is 15 ha, and the drainage coefficient is 8 mm/day?

Q.25.11 Calculate the equivalent depth and spacing of drains to keep a constant water table 1.25 m below the ground surface if the hydraulic conductivity is 0.5 m/day, the depth to the impervious layer is 2.5 m, the drainage coefficient is 0.005 m/day, and the height of the water table at the midplane is 0.5 m above the center of 0.15 m drains.

Q.25.12 Determine the diameter of corrugated plastic tubing and flow rate for a falling water table if the hydraulic conductivity is 0.20 m/day, the length of the tile drain is 120 m, and the drain spacing is 25 m. The drain slope is 0.0021. The maximum height of the water table above the drain at the midpoint is 0.7 m. The drain is at the impermeable layer.

Q.25.13 Using the data from Q.25.12, compute the diameter of corrugated plastic tubing and flow rate if the drain lies above the impermeable layer at 1.5 m. The water table drops by 0.25 m after 1 day.

Q.25.14 Determine the flow rate and diameter of concrete tile with surface ponding of 0.06 m, the saturated hydraulic conductivity is 0.25 m/day, the depth from the soil surface to the center of the drain is 0.45 m, and the outside radius of the drain is 0.24 m, assuming the tile is running full. The drain slope is 0.0022 and the length of the drain is 150 m.

REFERENCES

Hooghoudt, S. M. 1940. Bijdragen tot de kennis van eenige natuurkundige grootheden van den grond, 7, Algemeene beschouwing van het probleem van de detail ontwatering en de infiltratie door middel van parallel loopende drains, greppels, slooten, en kanalen. *Versl. Landbouwk Ond*, 46(5): 15–707.

Schwab, G. O., Fangmeier, D. D., Elliot, W. J., and Fervert, R. K. (1993). *Soil and Water Conservation Engineering*, 4th ed. New York: Wiley.

Toksöz, S., and Kirkham, D. (1961). Graphical solution and interpretation of a new drain-spacing formula. *Journal of Geophysical Research*, 66(2): 509–516.

US Bureau of Reclamation (1984). *Drainage Manual*, 2nd ed. Washington, DC: US Department of Interior, Bureau of Reclamation.

US Bureau of Reclamation (1993). *Drainage Manual*, 3rd ed. Washington, DC: US Department of Interior, Bureau of Reclamation.

USDA-NRCS (2001). Water management (drainage). In *National Engineering Handbook*. Washington, DC: USDA-NRCS.

US Department of Commerce (1981). *Bureau of Census, 1978 Census of Agricultural*. Washington, DC: US Department of Commerce.

van Schilfgaarde, J. (1963). Design of tile drainage of falling water tables. *Journal of Irrigation and Drainage Engineering, ASCE*, 89(IR-2): 1–11.

van Schilfgaarde, J. (ed.) (1974). *Drainage for Agriculture*. Madison, WI: American Society of Agronomy.

Visser, W. C. (1954). Tile drainage in the Netherlands: a summary of the addresses delivered on the tile drainage day. *Netherlands Journal of Agricultural Science*, 2(2): 69–87.

26 Farm Irrigation System Design

26.1 INTRODUCTION

Design of a farm irrigation system entails both technical and non-technical considerations. It is an integration of principles borrowed from agriculture, meteorology, hydrology, hydraulics, irrigation, and drainage engineering, as well as economic, environmental, and management sciences. A farm irrigation system is designed with the primary objective of maximizing the net benefit or crop production with a minimum water supply and energy expenditure. The system may also be applied to perform other tasks, such as protection of crops from frost damage, cooling of soil and crops, control of wind erosion, application of chemicals, land application of wastes, providing water for seed germination, and delaying fruit and bud development. These tasks can be of critical importance in achieving the primary objective. A short discussion of these tasks is given in the following section. This chapter provides a snapshot of steps involved in designing a farm irrigation system.

26.2 OTHER ADVANTAGES OF IRRIGATION SYSTEMS

When nights are clear and cool, it is not uncommon for radiation frost to occur, as commonly witnessed in northern India during winter. Plants radiate energy to a cold sky and are cooled 0.5–2 °C below the ambient air temperature and may be damaged, especially if several such cool nights occur in succession. Sprinklers are particularly useful for protecting plants from frost damage.

During hot days transpiration rate may be high and soil water content may decline quickly; it may therefore be desirable to cool the soil and plants to reduce plant water deficit and stomatal closure. It has been reported that cooling not only increases yield but also improves the quality of products such as almonds, walnuts, tomatoes, cucumbers, potatoes, beans, sugar beets, cotton, flowers, strawberries, cranberries, apples, grapes, apples, prunes, and cherries.

In many cases, irrigation water and chemicals, such as fertilizers, pesticides, herbicides, weedicides, desiccants, and defoliants, are applied together. Such an application results in savings in labor and equipment, better timing, and greater flexibility of farm operations. However, uniform application of water is needed for the application of chemicals.

Liquid wastes are generated by cities, towns, farms, and industrial plants. These wastes may contain suspended or dissolved materials, either organic or inorganic. These wastes are spread over the land by irrigation systems, where water is cleaned by filtration via percolation through the soil. If directly applied to land they may be harmful to plants, fungi, bacteria, and other useful flora.

Seedbeds may need control of temperature and salinity. Irrigation systems are used to move the salts down below the root zone and bring the temperature down to prevent burning off of young seedlings. This is often done in furrow irrigation and in greenhouses.

Deciduous fruit-tree blossoming may sometimes be impacted by frost. Irrigation systems are then employed to delay blossoming by applying water. Likewise, the cooling of buds slows bud growth. Thus, frost damage can be prevented.

26.3 FARM IRRIGATION SYSTEMS

Supply of irrigation water to farms involves diversion, transport or conveyance, and direct application, which can be accomplished by different methods.

26.3.1 Methods of Diversion

The source of water is either surface water or groundwater, and this is diverted either by gravity, pumping, or both. Surface water is first stored in a reservoir and then diverted into a farm canal, as discussed in Chapter 8. The diversion is usually under gravity. If the source of water is groundwater, then it is pumped and diverted into a farm channel. Pumps are discussed in Chapter 10.

26.3.2 Methods of Conveyance

The source of water is connected to the farm by a system of conveyance, which can be either an open channel or a pipe. Pipes can be open or closed. Open pipes are like open channels, where water flows under gravity. Closed pipes are pressurized, where water flows under pressure provided by a pumping system. Open-channel hydraulics is discussed in Chapter 8, and pipeline hydraulics is discussed in Chapter 9.

26.3.3 Methods of Application

Water is spread over the cropland using the basin, border, furrow, sprinkler, or trickle irrigation method. The basin irrigation method is described in Chapter 18, the border method in Chapter 19, the furrow method in Chapter 20, the sprinkle method in Chapter 21, and the trickle method in Chapter 22. Doneen and Westcot (1984) presented a guide for selecting a method of irrigation, as given in Table 26.1.

26.3.4 Flow Measurement and Regulation

Flow measurement and regulation are vital to the operation, maintenance, and management of farm irrigation systems. For example, to ensure that all parts of the farm are irrigated, flow may need to be adjusted. Likewise, flow measurement allows detection of clogged pipes and channels, plugged screens, worn pumps, and leaking channels and pipes. Flow measurement is discussed in Chapter 11. Flow regulation requires the employment of checks, check structures, division boxes, and gates on open channels, and various types of valves in pipelines.

26.4 DESIGN OF IRRIGATION SYSTEMS

The design of a farm irrigation system depends on different factors that must be optimally satisfied. The first set of factors relates to physical characteristics and the second set to socioeconomic characteristics. The physical characteristics include climate, soils, topography, hydrogeology, crops, rivers, streams, roads, and buildings, whereas socioeconomic characteristics include costs of water, energy, fertilizers, herbicides, weedicides, land, and seeds; markets; landowner's culture and financial situation; social and community structure; and crop prices. Data on these characteristics are needed. The main steps for farm irrigation system design can be stated as follows:

1. collection of data;
2. identification and evaluation of water sources;
3. evaluation of water quality;
4. determination of soil type and its characteristics;
5. identification of crops for irrigation;
6. determination of design daily irrigation requirements;
7. design of alternative systems for farms;
8. evaluation of the performance of alternative irrigation systems;
9. estimation of the annual cost of alternative system designs; and
10. selection of the most suitable system design.

26.4.1 Data Requirements

Collection of data needed is fundamental for the success of design, and every effort should be made to assemble accurate and reliable data. The data required for the design of an irrigation system are presented in Table 26.2. Different types of data are collected from different sources. Some of the data are easily available, even on the web, but some require a lot of time and effort. In the United States, climate data can be easily obtained from the National Oceanic and Atmospheric Administration (NOAA) of the US Department of Commerce, whereas data on soils and crops can be obtained from the Natural Resources Conservation Service (NRCS) of the US Department of Agriculture (USDA). Data on water supply and water quality can be obtained from the US Geological Survey (USGS). Other types of data can be obtained from state and local agencies, and other branches of the USDA, as well as from individuals.

26.4.2 Identification and Evaluation of Water Sources

Irrigated agriculture is vital for food security. If the source of water is surface water then it is normally developed by either a government agency or a cooperative, and most individual farmers obtain water from this source – at a cost, of course. If the source of water is groundwater, then it is developed by individual farmers or groups of farmers. There are, however, legal, administrative, and socioeconomic issues that have to be considered. Chapters 3 and 4 discuss surface and groundwater sources.

26.4.3 Evaluation of Water Quality

The quality of irrigation water is basic to crop production and the quality of produce. It is important that the water available is suitable for irrigation. Different crops need different types of water quality. Water quality is discussed in Chapter 5.

26.4.4 Determination of Soil Characteristics

Soil type, texture, structure, depth, organic content, salinity, steady infiltration rate, and water-holding capacity are needed for determining the type of crop that should be grown and the yield that can be expected. Soil characteristics are discussed in Chapter 6.

26.4.5 Identification of Crops

Some crops grow best in certain types of soil in a given climate. In order to get the best yield, crops should be selected for the given combination of climatic conditions, soils available, and the availability and quality of water. Crops are discussed in Chapter 7.

26.4.6 Design Daily Irrigation Requirements

The design daily irrigation requirement (DDIR) of a crop depends on the evapotranspiration (ET) and land

Table 26.1 *Guide for selecting a method of irrigation*

Method of irrigation	Topography	Crops	Remarks
Widely spaced borders	Land slopes capable of being graded to less than 1% slope and preferably 0.2%	Alfalfa and other deep-rooted close-growing crops and orchards	The most desirable surface method for irrigating close-growing crops where topographical conditions are favorable. Even grade in the direction of irrigation is required on flat land and is desirable but not essential on slopes of more than 0.5%. Grade changes should be slight and reverse grades must be avoided. Cross-slope is permissible when confined to differences in elevation between border strips of 6–9 cm.
Closely spaced borders	Land slopes capable of being graded to 4% or less and preferably less than 1%	Pastures	Especially adapted to shallow soils underlain by claypan or soils that have a lower water intake rate. Even grade in the direction of irrigation is desirable but not essential. Sharp grade changes and reverse grades should be smoothed out. Cross-slope is permissible when confined to differences in elevation between borders of 6–9 cm. Since the border strips may have less width, a greater total cross-slope is permissible than for border-irrigated alfalfa.
Check back and cross-furrows	Land slopes capable of being graded to 0.2% slope or less	Fruit	This method is especially designed to obtain adequate distribution and penetration of moisture in soils with low water intake rates.
Corrugations	Land slopes capable of being graded to slopes between 0.5% and 12%	Alfalfa, pasture, and grain	This method is especially adapted to steep land and small irrigation streams. An even grade in the direction of irrigation is desirable but not essential. Sharp grade changes and reverse grades should at least be smoothed out. Due to the tendency of corrugations to clog and overflow and cause serious erosion, cross-slopes should be avoided as much as possible.
Graded contour furrows	Variable land slopes of 2–25% but preferably less	Row crops and fruit	Especially adapted to row crops on steep land, though hazardous due to possible erosion from heavy rainfall. Unsuitable for rodent-infested fields or soils that crack excessively. Actual grade in the direction of irrigation is 0.5–1.5%. No grading required beyond filling gullies and removal of abrupt ridges.
Contour ditches	Irregular slopes up to 12%	Hay, pasture, and grain	Especially adapted to foothill conditions. Requires little or no surface grading.
Rectangular checks (levees)	Land slopes capable of being graded to single or multiple tree basins will be leveled within 6 cm	Orchards	Especially adapted to soils that have either a relatively high or low water intake rate. May require considerable grading.
Contour levee	Slightly irregular land slopes of less than 1%	Fruit, rice, grain, and forage crops	Reduces the need to grade land. Frequently employed to avoid altogether the necessity of grading. Adapted best to soils that have either a high or low intake rate.
Portable pipes	Irregular slopes up to 12%	Hay, pasture, and grain	Especially adapted to foothill conditions. Requires little or no surface grading.
Subirrigation	Smooth flat	Shallow-rooted crops such as potatoes or grass	Requires a water table, very permeable subsoil conditions, and precise leveling. Very few areas adapted to this method.

Table 26.1 (*cont.*)

Method of irrigation	Topography	Crops	Remarks
Sprinkler	Undulating 1–35% slope	All crops	High operation and maintenance costs. Good for rough or very sandy lands in areas of high production and good markets. Good method where power costs are low. May be the only practical method in areas of steep or rough topography. Good for high-rainfall areas where only a small supplemental water supply is needed.
Contour bench terraces	Sloping land – best for slopes under 3% but useful up to 6%	Any crop but particularly suited to cultivated crops	Considerable loss of productive land due to berms. Requires extensive drop structures for water erosion control.
Subirrigation (installed pipes)	Flat to uniform slopes up to 1%; surface should be smooth	Any crop; rows of high-value crops usually used	Requires installation of perforated plastic pipe in root zone at narrow spacings. Some difficulties in roots plugging the perforations. Also a problem as to correct spacing. Field trials on different soils are needed.
Localized (drip, trickle, etc.)	Any topographic condition for row crop farming	Any crops or fruit	Perforated pipe on the soil surface drips water at the base of individual vegetable plants or around fruit trees. Has been successfully used with saline irrigation water where irrigation frequency is high and the soil water salinity is nearly that of the applied water.

After Doneen and Westcot, 1984.

Table 26.2 *Main data needed for the design of a farm irrigation system*

Data	Specific type
Topography	Elevation map, topographic features, location of streams, and hydraulic structures.
Climate	Several years of temperature, relative humidity, wind, solar radiation, sunshine hours, and vapor pressure. Pan evaporation data may also be available. The type of data needed will depend on the method of estimating evapotranspiration.
Water supply	Volume of available water and rate of supply for surface water sources; distance from the farm; type of aquifer and aquifer properties for groundwater sources.
Water quality	Physical and chemical properties of water.
Soils	Soil characteristics, such as soil type, spatial distribution, infiltration rate, depth, drainage requirement, salinity, and erodibility.
Crops	Types of crops and areal distribution, farming practices, crop yield and prices, storage and distribution, and markets.
Energy source	Availability, type, and cost.
Capital and labor	Availability of finances; availability, cost, and skill of labor; equipment and maintenance requirement, and proximity and access to repair facility.
Other	Location of roads, buildings, trees, and drainage ways.

preparation. Of course, ET depends on the crop, soil, and climatic conditions. DDIR defines the rate at which the irrigation system must supply water to the crop to achieve the desired level of irrigation. To determine DDIR, the daily irrigation requirement should be determined for several years. Then, allowable depletion (AD) between irrigations and the minimum irrigation interval (IT_{min}) during the irrigation season should be determined. Then, DDIR is the ratio of AD to IT_{min}.

26.4.7 Alternative Irrigation System Designs

There are several ways by which a farm can be irrigated. These several ways or systems depend on the method of

Table 26.3 *Comparison of irrigation systems in relation to size and situation factors*

Size and situation factors	Redesigned surface system	Level basin	Intermittent mechanical movement	Continuous mechanical movement	Solid set and permanent	Emitters and porous tubes
Infiltration rate	Moderate to low	Moderate	All	Medium to high	All	All
Topography	Moderate slopes	Small slopes	Level to rolling	Level to rolling	Level to rolling	All
Crops	All	All	Generally shorter crops	All but trees and vineyards	All	High value required
Water supply	Large streams	Very large streams	Small streams	Small streams	Small streams	Small streams
Water quality	All but very high salts	All	Salty water may harm plants	Salty water may harm plants	Salty water may harm plants	All can potentially use high-salt waters
Efficiency	Average 60–70%	Average 80%	Average 70–80%	Average 80%	Average 70–80%	Average 80–90%
Labor requirement	High training required	Low, some training	Moderate, some training	Low, some training	Low to seasonal high training	Low to high, some training
Capital requirement	Low to moderate	Moderate	Moderate	Moderate	High	High
Energy requirement	Low	Low	Moderate to high	Moderate to high	Moderate	Low to moderate
Management skill	Moderate	Moderate	Moderate	Moderate to high	Moderate	High
Machinery operation	Medium to long fields	Short fields	Moderate field length, small interference	Some interference, circular fields	Some interference	May have considerable interference
Duration of use	Short to long	Long	Short to medium	Short to medium	Long term	Long term, but durability unknown
Weather	All	All	Poor in windy conditions	Better in windy conditions than other sprinklers	Windy conditions reduce performance; good for cooling	All
Chemical applications	Fair	Good	Good	Good	Good	Very good

application, conveyance system, diversion system, landowner's preference, financial situation, flexibility, prospects for future expansion, and associated costs. For each system these should be clearly laid out. Schwab et al. (1981) have provided a list of major factors that affect the selection of application method, as shown in Table 26.3 (Schwab et al., 1981).

26.4.8 Performance of Alternative Irrigation Systems

Water is a scarce commodity and its value is increasing with time. It is important that water is not wasted – that is, deep percolation, runoff, evaporation, and operational losses are controlled as much as possible. The irrigation system performance will vary from one system to another and is evaluated by the adequacy and uniformity of application in each field, as well as by the efficiency of water diversion, conveyance, and application.

26.4.9 Annual Cost of Alternative Systems

The annual cost of owning and operating each alternative system should be determined. It is used to determine the economic soundness of each system and to make a decision on selection. The financing agency may need this information before making a decision about whether or not to finance the system.

26.4.10 Selection of the System

The selection of the most suitable system depends on the needs, desires, finances, technical details, and economic analysis. All this information has to be cast in the context of the profit that can be generated by increasing crop yield while maintaining soil quality and environmental sustainability.

REFERENCES

Doneen, L. D., and Westcot, D. W. (1984). Irrigation practice and water management. FAO irrigation and drainage paper 1 (revised).

Schwab, G. O., Frevert, R. K., Edminster, T. W., and Barnes, K. K. (1981). *Soil and Water Conservation Engineering*. New York: Wiley.

Part VII Irrigation Operation and Management

27 Irrigation Scheduling

Notation

AD	allowable depletion (mm, in.)
AW	available water to z_r (percentage)
AWC	available water capacity or available water-holding capacity of the soil (in./in., in./ft, m/m, mm/m)
d_e	effective depth of water applied per irrigation (mm, in.)
d_{ep}	planned effective depth of irrigation water (mm, in.)
D	estimated depletion to the date of irrigation (mm, in.)
D_0	current soil water depletion (mm, in.)
D_{ag}	days after germination (days)
DP_i	deep percolation at time i (mm, in.)
D_{rz}	effective root zone depth that increases during the growing season and reaches a maximum depth (m, ft)
D_{tm}	days from germination to maximum effective cover (days)
E	irrigation efficiency (percentage)
ED	earliest date of irrigation (days)
ET_c	crop evapotranspiration (mm/day)
ET_{ci}	crop evapotranspiration at time i (mm/day, in./day)
f_{dmax}	maximum allowable fraction depletion to maintain maximum yields of crops (fraction)
f_r	fraction of available water remaining (fraction)
f_{rmin}	minimum fraction of available water remaining to maintain maximum yields of crops (fraction)
FC	field capacity (in./in., in./ft, m/m, mm/m)
G_{ag}	growing degree days (days)
G_{tm}	growing degree days to the maximum effective cover (days)
I_{nl}	net irrigation depth after considering leaching requirement (L)
LD	latest date by which irrigation should occur before the soil moisture reservoir reaches a critical limit (days)
LR	leaching requirement (fraction)
MAD	management allowed depletion (fraction)
MB	minimum allowable soil water, or minimum balance (mm, in.)
N	number of days to the next irrigation
P_{ei}	effective precipitation on day i (mm, in.)
PWP	permanent wilting point (in./in., in./ft, m/m, mm/m)
r_a	rainfall allowance (mm, in.)
R_d	root depth (m, ft)
R_{dmax}	maximum effective root depth (m, ft)
R_{dmin}	minimum root depth for young plants (m, ft)
SMD_i	total soil moisture depletion in the root zone at day i in the water balance approach (mm, in.)
SWD	soil water deficit (mm, in.)
t_i	thickness of the soil layer i (m, ft)
T	time interval between irrigations (days)
TAW	total available water (mm, in.)
T_{max}	maximum time interval between irrigations (days)
U_{fi}	upward flow of groundwater on day i due to capillary rise or ground water contribution (mm, in.)
W	water content (mm)
W_0	water content when $t = 1$ day
WB	available water balance (mm, in.)
W_d	daily drainage from the root zone or upward movement from the saturated zone (with negative sign) or cumulative drainage (L)
W_I	amount of irrigation water to be applied (mm, in.)
z_c	effective height of the capillary fringe above the water table (m, ft)
z_w	depth to water table (m, ft)
θ_{i-1}	initial soil moisture content (L/L)
θ_{fc}	moisture content at field capacity (L/L)
θ_v	current soil water content (L/L)
θ_{wp}	soil water content at the permanent wilting point (L/L)

27.1 INTRODUCTION

Irrigation scheduling is a fundamental component of irrigation management and is vital for optimum agricultural production. It varies with the type of crop, soil, climate, irrigation method, and agricultural practices. Proper irrigation scheduling is essential to achieve the objectives of irrigation that entails the efficient use of water, energy, fertilizer, and labor for optimum crop production without undue environmental impact. Hence, irrigation management significantly depends on irrigation scheduling, which answers essentially two questions: When to irrigate? And how much to irrigate? When to irrigate includes frequency and time interval of irrigation; how much to irrigate includes the duration and amount of irrigation. Thus, irrigation scheduling can be defined as the determination of the amount and timing of irrigation application that meets a specific management objective. Management objectives may include maximum yield or biomass production per unit area or per unit water applied, maximizing net benefits, minimizing energy requirements, maintaining plant life, crop quality, frost protection or crop cooling, maximizing economic return, salinity control, functional value of plants (i.e., athletic field), aesthetic value (i.e., keeping plants healthy), minimizing environmental impact, and maintaining minimum streamflow. In any event, the water supplied by irrigation replaces the soil water consumed by plants and the amount supplied should lead to uniformity and efficiency of application and leaching if needed. The time of application of this water depends on the effective rooting depth of the crop, water-holding capacity of the soil, crop consumptive use rate, crop sensitivity to water deficit, water quality and crop tolerance to salinity, type and capacity of irrigation system, and water supply and delivery schedule. There are different ways to determine the timing and amount of water application. This chapter discusses some of the commonly used methods of irrigation scheduling.

27.2 FACTORS AFFECTING IRRIGATION SCHEDULING

Irrigation scheduling depends on soil type, crop type, stage of crop growth, season, climate, intercultural operations, fertilizer application, irrigation system type, water supply, and irrigation management. Management considers the maximization of net return and crop yield, minimization of irrigation cost, and optimum distribution of water in the field. Irrigation scheduling also affects irrigation efficiency and uniformity. Further, different crops have different degrees of sensitivity to water stress. Obviously, some crops are more sensitive than others. The sensitivity of various crops to water shortages is shown in Table 27.1 (Brouwer et al., 1989). The crop sensitivity is not the same throughout the growing season and is much higher during certain growth stages, and that is when irrigation should necessarily be done in order to minimize the reduction in crop yield. Table 27.2 shows the periods of growth stage of various crops that are more sensitive to water shortage (Brouwer et al., 1989).

Table 27.1 *Sensitivity of various field crops to water shortage*

Sensitivity			
Low	Low–medium	Medium–high	High
Crops			
Cassava	Alfalfa	Beans	Banana
Cotton	Citrus	Cabbage	Fresh greens
Millet	Grape	Maize	Vegetables
Pigeon pea	Groundnuts	Onion	Paddy rice
Sorghum	Soybean	Peas	Potato
	Sugar beet	Pepper	Sugarcane
	Sunflower	Tomato	
	Wheat	(Water) melon	

After Brouwer et al., 1989.

Table 27.2 *Periods sensitive to water shortages*

Crop	Sensitive period
Alfalfa	Just after cutting
Alfalfa (for seed prod.)	Flowering
Banana	Throughout
Bean	Flowering and pod filling
Cabbage	Head enlargement and ripening
Citrus	Flowering and fruit setting more than fruit enlargement
Cotton	Flowering and boll formation
Grape	Vegetative period and flowering more than fruit filling
Groundnut	Flowering and pod setting
Maize	Flowering and grain filling
Olive	Just prior to flowering and yield formation
Onion	Bulb enlargement
Onion (for seed prod.)	Flowering
Pea, fresh	Flowering and yield formation
Pea, dry	Ripening
Pepper	Throughout
Pineapple	Vegetative period
Potato	Stolonization and tuber initiation
Rice	Head development and flowering
Sorghum	Flowering and yield formation
Soybean	Flowering and yield formation
Sugar beet	First month after emergence
Sugarcane	Vegetative period (tillering and stem elongation)
Sunflower	Flowering more than yield formation
Tobacco	Period of rapid growth
Tomato	Flowering more than yield formation
Watermelon	Flowering and fruit filling
Wheat	Flowering more than yield formation

After Brouwer et al., 1989.

Table 27.3 *Soil moisture-based indicators of when to irrigate*

Methods and observed/ measured parameters	Required instrument or procedure	Principal advantages	Principal disadvantages
Feel and appearance (soil moisture)	Hand probe	Simple	Time-consuming, approximate, requires interpretative skills
Gravimetric method (soil moisture)	Sample cans, soil augur, scale, and oven	Simple and accurate; time gap between sample and results	Time-consuming and destructive methods; repetitive measurement at the same location and depth are not possible
Tensiometer (soil matric potential)	Tensiometer; gauge	Measures fundamental parameter affecting soil water flux; good accuracy; instantaneous reading of soil moisture tension	Requires careful installation, calibration, and frequent readings and maintenance; breaks at tensions above 0.7 atm
Electrical resistance (electrical conductivity)	Porous block	Provides indirect measure of soil water content; can be used for remote reading	Affected by soil salinity; not sensitive at low tensions and also in coarse-textured soils; short block life
Neutron scattering (soil moisture)	Neutron probe and access tubes	Gives volumetric soil moisture, repetitive measurements at same depth and location are possible; quick and accurate	Expensive and requires special handling and storage; calibration is affected by changes in soil organic matter
TDR (soil moisture)	TDR probe	Accurate, can be automated to continuously take measurements	Expensive; monitors relatively small volume; requires careful installation, particularly in gravelly and heavy soils

After Stegman et al., 1980. Copyright © 1980 by American Society of Agricultural and Biological Engineers, St. Joseph, Michigan. Reprinted with permission from ASABE.

27.3 FULL OR DEFICIT IRRIGATION

Whether to do full irrigation or deficit irrigation depends on the source of water supply. Full irrigation is done if there is enough water available and cost of irrigation is low. In this case, crops are not subject to stress. Deficit irrigation is done when there is scarcity in the water supply. In this case plants will experience stress. However, deficit irrigation is not recommended throughout the growing season. Adequate irrigation must be done during critical stages of plant growth to maximize crop water use. Deficit irrigation can be practiced by balancing crop production cost and revenue loss due to decreased crop yield.

27.4 METHODS OF IRRIGATION SCHEDULING

Irrigation scheduling depends on the state of soil moisture and plants. Three general approaches to irrigation scheduling include: (1) maintaining soil water balance within desired limits, which depends on the capacity of the soil moisture reservoir; (2) use of plant status indicators to signal the need for water, which depends on crop water use (ET_c) and soil water; and (3) irrigation based on the calendar date or other fixed schedule. There are a multitude of methods to evaluate the status of moisture in the soil and plants. The status of soil moisture is judged in many different ways and some of these are qualitative, such as feel and appearance. Others are quantitative, employing gravimetric sampling, tensiometers, gypsum blocks, neutron probes, dielectric and electromagnetic properties of soil, and remote sensing. Irrigation scheduling, based on the status of soil moisture, assumes that soil moisture is the determinant of the ability of plants to extract water for growth. It determines when to irrigate and how much to irrigate; the soil moisture in the root zone is raised to the field capacity via irrigation. Stegman et al. (1980) summarized these methods (Table 27.3).

The status of plants can be monitored by evaluating stress (trunk or branch diameter change, leaf water potential, sap flow) or by canopy temperature using radiometers and infrared devices, such as remote sensing. Plant stress is overcome using irrigation, but it does not indicate how much to irrigate. Plant-based methods are summarized by Stegman et al. (1980) (Table 27.4). More common is the water balance approach, which yields information on the amount of water used by plants and determines when to irrigate and how much to irrigate. Plant-based irrigation scheduling methods can be used in conjunction with soil moisture-based and water balance methods to determine the synergy between plant response and scheduling methods. Table 27.5 summarizes different methods and their advantages and disadvantages (Ihuoma and Madramootoo, 2017).

Table 27.4 *Plant-based indicators of when to irrigate*

Methods and observed/measured parameters	Required instrument or procedure	Principal advantages	Principal disadvantages
Appearance	Eye	Simple	Yield potential is affected before color and other changes are observed; approximate, requires interpretative skills
Leaf temperature	Non-contact thermometers	Can be sensed remotely	Application methods not well developed
Leaf water potential	Pressure chamber or thermocouple psychrometer	Indicates integrated effect of aerial and soil environment on degree of plant hydration; is correlated with metabolic processes; a fundamental parameter affecting water flux	Subject to large durational fluctuations; time-consuming; requires sampling skills and data are not easily interpreted
Stomatal resistance	Diffusion porometer	Measures stomatal openings	Same as for leaf water potential

After Stegman et al., 1980.

Table 27.5 *Summary of the main classes of monitoring crop water stress for irrigation scheduling, indicating their main advantages and disadvantages*

Methods	Description	Advantages	Disadvantages
I. Soil water measurement			
(a) Gravimetric method	Soil is weighed, oven-dried, and reweighed to estimate water lost from the plant–soil system	Reliable and serves as a guide to the amount of water to apply during irrigation	Labor-intensive, destructive, and time-consuming
(b) Soil moisture sensors			
(b1) Neutron probe	Based on the emission of high-energy neutrons by a radioactive source into the soil	Fast, non-destructive, and repetitive	Requires adequate operator training, storage, licensing, and inspection due to its radioactive source
(b2) Time-domain and frequency-domain reflectometer (TDR and FDR)	Based on the difference between the dielectric constant of water and soil	Precise and easy to apply in practice; estimates soil water levels at different depths along with the soil profile; readings can be logged automatically	Several sensors are required for the entire field; high cost of installation of sensors
(b3) Tensiometers	Measure soil water potential	Easy to use for irrigation scheduling	Useful in coarse-textured soils or high-frequency irrigation only; used for a narrow range of available soil water
II. Soil water balance approach	Indirect estimate of soil moisture status based on soil water balance calculations	A good indicator of the amount of irrigation water and easy to apply	Not very accurate and requires calibration with soil measurements; requires estimation of evaporation, rainfall, and irrigation events; errors are cumulative, so regular recalibration needed
III. Plant-based approaches			
(a) Stomatal conductance	Indirect indicator of plant water stress by measuring the stomata opening	A good measure of plant water status; can be used as the benchmark for most research studies	Labor-intensive and unsuitable for automation and commercial application; not very accurate for anisohydric crops

Table 27.5 (*cont.*)

Methods	Description	Advantages	Disadvantages
(b) Leaf water potential	Direct measurement of leaf water content	Widely accepted reference technique	Slow, destructive, and unsuitable for isohydric crops
(c) Relative water content	Direct measurement of leaf water status	Good indicator of plant water status; requires less sophisticated equipment	Destructive and time-consuming
(d) Sap flow measurement	Measures the rate of transpiration through heat pulse	Sensitive to stomatal closure and water deficits; adapted for automated recording and controlling of irrigation systems	Needs calibration for each plant and is difficult to replicate; requires complex instrumentation and expertise
IV. Remote sensing methods			
(a) Infrared thermometry	Measures canopy temperature, which increases as a result of water stress	Reliable and non-destructive	Based on only a few point measurements; does not account for soil and crop heterogeneity
(b) Spectral vegetation indices			
(b1) Structural indices	Measures reflectance indices within the vegetation indices and near-infrared spectral range to indicate canopy changes due to water stress	Non-destructive, with high temporal and spectral resolutions	Requisite image analysis is still a challenging task; precision reduces from leaf scale to canopy scale
(b2) Xanthophyll indices	Measures PRI and PRI_{norm}, which are sensitive to the epoxidation state of the xanthophyll cycle pigments	Account for physiological changes in photosynthetic pigment changes due to water stress	More work is needed to convert raw imagery to a user-friendly irrigation application
(b3) Water indices	Measures the reflectance in the near-infrared region to represent canopy moisture content	Rapid and non-destructive measurement of leaf water content	Problems of scaling up to canopy level

After Ihuoma and Madramootoo, 2017.

27.4.1 Plant Response to ET_c and Soil Water

There is a linear relationship between relative growth and/or yield and ET_c, as shown in Figure 27.1. If the management objective is maximum yield, then the value of ET_c can be determined for the maximum yield. Also, there is a relation between relative growth and/or yield and fraction of available water remaining, f_r, as shown in Figure 27.2 (Stegman, 1983) wherein f_{dmax} (maximum allowable fraction depletion to maintain maximum yield) depends on plant species, genotype, and weather conditions. f_{rmin} is the minimum fraction of available water remaining to maintain maximum crop yield, where $f_{dmax} = 1 - f_{rmin}$. The weather conditions influence ET_c for a given day. The values of f_{dmax} range from 0.18 to 0.88, depending on how plants respond to soil water deficit and the maximum ET_c for a given day. Table 27.6 gives an estimated maximum allowable fraction depletion to maintain maximum yields of crops grouped according to sensitivity (Doorenbos and Kassam, 1979). The maximum allowable fraction depletion can also be found in Table 13.3.

Table 27.6 *Estimated maximum allowable fraction depletion to maintain maximum yields of crops grouped according to sensitivity*

		Maximum ET_c (in./day [mm/day])						
		0.08 (2)	0.12 (3)	0.16 (4)	0.20 (5)	0.24 (6)	0.28 (7)	0.31 (8)
Crop group		f_{dmax} to maintain maximum evapotranspiration						
1	Onion, pepper, potato	0.50	0.43	0.35	0.30	0.25	0.23	0.20
2	Banana, cabbage, pea, tomato	0.68	0.58	0.45	0.40	0.35	0.33	0.28
3	Alfalfa, citrus, groundnut, pineapple, field beans	0.80	0.70	0.60	0.50	0.45	0.43	0.38
4	Sunflower, watermelon, wheat, cotton, sorghum, olive, grape, corn, soybean, tobacco	0.88	0.80	0.70	0.60	0.55	0.50	0.45

After Doorenbos and Kassam, 1979.

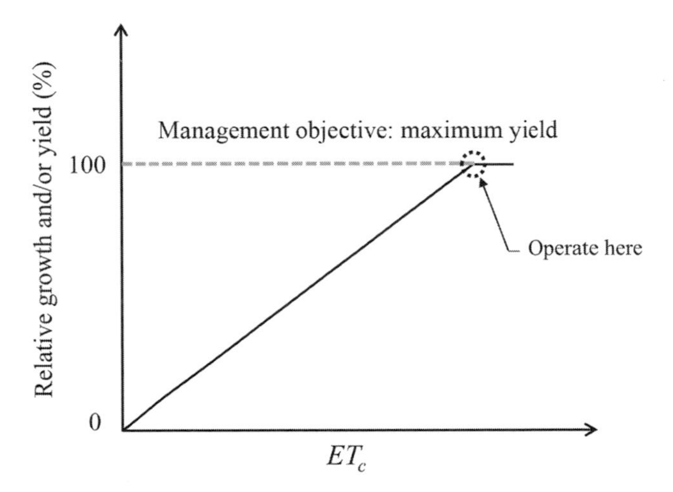

Figure 27.1 Relationship between yield and ET_c.

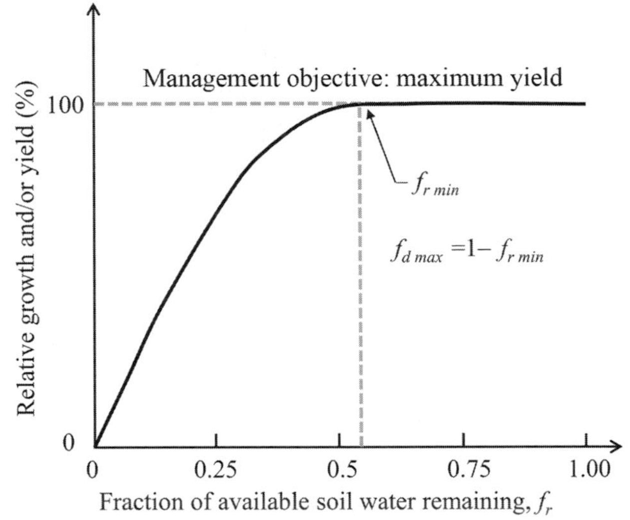

Figure 27.2 Relationship between available water and yield.

Example 27.1: Consider a corn crop with ET_c of 0.28 in./day (7.1 mm/day). Compute the value of f_{dmax} to maintain the maximum evapotranspiration. Now consider onion with the same value of ET_c of 0.28 in./day. Then, compute the value of f_{dmax} to maintain the maximum evapotranspiration. Also, compute f_{rmin} in both cases. Now consider ET_c of 0.20 in./day (5.1 mm/day) for both corn and onion. Compute f_{dmax} and f_{rmin} for both corn and onion crops.

Solution: Given $ET_c = 0.28$ in./day, from Table 27.6 corn is in group 4 so the value of f_{dmax} corresponding to the given ET_c is 0.50. This gives f_{rmin} as 0.50 ($= 1 - f_{dmax} = 1 - 0.5$).

Onion falls in group 1, so the value of f_{dmax} corresponding to the given ET_c is 0.23. This gives f_{rmin} as $1.0 - 0.23 = 0.77$.

Again, given $ET_c = 0.20$ in./day, from Table 27.6 for corn $f_{dmax} = 0.6$ and $f_{rmin} = 0.4$. Likewise, for onion, $f_{dmax} = 0.3$ and hence $f_{rmin} = 0.70$.

This example shows that if the weather has a relatively low ET_c, a high percentage of crop root zone can be depleted before stress occurs. Conversely, on days with high ET_c, less soil water depletion is allowed before plants undergo water stress.

Table 27.7 *Approximate range of available water in different soil textures*

Soil texture	Available water (in./ft.) Range	Available water (in./ft.) Typical	Available water (cm/m) Range	Available water (cm/m) Typical
Very coarse sand	0.4–0.8	0.5	3–6	4
Coarse sand, fine sand, and loamy sand	0.8–1.3	1.0	6–10	8
Sandy loam	1.3–1.8	1.5	10–15	13
Very fine sandy loam, loam, and silt loam	1.5–2.3	2.0	13–20	17
Clay loam, silty clay loam, sandy clay loam	1.8–2.5	2.2	15–21	18
Sandy clay, silty clay, clay	1.4–2.5	2.3	13–21	19
Peat and muck soils	2.0–3.0	2.5	17–25	21

After USDA-NRCS, 2016.

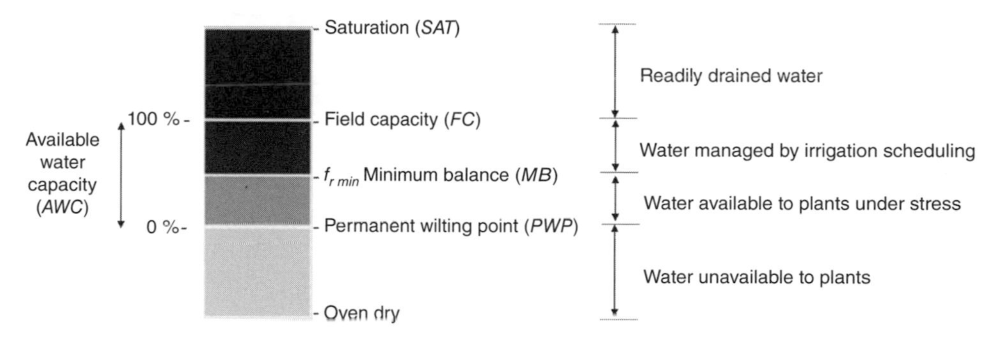

Figure 27.3 Soil water reservoir.

27.4.2 Capacity of Soil Moisture Reservoir

Soil types are classified into two groups: (1) medium- and fine-textured soils, including silt loams (more readily managed and substantial water-holding capacity of 4.5–6 inches in the top 3 ft [13–17 cm in the top 1 m]); and (2) coarse-textured soils, including fine sands and loamy sands (more difficult to manage and relatively small water-holding capacity of less than 4.5 inches in the top 3 ft [13 cm in the top 1 m]). The approximate range of available water is given in Table 27.7.

The total available water (*TAW*) in the root zone can be defined as the product of crop root zone depth (R_d) and available water-holding capacity of the soil (*AWC*):

$$TAW = R_d \times AWC. \tag{27.1}$$

For layered soils, *TAW* can be expressed as

$$TAW = \sum_{i=1}^{N} (AWC_i)t_i, \tag{27.2}$$

where AWC_i is the available water-holding capacity of soil layer i, and t_i is the thickness of soil layer i. The allowable depletion (*AD*) is given by

$$AD = f_{dmax}(TAW). \tag{27.3}$$

Likewise, the minimum allowable soil water, or minimum balance (*MB*), is given by

$$MB = f_{rmin}(TAW). \tag{27.4}$$

It may be noted that

$$MB = TAW - AD \tag{27.5}$$

or

$$TAW = MB + AD. \tag{27.6}$$

This shows that the size of the soil water reservoir is dependent on both the soil and root zone depth. Figures 27.3–27.5 show the soil water with reservoir analog. *AWC* is the difference between field capacity (*FC*) and the permanent wilting point (*PWP*). The water between saturation (*SAT*) and *FC* is readily drained due to gravity. The water managed by irrigation scheduling is the difference between *FC* and f_{rmin} (*MB*). Below *MB*, water is still available to the plant until it reaches *PWP*, but plants will be under water stress. When the soil moisture is under *PWP*, water is unavailable to plants.

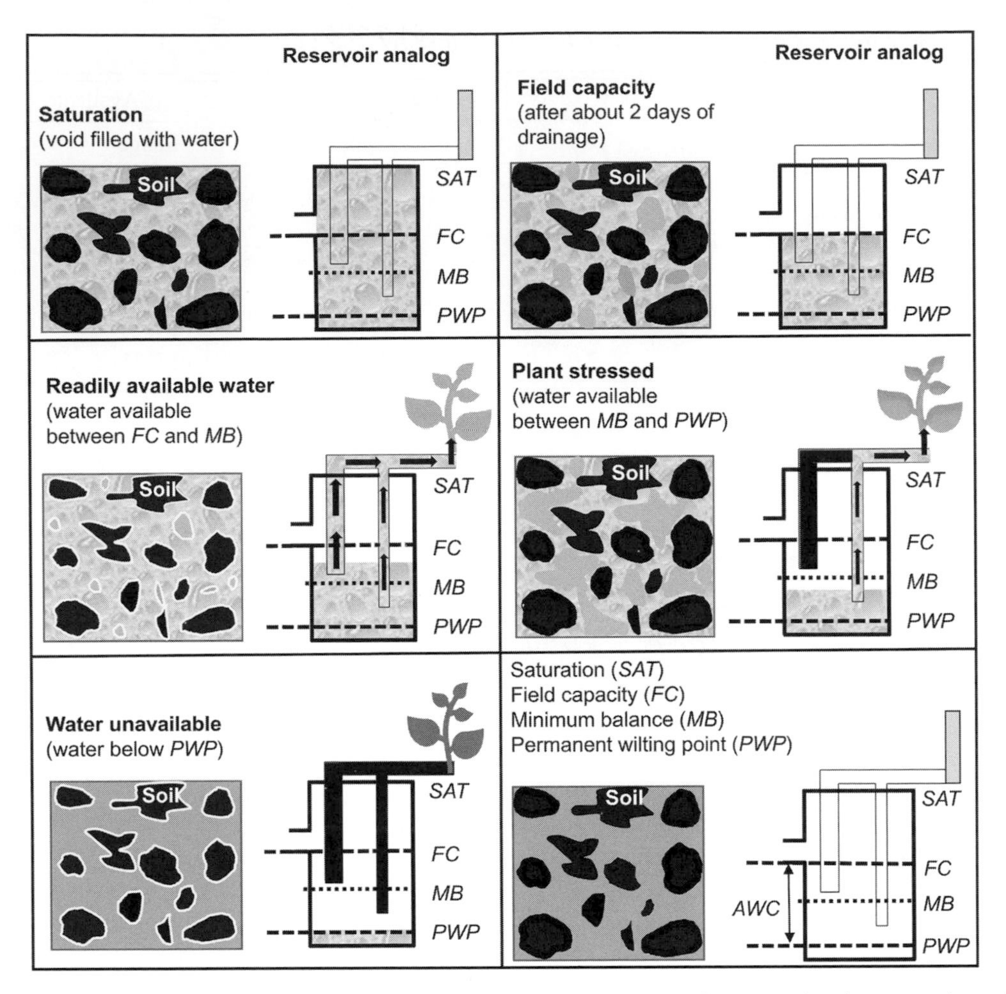

Figure 27.4 Soil water with reservoir analog. A black and white version of this figure will appear in some formats. For the color version, please refer to the plate section.

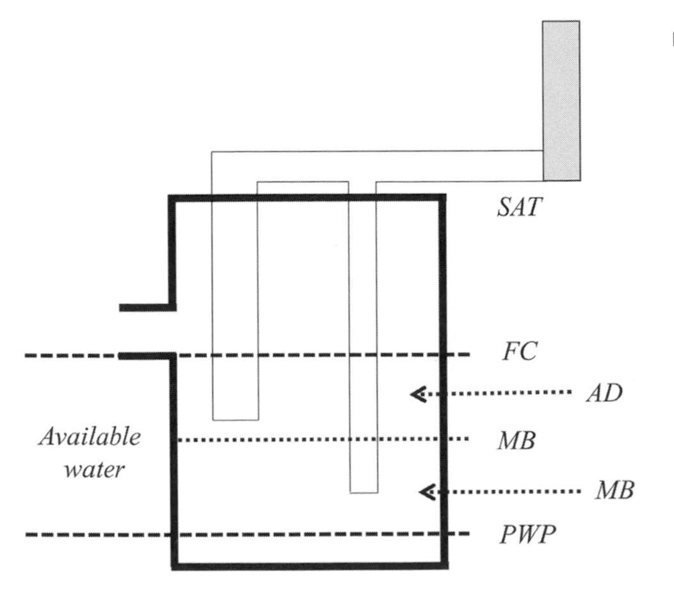

Figure 27.5 Reservoir analog.

Example 27.2: The root zone depth (R_d) is given as 3 feet (0.91 m), AWC equals 0.17 in./in., and f_{dmax} is 0.45. Determine the total allowable water (TAW), allowable depletion (AD), and minimum balance (MB).

Solution:

$$TAW = R_d \times AWC = 3 \text{ ft} \times 0.17 \frac{\text{in.}}{\text{in.}} \times \frac{12 \text{ in.}}{1 \text{ ft}}$$
$$= 6.12 \text{ in. } (15.5 \text{ cm}),$$

$$AD = f_{dmax} \times TAW = 0.45 \times 6.12 \text{ in.} = 2.75 \text{ in. } (7.0 \text{ cm}),$$

$$MB = f_{rmin} \times TAW = (1 - 0.45) \times 6.12 \text{ in.}$$
$$= 3.37 \text{ in. } (8.6 \text{ cm}).$$

It may be noted that $AD + MB = TAW$, as shown in eq. (27.6).

The crop should be irrigated when or before 2.75 in. (7.0 cm) of water has been depleted from the crop root zone. If AD is reached (i.e., $SWD = AD$ at the time of irrigation), the maximum amount of water that the root zone would hold without exceeding the field capacity would be 2.75 in. (7.0 cm).

27.4.3 Plant Root Zone Depth

For annual crops the root depth prior to the date of maximum rooting is described by a linear equation:

$$R_d = R_{dmin} + (R_{dmax} - R_{dmin}) \times R_f, \tag{27.7}$$

where R_d is the root depth (m, ft), R_{dmin} is the minimum root depth for young plants (m, ft), R_{dmax} is the maximum effective root depth (m, ft), and R_f is the root growth factor (fraction). The development of a corn plant's root zone is shown in Figure 27.6. The minimum root depth for seedlings is normally considered to be 4–6 in. (10–15 cm). The actual initial depth may deviate slightly from this value, but the error will have little effect on the soil water balance or irrigation scheduling.

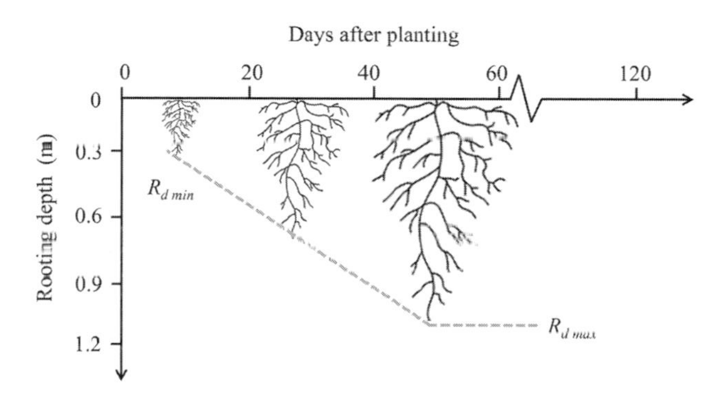

Figure 27.6 Development of a corn plant's root zone.

The root growth factor, which describes the rate of growth during the season, can be computed as

$$R_f = \frac{D_{ag}}{D_{tm}} \text{ or } R_f = \frac{G_{ag}}{G_{tm}}, \tag{27.8}$$

where D_{ag} is the days after germination, D_{tm} is the days from germination to maximum effective cover, G_{ag} is the growing degree days, and G_{tm} is the growing degree days to the maximum effective cover.

The range of the maximum effective rooting depths for different crops was provided in Table 13.3. The maximum effective rooting depths for fully grown plants are listed in Table 27.8.

Table 27.8 *Range of maximum effective rooting depths for fully grown plants*

Crop	Maximum effective depth (ft [m])	Crop	Maximum effective depth (ft [m])
Alfalfa (hay/seed)	3.0–10.0 (1.0–3.0 m)	Olive	1.6–3.3 (0.5–1.0 m)
Banana	1.3–2.6 (0.4–0.8 m)	Onion	1.0–2.0 (0.3–0.6 m)
Barley	3.3–4.3 (1.0–1.3 m)	Palm tree	2.3–3.6 (2.7–1.1 m)
Beans	1.3–2.6 (0.4–0.8 m)	Pea	2.0–3.3 (0.6–1.0 m)
Cabbage	2.0–3.3 (0.6–1.0 m)	Pineapple	1.0–2.0 (0.3–0.6 m)
Carrot	1.6–3.3 (0.5–1.0 m)	Potato	1.3–2.0 (0.4–0.6 m)
Celery	1.0–1.7 (0.3–0.5 m)	Safflower	3.3–6.6 (1.0–2.0 m)
Citrus	3.3–5.9 (1.0–1.8 m)	Sorghum	3.3–6.6 (1.0–2.0 m)
Corn	3.3–5.6 (1.0–1.7 m)	Soybean	2.0–4.3 (0.6–1.3 m)
Cotton	3.3–5.6 (1.0–1.7 m)	Shrub	2.0 (0.6 m)
Cucumber	2.3–4.0 (0.7–1.2 m)	Turfgrass	1.6–3.3 (0.5–1.0 m)

After Allen et al., 1998.

Example 27.3: Determine the root zone depth for corn at early tassel assuming that the depth at germination is 6 in. (0.15 m), the maximum root depth is 4 ft (1.2 m), full depth occurs 90 days after germination, and early tassel occurs 50 days after germination.

Solution:

$$R_f = \frac{D_{ag}}{D_{tm}} = \frac{50 \text{ days}}{90 \text{ days}} = 0.56$$

Given the minimum root depth $R_{dmin} = 6$ in. $= 0.5$ ft, the maximum root depth $R_{dmax} = 4$ ft.

$$\begin{aligned} R_d &= R_{dmin} + (R_{dmax} - R_{dmin}) \times R_f \\ &= 0.5 \text{ ft} + (4 - 0.5) \text{ ft} \times 0.56 \\ &= 2.46 \text{ ft.} \end{aligned}$$

The root zone depths for various stages of development are given in Table 27.9. Plants do not extract water uniformly throughout the rooting depth. Usually there is more water extracted from shallow depths and less from greater depths. An approximation of this variation is shown in Figure 27.7. The average moisture extraction from the plant root zone follows a 4–3–2–1 rule, as shown in Figure 27.8.

This concept applies only if the root zone depth is completely refilled to field capacity during irrigation. If the root zone depth is not completely refilled to field capacity then more water will be obtained from shallower depths. Under these conditions there is usually a sandwiched layer of drier soil between the upper and lower parts of the root zone.

27.4.4 Soil Water Maintenance

With soil water maintenance, the plant's water needs are assumed to be met as long as the soil water is maintained between *TAW* and *MB*. An important variable in irrigation scheduling is *AD*. The time interval between two irrigations is controlled by the soil water deficit and evapotranspiration. The maximum time interval between irrigations, T_{max}, is

$$T_{max} = \frac{AD}{ET_c}, \tag{27.9}$$

where ET_c is the crop evapotranspiration.

Table 27.9 *Example root zone information for various annual crops*

	Corn	Grain sorghum	Soybean	Winter wheat	Cotton
Growing stage	Assumed root depth (m)				
Initial	0.3	0.30	0.3	0.3	0.3
Crop development	>>	>>	>>	>>	>>
Midseason	>>	>>	>>	>>	>>
Late season	1.0	1.4	1.4	1.0	1.4

>> means that the root depth increases linearly during this stage.
Data from the FAO: www.fao.org/land-water/databases-and-software/crop-information.

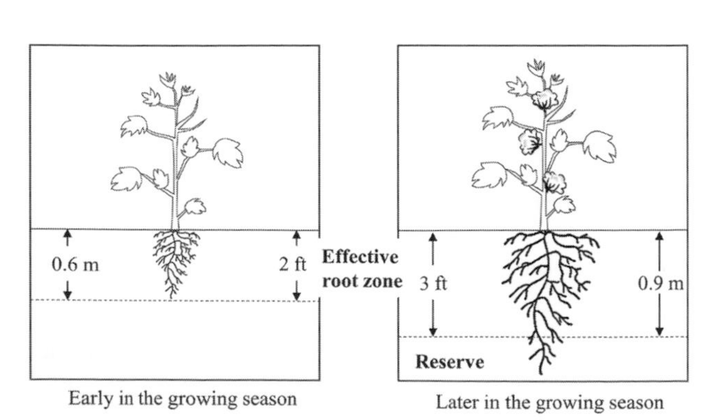

Figure 27.7 Effective root zone early and later in the growing season.

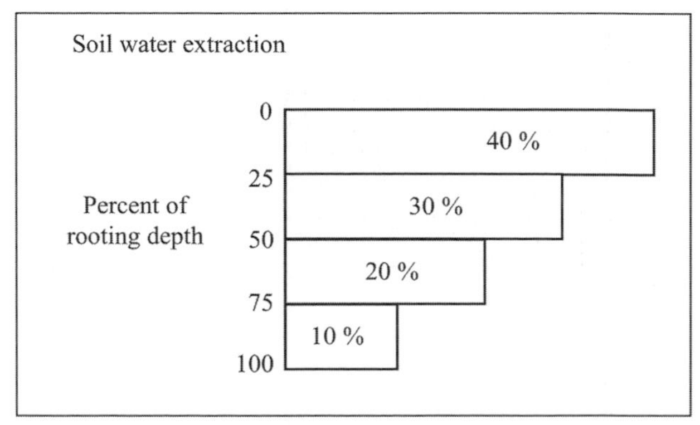

Figure 27.8 Average moisture extraction from the plant root zone (4–3–2–1 rule).

Example 27.4: What would be the value of T_{max} if the allowed deficit was 2 in. (50.8 mm) and if ET_c was 0.3 in./day (7.6 mm/day)?

Solution:

$$T_{max} = \frac{AD}{ET_c} = \frac{2\text{ in.}}{0.31\ \frac{\text{in.}}{\text{day}}} = 6.5\text{ days} \approx 6\text{ days}.$$

The calculated irrigation interval can be rounded off to the nearest integer number of days. In some cases it is also acceptable to round up if the value is very close to the next highest integer number. The irrigation interval can be fractional if an automated sprinkler system is used. The range of irrigation intervals for agricultural crops is usually 0.25–8 days. The root zone depth, crop evapotranspiration, and the available water capacity of the soil influence the frequency and amount of irrigation. A shallow root zone requires more frequent irrigation, but light application. A coarse-textured soil with a lower available water capacity requires lighter and more frequent irrigation. Medium-textured soils combined with deep root zones allow for less frequent irrigation and greater water application.

The irrigation time interval does not have to be equal to the value of T_{max} and can be less. It is controlled by ET_c and the effective depth of water application:

$$T = \frac{d_e}{ET_o}, \tag{27.10}$$

where T is the time interval between irrigations, d_e is the effective depth of water applied per irrigation, and ET_c is the crop evapotranspiration.

Many of the modern automatic irrigation methods are managed to apply light and frequent irrigations even when the root zones are deep and AWC is large. For example, a center-pivot irrigation system might be managed to apply an effective depth of 0.8 in. (20 mm) even if AD is much larger. Suppose $SWD = AD = 0.7$ in. (18 mm) on the day of irrigation. The effective application of 0.8 in. (20 mm) is okay as long as the irrigation frequency is adjusted accordingly. Using the earlier example in which ET_c is 0.3 in./day, the appropriate interval would be:

$$T = \frac{0.8\text{ in.}}{0.3\text{ in./day}} = 2.7\text{ days} \cong 2\text{ days}.$$

The latest date (LD) is the date by which irrigation should occur before the soil moisture reservoir reaches a critical limit (AD), as described in Figure 27.9.

The basic goals of irrigation management are that the deficit does not exceed AD before water is applied and infiltration does not exceed SWD. To avoid exceeding AD, irrigation should occur on or before LD, which is defined as

$$LD = \frac{AD - SWD}{ET_c}, \tag{27.11}$$

or using the balance approach:

$$LD = \frac{WB - MB}{ET_c}, \tag{27.12}$$

where WB is the available water balance and MB is the minimum balance.

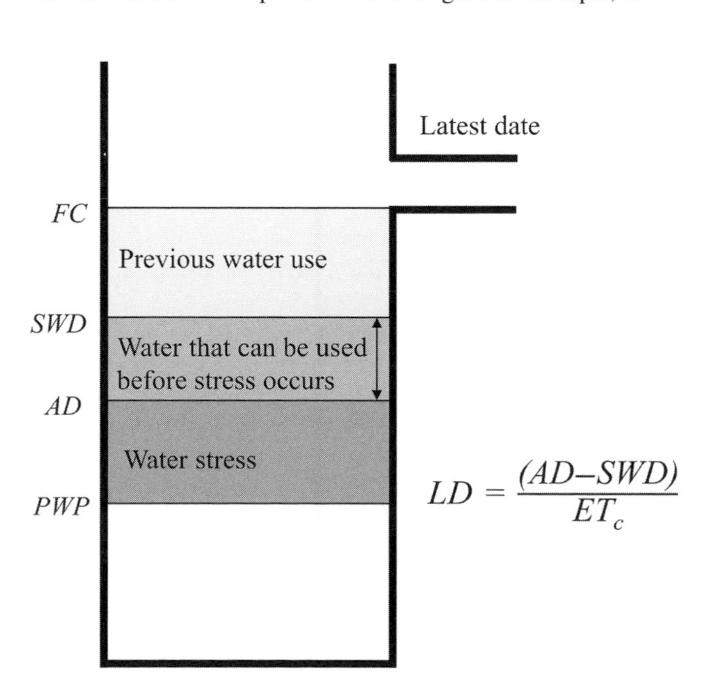

Figure 27.9 Illustration of the latest day (LD) concept.

For uniform (non-layered) soils, WB can be written as

$$WB = \left(\theta_v - \theta_{wp}\right) R_d, \tag{27.13}$$

where θ_v is the current soil water content, θ_{wp} is the soil water content at the permanent wilting point, and R_d is the root depth.

Example 27.5: Field beans (crop group 3; see Table 27.6) are being grown on a fine sandy loam soil $AWC = 0.13$ in./in. The appearance and feel method of soil moisture measurement indicates that the average $f_r = 0.8$ in the root zone. Determine the latest date of irrigation (LD). Assume the root zone depth (R_d) is 24 in. (0.61 m) and ET_c is 0.31 in./day (7.9 mm/day).

Solution: Given $AWC = 0.13$ in./in., $R_d = 24$ in. (0.61 m), the total available water:

$$TAW = R_d \times AWC = 24 \text{ in.} \times 0.13 \frac{\text{in.}}{\text{in.}}$$
$$= 3.12 \text{ in. } (79.2 \text{ mm}).$$

Given $ET_c = 0.31$ in./day, for field bean, $f_{dmax} = 0.38$ (Table 27.6). AD is:

$$AD = f_{dmax} \times TAW = 0.38 \times 3.12 \text{ in.}$$
$$= 1.19 \text{ in. } (30.2 \text{ mm}).$$

Given the fraction of available soil water remaining $f_r = 0.8$, the soil water deficit is:

$$SWD = TAW \times (1 - f_r) = 3.12 \text{ in.} \times (1 - 0.8)$$
$$= 0.62 \text{ in. } (15.7 \text{ mm}).$$

$$LD = \frac{AD - SWD}{ET_c} = \frac{1.19 \text{ in.} - 0.62 \text{ in.}}{0.31 \frac{\text{in.}}{\text{day}}} = 1.84 \cong 2 \text{ days.}$$

Example 27.6: Provide an alternative solution (balance approach) to Example 27.5.

Solution: From Example 27.5, one gets $TAW = 3.12$ in. and $f_{dmax} = 0.38$. Then, the minimum allowable fraction depletion is:

$$f_{rmin} = 1 - f_{dmax} = 1 - 0.38 = 0.62,$$

$$MB = f_{rmin} \times TAW = 0.62 \times 3.12 \text{ in.} = 1.93 \text{ in. } (49.0 \text{ mm}),$$

$$WB = TAW \times (f_r) = 3.12 \text{ in.} \times (0.8) = 2.50 \text{ in. } (63.5 \text{ mm}),$$

$$LD = \frac{WB - MB}{ET_c} = \frac{2.50 \text{ in.} - 1.93 \text{ in.}}{0.31 \frac{\text{in.}}{\text{day}}} = 1.84 \cong 2 \text{ days.}$$

The system should water this location in the irrigated area within two days to prevent plant stress. If it takes three days to get there, irrigation will be late by one day. Usually, a beginning or starting position and an ending or stop position are designated within the irrigated area. A record should be kept of each position so that irrigation occurs before AD is exceeded at either position.

Another goal is not to irrigate too soon, called the earliest date (ED), as described in Figure 27.10. There must be room to store the planned effective depth of irrigation water (d_{ep}). In addition, in humid and semi-humid regions it is good to allow room in the soil profile for storing rainfall that might occur immediately following irrigation. This is called rainfall allowance, r_a.

ED is calculated as:

$$ED = \frac{r_a + d_{ep} - SWD}{ET_c}. \tag{27.14}$$

This concept is illustrated in Figure 27.10. ED can also be calculated as

$$ED = \frac{r_a + d_{ep} - (TAW - WB)}{ET_c}. \tag{27.15}$$

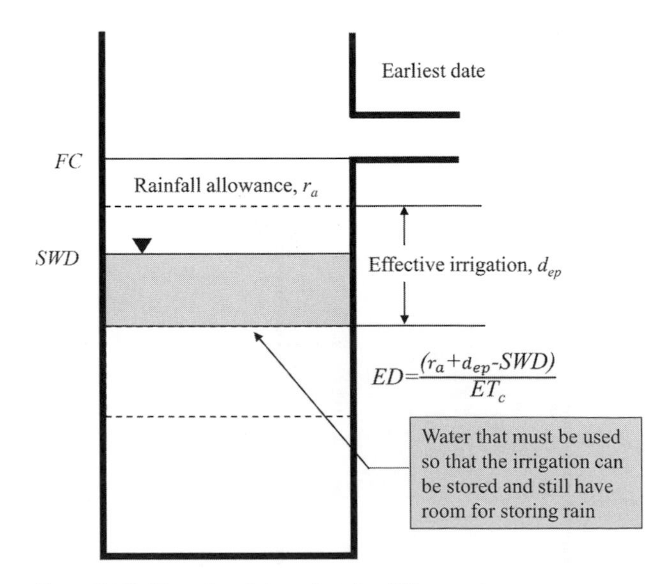

Figure 27.10 Illustration of the earliest date (ED) concept.

Example 27.7: Suppose $d_{ep}=0.5$ in. (12.7 mm), $r_a=0.4$ in. (10.2 mm), $ET_c=0.3$ in./day (7.6 mm/day), and $SWD=0.6$ in. (15.24 mm). Determine the earliest date that one should irrigate.

Solution:

$$ED=\frac{r_a+d_{ep}-SWD}{ET_c}=\frac{0.4+0.5-0.6}{0.3}=1\text{ day}.$$

From the previous example, since *LD* was two days, the irrigation should occur either one or two days from now.

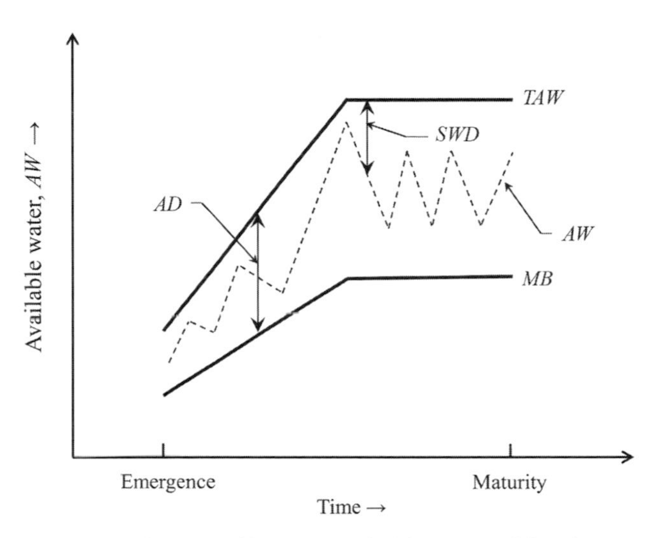

Figure 27.11 Illustration of key irrigation scheduling terms and their changes with time for annual crops.

Figure 27.11 illustrates key irrigation scheduling terms and their changes with time for annual crops. An example of irrigation scheduling using a center-pivot irrigation system is shown in Figure 27.12, and the typical outputs of the irrigation schedule are listed in Table 27.10. The depletion updated table includes evapotranspiration (water used), irrigation and rain, and soil water depletion measured in the field at the start and stop position on the observed date. The forecast depletion table provides information required to schedule irrigation, but still requires judgment. For example, irrigation could be applied when a negative sign (–) is used up. An asterisk (*) indicates that water depletion exceeds the allowable depletion. The schedule table shows a specific irrigation schedule.

27.4.5 Checkbook Accounting Method

One of the important variables in the *LD* and *ED* calculations is *SWD*. *SWD* can be determined either by direct measurement or by checkbook accounting. The checkbook

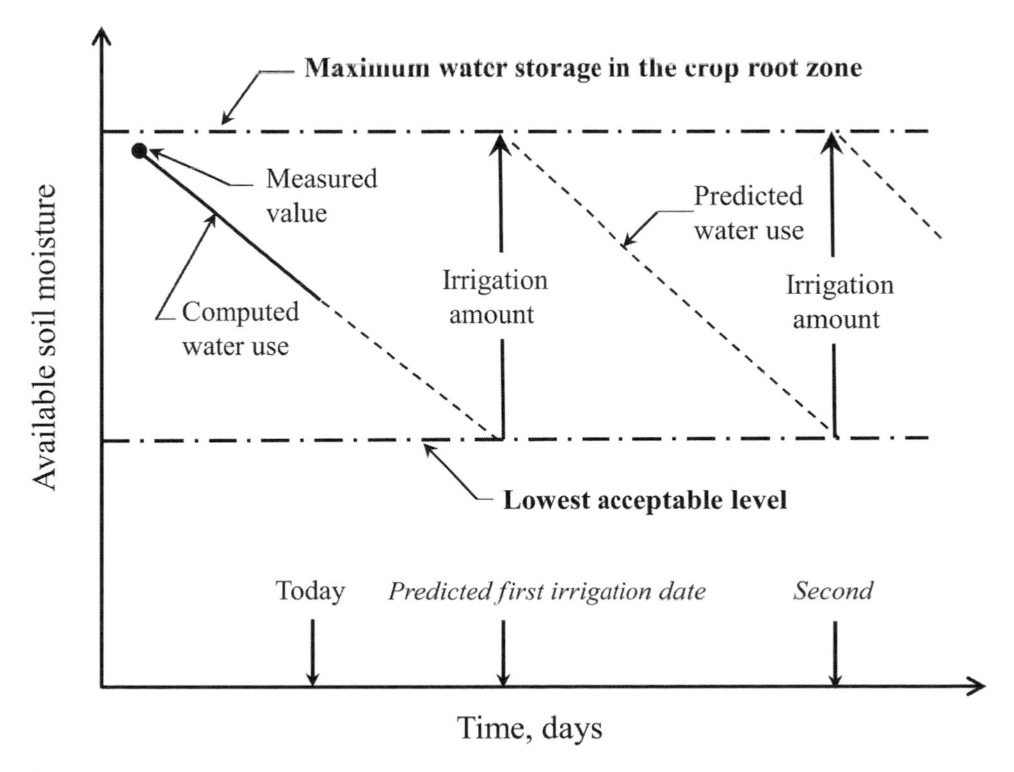

Figure 27.12 Irrigation scheduling: center-pivot system.

Table 27.10 *Example output of irrigation scheduling: center-pivot system*

Depletion updated

Field example. Crop: corn/Date: July 30

Day	Water used (mm)	Irrigate and rain (mm)	Application	Depletion (mm)	
				Start position	Stop position
July 23	6.0			6.0	10.2
July 24	5.9	17.8	Irrigate	0.0	0.0
July 25	5.6	15.2	Rain	0.0	0.0
July 26	1.2			1.2	0.0
July 27	5.6			6.8	5.6
July 28	5.0			11.8	10.6
July 29	7.6			19.5	18.2

Forecast depletion table

Maximum useful irrigation amounts

Day	Start position (mm)	Stop position (mm)
July 30	26.0 (–)	24.8 (–)
July 31	32.5 (–)	31.2 (–)
Aug 1	38.9 (–)	37.7 (–)
Aug 2	45.3 (–)	44.1 (–)
Aug 3	51.7 (–)	50.5 (–)
Aug 4	58.1 (–)	56.9 (–)
Aug 5	64.4 (–)	63.2 (–)
Aug 6	70.7 (–)	69.5 (–)
Aug 7	76.9 (*)	75.7 (*)
Aug 8	83.0 (*)	81.8 (*)

(–) indicates that the forecasted depletion exceeds the average application depth of 20.3 mm.
(*) indicates that the forecasted depletion exceeds the allowable depletion of 73.2 mm.

Schedule

If 20.3 mm are applied in 60 h, the starting times are:

Amount of rain (mm)	Earliest date	Latest date
0	July 30	Aug 5
5	July 30	Aug 6
10	July 31	Aug 7
15	Aug 1	Aug 8
20	Aug 2	Aug 8

Assuming the system was started on July 30, the next starting times are:

Amount of rain (mm)	Earliest date	Latest date
0	Aug 2	Aug 8
5	Aug 3	Aug 9
10	Aug 3	Aug 10
15	Aug 4	Aug 11
20	Aug 5	Aug 12

Assuming the system was started on Aug 2, the next starting times are:

Amount of rain (mm)	Earliest date	Latest date
0	Aug 5	Aug 12
5	Aug 6	Aug 13
10	Aug 7	Aug 14
15	Aug 7	Aug 15
20	Aug 8	Aug 16

Adapted from Harrington and Heermann, 1981.

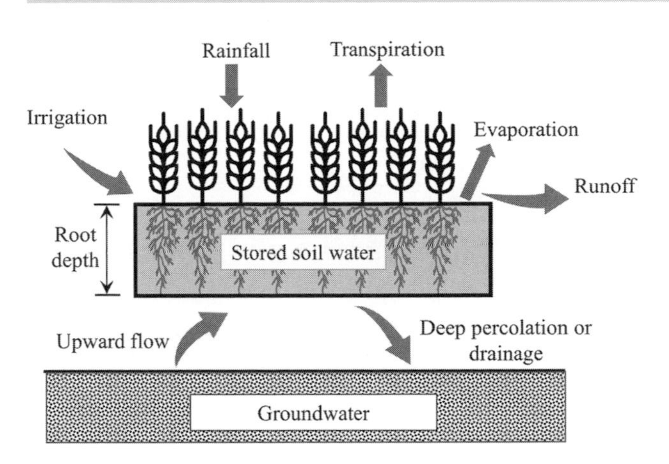

Figure 27.13 Soil water balance.

Figure 27.14 Additions and subtractions from the plant root zone.

accounting or water balance approach (Figure 27.13) can be used to schedule irrigation. This approach accounts for all the additions and withdrawals to and from the root zone (Figure 27.14) (a similar figure can be found in Cassel, 1984).

SWD can be calculated as

$$SWD_i = SWD_{i-1} + ET_{ci} - d_{ei} - P_{ei} - U_{fi}, \tag{27.16}$$

where SWD_i is *SWD* on day i, SWD_{i-1} is the *SWD* on day $i-1$, ET_{ci} is the ET_c on day i, d_{ei} is effective irrigation depth on day i, P_{ei-1} is the effective precipitation on day i, and U_{fi-1} is the upward flow of groundwater on day i.

The water balance equation can be expressed as

$$WB_i = WB_{i-1} + ET_{ci} - d_{ei} - P_{ei} - U_{fi}, \tag{27.17}$$

where WB_i is the *WB* on day i, and WB_{i-1} is the *WB* on day $i-1$. It may be noted that runoff and deep percolation are not considered in eqs (27.16) and (27.17). This is because effective precipitation and effective irrigation account for them. If the infiltrated depth of water from precipitation and irrigation exceeds *SWD*, then the effective depth equals *SWD*.

Using the water balance is analogous to keeping the balance in the checkbook. *WB* is the balance, irrigation and rainfall are the deposits, and ET_c is the withdrawal. If *WB* becomes less than the minimum allowable balance (*MB*), a penalty is paid in terms of yield loss.

An initial estimate of *SWD* or *WB* is required to start the process. This may be a measured value using the soil moisture measuring techniques, or the process can be started following a wet period or through irrigation when soils can be assumed to be at or near field capacity. Crop evapotranspiration can be calculated from weather data. A question that often arises is: What should be used as the forecast ET_c in the *WB* calculations? The forecast ET_c can be based on long-term averages for an area or it can be based on the past few days. If the weather is forecast to be similar to that which just occurred, then it can be assumed that the forecast ET_c is equal to the ET_c of the prior few days.

For upward flow, when a water table exists close to or near the root zone, crops may extract water from the capillary fringe, or water may flow upward into the crop root zone. Upward flow can be significant in areas where the required irrigation rate is low due to rain or mild climate. In areas where saline irrigation water is used, leaching is necessary to remove salts from the crop root zone. Generally, that water should not be counted on for crop water use.

For annual crops where the root zone depth expands with time, the *WB* and *SWD* calculations should consider the soil water conditions that the roots are growing into.

Example 27.8: Corn is being grown on a salt loam soil in two different locations with the following variables: $f_{dmax} = 0.45$, $R_d = 2.5$ ft, $AWC = 0.20$ in./in., rainfall allowance $r_a = 0.5$ in., depth to the water table is 10 ft, and the planned irrigation depth (d_{ep}) is 1.1 in. *SWD* at the start of June 25 at location 1 was 2.2 in. and at location 2 it was 0.8 in. The crop ET_c and rainfall are given as follows:

Date	Actual ET_c (in. [mm])	Forecast ET_c (in. [mm])	P_e (in. [mm])
June 25	0.20 (5.08)	0.18 (4.57)	0.00
June 26	0.21 (5.33)	0.18 (4.57)	0.00
June 27	0.13 (3.30)	0.18 (4.57)	0.30 (7.62)
June 28	0.17 (4.32)	0.18 (4.57)	0.00

Determine the earliest irrigation date (*ED*) and the latest irrigation date (*LD*) for each location for each of the given dates.

Solution:

$$TAW = R_d \times AWC$$
$$= (2.5\text{ ft}) \times (0.2\text{ in./in.}) \times (12\text{ in./ft})$$
$$= 6.0\text{ in. } (152.4\text{ mm}),$$

$$AD = f_{dmax} \times TAW = 0.45 \times (6.0\text{ in.})$$
$$= 2.7\text{ in. } (68.6\text{ mm}).$$

Water balance at Location 1:

June 25

Given the depth between the groundwater and root zone is 10 ft − 2.5 ft = 7.5 ft, $U_f = 0$ from Figure 27.15, *SWD* at the end of day:

$$SWD_i = SWD_{i-1} + ET_{ci} - d_{ei} - P_{ei} - U_{fi}$$
$$= 2.2 + 0.20 - 0 - 0 - 0 = 2.40 \text{ in. } (61.0 \text{ mm}).$$

June 26

$$SWD_i = SWD_{i-1} + ET_{ci} - d_{ei} - P_{ei} - U_{fi}$$
$$= 2.4 + 0.21 - 0 - 0 - 0 = 2.61 \text{ in. } (66.3 \text{ mm}).$$

June 27

Since *AD* is 2.7 in., irrigation is required today, and $d_e = d_{ep} = 1.1$ in., at the end of June 27,

$$SWD_i = SWD_{i-1} + ET_{ci} - d_{ei} - P_{ei} - U_{fi}$$
$$= 2.61 + 0.13 - 1.1 - 0.3 - 0$$
$$= 1.34 \text{ in. } (34.0 \text{ mm}).$$

June 28

$$SWD_i = SWD_{i-1} + ET_{ci} - d_{ei} - P_{ei} - U_{fi}$$
$$= 1.34 + 0.17 - 0 - 0 - 0 = 1.51 \text{ in. } (38.4 \text{ mm}).$$

***ED* calculation:**

Given $r_a = 0.50$ in., $d_{ep} = 1.1$ in., eq. (27.14) is used to calculate *ED*:

$$ED = \frac{r_a + d_{ep} - SWD}{ET_c(\text{forecast})}.$$

June 25: $ED = \frac{0.50 + 1.10 - 2.40}{0.18} = -4.4$ days.

June 26: $ED = \frac{0.50 + 1.10 - 2.61}{0.18} = -5.6$ days.

June 27: $ED = \frac{0.50 + 1.10 - 1.34}{0.18} = 1.4$ days (use 1 day).

June 28: $ED = \frac{0.50 + 1.10 - 1.51}{0.18} = 0.5$ day (use 0 day).

***LD* calculation:**

Since $AD = 2.7$ in., eq. (27.11) is used to calculate *LD*:

$$LD = \frac{AD - SWD}{ET_c(\text{forecast})}$$

June 25: $LD = \frac{2.70 - 2.40}{0.18} = 1.7$ days (use 1 day)

June 26: $LD = \frac{2.70 - 2.61}{0.18} = 0.5$ day (use 0 day).

June 27: $LD = \frac{2.70 - 1.34}{0.18} = 7.6$ days (use 7 days).

June 28: $LD = \frac{2.70 - 1.51}{0.18} = 6.6$ days (use 6 days).

The results are summarized in the following table.

Date	Actual ET_c (in. [mm])	Forecast ET_c (in. [mm])	P_e (in. [mm])	*SWD* (in.)	d_{ep}	*ED*	*LD*
June 25	0.20 (5.08)	0.18 (4.6)	0.00	2.20 (55.9) 2.40 (61.0)	0.00	××	1 day
June 26	0.21 (5.33)	0.18 (4.6)	0.00	2.40 (61.0) 2.61 (66.3)	0.00	××	0 day
June 27	0.13 (3.3)	0.18 (4.6)	0.30 (7.6)	2.61 (66.3) 1.34 (34.0)	1.10 (27.9)	1 day	7 days
June 28	0.17 (4.3)	0.18 (4.6)	0.00	1.34 (34.0) 1.51 (38.4)	0.00	0 day	6 days

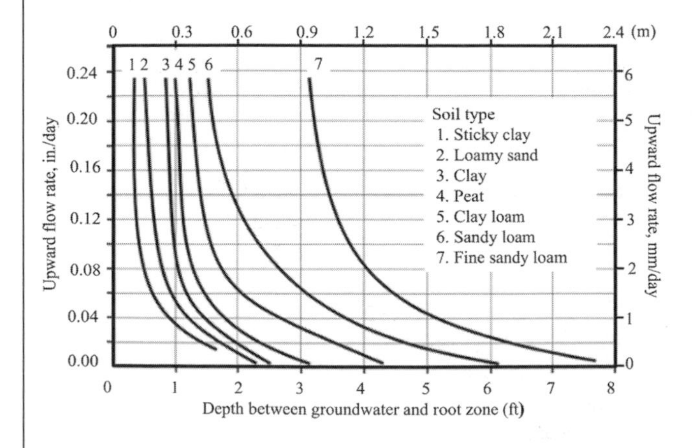

Figure 27.15 Upward flow of water from a groundwater table. (Adapted from Doorenbos and Pruitt, 1977.)

Water balance at Location 2:

June 25

Given the depth between the groundwater and root zone is 10 ft – 2.5 ft = 7.5 ft, $U_f = 0$ from Figure 27.15, *SWD* at the end of day is:

$$SWD_i = SWD_{i-1} + ET_{ci} - d_{ei} - P_{ei} - U_{fi}$$
$$= 0.80 + 0.20 - 0 - 0 - 0 = 1.00 \text{ in. } (25.4 \text{ mm}).$$

June 26

$$SWD_i = SWD_{i-1} + ET_{ci} - d_{ei} - P_{ei} - U_{fi}$$
$$= 1.00 + 0.21 - 0 - 0 - 0 = 1.21 \text{ in. } (30.7 \text{ mm}).$$

June 27

$$SWD_i = SWD_{i-1} + ET_{ci} - d_{ei} - P_{ei} - U_{fi}$$
$$= 1.21 + 0.13 - 0 - 0.3 - 0 = 1.04 \text{ in. } (26.4 \text{ mm}).$$

June 28

$$SWD_i = SWD_{i-1} + ET_{ci} - d_{ei} - P_{ei} - U_{fi}$$
$$= 1.04 + 0.17 - 0 - 0 - 0 = 1.21 \text{ in. } (30.7 \text{ mm}).$$

***ED* calculation:**

Given $r_a = 0.5$ in., $d_{ep} = 1.1$ in., eq. (27.14) is used to calculate *ED*:

$$ED = \frac{r_a + d_{ep} - SWD}{ET_c(\text{forecast})}$$

June 25: $ED = \frac{0.50 + 1.10 - 1.00}{0.18} = 3.3$ days (use 3 days).

June 26: $ED = \frac{0.50 + 1.10 - 1.21}{0.18} = 2.2$ days (use 2 days).

June 27: $ED = \frac{0.50 + 1.10 - 1.04}{0.18} = 3.1$ days (use 3 days).

June 28: $ED = \frac{0.50 + 1.10 - 1.21}{0.18} = 2.2$ days (use 2 days).

***LD* calculation:**

Since $AD = 2.7$ in., eq. (27.11) is used to calculate *LD*:

$$LD = \frac{AD - SWD}{ET_c(\text{forecast})}$$

June 25: $LD = \frac{2.7 - 1.00}{0.18} = 9.4$ days (use 9 days).

June 26: $LD = \frac{2.7 - 1.21}{0.18} = 8.3$ days (use 8 days).

June 27: $LD = \frac{2.7 - 1.04}{0.18} = 9.2$ days (use 9 days).

June 28: $LD = \frac{2.7 - 1.21}{0.18} = 8.3$ days (use 8 days).

The results are summarized in the following table.

Date	*Actual* ET_c (in. [mm])	*Forecast ET* (in. [mm])	P_e (in. [mm])	*SWD* in. [mm])	d_{ep} in. [mm])	*ED*	*LD*
June 25	0.20 (5.1)	0.18 (4.6)	0.00 (0)	0.80 (20.3) 1.00 (25.4)	0.00 (0)	3 days	9 days
June 26	0.21 (5.3)	0.18 (4.6)	0.00 (0)	1.00 (25.4) 1.21 (30.7)	0.00 (0)	2 days	8 days
June 27	0.13 (3.3)	0.18 (4.6)	0.30 (7.6)	1.21 (30.7) 1.04 (26.4)	0.00 (0)	3 days	9 days
June 28	0.17 (4.3)	0.18 (4.6)	0.00 (0)	1.04 (26.4) 1.21 (30.7)	0.00 (0)	2 days	8 days

Two cases using checkbook accounting are shown in Figures 27.16 and 27.17. The programmed soil moisture depletion using checkbook accounting was compared with the measured data in the field. Figure 27.16: In this case, f_r is maintained between 0.4 and 0.7 throughout the growing season. Figure 27.17: In this case f_r is allowed to decline over the season.

27.4.6 Simplified Application of ET_c Data

A major limitation of the checkbook accounting method is sometimes the lack of reliable real-time ET_r data. Weather data are used to calculate reference crop ET_r and crop ET_c on a continuous basis. With the easier availability of *ET* data now, the biggest problem is the estimation of effective irrigation depths. Water measurement is the key to this management problem. Once water is measured, effective depths can be determined. The checkbook accounting procedure requires daily record keeping. Computer software can be used to perform calculations, but daily record keeping is necessary.

One way that irrigators apply ET_c data without daily recording is to simply irrigate when the effective depth has been consumed. For example, if ET_c is 0.25 in./day and the effective depth is 0.75 in. per application, then irrigation must be applied every three days. Thus, the water manager is reacting to the amount of water applied and ET_c. To adjust for rainfall, irrigation can be delayed in accordance with how long it will take to consume the rainfall. If 0.5 in. of rainfall

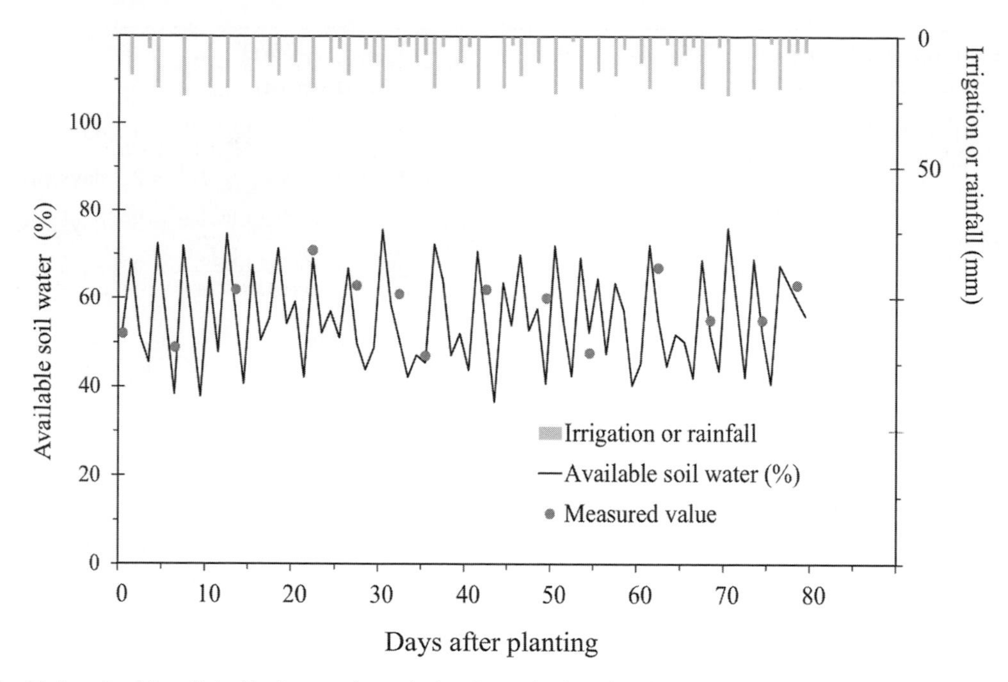

Figure 27.16 Graphical results of the soil checkbook accounting method. Soil water levels are kept between 0.4 and 0.7 throughout the growing season.

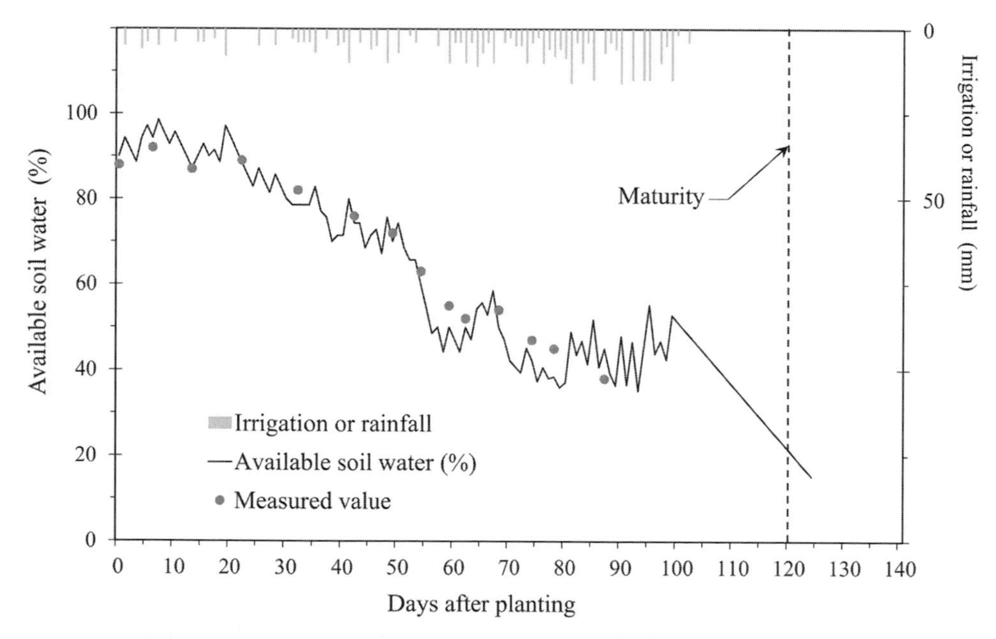

Figure 27.17 Graphical results of the soil checkbook accounting method. Soil water levels are allowed to decline over the season.

occurs, the irrigation should be delayed by two days, assuming one was not behind with irrigation before the rainfall occurred and all the rainfall was effective in satisfying ET_c.

Another simple checkbook accounting approach is to adjust the effective depth of application according to the amount of ET_c and rainfall that occurred over some pre-established time interval. Suppose weekly irrigations are desired. Then, the ET_c and effective precipitation is summed for the time interval. The effective irrigation is calculated as

$$d_{ep} = \sum ET_c - \sum P_e. \tag{27.18}$$

Example 27.9: The daily ET for a turf grass and the effective precipitation (P_e) for a one-week period is given below. How much effective irrigation water is needed to make up the balance between ET_c and rainfall?

Day	ET_c (in. [mm])	P_e (in. [mm])
1	0.20 (5.08)	–
2	0.30 (7.62)	0.50 (12.70)
3	0.15 (3.81)	–
4	0.25 (6.35)	0.20 (5.08)
5	0.20 (5.08)	–
6	0.25 (6.35)	–
7	0.25 (6.35)	–
Total	1.60 (40.64)	0.70 (17.78)

Solution: $d_{ep} = \sum ET_c - \sum P_e$

The effective irrigation is calculated as $1.60 - 0.70 =$ 0.90 in. (22.86 mm).

A drawback to the two simpler checkbook accounting approaches is that the water applications lag behind the time of water use. On soils with relatively low available water-holding capacities and/or plants with shallow root zones, the lag may cause some moisture stress before water is applied.

27.4.7 Water Balance Approach

The purpose of water balance is to calculate the soil moisture depletion in the root zone, which is needed for irrigation scheduling. The soil water balance at any given time can be expressed as the algebraic sum of the supply of water to the soil by irrigation and precipitation, and the loss of water by crop evapotranspiration and deep percolation, as well as the contribution from groundwater. The water balance can thus be expressed as

$$SMD_i = SMD_{i-1} + ET_{ci} + DP_i - d_{ei} - P_{ei} + U_{fi}, \quad (27.19)$$

where subscript i refers to the time index, SMD_i is the total soil moisture depletion in the root zone defined as the difference between total soil moisture stored in the root zone at field capacity and the moisture status at time i; ET_{ci} is the crop evapotranspiration at time i; DP_i is the deep percolation at time i; d_{ei} is the effective depth of water applied per irrigation; P_{ei} is the effective precipitation; and U_{fi} is the upward flow of groundwater on day i due to capillary rise or groundwater contribution.

Each term in eq. (27.19) is either computed or measured. Crop evapotranspiration is computed using one of the methods discussed in Chapters 14 and 15. Precipitation is measured and the effective amount is then computed. The amount of irrigation water is known from measurements. The contribution of groundwater is computed and if the groundwater table is too far below, it can be assumed negligible. Deep percolation is computed as discussed in Chapter 13. The soil moisture depletion at the beginning $(i - 1)$ can be determined from the measured value of moisture content and effective root zone depth as:

$$SMD_{i-1} = \left(\theta_{fc} - \theta_{i-1}\right) D_{rz}, \quad (27.20)$$

where SMD_{i-1} is the soil moisture depletion, D_{rz} is the effective root zone depth that increases during the growing season and reaches a maximum depth, θ_{fc} is the moisture content at field capacity, and θ_{i-1} is the initial soil moisture content.

Once SMD_i is known, irrigation scheduling can be determined based on several criteria, such as fixed depth, fixed interval, or MAD. If the fixed depth criterion is used, then irrigation is required when SMD becomes equal to the depth of irrigation. If the fixed time interval is used then the estimated irrigation requirement is equal to SMD at the end of the interval. For the MAD criterion, the AD and TAW must be determined. TAW is the difference between water content at field capacity and permanent wilting point. In this case, both the day of irrigation and the depth are calculated. Now, AD can be computed as:

$$AD = TAW \times MAD. \quad (27.21)$$

It should be noted that MAD is the management allowed depletion, which is defined as the fraction of TAW that can be safely removed from the soil to meet the daily ET demand on day i. MAD values for different crops can be found in Table 13.3. When SMD reaches AD on day i, the field is irrigated. In this case, the irrigation depth required equals SMD.

For drip-irrigated crops, irrigation can be scheduled based on actual crop evapotranspiration because drip systems apply low-volume irrigation at frequent intervals. Considering leaching fraction, the irrigation depth can be computed as:

$$I_{nl} = \frac{SMD}{1 - LR}, \quad (27.22)$$

where I_{nl} is the net irrigation depth after considering leaching requirement, SMD is the soil moisture depletion, and LR is the leaching requirement expressed as a fraction. The gross irrigation depth can then be determined by substituting I_{nl} in place of d_{ei} in eq. (27.19).

27.4.8 USDA Water Balance Technique

The USDA-SCS (1970) developed a widely used water balance-based computer program for irrigation scheduling, which was modified by Jensen et al. (1971). The program computes the soil water depletion as

$$D=\sum_{i=1}^{n}(ET_c - P_e - d_e + W_d)_i, \qquad (27.23)$$

where D is the water depletion in the root zone, P_e is the effective daily precipitation or precipitation minus runoff, d_e is the daily net irrigation, and W_d is the daily drainage from the root zone or upward movement from the saturated zone (with negative sign), and i is the daily time increment. After irrigation, D should be 0, unless sufficient water is not applied.

Ogata and Richards (1957) expressed the water content from a draining soil empirically as

$$W = cW_0 t^{-m}, \qquad (27.24)$$

where W is the water content in the soil (mm), W_0 is the water content when $t = 1$ day, t is the time in days after irrigation has ceased, m is a constant for a given soil, and c is a constant (dimensional). The drainage rate can be calculated by differentiating eq. (27.24):

$$\frac{dW}{dt} = -cmW_0 t^{-m-1}. \qquad (27.25)$$

Equation (27.25) can be cast, with the substitution of eq. (27.24), as

$$\frac{dW}{dt} = -mW\left(\frac{W}{cW_0}\right)^{\frac{1}{m}}. \qquad (27.26)$$

Jensen (1972) approximated the cumulative daily drainage as

$$W_d = \sum_{i=1}^{\infty} m(W_{i-1} - ET_i)\left(\frac{W_{i-1} - ET_i}{W_0}\right)^{\frac{1}{m}}, \qquad (27.27)$$

in which W_d is the cumulative drainage, ET is the daily evapotranspiration, and i is the ith day. The value of m varies from 0.1 to 0.15 for a range of soil textures (Miller and Aarstad, 1971).

For upward flow or negative drainage, Jensen (1972) suggested

$$W_d = \left(1 - \frac{AW - a_c}{100 - a_c}\right)\left(\frac{z_c}{z_w - R_d}\right)^n ET, \qquad (27.28)$$

where z_c is the effective height of the capillary fringe above the water table, z_w is the depth to the water table, R_d is the root zone depth, AW is the available water to R_d in percent, a_c is a constant having a value of about 25, and n is a constant for a given soil varying between 1 and 3.

The days to the next irrigation are calculated as

$$N = \frac{D_0 - D}{E\left(\frac{dD}{dt}\right)}, \qquad (27.29)$$

where N is the number of days to the next irrigation, D_0 is the current soil water depletion, D is the estimated depletion to the date of irrigation, and $E(dD/dt)$ is the expected (average) rate of soil water depletion until the next irrigation is required. The $E(dD/dt)$ average depletion rate can be calculated from the mean ET rate for the previous 3–5 days.

Now the amount of water to be applied by irrigation can be computed as

$$W_I = \frac{D_0}{E} \times 100; \; D_0 > D \qquad (27.30)$$

or

$$W_I = \frac{D}{E} \times 100; \; D > D_0, \qquad (27.31)$$

where W_I is the amount of irrigation water to be applied and E is the irrigation efficiency as a percentage.

Israelson and Hansen (1962) provided a guide for judging the amount of soil moisture that has been removed from the soil, as shown in Table 27.11.

27.4.9 Soil Water Measurement Method

An alternative or supplement to the checkbook accounting method is to measure soil water directly for irrigation scheduling. In concept it is simple. Rather than predicting or calculating *SWD*, *SWD* is inferred from measurements of soil moisture or soil water tension. Once *SWD* or *AW* is determined, *ED* and *LD* can be calculated. The soil water content must be measured throughout the root zone. Techniques such as feel and appearance, gravimetric sampling, neutron scattering, and TDR can be used to measure soil water content directly. Water content can then be used to calculate *LD* and *ED*. When soil water potential (soil water tension) is measured, as with the tensiometers or electrical resistance blocks, a soil water release curve is needed to convert tension to volumetric water content.

An alternative to converting soil water tension to water content is to monitor the soil water sensors frequently and irrigate when the soil water tension has reached a threshold level. In fact, manufacturers of soil water sensing equipment often provide users with guidelines for various crops and soil textures. They are usually based on sensing near the center of the root zone and then estimating a tension that corresponds to f_{dmax} for each soil texture. One problem with this approach is that it is difficult to predict ahead to determine *LD*. This can be overcome with frequent monitoring and graphical extrapolation. A limitation of the threshold level method is that the irrigator does not know how much water the soil can hold during irrigation. Table 27.12 lists threshold soil water tensions of different soils for irrigation scheduling. A graphical method of predicting the date of irrigation using the threshold level method is shown in Figure 27.18.

An important, and often frustrating, consideration is the number of locations that must be sampled to reliably estimate the average soil water condition within the area of interest. Not only must the spatial variability of the soil be

Table 27.11 *Guide for judging the amount of moisture that has been removed from the soil*

Soil moisture deficiency	Feel or appearance of soil and moisture deficiency in cm per meter of soil			
	Coarse texture	Moderately coarse texture	Medium texture	Fine and very fine texture
0% (field capacity)	Upon squeezing, no free water appears on soil, but a wet outline of the ball is left on the hand. 0.0	Upon squeezing, no free water appears but a wet outline of the ball is left on the hand. 0.0	Upon squeezing, no free water appears but a wet outline of the ball is left on the hand. 0.0	Upon squeezing, no free water appears but a wet outline of the ball is left on the hand. 0.0
0–25%	Tends to stick together slightly. Sometimes forms a very weak ball under pressure. 0.0–1.7	Forms a weak ball, breaks easily, will not stick. 0.0–3.4	Forms a ball, is very pliable, sticks readily if relatively high in clay. 0.0–4.2	Easily ribbons out between fingers, has sticky feeling. 0.0–5.0
25–50%	Appears to be dry, will not form a ball with pressure. 1.7–4.2	Tends to ball under pressure but seldom holds together. 3.4–6.7	Forms a ball that is somewhat plastic, will stick slightly with pressure. 4.2–8.3	Forms a ball, ribbons out between thumb and forefinger. 5.0–10.0
50–75%	Appears to be dry, will not form a ball with pressure.* 4.2–6.7	Appears to be dry, will not form a ball.* 6.7–10.0	Somewhat crumbly but holds together from pressure. 8.3–12.5	Somewhat pliable, will ball under pressure.* 10.0–15.8
75–100% (100% is permanent wilting)	Dry, loose, single-grained, flows through fingers. 6.7–8.3	Dry, loose, flows through fingers. 10.0–12.5	Powdery, dry, sometimes slightly crusted but easily broken down into powdery condition. 12.5–16.7	Hard, baked, cracked, sometimes has loose crumbs on the surface. 15.8–20.8

* The ball is formed by squeezing a handful of soil very firmly.
After Israelson and Hansen, 1962. Copyright © 1962 by John Wiley & Sons, New York.

Table 27.12 *Threshold soil water tensions for irrigation scheduling*

Texture	Threshold tension (cb)
Fine sand	30
Loamy sand	40
Sandy loam	50
Fine sandy loam	55
Silt loam	60
Clay loam	70

After USDA-NRCS, 1997.

considered, but also the variability of the water application of the irrigation system. A minimum of four locations should be sampled in a large irrigated area that has relatively uniform soils. Using the checkbook accounting method in conjunction with the soil water sensing method reduces the number of locations that must be sampled.

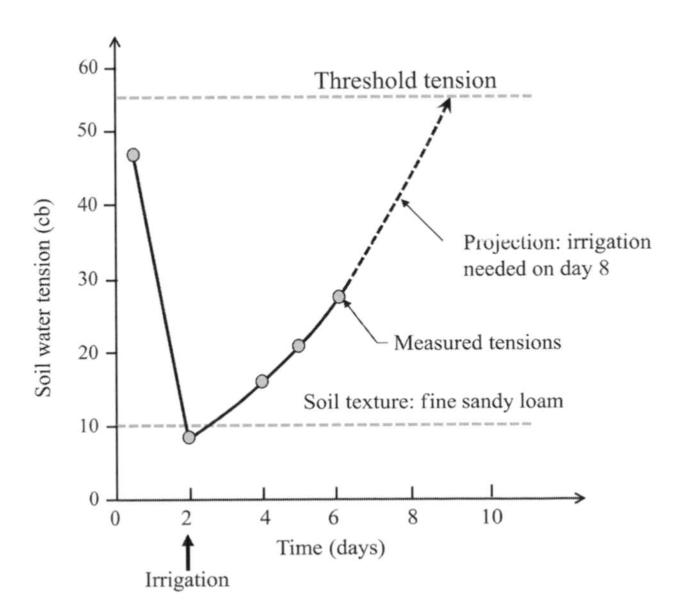

Figure 27.18 Graphical method for predicting date of irrigation.

27.4.10 Plant Status Indicators

Plant status indicators include leaf water potential, foliage–air temperature difference, other plant indicators, and stage of plant development. The techniques of scheduling irrigation, based on the management of soil water reservoir, do not directly evaluate all the factors influencing irrigation management. Plant status indicators integrate all of the factors (i.e., soil water conditions, atmospheric demand for water, and plant characteristics). All three factors are taken into account by selecting the appropriate f_{dmax} given in Table 27.6. Nevertheless, direct measures or indicators of plant water status are often suggested for irrigation scheduling.

The leaf water potential, a measure of the energy status of water in plant leaves, is an indicator of the water status of a plant. The relationship between various water potentials in the soil–plant–atmosphere continuum is shown in Figure 27.19.

Stegman (1983) found that threshold levels of leaf water potential ranged from –12 to –12.5 bar for corn (mid-afternoon readings). The thresholds were dependent on the ambient air temperatures, much like the term f_{dmax} is dependent on ET_c. While leaf water potential is a direct measure of plant water status, using it as a scheduling tool has some limitations. Threshold levels must be developed for the plants in question. Also, like using threshold levels of soil water potential, it takes predictability. Third, measuring leaf water potential is time-consuming and must be done during a narrow time window around midday. Further, a large number of samples is necessary for an accurate estimation of the mean.

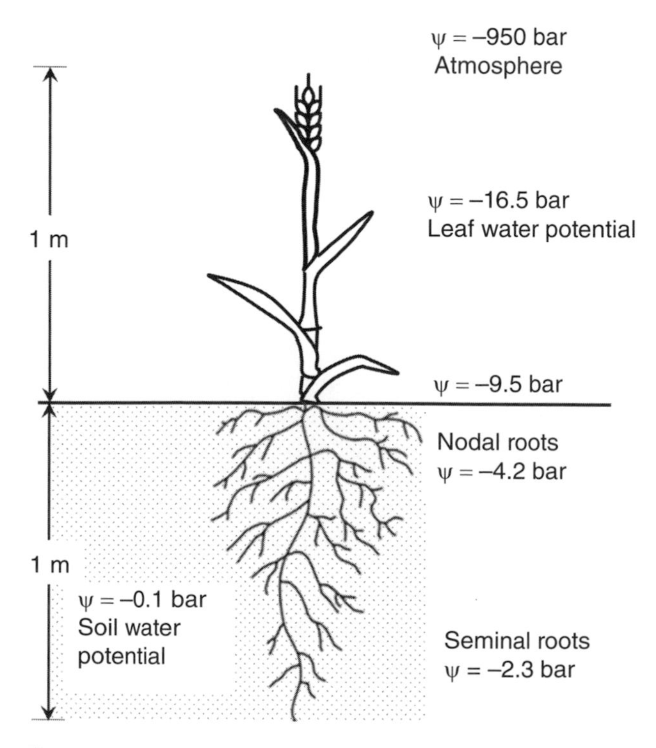

Figure 27.19 Water potential at different points for water transport through a wheat plant.

Since evaporation of water is a cooling process, the foliage of well-watered plants is usually cooler than the surrounding air, especially in arid climates. The temperature difference (*DT*) depends on both the plant water status and the vapor pressure deficit (*VPD*) of the air. Plant stress can be related to *DT* and *VPD*, and *DT* may be a very useful tool for scheduling because it is easy to measure with hand-held infrared thermometers for measuring leaf temperature. Some difficulties of the method, however, include the effects of bare soil, threshold values, and effect of wind and cloud cover and their relationship to irrigation management.

Other measures of plant stress include color, leaf rolling, and wilting. With many crops, stressed plants often turn a darker color if soil water stress occurs. This is particularly evident in turfgrass. Bluegrass, for example, will turn a blue color when under stress. While all of these visual techniques are useful, many often appear too late to be useful for water management. Large economic losses may have already occurred. Prediction is still a problem for these techniques.

It is often heard that irrigation scheduling at critical stages of plant growth is a good way of irrigation scheduling. While the method has merit, local calibration is necessary to account for soil–crop–climate conditions.

27.5 IRRIGATION MODELS

A number of mathematical models have been developed for scheduling irrigation (Lamacq and Wallender, 1994, Singh et al., 1995; George et al., 2000). Some of the models employ stochastic methods for irrigation scheduling by forecasting irrigation dates and amounts (Villalobos and Fereres, 1989; Raghuwanshi, 1994). For example, George et al. (2000) developed a user-friendly irrigation scheduling model for field crops. In irrigation scheduling it is important that soil–plant–water–atmosphere interactions are adequately accounted for. This therefore suggests that the scheduling models are integrated with recommendations for irrigation, fertilization, crop growth, and plant protection.

27.6 REAL-TIME IRRIGATION SCHEDULING

Real-time irrigation (RTI) scheduling entails measuring soil moisture on a continuous basis using a suitable monitoring device, employing a short-term weather forecasting capability, and determining an appropriate schedule for irrigation using a decision-support system that uses field moisture status, weather forecast, and crop cultural practices (Malano et al., 1996). Ideally, no part of the field has too much or too little irrigation. In cooperation with Michigan State University the USDA Soil Conservation Service (SCS) developed a computerized irrigation scheduling program utilizing real-time (daily) local weather station data to calculate a daily soil water balance for the crop root zone (Shayya and Bralts, 1988). The PUTU-any crop model was developed to provide weekly irrigation scheduling advice to farmers for any crop in South Africa (De Jager and Singels, 1990; De Jager, 1992; De Jager et al., 2001). Annandale et al. (1999) developed a crop model, SWB, for RTI scheduling. Singels and Smith (2006) developed a weather-based irrigation scheduling advice service (MyCanesim system) for sugarcane growers. Humbuckle et al. (2009) used a remote sensing method for assessing within-field crop health variations and linked it to reference ET values from nearby weather stations to provide field-specific scheduling information.

However, in developing countries, farm holdings are small and continuous measurement of soil moisture is beyond the reach of farmers because of the costs involved. Further, farmers usually do not have the expertise to integrate different systems. In the above discussion on irrigation scheduling the focus is on irrigation scheduling without regard to constraints on water delivery. In many cases, a large part of the irrigated area receives water from irrigation projects through a network of canals, and their operation determines the water delivery pattern to agricultural fields.

27.7 CANAL IRRIGATION SCHEDULING

In general, canals are operated using either a rigid (predetermined) or flexible (modifiable) schedule. In comparison with free access of a flexible system, which is conducive to misuse, a rotation (rigid) schedule sometimes improves irrigation efficiencies in areas where farm operators are not exposed to new irrigation technologies. In any schedule, the goal is to provide water at the right time in the right amount and at the right place.

27.7.1 Rigid Schedule

A major advantage of a rotation schedule is that it requires relatively low capital investment in canals or distribution pipelines and low management. The disadvantages are that it may lead to less than proper timing, over-irrigation and runoff, inadequate use of effective rainfall, conflicts over water, and may impede good farm management (Merriam and Keller, 1978). Further, without on-farm water storage, this schedule forces farm operators to operate at low-water-use efficiency. The only communication between farm operators and water agencies is the posting of the schedule. It can sometimes be made more effective if some flexibility can be introduced in water delivery. Typical flow rates of rigid-schedule systems are 15–30 L/s; typical frequencies are 7 to 14 days; and typical durations are 8, 12, or 24 hours.

27.7.2 Flexible Schedule

A demand (flexible) schedule involves a water supply that is flexible in terms of frequency, rate, and duration. Flexibility is essential for optimizing farm operations and maintaining sustainable irrigated agriculture. The flexible system has several characteristics: (1) It is sufficiently large to satisfy the demands of all users at any one time. (2) Requests must be made 1–4 days in advance of the desired delivery date. (3) The maximum rate is set fairly high to use labor economically on the farm. (4) It requires operational spillage, storage, or automation of the supply. (5) It enables the farm operator to irrigate each crop when needed, and to use a stream size. The disadvantage is that it is often too expensive, especially if a high rate or complete time flexibility is offered. Nearly all large projects in the northwestern United States operate on some type of arranged schedule.

27.7.3 Warabandi

In the Warabandi system the water is distributed by turns based on a predetermined irrigation schedule (day, time, and duration) in proportion to holdings in the outlet command. The irrigation time cycle is divided equitably. The time interval between two turns is always constant and is repeated at regular intervals. A seven-day rotation has been found to be suitable (Ajmera and Shrivastava, 2013), and watering weekly is also suited to crop water requirements. The distribution of water is a two tier operation. The upper tier is operated by the state agency (i.e., all distribution systems are operated with their full discharge). The distribution of water below the outlet is the lower tier, which is managed by organizations of farmers with their land under that outlet. Because both the state and farmers have their well-defined roles in irrigation water management, the system has greater social acceptance and there are inbuilt mechanisms for equity and efficiency.

27.7.4 Shejpali

This system was introduced to counter the tail-end problem by giving priority to tail-end farmers and serving head-end farmers last in each rotation. Each farmer is required to apply for irrigation water for a particular crop in a particular area each season. The state agency then sanctions some or all of the applications. Once sanctioned, the agency is obligated to deliver sufficient water to each farmer to bring their crop to harvest. In practice, the system is too rigid and difficult to enforce. It is practiced primarily in Maharashtra and Gujarat, India.

QUESTIONS

Q.27.1 Daily soil moisture measurements for a farm in the month of August are given as:

Date	θ (%)
10	30.2
11	29.5
12	28.2
13	27.3
14	26.8
15	27.4
17	24.5

It is required that the soil moisture content must be 25% or greater. What should be the next date of irrigation?

Q.27.2 Determine the next date of irrigation if the following data in the month of May are given:

Day	ET_r (in.)	P_e (in.)	θ (%) (at the end of the day)
15	0.45	0.0	24.4
16	0.42	0.0	23.9
17	0.44	0.0	22.8
18	0.39	0.04	22.1
19	0.38	0.05	21.7
20	0.35	0.06	20.5
21	0.39	0.01	20.1
22	0.40	0.0	19.5

The soil water content on the morning of May 15 was measured as 25% by volume, and the moisture content should not be less than 20%. The root zone depth is 30 in. below the surface.

Q.27.3 Using the data from Q.27.2, how much water should be applied if irrigation efficiency is 75% and the field capacity is 27.5% on the day of irrigation?

Q.27.4 A 5-ha field is being irrigated with an inflow discharge of 3500 L/min. The soil moisture content of the 1-m deep soil prior to the start of irrigation was 15% by volume. The field capacity of the soil is 25% by volume. What will be the depth of water that can be applied without deep percolation and what will be the duration of irrigation without deep percolation?

Q.27.5 The following climatic data are given for a farm for which the soil is at field capacity on the morning of day 1, the depth of water in the soil at field capacity is 10 cm, the depth of water at saturation is 12 cm, and the readily available water is 5 cm.

Day (June)	ET_r (mm)	Precipitation (mm)
1	9.5	0
2	8.5	0
3	6.8	5
4	10.2	0
5	4.5	15
6	9.5	0
7	8.5	4
8	6.5	8
9	8.2	0
10	5.5	12.5
11	7.9	0
12	11.5	0

What day will irrigation be needed? What will be the depth of irrigation? What will be the total depths of irrigation and deep percolation at the end of the 12-day period? What will be the depth of water remaining in the soil at the end of the 12-day period?

Q.27.6 The following information is given for a wheat field for which the soil is 80-cm deep sandy loam, field capacity is 25%, the available soil water content is 0.15 m/m of soil depth, the average minimum relative humidity is 40%, wind velocity is 1.5 m/s, and the soil water content in the morning of day 1 is 20%. The wheat is at the midseason stage. Compute the total effective precipitation, date and amount of each irrigation, total depth of irrigation water applied during the 12-day period, and the total depth of water used by wheat during the 12-day period.

Days	ET_r (mm)	Precipitation (mm)
1	8.5	10.0
2	9.2	0.0
3	8.9	5.0
4	10.5	0.0
5	11.2	0.0
6	10.3	0.0
7	9.8	0.0
8	9.5	4.0
9	8.8	12.0
10	8.2	10.0
11	9.5	4.0
12	8.5	8.5

Q.27.7 The following monthly data are given for a wheat field for which the field capacity is 30%, wilting point is 15%, and the soil is 65 cm deep. The farm is located at 45° N latitude. Wheat is planted on April 1.

Month	Temperature (°C)	Precipitation (mm)	n/N	u_2 (m/s)	RH_{min} (%)
April	15.0	30.0	0.75	2.2	35
May	18.0	20.0	0.70	2.0	32
June	22.0	15.5	0.85	1.9	28
July	24.0	10.5	0.80	1.8	25
August	15.0	12.0	0.78	1.5	30

What will be the number of irrigations required during the growing season? It is assumed that the soil is at field capacity on April 1.

Q.27.8 What will be the irrigation interval for a crop with effective root zone depth of 1 m, field capacity of 20%, permissible depletion of 8%, and crop evapotranspiration of 250 mm/month?

Q.27.9 For the data in Q.27.8, the effective precipitation is given as 40 mm. What will be the irrigation interval then?

Q.27.10 Develop an irrigation schedule using the management allowed depletion for a corn crop being grown in a sandy loam soil. The crop is fully developed. At the start of the midseason of crop stage development the soil moisture is at field capacity. It is known that the groundwater table is 4 m below the land surface, field capacity is 0.35 m/m or 35%, wilting point is 0.15 m/m or 15%, the effective root zone depth is 60 cm, and crop coefficient K_{cmid} is 1.1.

The data on reference crop evapotranspiration and precipitation are given by a 12-day period as follows:

Day	ET_r (mm)	Precipitation (mm)
1	6.5	5
2	7	0
3	7.5	0
4	6.8	4
5	6.4	5
6	6.3	8
7	7	0
8	7.3	0
9	7.2	0
10	6.9	0
11	6.5	7.5
12	7.4	0

(1) Determine the crop evapotranspiration. (2) Calculate the maximum allowable fraction depletion f_{dmax} to maintain the maximum evapotranspiration. (3) Calculate the allowable deficit (*AD*). (4) Determine the upward flow of water (U_f). (5) Calculate the daily soil water deficit (*SWD*). (6) Determine the irrigation date and amount. (7) If a fixed irrigation interval of one week is used, determine the irrigation amount on day 7. What will *SWD* be on day 12?

Q.27.11 Determine the root zone depth for cotton at 50 days after germination, assuming that the depth at germination is 0.3 m, the maximum root depth is 1.4 m, and full depth occurs 110 days after germination.

Q.27.12 A crop has an effective root zone depth of 1 m, the field soil has a field capacity of 0.27 m/m, permanent wilting point of 0.12 m/m, the maximum allowable fraction depletion $f_{dmax} = 0.55$, and crop evapotranspiration is 250 mm/month (assume 30 days per month).

1. What will be the maximum irrigation interval?
2. If the appearance and feel method of soil moisture measurement indicates that the average fraction of available soil water remaining in the root zone $f_r = 0.7$, determine the latest date of irrigation (*LD*).

REFERENCES

Ajmera, S., and Srivastava, R. K. (2013). Water distribution schedule under Warabandi system considering seepage losses for an irrigation project: a case study. *International Journal of Innovations in Engineering and Technology*, 2(4).

Allen, R. G., Pereira, L. S., Raes, D., and Smith, M. (1998). Crop evapotranspiration (guidelines for computing crop water requirements). FAO irrigation and drainage paper 56.

Annandale, J. G., Steyn, J. M., Benade, N., Jovanovic, N. Z., and Soundy, P. (1999). Facilitating irrigation scheduling by means of the soil water balance model. WRC report 753/1/99.

Brouwer, C., Prins, K., and Heibloem, M. (1989). *Irrigation Water Management: Irrigation Scheduling – FAO Training Manual No. 4*. Rome: FAO.

Cassel, D. K. (1984). Irrigation scheduling. *Crop and Soil Magazine*, 36(6): 15–18.

De Jager, J. M. (1992). *The PUTU System*. Bloemfontein: Department of Agrometeorology, University of the Orange Free State.

De Jager, J. M., and Singels, A. (1990). Using expected gross margin uncertainty to delimit good and poor areas for maize in the semi-arid regions of South Africa. *Applied Plant Science*, 4: 25–29.

De Jager, J. M., Mottram, R., and Kennedy J. A. (2001). Research on a computerised weather-based irrigation water management system. WRC report 581/01/01.

Doorenbos, J., and Kassam, A. H. (1979). Yield response to water. FAO irrigation and drainage paper 33.

Doorenbos, J., and Pruitt, W. O. (1977). Guidelines for predicting crop water requirements, FAO irrigation and drainage paper 24.

George, B. A., Shende, S. A., and Raghuwanshi, N. S. (2000). Development and testing of an irrigation scheduling model. *Agricultural Water Management*, 46(2): 121–136.

Harrington, G. J., and Heermann, D. F. (1981). State of the art irrigation scheduling computer program. In Irrigation Scheduling for Water and Energy Conservation in the 80s. Proceedings of the American Society of Agricultural Engineers Irrigation Scheduling Conference. St. Joseph, MI: ASAE, pp.171–178.

Hornbuckle, J. W., Car, N. J., Christen, E. W., Stein, T. M., and Williamson, B. (2009). *IrriSatSMS Irrigation Water Management by Satellite and SMS: A Utilisation Framework*. Griffith, NSW, CSIRO Land & Water.

Ihuoma, S. O., and Madramootoo, C. A. (2017). Recent advances in crop water stress detection. *Computers and Electronics in Agriculture*, 55(141): 267–275.

Israelson, O. W., and Hansen, V. E. (1962). *Irrigation Principles and Practices*, 3rd edition. New York: Wiley.

Jensen, M. E. (1972). Programming irrigation for greater efficiency. In *Optimizing the Soil Physical Environment toward Greater Crop Yields*, edited by D. Hillel. New York: Academic Press, pp. 133–161.

Jensen, M. E., Wright, J. L., and Pratt, B. J. (1971). Estimating soil moisture depletion from climate, crop and soil data. *Transactions of the ASAE*, 14(5): 954–959.

Lamacq. S., and Wallender, W. W. (1994) Soil water model for evaluating water delivery flexibility. *Journal of Irrigation and Drainage Engineering*, 120(4): 756–774.

Malano, H. M., Turral, H. N., and Wood, W. L. (1996). Surface irrigation management in real time in South Eastern Australia: irrigation scheduling and field application. In *Irrigation Scheduling: From Theory to Practice*, edited by M. Smith et al. Rome: ICID and FAO, pp. 105–118.

Merriam, J. L., and Keller, J. (1978). *Farm Irrigation System Evaluation: A Guide for Management*. Logan, UT: Utah State University.

Miller, D. E., and Aarstad, J. S. (1971). Available water as related to evapotranspiration rates and deep drainage. *Soil Science Society of America Proceedings*, 35: 131–134.

Ogata, G., and Richards, L. A. (1957). Water content changes following irrigation of bare soil that is protected from evaporation. *Soil Science Society of America Proceedings*, 21: 355–356.

Raghuwanshi, N. S. (1994). *Stochastic Scheduling, Simulation and Optimization of Furrow Irrigation*. Davis, CA: University of California.

Shayya, W. H., and Bralts, V. F. (1988). *Guide to Microcomputer Irrigation Scheduler SCS Version 1.00*. East Lansing, MI: Department of Agricultural Engineering, Michigan State University.

Singels, A., and Smith, M. T. (2006). Provision of irrigation scheduling advice to small-scale sugarcane farmers using a web based crop model and cellular technology: a South African case study. *Irrigation and Drainage*, 55: 363–372.

Singh, B., Boivin, J., Kirkpatrick, G., and Hum, B. (1995). Automatic Irrigation Scheduling System (AISSUM): principles and applications. *Journal of Irrigation and Drainage Engineering, ASCE*, 121(1): 43–56.

Stegman, E. C. (1983). Irrigation scheduling: applied timing criteria. In *Advances in Irrigation*, vol 2, edited by D. Hillel. New York: Academic Press, pp. 1–30.

Stegman, E. C., Musick, J. T., and Stewart, J. I. (1980). Irrigation water management. In *Design and Operation of Farm Systems*, edited by M. E. Jensen. St. Joseph, MI: ASAE.

USDA-NRCS (1997). Irrigation water management. In *National Engineering Handbook*. Washington, DC: USDA.

USDA-NRCS (2016). Sprinkler irrigation. In *National Engineering Handbook*. Washington, DC: USDA.

USDA-SCS. (1970). *Irrigation Water Requirements*. Washington, DC: USDA.

Villalobos, F. J., and Fereres, E. (1989). A simplified model for irrigation scheduling under variable rainfall. *Transactions of the American Society of Agricultural Engineering*, 32(1): 181–188.

28 Environmental Impact

28.1 INTRODUCTION

In the United States, irrigated agriculture is the largest user of water. In 2012, irrigated land represented approximately 56 million acres (or 7.6% of all US cropland and pastureland). About 75% of the irrigated acreage is in the 17 western states. The amount of water diverted to irrigate 56 million acres (22.7 million ha) is about 260 million acre-feet (3.2×10^{11} m^3). It may be noted that irrigated agriculture accounts for 47% of the total diversions, 82% of the total water consumptively used, and 32% of total return flows. Under current management and operation, irrigation returns 92 million acre-feet (1.1×10^{11} m^3) of gross diversions to streams (USDA-SCS, 1976).

Most of the irrigated land in the United States is in the 17 western states, excluding Alaska and Hawaii. Table 28.1 shows irrigation return flow in nine western water resource regions, which amounts to 87 million acre-feet (Boone, 1976). Other states having more than 500,000 acres (2.0×10^5 ha) of irrigated land are Arkansas, Florida, and Louisiana. Table 28.2 provides information on the irrigation water budget for the western states (Boone, 1976). Only about 52% of water diverted to farms is used for growing crops; the remaining 48% is used up by plants with low economic value, percolates into non-recoverable groundwater aquifers, evaporates from poorly drained areas, or returns to streams. Return flows, either surface or subsurface, transport much of the sediment, salts, and other pollutants to the receiving streams. Boone (1976) has provided a good discussion of return flow, and the discussion here draws on his work. Figure 28.1 shows irrigation water requirements for the United States.

Irrigation of crops and drainage of excess water have both positive and negative environmental consequences. Irrigation return flows degrade the quality of the receiving streamflow as they transport pollutants. Although return flows cannot be entirely eliminated, they can be reduced by appropriate water management and improved conveyance and delivery systems. This chapter provides a snapshot of the environmental impact caused by irrigation and drainage of agricultural lands, and the importance of return flows and the pollutants transported by them.

28.2 IMPORTANCE OF IRRIGATION RETURN FLOWS

The amount and quality of return flow are impacted by the management of irrigation water applied to crops, and in turn the transport of pollutants to receiving streams. However, the management of irrigation water is not uniquely defined. Often it is expressed in terms of irrigation efficiency, which itself is defined in many different ways, as discussed in Chapter 18. From the perspective of return flow, irrigation efficiency can be computed as a percentage of total withdrawal, as illustrated in Table 28.3 (Boone, 1976). If the average on-farm efficiency is used, then it can be calculated from the values in Table 28.3 by dividing the percentage of farm delivery by the percentage of crop consumptive use.

28.3 IMPACT ON WATER QUALITY

Irrigation return flows degrade water quality in a variety of ways, as shown in Table 28.4 (Boone, 1976). They increase levels of suspended and dissolved solids, nutrients, pesticides, weedicides, and the temperature of the receiving waters. Pollution of receiving streams by irrigation is now second to that by industrial and municipal waste, and is perhaps higher in developing countries. Figure 28.2 shows a schematic of pollution due to irrigation return flow.

28.3.1 Salinity

Dissolved mineral content is commonly referred to as salinity. It is a problem throughout the western United States, except in the Columbia River basin. Many rivers in arid and semi-arid regions of the western United States do not have salinity problems in their headwaters, but show increasing salinity as they traverse downstream toward their outlets, especially when irrigated agriculture uses a large part of their water and there are intervening sources of salt. Examples of such rivers are the Colorado River, which supplies water to parts of seven states and some parts of Mexico; the Rio Grande and the Pecos River in New Mexico and Texas; the San Joaquin River in California; the closed

Table 28.1 *Irrigation return flows*

Region	Irrigated area Million acres (million ha)	Gross diversions Million acre-feet (10^9 m^3)	Net depletions Million acre-feet (10^9 m^3)	Return flow Million acre-feet (10^9 m^3)	%
Missouri	7.90 (3.20)	35.9 (44.28)	14.83 (18.29)	21.07 (25.99)	59
Arkansas –White Red	3.97 (1.61)	11.61 (14.32)	8.18 (10.09)	3.53 (4.35)	30
Texas – Gulf	4.53 (1.83)	13.47 (16.61)	11.05 (13.63)	2.42 (2.99)	18
Rio Grande	1.96 (0.79)	7.07 (8.72)	4.83 (5.96)	2.24 (2.76)	32
Upper Colorado	1.37 (0.55)	7.92 (9.77)	2.72 (3.36)	5.20 (6.41)	66
Lower Colorado	1.33 (0.50)	9.21 (11.36)	4.64 (5.72)	4.57 (5.64)	50
Great Basin	1.75 (0.71)	8.88 (10.95)	4.14 (5.11)	4.74 (5.85)	53
Columbia	5.75 (2.33)	45.36 (55.95)	14.36 (17.71)	31.00 (38.24)	68
California	8.95 (3.62)	41.88 (51.66)	29.27 (36.1)	12.61 (15.55)	30
West subtotal	37.51 (15.18)	181.30 (223.63)	94.02 (115.97)	87.38 (107.78)	48
Other regions	4.74 (1.92)	13.32 (16.43)	9.00 (11.10)	4.32 (5.33)	32
United States	42.25 (17.10)	194.62 (240.06)	103.02 (127.07)	91.7 (113.11)	47

After Boone, 1976.

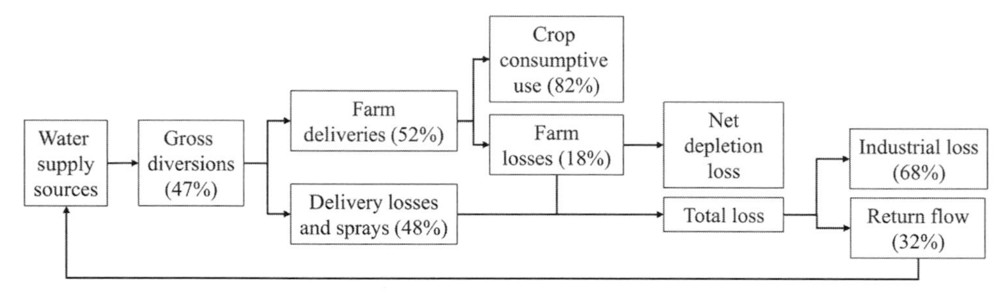

Figure 28.1 Irrigation water requirements of the United States. (Data from Boone, 1976.)

system of the Great Basin; the South Platte in Colorado; the Arkansas and Red Rivers in Colorado, Kansas, Oklahoma, and Texas; and the Brazos and Colorado Rivers in Texas. Many of the rivers have their headwaters in high-altitude forest and mountainous areas, and their source of runoff is snowmelt, having excellent quality of water. These arid and semi-arid regions at their lower altitudes have soils that contain large amounts of salts, and their surface waters have therefore large concentrations of salts and sediments. Thus, salinity is a result of interactions of water with soil and other geologic formations, as well as human activities.

Salinity of return flow is also impacted by irrigation water in other ways. For example, the water used by plants through evapotranspiration is essentially salt-free. The remainder of water percolating below the root zone will therefore have a larger concentration of salts, increasing the salinity of drainage water. Further, the water percolating through the soil may pick up additional salts by dissolving weathered minerals and previously precipitated salts.

High salinity affects the use of water for irrigation as well as for domestic and municipal use without expensive treatment. It reduces crop yields and increases operational costs. However, the increase in salinity is a result of natural irrigation processes and can be reduced by system improvements and improved water management, although it cannot be entirely eliminated.

28.3.2 Nutrients

Nutrients being transported by irrigation return flows enrich receiving streams. Enrichment of streams causes accelerated eutrophication of lakes and reservoirs, depressed oxygen concentrations, impairment of fisheries, taste, and odor in drinking water supplies, interference in waste treatment processes, and impairment of navigation. There are many areas of the world where enrichment has become disproportionally high, such as Snake River and San Joaquin River in the western United States. In certain areas phosphate levels are high, which combines with nitrogen due to irrigation return flow to stimulate nuisance aquatic growth.

28.3.3 Pesticides

Pesticide pollution due to irrigated agriculture is caused in different ways. First, there is a direct application to crops to

Table 28.2 *Irrigation water requirements*

State	Irrigated land	Withdrawals or diversions	Off-farm conveyance loss	Farm delivery	Consumptive use: Crop and pastures	Consumptive use: Associated losses	Return flow
	'000 acres (10^3 ha)	Thousand acre-feet (10^6 m^3)					
Arizona	1,250 (506)	8,690 (10,719)	1,910 (2,356)	6,780 (8,363)	3,840 (4,737)	570 (703)	4,280 (5,279)
California	8,730 (3,533)	40,520 (49,981)	7,980 (9,843)	32,540 (40,137)	21,280 (26,248)	7,590 (9,362)	11,650 (14,370)
Colorado	2,910 (1,178)	12,970 (15,998)	3,760 (4,638)	9,210 (11,360)	4,320 (5,329)	980 (1,209)	7,670 (9,461)
Idaho	2,900 (1,174)	28,690 (35,389)	7,780 (9,596)	20,910 (25,792)	5,530 (6,821)	1,390 (1,715)	21,770 (26,853)
Kansas	1,590 (643)	3,850 (4,749)	80 (99)	3,770 (4,650)	1,890 (2,331)	490 (604)	1,470 (1,813)
Montana	1,900 (769)	13,530 (16,689)	6,780 (8,363)	6,750 (8,326)	2,890 (3,565)	790 (974)	9,850 (12,150)
Nebraska	3,330 (1,348)	10,170 (12,544)	1,440 (1,776)	8,730 (10,768)	5,020 (6,192)	1,460 (1,801)	3,690 (4,552)
Nevada	830 (336)	4,000 (4,934)	270 (333)	3,730 (4,601)	1,670 (2,060)	610 (752)	1,720 (2,122)
New Mexico	870 (352)	4,250 (5,242)	850 (1,048)	3,400 (4,194)	1,680 (2,072)	690 (851)	1,880 (2,319)
North Dakota	80 (32)	280 (345)	90 (111)	190 (234)	100 (123)	40 (49)	140 (173)
Oklahoma	530 (214)	1,680 (2,072)	100 (123)	1,580 (1,949)	860 (1,061)	460 (567)	360 (444)
Oregon	1,600 (647)	9,730 (12,002)	3,290 (4,058)	6,440 (7,944)	2,580 (3,182)	790 (974)	6,360 (7,845)
South Dakota	150 (61)	450 (555)	60 (74)	390 (481)	190 (234)	150 (185)	110 (136)
Texas	6,930 (2,804)	20,160 (24,867)	810 (999)	19,350 (23,868)	12,870 (15,875)	4,000 (4,934)	3,290 (4,058)
Utah	1,070 (433)	5,750 (7,093)	1,280 (1,579)	4,470 (5,514)	1,940 (2,393)	460 (567)	3,350 (4,132)
Washington	1,220 (494)	6,710 (8,277)	1,930 (2,381)	4,780 (5,896)	2,760 (3,404)	1,240 (1,530)	2,710 (3,343)
Wyoming	1,550 (627)	9,650 (11,903)	2,990 (3,688)	6,660 (8,215)	2,310 (2,849)	430 (530)	6,910 (8,523)
Total	37,440 (15,151)	181,080 (223,359)	41,400 (51,066)	139,680 (172,292)	71,730 (88,478)	22,140 (27,309)	87,210 (107,572)

After Boone, 1976.

Table 28.3 *Relative irrigation water rates and efficiencies*

	Withdrawals or diversions	Off-farm conveyance	Farm delivery	Consumptive use			Return flow	
				Crops and pastures		Associated losses		
State	acre-feet/acre (m)	%	%	%	Acre-feet/ acre (m)	%	%	Acre-feet/ acre (m)
Arizona	7 (2.1)	21	79	44	3.1 (0.9)	7	49	3.9 (1.2)
California	4.6 (1.4)	20	80	52	2.4 (0.7)	19	29	2.2 (0.7)
Colorado	4.5 (1.4)	29	71	33	1.5 (0.5)	8	59	3 (0.9)
Idaho	9.9 (3)	27	73	19	1.9 (0.6)	5	76	8 (2.4)
Kansas	2.4 (0.7)	2	98	49	1.2 (0.4)	13	38	1.2 (0.4)
Montana	7.1 (2.2)	50	50	21	1.5 (0.5)	6	73	5.6 (1.7)
Nebraska	3.1 (0.9)	14	86	50	1.5 (0.5)	14	36	1.6 (0.5)
Nevada	4.9 (1.5)	7	93	42	2 (0.6)	15	43	2.9 (0.9)
New Mexico	4.9 (1.5)	20	80	40	1.9 (0.6)	16	44	3 (0.9)
North Dakota	3.6 (1.1)	33	67	35	1.2 (0.4)	15	50	2.4 (0.7)
Oklahoma	3.2 (1)	6	94	52	1.7 (0.5)	27	21	1.5 (0.5)
Oregon	6.1 (1.9)	34	66	27	1.6 (0.5)	8	65	4.5 (1.4)
South Dakota	3 (0.9)	14	86	42	1.3 (0.4)	33	25	1.7 (0.5)
Texas	2.9 (0.9)	4	96	64	1.9 (0.6)	20	16	1 (0.3)
Utah	5.4 (1.6)	22	78	34	1.8 (0.5)	8	58	3.6 (1.1)
Washington	5.5 (1.7)	29	71	42	2.3 (0.7)	18	40	3.2 (1)
Wyoming	6.2 (1.9)	31	69	24	1.5 (0.5)	4	72	4.7 (1.4)
Average	4.8 (1.5)	23	77	40	1.9 (0.6)	12	48	2.3 (0.7)
Range	2.4~9.9 (0.7–3.0)	2~50	50~98	16~64	1.2~3.1 (0.4-0.9)	4~33	16~76	1.0~8.0 (0.3–2.4)

After Boone, 1976.

control weeds and aquatic insects and pests. Second, there is aerial spraying of pesticides, part of which reaches canals, streams, lakes, and drains as a result of drift and overspray. Third, there is storm runoff from irrigation fields. Fourth, there is dumping of equipment clean-up solutions and excess mixes. Pesticides should not be released in toxic amounts.

28.3.4 Temperature

The temperature of receiving water may change due to irrigation return flow. Normally, surface return flow may be exposed to elevated ambient temperatures. When applied in sufficient quantities, it may increase streamflow temperature. However, subsurface return flow temperature may not be raised enough to increase receiving water temperature.

28.3.5 Solids

Suspended and settleable solids occur in irrigation return flow because of sediment from erosion of irrigated fields, canals, and drains; presence of solids in irrigation water; and sediment from erosion due to storm runoff. These solids are a loss in soil productivity and cause silting of streambeds, reservoirs, lakes, and estuaries.

28.4 MEASURES FOR REDUCING RETURN FLOWS

Some return flow cannot be avoided. For example, high rainfall following the application of chemical fertilizers causes runoff and surface, as well as subsurface, return flow. The same applies if there is heavy rainfall after pesticide application. Nevertheless, some return flow can be reduced by water management practices, which are briefly discussed here.

28.4.1 Water Delivery

A significant portion of water (about 20–25%) is lost during conveyance between the source of water and farmland. Of course, this loss occurs for irrigation, otherwise this lost water, at least in part, may return to the stream. Some of it is caused by non-crop plant growth, evaporation, and deep percolation. This also reduces irrigation efficiency. When this water returns to the stream, it brings with it sediment and salts. It is therefore advisable to reduce this loss by such practices as lining canals, removing leakages, and controlling weed growth.

Table 28.4 *Irrigation return flow*

Quality factors	Irrigation return flow	
	Surface	Subsurface drainage
Salts (TDS)	Not greatly different from source	Concentration increases usually 2–7 times Depends on amount in the supply, number of times reused, the amount of residual salts being removed, and the amount from non-agricultural sources
Sodium and chloride ions	Relatively unchanged	Both proportion and concentration likely to increase
Nitrate	More likely a slight increase than a decrease and highly variable	Likely to decrease if the control of irrigation water is high and increase if amounts are low Greatest hazard from heavily fertilized porous soils that are over-irrigated
Phosphate	Content may increase, bur closely correlated with erosion of fertile top soil	Decrease if considerable in source Not likely to greatly increase
Pesticides	Highly variable content Surface waters subject to pollution Likely associated with amount of erosion	A reduction in many instances Concentrations likely to be low
Pathogens and other organisms	Variable and may increase or decrease	Low content with a likely reduction in almost all pathogens Other organisms may increase or decrease
Sediments and colloids organics	Often more than in source but may be less variable Manures, debris, etc., likely to increase	Little or no sediment or colloidal materials in the flow Most oxidizable and degradable materials decrease
Heavy metals	Kinds and amounts are variable Likely to be greater than in subsurface flow	Most likely to decrease in concentration
Sewage effluents	Not greatly changed except by filtering and oxidation effect of crops if sprinkled	Concentration of all pollutants reduced except common soluble salts

After Boone, 1976. Copyright © 1976 by ASCE. Reprinted with permission from ASCE.

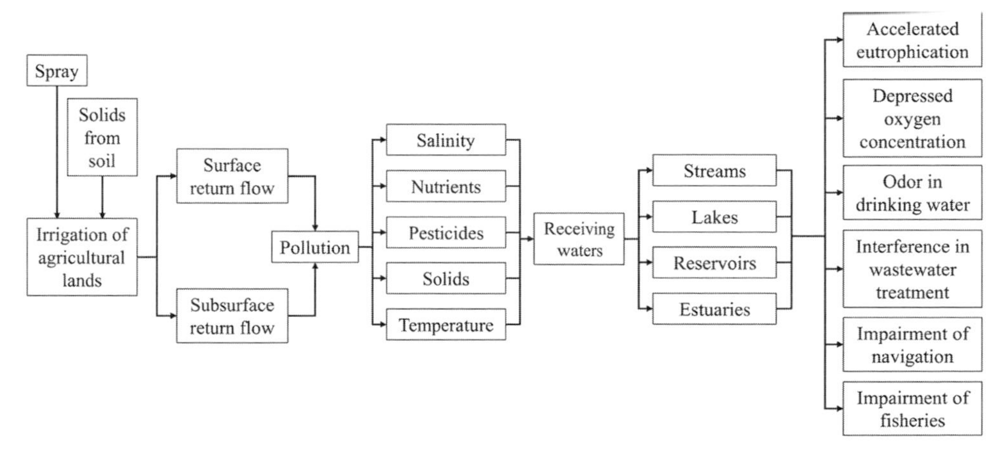

Figure 28.2 A schematic of pollution due to irrigation return flow.

28.4.2 Farm Water Management

The main point here is to improve irrigation efficiency, which can be accomplished by: (1) improving on-farm irrigation systems; (2) application of water to meet actual crop requirements; and (3) irrigation scheduling when needed. Gravity surface irrigation systems can be improved by land leveling and the desired amount of water application. However, one must pay attention to salt balance in the root zone.

28.5 IMPACT OF DRAINAGE

There are two basic points that should be kept in mind. First, well-drained lands are not always available and artificial drainage must be provided, which changes the natural flow regime but improves the soil environment for crop production. Second, drainage systems provide pathways for the transport of pollutants and do not degrade the environment themselves. It is actually human activities that are detrimental to environmental quality.

Agricultural drainage changes the conditions of the receiving stream. Surface drainage eliminates the surface storage capacity of the land, accelerates runoff, and increases peak flows, thus altering runoff volumes and distribution. It shortens the time of runoff concentration and heightens the runoff peak, especially when rainstorms are intense. It is also possible that surface drainage may reduce peak flow if drainage measures in different parts of a large watershed yield a better distribution of the contributions from these parts. Subsurface drainage does not eliminate the natural storage capacity of the land, but changes surface runoff into delayed subsurface runoff.

Drainage changes the concentration of dissolved and suspended materials in the receiving stream waters. Soil erosion and erosion from the bottom and banks of canals, ditches, and streams alter the concentration of suspended sediment. The silting of stream beds, lakes, and reservoirs reduces their capacities, increases maintenance costs, and enhances flood risk. The degradation of water quality limits the use of water for domestic, industrial, and agricultural purposes, harms fish, and increases the cost of water treatment. Field drainage causes downstream pollution. Loss of clay and silt particles, which are the most active fraction of any soil, causes long-term damage to soil productivity. Leaching of various substances from soil may have both negative and positive consequences. It may be destructive to soil fertility but may be beneficial if excess salts are removed.

REFERENCES

Boone, S. G. (1976). Problems of irrigation return flows. In *Environmental Aspects of Irrigation and Drainage, Proceedings of Specialty Conference Conducted by the Irrigation and Drainage Division of the American Society of Civil Engineers*, University of Ottawa, Canada, June 21–23, pp. 673–689.

USDA-SCS (1976). *Crop Consumptive Irrigation Requirements and Irrigation Efficiency Coefficients for the United States*. Washington, DC: USDA.

29 Economic Analysis

Notation			
A	annual revenues ($)	SFF	sinking fund factor
CRF	capital-recovery factor	SPCAF	single-payment compound-amount factor
F	future worth ($)	SPPWF	single-payment present-worth factor
G	gradient amount ($)	SPWF	series present-worth factor
i	interest rate (fraction)	UGSCAF	uniform gradient series compound-amount factor
N	investment period (years)	UGSPWF	uniform gradient series present-worth factor
P	present investment ($)		
SCAF	series compound-amount factor		

29.1 INTRODUCTION

Irrigated agriculture is vital for food security. Irrigation systems that make it possible are usually designed with the long-term objective that they are economically sustainable, although that is not always the case in many developing countries. Irrigation systems vary widely from country to country and from area to area in the same country. For example, in the United States, canal irrigation is common in California, Arizona, New Mexico, and Texas. In India and Pakistan, large networks of canals have been built and canal irrigation is the common mode of irrigation. However, the ownership and the governance of these canal irrigation systems in India and Pakistan are quite different from those in the United States. In south Texas, where fruits and vegetables are predominantly grown, sprinkler and trickle irrigation systems are quite common, whereas in east Texas canal irrigation is common. Economic analysis greatly depends on the type of irrigation systems and who owns them and how they are operated.

There is a broad range of scales of irrigation systems across the globe. In China, India, and Pakistan, large canal irrigation systems have been constructed by governments to bring water from rivers and reservoirs to farms. However, irrigation by water delivered from canal outlets to farms is done by the farmers themselves. Most of the farm holdings are small and individual farmers do not do any systematic economic analysis when irrigating their lands. For many farmers, agriculture is for subsistence and is not regarded so much as a business. Large farms in the United States, Canada, Australia, and Brazil are quite common and farmers design irrigation systems using economic analysis, because irrigated agriculture for them is a business. This indicates that the goal of agriculture can vary from subsistence to business. Therefore, economic analysis also varies. This chapter visits fundamental concepts needed for analyzing benefits and costs, which in the long term define the benefit–cost (BC) ratio.

29.2 BENEFITS AND COSTS

The primary benefit of irrigation is increased crop production. Under appropriate climatic conditions, irrigation enables more than one crop each year. In India, Bangladesh, and Pakistan, many farmers are able to grow three crops per year. In Texas two crops per year is not uncommon. Irrigation also helps convert wasteland into productive land, and the value of land is substantial. In sprinkler and drip irrigation systems, fertilizer can also be applied along with irrigation water. Irrigation also helps improve the quality of fruits, vegetables, and other crops. Protection of crops from frost and some diseases is enabled by irrigation. Also, the land is protected from wind erosion, and its productivity is maintained. Further, many small farmers survive and are able to sustain their families because of irrigation, and some farmers are able to enhance their well-being because of irrigation. Irrigation also generates employment, which in turn helps people with their livelihoods. From the viewpoint of economic analysis, some of the benefits can be easily quantified, but others cannot. For example, the increased crop is easy to quantify but increased well-being is not.

Costs of irrigation systems vary from system to system. For example, for sprinkler and trickle irrigation systems there are costs of the system, water delivery, operation, maintenance, labor, and replacement of parts. In the case of canal irrigation systems, there are costs of construction, equipment, operation, maintenance, and farm irrigation methods. In developing countries, canals are constructed by the government as the cost is prohibitively high for farmers. In the United States there are irrigation districts that are responsible for such systems. Then, there is the cost of water itself. If proper drainage is not provided, irrigation leads to alkalinization and salinization, and hence loss of productive lands. Irrigation also leads to bacterial and viral activity and vector-borne diseases, which are a cost to human health. It also leads to weed growth, whose removal costs money. Likewise, irrigated agriculture promotes the application of fertilizers and, in turn, leads to non-point source pollution. Some of the costs can be easily quantified, but others cannot.

However, in the real world, the situation is far more complicated because of a number of factors and uncertainties, and costs and benefits cannot be quantified every year. To illustrate, in 2018 in northern India a bumper harvest was expected. Just before the harvest, unexpectedly high rainfall accompanied by tornado-like winds through the Indian plains damaged nearly 80% of crops. To make matters worse, when the little left was harvested it was blown away by high winds and farmers were left with very little. In this case, irrigation did its job but the farmers got virtually nothing. The question then arises: How would one compute the benefit in this case? The vagaries of nature can complicate economic analysis. In many cases in developing counties, wild animals can cause significant damage to crops. Untimely rainfall and flooding, high winds, fog, extremely high and extremely low temperatures, and unexpected diseases cause huge damage to crops. These events are hard to predict, especially under the specter of climate change.

29.3 ECONOMIC ANALYSIS

Economic analysis entails quantifying benefits and costs and comparing them for the purpose of evaluating the long-term sustainability of irrigation systems. Unless irrigation is vital for survival, which is often the case for small farmers in developing countries, an irrigation system must be economically cost-effective; that is, in the long term it must pay for itself and produce benefit or it will not be implemented. Its BC ratio must be greater than 1. Using economic analysis, different irrigation systems and their designs can be evaluated, and the one which is most cost-effective can be selected.

29.3.1 Cash-flow Diagram

In doing economic analysis, benefits and costs are computed each year for the duration of the irrigation system's life. When these are graphed, they lead to what is termed a cash-flow diagram. To illustrate, a farmer would like to have a sprinkler system on his farm. It is assumed that the source of water supply is a water reservoir and the water level in the reservoir is high enough so no pumping will be necessary, but some outlets will be necessary. Initially, the farmer will have to design either a pipeline or an open channel with the necessary accessories to bringing the water to his farm. In either case, it will cost money. For a sprinkler system, a pumping system and its associated accessories will be needed. Then, the sprinkler system, involving the main and submains, laterals, and sprinklers and accessories, will be necessary. The farmer will have to invest money over a period of time, say one year, to build the system and put it into operation. Each year there will be costs associated with operation and maintenance and unexpected occurrences. Assuming the entire farm is irrigated with the sprinkler system, the increase in crop yield is realized from the second year onward. It is assumed that the sprinkler irrigation system is designed for a period of 30 years, with sprinklers being replaced every 10 years. Figure 29.1 shows a cash-flow diagram for such a simplified situation. Similar

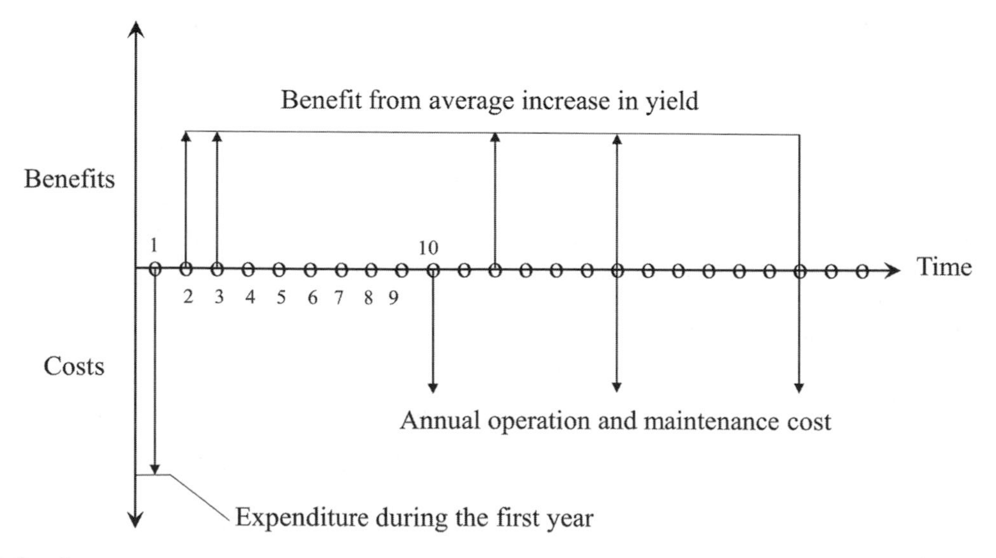

Figure 29.1 Cash-flow diagram.

cash-flow diagrams can be constructed for other irrigation systems. Then, diagrams can be compared to get an idea about different systems and to evaluate whether they meet the project objectives. Because different economic factors are involved in different systems, cash-flow diagrams are suitable for selecting the best system or best design. These factors are now discussed.

29.3.2 Single-Payment Factors

The question we ask is: What is the future worth of an investment made initially? The other question is: What investment should be made at an interest rate of i in order to have a given future worth? The future is defined by a number of years, say N. Thus, we have four terms: present investment P, future worth F, i, and N. The first question is answered by what is called the single-payment compound-amount factor (SPCAF) and the second question is answered by the single-payment present-worth factor (SPPWF).

The SPCAF involves computing the factor (F/P). Here, the amount P is invested initially at an interest rate of i for a period of N years. At the end of one year the amount P will increase by iP. Thus, at the end of the first year, the total amount will be $P+iP$. At the end of the second year, this amount will increase by $i(P+iP)$. Thus, the total amount will be $P+iP+i(P+iP)=P+2iP+i^2P=P(1+2i+i^2)$. Likewise, at the end of the third year, the total amount will be $P(1+2i+i^2)+iP(1+2i+i^2)=P+3iP+3Pi^2+Pi^3=P(1+3i+3i^2+i^3)$. In a similar manner, at the end of N years F will be:

$$F=P\left(1+Ni+\frac{N(N-1)}{2!}i^2+\frac{N(N-1)(N-2)}{3!}i^3+\cdots+i^N\right)$$
$$=P(1+i)^N \quad \text{or} \quad \frac{F}{P}=(1+i)^N. \tag{29.1}$$

This compounding of interest follows a binomial series which leads to eq. (29.1). Figure 29.2 shows a plot of eq. (29.1) for different values of P and interest rates. From this figure, the value of F can be obtained for a given value of P and interest i. This has applications in irrigation engineering. Suppose a small farmer buys a pumping system for irrigation by borrowing \$5000 at an interest rate of 10%. He would have to pay the money back with interest at the end of five years. Hence, he would like to know how much money he would have to pay. This means that he must make more

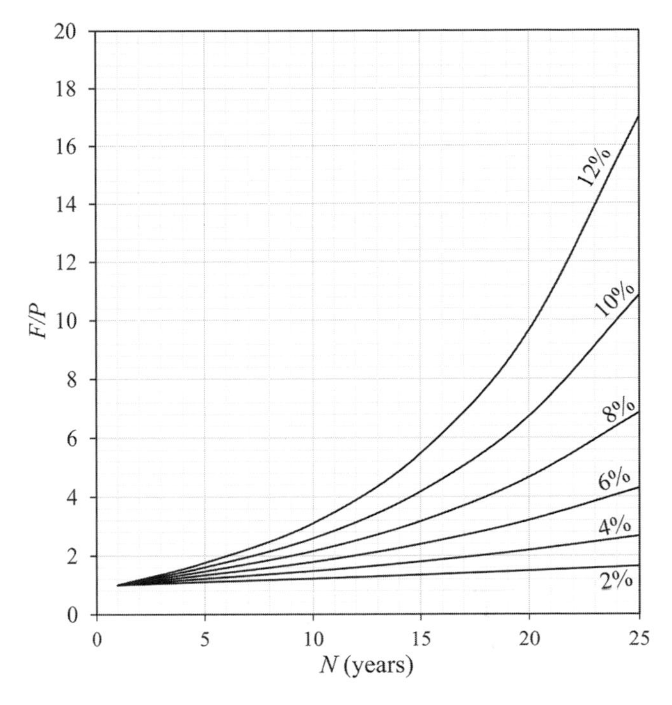

Figure 29.2 A plot of F/P versus N (eq. 29.1)].

than this amount by the end of five years. In this case, $P=\$5000$, $i=0.10$, and $F=\$8053$. His accumulated benefit due to the pumping system should be more than \$8053.

In the SPPWF, the value of F is given and the value of interest i is given, and one wants to determine P that should be invested. This is the inverse of SPCAF. Therefore, the factor P/F is determined in this case as

$$\frac{P}{F}=\frac{1}{(1+i)^N}. \tag{29.2}$$

Figure 29.3 plots P as a function of F for different values of interest rates. This has an application in farm irrigation. Suppose a farmer sets a target of earning \$100,000 at the end of five years. He can invest money at an annual rate of return of 10%. The question he asks himself is how much money he should invest. In this case, $i=0.1$ and $F=\$100{,}000$; P is to be determined, which equals \$62,092. In the context of irrigation, if a farmer invests \$62,092 at an interest rate of 10% to build an irrigation system, then they should be able to make \$100,000 at the end of five years in order to justify his investment.

Example 29.1: A farmer borrows \$10,000 from a bank at an interest rate of 10% that he must pay back in a lump sum. How much money will he have to pay the bank after 10 years?

Solution: $P=\$10{,}000$, $N=10$, and $i=0.1$.

From eq. (29.2),

$$\frac{P}{F}=\frac{1}{(1+i)^N},$$

$$F=P(1+i)^N=\$10{,}000\times(1+0.1)^{10}=\$25{,}937.$$

The farmer must pay \$25,937 at the end of 10 years.

Or, one can use Figure 29.2: $F/P=2.59$, given that $N=10$ and $i=0$. Then $F=2.59P=2.59\times\$10{,}000=\$25{,}900$.

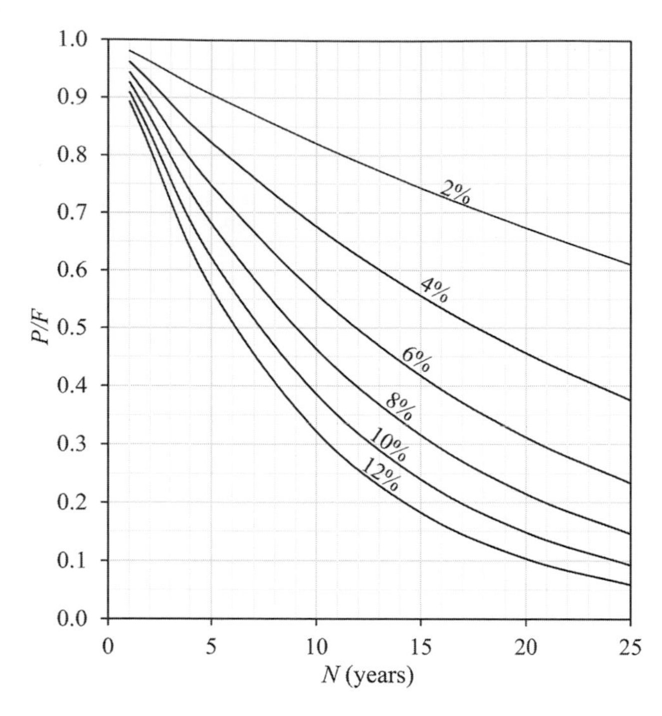

Figure 29.3 A plot of P/F versus N (eq. 29.2)].

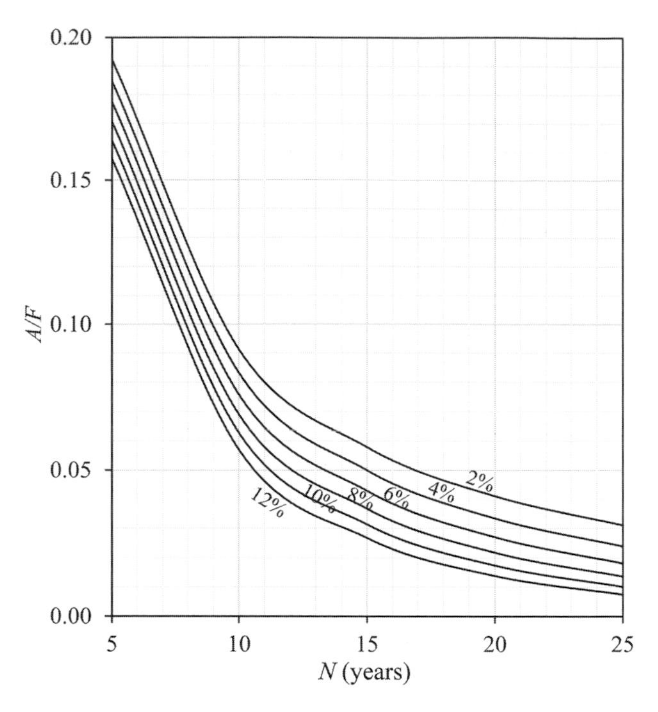

Figure 29.4 A plot of A/F versus N (eq. 29.3).

29.3.3 Uniform Annual Series Factors

In economic analysis, present or future benefits and costs are related to their equivalent value on an annual basis given an interest rate. In this case, several questions arise: (1) What should be the annual investment at a given interest rate (i) over a period of N years to produce a future amount F? (2) How much annual amount can one have when the present amount P is invested at an annual rate of i over a period of N years? (3) What will be the future amount F after N years of annual investment (A) at an interest rate of i? (4) What should be the present P that must be invested at an annual interest rate of i to be able to withdraw an annual amount A for N years?

To answer the first question, one computes what is called the sinking fund factor (SFF) (A/F). Here, F, i, and N are given and A is unknown. The SFF is given as

$$\frac{A}{F} = \frac{i}{(1+i)^N - 1}. \tag{29.3}$$

Equation (29.3) is graphed in Figure 29.4, from which, given the values of F, i, and N, one can obtain A. If a farmer invests an amount of A annually at an interest rate of i over a period of N years, then he will have invested an amount equivalent to F. He must therefore be able to generate revenues greater than F.

The second question is answered by considering what is called capital-recovery factor (CRF) (A/P). Here, the present amount P, annual interest rate i, and the period of investment N are given and the annual amount A is unknown. The CRF can be computed as

$$\frac{A}{P} = \frac{A}{F}\frac{F}{P}. \tag{29.4}$$

Inserting eq. (29.3) for A/F and eq. (29.1) for F/P, one obtains

$$\frac{A}{P} = \frac{i(1+i)^N}{(1+i)^N - 1}. \tag{29.5}$$

Equation (29.5) is graphed in Figure 29.5. If a farmer invests \$5000 at an annual interest rate of 10% for five years, then he must be able to generate annual revenues of \$1319.

For the third question, the future amount F is to be determined for known values of annual investment A, interest rate i, and period of investment N. This question is answered by computing what is called series compound-amount factor (SCAF) (F/A), defined as

$$\frac{F}{A} = \frac{(1+i)^N - 1}{i}. \tag{29.6}$$

Equation (29.6) is the inverse of eq. (29.3). This equation is graphed in Figure 29.6.

The fourth question is answered by considering the series present-worth factor (SPWF) (P/A). Here, P is unknown and annual investment A, interest rate i, and period of investment N are known. Hence, the SPWF can be computed as

$$\frac{P}{A} = \frac{P}{F}\frac{F}{A}. \tag{29.7}$$

Inserting eq. (29.2) for (P/F) and eq. (29.6) for F/A in eq. (29.7), one obtains

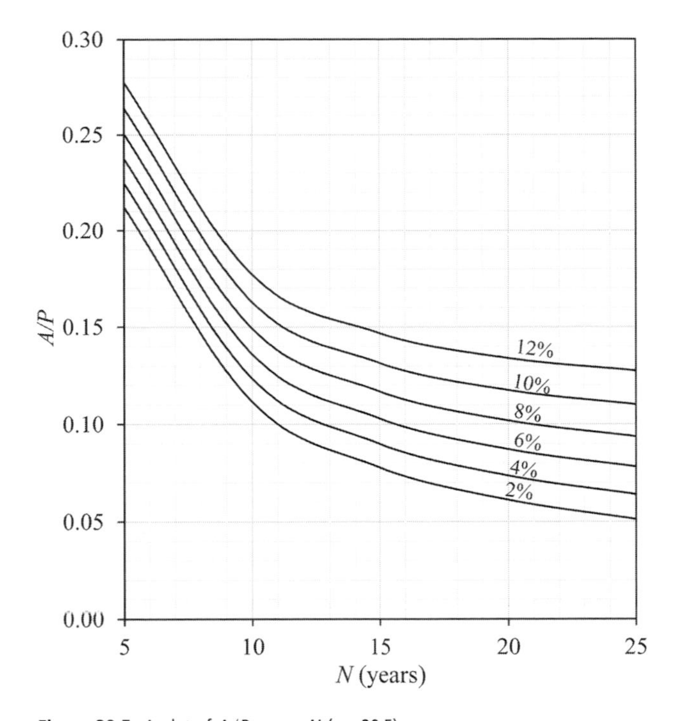

Figure 29.5 A plot of A/P versus N (eq. 29.5).

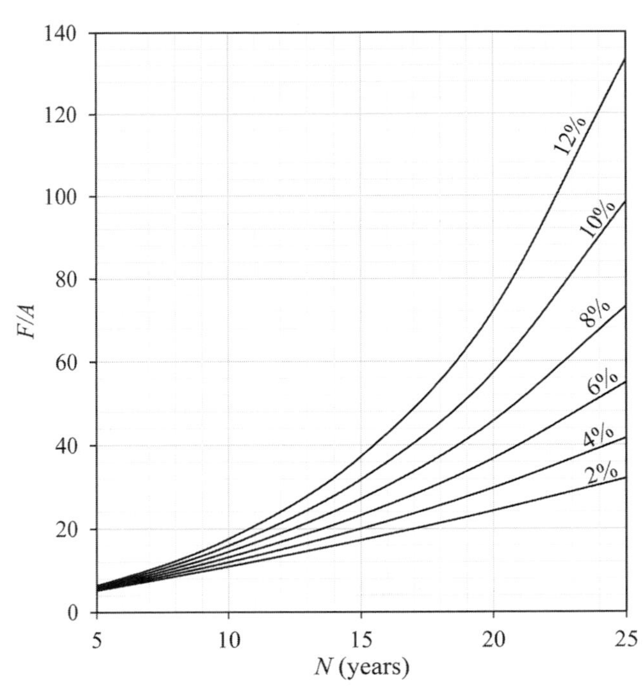

Figure 29.6 A plot of F/A versus N (eq. 29.6).

$$\frac{P}{A} = \frac{(1+i)^N - 1}{i(1+i)^N}. \tag{29.8}$$

Equation (29.8) is graphed in Figure 29.7. This shows the amount P that must be invested at an interest rate of i to receive an annual return of A for N years.

Example 29.2: A farmer buys a machine by borrowing \$15,000 from the bank at an interest rate of 12% for a period of 15 years. The machine will have no value at the end of this period. What will be the equivalent annual cost to the farmer, which will be the revenue that he must generate because of this investment?

Solution: $P = \$15{,}000$, $i = 0.12$, and $N = 15$, and A is to be computed.

Therefore,

$$\frac{A}{P} = \frac{i(1+i)^N}{(1+i)^N - 1} = \frac{0.12(1+0.12)^{15}}{(1+0.12)^{15} - 1} = 0.1468,$$

$$A = P \times 0.1468 = 15{,}000 \times 0.1468 = 2202.$$

The revenue to be generated by this investment is \$2202.

Example 29.3: For the data in Example 29.2, what will be the equivalent annual cost if the machine has a salvage value of 20% of the original value?

Solution: In this case, $F = 0.2 \times \$15{,}000 = \$3000$.

Then, the annual value of this amount must be deducted. Therefore,

$$A = \frac{i(1+i)^N}{(1+i)^N - 1} P - \frac{i}{(1+i)^N - 1} F$$

$$= \frac{0.12(1+0.12)^{15}}{(1+0.12)^{15} - 1} \times \$15{,}000 - \frac{0.12}{(1+0.12)^{15} - 1} \times \$3000$$

$$= 0.1468 \times \$15{,}000 - 0.0268 \times \$3000 = \$2122.$$

The annual equivalent cost to the farmer is \$2122, which he must generate to pay for the machine.

Example 29.4: A farmer wants to install an irrigation system that will cost \$100,000. The system will have an economic life of 25 years and a salvage value of 10% of the original cost. He also has to invest in bringing water to the irrigation system by installing a pumping system costing \$10,000, which he also borrows from the bank for a period of 10 years at the same interest rate. The salvage value of the pumping system after 10 years will be 25% of the original cost. Compute the annual cost of the total system for the first 10 years.

Solution: For the irrigation system, $F = \$100{,}000 \times 0.10 = \$10{,}000$.

Then,

$$A = \frac{i(1+i)^N}{(1+i)^N - 1} P - \frac{i}{(1+i)^N - 1} F$$

$$= \frac{0.10(1+0.10)^{25}}{(1+0.10)^{25} - 1} \times \$100{,}000 - \frac{0.10}{(1+0.10)^{25} - 1} \times \$10{,}000$$

$$= 0.1102 \times \$100{,}000 - 0.0102 \times \$10{,}000 = \$10{,}915.$$

For the pumping system, $F = \$0.25 \times 10{,}000 = \$2{,}500$.

Then,

$$A = \frac{i(1+i)^N}{(1+i)^N - 1} P - \frac{i}{(1+i)^N - 1} F$$

$$= \frac{0.10(1+0.10)^{10}}{(1+0.10)^{10} - 1} \times \$10{,}000 - \frac{0.10}{(1+0.10)^{10} - 1} \times \$2500$$

$$= 0.162745 \times \$10{,}000 - 0.062745 \times \$2500 = \$1471.$$

For the total system,

$$\text{Total} = \$10{,}915 + \$1471 = \$12{,}386.$$

Example 29.5: For the data in Example 29.4, the farmer has to buy a replacement pump for \$12,000 after 10 years, which will last 15 years and will have a salvage value of 30% of the original cost. Compute the new annual cost for the total system for the full 25 years of use.

Solution: First, the net present worth costs of the two pumps are computed. For the first pump is

$$P_1 = \$10{,}000 - \frac{1}{(1+0.1)^{10}} \times \$2500 = \$9036.$$

For the second pump, $F = 0.30 \times \$12{,}000 = \$3600$.

The present worth cost of the second pump is

$$P_2 = \$12{,}000 \times \frac{1}{(1+0.1)^{10}} - \$3600 \times \frac{1}{(1+0.1)^{25}} = \$4294.$$

The total net present worth cost of the two pumps is

$$P_T = P_1 + P_2 = \$9036 + \$4294 = \$13{,}330.$$

Now the equivalent annual cost of the pumps over 25 years is

$$A = P_T \times \frac{i(1+i)^N}{(1+i)^N - 1} = \$13{,}330 \times \frac{0.1 \times (1+0.1)^{25}}{(1+0.1)^{25} - 1} = \$1469.$$

The total new annual cost of the pumping system and the irrigation system over 25 years is

$$\text{Total} = \$10{,}915 + \$1469 = \$12{,}384.$$

29.3.4 Uniform Gradient Series Factors

When annual cash flows are not equal but increase or decrease by a constant amount each year, uniform gradient series factors are applied. Let the gradient amount be denoted by G. Three questions arise when the gradient series is uniformly increasing, as shown in Figure 29.8. In this case, one has four elements: F, P, i, and G. First, what will be the future amount F after N years if the gradient amount G is invested annually at an interest rate i? Second, how much should the present P be invested at an interest rate i to receive the annual gradient amount G? Third, what will be the equivalent annual series for a given gradient series? These questions can also be addressed if the gradient series is decreasing. The first question is answered using the uniform gradient series compound-amount factor (UGSCAF), and the second question by uniform gradient series present-worth factor (UGSPWF).

29.3.4.1 Uniform Gradient Series Compound-Amount Factor

The UGSCAF, denoted by F/G, allows us to compute the future amount F when the gradient amount G is invested at an interest rate of i. To determine F, one can apply the SPCAF to each gradient amount, and then sum to get the accumulated amount after the last investment. The first investment will become $G(1+i)^{N-1}$, and the last investment will be NG. In this manner, one gets (James and Lee, 1971):

$$\frac{F}{G} = N + (N-1)(1+i) + \cdots + 2(1+i)^{N-2} + (1+i)^{N-1}. \tag{29.9}$$

Equation (29.9) can be simplified as follows. Multiplying it by $(1 + i)$, one obtains

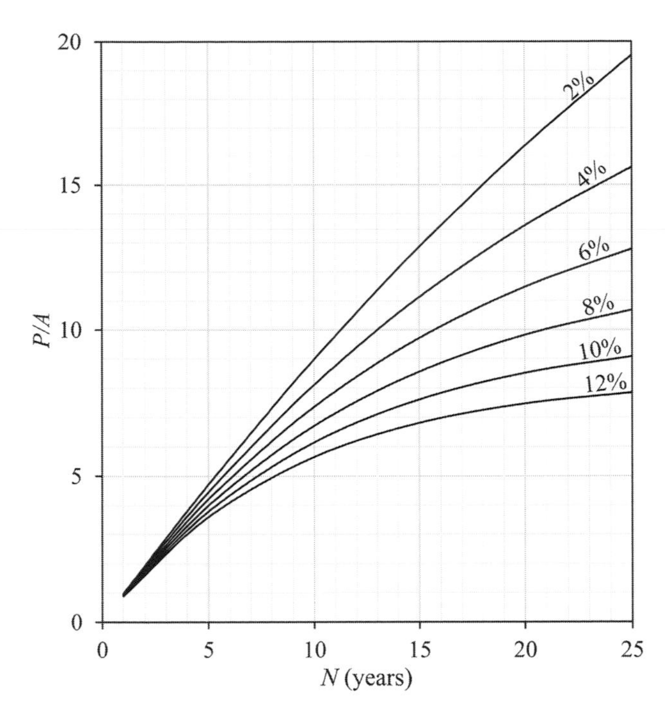

Figure 29.7 A plot of P/A versus N (eq. 29.8).

$$\frac{F}{G}(1+i) = N(1+i) + (N-1)(1+i)^2 + \cdots + 2(1+i)^{N-1} + (1+i)^N. \tag{29.10}$$

Subtracting eq. (29.9) from eq. (29.10), the result is

$$\frac{F}{G}i = -N + (1+i) + \cdots + (1+i)^{N-2} + (1+i)^{N-1} + (1+i)^N. \tag{29.11}$$

Multiplying eq. (29.11) by (1 + i), one obtains

$$\frac{F}{G}i(1+i) = -N(1+i) + (1+i)^2 + \cdots + (1+i)^N + (1+i)^{N+1}. \tag{29.12}$$

Subtracting eq. (29.11) from eq. (29.12), the result is

$$\frac{F}{G}i^2 = N - N(1+i) - (1+i) + (1+i)^{N+1}. \tag{29.13}$$

Therefore,

$$\frac{F}{G} = \frac{-(1+iN+i) + (1+i)^{N+1}}{i^2}. \tag{29.14}$$

Equation (29.14) shows that F can be obtained by knowing G, i, and N.

29.3.4.2 Uniform Gradient Series Present-Worth Factor

The UGSPWF is defined as P/G. This can be expressed as

$$\frac{P}{G} = \frac{P}{F}\frac{F}{G}. \tag{29.15}$$

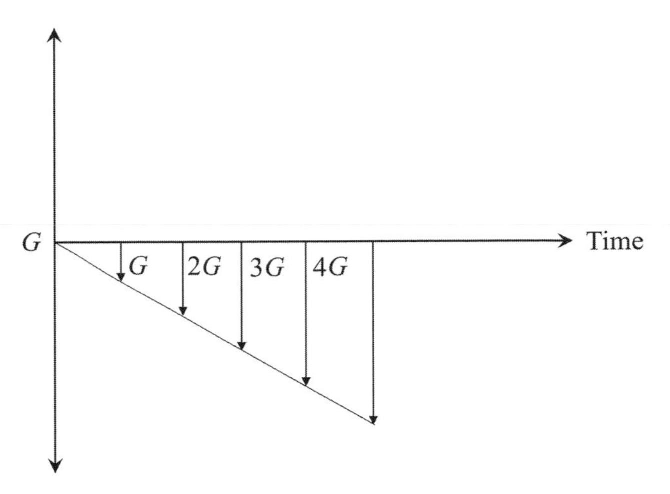

Figure 29.8 Uniform gradient series factors (uniformly increasing).

The ratio P/F is the SPPWF given by eq. (29.2) and the ratio F/G is the UGSCAF given by eq. (29.14). Using these equations, one obtains

$$\frac{P}{G} = \frac{(1+i)^{N+1} - (1+Ni+i)}{i^2(1+i)^N}. \tag{29.16}$$

Equation (29.16) gives the present amount P for given values of G, i, and N.

29.3.4.3 Equivalent Annual Series

The gradient series can be transformed to equivalent annual series (A) by noting

$$\frac{A}{G} = \frac{A}{P}\frac{P}{G}. \tag{29.17}$$

The ratio A/P is given by eq. (29.5) and P/G is given by eq. (29.16). Therefore,

$$\frac{A}{G} = \frac{i(1+i)^N}{(1+i)^N - 1}\,\frac{(1+i)^{N+1} - (1+iN+i)}{i^2(1+i)^N}. \tag{29.18}$$

29.3.4.4 Uniformly Decreasing Gradient Series

In the case of uniformly decreasing gradient series, one can subtract a uniformly increasing series from a uniform annual series, as illustrated in Figure 29.8. In practice, there is a uniformly increasing gradient series for a certain period of time and thereafter there is a uniformly decreasing gradient series. As shown for a cash-flow diagram in Figure 29.9, there is an increasing uniform gradient series for 15 years denoted by curve a, a uniform gradient series for 15 years denoted by b, and a decreasing gradient series denoted by c. The enveloping curve in this case is shown by $a + b - c$.

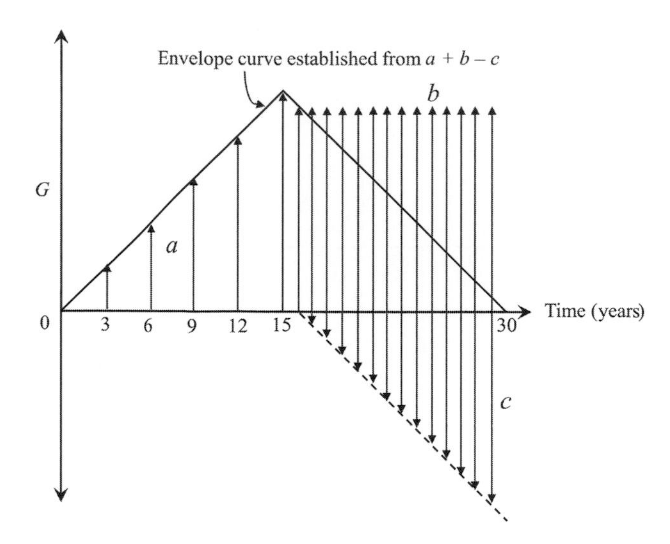

Figure 29.9 Uniform gradient series factors (uniformly decreasing).

QUESTIONS

Q.29.1 A farmer wants to make a decision on purchasing pumping equipment that should last 20 years. He has two choices. First, he can buy equipment for $12,000. Second, he can buy the equipment for $8000 that will last 12 years and buy another set of equipment for $7000 that will last 8 years. The interest rate is 10% and operation costs and other costs are the same. It is assumed that the salvage value will be negligible in both cases. Which option should the farmer choose?

Q.29.2 For the data in Q.29.1, what will be the choice of the farmer if the salvage value of the $12,000 pumping equipment is 20% of the original value, and the salvage value of the pumping system ($8000) after 12 years will be 15% of the original cost, and the cost of the pumping system ($7000) after 8 years will be 10% of the original cost?

Q.29.3 Equipment is bought for irrigation by borrowing $8000 from a bank, at an interest rate of 10%. How much money will be paid in a lump sum after 5 years.

Q.29.4 A farmer wants to generate a rainy-day fund. He deposits $10,000 each year in the fund and the money is invested in a safe mutual fund that earns 8% per year, compounded annually. What will be the amount available at the end of 10 years?

Q.29.5 A farmer has a 10-year annuity for which he pays $5000 at the end of each year. What will be the present value of the annuity if discounted at 8%?

Q.29.6 Compute the compounding factor for the first 10 years if the interest rate is 3%, 5%, 8%, and 10%.

Q.29.7 A farmer wants to install an irrigation system that will last 25 years. He plans to buy the first irrigation system for $10,000, which will last 15 years, and buy another one for $8000. which will last 10 years. The interest rate is 10%. It is assumed that the salvage value will be 10% of the original cost in both irrigation systems. Compute the equivalent annual cost of the irrigation system over 25 years.

REFERENCE

James, L. D., and Lee, R. R. (1971). *Economics of Water Resources Planning*. New York: McGraw-Hill.

30 Irrigation Management

30.1 INTRODUCTION

Irrigation management entails management of water, water structures, and machinery, and governance organization (Stegun et al., 1980). Management of water structures involves design, construction, control, operation, and maintenance. The organization deals with the governance of the entire irrigation enterprise, involving individuals or groups managing the irrigation system. It includes decision-making, resource mobilization, communication, and conflict management. When these three management components and their elements are considered together, they form a three-dimensional matrix consisting of management activities (Uphoff, 1985). The matrix helps facilitate comparison of management approaches across climatic, cultural, and national boundaries. Uphoff (1985) presented an analytical framework for analyzing irrigation organizations that comprise physical system, social structure, cultural precedents, biological circumstances, administrative structure, and economic opportunities for crop production. There are many aspects to irrigation management.

When managing a large irrigation system, it is important to recognize that irrigation is an interdisciplinary field and its operation and management is even more interdisciplinary. This is seen from the involvement of different disciplines in understanding and development of different components. For example, the areas involved in planning, designing, and developing a source system may include climatology, hydrology, hydraulics, geotechniques, structures, and geology. The conveyance system involves hydraulics, hydrology, geotechniques, and structures. The delivery system involves climatology, agricultural science, hydrology, hydraulics, geotechniques, and structures. The receiving system involves agricultural sciences, climatology, hydrology, hydraulics, remote sensing, and hydrometry. In multiple-purpose projects, the water in the system is used for purposes other than irrigation. The design of the system and its operation and maintenance will vary considerably, depending on the number of other uses and priorities for use.

Another point to be noted is that operation and management of an irrigation system depends on scale. The spatial scale reflects the farm size. This is primarily measured in the horizontal plane. Irrigation systems can be classified as small, medium, and large. The timescale is associated with the operation of the system. The timescale can be daily, weekly, monthly, seasonal, or yearly. The issues of maintenance and operation are associated with the temporal scale. This chapter provides a snapshot of these different aspects.

30.2 IRRIGATION SYSTEM

Although irrigation is defined in terms of the delivery of water to the land or farm, the system that makes the delivery possible is quite complicated. The system may also include all of the physical improvements necessary to transfer the water in time and space from the source to the farm. Therefore, management and operation of an irrigation system requires an understanding of the system in full detail.

An irrigation system may be composed of four subsystems: (1) source subsystem; (2) conveyance subsystem; (3) delivery subsystem; and (4) receiver subsystem. There are five types of activities related to these systems: (1) acquisition (adequate and assured water supply and allocation) is related to the source subsystem and conveyance subsystem; (2) distribution of water is related to the delivery subsystem; and (3) application and movement of water after application, (4) drainage of excess water, and (5) quality of water are related to the receiver subsystem. These subsystems and activities are related to water and structures, including equipment. Thus, irrigation management entails not only management of these subsystems but also an organized governance structure that does the management, simply referred to as organization.

The management of an irrigation system depends on the type of the system or its subsystems. Irrigation systems can be classified as: (1) source–conveyance–delivery–receiver system; (2) source–conveyance–receiver system; (3) source–delivery–receiver system; and (4) source–receiver system. All four subsystems or components constitute an irrigation system, such as a canal irrigation system. The subsystems are the most versatile and are usually designed for agricultural irrigation. In the case of a source–conveyance–receiver system, there is no delivery system. The water is conveyed directly from the conveyance system to the farm or the receiver system. Examples are farmlands located along rivers, canals, aqueducts, etc. Thus, water is received by the receiver system directly. Another example is an artesian well, where water is withdrawn from underground or an aquifer. In the case of source–delivery systems, water is delivered directly from the source system to the receiver system. Examples are water withdrawn from a reservoir to the farm or water delivered from a well. In the case of a source–receiver system, the water is delivered to the farm directly from the source. This happens when water is pumped from a reservoir to the farm. Not all irrigation systems involve all subsystems.

30.3 WATER MANAGEMENT

Although the ultimate objective of water management is the use of water, it is perceived differently by different people (Hargreaves and Merkley, 1998). For example, the engineer is primarily concerned with the efficient use of water; the agronomist is concerned with the use of water for crop production and pest control; the farmer is interested in water availability for maximum crop production; the economist deals with water from the vantage point of profitability, markets, and economic benefits; the environmentalist focuses on the minimal damage to the environment, such as deterioration of water quality, flooding, loss of wetlands; the ecologist wants to know the impact on human health and bacterial and viral activity; the sociologist emphasizes equity in water distribution and adherence to traditional practices; and the politician is concerned with water from the viewpoint of the promotion of rural well-being and limiting the migration to cities.

The engineer is responsible for addressing technical issues in both time and space, which occupy center-stage in the entire irrigation enterprise. The spatial scale can vary from the farm scale to the watershed to the river basin scale, and even beyond. Likewise, the timescale can vary from day to week to month to season to year and even beyond. The ultimate objective is to increase production without detrimental effects. Often, detrimental effects cannot be entirely eliminated, but the goal is to minimize them. Consider, for example, the case of a multipurpose reservoir that supplies water for irrigation, is the source of domestic and industrial water supplies, provides flood control, generates electricity, and is used for recreation and tourism. In the event of a drought, the reservoir may not be able to satisfy all objectives, and a decision has to be made about how to allocate water for different purposes. In so doing, political, economic, social, and environmental factors should be considered, and there is no unique water allocation strategy. The strategy to be adopted will depend on priorities and the minimization of damage. The priorities may include domestic and industrial water supply and limited irrigation water supply. On the other hand, if there is a flood, a very different situation may arise when excess water has to be discharged. The question is which sector should bear the brunt of flooding? Public safety may take precedence over all other priorities. Further, peak agricultural water demand and peak streamflow may not coincide, meaning the reservoir should have enough water in storage to meet agricultural demand at the peak time, but this might necessitate curtailing supply to other sectors. The question is: Will that be feasible, and if yes then how to do it?

From the standpoint of the farmer and the agronomist, the question arises as to the particular crop that will bring more profit. The selection of a particular crop depends on the climate, water supply, water quality, soil, labor, traditional farming practices, social culture, level of technology, and market demand. This may call for crop-by-crop analysis in tandem with the farmer's preferences, which may lead to an optimal crop mix.

Farm size is another factor that determines the level of water management. In countries like India, Pakistan, China, Japan, and South Korea, farm sizes are usually very small – only a few acres – barring a few exceptions. These sizes may be referred to as subsistence farm sizes. Such small farms are managed by individual families and there is hardly any well-thought-out management strategy. On the other hand, in countries like Australia, Canada, and the United States, there are many farmers who own thousands of acres, and farming for them is like any other business. In that case, farmers are like businessmen who have a management structure, generate employment, and employ a high level of technology. They go for economic and environmental analyses.

Agricultural lands are part of a watershed, and their irrigation is intimately connected with watershed management. Irrigation may often improve watershed conditions. Irrigated agriculture is found to be more profitable than grazing and ranching. But irrigation also impacts the environment and may deteriorate stream water quality. Soil conservation, stream restoration, land reclamation, and irrigated agriculture are interconnected, and irrigation of agricultural lands has to be practiced within the overall context of watershed management.

30.4 MANAGEMENT OF WATER STRUCTURES

Management of water structures primarily deals with operation and maintenance. Operation involves the delivery of water by the structures without unexpected interruption. These structures relate to acquisition, allocation, distribution, and drainage and recycling of water, and should be operated as per their objective. Maintenance involves upkeep of these structures so that their life is sustained and they remain operational and perform their intended functions.

30.5 ORGANIZATION AND MANAGEMENT

The organizational component entails the governing structure, management of people, mobilization of resources, allocation and distribution of resources, conflict resolution, and fiscal health. Irrigation organizations vary from country to country and from one part of a country to another. In developing countries, irrigation facilities such as canals, reservoirs, and structures are planned, designed, operated, maintained, and managed by government organizations, and individual farmers have virtually no role, even though the facilities are all meant for the farmers. The organizations supply water to farmers at subsidized cost. Farmers are responsible for maintaining irrigation channels that bring water to the farm.

Operation, management, and maintenance of irrigation systems are the responsibility of irrigation organization. The success or failure of an irrigation system is generally believed to be related to management. Management, in the establishment of irrigation and drainage projects, includes planning, organizing, equipping, and staffing for the direction and control of a complex business venture. Many factors are involved in what is actually the management of a natural resource, water, and the effective use and conservation of that resource, utilizing efficient operation and maintenance.

The management of an irrigation and drainage system not only depends on people, as do many business enterprises, but the equipment, procedures, practices, and crops to be grown are important considerations. In arid and semi-arid areas, where irrigation and drainage are essential to the people, a strong, dependable, and capable administrative organization is indispensable.

Irrigation management covers (1) all land within the area of jurisdiction; (2) the water to be used in the system (including water released from the area); (3) the physical system of improvements required to move and control water from the source or sources to the farm (including those necessary to take care of unwanted water); (4) people and equipment necessary to operate and manage the system; (5) people within the area; and (6) improvements and controls to provide for the use of water for other purposes (e.g., hydropower, navigation, and flood control). The management is limited by legal considerations, economics, and ethical norms. Normally it does not include activities within the farm areas except those directly affecting the operation of the system (Early, 1990).

30.5.1 Objectives

The objectives of irrigation system operation and management are: (1) to deliver irrigated water equitably to all water users at the lowest possible cost, commensurate with the type of service for which the water users are willing to pay and consistent with sound management practices and conservation of water; (2) to maintain the total irrigation system, and drainage system if present, to the degree necessary to preserve the capacity and condition of the facilities in a manner to avoid undue depreciation, and to store and deliver irrigation water as required by the water users; (3) to keep the system seepage and operational waste to a minimum; and (4) to be constantly alert to technical advances to adopt new practices that will improve system operating efficiency. To these can be added salinity control, frost control, plant and soil cooling, wind erosion control, herbicide and insecticide applications, and plant disease control.

Operation and management of an irrigation system includes all of the various items related to operation and management of all of the facilities associated with the delivery of water to the farm. It must be emphasized that management implies more than just the business aspects, but must also include the concept of maintaining permanent agriculture. By implication, this embraces any operation having to do with the management of land. This larger perspective of management must always be kept in mind by management and operational staff. Emphasis must be placed on business management and the operation and maintenance of physical facilities of the irrigation system if a successful and sustainable organization is to be accomplished.

Many irrigation systems are extensive and complicated. It is anticipated that the construction of facilities for irrigation of agricultural lands will decrease in the future. Thus, there are many real unsolved problems of irrigation in the field of technical management of irrigated areas rather than in design and construction. Social, political, ecological, and environmental factors will become more acute as population increases and management of irrigation systems will become more complicated. This will be more pronounced in the older systems, where all problems may not have been considered in depth during the planning, design, and construction stages.

An understanding of the operation and management of irrigation systems is vital for successful planning, design, and construction of such systems. Many of the difficulties in the operation and management of systems can be traced to the lack of proper consideration of these matters in the earlier stages of building the systems.

30.5.2 Aids for Operation and Management

The aids provided by the planner include (1) the concept of organization and (2) the aims and objectives. Those provided by the designer include (1) the design criteria, (2) the designer's operating instructions, and (3) maps and drawings. Those provided by the construction agency include (1) maps and final drawings, (2) information on sources of equipment, lists of spare parts, etc., and (3) miscellaneous information on geology, groundwater, etc., encountered and photographs of the construction.

30.5.3 Resource Mobilization

Resources are needed for managing irrigation systems. These include funds, manpower, materials, machinery and equipment, information, and data. Finances are partly generated by farmer participation and partly by obtaining funds from the government, and by loans. It is important that the managers and the farmers are in constant communication, and that the managers involve the public so that people understand the importance of irrigation systems and feel that the system belongs to them and is for their benefit, even if they are not the owners.

30.5.4 Communication

Communication is key to any organization. Open channels of communication between management and farmers are of utmost importance. Farmers should be free to communicate any problems regarding the delivery of water and the operation of the irrigation system to the management. Likewise, the management should keep farmers informed on a periodic basis regarding the expenditure of funds for operation, maintenance, rehabilitation, and purchases. Farmers' voices should be heard in terms of the hiring and firing of personnel.

30.5.5 Conflict Resolution

Conflicts are a natural occurrence, especially when people and resources are involved. They occur in several areas, including people, resources, and management. Water is usually in short supply and allocation of this resource becomes a challenge. All people in the same organization do not always see things in the same way, and will have different opinions and styles. Farmers and management may not always see things in the same way. The best way to manage conflicts is to establish rules and procedures, and to follow them consistently. However, these rules and procedures should be adopted with the approval of the farmers, and farmers should take ownership of them. Allowances should be made for exigencies, and extenuating circumstances should be dealt with on a case-by-case basis, with the approval of stakeholders.

30.6 WATER LAWS AND RIGHTS

The management must be aware of and abide by the laws of the land. Laws and water rights may vary from state to state and may be different for surface water and groundwater (Hargreaves and Merkley, 1998). Further, these rights and laws may change with time, as needs change. In many countries, surface water is owned by the government and there are no laws for groundwater: farmers may extract groundwater as and when needed. In India and China groundwater depletion has reached alarmingly high levels because there are no controls on its extraction.

REFERENCES

Early, A. C. (1990). Irrigation management in the Poudre Valley of northern Colorado. In *Management of Farm Irrigation Systems*, edited by G. J. A. Howell, and K. H. Solomon. St. Joseph, MI: ASAE, pp. 985–1015.

Hargreaves, G. H., and Merkley, G. P. (1998). *Irrigation Fundamentals*. Highlands Ranch, CO: Water Resources Publications.

Stegun, E. C., Musick, J. T., and Stewart, J. I. (1980). Irrigation water management. In *Design and Operation of Farm Irrigation Systems*, edited by M. E. Jensen. St. Joseph, MI: ASAE.

Uphoff, N. T. (1985). Community response to irrigation. Cornell University Irrigation Studies Group paper 1.

31 Irrigated Agriculture for Food Security

31.1 FOOD SECURITY

Food security entails three dimensions: (1) availability of sufficient food of appropriate quality and quantity at all times; (2) access to adequate resources to acquire sufficient and appropriately nutritious food in quantity; and (3) utilization of food through adequate diet, clean water, sanitation, and healthcare for nutritional well-being (Schmidhuber and Tubiello, 2007; CFS, 2012). Thus, food security can be defined as when all people, at all times, have physical, social, and economic access to sufficient safe and nutritious food to meet their dietary needs and food preference for an active and healthy life (Shaw, 2007). The factors that affect food security include population, rising standard of living, decreasing agricultural land, food wastage, climate change, natural calamities, water security, and energy security. These factors will be discussed in this chapter.

31.2 NUTRITIONAL SECURITY

What people eat and drink to maintain life and growth defines food, but nutrition encompasses those aspects that relate to health services, healthy environment, and caring practices (Pangaribowo et al., 2013). Nutritional security entails four dimensions: (1) adequacy of diet, (2) consumption of food, (3) immunity to and recovery from disease, and (4) adequacy of health. Thus, nutritional security is defined by a nutritionally adequate diet and the consumption of food such that adequate growth, resistance to or recovery from disease, pregnancy, lactation, and physical work are maintained (Frankenberger et al., 1997). UNICEF developed a framework for food and nutrition security (FNS) by recognizing the basic underlying and immediate causes of undernutrition.

Basic causes are associated with socioeconomic and political conditions that impact undernutrition, and are required for macroeconomic stability, economic growth and its distribution, public expenditure, and governance and quality of institutions affecting FNS (Ecker and Breisinger, 2012). Government policies can mitigate undernutrition by means of changing the social, economic, and political context (Pangaribowo et al., 2013).

At the individual human level, the immediate cause of nutritional status is dietary intake, which should meet a certain threshold in terms of quality and quantity, and should have a balance of carbohydrates, proteins, and fats (macronutrients), and vitamins and minerals (micronutrients). Other factors affecting individual nutritional status are household insecurity (in terms of availability and access), inadequate care, lack of health services, and unhealthy environments. The factors that emphasize the importance of child-feeding and health-seeking behaviors, support for mothers during pregnancy and lactation, and mothers' autonomy in household decision-making (Smith and Haddad, 2000) are shown in Figure 31.1 (Black et al., 2008).

In developing countries, infectious diseases, such as diarrheal diseases, and respiratory infections are major nutritional health problems, and the quality of health services might explain nutritional status (Strauss, 1990). Diarrheal diseases are associated with water and sanitation conditions (Pangaribowo et al., 2013), and water and sanitation improvements have a significant impact on the population and its health (Smith and Hadadd, 2000).

31.3 POPULATION

Population has a direct impact on food security. It is therefore pertinent to look at the timeline of population increase.

31.3.1 Global Population

The world population was around 7.7 billion in 2019 and is continuing to grow, but at a decreasing rate. The population growth rate reached its peak in 1965–1970, at around 2.1% per year. Since then, the population growth rate has shown a decreasing trend and decreased to 1.1% from 2015 to 2020, and is projected to continue to decrease from 2020 through 2100 (Figure 31.2). The UN Population Division projects the population will reach 8.5 billion by 2030, 9.7 billion by 2050, and 10.9 billion toward the end of the century, based on the medium-variance projection (Figure 31.2). Of course, there exists a large uncertainty in the population projection, and the projected population varies between 8.5 and 8.6 billion in 2030, with a 95% confidence interval. In the eighteenth century, the global population was around 1 billion, which rose to 1.6 billion by 1900, 2.5 billion by 1950, and 6.4 billion by 2000, as shown in Figure 31.2.

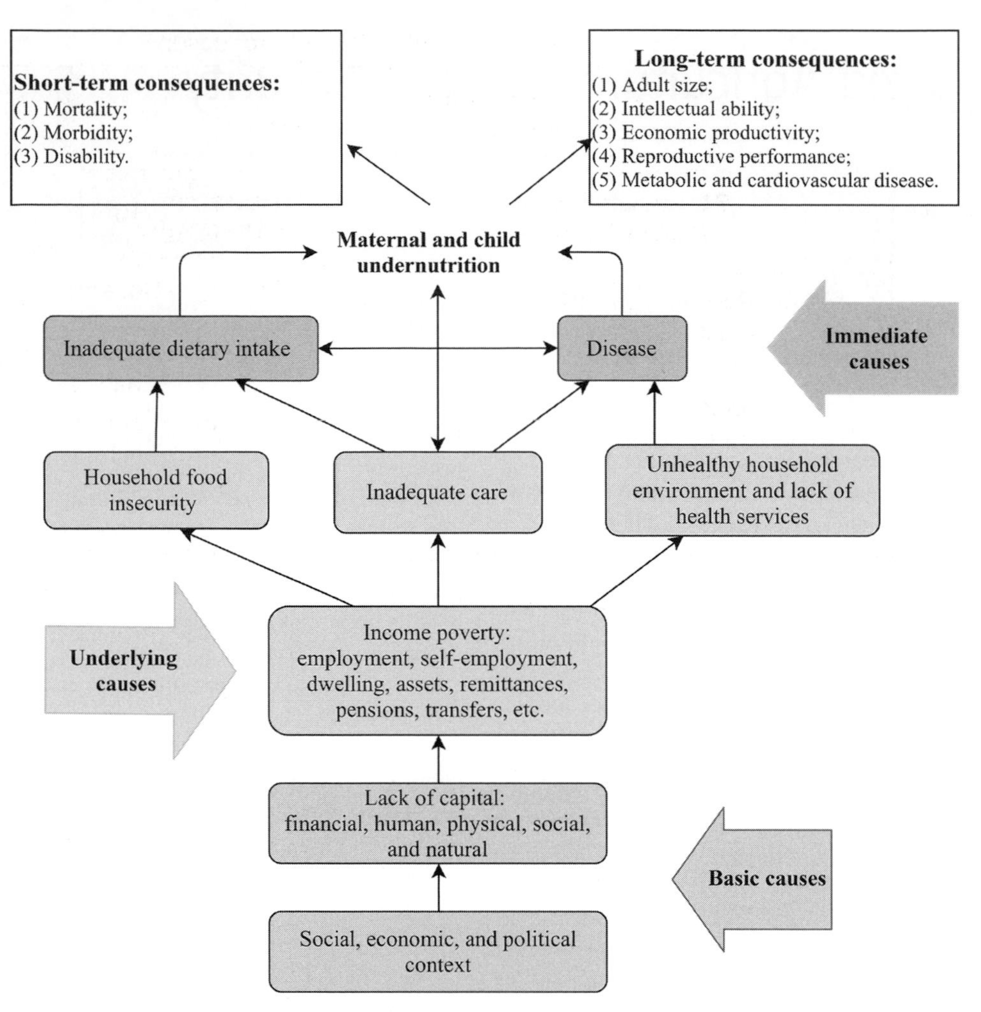

Figure 31.1 Conceptual framework of undernutrition. (After Black et al., 2008. Copyright © 2008. Reprinted with permission from Elsevier.) A black and white version of this figure will appear in some formats. For the color version, please refer to the plate section.

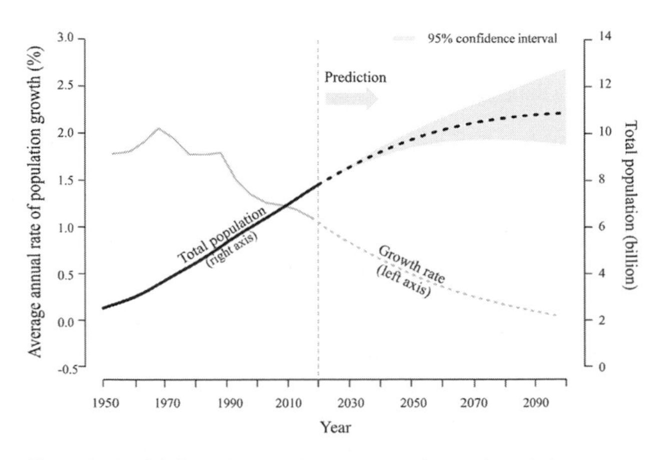

Figure 31.2 Global population and average annual rate of population growth (%) as a function of time. The projections are based on total fertility and life expectancy at birth. The gray shaded area represents 95% prediction intervals, 2020–2100. (Data from United Nations, DESA, Population Division, World Population Prospects 2019. http://population.un.org/wpp.) A black and white version of this figure will appear in some formats. For the color version, please refer to the plate section.

31.3.2 Countrywise Population

The population growth rates vary differently across countries. Several of the more populous countries contribute to much of the projected changes between 2019 and 2050 – for example, the ranking of the absolute increase in population are India, Nigeria, Pakistan, Congo, Ethiopia, Tanzania, Indonesia, Egypt, and the United States. Most of the countries in sub-Saharan Africa will see rapid population increase, and the population of total sub-Saharan Africa is projected to increase by 99% by 2050. Oceania, Northern Africa, and Western Asia would also increase very rapidly (nearly 50%). Eastern and Southeastern Asia, Europe, and Northern America are projected to grow slowly, and many countries in these regions are already experiencing decreasing populations. China and India are the top two populous countries in 2019, representing nearly 37% of the global population. Both countries experienced rapid population growth from 1950 to 2020, and China will reach its peak population size in 2031 and see a slight decrease from then,

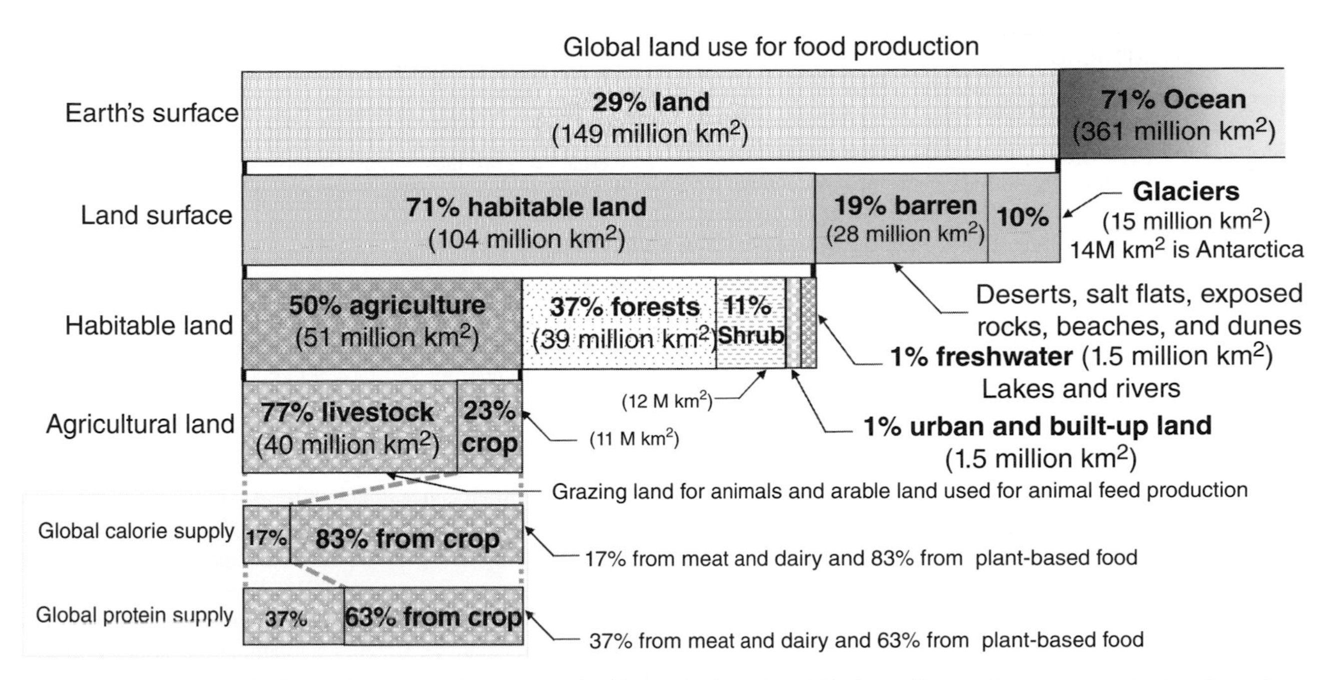

Figure 31.3 Global land use for food production. (Data from FAO, 2016.) A black and white version of this figure will appear in some formats. For the color version, please refer to the plate section.

while India is projected to increase until reaching its peak in 2058. India will surpass China and become the most populous country around 2027. Nigeria and Pakistan are projected to increase rapidly in the next several decades, and Nigeria will surpass the United States and become the third most populous country in 2047.

31.4 DISTRIBUTION OF LAND

Global food production depends on agricultural land, as shown in Figure 31.3 (a similar figure can be found in https://ourworldindata.org/agricultural-land-by-global-diets). About 29% of the Earth's surface is land, of which about 34% is agricultural land. Only 23% of the agricultural land is used for crop production. Although only a small proportion of land is used for crop production, it provides 83% of the global calorie supply.

Global cropland area has increased by 23% over the last 60 years (1961–2018), mostly at the expense of other habitable land (e.g., forest, wetland, and grassland) (FAO, 2013). At the same time, the global agricultural land area saw a rapid increase from 1961 to 2001, and then gradually decreased after peaking in 2001. Therefore, further expansion of cropland is limited. Agricultural land varies from country to country and from one part of a country to another within the same country.

31.4.1 Agricultural Land

Asia accounts for 37% of the total cropland, followed by the Americas (24%), Europe (19%), Africa (18%), and Oceania (2%). India has the largest cropland area (169.4 million ha), followed by the United States, China, Russia, Brazil, Indonesia, Nigeria, Argentina, Canada, and the Ukraine. Since 2001 most countries have had relatively stable amounts of agricultural land. However, Australia and the United States, which rank third and second largest in terms of agricultural land, exhibit decreasing trends. Cropland in the key countries has witnessed a limited increase since 2001 (e.g., there was a total increase of 74 million ha from 2001 to 2018). Globally, agricultural land/cropland per capita shows a decreasing trend. The agricultural land/cropland per capita in major countries also has had a decreasing trend. China and India have relatively high amounts of cropland, but the cropland per capita is only 46% and 61% of the global mean, respectively.

31.4.2 Irrigated Land

Food security depends heavily on irrigated land. Currently, cropland that is equipped with irrigation accounts for 22% of the total cropland. From 1961 to 2018 the amount of global irrigated land doubled. About 70.3% of the total irrigated land is in Asia, followed by the Americas (16%), Africa (5%), Europe (8%), and Oceania (1%). Rainfed agriculture is the dominant agricultural system worldwide. However, Bangladesh (60%), Pakistan (56%), and India (39%) have relatively higher proportions of irrigated land. China has the largest irrigated land area (74 million ha), followed by India, the United States, Pakistan, Iran, Brazil, Mexico, Indonesia, Thailand, and Bangladesh. China and India are the top two countries, and their irrigated areas have increased rapidly since 2001.

31.5 FOOD PRODUCTION

31.5.1 Global Food Production

Global food production for the past 50 years is shown in Figure 31.4, obtained by summing up the production of all crop types in billion tons per year. Global food production almost tripled from 1969 to 2019, increasing from just over 7.3 billion tons in 1969 to 20.9 billion tons in 2019. The annual increase rate is relatively stable (around 2%) from 1969 to 2004. After 2004, global food production increased slightly faster, with an annual mean increase rate of 2.5%. Cereal crop production has the most dramatic increase, from 0.9 billion tons in 1969 to 2.8 billion tons in 2019.

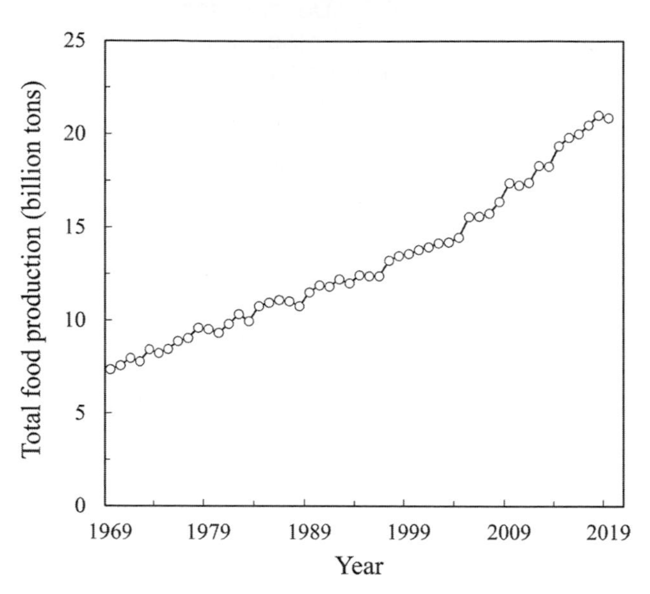

Figure 31.4 Global food production over the period 1969–2019. (Data from FAOSTAT, 2020.)

31.5.2 Countrywise Food Production

Overall, global per capita food production has been increasing consistently for the past 50 years, but the trend varies across different countries. Figure 31.5 shows the per capita food production of six countries. The per capita food production of Canada and the United States has been approximately constant since the 1980s. Australia even shows a decreasing trend since 2005. These three countries have relatively higher per capita food production, varying between 7.1 and 7.8 tons in 2018. Brazil, China, and India all experienced a significant increase over the same 50-year period. For example, Brazil's per capita food production was only 3 tons in 1969, but the value has been comparable to Canada since 2010. Although the increase of per capita food production in China and India is significant, they still have relatively lower per capita production ability than countries in Europe and Oceania due to their large population sizes.

Calorie supply per capita is the amount of food available for consumption (crop equivalent), measured in kilocalories per capita per day, as shown in Figure 31.6. This figure is reached by dividing the total available food supply for

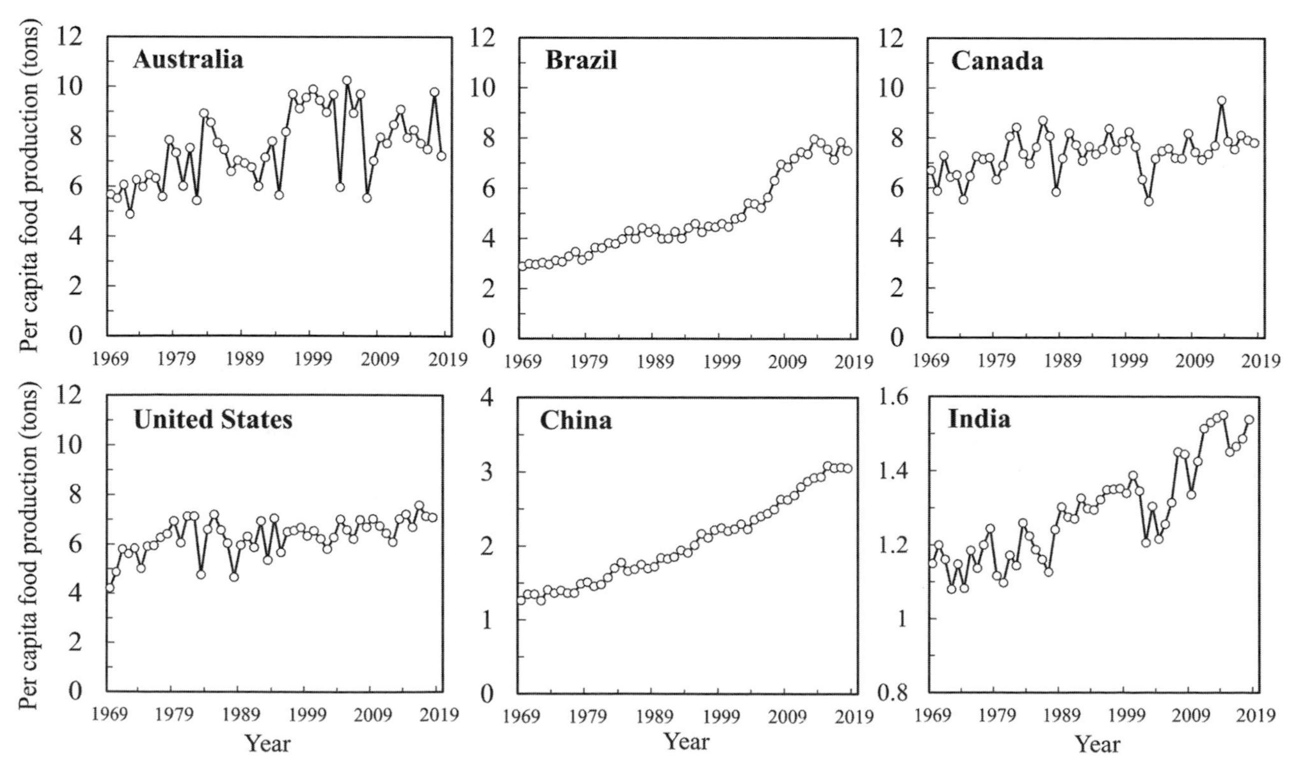

Figure 31.5 Countrywise per capita food production over the period 1969–2019. (Data from FAOSTAT, 2020.)

human consumption by the global population. Global calorie supply per capita has a similar trend as per capita food production. The global calorie supply per capita consistently increased from 2200 kilocalories per capita per day in 1961 to 2884 kilocalories per capita per day in 2016. Countrywise per capita calorie supply trends show considerable variability (Figure 31.7). Calorie supply across Oceania and Europe has been nearly constant from 1961 to 2013, while in other countries calorie supply has increased. The increase is significant in countries in Asia, Africa, and South America (e.g., China, India, Nigeria, and Brazil). In North America, calorie supply has nearly plateaued since 2000. In terms of food supply, the gap between developing and developed countries is constantly decreasing now. In 2013, the calorie supply per capita in most countries in Europe, North America, and Oceania was greater than 3100 kilocalories per capita per day, and the values in countries in Asia, South America, and sub-Saharan Africa were in the range of 2300–3000 kilocalories per capita per day.

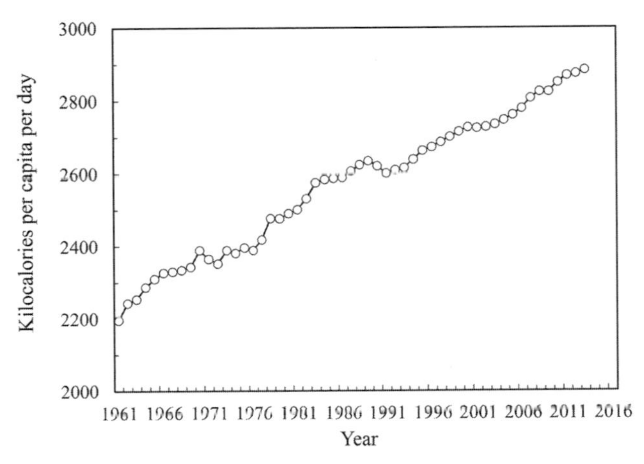

Figure 31.6 Global amount of food available for consumption (1961–2013). (Data from FAOSTAT, 2020.)

31.5.3 Countrywise per Capita Food Production/Supply versus Agricultural Land

Over the last 50 years the global per capita food production increase rate has exceeded the agricultural land extension rate, indicating an increasing trend in crop productivity. For example, global food production tripled, and per capita food production increased by 40% during this period, with only a 23% increase in agricultural land. Countrywise per capita food production/supply versus agricultural land shows different trends. From 1970 to 1990, although agricultural land showed a decreasing trend across Europe, Oceania, and North America, both their per capita food production and supply increased significantly due to the increased crop productivity. The crop productivity improvement in these countries is attributed to the adoption of new crop varieties and the application of irrigation and fertilizer. During this period, per capita food production/supply in East and South Asia and South America also increased very rapidly, which was mainly due to the expansion of cropland – agricultural land in China, Thailand, and Brazil increased by 25–40% from 1970 to 1990. Most countries in Africa had a relatively

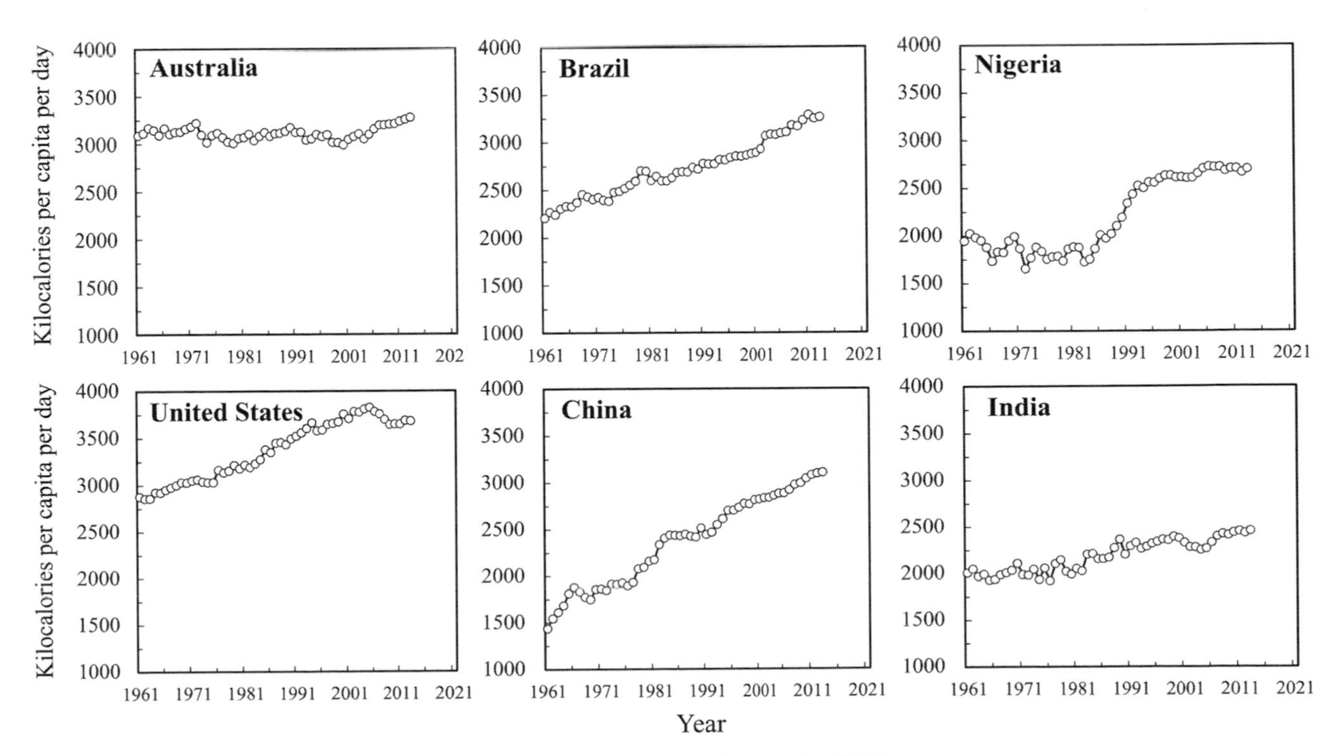

Figure 31.7 Countrywise per capita food production over the period 1961–2013. (Data from FAOSTAT, 2020.)

Table 31.1 *Irrigated cropland and its percentage and irrigation water sources*

	Equipped area (million ha)		Percentage of cropland		Groundwater irrigation	
Regions	1961	2006	1961	2006	Area (million ha)	Percentage of total irrigated area
Africa	**7.4**	**13.6**	**4.4**	**5.4**	**2.5**	**18.5**
Northern Africa	3.9	6.4	17.1	22.7	2.1	32.8
Sub-Saharan Africa	3.5	7.2	2.4	3.2	0.4	5.8
Americas	**22.6**	**48.9**	**6.7**	**12.4**	**21.6**	**44.1**
Northern America	17.4	35.5	6.7	14	19.1	54
Central America and Caribbean	0.6	1.9	5.5	12.5	0.7	36.3
Southern America	4.7	11.6	6.8	9.1	1.7	14.9
Asia	**95.6**	**211.8**	**19.6**	**39.1**	**80.6**	**38**
Western Asia	9.6	23.6	16.2	36.6	10.8	46
Central Asia	7.2	14.7	13.4	37.2	1.1	7.8
South Asia	36.3	85.1	19.1	41.7	48.3	56.7
East Asia	34.5	67.6	29.7	51	19.3	28.6
Southeast Asia	8	20.8	11.7	22.5	1	4.7
Europe	**12.3**	**22.7**	**3.6**	**7.7**	**7.3**	**32.4**
Western and Central Europe	8.7	17.8	5.8	14.2	6.9	38.6
Eastern Europe and Russia	3.6	4.9	1.9	2.9	0.5	10.1
Oceania	**1.1**	**4**	**3.2**	**8.7**	**0.9**	**23.9**
Australia and New Zealand	1.1	4	3.2	8.8	0.9	24
Pacific Islands	0.001	0.004	0.2	0.6	0	18.7
World	**139**	**300.9**	**10.2**	**19.7**	**112.9**	**37.5**
High income	26.7	54	6.9	14.7	26.5	49.1
Middle income	66.6	137.9	10.5	19.3	36.1	26.1
Low income	45.8	108.9	13.1	24.5	50.3	46.2
Low income food deficit	82.5	187.6	16.6	29.2	71.9	38.3
Least developed	6.1	17.5	5.2	10.1	5	28.8

After FAO, 2013.

low per capita food production and constant agricultural land during this period.

After 1990, despite the limited increase in agricultural land, South and East Asia produced the most rapid per capita food production/supply increase, with 77% of the production increase attributable to increased cropping intensity and improved crop productivity arising from irrigation, fertilizer, and the adoption of new crop varieties (Bruinsma, 2003). The expansion of agricultural land is still the primary factor in per capita food production in some countries in sub-Saharan African and South America. The application of irrigation, fertilizer, and the adoption of new crop varieties lagged in sub-Saharan African. Therefore, the most significant growth of food production in the future is expected to be in sub-Saharan Africa.

31.5.4 Food Production from Irrigated and Non-irrigated Lands

The global cropland area increased by 159 million ha from 1961 to 2009, which is attributed to the net increase in irrigated cropland – irrigated land had a net increase of 162 million ha from 1961 to 2009. The amount of irrigated cropland has more than doubled during this period. Rainfed cropland, however, shows a slightly decreasing trend. Simultaneously, the cropland per capita has seen a rapid decrease since 1961, reducing from 0.45 ha to 0.22 ha per capita (FAO, 2013). This indicates that less cropland is needed to feed each person, and that irrigated land has significantly increased crop production. More than 70% of irrigated areas are located in Asia (e.g., China, northern India, Bangladesh, Indonesia, Pakistan, Thailand, and Vietnam), where some 90% of the world rice production area is located. Large irrigated cropland is also located in some regions of the United States (e.g., the Great Plains and California) and in Africa (e.g., Egypt, which relies almost exclusively on irrigation for agricultural production). The details of irrigated cropland and the irrigation water sources in different regions are shown in Table 31.1.

In 2007 global irrigated cropland was estimated at 17%, but it produced 40% of the world's food, and the remaining

Table 31.2 *Food production and percentage of cropland from irrigated and rainfed agriculture in 2007*

Cropland	Percentage of cropland	Food production (million tons)	Food production (%)
Irrigated cropland	17	6,546	40
Rainfed cropland	83	9,820	60
Total	100	16,366	100

Table 31.3 *Cereals production, yields per area from irrigated and rainfed agriculture, potential loss (%) without irrigation in irrigated and total cropland (1998–2002)*

	Total crop production		Crop yields per area		Potential loss without irrigation (%)	
Regions	All cropland (Tg/yr[1])	Irrigated land (%)	Rainfed (Mg/km^2)	Irrigated (Mg/km^2)	Irrigated cropland	All cropland
Eastern Africa	22	16.8	126	221	51.6	8.7
Middle Africa	4	3	82	206	53.9	1.6
Northern Africa	31	68.9	69	540	95.6	66
Southern Africa	11	11.9	215	335	42.6	5.1
Western Africa	32	3.1	94	272	75.8	2.3
Africa	**100**	**27.6**	**102**	**428**	**86.5**	**23.9**
Caribbean	2	49.3	152	314	38.5	18.9
Central America	32	34.1	228	320	37.7	12.9
Northern America	383	17	472	844	54.3	9.2
South America	101	15.9	289	463	48.3	7.7
Americas	**518**	**17.9**	**398**	**625**	**51.2**	**9.2**
Central Asia	21	30.6	114	282	47.5	14.9
Eastern Asia	545	78.2	389	549	35.1	27.5
Southeastern Asia	169	50.2	272	419	38.8	19.5
Southern Asia	326	70.4	150	305	64	45.1
Western Asia	42	34.7	179	283	61.9	21.6
Asia	**1104**	**69**	**221**	**422**	**44.9**	**31**
Eastern Europe	151	4	226	303	36.3	1.5
Northern Europe	42	1.8	527	625	0.1	<0.05
Southern Europe	58	33.3	306	865	49.3	16.4
Western Europe	102	5.5	670	942	21.8	1.2
Europe and Russia	**353**	**8.9**	**325**	**643**	**40.8**	**3.7**
Oceania	**33**	**6.5**	**190**	**623**	**65**	**4.2**
World	**2108**	**43.5**	**266**	**442**	**46.7**	**20.3**

83% of cropland was unirrigated or rainfed, as shown in Table 31.2. Table 31.3 shows cereals production and yields per area from irrigated and rainfed agriculture in different continents during 1998–2002 using the global crop model (Siebert and Doll, 2010). Globally, the total irrigated cereal product accounted for 43.5% of the total cereal production from 1998 to 2002. In specific regions in east China, northern India, and the central United States, more than 70% of cereals are from irrigated land.

Irrigated land can significantly increase crop yields (Table 31.3) – the average global cereals yield from irrigated cropland (442 Mg km^{-2}) is 1.7 times that from rainfed cropland (266 Mg km^{-2}). Siebert and Doll (2010) estimated the potential loss without irrigation in total cropland and current cropland equipped with irrigation using the global crop water model during the period 1998–2002 (Table 31.3). Globally, 47% of loss is expected in irrigated cropland, and Africa will suffer an 87% loss in irrigated cropland, followed by Oceania (65%), the Americas (51%), Asia (45%), and Europe (41%). For total cropland, the global production is projected to have a 20% loss, and Asia will suffer the most loss (31%), followed by Africa (24%), the Americas (9%), Oceania (4%), and Europe (4%).

31.6 IMPACT OF CLIMATE CHANGE

31.6.1 Impact on Global Food Production

Climate change has a major impact on crop production (Piao et al., 2010). Connor et al. (2011) state that agriculture needs to adapt production systems to changing climate and to mitigate anthropogenic global warming by both reducing emissions of greenhouse gases (GHGs) from agriculture and by developing systems to sequester CO_2 emission from urban, industrial, and transport activities. The effects on crop yield may be positive or negative, caused by rising CO_2 concentrations, higher temperatures, changing precipitation patterns, changing water availability, increasing frequency of weather extremes (i.e., floods, heavy storms, and droughts), climate-induced soil erosion, and sea-level rise (Lotze-Campen, 2011).

Temperature is one of the primary factors that affect the rate of plant development and productivity (Hatfield and Prueger, 2015). Under global warming and climate change, higher temperatures are being observed across the globe, but with significant regional and seasonal variations (Lotze-Campen, 2011). Meehl et al. (2007) stated that daily minimum temperatures would increase more rapidly than daily maximum temperatures and there would be a greater likelihood of extreme events having detrimental effects on grain yield. Rising temperatures may lengthen the growing season by 1.2–3.6 days per decade (Gitay et al., 2001). An expansion of suitable crop areas may be possible in the Russian Federation, North America, North Europe, and Northeast Asia, but significant losses may occur in Africa due to heat and water stress and an expansion of arid and semi-arid regions (Fischer et al., 2005).

Rising temperatures are projected in the four crop-growing areas under four representative concentration pathway (RCP) scenarios of increasing greenhouse gas concentrations during the period 2071–2100. Compared with the baseline (1981–2010), temperature increases range from about 1 to 4 °C under RCP 2.6 and 8.5 scenarios. The principal cause of rising atmospheric CO_2 is the combustion of fossil fuels and land-use change (Connor et al., 2011). Plants growing in atmospheric CO_2 tend to increase the rate of net photosynthesis (Rosenzweig and Iglesias, 1998), and CO_2 reduces transpiration per unit leaf area. Thus, high levels of CO_2 improve water-use efficiency, which is the ratio of crop biomass accumulation or yield to the amount of water used in evapotranspiration. Free air carbon enrichment increases productivity in the range of 15–25% for C_3 crops (e.g., wheat, rice, and soybean) and 5–10% for C_4 crops (e.g., maize, sorghum, and sugarcane).

The median yield change under RCP 8.5 at the end of the century (2070–2099) in comparison to the baseline (1980–2010) under all global gridded crop models shows that the tropics are projected to see more severe climate impacts since the effects of CO_2 fertilization cannot compensate for the side effect of increased demand for irrigation water. For maize and wheat, increased production is projected at high latitudes and decreased production is projected at low latitudes, with a general agreement among different models. Rice and soybean also show increased production in mid- and high-latitude areas, but are less consistent in the tropical regions.

31.6.2 Impact on Water Security

Climate change has perturbed the hydrologic cycle and has affected both water availability and water demand. The space–time distribution of rainfall is changing and some regions are experiencing less rain and more droughts; others are seeing more rains, flooding, and landslides. Impacts of climate change on water security are dramatic, and for some regions likely beyond the adaptation limits for natural and human systems (Jaramillo and Nazemi, 2018).

In 1955 only 7 countries were found to be under water-stressed conditions. In 1990 this rose to 20 countries and it is expected that by the year 2025 another 10–15 countries will be added to this list. It is further predicted that by 2050, two-thirds of the world population may face water-stressed conditions (Misra, 2014).

Under climate change, higher temperatures are increasing evapotranspiration from vegetation, land, surface water, and oceans. A warmer atmosphere is holding more water, and the global mean temperature has increased by 0.74 °C during the last 100 years (Misra, 2014). As air holds more water, more precipitation is leading to increased flooding. The dangers of a stormier world are more easily appreciated by the public than the dangers of changes in temperature and rainfall patterns, if only because they are transmitted through media images of death and destruction following events like the devastation of New Orleans by Hurricane Katrina in 2005 and the floods in Bangladesh in 2007 caused by Cyclone Sidr, which took 3400 lives and left nearly one million people homeless (Sadoff and Muller, 2009).

A warmer climate also translates to having more precipitation in the form of rain and less as snow. Snow represents natural water storage, valuable for later irrigation seasons. At the same time, areas like the Southwest United States will experience less precipitation because of climate change, leading to longer and more severe drought periods (Ortiz-Partida, 2019). In addition, rainy seasons will become shorter, creating more days when irrigation is needed and therefore increasing water demands. Warmer water in streams and rivers has an impact on metabolism, life cycle, and behavior of aquatic species. These cumulative impacts on water resources make water availability harder to predict and manage. This is intensifying problems for areas that are already experiencing such impacts and extending water stress into new places that will need to learn and adapt (Ortiz-Partida, 2019).

It is expected that climate change characterized by increased temperatures will continue. Therefore, it is

Table 31.4 *Impact of climate change on water quality*

Parameter	Impact	References
Temperature	River water temperatures are in close equilibrium with air temperature and, as air temperatures rise, so will river temperatures.	Whitehead et al., 2009
pH	pH is affected by the solubility of the salts at a higher temperature. It can also modify the metabolic processes of microorganisms.	Bergara et al., 2007
BOD (biological oxygen demand) and COD (chemical oxygen demand)	Reduced dilution effects will impact organic pollutant concentrations with increased BOD and COD.	Whitehead et al., 2009
DO (dissolved oxygen)	When the temperature rises, the water's ability to retain dissolved oxygen is reduced.	Bergara et al., 2007
Nutrients	In general, an increase in the temperature and a decrease in precipitation will promote the mineralization and salinization of water. The increased availability of nutrients translates into greater eutrophication.	Xia et al., 2015
Toxic compounds	Increasing water temperatures generally increase the toxicity of metals in aquatic ecosystems. Ammoniacal nitrogen changes its ionization with pH at 8.0. It is 65% more toxic than at neutral pH.	García-López et al., 2011
Biomass	Among the consequences of eutrophication is the greater probability of the presence of cyanobacteria blooms.	García-López et al., 2011

essential to pay attention to the issues of water security under global warming, to enhance scientific assessment, mitigation, and adaptation of water security to climate change, and to provide scientific support.

Extreme climate events, such as heat waves, heavy storms, floods, or droughts, may damage crops in specific development stages. A substantial and widespread increase in the number of heavy rainfall events is expected, even in regions where the total precipitation amount decreases. Bouts of heavy rainfall are very likely in Southern and Eastern Asia and in Northern Europe, which are major agricultural production areas (Solomon et al., 2007). Observations show an increase in frequency and duration of warm weather extremes. In many regions, especially in the tropics and subtropics, droughts have been more prolonged and more extensive since the 1970s because of higher temperatures and reduced precipitation. In arid and semi-arid regions higher rainfall intensity will increase the risk of soil erosion and salinization.

31.6.3 Impact on Water Quality

Climate change can alter water quality and water ecosystems through biochemical processes (Dalla Valle et al., 2007; Delpla et al., 2009). Specific changes vary among different regions and water bodies (Whitehead et al., 2009). The relationship between climate change and water quality in water bodies depends on hydrodynamics and biochemical processes (Mooij et al., 2005; Delpla et al., 2009). As shown in Table 31.4, water quality can be impacted in many ways: physical, including temperature and turbidity; chemical, including pH and chemical concentrations, and biological, including biodiversity and species abundance, across the entire food web from microbial pools and macrophytes up to fish (Ducharne, 2008).

Some studies have reported the impacts of droughts on water quality (Caruso, 2002; Evans et al., 2005; Monteith et al., 2007; Ducharne, 2008), which mainly include increased pollutant concentrations, enhanced nitrogen mineralization, and delayed recovery from acidification. During the drought period, lower flows can weaken the dilution effects of some pollutants (Zwolsman and Van Bokhoven et al., 2007; Elsdon et al., 2009).

31.7 WATER REQUIREMENTS

31.7.1 Global Water Requirements for Irrigated Agriculture

The amount of water required to produce crops varies by crop and region, depending on climate, mode of cultivation (rainfed or irrigated, high-input or low-input agriculture), crop variety and length of the growing season, and crop yields. It is estimated that 7130 km^3 are consumed by crops globally, including feed and food crops. Table 31.5 shows crop water consumption of different crops in 2000.

Table 31.5 *Crop water consumption (km^3) in 2000*

Region	Crops						
	Total cereals	Roots and tubers	Sugar	Vegetables and fruits	Soybeans	Other	Total
Sub-Saharan Africa	557	154	25	26	7	312	1071
East Asia	960	99	67	172	68	325	1661
South Asia	896	18	135	84	37	335	1505
Central Asia and Eastern Europe	525	44	14	7	4	193	772
Latin America	336	29	163	35	176	169	895
Middle East and North Africa	166	4	6	32	1	30	225
OECD countries	640	12	24	15	134	181	990
World	4089	363	434	370	427	1547	7130

After de Fraiture et al., 2007.

Table 31.6 *Projected increases for water withdrawals for irrigation (km^3)*

Source	1995	2025	Increase (%) 1995–2025
Shiklomanov (2000)	2488	3097	24
Seckler et al. (2000)	2469	2915	18
Faurès et al. (2002)	2128	2420[a]	14

[a] This estimate uses 2030 as the projection year and covers projections for developing countries only, constituting 75–80% of global withdrawals.
After Molden and de Fraiture, 2004.

Table 31.6 presents some best estimates on water withdrawals for irrigated agriculture in 2025.

31.7.2 Countrywise Water Availability

Water withdrawal is defined as the quantity of freshwater taken from groundwater or surface water sources for use in agricultural, industrial, or domestic purposes. Water levels vary significantly across the world. The large variance in levels of water withdrawal depends on different factors, such as latitude, climate, and agriculture or industrial sector. Areas equipped for full-control irrigation for different irrigation methods countrywise are given in Table 31.7.

31.7.3 Countrywise Crop Water Requirement

Table 31.8 shows the observed annual crop water requirement of some major countries. India has the largest crop water requirement, followed by China, the United States, and Pakistan.

Table 31.7 *Areas equipped for full-control irrigation for different irrigation methods of some major countries in 2017*

Country	Percentage of different methods			Total areas
	Surface	Sprinkler	Localized	('000 ha)
Australia	71.9	20.6	7.5	2,546
Brazil	30.5	20.3	49.2	5,066
Canada	46.2	53.3	0.5	1,281
China	94.3	4.5	1.2	69,863
Egypt	75.6	11.4	13.0	3,823
Germany	2.1	97.0	1.0	676
India	96.8	2.3	0.9	70,400
Indonesia	100.0	0.0	0.0	6,722
Iran	91.4	3.4	5.2	8,700
Israel	0.0	26.2	73.8	229
Italy	37.4	38.6	24.0	4,124
Japan	80.4	17.2	2.4	2,500
Mexico	91.7	5.7	2.6	6,305
Nigeria	100.0	0.0	0.0	238
Pakistan	100.0	0.0	0.0	19,270
South Africa	23.1	55.1	21.9	1,670
United Kingdom	51.3	46.1	2.6	208
United States	45.2	47.9	6.9	26,708
Vietnam	100.0	0.0	0.0	4,585

Data from FAO database (FAOSTAT, 2020).

31.8 ENERGY REQUIREMENT FOR IRRIGATED AGRICULTURE

Energy inputs for crop production can be direct, such as fuel sources used to operate farm machinery and pumps, or indirect, such as energy used to produce equipment and other goods and services that are used on the farm (Pimentel, 1992). Irrigation is a primary direct energy

Table 31.8 *Total crop water requirement of some major countries*

Country	Year	Crop water requirement (km^3/year)	Country	Year	Crop water requirement (km^3/year)
India	2010	688.0	Thailand	2005	51.8
China	2015	385.2	Brazil	2005	31.7
United States	2010	175.1	Bangladesh	2010	31.5
Pakistan	2010	172.4	Turkey	2005	29.6
Indonesia	2000	92.8	Chile	2005	29.4
Iran	2005	86.0	Spain	2010	25.5
Vietnam	2005	77.8	Russia	2000	13.2
Philippines	2010	67.1	Italy	2005	12.9
Mexico	2005	60.6	Australia	2015	10.6
Egypt	2000	59.0	Nigeria	2010	5.5
Japan	1995	58.9	France	2005	3.9
Uzbekistan	1995	54.4	Germany	2010	1.0
Iraq	2000	52.0	United Kingdom	2010	0.2

Data from FAOSTAT, 2020.

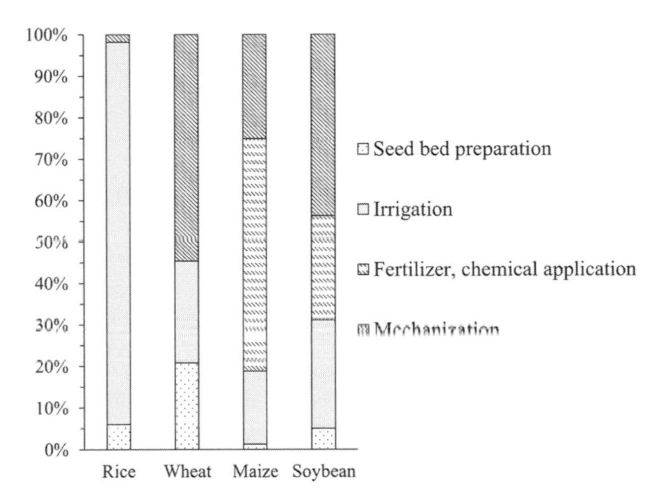

Figure 31.8 Proportions of different operations on direct energy use for the cultivation of different crops. (Rice and wheat data from Pathak and Bining, 1985; maize and soybean data from Alluvione et al., 2011.)

consumer on the farm (Naylor, 1996). For example, energy used for on-farm pumping is between 18% and 48% of direct energy used for crop production (Hodges et al., 1994; Alluvione et al., 2011). Figure 31.8 compares direct energy use of various farm operations for the cultivation of different crops. For some high water requirement crops, such as rice, irrigation can account for 92% of the total direct energy use. Therefore, any changes in irrigation technology used on the farm are expected to impact energy consumption, that is, energy consumption and irrigation are intertwined (Plappally and Lienhard, 2012), and this relationship is called the energy–irrigation nexus, electricity–irrigation nexus, or diesel–irrigation nexus. Energy consumption in irrigation depends on pumps and the pumping system, irrigation area, field characteristics, crop variety and crop pattern, climate, method of irrigation, irrigation scheduling, and energy cost (Lal, 2004; Martin et al., 2011). Brown and Elliot (2005) pointed out that irrigation pumping has the most considerable energy-saving potential in farm systems. Pathak and Bining (1985) reported that 50% of energy could be saved by improving irrigation technology and water management practices. For example, pressurized microirrigation systems require less energy consumption if operating pressures and pumping volumes are reduced (Hodges et al., 1994). When surface water is used for irrigation, it requires less energy than when groundwater is used.

With increasing demand due to growing populations and competition for water use from other sectors, complicated by climate change, irrigated agriculture nowadays is under greater pressure to reduce water use by adopting high water-use efficiency irrigation systems. Sprinkler and drip irrigation systems have the potential to enhance irrigation efficiency compared with conventional surface irrigation, e.g., border, basin, and furrow irrigation. However, they are more energy-intensive, which may increase the energy use and greenhouse gas emissions from irrigated agriculture. Proper system design and management are required to increase both water and energy use efficiency in irrigated agriculture.

Irrigation involves both high-pressure sprinkler systems to pump large volumes of water and low-pressure drip irrigation systems. Figure 31.9 provides the reported energy requirements for irrigating a farm using different sprinkler and drip irrigation systems. Energy intensity in mobile sprinkler irrigation is comparatively higher, due to its high pressure, than in stationary and row sprinkler systems. Row sprinkler systems with movable hose have a relatively lower pressure, which allows reducing energy consumption per unit of water. Also, the use of tubes and nozzles in sprinkler systems can reduce energy consumption, with energy requirements comparable to those of drip irrigation systems.

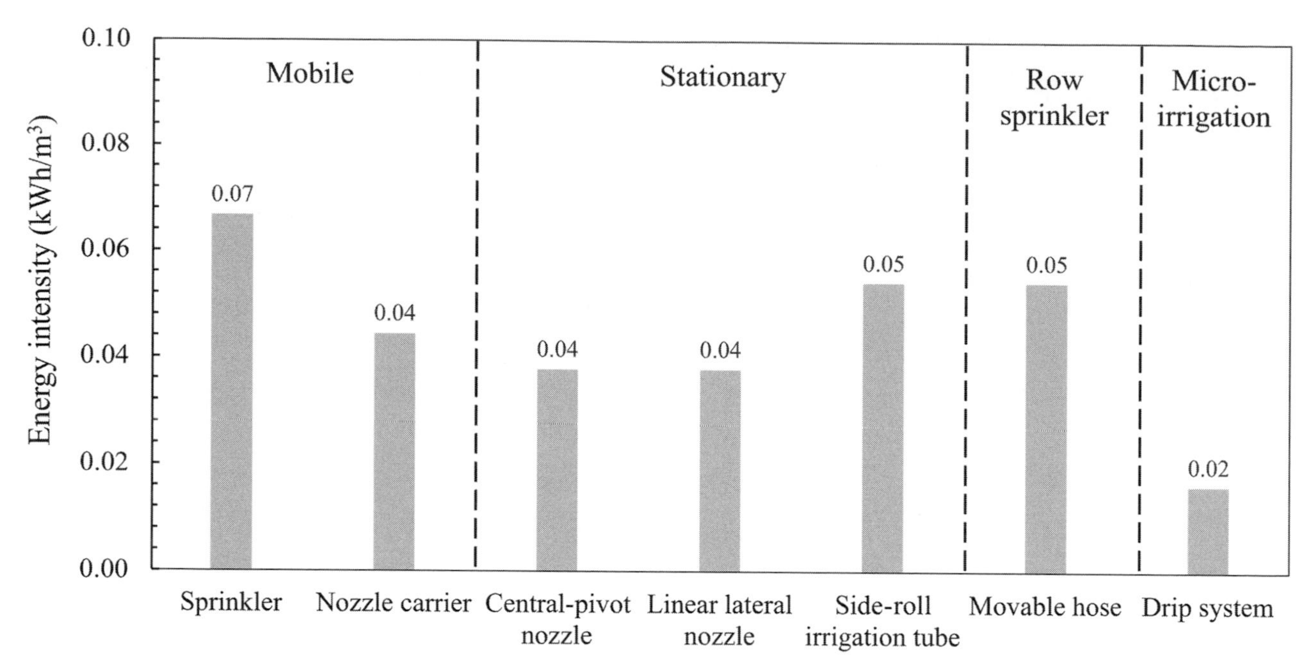

Figure 31.9 Energy requirements in different irrigation systems. (Data from Chiaramonti et al., 2000.)

REFERENCES

Alluvione, F., Moretti, B., Sacco, D., and Grignani, C. (2011). EUE (energy use efficiency) of cropping systems for a sustainable agriculture. *Energy*, 36: 4468–4481.

AQUASTAT (1999). FAO's Information System on Water and Agriculture. Software. https://lccn.loc.gov/2004616053.

Bergara, J., Millain, R., Painemilla, V., et al. (2007). *Efectos del cambio climático en la calidad del agua del río Chimehuín.* Patagonia: The Globe Program.

Black, R. E., Allen, L. H., Bhutta, Z. A., et al. (2008). Maternal and child undernutrition: global and regional exposures and health consequences. *Lancet*, 371(9608): 243–260.

Brown, E., and Elliot, R. N. (2005). Potential energy efficiency savings in the agriculture sector. American Council for an Energy-Efficient Economy. www.aceee.org/pubs/ie053.htm.

Bruinsma, J. (2003). *World Agriculture: Towards 2015/2030: An FAO Perspective*. London: Earthscan.

Caruso, B. (2002). Temporal and spatial patterns of extreme low flows and effects on stream ecosystems in Otago, New Zealand. *Journal of Hydrology*, 257(1–4): 115–133.

CFS (2012). *Coming to Terms with Food Security, Nutrition Security, Food Security and Nutrition*, Rome: FAO. www.fao.org/docrep/meeting/026/MD776E.pdf

Chiaramonti, D., Grimm, H., Bassam, N. E., and Cendagorta, M. (2000). Energy crops and bioenergy for rescuing deserting coastal area by desalination: feasibility study. *Bioresource Technology*, 72: 131–146.

Connor, D. J., Loomis, R. S., and Cassman, K. G. (2011). *Crop Ecology: Productivity and Management in Agricultural Systems*. Cambridge: Cambridge University Press.

Dalla Valle, M., Codato, E., and Marcomini, A. (2007). Climate change influence on POPs distribution and fate: a case study. *Chemosphere*, 67(7): 1287–1295.

Delpla, I., Jung, V. A., Baures, E., Clement, M., and Thomas, O. (2009). Impacts of climate change on surface water quality in relation to drinking water production. *Environment International*, 35(8): 1225–1233.

de Fraiture, C., Wichelns, D., Rockstrom, J., et al. (2007). Looking ahead to 2050: scenarios of alternative investment approaches. In *Water for Food, Water for Life: A Comprehensive Assessment of Water Management in Agriculture*, edited by D. Molden. London: Earthscan and International Water Management Institute (IWMI), pp. 91–145.

Ducharne, A. (2008). Importance of stream temperature to climate change impact on water quality. *Hydrology and Earth System Sciences*, 12: 797–810.

Ecker, O., and Breisinger, C. (2012). The food security system: a new conceptual framework. International Food Policy Research Institute (IFPRI) paper 1166.

Elsdon, T. S., DeBruin, M. A. N. B., Diepen, J. N., and Gillandersa, M. B. (2009). Extensive drought negates human influence on nutrients and water quality in estuaries. *Science of the Total Environment*, 407(8): 3033– 3043.

Evans, C. D., Monteith, D. T., and Cooper, D. M. (2005). Long-term increases in surface water dissolved organic 542 carbon: observations, possible causes and environmental impacts. *Environmental Pollution*, 137(1): 55–71.

FAO (2013). *The State of the World's Land and Water Resources for Food and Agriculture: Managing Systems at Risk.* London: Routledge.

FAO (2016). AQUASTAT Main Database.

FAOSTAT (2020). Database. www.fao.org/faostat/en/#data.

Faurès, J. M., Hoogeveen, J., and Bruinsma, J. (2002). *The FAO Irrigated Area Forecast for 2030*. Rome: FAO.

Fischer, G., Shah, M., Tubiello, F. N., and van Velhuizen, H. (2005) Socio-economic and climate change impacts on agriculture: an integrated assessment, 1990–2080. *Philosophical Transactions of the Royal Society B*, 360: 2067–2073.

Frankenberger, T. R., Oshaug, A., and Smith, L. C. (1997). *A Definition of Nutrition Security*. Mimeo.

García-López, A., Ramírez-Salinas, N., Vázquez-Bustos, C., and Leal-Ascencio, M. (2011). Impacto del cambio climático en la calidad del agua en México. IMTA/SEMARNAT.

Gitay, H., Brown, S., Easterlin, W., and Jallow, B. (2001). Ecosystems and their goods and services. In *Climate Change 2001: Impacts, Adaptation and Vulnerability. Working Group II of the Intergovernmental Panel on Climate Change*. Cambridge: Cambridge University Press.

Hatfield, J. L., and Prueger, J. H. (2015). Temperature extremes: effect on plant growth and development. *Weather and Climate Extremes*, 10: 4–10.

Hodges, A. W., Lynne, G. D., Rahmani, M., and Casey, C. F. (1994). *Adoption of Energy and Water-Conserving Irrigation Technologies in Florida*. Gainesville, FL: University of Florida.

Jaramillo, P., and Nazemi, A. (2018). Assessing urban water security under changing climate: challenges and ways forward. *Sustainable Cities and Society*, 41: 907–918.

Lal, R. (2004). Carbon emission from farm operations. *Environment International*, 30: 981–990.

Lotze-Campen, H. (2011). Climate change, population growth, and crop production: an overview. In *Crop Adaptation to Climate Change*, edited by S. Yadav, R. Redden, J. Hatfield, et al. Chichester: Wiley, pp. 1–11.

Martin, D. L., Dorn, T. W., Melvin, S. R., Corr, A. J., and Kranz, W. L. (2011). Evaluating energy use for pumping irrigation water. In the *23rd Annual Central Plains Irrigation Conference*, pp. 104–116.

Meehl, G. A., Stocker, T. F., Collins, W. D., et al. (2007). *Global Climate Projections*. New York: Academic Press.

Misra, A. K. (2014). Climate change and challenges of water and food security. *International Journal of Sustainable Built Environment*, 3(1): 153–165.

Molden, D., and de Fraiture, C. (2004). Investing in water for food, ecosystems and livelihoods. International Water Management Institute discussion draft.

Monteith, D. T., Stoddard, L. J., Evans, D. C., et al. (2007). Dissolved organic carbon trends resulting from changes in atmospheric deposition chemistry. *Nature*, 450(7169): 537–540.

Mooij, W. M., Hülsmann, S., De Senerpont Domis, L. N., et al. (2005). The impact of climate change on lakes in the Netherlands: a review. *Aquatic Ecology*, 29: 381–400.

Naylor, R. L. (1996). Energy and resource constraints on intensive agricultural production. *Annual Review of Energy and the Environment*, 21: 99–123.

Ortiz-Partida, J. (2019). The world is in a water crisis and climate change is making it worse [Blog post]. https://blog.ucsusa.org/pablo-ortiz/the-world-is-in-a-water-crisis-and-climate-change-is-making-it-worse/

Pangaribowo, E. H., Gerber, N., and Torero, M. (2013). Food and nutrition security indicators: a review. ZEF working paper 108.

Pathak, B. S., and Bining, A. S. (1985). Energy use pattern and potential for energy saving in rice–wheat cultivation. *Energy in Agriculture*, 4: 271–278.

Piao, S., Ciais, P., Huang, Y., et al. (2010). The impacts of climate change on water resources and agriculture in China. *Nature*, 467 (7311): 43–51.

Pimentel, D. (1992). Energy inputs in production agriculture. In *Energy in Farm Production*, edited by R. C. Fluck. Amsterdam: Elsevier.

Plappally, A. K., and Lienhard, V. (2012). Energy requirements for water production, treatment, end use, reclamation, and disposal. *Renewable and Sustainable Energy Reviews*, 16(7): 4818–4848.

Rosenzweig, C., and Iglesias, A. (1998). The use of crop models for international climate change impact assessment. In *Understanding Options for Agricultural Production*. Dordrecht: Springer, pp. 267–292.

Sadoff, C., and Muller, M. (2009). *Water Management, Water Security and Climate Change Adaptation: Early Impacts and Essential Responses*. Stockholm: Global Water Partnership.

Schmidhuber, J., and Tubiello, F. N. (2007). Global food security under climate change. *Proceedings of the National Academy of Sciences*, 104(50): 19703–19708.

Seckler, D., Molden, D., Amarasinghe, U., and De Fraiture, C. (2000). *Water Issues for 2025: A Research Perspective*. Colombo: International Water Management Institute.

Shaw, D. J. (2007). World Food Summit, 1996. In *World Food Security*. London: Palgrave Macmillan, pp. 347–360.

Shiklomanov, I. A. (2000). Appraisal and assessment of world water resources. *Water International*, 25(1): 11–32.

Siebert, S., and Doll, P. (2010). Quantifying blue and green virtual water contents in global crop production as well as potential production losses without irrigation. *Journal of Hydrology*, 384 (3–4): 198–217.

Smith, L. C., and Haddad, L. J. (2000). *Explaining Child Malnutrition in Developing Countries: A Cross-Country Analysis*. Washington, DC: IFPRI.

Solomon, S., Manning, M., Marquis, M., and Qin, D. (2007). *Climate Change 2007 – The Physical Science Basis: Working group I Contribution to the Fourth Assessment Report of the IPCC*. Cambridge: Cambridge University Press.

Strauss, J. (1990). Households, communities, and preschool children's nutrition outcomes: evidence from rural Côte d'Ivoire. *Economic Development and Cultural Change*, 38(2): 231–261.

Whitehead, P. G., Wilby, R. L., Battarbee, R. W., Kernan, M., and Wade, A. J. (2009). A review of the potential impacts of climate change on surface water quality. *Hydrological Sciences Journal*, 54(1): 101–123.

Xia, X., Wu, Q., Mou, X. L., and Lai, Y. (2015). Potential impacts of climate change on the water quality of different water bodies. *Journal of Environmental Informatics*, 25: 85–98.

Zwolsman, J., and Van Bokhoven, A. (2007). Impact of summer droughts on water quality of the Rhine River: a preview of climate change? *Water Science and Technology*, 56(4): 45–55.

Author Index

Subject Index